Praise for Roger Penrose's

THE ROAD TO REALITY

"Provocative. . . . Delivers a sweeping overview of modern physics." —*Discover*

"A truly remarkable book. . . . Penrose does much to reveal the beauty and subtlety that connects nature and the human imagination, demonstrating that the quest to understand the reality of our physical world, and the extent and limits of our mental capacities, is an awesome, never-ending journey rather than a one-way cul-de-sac." —*The Sunday Times* (London)

"A philosophical work of art and a mathematical menagerie. . . . Some of the finest explanations of the theories of physical phenomena to be found anywhere." —*The Orlando Sentinel*

"Stimulating . . . genuinely magnificent." —*Scotland on Sunday*

"Science needs more people like Penrose, willing and able to point out the flaws in fashionable models from a position of authority and to signpost alternative roads to follow." —*The Independent* (London)

Roger Penrose

THE ROAD TO REALITY

Roger Penrose is Emeritus Rouse Ball Professor of Mathematics at Oxford University. He has received a number of prizes and awards, including the 1988 Wolf Prize in physics, which he shared with Stephen Hawking for their joint contribution to our understanding of the universe. His books include *The Emperor's New Mind*, *Shadows of the Mind*, and *The Nature of Space and Time*, which he wrote with Hawking. He has lectured extensively at universities throughout America. He lives in Oxford.

ALSO BY ROGER PENROSE

The Emperor's New Mind:
Concerning Computers, Minds, and the Laws of Physics

Shadows of the Mind:
A Search for the Missing Science of Consciousness

The Nature of Space and Time (with Stephen Hawking)

THE ROAD TO REALITY

THE ROAD TO
REALITY

A Complete Guide to
the Laws of the Universe

ROGER PENROSE

VINTAGE BOOKS
A Division of Random House, Inc.
New York

FIRST VINTAGE BOOKS EDITION, JANUARY 2007

The Library of Congress has cataloged the Knopf edition as follows:
Penrose, Roger.
The road to reality : a complete guide to the laws of the universe /
Roger Penrose. —1st American ed.
p. cm.
Includes bibliographical references and index.
1. Mathematical physics. 2. Physical laws. I. Title.
QC20.P366 2005
530.1—dc22
2004061543

Vintage ISBN: 978-0-679-77631-4

www.vintagebooks.com

Printed in the United States of America
10 9 8 7 6 5 4 3 2 1

Contents

I dedicate this book to the memory of

DENNIS SCIAMA

who showed me the excitement of physics

Preface

THE purpose of this book is to convey to the reader some feeling for what is surely one of the most important and exciting voyages of discovery that humanity has embarked upon. This is the search for the underlying principles that govern the behaviour of our universe. It is a voyage that has lasted for more than two-and-a-half millennia, so it should not surprise us that substantial progress has at last been made. But this journey has proved to be a profoundly difficult one, and real understanding has, for the most part, come but slowly. This inherent difficulty has led us in many false directions; hence we should learn caution. Yet the 20th century has delivered us extraordinary new insights—some so impressive that many scientists of today have voiced the opinion that we may be close to a basic understanding of *all* the underlying principles of physics. In my descriptions of the current fundamental theories, the 20th century having now drawn to its close, I shall try to take a more sober view. Not all my opinions may be welcomed by these 'optimists', but I expect further changes of direction greater even than those of the last century.

The reader will find that in this book I have not shied away from presenting mathematical formulae, despite dire warnings of the severe reduction in readership that this will entail. I have thought seriously about this question, and have come to the conclusion that what I have to say cannot reasonably be conveyed without a certain amount of mathematical notation and the exploration of genuine mathematical concepts. The understanding that we have of the principles that actually underlie the behaviour of our physical world indeed depends upon some appreciation of its mathematics. Some people might take this as a cause for despair, as they will have formed the belief that they have no capacity for mathematics, no matter at how elementary a level. How could it be possible, they might well argue, for them to comprehend the research going on at the cutting edge of physical theory if they cannot even master the manipulation of *fractions*? Well, I certainly see the difficulty.

Yet I am an optimist in matters of conveying understanding. Perhaps I am an incurable optimist. I wonder whether those readers who cannot manipulate fractions—or those who claim that they cannot manipulate fractions—are not deluding themselves at least a little, and that a good proportion of them actually have a potential in this direction that they are not aware of. No doubt there are some who, when confronted with a line of mathematical symbols, however simply presented, can see only the stern face of a parent or teacher who tried to force into them a non-comprehending parrot-like apparent competence—a duty, and a duty alone—and no hint of the magic or beauty of the subject might be allowed to come through. Perhaps for some it is too late; but, as I say, I am an optimist and I believe that there are many out there, even among those who could never master the manipulation of fractions, who have the capacity to catch some glimpse of a wonderful world that I believe must be, to a significant degree, genuinely accessible to them.

One of my mother's closest friends, when she was a young girl, was among those who could not grasp fractions. This lady once told me so herself after she had retired from a successful career as a ballet dancer. I was still young, not yet fully launched in my activities as a mathematician, but was recognized as someone who enjoyed working in that subject. 'It's all that cancelling', she said to me, 'I could just never get the hang of cancelling.' She was an elegant and highly intelligent woman, and there is no doubt in my mind that the mental qualities that are required in comprehending the sophisticated choreography that is central to ballet are in no way inferior to those which must be brought to bear on a mathematical problem. So, grossly overestimating my expositional abilities, I attempted, as others had done before, to explain to her the simplicity and logical nature of the procedure of 'cancelling'.

I believe that my efforts were as unsuccessful as were those of others. (Incidentally, her father had been a prominent scientist, and a Fellow of the Royal Society, so she must have had a background adequate for the comprehension of scientific matters. Perhaps the 'stern face' could have been a factor here, I do not know.) But on reflection, I now wonder whether she, and many others like her, did not have a more rational hang-up—one that with all my mathematical glibness I had not noticed. There is, indeed, a profound issue that one comes up against again and again in mathematics and in mathematical physics, which one first encounters in the seemingly innocent operation of cancelling a common factor from the numerator and denominator of an ordinary numerical fraction.

Those for whom the action of cancelling has become second nature, because of repeated familiarity with such operations, may find themselves insensitive to a difficulty that actually lurks behind this seemingly simple procedure. Perhaps many of those who find cancelling mysterious are seeing a certain profound issue more deeply than those of us who press onwards in a cavalier way, seeming to ignore it. What issue is this? It concerns the very way in which mathematicians can provide an existence to their mathematical entities and how such entities may relate to physical reality.

I recall that when at school, at the age of about 11, I was somewhat taken aback when the teacher asked the class what a fraction (such as $\frac{3}{8}$) actually is! Various suggestions came forth concerning the dividing up of pieces of pie and the like, but these were rejected by the teacher on the (valid) grounds that they merely referred to imprecise physical situations to which the precise mathematical notion of a fraction was to be *applied*; they did not tell us what that clear-cut mathematical notion actually *is*. Other suggestions came forward, such as $\frac{3}{8}$ is 'something with a 3 at the top and an 8 at the bottom with a horizontal line in between' and I was distinctly surprised to find that the teacher seemed to be taking these suggestions seriously! I do not clearly recall how the matter was finally resolved, but with the hindsight gained from my much later experiences as a mathematics undergraduate, I guess my schoolteacher was making a brave attempt at telling us the definition of a fraction in terms of the ubiquitous mathematical notion of an *equivalence class*.

What is this notion? How can it be applied in the case of a fraction and tell us what a fraction actually is? Let us start with my classmate's 'something with a 3 at the top and an 8 on the bottom'. Basically, this is suggesting to us that a fraction is specified by an ordered pair of whole numbers, in this case the numbers 3 and 8. But we clearly cannot regard the fraction as *being* such an ordered pair because, for example, the fraction $\frac{6}{16}$ is the same number as the fraction $\frac{3}{8}$, whereas the pair (6, 16) is certainly not the same as the pair (3, 8). This is only an issue of cancelling; for we can write $\frac{6}{16}$ as $\frac{3\times 2}{8\times 2}$ and then cancel the 2 from the top and the bottom to get $\frac{3}{8}$. Why are we allowed to do this and thereby, in some sense, 'equate' the pair (6, 16) with the pair (3, 8)? The mathematician's answer—which may well sound like a cop-out—has the cancelling rule just built in to the definition of a fraction: a pair of whole numbers $(a \times n, b \times n)$ is deemed to represent the same fraction as the pair (a, b) whenever n is any non-zero whole number (and where we should not allow b to be zero either).

But even this does not tell us what a fraction is; it merely tells us something about the way in which we represent fractions. What *is* a fraction, then? According to the mathematician's "equivalence class" notion, the fraction $\frac{3}{8}$, for example, simply is the infinite collection of all pairs

$$(3, 8), (-3, -8), (6, 16), (-6, -16), (9, 24), (-9, -24), (12, 32), \ldots,$$

where each pair can be obtained from each of the other pairs in the list by repeated application of the above cancellation rule.* We also need definitions telling us how to add, subtract, and multiply such infinite collections of pairs of whole numbers, where the normal rules of algebra hold, and how to identify the whole numbers themselves as particular types of fraction.

This definition covers all that we mathematically need of fractions (such as $\frac{1}{2}$ being a number that, when added to itself, gives the number 1, etc.), and the operation of cancelling is, as we have seen, built into the definition. Yet it seems all very formal and we may indeed wonder whether it really captures the intuitive notion of what a fraction is. Although this ubiquitous equivalence class procedure, of which the above illustration is just a particular instance, is very powerful as a pure-mathematical tool for establishing consistency and mathematical existence, it can provide us with very top-heavy-looking entities. It hardly conveys to us the intuitive notion of what $\frac{3}{8}$ is, for example! No wonder my mother's friend was confused.

In my descriptions of mathematical notions, I shall try to avoid, as far as I can, the kind of mathematical pedantry that leads us to define a fraction in terms of an 'infinite class of pairs' even though it certainly has its value in mathematical rigour and precision. In my descriptions here I shall be more concerned with conveying the idea—and the beauty and the magic—inherent in many important mathematical notions. The idea of a fraction such as $\frac{3}{8}$ is simply that it is some kind of an entity which has the property that, when added to itself 8 times in all, gives 3. The magic is that the idea of a fraction actually works despite the fact that we do not really directly experience things in the physical world that are exactly quantified by fractions—pieces of pie leading only to approximations. (This is quite unlike the case of natural numbers, such as 1, 2, 3, which do precisely quantify numerous entities of our direct experience.) One way to see that fractions do make consistent sense is, indeed, to use the 'definition' in terms of infinite collections of pairs of integers (whole numbers), as indicated above. But that does not mean that $\frac{3}{8}$ actually *is* such a collection. It is better to think of $\frac{3}{8}$ as being an entity with some kind of (Platonic) existence of its own, and that the infinite collection of pairs is merely one way of our coming to terms with the consistency of this type of entity. With familiarity, we begin to believe that we can easily grasp a notion like $\frac{3}{8}$ as something that has its own kind of existence, and the idea of an 'infinite collection of pairs' is merely a pedantic device—a device that quickly recedes from our imaginations once we have grasped it. Much of mathematics is like that.

* This is called an 'equivalence class' because it actually is a class of entities (the entities, in this particular case, being pairs of whole numbers), each member of which is deemed to be equivalent, in a specified sense, to each of the other members.

To mathematicians (at least to most of them, as far as I can make out), mathematics is not just a cultural activity that we have ourselves created, but it has a life of its own, and much of it finds an amazing harmony with the physical universe. We cannot get any deep understanding of the laws that govern the physical world without entering the world of mathematics. In particular, the above notion of an equivalence class is relevant not only to a great deal of important (but confusing) mathematics, but a great deal of important (and confusing) physics as well, such as Einstein's general theory of relativity and the 'gauge theory' principles that describe the forces of Nature according to modern particle physics. In modern physics, one cannot avoid facing up to the subtleties of much sophisticated mathematics. It is for this reason that I have spent the first 16 chapters of this work directly on the description of mathematical ideas.

What words of advice can I give to the reader for coping with this? There are four different levels at which this book can be read. Perhaps you are a reader, at one end of the scale, who simply turns off whenever a mathematical formula presents itself (and some such readers may have difficulty with coming to terms with fractions). If so, I believe that there is still a good deal that you can gain from this book by simply skipping all the formulae and just reading the words. I guess this would be much like the way I sometimes used to browse through the chess magazines lying scattered in our home when I was growing up. Chess was a big part of the lives of my brothers and parents, but I took very little interest, except that I enjoyed reading about the exploits of those exceptional and often strange characters who devoted themselves to this game. I gained something from reading about the brilliance of moves that they frequently made, even though I did not understand them, and I made no attempt to follow through the notations for the various positions. Yet I found this to be an enjoyable and illuminating activity that could hold my attention. Likewise, I hope that the mathematical accounts I give here may convey something of interest even to some profoundly non-mathematical readers if they, through bravery or curiosity, choose to join me in my journey of investigation of the mathematical and physical ideas that appear to underlie our physical universe. Do not be afraid to skip equations (I do this frequently myself) and, if you wish, whole chapters or parts of chapters, when they begin to get a mite too turgid! There is a great variety in the difficulty and technicality of the material, and something elsewhere may be more to your liking. You may choose merely to dip in and browse. My hope is that the extensive cross-referencing may sufficiently illuminate unfamiliar notions, so it should be possible to track down needed concepts and notation by turning back to earlier unread sections for clarification.

At a second level, you may be a reader who is prepared to peruse mathematical formulae, whenever such is presented, but you may not have the inclination (or the time) to verify for yourself the assertions that I shall be making. The confirmations of many of these assertions constitute the solutions of the exercises that I have scattered about the mathematical portions of the book. I have indicated three levels of difficulty by the icons –

 very straight forward

 needs a bit of thought

 not to be undertaken lightly.

It is perfectly reasonable to take these on trust, if you wish, and there is no loss of continuity if you choose to take this position.

If, on the other hand, you are a reader who does wish to gain a facility with these various (important) mathematical notions, but for whom the ideas that I am describing are not all familiar, I hope that working through these exercises will provide a significant aid towards accumulating such skills. It is always the case, with mathematics, that a little direct experience of thinking over things on your own can provide a much deeper understanding than merely reading about them. (If you need the solutions, see the website www.roadsolutions.ox.ac.uk.)

Finally, perhaps you are already an expert, in which case you should have no difficulty with the mathematics (most of which will be very familiar to you) and you may have no wish to waste time with the exercises. Yet you may find that there is something to be gained from my own perspective on a number of topics, which are likely to be somewhat different (sometimes very different) from the usual ones. You may have some curiosity as to my opinions relating to a number of modern theories (e.g. supersymmetry, inflationary cosmology, the nature of the Big Bang, black holes, string theory or M-theory, loop variables in quantum gravity, twistor theory, and even the very foundations of quantum theory). No doubt you will find much to disagree with me on many of these topics. But controversy is an important part of the development of science, so I have no regrets about presenting views that may be taken to be partly at odds with some of the mainstream activities of modern theoretical physics.

It may be said that this book is really about the relation between mathematics and physics, and how the interplay between the two strongly influences those drives that underlie our searches for a better theory of the universe. In many modern developments, an essential ingredient of these drives comes from the judgement of mathematical beauty, depth, and sophistication. It is clear that such mathematical influences can be vitally important, as with some of the most impressively successful achievements of 20th-century physics: Dirac's equation for the electron, the general framework of quantum mechanics, and Einstein's general relativity. But in all these cases, physical considerations—ultimately observational ones—have provided the overriding criteria for acceptance. In many of the modern ideas for fundamentally advancing our understanding of the laws of the universe, adequate physical criteria—i.e. experimental data, or even the possibility of experimental investigation—are not available. Thus we may question whether the accessible mathematical desiderata are sufficient to enable us to estimate the chances of success of these ideas. The question is a delicate one, and I shall try to raise issues here that I do not believe have been sufficiently discussed elsewhere.

Although, in places, I shall present opinions that may be regarded as contentious, I have taken pains to make it clear to the reader when I am actually taking such liberties. Accordingly, this book may indeed be used as a genuine guide to the central ideas (and wonders) of modern physics. It is appropriate to use it in educational classes as an honest introduction to modern physics—as that subject is understood, as we move forward into the early years of the third millennium.

Acknowledgements

IT is inevitable, for a book of this length, which has taken me about eight years to complete, that there will be a great many to whom I owe my thanks. It is almost as inevitable that there will be a number among them, whose valuable contributions will go unattributed, owing to congenital disorganization and forgetfulness on my part. Let me first express my special thanks—and also apologies—to such people: who have given me their generous help but whose names do not now come to mind. But for various specific pieces of information and assistance that I can more clearly pinpoint, I thank Michael Atiyah, John Baez, Michael Berry, Dorje Brody, Robert Bryant, Hong-Mo Chan, Joy Christian, Andrew Duggins, Maciej Dunajski, Freeman Dyson, Artur Ekert, David Fowler, Margaret Gleason, Jeremy Gray, Stuart Hameroff, Keith Hannabuss, Lucien Hardy, Jim Hartle, Tom Hawkins, Nigel Hitchin, Andrew Hodges, Dipankar Home, Jim Howie, Chris Isham, Ted Jacobson, Bernard Kay, William Marshall, Lionel Mason, Charles Misner, Tristan Needham, Stelios Negrepontis, Sarah Jones Nelson, Ezra (Ted) Newman, Charles Oakley, Daniel Oi, Robert Osserman, Don Page, Oliver Penrose, Alan Rendall, Wolfgang Rindler, Engelbert Schücking, Bernard Schutz, Joseph Silk, Christoph Simon, George Sparling, John Stachel, Henry Stapp, Richard Thomas, Gerard 't Hooft, Paul Tod, James Vickers, Robert Wald, Rainer Weiss, Ronny Wells, Gerald Westheimer, John Wheeler, Nick Woodhouse, and Anton Zeilinger. Particular thanks go to Lee Smolin, Kelly Stelle, and Lane Hughston for numerous and varied points of assistance. I am especially indebted to Florence Tsou (Sheung Tsun) for immense help on matters of particle physics, to Fay Dowker for her assistance and judgement concerning various matters, most notably the presentation of certain quantum-mechanical issues, to Subir Sarkar for valuable information concerning cosmological data and the interpretation thereof, to Vahe Gurzadyan likewise, and for some advance information about his cosmological findings concerning the overall geometry of the universe, and particularly to Abhay Ashtekar, for his comprehensive information about loop-variable theory and also various detailed matters concerning string theory.

I thank the National Science Foundation for support under grants PHY 93-96246 and 00-90091, and the Leverhulme Foundation for the award of a two-year Leverhulme Emeritus Fellowship, during 2000–2002. Part-time appointments at Gresham College, London (1998–2001) and The Center for Gravitational Physics and Geometry at Penn State University, Pennsylvania, USA have been immensely valuable to me in the writing of this book, as has the secretarial assistance (most particularly Ruth Preston) and office space at the Mathematical Institute, Oxford University.

Special assistance on the editorial side has also been invaluable, under difficult timetabling constraints, and with an author of erratic working habits. Eddie Mizzi's early editorial help was vital in initiating the process of converting my chaotic writings into an actual book, and Richard Lawrence, with his expert efficiency and his patient, sensitive persistence, has been a crucial factor in bringing this project to completion. Having to fit in with such complicated reworking, John Holmes has done sterling work in providing a fine index. And I am particularly grateful to William Shaw for coming to our assistance at a late stage to produce excellent computer graphics (Figs. 1.2 and 2.19, and the implementation of the transformation involved in Figs. 2.16 and 2.19), used here for the Mandelbrot set and the hyperbolic plane. But all the thanks that I can give to Jacob Foster, for his Herculean achievement in sorting out and obtaining references for me and for checking over the entire manuscript in a remarkably brief time and filling in innumerable holes, can in no way do justice to the magnitude of his assistance. His personal imprint on a huge number of the end-notes gives those a special quality. Of course, none of the people I thank here are to blame for the errors and omissions that remain, the sole responsibility for that lying with me.

Special gratitude is expressed to The M.C. Escher Company, Holland for permission to reproduce Escher works in Figs. 2.11, 2.12, 2.16, and 2.22, and particularly to allow the modifications of Fig. 2.11 that are used in Figs. 2.12 and 2.16, the latter being an explicit mathematical transformation. All the Escher works used in this book are copyright (2004) The M.C. Escher Company. Thanks go also to the Institute of Theoretical Physics, University of Heidelberg and to Charles H. Lineweaver for permission to reproduce the respective graphs in Figs. 27.19 and 28.19.

Finally, my unbounded gratitude goes to my beloved wife Vanessa, not merely for supplying computer graphics for me on instant demand (Figs. 4.1, 4.2, 5.7, 6.2–6.8, 8.15, 9.1, 9.2, 9.8, 9.12, 21.3b, 21.10, 27.5, 27.14, 27.15, and the polyhedra in Fig. 1.1), but for her continued love and care, and her deep understanding and sensitivity, despite the seemingly endless years of having a husband who is mentally only half present. And Max, also, who in his entire life has had the chance to know me only in such a distracted state, gets my warmest gratitude—not just for slowing down the writing of this book (so that it could stretch its life, so as to contain at least two important pieces of information that it would not have done otherwise)—but for the continual good cheer and optimism that he exudes, which has helped to keep me going in good spirits. After all, it is through the renewal of life, such as he himself represents, that the new sources of ideas and insights needed for genuine future progress will come, in the search for those deeper laws that *actually* govern the universe in which we live.

Notation

(Not to be read until you are familiar with the concepts, but perhaps find the fonts confusing!)

I have tried to be reasonably consistent in the use of particular fonts in this book, but as not all of this is standard, it may be helpful to the reader to have the major usage that I have adopted made explicit.

Italic lightface (Greek or Latin) letters, such as in $w^2, p^n, \log z$, $\cos\theta$, $\mathrm{e}^{\mathrm{i}\theta}$, or e^x are used in the conventional way for mathematical variables which are numerical or scalar quantities; but established numerical constants, such as e, i, or π or established functions such as sin, cos, or log are denoted by upright letters. Standard physical constants such as c, G, h, $\hbar$, g, or k are italic, however.

A vector or tensor quantity, when being thought of in its (abstract) entirety, is denoted by a boldface italic letter, such as $\boldsymbol{R}$ for the Riemann curvature tensor, while its set of components might be written with lightface italic letters (both for the kernel symbol its indices) as R_{abcd}. In accordance with the abstract-index notation, introduced here in §12.8, the quantity R_{abcd} may alternatively stand for the entire tensor $\boldsymbol{R}$, if this interpretation is appropriate, and this should be clear from the text. Abstract linear transformations are kinds of tensors, and boldface italic letters such as $\boldsymbol{T}$ are used for such entities also. The abstract-index form $T^a{}_b$ is also used here for an abstract linear transformation, where appropriate, the staggering of the indices making clear the precise connection with the ordering of matrix multiplication. Thus, the (abstract-)index expression $S^a{}_b T^b{}_c$ stands for the product $\boldsymbol{ST}$ of linear transformations. As with general tensors, the symbols $S^a{}_b$ and $T^b{}_c$ could alternatively (according to context or explicit specification in the text) stand for the corresponding arrays of components—these being *matrices*—for which the corresponding bold upright letters **S** and **T** can also be used. In that case, **ST** denotes the corresponding matrix product. This 'ambivalent' interpretation of symbols such as R_{abcd} or $S^a{}_b$ (either standing for the array of components or for the abstract tensor itself) should not cause confusion, as the algebraic (or differential) relations that these symbols are subject to are identical for both interpretations. A third notation for such quantities—the *diagrammatic* notation—is also sometimes used here, and is described in Figs. 12.17, 12.18, 14.6, 14.7, 14.21, 19.1 and elsewhere in the book.

There are places in this book where I need to distinguish the 4-dimensional spacetime entities of relativity theory from the corresponding ordinary 3-dimensional purely spatial entities. Thus, while a boldface italic notation might be used, as above, such as $\boldsymbol{p}$ or $\boldsymbol{x}$, for the 4-momentum or 4-position, respectively, the corresponding 3-dimensional purely spatial entities would be denoted by the corresponding upright bold letters $\mathbf{p}$ or $\mathbf{x}$. By analogy with the notation $\mathbf{T}$ for a matrix, above, as opposed to $\boldsymbol{T}$ for an abstract linear transformation, the quantities $\mathbf{p}$ and $\mathbf{x}$ would tend to be thought of as 'standing for' the three spatial components, in each case, whereas $\boldsymbol{p}$ and $\boldsymbol{x}$ might be viewed as having a more abstract component-free interpretation (although I shall not be particularly strict about this). The Euclidean 'length' of a 3-vector quantity $\mathbf{a} = (a_1,a_2,a_3)$ may be written a, where $a^2 = a_1^2 + a_2^2 + a_3^2$, and the scalar product of $\mathbf{a}$ with $\mathbf{b} = (b_1,b_2,b_3)$, written $\mathbf{a} \bullet \mathbf{b} = a_1b_1 + a_2b_2 + a_3b_3$. This 'dot' notation for scalar products applies also in the general n-dimensional context, for the scalar (or inner) product $\boldsymbol{\alpha} \bullet \boldsymbol{\xi}$ of an abstract covector $\boldsymbol{\alpha}$ with a vector $\boldsymbol{\xi}$.

A notational complication arises with quantum mechanics, however, since physical quantities, in that subject, tend to be represented as linear operators. I do not adopt what is a quite standard procedure in this context, of putting 'hats' (circumflexes) on the letters representing the quantum-operator versions of the familiar classical quantities, as I believe that this leads to an unnecessary cluttering of symbols. (Instead, I shall tend to adopt a philosophical standpoint that the classical and quantum entities are really the 'same'—and so it is fair to use the same symbols for each—except that in the classical case one is justified in ignoring quantities of the order of $\hbar$, so that the classical commutation properties $ab = ba$ can hold, whereas in quantum mechanics, ab might differ from ba by something of order $\hbar$.) For consistency with the above, such linear operators would seem to have to be denoted by italic bold letters (like $\boldsymbol{T}$), but that would nullify the philosophy and the distinctions called for in the preceding paragraph. Accordingly, with regard to specific quantities, such as the momentum $\mathbf{p}$ or $\boldsymbol{p}$, or the position $\mathbf{x}$ or $\boldsymbol{x}$, I shall tend to use the same notation as in the classical case, in line with what has been said earlier in this paragraph. But for less specific quantum operators, bold italic letters such as $\boldsymbol{Q}$ will tend to be used.

The shell letters $\mathbb{N}$, $\mathbb{Z}$, $\mathbb{R}$, $\mathbb{C}$, and $\mathbb{F}_q$, respectively, for the system of natural numbers (i.e. non-negative integers), integers, real numbers, complex numbers, and the finite field with q elements (q being some power of a prime number, see §16.1), are now standard in mathematics, as are the corresponding $\mathbb{N}^n$, $\mathbb{Z}^n$, $\mathbb{R}^n$, $\mathbb{C}^n$, $\mathbb{F}_q^n$, for the systems of ordered n-tuples of such numbers. These are canonical mathematical entities in standard use. In this book (as is not all that uncommon), this notation is extended to some other standard mathematical structures such as Euclidean 3-space $\mathbb{E}^3$ or, more generally, Euclidean n-space $\mathbb{E}^n$. In frequent use in this book is the standard flat 4-dimensional Minkowski spacetime, which is itself a kind of 'pseudo-' Euclidean space, so I use the shell letter $\mathbb{M}$ for this space (with $\mathbb{M}^n$ to denote the n-dimensional version—a 'Lorentzian' spacetime with 1 time and $(n-1)$ space dimensions). Sometimes I use $\mathbb{C}$ as an adjective, to denote 'complexified', so that we might consider the complex Euclidean 4-space, for example, denoted by $\mathbb{C}\mathbb{E}^n$. The shell letter $\mathbb{P}$ can also be used as an adjective, to denote 'projective' (see §15.6), or as a noun, with $\mathbb{P}^n$ denoting projective n-space (or I use $\mathbb{R}\mathbb{P}^n$ or $\mathbb{C}\mathbb{P}^n$ if it is to be made clear that we are concerned with real or complex projective n-space, respectively). In twistor theory (Chapter 33), there is the complex 4-space $\mathbb{T}$, which is related to $\mathbb{M}$ (or its complexification $\mathbb{C}\mathbb{M}$) in a canonical way, and there is also the projective version $\mathbb{P}\mathbb{T}$. In this theory, there is also a space $\mathbb{N}$ of *null* twistors (the double duty that this letter serves causing no conflict here), and its projective version $\mathbb{P}\mathbb{N}$.

The adjectival role of the shell letter $\mathbb{C}$ should not be confused with that of the lightface sans serif C, which here stands for 'complex conjugate of' (as used in §13.1,2). This is basically similar to another use of C in particle physics, namely *charge conjugation*, which is the operation which interchanges each particle with its antiparticle (see Chapters 25, 30). This operation is usually considered in conjunction with two other basic particle-physics operations, namely P for *parity* which refers to the operation of reflection in a mirror, and T, which refers to *time-reversal*. Sans serif letters which are bold serve a different purpose here, labelling *vector spaces*, the letters **V**, **W**, and **H**, being most frequently used for this purpose. The use of **H**, is specific to the Hilbert spaces of quantum mechanics, and $\mathbf{H}^n$ would stand for a Hilbert space of n complex dimensions. Vector spaces are, in a clear sense, flat. Spaces which are (or could be) *curved* are denoted by script letters, such as $\mathcal{M}$, $\mathcal{S}$, or $\mathcal{T}$, where there is a special use for the particular script font $\mathscr{I}$ to denote *null infinity*. In addition, I follow a fairly common convention to use script letters for Lagrangians ($\mathcal{L}$) and Hamiltonians ($\mathcal{H}$), in view of their very special status in physical theory.

Prologue

Am-tep was the King's chief craftsman, an artist of consummate skills. It was night, and he lay sleeping on his workshop couch, tired after a handsomely productive evening's work. But his sleep was restless—perhaps from an intangible tension that had seemed to be in the air. Indeed, he was not certain that he was asleep at all when it happened. Daytime had come—quite suddenly—when his bones told him that surely it must still be night.

He stood up abruptly. Something was odd. The dawn's light could not be in the north; yet the red light shone alarmingly through his broad window that looked out northwards over the sea. He moved to the window and stared out, incredulous in amazement. The Sun had never before risen in the north! In his dazed state, it took him a few moments to realize that this could not possibly be the Sun. It was a distant shaft of a deep fiery red light that beamed vertically upwards from the water into the heavens.

As he stood there, a dark cloud became apparent at the head of the beam, giving the whole structure the appearance of a distant giant parasol, glowing evilly, with a smoky flaming staff. The parasol's hood began to spread and darken—a daemon from the underworld. The night had been clear, but now the stars disappeared one by one, swallowed up behind this advancing monstrous creature from Hell.

Though terror must have been his natural reaction, he did not move, transfixed for several minutes by the scene's perfect symmetry and awesome beauty. But then the terrible cloud began to bend slightly to the east, caught up by the prevailing winds. Perhaps he gained some comfort from this and the spell was momentarily broken. But apprehension at once returned to him as he seemed to sense a strange disturbance in the ground beneath, accompanied by ominous-sounding rumblings of a nature quite unfamiliar to him. He began to wonder what it was that could have caused this fury. Never before had he witnessed a God's anger of such magnitude.

His first reaction was to blame himself for the design on the sacrificial cup that he had just completed—he had worried about it at the time. Had his depiction of the Bull-God not been sufficiently fearsome? Had that god been offended? But the absurdity of this thought soon struck him. The fury he had just witnessed could not have been the result of such a trivial action, and was surely not aimed at him specifically. But he knew that there would be trouble at the Great Palace. The Priest-King would waste no time in attempting to appease this Daemon-God. There would be sacrifices. The traditional offerings of fruits or even animals would not suffice to pacify an anger of this magnitude. The sacrifices would have to be human.

Quite suddenly, and to his utter surprise, he was blown backwards across the room by an impulsive blast of air followed by a violent wind. The noise was so extreme that he was momentarily deafened. Many of his beautifully adorned pots were whisked from their shelves and smashed to pieces against the wall behind. As he lay on the floor in a far corner of the room where he had been swept away by the blast, he began to recover his senses, and saw that the room was in turmoil. He was horrified to see one of his favourite great urns shattered to small pieces, and the wonderfully detailed designs, which he had so carefully crafted, reduced to nothing.

Am-tep arose unsteadily from the floor and after a while again approached the window, this time with considerable trepidation, to re-examine that terrible scene across the sea. Now he thought he saw a disturbance, illuminated by that far-off furnace, coming towards him. This appeared to be a vast trough in the water, moving rapidly towards the shore, followed by a clifflike wall of wave. He again became transfixed, watching the approaching wave begin to acquire gigantic proportions. Eventually the disturbance reached the shore and the sea immediately before him drained away, leaving many ships stranded on the newly formed beach. Then the cliff-wave entered the vacated region and struck with a terrible violence. Without exception the ships were shattered, and many nearby houses instantly destroyed. Though the water rose to great heights in the air before him, his own house was spared, for it sat on high ground a good way from the sea.

The Great Palace too was spared. But Am-tep feared that worse might come, and he was right—though he knew not how right he was. He did know, however, that no ordinary human sacrifice of a slave could now be sufficient. Something more would be needed to pacify the tempestuous anger of this terrible God. His thoughts turned to his sons and daughters, and to his newly born grandson. Even they might not be safe.

Am-tep had been right to fear new human sacrifices. A young girl and a youth of good birth had been soon apprehended and taken to a nearby temple, high on the slopes of a mountain. The ensuing ritual was well under way when yet another catastrophe struck. The ground shook with devastating violence, whence the temple roof fell in, instantly killing all the priests and their intended sacrificial victims. As it happened, they would lie there in mid-ritual—entombed for over three-and-a-half millennia!

The devastation was frightful, but not final. Many on the island where Am-tep and his people lived survived the terrible earthquake, though the Great Palace was itself almost totally destroyed. Much would be rebuilt over the years. Even the Palace would recover much of its original splendour, constructed on the ruins of the old. Yet Am-tep had vowed to leave the island. His world had now changed irreparably.

In the world he knew, there had been a thousand years of peace, prosperity, and culture where the Earth-Goddess had reigned. Wonderful art had been allowed to flourish. There was much trade with neighbouring lands. The magnificent Great Palace was a huge luxurious labyrinth, a virtual city in itself, adorned by superb frescoes of animals and flowers. There was running water, excellent drainage, and flushed sewers. War was almost unknown and defences unnecessary. Now, Am-tep perceived the Earth-Goddess overthrown by a Being with entirely different values.

It was some years before Am-tep actually left the island, accompanied by his surviving family, on a ship rebuilt by his youngest son, who was a skilled carpenter and seaman. Am-tep's grandson had developed into an alert child, with an interest in everything in the world around. The voyage took some days, but the weather had been supremely calm. One clear night, Am-tep was explaining to his grandson about the patterns in the stars, when an odd thought overtook him: *The patterns of stars had been disturbed not one iota from what they were before the Catastrophe of the emergence of the terrible Daemon.*

Am-tep knew these patterns well, for he had a keen artist's eye. Surely, he thought, those tiny candles of light in the sky should have been blown at least a little from their positions by the violence of that night, just as his pots had been smashed and his great urn shattered. The Moon also had kept her face, just as before, and her route across the star-filled heavens had changed not one whit, as far as Am-tep could tell. For many moons after the Catastrophe, the skies had appeared different. There had been darkness and strange clouds, and the Moon and Sun had sometimes worn unusual colours. But this had now passed, and their motions seemed utterly undisturbed. The tiny stars, likewise, had been quite unmoved.

If the heavens had shown such little concern for the Catastrophe, having a stature far greater even than that terrible Daemon, Am-tep reasoned, why should the forces controlling the Daemon itself show concern for what the little people on the island had been doing, with their foolish rituals and human sacrifice? He felt embarrassed by his *own* foolish thoughts at the time, that the Daemon might be concerned by the mere patterns on his pots.

Yet Am-tep was still troubled by the question 'why?' What deep forces control the behaviour of the world, and why do they sometimes burst forth in violent and seemingly incomprehensible ways? He shared his questions with his grandson, but there were no answers.

...

A century passed by, and then a millennium, and still there were no answers.

...

Amphos the craftsman had lived all his life in the same small town as his father and his father's father before him, and his father's father's father before that. He made his living constructing beautifully decorated gold bracelets, earrings, ceremonial cups, and other fine products of his artistic skills. Such work had been the family trade for some forty generations—a line unbroken since Am-tep had settled there eleven hundred years before.

But it was not just artistic skills that had been passed down from generation to generation. Am-tep's questions troubled Amphos just as they had troubled Am-tep earlier. The great story of the Catastrophe that destroyed an ancient peaceful civilization had been handed down from father to son. Am-tep's perception of the Catastrophe had also survived with his descendants. Amphos, too, understood that the heavens had a magnitude and stature so great as to be quite unconcerned by that terrible event. Nevertheless, the event had had a catastrophic effect on the little people with their cities and their human sacrifices and insignificant religious rituals. Thus, by comparison, the event itself must have been the result of enormous forces quite unconcerned by those trivial actions of human beings. Yet the nature of those forces was as unknown in Amphos's day as it was to Am-tep.

Amphos had studied the structure of plants, insects and other small animals, and crystalline rocks. His keen eye for observation had served him well in his decorative designs. He took an interest in agriculture and was fascinated by the growth of wheat and other plants from grain. But none of this told him 'why?', and he felt unsatisfied. He believed that there was indeed reason underlying Nature's patterns, but he was in no way equipped to unravel those reasons.

One clear night, Amphos looked up at the heavens, and tried to make out from the patterns of stars the shapes of those heroes and heroines who formed constellations in the sky. To his humble artist's eye, those shapes made poor resemblances. He could himself have arranged the stars far more convincingly. He puzzled over why the Gods had not organized the stars in a more appropriate way? As they were, the arrangements seemed more like scattered grains randomly sowed by a farmer, rather than the deliberate design of a god. Then an odd thought overtook him: *Do not seek for reasons in the specific patterns of stars, or of other scattered arrangements of objects; look, instead, for a deeper universal order in the way that things behave.*

Amphos reasoned that we find order, after all, not in the patterns that scattered seeds form when they fall to the ground, but in the miraculous way that each of those seeds develops into a living plant having a superb structure, similar in great detail to one another. We would not try to seek the meaning in the precise arrangement of seeds sprinkled on the soil; yet, there must be meaning in the hidden mystery of the inner forces controlling the growth of each seed individually, so that each one follows essentially the same wonderful course. Nature's laws must indeed have a superbly organized precision for this to be possible.

Amphos became convinced that without precision in the underlying laws, there could be no order in the world, whereas much order is indeed perceived in the way that things behave. Moreover, there must be precision in our ways of thinking about these matters if we are not to be led seriously astray.

It so happened that word had reached Amphos of a sage who lived in another part of the land, and whose beliefs appeared to be in sympathy with those of Amphos. According to this sage, one could not rely on the teachings and traditions of the past. To be certain of one's beliefs, it was necessary to form precise conclusions by the use of unchallengeable reason. The nature of this precision had to be mathematical—ultimately dependent on the notion of *number* and its application to geometric forms. Accordingly, it must be number and geometry, not myth and superstition, that governed the behaviour of the world.

As Am-tep had done a century and a millennium before, Amphos took to the sea. He found his way to the city of Croton, where the sage and his brotherhood of 571 wise men and 28 wise women were in search of truth. After some time, Amphos was accepted into the brotherhood. The name of the sage was *Pythagoras*.

1
The roots of science

1.1 The quest for the forces that shape the world

WHAT laws govern our universe? How shall we know them? How may this knowledge help us to comprehend the world and hence guide its actions to our advantage?

Since the dawn of humanity, people have been deeply concerned by questions like these. At first, they had tried to make sense of those influences that do control the world by referring to the kind of understanding that was available from their own lives. They had imagined that whatever or whoever it was that controlled their surroundings would do so as they would themselves strive to control things: originally they had considered their destiny to be under the influence of beings acting very much in accordance with their own various familiar human drives. Such driving forces might be pride, love, ambition, anger, fear, revenge, passion, retribution, loyalty, or artistry. Accordingly, the course of natural events—such as sunshine, rain, storms, famine, illness, or pestilence—was to be understood in terms of the whims of gods or goddesses motivated by such human urges. And the only action perceived as influencing these events would be appeasement of the god-figures.

But gradually patterns of a different kind began to establish their reliability. The precision of the Sun's motion through the sky and its clear relation to the alternation of day with night provided the most obvious example; but also the Sun's positioning in relation to the heavenly orb of stars was seen to be closely associated with the change and relentless regularity of the seasons, and with the attendant clear-cut influence on the weather, and consequently on vegetation and animal behaviour. The motion of the Moon, also, appeared to be tightly controlled, and its phases determined by its geometrical relation to the Sun. At those locations on Earth where open oceans meet land, the tides were noticed to have a regularity closely governed by the position (and phase) of the Moon. Eventually, even the much more complicated apparent motions of the planets began to yield up their secrets, revealing an immense underlying precision and regularity. If the heavens were indeed controlled by the whims of gods, then these gods themselves seemed under the spell of exact mathematical laws.

Likewise, the laws controlling earthly phenomena—such as the daily and yearly changes in temperature, the ebb and flow of the oceans, and the growth of plants—being seen to be influenced by the heavens in this respect at least, shared the mathematical regularity that appeared to guide the gods. But this kind of relationship between heavenly bodies and earthly behaviour would sometimes be exaggerated or misunderstood and would assume an inappropriate importance, leading to the occult and mystical connotations of astrology. It took many centuries before the rigour of scientific understanding enabled the true influences of the heavens to be disentangled from purely suppositional and mystical ones. Yet it had been clear from the earliest times that such influences did indeed exist and that, accordingly, the mathematical laws of the heavens must have relevance also here on Earth.

Seemingly independently of this, there were perceived to be other regularities in the behaviour of earthly objects. One of these was the tendency for all things in one vicinity to move in the same downward direction, according to the influence that we now call *gravity*. Matter was observed to transform, sometimes, from one form into another, such as with the melting of ice or the dissolving of salt, but the total quantity of that matter appeared never to change, which reflects the law that we now refer to as *conservation of mass*. In addition, it was noticed that there are many material bodies with the important property that they retain their shapes, whence the idea of rigid spatial motion arose; and it became possible to understand spatial relationships in terms of a precise, well-defined geometry—the 3-dimensional geometry that we now call *Euclidean*. Moreover, the notion of a 'straight line' in this geometry turned out to be the same as that provided by rays of light (or lines of sight). There was a remarkable precision and beauty to these ideas, which held a considerable fascination for the ancients, just as it does for us today.

Yet, with regard to our everyday lives, the implications of this mathematical precision for the actions of the world often appeared unexciting and limited, despite the fact that the mathematics itself seemed to represent a deep truth. Accordingly, many people in ancient times would allow their imaginations to be carried away by their fascination with the subject and to take them far beyond the scope of what was appropriate. In astrology, for example, geometrical figures also often engendered mystical and occult connotations, such as with the supposed magical powers of pentagrams and heptagrams. And there was an entirely suppositional attempted association between Platonic solids and the basic elementary states of matter (see Fig. 1.1). It would not be for many centuries that the deeper understanding that we presently have, concerning the actual

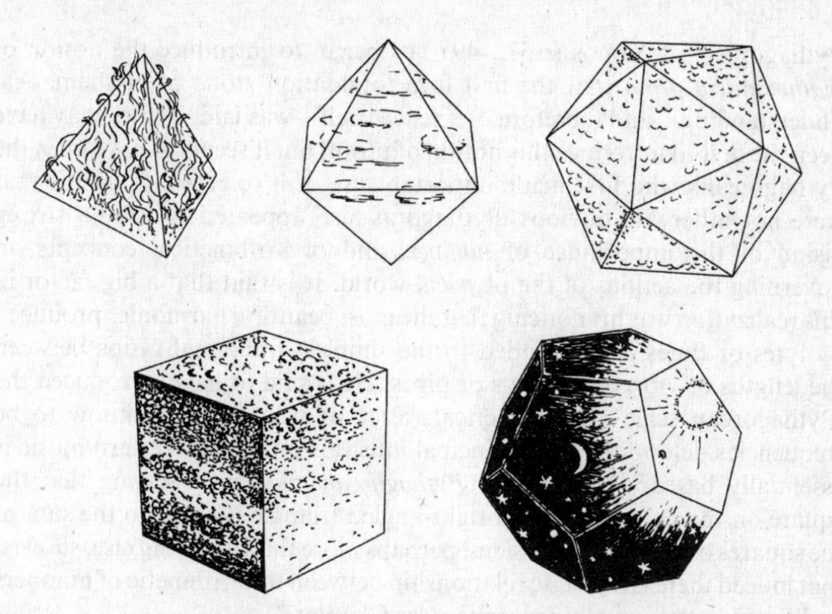

Fig. 1.1 A fanciful association, made by the ancient Greeks, between the five Platonic solids and the four 'elements' (fire, air, water, and earth), together with the heavenly firmament represented by the dodecahedron.

relationships between mass, gravity, geometry, planetary motion, and the behaviour of light, could come about.

1.2 Mathematical truth

The first steps towards an understanding of the real influences controlling Nature required a disentangling of the true from the purely suppositional. But the ancients needed to achieve something else first, before they would be in any position to do this reliably for their understanding of Nature. What they had to do first was to discover how to disentangle the true from the suppositional in *mathematics*. A procedure was required for telling whether a given mathematical assertion is or is not to be trusted as true. Until that preliminary issue could be settled in a reasonable way, there would be little hope of seriously addressing those more difficult problems concerning forces that control the behaviour of the world and whatever their relations might be to mathematical truth. This realization that the key to the understanding of Nature lay within an unassailable mathematics was perhaps the first major breakthrough in science.

Although mathematical truths of various kinds had been surmised since ancient Egyptian and Babylonian times, it was not until the great Greek philosophers Thales of Miletus (*c*.625–547 BC) and

Pythagoras[1*] of Samos (*c.*572–497 BC) began to introduce the notion of *mathematical proof* that the first firm foundation stone of mathematical understanding—and therefore of science itself—was laid. Thales may have been the first to introduce this notion of proof, but it seems to have been the Pythagoreans who first made important use of it to establish things that were not otherwise obvious. Pythagoras also appeared to have a strong vision of the importance of *number*, and of arithmetical concepts, in governing the actions of the physical world. It is said that a big factor in this realization was his noticing that the most beautiful harmonies produced by lyres or flutes corresponded to the simplest fractional ratios between the lengths of vibrating strings or pipes. He is said to have introduced the 'Pythagorean scale', the numerical ratios of what we now know to be frequencies determining the principal intervals on which Western music is essentially based.[2] The famous *Pythagorean theorem*, asserting that the square on the hypotenuse of a right-angled triangle is equal to the sum of the squares on the other two sides, perhaps more than anything else, showed that indeed there is a precise relationship between the arithmetic of numbers and the geometry of physical space (see Chapter 2).

He had a considerable band of followers—the *Pythagoreans*—situated in the city of Croton, in what is now southern Italy, but their influence on the outside world was hindered by the fact that the members of the Pythagorean brotherhood were all sworn to secrecy. Accordingly, almost all of their detailed conclusions have been lost. Nonetheless, some of these conclusions were leaked out, with unfortunate consequences for the 'moles'—on at least one occasion, death by drowning!

In the long run, the influence of the Pythagoreans on the progress of human thought has been enormous. For the first time, with mathematical proof, it was possible to make significant assertions of an unassailable nature, so that they would hold just as true even today as at the time that they were made, no matter how our knowledge of the world has progressed since then. The truly timeless nature of mathematics was beginning to be revealed.

But what is a mathematical proof? A proof, in mathematics, is an impeccable argument, using only the methods of pure logical reasoning, which enables one to infer the validity of a given mathematical assertion from the pre-established validity of other mathematical assertions, or from some particular primitive assertions—the *axioms*—whose validity is taken to be self-evident. Once such a mathematical assertion has been established in this way, it is referred to as a *theorem*.

Many of the theorems that the Pythagoreans were concerned with were geometrical in nature; others were assertions simply about numbers. Those

*Notes, indicated in the text by superscript numbers, are gathered at the ends of the chapter (in this case on p. 23).

that were concerned merely with numbers have a perfectly unambiguous validity today, just as they did in the time of Pythagoras. What about the *geometrical* theorems that the Pythagoreans had obtained using their procedures of mathematical proof? They too have a clear validity today, but now there is a complicating issue. It is an issue whose nature is more obvious to us from our modern vantage point than it was at that time of Pythagoras. The ancients knew of only one kind of geometry, namely that which we now refer to as *Euclidean geometry*, but now we know of many other types. Thus, in considering the geometrical theorems of ancient Greek times, it becomes important to specify that the notion of geometry being referred to is indeed Euclid's geometry. (I shall be more explicit about these issues in §2.4, where an important example of non-Euclidean geometry will be given.)

Euclidean geometry is a specific mathematical structure, with its own specific axioms (including some less assured assertions referred to as postulates), which provided an excellent approximation to a particular aspect of the physical world. That was the aspect of reality, well familiar to the ancient Greeks, which referred to the laws governing the geometry of rigid objects and their relations to other rigid objects, as they are moved around in 3-dimensional space. Certain of these properties were so familiar and self-consistent that they tended to become regarded as 'self-evident' mathematical truths and were taken as axioms (or postulates). As we shall be seeing in Chapters 17–19 and §§27.8,11, Einstein's general relativity—and even the Minkowskian spacetime of special relativity—provide geometries for the physical universe that are different from, and yet more accurate than, the geometry of Euclid, despite the fact that the Euclidean geometry of the ancients was already extraordinarily accurate. Thus, we must be careful, when considering geometrical assertions, whether to trust the 'axioms' as being, in any sense, actually *true*.

But what does 'true' mean, in this context? The difficulty was well appreciated by the great ancient Greek philosopher Plato, who lived in Athens from *c*.429 to 347 BC, about a century and a half after Pythagoras. Plato made it clear that the mathematical propositions—the things that could be regarded as unassailably true—referred not to actual physical objects (like the approximate squares, triangles, circles, spheres, and cubes that might be constructed from marks in the sand, or from wood or stone) but to certain idealized entities. He envisaged that these ideal entities inhabited a different world, distinct from the physical world. Today, we might refer to this world as the *Platonic world of mathematical forms*. Physical structures, such as squares, circles, or triangles cut from papyrus, or marked on a flat surface, or perhaps cubes, tetrahedra, or spheres carved from marble, might conform to these ideals very closely, but only approximately. The actual *mathematical* squares, cubes, circles, spheres, triangles, etc., would not be part of the physical world, but would be inhabitants of Plato's idealized mathematical world of forms.

1.3 Is Plato's mathematical world 'real'?

This was an extraordinary idea for its time, and it has turned out to be a very powerful one. But does the Platonic mathematical world actually exist, in any meaningful sense? Many people, including philosophers, might regard such a 'world' as a complete fiction—a product merely of our unrestrained imaginations. Yet the Platonic viewpoint is indeed an immensely valuable one. It tells us to be careful to distinguish the precise mathematical entities from the approximations that we see around us in the world of physical things. Moreover, it provides us with the blueprint according to which modern science has proceeded ever since. Scientists will put forward models of the world—or, rather, of certain aspects of the world—and these models may be tested against previous observation and against the results of carefully designed experiment. The models are deemed to be appropriate if they survive such rigorous examination and if, in addition, they are internally consistent structures. The important point about these models, for our present discussion, is that they are basically purely abstract *mathematical* models. The very question of the internal consistency of a scientific model, in particular, is one that requires that the model be precisely specified. The required precision demands that the model be a mathematical one, for otherwise one cannot be sure that these questions have well-defined answers.

If the model itself is to be assigned any kind of 'existence', then this existence is located within the Platonic world of mathematical forms. Of course, one might take a contrary viewpoint: namely that the model is itself to have existence only within our various *minds*, rather than to take Plato's world to be in any sense absolute and 'real'. Yet, there is something important to be gained in regarding mathematical structures as having a reality of their own. For our individual minds are notoriously imprecise, unreliable, and inconsistent in their judgements. The precision, reliability, and consistency that are required by our scientific theories demand something beyond any one of our individual (untrustworthy) minds. In mathematics, we find a far greater robustness than can be located in any particular mind. Does this not point to something outside ourselves, with a reality that lies beyond what each individual can achieve?

Nevertheless, one might still take the alternative view that the mathematical world has no independent existence, and consists merely of certain ideas which have been distilled from our various minds and which have been found to be totally trustworthy and are agreed by all.

Yet even this viewpoint seems to leave us far short of what is required. Do we mean 'agreed by all', for example, or 'agreed by those who are in their right minds', or 'agreed by all those who have a Ph.D. in mathematics' (not much use in Plato's day) and who have a right to venture an 'authoritative' opinion? There seems to be a danger of circularity here; for to judge whether or not someone is 'in his or her right mind' requires some external standard. So also does the meaning of 'authoritative', unless some standard of an unscientific nature such as 'majority opinion' were to be adopted (and it should be made clear that majority opinion, no matter how important it may be for democratic government, should in no way be used as the criterion for scientific acceptability). Mathematics itself indeed seems to have a robustness that goes far beyond what any individual mathematician is capable of perceiving. Those who work in this subject, whether they are actively engaged in mathematical research or just using results that have been obtained by others, usually feel that they are merely explorers in a world that lies far beyond themselves—a world which possesses an objectivity that transcends mere opinion, be that opinion their own or the surmise of others, no matter how expert those others might be.

It may be helpful if I put the case for the actual existence of the Platonic world in a different form. What I mean by this 'existence' is really just the objectivity of mathematical truth. Platonic existence, as I see it, refers to the existence of an objective external standard that is not dependent upon our individual opinions nor upon our particular culture. Such 'existence' could also refer to things other than mathematics, such as to morality or aesthetics (cf. §1.5), but I am here concerned just with mathematical objectivity, which seems to be a much clearer issue.

Let me illustrate this issue by considering one famous example of a mathematical truth, and relate it to the question of 'objectivity'. In 1637, Pierre de Fermat made his famous assertion now known as 'Fermat's Last Theorem' (that no positive *n*th power[3] of an integer, i.e. of a whole number, can be the sum of two other positive *n*th powers if n is an integer greater than 2), which he wrote down in the margin of his copy of the *Arithmetica*, a book written by the 3rd-century Greek mathematician Diophantos. In this margin, Fermat also noted: 'I have discovered a truly marvellous proof of this, which this margin is too narrow to contain.' Fermat's mathematical assertion remained unconfirmed for over 350 years, despite concerted efforts by numerous outstanding mathematicians. A proof was finally published in 1995 by Andrew Wiles (depending on the earlier work of various other mathematicians), and this proof has now been accepted as a valid argument by the mathematical community.

Now, do we take the view that Fermat's assertion was always true, long before Fermat actually made it, or is its validity a purely cultural matter, dependent upon whatever might be the subjective standards of the community of human mathematicians? Let us try to suppose that the validity of the Fermat assertion is in fact a subjective matter. Then it would not be an absurdity for some other mathematician X to have come up with an actual and specific counter-example to the Fermat assertion, so long as X had done this before the date of 1995.[4] In such a circumstance, the mathematical community would have to accept the correctness of X's counter-example. From then on, any effort on the part of Wiles to prove the Fermat assertion would have to be fruitless, for the reason that X had got his argument in first and, as a result, the Fermat assertion would now be false! Moreover, we could ask the further question as to whether, consequent upon the correctness of X's forthcoming counter-example, Fermat himself would necessarily have been mistaken in believing in the soundness of his 'truly marvellous proof', at the time that he wrote his marginal note. On the subjective view of mathematical truth, it could possibly have been the case that Fermat had a valid proof (which would have been accepted as such by his peers at the time, had he revealed it) and that it was Fermat's secretiveness that allowed the possibility of X later obtaining a counter-example! I think that virtually all mathematicians, irrespective of their professed attitudes to 'Platonism', would regard such possibilities as patently absurd.

Of course, it might still be the case that Wiles's argument in fact contains an error and that the Fermat assertion is indeed false. Or there could be a fundamental error in Wiles's argument but the Fermat assertion is true nevertheless. Or it might be that Wiles's argument is correct in its essentials while containing 'non-rigorous steps' that would not be up to the standard of some future rules of mathematical acceptability. But these issues do not address the point that I am getting at here. The issue is the objectivity of the Fermat assertion itself, not whether anyone's particular demonstration of it (or of its negation) might happen to be convincing to the mathematical community of any particular time.

It should perhaps be mentioned that, from the point of view of mathematical logic, the Fermat assertion is actually a mathematical statement of a particularly simple kind,[5] whose objectivity is especially apparent. Only a tiny minority[6] of mathematicians would regard the truth of such assertions as being in any way 'subjective'—although there might be some subjectivity about the types of argument that would be regarded as being convincing. However, there are other kinds of mathematical assertion whose truth could plausibly be regarded as being a 'matter of opinion'. Perhaps the best known of such assertions is the *axiom of choice*. It is not important for us, now, to know what the axiom of choice is. (I shall describe it in §16.3.) It is cited here only as an example. Most mathematicians would probably regard the axiom of choice as 'obviously true', while others may regard it as a somewhat questionable assertion which might even be false (and I am myself inclined, to some extent, towards this second viewpoint). Still others would take it as an assertion whose 'truth' is a mere matter of opinion or, rather, as something which can be taken one way or the other, depending upon which system of axioms and rules of procedure (a 'formal system'; see §16.6) one chooses to adhere to. Mathematicians who support this final viewpoint (but who accept the objectivity of the truth of particularly clear-cut mathematical statements, like the Fermat assertion discussed above) would be relatively weak Platonists. Those who adhere to objectivity with regard to the truth of the axiom of choice would be stronger Platonists.

I shall come back to the axiom of choice in §16.3, since it has some relevance to the mathematics underlying the behaviour of the physical world, despite the fact that it is not addressed much in physical theory. For the moment, it will be appropriate not to worry overly about this issue. If the axiom of choice can be settled one way or the other by some appropriate form of unassailable mathematical reasoning,[7] then its truth is indeed an entirely objective matter, and either it belongs to the Platonic world or its negation does, in the sense that I am interpreting this term 'Platonic world'. If the axiom of choice is, on the other hand, a mere matter of opinion or of arbitrary decision, then the Platonic world of absolute mathematical forms contains neither the axiom of choice nor its negation (although it could contain assertions of the form 'such-and-such follows from the axiom of choice' or 'the axiom of choice is a theorem according to the rules of such-and-such mathematical system').

The mathematical assertions that can belong to Plato's world are precisely those that are objectively true. Indeed, I would regard mathematical objectivity as really what mathematical Platonism is all about. To say that some mathematical assertion has a Platonic existence is merely to say that it is true in an objective sense. A similar comment applies to mathematical *notions*—such as the concept of the number 7, for example, or the rule of multiplication of integers, or the idea that some set contains infinitely many elements—all of which have a Platonic existence because they are objective notions. To my way of thinking, Platonic existence is simply a matter of objectivity and, accordingly, should certainly not be viewed as something 'mystical' or 'unscientific', despite the fact that some people regard it that way.

As with the axiom of choice, however, questions as to whether some particular proposal for a mathematical entity is or is not to be regarded as having objective existence can be delicate and sometimes technical. Despite this, we certainly need not be mathematicians to appreciate the general robustness of many mathematical concepts. In Fig. 1.2, I have depicted various small portions of that famous mathematical entity known

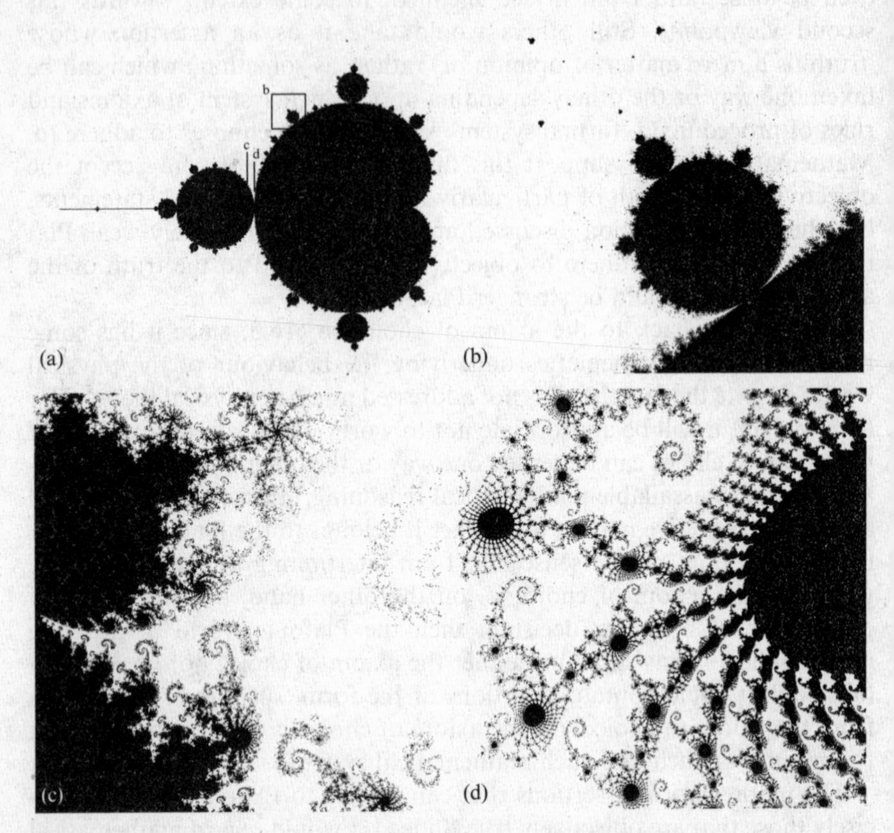

Fig. 1.2 (a) The Mandelbrot set. (b), (c), and (d) Some details, illustrating blow-ups of those regions correspondingly marked in Fig. 1.2a, magnified by respective linear factors 11.6, 168.9, and 1042 (and caps 300, 300, 200, 200; see Note 4.10).

as the *Mandelbrot set*. The set has an extraordinarily elaborate structure, but it is not of any human design. Remarkably, this structure is defined by a mathematical rule of particular simplicity. We shall come to this explicitly in §4.5, but it would distract us from our present purposes if I were to try to provide this rule in detail now.

The point that I wish to make is that no one, not even Benoit Mandelbrot himself when he first caught sight of the incredible complications in the fine details of the set, had any real preconception of the set's extraordinary richness. The Mandelbrot set was certainly no invention of any human mind. The set is just objectively there in the mathematics itself. If it has meaning to assign an actual existence to the Mandelbrot set, then that existence is not within our minds, for no one can fully comprehend the set's endless variety and unlimited complication. Nor can its existence lie within the multitude of computer printouts that begin to capture some of its incredible sophistication and detail, for at best those printouts capture but a shadow of an approximation to the set itself. Yet it has a robustness that is beyond any doubt; for the same structure is revealed—in all its perceivable details, to greater and greater fineness the more closely it is examined—independently of the mathematician or computer that examines it. Its existence can only be within the Platonic world of mathematical forms.

I am aware that there will still be many readers who find difficulty with assigning any kind of actual existence to mathematical structures. Let me make the request of such readers that they merely broaden their notion of what the term 'existence' can mean to them. The mathematical forms of Plato's world clearly do not have the same kind of existence as do ordinary physical objects such as tables and chairs. They do not have spatial locations; nor do they exist in time. Objective mathematical notions must be thought of as timeless entities and are not to be regarded as being conjured into existence at the moment that they are first humanly perceived. The particular swirls of the Mandelbrot set that are depicted in Fig. 1.2c or 1.2d did not attain their existence at the moment that they were first seen on a computer screen or printout. Nor did they come about when the general idea behind the Mandelbrot set was first humanly put forth—not actually first by Mandelbrot, as it happened, but by R. Brooks and J. P. Matelski, in 1981, or perhaps earlier. For certainly neither Brooks nor Matelski, nor initially even Mandelbrot himself, had any real conception of the elaborate detailed designs that we see in Fig. 1.2c and 1.2d. Those designs were already 'in existence' since the beginning of time, in the potential timeless sense that they would necessarily be revealed precisely in the form that we perceive them today, no matter at what time or in what location some perceiving being might have chosen to examine them.

1.4 Three worlds and three deep mysteries

Thus, mathematical existence is different not only from physical existence but also from an existence that is assigned by our mental perceptions. Yet there is a deep and mysterious connection with each of those other two forms of existence: the physical and the mental. In Fig. 1.3, I have schematically indicated all of these three forms of existence—the physical, the mental, and the Platonic mathematical—as entities belonging to three separate 'worlds', drawn schematically as spheres. The mysterious connections between the worlds are also indicated, where in drawing the diagram

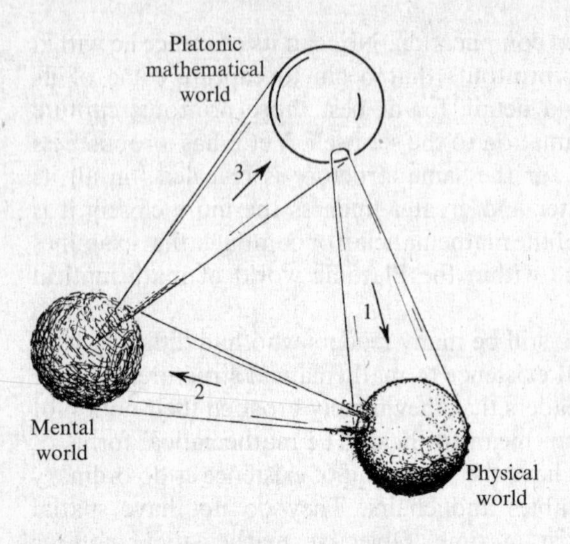

Fig. 1.3 Three 'worlds'—the Platonic mathematical, the physical, and the mental—and the three profound mysteries in the connections between them.

I have imposed upon the reader some of my beliefs, or prejudices, concerning these mysteries.

It may be noted, with regard to the *first* of these mysteries—relating the Platonic mathematical world to the physical world—that I am allowing that only a small part of the world of mathematics need have relevance to the workings of the physical world. It is certainly the case that the vast preponderance of the activities of pure mathematicians today has no obvious connection with physics, nor with any other science (cf. §34.9), although we may be frequently surprised by unexpected important applications. Likewise, in relation to the *second* mystery, whereby mentality comes about in association with certain physical structures (most specifically, healthy, wakeful human brains), I am not insisting that the majority of physical structures need induce mentality. While the brain of a cat may indeed evoke mental qualities, I am not requiring the same for a rock. Finally, for the *third* mystery, I regard it as self-evident that only a small fraction of our mental activity need be concerned with absolute mathematical truth! (More likely we are concerned with the multifarious irritations, pleasures, worries, excitements, and the like, that fill our daily lives.) These three facts are represented in the smallness of the base of the connection of each world with the next, the worlds being taken in a clockwise sense in the diagram. However, it is in the encompassing of each entire world within the scope of its connection with the world preceding it that I am revealing my prejudices.

Thus, according to Fig. 1.3, the entire physical world is depicted as being governed according to mathematical laws. We shall be seeing in later chapters that there is powerful (but incomplete) evidence in support of this contention. On this view, everything in the physical universe is indeed governed in completely precise detail by mathematical principles—perhaps by equations, such as those we shall be learning about in chapters to follow, or perhaps by some future mathematical notions fundamentally different from those which we would today label by the term 'equations'. If this is right, then even our own physical actions would be entirely subject to such ultimate mathematical control, where 'control' might still allow for some random behaviour governed by strict probabilistic principles.

Many people feel uncomfortable with contentions of this kind, and I must confess to having some unease with it myself. Nonetheless, my personal prejudices are indeed to favour a viewpoint of this general nature, since it is hard to see how any line can be drawn to separate physical actions under mathematical control from those which might lie beyond it. In my own view, the unease that many readers may share with me on this issue partly arises from a very limited notion of what 'mathematical control' might entail. Part of the purpose of this book is to touch upon, and to reveal to the reader, some of the extraordinary richness, power, and beauty that can spring forth once the right mathematical notions are hit upon.

In the Mandelbrot set alone, as illustrated in Fig. 1.2, we can begin to catch a glimpse of the scope and beauty inherent in such things. But even these structures inhabit a very limited corner of mathematics as a whole, where behaviour is governed by strict computational control. Beyond this corner is an incredible potential richness. How do I really feel about the possibility that all my actions, and those of my friends, are ultimately governed by mathematical principles of this kind? I can live with that. I would, indeed, prefer to have these actions controlled by something residing in some such aspect of Plato's fabulous mathematical world than to have them be subject to the kind of simplistic base motives, such as pleasure-seeking, personal greed, or aggressive violence, that many would argue to be the implications of a strictly scientific standpoint.

Yet, I can well imagine that a good many readers will still have difficulty in accepting that all actions in the universe could be entirely subject to mathematical laws. Likewise, many might object to two other prejudices of mine that are implicit in Fig. 1.3. They might feel, for example, that I am taking too hard-boiled a scientific attitude by drawing my diagram in a way that implies that all of mentality has its roots in physicality. This is indeed a prejudice, for while it is true that we have no reasonable scientific evidence for the existence of 'minds' that do not have a physical basis, we cannot be completely sure. Moreover, many of a religious persuasion would argue strongly for the possibility of physically independent minds and might appeal to what they regard as powerful evidence of a different kind from that which is revealed by ordinary science.

A further prejudice of mine is reflected in the fact that in Fig. 1.3 I have represented the entire Platonic world to be within the compass of mentality. This is intended to indicate that—at least in principle—there are no mathematical truths that are beyond the scope of reason. Of course, there are mathematical statements (even straightforward arithmetical addition sums) that are so vastly complicated that no one could have the mental fortitude to carry out the necessary reasoning. However, such things would be *potentially* within the scope of (human) mentality and would be consistent with the meaning of Fig. 1.3 as I have intended to represent it. One must, nevertheless, consider that there might be other mathematical statements that lie outside even the potential compass of reason, and these would violate the intention behind Fig. 1.3. (This matter will be considered at greater length in §16.6, where its relation to Gödel's famous incompleteness theorem will be discussed.)[8]

In Fig. 1.4, as a concession to those who do not share all my personal prejudices on these matters, I have redrawn the connections between the three worlds in order to allow for all three of these possible violations of my prejudices. Accordingly, the possibility of physical action beyond the scope of mathematical control is now taken into account. The diagram also allows for the belief that there might be mentality that is not rooted in physical structures. Finally, it permits the existence of true mathematical assertions whose truth is in principle inaccessible to reason and insight.

This extended picture presents further potential mysteries that lie even beyond those which I have allowed for in my own preferred picture of the world, as depicted in Fig. 1.3. In my opinion, the more tightly organized scientific viewpoint of Fig. 1.3 has mysteries enough. These mysteries are not removed by passing to the more relaxed scheme of Fig. 1.4. For it

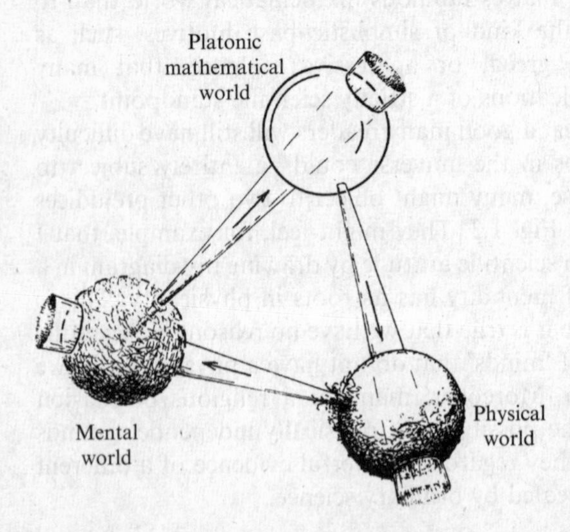

Fig. 1.4 A redrawing of Fig. 1.3 in which violations of three of the prejudices of the author are allowed for.

remains a deep puzzle why mathematical laws should apply to the world with such phenomenal precision. (We shall be glimpsing something of the extraordinary accuracy of the basic physical theories in §19.8, §26.7, and §27.13.) Moreover, it is not just the precision but also the subtle sophistication and mathematical beauty of these successful theories that is profoundly mysterious. There is also an undoubted deep mystery in how it can come to pass that appropriately organized physical material—and here I refer specifically to living human (or animal) brains—can somehow conjure up the mental quality of conscious awareness. Finally, there is also a mystery about how it is that we perceive mathematical truth. It is not just that our brains are programmed to 'calculate' in reliable ways. There is something much more profound than that in the insights that even the humblest among us possess when we appreciate, for example, the actual meanings of the terms 'zero', 'one', 'two', 'three', 'four', etc.[9]

Some of the issues that arise in connection with this third mystery will be our concern in the next chapter (and more explicitly in §§16.5,6) in relation to the notion of *mathematical proof*. But the main thrust of this book has to do with the first of these mysteries: the remarkable relationship between mathematics and the actual behaviour of the physical world. No proper appreciation of the extraordinary power of modern science can be achieved without at least some acquaintance with these mathematical ideas. No doubt, many readers may find themselves daunted by the prospect of having to come to terms with such mathematics in order to arrive at this appreciation. Yet, I have the optimistic belief that they may not find all these things to be so bad as they fear. Moreover, I hope that I may persuade many readers that, despite what she or he may have previously perceived, mathematics can be fun!

I shall not be especially concerned here with the second of the mysteries depicted in Figs. 1.3 and 1.4, namely the issue of how it is that mentality—most particularly conscious awareness—can come about in association with appropriate physical structures (although I shall touch upon this deep question in §34.7). There will be enough to keep us busy in exploring the physical universe and its associated mathematical laws. In addition, the issues concerning mentality are profoundly contentious, and it would distract from the purpose of this book if we were to get embroiled in them. Perhaps one comment will not be amiss here, however. This is that, in my own opinion, there is little chance that any deep understanding of the nature of the mind can come about without our first learning much more about the very basis of physical reality. As will become clear from the discussions that will be presented in later chapters, I believe that major revolutions are required in our physical understanding. Until these revolutions have come to pass, it is, in my view, greatly optimistic to expect that much real progress can be made in understanding the actual nature of mental processes.[10]

1.5 The Good, the True, and the Beautiful

In relation to this, there is a further set of issues raised by Figs. 1.3 and 1.4. I have taken Plato's notion of a 'world of ideal forms' only in the limited sense of mathematical forms. Mathematics is crucially concerned with the particular ideal of *Truth*. Plato himself would have insisted that there are two other fundamental absolute ideals, namely that of the *Beautiful* and of the *Good*. I am not at all averse to admitting to the existence of such ideals, and to allowing the Platonic world to be extended so as to contain absolutes of this nature.

Indeed, we shall later be encountering some of the remarkable interrelations between truth and beauty that both illuminate and confuse the issues of the discovery and acceptance of physical theories (see §§34.2,5,9 particularly; see also Fig. 34.1). Moreover, quite apart from the undoubted (though often ambiguous) role of beauty for the mathematics underlying the workings of the physical world, aesthetic criteria are fundamental to the development of mathematical ideas for their own sake, providing both the drive towards discovery and a powerful guide to truth. I would even surmise that an important element in the mathematician's common conviction that an external Platonic world actually has an existence independent of ourselves comes from the extraordinary unexpected hidden beauty that the ideas themselves so frequently reveal.

Of less obvious relevance here—but of clear importance in the broader context—is the question of an absolute ideal of morality: what is good and what is bad, and how do our minds perceive these values? Morality has a profound connection with the mental world, since it is so intimately related to the values assigned by conscious beings and, more importantly, to the very presence of consciousness itself. It is hard to see what morality might mean in the absence of sentient beings. As science and technology progress, an understanding of the physical circumstances under which mentality is manifested becomes more and more relevant. I believe that it is more important than ever, in today's technological culture, that scientific questions should not be divorced from their moral implications. But these issues would take us too far afield from the immediate scope of this book. We need to address the question of separating true from false before we can adequately attempt to apply such understanding to separate good from bad.

There is, finally, a further mystery concerning Fig. 1.3, which I have left to the last. I have deliberately drawn the figure so as to illustrate a paradox. How can it be that, in accordance with my own prejudices, each world appears to encompass the next one in its entirety? I do not regard this issue as a reason for abandoning my prejudices, but merely for demonstrating the presence of an even deeper mystery that transcends those which I have been pointing to above. There may be a sense in which the three worlds are not separate at all, but merely reflect, individually, aspects of a deeper truth about the world as a whole of which we have little conception at the present time. We have a long way to go before such matters can be properly illuminated.

I have allowed myself to stray too much from the issues that will concern us here. The main purpose of this chapter has been to emphasize the central importance that mathematics has in science, both ancient and modern. Let us now take a glimpse into Plato's world—at least into a relatively small but important part of that world, of particular relevance to the nature of physical reality.

Notes

Section 1.2

1.1. Unfortunately, almost nothing reliable is known about Pythagoras, his life, his followers, or of their work, apart from their very existence and the recognition by Pythagoras of the role of simple ratios in musical harmony. See Burkert (1972). Yet much of great importance is commonly attributed to the Pythagoreans. Accordingly, I shall use the term 'Pythagorean' simply as a label, with no implication intended as to historical accuracy.

1.2. This is the pure 'diatonic scale' in which the frequencies (in inverse proportion to the lengths of the vibrating elements) are in the ratios 24 : 27 : 30 : 32 : 36 : 40 : 45 : 48, giving many instances of simple ratios, which underlie harmonies that are pleasing to the ear. The 'white notes' of a modern piano are tuned (according to a compromise between Pythagorean purity of harmony and the facility of key changes) as approximations to these Pythagorean ratios, according to the *equal temperament* scale, with relative frequencies $1{:}\alpha^2{:}\alpha^4{:}\alpha^5{:}\alpha^7{:}\alpha^9{:}\alpha^{11}{:}\alpha^{12}$, where $\alpha = \sqrt[12]{2} = 1.05946\ldots$ (Note: α^5 means the fifth power of α, i.e. $\alpha \times \alpha \times \alpha \times \alpha \times \alpha$. The quantity $\sqrt[12]{2}$ is the twelfth root of 2, which is the number whose twelfth power is 2, i.e. $2^{1/12}$, so that $\alpha^{12} = 2$. See Note 1.3 and §5.2.)

Section 1.3

1.3. Recall from Note 1.2 that the *n*th power of a number is that number multiplied by itself *n* times. Thus, the third power of 5 is 125, written $5^3 = 125$; the fourth power of 3 is 81, written $3^4 = 81$; etc.

1.4. In fact, while Wiles was trying to fix a 'gap' in his proof of Fermat's Last Theorem which had become apparent after his initial presentation at Cambridge in June 1993, a rumour spread through the mathematical community that the mathematician Noam Elkies had found a counter-example to Fermat's assertion. Earlier, in 1988, Elkies had found a counter-example to Euler's conjecture—that there are no integer solutions to the equation $x^4 + y^4 + z^4 = w^4$—thereby proving it false. It was not implausible, therefore, that he had proved that Fermat's assertion also was false. However, the e-mail that started the rumour was dated 1 April and was revealed to be a spoof perpetrated by Henri Darmon; see Singh (1997), p. 293.

1.5. Technically it is a Π_1-sentence; see §16.6.

1.6. I realize that, in a sense, I am falling into my own trap by making such an assertion. The issue is not really whether the mathematicians taking such an extreme subjective view happen to constitute a tiny minority or not (and I have certainly not conducted a trustworthy survey among mathematicians on this point); the issue is whether such an extreme position is actually to be taken seriously. I leave it to the reader to judge.

1.7. Some readers may be aware of the results of Gödel and Cohen that the axiom of choice is independent of the more basic standard axioms of set theory (the Zermelo–Frankel axiom system). It should be made clear that the Gödel–Cohen argument does not in itself establish that the axiom of choice will never be settled one way or the other. This kind of point is stressed, for example, in the final section of Paul Cohen's book (Cohen 1966, Chap. 14, §13), except that, there, Cohen is more explicitly concerned with the *continuum hypothesis* than the axiom of choice; see §16.5.

Section 1.4

1.8. There is perhaps an irony here that a fully fledged anti-Platonist, who believes that mathematics is 'all in the mind' must also believe—so it seems—that there are no true mathematical statements that are in principle beyond reason. For example, if Fermat's Last Theorem had been inaccessible (in principle) to reason, then this anti-Platonist view would allow no validity either to its truth or to its falsity, such validity coming only through the mental act of perceiving some proof or disproof.

1.9. See e.g. Penrose (1997b).

1.10. My own views on the kind of change in our physical world-view that will be needed in order that conscious mentality may be accommodated are expressed in Penrose (1989, 1994, 1997a,1997b).

2
An ancient theorem and a modern question

2.1 The Pythagorean theorem

LET us consider the issue of geometry. What, indeed, are the different 'kinds of geometry' that were alluded to in the last chapter? To lead up to this issue, we shall return to our encounter with Pythagoras and consider that famous theorem that bears his name:[1] for any right-angled triangle, the square of the length of the hypotenuse (the side opposite the right angle) is equal to the sum of the squares of the lengths of the other two sides (Fig. 2.1). What reasons do we have for believing that this assertion is true? How, indeed, do we 'prove' the Pythagorean theorem? Many arguments are known. I wish to consider two such, chosen for their particular transparency, each of which has a different emphasis.

For the first, consider the pattern illustrated in Fig. 2.2. It is composed entirely of squares of two different sizes. It may be regarded as 'obvious' that this pattern can be continued indefinitely and that the entire plane is thereby covered in this regular repeating way, without gaps or overlaps, by squares of these two sizes. The repeating nature of this pattern is made manifest by the fact that if we mark the centres of the larger squares, they form the vertices of another system of squares, of a somewhat greater size than either, but tilted at an angle to the original ones (Fig. 2.3) and which alone will cover the entire plane. Each of these tilted squares is marked in exactly the same way, so that the markings on these squares fit together to

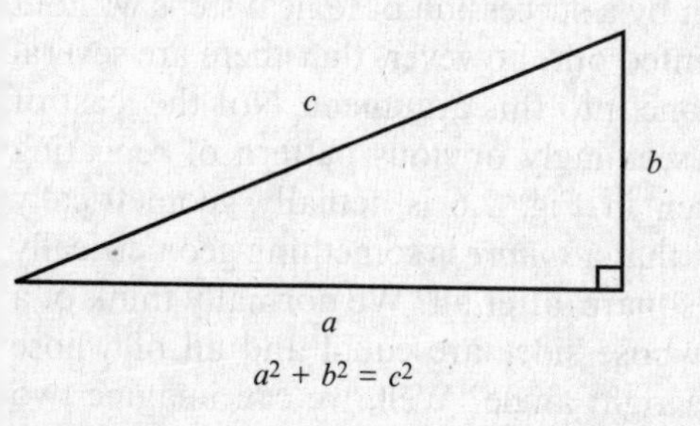

Fig. 2.1 The Pythagorean theorem: for any right-angled triangle, the squared length of the hypotenuse c is the sum of the squared lengths of the other two sides a and b.

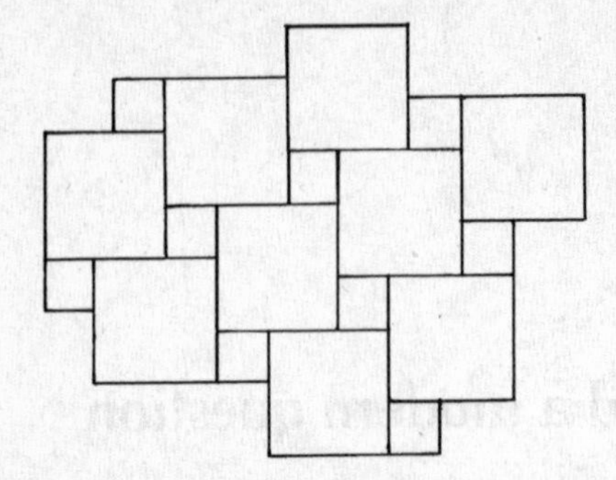

Fig. 2.2 A tessellation of the plane by squares of two different sizes.

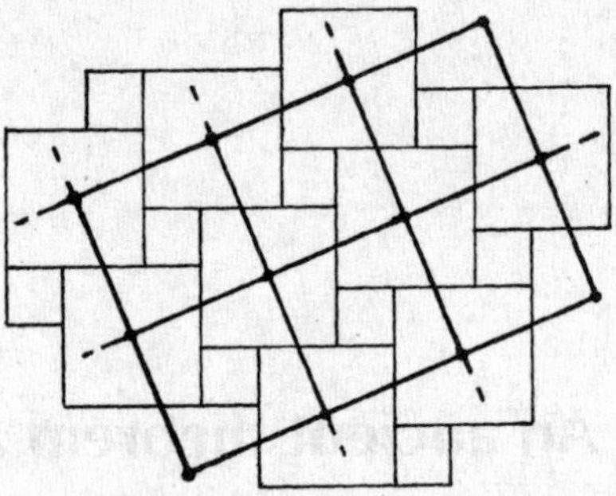

Fig. 2.3 The centres of the (say) larger squares form the vertices of a lattice of still larger squares, tilted at an angle.

form the original two-square pattern. The same would apply if, instead of taking the centres of the larger of the two squares of the original pattern, we chose any other point, together with its set of corresponding points throughout the pattern. The new pattern of tilted squares is just the same as before but moved along without rotation—i.e. by means of a motion referred to as a *translation*. For simplicity, we can now choose our starting point to be one of the corners in the original pattern (see Fig. 2.4).

It should be clear that the area of the tilted square must be equal to the sum of the areas of the two smaller squares—indeed the pieces into which the markings would subdivide this larger square can, for any starting point for the tilted squares, be moved around, without rotation, until they fit together to make the two smaller squares (e.g. Fig. 2.5). Moreover, it is evident from Fig. 2.4 that the edge-length of the large tilted square is the hypotenuse of a right-angled triangle whose two other sides have lengths equal to those of the two smaller squares. We have thus established the Pythagorean theorem: the square on the hypotenuse is equal to the sum of the squares on the other two sides.

The above argument does indeed provide the essentials of a simple proof of this theorem, and, moreover, it gives us some 'reason' for believing that the theorem has to be true, which might not be so obviously the case with some more formal argument given by a succession of logical steps without clear motivation. It should be pointed out, however, that there are several implicit assumptions that have gone into this argument. Not the least of these is the assumption that the seemingly obvious pattern of repeating squares shown in Fig. 2.2 or even in Fig. 2.6 is actually geometrically possible—or even, more critically, that a *square* is something geometrically possible! What do we mean by a 'square' after all? We normally think of a square as a plane figure, all of whose sides are equal and all of whose angles are right angles. What is a right angle? Well, we can imagine two

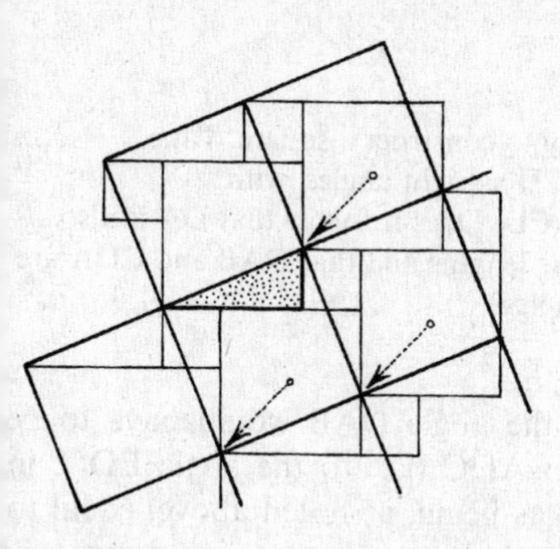

Fig. 2.4 The lattice of tilted squares can be shifted by a translation, here so that the vertices of the tilted lattice lie on vertices of the original two-square lattice, showing that the side-length of a tilted square is the hypotenuse of a right-angled triangle (shown shaded) whose other two side-lengths are those of the original two squares.

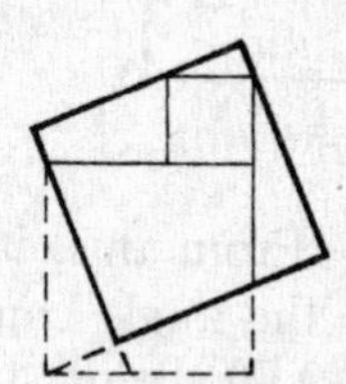

Fig. 2.5 For any particular starting point for the tilted square, such as that depicted, the tilted square is divided into pieces that fit together to make the two smaller squares.

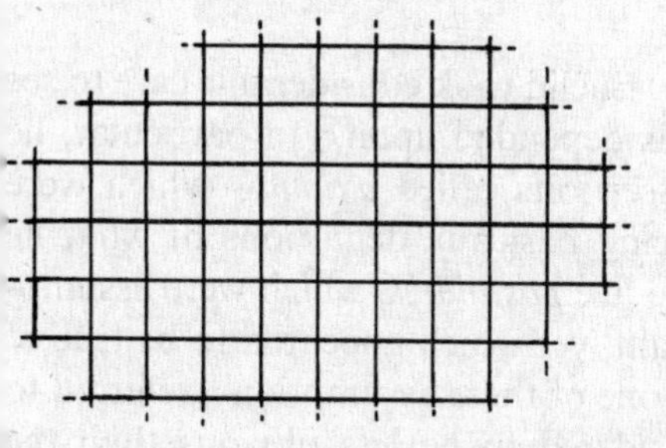

Fig. 2.6 The familiar lattice of equal squares. How do we know it exists?

straight lines crossing each other at some point, making four angles that are all equal. Each of these equal angles is then a right angle.

Let us now try to construct a square. Take three equal line segments AB, BC, and CD, where ABC and BCD are right angles, D and A being on the same side of the line BC, as in Fig. 2.7. The question arises: is AD the same length as the other three segments? Moreover, are the angles DAB and CDA also right angles? These angles should be equal to one another by a left–right symmetry in the figure, but are they actually right angles? This only seems obvious because of our familiarity with squares, or perhaps because we can recall from our schooldays some statement of Euclid that can be used to tell us that the sides BA and CD would have to be 'parallel' to each other, and some statement that any 'transversal' to a pair of parallels has to have corresponding angles equal, where it meets the two

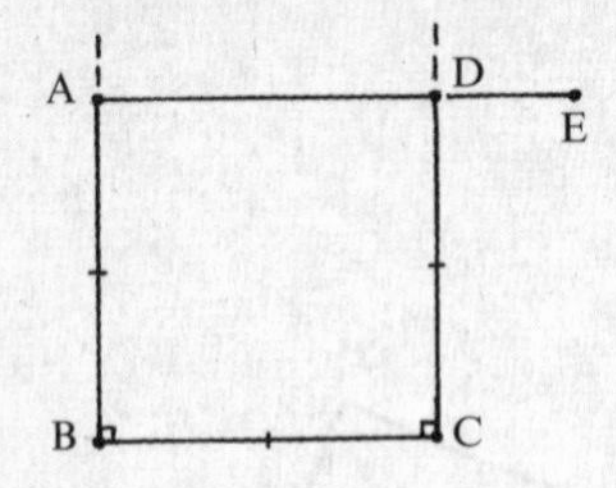

Fig. 2.7 Try to construct a square. Take ABC and BCD as right angles, with AB = BC = CD. Does it follow that DA is also equal to these lengths and that DAB and CDA are also right angles?

parallels. From this, it follows that the angle DAB would have to be equal to the angle complementary to ADC (i.e. to the angle EDC, in Fig. 2.7, ADE being straight) as well as being, as noted above, equal to the angle ADC. An angle (ADC) can only be equal to its complementary angle (EDC) if it is a right angle. We must also prove that the side AD has the same length as BC, but this now also follows, for example, from properties of transversals to the parallels BA and CD. So, it is indeed true that we can prove from this kind of Euclidean argument that squares, made up of right angles, actually do exist. But there is a deep issue hiding here.

2.2 Euclid's postulates

In building up his notion of geometry, Euclid took considerable care to see what assumptions his demonstrations depended upon.[2] In particular, he was careful to distinguish certain assertions called *axioms*—which were taken as self-evidently true, these being basically definitions of what he meant by points, lines, etc.—from the five *postulates*, which were assumptions whose validity seemed less certain, yet which appeared to be true of the geometry of our world. The final one of these assumptions, referred to as Euclid's fifth postulate, was considered to be less obvious than the others, and it was felt, for many centuries, that it ought to be possible to find a way of proving it from the other more evident postulates. Euclid's fifth postulate is commonly referred to as the *parallel postulate* and I shall follow this practice here.

Before discussing the parallel postulate, it is worth pointing out the nature of the other four of Euclid's postulates. The postulates are concerned with the geometry of the (Euclidean) plane, though Euclid also considered three-dimensional space later in his works. The basic elements of his plane geometry are points, straight lines, and circles. Here, I shall consider a 'straight line' (or simply a 'line') to be indefinitely extended in both directions; otherwise I refer to a 'line segment'. Euclid's *first* postulate effectively asserts that there is a (unique) straight line segment connecting any two points. His *second* postulate asserts the unlimited (continuous) extendibility of any straight line segment. His *third* postulate asserts the existence of a circle with any centre and with any value for its radius. Finally, his *fourth* postulate asserts the equality of all right angles.[3]

From a modern perspective, some of these postulates appear a little strange, particularly the fourth, but we must bear in mind the origin of the ideas underlying Euclid's geometry. Basically, he was concerned with the movement of idealized rigid bodies and the notion of *congruence* which was signalled when one such idealized rigid body was moved into coincidence with another. The equality of a right angle on one body with that on another had to do with the possibility of moving the one so that the lines forming its right angle would lie along the lines forming the right angle of the other. In effect, the fourth postulate is asserting the isotropy and homogeneity of space, so that a figure in one place could have the 'same' (i.e. congruent) geometrical shape as a figure in some other place. The second and third postulates express the idea that space is indefinitely extendible and without 'gaps' in it, whereas the first expresses the basic nature of a straight line segment. Although Euclid's way of looking at geometry was rather different from the way that we look at it today, his first four postulates basically encapsulated our present-day notion of a (two-dimensional) metric space with complete homogeneity and isotropy, and infinite in extent. In fact, such a picture seems to be in close accordance with the very large-scale spatial nature of the actual universe, according to modern cosmology, as we shall be coming to in §27.11 and §28.10.

What, then, is the nature of Euclid's fifth postulate, the parallel postulate? As Euclid essentially formulated this postulate, it asserts that if two straight line segments *a* and *b* in a plane both intersect another straight line *c* (so that *c* is what is called a *transversal* of *a* and *b*) such that the sum of the interior angles on the same side of *c* is less than two right angles, then *a* and *b*, when extended far enough on that side of *c*, will intersect somewhere (see Fig. 2.8a). An equivalent form of this postulate (sometimes referred to as *Playfair's axiom*) asserts that, for any straight line and for any point not on the line, there is a unique straight line through the point which is parallel to the line (see Fig. 2.8b). Here, 'parallel' lines would be two straight lines in the same plane that do not intersect each other (and recall that *my* 'lines' are fully extended entities, rather than Euclid's 'segments of lines').[2.1]

[2.1] Show that if Euclid's form of the parallel postulate holds, then Playfair's conclusion of the uniqueness of parallels must follow.

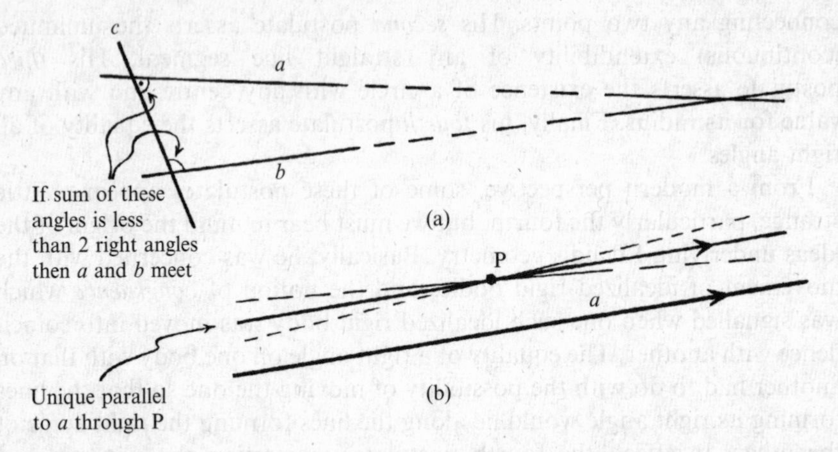

Fig. 2.8 (a) Euclid's parallel postulate. Lines *a* and *b* are transversals to a third line *c*, such that the interior angles where *a* and *b* meet *c* add to less than two right angles. Then *a* and *b* (assumed extended far enough) will ultimately intersect each other. (b) Playfair's (equivalent) axiom: if *a* is a line in a plane and P a point of the plane not on *a*, then there is just one line parallel to *a* through P, in the plane.

Once we have the parallel postulate, we can proceed to establish the property needed for the existence of a square. If a transversal to a pair of straight lines meets them so that the sum of the interior angles on one side of the transversal is two right angles, then one can show that the lines of the pair are indeed parallel. Moreover, it immediately follows that any other transversal of the pair has just the same angle property. This is basically just what we needed for the argument given above for the construction of our square. We see, indeed, that it is just the parallel postulate that we must use to show that our construction actually yields a square, with all its angles right angles and all its sides the same. Without the parallel postulate, we cannot establish that squares (in the normal sense where all their angles are right angles) actually exist.

It may seem to be merely a matter of mathematical pedantry to worry about precisely which assumptions are needed in order to provide a 'rigorous proof' of the existence of such an obvious thing as a square. Why should we really be concerned with such pedantic issues, when a 'square' is just that familiar figure that we all know about? Well, we shall be seeing shortly that Euclid actually showed some extraordinary perspicacity in worrying about such matters. Euclid's pedantry is related to a deep issue that has a great deal to say about the actual geometry of the universe, and in more than one way. In particular, it is not at all an obvious matter whether physical 'squares' exist on a cosmological scale in the actual universe. This is a matter for observation, and the evidence at the moment appears to be conflicting (see §2.7 and §28.10).

2.3 Similar-areas proof of the Pythagorean theorem

I shall return to the mathematical significance of *not* assuming the parallel postulate in the next section. The relevant physical issues will be re-examined in §18.4, §27.11, §28.10, and §34.4. But, before discussing such matters, it will be instructive to turn to the other proof of the Pythagorean theorem that I had promised above.

One of the simplest ways to see that the Pythagorean assertion is indeed true in Euclidean geometry is to consider the configuration consisting of the given right-angled triangle subdivided into two smaller triangles by dropping a perpendicular from the right angle to the hypotenuse (Fig. 2.9). There are now three triangles depicted: the original one and the two into which it has now been subdivided. Clearly the area of the original triangle is the sum of the areas of the two smaller ones.

Now, it is a simple matter to see that these three triangles are all *similar* to one another. This means that they are all the same *shape* (though of different sizes), i.e. obtained from one another by a uniform expansion or contraction, together with a rigid motion. This follows because each of the three triangles possesses exactly the same angles, in some order. Each of the two smaller triangles has an angle in common with the largest one and one of the angles of each triangle is a right angle. The third angle must also agree because the sum of the angles in any triangle is always the same. Now, it is a general property of similar plane figures that their areas are in proportion to the squares of their corresponding linear dimensions. For each triangle, we can take this linear dimension to be its longest side, i.e. its hypotenuse. We note that the hypotenuse of each of the smaller triangles is

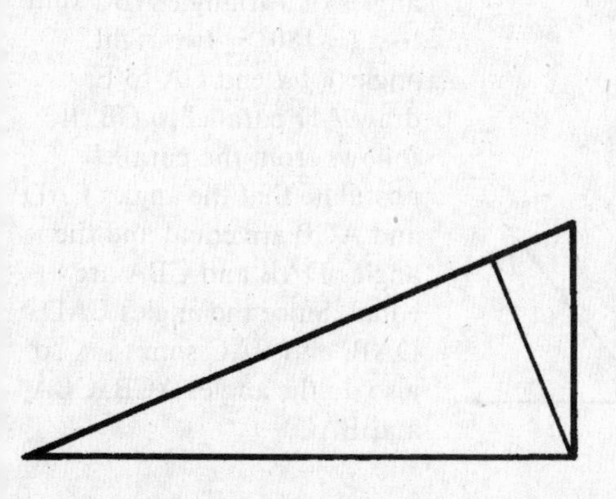

Fig. 2.9 Proof of the Pythagorean theorem using similar triangles. Take a right-angled triangle and drop a perpendicular from its right angle to its hypotenuse. The two triangles into which the original triangle is now divided have areas which sum to that of the original triangle. All three triangles are similar, so their areas are in proportion to the squares of their respective hypotenuses. The Pythagorean theorem follows.

the same as one of the (non-hypotenuse) sides of the original triangle. Thus, it follows at once (from the fact that the area of the original triangle is the sum of the areas of the other two) that the square on the hypotenuse on the original triangle is indeed the sum of the squares on the other two sides: *the Pythagorean theorem*!

There are, again, some particular assumptions in this argument that we shall need to examine. One important ingredient of the argument is the fact that the angles of a triangle always add up to the same value. (This value of this sum is of course 180°, but Euclid would have referred to it as 'two right angles'. The more modern 'natural' mathematical description is to say that the angles of a triangle, in Euclid's geometry, add up to π. This is to use radians for the absolute measure of angle, where the degree sign '°' counts as $\pi/180$, so we can write $180° = \pi$.) The usual proof is depicted in Fig. 2.10. We extend CA to E and draw a line AD, through A, which is parallel to CB. Then (as follows from the parallel postulate) the angles EAD and ACB are equal, and also DAB and CBA are equal. Since the angles EAD, DAB, and BAC add up to π (or to 180°, or to two right angles), so also must the three angles ACB, CBA, and BAC of the triangle—as was required to prove. But notice that the parallel postulate was used here.

This proof of the Pythagorean theorem also makes use of the fact that the areas of similar figures are in proportion to the squares of any linear measure of their sizes. (Here we chose the hypotenuse of each triangle to represent this linear measure.) This fact not only depends on the very existence of similar figures of different sizes—which for the triangles of Fig. 2.9 we established using the parallel postulate—but also on some more sophisticated issues that relate to how we actually define 'area' for non-rectangular shapes. These general matters are addressed in terms of the carrying out of limiting procedures, and I do not want to enter into

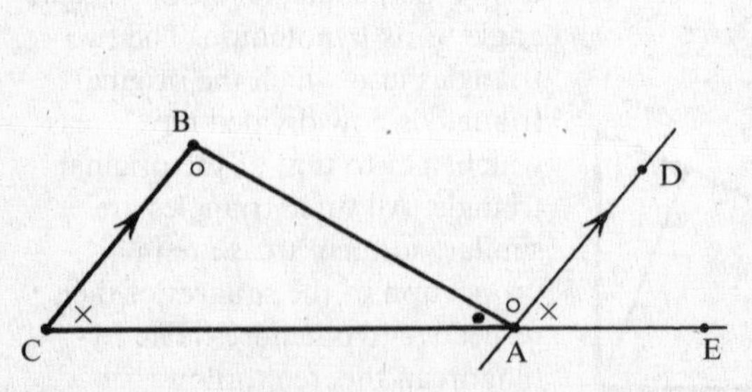

Fig. 2.10 Proof that the angles of a triangle ABC sum to π (= 180° = two right angles). Extend CA to E; draw AD parallel to CB. It follows from the parallel postulate that the angles EAD and ACB are equal and the angles DAB and CBA are equal. Since the angles EAD, DAB, and BAC sum to π, so also do the angles ACB, CBA, and BAC.

this kind of discussion just for the moment. It will take us into some deeper issues related to the kind of numbers that are used in geometry. The question will be returned to in §§3.1–3.

An important message of the discussion in the preceding sections is that the Pythagorean theorem seems to depend on the parallel postulate. Is this really so? Suppose the parallel postulate were false? Does that mean that the Pythagorean theorem might itself actually be false? Does such a possibility make any sense? Let us try to address the question of what would happen if the parallel postulate is indeed allowed to be taken to be false. We shall seem to be entering a mysterious make-belief world, where the geometry that we learned at school is turned all topsy-turvy. Indeed, but we shall find that there is also a deeper purpose here.

2.4 Hyperbolic geometry: conformal picture

Have a look at the picture in Fig. 2.11. It is a reproduction of one of M. C. Escher's woodcuts, called *Circle Limit I*. It actually provides us with a very accurate representation of a kind of geometry—called *hyperbolic* (or sometimes *Lobachevskian*) geometry—in which the parallel postulate is false, the Pythagorean theorem fails to hold, and the angles of a triangle do not add to π. Moreover, for a shape of a given size, there does not, in general, exist a similar shape of a larger size.

In Fig. 2.11, Escher has used a particular representation of hyperbolic geometry in which the entire 'universe' of the hyperbolic plane is 'squashed' into the interior of a circle in an ordinary Euclidean plane. The bounding circle represents 'infinity' for this hyperbolic universe. We can see that, in Escher's picture, the fish appear to get very crowded as they get close to this bounding circle. But we must think of this as an illusion. Imagine that you happened to be one of the fish. Then whether you are situated close to the rim of Escher's picture or close to its centre, the entire (hyperbolic) universe will look the same to you. The notion of 'distance' in this geometry does not agree with that of the Euclidean plane in terms of which it has been represented. As we look down upon Escher's picture from our Euclidean perspective, the fish near the bounding circle appear to us to be getting very tiny. But from the 'hyperbolic' perspective of the white or the black fish themselves, they think that they are exactly the same size and shape as those near the centre. Moreover, although from our outside Euclidean perspective they appear to get closer and closer to the bounding circle itself, from their own hyperbolic perspective that boundary always remains infinitely far away. Neither the bounding circle nor any of the 'Euclidean' space outside it has any existence for them. Their entire universe consists of what to us seems to lie strictly within the circle.

Fig. 2.11 M. C. Escher's woodcut *Circle Limit I*, illustrating the conformal representation of the hyperbolic plane.

In more mathematical terms, how is this picture of hyperbolic geometry constructed? Think of any circle in a Euclidean plane. The set of points lying in the interior of this circle is to represent the set of points in the entire hyperbolic plane. Straight lines, according to the hyperbolic geometry are to be represented as segments of Euclidean circles which meet the bounding circle *orthogonally*—which means at right angles. Now, it turns out that the hyperbolic notion of an *angle* between any two curves, at their point of intersection, is precisely the same as the Euclidean measure of angle between the two curves at the intersection point. A representation of this nature is called *conformal*. For this reason, the particular representation of hyperbolic geometry that Escher used is sometimes referred to as the *conformal model* of the hyperbolic plane. (It is also frequently referred to as the *Poincaré disc*. The dubious historical justification of this terminology will be discussed in §2.6.)

We are now in a position to see whether the angles of a triangle in hyperbolic geometry add up to π or not. A quick glance at Fig. 2.12 leads us to suspect that they do not and that they add up to something less. In fact, the sum of the angles of a triangle in hyperbolic geometry always falls short of π. We might regard that as a somewhat unpleasant feature of hyperbolic geometry, since we do not appear to get a 'neat' answer for the

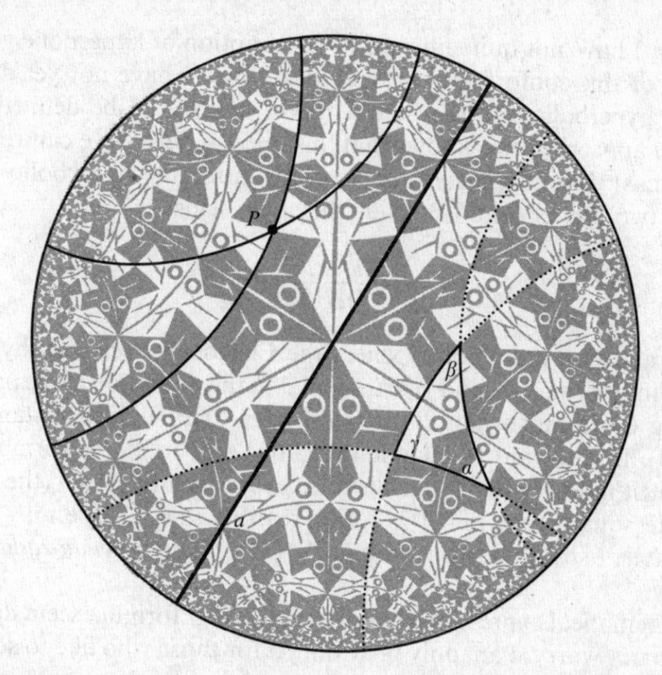

Fig. 2.12 The same Escher picture as Fig. 2.11, but with hyperbolic straight lines (Euclidean circles or lines meeting the bounding circle orthogonally) and a hyperbolic triangle illustrated. Hyperbolic angles agree with the Euclidean ones. The parallel postulate is evidently violated (lettering as in Fig. 2.8b) and the angles of a triangle sum to less than π.

sum of the angles of a triangle. However, there is actually something particularly elegant and remarkable about what does happen when we add up the angles of a hyperbolic triangle: the shortfall is always proportional to the area of the triangle. More explicitly, if the three angles of the triangle are α, β, and γ, then we have the formula (found by Johann Heinrich Lambert 1728–1777)

$$\pi - (\alpha + \beta + \gamma) = C\Delta,$$

where Δ is the area of the triangle and C is some constant. This constant depends on the 'units' that are chosen in which lengths and areas are to be measured. We can always scale things so that $C = 1$. It is, indeed, a remarkable fact that the area of a triangle can be so simply expressed in hyperbolic geometry. In Euclidean geometry, there is no way to express the area of a triangle simply in terms of its angles, and the expression for the area of a triangle in terms of its side-lengths is considerably more complicated.

In fact, I have not quite finished my description of hyperbolic geometry in terms of this conformal representation, since I have not yet described how the hyperbolic *distance* between two points is to be defined (and it would be appropriate to know what 'distance' is before we can really talk about areas). Let me give you an expression for the hyperbolic distance between two points A and B inside the circle. This is

$$\log \frac{QA \cdot PB}{QB \cdot PA},$$

where P and Q are the points where the Euclidean circle (i.e. hyperbolic straight line) through A and B orthogonal to the bounding circle *meets* this bounding circle and where 'QA', etc., refer to Euclidean distances (see Fig. 2.13). If you want to include the C of Lambert's area formula (with $C \neq 1$), just multiply the above distance expression by $C^{-1/2}$ (the reciprocal of the square root of C)^4.[2.2] For reasons that I hope may become clearer later, I shall refer to the quantity $C^{-1/2}$ as the *pseudo-radius* of the geometry.

If mathematical expressions like the above 'log' formula seem daunting, please do not worry. I am only providing it for those who like to see things explicitly. In any case, I am not going to explain why the expression works (e.g. why the shortest hyperbolic distance between two points, defined in this way, is actually measured along a hyperbolic straight line, or why the distances along a hyperbolic straight line 'add up' appropriately).[2.3] Also, I apologize for the 'log' (logarithm), but that is the way things are. In fact,

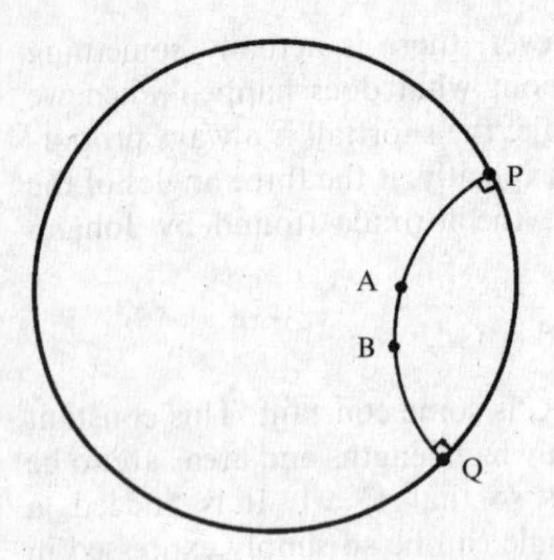

Fig. 2.13 In the conformal representation, the hyperbolic distance between A and B is log {QA.PB/QB.PA} where QA, etc. are Euclidean distances, P and Q being where the Euclidean circle through A and B, orthogonal to the bounding circle (hyperbolic line), meets this circle.

[2.2] Can you see a simple reason why?

[2.3] See if you can prove that, according to this formula, if A, B, and C are three successive points on a hyperbolic straight line, then the hyperbolic distances 'AB', etc. satisfy 'AB' + 'BC' = 'AC'. You may assume the general property of logarithms, $\log(ab) = \log a + \log b$ as described in §§5.2, 3.

this is a *natural logarithm* ('log to the base e') and I shall be having a good deal to say about it in §§5.2,3. We shall find that logarithms are really very beautiful and mysterious entities (as is the number e), as well as being important in many different contexts.

Hyperbolic geometry, with this definition of distance, turns out to have all the properties of Euclidean geometry apart from those which need the parallel postulate. We can construct triangles and other plane figures of different shapes and sizes, and we can move them around 'rigidly' (keeping their hyperbolic shapes and sizes from changing) with as much freedom as we can in Euclidean geometry, so that a natural notion of when two shapes are 'congruent' arises, just as in Euclidean geometry, where 'congruent' means 'can be moved around rigidly until they come into coincidence'. All the white fish in Escher's woodcut are indeed congruent to each other, according to this hyperbolic geometry, and so also are all the black fish.

2.5 Other representations of hyperbolic geometry

Of course, the white fish do not all look the same shape and size, but that is because we are viewing them from a Euclidean rather than a hyperbolic perspective. Escher's picture merely makes use of one particular Euclidean *representation* of hyperbolic geometry. Hyperbolic geometry itself is a more abstract thing which does not depend upon any particular Euclidean representation. However, such representations are indeed very helpful to us in that they provide ways of visualizing hyperbolic geometry by referring it to something that is more familiar and seemingly more 'concrete' to us, namely Euclidean geometry. Moreover, such representations make it clear that hyperbolic geometry is a consistent structure and that, consequently, the parallel postulate cannot be proved from the other laws of Euclidean geometry.

There are indeed other representations of hyperbolic geometry in terms of Euclidean geometry, which are distinct from the conformal one that Escher employed. One of these is that known as the *projective* model. Here, the entire hyperbolic plane is again depicted as the interior of a circle in a Euclidean plane, but the hyperbolic straight lines are now represented as straight Euclidean lines (rather than as circular arcs). There is, however, a price to pay for this apparent simplification, because the hyperbolic angles are now not the same as the Euclidean angles, and many people would regard this price as too high. For those readers who are interested, the hyperbolic distance between two points A and B in this representation is given by the expression (see Fig. 2.14)

$$\frac{1}{2}\log\frac{\mathrm{RA}\cdot\mathrm{SB}}{\mathrm{RB}\cdot\mathrm{SA}}$$

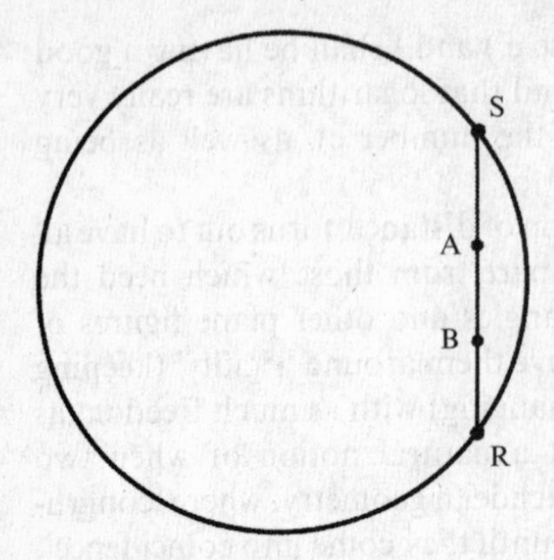

Fig. 2.14 In the projective representation, the formula for hyperbolic distance is now $\frac{1}{2}\log\{\mathrm{RA.SB/RB.SA}\}$, where R and S are the intersections of the Euclidean (i.e. hyperbolic) straight line AB with the bounding circle.

(taking $C = 1$, this being almost the same as the expression we had before, for the conformal representation), where R and S are the intersections of the extended straight line AB with the bounding circle. This representation of hyperbolic geometry, can be obtained from the conformal one by means of an expansion radially out from the centre by an amount given by

$$\frac{2R^2}{R^2 + r_c^2},$$

where R is the radius of the bounding circle and r_c is the Euclidean distance out from the centre of the bounding circle of a point in the conformal representation (see Fig. 2.15).[2.4] In Fig. 2.16, Escher's picture of Fig. 2.11 has been transformed from the conformal to the projective model using this formula. (Despite lost detail, Escher's precise artistry is still evident.)

There is a more directly geometrical way of relating the conformal and projective representations, via yet another clever representation of this same geometry. All three of these representations are due to the ingenious

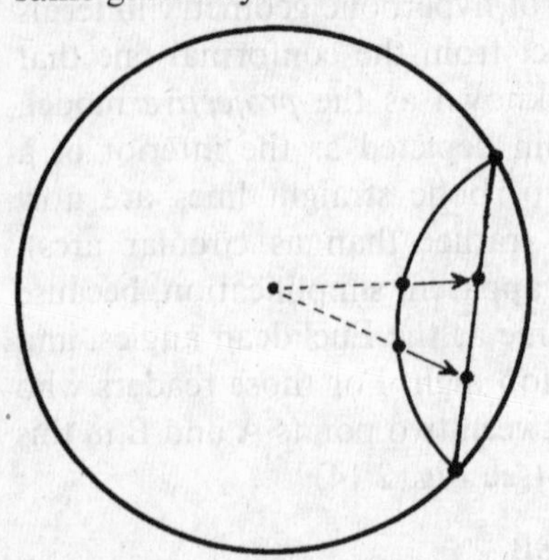

Fig. 2.15 To get from the conformal to the projective representation, expand out from the centre by a factor $2R^2/(R^2 + r_c^2)$, where R is the radius of the bounding circle and r_c is the Euclidean distance out of the point in the conformal representation.

[2.4] Show this. (*Hint*: You can use Beltrami's geometry, as illustrated in Fig. 2.17, if you wish.)

Fig. 2.16 Escher's picture of Fig. 2.11 transformed from the conformal to the projective representation.

Italian geometer Eugenio Beltrami (1835–1900). Consider a sphere S, whose equator coincides with the bounding circle of the projective representation of hyperbolic geometry given above. We are now going to find a representation of hyperbolic geometry on the *northern hemisphere* S^+ of S, which I shall call the *hemispheric* representation. See Fig. 2.17. To pass from the projective representation in the plane (considered as horizontal) to the new one on the sphere, we simply project vertically upwards (Fig. 2.17a). The straight lines in the plane, representing hyperbolic straight lines, are represented on S^+ by semicircles meeting the equator orthogonally. Now, to get from the representation on S^+ to the conformal representation on the plane, we project from the *south pole* (Fig. 2.17b). This is what is called *stereographic projection*, and it will play important roles later on in this book (see §8.3, §18.4, §22.9, §33.6). Two important properties of stereographic projection that we shall come to in §8.3 are that it is *conformal*, so that it preserves angles, and that it sends circles on the sphere to circles (or, exceptionally, to straight lines) on the plane.[2.5], [2.6]

[2.5] Assuming these two stated properties of stereographic projection, the conformal representation of hyperbolic geometry being as stated in §2.4, show that Beltami's hemispheric representation is conformal, with hyperbolic 'straight lines' as vertical semicircles.

[2.6] Can you see how to prove these two properties? (*Hint*: Show, in the case of circles, that the cone of projection is intersected by two planes of exactly opposite tilt.)

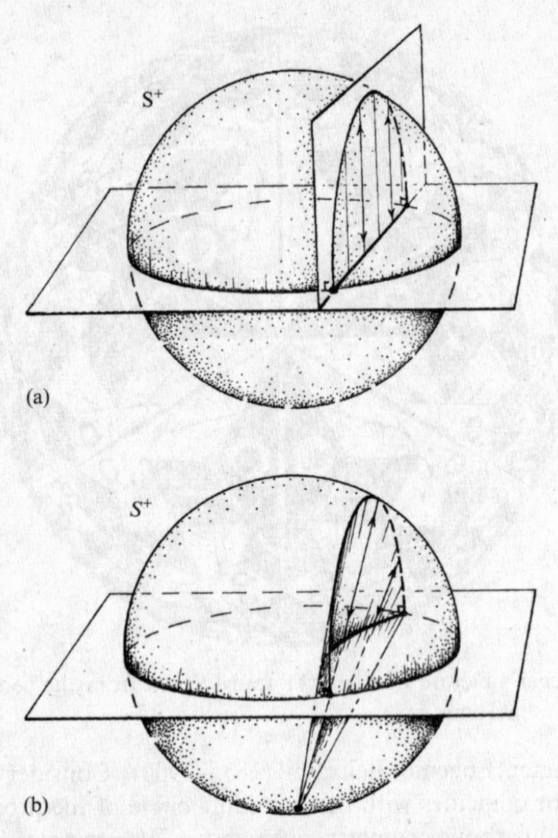

Fig. 2.17 Beltrami's geometry, relating three of his representations of hyperbolic geometry. (a) The hemispheric representation (conformal on the *northern hemisphere* S^+) projects vertically to the projective representation on the equatorial disc. (b) The hemispheric representation projects stereographically, from the *south pole* to the conformal representation on the equatorial disc.

The existence of various different models of hyperbolic geometry, expressed in terms of Euclidean space, serves to emphasize the fact that these are, indeed, merely 'Euclidean models' of hyperbolic geometry and are not to be taken as telling us what hyperbolic geometry actually *is*. Hyperbolic geometry has its own 'Platonic existence', just as does Euclidean geometry (see §1.3 and the Preface). No one of the models is to be taken as the 'correct' picturing of hyperbolic geometry at the expense of the others. The representations of it that we have been considering are very valuable as aids to our understanding, but only because the Euclidean framework is the one which we are more used to. For a sentient creature brought up with a direct experience of hyperbolic (rather than Euclidean) geometry, a model of Euclidean geometry in hyperbolic terms might seem the more natural way around. In §18.4, we shall encounter yet another model of hyperbolic geometry, this time in terms of the Minkowskian geometry of special relativity.

To end this section, let us return to the question of the existence of squares in hyperbolic geometry. Although squares whose angles are right angles do not exist in hyperbolic geometry, there are 'squares' of a more general type, whose angles are less than right angles. The easiest way to construct a square of this kind is to draw two straight lines intersecting at right angles at a point O. Our 'square' is now the quadrilateral whose four vertices are the intersections A, B, C, D (taken cyclicly) of these two lines with some circle with centre O. See Fig. 2.18. Because of the symmetry of the figure, the four sides of the resulting quadrilateral ABCD are all equal and all of its four angles must also be equal. But are these angles right angles? Not in hyperbolic geometry. In fact they can be any (positive) angle we like which is less than a right angle, but not equal to a right angle. The bigger the (hyperbolic) square (i.e. the larger the circle, in the above construction), the smaller will be its angles. In Fig. 2.19a, I have depicted a lattice of hyperbolic squares, using the conformal model, where there are five squares at each vertex point (instead of the Euclidean four), so the angle is $\frac{2}{5}\pi$, or 72°. In Fig. 2.19b, I have depicted the same lattice using the projective model. It will be seen that this does not allow the modifications that would be needed for the two-square lattice of Fig. 2.2.[2.7]

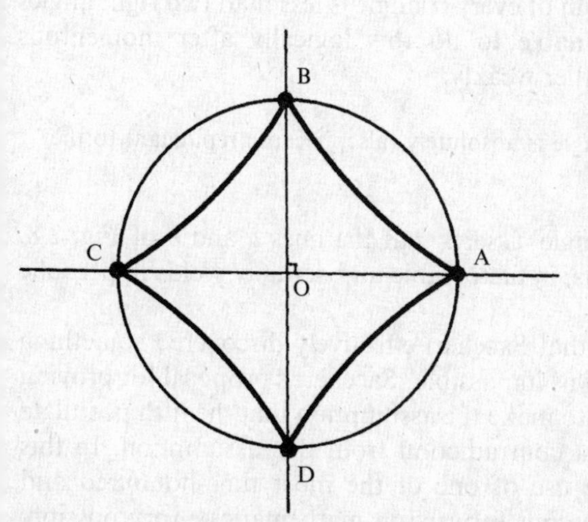

Fig. 2.18 A hyperbolic 'square' is a hyperbolic quadrilateral, whose vertices are the intersections A, B, C, D (taken cyclically) of two perpendicular hyperbolic straight lines through some point O with some circle centred at O. Because of symmetry, the four sides of ABCD as well as all the four angles are equal. These angles are not right angles, but can be equal to any given positive angle less than $\frac{1}{2}\pi$.

[2.7] See if you can do something similar, but with hyperbolic regular pentagons and squares.

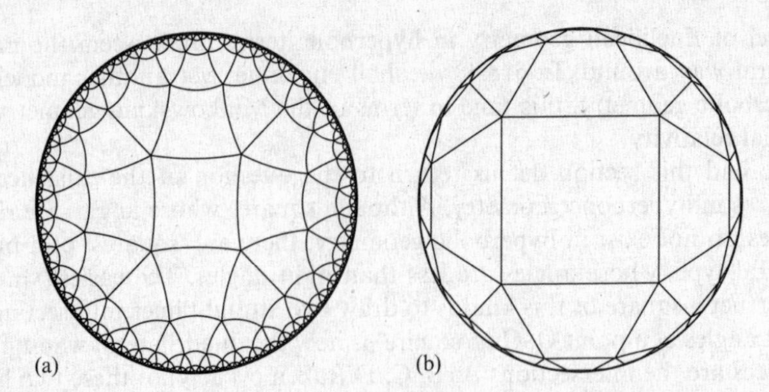

Fig. 2.19 A lattice of squares, in hyperbolic space, in which five squares meet at each vertex, so the angles of the square are $\frac{2\pi}{5}$, or 72°. (a) Conformal representation. (b) Projective representation.

2.6 Historical aspects of hyperbolic geometry

A few historical comments concerning the discovery of hyperbolic geometry are appropriate here. For centuries following the publication of Euclid's elements, in about 300 BC, various mathematicians attempted to prove the fifth postulate from the other axioms and postulates. These efforts reached their greatest heights with the heroic work by the Jesuit Girolamo Saccheri in 1733. It would seem that Saccheri himself must ultimately have thought his life's work a failure, constituting merely an unfulfilled attempt to *prove* the parallel postulate by showing that the hypothesis that the angle sum of every triangle is less than two right angles led to a contradiction. Unable to do this logically after momentous struggles, he concluded, rather weakly:

> The hypothesis of acute angle is absolutely false; because repugnant to the nature of the straight line.[5]

The hypothesis of 'acute angle' asserts that the lines *a* and *b* of Fig. 2.8. sometimes do not meet. It is, in fact, viable and actually yields hyperbolic geometry!

How did it come about that Saccheri effectively discovered something that he was trying to show was impossible? Saccheri's proposal for proving Euclid's fifth postulate was to make the assumption that the fifth postulate was false and then derive a contradiction from this assumption. In this way he proposed to make use of one of the most time-honoured and fruitful principles ever to be put forward in mathematics—very possibly first introduced by the Pythagoreans—called *proof by contradiction* (or *reductio ad absurdum*, to give it its Latin name). According to this procedure, in order to prove that some assertion is true, one first makes the supposition that the assertion in question is *false*, and one then argues from this that some contradiction ensues. Having found such a contradiction, one deduces that the assertion must be true after all.[6] Proof by contradiction provides a very powerful method of reasoning in mathematics, frequently applied today. A quotation from the distinguished mathematician G. H. Hardy is apposite here:

> *Reductio ad absurdum*, which Euclid loved so much, is one of a mathematician's finest weapons. It is a far finer gambit than any chess gambit: a chess player may offer the sacrifice of a pawn or even a piece, but a mathematician offers *the game*.[7]

We shall be seeing other uses of this important principle later (see §3.1 and §§16.4,6).

However, Saccheri failed in his attempt to find a contradiction. He was therefore not able to obtain a proof of the fifth postulate. But in striving for it he, in effect, found something far greater: a new geometry, different from that of Euclid—the geometry, discussed in §§2.4,5, that we now call *hyperbolic geometry*. From the assumption that Euclid's fifth postulate was false, he derived, instead of an actual contradiction, a host of strange-looking, barely believable, but interesting theorems. However, strange as these results appeared to be, none of them was actually a contradiction. As we now know, there was no chance that Saccheri would find a genuine contradiction in this way, for the reason that hyperbolic geometry *does* actually exist, in the mathematical sense that there is such a consistent structure. In the terminology of §1.3, hyperbolic geometry inhabits Plato's world of mathematical forms. (The issue of hyperbolic geometry's *physical* reality will be touched upon in §2.7 and §28.10.)

A little after Saccheri, the highly insightful mathematician Johann Heinrich Lambert (1728–1777) also derived a host of fascinating geometrical results from the assumption that Euclid's fifth postulate is false, including the beautiful result mentioned in §2.4 that gives the area of a hyperbolic triangle in terms of the sum of its angles. It appears that Lambert may well have formed the opinion, at least at some stage of his life, that a consistent geometry perhaps could be obtained from the denial of Euclid's fifth postulate. Lambert's tentative reason seems to have been that he could contemplate the theoretical possibility of the geometry on a 'sphere of imaginary radius', i.e. one for which the 'squared radius' is negative. Lambert's formula $\pi - (\alpha + \beta + \gamma) = C\Delta$ gives the area, Δ, of a hyperbolic triangle, where α, β, and γ are the angles of the triangle and where C is a constant ($-C$ being what we would now call the 'Gaussian curvature' of the hyperbolic plane). This formula looks basically the same as a previously known one due, in 1603, to Thomas Hariot (1560–1621), $\Delta = R^2(\alpha + \beta + \gamma - \pi)$, for the area Δ of a *spherical triangle*, drawn with great circle arcs[8] on a sphere of radius R (see Fig. 2.20).[2.8] To retrieve Lambert's formula, we have to put

$$C = -\frac{1}{R^2}.$$

But, in order to give the *positive* value of C, as would be needed for hyperbolic geometry, we require the sphere's radius to be 'imaginary' (i.e. to be the square root of a negative number). Note that the radius R is given by the imaginary quantity $(-C)^{-1/2}$. This explains the term 'pseudo-radius', introduced in §2.4, for the real quantity $C^{-1/2}$. In fact Lambert's procedure is perfectly justified from our more modern perspectives (see Chapter 4 and §18.4, Fig. 18.9), and it indicates great insight on his part to have foreseen this.

It is, however, the conventional standpoint (somewhat unfair, in my opinion) to deny Lambert the honour of having first constructed non-Euclidean geometry, and to consider that (about half a century later) the first person to have come to a clear acceptance of a fully consistent geometry, distinct from that of Euclid, in which the parallel postulate is false, was the great mathematician Carl Friedrich Gauss. Being an exceptionally cautious man, and being fearful of the controversy that such a revelation might cause, Gauss did not publish his findings, and kept them to himself.[9] Some 30 years after Gauss had begun working on it, hyperbolic

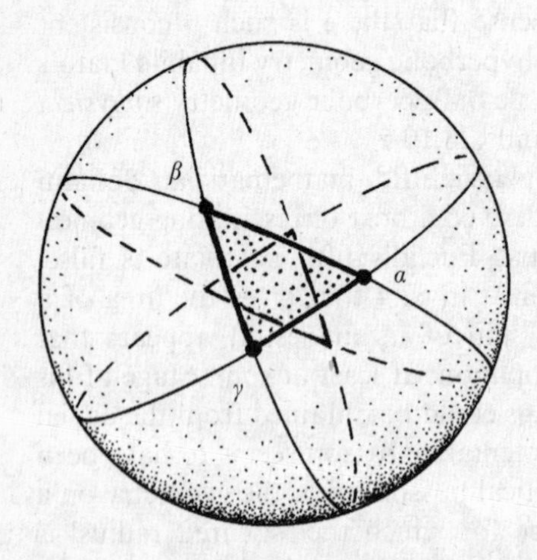

Fig. 2.20 Hariot's formula for the area of a *spherical triangle*, on a sphere of radius R, with angles α, β, γ, is $\Delta = R^2(\alpha + \beta + \gamma - \pi)$. Lambert's formula, for a hyperbolic triangle, has $C = -1/R^2$.

[2.8] Try to prove this spherical triangle formula, basically using only symmetry arguments and the fact that the total area of the sphere is $4\pi R^2$. *Hint*: Start with finding the area of a segment of a sphere bounded by two great circle arcs connecting a pair of antipodal points on the sphere; then cut and paste and use symmetry arguments. Keep Fig. 2.20 in mind.

geometry was independently rediscovered by various others, including the Hungarian János Bolyai (by 1829) and, most particularly, the Russian geometer Nicolai Ivanovich Lobachevsky in about 1826 (whence hyperbolic geometry is frequently called *Lobachevskian* geometry).

The specific projective and conformal realizations of hyperbolic geometry that I have described above were both found by Eugenio Beltrami, and published in 1868, together with some other elegant representations including the hemispherical one mentioned in §2.5. The conformal representation is, however, commonly referred to as the 'Poincaré model', because Poincaré's rediscovery of this representation in 1882 is better known than the original work of Beltrami (largely because of the important use that Poincaré made of this model).[10] Likewise, poor old Beltrami's projective representation is sometimes called the 'Klein representation'. It is not uncommon in mathematics that the name normally attached to a mathematical concept is not that of the original discoverer. At least, in this case, Poincaré did *re*discover the conformal representation (as did Klein the projective one in 1871). There are other instances in mathematics where the mathematician(s) whose name(s) are attached to a result did not even know of the result in question![11]

The representation of hyperbolic geometry that Beltrami is best known for is yet another one, which he found also in 1868. This represents the geometry on a certain surface known as a *pseudo-sphere* (see Fig. 2.21). This surface is obtained by rotating a *tractrix*, a curve first investigated by Isaac Newton in 1676, about its 'asymptote'. The asymptote is a straight line which the curve approaches, becoming asymptotically tangent to it as the curve recedes to infinity. Here, we are to imagine the asymptote to be drawn on a horizontal plane of rough texture. We are to think of a light, straight, stiff rod, at one end P of which is attached a heavy point-like weight, and the other end R moves along the asymptote. The point P then traces out a tractrix. Ferdinand Minding found, in 1839, that the pseudo-sphere has a constant

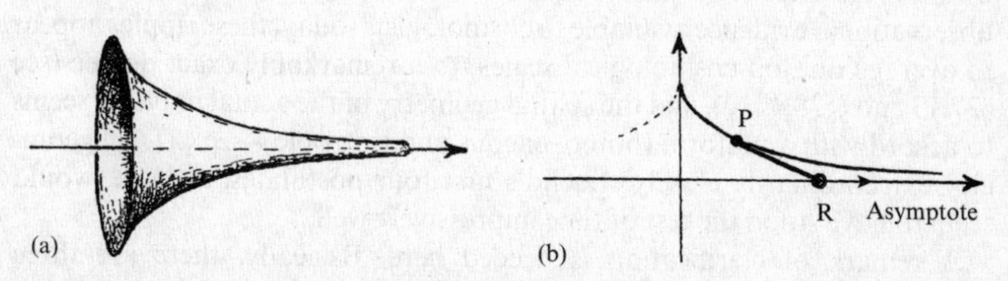

Fig. 2.21 (a) A *pseudo-sphere*. This is obtained by rotating, about its asymptote (b) a *tractrix*. To construct a tractrix, imagine its plane to be horizontal, over which is dragged a light, frictionless straight, stiff rod. One end of the rod is a point-like weight P with friction, and the other end R moves along the (straight) asymptote.

negative intrinsic geometry, and Beltrami used this fact to construct the first model of hyperbolic geometry. Beltrami's pseudo-sphere model seems to be the one that persuaded mathematicians of the consistency of plane hyperbolic geometry, since the measure of hyperbolic distance agrees with the Euclidean distance along the surface. However, it is a somewhat awkward model, because it represents hyperbolic geometry only locally, rather than presenting the entire geometry all at once, as do Beltrami's other models.

2.7 Relation to physical space

Hyperbolic geometry also works perfectly well in higher dimensions. Moreover, there are higher-dimensional versions of both the conformal and projective models. For three-dimensional hyperbolic geometry, instead of a bounding circle, we have a bounding sphere. The entire infinite three-dimensional hyperbolic geometry is represented by the interior of this finite Euclidean sphere. The rest is basically just as we had it before. In the conformal model, straight lines in this three-dimensional hyperbolic geometry are represented as Euclidean circles which meet the bounding sphere orthogonally; angles are given by the Euclidean measures, and distances are given by the same formula as in the two-dimensional case. In the projective model, the hyperbolic straight lines are Euclidean straight lines, and distances are again given by the same formula as in the two-dimensional case.

What about our actual universe on cosmological scales? Do we expect that its spatial geometry is Euclidean, or might it accord more closely with some other geometry, such as the remarkable hyperbolic geometry (but in three dimensions) that we have been examining in §§2.4–6. This is indeed a serious question. We know from Einstein's general relativity (which we shall come to in §17.9 and §19.6) that Euclid's geometry is only an (extraordinarily accurate) approximation to the actual geometry of physical space. This physical geometry is not even exactly uniform, having small ripples of irregularity owing to the presence of matter density. Yet, strikingly, according to the best observational evidence available to cosmologists today, these ripples appear to average out, on cosmological scales, to a remarkably exact degree (see §27.13 and §§28.4–10), and the spatial geometry of the actual universe seems to accord with a uniform (homogeneous and isotropic—see §27.11) geometry extraordinarily closely. Euclid's first four postulates, at least, would seem to have stood the test of time impressively well.

A remark of clarification is needed here. Basically, there are three types of geometry that would satisfy the conditions of homogeneity (every point the same) and isotropy (every direction the same), referred to as Euclidean, hyperbolic, and elliptic. Euclidean geometry is familiar to us (and has been for some 23 centuries). Hyperbolic geometry has been our main concern in this chapter. But what is elliptic geometry? Essentially, elliptic plane geometry is that satisfied by figures drawn on the surface of a sphere. It figured in the discussion of Lambert's approach to hyperbolic geometry in §2.6. See Fig. 2.22a,b,c,

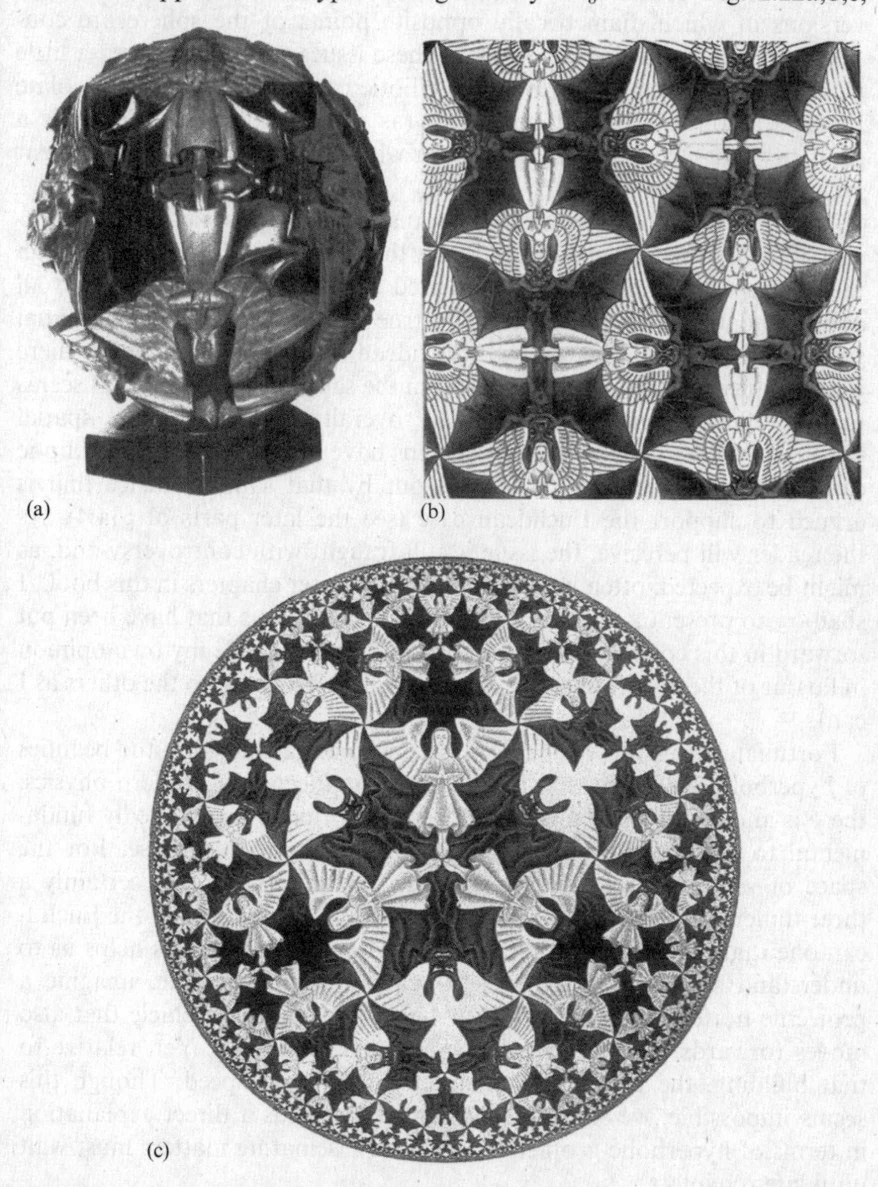

Fig. 2.22 The three basic kinds of uniform plane geometry, as illustrated by Escher using tessellations of angels and devils. (a) Elliptic case (positive curvature), (b) Euclidean case (zero curvature), and (c) Hyperbolic case (negative curvature)—in the conformal representation (Escher's *Circle Limit IV*, to be compared with Fig. 2.11).

for Escher's rendering of the elliptic, Euclidean, and hyperbolic cases, respectively, using a similar tessellation of angels and devils in all three cases, the third one providing an interesting alternative to Fig. 2.11. (There is also a three-dimensional version of elliptic geometry, and there are versions in which diametrically opposite points of the sphere are considered to represent the same point. These issues will be discussed a little more fully in §27.11.) However, the elliptic case could be said to violate Euclid's second and third postulates (as well as the first). For it is a geometry that is finite in extent (and for which more than one line segment joins a pair of points).

What, then, is the observational status of the large-scale spatial geometry of the universe? It is only fair to say that we do not yet know, although there have been recent widely publicized claims that Euclid was right all along, and his fifth postulate holds true also, so the averaged spatial geometry is indeed what we call 'Euclidean'.[12] On the other hand, there is also evidence (some of it coming from the same experiments) that seems to point fairly firmly to a *hyperbolic* overall geometry for the spatial universe.[13] Moreover, some theoreticians have long argued for the elliptic case, and this is certainly not ruled out by that same evidence that is argued to support the Euclidean case (see the later parts of §34.4). As the reader will perceive, the issue is still fraught with controversy and, as might be expected, often heated argument. In later chapters in this book, I shall try to present a good many of the considerations that have been put forward in this connection (and I do not attempt to hide my own opinion in favour of the hyperbolic case, while trying to be as fair to the others as I can).

Fortunately for those, such as myself, who are attracted to the beauties of hyperbolic geometry, and also to the magnificence of modern physics, there is another role for this superb geometry that is undisputedly fundamental to our modern understanding of the physical universe. For the space of *velocities*, according to modern relativity theory, is certainly a three-dimensional hyperbolic geometry (see §18.4), rather than the Euclidean one that would hold in the older Newtonian theory. This helps us to understand some of the puzzles of relativity. For example, imagine a projectile hurled forward, with near light speed, from a vehicle that also moves forwards with comparable speed past a building. Yet, relative to that building, the projectile can never exceed light speed. Though this seems impossible, we shall see in §18.4 that it finds a direct explanation in terms of hyperbolic geometry. But these fascinating matters must wait until later chapters.

What about the Pythagorean theorem, which we have seen to fail in hyperbolic geometry? Must we abandon this greatest of the specific Pythagorean gifts to posterity? Not at all, for hyperbolic geometry—and, indeed, all the 'Riemannian' geometries that generalize hyperbolic geometry in an irregularly curved way (forming the essential framework for Einstein's general theory of relativity; see §13.8, §14.7, §18.1, and §19.6)—depends vitally upon the Pythagorean theorem holding in the limit of small distances. Moreover, its enormous influence permeates other vast areas of mathematics and physics (e.g. the 'unitary' metric structure of quantum mechanics, see §22.3). Despite the fact that this theorem is, in a sense, superseded for 'large' distances, it remains central to the small-scale structure of geometry, finding a range of application that enormously exceeds that for which it was originally put forward.

Notes

Section 2.1

2.1. It is historically very unclear who actually first proved what we now refer to as the 'Pythagorean theorem', see Note 1.1. The ancient Egyptians and Babylonians seem to have known at least many instances of this theorem. The true role played by Pythagoras or his followers is largely surmise.

Section 2.2

2.2. Even with this amount of care, however, various hidden assumptions remained in Euclid's work, mainly to do with what we would now call 'topological' issues that would have seemed to be 'intuitively obvious' to Euclid and his contemporaries. These unmentioned assumptions were pointed out only centuries later, particularly by Hilbert at the end of the 19th century. I shall ignore these in what follows.

2.3. See e.g. Thomas (1939). Compare also Schutz (1997), who gives a nice axiomatic account of Minkowski's 4-dimensional spacetime geometry (§17.8, §18.1).

Section 2.4

2.4. The 'exponent' notation, such as $C^{-1/2}$, is frequently used in this book. As already referred to in Note 1.2, a^5 means $a \times a \times a \times a \times a$; correspondingly, for a positive integer n, the product of a with itself a total of n times is written a^n. This notation extends to negative exponents, so that a^{-1} is the reciprocal $1/a$ of a, and a^{-n} is the reciprocal $1/a^n$ of a^n, or equivalently $(a^{-1})^n$. In accordance with the more general discussion of §5.2, $a^{1/n}$, for a positive number a, is the 'nth root of a', which is the (positive) number satisfying $(a^{1/n})^n = a$ (see Note 1.2). Moreover, $a^{m/n}$ is the mth power of $a^{1/n}$.

Section 2.6

2.5. Saccheri (1733), Prop. XXXIII.

2.6. There is a standpoint known as *intuitionism*, which is held to by a (rather small) minority of mathematicians, in which the principle of 'proof by contradiction' is not accepted. The objection is that this principle can be *non-constructive* in that it sometimes leads to an assertion of the existence of some mathematical entity, without any actual construction for it having been provided. This has some relevance to the issues discussed in §16.6. See Heyting (1956).

2.7. Hardy (1940), p. 34.

2.8. Great circle arcs are the 'shortest' curves (geodesics) on the surface of a sphere; they lie on planes through the sphere's centre.

2.9. It is a matter of some dispute whether Gauss, who was professionally concerned with matters of geodesy, might actually have tried to ascertain whether there are measurable deviations from Euclidean geometry in physical space. Owing to his well-known reticence in matters of non-Euclidean geometry, it is unlikely that he would let it be known if he were in fact trying to do this, particularly since (as we now know) he would be bound to fail, owing to the smallness of the effect, according to modern theory. The present consensus seems to be that he was 'just doing geodesy', being concerned with the curvature of the Earth, and not of space. But I find it a little hard to believe that he would not also have been on the lookout for any significant discrepancy with Euclidean geometry; see Fauvel and Gray (1987), Gray (1979).

2.10. The so-called 'Poincaré half-plane' representation (with metric form $(dx^2 + dy^2)/y^2$; see §14.7) is also due to Beltrami; see Beltrami (1868). The constant negative curvature of the 'Poincaré metric' $4(dx^2 + dy^2)/(1 - x^2 - y^2)^2$ of Figs. 2.11–13 was actually noted by Riemann (1854)!

2.11. This appears to have applied even to the great Gauss himself (who had, on the other hand, very frequently anticipated other mathematicians' work). There is an important topological mathematical theorem now referred to as the 'Gauss–Bonnet theorem', which can be elegantly proved by use of the so-called 'Gauss map', but the theorem itself appears actually to be due to Blaschke and the elegant proof procedure just referred to was found by Olinde Rodrigues. It appears that neither the result nor the proof procedure were even known to Gauss or to Bonnet. There is a more elemental 'Gauss–Bonnet' theorem, correctly cited in several texts, see Willmore (1959), also Rindler (2001).

Section 2.7

2.12. The main evidence for the overall structure of the universe, as a whole comes from a detailed analysis of the *cosmic microwave background radiation* (CMB) that will be discussed in §§27.7,10,11,13, §§28.5,10, and §30.14. A basic reference is de Bernardis *et al.* (2000); for more accurate, more recent data, see Netterfield *et al.* (2002) (concerning BOOMERanG). See also Hanany *et al.* (2000) (concerning MAXIMA), Halverson *et al.* (2001) (concerning DASI), and Bennett *et al.* (2003) (concerning WMAP).

2.13. See Gurzadyan and Torres (1997) and Gurzadyan and Kocharyan (1994) for the theoretical underpinnings, and Gurzadyan and Kocharyan (1992) (for COBE data) and Gurzadyan *et al.* (2002, 2003) (for BOOMERanG data and (2004) for WMAP data) for the corresponding analysis of the actual CMB data.

3
Kinds of number in the physical world

3.1 A Pythagorean catastrophe?

LET us now return to the issue of proof by contradiction, the very principle that Saccheri tried hard to use in his attempted proof of Euclid's fifth postulate. There are many instances in classical mathematics where the principle *has* been successfully applied. One of the most famous of these dates back to the Pythagoreans, and it settled a mathematical issue in a way which greatly troubled them. This was the following. Can one find a rational number (i.e. a fraction) whose square is precisely the number 2? The answer turns out to be no, and the mathematical assertion that I shall demonstrate shortly is, indeed, that there is no such rational number.

Why were the Pythagoreans so troubled by this discovery? Recall that a fraction—that is, a rational number—is something that can be expressed as the ratio a/b of two integers (or whole numbers) a and b, with b non-zero. (See the Preface for a discussion of the definition of a fraction.) The Pythagoreans had originally hoped that all their geometry could be expressed in terms of lengths that could be measured in terms of rational numbers. Rational numbers are rather simple quantities, being describable and understood in simple finite terms; yet they can be used to specify distances that are as small as we please or as large as we please. If all geometry could be done with rationals, then this would make things relatively simple and easily comprehensible. The notion of an 'irrational' number, on the other hand, requires infinite processes, and this had presented considerable difficulties for the ancients (and with good reason). Why is there a difficulty in the fact that there is no rational number that squares to 2? This comes from the Pythagorean theorem itself. If, in Euclidean geometry, we have a square whose side length is unity, then its diagonal length is a number whose square is $1^2 + 1^2 = 2$ (see Fig. 3.1). It would indeed be catastrophic for geometry if there were no actual number that could describe the length of the diagonal of a square. The Pythagoreans tried, at first, to make do with a notion of 'actual number' that could be described simply in terms of ratios of whole numbers. Let us see why this will not work.

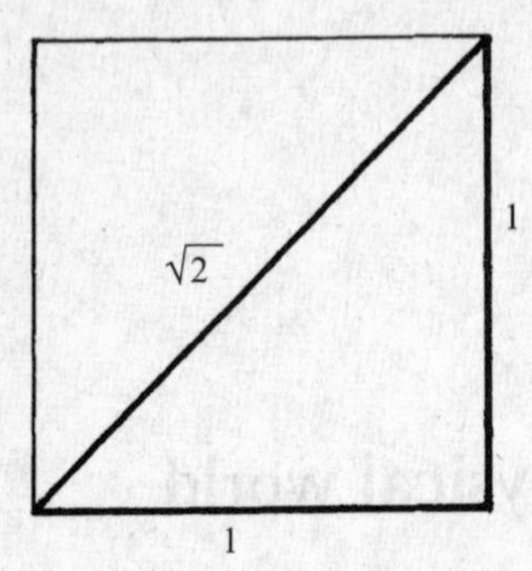

Fig. 3.1 A square of unit side-length has diagonal $\sqrt{2}$, by the Pythagorean theorem.

The issue is to see why the equation

$$\left(\frac{a}{b}\right)^2 = 2$$

has no solution for integers a and b, where we take these integers to be positive. We shall use proof by contradiction to prove that no such a and b can exist. We therefore try to suppose, on the contrary, that such an a and b *do* exist. Multiplying the above equation by b^2 on both sides, we find that it becomes

$$a^2 = 2b^2$$

and we clearly conclude[1] that $a^2 > b^2 > 0$. Now the right-hand side, $2b^2$, of the above equation is even, whence a must be even (not odd, since the square of any odd number is odd). Hence $a = 2c$, for some positive integer c. Substituting $2c$ for a in the above equation, and squaring it out, we obtain

$$4c^2 = 2b^2,$$

that is, dividing both sides by 2,

$$b^2 = 2c^2,$$

and we conclude $b^2 > c^2 > 0$. Now, this is precisely the same equation that we had displayed before, except that b now replaces a, and c replaces b. Note that the corresponding integers are now smaller than they were before. We can now repeat the argument again and again, obtaining an unending sequence of equations

$$a^2 = 2b^2, b^2 = 2c^2, c^2 = 2d^2, d^2 = 2e^2, \ldots,$$

where

$$a^2 > b^2 > c^2 > d^2 > e^2 > \ldots,$$

all of these integers being positive. But any decreasing sequence of positive integers must come to an end, contradicting the fact that this sequence is unending. This provides us with a contradiction to what has been supposed, namely that there is a rational number which squares to 2. It follows that there is no such rational number—as was required to prove.[2]

Certain points should be remarked upon in the above argument. In the first place, in accordance with the normal procedures of mathematical proof, certain properties of numbers have been appealed to in the argument that were taken as either 'obvious' or having been previously established. For example, we made use of the fact that the square of an odd number is always odd and, moreover, that if an integer is not odd then it is even. We also used the fundamental fact that every strictly decreasing sequence of positive integers must come to an end.

One reason that it can be important to identify the precise assumptions that go into a proof—even though some of these assumptions could be perfectly 'obvious' things—is that mathematicians are frequently interested in other kinds of entity than those with which the proof might be originally concerned. If these other entities satisfy the same assumptions, then the proof will still go through and the assertion that had been proved will be seen to have a greater generality than originally perceived, since it will apply to these other entities also. On the other hand, if some of the needed assumptions fail to hold for these alternative entities, then the assertion that may turn out to be false for these entities. (For example, it is important to realize that the parallel postulate was used in the proofs of the Pythagorean theorem given in §2.2, for the theorem is actually false for hyperbolic geometry.)

In the above argument, the original entities are integers and we are concerned with those numbers—the rational numbers—that are constructed as quotients of integers. With such numbers it is indeed the case that none of them squares to 2. But there are other kinds of number than merely integers and rationals. Indeed, the need for a square root of 2 forced the ancient Greeks, very much against their wills at the time, to proceed outside the confines of integers and rational numbers—the only kinds of number that they had previously been prepared to accept. The kind of number that they found themselves driven to was what we now call a 'real number': a number that we now express in terms of an unending decimal expansion (although such a representation was not available to the ancient Greeks). In fact, 2 does indeed have a real-number square root, namely (as we would now write it)

$$\sqrt{2} = 1.414\,213\,562\,373\,095\,048\,801\,688\,72\ldots.$$

We shall consider the *physical* status of such 'real' numbers more closely in the next section.

As a curiosity, we may ask why the above proof of the non-existence of a square root of 2 fails for real numbers (or for real-number ratios, which amounts to the same thing). What happens if we replace 'integer' by 'real number' throughout the argument? The basic difference is that it is not true that any strictly decreasing sequence of positive reals (or even of fractions) must come to an end, and the argument breaks down at that point.[3] (Consider the unending sequence $1, \frac{1}{2}, \frac{1}{4}, \frac{1}{8}, \frac{1}{16}, \frac{1}{32}, \ldots$, for example.) One might worry what an 'odd' and 'even' real number would be in this context. In fact the argument encounters no difficulty at that stage because *all* real numbers would have to count as 'even', since for any real a there is always a real c such that $a = 2c$, division by 2 being always possible for reals.

3.2 The real-number system

Thus it was that the Greeks were forced into the realization that rational numbers are not enough, if the ideas of (Euclid's) geometry are to be properly developed. Nowadays, we do not worry unduly if a certain geometrical quantity cannot be measured simply in terms of rational numbers alone. This is because the notion of a 'real number' is very familiar to us. Although our pocket calculators express numbers in terms of only a finite number of digits, we readily accept that this is an approximation forced upon us by the fact that the calculator is a finite object. We are prepared to allow that the ideal (Platonic) mathematical number could certainly require that the decimal expansion continues indefinitely. This applies, of course, even to the decimal representation of most fractions, such as

$$\tfrac{1}{3} = 0.333\,333\,333\ldots,$$
$$\tfrac{29}{12} = 2.416\,666\,666\ldots,$$
$$\tfrac{9}{7} = 1.285\,714\,285\,714\,285,$$
$$\tfrac{237}{148} = 1.601\,351\,351\,35\ldots.$$

For a fraction, the decimal expanson is always *ultimately periodic*, which is to say that after a certain point the infinite sequence of digits consists of some finite sequence repeated indefinitely. In the above examples the repeated sequences are, respectively, 3, 6, 285714, and 135.

Decimal expansions were not available to the ancient Greeks, but they had their own ways of coming to terms with irrational numbers. In effect, what they adopted was a system of representing numbers in terms of what are now called *continued fractions*. There is no need to go into this in full detail here, but some brief comments are appropriate. A continued fraction[4] is a finite or infinite expression $a+(b+(c+(d+\cdots)^{-1})^{-1})^{-1}$, where $a, b, c, d, \ldots$ are positive integers:

$$a+\cfrac{1}{b+\cfrac{1}{c+\cfrac{1}{d+\cdots}}}$$

Any rational number larger than 1 can be written as a *terminating* such expression (where to avoid ambiguity we normally require the final integer to be greater than 1), e.g. $52/9 = 5+(1+(3+(2)^{-1})^{-1})^{-1}$:

$$\frac{52}{9}=5+\cfrac{1}{1+\cfrac{1}{3+\cfrac{1}{2}}}$$

and, to represent a positive rational less than 1, we just allow the first integer in the expression to be zero. To express a real number, which is not rational, we simply[3.1] allow the continued-fraction expression to run on forever, some examples being[5]

$$\sqrt{2}=1+(2+(2+(2+(2+\cdots)^{-1})^{-1})^{-1})^{-1},$$

$$7-\sqrt{3}=5+(3+(1+(2+(1+(2+(1+(2+\cdots)^{-1})^{-1})^{-1})^{-1})^{-1})^{-1})^{-1},$$

$$\pi=3+(7+(15+(1+(292+(1+(1+(1+(2+\cdots)^{-1})^{-1})^{-1})^{-1})^{-1})^{-1})^{-1})^{-1}.$$

In the first two of these infinite examples, the sequences of natural numbers that appear—namely 1, 2, 2, 2, 2, ... in the first case and 5, 3, 1, 2, 1, 2, 1, 2, ... in the second—have the property that they are ultimately periodic (the 2 repeating indefinitely in the first case and the sequence 1, 2 repeating indefinitely in the second).[3.2] Recall that, as

---

[3.1] Experiment with your pocket calculator (assuming you have '$\sqrt{\ }$' and 'x^{-1}' keys) to obtain these expansions to the accuracy available. Take $\pi = 3.141\,592\,653\,589\,793\ldots$ (*Hint*: Keep taking note of the integer part of each number, subtracting it off, and then forming the reciprocal of the remainder.)

[3.2] Assuming this eventual periodicity of these two continued-fraction expressions, show that the numbers they represent must be the quantities on the left. (*Hint*: Find a quadratic equation that must be satisfied by this quantity, and refer to Note 3.6.)

already noted above, in the familiar decimal notation, it is the *rational* numbers that have (finite or) ultimately periodic expressions. We may regard it as a strength of the Greek 'continued-fraction' representation, on the other hand, that the rational numbers now always have a finite description. A natural question to ask, in this context, is: which numbers have an *ultimately periodic* continued-fraction representation? It is a remarkable theorem, first proved, to our knowledge, by the great 18th-century mathematician Joseph L. Lagrange (whose most important other ideas we shall encounter later, particularly in Chapter 20) that the numbers whose representation in terms of continued fractions are ultimately periodic are what are called *quadratic irrationals*.[6]

What is a quadratic irrational and what is its importance for Greek geometry? It is a number that can be written in the form

$$a + \sqrt{b},$$

where a and b are fractions, and where b is not a perfect square. Such numbers are important in Euclidean geometry because they are the most immediate irrational numbers that are encountered in ruler-and-compass constructions. (Recall the Pythagorean theorem, which in §3.1 first led us to consider the problem of $\sqrt{2}$, and other simple constructions of Euclidean lengths directly lead us to other numbers of the above form.)

Particular examples of quadratic irrationals are those cases where $a = 0$ and b is a (non-square) natural number, or rational greater than 1, e.g.

$$\sqrt{2}, \sqrt{3}, \sqrt{5}, \sqrt{6}, \sqrt{7}, \sqrt{8}, \sqrt{10}, \sqrt{11}, \ldots.$$

The continued-fraction representation of such a number is particularly striking. The sequence of natural numbers that defines it as a continued fraction has a curious characteristic property. It starts with some number A, then it is immediately followed by a 'palindromic' sequence (i.e. one which reads the same backwards), $B, C, D, \ldots, D, C, B$, followed by $2A$, after which the sequence $B, C, D, \ldots, D, C, B, 2A$ repeats itself indefinitely. The number $\sqrt{14}$ is a good example, for which the sequence is

$$3, 1, 2, 1, 6, 1, 2, 1, 6, 1, 2, 1, 6, 1, 2, 1, 6, \ldots.$$

Here $A = 3$ and the palindromic sequence $B, C, D, \ldots, D, C, B$ is just the three-term sequence 1, 2, 1.

How much of this was known to the ancient Greeks? It seems very likely that they knew quite a lot—very possibly *all* the things that I have described above (including Lagrange's theorem), although they may well have lacked rigorous proofs for everything. Plato's contemporary

Theaetetos seems to have established much of this. There appears even to be some evidence of this knowledge (including the repeating palindromic sequences referred to above) revealed in Plato's dialectics.[7]

Although incorporating the quadratic irrationals gets us some way towards numbers adequate for Euclidean geometry, it does not do all that is needed. In the tenth (and most difficult) book of Euclid, numbers like $\sqrt{a+\sqrt{b}}$ are considered (with a and b positive rationals). These are *not* generally quadratic irrationals, but they occur, nevertheless, in ruler-and-compass constructions. Numbers sufficient for such geometric constructions would be those that can be built up from natural numbers by repeated use of the operations of addition, subtraction, multiplication, division, and the taking of square roots. But operating exclusively with such numbers gets extremely complicated, and these numbers are still too limited for considerations of Euclidean geometry that go beyond ruler-and-compass constructions. It is much more satisfactory to take the bold step—and how bold a step this actually is will be indicated in §§16.3–5—of allowing infinite continued-fraction expressions that are completely general. This provided the Greeks with a way of describing numbers that do turn out to be adequate for Euclidean geometry.

These numbers are indeed, in modern terminology, the so-called 'real numbers'. Although a fully satisfactory definition of such numbers is not regarded as having been found until the 19th century (with the work of Dedekind, Cantor, and others), the great ancient Greek mathematician and astronomer Eudoxos, who had been one of Plato's students, had obtained the essential ideas already in the 4th century BC. A few words about Eudoxos's ideas are appropriate here.

First, we note that the numbers in Euclidean geometry can be expressed in terms of *ratios* of lengths, rather than directly in terms of lengths. In this way, no specific unit of length (such as 'inch' or Greek 'dactylos') was needed. Moreover, with ratios of lengths, there would be no restriction as to how many such ratios might be multiplied together (obviating the apparent need for higher-dimensional 'hypervolumes' when more than three lengths are multiplied together). The first step in the Eudoxan theory was to supply a criterion as to when a length ratio $a : b$ would be *greater* than another such ratio $c : d$. This criterion is that some positive integers M and N exist such that the length a added to itself M times exceeds b added to itself N times, while also d added to itself N times exceeds c added to itself M times.[3.3] A corresponding criterion holds expressing the condition that the ratio $a : b$ be *less* than the ratio $c : d$. The condition for equality of these ratios would be that neither of these criteria hold. With this ingenious notion of 'equality' of such ratios, Eudoxos had, in effect, an

[3.3] Can you see why this works?

abstract concept of a 'real number' in terms of length ratios. He also provided rules for the sum and product of such real numbers.[3.4]

There was a basic difference in viewpoint, however, between the Greek notion of a real number and the modern one, because the Greeks regarded the number system as basically 'given' to us, in terms of the notion of *distance* in physical space, so the problem was to try to ascertain how these 'distance' measures actually behaved. For 'space' may well have had the appearance of being itself a Platonic absolute even though actual physical objects existing in this space would inevitably fall short of the Platonic ideal.[8] (However, we shall be seeing in §17.9 and §§19.6,8 how Einstein's general theory of relativity has now changed this perspective on space and matter in a fundamental way.)

A physical object such as a square drawn in the sand or a cube hewn from marble might have been regarded by the ancient Greeks as a reasonable or sometimes an excellent approximation to the Platonic geometrical ideal. Yet any such object would nevertheless provide a mere approximation. Lying behind such approximations to the Platonic forms—so it would have appeared—would be space itself: an entity of such abstract or notional existence that it could well have been regarded as a direct realization of a Platonic reality. The measure of distance in this ideal geometry would be something to *ascertain*; accordingly, it would be appropriate to try to extract this ideal notion of real number from a geometry of a Euclidean space that was assumed to be *given*. In effect, this is what Eudoxos succeeded in doing.

By the 19th and 20th centuries, however, the view had emerged that the mathematical notion of number should stand separately from the nature of physical space. Since mathematically consistent geometries other than that of Euclid had been shown to exist, this rendered it inappropriate to insist that the mathematical notion of 'geometry' should be necessarily extracted from the supposed nature of 'actual' physical space. Moreover, it could be very difficult, if not impossible, to ascertain the detailed nature of this supposed underlying 'Platonic physical geometry' in terms of the behaviour of imperfect physical objects. In order to know the nature of the numbers according to which 'geometrical distance' is to be defined, for example, it would be necessary to know what happens both at indefinitely tiny and indefinitely large distances. Even today, these questions are without clear-cut resolution (and I shall be addressing them again in later chapters). Thus, it was far more appropriate to develop the nature of number in a way that does not directly refer to physical measures. Accordingly, Richard Dedekind and Georg Cantor developed their ideas of what real numbers 'are' by use of notions that do not directly refer to geometry.

[3.4] Can you see how to formulate these?

Dedekind's definition of a real number is in terms of infinite sets of rational numbers. Basically, we think of the rational numbers, both positive and negative (and zero), to be arranged in order of size. We can imagine that this ordering takes place from left to right, where we think of the negative rationals as being displayed going off indefinitely to the left, with 0 in the middle, and the positive rationals displayed going off indefinitely to the right. (This is just for visualization purposes; in fact Dedekind's procedure is entirely abstract.) Dedekind imagines a 'cut' which divides this display neatly in two, with those to the left of the cut being all smaller than those to the right. When the 'knife-edge' of the cut does not 'hit' an actual rational number but falls between them, we say that it defines an *irrational* real number. More correctly, this occurs when those to the left have no actual largest member and those to the right, no actual smallest one. When the system of 'irrationals', as defined in terms of such cuts, is adjoined to the system of rational numbers that we already have, then the complete family of *real numbers* is obtained.

Dedekind's procedure leads, by means of simple definitions, directly to the laws of addition, subtraction, multiplication, and division for real numbers. Moreover, it enables one to go further and define *limits*, whereby such things as the infinite continued fraction that we saw before

$$1+(2+(2+(2+(2+\cdots)^{-1})^{-1})^{-1})^{-1}$$

or the infinite sum

$$1-\frac{1}{3}+\frac{1}{5}-\frac{1}{7}+\frac{1}{9}-\cdots$$

may be assigned real-number meanings. In fact, the first gives us the irrational number $\sqrt{2}$, and the second, $\frac{1}{4}\pi$. The ability to take limits is fundamental for many mathematical notions, and it is this that gives the real numbers their particular strengths.[9] (The reader may recall that the need for 'limiting procedures' was a requirement for the general definition of areas, as was indicated in §2.3.)

3.3 Real numbers in the physical world

There is a profound issue that is being touched upon here. In the development of mathematical ideas, one important initial driving force has always been to find mathematical structures that accurately mirror the behaviour of the physical world. But it is normally not possible to examine the physical world itself in such precise detail that appropriately clear-cut mathematical notions can be abstracted directly from it. Instead, progress is made because mathematical notions tend to have a 'momentum' of their own that appears to spring almost entirely from within the subject itself. Mathematical ideas develop, and various kinds of problem seem to arise naturally. Some of these (as was the case with the problem of finding the length of the diagonal of a square) can lead to an essential extension of the original mathematical concepts in terms of which the problem had been formulated. Such extensions may seem to be forced upon us, or they may arise in ways that appear to be matters of convenience, consistency, or mathematical elegance. Accordingly, the development of mathematics may seem to diverge from what it had been set up to achieve, namely simply to reflect physical behaviour. Yet, in many instances, this drive for mathematical consistency and elegance takes us to mathematical structures and concepts which turn out to mirror the physical world in a much deeper and more broad-ranging way than those that we started with. It is as though Nature herself is guided by the same kind of criteria of consistency and elegance as those that guide human mathematical thought.

An example of this is the real-number system itself. We have no direct evidence from Nature that there is a physical notion of 'distance' that extends to arbitrarily large scales; still less is there evidence that such a notion can be applied on the indefinitely tiny level. Indeed, there is no evidence that 'points in space' actually exist in accordance with a geometry that precisely makes use of real-number distances. In Euclid's day, there was scant evidence to support even the contention that such Euclidean 'distances' extended outwards beyond, say, about 10^{12} metres,[10] or inwards to as little as 10^{-5} metres. Yet, having been driven mathematically by the consistency and elegance of the real-number system, all of our broad-ranging and successful physical theories to date have, without exception, still clung to this ancient notion of 'real number'. Although there might appear to have been little justification for doing this from the evidence that was available in Euclid's day, our faith in the real-number system appears to have been rewarded. For our successful modern theories of cosmology now allow us to extend the range of our real-number distances out to about 10^{26} metres or more, while the accuracy of our theories of particle physics extends this range inwards to 10^{-17} metres or less. (The only scale at which it has been seriously proposed that a change might come about is some 18 orders of magnitude smaller even than that, namely 10^{-35} metres, which is the 'Planck scale' of quantum gravity that will feature strongly in some of our later discussions; e.g. §§31.1,6–12,14 and §32.7.) It may be regarded as a remarkable justification of our use of mathematical idealizations that the range of validity of the real-number system has extended from the total of about 10^{17}, from the smallest to the largest, that seemed appropriate in Euclid's day to at least the 10^{43} that our theories directly employ today, this representing a stupendous increase by a factor of some 10^{26}.

There is a good deal more to the physical validity of the real-number system than this. In the first place, we must consider that areas and volumes are also quantities for which real-number measures are accurately appropriate. A volume measure is the cube of a distance measure (and an area is the square of a distance). Accordingly, in the case of volumes, we may consider that it is the cube of the above range that is relevant. For Euclid's time, this would give us a range of about $(10^{17})^3 = 10^{51}$; for today's theories, at least $(10^{43})^3 = 10^{129}$. Moreover, there are other physical measures that require real-number descriptions, according to our presently successful theories. The most noteworthy of these is time. According to relativity theory, this needs to be adjoined to space to provide us with *spacetime* (which is the subject of our deliberations in Chapter 17). Spacetime volumes are four-dimensional, and it might well be considered that the temporal range (of again about 10^{43} or more in total range, in our well-tested theories) should also be incorporated into our considerations, giving a total of something like at least 10^{172}. We shall see some far larger real numbers even than this coming into our later considerations (see §27.13 and §28.7), although it is not really clear in some cases that the use of real numbers (rather than, say, integers) is essential.

More importantly for physical theory, from Archimedes, through Galileo and Newton, to Maxwell, Einstein, Schrödinger, Dirac, and the rest, a crucial role for the real-number system has been that it provides a necessary framework for the standard formulation of the *calculus* (see Chapter 6). All successful dynamical theories have required notions of the calculus for their formulations. Now, the conventional approach to calculus requires the *infinitesimal* nature of the reals to be what it is. That is to say, on the small end of the scale, it is the entire range of the real numbers that is in principle being made use of. The ideas of calculus underlie other physical notions, such as velocity, momentum, and energy. Consequently, the real-number system enters our successful physical theories in a fundamental way for our description of all these quantities also. Here, as mentioned earlier in connection with areas, in §2.3 and §3.2, the infinitesimal limit of small-scale structure of the real-number system is being called upon.

Yet we may still ask whether the real-number system is really 'correct' for the description of physical reality at its deepest levels. When quantum-mechanical ideas were beginning to be introduced early in the 20th century, there was the feeling that perhaps we were now beginning to witness a discrete or granular nature to the physical world at its smallest scales.[11] Energy could apparently exist only in discrete bundles—or 'quanta'—and the physical quantities of 'action' and 'spin' seemed to occur only in discrete multiples of a fundamental unit (see §§20.1,5 for the classical concept of *action* and §26.6 for its quantum counterpart; see §§22.8–12 for *spin*). Accordingly, various physicists attempted to build up an alternative picture of the world in which discrete processes governed all actions at the tiniest levels.

However, as we now understand quantum mechanics, that theory does not force us (nor even lead us) to the view that there is a discrete or granular nature to space, time, or energy at its tiniest levels (see Chapters 21 and 22, particularly the last sentence of §22.13). Nevertheless, the idea has remained with us that there may indeed be, at root, such a fundamental discreteness to Nature, despite the fact that quantum mechanics, in its standard formulation, certainly does not imply this. For example, the great quantum physicist Erwin Schrödinger was among the first to propose that a change to some form of fundamental spatial discreteness might actually be necessary:[12]

> The idea of a *continuous range*, so familiar to mathematicians in our days, is something quite exorbitant, an enormous extrapolation of what is accessible to us.

He related this proposal to some early Greek thinking concerning the discreteness of Nature. Einstein, also, suggested, in his last published words, that a discretely based ('algebraic') theory might be the way forward for the future physics:[13]

> One can give good reasons why reality cannot be represented as a continuous field.... Quantum phenomena... must lead to an attempt to find a purely algebraic theory for the description of reality. But nobody knows how to obtain the basis of such a theory.[14]

Others[15] also have pursued ideas of this kind; see §33.1. In the late 1950s, I myself tried this sort of thing, coming up with a scheme that I referred to as the theory of 'spin networks', in which the discrete nature of quantum-mechanical *spin* is taken as the fundamental building block for a *combinatorial* (i.e. discrete rather than real-number-based) approach to physics. (This scheme will be briefly described in §32.6.) Although my own ideas along this particular direction did not develop to a comprehensive theory (but, to some extent, became later transmogrified into 'twistor theory'; see §33.2), the theory of spin networks has now been imported, by others, into one of the major programmes for attacking the fundamental problem of *quantum gravity*.[16] I shall give brief descriptions of these various ideas in Chapter 32. Nevertheless, as tried and tested physical theory stands today—as it has for the past 24 centuries—real numbers still form a fundamental ingredient of our understanding of the physical world.

3.4 Do natural numbers need the physical world?

In the above description, in §3.2, of the Dedekind approach to the real-number system, I have presupposed that the *rational numbers* are already taken as 'understood'. In fact, it is not a difficult step from the integers to the rationals; rationals are just ratios of integers (see the Preface). What about the integers themselves, then? Are these rooted in physical ideas? The discrete approaches to physics that were referred to in the previous two paragraphs certainly depend upon our notion of *natural number* (i.e. 'counting number') and its extension, by the inclusion of the negative numbers, to the integers. Negative numbers were not considered, by the Greeks, to be actual 'numbers', so let us continue our considerations by first asking about the physical status of the natural numbers themselves.

The *natural numbers* are the quantities that we now denote by 0, 1, 2, 3, 4, etc., i.e. they are the non-negative whole numbers. (The modern procedure is to include 0 in this list, which is an appropriate thing to do from the mathematical point of view, although the ancient Greeks appear not to have recognized 'zero' as an actual number. This had to wait for the Hindu mathematicians of India, starting with Brahmagupta in 7th century and followed up by Mahavira and Bhaskara in the 9th and 12th century, respectively.) The role of the natural numbers is clear and unambiguous. They are indeed the most elementary 'counting numbers', which have a basic role whatever the laws of geometry or physics might be. Natural numbers are subject to certain familiar operations, most particularly the operations of *addition* (such as $37 + 79 = 116$) and *multiplication* (e.g. $37 \times 79 = 2923$), which enable pairs of natural numbers to be combined together to produce new natural numbers. These operations are independent of the nature of the geometry of the world.

We can, however, raise the question of whether the natural numbers themselves have a meaning or indeed existence independent of the actual nature of the physical world. Perhaps our notion of natural numbers depends upon there being, in our universe, reasonably well-defined discrete objects that persist in time. Natural numbers initially arise when we wish to count things, after all. But this seems to depend upon there actually being persistent distinguishable 'things' in the universe which are available to be 'counted'. Suppose, on the other hand, our universe were such that numbers of objects had a tendency to keep changing. Would natural numbers actually be 'natural' concepts in such a universe? Moreover, perhaps the universe actually contains only a finite number of 'things', in which case the 'natural' numbers might themselves come to an end at some point! We can even envisage a universe which consists only of an amorphous featureless substance, for which the very notion of numerical quantification might seem intrinsically inappropriate. Would the notion of 'natural number' be at all relevant for the description of universes of this kind?

Even though it might well be the case that inhabitants of such a universe would find our present mathematical concept of a 'natural number' difficult to come upon, it is hard to imagine that there would not still be an important role for such fundamental entities. There are various ways in which natural numbers can be introduced in pure mathematics, and these do not seem to depend upon the actual nature of the physical universe at all. Basically, it is the notion of a 'set' which needs to be brought into play, this being an abstraction that does not appear to be concerned, in any essential way, with the specific structure of the physical universe. In fact, there are certain definite subtleties concerning this question, and I shall return to that issue later (in §16.5). For the moment, it will be convenient to ignore such subtleties.

Let us consider one way (developed by Cantor from ideas of Giuseppe Peano, and promoted by the distinguished mathematician John von Neumann) that natural numbers can be introduced merely using the abstract notion of set. It also leads on to what are called 'ordinal numbers'. The simplest set of all is referred to as the 'null set' or the 'empty set', and it is characterized by the fact that it contains no members whatever! The empty set is usually denoted by the symbol $\varnothing$, and we can write this definition

$$\varnothing = \{\ \},$$

where the curly brackets delineate a *set*, the specific set under consideration having, as its members, the quantities indicated within the brackets. In this case, there is nothing within the brackets, so the set being described is indeed the empty set. Let us associate $\varnothing$ with the natural number 0. We can now proceed further and define the set whose only member is $\varnothing$; i.e. the set $\{\varnothing\}$. It is important to realize that $\{\varnothing\}$ is not the same set as the empty set $\varnothing$. The set $\{\varnothing\}$ has *one* member (namely $\varnothing$), whereas $\varnothing$ itself has none at all. Let us associate $\{\varnothing\}$ with the natural number 1. We next define the set whose two members are the two sets that we just encountered, namely $\varnothing$ and $\{\varnothing\}$, so this new set is $\{\varnothing, \{\varnothing\}\}$, which is to be associated with the natural number 2. Then we associate with 3 the collection of all the three entities that we have encountered up to this point, namely the set $\{\varnothing, \{\varnothing\}, \{\varnothing, \{\varnothing\}\}\}$, and with 4 the set $\{\varnothing, \{\varnothing\}, \{\varnothing, \{\varnothing\}\}, \{\varnothing, \{\varnothing\}, \{\varnothing, \{\varnothing\}\}\}\}$, whose members are again the sets that we have encountered previously, and so on. This may not be how we usually think of natural numbers, as a matter of definition, but it is one of the ways that mathematicians can come to the concept. (Compare this with the discussion in the Preface.) Moreover, it shows us, at least, that things like the natural numbers[17] can be conjured literally out of nothing, merely by employing the abstract notion of 'set'. We get an infinite sequence of abstract

(Platonic) mathematical entities—sets containing, respectively, zero, one, two, three, etc., elements, one set for each of the natural numbers, quite independently of the actual physical nature of the universe. In Fig.1.3 we envisaged a kind of independent 'existence' for Platonic mathematical notions—in this case, the natural numbers themselves—yet this 'existence' can seemingly be conjured up by, and certainly accessed by, the mere exercise of our mental imaginations, without any reference to the details of the nature of the physical universe. Dedekind's construction, moreover, shows how this 'purely mental' kind of procedure can be carried further, enabling us to 'construct' the entire system of real numbers,[18] still without any reference to the actual physical nature of the world. Yet, as indicated above, 'real numbers' indeed seem to have a direct relevance to the real structure of the world—illustrating the very mysterious nature of the 'first mystery' depicted in Fig.1.3.

3.5 Discrete numbers in the physical world

But I am getting slightly ahead of myself. We may recall that Dedekind's construction really made use of sets of *rational* numbers, not of natural numbers directly. As indicated above, it is not hard to 'define' what we mean by a rational number once we have the notion of natural number. But, as an intermediate step, it is appropriate to define the notion of an *integer*, which is a natural number or the *negative* of a natural number (the number zero being its own negative). In a formal sense, there is no difficulty in giving a mathematical definition of 'negative': roughly speaking we just attach a 'sign', written as '–', to each natural number (except 0) and define all the arithmetical rules of addition, subtraction, multiplication, and division (except by 0) consistently. This does not address the question of the 'physical meaning' of a negative number, however. What might it mean to say that there are minus three cows in a field, for example?

I think that it is clear that, unlike the natural numbers themselves, there is no evident physical content to the notion of a negative number of physical objects. Negative integers certainly have an extremely valuable organizational role, such as with bank balances and other financial transactions. But do they have direct relevance to the *physical* world? When I say 'direct relevance' here, I am not referring to circumstances where it would appear that it is negative real numbers that are the relevant measures, such as when a distance measured in one direction counts as positive while that measured in the opposite direction would count as negative (or the same thing with regard to time, in which times extending into the past might count as negative). I am referring, instead, to numbers that are *scalar* quantities, in the sense that there is no directional (or temporal) aspect to the quantity in question. In these circumstances it appears to be the case that it is the system of integers, both positive and negative, that has direct physical relevance.

It is a remarkable fact that only in about the last hundred years has it become apparent that the system of integers does indeed seem to have such direct physical relevance. The first example of a physical quantity which seems to be appropriately quantified by integers is *electric charge*.[19] As far as is known (although there is as yet no complete theoretical justification of this fact), the electric charge of any discrete isolated body is indeed quantified in terms of integral multiples, positive, negative, or zero, of one particular value, namely the charge on the proton (or on the electron, which is the negative of that of the proton).[20] It is now believed that protons are composite objects built up, in a sense, from smaller entities referred to as 'quarks' (and additional chargeless entities called 'gluons'). There are three quarks to each proton, the quarks having electric charges with respective values $\frac{2}{3}$, $\frac{2}{3}$, $-\frac{1}{3}$. These constituent charges add up to give the total value 1 for the proton. If quarks are fundamental entities, then the basic charge unit is one third of that which we seemed to have before. Nevertheless, it is still true that electric charge is measured in terms of integers, but now it is integer multiples of one third of a proton charge. (The role of quarks and gluons in modern particle physics will be discussed in §§25.3–7.)

Electric charge is just one instance of what is called an *additive quantum number*. Quantum numbers are quantities that serve to characterize the particles of Nature. Such a quantum number, which I shall here take to be a real number of some kind, is 'additive' if, in order to derive its value for a composite entity, we simply add up the individual values for the constituent particles—taking due account of the signs, of course, as with the above-mentioned case of the proton and its constituent quarks. It is a very striking fact, according to the state of our present physical knowledge, that all known additive quantum numbers[21] are indeed quantified in terms of the system of integers, not general real numbers, and not simply natural numbers—so that the negative values actually do occur.

In fact, according to 20th-century physics, there is now a certain sense in which it *is* meaningful to refer to a negative number of physical entities. The great physicist Paul Dirac put forward, in 1929–31, his theory of antiparticles, according to which (as it was later understood), for each type of particle, there is also a corresponding *antiparticle* for which each additive quantum number has precisely the negative of the value that it has for the original particle; see §§24.2,8. Thus, the system of integers (with negatives included) does indeed appear to have a clear relevance to the physical universe—a physical relevance that has become apparent only in the 20th century, despite those many centuries for which integers have found great value in mathematics, commerce, and many other human activities.

One important qualification should be made at this juncture, however. Although it is true that, in a sense, an antiproton is a negative proton, it is not really 'minus one proton'. The reason is that the sign reversal refers only to *additive* quantum numbers, whereas the notion of *mass* is not additive in modern physical theory. This issue will be explained in a bit more detail in §18.7. 'Minus one proton' would have to be an antiproton whose mass is the negative of the mass value of an ordinary proton. But the mass of an actual physical particle is not allowed to be negative. An antiproton has the same mass as an ordinary proton, which is a positive mass. We shall be seeing later that, according to the ideas of quantum field theory, there are things called 'virtual' particles for which the mass (or, more correctly, energy) can be negative. 'Minus one proton' would really be a virtual antiproton. But a virtual particle does not have an independent existence as an 'actual particle'.

Let us now ask the corresponding question about the rational numbers. Has this system of numbers found any direct relevance to the physical universe? As far as is known, this does not appear to be the case, at least as far as conventional theory is concerned. There are some physical curiosities[22] in which the family of rational numbers does play its part, but it would be hard to maintain that these reveal any fundamental physical role for rational numbers. On the other hand, it may be that there is a particular role for the rationals in fundamental quantum-mechanical probabilities (a rational probability possibly representing a choice between alternatives, each of which involves just a finite number of possibilities). This kind of thing plays a role in the theory of spin networks, as will be briefly described in §32.6. As of now, the proper status of these ideas is unclear.

Yet, there are other kinds of number which, according to accepted theory, do appear to play a fundamental role in the workings of the universe. The most important and striking of these are the *complex numbers*, in which the seemingly mystical quantity $\sqrt{-1}$, usually denoted by 'i', is introduced and adjoined to the real-number system. First encountered in the 16th century, but treated for hundreds of years with distrust, the mathematical utility of complex numbers gradually impressed the mathematical community to a greater and greater degree, until complex numbers became an indispensable, even magical, ingredient of our mathematical thinking. Yet we now find that they are fundamental not just to mathematics: these strange numbers also play an extraordinary and very basic role in the operation of the physical universe at its tiniest scales. This is a cause for wonder, and it is an even more striking instance of the convergence between mathematical ideas and the deeper workings of the physical universe than is the system of real numbers that we have been considering in this section. Let us come to these remarkable numbers next.

Notes

Section 3.1

3.1. The notations $>$, $<$, $\geqslant$, $\leqslant$, frequently used in this book, respectively stand for 'is greater than', 'is less than', 'is greater than or equal to', and 'is less than or equal to' (made appropriately grammatical).

3.2. Some readers might be aware of an apparently shorter argument which starts by demanding that a/b be 'in its lowest terms' (i.e. that a and b have no common factor). However, this assumes that such a lowest-terms expression always exists, which, though perfectly true, needs to be shown. Finding a lowest-term expression for a given fraction A/B (implicitly or explicitly—say using the procedure known as Euclid's algorithm; see, for example, Hardy and Wright 1945, p. 134; Davenport 1952, p. 26; Littlewood 1949, Chap. 4; and Penrose 1989, Chap. 2) involves reasoning similar to that given in the text, but more complicated.

3.3. One might well object that it is somewhat curious to use real numbers in the above proof, since the 'real rationals' (i.e. quotients of reals) would simply be real numbers all over again. This does not invalidate what has just been said, however. It may be remarked that it is as well that a and b were taken to be integers, in the original argument, and not themselves taken to be rationals. For, if a and b were merely rational, then the argument would fail at the 'decreasing sequence' part, even though the result itself would still be true.

Section 3.2

3.4. At a casual glance, expressions like $a+(b+(c+(d+\cdots)^{-1})^{-1})^{-1}$ may look rather odd. However, they are very natural in the context of ancient Greek thinking (although the Greeks did not use this particular notation). The procedure of *Euclid's algorithm* was referred to in Note 3.2 in the context of finding the lowest-term form of a fraction. Euclid's algorithm (when unravelled) leads precisely to such a continued fraction expression. The Greeks would apply this same procedure to the ratio of two geometrical lengths. In the most general case, the result would be an *infinite* continued fraction, of the kind considered here.

3.5. For more information (with proofs) concerning continued fractions, see the elegant account given in Chapter 4 of Davenport (1952). It may be remarked that in certain respects the continued-fraction representation of real numbers is deeper and more interesting than the normal one in terms of decimal expansions, finding applications in many different areas of modern mathematics, including the hyperbolic geometry discussed in §§2.4,5. On the other hand, continued fractions are not at all well suited for (most) practical calculation, the conventional decimal representation being far easier to use.

3.6. Quadratic irrationals are so called because they arise in the solution of a general quadratic equation

$$Ax^2+Bx+C=0,$$

with A non-zero, the solutions being

$$-\frac{B}{2A}+\sqrt{\left(\frac{B}{2A}\right)^2-\frac{C}{A}} \text{ and } -\frac{B}{2A}-\sqrt{\left(\frac{B}{2A}\right)^2-\frac{C}{A}}$$

where, to keep within the realm of real numbers, we must have B^2 greater than $4AC$. When A, B, and C are integers or rational numbers, and where there is no rational solution to the equation, the solutions are indeed quadratic irrationals.

3.7. Professor Stelios Negrepontis informs me that this evidence is to be found in the Platonic dialogue the Statesman (= Politikos), the third in the 'trilogy' the Theaetetos-the Sophist-the Politikos. See Negrepontis (2000).

3.8. See Sorabji (1984, 1988) for an account of ancient Greek thinking on the nature of space.

3.9. See Hardy (1914); Conway (1976); Burkill (1962).

Section 3.3

3.10. The scientific notation '10^{12}' for a 'million million' makes use of *exponents*, as described in Notes 1.2 and 2.4. In this book, I shall tend to avoid verbal terms such as 'million', and especially 'billion', in preference to this much clearer scientific notation. The word 'billion' is particularly confusing, as in American usage—now commonly adopted also in the UK—'billion' refers to 10^9, whereas, in the older (more logical) UK usage, in agreement with most other European languages, it refers to 10^{12}. Negative exponents, such as in 10^{-6} (which refers to 'one millionth'), are also used here in accordance with the normal scientific notation.

The distance 10^{12} metres is about 7 times the Earth–Sun separation.This is roughly the distance of the planet Jupiter, although that distance was not known in Euclid's day and would have been guessed to be rather smaller.

3.11. See, for example, Russell (1927), Chap. 4.

3.12. Schrödinger (1952), pp. 30–1.

3.13. See Stachel (1995).

3.14. Einstein (1955), p. 166.

3.15. See e.g. Snyder (1947); Schild (1949); and Ahmavaara (1965).

3.16. See Ashtekar (1986); Ashtekar and Lewandowski (2004); Smolin (1998, 2001); Rovelli (1998, 2003).

Section 3.4

3.17. The notion of 'ordinal number', that is implied here in the finite case, extends also to *infinite* ordinal numbers, the smallest being Cantor's 'ω', which is the ordered collection of *all* finite ordinals.

3.18. This notion of 'construct' should not be taken in too strong a sense, however. We shall be finding in §16.6 that there are certain real numbers (in fact most of them) that are inaccessible by any computational procedure.

Section 3.5

3.19. The Irish physicist George Johnstone Stoney was the first, in 1874, to give a (crude) estimate of the basic electric charge, and, in 1891, coined the term 'electron' for this fundamental unit. In 1909, the American physicist Robert Andrews Millikan designed his famous 'oil-drop' experiment, which precisely showed that the charge on electrically charged bodies (the oil drops, in his experiment) came in integer multiples of a well-defined value—the electron charge.

3.20. In 1959, R. A. Lyttleton and H. Bondi proposed that a slight difference in the proton and (minus) the electron charges, of the order of one part in 10^{18} might account for the expansion of the universe, (for which, see §§27.11,13, and Chapter 28). See Lyttleton and Bondi (1959). Unfortunately, for this theory, such a discrepancy was soon disproved in several experiments. Nevertheless, this idea provided an excellent example of creative thinking.

3.21. I am here distinguishing the 'additive' quantum numbers from the numbers that physicists call 'multiplicative', which we shall come to in §5.5.

3.22. For example, in the 'fractional quantum Hall effect', one finds rational numbers playing a key role; see, for example, Fröhlich and Pedrini (2000).

4
Magical complex numbers

4.1 The magic number 'i'

How is it that -1 can have a square root? The square of a positive number is always positive, and the square of a negative number is again positive (and the square of 0 is just 0 again, so that is hardly of use to us here). It seems impossible that we can find a number whose square is actually negative. Yet, this is the kind of situation that we have seen before, when we ascertained that 2 has no square root within the system of rational numbers. In that case we resolved the situation by extending our system of numbers from the rationals to a larger system, and we settled on the system of reals. Perhaps the same trick will work again.

Indeed it will. In fact what we have to do is something much easier and far less drastic than the passage from the rationals to the reals. (Raphael Bombelli introduced the procedure in 1572 in his work *L'Algebra*, following Gerolamo Cardano's original encounters with complex numbers in his *Ars Magna* of 1545.) All we need do is introduce a single quantity, called 'i', which is to square to -1, and adjoin it to the system of reals, allowing combinations of i with real numbers to form expressions such as

$$a + \mathrm{i}b,$$

where a and b are arbitrary real numbers. Any such combination is called a *complex number*. It is easy to see how to add complex numbers:

$$(a + \mathrm{i}b) + (c + \mathrm{i}d) = (a + c) + \mathrm{i}(b + d)$$

which is of the same form as before (with the real numbers $a + c$ and $b + d$ taking the place of the a and b that we had in our original expression). What about multiplication? This is almost as easy. Let us find the product of $a + \mathrm{i}b$ with $c + \mathrm{i}d$. We first simply multiply these factors, expanding the expression using the ordinary rules of algebra:[1]

$$\begin{aligned}(a + \mathrm{i}b)(c + \mathrm{i}d) &= ac + \mathrm{i}bc + a\mathrm{i}d + \mathrm{i}b\mathrm{i}d \\ &= ac + \mathrm{i}(bc + ad) + \mathrm{i}^2 bd.\end{aligned}$$

But $i^2 = -1$, so we can rewrite this as

$$(a + ib)(c + id) = (ac - bd) + i(bc + ad),$$

which is again of the same form as our original $a + ib$, but with $ac - bd$ taking the place of a and $bc + ad$ taking the place of b.

It is easy enough to subtract two complex numbers, but what about division? Recall that in the ordinary arithmetic we are allowed to divide by any real number that is not zero. Now let us try to divide the complex number $a + ib$ by the complex number $c + id$. We must take the latter to be non-zero, which means that the real numbers c and d cannot both be zero. Hence $c^2 + d^2 > 0$, and therefore $c^2 + d^2 \neq 0$, so we are allowed to divide by $c^2 + d^2$. It is a direct exercise[4.1] to check (multiplying both sides of the expression below by $c + id$) that

$$\frac{(a + ib)}{(c + id)} = \frac{ac + bd}{c^2 + d^2} + i\frac{bc - ad}{c^2 + d^2}.$$

This is of the same general form as before, so it is again a complex number.

When we get used to playing with these complex numbers, we cease to think of $a + ib$ as a *pair* of things, namely the two real numbers a and b, but we think of $a + ib$ as an entire thing on its own, and we could use a single letter, say z, to denote the whole complex number $z = a + ib$. It may be checked that all the normal rules of algebra are satisfied by complex numbers.[4.2] In fact, all this is a good deal more straightforward than checking everything for real numbers. (For that check, we imagine that we had previously convinced ourselves that the rules of algebra are satisfied for fractions, and then we have to use Dedekind's 'cuts' to show that the rules still work for real numbers.) From this point of view, it seems rather extraordinary that complex numbers were viewed with suspicion for so long, whereas the much more complicated extension from the rationals to the reals had, after ancient Greek times, been generally accepted without question.

Presumably this suspicion arose because people could not 'see' the complex numbers as being presented to them in any obvious way by the physical world. In the case of the real numbers, it had seemed that distances, times, and other physical quantities were providing the reality that such numbers required; yet the complex numbers had appeared to be merely invented entities, called forth from the imaginations of mathemat-

[4.1] Do this. (Alternatively, can you check it by multiplying both sides by $(c - id)$?)

[4.2] Check this, the relevant rules being $w + z = z + w$, $w + (u + z) = (w + u) + z$, $wz = zw$, $w(uz) = (wu)z$, $w(u + z) = wu + wz$, $w + 0 = w$, $w1 = w$.

icians who desired numbers with a greater scope than the ones that they had known before. But we should recall from §3.3 that the connection the mathematical real numbers have with those physical concepts of length or time is not as clear as we had imagined it to be. We cannot directly see the minute details of a Dedekind cut, nor is it clear that arbitrarily great or arbitrarily tiny times or lengths actually exist in nature. One could say that the so-called 'real numbers' are as much a product of mathematicians' imaginations as are the complex numbers. Yet we shall find that complex numbers, as much as reals, and perhaps even more, find a unity with nature that is truly remarkable. It is as though Nature herself is as impressed by the scope and consistency of the complex-number system as we are ourselves, and has entrusted to these numbers the precise operations of her world at its minutest scales. In Chapters 21–23, we shall be seeing, in detail, how this works.

Moreover, to refer just to the scope and to the consistency of complex numbers does not do justice to this system. There is something more which, in my view, can only be referred to as 'magic'. In the remainder of this chapter, and in the next, I shall endeavour to convey to the reader something of the flavour of this magic. Then, in Chapters 7–9, we shall again witness this complex-number magic in some of its most striking and unexpected manifestations.

Over the four centuries that complex numbers have been known, a great many magical qualities have been gradually revealed. Yet this is a magic that had been perceived to lie within mathematics, and it indeed provided a utility and a depth of mathematical insight that could not be achieved by use of the reals alone. There had not been any reason to expect that the physical world should be concerned with it. And for some 350 years from the time that these numbers were introduced through the works of Cardano and Bombelli, it was purely through their mathematical role that the magic of the complex-number system was perceived. It would, no doubt, have come as a great surprise to all those who had voiced their suspicion of complex numbers to find that, according to the physics of the latter three-quarters of the 20th century, the laws governing the behaviour of the world, at its tiniest scales, is fundamentally governed by the complex-number system.

These matters will be central to some of the later parts of this book (particularly in Chapters 21–23, 26, and 31–33). For the moment, let us concentrate on some of the *mathematical* magic of complex numbers, leaving their physical magic until later. Recall that all we have done is to demand that -1 have a square root, together with demanding that the normal laws of arithmetic be retained, and we have ascertained that these demands can be satisfied consistently. This seems like a fairly simple thing to have done. But now for the magic!

4.2 Solving equations with complex numbers

In what follows, I shall find it necessary to introduce somewhat more mathematical notation than previously. I apologize for this. However, it is hardly possible to convey serious mathematical ideas without the use of a certain amount of notation. I appreciate that there will be many readers who are uncomfortable with these things. My advice to such readers is basically just to read the words and not to bother too much about trying to understand the equations. At least, just skim over the various formulae and press on. There will, indeed, be quite a number of serious mathematical expressions scattered about this book, particularly in some of the later chapters. My guess is that certain aspects of understanding will eventually begin to come through even if you make little attempt to understand what all the expressions actually mean in detail. I hope so, because the magic of complex numbers, in particular, is a miracle well worth appreciating. If you can cope with the mathematical notation, then so much the better.

First of all, we may ask whether other numbers have square roots. What about -2, for example? That's easy. The complex number $i\sqrt{2}$ certainly squares to -2, and so also does $-i\sqrt{2}$. Moreover, for any positive real number a, the complex number $i\sqrt{a}$ squares to $-a$, and $-i\sqrt{a}$ does also. There is no real magic here. But what about the general complex number $a + ib$ (where a and b are real)? We find that the complex number

$$\sqrt{\frac{1}{2}\left(a+\sqrt{a^2+b^2}\right)}+i\sqrt{\frac{1}{2}\left(-a+\sqrt{a^2+b^2}\right)}$$

squares to $a + ib$ (and so does its negative).[4.3] Thus, we see that, even though we only adjoined a square root for a single quantity (namely -1), we find that every number in the resulting system now automatically has a square root! This is quite different from what happened in the passage from the rationals to the reals. In that case, the mere introduction of the quantity $\sqrt{2}$ into the system of rationals would have got us almost nowhere.

But this is just the very beginning. We can ask about cube roots, fifth roots, 999th roots, πth roots—or even i-th roots. We find, miraculously, that whatever complex root we choose and whatever complex number we apply it to (excluding 0), there is always a complex-number solution to this problem. (In fact, there will normally be a number of different solutions to the problem, as we shall be seeing shortly. We noted above that for square roots we get two solutions, the negative of the square root of a complex number z being also a square root of z. For higher roots there are more solutions; see §5.4.)

[4.3] Check this.

We are still barely scratching the surface of complex-number magic. What I have just asserted above is really quite simple to establish (once we have the notion of a *logarithm* of a complex number, as we shall shortly, in Chapter 5). Somewhat more remarkable is the so-called 'fundamental theorem of algebra' which, in effect, asserts that any polynomial equation, such as

$$1 - z + z^4 = 0$$

or

$$\pi + \mathrm{i}z - \sqrt{417}z^3 + z^{999} = 0,$$

must have complex-number solutions. More explicitly, there will always be a solution (normally several different ones) to any equation of the form

$$a_0 + a_1 z + a_2 z^2 + a_3 z^3 + \cdots + a_n z^n = 0,$$

where $a_0, a_1, a_2, a_3, \ldots, a_n$ are given complex numbers with the a_n taken as non-zero.[2] (Here n can be any positive integer that we care to choose, as big as we like.) For comparison, we may recall that i was introduced, in effect, simply to provide a solution to the one particular equation

$$1 + z^2 = 0.$$

We get all the rest *free*!

Before proceeding further, it is worth mentioning the problem that Cardano had been concerned with, from around 1539, when he first encountered complex numbers and caught a hint of another aspect of their attendant magical properties. This problem was, in effect, to find an expression for the general solution of a (real) *cubic* equation (i.e. $n = 3$ in the above). Cardano found that the general cubic could be reduced to the form

$$x^3 = 3px + 2q$$

by a simple transformation. Here p and q are to be real numbers, and I have reverted to the use of x in the equation, rather than z, to indicate that we are now concerned with real-number solutions rather than complex ones. Cardano's complete solution (as published in his 1545 book *Ars Magna*) seems to have been developed from an earlier partial solution that he had learnt in 1539 from Niccolò Fontana ('Tartaglia'), although this partial solution (and perhaps even the complete solution) had been found earlier (before 1526) by Scipione del Ferro.[3] The (del Ferro–)Cardano solution was essentially the following (written in modern notation):

$$x = (q + w)^{\frac{1}{3}} + (q - w)^{\frac{1}{3}},$$

where

$$w = (q^2 - p^3)^{\frac{1}{2}}.$$

Now this equation presents no fundamental problem within the system of real numbers if

$$q^2 \geqslant p^3.$$

In this case there is just one real solution x to the equation, and it is indeed correctly given by the (del Ferro–)Cardano formula, as given above. But if

$$q^2 < p^3,$$

the so-called *irreducible case*, then, although there are now *three* real solutions, the formula involves the square root of the *negative* number $q^2 - p^3$ and so it cannot be used without bringing in complex numbers. In fact, as Bombelli later showed (in Chapter 2 of his *L'Algebra* of 1572), if we do allow ourselves to admit complex numbers, then all three real solutions are indeed correctly expressed by the formula.[4] (This makes sense because the expression provides us with two complex numbers added together, where the parts involving i cancel out in the sum, giving a real-number answer.[5]) What is mysterious about this is that even though it would seem that the problem has nothing to do with complex numbers—the equation having real coefficients and all its solutions being real (in this 'irreducible' case)—we need to journey through this seemingly alien territory of the complex-number world in order that the formula may allow us to return with our purely real-number solutions. Had we restricted ourselves to the straight and narrow 'real' path, we should have returned empty-handed. (Ironically, *complex* solutions to the original equation can only come about in those cases when the formula does *not* necessarily involve this complex journey.)

4.3 Convergence of power series

Despite these remarkable facts, we have still not got very far into complex-number magic. There is much more to come! For example, one area where complex numbers are invaluable is in providing an understanding of the behaviour of what are called *power series*. A power series is an *infinite* sum of the form

$$a_0 + a_1x + a_2x^2 + a_3x^3 + \cdots.$$

Because this sum involves an infinite number of terms, it may be the case that the series *diverges*, which is to say that it does not settle down to a particular finite value as we add up more and more of its terms. For an example, consider the series

$$1 + x^2 + x^4 + x^6 + x^8 + \cdots$$

(where I have taken $a_0 = 1, a_1 = 0, a_2 = 1, a_3 = 0, a_4 = 1, a_5 = 0, a_6 = 1, \ldots$). If we put $x = 1$, then, adding the terms successively, we get

$$1, \quad 1+1=2, \quad 1+1+1=3,$$
$$1+1+1+1=4, \quad 1+1+1+1+1=5, \quad \text{etc.,}$$

and we see that the series has no chance of settling down to a particular finite value, that is, it is *divergent*. Things are even worse if we try $x = 2$, for example, since now the individual terms are getting bigger, and adding terms successively we get

$$1, \quad 1+4=5, \quad 1+4+16=21, \quad 1+4+16+64=85, \quad \text{etc.,}$$

which clearly diverges. On the other hand, if we put $x = \frac{1}{2}$, then we get

$$1, \quad 1+\tfrac{1}{4}=\tfrac{5}{4}, \quad 1+\tfrac{1}{4}+\tfrac{1}{16}=\tfrac{21}{16}, \quad 1+\tfrac{1}{4}+\tfrac{1}{16}+\tfrac{1}{64}=\tfrac{85}{64}, \quad \text{etc.,}$$

and it turns out that these numbers become closer and closer to the limiting value $\frac{4}{3}$, so the series is now *convergent*.

With this series, it is not hard to appreciate, in a sense, an underlying reason why the series cannot help but diverge for $x = 1$ and $x = 2$, while converging for $x = \frac{1}{2}$ to give the answer $\frac{4}{3}$. For we can explicitly write down the 'answer' to the sum of the entire series, finding[4.4]

$$1 + x^2 + x^4 + x^6 + x^8 + \cdots = (1 - x^2)^{-1}.$$

When we substitute $x = 1$, we find that this answer is $(1 - 1^2)^{-1} = 0^{-1}$, which is 'infinity',[6] and this provides us with an understanding of why the series has to diverge for that value of x. When we substitute $x = \frac{1}{2}$, the answer is $(1 - \frac{1}{4})^{-1} = \frac{4}{3}$, and the series actually converges to this particular value, as stated above.

This all seems very sensible. But what about $x = 2$? Now there is an 'answer' given by the explicit formula, namely $(1 - 4)^{-1} = -\frac{1}{3}$, although we do not seem to get this value simply by adding up the terms of the series. We could hardly get this answer because we are just adding together positive quantities, whereas $-\frac{1}{3}$ is negative. The reason that the series diverges is that, when $x = 2$, each term is actually bigger than the corresponding term was when $x = 1$, so that divergence for $x = 2$ follows, logically, from the divergence for $x = 1$. In the case of $x = 2$, it is not that the 'answer' is really infinite, but that we cannot reach this answer by attempting to sum the series directly. In Fig. 4.1, I have plotted the partial sums of the series (i.e. the sums up to some finite number of terms), successively up to four terms, together with the 'answer' $(1 - x^2)^{-1}$

[4.4] Can you see how to check this expression, in a formal algebraic way?

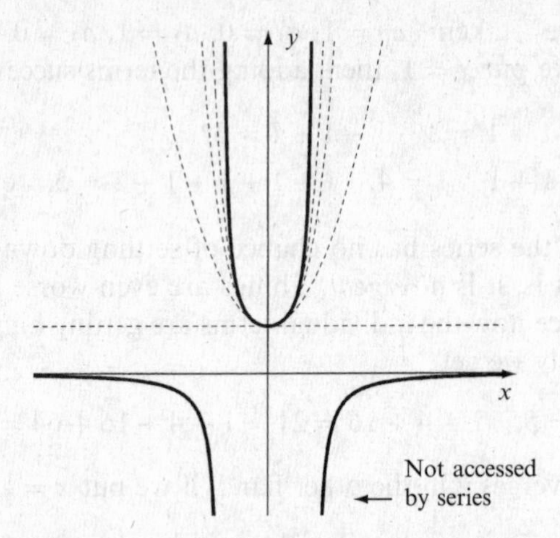

Fig. 4.1 The respective partial sums, 1, $1+x^2$, $1+x^2+x^4$, $1+x^2+x^4+x^6$ (dashed lines) of the series for $(1-x^2)^{-1}$ are plotted, illustrating the convergence of the series to $(1-x^2)^{-1}$ for $|x|<1$ and divergence for $|x|>1$.

and we can see that, provided x lies strictly[7] between the values -1 and $+1$, the curves depicting these partial sums do indeed converge on this answer, namely $(1-x^2)^{-1}$, as we expect. But outside this range, the series simply diverges and does not actually reach any finite value at all.

As a slight digression, it will be helpful to address a certain issue here that will be of importance to us later. Let us ask the following question: does the equation that we obtain by putting $x=2$ in the above expression, namely

$$1+2^2+2^4+2^6+2^8+\cdots=(1-2^2)^{-1}=-\frac{1}{3},$$

actually make any sense? The great 18th-century mathematician Leonhard Euler often wrote down equations like this, and it has become fashionable to poke gentle fun at him for holding to such absurdities, while one might excuse him on the grounds that in those early days nothing was properly understood about matters of 'convergence' of series and the like. Indeed, it is true that the rigorous mathematical treatment of series did not come about until the late 18th and early 19th century, through the work of Augustin Cauchy and others. Moreover, according to this rigorous treatment, the above equation would be officially classified as 'nonsense'. Yet, I think that it is important to appreciate that, in the appropriate sense, Euler really knew what he was doing when he wrote down apparent absurdities of this nature, and that there are senses according to which the above equation must be regarded as 'correct'.

In mathematics, it is indeed imperative to be absolutely clear that one's equations make strict and accurate sense. However, it is equally important not to be insensitive to 'things going on behind the scenes' which may ultimately lead to deeper insights. It is easy to lose sight of such things by adhering too rigidly to what appears to be strictly logical, such as the fact that the sum of the positive terms $1 + 4 + 16 + 64 + 256 + \cdots$ cannot possibly be $-\frac{1}{3}$. For a pertinent example, let us recall the logical absurdity of finding a real solution to the equation $x^2 + 1 = 0$. There is no solution; yet, if we leave it at that, we miss all the profound insights provided by the introduction of complex numbers. A similar remark applies to the absurdity of a rational solution to $x^2 = 2$. In fact, it is perfectly possible to give a mathematical sense to the answer '$-\frac{1}{3}$' to the above infinite series, but one must be careful about the rules telling us what is allowed and what is not allowed. It is not my purpose to discuss such matters in detail here,[8] but it may be pointed out that in modern physics, particularly in the area of quantum field theory, divergent series of this nature are frequently encountered (see particularly §§26.7,9 and §31.13). It is a very delicate matter to decide whether the 'answers' that are obtained in this way are actually meaningful and, moreover, actually correct. Sometimes extremely accurate answers are indeed obtained by manipulating such divergent expressions and are occasionally strikingly confirmed by comparison with actual physical experiment. On the other hand, one is often not so lucky. These delicate issues have important roles to play in current physical theories and are very relevant for our attempts to assess them. The point of immediate relevance to us here is that the 'sense' that one may be able to attribute to such apparently meaningless expressions frequently depends, in an essential way, upon the properties of complex numbers.

Let us now return to the issue of the convergence of series, and try to see how complex numbers fit into the picture. For this, let us consider a function just slightly different from $(1 - x^2)^{-1}$, namely $(1 + x^2)^{-1}$, and try to see whether it has a sensible power series expansion. There would seem to be a better chance of complete convergence now, because $(1 + x^2)^{-1}$ remains smooth and finite over the entire range of real numbers. There is, indeed, a simple-looking power series for $(1 + x^2)^{-1}$, only slightly different from the one that we had before, namely

$$1 - x^2 + x^4 - x^6 + x^8 - \cdots = (1 + x^2)^{-1},$$

the difference being merely a change of sign in alternate terms.[4.5] In Fig. 4.2, I have plotted the partial sums of the series, successively up to five terms, similarly to before, together with this answer $(1 + x^2)^{-1}$. What seems surprising is that the partial sums still only converge on the answer

---

[4.5] Can you see an elementary reason for this simple relationship between the two series?

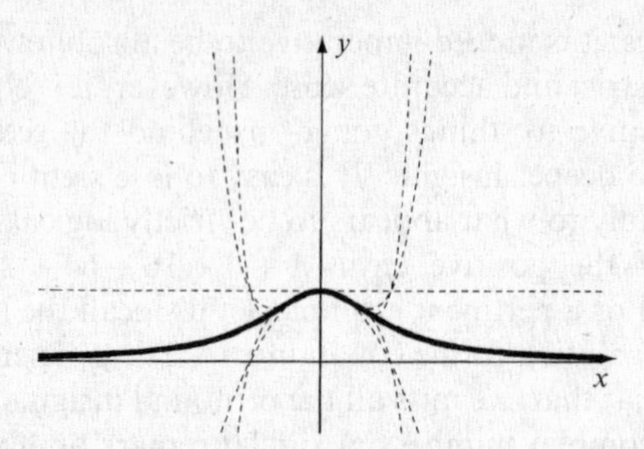

Fig. 4.2 The partial sums, 1, $1 - x^2$, $1 - x^2 + x^4$, $1 - x^2 + x^4 - x^6$, $1 - x^2 + x^4 - x^6 + x^8$, of the series for $(1 + x^2)^{-1}$ are likewise plotted, and again there is convergence for $|x| < 1$ and divergence for $|x| > 1$, despite the fact that the function is perfectly well behaved at $x = \pm 1$.

in the range strictly between values -1 and $+1$. We appear to be getting a divergence outside this range, even though the answer does not go to infinity at all, unlike in our previous case. We can test this explicitly using the same three values $x = 1$, $x = 2$, $x = \frac{1}{2}$ that we used before, finding that, as before, convergence occurs only in the case $x = \frac{1}{2}$, where the answer comes out correctly with the limiting value $\frac{4}{5}$ for the sum of the entire series:

$$x = 1: \quad 1, 0, 1, 0, 1, 0, 1, \text{etc.},$$

$$x = 2: \quad 1, -3, 13, -51, 205, -819, \text{etc.},$$

$$x = \tfrac{1}{2}: \quad 1, \tfrac{3}{4}, \tfrac{13}{16}, \tfrac{51}{64}, \tfrac{205}{256}, \tfrac{819}{1024}, \text{etc.}$$

We note that the 'divergence' in the first case is simply a failure of the partial sums of the series ever to settle down, although they do not actually diverge to infinity.

Thus, in terms of real numbers alone, there is a puzzling discrepancy between actually summing the series and passing directly to the 'answer' that the sum to infinity of the series is supposed to represent. The partial sums simply 'take off' (or, rather, flap wildly up and down) just at the same places (namely $x = \pm 1$) as where trouble arose in the previous case, although now the supposed answer to the infinite sum, namely $(1 + x^2)^{-1}$, does not exhibit any noticeable feature at these places at all. The resolution of the mystery is to be found if we examine *complex* values of this function rather than restricting our attention to real ones.

4.4 Caspar Wessel's complex plane

In order to see what is going on here, it will be important to use the now-standard *geometrical* representation of complex numbers in the Euclidean plane. Caspar Wessel in 1797, Jean Robert Argand in 1806, John Warren in 1828, and Carl Friedrich Gauss well before 1831, all independently, came up with the idea of the *complex plane* (see Fig. 4.3), in which they gave clear geometrical interpretations of the operations of addition and multiplication of complex numbers. In Fig. 4.3, I have used standard Cartesian axes, with the x-axis going off to the right horizontally and the y-axis going vertically upwards. The complex number

$$z = x + \mathrm{i}y$$

is represented as the point with Cartesian coordinates (x, y) in the plane.

We are now to think of a real number x as a particular case of the complex number $z = x + \mathrm{i}y$ where $y = 0$. Thus we are thinking of the x-axis in our diagram as representing the *real line* (i.e. the totality of real numbers, linearly ordered along a straight line). The complex plane, therefore, gives us a direct pictorial representation of how the system of real numbers extends outwards to become the entire system of complex numbers. This real line is frequently referred to as the 'real axis' in the complex plane. The y-axis is, correspondingly, referred to as the 'imaginary axis'. It consists of all real multiples of i.

Let us now return to our two functions that we have been trying to represent in terms of power series. We took these as functions of the real variable x, namely $(1 - x^2)^{-1}$ and $(1 + x^2)^{-1}$, but now we are going to extend these functions so that they apply to a complex variable z. There

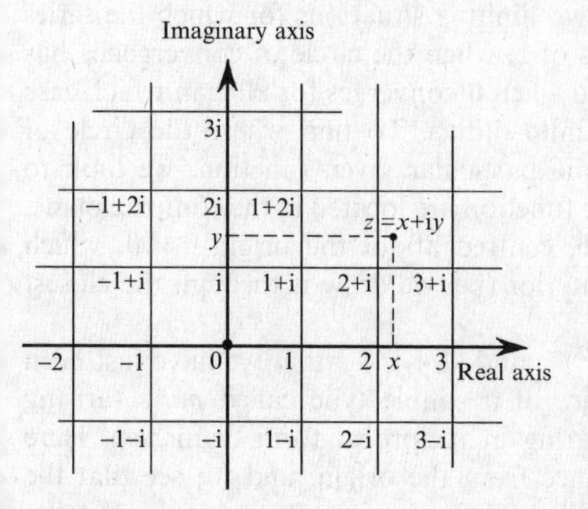

Fig. 4.3 The complex plane of $z = x + \mathrm{i}y$. In Cartesian coordinates (x, y), the x-axis horizontally to the right is the *real* axis; the y-axis vertically upwards is the *imaginary* axis.

is no problem about doing this, and we simply write these extended functions as $(1-z^2)^{-1}$ and $(1+z^2)^{-1}$, respectively. In the case of the first real function $(1-x^2)^{-1}$, we were able to recognize where the 'divergence' trouble starts, because the function is *singular* (in the sense of becoming infinite) at the two places $x = -1$ and $x = +1$; but, with $(1+x^2)^{-1}$, we saw no singularity at these places and, indeed, no real singularities at all. However, in terms of the complex variable z, we see that these two functions are much more on a par with one another. We have noted the singularities of $(1-z^2)^{-1}$ at two points $z = \pm 1$, of unit distance from the origin along the real axis; but now we see that $(1+z^2)^{-1}$ also has singularities, namely at the two places $z = \pm \mathrm{i}$ (since then $1 + z^2 = 0$), these being the two points of unit distance from the origin on the imaginary axis.

But what do these complex singularities have to do with the question of convergence or divergence of the corresponding power series? There is a striking answer to this question. We are now thinking of our power series as functions of the complex variable z, rather than the real variable x, and we can ask for those locations of z in the complex plane for which the series converges and those for which it diverges. The remarkable general answer,[9] for any power series whatever

$$a_0 + a_1 z + a_2 z^2 + a_3 z^3 + \cdots,$$

is that there is some circle in the complex plane, centred at 0, called the *circle of convergence*, with the property that if the complex number z lies strictly inside the circle then the series converges for that value of z, whereas if z lies strictly outside the circle then the series diverges for that value of z. (Whether or not the series converges when z lies actually on the circle is a somewhat delicate issue that will not concern us here, although it has relevance to the issues that we shall come to in §§9.6,7.) In this statement, I am including the two limiting situations for which the series diverges for all non-zero values of z, when the circle of convergence has shrunk down to zero radius, and when it converges for all z, in which case the circle has expanded to infinite radius. To find where the circle of convergence actually is for some particular given function, we look to see where the singularities of the function are located in the complex plane, and we draw the largest circle, centred about the origin $z = 0$, which contains no singularity in its interior (i.e. we draw it through the closest singularity to the origin).

In the particular cases $(1-z^2)^{-1}$ and $(1+z^2)^{-1}$ that we have just been considering, the singularities are of a simple type called *poles* (arising where some polynomial, appearing in reciprocal form, vanishes). Here these poles all lie at unit distance from the origin, and we see that the

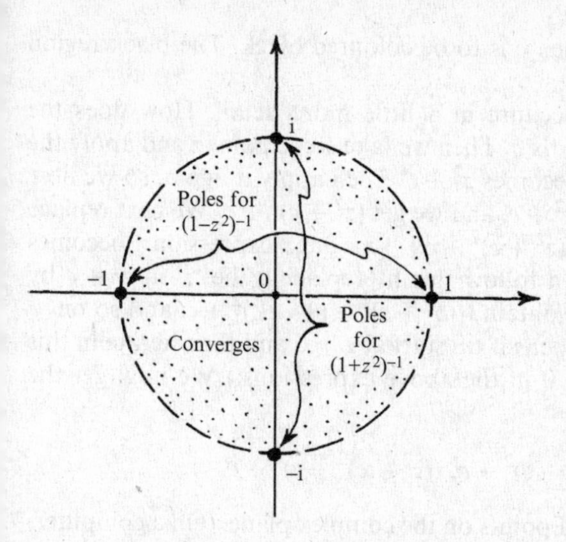

Fig. 4.4 In the complex plane, the functions $(1 - z^2)^{-1}$ and $(1 + z^2)^{-1}$ have the same circle of convergence, there being poles for the former at $z = \pm 1$ and poles for the latter at $z = \pm$ i, all having the same (unit) distance from the origin.

circle of convergence is, in both cases, just the unit circle about the origin. The places where this circle meets the real axis are the same in each case, namely the two points $z = \pm 1$ (see Fig. 4.4). This explains why the two functions converge and diverge in the same regions—a fact that is not manifest from their properties simply as functions of real variables. Thus, complex numbers supply us with deep insights into the behaviour of power series that are simply not available from the consideration of their real-variable structure.

4.5 How to construct the Mandelbrot set

To end this chapter, let us look at another type of convergence/divergence issue. It is the one that underlies the construction of that extraordinary configuration, referred to in §1.3 and depicted in Fig. 1.2, known as the *Mandelbrot set*. In fact, this is just a subset of Wessel's complex plane which can be defined in a surprisingly simple way, considering the extreme complication of this set. All we need to do is examine repeated applications of the replacement

$$z \mapsto z^2 + c,$$

where c is some chosen complex number. We think of c as a point in the complex plane and start with $z = 0$. Then we *iterate* this transformation (i.e. repeatedly apply it again and again) and see how the point z in the plane behaves. If it wanders off to infinity, then the point c is to be coloured white. If z wanders around in some restricted region without ever receding to infinity, then c is to be coloured black. The black region gives us the Mandelbrot set.

Let us describe this procedure in a little more detail. How does the iteration proceed? First, we fix c. Then we take some point z and apply the transformation, so that z becomes $z^2 + c$. Then apply it again, so we now replace the 'z' in $z^2 + c$ by $z^2 + c$, and we get $(z^2 + c)^2 + c$. We next replace the 'z' in $z^2 + c$ by $(z^2 + c)^2 + c$, so our expression becomes $((z^2 + c)^2 + c)^2 + c$. We then follow this by replacing the 'z' in $z^2 + c$ by $((z^2 + c)^2 + c)^2 + c$, and we obtain $(((z^2 + c)^2 + c)^2 + c)^2 + c$, and so on.

Let us now see what happens if we start at $z = 0$ and then iterate in this way. (We can just put $z = 0$ in the above expressions.) We now get the sequence

$$0,\ c,\ c^2 + c,\ (c^2 + c)^2 + c,\ ((c^2 + c)^2 + c)^2 + c,\ \ldots .$$

This gives us a succession of points on the complex plane. (On a computer, one would just work these things out purely numerically, for each individual choice of the complex number c, rather than using the above algebraic expressions. It is computationally much 'cheaper' just to do the arithmetic afresh each time.) Now, for any given value of c, one of two things can happen: (i) points of the sequence eventually recede to greater and greater distances from the origin, that is, the sequence is *unbounded*, or (ii) every one of the points lies within some fixed distance from the origin (i.e. within some circle about the origin) in the complex plane, that is, the sequence is *bounded*. The white regions of Fig. 1.2a are the locations of c that give an unbounded sequence (i), whereas the black regions are the locations of c where it is the bounded case (ii) that holds, the Mandelbrot set itself being the entire black region.[10]

The complication of the Mandelbrot set arises from the fact that there are many different and often highly involved ways in which the iterated sequence can remain bounded. There can be elaborate combinations of cycles and 'almost' cycles of various kinds, dotting around the plane in various intricate ways—but it would take us too far afield to try to understand in any detail how the extraordinary complication of this set comes about, and where subtle issues of complex analysis and number theory are involved. The interested reader may care to consult Peitgen and Reichter (1986) and Peitgen and Saupe (1988) for further information and pictures (see also Douady and Hubbard 1985).

Notes

Section 4.1

4.1. See Exercise [4.2] for these rules.

Section 4.2

4.2. It is a direct consequence[4.6] that any complex polynomial in the single variable z factorizes into linear factors,

$$a_0 + a_1 z + a_2 z^2 + \cdots + a_n z^n = a_n(z - b_1)(z - b_2)\cdots(z - b_n),$$

and it is *this* statement that is normally termed 'the fundamental theorem of algebra'.

4.3. As the story goes, Tartaglia had revealed his partial solution to Cardano only after Cardano had been sworn to secrecy. Accordingly, Cardano could not publish his more general solution without breaking this oath. However, on a subsequent trip to Bologna, in 1543, Cardano examined del Ferro's posthumous papers and satisfied himself of del Ferro's actual priority. He considered that this freed him to publish all these results (with due acknowledgement both to Tartaglia and del Ferro) in *Ars Magna* in 1545. Tartaglia disagreed, and the dispute had very bitter consequences (see Wykes 1969).

4.4. For more information, see van der Waerden (1985).

4.5. The reason for this is that we are adding together two numbers which are *complex conjugates* of each other (see §10.1) and such a sum is always a real number.

Section 4.3

4.6. Recall from Note 2.4 that 0^{-1} should mean $\frac{1}{0}$, i.e. 'one divided by zero'. It is a convenient 'shorthand' to express the 'result' of this illegal operation '$0^{-1} = \infty$'.

4.7. 'Strictly' means that the end-values are not included in the range.

4.8. For further information, see, for example, Hardy (1949).

Section 4.4

4.9. See e.g. Priestley (2003), p.71—referred to as 'radius of convergence'—and Needham (1997), pp. 67, 264.

Section 4.5

4.10. In computer-produced pictures of the Mandelbrot set (such as Fig. 1.2), one cannot, of course, compute indefinitely to ensure that an apparently bounded sequence is *actually* bounded. It is usual to 'cap' the iteration at some suitable number of steps. Simply adopting a larger cap does not, however, necessarily improve the accurate appearance of a picture, because the filaments tend to get lost.

[4.6] Show this. (*Hint*: Show that no remainder survives if this polynomial is 'divided' by $z - b$ whenever $z = b$ solves the given equation.)

5
Geometry of logarithms, powers, and roots

5.1 Geometry of complex algebra

THE aspects of complex-number magic discussed at the end of the previous chapter involve many subtleties, so let us pull back a little and look at some more elementary, though equally enigmatic and important, pieces of magic. First, let us see how the rules for addition and multiplication that we encountered in §4.1 are geometrically represented in the complex plane. We can exhibit these as the *parallelogram* law and the *similar-triangle* law, respectively, depicted in Fig. 5.1a,b. Specifically, for two general complex numbers w and z, the points representing $w + z$ and wz are determined by the respective assertions:

the points 0, w, w + z, z are the vertices of a parallelogram

and

the triangles with vertices 0, 1, *w and* 0, *z*, *wz are similar*.

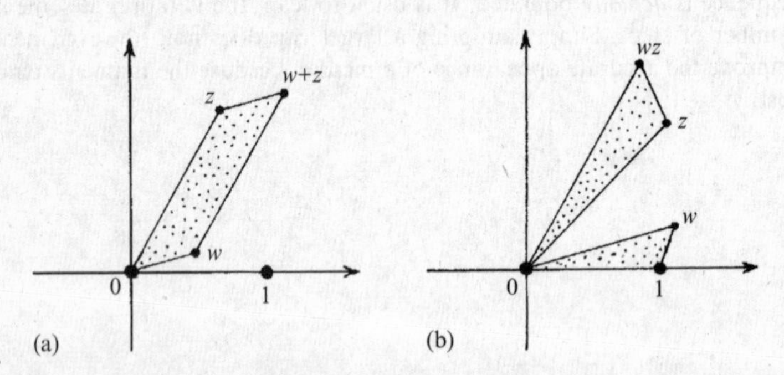

Fig. 5.1 Geometrical description of the basic laws of complex-number algebra. (a) Parallelogram law of addition: 0, w, $w + z$, z give the vertices of a parallelogram. (b) Similar-triangle law of multiplication: the triangles with vertices 0, 1, w and 0, z, wz are similar.

(Normal conventions about orderings and orientations are being adopted here. By this, I mean that we go around the parallelogram cyclicly, so the line segment from w to $w+z$ is parallel to that from 0 to z, etc.; moreover, there is to be no 'reflection' involved in the similarity relation between the two triangles. Also, there are special cases where the triangles or parallelogram degenerate in various ways.[5.1]) The interested reader may care to check these rules by trigonometry and direct computation.[5.2] However, there is another way of looking at these things which avoids detailed computation and yields greater insights.

Let us consider addition and multiplication in terms of different *maps* (or 'transformations') that send the entire complex plane to itself. Any given complex number w defines an 'addition map' and a 'multiplication map', these being the operations which, when applied to an arbitrary complex number z, will add w to z and take the product of w with z, respectively, that is,

$$z \mapsto w + z \text{ and } z \mapsto wz.$$

It is easy to see that the addition map simply slides the complex plane along without rotation or change of size or shape—an example of a *translation* (see §2.1)—displacing the origin 0 to the point w; see Fig. 5.2a. The parallelogram law is basically a restatement of this. But what about the multiplication map? This provides a transformation which leaves the origin fixed and preserves shapes—sending 1 to the point w. In the general case it combines a (non-reflective) rotation with a uniform expansion (or

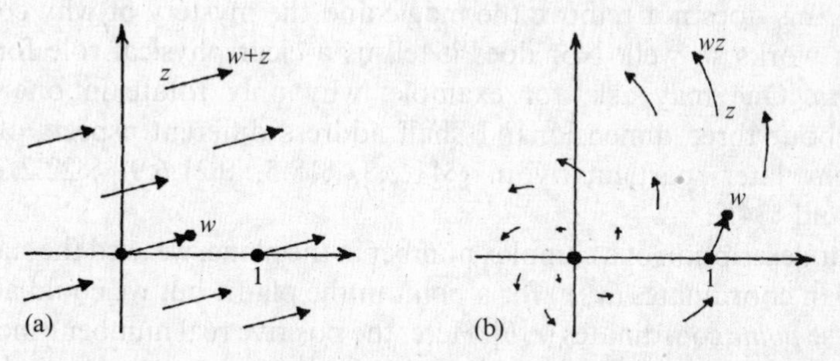

Fig. 5.2 (a) The addition map '$+w$' provides a translation of the complex plane, sending 0 to w. (b) The multiplication map '$\times w$' provides a rotation and expansion (or contraction) of the complex plane about 0, sending 1 to w.

[5.1] Examine the various possibilities.

[5.2] Do this.

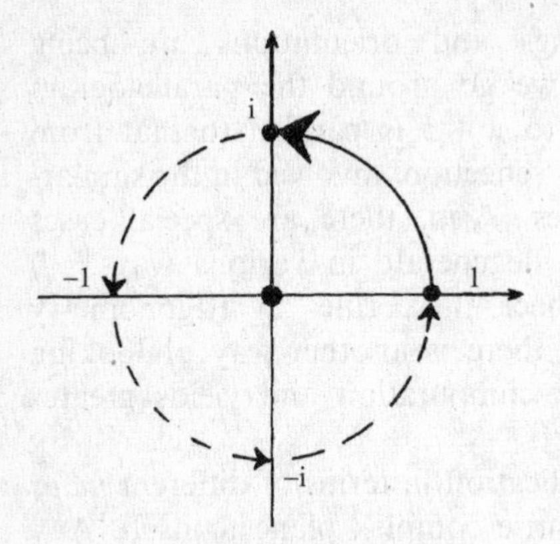

Fig. 5.3 The particular operation 'multiply by i' is realized, in the complex plane, as the geometrical transformation 'rotate through right angle'. The 'mysterious' equation $i^2 = -1$ is rendered visual.

contraction); see Fig. 5.2b.[5.3] The similar-triangle law effectively exhibits this. This map will have particular significance for us in §8.2.

In the particular case $w = i$, the multiplication map is simply a right-handed (i.e. anticlockwise) rotation through a right angle ($\frac{1}{2}\pi$). If we apply this operation twice, we get a rotation through π, which is simply a reflection in the origin; in other words, this is the multiplication map that sends each complex number z to its negative. This provides us with a graphic realization of the 'mysterious' equation $i^2 = -1$ (Fig. 5.3). The operation 'multiply by i' is realized as the geometrical transformation 'rotate through a right angle'. When viewed in this way, it does not seem so mysterious that the 'square' of this operation (i.e. doing it twice) should give the same effect as the operation of 'taking the negative'. Of course, this does not remove the magic and the mystery of why complex algebra works so well. Nor does it tell us a clear physical role for these numbers. One may ask, for example: why only rotate in one plane; what about three dimensions? I shall address different aspects of these questions later, particularly in §§11.2,3, §18.5, §§21.6,9, §§22.2,3,8–10, §33.2, and §34.8.

In our description of a complex number in the plane, we used the standard Cartesian coordinates (x, y) for a point in the plane, but we could alternatively use *polar* coordinates $[r, \theta]$. Here, the positive real number r measures the *distance* from the origin and the angle θ measures the *angle* that the line from the origin to the point z makes with the real axis, measured in an

---

[5.3] Try to show this without detailed calculation, and without trigonometry. (*Hint*: This is a consequence of the 'distributive law' $w(z_1 + z_2) = wz_1 + wz_2$, which shows that the 'linear' structure of the complex plane is preserved, and $w(iz) = i(wz)$, which shows that rotation through a right angle is preserved; i.e. right angles are preserved.)

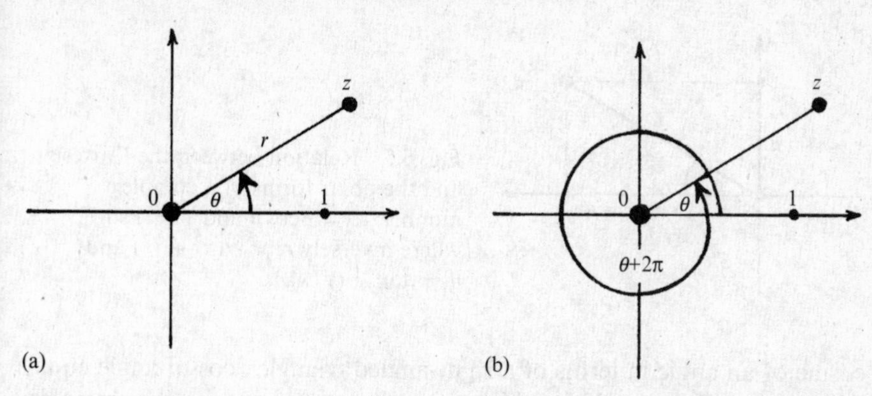

Fig. 5.4 (a) Passing from Cartesian (x, y) to polar $[r, \theta]$, where the *modulus* $r = |z|$ is the *distance* from the origin and the *argument* θ is the angle that the line from the origin to z makes with real axis, measured anticlockwise. (b) If we do not insist $-\pi < \theta \leq \pi$, we can allow z to wind around origin many times, adding any integer multiple of 2π to θ.

anticlockwise direction; see Fig. 5.4a. The quantity r is referred to as the *modulus* of the complex number z, which we sometimes write as

$$r = |z|,$$

and θ as its *argument* (or, in quantum theory, sometimes as its *phase*). For $z = 0$, we do not need to bother with θ, but we can still define r to be the distance from the origin, which in this case simply gives $r = 0$.

We could, for definiteness, insist that θ lie in a particular range, such as $-\pi < \theta \leq \pi$ (which is a standard convention). Alternatively, we may just think of the argument as something with the ambiguity that we are allowed to add integer multiples of 2π to it without affecting anything. This is just a matter of allowing us to wind around the origin as many times as we like, in either direction, when measuring the angle (see Fig. 5.4b). (This second point of view is actually the more profound one, and it will have implications for us shortly.) We see from Fig. 5.5 and basic trigonometry that

$$x = r\cos\theta \text{ and } y = r\sin\theta,$$

and, inversely, that

$$r = \sqrt{x^2 + y^2} \text{ and } \theta = \tan^{-1}\frac{y}{x},$$

where $\theta = \tan^{-1}(y/x)$ means some specific value of the many-valued function $\tan^{-1}$. (For those readers who have forgotten all their trigonometry, the first two formulae just re-express the definitions of the sine and

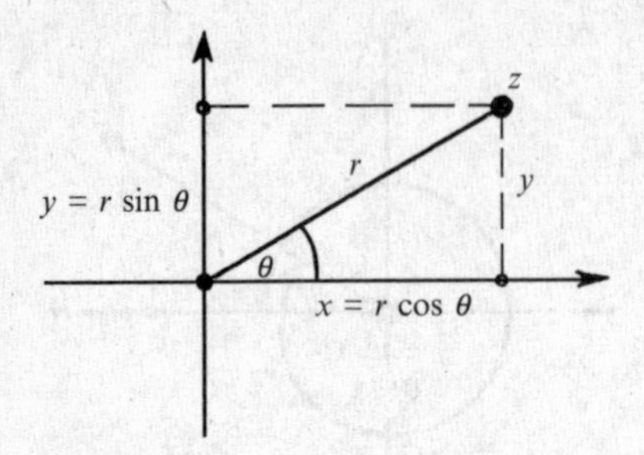

Fig. 5.5 Relation between the Cartesian and the polar forms of a complex number: $x = r\cos\theta$ and $y = r\sin\theta$, where inversely $r = \sqrt{(x^2 + y^2)}$ and $\theta = \tan^{-1}(y/x)$.

cosine of an angle in terms of a right-angled triangle: 'cos of angle equals adjacent over hypotenuse' and 'sin of angle equals opposite over hypotenuse', r being the hypotenuse; the second two express the Pythagorean theorem and, in inverse form, 'tan of angle equals opposite over adjacent'. One should also note that $\tan^{-1}$ is the *inverse function* of tan, not the reciprocal, so the above equation $\theta = \tan^{-1}(y/x)$ stands for $\tan\theta = y/x$. Finally, there is the ambiguity in $\tan^{-1}$ that any integer multiple of 2π can be added to θ and the relation will still hold.)[1]

5.2 The idea of the complex logarithm

Now, the 'similar-triangle law' of multiplication of two complex numbers, as illustrated in Fig. 5.1b, can be re-expressed in terms of the fact that when we multiply two complex numbers we add their arguments and multiply their moduli.[5.4] Note the remarkable fact here that, as far as the rule for the arguments is concerned, we have *converted multiplication into addition*. This fact is the basis of the use of *logarithms* (the logarithm of the product of two numbers is equal to the sum of their logarithms: $\log ab = \log a + \log b$), as is exhibited by the slide-rule (Fig. 5.6), and this property had fundamental importance to computational practice in earlier times.[2] Now we use electronic calculators to do our multiplication for us. Although this is far faster and more accurate than the use of a slide-rule or log tables, we lose something very significant for our understanding if we gain no direct experience of the beautiful and deeply important logarithmic operation. We shall see that logarithms have a profound rôle to play in relation to complex numbers. Indeed, the argument of a complex number really *is* a logarithm, in a certain clear sense. We shall try to understand how this comes about.

Also, recall the assertion in §4.2 that the taking of roots for complex numbers is basically a matter of understanding complex logarithms. We

[5.4] Spell this out.

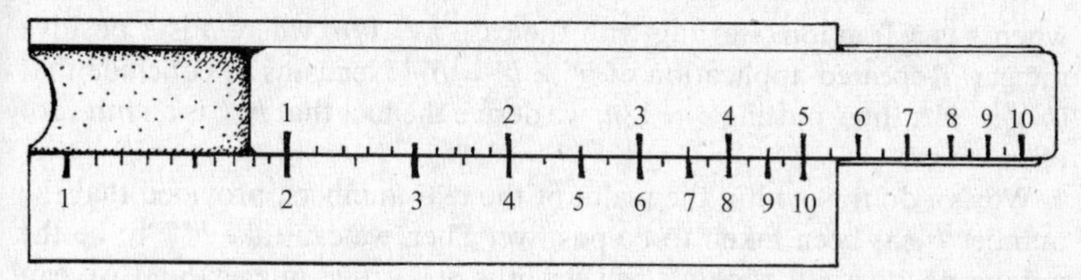

Fig. 5.6 Slide rules display numbers on a logarithmic scale, thereby enabling multiplication to be expressed by the adding of distances, in accordance with the formula $\log_b(p \times q) = \log_b p + \log_b q$. (Multiplication by 2 is illustrated.)

shall find that there are some striking relations between complex logarithms and trigonometry. Let us try to see how all these things come together.

First, recall something about ordinary logarithms. A logarithm is the reverse of 'raising a number to a power', or of *exponentiation*. 'Raising to a power' *is* an operation that converts addition into multiplication. Why is this? Take any (non-zero) number b. Then note the formula (converting addition into multiplication)

$$b^{m+n} = b^m \times b^n,$$

which is obvious if m and n are positive integers, because each side just represents $m + n$ instances of the number b, all multiplied together. What we have to do is to find a way of generalizing this so that m and n do not have to be positive integers, but can be any complex numbers whatever. For this, we need to find the right definition of 'b raised to the power z', for complex z, and we want the same formula as the above, namely $b^{w+z} = b^w \times b^z$, to hold when the exponents w and z are complex.

In fact, the procedure for doing this mirrors, to some extent, the very history of generalizing, step by step, from the positive integers to the complex numbers, as was done, starting from Pythagoras, via the work of Eudoxos, through Brahmagupta, until the time of Cardano and Bombelli (and later), as was indicated in §3.4, §4.1. First, the notion of 'b^z' is initially understood, when z is a positive integer, as simply $b \times b \times \cdots \times b$, with z b's multiplied together; in particular, $b^1 = b$. Then (following the lead of Brahmagupta) we allow z to be zero, realizing that to preserve $b^{w+z} = b^w \times b^z$ we need to define $b^0 = 1$. Next we allow z to be negative, and realize, for the same reason, that for the case $z = -1$ we must define b^{-1} to be the reciprocal of b (i.e. $1/b$), and that b^{-n}, for a natural number n, must be the nth power of b^{-1}. We then try to generalize to the situations when z is a fraction, starting with the case $z = 1/n$, where n is a positive integer. Repeated application of $b^w \times b^z = b^{w+z}$ leads us to conclude that $(b^z)^n = b^{zn}$; thus, putting $z = 1/n$, we derive the fact that $b^{1/n}$ is an nth root of b.

We can do this within the realm of the real numbers, provided that the number b has been taken to be positive. Then we can take $b^{1/n}$ to be the unique positive nth root of b (when n is a positive integer) and we can continue with defining b^z uniquely for any rational number $z = m/n$ to be the mth power of the nth root of b and thence (using a limiting process) for any real number z. However, if b is allowed to be negative, then we hit a snag at $z = \frac{1}{2}$, since $\sqrt{b}$ then requires the introduction of i and we are down the slippery slope to the complex numbers. At the bottom of that slope we find our magical complex world, so let us brace ourselves and go all the way down.

We require a definition of b^p such that, for all complex numbers p, q, and b (with $b \neq 0$), we have

$$b^{p+q} = b^p \times b^q.$$

We could then hope to define the *logarithm to the base b* (the operation denoted by '$\log_b$') as the inverse of the function defined by $f(z) = b^z$, that is,

$$z = \log_b w \quad \text{if} \quad w = b^z.$$

Then we should expect

$$\log_b (p \times q) = \log_b p + \log_b q,$$

so this notion of logarithm would indeed convert multiplication into addition.

5.3 Multiple valuedness, natural logarithms

Although this is basically correct, there are certain technical difficulties about doing this (which we shall see how to deal with shortly). In the first place, b^z is 'many valued'. That is to say, there are many different answers, in general, to the meaning of 'b^z'. There is also an additional many-valuedness to $\log_b w$. We have seen the many-valuedness of b^z already with fractional values of z. For example, if $z = \frac{1}{2}$, then 'b^z' ought to mean 'some quantity t which squares to b', since we require $t^2 = t \times t = b^{\frac{1}{2}} \times b^{\frac{1}{2}} = b^{\frac{1}{2}+\frac{1}{2}} = b^1 = b$. If some number t satisfies this property, then $-t$ will do so also (since $(-t) \times (-t) = t^2 = b$). Assuming that $b \neq 0$, we have two distinct answers for $b^{1/2}$ (normally written $\pm\sqrt{b}$). More generally, we have n distinct complex answers for $b^{1/n}$, when n is a positive integer: 1, 2, 3, 4, 5, ... In fact, we have some finite number of answers whenever n is a (non-zero) rational number. If n is irrational, then we have an infinite number of answers, as we shall be seeing shortly.

Let us try to see how we can cope with these ambiguities. We shall start by making a particular choice of b, above, namely the fundamental number 'e', referred to as the *base of natural logarithms*. This will reduce our ambiguity problem. We have, as a definition of e:

$$\mathrm{e} = 1 + \frac{1}{1!} + \frac{1}{2!} + \frac{1}{3!} + \frac{1}{4!} + \cdots = 2.718\,281\,828\,5\ldots,$$

where the exclamation points denote *factorials*, i.e.

$$n! = 1 \times 2 \times 3 \times 4 \times \cdots \times n,$$

so that $1! = 1$, $2! = 2$, $3! = 6$, etc. The function defined by $f(z) = \mathrm{e}^z$ is referred to as the *exponential* function and sometimes written 'exp'; it may be thought of as 'e raised to the power z' when acting on z, this 'power' being defined by the following simple modification of the above series for e:

$$\mathrm{e}^z = 1 + \frac{z}{1!} + \frac{z^2}{2!} + \frac{z^3}{3!} + \frac{z^4}{4!} + \cdots.$$

This important power series actually converges for all values of z (so it has an infinite circle of convergence; see §4.4). The infinite sum makes a particular choice for the ambiguity in 'b^z' when $b = \mathrm{e}$. For example, if $z = \frac{1}{2}$, then the series gives us the particular positive quantity $+\sqrt{\mathrm{e}}$ rather than $-\sqrt{\mathrm{e}}$. The fact that $z = \frac{1}{2}$ actually gives a quantity $\mathrm{e}^{1/2}$ that squares to e follows from the fact that e^z, as defined by this series,[5.5] indeed always has the required 'addition-to-multiplication' property

$$\mathrm{e}^{a+b} = \mathrm{e}^a \mathrm{e}^b,$$

so that $\left(\mathrm{e}^{\frac{1}{2}}\right)^2 = \mathrm{e}^{\frac{1}{2}}\mathrm{e}^{\frac{1}{2}} = \mathrm{e}^{\frac{1}{2}+\frac{1}{2}} = \mathrm{e}^1 = \mathrm{e}$.

Let us try to use this definition of e^z to provide us with an unambiguous logarithm, defined as the *inverse* of the exponential function:

$$z = \log w \quad \text{if } w = \mathrm{e}^z.$$

This is referred to as the *natural* logarithm (and I shall write the function simply as 'log' without a base symbol).[3] From the above addition-to-multiplication property, we anticipate a 'multiplication-to-addition' rule:

[5.5] Check this directly from the series. (*Hint*: The 'binomial theorem' for integer exponents asserts that the coefficient of $a^p b^q$ in $(a+b)^n$ is $n!/p!q!$.)

$$\log ab = \log a + \log b.$$

It is not immediately obvious that such an inverse to e^z will necessarily exist. However, it turns out in fact that, for any complex number w, apart from 0, there always does exist z such that $w = e^z$, so we can define $\log w = z$. But there is a catch here: there is more than one answer.

How do we express these answers? If $[r, \theta]$ is the polar representation of w, then we can write its logarithm z in ordinary Cartesian form $(z = x + iy)$ as

$$z = \log r + i\theta,$$

where $\log r$ is the ordinary natural logarithm of a positive real number—the inverse of the real exponential. Why? It is intuitively clear from Fig. 5.7 that such a real logarithm function exists. In Fig. 5.7a we have the graph of $r = e^x$. We just flip the axes over to get the graph of the inverse function $x = \log r$, as in Fig. 5.7b. It is not so surprising that the real part of $z = \log w$ is just an ordinary real logarithm. What is somewhat more remarkable[4] is that the imaginary part of z is just the angle θ that is the argument of the complex number w. This fact makes explicit my earlier comment that the argument of a complex number is really just a form of logarithm.

Recall that there is an ambiguity in the definition of the argument of a complex number. We can add any integer multiple of 2π to θ, and this will do just as well (recall Fig. 5.4b). Accordingly, there are many different solutions z for a given choice of w in the relation $w = e^z$. If we take one such z, then $z + 2\pi in$ is another possible solution, where n is any integer that we care to choose. Thus, the logarithm of w is ambiguous up to the

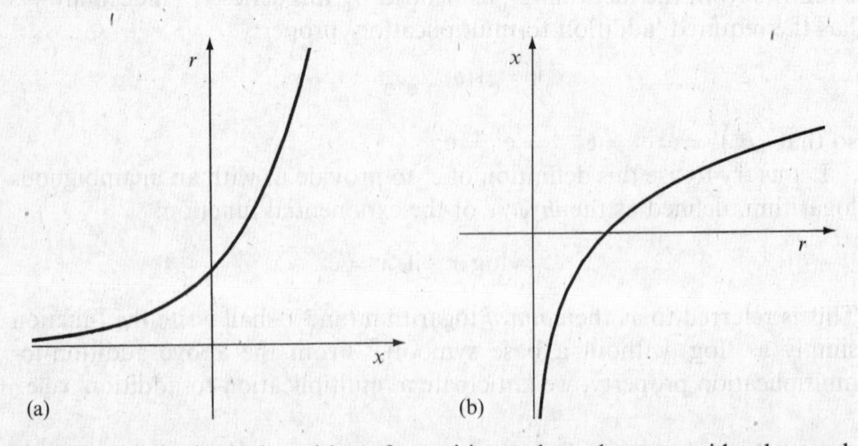

Fig. 5.7 To obtain the logarithm of a positive real number r, consider the graph (a) of $r = e^x$. All positive values of r are reached, so flipping the picture over, we get the graph (b) of the inverse function $x = \log r$ for positve r.

addition of any integer multiple of $2\pi i$. We must bear this in mind with expressions such as $\log ab = \log a + \log b$, making sure that the appropriately corresponding choices of logarithm are made.

This feature of the complex logarithm seems, at this stage, to be just an awkward irritation. However, we shall be seeing in §7.2 that it is absolutely central to some of the most powerful, useful, and magical properties of complex numbers. Complex analysis depends crucially upon it. For the moment, let us just try to appreciate the nature of the ambiguity.

Another way of understanding this ambiguity in $\log w$ is to note the striking formula

$$e^{2\pi i} = 1,$$

whence $e^{z+2\pi i} = e^z = w$, etc., showing that $z + 2\pi i$ is just as good a logarithm of w as z is (and then we can repeat this as many times as we like). The above formula is closely related to the famous *Euler formula*

$$e^{\pi i} + 1 = 0$$

(which relates the five fundamental numbers 0, 1, i, π, and e in one almost mystical expression).[5.6]

We can best understand these properties if we take the exponential of the expression $z = \log r + i\theta$ to obtain

$$w = e^z = e^{\log r + i\theta} = e^{\log r}e^{i\theta} = re^{i\theta}.$$

This shows that the polar form of any complex number w, which I had previously been denoting by $[r, \theta]$, can more revealingly be written as

$$w = re^{i\theta}.$$

In this form, it is evident that, if we multiply two complex numbers, we take the product of their moduli and the sum of their arguments ($re^{i\theta}se^{i\phi} = rse^{i(\theta+\phi)}$, so r and s are multiplied, whereas θ and ϕ are added—bearing in mind that subtracting 2π from $\theta + \phi$ makes no difference), as is implicit in the similar-triangle law of Fig. 5.1b. I shall henceforth drop the notation $[r, \theta]$, and use the above displayed expression instead. Note that if $r = 1$ and $\theta = \pi$ then we get -1 and recover Euler's famous $e^{\pi i} + 1 = 0$ above, using the geometry of Fig. 5.4a; if $r = 1$ and $\theta = 2\pi$, then we get $+1$ and recover $e^{2\pi i} = 1$.

The circle with $r = 1$ is called the *unit circle* in the complex plane (see Fig. 5.8). This is given by $w = e^{i\theta}$ for real θ, according to the above expression. Comparing that expression with the earlier ones $x = r\cos\theta$ and $y = r\sin\theta$ given above, for the real and imaginary parts of what is

---

[5.6] Show from this that $z + \pi i$ is a logarithm of $-w$.

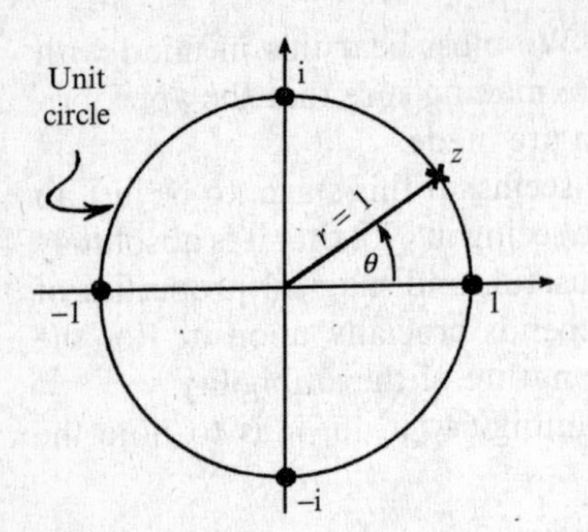

Fig. 5.8 The unit circle, consisting of unit-modulus complex numbers. The Cotes–Euler formula gives these as $e^{i\theta} = \cos\theta + i\sin\theta$ for real θ.

now the quantity $w = x + iy$, we obtain the prolific '(Cotes–) Euler formula'[5]

$$e^{i\theta} = \cos\theta + i\sin\theta,$$

which basically encapsulates the essentials of trigonometry in the much simpler properties of complex exponential functions.

Let us see how this works in elementary cases. In particular, the basic relation $e^{a+b} = e^a e^b$, when expanded out in terms of real and imaginary parts, immediately yields[5.7] the much more complicated-looking expressions (no doubt depressingly familiar to some readers)

$$\cos(a+b) = \cos a \cos b - \sin a \sin b,$$

$$\sin(a+b) = \sin a \cos b + \cos a \sin b.$$

Likewise, expanding out $e^{3i\theta} = (e^{i\theta})^3$, for example, quickly yields[6,[5.8]]

$$\cos 3\theta = \cos^3\theta - 3\cos\theta \sin^2\theta,$$

$$\sin 3\theta = 3\sin\theta \cos^2\theta - \sin^3\theta.$$

There is indeed a magic about the direct way that such somewhat complicated formulae spring from simple complex-number expressions.

5.4 Complex powers

Let us now return to the question of defining w^z (or b^z, as previously written). We can achieve such a thing by writing

$$w^z = e^{z \log w}$$

[5.7] Check this.

[5.8] Do it.

(since we expect $e^{z \log w} = (e^{\log w})^z$ and $e^{\log w} = w$). But we note that, because of the ambiguity in $\log w$, we can add any integer multiple of $2\pi i$ to $\log w$ to obtain another allowable answer. This means that we can multiply or divide any particular choice of w^z by $e^{z \cdot 2\pi i}$ any number of times and we still get an allowable 'w^z'. It is amusing to see the configuration of points in the complex plane that this gives in the general case. This is illustrated in Fig. 5.9. The points lie at the intersections of two equiangular spirals. (An equiangular—or *logarithmic*—spiral is a curve in the plane that makes a constant angle with the straight lines radiating from a point in the plane.)[5.9]

This ambiguity leads us into all sorts of problems if we are not careful.[5.10] The best way of avoiding these problems appears to be to adopt the rule that the notation w^z is used only when a *particular choice* of $\log w$ has been *specified*. (In the special case of e^z, the tacit convention is always to take the particular choice $\log e = 1$. Then the standard notation e^z is consistent with our more general w^z.) Once this choice of $\log w$ is specified, then w^z is unambiguously defined for all values of z.

It may be remarked at this point that we also need a specification of $\log b$ if we are to define the 'logarithm to the base b' referred to earlier in this section (the function denoted by '$\log_b$'), because we need an unambiguous $w = b^z$ to define $z = \log_b w$. Even so, $\log_b w$ will of course be many-valued (as was $\log w$), where we can add to $\log_b w$ any integer multiple of $2\pi i / \log b$.[5.11]

One curiosity that has greatly intrigued some mathematicians in the past is the quantity i^i. This might have seemed to be 'as imaginary as one could get'. However, we find the *real* answer

$$i^i = e^{i \log i} = e^{i \cdot \frac{1}{2}\pi i} = e^{-\pi/2} = 0.207\,879\,576\ldots,$$

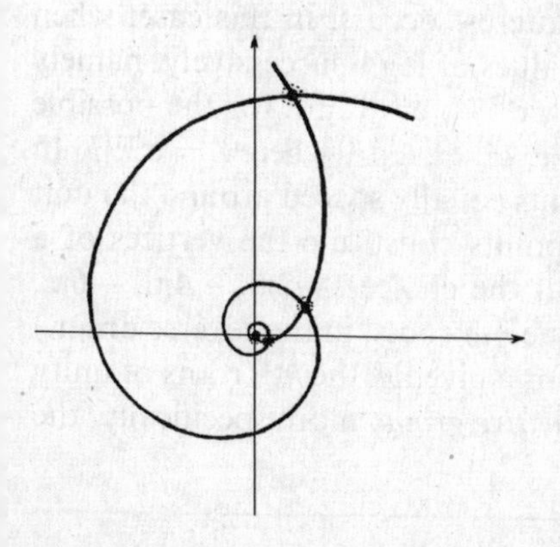

Fig. 5.9 The different values of $w^z (= e^{z \log w})$. Any integer multiple of $2\pi i$ can be added to $\log w$, which multiplies or divides w^z by $e^{z 2\pi i}$ an integer number of times. In the general case, these are represented in the complex plane as the intersections of two equiangular spirals (each making a constant angle with straight lines through the origin).

[5.9] Show this. How many ways? Also find all special cases.

[5.10] Resolve this 'paradox': $e = e^{1+2\pi i}$, so $e = (e^{1+2\pi i})^{1+2\pi i} = e^{1+4\pi i - 4\pi^2} = e^{1-4\pi^2}$.

[5.11] Show this.

by specifying $\log i = \frac{1}{2}\pi i$.[5.12] There are also many other answers, given by the other specifications of $\log i$. These are obtained by multiplying the above quantity by $e^{2\pi n}$, where n is any integer (or, equivalently, by raising the above quantity to any power of the form $4n+1$, where n is an integer—positive or negative[5.13]). It is striking that *all* the values of i^i are in fact real numbers.

Let us see how the notation w^z works for $z = \frac{1}{2}$. We expect to be able to represent the two quantities $\pm\sqrt{w}$ as '$w^{1/2}$' in some sense. In fact we get these two quantities simply by first specifying one value for $\log w$ and then specifying another one, where we add $2\pi i$ to the first one to get the second one. This results in a change of sign in $w^{1/2}$ (because of the Euler formula $e^{\pi i} = -1$). In a similar way, we can generate all n solutions $z^n = w$ when n is 3, 4, 5, ... as the quantity $w^{1/n}$, when successively different values of the $\log w$ are specified.[5.14] More generally, we can return to the question of zth roots of a non-zero complex number w, where z is any non-zero complex number, that was alluded to in §4.2. We can express such a zth root as the expression $w^{1/z}$, and we generally get an infinite number of alternative values for this, depending upon which choice of $\log w$ is specified. With the right specified choice for $\log w^{1/z}$, namely that given by $(\log w)/z$, we indeed get $\left(w^{1/z}\right)^z = w$. We note, more generally, that

$$(w^a)^b = w^{ab},$$

where once we have made a specification of $\log w$ (for the right-hand side), we must (for the left-hand side) specify $\log w^a$ to be $a \log w$.[5.15]

When $z = n$ is a positive integer, things are much simpler, and we get just n roots. A situation of particular interest occurs, in this case, when $w = 1$. Then, specifying some possible values of $\log 1$ successively, namely $0, 2\pi i, 4\pi i, 6\pi i, \ldots$, we get $1 = e^0, e^{2\pi i/n}, e^{4\pi i/n}, e^{6\pi i/n}, \ldots$ for the possible values of $1^{1/n}$. We can write these as $1, \epsilon, \epsilon^2, \epsilon^3, \ldots$, where $\epsilon = e^{2\pi i/n}$. In terms of the complex plane, we get n points equally spaced around the unit circle, called *nth roots of unity*. These points constitute the vertices of a regular n-gon (see Fig. 5.10). (Note that the choices, $-2\pi i$, $-4\pi i$, $-6\pi i$, etc., for $\log 1$ would merely yield the same nth roots, in the reverse order.)

It is of some interest to observe that, for a given n, the nth roots of unity constitute what is called a *finite multiplicative group*, more specifically, the

---

[5.12] Why is this an allowable specification?

[5.13] Show why this works.

[5.14] Spell this out.

[5.15] Show this.

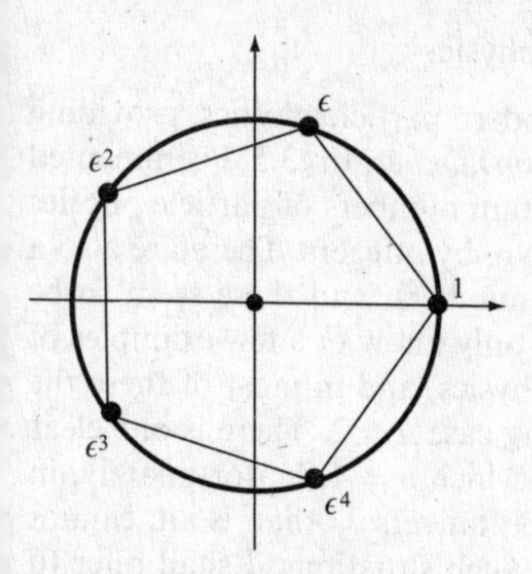

Fig. 5.10 The nth roots of unity $e^{2\pi ri/n}(r = 1, 2, \ldots, n)$, equally spaced around the unit circle, provide the vertices of a regular n-gon. Here $n = 5$.

cyclic group $\mathbb{Z}_n$ (see §13.1). We have n quantities with the property that we can multiply any two of them together and get another one. We can also divide one by another to get a third. As an example, consider the case $n = 3$. Now we get three elements 1, ω, and ω^2, where $\omega = e^{2\pi i/3}$ (so $\omega^3 = 1$ and $\omega^{-1} = \omega^2$). We have the following simple multiplication and division tables for these numbers:

$\times$	1	ω	ω^2
1	1	ω	ω^2
ω	ω	ω^2	1
ω^2	ω^2	1	ω

$\div$	1	ω	ω^2
1	1	ω^2	ω
ω	ω	1	ω^2
ω^2	ω^2	ω	1

In the complex plane, these particular numbers are represented as the vertices of an equilateral triangle. Multiplication by ω rotates the triangle through $\frac{2}{3}\pi$ (i.e. 120°) in an anticlockwise sense, and multiplication by ω^2 turns it through $\frac{2}{3}\pi$ in a clockwise sense; for division, the rotation is in the opposite direction (see Fig. 5.11).

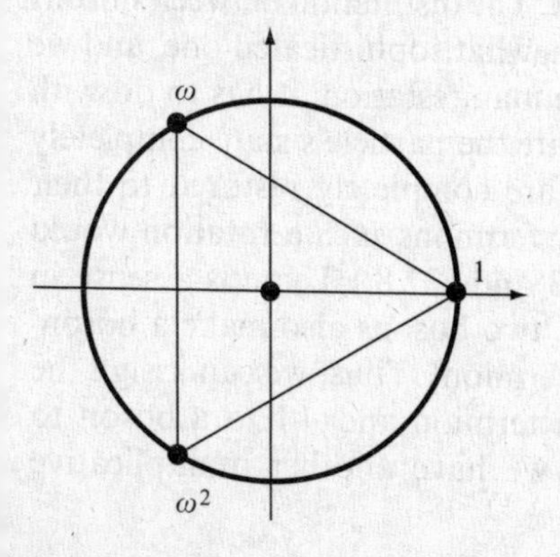

Fig. 5.11 Equilateral triangle of cube roots 1, ω, and ω^2 of unity. Multiplication by ω rotates through 120° anticlockwise, and by ω^2, clockwise.

5.5 Some relations to modern particle physics

Numbers such as these have interest in modern particle physics, providing the possible cases of a *multiplicative quantum number*. In §3.5, I commented on the fact that the additive (scalar) quantum numbers of particle physics are invariably quantified, as far as is known, by integers. There are also a few examples of multiplicative quantum numbers, and these seem to be quantified in terms of nth roots of unity. I only know of a few examples of such quantities in conventional particle physics, and in most of these the situation is the comparatively uninteresting case $n = 2$. There is one clear case where $n = 3$ and possibly a case for which $n = 4$. Unfortunately, in most cases, the quantum number is not universal, that is, it cannot consistently be applied to all particles. In such situations, I shall refer to the quantum number as being only *approximate*.

The quantity called *parity* is an (approximate) multiplicative quantum number with $n = 2$. (There are also other approximate quantities for which $n = 2$, similar in many respects to parity, such as *g-parity*. I shall not discuss these here.) The notion of parity for a composite system is built up (multiplicatively) from those of its basic constituent particles. For such a constituent particle, its parity can be even, in which case, the mirror reflection of the particle is the same as the particle itself (in an appropriate sense); alternatively, its parity can be odd, in which case its mirror reflection is what is called its antiparticle (see §3.5, §§24.1–3,8 and §26.4). Since the notion of mirror reflection, or of taking the antiparticle, is something that 'squares to unity', (i.e., doing it twice gets us back to where we started), the quantum number—let us call it ϵ—has to have the property $\epsilon^2 = 1$, so it must be an 'nth root of unity', with $n = 2$ (i.e. $\epsilon = +1$ or $\epsilon = -1$). This notion is only approximate, because parity is not a conserved quantity with respect to what are called 'weak interactions' and, indeed, there may not be a well-defined parity for certain particles because of this (see §§25.3,4).

Moreover, the notion of parity applies, in normal descriptions, only to the family of particles known as *bosons*. The remaining particles belong to another family and are known as *fermions*. The distinction between bosons and fermions is a very important but somewhat sophisticated one, and we shall come to it later, in §§23.7,8. (In one manifestation, it has to do with what happens when we continuously rotate the particle's state completely by 2π (i.e. through 360°). Only bosons are completely restored to their original states under such a rotation. For fermions such a rotation would have to be done twice for this. See §11.3 and §22.8.) There is a sense in which 'two fermions make a boson' and 'two bosons also make a boson' whereas 'a boson and a fermion make a fermion'. Thus, we can assign the multiplicative quantum number -1 to a fermion and $+1$ to a boson to describe its fermion/boson nature, and we have another multiplicative quantum number with $n = 2$. As far as is known, this quantity is an *exact* multiplicative quantum number.

It seems to me that there is also a parity notion that can be applied to fermions, although this does not seem to be a conventional terminology. This must be combined with the fermion/boson quantum number to give a combined multiplicative quantum number with $n = 4$. For a fermion, the parity value would have to be $+i$ or $-i$, and its double mirror reflection would have the effect of a 2π rotation. For a boson, the parity value would be ± 1, as before.

The multiplicative quantum number with $n = 3$ that I have referred to is what I shall call *quarkiness*. (This is not a standard terminology, nor is it usual to refer to this concept as a quantum number at all, but it does encapsulate an important aspect of our present-day understanding of particle physics.) In §3.5, I referred to the modern viewpoint that the 'strongly interacting' particles known as *hadrons* (protons, neutrons, π-mesons, etc.) are taken to be composed of *quarks* (see §25.6). These quarks have values for their electric charge which are not integer multiples of the electron's charge, but which are integer multiples of one-third of this charge. However, quarks cannot exist as separate individual particles, and their composites can exist as separate individuals only if their combined charges add up to an integer, in units of the electron's charge. Let q be the value of the electric charge measured in negative units of that of the electron (so that for the electron itself we have $q = -1$, the electron's charge being counted as negative in the normal conventions). For quarks, we have $q = \frac{2}{3}$ or $-\frac{1}{3}$; for antiquarks, $q = \frac{1}{3}$ or $-\frac{2}{3}$. Thus, if we take for the quarkiness the multiplicative quantum number $e^{-2q\pi i}$, we find that it takes values 1, ω, and ω^2. For a quark the quarkiness is ω, and for an antiquark it is ω^2. A particle that can exist separately on its own only if its quarkiness is 1. In accordance with §5.4, the degrees of quarkiness constitute the cyclic group $\mathbb{Z}_3$. (In §16.1, we shall see how, with an additional element '0' and a notion of addition, this group can be extended to the *finite field* $\mathbb{F}_4$.)

In this section and in the previous one, I have exhibited some of the mathematical aspects of the magic of complex numbers and have hinted at just a very few of their applications. But I have not yet mentioned those aspects of complex numbers (to be given in Chapter 7) that I myself found to be the most magical of all when I learned about them as a mathematics undergraduate. In later years, I have come across yet more striking aspects of this magic, and one of these (described at the end of Chapter 9) is strangely complementary to the one which most impressed me as an undergraduate. These things, however, depend upon certain basic notions of the *calculus*, so, in order to convey something of this magic to the reader, it will be necessary first to say something about these basic notions. There is, of course, an additional reason for doing this. Calculus is absolutely essential for a proper understanding of physics!

Notes

Section 5.1

5.1. The trigonometrical functions $\cot\theta = \cos\theta/\sin\theta = (\tan\theta)^{-1}$, $\sec\theta = (\cos\theta)^{-1}$, and $\operatorname{cosec}\theta = (\sin\theta)^{-1}$ should also be noted, as should the 'hyperbolic' versions of the trigonometrical functions, $\sinh t = \frac{1}{2}(e^t - e^{-t})$, $\cosh t = \frac{1}{2}(e^t + e^{-t})$, $\tanh t = \sinh t/\cosh t$, etc. Note also that the inverses of these operations are denoted by $\cot^{-1}$, $\sinh^{-1}$, etc., as with the '$\tan^{-1}(y/x)$' of §5.1.

Section 5.2

5.2. Logarithms were introduced in 1614 by John Neper (Napier) and made practical by Henry Briggs in 1624.

Section 5.3

5.3. The natural logarithm is also commonly written as 'ln'.

5.4. From what has been established so far here, we cannot infer that '$i\theta$' in the formula $z = \log r + i\theta$ should not be a real multiple of $i\theta$. This needs calculus.

5.5. Roger Cotes (1714) had the equivalent formula $\log(\cos\theta + i\sin\theta) = i\theta$. Euler's $e^{i\theta} = \cos\theta + i\sin\theta$ seems to have first appeared 30 years later (see Euler 1748).

5.6. I am using the convenient (but somewhat illogical) notation $\cos^3\theta$ for $(\cos\theta)^3$, etc., here. The notational inconsistency with (the more logical) $\cos^{-1}\theta$ should be noted, the latter being commonly also denoted as $\arccos\theta$. The formula $\sin n\theta + i\cos n\theta = (\sin\theta + i\cos\theta)^n$ is sometimes known as 'De Moivre's theorem'. Abraham De Moivre, a contemporary of Roger Cotes (see above endnote), seems also to have been a co-discoverer of $e^{i\theta} = \sin\theta + i\cos\theta$. For more fascinating information about complex numbers and their history, see Nahin (1998).

6
Real-number calculus

6.1 What makes an honest function?

Calculus—or, according to its more sophisticated name, *mathematical analysis*—is built from two basic ingredients: *differentiation* and *integration*. Differentiation is concerned with velocities, accelerations, the slopes and curvature of curves and surfaces, and the like. These are rates at which things change, and they are quantities defined *locally*, in terms of structure or behaviour in the tiniest neighbourhoods of single points. Integration, on the other hand, is concerned with areas and volumes, with centres of gravity, and with many other things of that general nature. These are things which involve measures of *totality* in one form or another, and they are not defined merely by what is going on in the local or infinitesimal neighbourhoods of individual points. The remarkable fact, referred to as the *fundamental theorem of calculus*, is that each one of these ingredients is essentially just the *inverse* of the other. It is largely this fact that enables these two important domains of mathematical study to combine together and to provide a powerful body of understanding and of calculational technique.

This subject of mathematical analysis, as it was originated in the 17th century by Fermat, Newton, and Leibniz, with ideas that hark back to Archimedes in about the 3rd century BC, is called 'calculus' because it indeed provides such a body of calculational technique, whereby problems that would otherwise be conceptually difficult to tackle can frequently be solved 'automatically', merely by the following of a few relatively simple rules that can often be applied without the exertion of a great deal of penetrating thought. Yet there is a striking contrast between the operations of differentiation and integration, in this calculus, with regard to which is the 'easy' one and which is the 'difficult' one. When it is a matter of applying the operations to explicit formulae involving known functions, it is differentiation which is 'easy' and integration 'difficult', and in many cases the latter may not be possible to carry out at all in an explicit way. On the other hand, when functions are not given in terms of formulae, but are provided in the form of tabulated lists of numerical data, then it is integration which is 'easy' and differentiation 'difficult', and the latter may not, strictly speaking, be possible at all in the ordinary way. Numerical techniques are generally concerned with approximations, but there is also a close analogue of this aspect of things in the exact theory, and again it is integration which can be performed in circumstances where differentiation cannot. Let us try to understand some of this. The issues have to do, in fact, with what one actually means by a 'function'.

To Euler, and the other mathematicians of the 17th and 18th centuries, a 'function' would have meant something that one could write down explicitly, like x^2 or $\sin x$ or $\log(3 - x + e^x)$, or perhaps something defined by some formula involving an integration or maybe by an explicitly given power series. Nowadays, one prefers to think in terms of 'mappings', whereby some array A of numbers (or of more general entities) called the *domain* of the function is 'mapped' to some other array B, called the *target* of the function (see Fig. 6.1). The essential point of this is that the function would assign a member of the target B to each member of the domain A. (Think of the function as 'examining' a number that belongs to A and then, depending solely upon which number it finds, it would produce a definite number belonging to B.) This kind of function can be just a 'look-up table'. There would be no requirement that there be a reasonable-looking 'formula' which expresses the action of the function in a manifestly explicit way.

Let us consider some examples. In Fig. 6.2, I have drawn the graphs of three simple functions[1], namely those given by x^2, $|x|$, and $\theta(x)$. In each case, the domain and target spaces are both to be the totality of *real numbers*, this totality being normally represented by the symbol $\mathbb{R}$. The function that I am denoting by 'x^2' simply takes the square of the real number that it is examining. The function denoted by '$|x|$' (called the *absolute value*) just yields x if x is non-negative, but gives $-x$ if x is negative; thus $|x|$ itself is never negative. The function '$\theta(x)$' is 0 if x is negative, and 1 if x is positive; it is usual also to define $\theta(0) = \frac{1}{2}$. (This function is called the Heaviside *step function*; see §21.1 for another important mathematical influence of Oliver Heaviside, who is perhaps better known for first postulating the Earth's atmospheric 'Heaviside layer', so vital to radio transmission.) Each of these is a perfectly good

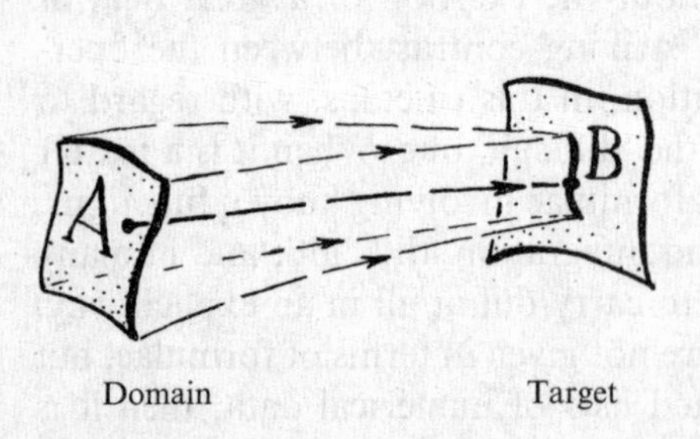

Fig. 6.1 A *function* as a 'mapping', whereby its *domain* (some array A of numbers or of other entities) is 'mapped' to its *target* (some other array B). Every element of A is assigned some particular value in B, though different elements of A may attain the same value and some values of B may not be reached.

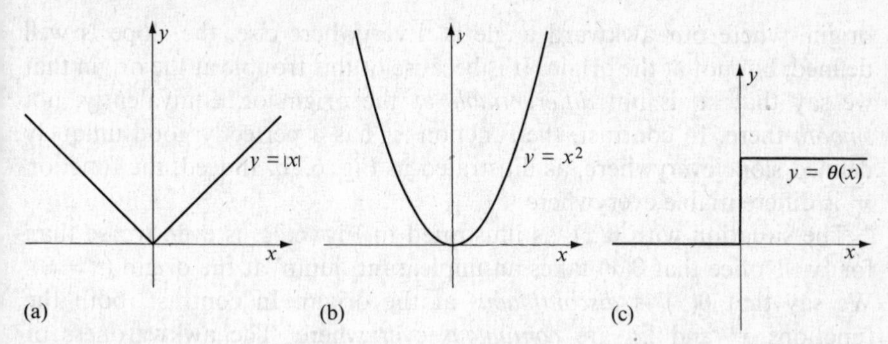

Fig. 6.2 Graphs of (a) $|x|$, (b) x^2, and (c) $\theta(x)$; the domain and target being the system of real numbers in each case.

function in this modern sense of the term, but Euler[2] would have had difficulty in accepting $|x|$ or $\theta(x)$ as a 'function' in *his* sense of the term.

Why might this be? One possibility is to think that the trouble with $|x|$ and $\theta(x)$ is that there is too much of the following sort of thing: 'if x is such-and-such then take so-and-so, whereas if x is ...', and there is no 'nice formula' for the function. However, this is a bit vague, and in any case we could wonder what is really wrong with $|x|$ being counted as a formula. Moreover, once we have accepted $|x|$, we could write[6.1] a *formula* for $\theta(x)$:

$$\theta(x) = \frac{|x| + x}{2x}$$

(although we might wonder if there is a good sense in which this gets the right value for $\theta(0)$, since the formula just gives 0/0). More to the point is that the trouble with $|x|$ is that it is not 'smooth', rather than that its explicit expression is not 'nice'. We see this in the 'angle' in the middle of Fig. 6.2a. The presence of this angle is what prevents $|x|$ from having a well-defined *slope* at $x = 0$. Let us next try to come to terms with this notion.

6.2 Slopes of functions

As remarked above, one of the things with which differential calculus is concerned is, indeed, the finding of 'slopes'. We see clearly from the graph of $|x|$, as shown in Fig. 6.2a, that it does not have a unique slope at the

[6.1] Show this (ignoring $x = 0$).

origin, where our awkward angle is. Everywhere else, the slope is well defined, but not at the origin. It is because of this trouble at the origin that we say that $|x|$ is not *differentiable* at the origin or, equivalently, not *smooth* there. In contrast, the function x^2 has a perfectly good uniquely defined slope everywhere, as illustrated in Fig. 6.2b. Indeed, the function x^2 is differentiable everywhere.

The situation with $\theta(x)$, as illustrated in Fig. 6.2c, is even worse than for $|x|$. Notice that $\theta(x)$ takes an unpleasant 'jump' at the origin ($x = 0$). We say that $\theta(x)$ is *discontinuous* at the origin. In contrast, both the functions x^2 and $|x|$ are *continuous* everywhere. The awkwardness of $|x|$ at the origin is not a failure of continuity but of differentiability. (Although the failure of continuity and of smoothness are different things, they are actually interconnected concepts, as we shall be seeing shortly.)

Neither of these failings would have pleased Euler, presumably, and they seem to provide reasons why $|x|$ and $\theta(x)$ might not be regarded as 'proper' functions. But now consider the two functions illustrated in Fig. 6.3. The first, x^3, would be acceptable by anyone's criteria; but what about the second, which can be defined by the expression $x|x|$, and which illustrates the function that is x^2 when x is non-negative and $-x^2$ when x is negative? To the eye, the two graphs look rather similar to each other and certainly 'smooth'. Indeed, they both have a perfectly good value for the 'slope' at the origin, namely zero (which means that the curves have a horizontal slope there) and are, indeed, 'differentiable' everywhere, in the most direct sense of that word. Yet, $x|x|$ certainly does not seem to be the 'nice' sort of function that would have satisfied Euler.

One thing that is 'wrong' with $x|x|$ is that it does not have a well-defined *curvature* at the origin, and the notion of curvature is certainly something that the differential calculus is concerned with. In fact, 'curvature' is something that involves what are called 'second derivatives', which

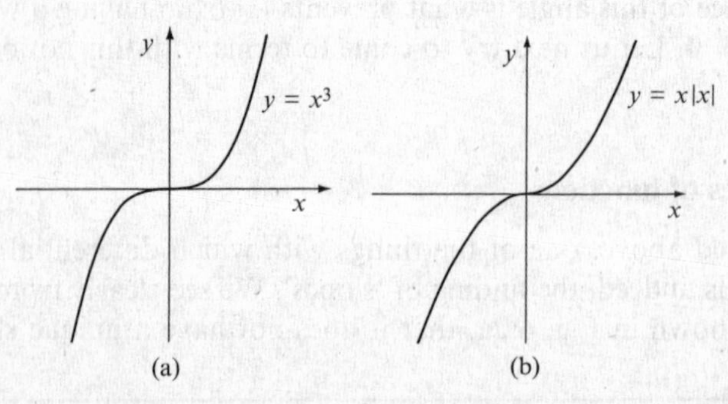

Fig. 6.3 Graphs of (a) x^3 and of (b) $x|x|$ (i.e. x^2 if $x \geq 0$ and $-x^2$ if $x < 0$).

means doing the differentiation twice. Indeed, we say that the function $x|x|$ is not *twice differentiable* at the origin. We shall come to second and higher derivatives in §6.3.

In order to start to understand these things, we shall need to see what the operation of differentiation really does. For this, we need to know how a *slope* is measured. This is illustrated in Fig. 6.4. I have depicted a fairly representative-looking function, which I shall call $f(x)$. The curve in Fig. 6.4a depicts the relation $y = f(x)$, where the value of the coordinate y measures the height and the value of x measures horizontal displacement, as is usual in a Cartesian description. I have indicated the slope of the curve at one particular point p, as the increment in the y coordinate divided by the increment in the x coordinate, as we proceed along the *tangent line* to the curve, touching it at the point p. (The technical definition of 'tangent line' depends upon the appropriate limiting procedures, but it is not my purpose here to provide these technicalities. I hope that the reader will find my intuitive descriptions adequate for our immediate purposes.[3]) The standard notation for the value of this slope is $\mathrm{d}y/\mathrm{d}x$ (and pronounced 'dy by dx'). We can think of 'dy' as a very tiny increase in the value of y along the curve and of 'dx' as the corresponding tiny increase in the value of x. (Here, technical correctness would require us to go to the 'limit', as these tiny increases each get reduced to zero.)

We can now consider another curve, which plots (against x) this slope at each point p, for the various possible choices of x-coordinate; see Fig. 6.4b. Again, I am using a Cartesian description, but now it is $\mathrm{d}y/\mathrm{d}x$ that is plotted vertically, rather than y. The horizontal displacement is still measured by x. The function that is being plotted here is commonly called $f'(x)$, and we can write $\mathrm{d}y/\mathrm{d}x = f'(x)$. We call $\mathrm{d}y/\mathrm{d}x$ the *derivative of y with respect to x*, and we say that the function $f'(x)$ is the *derivative*[4] *of* $f(x)$.

6.3 Higher derivatives; C^∞-smooth functions

Now let us see what happens when we take a *second* derivative. This means that we are now looking at the slope-function for the new curve of Fig. 6.4b, which plots $u = f'(x)$, where u now stands for $\mathrm{d}y/\mathrm{d}x$. In Fig. 6.4c, I have plotted this 'second-order' slope function, which is the graph of $\mathrm{d}u/\mathrm{d}x$ against x, in the same kind of way as I did before for $\mathrm{d}y/\mathrm{d}x$, so the value of $\mathrm{d}u/\mathrm{d}x$ now provides us with the slope of the second curve $u = f'(x)$. This gives us what is called the second derivative of the original function $f(x)$, and this is commonly written $f''(x)$. When we substitute $\mathrm{d}y/\mathrm{d}x$ for u in the quantity $\mathrm{d}u/\mathrm{d}x$, we get the *second derivative* of y with respect to x, which is

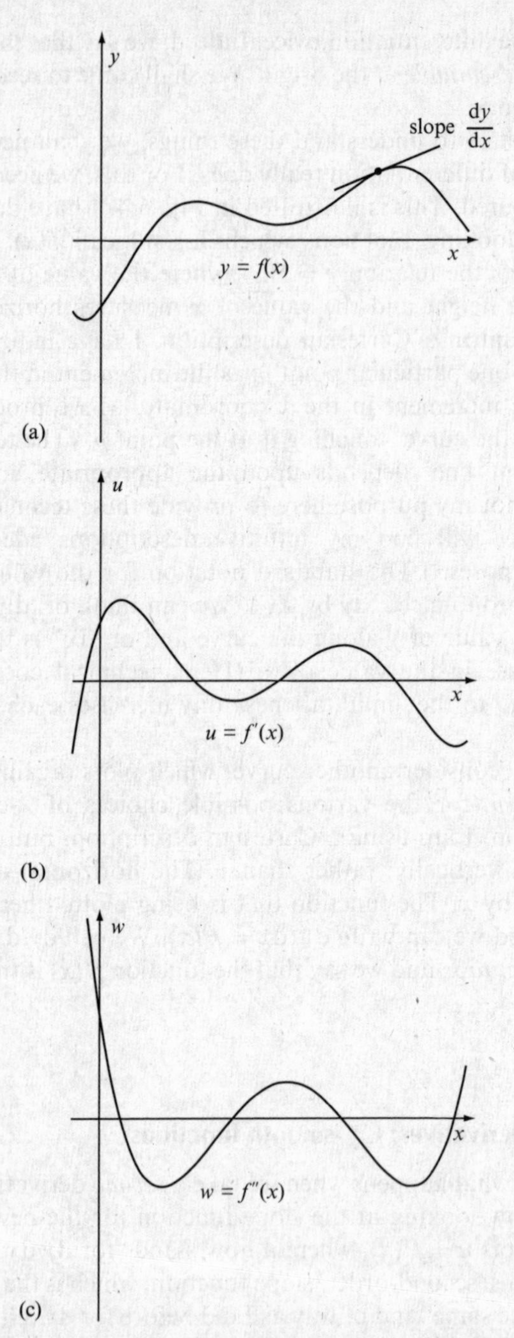

Fig. 6.4 Cartesian plot of (a) $y = f(x)$, (b) the derivative $u = f'(x)$ $(= dy/dx)$, and (c) the second derivative $f''(x) = d^2y/dx^2$. (Note that $f(x)$ has horizontal slope just where $f'(x)$ meets the x-axis, and it has an inflection point where $f''(x)$ meets the x-axis.)

(slightly illogically) written d^2y/dx^2 (and pronounced 'd-two-y by dx-squared').

Notice that the values of x where the original function $f(x)$ has a horizontal slope are just the values of x where $f'(x)$ meets the x-axis (so dy/dx vanishes for those x-values). The places where $f(x)$ acquires a (local) maximum or minimum occur at such locations, which is important when we are interested in finding the (locally) greatest and smallest values of a function. What about the places where the second derivative $f''(x)$ meets the x-axis? These occur where the *curvature* of $f(x)$ vanishes. In general, these points are where the direction in which the curve $y = f(x)$ 'bends' changes from one side of the curve to the other, at a place called a *point of inflection*. (In fact, it would not be correct to say that $f''(x)$ actually 'measures' the curvature of the curve defined by $y = f(x)$, in general; the actual curvature is given by a more complicated expression[5] than $f''(x)$, but it involves $f''(x)$, and the curvature vanishes whenever $f''(x)$ vanishes.)

Let us next consider our two (superficially) similar-looking functions x^3 and $x|x|$, considered above. In Fig. 6.5a,b,c, I have plotted x^3 and its first and second derivatives, as I did with the function $f(x)$ in Fig. 6.4, and, in Fig. 6.5d,e,f, I have done the same with $x|x|$. In the case of x^3, we see that

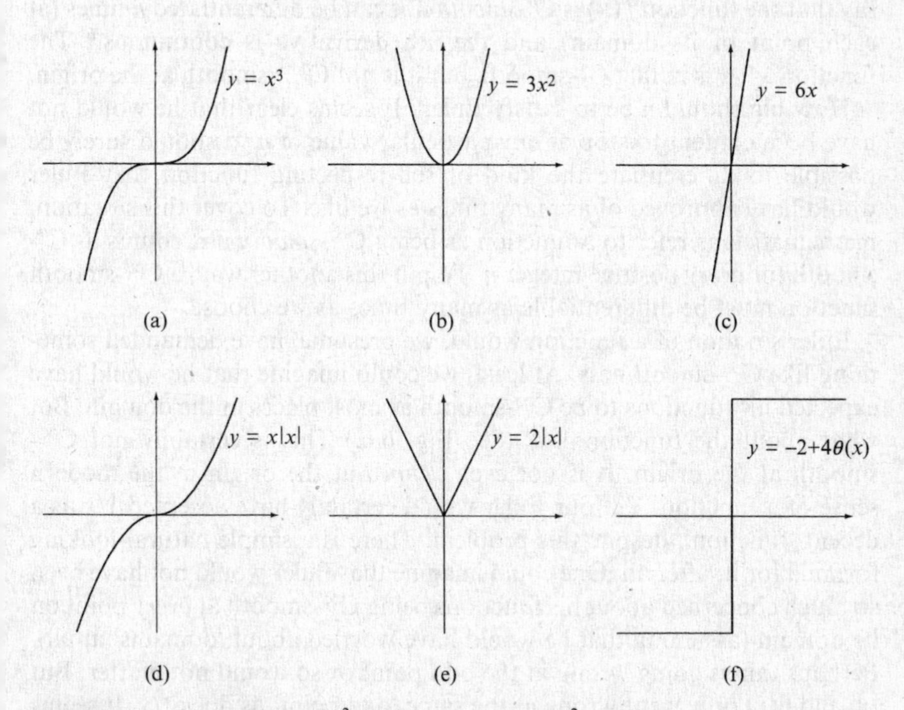

Fig. 6.5 (a), (b), (c) Plots of x^3, its first derivative $3x^2$, and its second derivative $6x$, respectively. (d), (e), (f) Plots of $x|x|$, its first derivative $2|x|$, and the second derivative $-2 + 4\theta(x)$, respectively.

there are no problems with continuity or smoothness with either the first or second derivative. In fact the first derivative is $3x^2$ and the second is $6x$, neither of which would have given Euler a moment of worry. (We shall see how to obtain these explicit expressions shortly.) However, in the case of $x|x|$, we find something very much like the 'angle' of Fig. 6.2a for the first derivative, and a 'step function' behaviour for the second derivative, very similar to Fig. 6.2c. We have failure of smoothness for the first derivative and failure of continuity for the second. Euler would not have cared for this at all. This first derivative is actually $2|x|$ and the second derivative is $-2 + 4\theta(x)$. (My more pedantic readers might complain that I should not so glibly write down a 'derivative' for $2|x|$, which is not actually differentiable at the origin. True, but this is just a quibble: full justification of this can be achieved using the notions that will be introduced at the end of Chapter 9.)

We can easily imagine that functions can be constructed for which such failure of smoothness or of continuity does not show up until many derivatives have been calculated. Indeed, functions of the form $x^n|x|$ will do the trick, where we can take n to be a positive integer which can be as large as we like. The mathematical terminology for this sort of thing is to say that the function $f(x)$ is C^n-*smooth* if it can be differentiated n times (at each point of its domain) and the nth derivative is continuous.[6] The function $x^n|x|$ is in fact C^n-smooth, but it is not C^{n+1}-smooth at the origin.

How big should n be to satisfy Euler? It seems clear that he would not have been content to stop at *any* particular value of n. It should surely be possible to differentiate the kind of self-respecting function that Euler would have approved of as many times as we like. To cover this situation, mathematicians refer to a function as being C^∞-*smooth* if it counts as C^n-smooth for *every* positive integer n. To put this another way, a C^∞-smooth function must be differentiable as many times as we choose.

Euler's notion of a function would, we presume, have demanded something like C^∞-smoothness. At least, we could imagine that he would have expected his functions to be C^∞-smooth at most places in the domain. But what about the function $1/x$? (See Fig. 6.6.) This is certainly not C^∞-smooth at the origin. It is not even *defined* at the origin in the modern sense of a function. Yet our Euler would certainly have accepted $1/x$ as a decent 'function', despite this problem. There is a simple natural-looking formula for it, after all. One could imagine that Euler would not have been so much concerned about his functions being C^∞-smooth at *every* point on its domain (assuming that he would have worried about 'domains' at all). Perhaps things going wrong at the odd point or so would not matter. But $|x|$ and $\theta(x)$ only went wrong at the same 'odd point' as does $1/x$. It seems that, despite all our efforts, we still have not captured the 'Eulerian' notion of a function that we have been striving for.

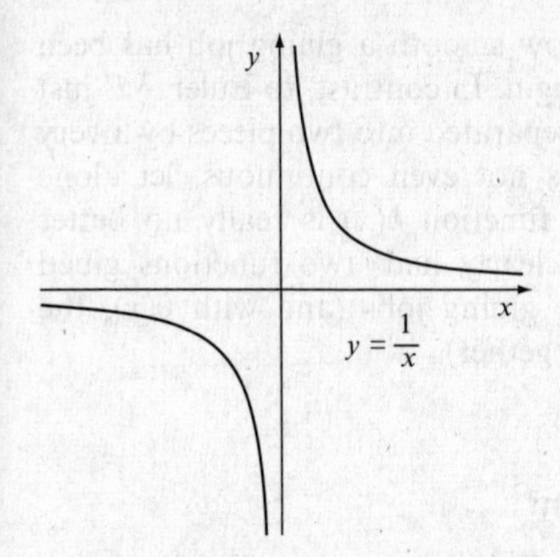

Fig. 6.6 Plot of $\frac{1}{x}$.

Let us take another example. Consider the function $h(x)$, defined by the rules

$$h(x) = \begin{cases} 0 & \text{if } x \leqslant 0, \\ e^{-1/x} & \text{if } x > 0. \end{cases}$$

The graph of this function is depicted in Fig. 6.7. This certainly looks like a smooth function. In fact it is *very* smooth. It is C^∞-smooth over the entire domain of real numbers. (Proving this is the sort of thing that one does in a mathematics undergraduate course. I remember having to tackle this one when I was an undergraduate myself.[6.2]) Despite its utter smoothness, one can certainly imagine Euler turning up his nose at a function defined in this kind of a way. It is clearly not just 'one function', in Euler's sense. It is 'two

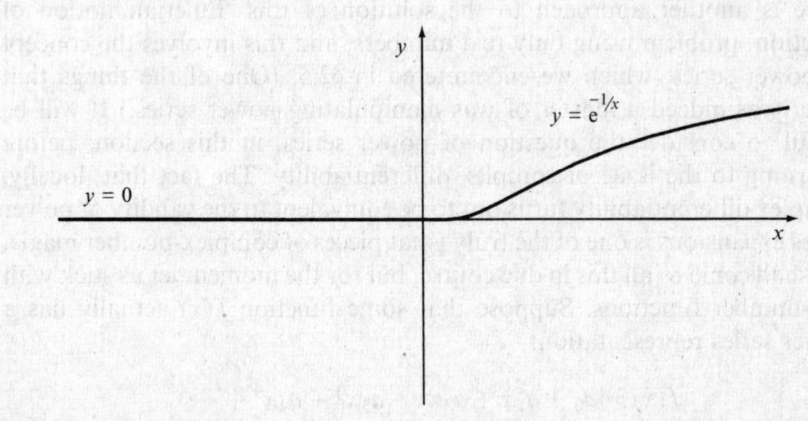

Fig. 6.7 Plot of $y = h(x)$ ($= 0$ if $x \leq 0$ and $= e^{-1/x}$ if $x > 0$), which is C^∞-smooth.

[6.2] Have a go at proving this if you have the background.

functions stuck together', no matter how smooth a gluing job has been done to paste over the 'glitch' at the origin. In contrast, to Euler, $\frac{1}{x}$ *is* just one function, despite the fact that it is separated into two pieces by a very nasty 'spike' at the origin, where it is not even continuous, let alone smooth (Fig. 6.6). To our Euler, the function $h(x)$ is really no better than $|x|$ or $\theta(x)$. In those cases, we clearly had 'two functions glued together', though with much shoddier gluing jobs (and with $\theta(x)$, the glued bits seem to have come apart altogether).

6.4 The 'Eulerian' notion of a function?

How are we to come to terms with this 'Eulerian' notion of having just a single function as opposed to a patchwork of separate functions? As the example of $h(x)$ clearly shows, C^∞-smoothness is not enough. It turns out that there are actually two completely different-looking approaches to resolving this issue. One of these uses complex numbers, and it is deceptively simple to state, though momentous in its implications. We simply demand that our function $f(x)$ be extendable to a function $f(z)$ of the complex variable z so that $f(z)$ is smooth in the sense that it is merely required to be *once* differentiable with respect to the complex variable z. (Thus $f(z)$ is, in the complex sense, a kind of C^1-function.) It is an extraordinary display of genuine magic that we do not need more than this. If $f(z)$ can be differentiated once with respect to the complex parameter z, then it can be differentiated as many times as we like!

I shall return to the matter of complex calculus in the next chapter. But there is another approach to the solution of this 'Eulerian notion of function' problem using only real numbers, and this involves the concept of power series, which we encountered in §2.5. (One of the things that Euler was indeed a master of was manipulating power series.) It will be useful to consider the question of power series, in this section, before returning to the issue of complex differentiability. The fact that, locally, complex differentiability turns out to be equivalent to the validity of power series expansions is one of the truly great pieces of complex-number magic.

I shall come to all this in due course, but for the moment let us stick with real-number functions. Suppose that some function $f(x)$ actually has a power series representation:

$$f(x) = a_0 + a_1 x + a_2 x^2 + a_3 x^3 + a_4 x^4 + \cdots.$$

Now, there are methods of finding out, from $f(x)$, what the coefficients $a_0, a_1, a_2, a_3, a_4, \ldots$ must be. For such an expansion to exist, it is necessary (although not sufficient, as we shall shortly see) that $f(x)$ be C^∞-smooth, so we shall have new functions $f'(x), f''(x), f'''(x), f''''(x), \ldots$, etc., which are the first, second, third, fourth, etc., derivatives of $f(x)$, respectively. In fact, we shall be concerned with the values of these functions only at the origin ($x = 0$), and we need the C^∞-smoothness of $f(x)$ only there. The result (sometimes called *Maclaurin's series*[7]) is that if $f(x)$ has such a power series expansion, then[6.3]

$$a_0 = f(0),\ a_1 = \frac{f'(0)}{1!},\ a_2 = \frac{f''(0)}{2!},\ a_3 = \frac{f'''(0)}{3!},\ a_4 = \frac{f''''(0)}{4!}, \ldots.$$

(Recall, from §5.3, that $n! = 1 \times 2 \times \ldots \times n$.) But what about the other way around? If the a's are given in this way, does it follow that the sum actually gives us $f(x)$ (in some interval encompassing the origin)?

Let us return to our seemingly seamless $h(x)$. Perhaps we can spot a flaw at the joining point ($x = 0$) using this idea. We try to see whether $h(x)$ actually has a power series expansion. Taking $f(x) = h(x)$ in the above, we consider the various coefficients $a_0, a_1, a_2, a_3, a_4, \ldots$, noticing that they all have to vanish, because the series has to agree with the value $h(x) = 0$, whenever x is just to the left of the origin. In fact, we find that they all vanish also for $e^{-1/x}$, which is basically the reason why $h(x)$ is C^∞-smooth at the origin, with all derivatives coming from the two sides matching each other. But this also tells us that there is no way that the power series can work, because all the terms are zero (see Exercise [6.2]) and therefore do not actually sum to $e^{-1/x}$. Thus there *is* a flaw at the join at $x = 0$: the function $h(x)$ cannot be expressed as a power series. We say that $h(x)$ is not *analytic* at $x = 0$.

In the above discussion, I have really been referring to what would be called a power series expansion *about the origin*. A similar discussion would apply to any other point of the real-number domain of the function. But then we have to 'shift the origin' to some other particular point, defined by the real number p in the domain, which means replacing x by $x - p$ in the above power series expansion, to obtain

$$f(x) = a_0 + a_1(x-p) + a_2(x-p)^2 + a_3(x-p)^3 + \cdots,$$

where now

$$a_0 = f(p),\ a_1 = \frac{f'(p)}{1!},\ a_2 = \frac{f''(p)}{2!},\ a_3 = \frac{f'''(p)}{3!}, \ldots.$$

This is called a power series expansion *about p*. The function $f(x)$ is called *analytic at p* if it can be expressed as such a power series expression in some interval encompassing $x = p$. If $f(x)$ is analytic at all points of its domain, we

[6.3] Show this, using rules given towards end of §6.5.

just call it an *analytic function* or, equivalently, a C^ω-smooth function. Analytic functions are, in a clear sense, even 'smoother' than C^∞-smooth functions. In addition, they have the property that it is not possible to get away with gluing two 'different' analytic functions together, in the manner of the examples $\theta(x)$, $|x|$, $x|x|$, $x^n|x|$, or $h(x)$, given above. Euler would have been pleased with analytic functions. These are 'honest' functions indeed!

However, all these power series are awkward things to be carrying around, even if only in the imagination. The 'complex' way of looking at things turns out to be enormously more economical. Moreover, it gives us a greater depth of understanding. For example, the function $\frac{1}{x}$ is not analytic at $x = 0$; yet it is still 'one function'.[6.4] The 'power series philosophy' does not directly tell us this. But from the point of view of complex numbers, $\frac{1}{x}$ is clearly just one function, as we shall be seeing.

6.5 The rules of differentiation

Before discussing these matters, it will be useful to say a little about the wonderful rules that the differential calculus actually provides us with—rules that enable us to differentiate functions almost without really thinking at all, but only after months of practice, of course! These rules enable us to see how to write down the derivative of many functions directly, particularly when they are represented in terms of power series.

Recall that, as a passing comment, I remarked above that the derivative of x^3 is $3x^2$. This is a particular case of a simple but important formula: the derivative of x^n is nx^{n-1}, which we can write

$$\frac{\mathrm{d}(x^n)}{\mathrm{d}x} = nx^{n-1}.$$

(It would distract us too much, here, for me to explain why this formula holds. It is not really hard to show, and the interested reader can find all that is required in any elementary textbook on calculus.[8] Incidentally, n need not be an integer.) We can also express[9] this equation ('multiplying through by $\mathrm{d}x$') by the convenient formula

$$\mathrm{d}(x^n) = nx^{n-1}\mathrm{d}x.$$

There is not much more that we need to know about differentiating power series. There are basically two other things. First, the derivative of a sum of functions is the sum of the derivatives of the functions:

$$\mathrm{d}[f(x) + g(x)] = \mathrm{d}f(x) + \mathrm{d}g(x).$$

[6.4] Consider the 'one function' e^{-1/x^2}. Show that it is C^∞, but not analytic at the origin.

This then extends to a sum of any finite number of functions.[10] Second, the derivative of a constant times a function is the constant times the derivative of that function:

$$\mathrm{d}\{a\,f(x)\} = a\,\mathrm{d}f(x).$$

By a 'constant' I mean a number that does not vary with x. The *coefficients* $a_0, a_1, a_2, a_3, \ldots$ in the power series are constants. With these rules, we can directly differentiate *any* power series.[6.5]

Another way of expressing the constancy of a is

$$\mathrm{d}a = 0.$$

Bearing this in mind, we find that the rule given immediately above this one is really a special case (with $g(x) = a$) of the 'Leibniz law':

$$\mathrm{d}\{f(x)\,g(x)\} = f(x)\,\mathrm{d}g(x) + g(x)\,\mathrm{d}f(x)$$

(and $\mathrm{d}(x^n)/\mathrm{d}x = nx^{n-1}$, for any natural number n, can also be derived from the Leibniz law[6.6]). A useful further law is

$$\mathrm{d}\{f(g(x))\} = f'(g(x))g'(x)\mathrm{d}x.$$

From the last two and the first, putting $f(x)[g(x)]^{-1}$ into the Leibniz law, we can deduce[6.7]

$$\mathrm{d}\left(\frac{f(x)}{g(x)}\right) = \frac{g(x)\,\mathrm{d}f(x) - f(x)\,\mathrm{d}g(x)}{g(x)^2}.$$

Armed with these few rules (and loads and loads of practice), one can become an 'expert at differentiation' without needing to have much in the way of actual *understanding* of why the rules work! This is the power of a good calculus.[6.8] Moreover, with the knowledge of the derivatives of just a few special functions,[6.9] one can become even more of an expert. Just so that the uninitiated reader can become an 'instant member' of the club of expert differentiators, let me provide the main examples:[11,][6.10]

[6.5] Using the power series for e^x given in §5.3, show that $\mathrm{de}^x = \mathrm{e}^x\mathrm{d}x$.

[6.6] Establish this.

[6.7] Derive this.

[6.8] Work out $\mathrm{d}y/\mathrm{d}x$ for $y = (1 - x^2)^4$, $y = (1 + x)/(1 - x)$.

[6.9] With a constant, work out $\mathrm{d}(\log_a x)$, $\mathrm{d}(\log_x a)$, $\mathrm{d}(x^x)$.

[6.10] For the first, see Exercise [6.5]; derive the second from $\mathrm{d}(\mathrm{e}^{\log x})$; the third and fourth from $\mathrm{de}^{\mathrm{i}x}$, assuming that the complex quantities work like real ones; and derive the rest from the earlier ones, using $\mathrm{d}(\sin(\sin^{-1} x))$, etc., and noting that $\cos^2 x + \sin^2 x = 1$.

$$\begin{aligned}
\mathrm{d}(\mathrm{e}^x) &= \mathrm{e}^x\,\mathrm{d}x,\\
\mathrm{d}(\log x) &= \frac{\mathrm{d}x}{x},\\
\mathrm{d}(\sin x) &= \cos x\,\mathrm{d}x,\\
\mathrm{d}(\cos x) &= -\sin x\,\mathrm{d}x,\\
\mathrm{d}(\tan x) &= \frac{\mathrm{d}x}{\cos^2 x},\\
\mathrm{d}(\sin^{-1} x) &= \frac{\mathrm{d}x}{\sqrt{1-x^2}},\\
\mathrm{d}(\cos^{-1} x) &= \frac{-\mathrm{d}x}{\sqrt{1-x^2}},\\
\mathrm{d}(\tan^{-1} x) &= \frac{\mathrm{d}x}{1+x^2}.
\end{aligned}$$

This illustrates the point referred to at the beginning of this section that, when we are given explicit formulae, the operation of differentiation is 'easy'. Of course, I do not mean by this that this is something that you could do in your sleep. Indeed, in particular examples, it may turn out that the expressions get very complicated indeed. When I say 'easy', I just mean that there is an explicit computational procedure for carrying out differentiation. If we know how to differentiate each of the ingredients in an expression, then the procedures of calculus, as given above, tell us how to go about differentiating the entire expression. 'Easy', here, really means something that could be readily put on a computer. But things are very different if we try to go in the reverse direction.

6.6 Integration

As stated at the beginning of the chapter, *integration* is the reverse of differentiation. What this amounts to is trying to find a function $g(x)$ for which $g'(x) = f(x)$, i.e. finding a solution $y = g(x)$ to the equation $\mathrm{d}y/\mathrm{d}x = f(x)$. Another way of putting this is that, instead of moving down the picture in Fig. 6.4 (or Fig. 6.5), we try to work our way upwards. The beauty of the 'fundamental theorem of calculus' is that this procedure is telling us how to work out areas under each successive curve. Have a look at Fig. 6.8. Recall that the bottom curve $u = f(x)$ can be obtained from the top curve $y = g(x)$ because it plots the slopes of that curve, $f(x)$ being the derivative of $g(x)$. This is just what we had before. But now let us start with the bottom curve. We find that the top curve simply maps out the areas beneath the bottom curve. A little more explicitly: if we take two vertical lines in the bottom picture given by $x = a$ and $x = b$, respectively, then the area bounded by these two lines, the x-axis, and the curve itself, will be the difference between the heights of the top curve at those two x-values. Of course, in matters such as this, we must

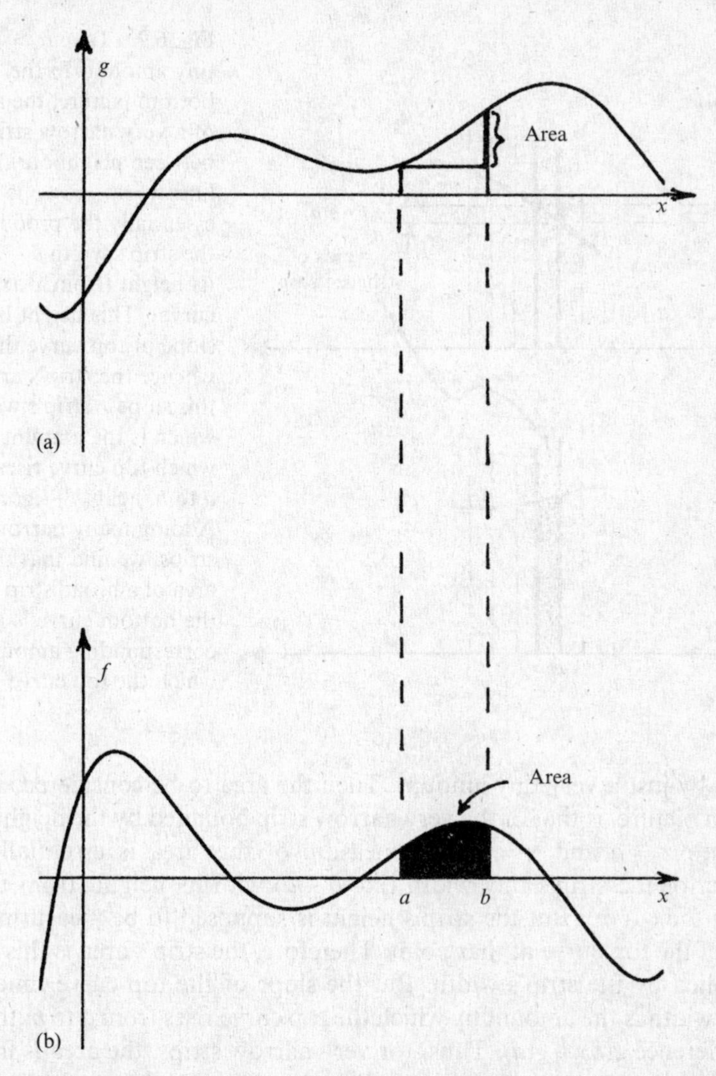

Fig. 6.8 Fundamental theorem of calculus: re-interpret Fig. 6.4a,b, proceeding upwards rather than downwards. Top curve (a) plots areas under bottom curve (b), where the area bounded by two vertical lines $x = a$ and $x = b$, the x-axis, and the bottom curve is the difference, $g(b) - g(a)$, of heights of the top curve at those two x-values (signs taken into account).

be careful about 'signs'. In regions where the bottom curve dips below the x-axis, the areas count negatively. Moreover, in the picture, I have taken $a < b$ and the 'difference between the heights' of the top curve in the form $g(b) - g(a)$. Signs would be reversed if $a > b$.

In Fig. 6.9, I have tried to make it intuitively believable why there is this inverse relationship between slopes and areas. We imagine b to be greater

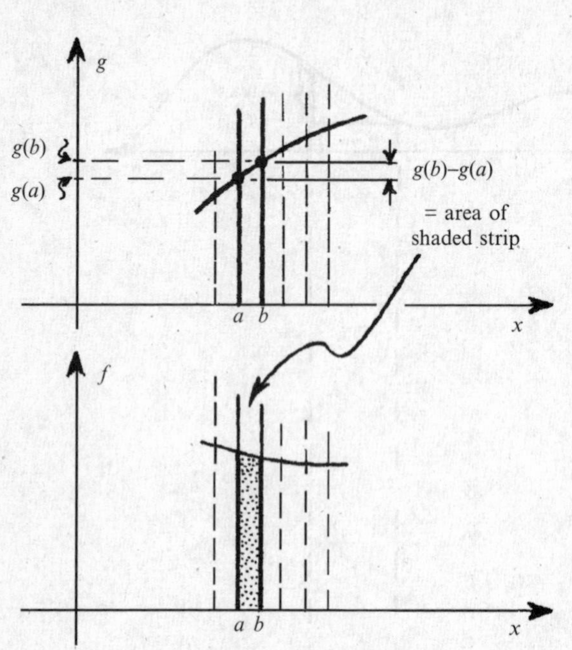

Fig. 6.9 Take $b > a$ by a tiny amount. In the bottom picture, the area of a very narrow strip between neighbouring lines $x = a$, $x = b$ is essentially the product of the strip's width $b - a$ with its height (from x-axis to curve). This height is the slope of top curve there, whence the strip's area is this slope × strip's width, which is the amount by which top curve rises from a to b, i.e. $g(b) - g(a)$. Adding many narrow strips, we find that the area of a broad strip under the bottom curve is the corresponding amount by which the top curve rises.

than a by just a very tiny amount. Then the area to be considered, in the bottom picture, is that of the very narrow strip bounded by the neighbouring lines $x = a$ and $x = b$. The measure of this area is essentially the product of the strip's tiny width (i.e. $b - a$) with its height (from the x-axis to the curve). But the strip's height is supposed to be measuring the slope of the top curve at that point. Therefore, the strip's area is this slope multiplied by the strip's width. But the slope of the top curve times the strip's width is the amount by which the top curve rises from a to b, that is, the difference $g(b) - g(a)$. Thus, for very narrow strips, the area is indeed measured by this stated difference. Broad strips are taken to be built up from large numbers of narrow strips, and we get the total area by measuring how much the top curve rises over the entire interval.

There is a significant point that I should bring out here. In the passage from the bottom curve to the top curve there is a non-uniqueness about how high the whole top curve is to be placed. We are only concerned with *differences* between heights on the top curve, so sliding the whole curve up or down by some constant amount will not make any difference. This is clear from the 'slope' interpretation too, since the slope at different points on the top curve will be just the same as before if we slide it up or down. What this amounts to, in our calculus, is that if we add a constant C to $g(x)$, then the resulting function still differentiates to $f(x)$:

$$\mathrm{d}(g(x) + C) = \mathrm{d}g(x) + \mathrm{d}C = f(x)\,\mathrm{d}x + 0 = f(x)\,\mathrm{d}x.$$

Such a function $g(x)$, or equivalently $g(x) + C$ for some arbitrary constant C, is called an *indefinite integral* of $f(x)$, and we write

$$\int f(x)\,\mathrm{d}x = g(x) + \text{const.}$$

This is just another way of expressing the relation $\mathrm{d}[g(x) + \text{const.}] = f(x)\mathrm{d}x$, so we just think of the '$\int$' sign as the inverse of the 'd' symbol. If we want the specific area between $x = a$ and $x = b$, then we want what is called the *definite integral*, and we write

$$\int_a^b f(x)\,\mathrm{d}x = g(b) - g(a).$$

If we know the function $f(x)$ and we wish to obtain its integral $g(x)$, we do not have nearly such straightforward rules for obtaining it as we did for differentiation. A great many tricks are known, a variety of which can be found in standard textbooks and computer packages, but these do not suffice to handle all cases. In fact, we frequently find that the family of explicit standard functions that we had been using previously has to be broadened, and that new functions have to be 'invented' in order to express the results of the integration. We have, in effect, seen this already in the special examples given above. Suppose that we were familiar just with functions made up of combinations of powers of x. For a general power x^n, we can integrate it to get $x^{n+1}/(n+1)$. (This is just using our formula above, in §6.5, with $n+1$ for n: $\mathrm{d}(x^{n+1})/\mathrm{d}x = (n+1)x^n$.) Everything is fine until we worry about what to do with the case $n = -1$. Then the supposed answer $x^{n+1}/(n+1)$ has zero in the denominator, so this won't work. How, then, do we integrate x^{-1}? Well, we notice that, by the greatest of good fortune, there is the formula $\mathrm{d}(\log x) = x^{-1}\mathrm{d}x$ sitting in our list in §6.5. So the answer is $\log x + \text{const.}$

This time we were lucky! It just happened that we had been studying the logarithm function before for a different reason, and we knew about some of its properties. But on other occasions, we might well find that there is no function that we had previously known about in terms of which we can express our answer. Indeed, integrals frequently provide the appropriate means whereby new functions are *defined*. It is in this sense that explicit integration is 'difficult'.

On the other hand, if we are not so interested in explicit expressions, but are concerned with questions of *existence* of functions that are the derivatives or integrals of given functions, then the boot is on the other foot. Integration is now the operation that works smoothly, and differentiation causes the problems. The same applies when performing these operations with numerical data. Basically, the problem with differentiation is that it depends very critically on the fine details of the function to be differentiated. This can present a problem if we do not have an explicit expression for the function to be differentiated. Integration, on the other hand, is relatively insensitive to such matters, being concerned with the broad overall nature of the function to be integrated. In fact, any continuous function (a C^0-function) whose domain is a 'closed' interval $a \leqslant x \leqslant b$ can be integrated,[12] the result being C^1 (i.e. C^1-smooth). This can be integrated again, the result being C^2, and then again, giving a C^3-smooth function, and so on. Integration makes the functions smoother and smoother, and we can keep on going with this indefinitely. Differentiation, on the other hand just makes things worse, and it may come to an end at a certain point, where the function becomes 'non-differentiable'.

Yet, there are approaches to these issues that enable the process of differentiation to be continued indefinitely also. I have hinted at this already, when I allowed myself to differentiate the function $|x|$ to obtain $\theta(x)$, even though $|x|$ is 'not differentiable'. We could attempt to go further and differentiate $\theta(x)$ also, despite the fact that it has an infinite slope at the origin. The 'answer' is what is called the Dirac[13] *delta function*—an entity of considerable importance in the mathematics of quantum mechanics. The delta function is not really a function at all, in the ordinary (modern) sense of 'function' which maps domains to target spaces. There is no 'value' for the delta function at the origin (which could only have been *infinity* there) and it is zero elsewhere. Yet the delta function does finds a clear mathematical definition within various broader classes of mathematical entities, the best known being *distributions*.

For this, we need to extend our notion of C^n-functions to cases where n can be a negative integer. The function $\theta(x)$ is then a C^{-1}-function and the delta function is C^{-2}. Each time we differentiate, we must decrease the differentiability class by unity (i.e. the class becomes more negative by one unit). It would seem that we are getting farther and farther from Euler's notion of a 'decent function' with all this and that he would tell us to have no truck with such things, were it not for the fact that they seem to be useful. Yet, we shall be finding, in due course, that it is here that complex numbers astound us with an irony—an irony that is expressed in one of their finest magical feats of all! We shall have to wait until the end of Chapter 9 to witness this feat, for it is not something that I can properly describe just yet. The reader must bear with me for a while, for the ground needs first to be made ready, paved with other superbly magical ingredients.

Notes

Section 6.1

6.1. I am adopting a slight 'abuse of notation' here, as technically x^2, for instance, denotes the *value* of the function rather than the function. The function itself maps x to x^2 and might be denoted by $x \mapsto x^2$, or by $\lambda x[x^2]$ according to Alonzo Church's (1941) *lambda calculus*; see Chapter 2 of Penrose (1989).

6.2. In this section, I shall frequently refer to what Euler's beliefs might well have been with regard to the notion of a function. However, I should make clear here that the 'Euler' that I am referring to is really a hypothetical or idealized individual. I have no direct information about what the real Leonhard Euler's views were in any particular case. But the views that I am attributing to my 'Euler' do not appear to be out of line with the kind of views that the real Euler might well have expressed. For more information about Euler, see Boyer (1968); Thiele (1982); Dunham (1999).

Section 6.2

6.3. For details, see Burkill (1962).

6.4. Strictly, it is the function f' that is the derivative of the function f; we cannot obtain the value of f' at x simply from the value of f at x. See Note 6.1.

Section 6.3

6.5. Viz., $f''(x)/[1+f'(x)^2]^{3/2}$.

6.6. In fact, this implies that all the derivatives up to and including the nth must be continuous, because the technical definition of differentiability requires continuity.

Section 6.4

6.7. Traditionally, this power series expansion about the origin is known (with little historical justification) as Maclaurin's series; the more general result about the point p (see later in the section) is attributed to Brook Taylor (1685–1731).

Section 6.5

6.8. See Edwards and Penney (2002).

6.9. For the moment, just treat the following expressions formally, or else mentally 'divide back through by dx' if this makes you happier. The notation that I am using here is consistent with that of differential forms, which will be discussed in §§12.3–6.

6.10. However, there is a technical subtlety about applying this law to the sum of the infinite number of terms that we need for a power series. This subtlety can be ignored for values of x strictly within the circle of convergence; see §4.4. See Priestley (2003).

6.11. Recall from §5.1 that $\sin^{-1}$, $\cos^{-1}$, and $\tan^{-1}$ are the inverse functions of $\sin, \cos$, and $\tan$, respectively. Thus $\sin(\sin^{-1} x) = x$, etc. We must bear in mind that these inverse functions are 'many-valued functions', however, and it is usual to select the values for which $-\frac{\pi}{2} \leqslant \sin^{-1}x \leqslant \frac{\pi}{2}$, $0 \leqslant \cos^{-1}x \leqslant \pi$, and $-\frac{\pi}{2} < \tan^{-1}x < \frac{\pi}{2}$.

Section 6.6

6.12. The significant requirement on the domain is that it be what is called *compact*; see §12.6. Finite intervals of the real line including their end-points are indeed compact.

6.13. Apparently, Oliver Heaviside had also conceived the 'delta function' many years before Dirac.

7
Complex-number calculus

7.1 Complex smoothness; holomorphic functions

How are we to understand the notion of differentiation when this is applied to a *complex* function $f(z)$? It is certainly not appropriate, in this book, that I attempt to address this issue in full detail.[1] I did not even properly address such details, in §6.2, for a real function. But at least I can attempt to convey the gist of what is involved. The following is a very rapid outline of the essential argument to show what complex differentiability achieves. Afterwards I shall be a little more explicit about some of its surprising ingredients.

Basically, for complex differentiation, we require that there be a notion of 'slope' of the complex curve $w = f(z)$ at any point z in the function's domain. (The function $f(z)$ and the variable z are now both allowed to take complex values.) For this notion of 'slope' to make consistent sense, as we move the variable z around slightly in different directions in z's complex plane, it is necessary for $f(z)$ to satisfy a certain pair of equations called the *Cauchy–Riemann* equations[2] (involving the derivatives of the real and imaginary parts of $f(z)$, taken with respect to the real and imaginary parts of z; see §10.5). These equations establish for us something rather remarkable about complex integration—something which then enables a new notion of integration to be defined, called *contour integration*. A beautiful formula can then be given, in terms of this contour integration, for the nth derivative of $f(z)$. Thus, once we have the first derivative, we get all higher derivatives *free*.

We next use this formula to provide us with the coefficients of a proposed Taylor series for $f(z)$, which we have to show actually converges to $f(z)$. Having achieved this, we have a Taylor series expression for $f(z)$ that works inside any circle in the complex z-plane throughout which $f(z)$ is defined and differentiable. The magical fact thus arises, that any complex function that is complex-smooth is necessarily analytic!

Accordingly, there is no problem, in complex analysis, in recognizing the limitations of the 'gluing jobs' in certain C^∞-functions, such as the '$h(x)$' defined in the previous chapter. The power of complex smoothness would surely have delighted Euler. (Unfortunately for the real Leonhard Euler, the astounding power of this complex smoothness was appreciated too late for him, as it was first found by Augustin Cauchy in 1821, some 38 years after Euler's death.) We see that complex smoothness provides a much more economical way of expressing what is required for our 'Eulerian' notion of a function than does the existence of power series expansions. But there is also another advantage in looking at such functions from the complex point of view. Recall our troublesome '$1/x$' that seemed to be 'just one function' despite the fact that the real curve $y = 1/x$ consists of two separate pieces which are not joined 'analytically' to each other through real values of x. From the complex perspective, we see clearly that $1/z$ is indeed a single function. The one place where the function 'goes wrong' in the complex plane is the origin $z = 0$. If we remove this one point from the complex plane, we still get a connected region. The part of the real line for which $x < 0$ is connected to the part for which $x > 0$ through the complex plane. Thus, $1/z$ is indeed one connected complex function, this being quite different from the real-number situation.

Functions that are complex-smooth (complex-analytic) in this sense are called *holomorphic*. Holomorphic functions will play a vital part in many of our later deliberations. We shall see their importance in connection with conformal mappings and Riemann surfaces in Chapter 8, and with Fourier series (fundamental to the theory of vibrations) in Chapter 9. They have important roles to play in quantum theory and in quantum field theory (as we shall see in §24.3 and §26.3). They are also fundamental to some approaches to the developing of new physical theories (particularly twistor theory—see Chapter 33—and they also have a significant part to play in string theory; see §§31.5,13,14).

7.2 Contour integration

Although this is not the place to spell out all the details of the mathematical arguments indicated in §7.1, it will nevertheless be illuminating to elaborate upon the above outline. In particular, it will be of benefit to have an account of contour integration here, which will provide the reader with some understanding of the way in which contour integration can be used to establish what is needed for the requirements of §7.1. First let us recall the notation for a definite integral that was given, in the previous chapter, for a real variable x, and now think of it as applying to a complex variable z:

$$\int_a^b f(z)\mathrm{d}z = g(b) - g(a),$$

where $g'(z) = f(z)$. In the real case, the integral is taken from one point a on the real line to another point b on that line. There is only one way to get from a to b along the real line. Now think of it as a complex formula. Here we have a and b as two points on the complex plane instead. Now, we do not just have one route from a to b, but we could draw lots of different paths connecting a to b. What the Cauchy–Riemann equations tell us is that if we do our integration along one such path[3] then we get the same answer as along any other such path that can be obtained from the first by continuous deformation within the domain of the function. (See Fig. 7.1. This property is a consequence of a simple case of the 'fundamental theorem of exterior calculus', described in §12.6.) For some functions, $1/z$ being a case in point, the domain has a 'hole' in it (the hole being $z = 0$ in the case of $1/z$), so there may be several essentially different ways of getting from a to b. Here 'essentially different' refers to the fact that one of the paths cannot be continuously deformed into another while remaining in the domain of the function. In such cases, the value of the integral from a to b may give a different answer for the various paths.

One point of clarification (or, rather, of correction) should be made here. When I talk about one path being continuously deformed into another, I am referring to what mathematicians call *homologous* deformations, not *homotopic* ones. With a homologous deformation, it is legitimate for parts of paths to cancel one another out, provided that those portions are being traversed in opposite directions. See Fig. 7.2 for an example of this sort of allowable deformation. Two paths that are deformable one into the other in this way are said to *belong to the same homology class*. By contrast, homotopic deformations do not permit this kind of cancellation. Paths deformable one into another, where such cancellation is not permitted, belong to the same *homotopy class*. Homotopic curves are always homologous, but not necessarily the other way around. Both homotopy and homology are to do with equivalence under continuous motions. Thus they are part of the

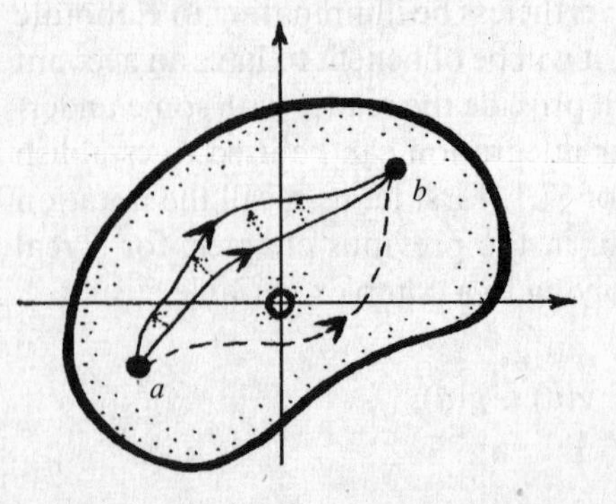

Fig. 7.1 Different paths from a to b. Integrating a holomorphic function f along one path yields the same answer as along any other path obtainable from it by continuous deformation within f's domain. For some functions, the domain has a 'hole' in it (e.g. $z = 0$, for $1/z$), obstructing certain deformations, so different answers may be obtained.

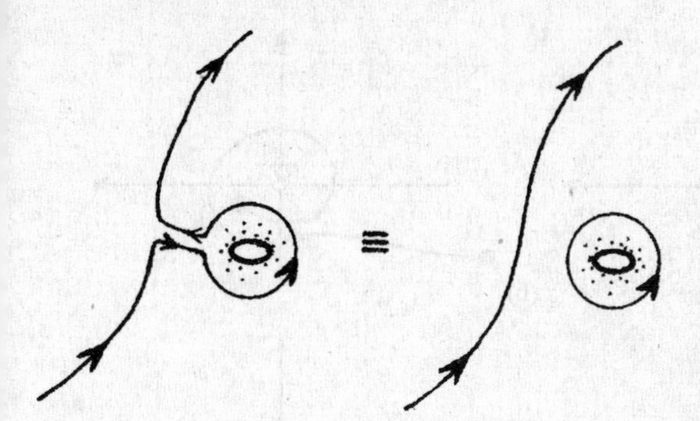

Fig. 7.2 With a homologous deformation, parts of paths cancel each other, if traversed in opposite directions. Sometimes this gives rise to separated loops.

subject of *topology*. We shall be seeing different aspects of topology playing important roles in other areas later.

The function $f(z) = 1/z$ is in fact one for which different answers *are* obtained when the paths are not homologous. We can see why this must be so from what we already know about logarithms. Towards the end of the previous chapter, it was noted that $\log z$ is an indefinite integral of $1/z$. (In fact, this was only stated for a real variable x, but the same reasoning that obtains the real answer will also obtain the corresponding complex answer. This is a general principle, applying to our other explicit formulae also.) We therefore have

$$\int_a^b \frac{\mathrm{d}z}{z} = \log b - \log a.$$

But recall, from §5.3, that there are different alternative 'answers' to a complex logarithm. More to the point is that we can get continuously from one answer to another. To illustrate this, let us keep a fixed and allow b to vary. In fact, we are going to allow b to circle continuously once around the origin in a positive (i.e. anticlockwise) sense (see Fig. 7.3a), restoring it to its original position. Remember, from §5.3, that the imaginary part of $\log b$ is simply its *argument* (i.e. the angle that b makes with the positive real axis, measured in the positive sense; see Fig. 5.4b). This argument increases precisely by 2π in the course of this motion, so we find that $\log b$ has increased by $2\pi\mathrm{i}$ (see Fig. 7.3b). Thus, the value of our integral is increased by $2\pi\mathrm{i}$ when the path over which the integral is performed winds once more (in the positive sense) about the origin.

We can rephrase this result in terms of *closed contours*, the existence of which is a characteristic and powerful feature of complex analysis. Let us consider the difference between the second and the first of our two paths, that is to say, we traverse the second path first and then we traverse the first path in the reverse direction (Fig. 7.3c). We consider this difference in the homologous sense, so we can cancel out portions that 'double back' and straighten out the rest, in a continuous fashion. The result is a closed

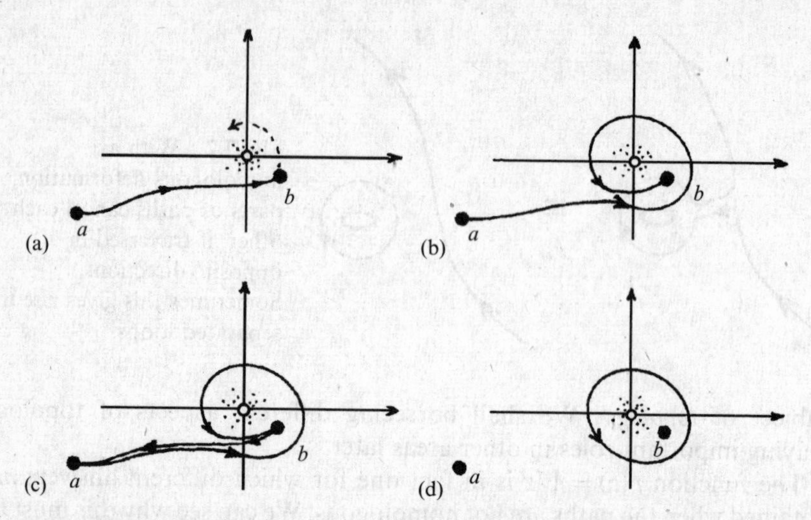

Fig. 7.3 (a) Integrating z^{-1} dz from a to b gives $\log b - \log a$. (b) Keep a fixed, and allow b to circle once anticlockwise about the origin, increasing $\log b$ in the answer by $2\pi i$. (c) Then return to a backwards along original route. (d) When the part of the path is cancelled from a, we are left with an anticlockwise closed contour integral $\oint z^{-1}\,dz = 2\pi i$.

path—or *contour*—that loops just once about the origin (see Fig. 7.3d), and it is not concerned with the location of either a or b. This gives an example of a (closed) contour integral, usually written with the symbol $\oint$, and we find, in this example,[7.1]

$$\oint \frac{dz}{z} = 2\pi i.$$

Of course, when using this symbol, we must be careful to make clear which actual contour is being used—or, rather, which homology class of contour is being used. If our contour had wound around twice (in the positive sense), then we would get the answer $4\pi i$. If it had wound once around the origin in the opposite direction (i.e. clockwise), then the answer would have been $-2\pi i$.

It is interesting that this property of getting a non-trivial answer with such a closed contour depends crucially on the multivaluedness of the complex logarithm, a feature which might have seemed to be just an awkwardness in the definition of a logarithm. We shall see in a moment that this is not just a curiosity. The power of complex analysis, in effect,

[7.1] Explain why $\oint z^n dz = 0$ when n is an integer other than -1.

depends critically upon it. In the following two paragraphs, I shall outline some of the implications of this sort of thing. I hope that non-mathematical readers can get something of value from the discussion. I believe that it conveys something that is both genuine and surprising in the nature of mathematical argument.

7.3 Power series from complex smoothness

The above displayed expression is a particular case (for the constant function $f(z) = 2\pi i$) of the famous *Cauchy formula* which expresses the value of a holomorphic function at the origin in terms of an integral around a contour surrounding the origin:[4]

$$\frac{1}{2\pi i}\oint \frac{f(z)}{z}\,dz = f(0).$$

Here, $f(z)$ is holomorphic at the origin (i.e. complex-smooth throughout some region encompassing the origin), and the contour is some loop just surrounding the origin—or it could be any loop homologous to that one, in the domain of the function with the origin removed. Thus, we have the remarkable fact that what the function is doing *at* the origin is completely fixed by what it is doing at a set of points *surrounding* the origin. (Cauchy's formula is basically a consequence of the Cauchy–Riemann equations, together with the above expression $\oint z^{-1}dz = 2\pi i$, taken in the limit of small loops; but it would not be appropriate for me to go into the details of all this here.)

If, instead of using $1/z$ in Cauchy's formula, we use $1/z^{n+1}$, where n is some positive integer, we get a 'higher-order' version of the Cauchy formula, yielding what turns out to be the nth derivative $f^{(n)}(z)$ of $f(z)$ at the origin:

$$\frac{n!}{2\pi i}\oint \frac{f(z)}{z^{n+1}}\,dz = f^{(n)}(0).$$

(Recall $n!$ from §5.3.) We can see that this formula 'has to be the right answer' by examining the power series for $f(z)$,[7.2] but it would be begging the question to *use* this fact, because we do not yet know that the power series expansion exists, or even that the nth derivative of f exists. All that we know at this stage is that $f(z)$ is complex-smooth, without knowing that it can be differentiated more than once. However, we simply use this formula as providing the *definition* of the nth derivative at the origin. We can then incorporate this 'definition' into the Maclaurin formula $a_n = f^{(n)}(0)/n!$ for the coefficients in the power series (see §6.4)

[7.2] Show this simply by substituting the Maclaurin series for $f(z)$ into the integral.

$$a_0 + a_1 z + a_2 z^2 + a_3 z^3 + a_4 z^4 + \cdots,$$

and with a bit of work we can prove that this series actually does sum to $f(z)$ in some region encompassing the origin. Consequently, the function has an actual nth derivative at the origin as given by the formula.[7.3] This contains the essence of the argument showing that complex smoothness in a region surrounding the origin indeed implies that the function is actually (complex-) *analytic* at the origin (i.e. holomorphic).

Of course, there is nothing special about the origin in all this. We can equally well talk about power series about any other point p in the complex plane and use Taylor's series, as we did in §6.4. For this, we simply displace the origin to the point p to obtain Cauchy's formula in the 'origin-shifted' form

$$\frac{1}{2\pi \mathrm{i}} \oint \frac{f(z)}{(z-p)} \mathrm{d}z = f(p),$$

and also the nth-derivative expression

$$\frac{n!}{2\pi \mathrm{i}} \oint \frac{f(z)}{(z-p)^{n+1}} \mathrm{d}z = f^{(n)}(p),$$

where now the contour surrounds the point p in the complex plane. Thus, complex smoothness implies analyticity (holomorphicity) at *every* point of the domain.

I have chosen to demonstrate the basics of the argument that, locally, complex smoothness implies analyticity, rather than simply request that the reader take the result on trust, because it is a wonderful example of the way that mathematicians can often obtain their results. Neither the premise ($f(z)$ is complex-smooth) nor the conclusion ($f(z)$ is analytic) contains a hint of the notion of contour integration or of the multivaluedness of a complex logarithm. Yet, these ingredients provide the essential clues to the true route to finding the answer. It is difficult to see how any 'direct' argument (whatever that might be) could have achieved this. The key is mathematical playfulness. The enticing nature of the complex logarithm itself is what beguiles us into studying its properties. This intrinsic appeal is apparently independent of any applications that the logarithm might have in other areas. The same, to an even greater degree, can be said for contour integration. There is an extraordinary elegance in the basic conception, where topological freedom combines with explicit expressions

---

[7.3] Show all this at least at the level of formal expressions; don't worry about the rigorous justification. *Hint*: Look at the origin-shifted Cauchy formula.

with exquisite precision.[7.4] But it is not merely elegance: contour integration also provides a very powerful and useful mathematical technique in many different areas, containing much complex-number magic. In particular, it leads to surprising ways of evaluating definite integrals and explicitly summing various infinite series.[7.5],[7.6] It also finds many other applications in physics and engineering, as well as in other areas of mathematics. Euler would have revelled in it all!

7.4 Analytic continuation

We now have the remarkable result that complex smoothness throughout some region is equivalent to the existence of a power series expansion about any point in the region. However, I should make it a little clearer what a 'region' is to mean in this context. Technically, I mean what mathematicians call an *open* region. We can express this by saying that if a point a is in the region then there is a circle centred at a whose interior is also contained in the region. This may not be very intuitive, so let me give some examples. A single point is not an open region, nor is an ordinary curve. But the *interior* of the unit circle in the complex plane, that is, the set of points whose distance from the origin is strictly less than unity, is an open region. This is because any point strictly inside the circle, no matter how close it is to the circumference, can be surrounded by a much smaller circle whose interior still lies strictly within the unit circle (see Fig. 7.4). On the other hand, the *closed disc*, consisting of points whose distance from the origin is either less than *or equal* to unity, is not an open region, because the circumference is now included, and a point on the circumference does not have the property that there is a circle centred at that point whose interior is contained within the region.

[7.4] The function $f(z)$ is holomorphic everywhere on a closed contour Γ, and also within Γ except at a finite set of points where f has poles. Recall from §4.4 that a *pole* of order n at $z = \alpha$ occurs where $f(z)$ is of the form $h(z)/(z-\alpha)^n$, where $h(z)$ is regular at α. Show that $\oint_\Gamma f(z)\mathrm{d}z = 2\pi i \times$ {sum of the residues at these poles}, where the *residue* at the pole α is $h^{(n-1)}(\alpha)/(n-1)!$

[7.5] Show that $\int_0^\infty x^{-1} \sin x \, \mathrm{d}x = \frac{\pi}{2}$ by integrating $z^{-1}e^{iz}$ around a closed contour Γ consisting of two portions of the real axis, from $-R$ to $-\epsilon$ and from ϵ to R (with $R > \epsilon > 0$) and two connecting semi-circular arcs in the upper half-plane, of respective radii ϵ and R. Then let $\epsilon \to 0$ and $R \to \infty$.

[7.6] Show that $1 + \frac{1}{2^2} + \frac{1}{3^2} + \frac{1}{4^2} + \cdots = \frac{\pi^2}{6}$ by integrating $f(z) = z^{-2} \cot \pi z$ (see Note 5.1) around a large contour, say a square of side-length $2N + 1$ centred at the origin (N being a large integer), and then letting $N \to \infty$. (*Hint*: Use Exercise [7.4], finding the poles of $f(z)$ and their residues. Try to show why the integral of $f(z)$ around Γ approaches the limiting value 0 as $N \to \infty$.)

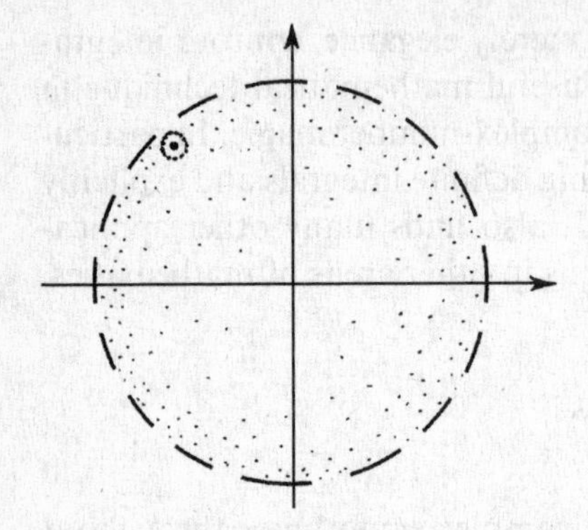

Fig. 7.4 The open unit disc $|x| < 1$. Any point strictly inside, no matter how close to the circumference, is surrounded by much smaller circle whose interior still lies strictly within unit circle. On the other hand, for the closed disc $|x| \leq 1$, this fails for points on the boundary.

Let us now consider the *domain*[5] D of some holomorphic function $f(z)$, where we take D to be an open region. At every point of D, the function $f(z)$ is to be complex-smooth. Thus, in accordance with the above, if we select any point p in D, then we have a convergent power series about p that represents $f(z)$ in a suitable region containing p. How big is this 'suitable region'? It will tend to be the case that, for a particular p, the power series will not work for the whole of D. Recall the *circle of convergence* described in §4.4. This would be some circle centred at p (infinite radius permitted) such that for points strictly within this circle the power series will converge, but for points z strictly outside the circle it will not. Suppose that $f(z)$ has a *singularity* at some point q, namely a point that the function $f(z)$ cannot be extended to while remaining complex-smooth. (For example, the origin $q = 0$ is a singularity of the function $f(z) = 1/z$; see §7.1. A singularity is sometimes referred to as a 'singular point' of the function. A *regular* point is just a place where the function is non-singular, and hence holomorphic.) Then the circle of convergence cannot be so large that it contains q in its interior. We therefore have a patchwork of circles of convergence (usually infinite in number) which together cover the whole of D, while generally no single circle will cover it. The case $f(z) = 1/z$ illustrates the issue (see Fig. 7.5). Here the domain D is the complex plane with the origin removed. If we select a point p in D, we find that the circle of convergence is the circle centred at p passing through the origin.[7.7] We need an infinite number of such circles to cover the entire region D.

This leads us to the important issue of *analytic continuation*. Suppose that we are given some function $f(z)$, holomorphic in some domain D, and we consider the question: can we extend D to a larger region D' so that $f(z)$ also extends holomorphically to D'? For example, $f(z)$ might have been given to us in the form of a power series, convergent within its particular circle of convergence, and we might wish to extend $f(z)$ outside that circle.

[7.7] What is the power series, taken about the point p, for $f(z) = 1/z$?

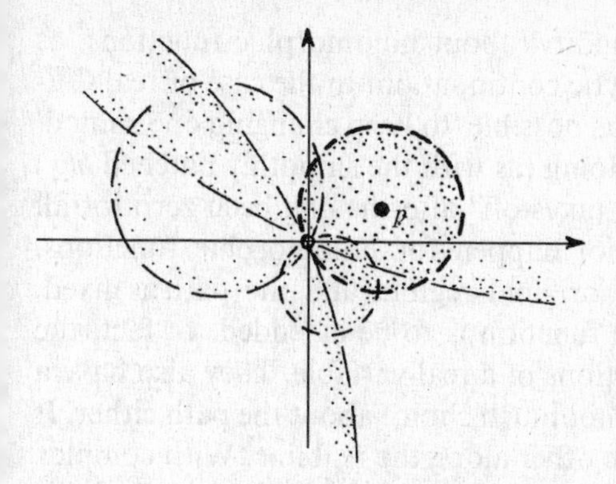

Fig. 7.5 For $f(z) = 1/z$, the domain D is complex plane with the origin removed. The circle of convergence about any point p in D is centred at p and passes through the origin. To cover the whole of D we need a patchwork (infinite) of such circles.

Frequently this is possible. In §4.4, we considered the series $1 - z^2 + z^4 - z^6 + \cdots$, which has the unit circle as its circle of convergence; yet it has the natural extension to the function $(1 + z^2)^{-1}$, which is holomorphic over the entire complex plane with only the two points $+i$ and $-i$ removed. Thus, in this case, the function can indeed be analytically extended far beyond the domain over which it was initially given.

Here, we were able to write down an explicit formula for the function, but in other cases this may not be so easy. Nevertheless, there is a general procedure according to which analytic continuation may frequently be carried out. We can imagine starting in some small region where a locally valid power series expression for the holomorphic function $f(z)$ is known. We might then go wandering off along some path, continuing the function as we go by the repeated use of power series based at different points. For this, we would use a sequence of points along the path and take a succession of power series expressions successively about each of these points in turn. This will fix the continuation provided that the interiors of the successive circles of convergence can be made to overlap (see Fig. 7.6). When this procedure can be carried out consistently, the resulting function is uniquely determined by the values of the function in the initial region and on the path along which it is being continued.

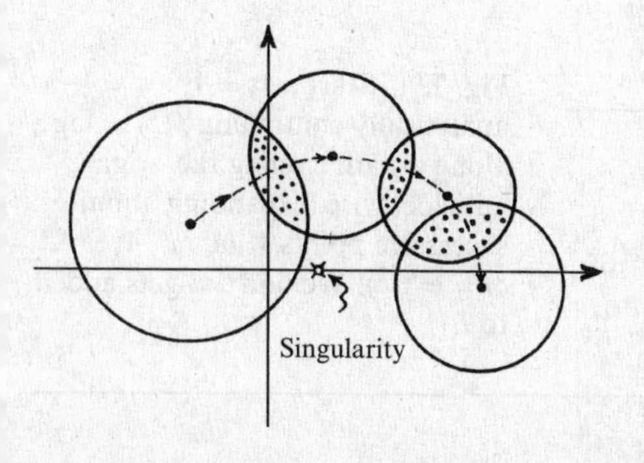

Fig. 7.6 A holomorphic function can be analytically continued, using a succession of power series expressions about a sequence of points. This proceeds uniquely along the connecting path, assuming successive circles of convergence overlap.

There is thus a remarkable 'rigidity' about holomorphic functions, as manifested in this process of analytic continuation. In the case of real C^∞-functions, on the other hand, it was possible 'to keep changing one's mind' about what the function is to be doing (as with the smoothly patched $h(x)$ of §6.3, Fig. 6.7, which suddenly 'takes off' after having been zero for all negative values of x). This cannot happen for holomorphic functions. Once the function is fixed in its original region, and the path is fixed, there is no choice about how the function is to be extended. In fact, the same is true for real-analytic functions of a real variable. They also have a similar 'rigidity', but now there is not much choice about the path either. It can only be in one direction or the other along the real line. With complex functions, analytic continuation can be more interesting because of this freedom of the path within a two-dimensional plane.

To illustrate, consider our old friend $\log z$. It certainly has no power series expansion about the origin, as it has a singularity there. But if we like, we can expand it about the point $p = 1$, say, to obtain the series[7.8]

$$\log z = (z-1) - \frac{1}{2}(z-1)^2 + \frac{1}{3}(z-1)^3 - \frac{1}{4}(z-1)^4 + \cdots.$$

The circle of convergence is the circle of unit radius centred at $z = 1$. Let us imagine performing an analytic continuation along a path that circles the origin in an anticlockwise direction. We could, if we choose, use power series taken about the successive points 1, ω, ω^2, and back to 1, thus returning to our starting point having encircled the origin once (Fig. 7.7). Here I have used the three cube roots of unity, regularly placed around the unit circle, namely 1, $\omega = e^{2\pi i/3}$, and $\omega^2 = e^{4\pi i/3}$, as discussed at the end of §5.4, and the route around the origin can be taken as an equilateral

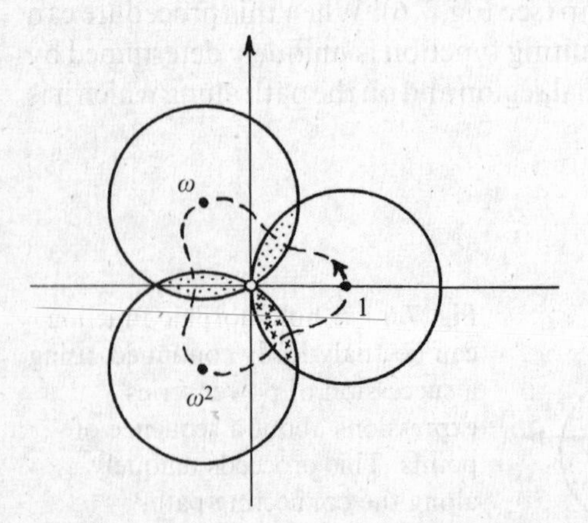

Fig. 7.7 Start at $z = 1$, analytically continuing $f(z) = \log z$ along a path circling the origin anticlockwise (expanding about successive points 1, ω, ω^2, 1; $\omega = e^{2\pi i/3}$). We find $2\pi i$ gets added to f.

[7.8] Derive this series.

triangle. Alternatively, I could have used 1, i, -1, $-$i, 1, which is slightly less economical. In any case, there is no need to work out the power series, since we already know the explicit answer for the function itself, namely $\log z$. The problem, of course, is that when we have gone once around the origin, uniquely following the function as it goes, we find that we have uniquely extended it to a value different from the one that we started with. Somehow, 2πi has got added to the function as we went around. Had we chosen to proceed around the origin in the opposite direction, then we should have found that 2πi would have been subtracted from the function that we started from. Thus, the uniqueness of analytic continuation can be quite a subtle thing, and it can definitely depend upon the path taken. For 'many-valued' functions more complicated than $\log z$, we can get something much more elaborate than just adding a constant (like 2πi) to the function.

As an aside, it is worth pointing out that the notion of analytic continuation need not refer particularly to power series, despite the fact that I have found it useful to employ them in some of my descriptions. For example, there is another class of series that has great significance in number theory, namely those called *Dirichlet series*. The most important of these is the *(Euler–)Riemann zeta function*,[6] defined by the infinite sum[7]

$$\zeta(z) = 1^{-z} + 2^{-z} + 3^{-z} + 4^{-z} + 5^{-z} + \cdots,$$

which converges to the holomorphic function denoted by $\zeta(z)$ when the real part of z is greater than 1. Analytic continuation of this function defines it uniquely (and 'single-valuedly') on the whole of the complex plane but with the point $z = 1$ removed. Perhaps the most important unsolved mathematical problem today is the *Riemann hypothesis*, which is concerned with the *zeros* of this analytically extended zeta function, that is, with the solutions of $\zeta(z) = 0$. It is relatively easy to show that $\zeta(z)$ becomes zero for $z = -2, -4, -6, \ldots$; these are the real zeros. The Riemann hypothesis asserts that all the remaining zeros lie on the line $\mathrm{Re}(z) = \frac{1}{2}$, that is, $\zeta(z)$ becomes zero (unless z is a negative even integer) only when the real part of z is equal to $\frac{1}{2}$. All numerical evidence to date supports this hypothesis, but its actual truth is unknown. It has fundamental implications for the theory of prime numbers.[8]

Notes

Section 7.1

7.1. To those readers wishing to explore these fascinating matters in greater geometric detail, I strongly recommend Needham (1997).

7.2. I shall give them in §10.5, after the notion of *partial derivative* has been introduced.

Section 7.2

7.3. More explicitly, integration of f 'along' a path given by $z = p(t)$ (where p is a smooth complex-valued function p of a real parameter t) can be expressed as the definite integral $\int_u^v f(p(t))p'(t)\mathrm{d}t = \int_a^b f(z)\mathrm{d}z$, where $p(u)$ is the initial point a of the path and $p(v)$ is its final point b.

Section 7.3

7.4. A 'reason' that Cauchy's formula must be true is that for a small loop around the origin, $f(z)$ may actually be treated as the *constant* value $f(0)$ and then the situation reduces to that studied in §7.2.

7.5. It is one of the irritations of the terminology of this subject that the term 'domain' has two distinct meanings. The one that is *not* intended here is simply any 'connected open region in the complex plane'. Here, as before (see §6.1), I mean the region in the complex plane where the function f is defined, which in general need not be necessarily open or connected (though here it is taken to be open).

7.6. The zeta function was first considered by Euler, but it is normally named after Riemann, in view of his fundamental work involving the extension of this function to the complex plane.

7.7. Note the curious 'upside-down' relation between this series and an ordinary power series, namely for $(-z) + (-z)^2 + (-z)^3 + \cdots = -z(1+z)^{-1}$.

7.8. For further information on the ζ-function and Riemann hypothesis, see Apostol (1976); Priestley (2003). For popular accounts, see Derbyshire (2003); du Sautoy (2003); Sabbagh (2003); Devlin (1988, 2002).

8
Riemann surfaces and complex mappings

8.1 The idea of a Riemann surface

THERE is a way of understanding what is going on with this analytic continuation of the logarithm function—or of any other 'many-valued function'—in terms of what are called *Riemann surfaces*. Riemann's idea was to think of such functions as being defined on a domain which is not simply a subset of the complex plane, but as a many-sheeted region. In the case of $\log z$, we can picture this as a kind of spiral ramp flattened down vertically to the complex plane. I have tried to indicate this in Fig. 8.1. The logarithm function is single-valued on this winding many-sheeted version of the complex plane because each time we go around the origin, and $2\pi i$ has to be added to the logarithm, we find ourselves on another sheet of the domain. There is no conflict between the different values of the logarithm now, because its domain is this more extended winding space—an example of a Riemann surface—a space subtly different from the complex plane itself.

Bernhard Riemann, who introduced this idea, was one of the very greatest of mathematicians, and in his short life (1826–66) he put forward a multitude of mathematical ideas that have profoundly altered the course of mathematical thought on this planet. We shall encounter some of his

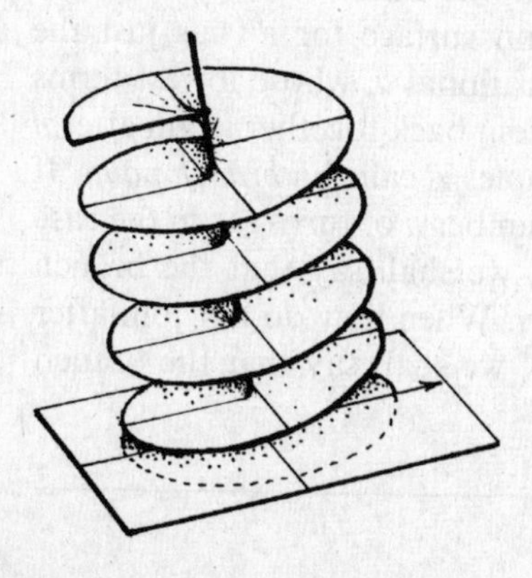

Fig. 8.1 The Riemann surface for $\log z$, pictured as a spiral ramp flattened down vertically.

other contributions later in this book, such as that which underlies Einstein's general theory of relativity (and one very important contribution of Riemann's, of a different kind, was referred to at the end of Chapter 7). Before Riemann introduced the notion of what is now called a 'Riemann surface', mathematicians had been at odds about how to treat these so-called 'many-valued functions', of which the logarithm is one of the simplest examples. In order to be rigorous, many had felt the need to regard these functions in a way that I would personally consider distasteful. (Incidentally, this was still the way that I was taught to regard them myself while at university, despite this being nearly a century after Riemann's epoch-making paper on the subject.) In particular, the domain of the logarithm function would be 'cut' in some arbitrary way, by a line out from the origin to infinity. To my way of thinking, this was a brutal mutilation of a sublime mathematical structure. Riemann taught us we must think of things differently. Holomorphic functions rest uncomfortably with the now usual notion of a 'function', which maps from a fixed domain to a definite target space. As we have seen, with analytic continuation, a holomorphic function 'has a mind of its own' and decides itself what its domain should be, irrespective of the region of the complex plane which we ourselves may have initially allotted to it. While we may regard the function's domain to be represented by the Riemann surface associated with the function, the domain is not given ahead of time; it is the explicit form of the function itself that tells us which Riemann surface the domain actually is.

We shall be encountering various other kinds of Riemann surface shortly. This beautiful concept plays an important role in some of the modern attempts to find a new basis for mathematical physics—most notably in string theory (§§31.5,13) but also in twistor theory (§§33.2,10). In fact, the Riemann surface for $\log z$ is one of the simplest of such surfaces. It gives us merely a hint of what is in store for us. The function z^a perhaps is marginally more interesting than $\log z$ with regard to its Riemann surface, but only when the complex number a is a rational number. When a is irrational, the Riemann surface for z^a has just the same structure as that for $\log z$, but for a rational a, whose lowest-terms expression is $a = m/n$, the spiralling sheets join back together again after n turns.[8.1] The origin $z = 0$ in all these examples is called a *branch point*. If the sheets join back together after a finite number n of turns (as in the case $z^{m/n}$, m and n having no common factor), we shall say that the branch point has *finite order*, or that it is of *order n*. When they do not join after any number of turns (as in the case $\log z$), we shall say that the branch point has *infinite order*.

[8.1] Explain why.

Expressions like $(1-z^3)^{1/2}$ give us more food for thought. Here the function has three branch points, at $z=1$, $z=\omega$, and $z=\omega^2$ (where $\omega = e^{2\pi i/3}$; see §5.4, §7.4), so $1-z^3=0$, and there is another 'branch point at infinity'. As we circle by one complete turn, around each individual branch point, staying in its immediate neighbourhood (and for 'infinity' this just means going around a very large circle), we find that the function changes sign, and, circling it again, the function goes back to its original value. Thus, we see that the branch points all have order 2. We have two sheets to the Riemann surface, patched together in the way that I have tried to indicate in Fig. 8.2a. In Fig. 8.2b, I have attempted to show, using some topological contortions, that the Riemann surface actually has the topology of a *torus*, which is topologically the surface of a bagel (or of an American donut), but with four tiny holes in it corresponding to the branch points themselves. In fact, the holes can be filled in unambiguously

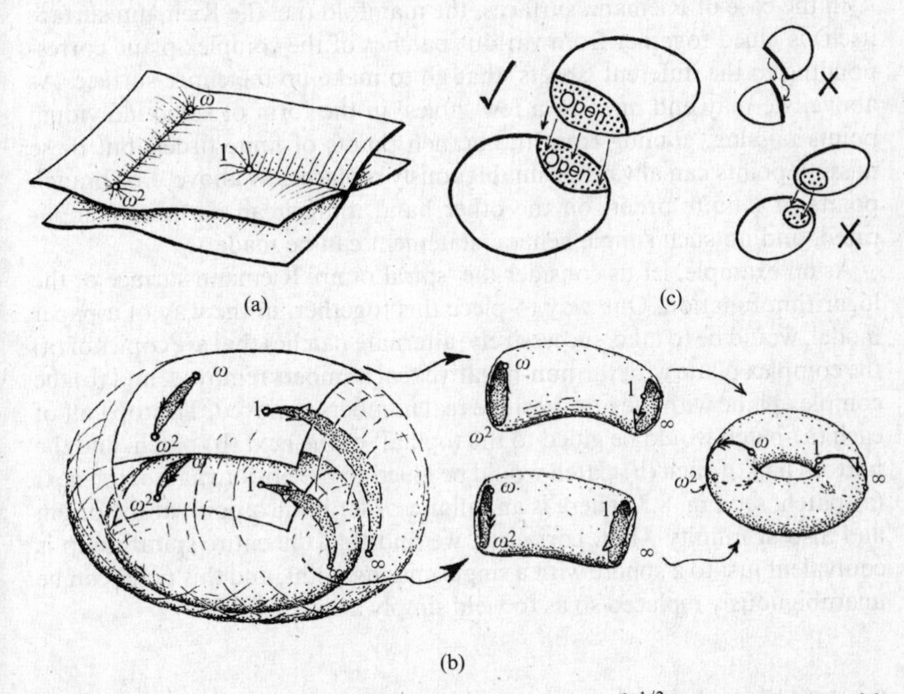

Fig. 8.2 (a) Constructing the Riemann surface for $(1-z^3)^{1/2}$ from two sheets, with branch points of order 2 at 1, ω, ω^2 (and also ∞). (b) To see that the Riemann surface for $(1-z^3)^{1/2}$ is topologically a torus, imagine the planes of (a) as two Riemann spheres with slits cut from ω to ω^2 and from 1 to ∞, identified along matching arrows. These are topological cylinders glued correspondingly, giving a torus. (c) To construct a Riemann surface (or a manifold generally) we can glue together patches of coordinate space—here open portions of the complex plane. There must be (open-set) overlaps between patches (and when joined there must be no 'non-Hausdorff branching', as in the final case above; see Fig. 12.5b, §12.2).

(with four single points), and the resulting Riemann surface then has exactly the topology of a torus.[8.2]

Riemann's surfaces provided the first instances of the general notion of a *manifold*, which is a space that can be thought of as 'curved' in various ways, but where, *locally* (i.e. in a small enough neighbourhood of any of its points), it looks like a piece of ordinary Euclidean space. We shall be encountering manifolds more seriously in Chapters 10 and 12. The notion of a manifold is crucial in many different areas of modern physics. Most strikingly, it forms an essential part of Einstein's general relativity. Manifolds may be thought of as being glued together from a number of different patches, where the gluing job really is seamless, unlike the situation with the function $h(x)$ at the end of §6.3. The seamless nature of the patching is achieved by making sure that there is always an appropriate (open-set) overlap between one patch and the next (see Fig. 8.2c and also §12.2, Fig. 12.5).

In the case of Riemann surfaces, the manifold (i.e. the Riemann surface itself) is glued together from various patches of the complex plane corresponding to the different 'sheets' that go to make up the entire surface. As above, we may end up with a few 'holes' in the form of some individual points missing, coming from the branch points of finite order, but these missing points can always be unabiguously replaced, as above. For branch points of infinite order, on the other hand, things can be more complicated, and no such simple general statement can be made.

As an example, let us consider the 'spiral ramp' Riemann surface of the logarithm function. One way to piece this together, in the way of a paper model, would be to take, successively, alternate patches that are copies of (a) the complex plane with the non-negative real numbers removed, and (b) the complex plane with the non-positive real numbers removed. The top half of each (a)-patch would be glued to the top half of the next (b)-patch, and the bottom half of each (b)-patch would be glued to the bottom half of the next (a)-patch; see Fig. 8.3. There is an infinite-order branch point at the origin and also at infinity—but, curiously, we find that the entire spiral ramp is equivalent just to a sphere with a single missing point, and this point can be unambiguously replaced so as to yield simply a sphere.[8.3]

8.2 Conformal mappings

When piecing together a manifold, we have to consider what local structure has to be preserved from one patch to the next. Normally, one deals with real manifolds, and the different patches are pieces of Euclidean space

[8.2] Now try $(1 - z^4)^{1/2}$.

[8.3] Can you see how this comes about? (*Hint*: Think of the Riemann sphere of the variable $w(= \log z)$; see §8.3.)

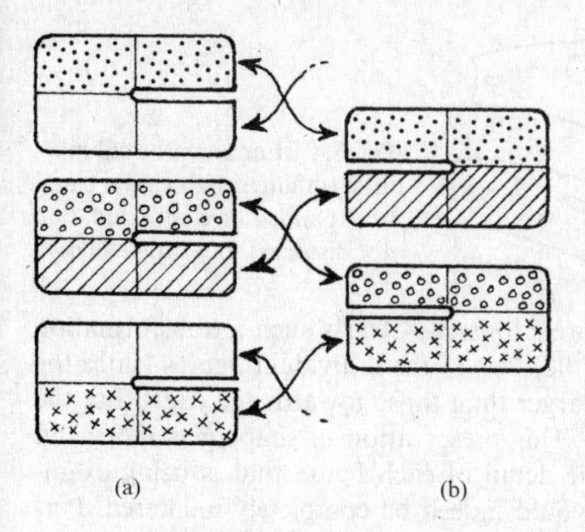

Fig. 8.3 We can construct the Riemann surface for $\log z$ by taking alternate patches of (a) the complex plane with the non-negative real axis removed, and (b) the complex plane with the non-positive real axis removed. The top half each (a)-patch is glued to the top half of the next (b)-patch, and the bottom half of each (b)-patch glued to the bottom half of the next (a)-patch.

(of some fixed dimension) that are glued together along various (open) overlap regions. The local structure to be matched from one patch to the next is normally just a matter of preserving continuity or smoothness. This issue will be discussed in §10.2. In the case of Riemann surfaces, however, we are concerned with *complex* smoothness, and we recall, from §7.1, that this is a more sophisticated matter, involving what are called the *Cauchy–Riemann* equations. Although we have not seen them explicitly yet (we shall be coming to them in §10.5), it will be appropriate now to understand the geometrical meaning of the structure that is encoded in these equations. It is a structure of remarkable elegance, flexibility, and power, leading to mathematical concepts with a great range of application.

The notion is that of *conformal geometry*. Roughly speaking, in conformal geometry, we are interested in shape but not size, this referring to shape on the infinitesimal scale. In a conformal map from one (open) region of the plane to another, shapes of finite size are generally distorted, but *infinitesimal* shapes are preserved. We can think of this applying to small (infinitesimal) circles drawn on the plane. In a conformal map, these little circles can be expanded or contracted, but they are not distorted into little ellipses. See Fig. 8.4.

To get some understanding of what a conformal transformation can be like, look at M. C. Escher's picture, given in Fig. 2.11, which provides a conformal representation of the hyperbolic plane in the Euclidean plane, as described in §2.4 (Beltrami's 'Poincaré disc'). The hyperbolic plane is very symmetrical. In particular, there are transformations which take the figures in the central region of Escher's picture to corresponding very tiny figures that lie just inside the bounding circle. We can represent such a transformation as a conformal motion of the Euclidean plane that takes

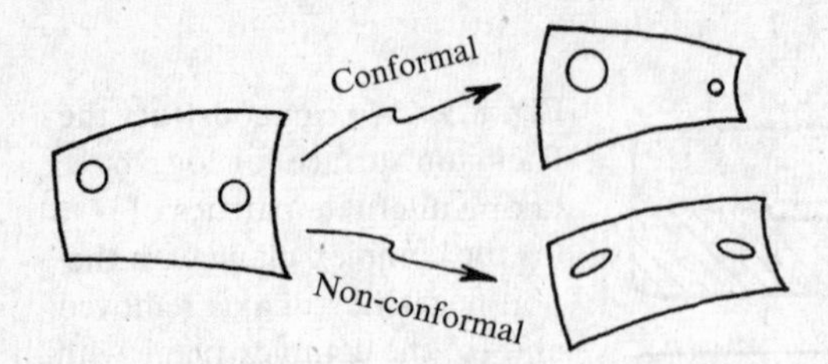

Fig. 8.4 For a conformal map, little (infinitesimal) circles can be expanded or contracted, but not distorted into little ellipses.

the interior of the bounding circle to itself. Clearly such a transformation would not generally preserve the sizes of the individual figures (since the ones in the middle are much larger than those towards the edge), but the shapes are roughly preserved. This preservation of shape gets more and more accurate, the smaller the detail of each figure that is being examined, so *infinitesimal* shapes would indeed be completely unaltered. Perhaps the reader would find a slightly different characterization more helpful: *angles* between curves are unaltered by conformal transformation. This characterizes the conformal nature of a transformation.

What does this conformal property have to do with the complex smoothness (holomorphicity) of some function $f(z)$? We shall try to obtain an intuitive idea of the geometric content of complex smoothness. Let us return to the 'mapping' viewpoint of a function f and think of the relation $w = f(z)$ as providing a mapping of a certain region in z's complex plane (the domain of the function f) into w's complex plane (the target); see Fig. 8.5. We ask the question: what local geometrical property characterizes this mapping as being holomorphic? There is a striking answer. Holomorphicity of f is indeed equivalent to the map being *conformal* and *non-reflective* (non-reflective—or *orientation-preserving*—meaning that the small shapes preserved in the transformation are not reflected, i.e. not 'turned over'; see end of §12.6).

The notion of 'smoothness' in our transformation $w = f(z)$ refers to how the transformation acts in the infinitesimal limit. Think of the real case first, and let us re-examine our real function $f(x)$ of §6.2, where the graph of $y = f(x)$ is illustrated in Fig. 6.4. The function f is *smooth* at

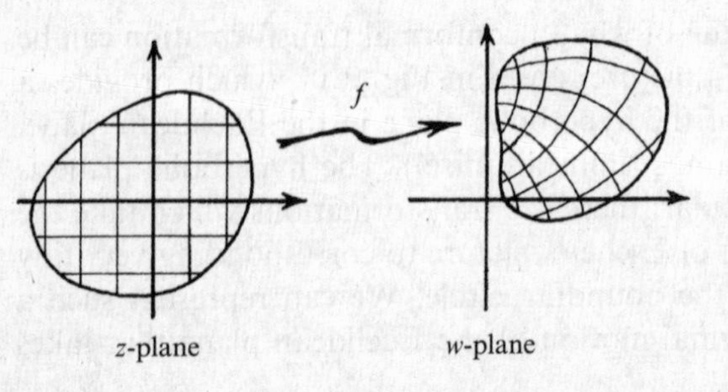

Fig. 8.5 The map $w = f(z)$ has domain an open region in the complex z-plane and target an open region in the complex w-plane. Holomorphicity of f is equivalent to this being conformal and non-reflective.

some point if the graph has a well-defined tangent at that point. We can picture the tangent by imagining that a larger and larger magnification is applied to the curve at that point, and, so long as it is smooth, the curve looks more and more like a straight line through that point as the magnification increases, becoming identical with the tangent line in the limit of infinite magnification. The situation with complex smoothness is similar, but now we apply the idea to the map from the z-plane to the w-plane. To examine the infinitesimal nature of this map, let us try to picture the immediate neighbourhood of a point z, in one plane, mapping this to the immediate neighbourhood of w in the other plane. To examine the immediate neighbourhood of the point, we imagine magnifying the neighbourhood of z by a huge factor and the corresponding neighbourhood of w by the same huge factor. In the limit, the map from the expanded neighbourhood of z to the expanded neighbourhood of w will be simply a linear transformation of the plane, but, if it is to be holomorphic, this must basically be one of the transformations studied in §5.1. From this it follows (by a little consideration) that, in the general case, the transformation from z's neighbourhood to w's neighbourhood simply combines a rotation with a uniform expansion (or contraction); see Fig. 5.2b. That is to say, small shapes (or angles) are preserved, without reflection, showing that the map is indeed conformal and non-reflective.

Let us look at a few simple examples. The very particular situations of the maps provided by the adding of a constant b to z or of multiplying z by a constant a, as considered already in §5.1 (see Fig. 5.2), are obviously holomorphic ($z + b$ and az being clearly differentiable) and are also obviously conformal. These are particular instances of the general case of the combined (inhomogeneous-linear) transformation

$$w = az + b.$$

Such transformations provide the Euclidean motions of the plane (without reflection), combined with uniform expansions (or contractions). In fact, they are the only (non-reflective) conformal maps of the entire complex z-plane to the entire complex w-plane. Moreover, they have the very special property that actual circles—not just infinitesimal circles—are mapped to actual circles, and also straight lines are mapped to straight lines.

Another simple holomorphic function is the *reciprocal* function,

$$w = z^{-1},$$

which maps the complex plane with the origin removed to the complex plane with the origin removed. Strikingly, this transformation also maps actual circles to actual circles[8.4] (where we think of straight lines as being

[8.4] Show this.

particular cases of circles—of infinite radius). This transformation, together with a reflection in the real axis, is what is called an *inversion*. Combining this with the inhomogeneous linear maps just considered, we get the more general transformation[8.5]

$$w = \frac{az + b}{cz + d},$$

called a *bilinear* or *Möbius* transformation. From what has been said above, these transformations must also map circles to circles (straight lines again being regarded as special circles). This Möbius transformation actually maps the entire complex plane with the point $-d/c$ removed to the entire complex plane with a/c removed—where, for the transformation to give a non-trivial mapping at all, we must have $ad \neq bc$ (so that the numerator is not a fixed multiple of the denominator).

Note that the point removed from the z-plane is that value ($z = -d/c$) which would give '$w = \infty$'; correspondingly, the point removed from the w-plane is that value ($w = a/c$) which would be achieved by '$z = \infty$'. In fact, the whole transformation would make more global sense if we were to incorporate a quantity '∞' into both the domain and target. This is one way of thinking about the simplest (compact) Riemann surface of all: the *Riemann sphere*, which we come to next.

8.3 The Riemann sphere

Simply adjoining an extra point called '∞' to the complex plane does not make it completely clear that the required seamless structure holds in the neighbourhood of ∞, the same as everywhere else. The way that we can address this issue is to regard the sphere to be constructed from two 'coordinate patches', one of which is the z-plane and the other the w-plane. All but two points of the sphere are assigned both a z-coordinate and a w-coordinate (related by the Möbius transformation above). But one point has only a z-coordinate (where w would be 'infinity') and another has only a w-coordinate (where z would be 'infinity'). We use either z or w or both in order to define the needed conformal structure and, where we use both, we get the same conformal structure using either, because the relation between the two coordinates is holomorphic.

In fact, for this, we do not need such a complicated transformation between z and w as the general Möbius transformation. It suffices to consider the particularly simple Möbius transformation given by

[8.5] Verify that the sequence of transformations $z \mapsto Az + B$, $z \mapsto z^{-1}$, $z \mapsto Cz + D$ indeed leads to a bilinear map.

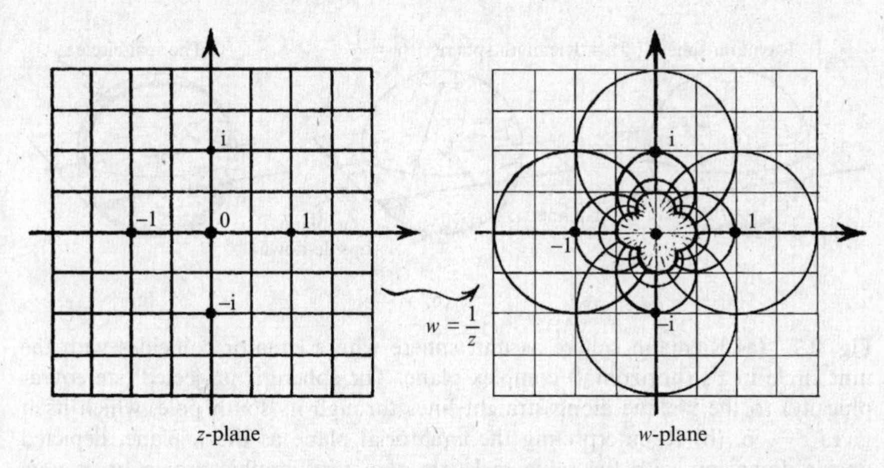

Fig. 8.6 Patching the Riemann sphere from the complex z- and w-planes, via $w = 1/z$, $z = 1/w$. (Here, the z grid lines are shown also in the w-plane.) The overlap regions exclude only the origins, $z = 0$ and $w = 0$ each giving '∞' in the opposite patch.

$$w = \frac{1}{z}, \quad z = \frac{1}{w},$$

where $z = 0$ and $w = 0$, would each give ∞ in the opposite patch. I have indicated in Fig. 8.6 how this transformation maps the real and imaginary coordinate lines of z.

All this defines the Riemann sphere in a rather abstract way. We can see more clearly the reason that the Riemann sphere is called a 'sphere' by employing the geometry illustrated in Fig. 8.7a. I have taken the z-plane to represent the *equatorial plane* of this geometrical sphere. The points of the sphere are mapped to the points of the plane by what is called *stereographic projection* from the south pole. This just means that I draw a straight line in the Euclidean 3-space (within which we imagine everything to be taking place) from the south pole through the point z in the plane. Where this line meets the sphere again is the point on the sphere that the complex number z represents. There is one additional point on the sphere, namely the south pole itself, and this represents $z = \infty$. To see how w fits into this picture, we imagine its complex plane to be inserted upside down (with $w = 1, \mathrm{i}, -1, -\mathrm{i}$ matching $z = 1, -\mathrm{i}, -1, \mathrm{i}$, respectively), and we now project stereographically from the *north* pole (Fig. 8.7b).[8.6] An important and beautiful property of stereographic projection is that it maps circles on the sphere to circles (or straight lines) on the plane.[1]

[8.6] Check that these two stereographic projections are related by $w = z^{-1}$.

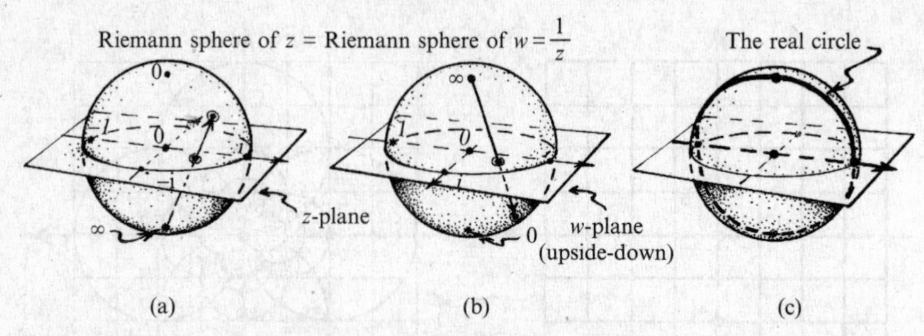

Fig. 8.7 (a) Riemann sphere as unit sphere whose equator coincides with the unit circle in z's (horizontal) complex plane. The sphere is projected (stereographically) to the z-plane along straight lines through its south pole, which itself gives $z = \infty$. (b) Re-interpreting the equatorial plane as the w-plane, depicted upside down but with the same real axis, the stereographic projection is now from the north pole ($w = \infty$), where $w = 1/z$. (c) The real axis is a great circle on this Riemann sphere, like the unit circle but drawn vertically rather than horizontally.

Hence, bilinear (Möbius) transformations send circles to circles on the Riemann sphere. This remarkable fact has a significance for relativity theory that we shall come to in §18.5 (and it has deep relevance to spinor and twistor theory; see §22.8, §24.7, §§33.2,4).

We notice that, from the point of view of the Riemann sphere, the real axis is 'just another circle', not essentially different from the unit circle, but drawn vertically rather than horizontally (Fig. 8.7c). One is obtained from the other by a rotation. A rotation is certainly conformal, so it is given by a holomorphic map of the sphere to itself. In fact every (non-reflective) conformal map which takes the entire Riemann sphere to itself is achieved by a bilinear (i.e. Möbius) transformation. The particular rotation that we are concerned with can be exhibited explicitly as a relation between the Riemann spheres of the complex parameters z and t given by the bilinear correspondence[8.7]

$$t = \frac{z-1}{\mathrm{i}z+\mathrm{i}}, \quad z = \frac{-t+\mathrm{i}}{t+\mathrm{i}}.$$

In Fig. 8.8, I have plotted this correspondence in terms of the complex planes of t and z, where I have specifically marked how the upper half-plane of t, bounded by its real axis, is mapped to the unit disc of z, bounded by its unit circle. This particular transformation will have importance for us in the next chapter.

[8.7] Show this.

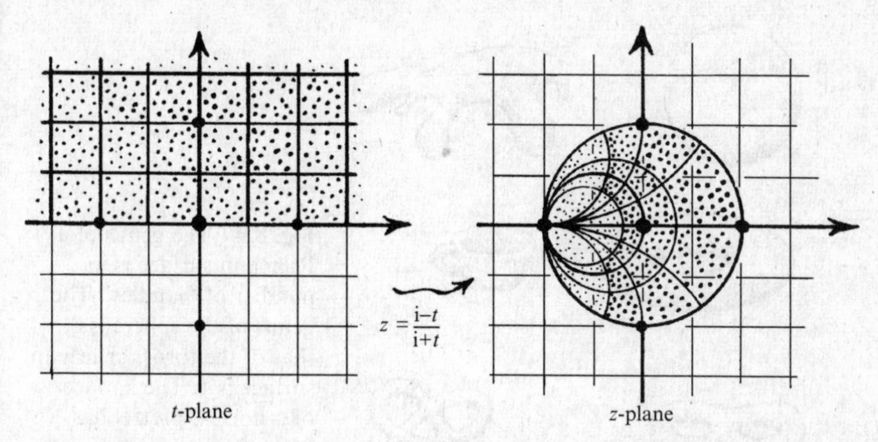

Fig. 8.8 The correspondence $t = (z-1)/(iz+i)$, $z = (-t+i)/(t+i)$ in terms of the complex planes of t and z. The upper half-plane of t, bounded by its real axis, is mapped to the unit disc of z, bounded by its unit circle.

The Riemann sphere is the simplest of the *compact*—or '*closed*'—Riemann surfaces.[2] See §12.6 for the notion of 'compact'. By contrast, the 'spiral ramp' Riemann surface of the logarithm function, as I have described it, is *non-compact*. In the case of the Riemann surface of $(1-z^3)^{1/2}$, we need to fill the four holes arising from the branch points to make it compact (and it is non-compact if we do not do this), but this 'compactification' is the usual thing to do. As remarked earlier, this 'hole-filling' is always possible with a branch point of finite order. As we saw at the end of §8.1, for the logarithm we can actually fill the branch points at the origin and at infinity, both together, with a single point, to obtain the Riemann sphere as the compactification. In fact, there is a complete classification of compact Riemann sufaces (achieved by Riemann himself), which is important in many areas (including string theory). I shall briefly outline this classification next.

8.4 The genus of a compact Riemann surface

The first stage is to classify the surfaces according to their *topology*, that is to say, according to that aspect of things preserved by continuous transformations. The topological classification of compact 2-dimensional orientable (see end of §12.6) surfaces is really very simple. It is given by a single natural number called the *genus* of the surface. Roughly speaking, all we have to do is count the number of 'handles' that the surface has. In the case of the sphere the genus is 0, whereas for the torus it is 1. The surface of an ordinary teacup also has genus 1 (one handle!), so it is topologically the

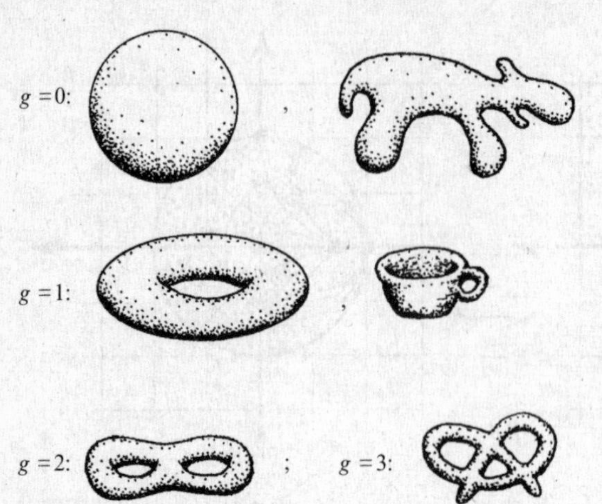

Fig. 8.9 The genus of a Riemann surface is its number of 'handles'. The genus of the sphere is 0, that of the torus, or teacup surface is 1. The surface of a normal pretzel has genus 3.

same as a torus. The surface of a normal pretzel has genus 3. See Fig. 8.9 for several examples.

The genus does not in itself fix the Riemann surface, however, except for genus 0. We also need to know certain complex parameters known as *moduli.* Let me illustrate this issue in the case of the torus (genus 1). An easy way to construct a Riemann surface of genus 1 is to take a region of the complex plane bounded by a parallelogram, say with vertices $0, 1, 1+p, p$ (described cyclicly). See Fig. 8.10. Now we must imagine that opposite edges of the parallelogram are glued together, that is, the edge from 0 to 1 is glued to that from p to $1+p$, and the edge from 0 to p is glued to that from 1 to $1+p$. (We could always find other patches to cover the seams, if we like.) The resulting Riemann surface is indeed topologically a torus. Now, it turns out that, for differing values of p, the resulting surfaces are generally *inequivalent* to each other; that is to say, it is not possible to transform one into another by means of a holomorphic mapping. (There are certain discrete equivalences, however, such as those arising when p is replaced by $1+p$, by $-p$, or by $1/p$.)[8.8] It can be made intuitively plausible that not all Riemann surfaces with the same topology

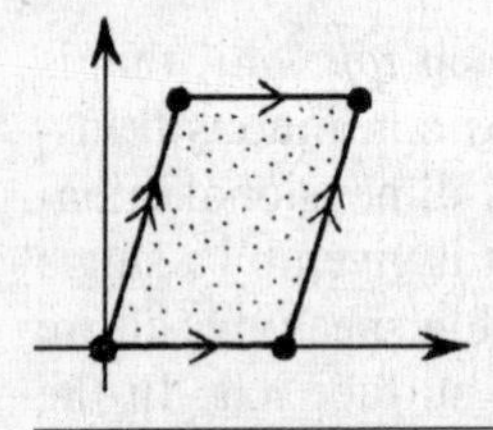

Fig. 8.10 To construct a Riemann surface of genus 1, take a region of the complex plane bounded by a parallelogram, vertices $0, 1, 1+p, p$ (cyclicly), with opposite edges identified. The quantity p provides a modulus for the Riemann surface.

[8.8] Show that these replacements give holomorphically equivalent spaces. Find all the special values of p where these equivalences lead to additional discrete symmetries of the Riemann surface.

Fig. 8.11 Two inequivalent torus-topology Riemann surfaces.

can be equivalent, by considering the two cases illustrated in Fig. 8.11. In one case I have chosen a very tiny value of p, and we have a very stringy looking torus, and in the other case I have chosen p close to i, where the torus is nice and fat. Intuitively, it seems pretty clear that there can be no conformal equivalence between the two, and indeed there is none.

There is just this one complex modulus p in the case of genus 1, but for genus 2 we find that there are three. To construct a Riemann surface of genus 2 by pasting together a shape, in the manner of the parallelogram that we used for genus 1, we could construct the shape from a piece of the hyperbolic plane; see Fig. 8.12. The same would hold for any higher genus. The number m of complex moduli for genus g, where $g \geqslant 2$, is $m = 3g - 3$.

One might regard it as a little strange that the formula $3g - 3$ for the number of moduli works for all values of the genus $g = 2, 3, 4, 5, \dots$ but it fails for $g = 0$ or 1. There is actually a 'reason' for this, which has to do with the number s of complex parameters that are needed to specify the different continuous (holomorphic) *self-transformations* of the Riemann surface. For $g \geqslant 2$, there are no such continuous self-transformations (although there can be discrete ones), so $s = 0$. However, for $g = 1$, the complex plane of the parallelogram of Fig. 8.10 can be translated (moved rigidly without rotation) in any direction in the plane. The amount (and direction) of this displacement can be specified by a single complex parameter a, the translation being achieved by $z \mapsto z + a$, so $s = 1$ when $g = 1$. In the case of the sphere (genus 0), the self-transformations are achieved by the bilinear transformations described above, namely $z \mapsto (az + b)/(cz + d)$.

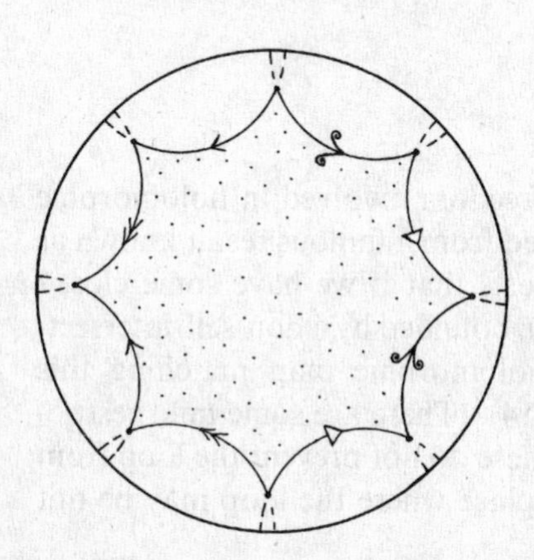

Fig. 8.12 An octagonal region of the hyperbolic plane, in the conformal representation of Fig. 2.12, with identifications to yield a genus-2 Riemann surface.

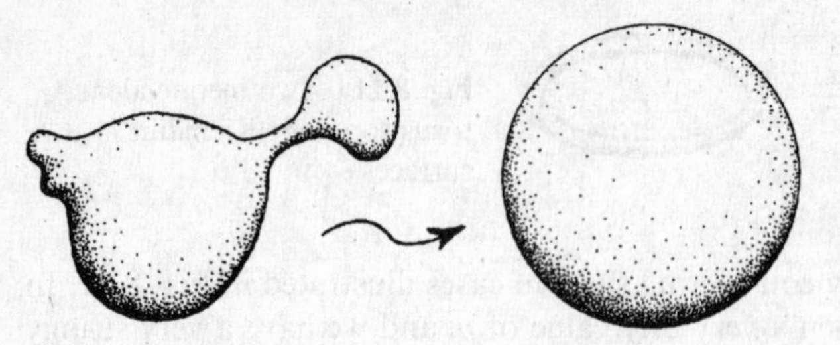

Fig. 8.13 Every $g = 0$ metric geometry is conformally identical to that of the standard ('round') unit sphere.

Here, the freedom is given by the three[3] independent ratios $a : b : c : d$. Thus, in the case $g = 0$, we have $s = 3$. Hence, in all cases, the difference $m - s$ between the number of complex moduli and the number of complex parameters required to specify a self-transformation satisfies

$$m - s = 3g - 3.$$

(This formula is related to some deeper issues that are beyond the scope of this book.[4])

It is clear that there is some considerable freedom, within the family of conformal (holomorphic) transformations, for altering the apparent 'shape' of a Riemann surface, while keeping its structure as a Riemann surface unaltered. In the case of spherical topology, for example, many different metrical geometries are possible (as is illustrated in Fig. 8.13); yet these are all conformally identical to the standard ('round') unit sphere. (I shall be more explicit about the notion of 'metric' in §14.7.) Moreover, for higher genus, the seemingly large amount of freedom in the 'shape' of the surface can all be reduced down to the finite number of complex moduli given by the above formulae. But there is still some overall information in the shape of the surface that cannot be eliminated by the use of this conformal freedom, namely that which is defined by the moduli themselves. Exactly how much can be achieved globally by the use of such freedom is quite a subtle matter.

8.5 The Riemann mapping theorem

Some appreciation of the considerable freedom involved in holomorphic transformations can, however, be obtained from a famous result known as the *Riemann mapping theorem*. This asserts that if we have some closed region in the complex plane (see Note 8.2), bounded by a non-self-intersecting closed loop, then there exists a holomorphic map matching this region to the closed unit disc (see Fig. 8.14). (There are some mild restrictions on the 'tameness' of the loop, but these do not prevent the loop from having corners or other worse kinds of place where the loop may be not

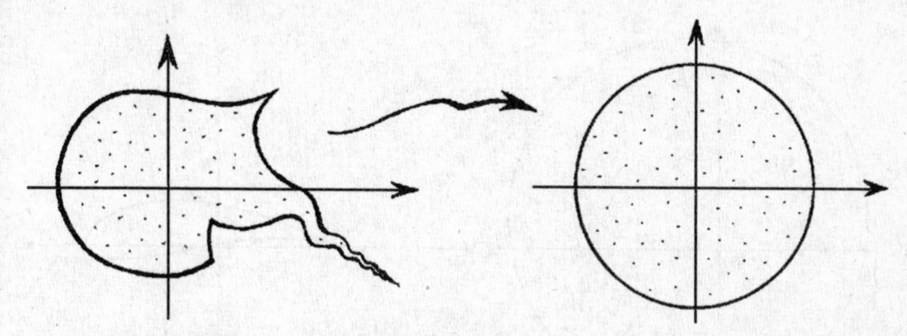

Fig. 8.14 The Riemann mapping theorem asserts that any open region in the complex plane, bounded by a simple closed (not necessarily smooth) loop, can be mapped holomorphically to the interior of the unit circle, the boundary being also mapped accordingly.

differentiable, as is illustrated in the particular example of Fig. 8.14.) One can go further than this and select, in a quite arbitrary way, three distinct points a, b, c on the loop, and insist that they be taken by the map to three specified points a', b', c' on the unit circle (say $a' = 1$, $b' = \omega$, $c' = \omega^2$), the only restriction being that the cyclic ordering of the points a, b, c, around the loop agrees with that of a', b', c' around the unit circle. Furthermore, the map is then determined uniquely. Another way of specifying the map uniquely would be to choose just one point a on the loop and one additional point j inside it, and then to insist that a maps to a specific point a' on the unit circle (say $a' = 1$) and j maps to a specific point j' inside the unit circle (say $j' = 0$).

Now, let us imagine that we are applying the Riemann mapping theorem on the *Riemann sphere*, rather than on the complex plane. From the point of view of the Riemann sphere, the 'inside' of a closed loop is on the same footing as the 'outside' of the loop (just look at the sphere from the other side), so the theorem can be applied equally well to the outside as to the inside of the loop. Thus, there is an 'inverted' form of the Riemann mapping theorem which asserts that the outside of a loop in the complex plane can be mapped to the outside of the unit circle and uniqueness is now ensured by the simple requirement that one specified point a on the loop maps to one specified point a' on the unit circle (say $a' = 1$), where now ∞ takes over the role of j and j' in the description provided at the end of the above paragraph.[5]

Often such desired maps can be achieved explicitly, and one of the reasons that such maps might indeed be desired is that they can provide solutions to physical problems of interest, for example to the flow of air past an aerofoil shape (in the idealized situation where the flow is what is called 'non-viscous', 'incompressible', and 'irrotational'). I remember being very struck by such things when I was an undergraduate mathematics student, most particularly by what is known as the Zhoukowski (or Joukowski)

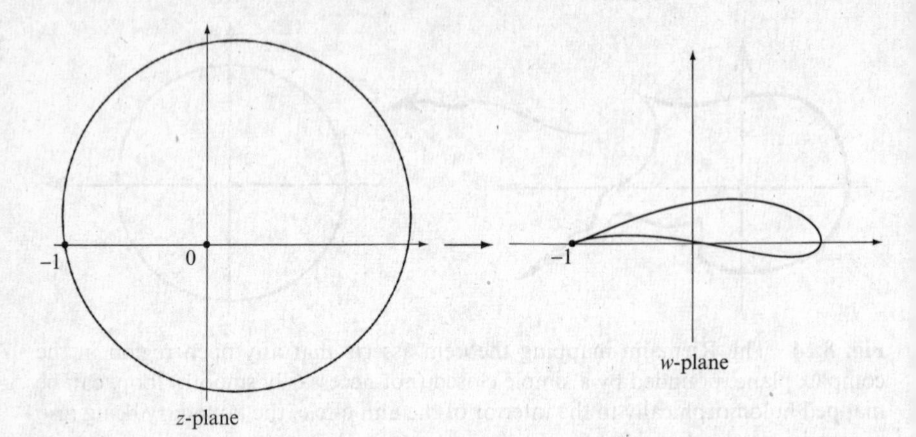

Fig. 8.15 Zhoukowski's transformation $w = \frac{1}{2}(z + 1/z)$ takes the exterior of a circle through $z = -1$ to an aerofoil cross-section, enabling the airflow pattern about the latter to be calculated.

aerofoil transformation, illustrated in Fig. 8.15, which can be given explicitly by the effect of the transformation

$$w = \tfrac{1}{2}\left(z + \frac{1}{z}\right),$$

on a suitable circle passing through the point $z = -1$. This shape indeed closely resembles a cross-section through the wing of an aeroplane of the 1930s, so that the (idealized) airflow around it can be directly obtained from that around a 'wing' of circular cross-section—which, in turn, is obtained by another such holomorphic transformation. (I was once told that the reason that such a shape was so commonly used for aeroplane wings was merely that then one could study it mathematically by just employing the Zhoukowski transformation. I hope that this is not true!)

Of course, there are specific assumptions and simplifications involved in applications such as these. Not only are the assumptions of zero viscosity and incompressible, irrotational flow mere convenient simplifications, but there is also the very drastic simplification that the flow can be regarded as the same all along the length of the wing, so that an essentially three-dimensional problem can be reduced to one entirely in two dimensions. It is clear that for a completely realistic computation of the flow around an aeroplane wing, a far more complicated mathematical treatment would be needed. There is no reason to expect that, in a more realistic treatment, we could get away with anything approaching such a direct and elegant use of holomorphic functions as we have with the Zhoukowski transformation.

It could, indeed, be argued that there is a strong element of good fortune in finding such an attractive application of complex numbers to a problem which had a distinctive importance in the real world. Air, of course, consists of enormous numbers of individual fundamental particles (in fact, about 10^{20} of them in a cubic centimetre), so airflow is something whose macroscopic description involves a considerable amount of averaging and approximation. There is no reason to expect that the mathematical equations of aerodynamics should reflect a great deal of the mathematics that is deeply involved in the physical laws that govern those individual particles.

In §4.1, I referred to the 'extraordinary and very basic role' that complex numbers actually play at the 'tiniest scales' of physical action, and there is indeed a holomorphic equation governing the behaviour of particles (see §21.2). However, for macroscopic systems, this 'complex structure' generally becomes completely buried, and it would appear that only in exceptional circumstances (such as in the airflow problem considered above) would complex numbers and holomorphic geometry find a natural utility. Yet there are circumstances where a basic underlying complex structure shows through even at the macroscopic level. This can sometimes be seen in Maxwell's electromagnetic theory and other wave phenomena. There is also a particularly striking example in relativity theory (see §18.5). In the following chapter, we shall see something of the remarkable way in which complex numbers and holomorphic functions can exert their magic from behind the scenes.

Notes

Section 8.3

8.1. See Exercise [2.5].

8.2. There is scope for terminological confusion in the use of the word 'closed' in the context of surfaces—or of the more general manifolds (n-surfaces) that will be considered in Chapter 12. For such a manifold, 'closed' means 'compact without boundary', rather than merely 'closed' in the topological sense, which is the complementary notion to 'open' as discussed in §7.4. (Topologically, a *closed* set is one that contains all its limit points. The *complement* of a closed set is an open one, and vice versa—where 'complement' of a set $\mathcal{S}$ within some ambient topological space $\mathcal{V}$ is the set of members of $\mathcal{V}$ which are not in $\mathcal{S}$.) There is additional confusion in that the term 'boundary', above, refers to a notion of 'manifold-with-boundary', which I do not discuss in this book. For the ordinary manifolds referred to in Chapter 12 (i.e. manifolds-without-boundary), the manifold notion of 'closed' (as opposed to the topological one) is equivalent to 'compact'. To avoid confusion, I shall normally just use the term 'compact', in this book, rather than 'closed'. Exceptions are the use of 'closed curve' for a real 1-manifold which is topologically a circle S^1 and 'closed universe' for a universe model which is *spatially* compact, that is, which contains a compact spacelike hypersurface; see Note 27.36.

Section 8.4

8.3. The transformation is unaffected if we multiply (rescale) each of a, b, c, d by the same non-zero complex number, but it changes if we alter any of them individually. This overall rescaling freedom reduces by one the number of independent parameters involved in the transformation, from four to three.

8.4. This may be thought of as the beginning of a long story whose climax is the very general and powerful Atiyah–Singer (1963) theorem.

Section 8.5

8.5. It should be noted that only for a loop that is an exact circle will the combination of both versions of the Riemann mapping theorem give us a complete smooth Riemann sphere.

9
Fourier decomposition and hyperfunctions

9.1 Fourier series

LET us return to the question, raised in §6.1, of what Euler and his contemporaries might have regarded as an acceptable notion of 'honest function'. In §7.1, we settled on the holomorphic (complex-analytic) functions as best satisfying what Euler might well have had in mind. Yet, most mathematicians today would regard such a notion of a 'function' as being unreasonably restrictive. Who is right? We shall be coming to a very remarkable answer to this question at the end of this chapter. But first let us try to understand what the issues are.

In the application of mathematics to problems of the physical world, it is a frequent requirement that there be a flexibility that neither the holomorphic functions nor their real counterparts—the analytic (i.e. C^{ω}-) functions—appear to possess. Because of the uniqueness of analytic continuation, as described in §7.4, the global behaviour of a holomorphic function defined throughout some connected open region $\mathcal{D}$ of the complex plane, is completely fixed, once it is known in some small open subregion of $\mathcal{D}$. Similarly, an analytic function of a real variable, defined on some connected segment $\mathcal{R}$ of the real line $\mathbb{R}$ is also completely fixed once the function is known in some small open subregion of $\mathcal{R}$. Such rigidity seems inappropriate for the realistic modelling of physical systems.

It would be particularly awkward when the propagation of *waves* is under consideration. Wave propagation, which includes the sending of signals via the electromagnetic vibrations of radio waves or light, gains much of its utility from the fact that information can be transmitted by such means. The whole point of signalling, after all, is that there must be the potential for sending a message that might be unexpected by the receiver. If the form of the signal has to be given by an analytic function, then there is not the possibility of 'changing one's mind' in the middle of the message. Any small part of the signal would completely fix the signal in its entirety for all time. Indeed, wave propagation is frequently studied in terms of the question as to how discontinuities, or other deviations from analyticity, will actually propagate.

Let us consider waves and ask how such things are described mathematically. One of the most effective ways of studying wave forms is through the procedure known as *Fourier analysis*. Joseph Fourier was a French mathematician who lived from 1768 until 1830. He had been concerned with the question of decomposing periodic vibrations into their component 'sine-wave' parts. In music, this is basically what is involved in representing some musical sound in terms of its constituent 'pure tones'. The term 'periodic' means that the pattern (say of physical displacements of the object which is vibrating) exactly repeats itself after some period of time, or it could refer to periodicity in space, like the repeating patterns in a crystal or on wallpaper or in waves in the open sea. Mathematically, we say that a function f (say[1] of a real variable χ) is *periodic* if, for all χ, it satisfies

$$f(\chi + l) = f(\chi),$$

where l is some fixed number referred to as the *period*. Thus, if we 'slide' the graph of $y = f(\chi)$ along the χ-axis by an amount l, it looks just the same as it did before (Fig. 9.1a). (The way in which Fourier handled functions that need *not* be periodic—by use of the Fourier *transform*—will be described in §9.4.)

The 'pure tones' are things like $\sin\chi$ or $\cos\chi$ (Fig. 9.1b). These have period 2π, since

$$\sin(\chi + 2\pi) = \sin\chi, \qquad \cos(\chi + 2\pi) = \cos\chi,$$

these relations being manifestations of the periodicity of the single complex quantity $e^{i\chi} = \cos\chi + i\sin\chi$,

$$e^{i(\chi+2\pi)} = e^{i\chi},$$

which we encountered in §5.3. If we want periodicity l, rather than 2π, then we can 'rescale' the χ as it appears in the function, and take $e^{i2\pi\chi/l}$ instead of $e^{i\chi}$. The real and imaginary parts $\cos(2\pi\chi/l)$ and $\sin(2\pi\chi/l)$ will correspondingly also have period l. But this is not the only possibility. Rather than oscillating just once, in the period l, the function could oscillate twice, three times, or indeed n times, where n is any positive integer (see Fig. 9.1c), so we find that each of

$$e^{i\cdot 2\pi n\chi/l}, \quad \sin\left(\frac{2\pi n\chi}{l}\right), \quad \cos\left(\frac{2\pi n\chi}{l}\right)$$

has period l (in addition to having also a smaller period l/n). In music, these expressions, for $n = 2, 3, 4, \ldots$, are referred to as *higher harmonics*.

One problem that Fourier addressed (and solved) was to find out how to express a general periodic function $f(\chi)$, of period l, as a sum of pure tones.

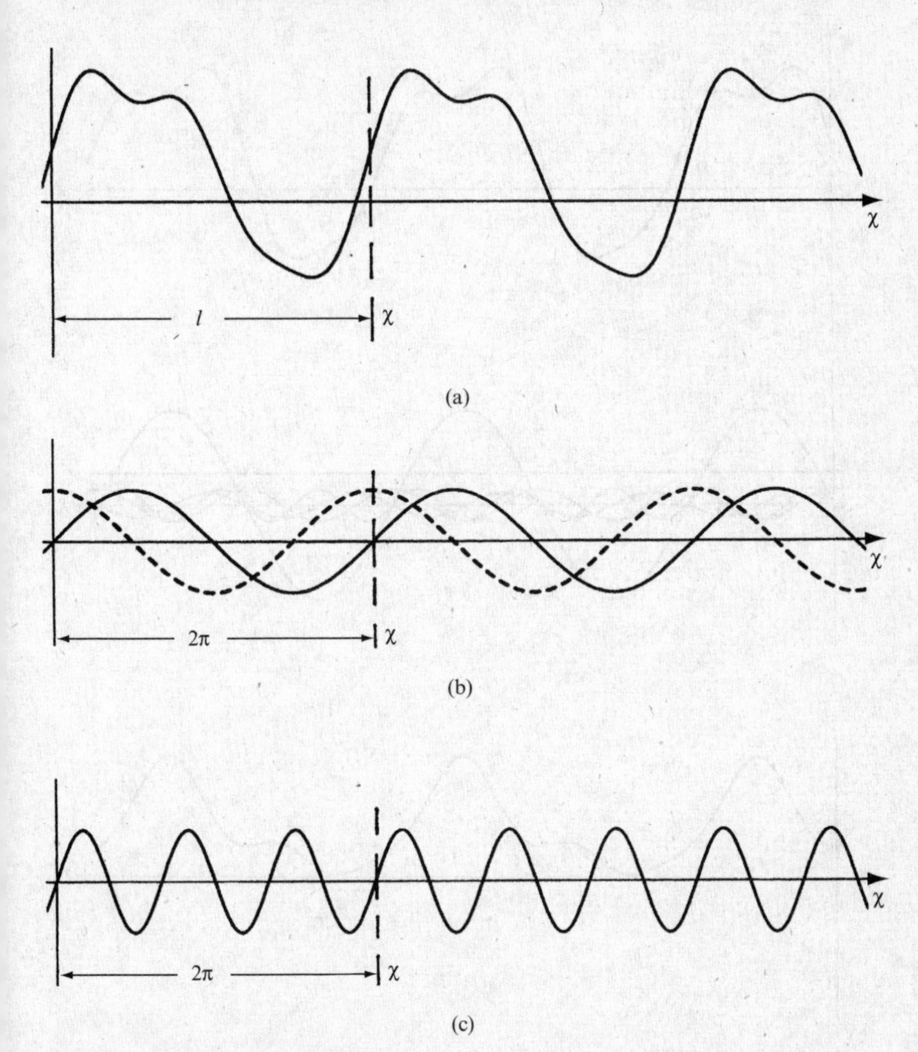

Fig. 9.1 Periodic functions. (a) $f(\chi)$ has period l if $f(\chi)=f(\chi+l)$ for all χ, meaning that if we slide the graph of $y=f(\chi)$ along the χ-axis by l, it looks just the same as before. (b) The basic 'pure tones' $\sin\chi$ or $\cos\chi$ (shown dotted) have period $l=2\pi$. (c) 'Higher harmonic' pure tones oscillate several times in the period l; they still have period l, while also having a shorter period ($\sin 3\chi$ is illustrated, having period $l=2\pi$ as well as the shorter period $2\pi/3$).

For each n, there will generally be a different magnitude of that pure tone's contribution to the total, and this will depend upon the wave form (i.e. upon the shape of the graph $y=f(\chi)$). Some simple examples are illustrated in Fig. 9.2. Usually, the number of different pure tones that contribute to $f(\chi)$ will be infinite, however. More specifically, what Fourier required was the

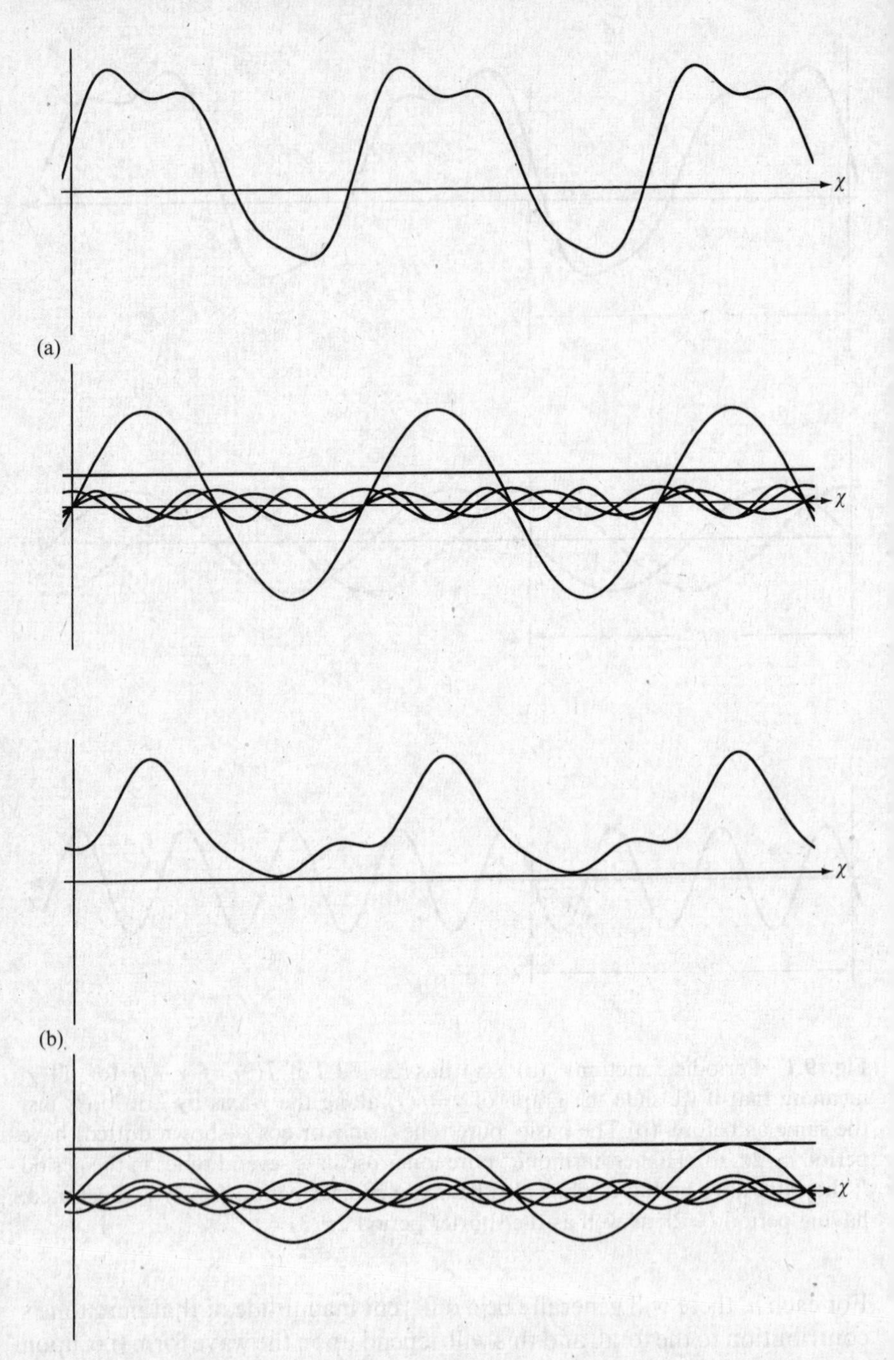

Fig. 9.2 Examples of Fourier decomposition of periodic functions. The wave form (shape of the graph) is determined by the Fourier coefficients. The functions and their individual Fourier components beneath. (a) $f(\chi) = \frac{2}{3} + 2\sin\chi + \frac{1}{3}\cos 2\chi + \frac{1}{4}\sin 2\chi + \frac{1}{3}\sin 3\chi$. (*b*) $f(\chi) = \frac{1}{2} + \sin\chi - \frac{1}{3}\cos 2\chi - \frac{1}{4}\sin 2\chi - \frac{1}{5}\sin 3\chi$.

collection of coefficients c, a_1, b_1, a_2, b_2, a_3, b_3, a_4, ... in the decomposition of $f(\chi)$ into its constituent pure tones, as given by the expression

$$f(\chi) = c + a_1 \cos \omega\chi + b_1 \sin \omega\chi + a_2 \cos 2\omega\chi + b_2 \sin 2\omega\chi + \\ a_3 \cos 3\omega\chi + b_3 \sin 3\omega\chi + \cdots,$$

where, in order to make the expressions look simpler, I have written them in terms of the *angular frequency* ω (nothing to do with the 'ω' of §§5.4,5, §8.1) given by $\omega = 2\pi/l$.

Some readers may well feel that this expression for $f(\chi)$ still looks unduly complicated—and such a reader is indeed correct. The formula actually looks a lot tidier if we incorporate the cos and sin terms together as complex exponentials $(e^{iA\chi} = \cos A\chi + i \sin A\chi)$, so that

$$f(\chi) = \cdots + \alpha_{-2}e^{-2i\omega\chi} + \alpha_{-1}e^{-i\omega\chi} + \alpha_0 + \alpha_1 e^{i\omega\chi} + \alpha_2 e^{2i\omega\chi} + \alpha_3 e^{3i\omega\chi} + \cdots,$$

where[2,][9.1]

$$a_n = \alpha_n + \alpha_{-n}, \qquad b_n = i\alpha_n - i\alpha_{-n}, \qquad c = \alpha_0$$

for $n = 1, 2, 3, 4, \ldots$. The expression looks even tidier if we put $z = e^{i\omega\chi}$, and define the function $F(z)$ to be just the same quantity as $f(\chi)$ but now expressed in terms of the new complex variable z. For then we get

$$F(z) = \cdots + \alpha_{-2}z^{-2} + \alpha_{-1}z^{-1} + \alpha_0 z^0 + \alpha_1 z^1 + \alpha_2 z^2 + \alpha_3 z^3 + \cdots,$$

where

$$F(z) = F(e^{i\omega\chi}) = f(\chi).$$

And we can make it look tidier still by using the summation sign $\sum$, which here means 'add together all the terms, for all integer values of r':

$$F(z) = \sum \alpha_r z^r.$$

This looks like a power series (see §4.3), except that there are *negative* as well as positive powers. It is called a *Laurent* series. We shall be seeing the importance of this expression in the next section.[9.2]

9.2 Functions on a circle

The Laurent series certainly gives us a very economical way of representing Fourier series. But this expression also suggests an interesting

[9.1] Show this.

[9.2] Show that when F is analytic on the unit circle the coefficients α_n, and hence the a_n, b_n, and c, can be obtained by use of the formula $\alpha_n = (2\pi i)^{-1} \oint z^{-n-1}F(z)\,dz$.

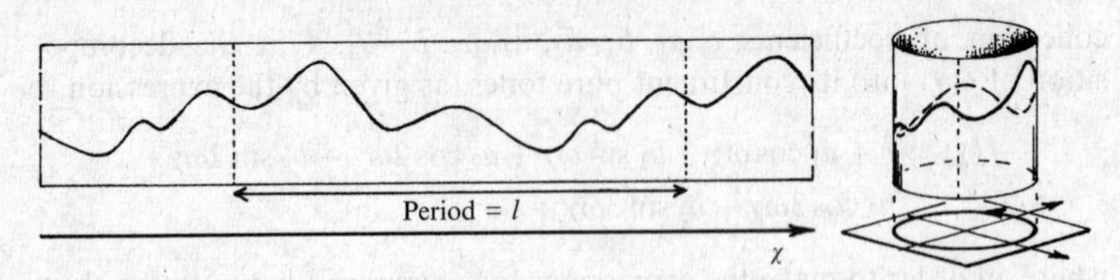

Fig. 9.3 A periodic function of a real variable χ may be thought of as defined on a circle of circumference l where we 'wrap up' the real axis of χ into the circle. With $l=2\pi$, we may take this circle as the unit circle in the complex plane.

alternative perspective on Fourier decomposition. Since a periodic function simply repeats itself endlessly, we may think of such a function (of a real variable χ) as being defined on a *circle* (Fig. 9.3), where the function's period l is the length of the circle's circumference, χ measuring distance around the circle. Rather than simply going off in a straight line, these distances now wrap around the circle, so that the periodicity is automatically taken into account.

For convenience (at least for the time being), I take this circle to be the unit circle in the complex plane, whose circumference is 2π, and I take the period l to be 2π. Accordingly,

$$\omega = 1, \quad \text{so } z = e^{i\chi}.$$

(For any other value of the period, all we need to do is to reinstate ω by rescaling the χ-variable appropriately.) The different cos and sin terms that represent the various 'pure tones' of the Fourier decomposition are now simply represented as positive or negative powers of z, namely $z^{\pm n}$ for the nth harmonics. On the unit circle, these powers just give us the oscillatory cos and sin terms that we require; see Fig. 9.4.

We now have this very tidy way of representing the Fourier decomposition of some periodic function $f(\chi)$. We think of $f(\chi) = F(z)$ as defined on the unit circle in the z-plane, with $z = e^{i\chi}$, and then the Fourier decomposition is just the Laurent series description of this function, in terms of a complex variable z. But the advantage is not just a matter of tidiness. This representation also provides us with deeper insights into the nature of Fourier series and of the kind of function that they can represent. More significantly for the eventual purpose of this book, it has important connections with quantum mechanics and, therefore, for our deeper understanding of Nature. This comes about through the magic of complex numbers, for we can also use our Laurent series expression when z lies away from the unit circle. It turns out that

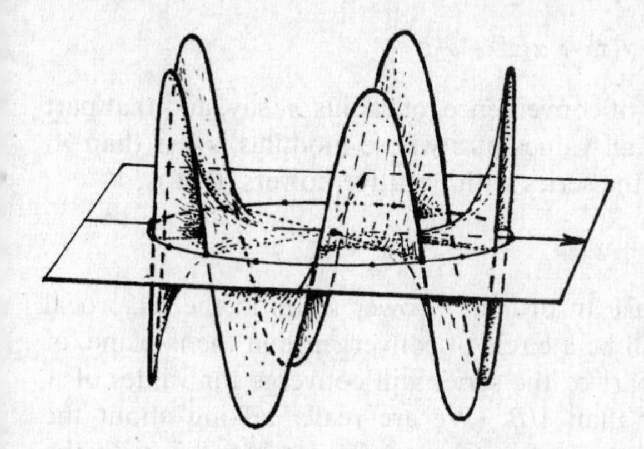

Fig. 9.4 On the unit circle, the real and imaginary parts of the function z^n appear as nth harmonic cos and sin waves (the real and imaginary parts of $e^{in\chi}$, respectively, where $z = e^{i\chi}$). Here, for $n = 5$, the real part of z^5 is plotted.

this series tells us something important about $F(z)$, for z lying *on* the unit circle, in terms of what the series does when z lies *off* the unit circle.

Now, let us recall (from §4.4) the notion of a circle of convergence, within which a power series converges and outside of which it diverges. There is a close analogue of this for a Laurent series: the *annulus of convergence*. This is the region lying strictly between two circles in the complex plane, both centred at the origin (see Fig. 9.5a). This is simple to understand once we have the notion of circle of convergence for an ordinary power series. The part of the series with positive powers,[3]

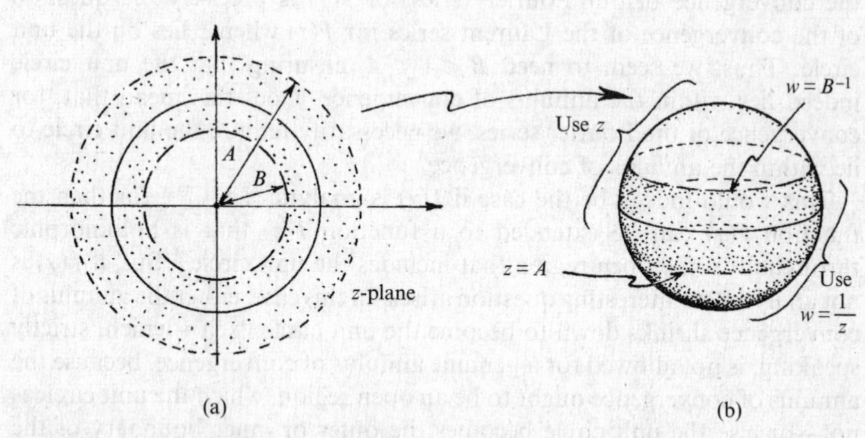

Fig. 9.5 (a) The annulus of convergence for a Laurent series $F(z) = F^+ + \alpha_0 + F^-$, where $F^+ = \cdots + \alpha_{-2}z^{-2} + \alpha_{-1}z^{-1}$, $F^- = \alpha_1 z^1 + \alpha_2 z^2 + \cdots$. The radius of convergence for F^- is A and, in terms of $w = z^{-1}$, for F^+ is B^{-1}. (b) The same, on the Riemann sphere (see Fig. 8.7), where z refers to the extended northern hemisphere and w $(= z^{-1})$ to the extended southern hemisphere.

$$F^- = \alpha_1 z^1 + \alpha_2 z^2 + \alpha_3 z^3 + \cdots,$$

will have an ordinary circle of convergence, of radius A, say, and that part of the series converges for all values of z whose modulus is less than A. With regard to the part of the series with negative powers, that is,

$$F^+ = \cdots + \alpha_{-3} z^{-3} + \alpha_{-2} z^{-2} + \alpha_{-1} z^{-1},$$

we can understand it as just an ordinary power series in the reciprocal variable $w = 1/z$. There will be a circle of convergence in the w-plane, of radius $1/B$, say, and that part of the series will converge for values of w whose modulus is smaller than $1/B$. (We are really talking about the Riemann sphere here, as described in Chapter 8—see Fig. 8.7, with the z-coordinate referring to one hemisphere and the w-coordinate referring to the other. See Fig. 9.5b. We shall explore the Riemann sphere aspect of this in the next section.) For values of z whose moduli are greater than B, therefore, the negative-power part of the series will converge. Provided that $B < A$, these two convergence regions will overlap, and we get the annulus of convergence for the entire Laurent series. Note that the whole Fourier or Laurent series for the function $f(\chi) = F(e^{i\chi}) = F(z)$ is given by

$$F(z) = F^+ + \alpha_0 + F^-,$$

where the additional constant term α_0 must be included.

In the present situation, we ask for convergence *on* the unit circle, since this is where we can have $z = e^{i\chi}$ for real values of χ, and the question of the convergence of our Fourier series for $f(\chi)$ is precisely the question of the convergence of the Laurent series for $F(z)$ when z lies on the unit circle. Thus, we seem to need $B < 1 < A$, ensuring that the unit circle indeed lies within the annulus of convergence. Does this mean that, for convergence of the Fourier series, we necessarily require the unit circle to lie within the annulus of convergence?

This would indeed be the case if $f(\chi)$ is analytic (i.e. C^ω); for then the function $f(\chi)$ can be extended to a function $F(z)$ that is holomorphic throughout some open region that includes the unit circle.[4] But, if $f(\chi)$ is not analytic, an interesting question arises. In this case, either the annulus of convergence shrinks down to become the unit circle itself—which, strictly speaking, is not allowed for a genuine annulus of convergence, because the annulus of convergence ought to be an open region, which the unit circle is not—or else the unit circle becomes the outer or inner boundary of the annulus of convergence. These questions will be important for us in §§9.6,7.

For the moment, let us not worry about what happens when $f(\chi)$ is not analytic, and consider the simpler situation that arises when $f(\chi)$ is analytic. Then we have the unit circle in the z-plane strictly contained within a genuine annulus of convergence for $F(z)$, this being bounded by circles (centred at the origin) of radii A and B, with $B < 1 < A$. The part of the Laurent series with positive powers, F^-, converges for points in the z-plane whose moduli are smaller than A and the part with negative powers, F^+, converges for points in the z-plane whose moduli are greater than B, so both converge within the annulus itself (and, in a very trivial sense, the constant term α_0 obviously 'converges' for all z). This provides us with a 'splitting' of the function $F(z)$ into two parts, one holomorphic inside the outer circle and the other holomorphic outside the inner circle, these being defined, respectively, by the series expressions for F^- and F^+.

There is a (mild) ambiguity about whether the constant term α_0 is to be included with F^- or with F^+ in this splitting. In fact, it is better just to live with this ambiguity. For there is a symmetry between F^- and F^+, which is made clearer if we adopt the Riemann sphere picture that was alluded to above (see Fig. 9.5b). This gives us a more complete picture of the situation, so let us explore this next.

9.3 Frequency splitting on the Riemann sphere

The coordinates z and w $(= 1/z)$ give us two patches covering the Riemann sphere. The unit circle becomes the equator of the sphere and the annulus is now just a 'collar' of the equator. We think of our splitting of $F(z)$ as expressing it as a sum of two parts, one of which extends holomorphically into the southern hemisphere—called the *positive-frequency* part of $F(z)$—as defined by $F^+(z)$, together with whatever portion of the constant term we choose to include, and the other, extending holomorphically into the northern hemisphere—called the *negative-frequency* part of $F(z)$—as defined by $F^-(z)$ and the remaining portion of the constant term. If we ignore the constant term, this splitting is uniquely determined by this holomorphicity requirement for the extension into one or other of the two hemispheres.[9.3]

It will be handy, from time to time, to refer to the 'inside' and the 'outside' of a circle (or other closed loop) drawn on the Riemann sphere by appealing to an *orientation* that is to be assigned to the circle. The standard orientation of the unit circle in the z-plane is given in terms of the direction of increase of the standard θ-coordinate, i.e. anticlockwise. If we reverse this orientation (e.g. replacing θ by $-\theta$), then we interchange positive with negative frequency. Our convention for a general closed loop is to be consistent with this. The orientation is anticlockwise if the 'clock face' is on the inside of the loop, so to speak, whereas it would be clockwise if the 'clock face' were to be placed on the outside of the loop. This serves to define the 'inside' and 'outside' of an oriented closed loop. Figure 9.6 should clarify the issue.

[9.3] Can you see why?

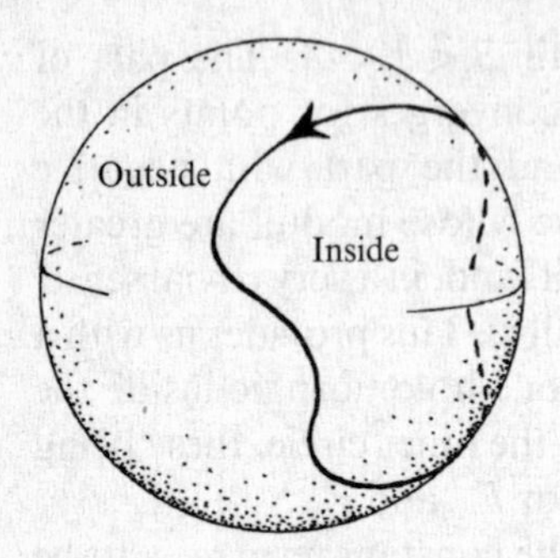

Fig. 9.6 An orientation assigned to a closed loop on the Riemann sphere defines its 'inside' and 'outside' as indicated: this orientation is anticlockwise for a 'clock face' inside the loop (and clockwise if outside).

This splitting of a function into its positive- and negative-frequency parts is a crucial ingredient of quantum theory, and most particularly of quantum field theory, as we shall be seeing in §24.3 and §§26.2–4. The particular formulation that I have given here is not quite the most usual way that this splitting is expressed, but it has some considerable advantages in a number of different contexts (particularly in twistor theory, for example; see §33.10). The usual formulation is not so concerned with holomorphic extensions as with the Fourier expansion directly. The positive-frequency components are those given by multiples of $\mathrm{e}^{-in\chi}$, where n is positive, as opposed to those given by multiples of $\mathrm{e}^{in\chi}$, which are negative-frequency components. A positive-frequency function is one composed entirely of positive-frequency components.

However, this description does not reveal the full generality of what is involved in this splitting. There are many holomorphic mappings of the Riemann sphere to itself which send each hemisphere to itself, but which do not preserve the north or south poles (i.e. the points $z = 0$ or $z = \infty$).[9.4] These preserve the positive/negative-frequency splitting but do not preserve the individual Fourier components $\mathrm{e}^{-in\chi}$ or $\mathrm{e}^{in\chi}$. Thus, the issue of the splitting into positive and negative frequencies (crucial to quantum theory) is a more general notion than the picking out of individual Fourier components.

In normal discussions of quantum mechanics, the positive/negative-frequency splitting refers to functions of *time* t, and we do not usually think of time as going round in a circle. But we can use a simple transformation to obtain the full range of t, from the 'past limit' $t = -\infty$ to the 'future limit' $t = \infty$, from a χ that goes once around the circle—here I take χ to range between the limits $\chi = -\pi$ and $\chi = \pi$ (so $z = \mathrm{e}^{i\chi}$ ranges round the unit circle in the complex plane, in an anticlockwise direction, from the point $z = -1$ and back to $z = -1$ again; see Fig. 9.7). Such a transformation is given by

---

[9.4] Which are these mappings, explicitly?

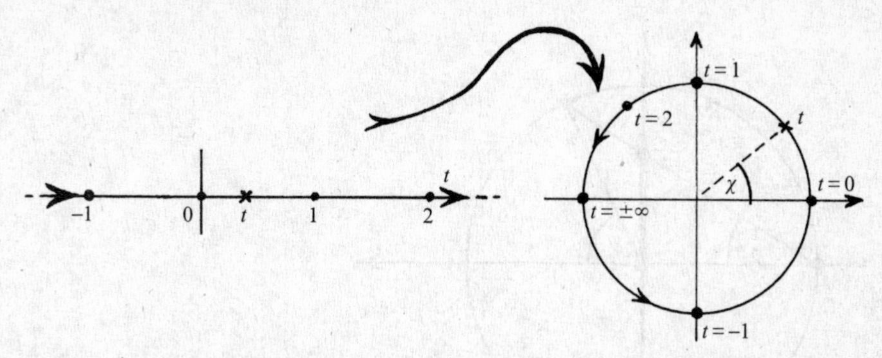

Fig. 9.7 In quantum mechanics, positive/negative-frequency splitting refers to functions of time t, not assumed periodic. The splitting of Fig. 9.5 can still be applied, for the full range of t (from $-\infty$ to $= +\infty$) if we use the transformation of relating t to $z(= e^{i\chi})$, where we go around unit circle, anticlockwise, from $z = -1$ and back to $z = -1$ again, so χ goes from $-\pi$ to π.

$$t = \tan \tfrac{1}{2}\chi.$$

The graph of this relationship is given in Fig. 9.8 and a simple geometrical description is provided in Fig. 9.9.

An advantage of this particular transformation is that it extends holomorphically to the entire Riemann sphere, this being a transformation that we already considered in §8.3 (see Fig. 8.8), which takes the unit circle (z-plane) into the real line (t-plane):[9.5]

$$t = \frac{z-1}{iz+i}, \quad z = \frac{-t+i}{t+i}.$$

The interior of the unit circle in the z-plane corresponds to the upper half-t-plane and the exterior of the z-unit circle corresponds to the lower half-t-plane. Hence, positive-frequency functions of t are those that extend holomorphically into the lower half-plane of t and negative-frequency ones, into the upper half-plane. (There is, however, a significant additional

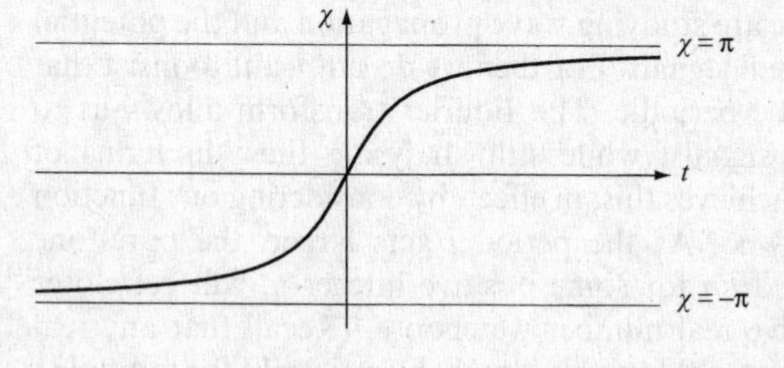

Fig. 9.8 Graph of $t = \tan \frac{1}{2}\chi$.

[9.5] Show that this gives the same t as above.

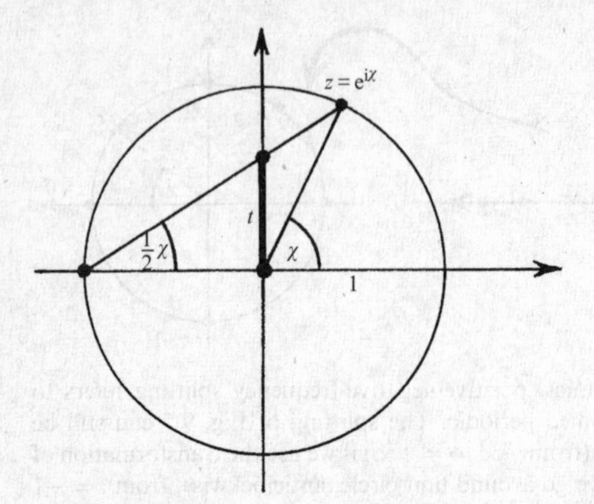

Fig. 9.9 Geometry of $t = \tan\frac{1}{2}\chi$.

technicality that we have to be careful about how we deal with the point '∞' of the t-plane; but this is handled appropriately if we always think in terms of the Riemann sphere, rather than simply the complex t-plane.)

In standard presentations, however, the notion of 'positive frequency' in terms of a time-coordinate t, is not usually stated in the particular way that I have just presented it here, but rather in terms of what is called the *Fourier transform* of $f(\chi)$. The answer is actually the same[5] as the one that I have given, but since Fourier transforms are of crucial significance for quantum mechanics in any case (and also in many other areas), it will be important to explain here what this transform actually is.

9.4 The Fourier transform

Basically, a Fourier transform is the limiting case of a Fourier series when the period l of our periodic function $f(\chi)$ is taken to get larger and larger until it becomes infinite. In this infinite limit, there is no restriction of periodicity on $f(\chi)$ at all: it is just an ordinary function.[6] This has considerable advantages when we are studying wave propagation and the potential for sending of 'unexpected' signals. For then we do not want to insist that the form of the signal be periodic. The Fourier transform allows us to consider such 'one-off' signals, while still analysing them in terms of periodic 'pure tones'. It achieves this, in effect, by considering our function $f(\chi)$ to have period $l \to \infty$. As the period l gets larger, the pure-tone harmonics, having period l/n for some positive integer n, will get closer and closer to any positive real number we choose. (Recall that any real number can be approximated arbitrarily closely by rationals, for example.) What this tells us is that any pure tone of any frequency whatever is now allowed as a Fourier component. Rather than having $f(\chi)$ expressed as a discrete sum of Fourier components, we now have $f(\chi)$ expressed as a continuous sum over all frequencies, which means that $f(\chi)$ is now expressed as an *integral* (see §6.6) with respect to the frequency.

Let us see, in outline, how this works. First, recall our 'tidiest' expression for the Fourier decomposition of a periodic function $f(\chi)$, of period l, as given above:

$$F(z) = \sum \alpha_r z^r, \quad \text{where } z = e^{i\omega\chi}$$

(the angular frequency ω being given by $\omega = 2\pi/l$). Let us take the period to be initially 2π, so $\omega = 1$. Now we are going to try to increase the period by some large integer factor N (whence $l = 2\pi N$), so the frequency is reduced by the same factor (i.e. $\omega = N^{-1}$). The oscillatory wave that used to be the fundamental pure tone now becomes the Nth harmonic with respect to this new lower frequency. A pure tone that used to be an nth harmonic would now be an (nN)th harmonic. When we take the limit as N approaches infinity, it becomes inappropriate to try to keep track of a particular oscillatory component by labelling it by its 'harmonic number' (i.e. by the number n), because this number keeps changing. That is to say, it is inappropriate to label this oscillatory component by the integer r in the above sum because a fixed value of r labels a particular harmonic ($r = \pm n$ for the nth harmonic), rather than keeping track of a particular tone frequency. Instead, it is r/N that keeps track of this frequency, and we need a new variable to label this. Bearing in mind the important use that Fourier transforms are due to be put to in later chapters (see §21.11 particularly), I shall call this variable 'p' which, in the limit when N tends to infinity, stands for the *momentum*[7] of some quantum-mechanical particle whose position is measured by χ. In this limit, one may also revert to the conventional use of x in place of χ, if desired, as we shall find that χ actually does become the real part of z in the limit in the following descriptions.

For finite N, I write

$$p = \frac{r}{N}.$$

In the limit as $N \to \infty$, the parameter p becomes a continuous variable and, since the 'coefficients α_r' in our sum will then depend on the continuous real-valued parameter p rather that on the discrete integer-valued parameter r, it is better to write the dependence of the coefficients α_r on r by using the standard type of functional notation, say $g(p)$, rather than just using a suffix (e.g. g_p), as in α_r. Effectively, we shall make the replacement

$$\alpha_r \mapsto g(p)$$

in our summation $\sum \alpha_r z^r$, but we must bear in mind that, as N gets larger, the number of actual terms lying within some small range of p-values gets larger (basically in proportion to N, because we are considering fractions n/N that lie in that range). Accordingly, the quantity $g(p)$ is really a measure of density, and it must be accompanied by the differential quantity dp in the limit as the summation $\sum$ becomes an integral $\int$. Finally, consider the term z^r in our sum $\sum \alpha_r z^r$. We have $z = e^{i\omega\chi}$, with $\omega = N^{-1}$; so $z = e^{i\chi/N}$. Thus $z^r = e^{ir\chi/N} = e^{i\chi p}$; so putting these things together, in the limit as $N \to \infty$, we get the expression

$$\sum \alpha_r z^r \to \int_{-\infty}^{\infty} g(p) e^{i\chi p} \mathrm{d}p$$

to represent our function $f(\chi)$. In fact it is usual to include a scaling factor of $(2\pi)^{-1/2}$ with the integral, for then there is the remarkable symmetry that the *inverse* relation, expressing $g(p)$ in terms of $f(\chi)$ has exactly the same form (apart from a minus sign) as that which expresses $f(\chi)$ in terms of $g(p)$:

$$f(\chi) = (2\pi)^{-1/2} \int_{-\infty}^{\infty} g(p) e^{i\chi p} \mathrm{d}p, \quad g(p) = (2\pi)^{-1/2} \int_{-\infty}^{\infty} f(\chi) e^{-i\chi p} \mathrm{d}\chi.$$

The functions $f(\chi)$ and $g(p)$ are called *Fourier transforms* of one another.[9.6]

9.5 Frequency splitting from the Fourier transform

A (complex) function $f(\chi)$, defined on the entire real line, is said to be of *positive frequency* if its Fourier transform $g(p)$ is zero for all $p \geqslant 0$. Thus, $f(\chi)$ is composed only of components of the form $e^{i\chi p}$ with $p < 0$. Euler might well have worried—see §6.1—about such a $g(p)$, which seems to be a blatant 'gluing job' between a non-zero function for $p < 0$ and simply zero for $p > 0$. Yet this seems to be representing a perfectly respectable 'holomorphic' property of $f(\chi)$. Another way of expressing this 'positive-frequency' condition is in terms of the holomorphic extendability of $f(\chi)$, as we did before for Fourier series. Now we think of the variable χ as labelling the points on the real axis (so we can take $\chi = x$ on this axis), where on the Riemann sphere this 'real axis' (including the point '$\chi = \infty$') is now the *real circle* (see Fig. 8.7c). This circle divides the sphere into two hemispheres, the 'outside' one being that which is the lower half-plane in the standard picture of the complex plane. The condition that $f(\chi)$ be of positive frequency is now that it extend holomorphically into this outside hemisphere.

There is one issue that requires some care, however, when we compare these two definitions of 'positive frequency'. This relates to the question of

---

[9.6] Show (in outline) how to obtain the expression for $g(p)$ in terms of $f(\chi)$ using a limiting form of the contour integral expression $\alpha_n = (2\pi i)^{-1} \oint z^{-n-1} F(z) \mathrm{d}z$ of Exercise [9.2].

how we treat the point $z = \infty$, since the function $f(\chi)$ will in general have some kind of singularity there. In fact, provided that we adopt the 'hyperfunctional' point of view that I shall be describing shortly (in §9.7), this singularity at $z = \infty$ presents us with no essential difficulty. With the appropriate point of view with regard to '$f(\infty)$', it turns out that the two definitions of positive frequency that I gave in the previous paragraph are in basic agreement with each other.[8]

For the interested reader, it may be helpful to examine, in terms of the Riemann sphere, some of the geometry that is involved in our limit of §9.4, taking us from Fourier series to Fourier transform. Let us return to the z-plane description that we had been considering earlier, for a function $f(\chi)$ of period 2π, where χ measures the arc length around a unit-radius circle. Suppose that we wish to change the period to values larger than 2π, in successively increasing steps, while retaining the interpretation of χ as a distance around a circle. We can achieve this by considering a sequence of larger and larger circles, but in order for the limiting procedure to make geometric sense we shall suppose that the circles are all touching each other at the starting point $\chi = 0$ (see Fig. 9.10a). For simplicity in what follows, let us choose this point to be the origin $z = 0$ (rather than $z = 1$), with all the circles lying in the lower half-plane. This makes our initial circle,

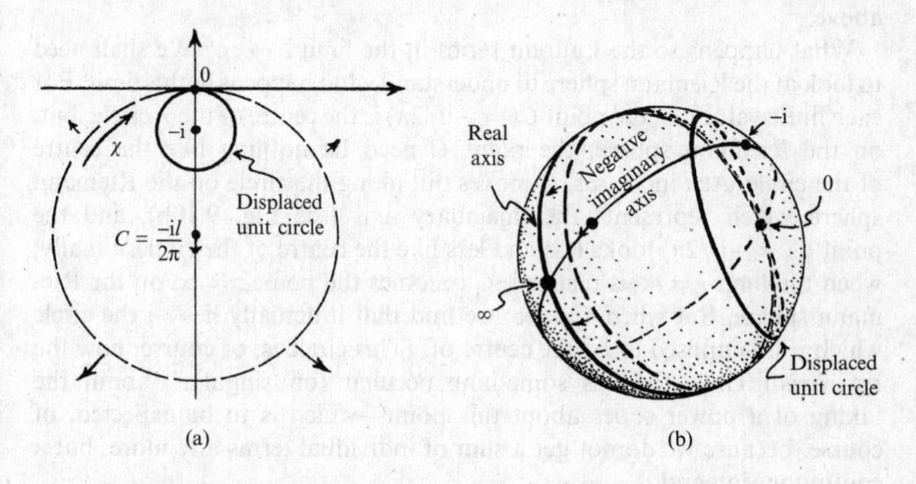

Fig. 9.10 Positive-frequency condition, as $l \to \infty$, where l is the period of $f(\chi)$. (a) Start with $l = 2\pi$, with f defined on the unit circle displaced to have its centre at $z = -\mathrm{i}$. For increasing l, the circle has radius l and centre at $C = -\mathrm{i}l/2\pi$. In each case χ measures arc length *clockwise*. Positive frequency is expressed as f being holomorphically extendible to the interior of the circle, and in the limit $l = \infty$, to the lower half-plane. (b) The same, on the Riemann sphere. For finite l, the Fourier series is obtained from a Laurent series about $z = -\mathrm{i}l/2\pi$, but on the sphere, this point is not the circle's centre, becoming the point ∞ (lying on it) in the limit $l = \infty$, where the Fourier series becomes the Fourier transform.

for period $l = 2\pi$, the unit circle centred at $z = -\mathrm{i}$, rather than at the origin. For a period $l > 2\pi$, the circle is centred at the point $C = -\mathrm{i}l/2\pi$ in the complex plane, and, in the limit as $l \to \infty$, we get the real axis itself (so $\chi = x$), the circle's 'centre' having moved off to infinity along the negative imaginary axis. In each case, we now take χ to measure arc length *clockwise* around the circle (or, in the limiting case, just positive distance along the real axis), with $\chi = 0$ at the origin. Since our circles now have a non-standard (i.e. clockwise) orientation, their 'outsides' are their *interiors* (see §9.3, Fig. 9.6), so our positive frequency condition refers to this interior. We now have the relation between χ and z expressed as[9.7]

$$z = \frac{\mathrm{i}l}{2\pi}\left(\mathrm{e}^{-\mathrm{i}\chi} - 1\right).$$

For finite l, we can express $f(\chi)$ as a Fourier series by referring to a Laurent series about the point $C = -\mathrm{i}l/2\pi$. We get the Fourier transform by taking the limit $l \to \infty$. For finite l, we obtain the condition of positive frequency as the holomorphic extendability of $f(\chi)$ into the *interior* of the relevant circle; in the limit $l \to \infty$, this becomes holomorphic extendability into the lower half-plane, in accordance with what has been stated above.

What happens to the Laurent series in the limit $l \to \infty$? We shall need to look at the Riemann sphere to understand what happens in this limit. For each finite value of l, the point $C(= -\mathrm{i}l/2\pi)$ is the centre of the χ-circle, but, on the Riemann sphere, the point C need be nothing like the centre of the circle. As l increases, C moves out along the circle on the Riemann sphere which represents the imaginary axis (see Fig. 9.10b), and the point $C(= -\mathrm{i}l/2\pi)$ looks less and less like the centre of the circle. Finally, when the limit $l = \infty$ is reached, C becomes the point $z = \infty$ on the Riemann sphere. But when $C = \infty$, we find that it actually lies *on* the circle which it is supposed to be the centre of! (This circle is, of course, now the real axis.) Thus, there is something peculiar (or 'singular') about the taking of a power series about this point—which is to be expected, of course, because we do not get a sum of individual terms any more, but a continuous integral.

9.6 What kind of function is appropriate?

Let us now return to the question posed at the beginning of this chapter, concerning the type of 'function' that is appropriate to use. We can raise

[9.7] Derive this expression.

the following issue: what kind of functions can we represent as Fourier transforms? It would seem to be inappropriate to restrict attention only to analytic (i.e to C^{ω}–) functions because, as we saw above, the Fourier transform $g(p)$ of a positive-frequency function $f(\chi)$—which can certainly be analytic—is a distinctly non-analytic 'gluing job' of a non-zero function to the zero function. The relation between a function and its Fourier transform is symmetrical, so it seems unreasonable to adopt such different standards for each. As a further point, it was noted above that the behaviour of $f(\chi)$ at the point $\chi = \infty$ is relevant to the issue of its positive/negative-frequency splitting, but only in very special circumstances would $f(\chi)$ actually be analytic (C^{ω}) at ∞ (since this would require a precise matching between the behaviour of $f(\chi)$ as $\chi \to +\infty$ and as $\chi \to -\infty$). In addition to all this, there is our initial *physical* motivation, referred to earlier, for studying Fourier transforms, namely that they allow us to treat signals which can transmit 'unexpected' (non-analytic) messages. Thus, we must return to the question which confronted us at the beginning of this chapter: what kind of function should we accept as being an 'honest' function?

We recall that, on the one hand, Euler and his contemporaries might indeed have probably settled for a holomorphic (or analytic) function as being the kind of thing that they had in mind for a respectable 'function'; yet, on the other hand, such functions seem unreasonably restrictive for many kinds of mathematical and physical problem, including those concerned with wave propagation, so a more general notion is needed. Is one of these points of view more 'correct' than the other? There is probably a strong prevailing opinion that supporters of the first viewpoint are 'old-fashioned', and that modern concepts lean heavily towards the second, so that holomorphic or analytic functions are just very special cases of the general notion of a 'function'. But is this necessarily the 'right' attitude to take? Let us try to put ourselves into an 18th-century frame of mind.

Enter Joseph Fourier early in the 19th century. Those who belonged to the 'analytic' ('Eulerian') school of thought would have received a nasty shock when Fourier showed that certain periodic functions, such as the square wave or saw tooth depicted in Fig. 9.11, have perfectly reasonable-looking Fourier representations! Fourier encountered a great deal of opposition from the mathematical establishment at the time. Many were reluctant to accept his conclusions. How could there be a 'formula' for the square-wave function, for example? Yet, as Fourier showed, the series

$$s(\chi) = \sin \chi + \tfrac{1}{3} \sin 3\chi + \tfrac{1}{5} \sin 5\chi + \tfrac{1}{7} \sin 7\chi + \cdots$$

actually sums to a square wave, taking this wave to oscillate between the constant values $\frac{1}{4}\pi$ and $-\frac{1}{4}\pi$ in the half-period π (see Fig. 9.12).

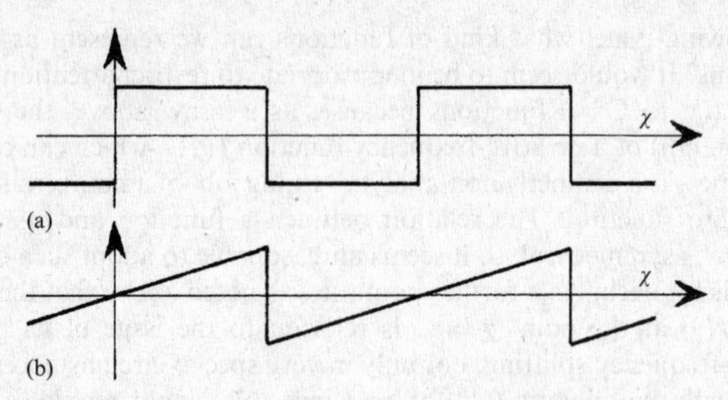

Fig. 9.11 Discontinuous periodic functions (with perfectly reasonable-looking Fourier representations): (a) Square wave (b) Saw tooth.

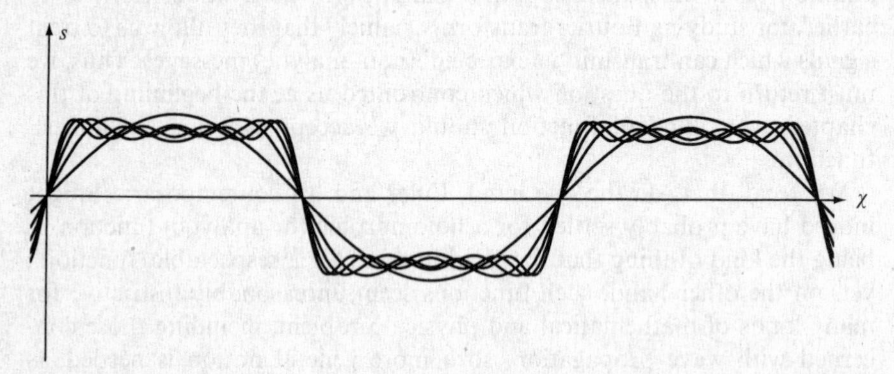

Fig. 9.12 Partial sums of the Fourier series $s(\chi) = \sin\chi + \frac{1}{3}\sin 3\chi + \frac{1}{5}\sin 5\chi + \frac{1}{7}\sin 7\chi + \frac{1}{9}\sin 9\chi + \cdots$, converging to a square wave (like that of Fig. 9.11a).

Let us consider the Laurent-series description for this, as given above. We have the rather elegant-looking expression[9.8]

$$2is(\chi) = \cdots - \tfrac{1}{5}z^{-5} - \tfrac{1}{3}z^{-3} - z^{-1} + z + \tfrac{1}{3}z^{3} + \tfrac{1}{5}z^{5} + \cdots,$$

where $z = e^{i\chi}$. In fact this is an example where the annulus of convergence shrinks down to the unit circle—with no actual open region left. However, we can still make sense of things in terms of holomorphic functions if we split the Laurent series into two halves, one with the positive powers, giving an ordinary power series in z, and one with the negative powers, giving a power series in z^{-1}. In fact, these are well-known series, and can be summed explicitly:[9.9]

[9.8] Show this.

[9.9] Do this, by taking advantage of a power series expansion for log z taken about $z = 1$, given towards the end of §7.4.

$$S^- = z + \tfrac{1}{3}z^3 + \tfrac{1}{5}z^5 + \cdots = \tfrac{1}{2}\log\left(\frac{1+z}{1-z}\right)$$

and

$$S^+ = \cdots - \tfrac{1}{5}z^{-5} - \tfrac{1}{3}z^{-3} - z^{-1} = -\tfrac{1}{2}\log\left(\frac{1+z^{-1}}{1-z^{-1}}\right),$$

giving $2\mathrm{i}s(\chi) = S^- + S^+$. A little rearrangement of these expressions leads to the conclusion that S^- and $-S^+$ differ only by $\pm\frac{1}{2}\mathrm{i}\pi$, telling us that $s(\chi) = \pm\frac{1}{4}\pi$.[9.10] But we need to look a little more closely to see why we actually get a square wave oscillating between these alternative values.

It is a little easier to appreciate what is going on if we apply the transformation $t = (z-1)/(\mathrm{i}z+\mathrm{i})$, given in §8.3, which takes the interior of the unit circle in the z-plane to the upper half-t-plane (as illustrated in Fig. 8.8). In terms of t, the quantity S^- now refers to this upper half-plane and S^+ to the lower half-plane, and we find (with possible $2\pi\mathrm{i}$ ambiguities in the logarithms)

$$S^- = -\tfrac{1}{2}\log t + \tfrac{1}{2}\log \mathrm{i}, \quad S^+ = \tfrac{1}{2}\log t + \tfrac{1}{2}\log \mathrm{i}.$$

Following the logarithms continuously from the respective starting points $t = \mathrm{i}$ (where $S^- = 0$) and $t = -\mathrm{i}$ (where $S^+ = 0$), we find that along the positive real t-axis we have $S^- + S^+ = +\frac{1}{2}\mathrm{i}\pi$, whereas along the negative real t-axis we have $S^- + S^+ = -\frac{1}{2}\mathrm{i}\pi$.[9.11] From this we deduce that along the top half of the unit circle in the z-plane we have $s(\chi) = +\frac{1}{4}\pi$, whereas along the bottom half we have $s(\chi) = -\frac{1}{4}\pi$. This shows that the Fourier series indeed sums to the square wave, just as Fourier had asserted.

What is the moral to be drawn from this example? We have seen that a particular (periodic) function that is not even continuous, let alone differentiable (in this case being a C^{-1}-function), can be represented as a perfectly sensible-looking Fourier series. Equivalently, when we think of the function as being defined on the unit circle, it can be represented as a reasonable-appearing Laurent series, although it is one for which the annulus of convergence has, in effect, shrunk down to the unit circle itself. The positive and the negative half of this Laurent series each sums to a perfectly good holomorphic function on half of the Riemann sphere. One is defined on one side of the unit circle, and the other is defined on the other side. We can think of the 'sum' of these two functions as giving the required square wave on the unit circle itself. It is because of the existence of branch singularities at the two points $z = \pm 1$ on

[9.10] Show this (assuming that $|s(\chi)| < 3\pi/2$).

[9.11] Show this.

the unit circle that the sum can 'jump' from one side to the other, giving the square wave that arises in this sum. These branch singularities also prevent the power series on the two sides from converging beyond the unit circle.

9.7 Hyperfunctions

This example is only a very special case, but it illustrates what we must do in general. Let us ask what is the most general type of function that can be defined on the unit circle (on the Riemann sphere) and represented as a 'sum' of some holomorphic function F^+ on the open region lying to one side of the circle and of another holomorphic function F^- on the open region lying to the other side, just as in the example that we have been considering. We shall find that the answer to this question leads us directly to an exotic but important notion referred to as a 'hyperfunction'.

In fact, it turns out to be more illuminating to think of f as being the 'difference' between F^- and $-F^+$. One reason for this is that, in the most general cases, there may be no analytic extension of either F^- or F^+ to the actual unit circle, so it is not clear what such a 'sum' could mean on the circle itself. However, we can think of the *difference* between F^- and $-F^+$ as representing the 'jump' between these two functions as their regions of definition come together at the unit circle.

This idea of a 'jump' between a holomorphic function on one side of a curve in the complex plane and another holomorphic function on the other—where neither holomorphic function need extend holomorphically over the curve itself—actually provides us with a new concept of a 'function' defined on the curve. This is, in effect, the definition of a *hyperfunction* on an (analytic) curve. It is a wonderful notion put forward by the Japanese mathematician Mikio Sato in 1958,[9] although, as we shall shortly be seeing, Sato's actual definition is considerably more elegant than just this.[10]

We do not need to think of a closed curve, like the entire unit circle, for the definition of a hyperfunction, but we can consider some part of a curve. Indeed, it is more usual to consider hyperfunctions as defined on some segment γ of the real line. We shall take γ to be the segment of the real line between a and b, where a and b are real numbers with $a < b$. A hyperfunction defined on γ is then the *jump* across γ, starting from a holomorphic function f on an open set $\mathcal{R}^-$ (having γ as its upper boundary) to a holomorphic function g on an open set $\mathcal{R}^+$ (having γ as its lower boundary) see Fig. 9.13.

Simply to refer to a 'jump' in this way does not give us much idea of what to do with such a thing (and it is not yet very mathematically precise). Sato's elegant resolution of these issues is to proceed in a rather

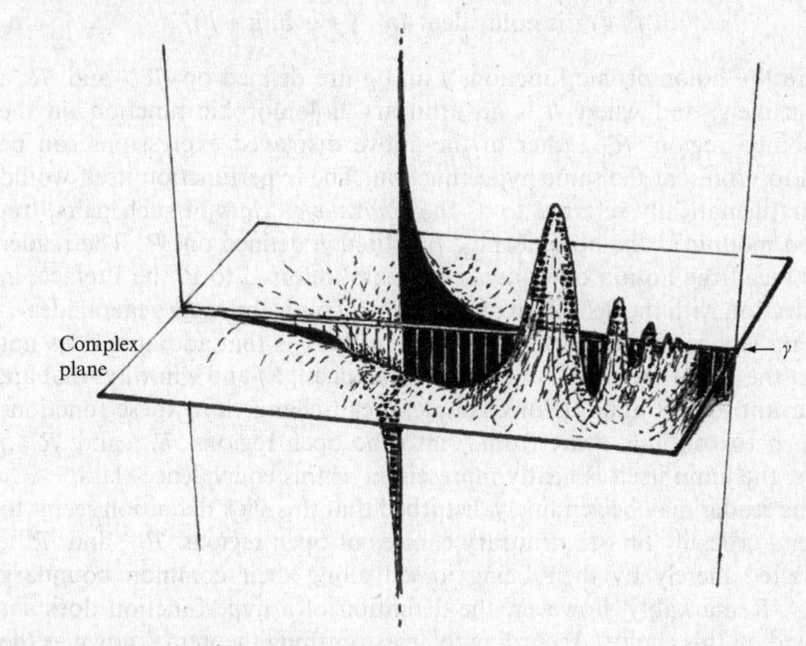

Fig. 9.13 A hyperfunction on a segment γ of the real axis expresses the 'jump' from a holomorphic function on one side of γ to one on the other.

formally algebraic way, which is actually extrordinarily simple. We merely represent this jump as the *pair* (f, g) of these holomorphic functions, but where we say that such a pair (f, g) is *equivalent* to another such pair (f_0, g_0) if the latter is obtained from the former by adding to both f and g the same holomorphic function h, where h is defined on the combined (open) region $\mathcal{R}$, which consists of $\mathcal{R}^-$ and $\mathcal{R}^+$ joined together along the curve segment γ; see Fig. 9.14. We can say

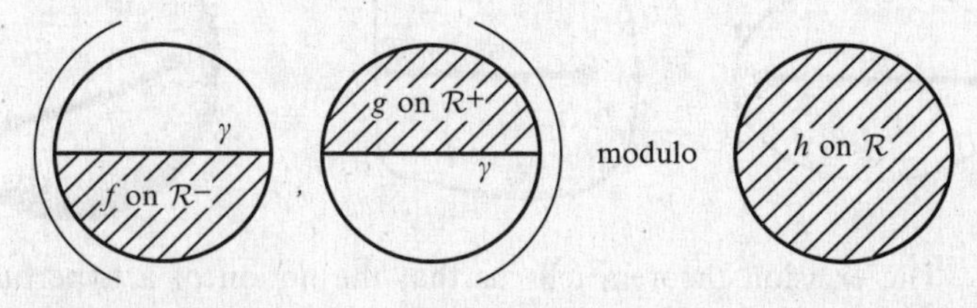

Fig. 9.14 A hyperfunction, on a segment γ of the real axis, is provided by a pair of holomorphic functions (f, g), with f defined on some open region $\mathcal{R}^-$, extending downwards from γ and g on an open region $\mathcal{R}^+$, extending upwards from γ. The actual hyperfunction, on γ, is (f, g) *modulo* quantities $(f + h, g + h)$, where h is holomorphic on the union $\mathcal{R}$ of $\mathcal{R}^-$, γ, and $\mathcal{R}^+$.

$$(f, g) \quad \text{is equivalent to} \quad (f+h, g+h),$$

where the holomorphic functions f and g are defined on $\mathcal{R}^-$ and $\mathcal{R}^+$, respectively, and where h is an arbitrary holomorphic function on the combined region $\mathcal{R}$. Either of the above displayed expressions can be used to represent the same hyperfunction. The hyperfunction itself would be mathematically referred to as the *equivalence class* of such pairs, 're-duced modulo'[11] the holomorphic functions h defined on $\mathcal{R}$. The reader may recall the notion of 'equivalence class' referred to in the Preface, in connection with the definition of a fraction. This is the same general idea—and no less confusing. The essential point here is that adding h does not affect the 'jump' between f and g, but h can change f and g in ways that are irrelevant to this jump. (For example, h can change how these functions happen to continue away from γ into the open regions $\mathcal{R}^-$ and $\mathcal{R}^+$.) Thus, the jump itself is neatly represented *as* this equivalence class.

The reader may be genuinely disturbed that this slick definition seems to depend crucially on our arbitrary choices of open regions $\mathcal{R}^-$ and $\mathcal{R}^+$, restricted merely by their being joined along their common boundary line γ. Remarkably, however, the definition of a hyperfunction does *not* depend on this choice. According to an astonishing theorem, known as the *excision theorem*, this notion of hyperfunction is actually quite independent of the particular choices of $\mathcal{R}^-$ and $\mathcal{R}^+$; see top three examples of Fig. 9.15.

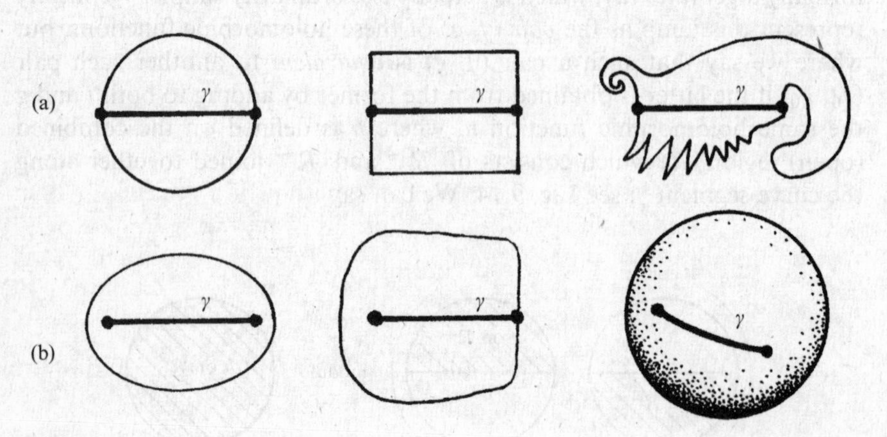

Fig. 9.15 The excision theorem tells us that the notion of a hyperfunction is independent of the choice of open region $\mathcal{R}$, so long as $\mathcal{R}$ contains the given curve γ. (a) The region $\mathcal{R} - \bar{\gamma}$ may consist of two separate pieces (so we get two distinct holomorphic functions f and g, as in Fig. 9.14) or (b) the region $\mathcal{R} - \bar{\gamma}$ may be a single connected piece, in which case f and g are simply two parts of the same holomorphic function.

In fact, the excision theorem gives us more than even this. We do not require that our open region $\mathcal{R}$ be divided into two (namely into $\mathcal{R}^-$ and $\mathcal{R}^+$) by the removal of γ. All we need is that the open region $\mathcal{R}$, in the complex plane, must contain the open[12] segment γ. It may be that $\mathcal{R} - \gamma$ (i.e. what is left of $\mathcal{R}$ when γ is removed from it[13]) consists of two separate pieces, just as we have been considering up to this point, but more generally the removal of γ from $\mathcal{R}$ may leave us with a single connected region, as illustrated in the bottom three examples of Fig. 9.15. In these cases, we must also remove any internal end-point a or b, of γ, so that we are left with an open set, which I refer to as $\mathcal{R} - \bar{\gamma}$. In this more general case, our hyperfunctions are defined as 'holomorphic functions on $\mathcal{R} - \bar{\gamma}$, reduced modulo holomorphic functions on $\mathcal{R}$'. It is quite remarkable that this very liberal choice of $\mathcal{R}$ makes no difference to the class of 'hyperfunctions' that is thereby defined.[9.12] The case when a and b both lie within $\mathcal{R}$ is useful for integrals of hyperfunctions, since then a closed contour in $\mathcal{R} - \bar{\gamma}$ can be used.

All this applies also to our previous case of a circle on the Riemann sphere. Here, there is some advantage in taking $\mathcal{R}$ to be the entire Riemann sphere, because then the functions that we have to 'mod out by' are the holomorphic functions that are global on the entire Riemann sphere, and there is a theorem which tells us that these functions are just constants. (These are actually the 'constants' α_0 that we chose not to worry about in §9.2.) Thus, modulo constants, a hyperfunction defined on a circle on the Riemann sphere is specified simply by one holomorphic function on the entire region on one side of the circle and another function on the other side. This gives the splitting of an arbitrary hyperfunction on the circle uniquely (modulo constants) into its positive- and negative-frequency parts.

Let us end by considering some basic properties of hyperfunctions. I shall use the notation $\langle\!| f, g |\!\rangle$ to denote the hyperfunction specified by the pair f and g defined holomorphically on $\mathcal{R}^-$ and $\mathcal{R}^+$, respectively (where I am reverting to the case where γ divides $\mathcal{R}$ into $\mathcal{R}^-$ and $\mathcal{R}^+$. Thus, if we have two different representations $\langle\!| f, g |\!\rangle$ and $\langle\!| f_0, g_0 |\!\rangle$ of the same hyperfunction, that is, $\langle\!| f, g |\!\rangle = \langle\!| f_0, g_0 |\!\rangle$, then $f - f_0$ and $g - g_0$ are both the *same* holomorphic function h defined on $\mathcal{R}$, but restricted to $\mathcal{R}^-$ and $\mathcal{R}^+$ respectively. It is then straightforward to express the *sum* of two hyperfunctions, the *derivative* of a hyperfunction, and the *product* of a hyperfunction with an analytic function q defined on γ:

[9.12] Why does 'holomorphic functions on $\mathcal{R} - \bar{\gamma}$, reduced modulo holomorphic functions on $\mathcal{R}$' become the definition of a hyperfunction that we had previously, when $\mathcal{R} - \bar{\gamma}$ splits into $\mathcal{R}^-$ and $\mathcal{R}^+$?

$$(\!|f, g|\!) + (\!|f_1, g_1|\!) = (\!|f + f_1, g + g_1|\!),$$

$$\frac{\mathrm{d}(\!|f, g|\!)}{\mathrm{d}z} = \left(\!\left|\frac{\mathrm{d}f}{\mathrm{d}z}, \frac{\mathrm{d}g}{\mathrm{d}z}\right|\!\right),$$

$$q(\!|f, g|\!) = = (\!|qf, qg|\!).$$

where, in the last expression, the analytic function q is extended holomorphically into a neighbourhood[14] of γ.[9.13] We can represent q itself as a hyperfunction by $q = (\!|q, 0|\!) = (\!|0, -q|\!)$, but there is no general product defined between two hyperfunctions. The lack of a product is not the fault of the hyperfunction approach to generalized functions. It is there with all approaches.[15] The fact that the Dirac delta function (referred to in §6.6; also see below) cannot be squared, for example, causes many quantum field theorists no end of trouble.

Some simple examples of hyperfunctional representations, in the case when $\gamma = \mathbb{R}$, and $\mathcal{R}^-$ and $\mathcal{R}^+$ are the upper and lower open complex half-planes, are the Heaviside step funtion $\theta(x)$ and the Dirac (-Heaviside) delta function $\delta(x)(= \mathrm{d}\theta(x)/\mathrm{d}x)$ (see §§6.1,6):

$$\theta(x) = \left(\!\left|\frac{1}{2\pi\mathrm{i}}\log z, \frac{1}{2\pi\mathrm{i}}\log z - 1\right|\!\right),$$

$$\delta(x) = \left(\!\left|\frac{1}{2\pi\mathrm{i}z}, \frac{1}{2\pi\mathrm{i}z}\right|\!\right),$$

where we take the branch of the logarithm for which $\log 1 = 0$. The integral of the hyperfunction $(\!|f, g|\!)$ over the entire real line can be expressed as the integral of f along a contour just below the real line minus the integral of g along a contour just above the real line (assuming these converge), both from left to right.[9.14] Note that the hyperfunction can be non-trivial even when f and g are analytic continuations of the same function.

How general are hyperfunctions? They certainly include all analytic functions. They also include discontinuous functions like $\theta(x)$ and the square wave (as our discussions above show), or other C^{-1}-functions obtained by adding such things together. In fact all C^{-1}-functions are examples of hyperfunctions. Moreover, since we can differentiate a hyperfunction to obtain another hyperfunction, and any C^{-2}-function can be obtained as the derivative of some C^{-1}-function, it follows that all C^{-2}-functions are also hyperfunctions. We have seen that this includes the

---

[9.13] There is a small subtlety here. Sort it out. *Hint*: Think carefully about the domains of definition.

[9.14] Check the standard property of the delta function that $\int q(x)\delta(x)\mathrm{d}x = q(0)$, in the case when $q(x)$ is analytic.

Dirac delta function. We can differentiate again, and then again. Indeed, any C^{-n}-function is a hyperfunction for any integer n whatever. What about the $C^{-\infty}$-functions, referred to as *distributions* (see §6.6). Yes, these also are all hyperfunctions.

The normal definition of a distribution[15] is as an element of what is called the *dual* space of the C^{∞}-smooth functions. The concept of a 'dual space' will be discussed in §12.3 (and §13.6). In fact, the dual (in an appropriate sense) of the space of C^{n}-functions is the space of C^{-2-n}-functions for any integer n, and this applies also to $n = \infty$, if we write $-2 - \infty = -\infty$ and $-2 + \infty = \infty$. Accordingly, the $C^{-\infty}$-functions are indeed dual to the C^{∞}-functions. What about the dual ($C^{-\omega}$) of the C^{ω}-functions? Indeed; with the appropriate definition of 'dual', these $C^{-\omega}$-functions are precisely the hyperfunctions!

We have come full circle. In trying to generalize the notion of 'function' as far as we can away from the apparently very restrictive notion of an 'analytic' or 'holomorphic' function—the type of function that would have made Euler happy—we have come round to the extremely general and flexible notion of a *hyperfunction*. But hyperfunctions are themselves defined, in a basically very simple way, in terms of the these very same 'Eulerian' holomorphic functions that we thought we had reluctantly abandoned. In my view, this is one of the supreme magical achievements of complex numbers.[16] If only Euler had been alive to appreciate this wondrous fact!

Notes

Section 9.1

9.1. I am using the greek letter χ ('chi') here, rather than an ordinary x, which might have seemed more natural, only because we need to distinguish this variable from the real part x of the complex number z, which will play an important part in what follows.

9.2. There is no requirement that $f(\chi)$ be real for real values of χ, that is, for the a_n,b_n, and c to be real numbers. It is perfectly legitimate to have complex functions of real variables. The condition that $f(\chi)$ be real is that α_{-n} be the complex conjugate of α_n. Complex conjugates will be discussed in §10.1.

Section 9.2

9.3. The odd-looking notational anomaly of using 'F^{-}' for the part of the series with positive powers and 'F^{+}' for the part with negative powers springs ultimately from a perhaps unfortunate sign convention that has become almost universal in the quantum-mechanical literature (see §§21.2,3 and §24.3). I apologize for this, but there is nothing that I can reasonably do about it!

9.4. It is a general principle that, for any C^{ω}-function f, defined on a real domain $\mathcal{R}$, it is possible to 'complexify' $\mathcal{R}$ to a slightly extended complex domain $\mathbb{C}\mathcal{R}$, called a 'complex thickening' of $\mathcal{R}$, containing $\mathcal{R}$ in its interior, such that f extends uniquely to a holomorphic function defined on $\mathbb{C}\mathcal{R}$.

9.5. See e.g. Bailey *et al.* (1982).

Section 9.4

9.6. On the other hand, it is usual to impose some requirement that $f(\chi)$ behaves 'reasonably' as χ tends to positive or negative infinity. This will not be of particular concern for us here and, in any case, with the approach that I am adopting, the normal requirements would be unnecessarily restrictive.

9.7. In quantum mechanics, there is also a constant quantity $\hbar$ introduced to fix the scaling of p appropriately, in relation to x (see §§21.2,11), but for the moment I am keeping things simple by taking $\hbar = 1$. In fact, $\hbar$ is Dirac's form of Planck's constant (i.e. $h/2\pi$, where h is Planck's original 'quantum of action'). The choice $\hbar = 1$ can always be made, by defining our basic units in a suitable way. See §27.10.

Section 9.5

9.8. See Bailey *et al.* (1982).

Section 9.7

9.9. See Sato (1958, 1959, 1960).

9.10. See also Bremermann (1965), although the term 'hyperfunction' is not used explicitly in this work.

9.11. Another aspect of the notion 'modulo' will be discussed in §16.1 (and compare Note 3.17).

9.12. Here 'open segment' simply refers to the fact that the actual end-points a and b are not included in γ, so that 'containing' γ does not imply the containing of a and b within $\mathcal{R}$.

9.13. This 'difference' between sets $\mathcal{R},\gamma$ is also commonly written $\mathcal{R} \setminus \gamma$.

9.14. The technical definition of 'neighbourhood of' is 'open set containing'.

9.15. For the more standard ('distribution') approach to the idea of 'generalized function', see Schwartz (1966); Friedlander (1982); Gel'fand and Shilov (1964); Trèves (1967); for an alternative proposal, useful in 'nonlinear' contexts, and which shifts the 'product non-existence problem' to a non-uniqueness problem—see Colombeau (1983, 1985) and Grosser *et al.* (2001).

9.16. There are also important interconnections between hyperfunctions and the holomorphic sheaf cohomology that will be discussed in §33.9. Such ideas play important roles in the theory of hyperfunctions on higher-dimensional surfaces, see Sato (1959, 1960) and Harvey (1966).

10
Surfaces

10.1 Complex dimensions and real dimensions

ONE of the most impressive achievements in the mathematics of the past two centuries is the development of various remarkable techniques that can handle non-flat spaces of various dimensions. It will be important for our purposes that I convey something of these ideas to the reader: for modern physics depends vitally upon them.

Up to this point, we have been considering spaces of only one dimension. The reader might well be puzzled by this remark, since the complex plane, the Riemann sphere, and various other Riemann surfaces have featured strongly in several of the previous chapters. However, in the context of holomorphic functions, these surfaces are really to be thought of as being, in essence, of only one dimension, this dimension being a *complex* dimension, as was indeed remarked upon in §8.2. The points of such a space are distinguished from one another (locally) by a single parameter, albeit a parameter that happens to be a complex number. Thus, these 'surfaces' are really to be thought of as *curves*, namely *complex* curves. Of course, one could split a complex number z into its real and imaginary parts (x, y), where $z = x + iy$, and think of x and y as being two independent real parameters. But the process of dividing a complex number up in this way is not something that belongs within the realm of holomorphic operations. So long as we are concerned only with holomorphic structures, as we have been up until now when considering our complex spaces, we must regard a single complex parameter as providing just a single dimension. This, at least, is the attitude of mind that I recommend should be adopted.

On the other hand, one may take an opposing position, namely that holomorphic operations constitute merely particular examples of more general operations, whereby x and y can, if desired, be split apart to be considered as separate independent parameters. The appropriate way of achieving this is via the notion of *complex conjugation*, which is a *non-holomorphic* operation. The complex conjugate of the complex number

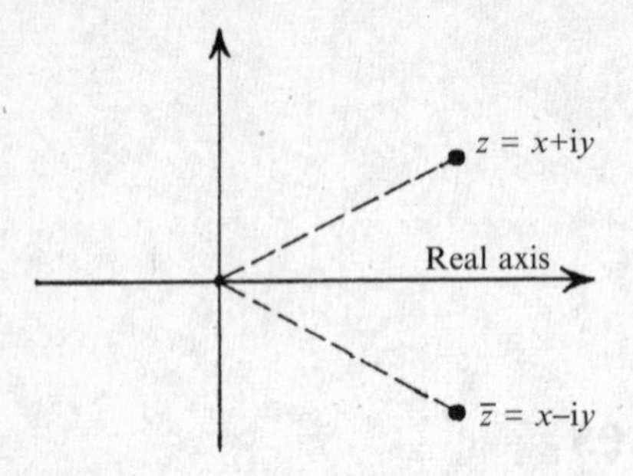

Fig. 10.1 The complex conjugate of $z = x + \mathrm{i}y$ (x, y real), is $\bar{z} = x - \mathrm{i}y$, obtained as a reflection of the z-plane in the real axis.

$z = x + \mathrm{i}y$, where x and y are real numbers, is the complex number $\bar{z}$ given by

$$\bar{z} = x - \mathrm{i}y.$$

In the complex z-plane, the operation of forming the complex conjugate of a complex number corresponds to a *reflection* of the plane in the real line (see Fig. 10.1). Recall from the discussion of §8.2 that holomorphic operations always preserve the orientation of the complex plane. If we wish to consider a conformal mapping of (a part of) the complex plane which reverses the orientation (such as turning the complex plane over on itself), then we need to include the operation of complex conjugation. But, when included with the other standard operations (adding, multiplying, taking a limit), complex conjugation also allows us to generalize our maps so that they need not be conformal at all. In fact, any map of a portion of the complex plane to another portion of the complex plane (let us say by a continuous transformation) can be achieved by bringing the operation of complex conjugation in with the other operations.

Let me elaborate on this comment. We may consider that holomorphic functions are those built up from the operations of addition and multiplication, as applied to complex numbers, together with the procedure of taking a limit (because these operations are sufficient for building up power series, an infinite sum being a limit of successive partial sums).[10.1] If we also incorporate the operation of complex conjugation, then we can generate general (say continuous) functions of x and y because we can express x and y individually by

$$x = \frac{z + \bar{z}}{2}, \qquad y = \frac{z - \bar{z}}{2\mathrm{i}}.$$

(Any continuous function of x and y can be built up from real numbers by sums, products, and limits.) I shall tend to use the notation $F(z, \bar{z})$, with $\bar{z}$ mentioned explicitly, when a non-holomorphic function of z is being considered. This serves to emphasize the fact that as soon as we move

[10.1] Explain why subtraction and division can be constructed from these.

outside the holomorphic realm, we must think of our functions as being defined on a 2-real-dimensional space, rather than on a space of a single complex dimension. Our function $F(z, \bar{z})$ can be considered, equally well, to be expressed in terms of the real and imaginary parts, x and y, of z, and we can write this function as $f(x, y)$, say. Then we have $f(x, y) = F(z, \bar{z})$, although, of course, f's explicit mathematical expression will in general be quite different from that of F. For example, if $F(z, \bar{z}) = z^2 + \bar{z}^2$, then $f(x, y) = 2x^2 - 2y^2$. As another example, we might consider $F(z, \bar{z}) = z\bar{z}$; then $f(x, y) = x^2 + y^2$, which is the square of the *modulus* $|z|$ of z, that is,[10.2]

$$z\bar{z} = |z|^2.$$

10.2 Smoothness, partial derivatives

Since, by considering functions of more than one variable, we are now beginning to venture into higher-dimensional spaces, some remarks are needed here concerning 'calculus' on such spaces. As we shall be seeing explicitly in the chapter following the next one, spaces—referred to as *manifolds*—can be of any dimension n, where n is a positive integer. (An n-dimensional manifold is often referred to simply as an n-manifold.) Einstein's general relativity uses a 4-manifold to describe spacetime, and many modern theories employ manifolds of higher dimension still. We shall explore general n-manifolds in Chapter 12, but for simplicity, in the present chapter, we just consider the situation of a real 2-manifold (or surface) $\mathcal{S}$. Then local (real) coordinates x and y can be used to label the different points of $\mathcal{S}$ (in some local region of $\mathcal{S}$). In fact, the discussion is very representative of the general n-dimensional case.

A 2-dimensional surface could, for example, be an ordinary plane or an ordinary sphere. But the surface is not to be thought of as a 'complex plane' or a 'Riemann sphere', because we shall not be concerned with assigning a structure to it as a complex space (i.e. with the attendant notion of 'holomorphic function' defined on the surface). Its only structure needs to be that of a *smooth manifold*. Geometrically, this means that we do not need to keep track of anything like a local conformal structure, as we did for our Riemann surfaces in §8.2, but we do need to be able to tell when a function defined on the space (i.e. a function whose domain is the space) is to be considered as 'smooth'.

For an intuitive notion of what a 'smooth' manifold is, think of a sphere as opposed to a cube (where, of course, in each case I am referring to the surface and not the interior). For an example of a smooth function

[10.2] Derive both of these.

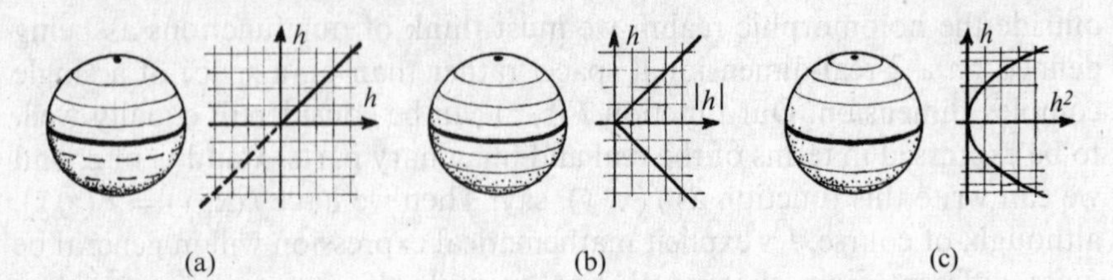

Fig. 10.2 Functions on a sphere $\mathcal{S}$, pictured as sitting in Euclidean 3-space, where h measures the distance above the equatorial plane. (a) The function h itself is smooth on $\mathcal{S}$ (negative values indicated by broken lines). (b) The modulus $|h|$ (see Fig. 6.2a) is not smooth along the equator. (c) The square h^2 is smooth all over $\mathcal{S}$.

on the sphere, we might think of a 'height function', say the distance above the equatorial plane (the sphere being pictured as sitting in ordinary Euclidean 3-space in the normal way, distances beneath the plane being counted negatively). See Fig. 10.2a. On the other hand, if our function is the *modulus* of this height function (see §6.1 and Fig. 10.2b), so that distances beneath the equator also count positively, then this function is not smooth along the equator. Yet, if we consider the *square* of the height function, then this function *is* smooth on the sphere (Fig. 10.2c). It is instructive to note that, in all these cases, the function is smooth at the north and south poles, despite the 'singular' appearance, at the poles, of the contour lines of constant height. The only instance of non-smoothness occurs in our second example, at the equator.

In order to understand what this means a little more precisely, let us introduce a system of *coordinates* on our surface $\mathcal{S}$. These coordinates need apply only locally, and we can imagine 'gluing' $\mathcal{S}$ together out of local pieces—*coordinate patches*—in a similar manner to our procedure for Riemann surfaces in §8.1. (For the sphere, for example, we do need more than one patch.) Within one patch, smooth coordinates label the different points; see Fig. 10.3. Our coordinates are to take real-number values, and let us call them x and y (without any suggestion intended that they ought to be combined together in the form of a complex number). Suppose, now,

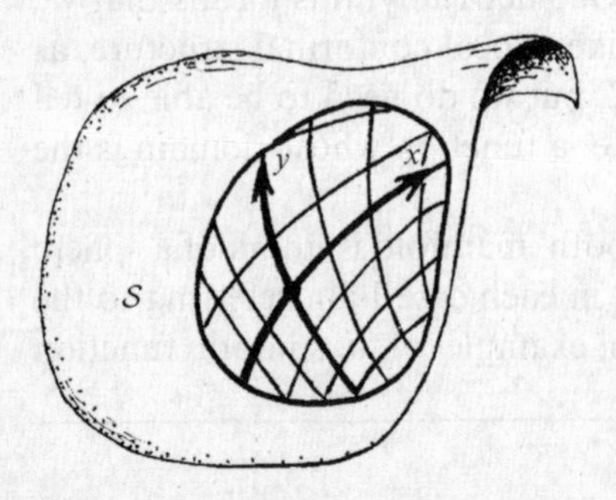

Fig. 10.3 Within one local patch, smooth (real-number) coordinates (x, y) label the points.

that we have some smooth function Φ defined on $\mathcal{S}$. In the modern mathematical terminology, Φ is a smooth *map* from $\mathcal{S}$ to the space of real numbers $\mathbb{R}$ (or complex numbers $\mathbb{C}$, in case Φ is to be a complex-valued function on $\mathcal{S}$) because Φ assigns to each point of $\mathcal{S}$ a real (or complex) number—i.e. Φ *maps* $\mathcal{S}$ to the real (or complex) numbers. Such a function is sometimes called a *scalar field* on $\mathcal{S}$. On a particular coordinate patch, the quantity Φ can be represented as a function of the two coordinates, let us say

$$\Phi = f(x, y),$$

where the *smoothness* of the quantity Φ is expressed as the *differentiability* of the function $f(x, y)$.

I have not yet explained what 'differentiability' is to mean for a function of more than one variable. Although intuitively clear, the precise definition is a little too technical for me to go into thoroughly here.[1] Some clarifying comments are nevertheless appropriate.

First of all, for f be differentiable, as a function of the pair of variables (x, y), it is certainly necessary that if we consider $f(x, y)$ in its capacity as a function of only the *one* variable x, where y is held to some constant value, then this function must be smooth (at least C^1), as a function of x, in the sense of functions of a *single* variable (see §6.3); moreover, if we consider $f(x, y)$ as a function of just the one variable y, where it is x that is now to be held constant, then it must be smooth (C^1) as a function of y. However, this is far from sufficient. There are many functions $f(x, y)$ which are separately smooth in x and in y, but for which it would be quite unreasonable to call smooth in the *pair* (x, y).[10.3] A sufficient additional requirement for smoothness is that the derivatives with respect to x and y separately are each *continuous* functions of the pair (x, y). Similar statements (of particular relevance to §4.3) would hold if we consider functions of more than two variables. We use the 'partial derivative' symbol ∂ to denote differentiation with respect to one variable, holding the other(s) fixed. The partial derivatives of $f(x, y)$ with respect to x and with respect to y, respectively, are written

---

[10.3] Consider the real function $f(x, y) = xy(x^2 + y^2)^{-N}$, in the respective cases $N = 2$, 1, and $\frac{1}{2}$. Show that in each case the function is differentiable (C^ω) with respect to x, for any fixed y-value (and that the same holds with the roles of x and y reversed). Nevertheless, f is not smooth as a function of the pair (x, y). Show this in the case $N = 2$ by demonstrating that the function is not even bounded in the neighbourhood of the origin (0, 0) (i.e. it takes arbitrarily large values there), in the case $N = 1$ by demonstrating that the function though bounded is not actually continuous as a function of (x, y), and in the case $N = \frac{1}{2}$ by showing that though the function is now continuous, it is not smooth along the line $x = y$. (*Hint*: Examine the values of each function along straight lines through the origin in the (x, y)-plane.) Some readers may find it illuminating to use a suitable 3-dimensional graph-plotting computer facility, if this is available—but this is by no means necessary.

$$\frac{\partial f}{\partial x} \text{ and } \frac{\partial f}{\partial y}.$$

(As an example, we note that if $f(x, y) = x^2 + xy^2 + y^3$, then $\partial f/\partial x = 2x + y^2$ and $\partial f/\partial y = 2xy + 3y^2$.) If these quantities exist and are continuous, then we say that Φ is a (C^1-)smooth function on the surface.

We can also consider higher orders of derivative, denoting the second partial derivative of f with respect to x and y, respectively, by

$$\frac{\partial^2 f}{\partial x^2} \text{ and } \frac{\partial^2 f}{\partial y^2}.$$

(Now we need C^2-smoothness, of course.) There is also a 'mixed' second derivative $\partial^2 f/\partial x\, \partial y$, which means $\partial(\partial f/\partial y)/\partial x$, namely the partial derivative, with respect to x, of the partial derivative of f with respect to y. We can also take this mixed derivative the other way around to get the quantity $\partial^2 f/\partial y\, \partial x$. In fact, it is a consequence of the (second) differentiability of f that these two quantities are equal:[10.4]

$$\frac{\partial^2 f}{\partial x\, \partial y} = \frac{\partial^2 f}{\partial y\, \partial x}.$$

(The full definition of C^2-smoothness, for a function of two variables, requires this.)[10.5] For higher derivatives (and higher-order smoothness), we have corresponding quantities:

$$\frac{\partial^3 f}{\partial x^3}, \quad \frac{\partial^3 f}{\partial x^2 \partial y} = \frac{\partial^3 f}{\partial x\, \partial y\, \partial x} = \frac{\partial^3 f}{\partial y\, \partial x^2}, \text{ etc.}$$

An important reason that I have been careful here to distinguish f from Φ, by using different letters (and I may be a good deal less 'careful' about this sort of thing later), is that we may want to consider a quantity Φ, defined on the surface, but expressed with respect to various different coordinate systems. The mathematical expression for the function $f(x, y)$ may well change from patch to patch, even though the value of the quantity Φ at any specific point of the surface 'covered' by those patches does not change. Most particularly, this can occur when we consider a region of overlap between different coordinate patches (see Fig. 10.4). If a second set of coordinates is denoted by (X, Y), then we have a new expression,

---

[10.4] Prove that the mixed second derivatives $\partial^2 f/\partial y \partial x$ and $\partial^2 f/\partial x \partial y$ are always equal if $f(x, y)$ is a polynomial. (*A polynomial* in x and y is an expression built up from x, y, and constants by use of addition and multiplication only.)

[10.5] Show that the mixed second derivatives of the function $f = xy(x^2 - y^2)/(x^2 + y^2)$ are unequal at the origin. Establish directly the lack of continuity in its second partial derivatives at the origin.

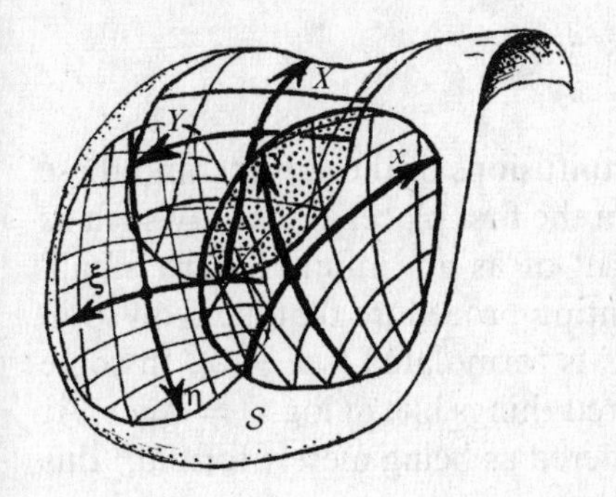

Fig. 10.4 To cover the whole of S we may have to 'glue' together several coordinate patches. A smooth function Φ on S would have a coordinate expression $\Phi = f(x, y)$ on one patch and $\Phi = F(X, Y)$ on another (with respective local coordinates (x, y), (X, Y)). On an overlap region $f(x, y) = F(X, Y)$, where X and Y are smooth functions of x and y.

$$\Phi = F(X, Y),$$

for the values of Φ on the new coordinate patch. On an overlap region between the two patches, we shall therefore have

$$F(X, Y) = f(x, y),$$

But, as indicated above, the particular expression that F represents, in terms of the quantities X and Y, will generally be quite different from the expression that f represents in terms of x and y. Indeed, X might be some complicated function of x and y on the overlap region and so might Y, and these functions would have to be incorporated in the passage from f to F.[10.6] Such functions, representing the coordinates of one system in terms of the coordinates of the other,

$$X = X(x, y) \quad \text{and} \quad Y = Y(x, y)$$

and their inverses

$$x = x(X, Y) \quad \text{and} \quad y = y(X, Y)$$

are called the *transition functions* that express the coordinate change from one patch to the other. These transition functions are to be smooth—let us, for simplicity, say C^∞-smooth—and this has the consequence that the 'smoothness' notion for the quantity Φ is independent of the choice of coordinates that are used in some patch overlap.

10.3 Vector fields and 1-forms

There is a notion of 'derivative' of a function that is independent of the coordinate choice. A standard notation for this, as applied to the function Φ defined on S, is $d\Phi$, where

[10.6] Find the form of $F(X,Y)$ explicitly when $f(x,y) = x^3 - y^3$, where $X = x - y$, $Y = xy$. *Hint*: What is $x^2 + xy + y^2$ in terms of X and Y; what does this have to do with f?

$$\mathrm{d}\Phi = \frac{\partial f}{\partial x}\mathrm{d}x + \frac{\partial f}{\partial y}\mathrm{d}y.$$

Here we begin to run into some of the confusions of the subject, and these take some while to get accustomed to. In the first place, a quantity such as '$\mathrm{d}\Phi$' or '$\mathrm{d}x$' initially tends to be thought of as an 'infinitesimally small' quantity, arising when we apply the limiting procedure that is involved in the calculus when the derivative '$\mathrm{d}y/\mathrm{d}x$' is formulated (see §6.2). In some of the expressions in §6.5, I also considered things like $\mathrm{d}(\log x) = \mathrm{d}x/x$. At that stage, these expressions were considered as being merely formal,[2] this last expression being thought of as just a convenient way ('multiplying through by $\mathrm{d}x$') of representing the 'more correct' expression $\mathrm{d}(\log x)/\mathrm{d}x = 1/x$. When I write '$\mathrm{d}\Phi$' in the displayed formula above, on the other hand, I mean a certain kind of geometrical entity that is called a *1-form* (although this is not the most general type of 1-form; see §10.4 below and §12.6), and this works for things like $\mathrm{d}(\log x) = \mathrm{d}x/x$, too. A 1-form is not an 'infinitesimal'; it has a somewhat different kind of interpretation, a type of interpretation that has grown in importance over the years, and I shall be coming to this in a moment. Remarkably, however, despite this significant change of interpretation of 'd', the formal mathematical expressions (such as those of §6.5)—provided that we do not try to divide by things like $\mathrm{d}x$—are not changed at all.

There is also another issue of potential confusion in the above displayed formula, which arises from the fact that I have used Φ on the left-hand side and f on the right. I did this mainly because of the warnings about the distinction between Φ and f that I issued above. The quantity Φ is a function whose domain is the manifold $\mathcal{S}$, whereas the domain of f is some (open) region in the (x, y)-plane that refers to a particular coordinate patch. If I am to apply the notion of 'partial derivative with respect to x', then I need to know what it means 'to hold the remaining variable y constant'. It is for this reason that f is used on the right, rather than Φ, because f 'knows' what the coordinates x and y are, whereas Φ doesn't. Even so, there is a confusion in this displayed formula, because the arguments of the functions are not mentioned. The Φ on the left is applied to a particular point p of the 2-manifold $\mathcal{S}$, while f is applied to the particular coordinate values (x, y) that the coordinate system assigns to the point p. Strictly speaking, this would have to be made explicit in order that the expression makes sense. However, it is a nuisance to have to keep saying this kind of thing, and it would be much more convenient to be able to write this formula as

$$\mathrm{d}\Phi = \frac{\partial \Phi}{\partial x}\mathrm{d}x + \frac{\partial \Phi}{\partial y}\mathrm{d}y,$$

or, in 'disembodied' operator form,

$$\mathrm{d} = \mathrm{d}x\frac{\partial}{\partial x} + \mathrm{d}y\frac{\partial}{\partial y}.$$

Indeed, I am going to try to make sense of these things. These formulae are instances of something referred to as the *chain rule*. As stated, they require meanings to be assigned to things like '$\partial\Phi/\partial x$' when Φ is some function defined on $\mathcal{S}$.

How are we to think of an operator, such as $\partial/\partial x$, as something that can be applied to a function, like Φ, that is defined on the manifold $\mathcal{S}$, rather than just to a function of the variables x and y? Let us first try to see what $\partial/\partial x$ means when we refer things to some other coordinate system (X, Y). The appropriate 'chain rule' formula now turns out to be

$$\frac{\partial}{\partial x} = \frac{\partial X}{\partial x}\frac{\partial}{\partial X} + \frac{\partial Y}{\partial x}\frac{\partial}{\partial Y}.$$

Thus, in terms of the (X, Y) system, we now have the more complicated-looking expression $(\partial X/\partial x)\partial/\partial X + (\partial Y/\partial x)\partial/\partial Y$ to represent exactly the same operation as the simple-looking $\partial/\partial x$ represents in the (x, y) system. This more complicated expression is a quantity $\boldsymbol{\xi}$, of the form

$$\boldsymbol{\xi} = A\frac{\partial}{\partial X} + B\frac{\partial}{\partial Y},$$

where A and B are (C^∞-) smooth functions of X and Y. In the particular case just given, with $\boldsymbol{\xi}$ representing $\partial/\partial x$ in the (x, y) system, we have $A = \partial X/\partial x$ and $B = \partial Y/\partial x$. But we can consider more general such quantities $\boldsymbol{\xi}$ for which A and B do not have these particular forms. Such a quantity $\boldsymbol{\xi}$ is called a *vector field* on $\mathcal{S}$ (in the (X, Y)-coordinate patch). We can rewrite $\boldsymbol{\xi}$ in the original (x, y) system, and find that $\boldsymbol{\xi}$ has just the same general form as in the (X, Y) system:

$$\boldsymbol{\xi} = a\frac{\partial}{\partial x} + b\frac{\partial}{\partial y}$$

(although the functions a and b are generally quite different from A and B).[10.7] This enables us to extend the vector field from the (X, Y)-patch to an overlapping (x, y)-patch. In this way, taking as many patches as we need, we can envisage extending the vector field $\boldsymbol{\xi}$ to the whole of $\mathcal{S}$.

All this has probably caused the reader great confusion! However, my purpose is not to confuse, but to find the right analytical form of a very basic geometrical notion. The differential operator $\boldsymbol{\xi}$, which we have called a 'vector field', with its (consequent) very specific way of transforming, as we pass from patch to patch, has a clear geometrical interpretation, as

[10.7] Find A and B in terms of a and b; by analogy, write down a and b in terms of A and B.

Fig. 10.5 The geometrical interpretation of a vector field $\boldsymbol{\xi}$ as a 'field of arrows' drawn on $\mathcal{S}$.

illustrated in Fig. 10.5. We are to visualize $\boldsymbol{\xi}$ as describing a 'field of little arrows' drawn on $\mathcal{S}$, although, at some places on $\mathcal{S}$, an arrow may shrink to a point, these being the places where $\boldsymbol{\xi}$ takes the value zero. (To get a good picture of a vector field, think of wind-flow charts on TV weather bulletins.) The arrows represent the directions in which the function upon which $\boldsymbol{\xi}$ acts is to be differentiated. Taking this function to be Φ, the action of $\boldsymbol{\xi}$ on Φ, namely $\boldsymbol{\xi}(\Phi) = a\,\partial\Phi/\partial x + b\,\partial\Phi/\partial y$, measures the rate of increase of Φ in the direction of the arrows; see Fig. 10.6. Also, the magnitude ('length') of the arrow has significance in determining the 'scale', in terms of which this increase is to be measured. A longer arrow gives a correspondingly greater measure of the rate of increase. More appropriately,

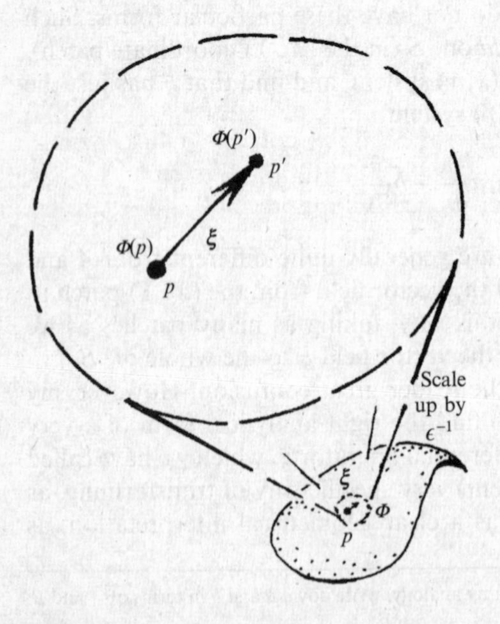

Fig. 10.6 The action of $\boldsymbol{\xi}$ on a scalar field Φ gives its rate of increase along the $\boldsymbol{\xi}$-arrows. Think of the arrows as *infinitesimal*, each connecting a point p of $\mathcal{S}$ ('tail' of the arrow) to a 'neighbouring' point p' of $\mathcal{S}$ ('head' of the arrow), pictured by applying a large magnification (by a factor ϵ^{-1}, where ϵ is small) to the neighbourhood of p. The difference $\Phi(p') - \Phi(p)$, divided by ϵ, is (in the limit $\epsilon \to 0$) the gradient $\boldsymbol{\xi}(\Phi)$ of Φ along $\boldsymbol{\xi}$.

we ought to think of all the arrows as being infinitesimal, each one connecting a point p of $\mathcal{S}$ (at the 'tail' of the arrow) with a 'neighbouring' point p' of $\mathcal{S}$ (at the 'head' of the arrow). To make this just a little more explicit, let us choose some small positive number ϵ as a measure of the separation, along the direction of $\boldsymbol{\xi}$, between two separate points p and p'. Then the difference $\Phi(p') - \Phi(p)$, divided by ϵ, gives us an approximation to the quantity $\boldsymbol{\xi}(\Phi)$. The smaller we choose ϵ to be, the better approximation we get. Finally, in the limit when p' approaches p (so $\epsilon \to 0$), we actually obtain $\boldsymbol{\xi}(\Phi)$, sometimes called the *gradient* (or slope) of Φ in the direction of $\boldsymbol{\xi}$.

In the particular case of the vector field $\partial/\partial x$, the arrows all point along the coordinate lines of constant y. This illustrates an issue that frequently leads to confusion with the standard mathematical notation '$\partial/\partial x$' for partial derivative. One might have thought that the expression '$\partial/\partial x$' referred most specifically to the quantity x. However, in a clear sense, it has more to do with the variable(s) that are not explicitly mentioned, here the variable y, than it has to do with x. The notation is particularly treacherous when one considers a change of coordinate variables, say from (x, y) to (X, Y), in which one of the coordinates remains the same. Consider, for example the very simple coordinate change

$$X = x, \quad Y = y + x.$$

Then we find[10.8]

$$\frac{\partial}{\partial X} = \frac{\partial}{\partial x} - \frac{\partial}{\partial y}, \quad \frac{\partial}{\partial Y} = \frac{\partial}{\partial y}.$$

Thus, we see that $\partial/\partial X$ is different from $\partial/\partial x$, despite the fact that X is the same as x—whereas, in this case, $\partial/\partial Y$ is the same as $\partial/\partial y$, even though Y differs from y. This is an instance of what my colleague Nick Woodhouse refers to as 'the second fundamental confusion of calculus'![3] It is *geometrically* clear, on the other hand, why $\partial/\partial X \neq \partial/\partial x$, since the corresponding 'arrows' point along different coordinate lines (Fig. 10.7).

We are now in a position to interpret the quantity $\mathrm{d}\Phi$. This is called the *gradient* (or exterior derivative) of Φ, and it carries the information of how Φ is varying in all possible directions along $\mathcal{S}$. A good geometrical way to think of $\mathrm{d}\Phi$ is in terms of a system of *contour lines* on $\mathcal{S}$. See Fig. 10.8a. We can think of $\mathcal{S}$ as being like an ordinary map (where by 'map' here I mean the thing made of stiff paper that you take with you when you go hiking, not the mathematical notion of 'map'), which might

[10.8] Derive this explicitly. *Hint*: You may use 'chain rule' expressions for $\partial/\partial X$ and $\partial/\partial Y$ that are the exact analogies of the expression for $\partial/\partial x$ that was displayed earlier.

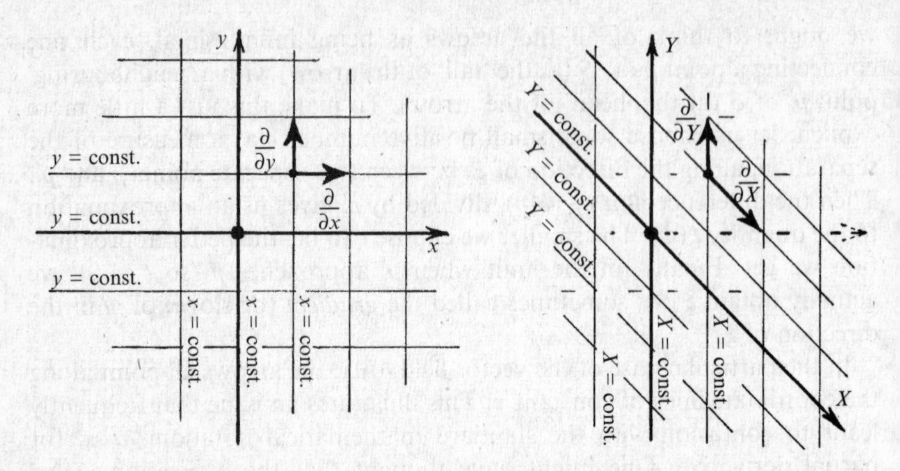

Fig. 10.7 Second fundamental confusion of calculus is illustrated: $\partial/\partial X \neq \partial/\partial x$ despite $X = x$, and $\partial/\partial Y = \partial/\partial y$ despite $Y \neq y$, for the coordinate change $X = x$, $Y = y + x$. The interpretation of partial differential operators as 'arrows' pointing along coordinate lines clarifies the geometry ($x =$ const. agree with $X =$ const., but $y =$ const. disagree with $Y =$ const.).

be a spherical globe, if we want to take into account that S might be a curved manifold. The function Φ might represent the height of the ground above sea level. Then $\mathrm{d}\Phi$ represents the slope of the ground as compared with the horizontal. The contour lines trace out places of equal height. At any one point p of S, the direction of the contour line tells us the direction along which the gradient vanishes (the 'axis of tilt' of the slope of the ground), so this is the direction of the arrow $\boldsymbol{\xi}$ at p for which $\boldsymbol{\xi}(\Phi) = 0$. We neither climb nor descend, when we follow a contour line. But if we cut across contour lines, then there will be an increase or decrease in Φ, and the rate at which this occurs, namely $\boldsymbol{\xi}(\Phi)$, will be measured by the crowding of the contour lines in the direction that we cross them. See Fig. 10.8b.

10.4 Components, scalar products

According to the expression

$$\boldsymbol{\xi} = a\frac{\partial}{\partial x} + b\frac{\partial}{\partial y},$$

the vector field $\boldsymbol{\xi}$ may be thought of as being composed of two parts, one being proportional to $\partial/\partial x$, which points along the lines of constant y, and the other, proportional to $\partial/\partial y$, which points along the lines of constant x.

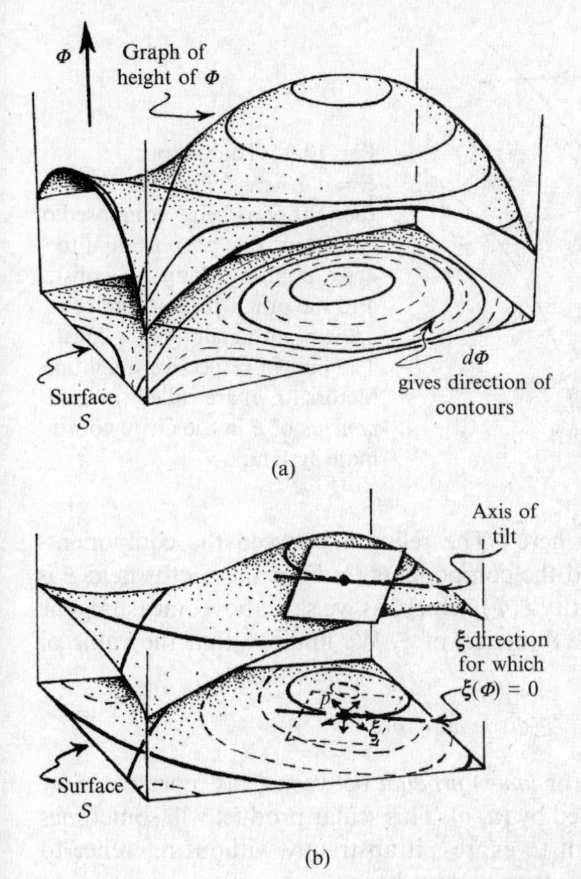

Fig. 10.8 We can geometrically picture the full gradient (exterior derivative) $\mathrm{d}\Phi$ of a scalar Φ in terms of a system of contour lines on $\mathcal{S}$. (a) The value Φ is here plotted vertically above $\mathcal{S}$, so the contour lines on $\mathcal{S}$ (constant Φ) describe constant height. (b) At any one point p of $\mathcal{S}$, the direction of the contour line tells us the direction along which the gradient vanishes (the 'axis of tilt' of the slope of the hill), i.e. the direction of the arrows $\boldsymbol{\xi}$ at p for which $\boldsymbol{\xi}(\Phi) = 0$. Cutting across contour lines gives an increase or decrease in Φ, $\boldsymbol{\xi}(\Phi)$ measuring the crowding of the lines in the direction of $\boldsymbol{\xi}$.

Thus, in the (x, y)-coordinate system, the pair of respective weighting factors (a, b) may be used to label $\boldsymbol{\xi}$. The numbers a and b are referred to as the *components* of $\boldsymbol{\xi}$ in this coordinate system; see Fig. 10.9. (Strictly speaking, the two 'components' of $\boldsymbol{\xi}$ would actually be the two vector fields $a\,\partial/\partial x$ and $b\,\partial/\partial y$ themselves, of which the vector field $\boldsymbol{\xi}$ is composed, as displayed in Fig. 10.9—and a similar remark would apply to the components of $\mathrm{d}\Phi$, below. However, the term 'component' has now acquired this meaning of 'coordinate label' in much mathematical literature, particularly in connection with the tensor calculus; see §12.8.)

Similarly, the quantity $\mathrm{d}\Phi$ (a '1-form') is composed of the two parts $\mathrm{d}x$ and $\mathrm{d}y$, according to the expression

$$\mathrm{d}\Phi = u\,\mathrm{d}x + v\,\mathrm{d}y$$

and so (u, v) may be used to label $\mathrm{d}\Phi$, and the numbers u and v are the *components* of $\mathrm{d}\Phi$ in this same coordinate system. (In fact, we have

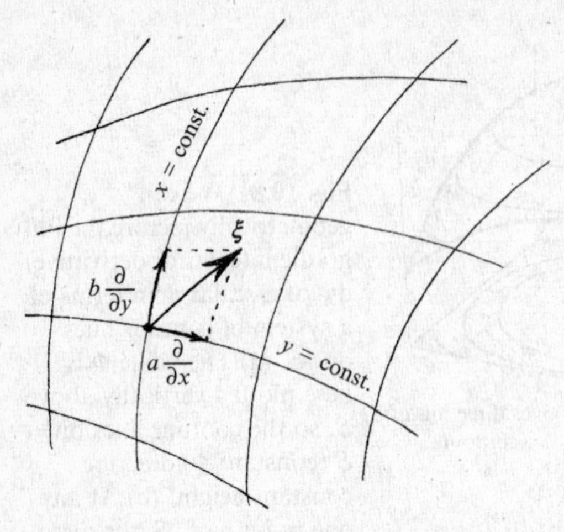

Fig. 10.9 The vector $\boldsymbol{\xi} = a\,\partial/\partial x + b\,\partial/\partial y$ may be thought of as being composed of two parts, one proportional to $\partial/\partial x$, pointing along $y = \text{const.}$, and the other, proportional to $\partial/\partial y$, pointing along $x = \text{const.}$ The pair of respective weighting factors (a, b) are called the *components* of $\boldsymbol{\xi}$ in the (x, y)-coordinate system.

$u = \partial\Phi/\partial x$ and $v = \partial\Phi/\partial y$ here.) The relation between the components (u, v) of the 1-form $\mathrm{d}\Phi$ and the components (a, b) of the vector field $\boldsymbol{\xi}$ is obtained through the quantity $\boldsymbol{\xi}(\Phi)$, which, as we saw above, measures the rate of increase of Φ in the direction of $\boldsymbol{\xi}$. We find[10.9] that the value of $\boldsymbol{\xi}(\Phi)$ is given by

$$\boldsymbol{\xi}(\Phi) = au + bv.$$

We call $au + bv$ the *scalar* (or *inner*) *product* between $\boldsymbol{\xi}$, as represented by (a, b), and $\mathrm{d}\Phi$, as represented by (u, v). This scalar product will sometimes be written $\mathrm{d}\Phi \bullet \boldsymbol{\xi}$ if we want to express it abstractly without reference to any particular coordinate system, and we have

$$\mathrm{d}\Phi \bullet \boldsymbol{\xi} = \boldsymbol{\xi}(\Phi).$$

The reason for having two different notations for the same thing, here, is that the operation expressed in $\mathrm{d}\Phi \bullet \boldsymbol{\xi}$ also applies to more general kinds of 1-form than those that can be expressed as $\mathrm{d}\Phi$ (see §12.3). If $\boldsymbol{\eta}$ is such a 1-form, then it has a scalar product with any vector field $\boldsymbol{\xi}$, which is written as $\boldsymbol{\eta} \bullet \boldsymbol{\xi}$.

In fact the definition of a 1-form is essentially that it is a quantity that can be combined with a vector field to form a 'scalar product' in this way. Thus, the fact that the quantity $\mathrm{d}\Phi$ is something that naturally forms a scalar product with vector fields is actually what characterizes it as a 1-form. (A 1-form is sometimes called a *covector*, depending on the context.) Technically, 1-forms (covectors) are *dual* to vector fields in this sense. This notion of a 'dual' object will be explored more fully in §12.3, where we shall see that

[10.9] Show this explicitly, using 'chain rule' expressions that we have seen earlier.

these ideas apply quite generally within a 'surface' of higher dimension (i.e. to an n-manifold). The geometrical meaning of a 1-form will also be filled out more fully in §§12.3–5, in the context of higher dimensions. For the moment, the family of contour lines itself will do, these lines representing the directions along which a $\boldsymbol{\xi}$-arrow must point if $\mathrm{d}\Phi \bullet \boldsymbol{\xi} = 0$ (i.e. if $\boldsymbol{\xi}(\Phi) = 0$).

10.5 The Cauchy–Riemann equations

But before making this leap to higher dimensions, which we shall be preparing ourselves for in the next chapter, let us return to the issue that we started with in this chapter: the property of a 2-dimensional surface that is needed in order that it can be reinterpreted as a complex 1-manifold. Essentially what is required is that we have a means of characterizing those complex-valued functions Φ which are holomorphic. The condition of holomorphicity is a local one, so that we can recognize it as something holding in each coordinate patch, and consistently on the overlaps between patches. On the (x, y)-patch, we require that Φ be holomorphic in the complex number $z = x + \mathrm{i}y$; on an overlapping (X, Y)-patch, holomorphic in $Z = X + \mathrm{i}Y$. The consistency between the two is ensured by the requirement that Z is a holomorphic function of z on the overlap and vice versa. (If Φ is holomorphic in z, and z is holomorphic in Z, then Φ must be holomorphic in Z, since a holomorphic function of a holomorphic function is again a holomorphic function.[10.10])

Now, how do we express the condition that Φ is holomorphic in z, in terms of the real and imaginary parts of Φ and z? These are the famous *Cauchy–Riemann* equations referred to in §7.1. But what are these equations explicitly? We can imagine Φ to be expressed as a function of z and $\bar{z}$ (since, as we saw at the beginning of this chapter, the real and imaginary parts of z, namely x and y, can be re-expressed in terms of z and $\bar{z}$ by using the expressions $x = (z + \bar{z})/2$ and $y = (z - \bar{z})/2\mathrm{i}$). We are required to express the condition that, in effect, Φ 'depends only on z' (i.e. that it is 'independent of $\bar{z}$').

What does this mean? Imagine that, instead of the complex conjugate pair of variables z and $\bar{z}$, we had a pair of independent real variables u and v, say, and we wished to express the fact that some quantity Ψ that is a function of u and v is in fact independent of v. This independence can be stated as

$$\frac{\partial \Psi}{\partial v} = 0$$

[10.10] Explain this from three different points of view: (a) intuitively, from general principles (how could a $\bar{z}$ appear?), (b) using the geometry of holomorphic maps described in §8.2, and (c) explicitly, using the chain rule and the Cauchy–Riemann equations that we are about to come to.

(because this equation tells us that, for each value of u, the quantity Ψ is constant in v; so Ψ is dependent only on u).[4] Accordingly, Φ being 'independent of $\bar{z}$' ought to be expressed as

$$\frac{\partial \Phi}{\partial \bar{z}} = 0,$$

and this does indeed express the holomorphicity of Φ (although the 'argument by analogy' that I have just given should not be taken as a proof of this fact)[5]. Using the chain rule, we can re-express this equation[10.11] in terms of partial derivatives in the (x, y)-system:

$$\frac{\partial \Phi}{\partial x} + \mathrm{i}\frac{\partial \Phi}{\partial y} = 0.$$

Writing Φ in terms of its real and imaginary parts,

$$\Phi = \alpha + \mathrm{i}\beta,$$

with α and β real, we obtain the *Cauchy–Riemann* equations[6,10.12]

$$\frac{\partial \alpha}{\partial x} = \frac{\partial \beta}{\partial y}, \qquad \frac{\partial \alpha}{\partial y} = -\frac{\partial \beta}{\partial x}.$$

Since, as remarked earlier, on an overlap between an (x, y)-coordinate patch and an (X, Y)-coordinate patch we require $Z = X + \mathrm{i}Y$ to be holomorphic in $z = x + \mathrm{i}y$, we also have the Cauchy–Riemann equations holding between (x, y) and (X, Y):

$$\frac{\partial X}{\partial x} = \frac{\partial Y}{\partial y}, \qquad \frac{\partial X}{\partial y} = -\frac{\partial Y}{\partial x}.$$

If this condition holds between any pair of coordinate patches, then we have assembled a Riemann surface $\mathcal{S}$. (These are the required analytic conditions that I skated over in §7.1.) Recall that such a surface can also be thought of as a complex 1-manifold. But, according to the present 'Cauchy–Riemann' way of looking at things, we think of $\mathcal{S}$ as being a real 2-manifold with the particular type of structure (namely that determined by the Cauchy–Riemann equations).

Whereas there is a certain 'purity' in trying to stick entirely to holomorphic operations (a philosophical perspective that will have importance for us later, in Chapter 33 and in §34.8) and in thinking of $\mathcal{S}$ as a 'curve', this alternative 'Cauchy–Riemann' standpoint is a powerful one in a

[10.11] Do this.

[10.12] Give a more direct derivation of the Cauchy–Riemann equations, from the definition of a derivative.

number of other contexts. For example, it allows us to prove results by appealing to many useful techniques in the *existence theory* of partial differential equations. Let me try to give a taste of this by appealing to an (important) example.

If the Cauchy–Riemann equations $\partial\alpha/\partial x = \partial\beta/\partial y$ and $\partial\alpha/\partial y = -\partial\beta/\partial x$ hold, then the quantities α and β each individually turn out to satisfy a particular equation (Laplace's equation). For we have[10.13]

$$\nabla^2\alpha = 0, \qquad \nabla^2\beta = 0,$$

where the second-order differential operator ∇^2, called the (2-dimensional) *Laplacian*, is defined by

$$\nabla^2 = \frac{\partial^2}{\partial x^2} + \frac{\partial^2}{\partial y^2}.$$

The Laplacian is important in many physical situations (see §21.2, §22.11, §§24.3–6). For example, if we have a soap film spanning a wire loop which deviates very slightly up and down from a horizontal plane, then the height of the film above the horizontal will be a solution of Laplace's equation (to a close approximation which gets better and better the smaller is this vertical deviation).[7] See Fig. 10.10. Laplace's equation (in three dimensions) also has a fundamental role to play in Newtonian gravitational theory (and in electrostatics; see Chapters 17 and 19) since it is the equation satisfied by a potential function determining the gravitational (or static electric) field in free space.

Solutions of the Cauchy–Riemann equations can be obtained from solutions of the 2-dimensional Laplace equation in a rather direct way. If we have any α satisfying $\nabla^2\alpha = 0$, then we can construct β by $\beta = \int (\partial\alpha/\partial x)\, dy$;

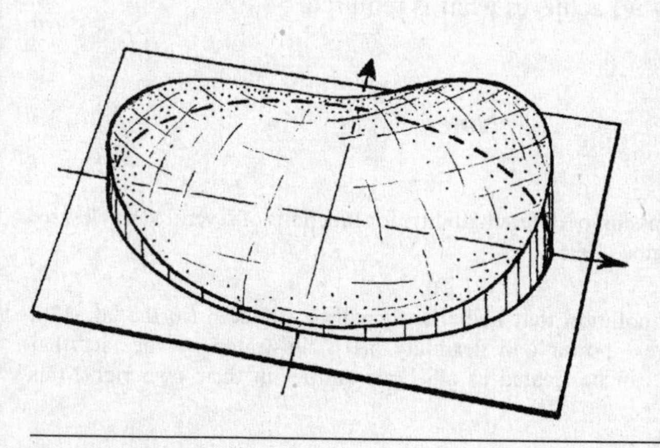

Fig. 10.10 A soap film spanning a wire loop which deviates only very slightly up and down from a horizontal plane. The height of the film above the horizontal gives a solution of Laplace's equation (to an approximation which gets better the smaller the vertical deviation).

[10.13] Show this.

we then find that both Cauchy–Riemann equations are consequently satisfied.[10.14] This fact can be used to demonstrate and illuminate some of the assertions made at the end of the previous chapter.

In particular, let us consider the remarkable fact, asserted at the end of §9.7, that any continuous function f defined on the unit circle in the complex plane can be represented as a hyperfunction. This assertion effectively states that any continuous f is the sum of two parts, one of which extends holomorphically into the interior of the unit circle and the other of which extends holomorphically into the exterior, where we now think of the complex plane completed to the Riemann sphere. This assertion is effectively equivalent (according to the discussion of §9.2) to the existence of a Fourier series representation of f, where f is regarded as a periodic function of a real variable. For simplicity, assume that f is real-valued. (The complex case follows by splitting f into real and imaginary parts.) Now, there are theorems that tell us that we can extend f continuously into the interior of the circle, where f satisfies $\nabla^2 f = 0$ inside the circle. (This fact is intuitively very plausible, because of the soap-film argument given above; see Fig. 10.10. Scaling f down appropriately to a new function ϵf, for some fixed small ϵ, we can imagine that our wire loop lies at the unit circle in the complex plane, deviating slightly[8] up and down vertically from it by the values of ϵf on the unit circle. The height of the spanning soap film provides ϵf and therefore f inside.) By the above prescription ($g = \int (\partial f/\partial x)\mathrm{d}y$), we can supply an imaginary part g to f, so that $f + \mathrm{i}g$ is holomorphic throughout the interior of the unit circle. This procedure also supplies an imaginary part g to f *on* the unit circle (generally in the form of a hyperfunction), so that $f + \mathrm{i}g$ is of negative frequency. We now repeat the procedure, applying it to the exterior of the unit circle (thought of as lying in the Riemann sphere), and find that $f - \mathrm{i}g$ extends there and is of positive frequency. The splitting $f = \frac{1}{2}(f + \mathrm{i}g) + \frac{1}{2}(f - \mathrm{i}g)$ achieves what is required.

Notes

Section 10.2

10.1 For a detailed discussion of differentiability, for functions of several variables, see Marsden and Tromba (1996).

Section 10.3

10.2 Although the 'dx' notation that Leibniz originally introduced (in the late 17th century) shows great power and flexibility, as is illustrated by the fact that quantities like dx can be treated as algebraic entities in their own right, this

[10.14] Show this.

does not extend to his 'd^2x' notation for second derivatives. Had he used a modification of this notation in which the second derivative of y with respect to x were written $(d^2y - d^2x\, dy/dx)/dx^2$ instead, then the quantity 'd^2x' would indeed behave in a consistent algebraic way (where 'dx^2' denotes $dxdx$, etc.). It is not clear how practical this would have been, owing to the complication of this expression, however.

10.3 The 'first fundamental confusion' has to do with the confusion between the use of f and Φ that we encountered in §10.2, particularly in relation to the taking of partial derivatives. See Woodhouse (1991).

Section 10.5

10.4 We must take this condition in a local sense only. For example, we can have a smooth function $\Phi(u, v)$ defined on a kidney-shaped region in the (u, v)-plane, within which $\partial\Phi/\partial v = 0$, but for which Φ is not fully consistent as a function of u.[10.15]

10.5 Although not the most rigorous route to the Cauchy–Riemann equations, this argument provides the underlying *reason* for their form.

10.6 In fact, Jean LeRond D'Alembert found these equations in 1752, long before Cauchy or Riemann (see Struik 1954, p. 219).

10.7 It turns out that the actual soap-film equation (to which the Laplace equation is an approximation) has a remarkable general solution, found by Weierstrass (1866), in terms of free holomorphic functions.

10.8 Since f is continuous on the circle, it must be *bounded* (i.e. its values lie between a fixed lower value and a fixed upper value). This follows from standard theorems, the circle being a compact space. (See §12.6 for the notion of 'compact' and Kahn 1995; Frankel 2001). We can then rescale f (multiplying it by a small constant ϵ), so that the upper and lower bounds are both very tiny. The soap film analogy then provides a reasonable plausibility argument for the existence of ϵf extended inside the circle, satisfying the Laplace equation. It is not a proof of course; see Strauss (1992) or Brown and Churchill (2004) for a more rigorous solution to this so-called, 'Dirichlet problem for a disc'.

[10.15] Spell this out in the case $\Phi(u, v) = \theta(v)h(u)$, where the functions θ and h are defined as in §§6.1,3. The kidney-shaped region must avoid the non-negative u-axis.

11
Hypercomplex numbers

11.1 The algebra of quaternions

How do we generalize all this to higher dimensions? I shall describe the standard (modern) procedure for studying n-manifolds in the next chapter, but it will be illuminating, for various other reasons, if I first acquaint the reader with certain earlier ideas aimed at the study of higher dimensions. These earlier ideas have acquired important direct relevance to some current activities in theoretical physics.

The beauty and power of complex analysis, such as with the above-mentioned property whereby solutions of the 2-dimensional Laplace equation—an equation of considerable physical importance—can be very simply represented in terms of holomorphic functions, led 19th-century mathematicians to seek 'generalized complex numbers', which could apply in a natural way to 3-dimensional space. The renowned Irish mathematician William Rowan Hamilton (1805–1865) was one who puzzled long and deeply over this matter. Eventually, on the 16 October 1843, while on a walk with his wife along the Royal Canal in Dublin, the answer came to him, and he was so excited by this discovery that he immediately carved his fundamental equations

$$\mathbf{i}^2 = \mathbf{j}^2 = \mathbf{k}^2 = \mathbf{ijk} = -1$$

on a stone of Dublin's Brougham Bridge.

Each of the three quantities **i**, **j**, and **k** is an independent 'square root of −1' (like the single i of complex numbers) and the general combination

$$q = t + u\mathbf{i} + v\mathbf{j} + w\mathbf{k},$$

where t, u, v, and w are real numbers, defines the general *quaternion*. These quantities satisfy all the normal laws of algebra bar one. The exception—and this was the true novelty[1] of Hamilton's entities—was the violation of the *commutative* law of multiplication. For Hamilton found that[11.1]

[11.1] Prove these directly from Hamilton's 'Brougham Bridge equations', assuming only the associative law $a(bc) = (ab)c$.

$$\mathbf{ij} = -\mathbf{ji}, \quad \mathbf{jk} = -\mathbf{kj}, \quad \mathbf{ki} = -\mathbf{ik},$$

which is in gross violation of the standard commutative law: $\boldsymbol{ab} = \boldsymbol{ba}$.

Quaternions still satisfy the commutative and associative laws of addition, the associative law of multiplication, and the distributive laws of multiplication over addition,[11.2] namely

$$\boldsymbol{a} + \boldsymbol{b} = \boldsymbol{b} + \boldsymbol{a},$$
$$\boldsymbol{a} + (\boldsymbol{b} + \boldsymbol{c}) = (\boldsymbol{a} + \boldsymbol{b}) + \boldsymbol{c},$$
$$\boldsymbol{a}(\boldsymbol{bc}) = (\boldsymbol{ab})\boldsymbol{c},$$
$$\boldsymbol{a}(\boldsymbol{b} + \boldsymbol{c}) = \boldsymbol{ab} + \boldsymbol{ac},$$
$$(\boldsymbol{a} + \boldsymbol{b})\boldsymbol{c} = \boldsymbol{ac} + \boldsymbol{bc},$$

together with the existence of additive and multiplicative 'identity elements' 0 and 1, such that

$$\boldsymbol{a} + 0 = \boldsymbol{a}, \quad 1\boldsymbol{a} = \boldsymbol{a}1 = \boldsymbol{a}.$$

These relations, if we exclude the last one, define what algebraists call a *ring*. (To my mind, the term 'ring' is totally non-intuitive—as is much of the terminology of abstract algebra—and I have no idea of its origins.) If we do include the last relation, we get what is called a *ring with identity*.

Quaternions also provide an example of what is called a *vector space* over the real numbers. In a vector space, we can add two elements (vectors[2]), $\boldsymbol{\xi}$ and $\boldsymbol{\eta}$, to form their sum $\boldsymbol{\xi} + \boldsymbol{\eta}$, where this sum is subject to commutativity and associativity

$$\boldsymbol{\xi} + \boldsymbol{\eta} = \boldsymbol{\eta} + \boldsymbol{\xi},$$
$$(\boldsymbol{\xi} + \boldsymbol{\eta}) + \boldsymbol{\zeta} = \boldsymbol{\xi} + (\boldsymbol{\eta} + \boldsymbol{\zeta}),$$

and we can multiply vectors by 'scalars' (here, just the real numbers f and g), where the following distributive and associative properties, etc., hold:

$$(f + g)\boldsymbol{\xi} = f\boldsymbol{\xi} + g\boldsymbol{\xi},$$
$$f(\boldsymbol{\xi} + \boldsymbol{\eta}) = f\boldsymbol{\xi} + f\boldsymbol{\eta},$$
$$f(g\boldsymbol{\xi}) = (fg)\boldsymbol{\xi},$$
$$1\boldsymbol{\xi} = \boldsymbol{\xi}.$$

Quaternions form a *4-dimensional* vector space over the reals, because there are just four independent 'basis' quantities 1, **i**, **j**, **k** that span the entire space of quaternions; that is, any quaternion can be expressed uniquely as a sum of real multiples of these basis elements. We shall be seeing many other examples of vector spaces later.

[11.2] Express the sum and product of two general quaternions so that all these indeed hold.

Quaternions also provide us with an example of what is called an *algebra* over the real numbers, because of the existence of a multiplication law, as described above. But what is remarkable about Hamilton's quaternions is that, in addition, we have an operation of *division* or, what amounts to the same thing, a (multiplicative) *inverse* $\boldsymbol{q}^{-1}$ for each non-zero quaternion $\boldsymbol{q}$. This inverse satisfies

$$\boldsymbol{q}^{-1}\boldsymbol{q} = \boldsymbol{q}\boldsymbol{q}^{-1} = 1,$$

giving the quaternions the structure of what is called a *division ring*, the inverse being explicitly

$$\boldsymbol{q}^{-1} = \bar{\boldsymbol{q}}(\boldsymbol{q}\bar{\boldsymbol{q}})^{-1},$$

where the (quaternionic) *conjugate* $\bar{\boldsymbol{q}}$ of $\boldsymbol{q}$ is defined by

$$\bar{\boldsymbol{q}} = t - u\mathbf{i} - v\mathbf{j} - w\mathbf{k},$$

with $\boldsymbol{q} = t + u\mathbf{i} + v\mathbf{j} + w\mathbf{k}$, as before. We find that

$$\boldsymbol{q}\bar{\boldsymbol{q}} = t^2 + u^2 + v^2 + w^2,$$

so that the *real number* $\boldsymbol{q}\bar{\boldsymbol{q}}$ cannot vanish unless $\boldsymbol{q} = 0$ (i.e. $t = u = v = w = 0$), so $(\boldsymbol{q}\bar{\boldsymbol{q}})^{-1}$ exists, whence $\boldsymbol{q}^{-1}$ is well defined provided that $\boldsymbol{q} \neq 0$.[11.3]

11.2 The physical role of quaternions?

This gives us a very beautiful algebraic structure and, apparently, the potential for a wonderful calculus finely tuned to the treatment of the physics and the geometry of our 3-dimensional physical space. Indeed, Hamilton himself devoted the remaining 22 years of his life attempting to develop such a calculus. However, from our present perspective, as we look back over the 19th and 20th centuries, we must still regard these heroic efforts as having resulted in relative failure. This is not to say that quaternions are mathematically (or even physically) unimportant. They certainly do have some very significant roles to play, and in a slightly indirect sense their influence has been enormous, through various types of generalization. But the original 'pure quaternions' still have not lived up to what must undoubtedly have initially seemed to be an extraordinary promise.

Why have they not? Is there perhaps a lesson for us to learn concerning modern attempts at finding the 'right' mathematics for the physical world?

[11.3] Check that this definition of $\boldsymbol{q}^{-1}$ actually works.

First, there is an obvious point. If we are to think of quaternions to be a higher-dimensional anologue of the complex numbers, the analogy is that the dimension has gone up not from 2 to 3 dimensions, but from 2 to 4. For, in each case, one of the dimensions is the 'real axis', which here corresponds to the 't' component in the above representation of $\boldsymbol{q}$ in terms of **i**, **j**, **k**. The temptation is strong to take this t to represent the *time*,[3] so that our quaternions would describe a four-dimensional *space-time*, rather than just space. We might think that this should be highly appropriate, from our 20th-century perspective, since a four-dimensional spacetime is central to modern relativity theory, as we shall be seeing in Chapter 17. But it turns out that quaternions are not really appropriate for the description of spacetime, largely for the reason that the 'quaternionically natural' *quadratic form* $\boldsymbol{q}\bar{\boldsymbol{q}} = t^2 + u^2 + v^2 + w^2$ has the 'incorrect signature' for relativity theory (a matter that we shall be coming to later; see §13.8, §18.1). Of course, Hamilton did not know about relativity, since he lived in the wrong century for that. In any case, there is a 'can of worms' here that I do not wish to get involved with just yet. I shall open it slowly later! (See §13.8, §§18.1–4, end of §22.11, §28.9, §31.13, §32.2.)

There is another reason, perhaps a more fundamental one, that quaternions are not really so mathematically 'nice' as they seem at first sight. They are relatively poor 'magicians'; and, certainly, they are no match for complex numbers in this regard. The reason appears to be that there is no satisfactory[4] quaternionic analogue of the notion of a holomorphic function. The basic reason for this is simple. We saw in the previous chapter that a holomorphic function of a complex variable z is characterized as being holomorphically 'independent' of the complex conjugate $\bar{z}$. But we find that, with quaternions, it is possible to express the quaternionic conjugate $\bar{\boldsymbol{q}}$ of $\boldsymbol{q}$ algebraically in terms of $\boldsymbol{q}$ and the constant quantities **i**, **j**, and **k** by use of the expression.[11.4]

$$\bar{\boldsymbol{q}} = -\frac{1}{2}(\boldsymbol{q} + \mathbf{i}\boldsymbol{q}\mathbf{i} + \mathbf{j}\boldsymbol{q}\mathbf{j} + \mathbf{k}\boldsymbol{q}\mathbf{k}).$$

If 'quaternionic-holomorphic' is to mean 'built up from quaternions by means of addition, multiplication, and the taking of limits', then $\bar{\boldsymbol{q}}$ has to count as a quaternionic-holomorphic function of $\boldsymbol{q}$, which rather spoils the whole idea.

Is it possible to find modifications of quaternions that might have more direct relevance to the physical world? We shall find that this is certainly true, but these all sacrifice the key property of quaternions, demonstrated above, that you can always divide by them (if non-zero). What about generalizations to higher dimensions? We shall be seeing shortly how

[11.4] Check this.

Clifford achieved this, and how this kind of generalization does have great importance for physics. But all these changes lead to the abandonment of the division-algebra property.

Are there generalizations of quaternions which preserve the division property? In fact, yes; but the first point to make is that there are theorems telling us that this is not possible unless we relax the rules of the algebra even further than our abandoning of the commutative law of multiplication. About two months after receiving a letter from Hamilton announcing the discovery of quaternions, in 1843, John Graves discovered that there exists a kind of 'double' quaternion—entities now referred to as *octonions*. These were rediscovered by Arthur Cayley in 1845. For octonians, the associative law $a(bc) = (ab)c$ is abandoned (although a remnant of this law is maintained in the form of the restricted identities $a(ab) = a^2b$ and $(ab)b = ab^2$). The beauty of this structure is that it is still a division algebra, although a non-associative one. (For each non-zero a, there is an a^{-1} such that $a^{-1}(ab) = b = (ba)a^{-1}$.) Octonions form an eight-dimensional non-associative division algebra. There are seven analogues of the **i**, **j**, and **k** of the quaternion algebra, which, together with 1, span the eight dimensions of the octonion algebra. The individual multiplication laws for these elements (analogues of $\mathbf{ij} = \mathbf{k} = -\mathbf{ji}$, etc.) are a little complicated and it is best that I postpone these until §16.2, where an elegant description will be given, illustrated in Fig. 16.3. Unhappily, there is no fully satisfactory generalization of the octonions to even higher dimensions if the division algebra property is to be retained, as follows from an algebraic result of Hurwitz (1898), which showed that the quaternionic (and octonionic) identity '$\boldsymbol{q}\bar{\boldsymbol{q}}$ = sum of squares' does not work for dimensions other than 1, 2, 4, 8. In fact, apart from these specific dimensions, there can be no algebra *at all* in which division is always possible (except by 0). This follows from a remarkable topological theorem[5] that we shall encounter in §15.4. The only division algebras are, indeed, the real numbers, the complex numbers, the quaternions, and the octonions.

If we are prepared to abandon the division property, then there *is* an important generalization of the notion of quaternions to higher dimensions, and it is a generalization that indeed has powerful implications in modern physics. This is the notion of a *Clifford algebra*, which was introduced[6] in 1878 by the brilliant but short-lived English mathematician William Kingdon Clifford (1845–1879). One may regard Clifford's algebra as actually having sprung from two sources, each of which was geared to the understanding of spaces of dimension higher than the two described by complex numbers. One of these sources was in fact the algebra of Hamilton's quaternions that we have been concerned with here; the other is an earlier important development, originally put forward[7] in 1844 and 1862 by a little-recognized German schoolmaster,

Hermann Grassmann (1809–1877). Grassmann algebras also have direct roles to play in modern theoretical physics. (In particular, the modern notion of *supersymmetry*—see §31.3—depends crucially upon them, supersymmetry being close to ubiquitous among modern attempts to develop the foundations of physics beyond the framework of its standard model.) It will be important for us to acquaint ourselves with both the Grassmann and Clifford algebras here, and we shall do so in §11.6 and §11.5, respectively.

Clifford (and Grassmann) algebras involve a new ingredient that comes from the higher dimensionality of the space under consideration. Before we can properly appreciate this point, it is best that we consider quaternions again, but from a somewhat different perspective—a geometrical one. This will lead us also into some other considerations that are of fundamental importance in modern physics.

11.3 Geometry of quaternions

Think of the basic quaternionic quantities **i**, **j**, **k** as referring to three mutually perpendicular (right-handed) axes in ordinary Euclidean 3-space (see Fig. 11.1). Now, we recall from §5.1 that the quantity i in ordinary complex-number theory can be interpreted in terms of the operation 'multiply by i' which, in its action on the complex plane, means 'rotate through a right angle about the origin, in the positive sense'. We might imagine that we could interpret the quaternion **i** in the same kind of way, but now as a rotation in 3 dimensions, in the positive sense (i.e. right-handed) about the **i**-axis (so the (**j**, **k**)-plane plays the role of the complex plane), where we would correspondingly think of **j** as representing a rotation (in the positive sense) about the **j**-axis, and **k** a rotation about the **k**-axis. However, if these rotations are indeed right-angle rotations, as was the case with complex numbers, then the product relations will not work, because if we follow the **i**-rotation by the **j**-rotation, we do not get (even a multiple of) the **k**-rotation.

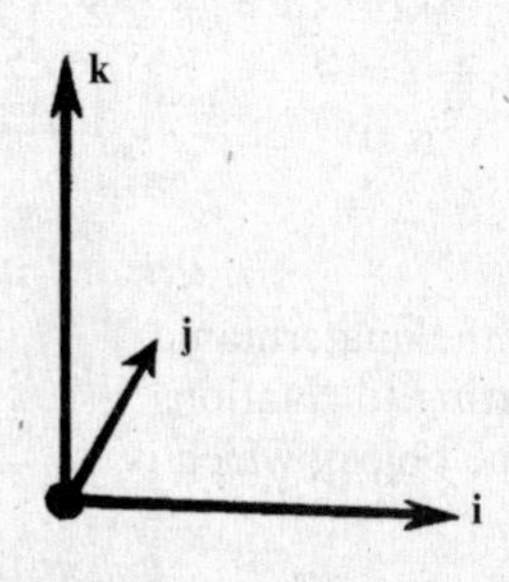

Fig. 11.1 The basic quaternions **i**, **j**, **k** refer to 3 mutually perpendicular (and right-handed) axes in ordinary Euclidean 3-space.

It is quite easy to see this explicitly by taking some ordinary object and physically rotating it. I suggest using a book. Lay the book flat on a horizontal table in front of you in the ordinary way, with the book closed, as though you were just about to open it to read it. Imagine the **k**-axis to be upwards, through the centre of the book, with the **i**-axis going off to the right and the **j**-axis going off directly away from you, both also through the centre. If we rotate the book through a right angle (in the right-handed sense) about **i** and then rotate it (in the right-handed sense) about **j**, we find that it ends up in a configuration (with its back spine upwards) that cannot be restored to its original state by any single rotation about **k**. (See Fig. 11.2.)

What we have to do to make things work is to rotate about *two* right angles (i.e. through 180°, or π). This seems an odd thing to do, as it is certainly not a direct analogy of the way that we understood the action of the complex number i. The main trouble would seem to be that if we apply this operation twice about the same axis, we get a rotation through 360° (or 2π), which simply restores the object (say our book) back to its original state, apparently representing $\mathbf{i}^2 = 1$, rather than $\mathbf{i}^2 = -1$. But here is where a wonderful new idea comes in. It is an idea of considerable subtlety and importance—a mathematical importance that is fundamental to the quantum physics of basic particles such as electrons, protons, and neutrons. As we shall be seeing in §23.7, ordinary solid matter could not exist without its consequences. The essential mathematical notion is that of a *spinor*.[8]

What is a spinor? Essentially, it is an object which turns into its negative when it undergoes a complete rotation through 2π. This may seem like an absurdity, because any classical object of ordinary experience is always returned to its original state under such a rotation, not to something else. To understand this curious property of spinors—or of what I shall refer to as *spinorial objects*—let us return to our book, lying on the table before us. We shall need some means of keeping track of how it has been rotated. We can do this by placing one end of a long belt firmly between the pages of the book and attaching the buckle rigidly to some fixed structure (say a

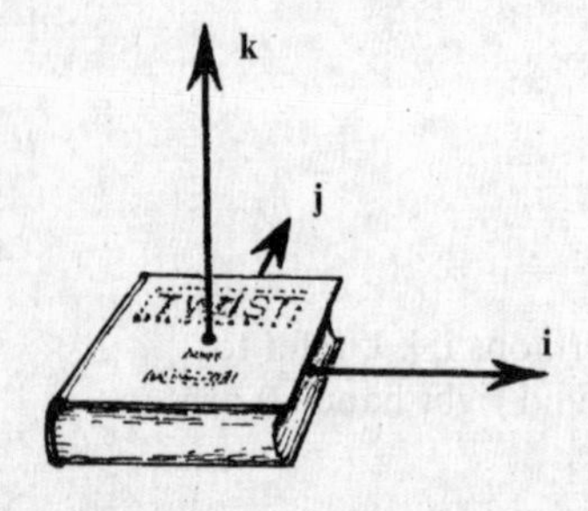

Fig. 11.2 We can think of the quaternionic operators **i**, **j**, and **k** as referring to rotations (through 180°, i.e. π) of some object, which is here taken to be a book.

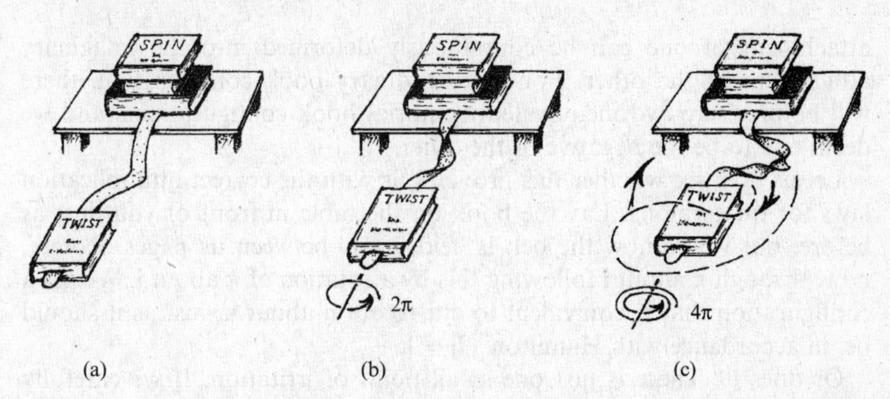

Fig. 11.3 A spinorial object, represented by the book of Fig. 11.2. An even number of 2π rotations is to be equivalent to no rotation, whereas an odd number of 2π rotations is not. (a) We keep track of the parity of the number of 2π rotations of the book by loosely attaching it, using a long belt, to some fixed object (here to a pile of books). (b) A rotation of our book through 2π twists the belt so that it cannot be undone without a further rotation. (c) A rotation of the book through 4π gives a twist that can be removed completely by looping the belt over the book.

pile of other books; see Fig. 11.3a). A rotation of the book through 2π twists the belt in a way that cannot be undone without further rotation of the book (Fig. 11.3b). But if we rotate the book through an additional angle of 2π, giving a total rotation through 4π, then we find, rather surprisingly, that the twist in the belt can be removed completely, simply by looping it over the book, keeping the book itself in the same position throughout the manoeuvre (Fig. 11.3c). Thus, the belt keeps track of the *parity* of the number of 2π rotations that the book undergoes, rather than totting up the entire number. That is to say, if we rotate the book through an even number of 2π rotations then the belt twist can be made to disappear completely, whereas if we rotate the book through an odd number of 2π rotations the belt inevitably remains twisted. This applies whatever rotation axis, or succession of different rotation axes, we choose to use.

Thus, to picture a spinorial object, we can think of an ordinary object in space, but where there is an imaginary flexible attachment to some fixed external structure, this imaginary attachment being represented by the belt that we have been just considering. The attachment may be moved around in any continuous way, but its ends must be kept fixed, one on the object itself and the other on the fixed external structure. The configuration of our 'spinorial book', so envisaged, is to be thought of as having such an imaginary attachment to some such fixed external structure, and two configurations of it are deemed to be equivalent only if the imaginary attachment of one can be continuously deformed into the imaginary attachment of the other. For every ordinary book configuration, there will be precisely two inequivalent spinorial book configurations, and we deem one to be the negative of the other.

Let us now see whether this provides us with the correct multiplication laws for quaternions. Lay the book on the table in front of you, just as before, but where now the belt is held firmly between its pages. Rotate, now, through π about **i** following this by a rotation of π about **j**. We get a configuration that is equivalent to a π rotation about **k**, just as it should be, in accordance with Hamilton's $\mathbf{ij} = \mathbf{k}$.

Or does it? There is just one small point of irritation. If we carefully insist that all these rotations are in the right-handed sense, then, keeping track of the belt twistings appropriately, we seem to get $\mathbf{ij} = -\mathbf{k}$, instead. This is not an important point, however, and it can be righted in a number of different ways. Either we can represent our quaternions by left-handed rotations through π instead of right-handed ones (in which case we do retrieve '$\mathbf{ij} = \mathbf{k}$') or we take our **i**, **j**, **k**-axes to have a left-handed orientation rather than a right-handed one. Or, best, we can adopt a convention of the ordering of multiplication of operators that is quite usual in mathematics, namely that the 'product $\boldsymbol{pq}$' represents $\boldsymbol{q}$ followed by $\boldsymbol{p}$, rather than $\boldsymbol{p}$ followed by $\boldsymbol{q}$.

In fact, there is a good reason for this odd-looking convention. This has to do with operators—such as things like $\partial/\partial x$—generally being understood to act on things written to the right of them. Thus, the operator $\boldsymbol{P}$ acting on Φ would be written $\boldsymbol{P}(\Phi)$, or simply $\boldsymbol{P}\Phi$. Accordingly, if we apply first $\boldsymbol{P}$ and then $\boldsymbol{Q}$ to Φ, we get $\boldsymbol{Q}(\boldsymbol{P}(\Phi))$ or simply $\boldsymbol{QP}\Phi$, which is $\boldsymbol{QP}$ acting on Φ.

My own way of resolving this awkward sign issue with quaternions will indeed be to take everything in the standard right-handed sense and to adopt this 'usual' reverse-order mathematical convention for the ordering of operators. It is now a simple matter for the reader to confirm that all of Hamilton's 'Brougham Bridge' equations $\mathbf{i}^2 = \mathbf{j}^2 = \mathbf{k}^2 = \mathbf{ijk} = -1$ are indeed satisfied by our 'spinorial book'. We bear in mind, of course, that **ijk** now stands for '**k** followed by **j** followed by **i**'.[9]

11.4 How to compose rotations

This curious property of rotation angles being twice what might have seemed geometrically appropriate can be demonstrated in another way. It is a particular feature of (proper, i.e. non-reflective) rotations in three dimensions that if we combine any number of them together then we always get a rotation about *some* axis. How can we find this axis in a simple geometrical way, and also the amount of this rotation? An elegant answer was found by Hamilton.[10] Let us see how this works. My presentation here will be a little different from that originally provided by Hamilton.

Recall that when we compose two different displacements that are simply *translations*, we can use the standard triangle law (equivalent to the parallelogram law illustrated in Fig. 5.1a) to get the answer. Thus, we can represent the first translation by a *vector* (by which I here mean an oriented line segment, the direction of the orientation being indicated by an arrow on the segment) and the second translation by another such vector, where the tail of the second vector is coincident with the head of the first. The vector stretching from the tail of the first vector to the head of the second represents the composition of the two translational motions. See Fig. 11.4a.

Can we do something similar for rotations? Remarkably, it turns out that we can. Think now of the 'vectors' as being oriented arcs of great circles drawn on a sphere—again depicted with an arrow to represent the orientation. (A great circle on a sphere is the intersection of the sphere with a plane through its centre.) We can imagine that such a 'vector arc' can be used to represent a rotation in the direction of the arrow. This rotation is to be about an axis, through the centre of the sphere, perpendicular to the plane of the great circle on which the arrow resides.

Can we think of the composition of two rotations, represented in this way, as being given by a 'triangle law' similar to the situation that we had for ordinary translations? Indeed we can; but there is a catch. The rotation that is to be represented by our 'vector arc' must be through an angle that is precisely *twice* the angle that is represented by the length of the arc. (For convenience, we can take the sphere to be of unit radius. Then the angle represented by the arc is simply the distance measured along the arc. For the 'triangle law' to hold, the angle through which the rotation is to take place must be twice this arc-length.) The reason that this works is illustrated in Fig. 11.4b. The curvilinear (spherical) triangle at the centre illustrates the 'triangle law' and the three external triangles are the respective reflections in its three vertices. The two initial rotations take one of these external triangles into a second one and then the second one into the third; the rotation that is the composition of the two takes the first into the third. We note that each of these rotations is through an angle which is precisely twice the corresponding arc-length of the original curvilinear triangle.[11.5] We shall be seeing a variant of this construction in relativistic physics, in §18.4 (Fig.18.13).

[11.5] In Hamilton's original version of this construction, the 'dual' spherical triangle to this one is used, whose vertices are where the sphere meets the three axes of rotation involved in the problem. Give a direct demonstration of how this works (perhaps 'dualizing' the argument given in the text), the amounts of the rotations being represented as twice the *angles* of this dual triangle.

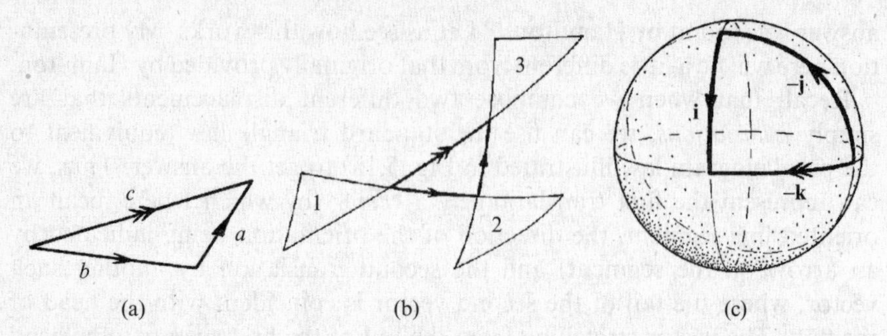

Fig. 11.4 (a) Translations in the Euclidean plane represented by oriented line segments. The double-arrowed segment represents the composition of the other two, by the triangle law. (b) For rotations in Euclidean 3-space, the segments are now great-circle arcs drawn on the unit sphere, each representing a rotation through *twice* the angle measured by the arc (about an axis perpendicular to its plane). To see why this works, reflect the triangle made by the arcs, in each vertex in turn. The first rotation takes triangle 1 into triangle 2, the second takes triangle 2 into triangle 3, and the composition takes triangle 1 into triangle 3. (c) The quaternionic relation $\mathbf{ij} = \mathbf{k}$ (in the form $\mathbf{i}(-\mathbf{j}) = -\mathbf{k}$), as a special case. The rotations are each through π, but represented by the half-angle $\frac{\pi}{2}$.

We can examine this in the particular situation that we considered above, and try to illustrate the quaternionic relation $\mathbf{ij} = \mathbf{k}$. The rotations described by $\mathbf{i}$, $\mathbf{j}$, and $\mathbf{k}$ are each through an angle π. Thus, we use arc-lengths that are just half this angle, namely $\frac{1}{2}\pi$, in order to depict the 'triangle law'. This is fully illustrated in Fig. 11.4c (in the form $\mathbf{i}(-\mathbf{j}) = -\mathbf{k}$, for clarity). We can also see the relation $\mathbf{i}^2 = -1$ as illustrated by the fact that a great circle arc, of length π, stretching from a point on the sphere to its *antipodal* point (depicting '-1') is essentially different from an arc of zero length or of length 2π, despite the fact that each represents a rotation of the sphere that restores it to its original position. The 'vector arc' description correctly represents the rotations of a 'spinorial object'.

11.5 Clifford algebras

To proceed to higher dimensions and to the idea of a Clifford algebra, we must consider what the analogue of a 'rotation about an axis' must be. In n dimensions, the basic such rotation has an 'axis' which is an $(n-2)$-dimensional space, rather than just the 1-dimensional line-axis that we get for ordinary 3-dimensional rotations. But apart from this, a rotation about an $(n-2)$-dimensional axis is similar to the familiar case of an ordinary 3-dimensional rotation about a 1-dimensional axis in that the rotation is completely determined by the direction of this axis and by the amount of the angle of the rotation. Again we have spinorial objects with the property that, if such an object is continuously rotated through the angle 2π, then it is not restored to its original state but to what we consider to be the 'negative' of that state. A rotation through 4π always does restore such an object to its original state.

There is, however, a 'new ingredient', alluded to above: that in dimension higher than 3, it is not true that the composition of basic rotations about $(n-2)$-dimensional axes will always again be a rotation about an $(n-2)$-dimensional axis. In these higher dimensions, general (compositions of) rotations cannot be so simply described. Such a (generalized) rotation may have an 'axis' (i.e. a space that is left undisturbed by the rotational motion) whose dimension can take a variety of different values. Thus, for a Clifford algebra in n dimensions, we need a hierarchy of different kinds of entity to represent such different kinds of rotation. In fact, it turns out to be better to start with something that is even more elementary than a rotation through π, namely a *reflection* in an $(n-1)$-dimensional (hyper)plane. A composition of two such reflections (with respect to two such planes that are perpendicular) provides a rotation through π, giving these previously basic π-rotations as 'secondary' entities, the primary entities being the reflections.[11.6]

We label these basic reflections $\boldsymbol{\gamma}_1, \boldsymbol{\gamma}_2, \boldsymbol{\gamma}_3, \ldots, \boldsymbol{\gamma}_n$, where $\boldsymbol{\gamma}_r$ reverses the rth coordinate axis, while leaving all the others alone. For the appropriate type of 'spinorial object', reflecting it twice in the same direction gives the negative of the object, so we have n quaternion-like relations,

$$\boldsymbol{\gamma}_1^2 = -1, \quad \boldsymbol{\gamma}_2^2 = -1, \quad \boldsymbol{\gamma}_3^2 = -1, \quad \ldots, \quad \boldsymbol{\gamma}_n^2 = -1,$$

satisfied by these primary reflections. The secondary entities, representing our original π-rotations, are products of pairs of distinct γ's, and these products have anticommutation properties (rather like quaternions):

$$\boldsymbol{\gamma}_p\boldsymbol{\gamma}_q = -\boldsymbol{\gamma}_q\boldsymbol{\gamma}_p \quad (p \neq q).$$

In the particular case of three dimensions ($n = 3$), we can define the three different 'second-order' quantities

$$\mathbf{i} = \boldsymbol{\gamma}_2\boldsymbol{\gamma}_3, \quad \mathbf{j} = \boldsymbol{\gamma}_3\boldsymbol{\gamma}_1, \quad \mathbf{k} = \boldsymbol{\gamma}_1\boldsymbol{\gamma}_2,$$

[11.6] Find the geometrical nature of the transformation, in Euclidean 3-space, which is the composition of two reflections in planes that are not perpendicular.

and it is readily checked that these three quantities **i**, **j**, and **k** satisfy the quaternion algebra laws (Hamilton's 'Brougham Bridge' equations).[11.7]

The general element of the Clifford algebra for an n-dimensional space is a sum of real-number multiples (i.e. a linear combination) of products of sets of distinct γ's. The first-order ('primary') entities are the n different individual quantities γ_p. The second-order ('secondary') entities are the $\frac{1}{2}n(n-1)$ independent products $\gamma_p\gamma_q$ (with $p < q$); there are $\frac{1}{6}n(n-1)(n-2)$ independent third-order entities $\gamma_p\gamma_q\gamma_r$ (with $p < q < r$), $\frac{1}{24}n(n-1)(n-2)(n-3)$ independent fourth-order entities, etc., and finally the single nth-order entity $\gamma_1\gamma_2\gamma_3\cdots\gamma_n$. Taking all these, together with the single zeroth-order entity 1, we get

$$1 + n + \frac{1}{2}n(n-1) + \frac{1}{6}n(n-1)(n-2) + \cdots + 1 = 2^n$$

entities in all,[11.8] and the general element of the Clifford algebra is a linear combination of these. Thus the elements of a Clifford algebra constitute a 2^n-dimensional algebra over the reals, in the sense described in §11.1. They form a ring with identity but, unlike quaternions, they do not form a division ring.

One reason that Clifford algebras are important is for their role in defining spinors. In physics, spinors made their appearance in Dirac's famous equation for the electron (Dirac 1928), the electron's state being a spinor quantity (see Chapter 24). A spinor may be thought of as an object upon which the elements of the Clifford algebra act as operators, such as with the basic reflections and rotations of a 'spinorial object' that we have been considering. The very notion of a 'spinorial object' is somewhat confusing and non-intuitive, and some people prefer to resort to a purely (Clifford-) algebraic[11] approach to their study. This certainly has its advantages, especially for a general and rigorous n-dimensional discussion; but I feel that it is important also not to lose sight of the geometry, and I have tried to emphasize this aspect of things here.

In n dimensions,[12] the full space of spinors (sometimes called *spin-space*) is $2^{n/2}$-dimensional if n is even, and $2^{(n-1)/2}$-dimensional if n is odd. When n is even, the space of spinors splits into two independent spaces (sometimes called the spaces of 'reduced spinors' or 'half-spinors'), each of which is $2^{(n-2)/2}$-dimensional; that is, each element of the full space is the sum of two elements—one from each of the two reduced spaces. A reflection in the (even) n-dimensional space converts one of these reduced spin-spaces into the other. The elements of one reduced spin-space have a certain 'chirality' or 'handedness'; those of the other have the opposite chirality. This appears

[11.7] Show this.

[11.8] Explain all this counting. *Hint*: Think of $(1+1)^n$.

to have deep importance in physics, where I here refer to the spinors for ordinary 4-dimensional spacetime. The two reduced spin-spaces are each 2-dimensional, one referring to right-handed entities and the other to left-handed ones. It seems that Nature assigns a different role to each of these two reduced spin-spaces, and it is through this fact that physical processes that are reflection non-invariant can emerge. It was, indeed, one of the most striking (and some would say 'shocking') unprecedented discoveries of 20th-century physics (theoretically predicted by Chen Ning Yang and Tsung Dao Lee, and experimentally confirmed by Chien-Shiung Wu and her group, in 1957) that there are actually fundamental processes in Nature which do not occur in their mirror-reflected form. I shall be returning to these foundational issues later (§§25.3,4, §32.2, §§33.4,7,11,14).

Spinors also have an important technical mathematical value in various different contexts[13] (see §§22.8–11, §§23.4,5, §§24.6,7, §§32.3,4, §§33.4,6,8,11), and they can be of practical use in certain types of computation. Because of the 'exponential' relation between the dimension of the spin-space ($2^{n/2}$, etc.) and the dimension n of the original space, it is not surprising that spinors are better practical tools when n is reasonably small. For ordinary 4-dimensional spacetime, for example, each reduced spin-space has dimension only 2, whereas for modern 11-dimensional 'M-theory' (see §31.14), the spin-space has 32 dimensions.

11.6 Grassmann algebras

Finally, let me turn to Grassmann algebra. From the point of view of the above discussion, we may think of Grassmann algebra as a kind of degenerate case of Clifford algebra, where we have basic anticommuting generating elements $\boldsymbol{\eta}_1, \boldsymbol{\eta}_2, \boldsymbol{\eta}_3, \ldots, \boldsymbol{\eta}_n$, similar to the $\boldsymbol{\gamma}_1, \boldsymbol{\gamma}_2, \boldsymbol{\gamma}_3, \ldots, \boldsymbol{\gamma}_n$ of the Clifford algebra, but where each η_s squares to *zero*, rather than to the -1 that we have in the Clifford case:

$$\boldsymbol{\eta}_1^2 = 0, \quad \boldsymbol{\eta}_2^2 = 0, \quad \ldots, \quad \boldsymbol{\eta}_n^2 = 0.$$

The anticommutation law

$$\boldsymbol{\eta}_p \boldsymbol{\eta}_q = -\boldsymbol{\eta}_q \boldsymbol{\eta}_p$$

holds as before, except that the Grassmann algebra is now more 'systematic' than the Clifford algebra, because we do not have to specify '$p \neq q$' in this equation. The case $\boldsymbol{\eta}_p \boldsymbol{\eta}_p = -\boldsymbol{\eta}_p \boldsymbol{\eta}_p$ simply re-expresses $\boldsymbol{\eta}_p^2 = 0$.

Indeed, Grassmann algebras are more primitive and universal than Clifford algebras, as they depend only upon a minimal amount of local structure. Basically, the point is that the Clifford algebra needs to 'know' what 'perpendicular' means, so that ordinary rotations can be built up out of reflections, whereas the notion of a 'rotation' is not part of what is described according to Grassmann algebras. To put this another way, the ordinary notions of 'Clifford algebra' and 'spinor' require that there be a *metric* on the space, whereas this is not necessary for a Grassmann algebra. (Metrics will be discussed in §13.8 and §14.7.)

What the Grassmann algebra is concerned with is the basic idea of a 'plane element' for different numbers of dimensions. Let us think of each of the basic quantities $\boldsymbol{\eta}_1, \boldsymbol{\eta}_2, \boldsymbol{\eta}_3, \ldots, \boldsymbol{\eta}_n$, as defining a *line element* or 'vector' (rather than a hyperplane of reflection) at the origin of coordinates in some n-dimensional space, each $\boldsymbol{\eta}$ being associated with one of the n different coordinate axes. (These can be 'oblique' axes, since Grassmann algebra is not concerned with orthogonality; see Fig. 11.5.) The general vector at the origin will be some combination

$$\boldsymbol{a} = a_1\boldsymbol{\eta}_1 + a_2\boldsymbol{\eta}_2 + \cdots + a_n\boldsymbol{\eta}_n,$$

where $a_1, a_2, \ldots, a_n$ are real numbers. (Alternatively the a_i could be complex numbers, in the case of a complex space; but the real and complex cases are similar in their algebraic treatment.) To describe the 2-dimensional plane element spanned by two such vectors $\boldsymbol{a}$ and $\boldsymbol{b}$, where

$$\boldsymbol{b} = b_1\boldsymbol{\eta}_1 + b_2\boldsymbol{\eta}_2 + \cdots + b_n\boldsymbol{\eta}_n,$$

we form the *Grassmann product* of $\boldsymbol{a}$ with $\boldsymbol{b}$. In order to avoid confusion with other forms of product, I shall henceforth adopt the (standard) notation $\boldsymbol{a} \wedge \boldsymbol{b}$ for this product (called the 'wedge product') rather than just using juxtaposition of symbols. Accordingly, what I previously wrote

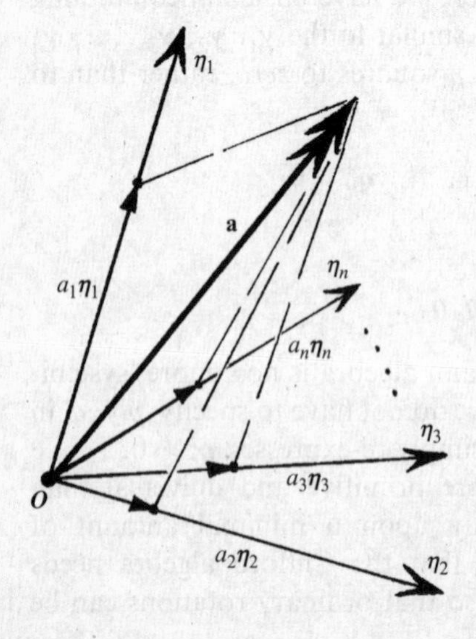

Fig. 11.5 Each basis element $\boldsymbol{\eta}_1, \boldsymbol{\eta}_2, \boldsymbol{\eta}_3, \ldots, \boldsymbol{\eta}_n$, of a Grassmann algebra defines a vector in n-dimensional space, at some origin-point $\boldsymbol{O}$. These vectors can be along the different coordinate axes (which can be 'oblique' axes; Grassmann algebra not being concerned with orthogonality). A general vector at $\boldsymbol{O}$ is a linear combination $\mathbf{a} = a_1\boldsymbol{\eta}_1 + a_2\boldsymbol{\eta}_2 + \cdots + a_n\boldsymbol{\eta}_n$.

as $\boldsymbol{\eta}_p\boldsymbol{\eta}_q$, I shall now denote by $\boldsymbol{\eta}_p \wedge \boldsymbol{\eta}_q$. The anticommutation law of these $\boldsymbol{\eta}$'s is now to be written

$$\boldsymbol{\eta}_p \wedge \boldsymbol{\eta}_q = -\boldsymbol{\eta}_q \wedge \boldsymbol{\eta}_p.$$

Adopting the distributive law (see §11.1) in defining the product $\boldsymbol{a} \wedge \boldsymbol{b}$, we consequently obtain the more general anticommutation property[11.9]

$$\boldsymbol{a} \wedge \boldsymbol{b} = -\boldsymbol{b} \wedge \boldsymbol{a}$$

for arbitrary vectors $\boldsymbol{a}$ and $\boldsymbol{b}$. The quantity $\boldsymbol{a} \wedge \boldsymbol{b}$ provides an algebraic representation of the plane element spanned by the vectors $\boldsymbol{a}$ and $\boldsymbol{b}$ (Fig. 11.6a). Note that this contains the information not only of an orientation for the plane element (since the sign of $\boldsymbol{a} \wedge \boldsymbol{b}$ has to do with which of $\boldsymbol{a}$ or $\boldsymbol{b}$ comes first), but also of a 'magnitude' assigned to the plane element.

We may ask how a quantity such as $\boldsymbol{a} \wedge \boldsymbol{b}$ is to be represented as a set of *components*, corresponding to the way that $\boldsymbol{a}$ may be represented as $(a_1, a_2, \ldots, a_n)$ and $\boldsymbol{b}$ as $(b_1, b_2, \ldots, b_n)$, these being the coefficients occurring when $\boldsymbol{a}$ and $\boldsymbol{b}$ are respectively presented as linear combinations of $\boldsymbol{\eta}_1, \boldsymbol{\eta}_2, \ldots, \boldsymbol{\eta}_n$. The quantity $\boldsymbol{a} \wedge \boldsymbol{b}$ may, correspondingly, be presented as a linear combination of $\boldsymbol{\eta}_1 \wedge \boldsymbol{\eta}_2$, $\boldsymbol{\eta}_1 \wedge \boldsymbol{\eta}_3$, etc., and we require the coefficients that arise. There is a certain choice of convention involved here because, for example, $\boldsymbol{\eta}_1 \wedge \boldsymbol{\eta}_2$ and $\boldsymbol{\eta}_2 \wedge \boldsymbol{\eta}_1$ are not independent (one being the negative of the other), so we may wish to single out one or the other of these. It turns out to be more systematic to include both terms and to divide the relevant coefficient equally between them. Then we find[11.10] the coefficients—that is, the *components*—of $\boldsymbol{a} \wedge \boldsymbol{b}$ to be the various quantities $a_{[p}b_{q]}$, where square brackets around indices denote *antisymmetrization*, defined by

$$A_{[pq]} = \frac{1}{2}\left(A_{pq} - A_{qp}\right),$$

whence

$$a_{[p}b_{q]} = \frac{1}{2}\left(a_p b_q - a_q b_p\right).$$

What about a 3-dimensional 'plane element'? Taking $\boldsymbol{a}$, $\boldsymbol{b}$, and $\boldsymbol{c}$ to be three independent vectors spanning this 3-element, we can form the *triple* Grassmann product $\boldsymbol{a} \wedge \boldsymbol{b} \wedge \boldsymbol{c}$ to represent this 3-element (again with an orientation and magnitude), finding the anticommutation properties

[11.9] Show this.

[11.10] Write out $\boldsymbol{a} \wedge \boldsymbol{b}$ fully in the case $n = 2$, to see how this comes about.

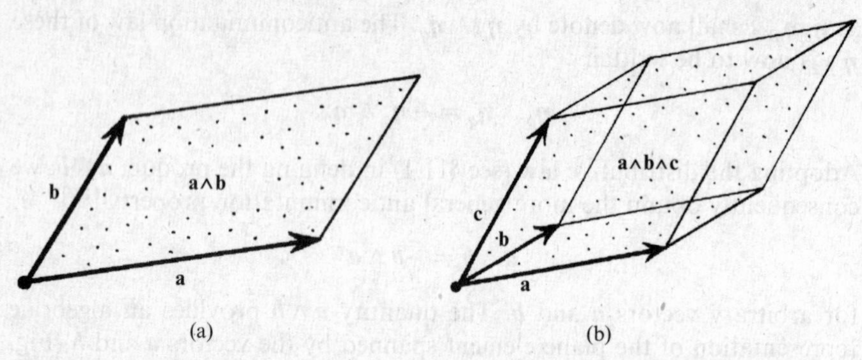

Fig. 11.6 (a) The quantity $\mathbf{a} \wedge \mathbf{b}$ represents the (oriented and scaled) plane-element spanned by independent vectors **a** and **b**. (b) The triple Grassmann product $\mathbf{a} \wedge \mathbf{b} \wedge \mathbf{c}$ represents the 3-element spanned by independent vectors **a, b** and **c**.

$$\boldsymbol{a} \wedge \boldsymbol{b} \wedge \boldsymbol{c} = \boldsymbol{b} \wedge \boldsymbol{c} \wedge \boldsymbol{a} = \boldsymbol{c} \wedge \boldsymbol{a} \wedge \boldsymbol{b} = -\boldsymbol{b} \wedge \boldsymbol{a} \wedge \boldsymbol{c} = -\boldsymbol{a} \wedge \boldsymbol{c} \wedge \boldsymbol{b} = -\boldsymbol{c} \wedge \boldsymbol{b} \wedge \boldsymbol{a}$$

(see Fig. 11.6b). The *components* of $\boldsymbol{a} \wedge \boldsymbol{b} \wedge \boldsymbol{c}$ are taken to be, in accordance with the above,

$$a_{[p}b_q c_{r]} = \frac{1}{6}\left(a_p b_q c_r + a_q b_r c_p + a_r b_p c_q - a_q b_p c_r - a_p b_r c_q - a_r b_q c_p\right),$$

the square brackets again denoting antisymmetrization, as illustrated by the expression on the right-hand side.

Similar expressions define general r-elements, where r ranges up to the dimension n of the entire space. The components of the rth-order wedge product are obtained by taking the antisymmetrized product of the components of the individual vectors.[11.11], [11.12] Indeed, Grassmann algebra provides a powerful means of describing the basic geometrical linear elements of arbitrary (finite) dimension.

The Grassmann algebra is a *graded* algebra in the sense that it contains rth-order elements (where r is the number of $\boldsymbol{\eta}$'s that are 'wedge-producted' together within the expression). The number r (where $r = 0, 1, 2, 3, \ldots, n$) is called the *grade* of the element of the Grassmann algebra. It should be noted, however, that the general element of the algebra of grade r need not be a simple wedge product (such as $\boldsymbol{a} \wedge \boldsymbol{b} \wedge \boldsymbol{c}$ in the case $r = 3$), but can be a sum of such expressions. Accordingly, there are many elements of the Grassmann algebra that do not directly describe

---

[11.11] Write down this expression explicitly in the case of a wedge product of four vectors.

[11.12] Show that the wedge product remains unaltered if $\boldsymbol{a}$ is replaced by $\boldsymbol{a}$ added to any multiple of any of the other vectors involved in the wedge product.

geometrical r-elements. A role for such 'non-geometrical' Grassmann elements will appear later (§12.7).

In general, if $\boldsymbol{P}$ is an element of grade p and $\boldsymbol{Q}$ is an element of grade q, we define their $(p+q)$-grade wedge product $\boldsymbol{P} \wedge \boldsymbol{Q}$ to have components $P_{[a...c}Q_{d...f]}$, where $P_{a...c}$ and $Q_{d...f}$ are the components $\boldsymbol{P}$ and $\boldsymbol{Q}$ respectively. Then we find[11.13], [11.14]

$$\boldsymbol{P} \wedge \boldsymbol{Q} = \begin{cases} +\boldsymbol{Q} \wedge \boldsymbol{P} & \text{if } p, \text{ or } q, \text{ or both, are even,} \\ -\boldsymbol{Q} \wedge \boldsymbol{P} & \text{if } p \text{ and } q \text{ are both odd.} \end{cases}$$

The sum of elements of a fixed grade r is again an element of grade r; we may also add together elements of different grades to obtain a 'mixed' quantity that does not have any particular grade. Such elements of the Grassmann algebra do not have such direct interpretations, however.

Notes

Section 11.1

11.1. According to Eduard and Klein (1898), Carl Friedrich Gauss had apparently already noted the multiplication law for quaternions in around 1820, but he had not published it (Gauss 1900). This, however, was disputed by Tait (1900) and Knott (1900). For further information, see Crowe (1967).

11.2. The term 'vector' has a spectrum of meanings. Here we require no association with the *differentiation* notion of a 'vector field', described in §10.3.

Section 11.2

11.3. It is not clear to me how seriously Hamilton himself may have yielded to this temptation. Prior to his discovery of quaternions, he had been interested in the algebraic treatment of the 'passage of time', and this could have had some influence on his preparedness to accept a fourth dimension in quaternionic algebra. See Crowe (1967), pp. 23–7.

11.4. Nevertheless, a fair amount of work has been directed at issue of quaternionic analogues of holomorphic notions and their value in physical theory. See Gürsey (1983); Adler (1995). One might regard the twistor expressions (§§33.8,9) for solving the massless free field equations as an appropriate 4-dimensional analogue of the holomorphic-function method of solution of the Laplace equation. This, however, uses complex analysis, not quaternionic. For a general reference on quaternions and octonions, see Conway and Smith (2003).

11.5. See Adams and Atiyah (1966).

11.6. See Clifford (1878). For modern references see Hestenes and Sobczyk (2001); Lounesto (1999).

[11.13] Show this.

[11.14] Deduce that $\boldsymbol{P} \wedge \boldsymbol{P} = 0$, if p is odd.

11.7. See Grassmann (1844, 1862); van der Waerden (1985), pp. 191–2; Crowe (1967), Chap. 3.

Section 11.3

11.8. We pronounce this as though it were spelt 'spinnor', not 'spynor'. See Note 24.12.

11.9. Although I do not know who first suggested this way of demonstrating quaternion multiplication, J. H. Conway used it in private demonstrations at the 1978 International Congress of Mathematicians in Helsinki—see also Newman (1942); Penrose and Rindler (1984), pp. 41–6.

Section 11.4

11.10. See Pars (1968).

Section 11.5

11.11. For an approach to many physical problems through Clifford algebra, see Lasenby *et al.* (2000) and references contained therein.

11.12. See Cartan (1966); Brauer and Weyl (1935); Penrose and Rindler (1986), Appendix; Harvey (1990); Budinich and Trautman (1988).

11.13. See Lounesto (1999); Cartan (1966); Crumeyrolle (1990); Chevalley (1954); Kamberov (2002) for a few examples.

12
Manifolds of n dimensions

12.1 Why study higher-dimensional manifolds?

LET us now come to the general procedure for building up higher-dimensional manifolds, where the dimension n can be any positive integer whatever (or even zero, if we allow ourselves to think of a single point as constituting a 0-manifold). This is an essential notion for almost all modern theories of basic physics. The reader might wonder why it is of interest, physically, to consider n-manifolds for which n is larger than 4, since ordinary spacetime has just four dimensions. In fact many modern theories, such as string theory, operate within a 'spacetime' whose dimension is much larger than 4. We shall be coming to this kind of thing later (§15.1, §§31.4,10–12,14–17), where we examine the physical plausibility of this general idea. But quite irrespective of the question of whether actual 'spacetime' might be appropriately described as an n-manifold, there are other quite different and very compelling reasons for considering n-manifolds generally in physics.

For example, the *configuration space* of an ordinary rigid body in Euclidean 3-space—by which I mean a space $\mathcal{C}$ whose different points represent the different physical locations of the body—is a non-Euclidean 6-manifold (see Fig. 12.1). Why of six dimensions? There are three dimensions (degrees of freedom) in the position of the centre of gravity and three more in the rotational orientation of the body.[12.1] Why non-Euclidean? There are many reasons, but a particularly striking one is that even its *topology* is different from that of Euclidean 6-space. This 'topological non-triviality' of $\mathcal{C}$ shows up simply in the 3-dimensional aspect of the space that refers to the *rotational* orientation of the body. Let us call this 3-space $\mathcal{R}$, so each point of $\mathcal{R}$ represents a particular rotational orientation of the body. Recall our consideration of rotations of a book in the previous chapter. We shall take our 'body' to be that book (which must, of course, remain unopened, for otherwise the configuration space would have many more dimensions corresponding to the movement of the pages).

[12.1] Explain this dimension count more explicitly.

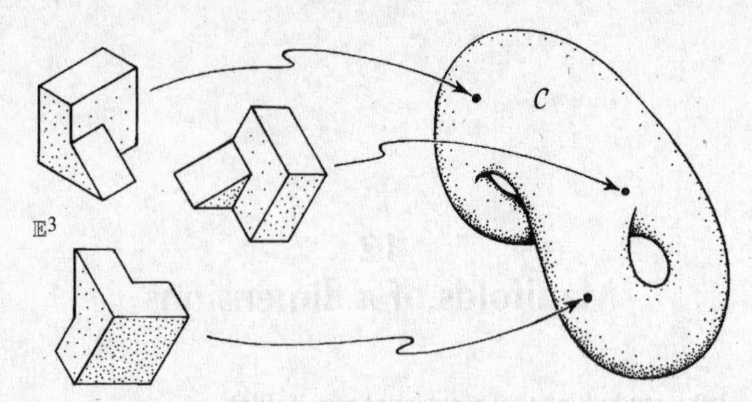

Fig. 12.1 Configuration space C, each of whose points represents a possible location of a given rigid body in Euclidean 3-space $\mathbb{E}^3$: C is a non-Euclidean 6-manifold.

How are we to recognize 'topological non-triviality'? We may imagine that this is not an easy matter for a 3- or 6-manifold. However, there are several mathematical procedures for ascertaining such things. Remember that in our examination of Riemann surfaces, as given in §8.4 (see Fig. 8.9), we considered various topologically non-trivial kinds of 2-surface. Apart from the (Riemann) sphere, the simplest such surface is the torus (surface of genus 1). How can we distinguish the torus from the sphere? One way is to consider closed loops on the surface. It is intuitively clear that there are loops that can be drawn on the torus for which there is no way to deform them continuously until they shrink away (down to a single point), whereas, on the sphere, every closed loop can be shrunk away in this manner (see Fig. 12.2). Loops on the Euclidean plane can also be all shrunk away. We say that the sphere and plane are *simply-connected* by virtue of this 'shrinkability' property. The torus (and surfaces of higher

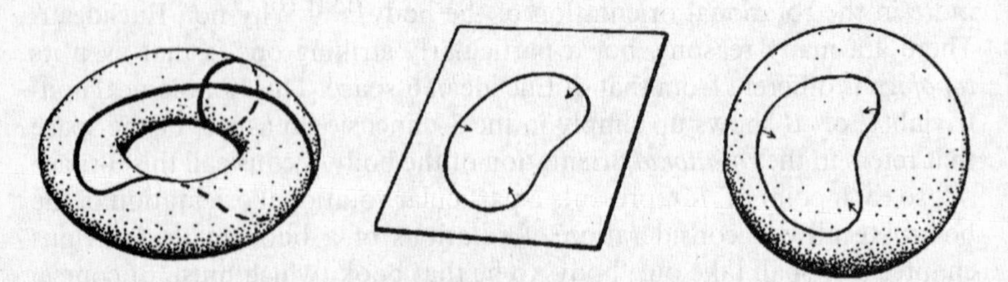

Fig. 12.2 Some loops on the torus cannot be shrunk away continuously (down to single point) while remaining in the surface, whereas on the plane or sphere, every closed loop can. Accordingly, the plane and sphere are simply-connected, but the torus (and surfaces of higher genus) are multiply-connected.

genus) are, on the other hand, *multiply-connected* because of the existence of non-shrinkable loops.[1] This provides us with one clear way, from within the surface itself, of distinguishing the torus (and surfaces of higher genus) from the sphere and from the plane.

We can apply the same idea to distinguish the topology of the 3-manifold $\mathcal{R}$ from the 'trivial' topology of Euclidean 3-space, or the topology of the 6-manifold $\mathcal{C}$ from that of 'trivial' Euclidean 6-space. Let us return to our 'book', which, as in §11.3, we picture as being attached to some fixed structure by an imaginary belt. Each individual rotational orientation of the book is to be represented by a corresponding point of $\mathcal{R}$. If we continuously rotate the book through 2π, so that it returns to its original rotational orientation, we find that this motion is represented, in $\mathcal{R}$, by a certain closed loop (see Fig. 12.3). Can we deform this closed loop in a continuous manner until it shrinks away (down to a single point)? Such a loop deformation would correspond to a gradual changing of our book rotation until it is no motion at all. But remember our imaginary belt attachment (which we can realize as an actual belt). Our 2π-rotation leaves the belt twisted; but this cannot be undone by a continuous belt motion while leaving the book unmoved. Now this 2π-twist must remain (or be transformed into an odd multiple of a 2π-twist) throughout the gradual deforming of the book rotation, so we conclude that it is impossible that the 2π-rotation can actually be continuously deformed to no rotation at all. Thus, correspondingly, there is no way that our chosen closed loop on $\mathcal{R}$ can be continuously deformed until it shrinks away. Accordingly, the 3-manifold $\mathcal{R}$ (and similarly the 6-manifold $\mathcal{C}$) must be multiply-connected and therefore topologically different from the simply-connected Euclidean 3-space (or 6-space).[2]

It may be noted that the multiple-connectivity of the spaces $\mathcal{R}$ and $\mathcal{C}$ is of a more interesting nature than that which occurs in the case of the

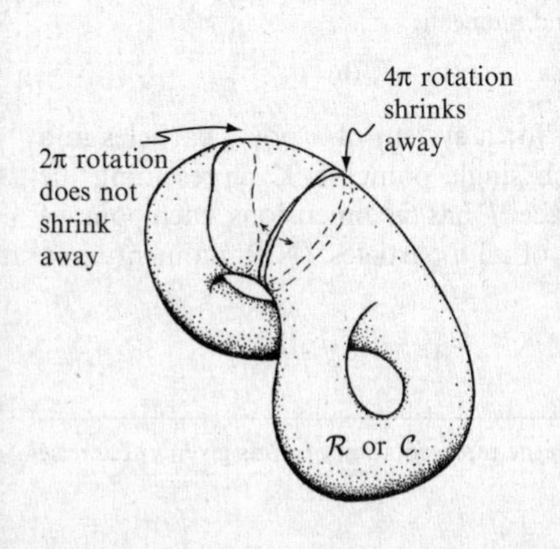

Fig. 12.3 The notion of multiple connectivity, as illustrated in Fig. 12.2, distinguishes the topology of the 3-manifold $\mathcal{R}$ (rotation space), or of the 6-manifold $\mathcal{C}$ (configuration space), from the 'trivial' topologies of Euclidean 3-space and 6-space. A loop on $\mathcal{R}$ or $\mathcal{C}$ representing a continuous rotation through 2π cannot be shrunk to a point, so $\mathcal{R}$ and $\mathcal{C}$ are multiply-connected. Yet, when traversed twice (representing a 4π-rotation) the loop does shrink to a point (topological torsion). See Fig. 11.3. (N.B. The 2-manifold depicted, being schematic only, does not actually have this last property.)

torus. For our loop that represents a 2π-rotation has the curious property that if we go around it twice (a 4π-rotation) then we obtain a loop which *can* now be deformed continuously to a point.[12.2] (This certainly does not happen for the torus.) This curious feature of loops in $\mathcal{R}$ and $\mathcal{C}$ is an instance of what is referred to as *topological torsion.*

We see from all this that it is of physical interest to study spaces, such as the 6-manifold $\mathcal{C}$, that are not only of dimension greater than that of ordinary spacetime but which also can have non-trivial topology. Moreover, such physically relevant spaces can have dimension enormously larger than 6. Very large-dimensional spaces can occur as configuration spaces, and also as what are called *phase spaces*, for systems involving large numbers of individual particles. The configuration space $\mathcal{K}$ of a gas, where the gas particles are described as individual points in 3-dimensional space, is of $3N$ dimensions, where N is the number of particles in the gas. Each point of $\mathcal{K}$ represents a gas configuration in which every particle's position is individually determined (Fig. 12.4a). In the case of the phase space $\mathcal{P}$ of the gas, we must keep track also of the *momentum* of each particle (which is the particle's velocity times its mass), this being a vector quantity (3 components for each particle), so that the overall dimension is $6N$. Thus, each single point of $\mathcal{P}$ represents not only the position of all the particles in the gas, but also of every individual particle's motion (Fig. 12.4b). For a thimbleful of ordinary air, there are could be some 10^{19} molecules,[3] so $\mathcal{P}$ has something like 60 000 000 000 000 000 000 dimensions! Phase spaces are particularly

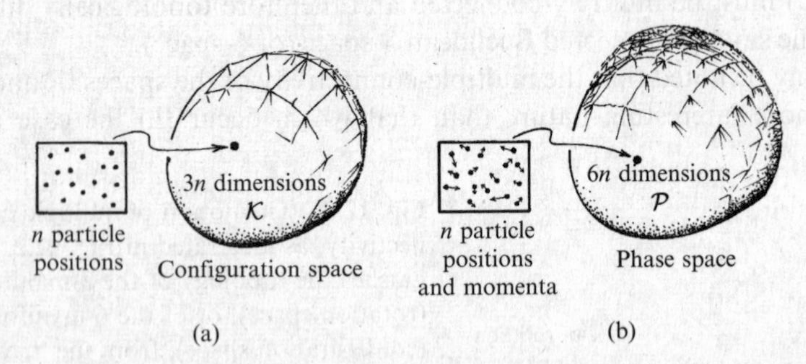

Fig. 12.4 (a) The configuration space $\mathcal{K}$, for a system of n point particles in a region of 3-space, has $3n$ dimensions, each single point of $\mathcal{K}$ representing the positions of all n particles. (b) The phase space $\mathcal{P}$ has $6n$ dimensions, each point of $\mathcal{P}$ representing the positions and momenta of all n particles. (N.B. momentum = velocity times mass.)

[12.2] Show how to do this, e.g. by appealing to the representation of $\mathcal{R}$ as given in Exercise [12.17].

useful in the study of the behaviour of (classical) physical systems involving many particles, so spaces of such large dimension can be physically very relevant.

12.2 Manifolds and coordinate patches

Let us now consider how the structure of an n-manifold may be treated mathematically. An n-manifold $\mathcal{M}$ can be constructed completely analogously to the way in which, in Chapters 8 and 10 (see §10.2), we constructed the surface $\mathcal{S}$ from a number of coordinate patches. However, now we need more coordinates in each patch than just a pair of numbers (x, y) or (X, Y). In fact we need n coordinates per patch, where n is a fixed number—the *dimension* of $\mathcal{M}$—which can be any positive integer. For this reason, it is convenient not to use a separate letter for each coordinate, but to distinguish our different coordinates

$$x^1, x^2, x^3, \ldots, x^n$$

by the use of an (upper) numerical index. Do not be confused here. These are not supposed to be different powers of a single quantity x, but separate independent real numbers. The reader might find it strange that I have apparently courted mystification, deliberately, by using an upper index rather than a lower one (e.g. $x_1, x_2, \ldots, x_n$), this leading to the inevitable confusion between, for instance, the coordinate x^3 and the cube of some quantity x. Confused readers are indeed justified in their confusion. I myself find it not only confusing but also, on occasion, genuinely irritating. For some historical reason, the standard conventions for classical *tensor analysis* (which we shall come to in a more serious way later in this chapter) have turned out this way around. These conventions involve tightly-knit rules governing the up/down placing of indices, and the consistent placing for the indices on the coordinates themselves has come out to be in the upper position. (These rules actually work well in practice, but it seems a great pity that the conventions had not been chosen the opposite way around. I am afraid that this is just something that we have to live with.)

How are we to picture our manifold $\mathcal{M}$? We think of it as 'glued together' from a number of coordinate patches, where each patch is an open region of $\mathbb{R}^n$. Here, $\mathbb{R}^n$ stands for the 'coordinate space' whose points are simply the n-tuples $(x^1, x^2, \ldots, x^n)$ of real numbers, where we may recall from §6.1 that $\mathbb{R}$ stands for the system of real numbers. In our gluing procedure, there will be *transition functions* that express the coordinates in one patch in terms of the coordinates in another, wherever in the manifold $\mathcal{M}$ we find one coordinate patch overlapping with another.

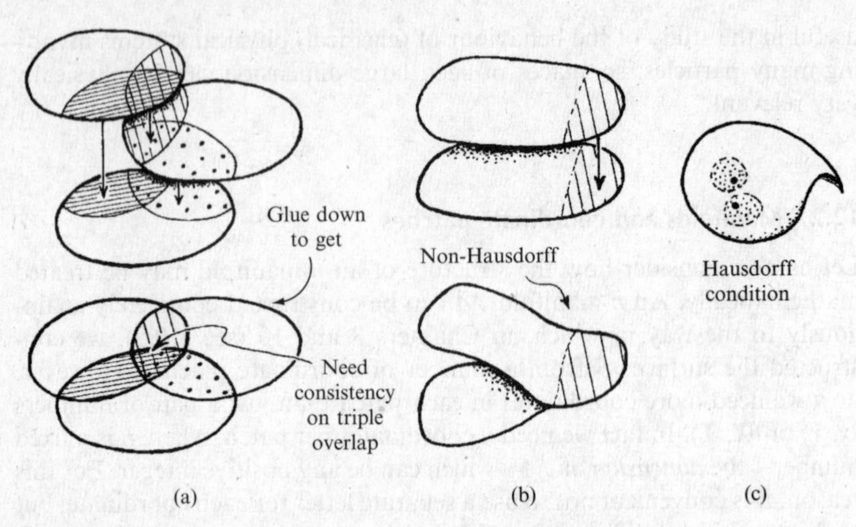

Fig. 12.5 (a) The transition functions that translate between coordinates in overlapping patches must satisfy a relation of consistency on every triple overlap. (b) The (open-set) overlap regions between pairs of patches must be appropriate; otherwise the 'branching' that characterizes a non-Haudorff space can occur. (c) A Hausdorff space is one with the property that any two distinct points possess neighbourhoods that do not overlap. (In (b), in order that the 'glued' part be an open set, its 'edge', where branching occurs, must remain separated, and it is along here that the Hausdorff condition fails.)

These transition functions must satisfy certain conditions among themselves to ensure the consistency of the whole procedure. The procedure is illustrated in Fig. 12.5a. But we must be careful, in order to produce the standard kind of manifold,[4] which is a *Hausdorff space*. (Non-Hausdorff manifolds can 'branch', in ways such as that indicated in Fig. 12.5b, see also Fig. 8.2c.) A Hausdorff space has the defining property that, for any two distinct points of the space, there are open sets containing each which do not intersect (Fig. 12.5c).

It is important to realize, however, that a manifold $\mathcal{M}$ is not to be thought of as 'knowing' where these individual patches are or what the particular coordinate values at some point might happen to be. A reasonable way to think of $\mathcal{M}$ is that it can be built up in some means, by the piecing together of a number of coordinate patches in this way, but then we choose to 'forget' the specific way in which these coordinate patches have been introduced. The manifold stands on its own as a mathematical structure, and the coordinates are just auxiliaries that can be reintroduced as a convenience when desired. However, the precise mathematical definition of a manifold (of which there are several alternatives) would be distracting for us here.[5]

12.3 Scalars, vectors, and covectors

As in §10.2, we have the notion of a *smooth function* Φ, defined on $\mathcal{M}$ (sometimes called a *scalar field* on $\mathcal{M}$) where Φ is defined, in any local coordinate patch, as a smooth function of the n coordinates in that patch. Here, 'smooth' will always be taken in the sense 'C^∞-smooth' (see §6.3), as this gives the most convenient theory. On each overlap between two patches, the coordinates on each patch are smooth functions of the coordinates on the other, so the smoothness of Φ in terms of one set of coordinates, on the overlap, implies its smoothness in terms of the other. In this way, the local ('patchwise') definition of smoothness of a scalar function Φ extends to the whole of $\mathcal{M}$, and we can speak simply of the *smoothness* of Φ on $\mathcal{M}$.

Next, we can define the notion of a *vector field* $\boldsymbol{\xi}$ on $\mathcal{M}$, which should be something with the geometrical interpretation as a family of 'arrows' on $\mathcal{M}$ (Fig. 10.5), where $\boldsymbol{\xi}$ is something which acts on any (smooth) scalar field Φ to produce another scalar field $\boldsymbol{\xi}(\Phi)$ in the manner of a differentiation operator. The interpretation of $\boldsymbol{\xi}(\Phi)$ is to be the 'rate of increase' of Φ in the direction indicated by the arrows that represent $\boldsymbol{\xi}$, just as for the 2-surfaces of §10.3. Being a 'differentiation operator', $\boldsymbol{\xi}$ satisfies certain characteristic algebraic relations (basically things that we have seen before in §6.5, namely $\mathrm{d}(f+g) = \mathrm{d}f + \mathrm{d}g$, $\mathrm{d}(fg) = f\mathrm{d}g + g\mathrm{d}f$, $\mathrm{d}a = 0$ if a is constant):

$$\boldsymbol{\xi}(\Phi + \Psi) = \boldsymbol{\xi}(\Phi) + \boldsymbol{\xi}(\Psi),$$
$$\boldsymbol{\xi}(\Phi\Psi) = \Phi\boldsymbol{\xi}(\Psi) + \Psi\boldsymbol{\xi}(\Phi),$$
$$\boldsymbol{\xi}(k) = 0 \text{ if } k \text{ is a constant.}$$

In fact, there is a theorem that tells us that these algebraic properties are *sufficient* to characterize $\boldsymbol{\xi}$ as a vector field.[6]

We can also use such purely algebraic means to define a *1-form* or, what is another name for the same thing, a *covector field*. (We shall be coming to the geometrical meaning of a covector shortly.) A covector field $\boldsymbol{\alpha}$ can be thought of as a map from vector fields to scalar fields, the action of $\boldsymbol{\alpha}$ on $\boldsymbol{\xi}$ being written $\boldsymbol{\alpha}\cdot\boldsymbol{\xi}$ (the *scalar product* of $\boldsymbol{\alpha}$ with $\boldsymbol{\xi}$; see §10.4), where, for any vector fields $\boldsymbol{\xi}$ and $\boldsymbol{\eta}$, and scalar field Φ we have *linearity*:

$$\boldsymbol{\alpha}\cdot(\boldsymbol{\xi} + \boldsymbol{\eta}) = \boldsymbol{\alpha}\cdot\boldsymbol{\xi} + \boldsymbol{\alpha}\cdot\boldsymbol{\eta},$$
$$\boldsymbol{\alpha}\cdot(\Phi\boldsymbol{\xi}) = \Phi(\boldsymbol{\alpha}\cdot\boldsymbol{\xi}).$$

These relations define covectors as *dual* objects to vectors (and this is what the prefix 'co' refers to). The relation between vectors and covectors turns out to be symmetrical, so we have corresponding expressions

$$(\boldsymbol{\alpha}+\boldsymbol{\beta})\bullet\boldsymbol{\xi}=\boldsymbol{\alpha}\bullet\boldsymbol{\xi}+\boldsymbol{\beta}\bullet\boldsymbol{\xi},$$
$$(\Phi\boldsymbol{\alpha})\bullet\boldsymbol{\xi}=\Phi(\boldsymbol{\alpha}\bullet\boldsymbol{\xi}),$$

leading to the definition of the sum of two covectors and the product of a covector by a scalar. When we take the dual of the space of covectors we get the original space of vectors, all over again. (In other words, a '*co*-covector' would be a vector.)

We can take these relations to be referring to entire fields or else merely to entities defined at a single point of $\mathcal{M}$. Vectors taken at a particular fixed point o constitute a *vector space*. (As described in §11.1, in a vector space, we can add elements $\boldsymbol{\xi}$ and $\boldsymbol{\eta}$, to form their sum $\boldsymbol{\xi}+\boldsymbol{\eta}$, with $\boldsymbol{\xi}+\boldsymbol{\eta}=\boldsymbol{\eta}+\boldsymbol{\xi}$ and $(\boldsymbol{\xi}+\boldsymbol{\eta})+\boldsymbol{\zeta}=\boldsymbol{\xi}+(\boldsymbol{\eta}+\boldsymbol{\zeta})$, and we can multiply them by scalars—here, real numbers f and g—where $(f+g)\boldsymbol{\xi}=f\boldsymbol{\xi}+g\boldsymbol{\xi}, f(\boldsymbol{\xi}+\boldsymbol{\eta})=f\boldsymbol{\xi}+f\boldsymbol{\eta}, f(g\boldsymbol{\xi})=(fg)\boldsymbol{\xi}$, $1\boldsymbol{\xi}=\boldsymbol{\xi}$.) We may regard this (flat) vector space as providing the structure of the manifold in the immediate neighbourhood of o (see Fig. 12.6). We call this vector space the *tangent space* T_o, to $\mathcal{M}$ at o. T_o may be intuitively understood as the limiting space that is arrived at when smaller and smaller neighbourhoods of o in $\mathcal{M}$ are examined at correspondingly greater and greater magnification. The immediate vicinity of o, in $\mathcal{M}$, thus appears to be infinitely 'stretched out' under this examination. In the limit, any 'curvature' of $\mathcal{M}$ would be 'ironed out flat' to give the *flat* structure of T_o. The vector space T_o has the (finite) dimension n, because we can find a set of *n basis elements*, namely the quantities $\partial/\partial x^1, \ldots, \partial/\partial x^n$, at the point o, pointing along coordinate axes, in terms of which any element of T_o can be uniquely linearly expressed (see also §13.5).

We can form the *dual* vector space to T_o (the space of covectors at o) in the way described above, and this is called the *cotangent space* T_o^* to $\mathcal{M}$ at o. A particular case of a covector *field* is the *gradient* (or exterior derivative) $\mathrm{d}\Phi$ of a scalar field Φ. (We have encountered this notation already, in

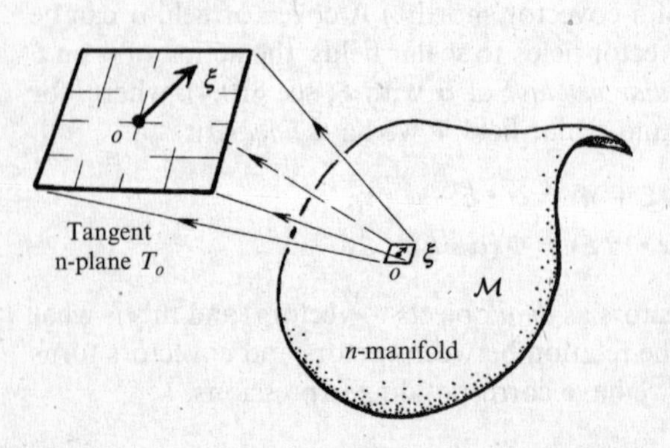

Fig. 12.6 The tangent space T_o, to an n-manifold $\mathcal{M}$ at a point o may be intuitively understood as the limiting space, when smaller and smaller neighbourhoods of o in $\mathcal{M}$ are examined at correspondingly greater and greater magnifications. (Compare Fig. 10.6.) The resulting T_o is flat: an n-dimensional vector space.

the 2-dimensional case, see §10.3). The covector $\mathrm{d}\Phi$ (with components $\partial\Phi/\partial x^1, \ldots \partial\Phi/\partial x^n$) has the defining property

$$\mathrm{d}\Phi \bullet \boldsymbol{\xi} = \boldsymbol{\xi}(\Phi).$$

(See also §10.4.)[12.3] Although not all covectors have the form $\mathrm{d}\Phi$, for some Φ, they can all be expressed in this way at any single point. We shall see in a moment why this does not extend to covector *fields*.

What is the geometrical difference between a covector and a vector? At each point of $\mathcal{M}$, a (non-zero) covector $\boldsymbol{\alpha}$ determines an $(n-1)$-*dimensional plane element*. The directions lying within this $(n-1)$-plane element are those determined by vectors $\boldsymbol{\xi}$ for which $\boldsymbol{\alpha} \bullet \boldsymbol{\xi} = 0$; see Fig. 12.7. In the particular case when $\boldsymbol{\alpha} = \mathrm{d}\Phi$, these $(n-1)$-plane elements are tangential to the family of $(n-1)$-dimensional surfaces[12.4] of constant Φ (which generalizes the notion of 'contour lines', as illustrated in Fig. 10.8a). However, in general the $(n-1)$-plane elements defined by a covector $\boldsymbol{\alpha}$ would twist around in a way that prevents them from consistently touching any such family of $(n-1)$-surfaces (see Fig. 12.8).[7]

In any particular coordinate patch, with coordinates $x^1, \ldots, x^n$, we can represent the vector (field) $\boldsymbol{\xi}$ by its set of *components* $(\xi^1, \xi^2, \ldots, \xi^n)$, these being the set of coefficients in the explicit representation of $\boldsymbol{\xi}$ in terms of partial differentiation operators

$$\boldsymbol{\xi} = \xi^1 \frac{\partial}{\partial x^1} + \xi^2 \frac{\partial}{\partial x^2} + \cdots + \xi^n \frac{\partial}{\partial x^n},$$

in the patch (see §10.4). For a vector at a particular point, $\xi^1, \ldots, \xi^n$ will just be n real numbers; for a vector field within some coordinate

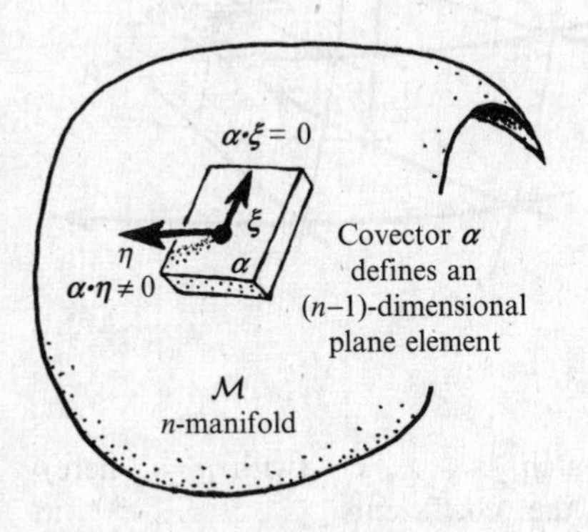

Fig. 12.7 A (non-zero) covector $\boldsymbol{\alpha}$ at a point of $\mathcal{M}$, determines an $(n-1)$-dimensional plane element there. The vectors $\boldsymbol{\xi}$ satisfying $\boldsymbol{\alpha} \bullet \boldsymbol{\xi} = 0$ define the directions within it.

[12.3] Show that '$\mathrm{d}\Phi$', defined in this way, indeed satisfies the 'linearity' requirements of a covector, as specified above.

[12.4] Why?

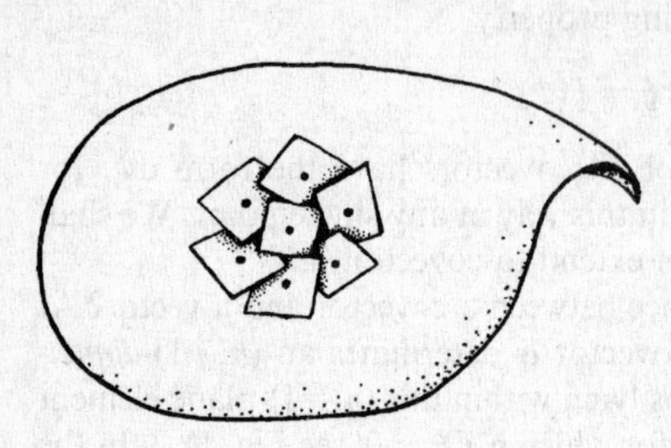

Fig. 12.8 The $(n-1)$-plane elements defined by a covector field $\boldsymbol{\alpha}$ would, in general, twist around in a way that prevents them from consistently touching a single family of $(n-1)$-surfaces—although in the particular case $\boldsymbol{\alpha} = \mathrm{d}\Phi$ (for a scalar field Φ), they would touch the surfaces $\Phi = \text{const.}$ (generalizing the 'contour lines' of Fig. 10.8).

patch, they will be n (smooth) functions of the coordinates $x^1, \ldots, x^n$ (and the reader is reminded that 'ξ^n' does *not* stand for 'the nth power of ξ', etc.). Recall that each of the operators '$\partial/\partial x^r$' stands for 'take the rate of change in the direction of the rth (local) coordinate axis'. The above expression for $\boldsymbol{\xi}$ simply expresses this vector (which, as an operator, we recall asserts 'take the rate of change in the $\boldsymbol{\xi}$-direction') as a linear combination of the vectors pointing along each of the coordinate axes (see Fig. 12.9a).

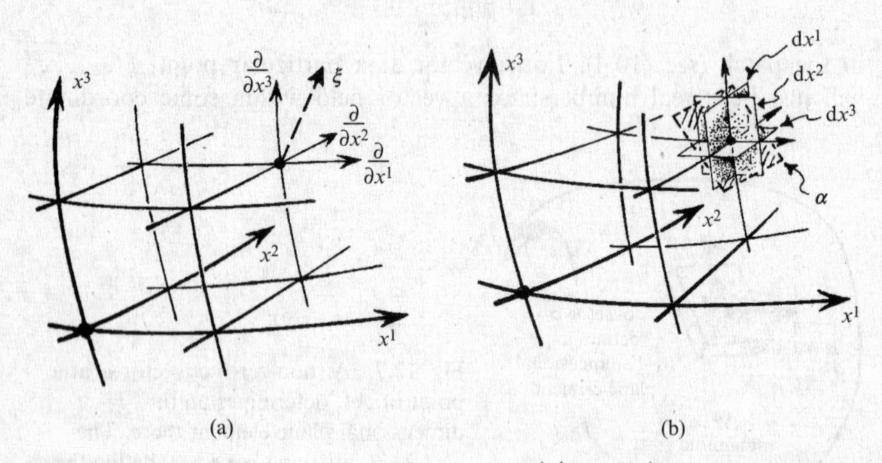

Fig. 12.9 Components in a coordinate patch $(x^1, \ldots, x^n)$ (with $n = 3$ here). (a) For a vector (field) $\boldsymbol{\xi}$, these are the coefficients $(\xi^1, \xi^2, \ldots, \xi^n)$ in $\boldsymbol{\xi} = \xi^1\partial/\partial x^1 + \xi^2\partial/\partial x^2 + \cdots + \xi^n\partial/\partial x^n$, where '$\partial/\partial x^r$' stands for 'rate of change along the rth (local) coordinate axis' (see also Fig. 10.9). (b) For a covector (field) $\boldsymbol{\alpha}$, these are the coefficients $(\alpha_1, \alpha_2, \ldots, \alpha_n)$ in $\boldsymbol{\alpha} = \alpha_1\mathrm{d}x^1 + \alpha_2\mathrm{d}x^2 + \cdots + \alpha_n\mathrm{d}x^n$, where $\mathrm{d}x^r$ stands for 'the gradient of x^r', and refers to the $(n-1)$-plane element spanned by the (local) coordinate axes except for the (local) x^r-axis.

In a similar way, a covector (field) $\boldsymbol{\alpha}$ is represented, in the coordinate patch, by a set of components $(\alpha_1, \alpha_2, \ldots, \alpha_n)$ in the patch, where now we write

$$\boldsymbol{\alpha} = \alpha_1 \mathrm{d}x^1 + \alpha_2 \mathrm{d}x^2 + \cdots + \alpha_n \mathrm{d}x^n,$$

expressing $\boldsymbol{\alpha}$ as a linear combination of the basic 1-forms (covectors)[8] $\mathrm{d}x^1, \mathrm{d}x^2, \ldots, \mathrm{d}x^n$. Geometrically, each $\mathrm{d}x^r$ refers to the $(n-1)$-plane element spanned by all the coordinate axes with the exception of the x^r-axis (see Fig. 12.9b).[12.5] The scalar product $\boldsymbol{\alpha} \bullet \boldsymbol{\xi}$ is given by the expression[12.6]

$$\boldsymbol{\alpha} \bullet \boldsymbol{\xi} = \alpha_1 \xi^1 + \alpha_2 \xi^2 + \cdots + \alpha_n \xi^n.$$

12.4 Grassmann products

Let us now consider the representation of plane elements of various other dimensions, using the idea of a *Grassmann product*, as defined in §11.6. A 2-plane element at a point of $\mathcal{M}$ (or a field of 2-plane elements over $\mathcal{M}$) will be represented by a quantity

$$\boldsymbol{\xi} \wedge \boldsymbol{\eta},$$

where $\boldsymbol{\xi}$ and $\boldsymbol{\eta}$ are two independent vectors (or vector fields) spanning the 2-plane(s) (see Figs. 11.6a and 12.10a). A quantity $\boldsymbol{\xi} \wedge \boldsymbol{\eta}$ is sometimes referred to as a (simple) *bivector*. Its components, in terms of those of $\boldsymbol{\xi}$ and $\boldsymbol{\eta}$, are the expressions

$$\xi^{[r}\eta^{s]} = \frac{1}{2}(\xi^r \eta^s - \xi^s \eta^r),$$

as described towards the end of the last chapter. A sum $\boldsymbol{\psi}$ of simple bivectors $\boldsymbol{\xi} \wedge \boldsymbol{\eta}$ is also called a bivector; its components ψ^{rs} have the characteristic property that they are *antisymmetric* in r and s, i.e. $\psi^{rs} = -\psi^{sr}$.

Similarly, a 3-plane element (or a field of such) would be represented by a simple *trivector*

[12.5] For example, show that $\mathrm{d}x^2$ has components $(0, 1, 0, \ldots, 0)$ and represents the tangent hyperplane elements to $x^2 =$ constant.

[12.6] Show, by use of the chain rule (see §10.3), that this expression for $\boldsymbol{\alpha} \bullet \boldsymbol{\xi}$ is consistent with $\mathrm{d}\Phi \bullet \boldsymbol{\xi} = \boldsymbol{\xi}(\Phi)$, in the particular case $\boldsymbol{\alpha} = \mathrm{d}\Phi$.

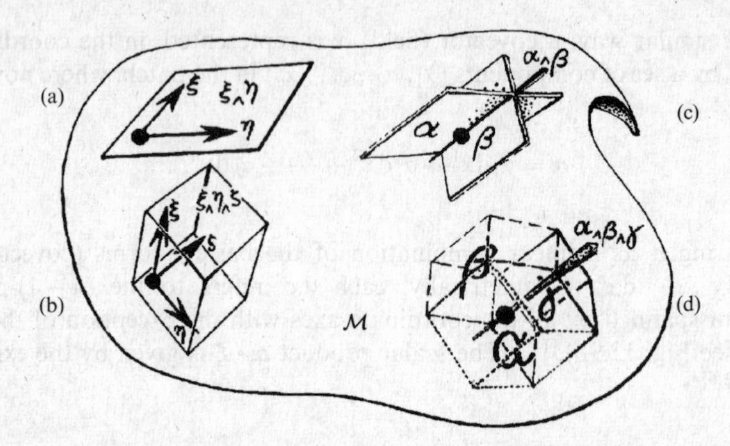

Fig. 12.10 (a) A 2-plane element at a point of $\mathcal{M}$, being spanned by independent vectors $\boldsymbol{\xi}$, $\boldsymbol{\eta}$, is described by the bivector $\boldsymbol{\xi} \wedge \boldsymbol{\eta}$. (b) Similarly, a 3-plane element spanned by $\boldsymbol{\xi}$, $\boldsymbol{\eta}$, $\boldsymbol{\zeta}$ is described by $\boldsymbol{\xi} \wedge \boldsymbol{\eta} \wedge \boldsymbol{\zeta}$. (c) Dually, an $(n-2)$-plane element, the intersection of two $(n-1)$-plane elements specified by 1-forms $\boldsymbol{\alpha}$, $\boldsymbol{\beta}$, is described by $\boldsymbol{\alpha} \wedge \boldsymbol{\beta}$. (d) The $(n-3)$-plane element of intersection of the three $(n-1)$-plane elements specified by $\boldsymbol{\alpha}$, $\boldsymbol{\beta}$, $\boldsymbol{\gamma}$, is described by $\boldsymbol{\alpha} \wedge \boldsymbol{\beta} \wedge \boldsymbol{\gamma}$.

$$\boldsymbol{\xi} \wedge \boldsymbol{\eta} \wedge \boldsymbol{\zeta},$$

where the vectors $\boldsymbol{\xi}$, $\boldsymbol{\eta}$, $\boldsymbol{\zeta}$ span the 3-plane (Figs. 11.6b and 12.10b), its components being

$$\xi^{[r}\eta^{s}\zeta^{t]} = \frac{1}{6}(\xi^{r}\eta^{s}\zeta^{t} + \xi^{s}\eta^{t}\zeta^{r} + \xi^{t}\eta^{r}\zeta^{s} - \xi^{r}\eta^{t}\zeta^{s} - \xi^{t}\eta^{s}\zeta^{r} - \xi^{s}\eta^{r}\zeta^{t}).$$

The general trivector $\boldsymbol{\tau}$ has completely antisymmetric components τ^{rst}, and would always be a sum of such simple trivectors. We can go on in a similar way to define 4-plane elements, represented by simple 4-vectors, and so on. The *general* n-vector has sets of components that are completely antisymmetric. It would always be expressible as a *sum* of simple n-vectors.

There is an issue arising here which may seem puzzling. It appears that we now have two different ways of representing an $(n-1)$-plane element, either as a 1-form (covector) or else as an $(n-1)$-vector quantity, obtained by 'wedging' together $n-1$ independent vectors spanning the $(n-1)$-plane. There is in fact a geometrical distinction between the quantities described in these two different ways, but it is a somewhat subtle one. The distinction is that the 1-form should be thought of as a kind of 'density', whereas the $(n-1)$-vector should not. In order to make this clearer, it will be helpful first to introduce the notion of a general p-form.

Essentially, we shall proceed just as for multivectors above, but starting with 1-forms rather than vectors. Given a number p of (independent) 1-forms $\boldsymbol{\alpha}, \boldsymbol{\beta}, \ldots, \boldsymbol{\delta}$, we can form their wedge product

$$\boldsymbol{\alpha} \wedge \boldsymbol{\beta} \wedge \cdots \wedge \boldsymbol{\delta},$$

this having components given by

$$\alpha_{[r}\beta_s \ldots \delta_{u]}$$

in a coordinate patch (using the general square-bracket-around-indices notation of §11.6). Such a quantity determines an $(n-p)$-plane element (or a field of such), this element being the *intersection* of the various $(n-1)$-plane elements determined by $\boldsymbol{\alpha}, \boldsymbol{\beta}, \ldots \boldsymbol{\delta}$ individually (Fig. 12.10c,d). This quantity is called a *simple p-form*. As was the case with p-vectors, the most general p-form is not expressible as a direct wedge product of covectors, however (except in the particular cases $p = 0, 1, n-1, n$), but is a sum of terms that are so expressible. In components, a general p-form $\boldsymbol{\varphi}$ is represented (in any coordinate patch) by a set of quantities

$$\varphi_{rs\ldots u}$$

(where each of $r, s, \ldots, u$ ranges over $1, \ldots, n$) which is *antisymmetrical* in its indices $r, s, \ldots, u$, these being p in number. As before, antisymmetry means that if we interchange any pair of index labels, we get a quantity that is precisely the negative of what we had before. In terms of our square-bracket notation (§11.6), we can express this antisymmetry property in the equation[12.7]

$$\varphi_{[rs\ldots u]} = \varphi_{rs\ldots u}.$$

It may also be remarked here that the $(p+q)$-form $\boldsymbol{\varphi} \wedge \boldsymbol{\chi}$, which is the wedge product of the p-form $\boldsymbol{\varphi}$ with a q-form $\boldsymbol{\chi}$, has components

$$\varphi_{[rs\ldots u}\chi_{jk\ldots m]},$$

the antisymmetrization being taken right across all the indices (where $\chi_{jk\ldots m}$ are the components of $\boldsymbol{\chi}$).[12.8] A similar notation applies for the wedge product of a p-vector with a q-vector.

12.5 Integrals of forms

Now let us return to the 'density' aspect of a p-form. Recall that, in ordinary physics, the *density* of an object is its mass per unit volume.

[12.7] Explain why this works.

[12.8] Justify the fact that $\boldsymbol{\varphi} \wedge \boldsymbol{\chi} = \boldsymbol{\alpha} \wedge \cdots \wedge \boldsymbol{\gamma} \wedge \boldsymbol{\lambda} \wedge \cdots \wedge \boldsymbol{\nu}$ where $\boldsymbol{\varphi} = \boldsymbol{\alpha} \wedge \cdots \wedge \boldsymbol{\gamma}$, $\boldsymbol{\chi} = \boldsymbol{\lambda} \wedge \cdots \wedge \boldsymbol{\nu}$.

This density is a property of the material of which the body is composed. We use this 'density' notion when we wish to evaluate the total mass of the object when we know its total volume and the nature of its material. Mathematically, what we would do would be to integrate its density over the volume that it occupies. Basically, the point about a density is that it is the appropriate kind of quantity that we can integrate over some region; it is the kind of quantity that we place after an integral sign. We should be a little careful here to distinguish integrals over spaces of different dimension, however. ('Mass per unit area' is a different kind of quantity from 'mass per unit volume', for example.) We shall find that a p-form is the appropriate quantity to integrate over a p-dimensional space.

Let us start with a 1-form. This is the simplest case. We are concerned with the integral of a quantity over a 1-dimensional manifold, that is, along some curve γ. Recall from §6.6 that ordinary (1-dimensional) integrals are things that are written

$$\int f(x)\,\mathrm{d}x,$$

where x is some real-valued quantity that we can take to be a parameter along the curve γ. We are to think of the quantity '$f(x)\,\mathrm{d}x$' as denoting a 1-form. The notation for 1-forms has, indeed, been carefully tailored to be consistent with the notation for ordinary integrals. This is a feature of the 20th-century calculus known as the *exterior calculus*, introduced by the outstanding French mathematician Élie Cartan (1869–1951), whom we shall encounter again in Chapters 13, 14, and 17, and it dovetails beautifully with the '$\mathrm{d}x$' notation introduced in the 17th century by Gottfried Wilhelm Leibniz (1646–1716). In Cartan's scheme we do not think of '$\mathrm{d}x$' as denoting an 'infinitesimal quantity', however, but as providing us with the appropriate kind of density (1-form) that one may integrate over a curve.

One of the beauties of this notation is that it automatically deals with any changes of variable that we may choose to invoke. If we change the parameter x to another one X, say, then the 1-form $\boldsymbol{\alpha} = f(x)\mathrm{d}x$ is deemed to remain the same—in the sense that $\int \boldsymbol{\alpha}$ remains the same—even though its explicit functional expression in terms of the given variable (x or X) will change.[12.9] We can also regard the 1-form $\boldsymbol{\alpha}$ as being defined throughout some larger-dimensional ambient space within which our curve resides. The parameter x or X could be taken to be one of the coordinates in a coordinate patch in this ambient space, where we are happy to change to a different coordinate when we pass to another coordinate patch. Everything takes care of itself. We can simply write this integral as

[12.9] Show this explicitly, explaining how to treat the limits, for a definite integral $\int_a^b \boldsymbol{\alpha}$.

$$\int \boldsymbol{\alpha} \quad \text{or} \quad \int_{\mathcal{R}} \boldsymbol{\alpha},$$

where $\mathcal{R}$ stands for some portion of the given curve γ, over which the integral is to be taken.

What about integrals over regions of higher dimension? For a 2-dimensional region, we need a 2-form after the integral sign.[9] This could be some quantity $f(x, y)\mathrm{d}x \wedge \mathrm{d}y$ (or a sum of things like this) and we can write

$$\int_{\mathcal{R}} f(x, y)\,\mathrm{d}x \wedge \mathrm{d}y = \int_{\mathcal{R}} \boldsymbol{\alpha}$$

(or a sum of such quantities), where $\mathcal{R}$ is now a 2-dimensional region over which the integral is to be performed, lying within some given 2-surface. Again, the parameters x and y, locally coordinatizing the surface, can be replaced by any other such pair, and the notation takes care of itself. This applies perfectly well if the 2-form inhabits some ambient higher-dimensional space within which the 2-region $\mathcal{R}$ resides. All this works also for 3-forms integrated over 3-dimensional regions or 4-forms integrated over 4-dimensional regions, etc. The wedge product in Cartan's differential-form notation (together with the exterior derivative of §12.6) takes care of everything if we choose to change our coordinates. (This eliminates the explicit mention of awkward quantities known as 'Jacobians', which would otherwise have to be brought in.)[12.10]

Recall, from §6.6, the *fundamental theorem of calculus*, which asserts, for 1-dimensional integrals, that integration is the inverse of differentiation, or, put another way, that

$$\int_a^b \frac{\mathrm{d}f(x)}{\mathrm{d}x}\,\mathrm{d}x = f(b) - f(a).$$

Is there a higher-dimensional analogue of this? There are, indeed, analogues for different dimensions that go under various names (Ostrogradski, Gauss, Green, Kelvin, Stokes, etc.), but the general result, essentially part of Cartan's exterior calculus of differential forms, will be called here 'the fundamental theorem of exterior calculus'.[10] This depends upon Cartan's general notion of *exterior derivative*, to which we now turn.

12.6 Exterior derivative

A 'coordinate-free' route to defining this important notion is to build up the exterior derivative axiomatically as the unique operator 'd', taking

---

[12.10] Let $G = \int_{-\infty}^{\infty} e^{-x^2} dx$. Explain why $G^2 = \int_{\mathbb{R}^2} e^{-(x^2+y^2)} dx_\wedge dy$ and evaluate this by changing to polar coordinates (r,θ). (See §5.1.) Hence prove $G = \sqrt{\pi}$.

p-forms to $(p+1)$-forms, for each $p = 0, 1, \ldots n-1$, which has the properties

$$\begin{aligned} \mathrm{d}(\boldsymbol{\alpha}+\boldsymbol{\beta}) &= \mathrm{d}\boldsymbol{\alpha}+\mathrm{d}\boldsymbol{\beta}, \\ \mathrm{d}(\boldsymbol{\alpha}\wedge\boldsymbol{\gamma}) &= \mathrm{d}\boldsymbol{\alpha}\wedge\boldsymbol{\gamma}+(-1)^p\boldsymbol{\alpha}\wedge\mathrm{d}\boldsymbol{\gamma}, \\ \mathrm{d}(\mathrm{d}\boldsymbol{\alpha}) &= 0, \end{aligned}$$

$\boldsymbol{\alpha}$ being a p-form, and where $\mathrm{d}\Phi$ has the same meaning ('gradient of Φ') for a 0-form (i.e. for a scalar) that it did in our earlier discussion (defined from $\mathrm{d}\Phi \cdot \boldsymbol{\xi} = \boldsymbol{\xi}(\Phi)$, the 'd' in $\mathrm{d}x$ also being this same operation). The final equation in the above list is frequently expressed simply as

$$\mathrm{d}^2 = 0,$$

which is a key property of the exterior derivative operator d. (We can perceive that the 'reason' for the awkward-looking term $(-1)^p$ in the second displayed equation is that the 'd' following it is really 'sitting in the wrong place', having to be 'pushed through' $\boldsymbol{\alpha}$, with its p antisymmetrical indices. This is made more manifest in the index expressions below.)[12.11]

A 1-form $\boldsymbol{\alpha}$ which is a *gradient* $\boldsymbol{\alpha} = \mathrm{d}\Phi$ must satisfy $\mathrm{d}\boldsymbol{\alpha} = 0$, by the above.[12.12] But not all 1-forms satisfy this relation. In fact, if a 1-form $\boldsymbol{\alpha}$ satisfies $\mathrm{d}\boldsymbol{\alpha} = 0$, then it follows that *locally* (i.e. in a sufficiently small open set containing any given point) it has the form $\boldsymbol{\alpha} = \mathrm{d}\Phi$ for some Φ. This is an instance of the important *Poincaré lemma*,[11],[12.13] which asserts that if a p-form $\boldsymbol{\beta}$ satisfies $\mathrm{d}\boldsymbol{\beta} = 0$, then *locally* $\boldsymbol{\beta}$ has the form $\boldsymbol{\beta} = \mathrm{d}\boldsymbol{\gamma}$, for some $(p-1)$-form $\boldsymbol{\gamma}$.

Exterior derivative is clarified, and made explicit, by the use of components. Consider a p-form $\boldsymbol{\alpha}$. In a coordinate patch, with coordinates $x^1, \ldots, x^n$, we have an antisymmetrical set of components $\alpha_{r\ldots t}$ ($= \alpha_{[r\ldots t]}$, where $r, \ldots, t$ are p in number; see §11.6) to represent $\boldsymbol{\alpha}$. We can write this representation

$$\boldsymbol{\alpha} = \sum \alpha_{r\ldots t}\, \mathrm{d}x^r \wedge \cdots \wedge \mathrm{d}x^t,$$

where the summation (indicated by the symbol $\sum$) is taken over all sets of p numbers $r, \ldots, t$, each running over the range $1, \ldots, n$. (Some people prefer to avoid a redundancy in this expression which arises because the antisymmetry in the wedge product leads to each non-zero term being repeated $p!$ times. However, the notation works much better if we simply live with this redundancy—which is my much preferred choice.) The *exterior derivative* of the p-form $\boldsymbol{\alpha}$ is a $(p+1)$-form that is written $\mathrm{d}\boldsymbol{\alpha}$, which has components

---

[12.11] Using the above relations, show that $\mathrm{d}(A\mathrm{d}x + B\mathrm{d}y) = (\partial B/\partial x - \partial A/\partial y)\mathrm{d}x \wedge \mathrm{d}y$.

[12.12] Why?

[12.13] Assuming the result of Exercise [12.11], prove the Poincaré lemma for $p = 1$.

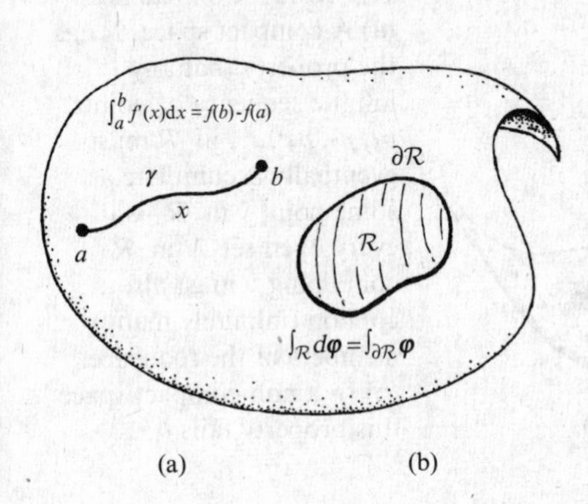

Fig. 12.11 The fundamental theorem of exterior calculus $\int_{\mathcal{R}} d\boldsymbol{\varphi} = \int_{\partial\mathcal{R}} \boldsymbol{\varphi}$. (a) The classical (17th century) case $\int_a^b f'(x)dx = f(b) - f(a)$, where $\boldsymbol{\varphi} = f(x)$ and $\mathcal{R}$ is the segment of a curve γ from a to b, parametrized by x, so $\partial\gamma$ consists of γ's end-points $x = a$ (counting negatively) and $x = b$ (positively). (b) The general case, for a p-form $\boldsymbol{\varphi}$, where $\mathcal{R}$ is a compact oriented $(p+1)$-dimensional region with p-dimensional boundary $\partial\mathcal{R}$.

$$(d\boldsymbol{\alpha})_{qr\ldots t} = \frac{\partial}{\partial x^{[q}} \alpha_{r\ldots t]},$$

(The notation looks a bit awkward here. The antisymmetrization—which is the key feature of the expression—extends across all $p+1$ indices, including the one on the derivative symbol.)[12.14],[12.15]

We are now in a position to write down the *fundamental theorem of exterior calculus*. This is expressed in the following very elegant (and powerful) formula for a p-form $\boldsymbol{\varphi}$ (see Fig. 12.11):

$$\int_{\mathcal{R}} d\boldsymbol{\varphi} = \int_{\partial\mathcal{R}} \boldsymbol{\varphi}.$$

Here $\mathcal{R}$ is some compact $(p+1)$-dimensional (oriented) region whose (oriented) p-dimensional boundary (consequently also compact) is denoted by $\partial\mathcal{R}$.

There are various words that I have employed here that I have not yet explained. For our purposes 'compact' means, intuitively, that the region $\mathcal{R}$ does not 'go off to infinity' and it does not have 'holes cut out of it' nor 'bits of its boundary removed'. More precisely, a *compact* region $\mathcal{R}$ is, for our purposes here,[12] a region with the property that any infinite

---

[12.14] Show directly that all the 'axioms' for exterior derivative are satisfied by this coordinate definition.

[12.15] Show that this coordinate definition gives the same quantity $d\boldsymbol{\alpha}$, whatever choice of coordinates is made, where the transformation of the components $\alpha_{r\ldots t}$ of a form is defined by the requirement that the form $\boldsymbol{\alpha}$ itself be unaltered by coordinate change. *Hint*: Show that this transformation is identical with the passive transformation of $\left[{0 \atop p}\right]$-valent tensor components, as given in §13.8.

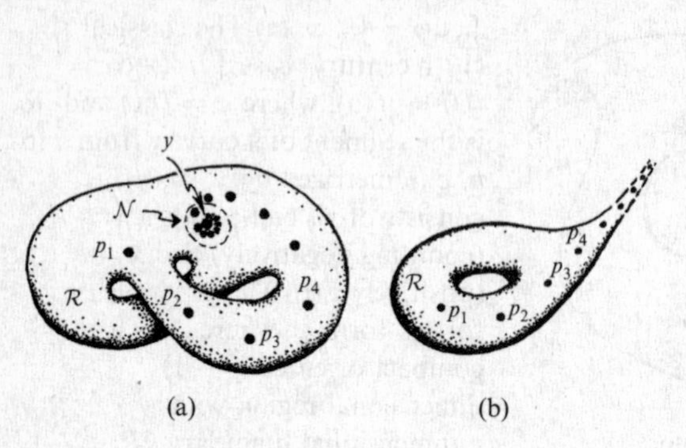

Fig. 12.12 Compactness. (a) A compact space $\mathcal{R}$ has the property that any infinite sequence of points $p_1, p_2, p_3, \ldots$ in $\mathcal{R}$ must eventually accumulate at some point y in $\mathcal{R}$—so every open set $\mathcal{N}$ in $\mathcal{R}$ containing y must also contain (infinitely many) members of the sequence. (b) In a non-compact space this property fails.

sequence of points lying in $\mathcal{R}$ must *accumulate* at some point within $\mathcal{R}$ (Fig. 12.12a). Here, an *accumulation point* y has the property that every open set in $\mathcal{R}$ (see §7.4) which contains y must also contain members of the infinite sequence (so that points of the sequence get closer and closer to y, without limit). The infinite Euclidean plane is not compact, but the surface of a sphere is, and so is the torus. So also is the set of points lying within or on the unit circle in the complex plane (closed unit disc); but if we remove the circle itself from the set, or even just the centre of the circle, then the resulting set is not compact. See Fig. 12.13.

The term 'oriented' refers to the assignment of a consistent 'handedness' at every point of $\mathcal{R}$ (Fig. 12.14). For a 0-manifold, or set of discrete *points*, the orientation simply assigns a 'positive' (+) or 'negative value' (−) to each point (Fig. 12.14a). For a 1-manifold, or *curve*, this orientation provides a 'direction' along the curve. This can be represented in a diagram by the placement of an 'arrow' on the curve to indicate this direction (Fig. 12.14b). For a 2-manifold, the orientation can be diagrammatically represented by a tiny circle or circular arc with an arrow on it (Fig. 12.14c); this indicates which rotation of a tangent vector at a point of the surface is considered to be in the 'positive' direction. For a 3-manifold the orientation specifies which triad of independent vectors at a point is to be regarded as 'right-handed' and which as 'left-handed' (recall §11.3 and Fig. 11.1). See Fig. 12.14d. Only for rather unusual spaces is it not possible to assign an orientation consistently. A ('non-orientable') example for which this cannot be done is the *Mobius strip*, as illustrated in Fig. 12.15.

The *boundary* $\partial\mathcal{R}$ of a (compact oriented) $(p+1)$-dimensional region $\mathcal{R}$ consists of those points of $\mathcal{R}$ that do not lie in its interior. If $\mathcal{R}$ is suitably

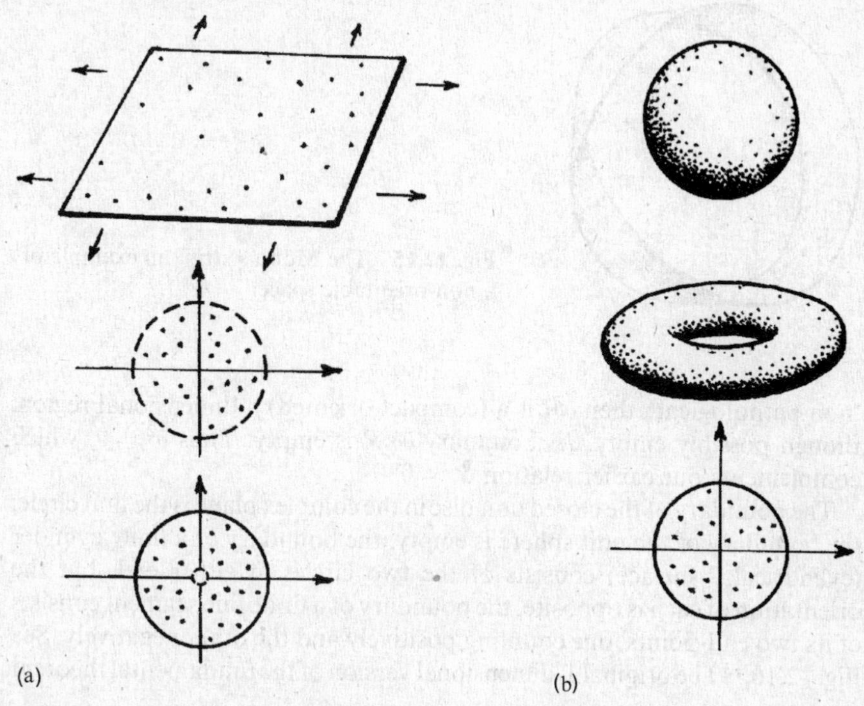

Fig. 12.13 (a) Some non-compact spaces: the infinite Euclidean plane, the open unit disc, and the closed disc with the centre removed. (b) Some compact spaces: the sphere, the torus, and the closed unit disc. (Solid boundary lines are part of the set; broken boundary lines are not.)

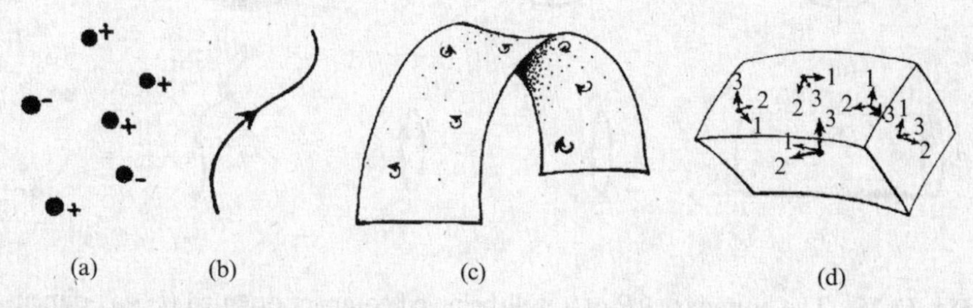

Fig. 12.14 Orientation. (a) A (multi-component) 0-manifold is a set of discrete points; the orientation simply assigns a 'positive' (+) or 'negative' (−) value to each. (b) For a 1-manifold, or curve, the orientation provides a 'direction' along the curve; represented in a diagram by the placement of an arrow on it. (c) For a 2-manifold, the orientation can be indicated by a tiny circular arc with an arrow on it, indicating the 'positive' direction of rotation of a tangent vector. (d) For a 3-manifold the orientation specifies which triads of independent vectors at a point are to be regarded as 'right-handed' (cf. Fig. 11.1).

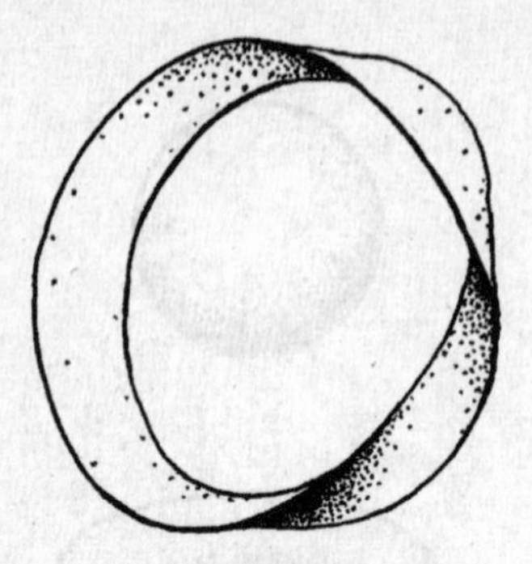

Fig. 12.15 The Möbius strip: an example of a non-orientable space.

'non-pathological', then $\partial\mathcal{R}$ is a (compact oriented) p-dimensional region, though possibly empty. *Its* boundary $\partial\partial\,\mathcal{R}$ is empty. Thus $\partial^2 = 0$, which complements our earlier relation $d^2 = 0$.

The boundary of the closed unit disc in the complex plane is the unit circle; the boundary of the unit sphere is empty, the boundary of a finite cylinder (cylindrical 2-surface) consists of the two circles at either end, but the orientation of each is opposite, the boundary of a finite line segment consists of its two end-points, one counting positively and the other negatively. See Fig. 12.16.[13] The original 1-dimensional version of the fundamental theorem

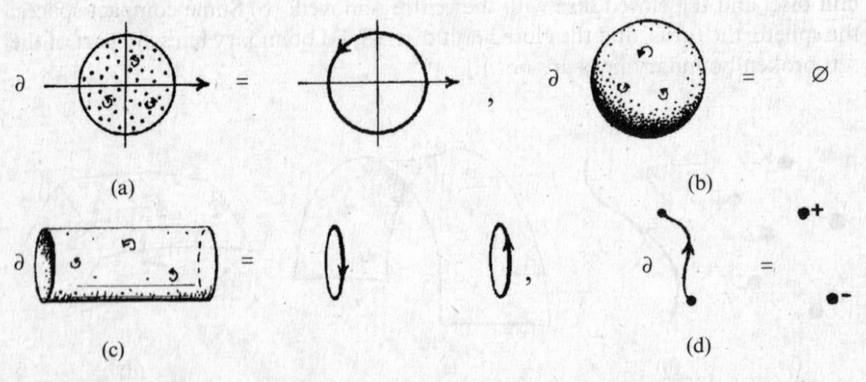

Fig. 12.16 The boundary $\partial\mathcal{R}$ of a well-behaved compact oriented $(p+1)$-dimensional region $\mathcal{R}$ is a (compact oriented) p-dimensional region (possibly empty), consisting of those points of $\mathcal{R}$ that do not lie in the $(p+1)$-dimensional interior. (a) The boundary of the closed unit disc (given by $|z| \leq 1$ in the complex plane $\mathbb{C}$) is the unit circle. (b) The boundary of the unit sphere is empty ($\varnothing$ denoting the empty set, see §3.4). (c) The boundary of a finite length of cylindrical surface consists of the two circles at either end, the orientation of each being opposite. (d) The boundary of a finite curve segment consists of two end-points, one positive and the other negative.

of calculus, as exhibited above, comes out as a special case of the fundamental theorem of exterior calculus, when $\mathcal{R}$ is taken to be such a line segment.

12.7 Volume element; summation convention

Let us now return to the distinction between—and the relation between—a p-form and an $(n-p)$-vector in an n-manifold $\mathcal{M}$. To understand this relationship, it is best to go first to the extreme case where $p = n$, so we are examining the relation between an n-form and a scalar field on $\mathcal{M}$. In the case of an n-form $\boldsymbol{\varepsilon}$, the associated n-surface element at a point o of $\mathcal{M}$ is just the entire tangent n-plane at o. The *measure* that $\boldsymbol{\varepsilon}$ provides is simply an n-density, with no directional properties at all. Such an n-density (assumed nowhere zero) is sometimes referred to as a *volume element* for the n-manifold $\mathcal{M}$. A volume element can be used to convert $(n-p)$-vectors to p-forms, and vice versa. (Sometimes there is a volume element assigned to a manifold, as part of its assigned 'structure'; in that case, the essential distinction between a p-form and an $(n-p)$-vector disappears.)

How can we use a volume element to convert an $(n-p)$-vector to a p-form? In terms of components, the n-form $\boldsymbol{\varepsilon}$ would be represented, in each coordinate patch, by a quantity with n antisymmetric lower indices:

$$\varepsilon_{r\dots w}.$$

(Some people might prefer to incorporate a factor $(n!)^{-1}$ into this; for '!' see §5.3. However, I shall not concern myself with the various awkward factorials that arise here, as they distract from the main ideas.) We can use the quantity $\varepsilon_{r\dots w}$ to convert the family of components $\psi^{u\dots w}$ of an $(n-p)$-vector $\boldsymbol{\psi}$ into the family of components $\alpha_{r\dots t}$ of a p-form $\boldsymbol{\alpha}$. We do this by taking advantage of the operations of *tensor algebra*, which we shall come to more fully in the next section. This algebra enables us to 'glue' the $n-p$ upper indices of $\psi^{u\dots w}$ to $n-p$ of the n lower indices of $\varepsilon_{r\dots w}$, leaving us with the p unattached lower indices that we need for $\alpha_{r\dots t}$. The 'gluing' operation that comes in here is what is referred to as tensor 'contraction' (or 'transvection'), and it enables each upper index to be paired off with a corresponding lower index, the two being 'summed over', so that both sets of indices are removed from the final expression.

The archetypical example of this is the *scalar product* (§12.3), which combines the components β_r of a covector $\boldsymbol{\beta}$ with the components ξ^r of a vector $\boldsymbol{\xi}$ by multiplying corresponding elements of the two sets of components together and then 'summing over' repeated indices to get

$$\boldsymbol{\beta}\cdot\boldsymbol{\xi} = \sum \beta_r \xi^r,$$

where the summation refers to the repeated index r (one up, one down). This summation procedure applies also with many-indexed quantities, and physicists find it exceedingly convenient to adopt a convention introduced by Einstein, referred to as the *summation convention*. What this convention amounts to is the omission of the actual summation signs, and it is assumed that a summation is taking place between a lower and an upper index whenever the same index letter appears in both positions in a term, the summation always being over the index values $1, \ldots, n$. Accordingly, the scalar product would now be written simply as

$$\boldsymbol{\beta} \cdot \boldsymbol{\xi} = \beta_r \xi^r.$$

Using this convention, we can write the procedure outlined above for expressing a p-form in terms of a corresponding $(n-p)$-vector and a volume form as

$$\alpha_{r \ldots t} \propto \varepsilon_{r \ldots tu \ldots w} \psi^{u \ldots w}$$

with contraction over the $n-p$ indices $u, \ldots, w$. Here, I am introducing the symbol '$\propto$', which stands for 'is proportional to', meaning that each side is a non-zero multiple of the other. This is so that our expressions do not get confusingly cluttered with complicated-looking factorials. We sometimes say that the $(n-p)$-vector $\boldsymbol{\psi}$ and the p-form $\boldsymbol{\alpha}$ are *dual*[14] to one another if this relation (up to proportionality) holds, in which case there will also be a corresponding inverse formula

$$\psi^{u \ldots w} \propto \alpha_{r \ldots t} \in^{r \ldots tu \ldots w}$$

for some suitable reciprocal volume form (n-vector) $\boldsymbol{\epsilon}$, often 'normalized' against $\boldsymbol{\varepsilon}$ according to

$$\boldsymbol{\varepsilon} \cdot \boldsymbol{\epsilon} = \varepsilon_{r \ldots w} \in^{r \ldots w} = n!$$

(although matters of normalization are not our main concern here).

These formulae are part of classical tensor algebra (see §12.8). This provides a powerful manipulative procedure (also extended to tensor calculus, of which we shall see more in Chapter 14), which gains much from the use of an index notation combined with Einstein's summation convention. The square-bracket notation for antisymmetrization (see §11.6) also plays a valuable role in this algebra, as does an additional round-bracket notation for *symmetrization*,

$$\psi^{(ab)} = \frac{1}{2}(\psi^{ab} + \psi^{ba}),$$

$$\psi^{(abc)} = \frac{1}{6}(\psi^{abc} + \psi^{acb} + \psi^{bca} + \psi^{bac} + \psi^{cab} + \psi^{cba}),$$

etc.,

in which all the minus signs defining the square bracket are replaced with plus signs.

As a further example of the value of the bracket notation, let us see how to write down the condition that a p-form $\boldsymbol{\alpha}$ or a q-vector $\boldsymbol{\psi}$ be *simple*, that is, the wedge product of p individual 1-forms or of q ordinary vectors. In terms of components, this condition turns out to be

$$\alpha_{[r\ldots t}\alpha_{u]v\ldots w} = 0 \quad \text{or} \quad \psi^{[r\ldots t}\psi^{u]v\ldots w} = 0,$$

where all indices of the first factor are 'skewed' with just one index of the second.[15] If $\boldsymbol{\alpha}$ and $\boldsymbol{\psi}$ happened to be dual to one another, then we could write either condition alternatively as

$$\psi^{r\ldots tu}\alpha_{uv\ldots w} = 0,$$

where a single index of $\boldsymbol{\psi}$ is contracted with a single index of $\boldsymbol{\alpha}$. The symmetry of this expression shows that the dual of a simple p-form is a simple $(n-p)$-vector and conversely.[12.16]

12.8 Tensors: abstract-index and diagrammatic notation

There is an issue that arises here which is sometimes seen as a conflict between the notations of the mathematician and the physicist. The two notations are exemplified by the two sides of the above equation, $\boldsymbol{\beta}\cdot\boldsymbol{\xi} = \beta_r\xi^r$. The mathematician's notation is manifestly independent of coordinates, and we see that the expression $\boldsymbol{\beta}\cdot\boldsymbol{\xi}$ (for which a notation such as $(\boldsymbol{\beta}, \boldsymbol{\xi})$ or $\langle\boldsymbol{\beta}, \boldsymbol{\xi}\rangle$ might be more common in the mathematical literature) makes no reference to any coordinate system, the scalar product operation being defined in entirely geometric/algebraic terms. The physicist's expression $\beta_r\xi^r$, on the other hand, refers explicitly to components in some coordinate system. These components would change when we move from coordinate patch to coordinate patch; moreover, the notation depends upon the 'objectionable' summation convention (which is in conflict with much standard mathematical usage). Yet, there is a great flexibility in the physicist's notation, particularly in the facility with which it can be used to construct new operations that do not come readily within the scope of the mathematician's specified operations. Somewhat complicated calculations (such as those that relate the last couple of displayed formulae above) are often almost unmanageable if one insists upon sticking to index-free expressions. Pure mathematicians often find themselves resorting to 'coordinate-patch' calculations

[12.16] Confirm the equivalence of all these conditions for simplicity; prove the sufficiency of $\alpha_{[rs}\alpha_{u]v} = 0$ in the case p = 2. (*Hint*: contract this expression with two vectors.)

(with some embarrassment!)—when some essential calculational ingredient is needed in an argument—and they rarely use the summation convention.

To me, this conflict is a largely artificial one, and it can be effectively circumvented by a shift in attitude. When a physicist employs a quantity 'ξ^a', she or he would normally have in mind the actual vector quantity that I have been denoting by $\boldsymbol{\xi}$, rather than its set of components in some arbitrarily chosen coordinate system. The same would apply to a quantity 'α_a', which would be thought of as an actual 1-form. In fact, this notion can be made completely rigorous within the framework of what has been referred to as the *abstract-index notation.*[16] In this scheme, the indices do *not* stand for one of $1, 2, \ldots, n$, referring to some coordinate system; instead they are just *abstract markers* in terms of which the algebra is formulated. This allows us to retain the practical advantages of the index notation without the conceptual drawback of having to refer, whether explicitly or not, to a coordinate system. Moreover, the abstract-index notation turns out to have numerous additional practical advantages, particularly in relation to spinor-based formalisms.[17]

Yet, the abstract-index notation still suffers from the visual problem that it can be hard to make out all-important details in a formula because the indices tend to be small and their precise arrangements awkward to ascertain. These difficulties can be eased by the introduction of yet another notation for tensor algebra that I shall next briefly describe. This is the *diagrammatic* notation.

First, we should know what a *tensor* actually is. In the index notation, a tensor is denoted by a quantity such as

$$Q_{a \ldots c}^{f \ldots h},$$

which can have q lower and p upper indices for any $p, q \geqslant 0$, and need have no special symmetries. We call this a tensor of *valence*[18] $[{p \atop q}]$ (or a $[{p \atop q}]$-valent tensor or just a $[{p \atop q}]$-tensor). Algebraically, this would represent a quantity $\boldsymbol{Q}$ which can be thought of as a function (of a particular kind known as *multilinear*[19]) of q vectors $\boldsymbol{A}, \ldots, \boldsymbol{C}$ and p covectors $\boldsymbol{F}, \ldots, \boldsymbol{H}$, where

$$\boldsymbol{Q}(\boldsymbol{A}, \ldots, \boldsymbol{C}; \boldsymbol{F}, \ldots, \boldsymbol{H}) = A^a \ldots C^c Q_{a \ldots c}^{f \ldots h} F_f \ldots H_h.$$

In the diagrammatic notation, the tensor $\boldsymbol{Q}$ would be represented as a distinctive symbol (say a rectangle or a triangle or an oval, according to convenience) to which are attached q lines extending downwards (the 'legs') and p lines extending upwards (the 'arms'). In any term of a tensor expression, the various elements that are multiplied together are drawn in some kind of juxtaposition, but not necessarily linearly ordered across the page. For any two indices that are contracted together, the lines must be connected, upper to lower. Some examples are illustrated in Figs. 12.17 and 12.18, including examples of various of the formulae that we have just

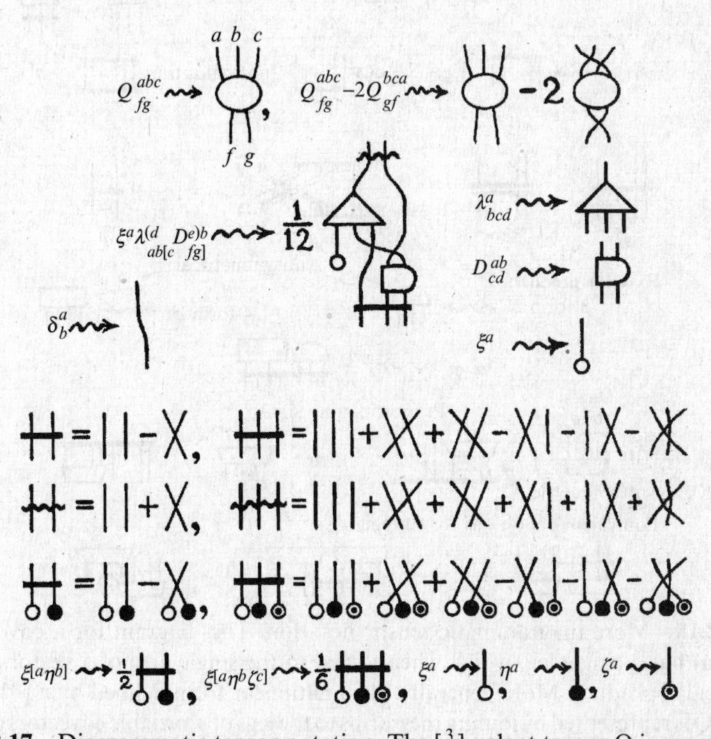

Fig. 12.17 Diagrammatic tensor notation. The $[{}^3_2]$-valent tensor Q is represented by an oval with 3 arms and 2 legs, where the general $[{}^p_q]$-valent tensor picture would have p arms and q legs. In an expression such as $Q^{abc}_{fg} - 2Q^{bca}_{gf}$, the diagrammatic notation uses positioning on the page of the ends of the arms and legs to keep track of which index is which, instead of employing individual index letters. Contractions of tensor indices are represented by the joining of an arm and a leg, as illustrated in the diagram for $\xi^a\lambda^{(d}_{ab[c}D^{e)b}_{fg]}$. This diagram also illustrates the use of a thick bar across index lines to denote antisymmetrization and a wiggly bar to represent symmetrization. The factor $\frac{1}{12}$ in the diagram results from the fact that (to facilitate calculations) the normal factorial denominator for symmetrizers and antisymmetrizers is omitted in the diagrammatic notation (so here we need $\frac{1}{2!} \times \frac{1}{3!} = \frac{1}{12}$). In the lower half of the figure, antisymmetrizers and symmetrizers are written out as 'disembodied' expressions (by use of the diagrammatic representation of the Kronecker delta δ^a_b that will be introduced in §13.3, Fig. 13.6c). This is then used to express the (multivector) wedge products $\boldsymbol{\xi} \wedge \boldsymbol{\eta}$ and $\boldsymbol{\xi} \wedge \boldsymbol{\eta} \wedge \boldsymbol{\zeta}$.

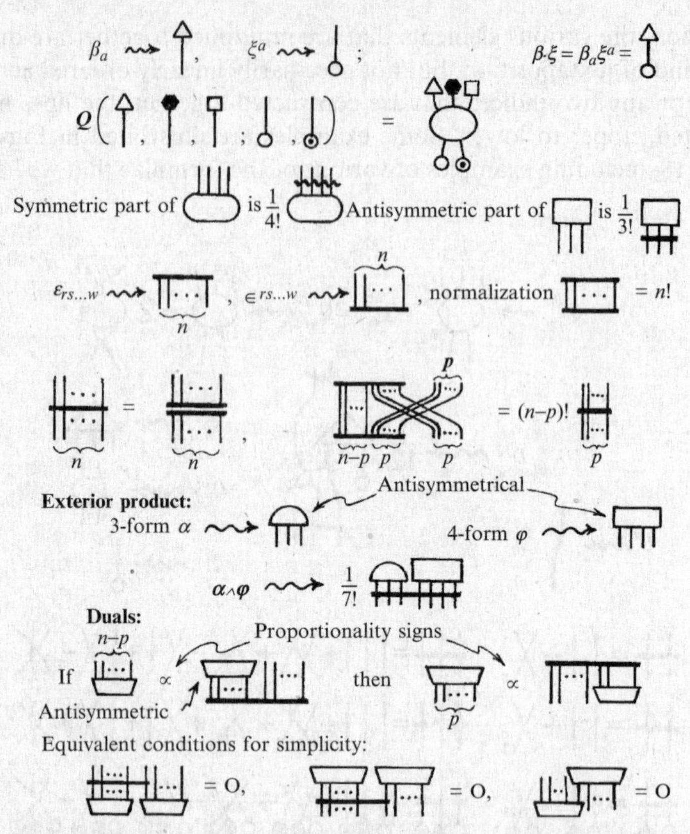

Fig. 12.18 More diagrammatic tensor notation. The diagram for a covector $\boldsymbol{\beta}$ (1-form) has a single leg, which when joined to the single arm of a vector $\boldsymbol{\xi}$ gives their scalar product. More generally, the multilinear form defined by a $[^p_q]$-valent tensor Q is represented by joining the p arms to the legs of p variable covectors and the q legs to the arms of q variable vectors (here $p = 3$ and $q = 2$). Symmetric and antisymmetric parts of general tensors can be expressed using the wiggly lines and thick bars of the operations of Fig. 12.17. Also, the bar notation combines with a related diagrammatic notation for the volume n-form $\varepsilon_{rs\ldots w}$ (for an n-dimensional space) and its dual n-vector $\epsilon^{rs\ldots w}$, normalized according to $\varepsilon_{rs\ldots w}\epsilon^{rs\ldots w} = n!$ Relations equivalent to $n!\delta^a_{[r}\delta^b_s\ldots\delta^f_{w]} = \epsilon^{ab\ldots f}\varepsilon_{rs\ldots w}$ (n antisymmetrized indices) and $\varepsilon_{a\ldots cu\ldots w}\,\epsilon^{a\ldots ce\ldots f} = p!(n-p)!\delta^e_{[u}\ldots\delta^f_{w]}$ (see § 13.3 and Fig. 13.6c) are also expressed. Exterior products of forms, the 'duality' between p-forms and $(n-p)$-vectors, and the conditions for 'simplicity' are then succinctly represented diagrammatically. (For exterior derivative diagrams, see Fig. 14.18.)

encountered. As part of this notation, a bar is drawn across index lines to denote antisymmetrization, mirroring the square-bracket notation of the index notation (although it proves to be convenient to adopt a different convention with regard to factorial multipliers). A 'wiggly' bar correspondingly mirrors symmetrization. Although the diagrammatic notation is hard to print, in the ordinary way, it can be enormously convenient in many handwritten calculations. I have been using it myself for over 50 years![20]

12.9 Complex manifolds

Finally, let us return to the issue of complex manifolds, as addressed in Chapter 10. When we think of a Riemann surface as being 1-dimensional, we are thinking solely in terms of holomorphic operations being performed on complex numbers. We can adopt precisely the same stance with higher-dimensional manifolds, considering our coordinates $x^1, \ldots, x^n$ now to be complex numbers $z^1, \ldots, z^n$ and our functions of them to be holomorphic functions. We again take our manifold to be 'glued together' from a number of coordinate patches, where each patch is now an open region of the coördinate space $\mathbb{C}^n$—the space whose points are the n-tuples $(z^1, z^2, \ldots, z^n)$ of complex numbers (and recall from §10.2 that '$\mathbb{C}$', by itself, stands for the system of complex numbers). The *transition functions* that express the coordinate transformations, when we move from coordinate patch to coordinate patch, are now to be given entirely by *holomorphic functions*. We can define holomorphic vector fields, covectors, p-forms, tensors, etc., in just the same way as we did above, in the case of a real n-manifold.

But then there is the alternative philosophical standpoint according to which we could express all our complex coordinates in terms of their real and imaginary parts $z^j = x^j + \mathrm{i}\, y^j$ (or, equivalently, include the notion of complex conjugation into our category of acceptable function, so that operations need no longer be exclusively holomorphic; see §10.1). Then, our 'complex n-manifold' is no longer viewed as being an n-dimensional space, but is thought of as being a real $2n$-manifold, instead. Of course, it is a $2n$-manifold with a very particular kind of local structure, referred to as a *complex structure*.

There are various ways of formulating this notion. Essentially, what is required is a higher-dimensional version of the Cauchy–Riemann equations (§10.5), but things are usually phrased somewhat differently from this. Let us think of the relation between complex vector fields and real vector fields on the manifold. We can think of a complex vector field $\boldsymbol{\zeta}$ as being represented in the form

$$\boldsymbol{\zeta} = \boldsymbol{\xi} + \mathrm{i}\boldsymbol{\eta},$$

where $\boldsymbol{\xi}$ and $\boldsymbol{\eta}$ are ordinary real vector fields on the $2n$-manifold. What the 'complex structure' does for us is to tell us how these real vector fields have to be related to each other and what differential equations they must satisfy in order that $\boldsymbol{\zeta}$ can qualify as 'holomorphic'. Now, consider the new complex vector field that arises when the complex field $\boldsymbol{\zeta}$ is multiplied by i. We see that, for consistency, we must have $\mathrm{i}\boldsymbol{\zeta} = -\boldsymbol{\eta} + \mathrm{i}\boldsymbol{\xi}$, so that the real vector field $\boldsymbol{\xi}$ is now replaced by $-\boldsymbol{\eta}$ and likewise $\boldsymbol{\eta}$ must be replaced by $\boldsymbol{\xi}$. The operation $\boldsymbol{J}$ which effects these replacements (i.e. $\boldsymbol{J}(\boldsymbol{\xi}) = -\boldsymbol{\eta}$ and $\boldsymbol{J}(\boldsymbol{\eta}) = \boldsymbol{\xi}$) is what is usually referred to as the 'complex structure'.

We note that if $\boldsymbol{J}$ is applied twice, it simply reverses the sign of what it acts on (since $\mathrm{i}^2 = -1$), so we can write

$$\boldsymbol{J}^2 = -1.$$

This condition alone defines what is referred to as an *almost complex structure*. To specialize this to an actual complex structure, so that a consistent notion of 'holomorphic' can arise for the manifold, a certain differential equation[21] in the quantity $\boldsymbol{J}$ must be satisfied. There is a remarkable theorem, the Newlander–Nirenberg theorem,[22] which tells us that this is *sufficient* (in addition to being necessary) for a $2n$-dimensional real manifold, with this $\boldsymbol{J}$-structure, to be reinterpreted as a complex n-manifold. This theorem allows us to move freely between the two philosophical standpoints with regard to complex manifolds.

Notes

Section 12.1

12.1. This 'shrinkability' is taken in the sense of *homotopy* (see §7.2, Fig. 7.2), so that 'cancellation' of oppositely oriented loop segments is not permitted; thus multiple-connectedness is part of homotopy theory. See Huggett and Jordan (2001); Sutherland (1975).

12.2. Strictly speaking this argument is incomplete, since I have presented no convincing reason that the 2π-twist of the belt cannot be continuously undone if the ends are held fixed.[12.17] See Penrose and Rindler (1984), pp. 41–4.

12.3. Here, we treat the molecules as point particles. The dimension of $\mathcal{P}$ would be considerably larger for molecules with internal or rotational degrees of freedom.

Section 12.2

12.4. The usual notion of 'manifold' presupposes that our space $\mathcal{M}$ is, in the first instance, a *topological space*. To assign a *topology* to a space $\mathcal{M}$ is to specify precisely which of its sets of points are to be called 'open' (cf. §7.4). The open sets

---

[12.17] By representing a rotation in ordinary 3-space as a vector pointing along the rotation axis of length equal to the angle of rotation, show that the topology of $\mathcal{R}$ can be described as a solid ball (of radius π) bounded by an ordinary sphere, where each point of the sphere is identified with its antipodal point. Give a direct argument to show why a closed loop representing a 2π-rotation cannot be continuously deformed to a point.

are to have the property that the intersection of any two of them is an open set and the union of any number of them (finite or infinite) is again an open set. In addition to the Hausdorff condition referred to in the text, it is usual to require that $\mathcal{M}$'s topology is restricted in certain other ways, most particularly that it satisfies a requirement called 'paracompactness'. For the meaning of this and other related terms, the interested reader is referred to Kelley (1965); Engelking (1968) or other standard text on general topology. But for our purposes here, it is sufficient to assume merely that $\mathcal{M}$ is constructed from a locally finite patchwork of open regions of $\mathbb{R}^n$, where 'locally finite' means that each patch is intersected by only finitely many other patches.

One final requirement that is sometimes made in the definition of a manifold is that it be *connected*, which means that it consists only of 'one piece' (which here can be taken to mean that it is not a disjoint union of two non-empty open sets). I shall not insist on this here; if connectness is required, then it will be stated explicitly (but disconnectedness will in any case be allowed only for a finite number of separate pieces).

12.5. See, for example, Kobayashi and Nomizu (1963); Hicks (1965); Lang (1972); Hawking and Ellis (1973). One interesting procedure for defining a manifold $\mathcal{M}$ is to reconstruct $\mathcal{M}$ itself simply from the commutative algebra of scalar fields defined on $\mathcal{M}$; see Chevalley 1946; Nomizu 1956; Penrose and Rindler (1984). This kind of idea generalizes to non-commutative algebras and leads to the 'non-commutative geometry' notion of Alain Connes (1994) which provides one of the modern approaches to a 'quantum spacetime geometry' (see §33.1).

Section 12.3

12.6. See Helgason (2001); Frankel (2001).

12.7. The general condition for the family of $(n-1)$-plane elements defined by a 1-form $\boldsymbol{\alpha}$ to touch a 1-parameter family of $(n-1)$-surfaces (so $\boldsymbol{\alpha} = \lambda \mathrm{d}\Phi$ for some scalar fields λ, Φ) is the *Frobenius condition* $\boldsymbol{\alpha} \wedge \mathrm{d}\boldsymbol{\alpha} = 0$; see Flanders (1963).

12.8. Confusion easily arises between the 'classical' idea that a thing like '$\mathrm{d}x^r$' should stand for an infinitesimal displacement (vector), whereas we here seem to be viewing it as a covector. In fact the notation *is* consistent, but it needs a clear head to see this! The quantity $\mathrm{d}x^r$ seems to have a vectorial character because of its upper index r, and this would indeed be the case if r is treated as an abstract index, in accordance with §12.8. On the other hand, if r is taken as a numerical index, say $r = 2$, then we do get a covector, namely $\mathrm{d}x^2$, the gradient of the scalar quantity $y = x^2$ ('x-two', not 'x squared'). But this depends upon the interpretation of 'd' as standing for the gradient rather than as denoting an infinitesimal, as it would have done in the classical tradition. In fact, if we treat *both* the r as abstract and the d as gradient, then '$\mathrm{d}x^r$' simply stands for the (abstract) Kronecker delta!

Section 12.5

12.9. This represents a shift in attitude from the 'infinitesimal' viewpoint with regard to quantities like '$\mathrm{d}x$'. Here, the anticommutation properties of '$\mathrm{d}x_\wedge \mathrm{d}y$' tell us that we are operating with densities with respect to oriented area measures.

12.10. A name suggested to me by N. M. J. Woodhouse. Sometimes this theorem is simply called *Stokes's theorem*. However, this seems particularly inappropriate since the only contribution made by Stokes was set in a (Cambridge) examination question he apparently got from William Thompson (Lord Kelvin).

Section 12.6

12.11. See Flanders (1963). (In that book, what I have called the 'Poincaré lemma' is referred to as the *converse* thereof.)

12.12. There is a more widely applicable definition of compactness of a topological space, which, however, is not so intuitive as that given in the text. A space $\mathcal{R}$ is compact if for every way that it can be expressed as a union of open sets, there is a finite collection of these sets whose union is still $\mathcal{R}$.

12.13. For more information on these matters, see Willmore (1959).

Section 12.7

12.14. This notion of 'dual' is rather different from that which has a covector be 'dual' to a vector, as decribed in §12.3. It is, however, closely connected with yet another concept of 'duality'—the *Hodge dual*. This plays a role in electromagnetism (see §19.2), and versions of it have importance in various approaches to quantum gravity (see §§31.5,14, §32.2, §§33.11,12) and particle physics (see §25.8). Unfortunately, this is only one place among many, where the limitations of mathematical terminology can cause confusion.

12.15. See Penrose and Rindler (1984), pp. 165, 166.

Section 12.8

12.16. See Penrose (1968a), pp. 135–41; Penrose and Rindler (1984), pp. 68–103; Penrose (1971).

12.17. See Penrose (1968a); Penrose and Rindler (1984, 1986); Penrose (1971) and O'Donnell (2003).

12.18. Sometimes the term *rank* is used for the value of $p+q$, but this is confusing because of a separate meaning for 'rank' in connection with matrices; see Note 13.10, §13.8.

12.19. This means separately linear in each of $\boldsymbol{A}, \ldots, \boldsymbol{C}; \boldsymbol{F}, \ldots, \boldsymbol{H}$; see also §§13.7–10.

12.20. See Penrose and Rindler (1984), Appendix; Penrose (1971b); Cvitanovič and Kennedy (1982).

Section 12.9

12.21. This is the vanishing of an expression called 'the Nijenhuis tensor constructed from $\boldsymbol{J}$', which we can express as $J^d_{[a}\partial J^c_{b]}/\partial x^d + J^c_d \partial J^d_{[a}/\partial x^{b]} = 0$.

12.22. Newlander and Nirenberg (1957).

13
Symmetry groups

13.1 Groups of transformations

SPACES that are symmetrical have a fundamental importance in modern physics. Why is this? It might be thought that completely exact symmetry is something that could arise only exceptionally, or perhaps just as some convenient approximation. Although a symmetrical object, such as a square or a sphere, has a precise existence as an idealized ('Platonic'; see §1.3) mathematical structure, any *physical* realization of such a thing would ordinarily be regarded as merely some kind of approximate representation of this Platonic ideal, therefore possessing no actual symmetry that can be regarded as exact. Yet, remarkably, according to the highly successful physical theories of the 20th century, all physical interactions (including gravity) act in accordance with an idea which, strictly speaking, depends crucially upon certain physical structures possessing a symmetry that, at a fundamental level of description, is indeed necessarily exact!

What is this idea? It is a concept that has come to be known as a 'gauge connection'. That name, as it stands, conveys little. But the idea is an important one, enabling us to find a subtle ('twisted') notion of differentiation that applies to general entities on a manifold (entities that are indeed more general than just those—the *p*-forms—which are subject to exterior differentiation, as described in Chapter 12). These matters will be the subject of the two chapters following this one; but as a prerequisite, we must first explore the basic notion of a *symmetry group*. This notion also has many other important areas of application in physics, chemistry, and crystallography, and also within many different areas of mathematics itself.

Let us take a simple example. What are the symmetries of a *square*? The question has two different answers depending upon whether or not we allow symmetries which reverse the orientation of the square (i.e. for which the square is turned over). Let us first consider the case in which these orientation-reversing symmetries are not allowed. Then the square's symmetries are generated from a single rotation through a right angle in the square's plane, repeated various numbers of times. For convenience, we can represent these motions in terms of complex numbers, as we did in

Chapter 5. We may, if we choose, think of the vertices of the square as occupying the points 1, i, −1, − i in the complex plane (Fig. 13.1a), and our basic rotation represented by multiplication by i (i.e. by 'i×'). The various *powers* of i represent all our rotations, there being four distinct ones in all:

$$\mathrm{i}^0 = 1, \quad \mathrm{i}^1 = \mathrm{i}, \quad \mathrm{i}^2 = -1, \quad \mathrm{i}^3 = -\mathrm{i}$$

(Fig. 13.1b). The fourth power $\mathrm{i}^4 = 1$ gets us back to the beginning, so we have no more elements. The product of any two of these four elements is again one of them.

These four elements provide us with a simple example of a *group*. This consists of a set of elements and a law of 'multiplication' defined between pairs of them (denoted by juxtaposition of symbols) for which the *associative* multiplication law holds

$$a(bc) = (ab)c,$$

where there is an *identity* element 1 satisfying

$$1a = a1 = a,$$

and where each element a has an *inverse* a^{-1}, such that[13.1]

$$a^{-1}a = aa^{-1} = 1.$$

The symmetry operations which take an object (not necessarily a square) into itself always satisfy these laws, called the *group axioms*.

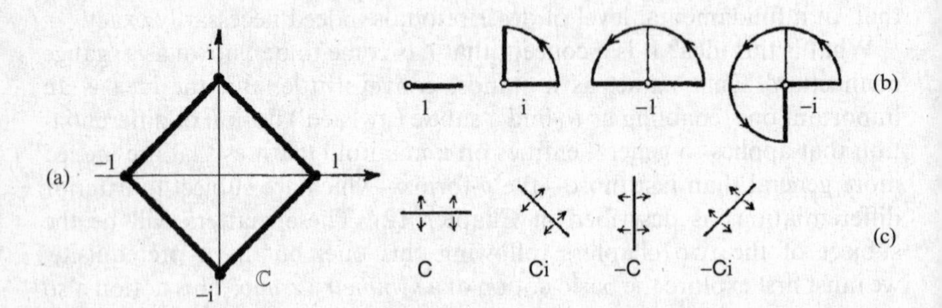

Fig. 13.1 Symmetry of a square. (a) We may represent the square's vertices by the points 1, i, − 1, − i in the complex plane ℂ. (b) The group of non-reflective symmetries are represented, in ℂ, as multiplication by $1 = \mathrm{i}^0$, $\mathrm{i} = \mathrm{i}^1$, $-1 = \mathrm{i}^2$, $-\mathrm{i} = \mathrm{i}^3$, respectively. (c) The reflective symmetries are given, in ℂ, by C (complex conjugation), Ci, − C, and − Ci.

[13.1] Show that if we just assume $1a = a$ and $a^{-1}a = 1$ for all a, together with associativity $a(bc) = (ab)c$, then $a1 = a$ and $aa^{-1} = 1$ can be *deduced*. (*Hint*: Of course a is not the only element asserted to have an inverse.) Show why, on the other hand, $a1 = a$, $a^{-1}a = 1$, and $a(bc) = (ab)c$ are insufficient.

Recall the conventions recommended in Chapter 11, where we think of b acting first and a afterwards, in the product ab. We can regard these as operations being performed upon some object appearing to the right. Thus, we could consider the motion, b, expressing a symmetry of an object $\boldsymbol{\Phi}$, as $\boldsymbol{\Phi} \mapsto b(\boldsymbol{\Phi})$, which we follow up by another such motion a, giving $b(\boldsymbol{\Phi}) \mapsto a(b(\boldsymbol{\Phi}))$. This results in the combined action $\boldsymbol{\Phi} \mapsto a(b(\boldsymbol{\Phi}))$, which we simply write $\boldsymbol{\Phi} \mapsto ab(\boldsymbol{\Phi})$, corresponding to the motion ab. The identity operation leaves the object alone (clearly always a symmetry) and the inverse is just the reverse operation of a given symmetry, moving the object back to where it came from.

In our particular example of non-reflective rotations of the square, we have the additional *commutative* property

$$ab = ba.$$

Groups that are commutative in this sense are called *Abelian*, after the tragically short-lived Norwegian mathematician Niels Henrik Abel.[1] Clearly any group that can be represented simply by the multiplication of complex numbers must be Abelian (since the multiplication of individual complex numbers always commutes). We saw other examples of this at the end of Chapter 5 when we considered the general case of a finite *cyclic* group $\mathbb{Z}_n$, generated by a single nth root of unity.[13.2]

Now let us allow the orientation-reversing *reflections* of our square. We can still use the above representation of the square in terms of complex numbers, but we shall need a new operation, which I denote by C, namely *complex conjugation*. (This flips the square over, about a horizontal line; see §10.1, Fig. 10.1.) We now find (see Fig. 13.1c) the 'multiplication laws'[13.3]

$$\mathsf{C}\mathrm{i} = (-\mathrm{i})\mathsf{C}, \quad \mathsf{C}(-1) = (-1)\mathsf{C}, \quad \mathsf{C}(-\mathrm{i}) = \mathrm{i}\mathsf{C}, \quad \mathsf{CC} = 1$$

(where[2] I shall henceforth write $(-\mathrm{i})\mathsf{C}$ as $-\mathrm{i}\mathsf{C}$, etc.). In fact, we can obtain the multiplication laws for the entire group just from the basic relations[13.4]

$$\mathrm{i}^4 = 1, \quad \mathsf{C}^2 = 1, \quad \mathsf{C}\mathrm{i} = \mathrm{i}^3\mathsf{C},$$

the group being non-Abelian, as is manifested in the last equation. The total number of distinct elements in a group is called its *order*. The order of this particular group is 8.

Now let us consider another simple example, namely the group of rotational symmetries of an ordinary sphere. As before, we can first consider the

---

[13.2] Explain why any vector space is an Abelian group—called an *additive* Abelian group—where the group 'multiplication' operation is the 'addition' operation of the vector space.

[13.3] Verify these relations (bearing in mind that Ci stands for 'the operation i×, followed by the operation C, etc.). (*Hint*: You can check the relations by just confirming their effects on 1 and i. Why?)

[13.4] Show this.

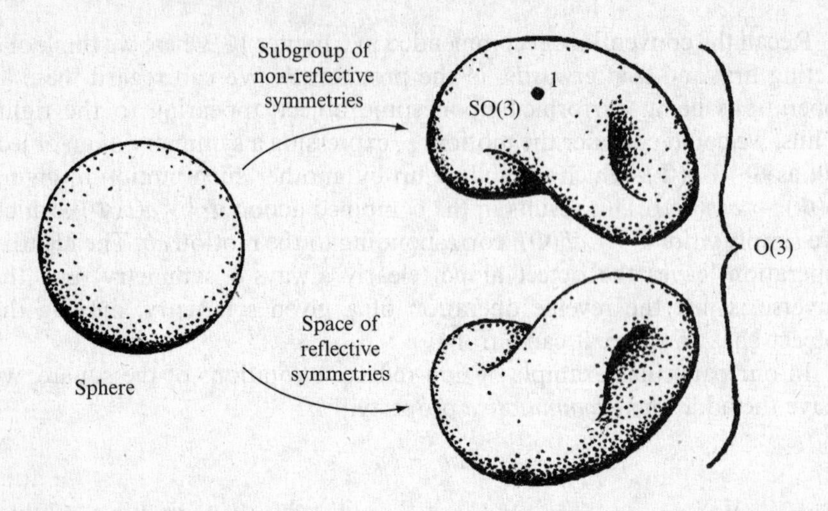

Fig. 13.2 Rotational symmetry of a sphere. The entire symmetry group, O(3), is a disconnected 3-manifold, consisting of two pieces. The component containing the identity element 1 is the (normal) subgroup SO(3) of non-reflective symmetries of the sphere. The remaining component is the 3-manifold of reflective symmetries.

case where reflections are excluded. This time, our symmetry group will have an infinite number of elements, because we can rotate through any angle about any axis direction in 3-space. The symmetry group actually constitutes a 3-dimensional space, namely the 3-manifold denoted by $\mathcal{R}$ in Chapter 12. Let me now give this group (3-manifold) its official name. It is called[3] SO(3), the non-reflective orthogonal group in 3 dimensions. If we now include the reflections, then we get a whole new set of symmetries—another 3-manifold's worth—which are disconnected from the first, namely those which involve a reversal of the orientation of the sphere. The entire family of group elements again constitutes a 3-manifold, but now it is a disconnected 3-manifold, consisting of two separate connected pieces (see Fig. 13.2). This entire group space is called O(3).

These two examples illustrate two of the most important categories of groups, the finite groups and the continuous groups (or *Lie* groups; see §13.6).[4] Although there is a great difference between these two types of group, there are many of the important properties of groups that are common to both.

13.2 Subgroups and simple groups

Of particular significance is the notion of a *subgroup* of a group. To exhibit a subgroup, we select some collection of elements within the group which themselves form a group, using the same multiplication and inversion operations as in the whole group. Subgroups are important in many modern theories of particle physics. It tends to be assumed that there is some fundamental symmetry of Nature that relates different kinds of particles to one another and also relates different particle interactions to one another. Yet one may not see this full group acting as a symmetry in any manifest way, finding, instead, that this symmetry is 'broken' down to some subgroup of the original group where the *subgroup* plays a manifest role as a symmetry. Thus, it is important to know what the possible subgroups of a putative 'fundamental' symmetry group actually are, in order that those symmetries that are indeed manifest in Nature might be able to be thought about as subgroups of this putative group. I shall be addressing questions of this kind in §§25.5–8, §26.11, and §28.1.

Let us examine some particular cases of subgroups, for the examples that we have been considering. The *non-reflective* symmetries of the square constitute a 4-element subgroup $\{1, \mathrm{i}, -1, -\mathrm{i}\}$ of the entire 8-element group of symmetries of the square. Likewise, the non-reflective rotation group SO(3) constitutes a subgroup of the entire group O(3). Another subgroup of the symmetries of the square consists of the four elements $\{1, -1, \mathsf{C}, -\mathsf{C}\}$; yet another has just the two elements $\{1, -1\}$.[13.5] Moreover there is always the 'trivial' subgroup consisting of the identity alone $\{1\}$ (and the whole group itself is, equally trivially, always a subgroup).

All the various subgroups that I have just described have a special property of particular importance. They are examples of what are called *normal* subgroups. The significance of a normal subgroup is that, in an appropriate sense, the action of any element of the whole group leaves a normal subgroup alone or, more technically, we say that each element of the whole group *commutes* with the normal subgroup. Let me be more explicit. Call the whole group $\mathcal{G}$ and the subgroup $\mathcal{S}$. If I select any particular element g of the group $\mathcal{G}$, then I can denote by $\mathcal{S}g$ the set consisting of all elements of $\mathcal{S}$ each individually multiplied by g on the right (what is called *postmultiplied* by g). Thus, in the case of the particular subgroup $\mathcal{S} = \{1, -1, \mathsf{C}, -\mathsf{C}\}$, of the symmetry group of the square, if we choose $g = \mathrm{i}$, then we obtain $\mathcal{S}\mathrm{i} = \{\mathrm{i}, -\mathrm{i}, \mathsf{C}\mathrm{i}, -\mathsf{C}\mathrm{i}\}$. Likewise, the notation $g\mathcal{S}$ will denote the set consisting of all elements of $\mathcal{S}$, each individually multiplied by g on the left (*premultiplied* by g). Thus, in our example, we now have $\mathrm{i}\mathcal{S} = \{\mathrm{i}, -\mathrm{i}, \mathrm{i}\mathsf{C}, -\mathrm{i}\mathsf{C}\}$. The condition for $\mathcal{S}$ to be a normal subgroup of $\mathcal{G}$ is that these two sets are the same, i.e.

$$\mathcal{S}g = g\mathcal{S}, \quad \text{for all } g \text{ in } \mathcal{G}.$$

In our particular example, we see that this is indeed the case (since $\mathsf{C}\mathrm{i} = -\mathrm{i}\mathsf{C}$ and $-\mathsf{C}\mathrm{i} = \mathrm{i}\mathsf{C}$), where we must bear in mind that the collection

---

[13.5] Verify that all these in this paragraph are subgroups (and bear in mind Note 13.2).

of things inside the curly brackets is to be taken as an *unordered* set (so that it does not matter that the elements $-$iC and iC appear in reverse order in the collection of elements, when $\mathcal{S}$i and i$\mathcal{S}$ are written out explicitly).

We can exhibit a *non*-normal subgroup of the group of symmetries of the square, as the subgroup of two elements $\{1, \mathrm{C}\}$. It is non-normal because $\{1, \mathrm{C}\}\mathrm{i} = \{\mathrm{i}, \mathrm{Ci}\}$ whereas $\mathrm{i}\{1, \mathrm{C}\} = \{\mathrm{i}, -\mathrm{Ci}\}$. Note that this subgroup arises as the new (reduced) symmetry group if we mark our square with a horizontal arrow pointing off to the right (see Fig. 13.3a). We can obtain another non-normal subgroup, namely $\{1, \mathrm{Ci}\}$ if we mark it, instead, with an arrow pointing diagonally down to the right (Fig. 13.3b).[13.6] In the case of O(3), there happens to be only one non-trivial normal subgroup,[13.7] namely SO(3), but there are many non-normal subgroups. Non-normal examples are obtained if we select some appropriate finite set of points on the sphere, and ask for the symmetries of the sphere with these points marked. If we mark just a single point, then the subgroup consists of rotations of the sphere about the axis joining the origin to this point (Fig. 13.3c). Alternatively, we could, for example, mark points that are the vertices of a regular polyhedron. Then the subgroup is finite, and consists of the symmetry group of that particular polyhedron (Fig. 13.3d).

One reason that normal subgroups are important is that, if a group $\mathcal{G}$ possesses a non-trivial normal subgroup, then we can break $\mathcal{G}$ down, in a sense, into smaller groups. Suppose that $\mathcal{S}$ is a normal subgroup of $\mathcal{G}$. Then the distinct sets $\mathcal{S}g$, where g runs through all the elements of $\mathcal{G}$, turn

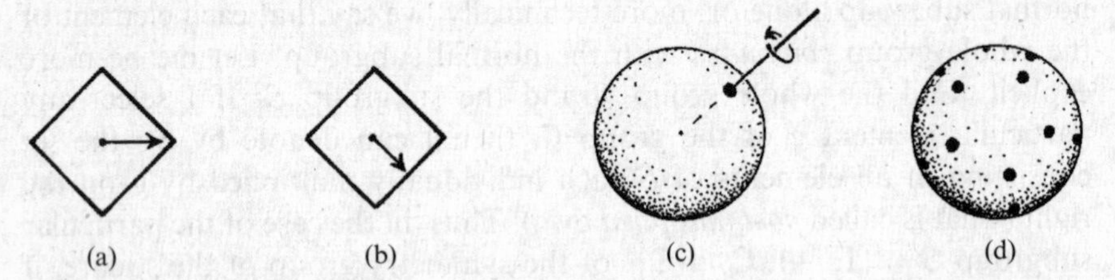

Fig. 13.3 (a) Marking the square of Fig. 13.1 with an arrow pointing to the right, reduces its symmetry group to a non-normal subgroup $\{1,\mathrm{C}\}$. (b) Marking it with an arrow pointing diagonally down to the right yields a different non-normal subgroup $\{1,\mathrm{Ci}\}$. (c) Marking the sphere of Fig. 13.2 with a single point reduces its symmetry to a (non-normal) O(2) subgroup of O(3): rotations about the axis joining the origin to this point. (d) If the sphere is marked with the vertices of a regular polyhedron (here a dodecahedron), its group of symmetries is a finite (non-normal) subgroup of O(3).

[13.6] Check these assertions, and find two more non-normal subgroups, showing that there are no further ones.

[13.7] Show this. (*Hint*: which *sets* of rotations can be rotation-invariant?)

out themselves to form a group. Note that for a given set $\mathcal{S}g$, the choice of g is generally not unique; we can have $\mathcal{S}g_1 = \mathcal{S}g_2$, for different elements g_1, g_2 of $\mathcal{G}$. The sets of the form $\mathcal{S}g$, for *any* subgroup $\mathcal{S}$, are called *cosets* of $\mathcal{G}$; but when $\mathcal{S}$ is normal, the cosets form a group. The reason for this is that if we have two such cosets $\mathcal{S}g$ and $\mathcal{S}h$ (g and h being elements of $\mathcal{G}$) then we can define the 'product' of $\mathcal{S}g$ with $\mathcal{S}h$ to be

$$(\mathcal{S}g)\,(\mathcal{S}h) = \mathcal{S}(gh),$$

and we find that all the group axioms are satisfied, provided that $\mathcal{S}$ is normal, essentially because the right-hand side is well defined, independently of which g and h were chosen in the representation of the cosets on the left-hand side of this equation.[13.8] The resulting group defined in this way is called the *factor group* of $\mathcal{G}$ by its normal subgroup $\mathcal{S}$. The factor group of $\mathcal{G}$ by $\mathcal{S}$ is written $\mathcal{G}/\mathcal{S}$. We can still write $\mathcal{G}/\mathcal{S}$ for the factor *space* (not a group) of distinct cosets $\mathcal{S}g$ even when $\mathcal{S}$ is not normal.[13.9]

Groups that possess no non-trivial normal subgroups at all are called *simple groups*. The group SO(3) is an example of a simple group. Simple groups are, in a clear sense, the basic building blocks of group theory. It is thus an important achievement of the 19th and 20th centuries in mathematics that all the finite simple groups and all the continuous simple groups are now known. In the continuous case (i.e. for Lie groups), this was a mathematical landmark, started by the highly influential German mathematician Wilhelm Killing (1847–1923), whose basic papers appeared in 1888–1890, and was essentially completed, in 1894, in one of the most important of mathematical papers ever written,[5] by the superb geometer and algebraist Élie Cartan (whom we have already encountered in Chapter 12, and whom we shall meet again in Chapter 17). This classification has continued to play a fundamental role in many areas of mathematics and physics, to the present day. It turns out that there are four families, known as A_m, B_m, C_m, D_m(for $m = 1, 2, 3, \ldots$), of respective dimension $m(m+2)$, $m(2m+1)$, $m(2m+1)$, $m(2m-1)$, called the *classical groups* (see end of §13.10) and five *exceptional groups* known as E_6, E_7, E_8, F_4, G_2, of respective dimension 78, 133, 248, 52, 14.

The classification of the finite simple groups is a more recent (and even more difficult) achievement, carried out over a great many years during the 20th century by a considerable number of mathematicians (with the aid of computers in more recent cases), being completed only in 1982.[6] Again there are some systematic families and a finite collection of *exceptional*

[13.8] Verify this and show that the axioms fail if $\mathcal{S}$ is not normal.

[13.9] Explain why the number of elements in $\mathcal{G}/\mathcal{S}$, for any finite subgroup $\mathcal{S}$ of a finite group $\mathcal{G}$, is the order of $\mathcal{G}$ divided by the order of $\mathcal{S}$.

finite simple groups. The largest of these exceptional groups is referred to as the *monster*, which is of order

$$= 808017424794512875886459904961710757005754368000000000.$$

$$= 2^{46}\times 3^{20}\times 5^{9}\times 7^{6}\times 11^{2}\times 13^{3}\times 17\times 19\times 23\times 29\times 31\times 41\times 47\times 59\times 71.$$

Exceptional groups appear to have a particular appeal for many modern theoretical physicists. The group E_8 features importantly in string theory (§31.14), while various people have expressed a hope that the huge but finite monster may feature in some future theory.[7]

The classification of the simple groups may be regarded as a major step towards the classification of groups generally since, as indicated above, general groups may be regarded as being built up out of simple groups (together with Abelian ones). In fact, this is not really the whole story because there is further information in how one simple group can build upon another. I do not propose to enter into the details of this matter here, but it is worth just mentioning the simplest way that this can happen. If $\mathcal{G}$ and $\mathcal{H}$ are any two groups, then they can be combined together to form what is called the *product group* $\mathcal{G}\times\mathcal{H}$, whose elements are simply *pairs* (g, h), where g belongs to $\mathcal{G}$ and h belongs to $\mathcal{H}$, the rule of group multiplication between elements (g_1, h_1) and (g_2, h_2), of $\mathcal{G}\times\mathcal{H}$, being defined as

$$(g_1, h_1)\,(g_2, h_2) = (g_1g_2,\, h_1h_2),$$

and it is very easy to verify that the group axioms are satisfied. Many of the groups that feature in particle physics are in fact product groups of simple groups (or elementary modifications of such).[13.10]

13.3 Linear transformations and matrices

In the general study of groups, there is a particular class of symmetry groups that have been found to play a central role. These are the groups of symmetries of vector spaces. The symmetries of a vector space are expressed by the *linear transformations* preserving the vector-space structure.

Recall from §11.1 and §12.3 that, in a vector space **V**, we have, defining its structure, a notion of addition of vectors and multiplication of vectors by numbers. We may take note of the fact that the geometrical picture of addition is obtained by use of the parallelogram law, while multiplication by a number is visualized as scaling the vector up (or down) by that number (Fig. 13.4). Here we are picturing it as a *real* number, but complex vector spaces are also allowed (and are particularly important in many

[13.10] Verify that $\mathcal{G}\times\mathcal{H}$ is a group, for any two groups $\mathcal{G}$ and $\mathcal{H}$, and that we can identify the factor group $(\mathcal{G}\times\mathcal{H})/\mathcal{G}$ with $\mathcal{H}$.

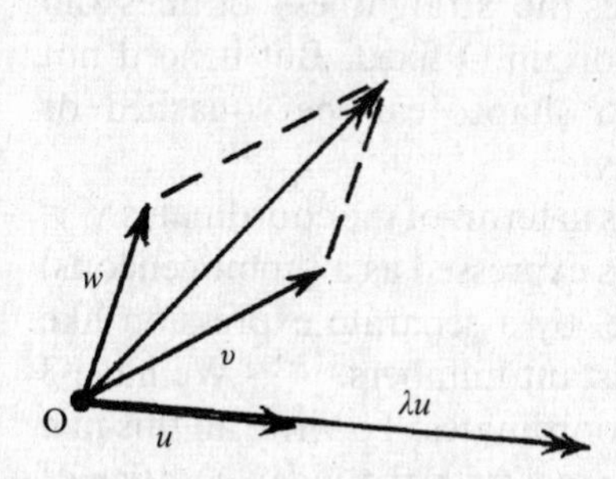

Fig. 13.4 A linear transformation preserves the vector-space structure of the space on which it acts. This structure is defined by the operations of addition (illustrated by the parallelogram law) and multiplication by a scalar λ (which could be a real number or, in the case of a complex vector space, a complex number). Such a transformation preserves the 'straightness' of lines and the notion of 'parallel', keeping the origin O fixed.

contexts, because of complex magic!), though hard to portray in a diagram. A linear transformation of **V** is a transformation that takes **V** to itself, preserving its structure, as defined by these basic vector-space notions. More generally, we can also consider linear transformations that take one vector space to another.

A linear transformation can be explicitly described using an array of numbers called a *matrix*. Matrices are important in many mathematical contexts. We shall examine these extremely useful entities with their elegant algebraic rules in this section (and in §§13.4,5). In fact, §§13.3–7 may be regarded as a rapid tutorial in matrix theory and its application to the theory of continuous groups. The notions described here are vital to a proper understanding of quantum theory, but readers already familiar with this material—or else who prefer a less detailed comprehension of quantum theory when we come to that—may prefer to skip these sections, at least for the time being.

To see what a linear transformation looks like, let us first consider the case of a 3-dimensional vector space and see its relevance to the rotation group O(3) (or SO(3)), discussed in §13.1, giving the symmetries of the sphere. We can think of this sphere as embedded in Euclidean 3-space $\mathbb{E}^3$ (this space being regarded as a vector space with respect to the origin O at the sphere's centre[8]) as the locus

$$x^2 + y^2 + z^2 = 1$$

in terms of ordinary Cartesian coordinates (x, y, z).[13.11] Rotations of the sphere are now expressed in terms of linear transformation of $\mathbb{E}^3$, but of a very particular type known as *orthogonal* which we shall be coming to in §13.8 (see also §13.1).

General linear transformations, however, would squash or stretch the sphere into an *ellipsoid*, as illustrated in Fig. 13.5. Geometrically,

[13.11] Show how this equation, giving the points of unit distance from O, follows from the Pythagorean theorem of §2.1.

a linear transformation is one that preserves the 'straightness' of lines and the notion of 'parallel' lines, keeping the origin O fixed. But it need not preserve right angles or other angles, so shapes can be squashed or stretched, in a uniform but anisotropic way.

How do we express linear transformations in terms of the coordinates x, y, z? The answer is that each new coordinate is expressed as a (homogeneous) *linear combination* of the original ones, i.e. by a separate expression like $\alpha x + \beta y + \gamma z$, where α, β, and γ are constant numbers.[13.12] We have 3 such expressions, one for each of the new coordinates. To write all this in a compact form, it will be useful to make contact with the *index notation* of Chapter 12. For this, we re-label the coordinates as (x^1, x^2, x^3), where

$$x^1 = x, \quad x^2 = y, \quad x^3 = z$$

(bearing in mind, again, that these upper indices do *not* denote powers see §12.2). A general point in our Euclidean 3-space has coordinates x^a, where $a = 1, 2, 3$. An advantage of using the index notation is that the discussion applies in any number of dimensions, so we can consider that a (and all our other index letters) run over $1, 2, \ldots, n$, where n is some fixed positive integer. In the case just considered, $n = 3$.

In the index notation, with Einstein's summation convention (§12.7), the general linear transformation now takes the form[9],[13.13]

$$x^a \mapsto T^a{}_b \, x^b.$$

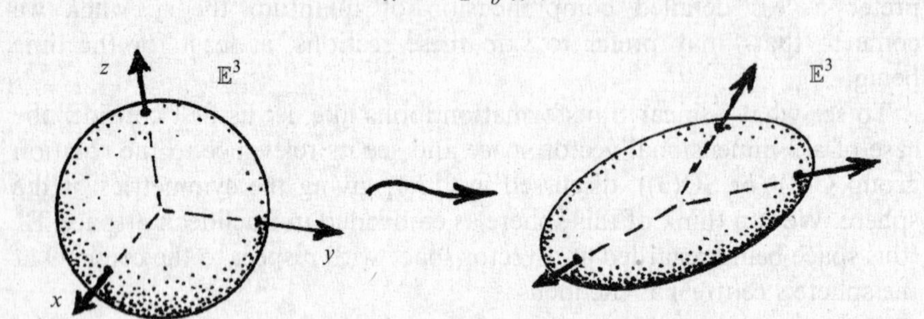

Fig. 13.5 A linear transformation acting on $\mathbb{E}^3$ (expressed in terms of Cartesian x, y, z coordinates) would generally squash or stretch the unit sphere $x^2 + y^2 + z^2 = 1$ into an ellipsoid. The orthogonal group O(3) consists of the linear transformations of $\mathbb{E}^3$ which preserve the unit sphere.

[13.12] Can you explain why? Just do this in the 2-dimensional case, for simplicity.

[13.13] Show this explicitly in the 3-dimensional case.

Calling this linear transformation $\boldsymbol{T}$, we see that $\boldsymbol{T}$ is determined by this set of *components* $T^a{}_b$. Such a set of *components* is referred to as an $n \times n$ *matrix*, usually set out as a *square*—or, in other contexts (see below) $m \times n$-rectangular—array of numbers. The above displayed equation, in the 3-dimensional case is then written

$$\begin{pmatrix} x^1 \\ x^2 \\ x^3 \end{pmatrix} \mapsto \begin{pmatrix} T^1{}_1 & T^1{}_2 & T^1{}_3 \\ T^2{}_1 & T^2{}_2 & T^2{}_3 \\ T^3{}_1 & T^3{}_2 & T^3{}_3 \end{pmatrix} \begin{pmatrix} x^1 \\ x^2 \\ x^3 \end{pmatrix},$$

this standing for three separate relations, starting with $x^1 \mapsto T^1{}_1 x^1 + T^1{}_2 x^2 + T^1{}_3 x^3$.[13.14]

We can also write this without indices or explicit coordinates, as $\boldsymbol{x} \mapsto \boldsymbol{T}\boldsymbol{x}$. If we prefer, we can adopt the *abstract–index* notation (§12.8) whereby '$x^a \mapsto T^a{}_b x^b$' is *not* a component expression, but actually represents this abstract transformation $\boldsymbol{x} \mapsto \boldsymbol{T}\boldsymbol{x}$. (When it is important whether an indexed expression is to be read abstractly or as components, this will be made clear by the wording.) Alternatively, we can use the *diagrammatic* notation, as depicted in Fig. 13.6a. In my descriptions, the matrix of numbers $(T^a{}_b)$ or the abstract linear transformation $\boldsymbol{T}$ will be used interchangeably when I am not concerned with the technical distinctions between these two concepts (the former depending upon a specific coordinate description of our vector space $\mathbf{V}$, the latter not).

Let us consider a second linear transformation $\boldsymbol{S}$, applied following the application of $\boldsymbol{T}$. The product $\boldsymbol{R}$ of the two, written $\boldsymbol{R} = \boldsymbol{S}\boldsymbol{T}$, would have a component (or abstract–index) description

$$R^a{}_c = S^a{}_b \, T^b{}_c$$

(summation convention for components!).[13.15] The diagrammatic form of the product $\boldsymbol{S}\boldsymbol{T}$ is given in Fig. 13.6b. Note that, in the diagrammatic notation, to form a successive product of linear transformations, we string

[13.14] Write this all out in full, explaining how this expresses $x^a \mapsto T^a{}_b x^b$.

[13.15] What is this relation between $\boldsymbol{R}$, $\boldsymbol{S}$, and $\boldsymbol{T}$, written out explicitly in terms of the elements of 3×3 square arrays of components. You may recognize this, the normal law for 'multiplication of matrices', if this is familiar to you.

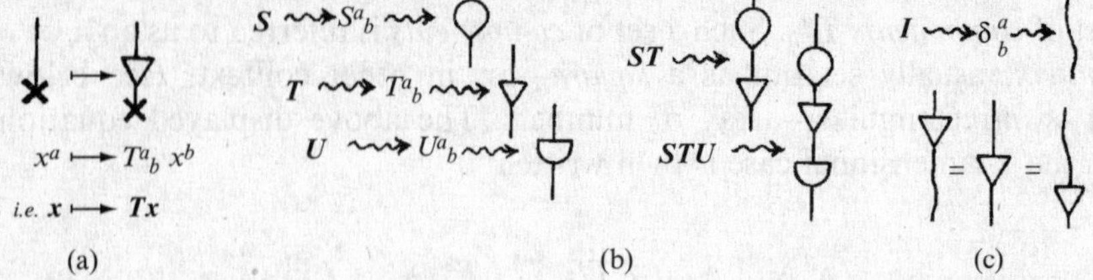

Fig. 13.6 (a) The linear transformation $x^a \mapsto T^a{}_b x^b$, or written without indices as $\boldsymbol{x} \mapsto \boldsymbol{Tx}$ (or read with the indices as abstract, as in §12.8), in diagrammatic form. (b) Diagrams for linear transformations $\boldsymbol{S}$, $\boldsymbol{T}$, $\boldsymbol{U}$, and their products $\boldsymbol{ST}$ and $\boldsymbol{STU}$. In a successive product, we string them in a line downwards. (c) The Kronecker delta δ^a_b, or identity transformation $\boldsymbol{I}$, is depicted as a 'disembodied' line, so relations $T^a{}_b\delta^b_c = T^a{}_c = \delta^a_b T^b{}_c$ become automatic in the notation (see also Fig. 12.17).

them in a line downwards. This happens to work out conveniently in the notation, but one could perfectly well adopt a different convention in which the connecting 'index lines' are drawn horizontally. (Then there would be a closer correspondence between algebraic and diagrammatic notations.)

The *identity* linear transformation $\boldsymbol{I}$ has components that are normally written δ^a_b (the *Kronecker delta*—the standard convention being that these indices are not normally staggered), for which

$$\delta^a_b = \begin{cases} 1 & \text{if } a = b, \\ 0 & \text{if } a \neq b, \end{cases}$$

and we have[13.16]

$$T^a{}_b\delta^b_c = T^a{}_c = \delta^a_b T^b{}_c$$

giving the algebraic relations $\boldsymbol{TI} = \boldsymbol{T} = \boldsymbol{IT}$. The square matrix of components δ^a_b has 1s down what is called the *main diagonal*, which extends from the top-left corner to bottom-right. In the case $n = 3$, this is

$$\begin{pmatrix} 1 & 0 & 0 \\ 0 & 1 & 0 \\ 0 & 0 & 1 \end{pmatrix}.$$

In the diagrammatic notation, we simply represent the Kronecker delta by a 'disembodied' line, and the above algebraic relations become automatic in the notation; see Fig. 13.6c.

[13.16] Verify.

Those linear transformations which map the entire vector space down to a region (subspace) of smaller dimension within that space are called *singular*.[10] An equivalent condition for $\boldsymbol{T}$ to be singular is the existence of a non-zero vector $\boldsymbol{v}$ such that[13.17]

$$\boldsymbol{T}\boldsymbol{v} = 0.$$

Provided that the transformation is non-singular, then it will have an inverse,[13.18] where the inverse of $\boldsymbol{T}$ is written $\boldsymbol{T}^{-1}$, so that

$$\boldsymbol{T}\boldsymbol{T}^{-1} = \boldsymbol{I} = \boldsymbol{T}^{-1}\boldsymbol{T},$$

as is required of an inverse. We can give the explicit expression for this inverse conveniently in the diagrammatic notation; see Fig. 13.7, where I have introduced the useful diagrams for the antisymmetrical (*Levi-Civita*) quantities $\varepsilon_{a...c}$ and $\in^{a...c}$ (with normalization $\varepsilon_{a\cdots c} \in^{a\cdots c} = n!$) that were introduced in §12.7 and Fig. 12.18.[13.19]

The algebra of matrices (initiated by the highly prolific English mathematician and lawyer Arthur Cayley in 1858)[11] finds a very broad range of application (e.g. statistics, engineering, crystallography, psychology, computing—not to mention quantum mechanics). This generalizes the algebra of quaternions and the Clifford and Grassmann algebras studied in §§11.3,5,6. I use bold-face *upright* letters ($\mathbf{A}$, $\mathbf{B}$, $\mathbf{C}$, . . .) for the arrays of components that constitute actual matrices (rather than abstract linear transformations, for which bold-face *italic* letters are being used).

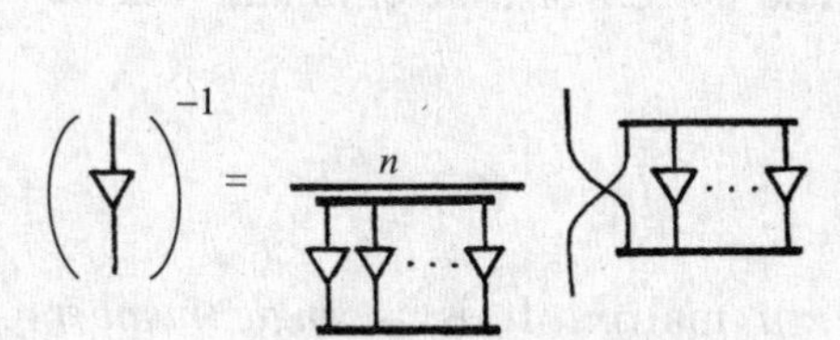

Fig. 13.7 The inverse $\boldsymbol{T}^{-1}$ of a non-singular $(n \times n)$ matrix $\boldsymbol{T}$ given here explicitly in diagrammatic form, using the diagrammatic form of the Levi-Civita antisymmetric quantities $\varepsilon_{a...c}$ and $\in^{a...c}$ (normalized by $\varepsilon_{a...c} \in^{a...c} = n!$) introduced in §12.7 and depicted in Fig. 12.18.

[13.17] Why? Show that this would happen, in particular, if the array of components has an entire column of 0s or two identical columns. Why does this also hold if there are two identical rows? *Hint*: For this last part, consider the determinant condition below.

[13.18] Show why, not using explicit expressions.

[13.19] Prove directly, using the diagrammatic relations given in Fig. 12.18, that this definition gives $\boldsymbol{T}\boldsymbol{T}^{-1} = \boldsymbol{I} = \boldsymbol{T}^{-1}\boldsymbol{T}$. *Hint*: see Fig. 13.8

Restricting attention to $n \times n$ matrices for fixed n, we have a system in which notions of addition and multiplication are defined, where the standard algebraic laws

$$\mathbf{A}+\mathbf{B}=\mathbf{B}+\mathbf{A}, \quad \mathbf{A}+(\mathbf{B}+\mathbf{C})=(\mathbf{A}+\mathbf{B})+\mathbf{C}, \quad \mathbf{A}(\mathbf{BC})=(\mathbf{AB})\mathbf{C},$$
$$\mathbf{A}(\mathbf{B}+\mathbf{C})=\mathbf{AB}+\mathbf{AC}, \quad (\mathbf{A}+\mathbf{B})\mathbf{C}=\mathbf{AC}+\mathbf{BC}$$

hold. (Each element of $\mathbf{A}+\mathbf{B}$ is simply the sum of the corresponding elements of $\mathbf{A}$ and $\mathbf{B}$.) However, we do not usually have the commutative law of multiplication, so that generally $\mathbf{AB} \neq \mathbf{BA}$. Moreover, as we have seen above, non-zero $n \times n$ matrices do not always have inverses.

It should be remarked that the algebra also extends to the *rectangular* cases of $m \times n$ matrices, where m need not be equal to n. However, addition is defined between an $m \times n$ matrix and a $p \times q$ matrix only when $m = p$ and $n = q$; multiplication is defined between them only when $n = p$, the result being an $m \times q$ matrix. This extended algebra subsumes products like the $\mathbf{Tx}$ considered above, where the 'column vector' $\mathbf{x}$ is thought of as being an $n \times 1$ matrix.[13.20]

The *general linear group* GL(n) is the group of symmetries of an n-dimensional vector space, and it is realized explicitly as the multiplicative group of $n \times n$ non-singular matrices. If we wish to emphasize that our vector space is *real*, and that the numbers appearing in our matrices are correspondingly real numbers, then we refer to this full linear group as GL(n, $\mathbb{R}$). We can also consider the complex case, and obtain the *complex* full linear group GL(n, $\mathbb{C}$). Each of these groups has a normal subgroup, written respectively SL(n, $\mathbb{R}$) and SL(n, $\mathbb{C}$)—or, more briefly when the underlying field (see §16.1) $\mathbb{R}$ or $\mathbb{C}$ is understood, SL(n)—called the *special* linear group. These are obtained by restricting the matrices to have their *determinants* equal to 1. The notion of a determinant will be explained next.

13.4 Determinants and traces

What is the determinant of an $n \times n$ matrix? It is a *single number* calculated from the elements of the matrix, which vanishes if and only if the matrix is singular. The diagrammatic notation conveniently describes the determinant explicitly; see Fig. 13.8a. The index-notation form of this is

$$\frac{1}{n!}\epsilon^{ab\ldots d}\,T^e{}_a T^f{}_b \ldots T^h{}_d \varepsilon_{ef\ldots h}$$

[13.20] Explain this, and give the full algebraic rules for rectangular matrices.

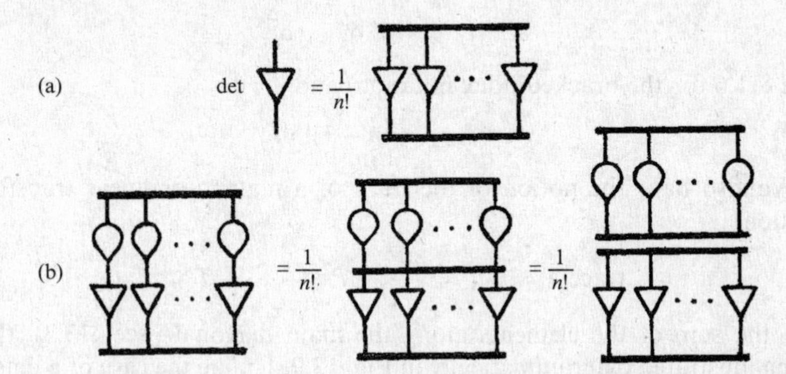

Fig. 13.8 (a) Diagrammatic notation for det $(T^a{}_b) = \det \boldsymbol{T} = |\boldsymbol{T}|$. (b) Diagrammatic proof that $\det(\boldsymbol{ST}) = \det \boldsymbol{S} \det \boldsymbol{T}$. The antisymmetrizing bar can be inserted in the middle term because there is already antisymmetry in the index lines that it crosses. See Figs. 12.17, 12.18.

where the quantities $\epsilon^{a...d}$ and $\varepsilon_{e...h}$ are antisymmetric (Levi-Civita) tensors, normalized according to

$$\epsilon^{a...d}\varepsilon_{a...d} = n!$$

for an n-dimensional space (and recall that $n! = 1 \times 2 \times 3 \times \cdots \times n$), where the indices $a, \ldots, d$ and $e, \ldots, h$ are each n in number.

We can refer to this determinant as det $(T^a{}_b)$ or det $\boldsymbol{T}$ (or sometimes $|\boldsymbol{T}|$ or as the array constituting the matrix but with vertical bars replacing the parentheses). In the particular cases of a 2×2 and a 3×3 matrix, the determinant is given by[13.21]

$$\det\begin{pmatrix} a & b \\ c & d \end{pmatrix} = ad - bc,$$

$$\det\begin{pmatrix} a & b & c \\ d & e & f \\ g & h & j \end{pmatrix} = aej - afh + bfg - bdj + cdh - ceg.$$

The determinant satisfies the important and rather remarkable relation

$$\det \mathbf{AB} = \det \mathbf{A} \det \mathbf{B},$$

which can be seen to be true quite neatly in the diagrammatic notation (Fig. 13.8b). The key ingredients are the formulae illustrated in Fig. 12.18[13.22] which, when written in the index notation, look like

[13.21] Derive these from the expression of Fig. 13.8a.

[13.22] Show why these hold.

$$\epsilon^{a\ldots c}\,\varepsilon_{f\ldots h} = n!\,\delta_f^{[a}\cdots\delta_h^{c]}$$

(see §11.6 for the bracket/index notation) and

$$\epsilon^{ab\ldots c}\varepsilon_{fb\ldots c} = (n-1)!\,\delta_f^a.$$

We also have the notion of the *trace* of a matrix (or linear transformation)

$$\text{trace}\,\mathbf{T} = T^a{}_a = T^1{}_1 + T^2{}_2 + \cdots + T^n{}_n$$

(i.e. the sum of the elements along the main diagonal—see §13.3), this being illustrated diagrammatically in Fig. 13.9. Unlike the case of a determinant, there is no particular relation between the trace of the product **AB** of two matrices and the traces of **A** and **B** individually. Instead, we have the relation[13.23]

$$\text{trace}\,(\mathbf{A}+\mathbf{B}) = \text{trace}\,\mathbf{A} + \text{trace}\,\mathbf{B}.$$

There is an important connection between the determinant and the trace which has to do with the determinant of an 'infinitesimal' linear transformation, given by an $n \times n$ matrix $\mathbf{I} + \varepsilon\mathbf{A}$ for which the number ε is considered to be 'infinitesimally small' so that we can ignore its square ε^2 (and also higher powers ε^3, ε^4, etc.). Then we find[13.24]

$$\det(\mathbf{I}+\varepsilon\mathbf{A}) = 1 + \varepsilon\,\text{trace}\,\mathbf{A}$$

(ignoring ε^2, etc.). In particular, infinitesimal elements of SL(n), i.e. elements of SL(n) representing infinitesimal rotations, being of *unit* determinant (as opposed to those of GL(n)), are characterized by the **A** in $\mathbf{I} + \varepsilon\mathbf{A}$ having zero trace. We shall be seeing the significance of this in §13.10. In fact the above formula can be extended to *finite* (that is, non-infinitesimal) linear transformations through the expression[13.25]

$$\det \mathrm{e}^{\mathbf{A}} = \mathrm{e}^{\text{trace}\,\mathbf{A}},$$

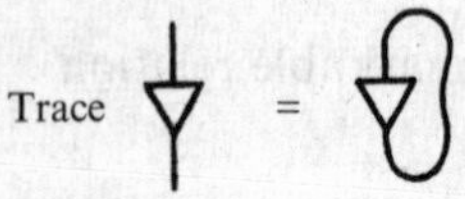

Trace [diagram] = [diagram]

Fig. 13.9 Diagrammatic notation for trace $\mathbf{T}(=T^a{}_a)$.

[13.23] Show this.

[13.24] Show this.

[13.25] Establish this expression. *Hint*: Use the 'canonical form' for a matrix in terms of its eigenvalues—as described in §13.5—assuming first that these eigenvalues are unequal (and see Exercise [13.27]). Then use a general argument to show that the equality of some eigenvalues cannot invalidate identities of this kind.

where 'e$^{\mathbf{A}}$' for matrices has just the same definition as it has for ordinary numbers (see §5.3), i.e.

$$e^{\mathbf{A}} = \mathbf{I} + \mathbf{A} + \frac{1}{2}\mathbf{A}^2 + \frac{1}{6}\mathbf{A}^3 + \frac{1}{24}\mathbf{A}^4 + \cdots.$$

We shall return to these issues in §13.6 and §14.6.

13.5 Eigenvalues and eigenvectors

Among the most important notions associated with linear transformations are what are called 'eigenvalues' and 'eigenvectors'. These are vital to quantum mechanics, as we shall be seeing in §21.5 and §§22.1,5, and to many other areas of mathematics and applications. An *eigenvector* of a linear transformation $\boldsymbol{T}$ is a non-zero complex vector $\boldsymbol{v}$ which $\boldsymbol{T}$ sends to a multiple of itself. That is to say, there is a complex number λ, the corresponding *eigenvalue*, for which

$$\boldsymbol{Tv} = \lambda\boldsymbol{v}, \text{ i.e. } T^a{}_b v^b = \lambda v^a.$$

We can also write this equation as $(\boldsymbol{T} - \lambda\boldsymbol{I})\boldsymbol{v} = 0$, so that, if λ is to be an eigenvalue of $\boldsymbol{T}$, the quantity $\boldsymbol{T} - \lambda\boldsymbol{I}$ must be *singular*. Conversely, if $\boldsymbol{T} - \lambda\boldsymbol{I}$ is singular, then λ is an eigenvalue of $\boldsymbol{T}$. Note that if $\boldsymbol{v}$ is an eigenvector, then so also is any non-zero complex multiple of $\boldsymbol{v}$. The complex 1-dimensional space of these multiples is unchanged by the transformation $\boldsymbol{T}$, a property which characterizes $\boldsymbol{v}$ as an eigenvector (Fig. 13.10).

From the above, we see that this condition for λ to be an eigenvalue of $\boldsymbol{T}$ is

$$\det(\boldsymbol{T} - \lambda\boldsymbol{I}) = 0.$$

Writing this out, we obtain a polynomial equation[13.26] of degree n in λ. By the 'fundamental theorem of algebra', §4.2, we can factorize the λ-polynomial $\det(\boldsymbol{T} - \lambda\boldsymbol{I})$ into linear factors. This reduces the above equation to

$$(\lambda_1 - \lambda)(\lambda_2 - \lambda)(\lambda_3 - \lambda)\ldots(\lambda_n - \lambda) = 0$$

where the complex numbers $\lambda_1, \lambda_2, \lambda_3, \ldots, \lambda_n$ are the various eigenvalues of $\boldsymbol{T}$. In particular cases, some of these factors may coincide, in which case we have a *multiple* eigenvalue. The multiplicity m of an eigenvalue λ_r is the number of times that the factor $\lambda_r - \lambda$ appears

[13.26] See if you can express the coefficients of this polynomial in diagrammatic form. Work them out for $n = 2$ and $n = 3$.

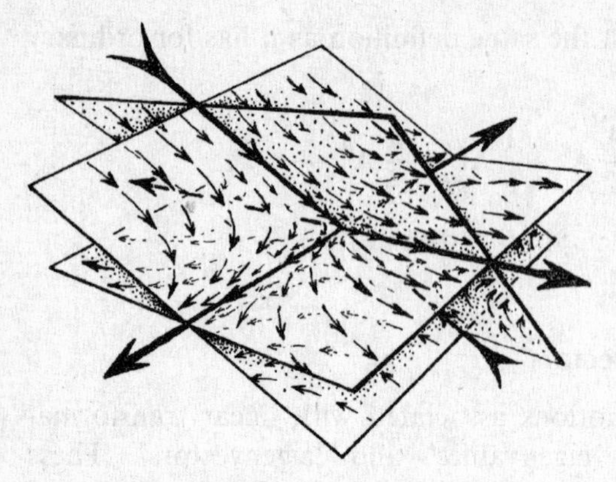

Fig. 13.10 The action of a linear transformation $\boldsymbol{T}$. Its eigenvectors always constitute linear spaces through the origin (here three lines). These spaces are unaltered by $\boldsymbol{T}$. (In this example, there are two (unequal) positive eigenvalues (outward pointing arrows) and one negative one (inward arrows).

in the above product. The total number of eigenvalues of $\boldsymbol{T}$, counted appropriately with multiplicities, is always equal to n, for an $n \times n$ matrix.[13.27]

For a particular eigenvalue λ of multiplicity r, the space of corresponding eigenvectors constitutes a linear space, of dimensionality d, where $1 \leq d \leq r$. For certain types of matrix, including the unitary, Hermitian, and normal matrices of most interest in quantum mechanics (see §13.9, §§22.4,6), we always have the maximum dimensionality $d = r$ (despite the fact that $d = 1$ is the most 'general' case, for given r). This is fortunate, because the (more general) cases for which $d < r$ are more difficult to handle. In quantum mechanics, eigenvalue multiplicities are referred to as *degeneracies* (cf. §§22.6,7).

A *basis* for an n-dimensional vector space **V** is an ordered set $\boldsymbol{e} = (\boldsymbol{e}_1, \ldots, \boldsymbol{e}_n)$ of n vectors $\boldsymbol{e}_1, \ldots, \boldsymbol{e}_n$ which are *linearly independent*, which means that there is no relation of the form $\alpha_1\boldsymbol{e}_1 + \cdots + \alpha_n\boldsymbol{e}_n = 0$ with $\alpha_1, \ldots, \alpha_n$ not all zero. Every element of **V** is then uniquely a linear combination of these basis elements.[13.28] In fact, this property is what characterizes a basis in the more general case when **V** can be infinite-dimensional, when the linear independence by itself is not sufficient.

Thus, given a basis $\boldsymbol{e} = (\boldsymbol{e}_1, \ldots, \boldsymbol{e}_n)$, *any* element $\boldsymbol{x}$ of **V** can be uniquely written

$$\begin{aligned}\boldsymbol{x} &= x^1\boldsymbol{e}_1 + x^2\boldsymbol{e}_2 + \cdots + x^n\boldsymbol{e}_n \\ &= x^j\boldsymbol{e}_j,\end{aligned}$$

[13.27] Show that $\det \boldsymbol{T} = \lambda_1\lambda_2\cdots\lambda_n$, $\operatorname{trace} \boldsymbol{T} = \lambda_1 + \lambda_2 + \cdots + \lambda_n$.

[13.28] Show this.

(the indices j *not* being abstract here) where $(x^1, x^2, \ldots, x^n)$ is the ordered set of *components* of $\boldsymbol{x}$ with respect to $\boldsymbol{e}$ (compare §12.3). A non-singular linear transformation $\boldsymbol{T}$ always sends a basis to another basis; moreover, if $\boldsymbol{e}$ and $\boldsymbol{f}$ are any two given bases, then there is a unique $\boldsymbol{T}$ sending each $\boldsymbol{e}_j$ to its corresponding $\boldsymbol{f}_j$:

$$\boldsymbol{T}\boldsymbol{e}_j = \boldsymbol{f}_j.$$

In terms of components taken with respect to $\boldsymbol{e}$, the components of $\boldsymbol{e}_j$ (the basis elements $\boldsymbol{e}_1, \boldsymbol{e}_2, \ldots, \boldsymbol{e}_n$ themselves) are, respectively, $(1, 0, 0, \ldots, 0)$, $(0, 1, 0, \ldots, 0)$, $\ldots$, $(0, 0, \ldots, 0, 1)$. In other words, the components of $\boldsymbol{e}_j$ are $(\delta^1_j, \delta^2_j, \delta^3_j, \ldots, \delta^n_j)$.[13.29] When all components are taken with respect to the $\boldsymbol{e}$ basis, we find that $\boldsymbol{T}$ is represented as the matrix $(T^i{}_j)$, where the components of $\boldsymbol{f}_j$ in the $\boldsymbol{e}$ basis would be[13.30]

$$(T^1{}_j, T^2{}_j, T^3{}_j, \ldots, T^n{}_j).$$

It should be recalled that the conceptual difference between a linear transformation and a matrix is that the latter refers to some basis-dependent presentation, whereas the former is abstract, not depending upon a basis.

Now, provided that each multiple eigenvalue of $\boldsymbol{T}$ (if there are any) satisfies $d = r$, i.e. its eigenspace dimensionality equals its multiplicity, it is possible to find a basis $(\boldsymbol{e}_1, \boldsymbol{e}_2, \ldots, \boldsymbol{e}_n)$ for $\mathbf{V}$, each of which is an eigenvector of $\boldsymbol{T}$.[13.31] Let the corresponding eigenvalues be $\lambda_1, \lambda_2, \ldots, \lambda_n$:

$$\boldsymbol{T}\boldsymbol{e}_1 = \lambda_1\boldsymbol{e}_1, \quad \boldsymbol{T}\boldsymbol{e}_2 = \lambda_2\boldsymbol{e}_2, \ldots, \quad \boldsymbol{T}\boldsymbol{e}_n = \lambda_n\boldsymbol{e}_n.$$

If, as above, $\boldsymbol{T}$ takes the $\boldsymbol{e}$ basis to the $\boldsymbol{f}$ basis, then the $\boldsymbol{f}$ basis elements are as above, so we have $\boldsymbol{f}_1 = \lambda_1\boldsymbol{e}_1, \boldsymbol{f}_2 = \lambda_2\boldsymbol{e}_2, \ldots, \boldsymbol{f}_n = \lambda_n\boldsymbol{e}_n$. It follows that $\boldsymbol{T}$, referred to the $\boldsymbol{e}$ basis, takes the *diagonal* matrix form

$$\begin{pmatrix} \lambda_1 & 0 & \cdots & 0 \\ 0 & \lambda_2 & \cdots & 0 \\ \vdots & \vdots & \ddots & \vdots \\ 0 & 0 & \cdots & \lambda_n \end{pmatrix},$$

that is $T^1_1 = \lambda_1, T^2_2 = \lambda_2, \ldots, T^n_n = \lambda_n$, the remaining components being zero. This *canonical form* for a linear transformation is very useful both conceptually and calculationally.[12]

[13.29] Explain this notation.

[13.30] Why? What are the components of $\boldsymbol{e}_i$ in the $\boldsymbol{f}$ basis?

[13.31] See if you can prove this. *Hint*: For each eigenvalue of multiplicity r, choose r linearly independent eigenvectors. Show that a linear relation between vectors of this entire collection leads to a contradiction when this relation is pre-multiplied by $\boldsymbol{T}$, successively.

13.6 Representation theory and Lie algebras

There is an important body of ideas (particularly significant for quantum theory) called the *representation theory* of groups. We saw a very simple example of a group representation in the discussion in §13.1, when we observed that the non-reflective symmetries of a square can be represented by complex numbers, the group multiplication being faithfully represented as actual multiplication of the complex numbers. However, nothing quite so simple can apply to non-Abelian groups, since the multiplication of complex numbers is commutative. On the other hand, linear transformations (or matrices) usually do not commute, so we may regard it as a reasonable prospect to represent non-Abelian groups in terms of them. Indeed, we already encountered this kind of thing at the beginning of §13.3, where we represented the rotation group O(3) in terms of linear transformations in three dimensions.

As we shall be seeing in Chapter 22, quantum mechanics is all to do with linear transformations. Moreover, various symmetry groups have crucial importance in modern particle physics, such as the rotation group O(3), the symmetry groups of relativity theory (Chapter 18), and the symmetries underlying particle interactions (Chapter 25). It is not surprising, therefore, that representations of these groups in particular, in terms of linear transformations, have fundamental roles to play in quantum theory.

It turns out that quantum theory (particularly the quantum field theory of Chapter 26) is frequently concerned with linear transformations of *infinite*-dimensional spaces. For simplicity, however, I shall phrase things here just for representations by linear transformations in the finite-dimensional case. Most of the ideas that we shall encounter apply also in the case of infinite-dimensional representations, although there are differences that can be important in some circumstances.

What is a group representation? Consider a group $\mathcal{G}$. Representation theory is concerned with finding a subgroup of GL(n) (i.e. a multiplicative group of $n \times n$ matrices) with the property that, for any element g in $\mathcal{G}$, there is a corresponding linear transformation $\boldsymbol{T}(g)$ (belonging to GL(n)) such that the multiplication law in $\mathcal{G}$ is preserved by the operations of GL(n), i.e. for any two elements g, h of $\mathcal{G}$, we have

$$\boldsymbol{T}(g)\boldsymbol{T}(h) = \boldsymbol{T}(gh).$$

The representation is called *faithful* if $\boldsymbol{T}(g)$ is different from $\boldsymbol{T}(h)$ whenever g is different from h. In this case we have an *identical copy* of the group $\mathcal{G}$, as a *subgroup* of GL(n).

In fact, every finite group has a faithful representation in GL(n, $\mathbb{R}$), where n is the order of $\mathcal{G}$,[13.32] and there are frequently many non-faithful representations. On the other hand, it is not quite true that every (finite-dimensional) continuous group has a faithful representation in some GL(n). However, if we are not worried about the global aspects of the group, then a representation *is* always (locally) possible.[13]

There is a beautiful theory, due to the profoundly original Norwegian mathematician Sophus Lie (1842–1899), which leads to a full treatment of the local theory of continuous groups. (Indeed, continuous groups are commonly called 'Lie groups'; see §13.1.) This theory depends upon a study of *infinitesimal* group elements.[14] These infinitesimal elements define a kind of algebra—referred to as a *Lie algebra*—which provides us with complete information as to the local structure of the group. Although the Lie algebra may not provide us with the full *global* structure of the group, this is normally considered to be a matter of lesser importance.

What is a Lie algebra? Suppose that we have a matrix (or linear transformation) $\boldsymbol{I} + \varepsilon\boldsymbol{A}$ to represent an 'infinitesimal' element a of some continuous group $\mathcal{G}$, where ε is taken as 'small' (compare end of §13.4). When we form the matrix product of $\boldsymbol{I} + \varepsilon\boldsymbol{A}$ and $\boldsymbol{I} + \varepsilon\boldsymbol{B}$ to represent the product ab of two such elements a and b, we obtain

$$\begin{aligned}(\boldsymbol{I} + \varepsilon\boldsymbol{A})\,(\boldsymbol{I} + \varepsilon\boldsymbol{B}) &= \boldsymbol{I} + \varepsilon(\boldsymbol{A} + \boldsymbol{B}) + \varepsilon^2\boldsymbol{AB}\\ &= \boldsymbol{I} + \varepsilon(\boldsymbol{A} + \boldsymbol{B})\end{aligned}$$

if we are allowed to ignore the quantity ε^2, as being 'too small to count'. In accordance with this, the matrix *sum* $\boldsymbol{A} + \boldsymbol{B}$ represents the *group product* ab of two infinitesimal elements a and b.

Indeed, the sum operation is part of the Lie algebra of the quantities $\boldsymbol{A}, \boldsymbol{B}, \ldots$. But the sum is commutative, whereas the group $\mathcal{G}$ could well be non-Abelian, so we do not capture much of the structure of the group if we consider only sums (in fact, only the dimension of $\mathcal{G}$). The non-Abelian nature of $\mathcal{G}$ is expressed in the *group commutators* which are the expressions[13.33]

$$a\,b\,a^{-1}\,b^{-1}.$$

[13.32] Show this. *Hint*: Label each column of the representing matrix by a separate element of the finite group $\mathcal{G}$, and also label each row by the corresponding group element. Place a 1 in any position in the matrix for which a certain relation holds (find it!) between the element of $\mathcal{G}$ labelling the row, that labelling the column, and the element of $\mathcal{G}$ that this particular matrix is representing. Place a 0 whenever this relation does not hold.

[13.33] Why is this expression just the *identity* group element when a and b commute?

Let us write this out in terms of $\boldsymbol{I}+\varepsilon\boldsymbol{A}$, etc., taking note of the power series expression $(\boldsymbol{I}+\varepsilon\boldsymbol{A})^{-1}=\boldsymbol{I}-\varepsilon\boldsymbol{A}+\varepsilon^2\boldsymbol{A}^2-\varepsilon^3\boldsymbol{A}^3+\cdots$ (this series being easily checked by multiplying both sides by $\boldsymbol{I}+\varepsilon\boldsymbol{A}$). Now it is ε^3 that we ignore as being 'too small to count', but we keep ε^2, whence[13.34]

$$(\boldsymbol{I}+\varepsilon\boldsymbol{A})\,(\boldsymbol{I}+\varepsilon\boldsymbol{B})\,(\boldsymbol{I}+\varepsilon\boldsymbol{A})^{-1}\,(\boldsymbol{I}+\varepsilon\boldsymbol{B})^{-1}$$
$$=(\boldsymbol{I}+\varepsilon\boldsymbol{A})\,(\boldsymbol{I}+\varepsilon\boldsymbol{B})\,(\boldsymbol{I}-\varepsilon\boldsymbol{A}+\varepsilon^2\boldsymbol{A}^2)\,(\boldsymbol{I}-\varepsilon\boldsymbol{B}+\varepsilon^2\boldsymbol{B}^2)$$
$$=\boldsymbol{I}+\varepsilon^2(\boldsymbol{AB}-\boldsymbol{BA})$$

This tells us that if we are to keep track of the precise way in which the group $\mathcal{G}$ is non-Abelian, we must take note of the 'commutators', or *Lie brackets*

$$[\boldsymbol{A},\boldsymbol{B}]=\boldsymbol{AB}-\boldsymbol{BA}.$$

The Lie algebra is now constructed by means of repeated application of the operations +, its inverse –, and the bracket operation [,], where it is customary also to allow the multiplication by ordinary numbers (which might be real or complex). The 'additive' aspect of the algebra has the usual vector-space structure (as with quaternions, in §11.1). In addition, Lie bracket satisfies distributivity, etc., namely

$$[\boldsymbol{A}+\boldsymbol{B},\boldsymbol{C}]=[\boldsymbol{A},\boldsymbol{C}]+[\boldsymbol{B},\boldsymbol{C}],\ [\lambda\boldsymbol{A},\boldsymbol{B}]=\lambda[\boldsymbol{A},\boldsymbol{B}],$$

the antisymmetry property

$$[\boldsymbol{A},\boldsymbol{B}]=-[\boldsymbol{B},\boldsymbol{A}],$$

(whence also $[\boldsymbol{A},\boldsymbol{C}+\boldsymbol{D}]=[\boldsymbol{A},\boldsymbol{C}]+[\boldsymbol{A},\boldsymbol{D}]$, $[\boldsymbol{A},\lambda\boldsymbol{B}]=\lambda[\boldsymbol{A},\boldsymbol{B}]$), and an elegant relation known as the *Jacobi identity*[13.35]

$$[\boldsymbol{A},[\boldsymbol{B},\boldsymbol{C}]]+[\boldsymbol{B},[\boldsymbol{C},\boldsymbol{A}]]+[\boldsymbol{C},[\boldsymbol{A},\boldsymbol{B}]]=0$$

(a more general form of which will be encountered in §14.6).

We can choose a basis $(\boldsymbol{E}_1,\boldsymbol{E}_2,\ldots,\boldsymbol{E}_N)$ for the vector space of our matrices $\boldsymbol{A},\boldsymbol{B},\boldsymbol{C},\ldots$ (where N is the dimension of the group $\mathcal{G}$, if the representation is faithful). Forming their various commutators $[\boldsymbol{E}_\alpha,\boldsymbol{E}_\beta]$, we express these in terms of the basis elements, to obtain relations (using the summation convention)

$$[\boldsymbol{E}_\alpha,\boldsymbol{E}_\beta]=\gamma_{\alpha\beta}{}^{\chi}\boldsymbol{E}_\chi.$$

[13.34] Spell out this 'order ε^2' calculation.

[13.35] Show all this.

The N^3 component quantities $\gamma_{\alpha\beta}{}^{\chi}$ are called *structure constants* for $\mathcal{G}$. They are not all independent because they satisfy (see §11.6 for bracket notation)

$$\gamma_{\alpha\beta}{}^{\chi} = -\gamma_{\beta\alpha}{}^{\chi}, \quad \gamma_{[\alpha\beta}{}^{\xi}\gamma_{\chi]\xi}{}^{\zeta} = 0,$$

by virtue of the above antisymmetry and Jacobi identity.[13.36] These relations are given in diagrammatic form in Fig. 13.11.

It is a remarkable fact that the structure of the Lie algebra for a faithful representation (basically, the knowledge of the structure constants $\gamma_{\alpha\beta}{}^{\chi}$) is sufficient to determine the precise local nature of the group $\mathcal{G}$. Here, 'local' means in a (sufficiently small) N-dimensional open region $\mathcal{N}$ surrounding the identity element $\boldsymbol{I}$ in the 'group manifold' $\tilde{\mathcal{G}}$ whose points represent the different elements of $\mathcal{G}$ (see Fig. 13.12). In fact, starting from a Lie group element $\boldsymbol{A}$, we can construct a corresponding actual *finite* (i.e. non-infinitesimal) group element by means of the 'exponentiation' operation $e^{\boldsymbol{A}}$ defined at the end of §13.4. (This will be considered a little more fully in §14.6.) Thus, the theory of representations of continuous groups by linear transformations (or by matrices) may be largely transferred to the study of representations of Lie algebras by such transformations—which, indeed, is the normal practice in physics.

This is particularly important in quantum mechanics, where the Lie algebra elements themselves, in a remarkable way, frequently have direct interpretations as physical quantities (such as angular momentum, when the group $\mathcal{G}$ is the rotation group, as we shall be seeing later in §22.8).

The Lie algebra matrices tend to be considerably simpler in structure than the corresponding Lie group matrices, being subject to linear rather

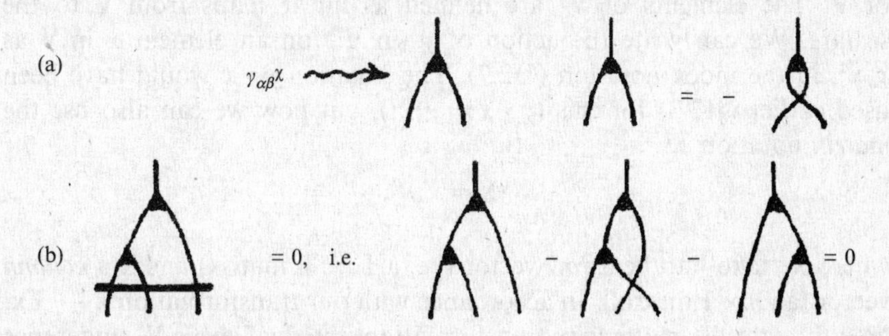

Fig. 13.11 (a) Structure constants $\gamma_{\alpha\beta}{}^{\chi}$ in diagrammatic form, depicting antisymmetry in α, β and (b) the Jacobi identity.

[13.36] Show this.

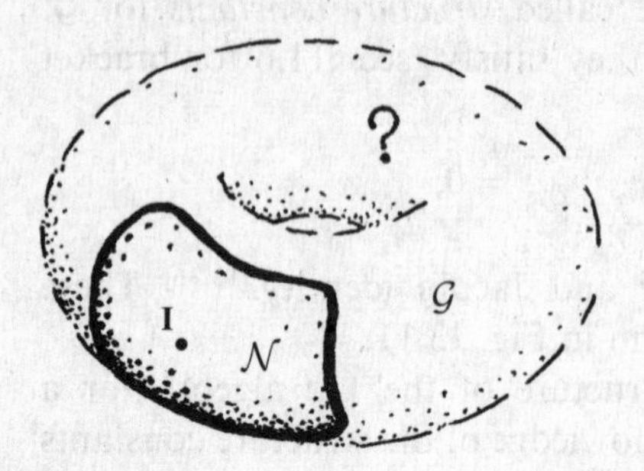

Fig. 13.12 The Lie algebra for a (faithful) representation of a Lie group $\mathcal{G}$ (basically, knowledge of the structure constants $\gamma_{\alpha\beta}{}^{\chi}$) determines the local structure of $\mathcal{G}$, i.e. it fixes the structure of $\mathcal{G}$ within some (sufficiently small) open region $\mathcal{N}$ surrounding the identity element **I**, but it does not tell us about the global nature of $\mathcal{G}$.

than nonlinear restrictions (see §13.10 for the case of the classical groups). This procedure is beloved of quantum physicists!

13.7 Tensor representation spaces; reducibility

There are ways of building up more elaborate representations of a group $\mathcal{G}$, starting from some particular one. How are we to do that? Suppose that $\mathcal{G}$ is represented by some family $\mathcal{T}$ of linear transformations, acting on an n-dimensional vector space **V**. Such a **V** is called a *representation space* for $\mathcal{G}$. Any element t of $\mathcal{G}$ is now represented by a corresponding linear transformation $\boldsymbol{T}$ in $\mathcal{T}$, where $\boldsymbol{T}$ effects $\boldsymbol{x} \mapsto \boldsymbol{Tx}$ for each $\boldsymbol{x}$ belonging to **V**. In the (abstract) index notation (§12.7) we write this $x^a \mapsto T^a{}_b x^b$, as in §13.3, or in diagrammatic form, as in Fig. 13.6a. Let us see how we can find other representation spaces for $\mathcal{G}$, starting from the given one **V**.

As a first example, recall, from §12.3, the definition of the *dual space* $\mathbf{V}^*$ of **V**. The elements of $\mathbf{V}^*$ are defined as linear maps from **V** to the scalars. We can write the action of $\boldsymbol{y}$ (in $\mathbf{V}^*$) on an element $\boldsymbol{x}$ in **V** as $y_a x^a$, in the index notation (§12.7). The notation $\boldsymbol{y} \bullet \boldsymbol{x}$ would have been used earlier (§12.3) for this ($\boldsymbol{y} \bullet \boldsymbol{x} = y_a x^a$), but now we can also use the *matrix* notation

$$\mathbf{yx} = y_a x^a,$$

where we take **y** to be a *row* vector (i.e. a $1 \times n$ matrix) and **x** a *column* vector (an $n \times 1$ matrix). In accordance with our transformation $\mathbf{x} \mapsto \boldsymbol{T}\mathbf{x}$, now thought of as a *matrix* transformation, the dual space $\mathbf{V}^*$ undergoes the linear transformation

$$\mathbf{y} \mapsto \mathbf{yS}, \quad \text{i.e.} \quad y_a \mapsto y_b S^b{}_a,$$

where **S** is the *inverse* of **T**:

$$\boldsymbol{S} = \boldsymbol{T}^{-1}, \quad \text{so} \quad S^a{}_b T^b{}_c = \delta^a_c,$$

since, if $\mathbf{x} \mapsto \mathbf{Tx}$, we need $\mathbf{y} \mapsto \mathbf{yT}^{-1}$ to ensure that $\mathbf{yx}$ is preserved by $\mapsto$.

The use of a row vector $\mathbf{y}$, in the above, gives us a non-standard multiplication ordering. It is more usual to write things the other way around, by employing the notation of the *transpose* $\mathbf{A}^\mathsf{T}$ of a matrix $\mathbf{A}$. The elements of the matrix $\mathbf{A}^\mathsf{T}$ are the same as those of $\mathbf{A}$, but with rows and columns interchanged. If $\mathbf{A}$ is square ($n \times n$), then so is $\mathbf{A}^\mathsf{T}$, its elements being those of $\mathbf{A}$ reflected in its main diagonal (see §13.3). If $\mathbf{A}$ is rectangular ($m \times n$), then $\mathbf{A}^\mathsf{T}$ is $n \times m$, correspondingly reflected. Thus $\mathbf{y}^\mathsf{T}$ is a standard column vector, and we can write the above $\mathbf{y} \mapsto \mathbf{yS}$ as

$$\mathbf{y}^\mathsf{T} \mapsto \mathbf{S}^\mathsf{T}\mathbf{y}^\mathsf{T},$$

since the transpose operation $^\mathsf{T}$ reverses the order of multiplication: $(\mathbf{AB})^\mathsf{T} = \mathbf{B}^\mathsf{T}\mathbf{A}^\mathsf{T}$. We thus see that the dual space $\mathbf{V}^*$, of any representation space $\mathbf{V}$ is itself a representation space of $\mathcal{G}$. Note that the inverse operation $^{-1}$ also reverses multiplication order, $(\mathbf{AB})^{-1} = \mathbf{B}^{-1}\mathbf{A}^{-1}$,[13.37] so the multiplication ordering needed for a representation is restored.

The same kinds of consideration apply to the various vector spaces of tensors constructed from $\mathbf{V}$; see §12.8. We recall that a tensor $\boldsymbol{Q}$ of valence $[{}^p_q]$ (over the vector space $\mathbf{V}$) has an index description as a quantity

$$Q^{f\ldots h}_{a\ldots c},$$

with q lower and p upper indices. We can add tensors to other tensors of the same valence and we can multiply them by scalars; tensors of fixed valence $[{}^p_q]$ form a vector space of dimension n^{p+q} (the total number of components).[13.38] Abstractly, we think of $\boldsymbol{Q}$ as belonging to a vector space that we refer to as the *tensor product*

$$\mathbf{V}^* \otimes \mathbf{V}^* \otimes \ldots \otimes \mathbf{V}^* \otimes \mathbf{V} \otimes \mathbf{V} \otimes \ldots \otimes \mathbf{V}$$

of q copies of the dual space $\mathbf{V}^*$ and p copies of $\mathbf{V}$ ($p, q \geq 0$). (We shall come to this notion of 'tensor product' a little more fully in §23.3.) Recall the abstract definition of a tensor, given in §12.8, as a multilinear function.

[13.37] Why?

[13.38] Why this number?

This will suffice for our purposes here (although there are certain subtleties in the case of an infinite-dimensional **V**, of relevance to the applications to many-particle quantum states, needed in §23.8).[15]

Whenever a linear transformation $x^a \mapsto T^a{}_b x^b$ is applied to **V**, this induces a corresponding linear transformation on the above tensor product space, given explicitly by[13.39]

$$Q_{a\ldots c}^{f\ldots h} \mapsto S^{a'}{}_a \ldots S^{c'}{}_c T^f{}_{f'} \ldots T^h{}_{h'} Q_{a'\ldots c'}^{f'\ldots h'}.$$

All these indices require good eyesight and careful scrutiny, in order to make sure of what is summed with what; so I recommend the diagrammatic notation, which is clearer, as illustrated in Fig. 13.13. We see that each lower index of $Q^{\cdots}_{\cdots}$ transforms by the inverse matrix $\mathbf{S} = \mathbf{T}^{-1}$ (or, rather, by $\mathbf{S}^{\mathsf{T}}$), as with y_a and each upper index by $\mathbf{T}$, as with x^a. Accordingly, the space of $[^p_q]$-valent tensors over **V** is also a representation space for $\mathcal{G}$, of dimension n^{p+q}.

These representation spaces are, however, likely to be what is called *reducible*. To illustrate this situation, consider the case of a $[^2_0]$-valent tensor Q^{ab}. Any such tensor can be split into its *symmetric* part $Q^{(ab)}$ and its *antisymmetric* part $Q^{[ab]}$ (§12.7 and §11.6):

$$Q^{ab} = Q^{(ab)} + Q^{[ab]},$$

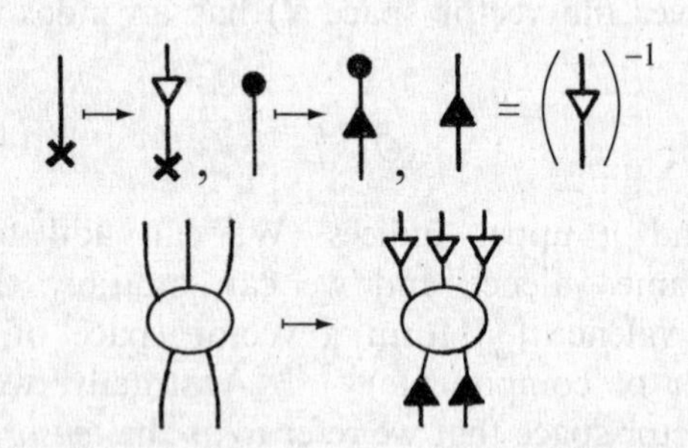

Fig. 13.13 The linear transformation $x^a \mapsto T^a{}_b x^b$, applied to $\mathbf{x}$ in the vector space **V** (with $\boldsymbol{T}$ depicted as a white triangle), extends to the dual space $\mathbf{V}^*$ by use of the inverse $\boldsymbol{S} = \boldsymbol{T}^{-1}$ (depicted as a black triangle) and thence to the spaces $\mathbf{V}^* \otimes \ldots \otimes \mathbf{V}^* \otimes \mathbf{V} \otimes \ldots \otimes \mathbf{V}$ of $[^p_q]$-valent tensors $\boldsymbol{Q}$. The case $p = 3$, $q = 2$ is illustrated, with $\boldsymbol{Q}$ shown as an oval with three arms and two legs undergoing $Q_{ab}{}^{cde} \mapsto S^{a'}{}_a S^{b'}{}_b T^c{}_{c'} T^d{}_{d'} T^e{}_{e'} Q_{a'b'}{}^{c'd'e'}$.

[13.39] Show this.

where

$$Q^{(ab)} = \tfrac{1}{2}(Q^{ab} + Q^{ba}), \quad Q^{[ab]} = \tfrac{1}{2}(Q^{ab} - Q^{ba}).$$

The dimension of the *symmetric* space $\mathbf{V}_+$ is $\frac{1}{2}n(n+1)$, and that of the *antisymmetric* space $\mathbf{V}_-$ is $\frac{1}{2}n(n-1)$.[13.40] It is not hard to see that, under the transformation $x^a \mapsto T^a{}_b x^b$, so that $Q^{ab} \mapsto T^a{}_c T^b{}_d Q^{cd}$, the symmetric and antisymmetric parts transform to tensors which are again, respectively, symmetric, and antisymmetric.[13.41] Accordingly, the spaces $\mathbf{V}_+$ and $\mathbf{V}_-$ are, separately, representation spaces for $\mathcal{G}$. By choosing a basis for $\mathbf{V}$ where the first $\frac{1}{2}n(n+1)$ basis elements are in $\mathbf{V}_+$ and the remaining $\frac{1}{2}n(n-1)$ are in $\mathbf{V}_-$, we obtain our representation with all matrices being of the $n^2 \times n^2$ 'block-diagonal' form

$$\begin{pmatrix} \mathbf{A} & \mathbf{O} \\ \mathbf{O} & \mathbf{B} \end{pmatrix},$$

where $\mathbf{A}$ stands for a $\frac{1}{2}n(n+1) \times \frac{1}{2}n(n+1)$ matrix and $\mathbf{B}$ for a $\frac{1}{2}n(n-1) \times \frac{1}{2}n(n-1)$ matrix, the two $\mathbf{O}$s standing for the appropriate rectangular blocks of zeros.

A representation of this form is referred to as the *direct sum* of the representation given by the $\mathbf{A}$ matrices and that given by the $\mathbf{B}$ matrices. The representation in terms of $\left[{2 \atop 0}\right]$-valent tensors is therefore *reducible*, in this sense.[13.42] The notion of 'direct sum' also extends to any number (perhaps infinite) of smaller representations.

In fact there is a more general meaning for the term 'reducible representation', namely one for which there is a choice of basis for which all the matrices of the representation can be put in the somewhat more complicated form

$$\begin{pmatrix} \mathbf{A} & \mathbf{C} \\ \mathbf{O} & \mathbf{B} \end{pmatrix},$$

where $\mathbf{A}$ is $p \times p$, $\mathbf{B}$ is $q \times q$, and $\mathbf{C}$ is $p \times q$, with $p, q \geq 1$ (for fixed p and q). Note that, if the representing matrices all have this form, then the $\mathbf{A}$ matrices and the $\mathbf{B}$ matrices each individually constitute a (smaller) representation of $\mathcal{G}$.[13.43] If the $\mathbf{C}$ matrices are all zero, we get the earlier case where the representation is the direct sum of these two smaller representations. A representation is called *irreducible* if it is not reducible (with C present or

---

[13.40] Show this.

[13.41] Explain this.

[13.42] Show that the representation space of $\left[{1 \atop 1}\right]$-valent tensors is also reducible. *Hint*: Split any such tensor into a 'trace-free' part and a 'trace' part.

[13.43] Confirm this.

not). A representation is called *completely reducible* if we never get the above situation (with non-zero C), so that it is a direct sum of irreducible representations.

There is an important class of continuous groups, known as *semi-simple* groups. This extensively studied class includes the simple groups referred to in §13.2. Compact semi-simple groups have the pleasing property that *all* their representations are completely reducible. (See §12.6, Fig. 12.12 for the definition of 'compact'.) It is sufficient to study *irreducible* representations of such a group, every representation being just a direct sum of these irreducible ones. In fact, every irreducible representation of such a group is finite-dimensional (which is not the case if we allow a semi-simple group to be non-compact, when representations that are *not* completely reducible can also occur).

What is a semi-simple group? Recall the 'structure constants' $\gamma_{\alpha\beta}{}^{\chi}$ of §13.6, which specify the Lie brackets and define the local structure of the group $\mathcal{G}$. There is a quantity of considerable importance known[16] as the '*Killing form*' κ that can be constructed from $\gamma_{\alpha\beta}{}^{\chi}$:[13.44]

$$\kappa_{\alpha\beta} = \gamma_{\alpha\zeta}{}^{\xi}\, \gamma_{\beta\xi}{}^{\zeta} = \kappa_{\beta\alpha}.$$

The diagrammatic form of this expression is given in Fig. 13.14. The condition for $\mathcal{G}$ to be semi-simple is that the matrix $\kappa_{\alpha\beta}$ be non-singular.

Some remarks are appropriate concerning the condition of compactness of a semi-simple group. For a given set of structure constants $\gamma_{\alpha\beta}{}^{\chi}$, assuming that we can take them to be real numbers, we could consider either the real or the complex Lie algebra obtained from them. In the complex case, we do not get a compact group $\mathcal{G}$, but we might do so in the real case. In fact, compactness occurs in the real case when $-\kappa_{\beta\alpha}$ is what is called *positive definite* (the meaning of which term we shall come to in §13.8). For fixed $\gamma_{\alpha\beta}{}^{\chi}$, in the case of a real group $\mathcal{G}$, we can always construct the *complexification* $\mathbb{C}\mathcal{G}$ (at least locally) of $\mathcal{G}$ which comes about merely by using the same $\gamma_{\alpha\beta}{}^{\chi}$, but with complex coefficients in the Lie algebra. However, different real groups $\mathcal{G}$ might sometimes give rise to the same[17] $\mathbb{C}\mathcal{G}$. These different real groups are called different *real forms* of the complex group. We shall be seeing important

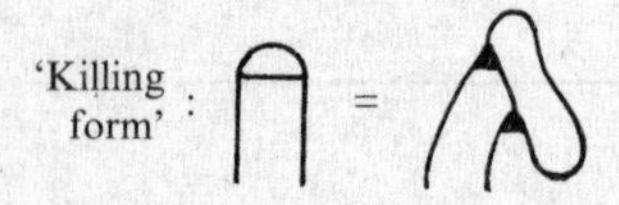

Fig. 13.14 The 'Killing form' $\kappa_{\alpha\beta}$ defined from the structure constants $\gamma_{\alpha\zeta}{}^{\xi}$ by $\kappa_{\alpha\beta} = \gamma_{\alpha\zeta}{}^{\xi}\gamma_{\beta\xi}{}^{\zeta}$.

[13.44] Why does $\kappa_{\alpha\beta} = \kappa_{\beta\alpha}$?

instances of this in later chapters, especially in §18.2, where the Euclidean motions in 4 dimensions and the Lorentz/Poincaré symmetries of special relativity are compared. It is a remarkable property of any complex semi-simple Lie group that it has exactly one real form $\mathcal{G}$ which is compact.

13.8 Orthogonal groups

Now let us return to the orthogonal group. We already saw at the beginning of §13.3 how to represent O(3) or SO(3) faithfully as linear transformations of a 3-dimensional real vector space, with ordinary Cartesian coordinates (x,y,z), where the sphere

$$x^2 + y^2 + z^2 = 1$$

is to be left invariant (the upper index 2 meaning the usual 'squared'). Let us write this equation in terms of the index notation (§12.7), so that we can generalize to n dimensions. The equation of our sphere can now be written

$$g_{ab}x^a x^b = 1,$$

which stands for $(x^1)^2 + \cdots + (x^n)^2 = 1$, the components g_{ab} being given by

$$g_{ab} = \begin{cases} 1 & \text{if } a = b, \\ 0 & \text{if } a \neq b. \end{cases}$$

In the diagrammatic notation, I recommend simply using a 'hoop' for g_{ab}, as indicated in Fig. 13.15a. I shall also use the notation g^{ab} (with the *same* explicit components as g_{ab}) for the *inverse* quantity ('inverted hoop' in Fig. 13.15a):

$$g_{ab}\, g^{bc} = \delta_a^c = g^{cb} g_{ba}.$$

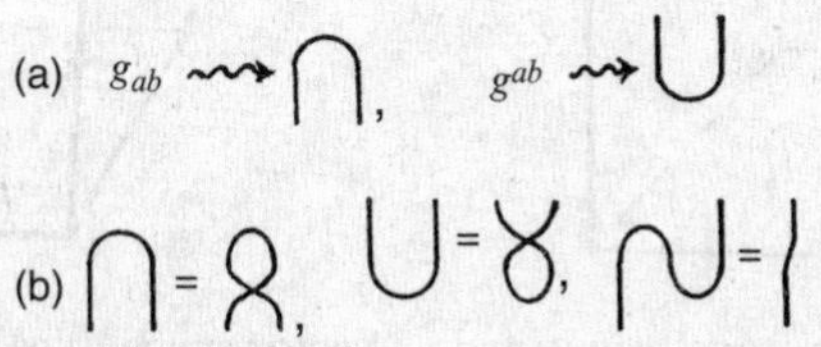

(a) g_{ab} ⇝ ∩, g^{ab} ⇝ ∪

(b) = , = , =

Fig. 13.15 (a) The metric g_{ab} and its inverse g^{ab} in the 'hoop' diagrammatic notation. (b) The relations $g_{ab} = g_{ba}$ (i.e. $\mathbf{g}^{\mathrm{T}} = \mathbf{g}$), $g^{ab} = g^{ba}$, and $g_{ab}g^{bc} = \delta_a^c$ in diagrammatic notation.

The puzzled reader might very reasonably ask why I have introduced two new notations, namely g_{ab} and g^{ab} for precisely the same matrix components that I denoted by δ^a_b in §13.3! The reason has to do with the consistency of the notation and with what happens when a linear transformation is applied to the *coordinates*, according to some replacement

$$x^a \mapsto t^a{}_b x^b,$$

$t^a{}_b$ being non-singular, so that it has an inverse $s^a{}_b$:

$$t^a{}_b s^b{}_c = \delta^a_c = s^a{}_b t^b{}_c.$$

This is formally the same as the type of linear transformation that we considered in §§13.3,7, but we are now thinking of it in a quite different way. In those sections, our linear transformation was thought of as *active*, so that the vector space **V** was viewed as being actually *moved* (over itself). Here we are thinking of the transformation as *passive* in that the objects under consideration—and, indeed, the vector space **V** itself—remain pointwise fixed, but the representations in terms of coordinates are changed. Another way of putting this is that the basis ($\boldsymbol{e}_1, \ldots, \boldsymbol{e}_n$) that we had previously been using (for the representation of vector/tensor quantities in terms of components[18]) is to be replaced by some other basis. See Fig. 13.16.

In direct correspondence with what we saw in §13.7 for the active transformation of a tensor, we find that the corresponding passive change in the components $Q^{a \ldots c}_{p \ldots r}$ of a tensor $\boldsymbol{Q}$ is given by[13.45]

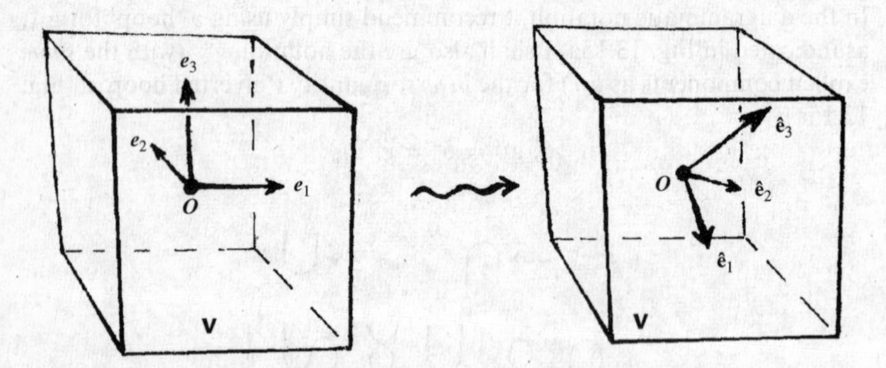

Fig. 13.16 A passive transformation in a vector space **V** leaves **V** pointwise fixed, but changes its coordinate description, i.e. the basis $\boldsymbol{e}_1, \boldsymbol{e}_2, \ldots, \boldsymbol{e}_n$ is replaced by some other basis (case $n = 3$ illustrated).

[13.45] Use Note 13.18 to establish this.

$$Q^{a\ldots c}_{p\ldots r} \mapsto t^a{}_d \cdots t^c{}_f \; Q^{d\ldots f}_{j\ldots l} \; s^j{}_p \ldots s^l{}_r.$$

Applying this to δ^a_b, we find that its components are completely unaltered,[13.46] whereas this is *not* the case for g_{ab}. Moreover, after a general such coordinate change, the components g^{ab} will be quite different from g_{ab} (inverse matrices). Thus, the reason for the additional symbols g^{ab} and g_{ab} is simply that they can only represent the same matrix of components as does δ^a_b in special types of coordinate system ('Cartesian' ones) and, in general, the components are just *different*. This has a particular importance for general relativity, where the coordinate system cannot normally be arranged to have this special (Cartesian) form.

A general coordinate change can make the matrix of components g_{ab} a more complicated although not completely general matrix. It retains the property of symmetry between a and b giving a *symmetric* matrix. The term 'symmetric' tells us that the square array of components is symmetrical about its main diagonal, i.e. $\mathbf{g}^{\mathrm{T}} = \mathbf{g}$ (using the 'transpose' notation of §13.7). In index-notation terms, this symmetry is expressed as either of the two equivalent[13.47] forms

$$g_{ab} = g_{ba}, \; g^{ab} = g^{ba},$$

and see Fig. 13.15b for the diagrammatic form of these relations.

What about going in the opposite direction? Can any non-singular $n \times n$ real symmetric matrix be reduced to the component form of a Kronecker delta? Not quite—not by a real linear transformation of coordinates. What it can be reduced to by such means is this same form except that there may be some terms 1 and some terms -1 along the main diagonal. The number, p, of these 1 terms and the number, q, of -1 terms is an *invariant*, which is to say we cannot get a different number by trying some other real linear transformation. This invariant (p, q) is called the *signature* of $\mathbf{g}$. (Sometimes it is $p - q$ that is called the signature; sometimes one just writes $+ \ldots + - \ldots -$ with the appropriate number of each sign.) In fact, this works also for a *singular* $\boldsymbol{g}$, but then we need some 0s along the main diagonal also and the number of 0s becomes part of the signature as well as the number of 1s and the number of -1s. If we only have 1s, so that $\mathbf{g}$ is non-singular and also $q = 0$, then we say that $\boldsymbol{g}$ is *positive-definite*. A non-singular $\mathbf{g}$ for which $p = 1$ and $q \neq 0$ (or $q = 1$ and $p \neq 0$) is called *Lorentzian*, in honour of the Dutch physicist H.A. Lorentz (1853–1928), whose important work in this connection provided one of the foundation stones of relativity theory; see §§17.6–9 and §§18.1–3.

[13.46] Why?

[13.47] Why equivalent?

An alternative characterization of a positive-definite matrix **A**, of considerable importance in certain other contexts (see §20.3, §24.3, §29.3) is that the real symmetric matrix **A** satisfy

$$\mathbf{x}^{\mathrm{T}}\mathbf{A}\mathbf{x} > 0$$

for all $\mathbf{x} \neq 0$. In index notation, this is: '$A_{ab}x^a x^b > 0$ unless the vector x^a vanishes'.[13.48] We say that A is *non-negative-definite* (or *positive-semi-definite*) if this holds but with $\geq$ in place of $>$ (so we now allow $\mathbf{x}^{\mathrm{T}}\mathbf{A}\mathbf{x} = 0$ for some non-zero $\mathbf{x}$).

Under appropriate circumstances, a symmetric non-singular $[{0 \atop 2}]$-tensor g_{ab}, is called a *metric*—or sometimes a *pseudometric* when $\boldsymbol{g}$ is not positive definite. This terminology applies if we are to use the quantity $\mathrm{d}s$, defined by its square $\mathrm{d}s^2 = g_{ab}\mathrm{d}x^a\mathrm{d}x^b$, as providing us with some notion of 'distance' along curves. We shall be seeing in §14.7 how this notion applies to curved manifolds (see §10.2, §§12.1,2), and in §17.8 how, in the Lorentzian case, it provides us with a 'distance' measure which is actually the *time* of relativity theory. We sometimes refer to the quantity

$$|\boldsymbol{v}| = (g_{ab}v^a v^b)^{\frac{1}{2}}$$

as the *length* of the vector $\boldsymbol{v}$, with index form v^a.

Let us return to the definition of the orthogonal group $\mathrm{O}(n)$. This is simply the group of linear transformations in n dimensions—called *orthogonal* transformations—that preserve a given positive-definite $\boldsymbol{g}$. 'Preserving' $\boldsymbol{g}$ means that an orthogonal transformation $\boldsymbol{T}$ has to satisfy

$$g_{ab}T^a{}_cT^b{}_d = g_{cd}.$$

This is an example of the (active) tensor transformation rule described in §13.7, as applied to g_{ab} (and see Fig. 13.17 for the diagrammatic form of this equation). Another way of saying this is that the metric form $\mathrm{d}s^2$ of the previous paragraph is unchanged by orthogonal transformations. We can, if we please, insist that the components g_{ab} be actually the Kronecker delta—this, in effect, providing the definition of O(3) given in §§13.1,3—but the group comes out the same[19] whatever positive-definite $n \times n$ array of g_{ab} we choose.[13.49]

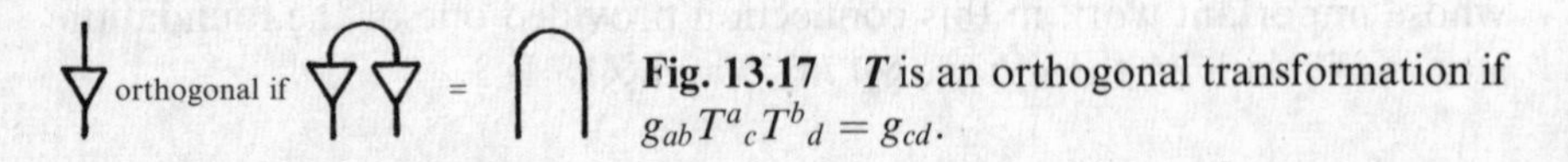

Fig. 13.17 $\boldsymbol{T}$ is an orthogonal transformation if $g_{ab}T^a{}_cT^b{}_d = g_{cd}$.

[13.48] Can you confirm this characterization?

[13.49] Explain why.

With the particular component realization of g_{ab} as the Kronecker delta, the matrices describing our orthogonal transformations are those satisfying[13.50]

$$T^{-1} = T^{\mathsf{T}},$$

called *orthogonal matrices*. The real orthogonal $n \times n$ matrices provide a concrete realization of the group O(n). To specialize to the non-reflective group SO(n), we require that the determinant be equal to unity:[13.51]

$$\det T = 1.$$

We can also consider the corresponding *pseudo-orthogonal* groups O(p, q) and SO(p, q) that are obtained when $\boldsymbol{g}$, though non-singular, is not necessarily positive definite, having the more general signature (p, q). The case when $p = 1$ and $q = 3$ (or equivalently $p = 3$ and $q = 1$), called the *Lorentz* group, plays a fundamental role in relativity theory, as indicated above. We shall also be finding (if we ignore time-reflections) that the Lorentz group is the same as the group of symmetries of the hyperbolic 3-space that was described in §2.7, and also (if we ignore space reflections) of the group of symmetries of the Riemann sphere, as achieved by the bilinear (Möbius) transformations as studied in §8.2. It will be better to delay the explanations of these remarkable facts until our investigation of the Minkowski spacetime geometry of special relativity theory (§§18.4,5). We shall also be seeing in §33.2 that these facts have a seminal significance for twistor theory.

How 'different' are the various groups O(p, q), for $p + q = n$, for fixed n? (The positive-definite and Lorentzian cases are contrasted, for $n = 2$ and $n = 3$, in Fig. 13.18.) They are closely related, all having the same dimension $\frac{1}{2}n(n-1)$; they are what are called *real forms* of one and the same complex group O(n, $\mathbb{C}$), the *complexification* of O(n). This complex group is defined in the same way as O(n) (= O(n, $\mathbb{R}$)), but where the linear transformations are allowed to be *complex*. Indeed, although I have phrased my considerations in this chapter in terms of real linear transformations, there is a parallel discussion where 'complex' replaces 'real' throughout. (Thus the coordinates x^a become complex and so do the components of our matrices.) The only essential difference, in what has been said above, arises with the concept of *signature*. There are complex linear coordinate transformations that can convert a -1 in a diagonal realization of g_{ab} into a $+1$ and *vice versa*,[13.52] so we do not now have a

---

[13.50] Explain this. What is T^{-1} in the pseudo-orthogonal cases (defined in the next paragraph)?

[13.51] Explain why this is equivalent to preserving the volume form given by $\varepsilon_{a...c}$, i.e. $\varepsilon_{a...c}T^a{}_p \ldots T^c{}_r = \varepsilon_{p...r}$? Moreover, why is the preservation of its sign sufficient?

[13.52] Why?

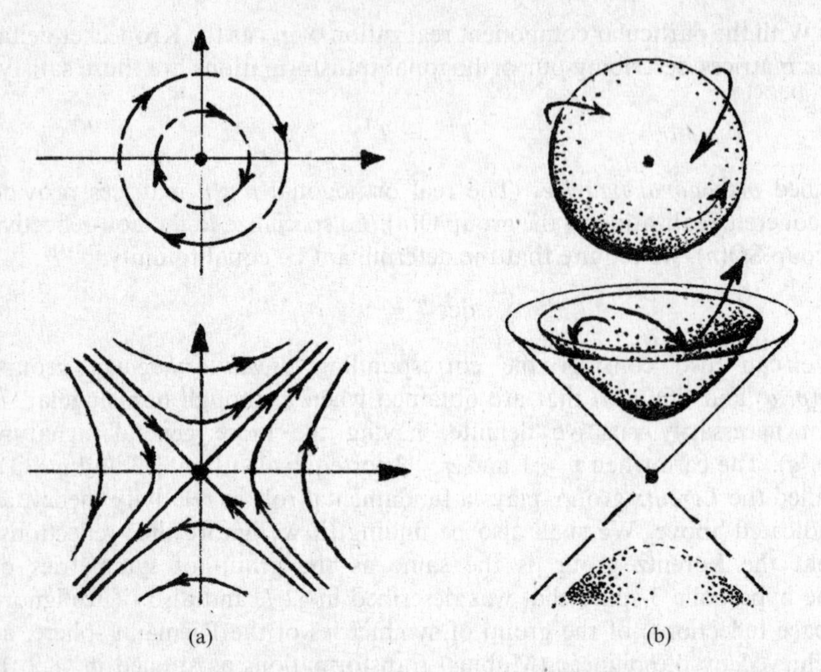

Fig. 13.18 (a) O(2,0) and O(1,1) are contrasted. (b) O(3,0) and O(1,2) are similarly contrasted, the 'unit sphere' being illustrated in each case. For O(1,2) (see §§2.4,5, §18.4), this 'sphere' is a hyperbolic plane (or two copies of such).

meaningful notion of signature. The only invariant[20] of $\boldsymbol{g}$, in the complex case, is what is called its *rank*, which is the number of non-zero terms in its diagonal realization. For a non-singular $\boldsymbol{g}$, the rank has to be maximal, i.e. n.

When is the difference between these various real forms important and when is it not? This can be a delicate question, but physicists are often rather cavalier about the distinctions, even though these can be important. The positive-definite case has the virtue that the group is compact, and much of the mathematics is easier for such situations (see §13.7). Sometimes people blithely carry over results from the compact case to the non-compact cases ($p \neq 0 \neq q$), but this is often not justified. (For example, in the compact case, one need only be concerned with representations that are finite-dimensional, but in the non-compact case additional infinite-dimensional representations arise.) On the other hand, there are other situations in which considerable insights can be obtained by ignoring the distinctions. (We may compare this with Lambert's discovery of the formula, in terms of angles, of the area of a hyperbolic triangle, given in §2.4. He obtained his formula by allowing his sphere to have an imaginary radius. This is similar to a signature change, which amounts to allowing some coordinates to have imaginary values. In §18.4, Fig. 18.9, I shall try to make the case that Lambert's approach to non-Euclidean geometry is perfectly justifiable.)

The different possible real forms of O(n, ℂ) are distinguished by certain sets of *inequalities* on the matrix elements (such as det $\boldsymbol{T} > 0$). A feature of *quantum theory* is that such inequalities are often *violated* in physical processes. For example, *imaginary* quantities can, in a sense, have a physically *real* significance in quantum mechanics, so the distinction between different signatures can become blurred. On the other hand, it is my impression that physicists are often somewhat less careful about these matters than they should be. Indeed, this question will have considerable relevance for us in our examination of a number of modern theories (§28.9, §31.13, §32.3). But more of this later. This is the 'can of worms' that I hinted at in §11.2!

13.9 Unitary groups

The group O(n, ℂ) provides us with *one* way in which the notion of a 'rotation group' can be generalized from the real numbers to the complex. But there is another way which, in certain contexts, has an even greater significance. This is the notion of a *unitary* group.

What does 'unitary' mean? The orthogonal group is concerned with the preservation of a *quadratic form*, which we can write equivalently as $g_{ab}x^a x^b$ or $\mathbf{x}^\mathsf{T}\mathbf{g}\mathbf{x}$. For a unitary group, we use *complex* linear transformations which preserve instead what is called a *Hermitian* form (after the important 19th century French mathematician Charles Hermite 1822–1901).

What is a Hermitian form? Let us first return to the orthogonal case. Rather than a quadratic form (in $\boldsymbol{x}$), we could equally have used the symmetric *bilinear* form (in $\boldsymbol{x}$ and $\boldsymbol{y}$)

$$\boldsymbol{g}(\boldsymbol{x}, \boldsymbol{y}) = g_{ab}x^a y^b = \mathbf{x}^\mathsf{T}\mathbf{g}\mathbf{y}.$$

This arises as a particular instance of the 'multilinear function' definition of a tensor given in §12.8, as applied to the $\left[\begin{smallmatrix}2\\0\end{smallmatrix}\right]$ tensor $\boldsymbol{g}$ (and putting $\boldsymbol{y} = \boldsymbol{x}$, we retrieve the quadratic form above). The symmetry of $\boldsymbol{g}$ would then be expressed as

$$\boldsymbol{g}(\boldsymbol{x}, \boldsymbol{y}) = \boldsymbol{g}(\boldsymbol{y}, \boldsymbol{x}),$$

and linearity in the second variable $\mathbf{y}$ as

$$\boldsymbol{g}(\boldsymbol{x}, \boldsymbol{y} + \boldsymbol{w}) = \boldsymbol{g}(\boldsymbol{x}, \boldsymbol{y}) + \boldsymbol{g}(\boldsymbol{x}, \boldsymbol{w}), \quad \boldsymbol{g}(\boldsymbol{x}, \lambda\boldsymbol{y}) = \lambda\boldsymbol{g}(\boldsymbol{x}, \boldsymbol{y}).$$

For *bilinearity*, we also require linearity in the *first* variable $\boldsymbol{x}$, but this now follows from the symmetry.

A *Hermitian form* $\boldsymbol{h}(\boldsymbol{x}, \boldsymbol{y})$ satisfies, instead, Hermitian symmetry

$$\boldsymbol{h}(\boldsymbol{x}, \boldsymbol{y}) = \overline{\boldsymbol{h}(\boldsymbol{y}, \boldsymbol{x})},$$

together with linearity in the second variable $\boldsymbol{y}$:

$$\boldsymbol{h}(\boldsymbol{x}, \boldsymbol{y} + \boldsymbol{w}) = \boldsymbol{h}(\boldsymbol{x}, \boldsymbol{y}) + \boldsymbol{h}(\boldsymbol{x}, \boldsymbol{w}), \quad \boldsymbol{h}(\boldsymbol{x}, \lambda \boldsymbol{y}) = \lambda \boldsymbol{h}(\boldsymbol{x}, \boldsymbol{y}).$$

The Hermitian symmetry now implies what is called *antilinearity* in the first variable:

$$\boldsymbol{h}(\boldsymbol{x} + \boldsymbol{w}, \boldsymbol{y}) = \boldsymbol{h}(\boldsymbol{x}, \boldsymbol{y}) + \boldsymbol{h}(\boldsymbol{w}, \boldsymbol{y}), \quad \boldsymbol{h}(\lambda \boldsymbol{x}, \boldsymbol{y}) = \bar{\lambda} \boldsymbol{h}(\boldsymbol{x}, \boldsymbol{y}).$$

Whereas an orthogonal group preserves a (non-singular) symmetric bilinear form, the complex linear transformations preserving a non-singular Hermitian form give us a unitary group.

What do such forms do for us? A (not necessarily symmetric) non-singular bilinear form $\boldsymbol{g}$ provides us with a means of identifying the vector space $\mathbf{V}$, to which $\boldsymbol{x}$ and $\boldsymbol{y}$ belong, with the dual space $\mathbf{V}^*$. Thus, if $\boldsymbol{v}$ belongs to $\mathbf{V}$, then $\boldsymbol{g}(\boldsymbol{v},\ )$ provides us with a linear map on $\mathbf{V}$, mapping the element $\boldsymbol{x}$ of $\mathbf{V}$ to the number $\boldsymbol{g}(\boldsymbol{v}, \boldsymbol{x})$. In other words, $\boldsymbol{g}(\boldsymbol{v},\ )$ is an element of $\mathbf{V}^*$ (see §12.3). In index form, this element of $\mathbf{V}^*$ is the covector $v^a g_{ab}$, which is customarily written with the same kernel letter $\boldsymbol{v}$, but with the index lowered (see also §14.7) by g_{ab}, according to

$$v_b = v^a g_{ab}.$$

The inverse of this operation is achieved by the raising of the index of v_a by use of the inverse metric $[{2 \atop 0}]$-tensor g^{ab}:

$$v^a = g^{ab} v_b.$$

We shall need the analogue of this in the Hermitian case. As before, each choice of element $\boldsymbol{v}$ from the vector space $\mathbf{V}$ provides us with an element $\boldsymbol{h}(\boldsymbol{v},\ )$ of the dual space $\mathbf{V}^*$. However, the difference is that now $\boldsymbol{h}(\boldsymbol{v},\ )$ depends antilinearly on $\boldsymbol{v}$ rather than linearly; thus $\boldsymbol{h}(\lambda\boldsymbol{v},\ ) = \bar{\lambda}\boldsymbol{h}(\boldsymbol{v},\ )$.

An equivalent way of saying this is that $\boldsymbol{h}(\boldsymbol{v},\ )$ is *linear* in $\bar{\boldsymbol{v}}$, this vector quantity $\bar{\boldsymbol{v}}$ being the 'complex conjugate' of $\boldsymbol{v}$. We consider these complex-conjugate vectors to constitute a separate vector space $\overline{\mathbf{V}}$. This viewpoint is particularly useful for the (abstract) index notation, where a separate 'alphabet' of indices is used, say $a', b', c', \ldots$, for these complex-conjugate elements, where contractions (summations) are not permitted between primed and unprimed indices. The operation of complex conjugation interchanges the primed with the unprimed indices. In the index notation, our Hermitian form is represented as an array of quantities $h_{a'b}$ with one (lower) index of each type, so

$$\boldsymbol{h}(\boldsymbol{x}, \boldsymbol{y}) = h_{a'b}\bar{x}^{a'}y^b$$

(with $\bar{x}^{a'}$ being the complex conjugate of the element x^a), where 'Hermiticity' is expressed as

$$h_{a'b} = \overline{h_{b'a}}.$$

The array of quantities $h_{a'b}$ allows us to lower or raise an index, but it now changes primed indices to unprimed ones, and vice versa, so it refers us to the dual of the complex-conjugate space:

$$\bar{v}_b = \bar{v}^{a'}h_{a'b}, \quad v_{a'} = h_{a'b}v^b.$$

For the inverses of these operations—where the Hermitian form is assumed non-singular (i.e. the matrix of components $h^{ab'}$ is non-singular)—we need the inverse $h^{ab'}$ of $h_{a'b}$

$$h^{ab'}h_{b'c} = \delta^a_c, \quad h_{a'b}h^{bc'} = \delta^{c'}_{a'},$$

whence[13.53]

$$\bar{v}^{a'} = \bar{v}_b h^{ba'}, \quad v^a = h^{ab'}v_{b'}.$$

Note that all primed indices can be eliminated using $h_{a'b}$ (and the corresponding inverse $h^{ab'}$) by virtue of the above relations, which can be applied index-by-index to any tensor quantity. The complex-conjugate space is thereby 'identified' with the dual space, instead of having to be a quite separate space.

The operation of 'complex conjugation'—usually called *Hermitian conjugation*—which incorporates this identification with the dual into the notion of complex conjugation (though not commonly written in the index notation) is of central importance to quantum mechanics, as well as to many other areas of mathematics and physics (such as twistor theory, see §33.5). In the quantum-mechanical literature this is often denoted by a *dagger* '†', but sometimes by an *asterisk* '∗'.

I prefer the asterisk, which is more usual in the mathematical literature, so I shall use this here. The asterisk is appropriate here because it interchanges the roles of the vector space **V** and its dual $\mathbf{V}^*$. A complex tensor of valence $[^p_q]$ (all primed indices having been eliminated, as above) is mapped by ∗ to a tensor of valence $[^q_p]$. Thus, upper indices become lower and lower indices become upper under the action of ∗. As applied to scalars, ∗ is simply the ordinary operation of complex conjugation. The operation ∗ is an equivalent notion to the Hermitian form $\boldsymbol{h}$ itself.

The most familiar Hermitian conjugation operation (which occurs when the components $h_{a'b}$ are taken to be the Kronecker delta) simply

[13.53] Verify these relations, explaining the notational consistency of $h^{ab'}$.

takes the complex conjugate of each component, reorganizing the components so as to read upper indices as lower ones and lower indices as upper ones. Accordingly, the matrix of components of a linear transformation is taken to the transpose of its complex conjugate (sometimes called the *conjugate transpose* of the matrix), so in the 2×2 case we have

$$\begin{pmatrix} a & b \\ c & d \end{pmatrix}^* = \begin{pmatrix} \bar{a} & \bar{c} \\ \bar{b} & \bar{d} \end{pmatrix}.$$

A *Hermitian matrix* is a matrix that is equal to its Hermitian conjugate in this sense. This concept, and the more general abstract *Hermitian operator*, are of great importance in quantum theory.

We note that $*$ is *antilinear* in the sense

$$(\boldsymbol{T} + \boldsymbol{U})^* = \boldsymbol{T}^* + \boldsymbol{U}^*,$$
$$(z\boldsymbol{T})^* = \bar{z}\boldsymbol{T}^*,$$

applied to tensors $\boldsymbol{T}$ and $\boldsymbol{U}$, both of the same valence, and for any complex number z. The action of $*$ must also preserve products of tensors but, because of the reversal of the index positions, it reverses the order of contractions; in particular, when $*$ is applied to linear transformations (regarded as tensors with one upper and one lower index), the order of multiplication is reversed:

$$(\boldsymbol{LM})^* = \boldsymbol{M}^*\boldsymbol{L}^*.$$

It is very handy, in the diagrammatic notation, to depict such a conjugation operation as reflection in a horizontal plane. This interchanges upper and lower indices, as required; see Fig. 13.19.

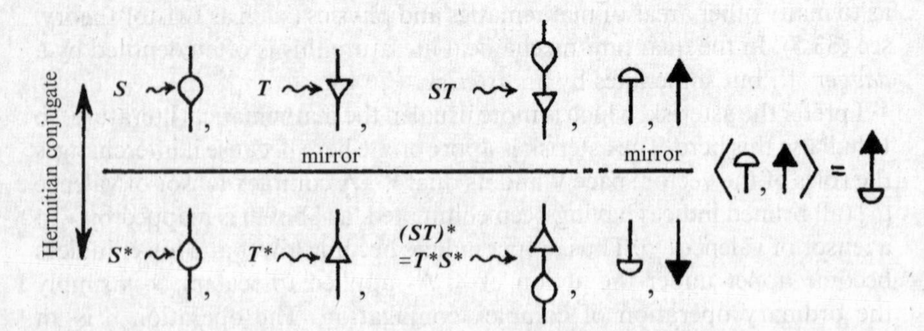

Fig. 13.19 The operation of Hermitian conjugation (*) conveniently depicted as reflection in a horizontal plane. This interchanges 'arms' with 'legs' and reverses the order of multiplication: $(\boldsymbol{ST})^* = \boldsymbol{T}^*\boldsymbol{S}^*$. The diagrammatic expression for the Hermitian scalar product $\langle \boldsymbol{v}|\boldsymbol{w}\rangle = \boldsymbol{v}^*\bullet\boldsymbol{w}$ is given (so that taking its complex conjugate would reflect the diagram on the far right upside-down).

The operation $*$ enables us to define a *Hermitian scalar product* between two elements $\boldsymbol{v}$ and $\boldsymbol{w}$, of V, namely the scalar product of the covector $\boldsymbol{v}^*$ with the vector $\boldsymbol{w}$ (the different notations being useful in different contexts):

$$\langle \boldsymbol{v} \mid \boldsymbol{w} \rangle = \boldsymbol{v}^* \bullet \boldsymbol{w} = \boldsymbol{h}(\boldsymbol{v}, \boldsymbol{w})$$

(and see Fig. 13.19), and we have

$$\langle \boldsymbol{v} \mid \boldsymbol{w} \rangle = \overline{\langle \boldsymbol{w} \mid \boldsymbol{v} \rangle}.$$

In the particular case $\boldsymbol{w} = \boldsymbol{v}$, we get the *norm* of $\boldsymbol{v}$, with respect to $*$:

$$\| \boldsymbol{v} \| = \langle \boldsymbol{v} \mid \boldsymbol{v} \rangle.$$

We can choose a *basis* $(\boldsymbol{e}_1, \boldsymbol{e}_2, \ldots, \boldsymbol{e}_n)$ for V, and then the components $h_{\bar{a}b}$ in this basis are simply the n^2 complex numbers

$$h_{\bar{a}b} = \boldsymbol{h}(\boldsymbol{e}_a, \boldsymbol{e}_b) = \langle \boldsymbol{e}_a \mid \boldsymbol{e}_b \rangle,$$

constituting the elements of a Hermitian matrix. The basis $(\boldsymbol{e}_1, \ldots, \boldsymbol{e}_n)$ is called *pseudo-orthonormal*, with respect to $*$, if

$$\langle \boldsymbol{e}_i \mid \boldsymbol{e}_j \rangle = \begin{cases} \pm 1 & \text{if } i = j \\ 0 & \text{if } i \neq j \end{cases};$$

in the case when all the $\pm$ signs are $+$, i.e. when each ± 1 is just 1, the basis is *orthonormal.*

A pseudo-orthonormal basis can always be found, but there are many choices. With respect to any such basis, the matrix $h_{\bar{a}b}$ is diagonal, with just 1s and -1s down the diagonal. The total number of 1s, p, always comes out the same, for a given $*$, independently of any particular choice of basis, and so also does the total number of -1s, q. This enables us to define the invariant notion of *signature* (p, q) for the operation $*$.

If $q = 0$, we say that $*$ is *positive-definite*. In this case,[21] the norm of any non-zero vector is always positive:[13.54]

$$\boldsymbol{v} \neq 0 \quad \textit{implies} \quad \| \boldsymbol{v} \| > 0.$$

Note that this notion of 'positive-definite' generalizes that of §13.8 to the complex case.

A linear transformation $\boldsymbol{T}$ whose inverse is $\boldsymbol{T}^*$, so that

$$\boldsymbol{T}^{-1} = \boldsymbol{T}^*, \text{ i.e. } \boldsymbol{T}\,\boldsymbol{T}^* = \boldsymbol{I} = \boldsymbol{T}^*\boldsymbol{T},$$

[13.54] Show this.

is called *unitary* in the case when $*$ is positive-definite, and *pseudo-unitary* in the other cases.[13.55] The term 'unitary matrix' refers to a matrix $\boldsymbol{T}$ satisfying the above relation when $*$ stands for the usual conjugate transpose operation, so that $\boldsymbol{T}^{-1} = \overline{\boldsymbol{T}}$.

The group of unitary transformations in n dimensions, or of $(n \times n)$ unitary matrices, is called the *unitary group* $\mathrm{U}(n)$. More generally, we get the pseudo-unitary group $\mathrm{U}(p, q)$ when $*$ has signature (p, q).[22] If the transformations have unit determinant, then we correspondingly obtain $\mathrm{SU}(n)$ and $\mathrm{SU}(p, q)$. Unitary transformations play an essential role in quantum mechanics (and they have great value also in many pure-mathematical contexts).

13.10 Symplectic groups

In the previous two sections, we encountered the orthogonal and unitary groups. These are examples of what are called *classical groups*, namely the simple Lie groups other than the exceptional ones; see §13.2. The list of classical groups is completed by the family of *symplectic* groups. Symplectic groups have great importance in *classical* physics, as we shall be seeing particularly in §20.4—and also in quantum physics, particularly in the infinite-dimensional case (§26.3).

What is a symplectic group? Let us return again to the notion of a bilinear form, but where instead of the symmetry ($\boldsymbol{g}(\boldsymbol{x}, \boldsymbol{y}) = \boldsymbol{g}(\boldsymbol{y}, \boldsymbol{x})$) required for defining the orthogonal group, we impose *antisymmetry*

$$\boldsymbol{s}(\boldsymbol{x}, \boldsymbol{y}) = -\boldsymbol{s}(\boldsymbol{y}, \boldsymbol{x}),$$

together with linearity

$$\boldsymbol{s}(\boldsymbol{x}, \boldsymbol{y} + \boldsymbol{w}) = \boldsymbol{s}(\boldsymbol{x}, \boldsymbol{y}) + \boldsymbol{s}(\boldsymbol{x}, \boldsymbol{w}), \quad \boldsymbol{s}(\boldsymbol{x}, \lambda \boldsymbol{y}) = \lambda \boldsymbol{s}(\boldsymbol{x}, \boldsymbol{y}),$$

where linearity in the first variable $\boldsymbol{x}$ now follows from the *anti*symmetry. We can write our antisymmetric form variously as

$$\boldsymbol{s}(\boldsymbol{x}, \boldsymbol{y}) = x^a s_{ab} y^b = \mathbf{x}^{\mathrm{T}}\mathbf{S}\mathbf{y},$$

just as in the symmetric case, but where s_{ab} is *antisymmetric*:

$$s_{ba} = -s_{ab} \quad \text{i.e.} \quad \mathbf{S}^{\mathrm{T}} = -\mathbf{S},$$

$\mathbf{S}$ being the matrix of components of s_{ab}. We require $\boldsymbol{S}$ to be non-singular. Then s_{ab} has an inverse s^{ab}, satisfying[23]

[13.55] Show that these transformations are precisely those which preserve the Hermitian correspondence between vectors $\boldsymbol{v}$ and covectors $\boldsymbol{v}^*$, and that they are those which preserve $h_{ab'}$.

$$s_{ab}s^{bc} = \delta_a^c = s^{cb}s_{ba},$$

where $s^{ab} = -s^{ba}$.

We note that, by analogy with a symmetric matrix, an antisymmetric matrix **S** equals *minus* its transpose. It is important to observe that an $n \times n$ antisymmetric matrix **S** can be non-singular only if n is even.[13.56] Here n is the dimension of the space **V** to which **x** and **y** belong, and we indeed take n to be even.

The elements **T** of GL(n) that preserve such a non-singular antisymmetric s_{ab} (or, equivalently, the bilinear form s), in the sense that

$$s_{ab}T^a{}_c\,T^b{}_d = s_{cd}, \quad \text{i.e. } \mathbf{T}^{\mathsf{T}}\mathbf{S}\,\mathbf{T} = \mathbf{S},$$

are called *symplectic*, and the group of these elements is called a *symplectic group* (a group of very considerable importance in classical mechanics, as we shall be seeing in §20.4). However, there is some confusion in the literature concerning this terminology. It is mathematically more accurate to define a (real) symplectic group as a real form of the *complex* symplectic group Sp($\frac{1}{2}n$, ℂ), which is the group of *complex* $T^a{}_b$ (or **T**) satisfying the above relation. The particular real form just defined is non-compact; but in accordance with the remarks at the end of §13.7—Sp($\frac{1}{2}n$, ℂ) being semi-simple—there is another real form of this complex group which *is* compact, and it is this that is normally referred to as the (real) symplectic group Sp($\frac{1}{2}n$).

How do we find these different real forms? In fact, as with the orthogonal groups, there is a notion of *signature* which is not so well known as in the cases of the orthogonal and unitary groups. The symplectic group of real transformations preserving s_{ab} would be the 'split-signature' case of signature ($\frac{1}{2}n$, $\frac{1}{2}n$). In the compact case, the symplectic group has signature (n, 0) or (0, n).

How is this signature defined? For each pair of natural numbers p and q such that $p + q = n$, we can define a corresponding 'real form' of the complex group Sp($\frac{1}{2}n$, ℂ) by taking only those elements which are also pseudo-unitary for signature (p, q)—i.e. which belong to U(p, q) (see §13.9). This gives[24] us the (pseudo-)symplectic group Sp(p, q). (Another way of saying this is to say that Sp(p, q) is the intersection of Sp($\frac{1}{2}n$, ℂ) with U(p, q).) In terms of the index notation, we can define Sp(p, q) to be the group of complex linear transformations $T^a{}_b$ that preserve both the antisymmetric s_{ab}, as above, and also a Hermitian matrix **H** of components $h_{a'b}$, in the sense that

$$\bar{T}^{a'}{}_{b'}T^a{}_b h_{a'a} = h_{b'b},$$

[13.56] Prove this.

where $\mathbf{H}$ has signature (p, q) (so we can find a pseudo-orthonormal basis for which $\mathbf{H}$ is diagonal with p entries 1 and q entries -1; see §13.9).[25] The compact *classical* symplectic group $\mathrm{Sp}(\frac{1}{2}n)$ is my $\mathrm{Sp}(n, 0)$ (or $\mathrm{Sp}(0, n)$), but the form of most importance in classical physics is $\mathrm{Sp}(\frac{1}{2}n, \frac{1}{2}n)$.[13.57]

As with the orthogonal and unitary groups, we can find choices of basis for which the components s_{ab} have a particularly simple form. We cannot now take this form to be diagonal, however, because the only antisymmetric diagonal matrix is zero! Instead, we can take the matrix of s_{ab} to consist of 2×2 blocks down the main diagonal, of the form

$$\begin{pmatrix} 0 & 1 \\ -1 & 0 \end{pmatrix}.$$

In the familiar split-signature case $\mathrm{Sp}(\frac{1}{2}n, \frac{1}{2}n)$, we can take the *real* linear transformations preserving this form. The general case $\mathrm{Sp}(p, q)$ is exhibited by taking, rather than real transformations, pseudo-unitary ones of signature (p, q).[13.58]

For various (small) values of p and q, some of the orthogonal, unitary, and symplectic groups are the same ('isomorphic') or at least locally the same ('locally isomorphic'), in the sense of having the same Lie algebras (cf. §13.6).[26] The most elementary example is the group SO(2), which describes the group of non-reflective symmetries of a circle, being the same as the unitary group U(1), the multiplicative group of unit-modulus complex numbers $e^{i\theta}$ (θ real).[13.59] Of a particular importance for physics is the fact that SU(2) and Sp(1) are the same, and are locally the same as SO(3) (being the twofold cover of this last group, in accordance with the twofold nature of the quaternionic representation of rotations in 3-space, as described in §11.3). This has great importance for the quantum physics of *spin* (§22.8). Of significance in relativity theory is the fact that $\mathrm{SL}(2, \mathbb{C})$, being the same as $\mathrm{Sp}(1, \mathbb{C})$, is locally the same as the non-reflective part of the Lorentz group O(1, 3) (again a twofold cover of it). We also find that SU(1, 1), Sp(1, 1), and SO(2, 1) are the same, and there are several other examples. Particularly noteworthy for twistor theory is the local identity between SU(2, 2) and the non-reflective part of the group O(2, 4) (see §33.3).

The Lie algebra of a symplectic group is obtained by looking for solutions $\mathbf{X}$ of the matrix equation

$$\mathbf{X}^{\mathrm{T}}\mathbf{S} + \mathbf{S}\,\mathbf{X} = 0, \quad \text{i.e. } \mathbf{S}\,\mathbf{X} = (\mathbf{S}\,\mathbf{X})^{\mathrm{T}},$$

[13.57] Find explicit descriptions of Sp(1) and Sp(1, 1) using this prescription. Can you see why the groups Sp(n, 0) are compact?

[13.58] Show why these two different descriptions for the case $p = q = \frac{1}{2}n$ are equivalent.

[13.59] Why are they the same?

so the infinitesimal transformation (Lie algebra element) **X** is simply $\mathbf{S}^{-1}$ times a symmetric $n \times n$ matrix. This enables the dimensionality $\frac{1}{2}n(n+1)$ of the symplectic group to be directly seen. Note that **X** has to be trace-free (i.e. trace $\mathbf{X} = 0$—see §13.4).[13.60] The Lie algebras for orthogonal and unitary groups are also readily obtained, in terms, respectively, of anti-symmetric matrices and pure-imaginary multiples of Hermitian matrices, the respective dimensions being $n(n-1)/2$ and n^2.[13.61]

We note from §13.4 that, for the transformations to have unit determinant, the trace of the infinitesimal element $\boldsymbol{X}$ must vanish. This is automatic in the symplectic case (noted above), and in the orthogonal case the infinitesimal elements all have unit determinant.[13.62] In the unitary case, restriction to SU(n) is one further condition (trace $\boldsymbol{X} = 0$), so the dimension of the group is reduced to $n^2 - 1$.

The *classical groups* referred to in §13.2, sometimes labelled A_m, B_m, C_m, D_m (for $m = 1, 2, 3, \ldots$), are simply the respective groups SU($m+1$), SO($2m+1$), Sp(m), and SO($2m$), that we have been examining in §§13.8–10, and we see from the above that they indeed have respective dimensions $m(m+2)$, $m(2m+1)$, $m(2m+1)$, and $m(2m-1)$, as asserted in §13.2. Thus, the reader has now had the opportunity to catch a significant glimpse of all the classical simple groups. As we have seen, such groups, and some of the various other 'real forms' (of their complexifications) play important roles in physics. We shall be gaining a little acquaintance with this in the next chapter. As mentioned at the beginning of this chapter, according to modern physics, all physical interactions are governed by 'gauge connections' which, technically, depend crucially on spaces having exact symmetries. However, we still need to know what a 'gauge theory' actually is. This will be revealed in Chapter 15.

Notes

Section 13.1

13.1. Abel was born in 1802 and died of consumption (tuberculosis) in 1829, aged 26. The more general non-Abelian ($ab \neq ba$) group theory was introduced by the even more tragically short-lived French mathematician Evariste Galois (1811–1832), who was killed in a duel before he reached 21, having been up the entire previous night feverishly writing down his revolutionary ideas involving the use of these groups to investigate the solubility of algebraic equations, now called *Galois theory*.

[13.60] Explain where the equation $\boldsymbol{X}^{\mathrm{T}}\boldsymbol{S} + \boldsymbol{S}\boldsymbol{X} = 0$ comes from and why $\boldsymbol{S}\boldsymbol{X} = (\boldsymbol{S}\boldsymbol{X})^{\mathrm{T}}$. Why does trace $\boldsymbol{X}$ vanish? Give the Lie algebra explicitly. Why is it of this dimension?

[13.61] Describe these Lie algebras and obtain these dimensions.

[13.62] Why, and what does this mean geometrically?

13.2. We should also take note that '$-$C' means 'take the complex conjugate, then multiply by -1', i.e. $-\mathsf{C} = (-1)\mathsf{C}$.

13.3. The S stands for 'special' (meaning 'of unit determinant') which, in the present context just tells us that orientation-reversing motions are excluded. The O stands for 'orthogonal' which has to do with the fact that the motions that it represents preserves the 'orthogonality' (i.e. the right-angled nature) of coordinate axes. The 3 stands for the fact that we are considering rotations in three dimensions.

13.4. There is a remarkable theorem that tells us that not only is every continuous group also *smooth* (i.e. C^0 implies C^1, in the notation of §§6.3,6, and even C^0 implies C^∞), but it is also *analytic* (i.e. C^0 implies C^ω). This famous result, which represented the solution of what had become known as 'Hilbert's 5th problem', was obtained by Andrew Mattei Gleason, Deane Montgomery, Leo Zippin, and Hidehiko Yamabe in 1953; see Montgomery and Zippin (1955). This justifies the use of power series in §13.6.

Section 13.2

13.5. See van der Waerden (1985), pp. 166–74.

13.6. See Devlin (1988).

13.7. See Conway and Norton (1972); Dolan (1996).

Section 13.3

13.8. We shall be seeing in §14.1 that a Euclidean space is an example of an *affine* space. If we select a particular point (origin) O, it becomes a vector space.

13.9. In many places in this book it will be convenient—and sometimes essential—to stagger the indices on a tensor-type symbol. In the case of a linear transformation, we need this to express the order of matrix multiplication.

13.10. This region is a vector space of dimension r (where $r < n$). We call r the *rank* of the matrix or linear transformation $\boldsymbol{T}$. A *non*-singular $n \times n$ matrix has rank n. (The concept of 'rank' applies also to rectangular matrices.) Compare Note 12.18.

13.11. For a history of the theory of matrices, see MacDuffee (1933).

Section 13.5

13.12. In those degenerate situations where the eigenvectors do not span the whole space (i.e., some d is less than the corresponding r), we can still find a canonical form, but we now allow 1s to appear just above the main diagonal, these residing just within square blocks whose diagonal terms are *equal* eigenvalues (*Jordan normal form*); see Anton and Busby (2003). Apparently Weierstrauss had (effectively) found this normal form in 1868, two years before Jordan; See Hawkins (1977).

Section 13.6

13.13. To illustrate this point, consider $\mathrm{SL}(n, \mathbb{R})$ (i.e. the unit-determinant elements of $\mathrm{GL}(n, \mathbb{R})$ itself). This group has a 'double cover' $\widetilde{\mathrm{SL}}(n, \mathbb{R})$ (provided that $n \geq 3$) which is obtained from $\mathrm{SL}(n, \mathbb{R})$ in basically the same way whereby we effectively found the double cover $\widetilde{\mathrm{SO}}(3)$ of SO(3) when we considered the rotations of a book, with belt attachment, in §11.3. Thus, $\widetilde{\mathrm{SO}}(3)$ is the group of (non-reflective) rotations of a *spinorial object* in ordinary 3-space. In the same way, we can consider 'spinorial objects' that are subject to the more general linear transformations that allow 'squashing' or 'stretching', as discussed in §13.3. In this way, we arrive at the group $\widetilde{\mathrm{SL}}(n, \mathbb{R})$, which is *locally* the same as $\mathrm{SL}(n, \mathbb{R})$, but which cannot, in fact, be faithfully represented in any $\mathrm{GL}(m)$. See Note 15.9.

13.14. This notion is well defined; cf. Note 13.4.

Section 13.7

13.15. See Thirring (1983).

13.16. Here, again, we have an instance of the capriciousness of the naming of mathematical concepts. Whereas many notions of great importance in this subject, to which Cartan's name is conventionally attached (e.g. 'Cartan subalgebra, Cartan integer') were originally due to Killing (see §13.2), what we refer to as the 'Killing form' is actually due to Cartan (and Hermann Weyl); see Hawkins (2000), section 6.2. However, the 'Killing vector' that we shall encounter in §30.6 *is* actually due to Killing (Hawkins 2000, note 20 on p. 128).

13.17. I am (deliberately) being mathematically a little sloppy in my use of the phrase 'the same' in this kind of context. The strict mathematical term is 'isomorphic'.

Section 13.8

13.18. I have not been very explicit about this procedure up to this point. A basis $\boldsymbol{e} = (\boldsymbol{e}_1, \ldots, \boldsymbol{e}_n)$ for $\mathbf{V}$ is associated with a *dual basis*—which is a basis $\boldsymbol{e}^* = (\boldsymbol{e}^1, \ldots, \boldsymbol{e}^n)$ for $\mathbf{V}^*$—with the property that $\boldsymbol{e}^i \bullet \boldsymbol{e}_j = \delta^i_j$. The components of a $[{p \atop q}]$-valent tensor $\boldsymbol{Q}$ are obtained by applying the multilinear function of §12.8 to the various collections of p dual basis elements and q basis elements: $Q^{f\ldots h}_{a\ldots c} = \boldsymbol{Q}(\boldsymbol{e}^f, \ldots, \boldsymbol{e}^h; \boldsymbol{e}_a, \ldots, \boldsymbol{e}_c)$.

13.19. See Note 13.3.

13.20. See Note 13.10. The reader may be puzzled about why the $T^a{}_b$ of §13.5 can have lots of invariants, namely all its *eigenvalues* $\lambda_1, \lambda_2, \lambda_3, \ldots, \lambda_n$, whereas g_{ab} does not. The answer lies simply in the difference in transformation behaviour implicit in the different index positioning.

Section 13.9

13.21. Note that, in the positive-definite case, $(\boldsymbol{e}_1^*, \boldsymbol{e}_2^*, \ldots, \boldsymbol{e}_n^*)$ is a *dual basis* to $(\boldsymbol{e}_1, \boldsymbol{e}_2, \ldots, \boldsymbol{e}_n)$, in the sense of Note 13.18.

13.22. The groups U(p, q), for fixed $p + q = n$, as well as GL(n, $\mathbb{R}$), all have the same complexification, namely GL(n, $\mathbb{C}$), and these can all be regarded as different real forms of this complex group.

Section 13.10

13.23. We can then use s_{ab} and s^{ab} to raise and lower indices of tensors, just as with g_{ab} and g^{ab}, so $v_a = s_{ab}v^b$, $v^a = s^{ab}v_b$ (see §13.8); but, because of the antisymmetry, we must be a little careful to make the ordering of the indices consistent. Those readers who are familiar with the 2-spinor calculus (see Penrose and Rindler 1984) may notice a slight notational discrepancy between our s_{ab} and the ε_{AB} of that calculus.

13.24. I am not aware of a standard terminology or notation for these various real forms, so the notation Sp(p, q) has been concocted for the present purposes.

13.25. In fact, every element of Sp($\frac{1}{2}n$, $\mathbb{C}$) has unit determinant, so we do not need an 'SSp($\frac{1}{2}n$)' by analogy with SO(n) and SU(n). The reason is that there is an expression (the 'Pfaffian') for Levi-Civita's $\varepsilon_{\ldots}$ in terms of the s_{ab}, which must be preserved whenever the s_{ab} are.

13.26. See Note 13.17.

14
Calculus on manifolds

14.1 Differentiation on a manifold?

In the previous chapter (in §§13.3,6–10), we saw how symmetry groups can act on vector spaces, represented by linear transformations of these spaces. For a specific group, we can think of the vector space as possessing some particular *structure* which is preserved by the transformations. This notion of 'structure' is an important one. For example, it could be a metric structure, in the case of the orthogonal group (§13.8), or a Hermitian structure, as is preserved by a unitary group (§13.9). As noted earlier, the representation theory of groups as actions on vector spaces has, in a general way, great importance in many areas of mathematics and physics, especially in quantum theory, where, as we shall see later (particularly in §22.3), vector spaces with a Hermitian (scalar-product) structure form the essential background for that theory.

However, a vector space is itself a very special type of space, and something much more general is needed for the mathematics of much of modern physics. Even Euclid's ancient geometry is not a vector space, because a vector space has to have a particular distinguished point, namely the *origin* (given by the zero vector), whereas in Euclidean geometry every point is on an equal footing. In fact, Euclidean space is an example of what is called an *affine* space. An affine space is like a vector space but we 'forget' the origin; in effect, it is a space in which there is a consistent notion of *parallelogram*.[14.1],[14.2] As soon as we specify a particular point as origin this allows us to define vector addition by the 'parallelogram law' (see §13.3, Fig.13.4).

[14.1] Let $[a, b; c, d]$ stand for the statement '*abdc* forms a parallelogram' (where a, b, d, and c are taken cyclicly, as in §5.1). Take as axioms (i) for any a, b, and c, there exists d such that $[a, b; c, d]$; (ii) if $[a, b; c, d]$, then $[b, a; d, c]$ and $[a, c; b, d]$; (iii) if $[a, b; c, d]$ and $[a, b; e, f]$, then $[c, d; e, f]$. Show that, when any chosen point is singled out and labelled as the origin, this algebraic structure reduces to that of a 'vector space', but without the 'scalar multiplication' operation, as given in §11.1—that is to say, we get the rules of an additive Abelian group; see Exercise [13.2].

[14.2] Can you see how to generalize this to the non-Abelian case?

The curved spacetime of Einstein's remarkable theory of general relativity is certainly more general than a vector space; it is a 4-manifold. Yet his notion of spacetime geometry does require some (local) structure—over and above just that of a smooth manifold (as studied in Chapter 12). Similarly, the configuration spaces or the phase spaces of physical systems (considered briefly in §12.1) also tend to possess local structures. How do we assign this needed structure? Such a local structure could provide a measure of 'distance' between points (in the case of a metric structure), or 'area' of a surface (as is specified in the case of a symplectic structure, cf. §13.10), or of 'angle' between curves (as with the conformal structure of a Riemann surface; see §8.2), etc. In all the examples just referred to, vector-space notions are what are needed to tell us what this local geometry is, the vector space in question being the n-dimensional *tangent space* $\mathcal{T}_p$ of a typical point p of the manifold $\mathcal{M}$ (where we may think of $\mathcal{T}_p$ as the immediate vicinity of p in $\mathcal{M}$ 'infinitely stretched out'; see Fig. 12.6).

Accordingly, the various group structures and tensor entities that we encountered in Chapter 13 can have a local relevance at the individual points of a manifold. We shall find that Einstein's curved spacetime indeed has a local structure that is given by a Lorentzian (pseudo)metric (§13.8) in each tangent space, whereas the phase spaces (cf. §12.1) of classical mechanics have local symplectic structures (§13.10). Both of these examples of manifolds with structure play vital roles in modern physical theory. But what form of calculus can be applied within such spaces?

As just remarked, the n-dimensional manifolds that we studied in Chapter 12 need only to be smooth, with no further local structure specified. In such an unstructured smooth manifold $\mathcal{M}$, there are relatively few meaningful calculus-based operations. Most importantly, we do not even have a general notion of differentiation that can be applied generally within $\mathcal{M}$.

I should clarify this point. In any particular coordinate patch, we could certainly simply differentiate the various quantities of interest with respect to each of the coordinates $x^1, x^2, \ldots, x^n$ in that patch, by use of the (partial) derivative operators $\partial/\partial x^1, \partial/\partial x^2, \ldots, \partial/\partial x^n$ (see §10.2). But in most cases, the answers would be geometrically meaningless, because they depend on the specific (arbitrary) choice of coordinates that has been made, and the answers would not generally match as we pass from one patch to another (cf. Fig. 10.7).

We did, however, take note of one important notion of differentiation, in §12.6, that actually does apply in a general smooth (unstructured) n-manifold—agreeing from one patch to the next—namely the *exterior derivative* of a differential form. Yet this operation is somewhat limited in its scope, as it applies only to p-forms and, moreover, does not give much information about how such a p-form is varying. Can we give a more complete notion of 'derivative' of some quantity on a general smooth manifold, say of a vector or tensor field? Such a notion would have to be defined independently of any particular coordinates that might happen to have been chosen to label points in some coordinate patch. It would, indeed, be good to have some kind of coordinate-independent calculus that can be applied to structures on manifolds, and which would enable us to express how a vector or tensor field varies as we move from place to place. But how can this be achieved?

14.2 Parallel transport

Recall from §10.3 and §12.3 that in the case of a *scalar* field Φ on a general smooth n-manifold $\mathcal{M}$, we were indeed able to provide an appropriate measure of its 'rate of change', namely the 1-form $\mathrm{d}\Phi$, where $\mathrm{d}\Phi = 0$ is the condition that Φ be constant (throughout connected regions of $\mathcal{M}$). However, this idea will not work for a general tensor quantity. It will not even work for a vector field $\boldsymbol{\xi}$. Why is this? One trouble is that in a general manifold we have no appropriate notion of $\boldsymbol{\xi}$ being constant (as we shall see in a moment), whereas any self-respecting differentiation ('gradient') operation that applies to $\boldsymbol{\xi}$ ought to have the property that its vanishing signals the constancy of $\boldsymbol{\xi}$ (as, indeed, $\mathrm{d}\Phi = 0$ signals the constancy of a scalar field Φ). More generally, we would expect that for a 'non-constant' $\boldsymbol{\xi}$, such a derivative operation ought to be measuring $\boldsymbol{\xi}$'s deviation from constancy.

Why is there a problem with this notion of vector 'constancy', on a general n-manifold $\mathcal{M}$? A constant vector field $\boldsymbol{\xi}$, in ordinary Euclidean space, should have the property that all the 'arrows' of its geometrical description are parallel to each other. Thus, some kind of notion of 'parallelism' would have to be part of $\mathcal{M}$'s structure. One might worry about this, bearing in mind the issue of Euclid's fifth postulate—the parallel postulate—that was central to the discussion of Chapter 2. Hyperbolic geometry, for example, does not admit vector fields that could unambiguously considered to be everywhere 'parallel'. In any case, a notion of 'parallelism' is not something that $\mathcal{M}$ would possess merely by virtue of its being a smooth manifold. In Fig. 14.1, the difficulty is illustrated in the case of a 2-manifold pieced together from two patches of Euclidean plane. The normal Euclidean notion of 'parallel' is not consistent from one patch to the next.

In order to gain some insights as to what kind of notion of parallelism is appropriate, it will be helpful for us first to examine the intrinsic geometry of an ordinary 2-dimensional sphere S^2. Let us choose a particular point p on S^2 (say, at the north pole, for definiteness) and a particular tangent vector $\boldsymbol{v}$

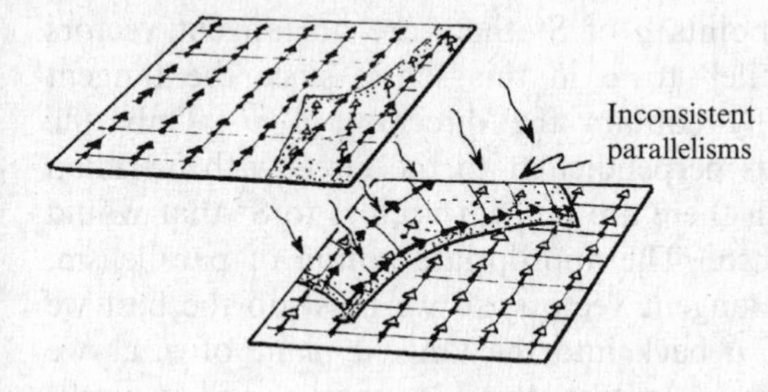

Fig. 14.1 The Euclidean notion of 'parallel' is likely to be inconsistent on the overlap between coordinate patches.

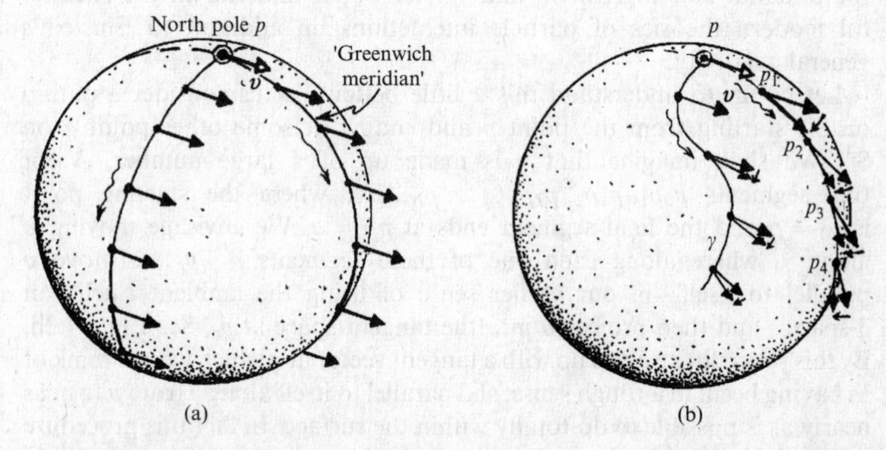

Fig. 14.2 Parallelism on the sphere S^2. Choose p at the north pole, with tangent vector $\boldsymbol{v}$ pointing along the Greenwich meridian. Which tangent vectors, at other points of S^2, are we to regard to being 'parallel' to $\boldsymbol{v}$? (a) The direct Euclidean notion of 'parallel', from the embedding of S^2 in $\mathbb{E}^3$, does not work because (except along the meridian perpendicular to the Greenwich meridian) the parallel $\boldsymbol{v}$s do not remain tangent to S^2. (b) Remedy this, moving $\boldsymbol{v}$ parallel along a given curve γ, by continually projecting back to tangency with the sphere. (Think of γ as made up of large number of tiny segments $p_0p_1, p_1p_2, p_2p_3, \ldots,$ projecting back at each stage. Then take the limit as the segments are made smaller and smaller.) This notion of *parallel transport* is indicated for the Greenwich meridian, but also for a general curve γ.

at p (say pointing along the Greenwich meridian; see Fig. 14.2a). Which other tangent vectors, at other points of S^2, are we to regard to being 'parallel' to $\boldsymbol{v}$? If we simply use the Euclidean notion of 'parallel' that is inherited from the standard embedding of S^2 in Euclidean 3-space, then we find that at most points q of S^2 there are no tangent vectors to S^2 at all that are 'parallel' to $\boldsymbol{v}$ in this sense, since the tangent plane at q does not usually contain the direction of $\boldsymbol{v}$. (Only the great circle through p that is perpendicular to the Greenwich meridian at p contains points at which there are tangent vectors to S^2 that would be 'parallel' to $\boldsymbol{v}$ in this sense.) The appropriate notion of parallelism, on S^2, should refer only to tangent vectors, so we must do the best we can to pull the direction of $\boldsymbol{v}$ back into the tangent plane of q, as we gradually move q away from p. In fact, this idea works, and it works beautifully, but there is now a new feature in that the notion of parallelism that we get is *dependent on the path* along which we move q away from p.[1] This path-dependence in the concept of 'parallelism' is the essential new ingredient, and versions of it underlie all the successful modern theories of particle interactions, in addition to Einstein's general relativity.

Let us try to understand this a little better. Let us consider a path γ on S^2, starting from the point p and ending at some other point q on S^2. We shall imagine that γ is made up of a large number, N, of tiny segments $p_0p_1, p_1p_2, p_2p_3, \ldots, p_{N-1}p_N$, where the starting point is $p_0 = p$ and the final segment ends at $p_N = q$. We envisage moving $\boldsymbol{v}$ along γ, where along each one of these segments $p_{r-1}p_r$ we move $\boldsymbol{v}$ parallel to itself—in our earlier sense of using the ambient Euclidean 3-space—and then project $\boldsymbol{v}$ into the tangent space at p_r. See Fig. 14.2b. By this procedure we end up with a tangent vector at q which we can think of as having been, in a rough sense, slid parallel to itself along γ from p to q, as nearly as is possible to do totally within the surface. In fact this procedure will depend slightly on how γ is approximated by the succession of segments, but it can be shown that in the limit, as the segments get smaller and smaller, we get a well-defined answer that does not depend upon the precise detailed way in which we break γ up into segments. This procedure is referred to as *parallel transport* of $\boldsymbol{v}$ along γ. In Fig. 14.3, I have indicated what this parallel transport would look like along five different paths (all great circles) starting at p.

What, then, is this path-dependence, referred to above? In Fig. 14.4, I have marked points p and q on S^2 and two paths from p to q, one of which is the direct great-circle route and the other of which consists of a pair of great-circle arcs jointed at the intermediate point r. From the geometry of Fig. 14.3, we see that parallel transport along these two paths (one having a corner on it, but this is not important) gives two quite different final results, differing from each other, in this case, by a right-angle rotation. Note that the discrepancy is just a rotation of the direction of the vector. There are general reasons that a notion of parallel transport defined in this particular way

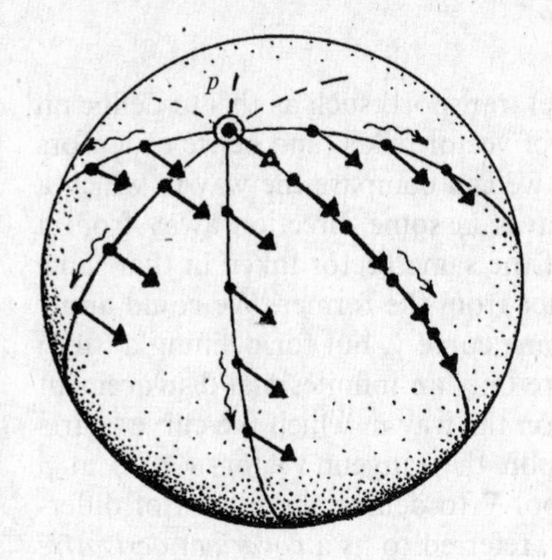

Fig. 14.3 Parallel transport of $\boldsymbol{v}$ along five different paths (all great circles).

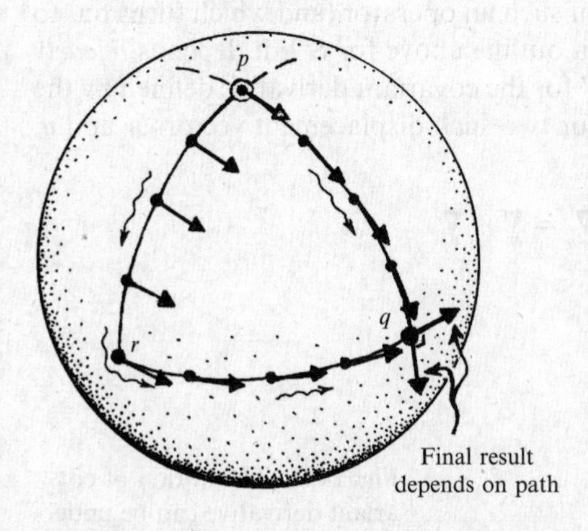

Fig. 14.4 Path dependence of parallel transport. This is illustrated using two distinct paths from p to q, one of which is a direct great-circle route, the other consisting of a pair of great-circle arcs jointed at an intermediate point r. Parallel transport along these two paths gives results at q differing by a right-angle rotation.

will always preserve the length of the vector. (However, there are other types of 'parallel transport' for which this is not the case. These issues will have importance for us in later sections (§14.7, §§15.7,8, §19.4.) We can see this angular discrepancy in an extreme form when our path γ is a closed loop (so that $p = q$), in which case there is likely to be a discrepancy between the initial and the final directions of the parallel-transported tangent vector. In fact, for an exact geometrical sphere of unit radius, this discrepancy is an angle of rotation which, when measured in radians, is precisely equal to the total area of the loop (with regions surrounded in the negative sense counting negatively).[14.3]

[14.3] See if you can confirm this assertion in the case of a spherical triangle (triangle on S^2 made up of great-circle arcs) where you may assume the Hariot's 1603 formula for the area of a spherical triangle given in §2.6.

14.3 Covariant derivative

How can we use a concept of 'parallel transport' such as this to define an appropriate notion of *differentiation* of vector fields (and hence of tensors generally)? The essential idea is that we can compare the way in which a vector (or tensor) field actually behaves in some direction away from a point p with the parallel transport of the same vector taken in that same direction from p, subtracting the latter from the former. We could apply this to a finite displacement along some curve γ, but for defining a (first) derivative of a vector field, we require only an infinitesimal displacement away from p, and this depends only on the way in which the curve 'starts out' from p; i.e. it depends only upon the tangent vector $\boldsymbol{w}$ of γ at p (Fig. 14.5). It is usual to use a symbol $\boldsymbol{\nabla}$ to denote the notion of differentiation, arising in this kind of way, referred to as a *covariant derivative operator* or simply a *connection*.

A fundamental requirement of such an operator (and which turns out to be true for the notion defined in outline above for S^2), it depends *linearly* on the vector $\boldsymbol{w}$. Thus, writing $\underset{\boldsymbol{w}}{\boldsymbol{\nabla}}$ for the covariant derivative defined by the displacement (direction) of $\boldsymbol{w}$, for two such displacement vectors $\boldsymbol{w}$ and $\boldsymbol{u}$, this must satisfy

$$\underset{\boldsymbol{w}+\boldsymbol{u}}{\boldsymbol{\nabla}} = \underset{\boldsymbol{w}}{\boldsymbol{\nabla}} + \underset{\boldsymbol{u}}{\boldsymbol{\nabla}},$$

and for a scalar multiplier λ:

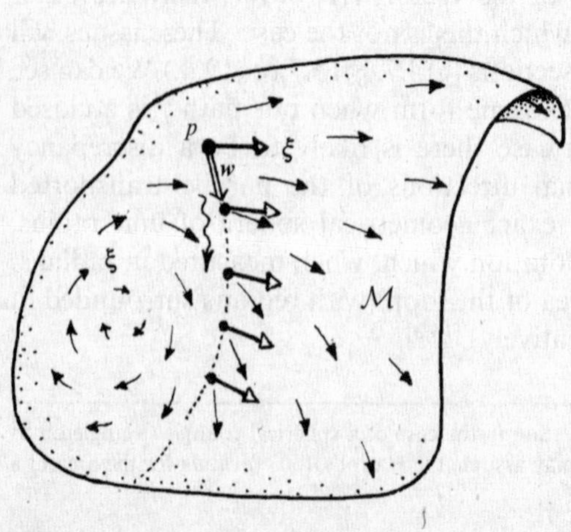

Fig. 14.5 The notion of covariant derivative can be understood in relation to parallel transport. The way in which a vector field $\boldsymbol{\xi}$ on $\mathcal{M}$ varies from point to point (black-headed arrows) is measured by its departure from that standard provided by parallel transport (white-headed arrows). This comparison can be made all along a curve γ, (starting at p), but for the covariant first derivative $\underset{\boldsymbol{w}}{\boldsymbol{\nabla}}$ at p we need to know only the tangent vector $\boldsymbol{w}$ to γ at p, which determines the covariant derivative $\underset{\boldsymbol{w}}{\boldsymbol{\nabla}}\boldsymbol{\xi}$ of $\boldsymbol{\xi}$ at p in the direction $\boldsymbol{w}$.

$$\underset{\lambda w}{\nabla} = \lambda \underset{w}{\nabla}.$$

It may seem that placing the vector symbol beneath the ∇ looks notationally awkward—as indeed it is! However, there is a genuine confusion between the mathematician's and the physicist's notation in the use of an expression such as '∇_w'. To our mathematician, this would be likely to denote the operation that I am using '$\underset{w}{\nabla}$' for here, whereas our physicist would be likely to interpret the w as an *index* and not as a vector field. In the physicist's notation, we would express the operator $\underset{w}{\nabla}$ as

$$\underset{w}{\nabla} = w^a \nabla_a,$$

and the above linearity simply reflects a consistency in the notation:

$$(w^a + u^a)\nabla_a = w^a \nabla_a + u^a \nabla_a \text{ and } (\lambda w^a)\nabla_a = \lambda(w^a \nabla_a).$$

The placing of a lower index on ∇ is consistent with its being a dual entity to a vector field (as is reflected in the above linearity; see §12.3), i.e. ∇ is a covector operator (meaning an operator of valence $\left[{0 \atop 1}\right]$). Thus, when ∇ acts on a vector field ξ (valence $\left[{1 \atop 0}\right]$), the resulting quantity $\nabla\xi$ is a $\left[{1 \atop 1}\right]$-valent tensor. This is made manifest in the index notation by the use of the notation $\nabla_a \xi^b$ for the component (or abstract–index) expression for the tensor $\nabla\xi$. In fact, there is a natural way to extend the scope of the operator ∇ from vectors to tensors of general valence, the action of ∇ on a $\left[{p \atop q}\right]$-valent tensor T yielding a $\left[{p \atop q+1}\right]$-valent tensor ∇T. The rules for achieving this can be conveniently expressed in the index notation, but there is an awkwardness in the mathematician's notation that we shall come to in a moment.

In its action on vector fields, ∇ satisfies the kind of rules that the differential operator d of §12.6 satisfies:

$$\nabla(\xi + \eta) = \nabla\xi + \nabla\eta$$

and the Leibniz law

$$\nabla(\lambda\xi) = \lambda\nabla\xi + \xi\nabla\lambda,$$

where ξ and η are vector fields and λ is a scalar field. As part of the normal reqirements of a connection, the action of ∇ on a scalar is to be identical with the action of the gradient (exterior derivative) d on that scalar:

$$\nabla\Phi = \mathrm{d}\Phi.$$

The extension of ∇ to a general tensor field is uniquely determined[14.4] by the following two natural requirements. The first is *additivity* (for tensors T and U of the same valence)

[14.4] Explain why unique. *Hint*: Consider the action of ∇ on $\alpha \bullet \xi$, etc.

$$\nabla(T + U) = \nabla T + \nabla U$$

and the second is that the appropriate form of Leibniz law holds. This Leibniz law is a little awkward to state, particularly in the mathematician's notation, which eschews indices. The rough form of this law (for tensors $\boldsymbol{T}$ and $\boldsymbol{U}$ of arbitrary valence) is

$$\nabla(\boldsymbol{T}\bullet\boldsymbol{U}) = (\nabla\boldsymbol{T})\bullet\boldsymbol{U} + \boldsymbol{T}\bullet\nabla\boldsymbol{U},$$

but this needs explanation. The dot • is to indicate some form of contracted product, where a set of upper and lower indices of $\boldsymbol{T}$ is contracted with a set of lower and upper indices of $\boldsymbol{U}$ (allowing that the sets could be vacuous, so that the product becomes an *outer* product, with no contractions at all). In the above formula, the contractions in both terms on the right-hand side are to mirror those on the left-hand side exactly, and the index letter on the ∇ is to be the same throughout the expression.

There is an especial awkwardness with the mathematician's notation—where indices are not referred to—in writing down the formula that expresses just what we mean by the tensor Leibniz law. This is slighly alleviated if we use $\underset{w}{\nabla}$ instead of ∇ since the $\boldsymbol{w}$ keeps track of the index on the ∇, and we can do something similar with the other indices if we wish, contracting each one with a vector or covector field (not acted on by ∇). In my own opinion, things are clearer with indices, but much more so in the diagrammatic notation where differentiation is denoted by drawing a ring around the quantity that is being differentiated. In Fig. 14.6, I have illustrated this with a representative example of the tensor Leibniz law.

All these properties would also be true of the 'coordinate derivative' operator $\partial/\partial x^a$ in place of ∇_a. In fact, in any one coordinate patch, we can use $\partial/\partial x^a$ to define a particular connection in that patch, which I shall call the *coordinate connection*. It is not a very interesting connection, since the coordinates are arbitrary. (It provides a notion of 'parallelism' in which all

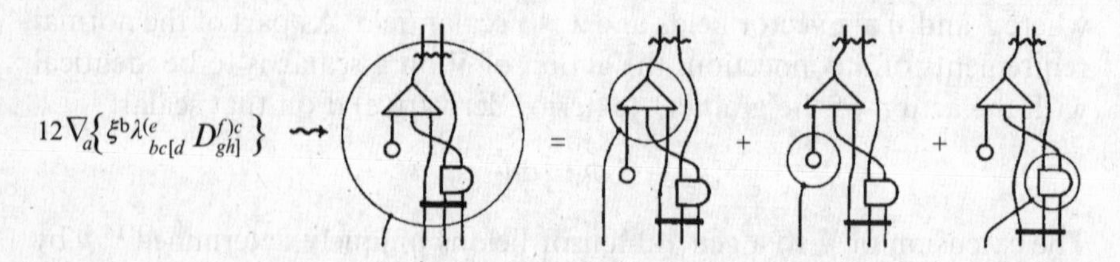

Fig. 14.6 In the diagrammatic notation, covariant differentiation is conveniently denoted by drawing a ring around the quantity being differentiated. This is illustrated here with example of the tensor Leibniz law applied to $\nabla_a\{\xi^b\lambda^{(e}_{bc[d}D^{f)c}_{gh]}\}$ (see Fig. 12.17). The (anti)symmetry factors give the '12'.

the coordinate lines count as 'parallel'.) On the overlap between two coordinate patches, the connection defined by the coordinates on one patch would usually not agree with that defined on the other (see Fig. 14.1). Although the coordinate connection is not 'interesting' (certainly not physically interesting), it is quite often useful in explicit expressions. The reason has to do with the fact that, if we take the difference between two connections, the action of this difference on some tensor quantity $\boldsymbol{T}$ can always be expressed entirely algebraically (i.e. without any differentiation) in terms of $\boldsymbol{T}$ and a certain tensor quantity $\boldsymbol{\Gamma}$ of valence $[{}^1_2]$.[14.5] This enables us to express the action of $\boldsymbol{\nabla}$ on any tensor $\boldsymbol{T}$ explicitly in terms of the coordinate derivatives[2] of the components $T^{a\ldots c}_{d\ldots f}$ together with some additional terms involving the components Γ^a_{bc}.[14.6]

14.4 Curvature and torsion

A coordinate connection is a rather special kind of connection in that, unlike the general case, it defines a parallelism that is independent of the path. This has to do with the fact (already noted in §10.2, in the form $\partial^2 f/\partial x \partial y = \partial^2 f/\partial y \partial x$) that coordinate derivative operators commute:

$$\frac{\partial^2}{\partial x^a \partial x^b} = \frac{\partial^2}{\partial x^b \partial x^a}.$$

Another way of saying this is that the quantity $\partial^2/\partial x^a \partial x^b$ is symmetric (in its indices ab). We shall be seeing what this has to do with the path independence of parallelism shortly. For a general connection $\boldsymbol{\nabla}$, this symmetry property does not hold for $\nabla_a \nabla_b$, its antisymmetric part $\nabla_{[a}\nabla_{b]}$ giving rise to two special tensors, one of valence $[{}^1_2]$ called the *torsion* tensor $\boldsymbol{\tau}$ and the other of valence $[{}^1_3]$ called the *curvature* tensor $\boldsymbol{R}$. Torsion is present when the action of $\nabla_{[a}\nabla_{b]}$ on a scalar quantity fails to vanish. In most physical theories, $\boldsymbol{\nabla}$ is

---

[14.5] See if you can show this, finding the expression explicitly. *Hints*: First look at the action of the difference between two connections on a vector field $\boldsymbol{\xi}$, giving the answer in the index form $\xi^c \Gamma^a_{bc}$; second, show that this difference of connections acting on a covector $\boldsymbol{\alpha}$ has the index form $-\alpha_c \Gamma^c_{ba}$; third, using the definition of a $[{}^p_q]$-valent tensor $\boldsymbol{T}$ as a multilinear function of q vectors on p covectors (cf. §12.8), find the general index expression for the difference between the connections acting on $\boldsymbol{T}$.

[14.6] As an application of this, take the two connections to be $\boldsymbol{\nabla}$ and the coordinate connection. Find a coordinate expression for the action of $\boldsymbol{\nabla}$ on any tensor, showing how to obtain the components Γ^a_{bc} explicitly from $\Gamma^a_{b1} = \nabla_b \delta^a_1, \ldots, \Gamma^a_{bn} = \nabla_b \delta^a_n$, i.e. in terms of the action of $\boldsymbol{\nabla}$ on each of the coordinate vectors $\delta^a_1, \ldots, \delta^a_n$. (Here a is a vector index, which may be thought of as an 'abstract index' in accordance with §12.8, so that 'δ^a_1' etc. indeed denote vectors and not simply sets of components, but n just denotes the dimension of the space. Note that the coordinate connection annihilates each of these coordinate vectors.)

taken to be *torsion-free*, i.e. $\boldsymbol{\tau} = 0$, and this certainly makes life easier. But there are some theories, such as supergravity and the Einstein–Cartan–Sciama–Kibble spin/torsion theories which employ a non-zero torsion that plays a significant physical role; see Note 19.10, §31.3. When torsion *is* present, its index expression $\tau_{ab}{}^{c}$, antisymmetric in *ab*, is defined by[14.7]

$$(\nabla_a \nabla_b - \nabla_b \nabla_a)\Phi = \tau_{ab}{}^{c} \nabla_c \Phi.$$

The *curvature* tensor $\boldsymbol{R}$, in the torsion-free case,[14.8] can be defined[3] by[14.9]

$$(\nabla_a \nabla_b - \nabla_b \nabla_a)\boldsymbol{\xi}^d = R_{abc}{}^{d} \boldsymbol{\xi}^c.$$

As is common in this subject, we run into daunting expressions with many little indices, so I recommend the diagrammatic version of these key expressions, e.g. Fig. 14.7a,b. In any case, I also recommend that indexed quantities be read, where appropriate, as tensors with abstract indices, as in §12.8 (Numerous different conventions exist in the literature about index orderings, signs, etc. I am imposing upon the reader the ones that I tend to use myself—at least in papers of which I am sole author!) The fact that $R_{abc}{}^{d}$ is antisymmetric in its first pair of indices *ab*, namely

$$R_{bac}{}^{d} = -R_{abc}{}^{d},$$

(see Fig. 14.7c) is evident from the corresponding antisymmetry of $\nabla_a \nabla_b - \nabla_b \nabla_a = 2\nabla_{[a} \nabla_{b]}$. We shall see the significance of this antisymmetry shortly. In the torsion-free case we have an additional symmetry relation[14.10] (Fig. 14.7d)

$$R_{[abc]}{}^{d} = 0, \quad \text{i.e. } R_{abc}{}^{d} + R_{bca}{}^{d} + R_{cab}{}^{d} = 0.$$

This relation is sometimes called 'the first Bianchi identity'. I shall call it the *Bianchi symmetry*. The term *Bianchi identity* (Fig. 14.7e) is normally reserved for the 'second' such identity which, in the absence of torsion, is[14.11]

[14.7] Explain why the right-hand side must have this general form; find the components τ^a_{bc} in terms of Γ^a_{bc}. See Exercise [14.6].

[14.8] Show what extra term is needed to make this expression consistent, when torsion is present.

[14.9] What is the corresponding expression for $\nabla_a \nabla_b - \nabla_b \nabla_a$ acting on a covector? Derive the expression for a general tensor of valence $\left[\begin{smallmatrix} p \\ q \end{smallmatrix}\right]$.

[14.10] First, explain the 'i.e.'; then derive this from the equation defining $R_{abc}{}^{d}$, above, by expanding out $\nabla_{[a} \nabla_b (\xi^d \nabla_{d]} \Phi)$. (Diagrams can help.)

[14.11] Derive this from the equation defining $R_{abc}{}^{d}$, above, by expanding out $\nabla_{[a} \nabla_b \nabla_{d]} \boldsymbol{\xi}^e$ in two ways. (Diagrams can again help.)

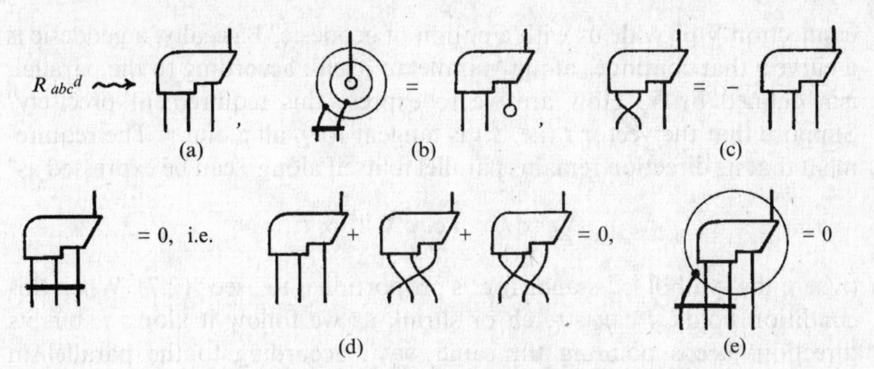

Fig. 14.7 (a) A convenient diagrammatic notation for the curvature tensor $R_{abc}{}^{d}$. (b) The Ricci identity $(\nabla_a\nabla_b - \nabla_b\nabla_a)\boldsymbol{\xi}^d = R_{abc}{}^{d}\boldsymbol{\xi}^c$. (c) The antisymmetry $R_{bac}{}^{d} = -R_{abc}{}^{d}$. (d) The Bianchi symmetry $R_{[abc]}{}^{d} = 0$, which reduces to $R_{abc}{}^{d} + R_{bca}{}^{d} + R_{cab}{}^{d} = 0$. (e) The Bianchi identity $\nabla_{[a}R_{bc]d}{}^{e} = 0$.

$$\nabla_{[a}R_{bc]d}{}^{e} = 0, \quad \text{i.e.} \nabla_a R_{bcd}{}^{e} + \nabla_b R_{cad}{}^{e} + \nabla_c R_{abd}{}^{e} = 0.$$

The Bianchi identity is the linchpin of the Einstein field equation, as we shall be seeing in §19.6.

Curvature is the essential quantity that expresses the path dependence of the connection (at least on the local scale). If we envisage transporting a vector around a small loop in the space $\mathcal{M}$, using the notion of parallel transport defined by $\boldsymbol{\nabla}$, then we find that it is $\boldsymbol{R}$ that measures how much that the vector has changed when we return to the starting point. It is easiest to think of the loop as an 'infinitesimal parallelogram' drawn in the space $\mathcal{M}$. (Such parallelograms adequately 'exist' when $\boldsymbol{\nabla}$ is torsion-free, as we shall see.) However, various notions here need clarification first.

14.5 Geodesics, parallelograms, and curvature

First, in order to build ourselves a parallelogram, let us consider the concept of a *geodesic*, as defined by the connection $\boldsymbol{\nabla}$. Geodesics are important to us for other reasons. They are the analogues of the straight lines of Euclidean geometry. In our example of the sphere S^2, considered above (Figs. 14.2–14.4), the geodesics are great circles on the sphere. More generally, for a curved surface in Euclidean space, the curves of minimum length (as would be taken up by a string stretched taut along the surface) are geodesics. We shall be seeing later (§17.9) that geodesics have a fundamental significance for Einstein's general relativity, representing the paths in spacetime that describe freely falling bodies. How does our connection $\boldsymbol{\nabla}$ provide us with a notion of geodesic? Basically, a geodesic is a curve γ that continues along 'parallel to itself', according to the parallelism defined by $\boldsymbol{\nabla}$. How are we to express this requirement precisely? Suppose that the vector $\boldsymbol{t}$ (i.e. t^a) is tangent to γ, all along γ. The requirement that its direction remains parallel to itself along γ can be expressed as[4]

$$\underset{t}{\boldsymbol{\nabla}}\boldsymbol{t} \propto \boldsymbol{t}, \quad \text{i.e. } t^a\boldsymbol{\nabla}_a t^b \propto t^b,$$

(where the symbol '$\propto$' stands for 'is proportional to'; see §12.7). When this condition holds, $\boldsymbol{t}$ can stretch or shrink as we follow it along γ, but its direction 'keeps pointing the same way', according to the parallelism notion defined by $\boldsymbol{\nabla}$. If we wish to assert that this 'stretching or shrinking' does not take place, so that the vector $\boldsymbol{t}$ itself remains constant along γ, then we demand the stronger condition that the tangent vector $\boldsymbol{t}$ be *parallel-transported* along γ, i.e. that

$$\underset{t}{\boldsymbol{\nabla}}\boldsymbol{t} = 0, \quad \text{i.e. } t^a\nabla_a t^b = 0,$$

holds all along γ, where the vector $\boldsymbol{t}$ (with index form t^a) is *tangent* to γ, along γ.

According to this stronger equation, not just the direction of $\boldsymbol{t}$, but also the 'scale' of $\boldsymbol{t}$ is kept constant along γ. What does this mean? The first thing to note is that any curve (not necessarily a geodesic), parameterized by an (appropriately smooth) coordinate u, is associated with a particular choice of scaling for its tangent vectors $\boldsymbol{t}$ along the curve. This is such that $\boldsymbol{t}$ stands for differentiation (d/du) with respect to u along the curve. We can write this condition, alternatively, as

$$\boldsymbol{t}(u) = 1$$

or as

$$\underset{t}{\boldsymbol{\nabla}}u = 1, \quad \text{i.e. } t^a\nabla_a u = 1$$

along the curve.[14.12]

In the case of a geodesic γ, the stronger choice of $\boldsymbol{t}$-scaling for which $\underset{t}{\boldsymbol{\nabla}}\boldsymbol{t} = 0$ is associated with a particular type of parameter u, known as an *affine* parameter[14.13] along γ. See Fig. 14.8. When we have an appropriate notion of 'distance' along curves, we can usually choose our affine param-

---

[14.12] Demonstrate the equivalence of all these conditions.

[14.13] Show that if u and v are two affine parameters on γ, with respect to two different choices of $\boldsymbol{t}$, then $v = Au + B$, where A and B are constant along γ.

eter to be this measure of distance. But affine parameters are more general. For example, in relativity theory, it turns out that we need such parameters for *light rays*, the appropriate 'distance measure' being useless here, because it is zero! (See §17.8 and §18.1.)

Let us now try to construct a parallelogram out of geodesics. Start at some point p in $\mathcal{M}$, and draw two geodesics λ and μ in $\mathcal{M}$ out from p, with respective tangent vectors $\boldsymbol{L}$ and $\boldsymbol{M}$ at p and respective affine parameters l and m. Choose some positive number ε and measure out an affine distance $l = \varepsilon$ along λ from p to reach the point q and also an affine distance $m = \varepsilon$ along μ from p to reach r; see Fig. 14.9a. (Intuitively, we may think of the geodesic segments pq and pr having the 'arrow lengths' of $\varepsilon\boldsymbol{L}$ and $\varepsilon\boldsymbol{M}$ respectively, for some small ε.) To complete the parallelogram, we need to move off from q along a new geodesic μ', in a direction which is 'parallel' to $\boldsymbol{M}$. To achieve this 'parallel' condition, we move $\boldsymbol{M}$ from p to q along λ by parallel transport (which means we require $\boldsymbol{M}$ to satisfy $\underset{L}{\nabla} M = 0$ along λ). Now, we try to locate the final vertex of the parallelogram at the point s which is measured out from q by an affine distance $m = \varepsilon$ along μ'. However, we could alternatively try to position this final vertex by proceeding the other way around: move out from r an affine distance $l = \varepsilon$ along λ' to a final point s' where the geodesic λ' starts off from r in the direction of $\boldsymbol{M}$ which has been carried from p to r along μ by parallel transport. For a thoroughly convincing parallelogram, we should require these alternative final vertices s and s' to be the same point ($s = s'$)!

However, except in very special cases (such as Euclidean geometry), these two points will be different. (Recall our attempts to construct a square in §2.1!) These points will not be 'very' different, in a certain sense,

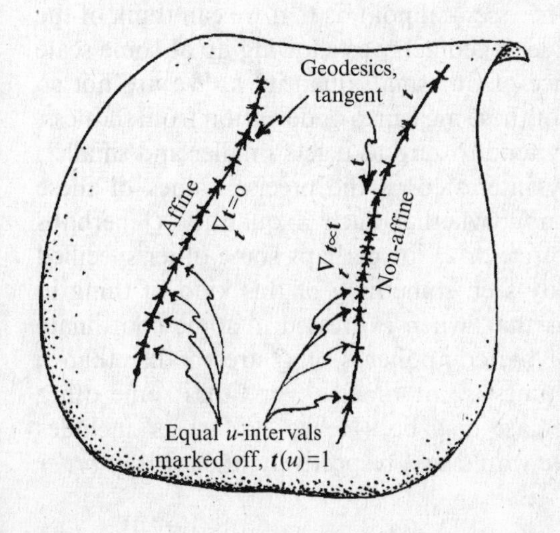

Fig. 14.8 For any (suitably smooth) parameter u defined along a curve γ, a field of tangent vectors $\boldsymbol{t}$ to γ is naturally associated with u so that, along γ, $\boldsymbol{t}$ stands for $\mathrm{d}/\mathrm{d}u$ (equivalently $\boldsymbol{t}(u) = 1$, or $t^a\nabla_a u = 1$). If γ is a geodesic, u is called an affine parameter if $\boldsymbol{t}$ is parallel-transported along γ, so $\underset{t}{\boldsymbol{\nabla}} \boldsymbol{t} = 0$ rather than just $\underset{t}{\boldsymbol{\nabla}} \boldsymbol{t} \propto \boldsymbol{t}$. An affine parameter is 'evenly spaced' along γ, according to $\boldsymbol{\nabla}$.

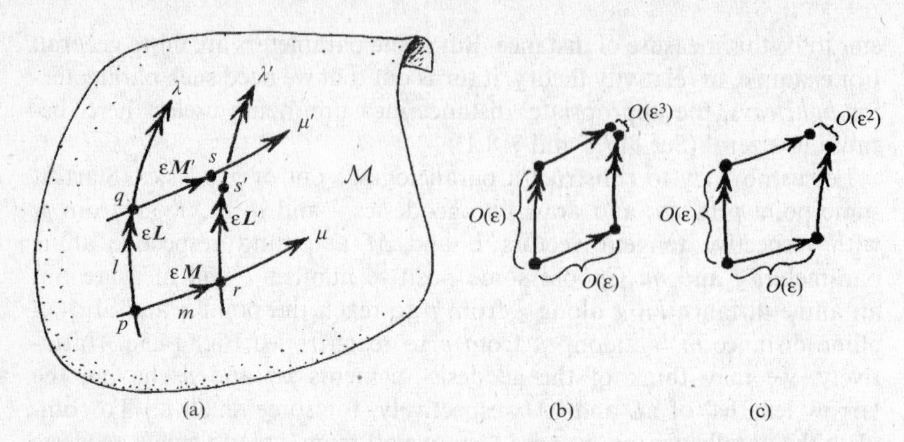

Fig. 14.9 (a) Try to make parallelogram out of geodesics. Take two geodesics λ, μ, through p, in $\mathcal{M}$, with respective tangent vectors $\boldsymbol{L}$, $\boldsymbol{M}$ at p and corresponding affine parameters l, m. Take q an affine distance $l = \varepsilon$ along λ from p, and r an affine distance $m = \varepsilon$ along μ from p (with $\varepsilon > 0$ a fixed small number). The geodesic segments pq and pr have respective 'arrow-lengths' $\varepsilon\boldsymbol{L}$, $\varepsilon\boldsymbol{M}$. To make the parallelogram, move $\boldsymbol{M}$ from p to q along λ by parallel transport ($\nabla_{\boldsymbol{L}}\boldsymbol{M} = 0$ along λ) giving us a neighbouring geodesic μ' to μ, extending from q to s along μ' by an affine distance ε along the new 'parallel' arrow $\varepsilon\boldsymbol{M}'$. Similarly, move $\boldsymbol{L}$ from p to r by parallel transport along μ, and extend from r to s' by a parallel arrow $\varepsilon\boldsymbol{L}'$ measured out from q an affine distance $m = \varepsilon$ along λ'. (b) Generally $s \neq s'$ and the parallelogram fails to close exactly, but this gap is only $O(\varepsilon^3)$ if the torsion $\boldsymbol{\tau}$ vanishes. (c) If there is a non-zero torsion $\boldsymbol{\tau}$, this will show up as an $O(\varepsilon^2)$ term.

if the vectors $\varepsilon\boldsymbol{L}$ and $\varepsilon\boldsymbol{M}$ are taken to be appropriately 'small'. But exactly how different they are has to do with the torsion $\boldsymbol{\tau}$. In order to understand this properly we need rather more in the way of calculus notions than I have provided up until now. The essential point is that we can think of the relevant deviations from Euclidean geometry as showing up at some scale that is dependent on the choice of our small quantity ε. We are not so concerned with the actual size of these measures of deviation from flatness, but with the rate at which they tend to zero as ε gets smaller and smaller. Thus, we are not particularly interested in the precise values of these quantities but we want to know whether such a quantity $\boldsymbol{Q}$ perhaps approaches zero as fast as ε, or ε^2, or ε^3, or perhaps some other specified function of ε. (We have already seen something of this kind of thing in §13.6.) Here 'as fast as' means that, when expressed in some coordinate system, the absolute values of the components of $\boldsymbol{Q}$ are smaller than a positive constant times ε, or times ε^2, or times ε^3, or times some other specified function of ε, as the case may be. (Hence 'as fast as' includes 'faster than'!) In these cases, we would say, respectively, that $\boldsymbol{Q}$ is *of order*

ε, or ε^2, or ε^3, etc., and we would write this $O(\varepsilon)$, or $O(\varepsilon^2)$, or $O(\varepsilon^3)$, etc. This is independent of the particular choice of coordinates, which is one reason that this notion of 'order of smallness' is a sensible and powerful notion. My description here has been very brief, and I refer the uninitiated interested reader to the literature concerning this remarkable and ubiquitous topic.[5] Intuitively, we just need to bear in mind that $O(\varepsilon^3)$ means very much smaller than $O(\varepsilon^2)$, which is itself much smaller than $O(\varepsilon)$, etc.

Let us return to our attempted parallelogram. The original vectors $\varepsilon\boldsymbol{L}$ and $\varepsilon\boldsymbol{M}$, at p, are both $O(\varepsilon)$, so the sides pq and pr are both $O(\varepsilon)$, and so also will be qs and rs'. How big do we expect the 'gap' ss' to be? The answer is that, if the connection is torsion-free, then ss' is always $O(\varepsilon^3)$. See Fig. 14.9b. In fact, this property characterizes the torsion-free condition completely. If a non-zero torsion $\boldsymbol{\tau}$ is present, then this will show up in (some) parallelograms, as an $O(\varepsilon^2)$ term. See Fig. 14.9c.[14.14] Sometimes we say (rather loosely) that the vanishing of torsion is the condition that parallelograms close (by which we mean 'close to order ε^2').

Suppose, now, that the torsion vanishes. Can we use our parallelogram to interpret curvature? Indeed we can. Let us suppose that we have a third vector $\mathbf{N}$ at p, and we carry this by parallel transport around our parallelogram from p to s, via q, and we compare this with transporting it from p to s', via r. (This comparison makes sense at order ε^2, when the torsion vanishes, because then the gap between s and s' is $O(\varepsilon^3)$ and can be ignored. When the torsion does not vanish, we have to worry about the additional torsion term; see Exercise [14.7].) We find the answer for the difference between the result of the pqs transport and the prs' transport to be

$$\varepsilon^2 L^a M^b N^c R_{abc}{}^d.$$

This provides us with a very direct geometrical interpretation of the curvature tensor $\boldsymbol{R}$; see Fig. 14.10. (An equivalent version of this interpretation is obtained if we think of transporting $\boldsymbol{N}$ all the way around the parallelogram, starting and ending at the same point p, where we ignore $O(\varepsilon^3)$ discrepancies in the vertices of the parallelogram. The difference between the starting and finishing values of $\boldsymbol{N}$ is again the above quantity $\varepsilon^2 L^a M^b N^c R_{abc}{}^d$.)

Recall the antisymmetry of $R_{abc}{}^d$ in ab. This means that the above expression is sensitive only to the antisymmetric part, $L^{[a}M^{b]}$, of L^aM^b, i.e. of the wedge product $\boldsymbol{L}\wedge\boldsymbol{M}$; see §11.6. Thus, it is the 2-plane element spanned by $\boldsymbol{L}$ and $\boldsymbol{M}$ at p that is of relevance. In the case when $\mathcal{M}$ is itself a

[14.14] Find this term.

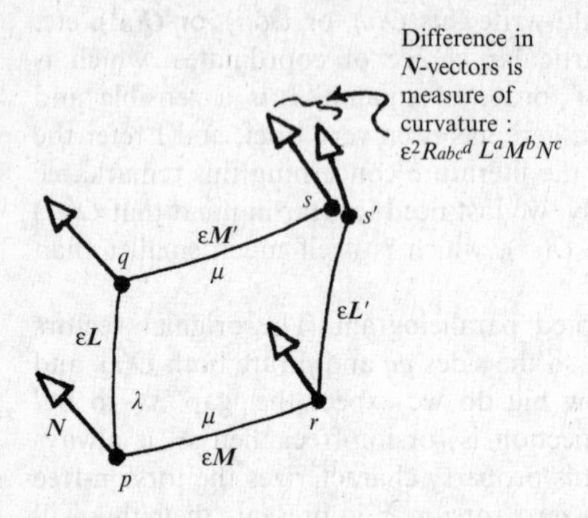

Fig. 14.10 Use the parallelogram to interpret curvature, when $\boldsymbol{\tau} = 0$. Carry a third vector $\boldsymbol{N}$, by parallel transport from p to s via q, comparing this with transporting it from p to s' via r. The $O(\varepsilon^2)$ term measuring the difference is $\varepsilon^2 L^a M^b N^c R_{abc}{}^d$, i.e. ε^2 $\boldsymbol{R}$ ($\boldsymbol{L}$, $\boldsymbol{M}$, $\boldsymbol{N}$), providing a direct geometrical interpretation of the curvature tensor $\boldsymbol{R}$.

2-surface, there is just one independent curvature component (since the 2-plane element has to be tangent to $\mathcal{M}$ at p). This component provides us with the *Gaussian curvature* of a 2-surface that I alluded to in §2.6, and which serves to distinguish the local geometries of sphere, Euclidean plane, and hyperbolic space. In higher dimensions, things are more complicated, as there are more components of curvature arising from the different possible choices of 2-plane element $\boldsymbol{L} \wedge \boldsymbol{M}$.

There is a particular version of this geometrical interpretation of curvature that has especial significance. This occurs if the vector $\boldsymbol{N}$ is chosen to be the same as $\boldsymbol{L}$. Then we can think of the sides pq and rs' of our parallelogram as being segments of two nearby geodesics γ and γ', respectively, and the vector $\boldsymbol{L}$ is tangent to these geodesics. The vector $\varepsilon\boldsymbol{M}$ at p measures the displacement of γ away from γ' at the point p. $\boldsymbol{M}$ is sometimes called a *connecting vector*. The geodesics γ and γ' start out parallel to each other (as compared at the two 'ends' of this connecting vector, i.e. along pr). Carrying the vector $\boldsymbol{L}$ $(=\boldsymbol{N})$ to s' by parallel transport along the second route prs' leaves it tangent to the geodesic γ' at the point s'. But if we take $\boldsymbol{L}$ to s by parallel transport along the first route pqs, then we arrive at the starting vector for another geodesic γ'' nearby to γ, where γ'' is starting out parallel to γ at the slightly 'later' point q. The $O(\varepsilon^2)$ difference between these two versions of $\boldsymbol{L}$ (one at s' and the other at s), namely $\varepsilon^2 L^a M^b L^c R_{abc}{}^d$, measures the 'relative acceleration' or 'geodesic deviation' of γ' away from γ. See Fig. 14.11. (This geodesic deviation is mathematically described by what is known as the *Jacobi equation*.) In Fig. 14.12, I have illustrated this

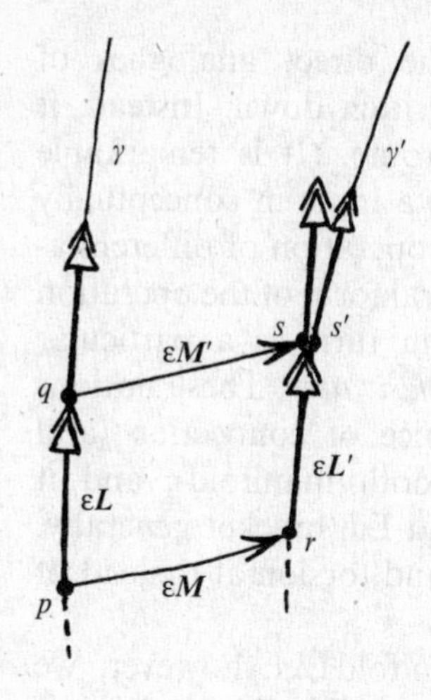

Fig. 14.11 Geodesic deviation: choose $\boldsymbol{N} = \boldsymbol{L}$ in the parallelogram of Fig. 14.10. The sides pq and rs' are segments of two neighbouring geodesics γ and γ' (γ being λ and γ' being λ') starting from p and r, respectively, with parallel-propagated tangent vectors $\boldsymbol{L}$ and $\boldsymbol{L}'$, the connecting vector at p being $\boldsymbol{M}$. The geodesic deviation between γ and γ' is measured by the difference between the results of parallel displacement of $\boldsymbol{L}$ along the routes prs' and pqs, which is basically $\varepsilon^2 L^a M^b L^c R_{abc}{}^d$.

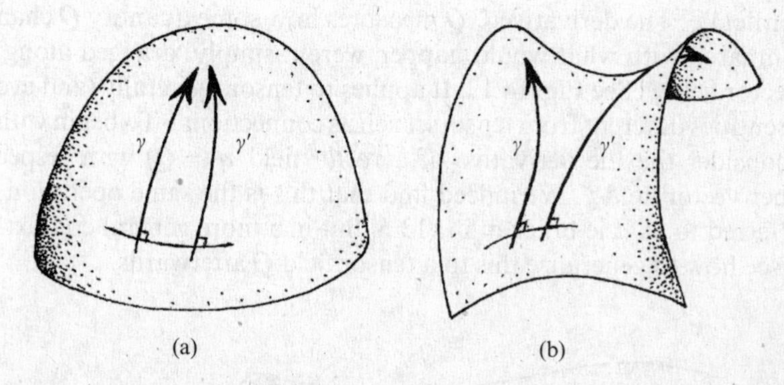

Fig. 14.12 Geodesic deviation when $\mathcal{M}$ is a 2-surface (a) of positive (Gaussian) curvature, when the geodesics γ, γ' bend towards each other, and (b) of negative curvature, when they bend apart.

geodesic deviation when $\mathcal{M}$ is a 2-surface of positive and negative (Gaussian) curvature, respectively. When the curvature is positive, the neighbouring geodesics, starting parallel, bend towards each other; when it is negative, they bend apart. We shall see the profound importance of this for Einstein's general relativity in §17.5 and §19.6.

14.6 Lie derivative

In the above discussion of the path dependence of parallelism, for a connection $\boldsymbol{\nabla}$, I have been expressing things using the physicist's index notation. In the mathematician's notation, the direct analogues of these particular expressions are not so easily written down. Instead, it becomes natural to follow a slightly different route. (It is remarkable how differences in notation can sometimes drive a topic in conceptually different directions!) This route involves another operation of differentiation, known as *Lie bracket*—which is a more general form of the operation of the same name introduced in §13.6. This, in turn, is a particular instance of an important concept known as *Lie derivative*. These notions are actually independent of any particular choice of connection (and therefore apply in a general unstructured smooth manifold), and it will be pertinent to discuss the Lie derivative and Lie bracket generally, before returning to their relevance to curvature and torsion at the end of this section.

For a Lie derivative to be defined on a manifold $\mathcal{M}$, however, we do require a vector field $\boldsymbol{\xi}$ to be pre-assigned on $\mathcal{M}$. The Lie derivative, written $\underset{\xi}{\pounds}$, is then an operation which is taken with respect to the vector field $\boldsymbol{\xi}$. The derivative $\pounds_\xi \boldsymbol{Q}$ measures how some quantity $\boldsymbol{Q}$ changes, as compared with what would happen were it simply 'dragged along', by the vector field $\boldsymbol{\xi}$. See Fig. 14.13. It applies to tensors generally (and even to some entities different from tensors, such as connections). To begin with, we just consider the Lie derivative of a *vector* field $\boldsymbol{\eta}$ $(=\boldsymbol{Q})$ with respect to another vector field $\boldsymbol{\xi}$. We indeed find that this is the same operation that we referred to as 'Lie bracket' in §13.6, but in a more general context. We shall see how to generalize this to a tensor field $\boldsymbol{Q}$ afterwards.

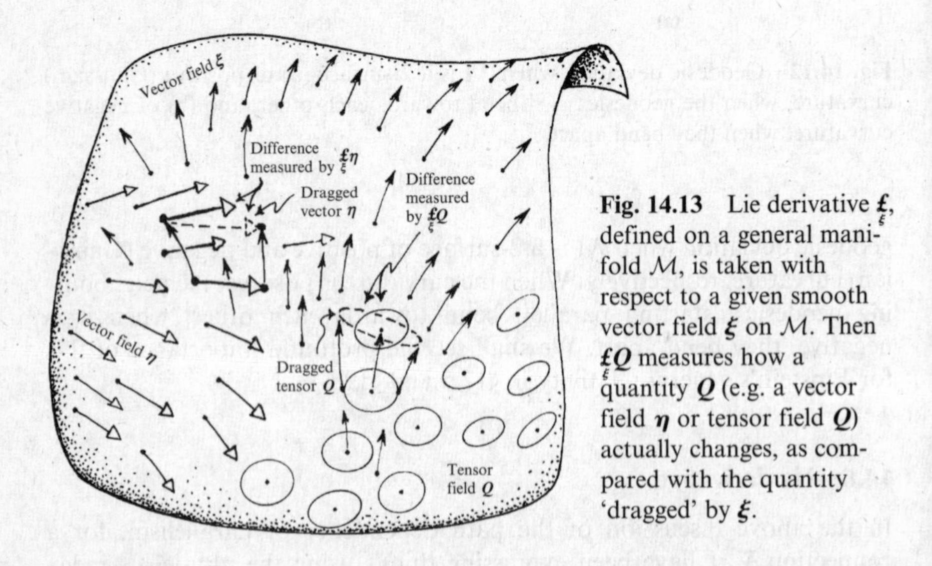

Fig. 14.13 Lie derivative $\underset{\xi}{\pounds}$, defined on a general manifold $\mathcal{M}$, is taken with respect to a given smooth vector field $\boldsymbol{\xi}$ on $\mathcal{M}$. Then $\underset{\xi}{\pounds}\boldsymbol{Q}$ measures how a quantity $\boldsymbol{Q}$ (e.g. a vector field $\boldsymbol{\eta}$ or tensor field $\boldsymbol{Q}$) actually changes, as compared with the quantity 'dragged' by $\boldsymbol{\xi}$.

Recall from §12.3 that a vector field can itself be interpreted as a differential operator acting on *scalar* fields Φ, Ψ, ... satisfying the three laws (i) $\boldsymbol{\xi}(\Phi + \Psi) = \boldsymbol{\xi}(\Phi) + \boldsymbol{\xi}(\Psi)$, (ii) $\boldsymbol{\xi}(\Phi\Psi) = \Psi\boldsymbol{\xi}(\Phi) + \Phi\boldsymbol{\xi}(\Psi)$, and (iii) $\boldsymbol{\xi}(k) = 0$ if k is a constant. It is a direct matter to show[14.15] that the operator $\boldsymbol{\omega}$, defined by

$$\boldsymbol{\omega}(\Phi) = \boldsymbol{\xi}(\boldsymbol{\eta}(\Phi)) - \boldsymbol{\eta}(\boldsymbol{\xi}(\Phi))$$

satisfies these same three laws, provided that $\boldsymbol{\xi}$ and both $\boldsymbol{\eta}$ do, so $\boldsymbol{\omega}$ must also be a vector field. The above *commutator* of the two operations $\boldsymbol{\xi}$ and $\boldsymbol{\eta}$ is frequently written (as in §13.6) in the *Lie bracket* notation

$$\boldsymbol{\omega} = \boldsymbol{\xi}\boldsymbol{\eta} - \boldsymbol{\eta}\zeta = [\boldsymbol{\xi}, \boldsymbol{\eta}].$$

The geometric meaning of the commutator between two vector fields $\boldsymbol{\xi}$ and $\boldsymbol{\eta}$ is illustrated in Fig. 14.14. We try to form a quadrilateral of 'arrows' made alternately from $\boldsymbol{\xi}$ and $\boldsymbol{\eta}$ (each taken to be $O(\varepsilon)$) and find that $\boldsymbol{\omega}$ measures the 'gap' (at order $O(\varepsilon^2)$). We can verify[14.16] that commutation satisfies the following relations

$$[\boldsymbol{\xi}, \boldsymbol{\eta}] = -[\boldsymbol{\eta}, \boldsymbol{\xi}], \qquad [\boldsymbol{\xi} + \boldsymbol{\eta}, \zeta] = [\boldsymbol{\xi}, \zeta] + [\boldsymbol{\eta}, \zeta],$$
$$[\boldsymbol{\xi}, [\boldsymbol{\eta}, \zeta]] + [\boldsymbol{\eta}, [\zeta, \boldsymbol{\xi}]] + [\zeta, [\boldsymbol{\xi}, \boldsymbol{\eta}]] = 0,$$

just as did the commutator of two infinitesimal elements of a Lie group, as we saw in §13.6.

How does our commutation operation, as defined above, relate to the algebra (§13.6) of infinitesimal elements of a Lie group? Let me digress briefly to explain this. We think of the group as a manifold $\mathcal{G}$ (called a

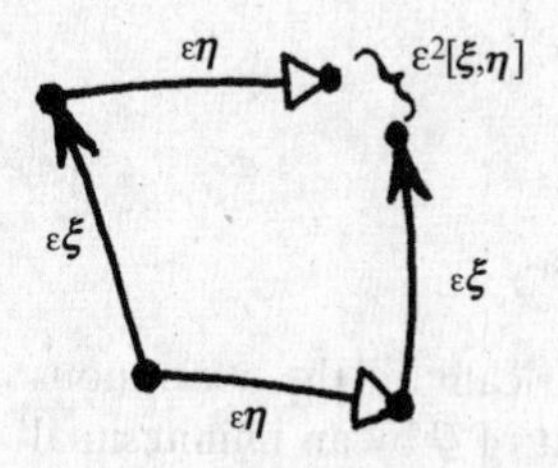

Fig. 14.14 The Lie bracket $[\boldsymbol{\xi},\boldsymbol{\eta}]$ ($= \pounds_{\boldsymbol{\xi}}\boldsymbol{\eta}$) between two vector fields $\boldsymbol{\xi}$, $\boldsymbol{\eta}$ measures the $O(\varepsilon^2)$ gap in an incomplete quadrilateral of $O(\varepsilon)$ 'arrows' made alternately from $\varepsilon\boldsymbol{\xi}$ and $\varepsilon\boldsymbol{\eta}$.

[14.15] Show it.

[14.16] Do it.

group manifold), whose points are the elements of our Lie group. More generally, we could think of any manifold $\mathcal{H}$ on which the elements act as smooth transformations (such as the sphere S^2 in the case of the rotation group $\mathcal{G} = SO(3)$, see Fig. 13.2). But, for now, we are primarily concerned with the group manifold $\mathcal{G}$, rather than the more general situation of $\mathcal{H}$, since we are interested in how the entire group $\mathcal{G}$ relates to the structure of its Lie algebra. The infinitesimal group elements are to be pictured as particular vector fields on $\mathcal{G}$ (or, indeed, $\mathcal{H}$). That is, we think of 'moving $\mathcal{G}$' infinitesimally along the relevant vector field $\boldsymbol{\xi}$ on $\mathcal{G}$, in order to express the transformation that corresponds to pre-multiplying each element of the group by the infinitesimal element represented by $\boldsymbol{\xi}$. See Fig. 14.15a.

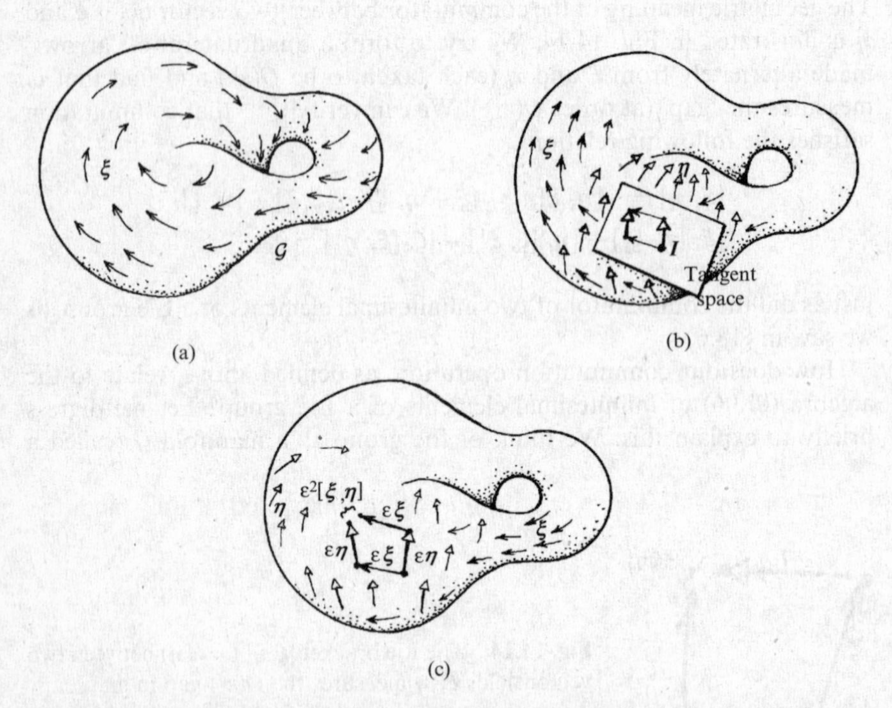

Fig. 14.15 Lie algebra operations, interpreted geometrically in the continuous group manifold $\mathcal{G}$. (a) Pre-multiplication of each element of $\mathcal{G}$ by an infinitesimal group element $\boldsymbol{\xi}$ (Lie algebra element) gives an infinitesimal shift of $\mathcal{G}$, i.e. a vector field $\boldsymbol{\xi}$ on $\mathcal{G}$. (b) To first order, the product of two such infinitesimal motions $\boldsymbol{\xi}$ and $\boldsymbol{\eta}$ just gives $\boldsymbol{\xi} + \boldsymbol{\eta}$, reflecting merely the structure of the tangent space (at $\boldsymbol{I}$). (c) The local group structure appears at second order, $\varepsilon^2[\boldsymbol{\xi}, \boldsymbol{\eta}]$, providing the $O(\varepsilon^2)$ gap in the 'parallelogram' with alternate sides $\varepsilon\boldsymbol{\xi}$ and $\varepsilon\boldsymbol{\eta}$ at $\boldsymbol{I}$.

Choosing a small positive quantity ε, we can think of $\varepsilon\boldsymbol{\xi}$ as being an $O(\varepsilon)$ motion of $\mathcal{G}$ along the vector field $\boldsymbol{\xi}$, the identity group element I corresponding to zero motion. The product of two such small group actions $\varepsilon\boldsymbol{\xi}$ and $\varepsilon\boldsymbol{\eta}$ is given, to $O(\varepsilon)$, by the *sum* $\varepsilon\boldsymbol{\xi}+\varepsilon\boldsymbol{\eta}$ of the two, so the 'arrows' representing $\varepsilon\boldsymbol{\xi}$ and $\varepsilon\boldsymbol{\eta}$ just add according to the parallelogram law (Fig. 14.15b). But this gives us little information about the structure of the group (only its *dimension*, in fact, as we are just revealing the additive structure of the tangent space at the identity element I of the group). To obtain the group structure, we need to go to $O(\varepsilon^2)$, and this is done, as in §13.6, by looking at the *commutator* $\boldsymbol{\xi\eta}-\boldsymbol{\eta\xi}=[\boldsymbol{\xi},\boldsymbol{\eta}]$. Now $\varepsilon^2[\boldsymbol{\xi},\boldsymbol{\eta}]$ corresponds to an $O(\varepsilon^2)$ gap in the 'parallelogram' whose initial sides are $\varepsilon\boldsymbol{\xi}$ and $\varepsilon\boldsymbol{\eta}$ at the origin I. The relevant notion of 'parallelism' comes from the group action, supplying the needed notion of 'parallel transport', which actually gives a connection with torsion but no curvature.[14.17] See Fig. 14.15c.

As was noted in §13.6, the Lie algebra of these vector fields provides the entire (local) structure of the group. The procedure whereby one obtains an ordinary *finite* (i.e. non-infinitesimal) group element x from a Lie algebra element $\boldsymbol{\xi}$ may be noted here. This is called *exponentiation* (cf. §5.3, §13.4):

$$x=\mathrm{e}^{\boldsymbol{\xi}}=\boldsymbol{I}+\boldsymbol{\xi}+\frac{1}{2}\boldsymbol{\xi}^2+\frac{1}{6}\boldsymbol{\xi}^3+\cdots.$$

Here $\boldsymbol{\xi}^2$ means 'the second derivative operator of applying $\boldsymbol{\xi}$ twice', etc. (and $\boldsymbol{I}$ is the identity operator). This is basically a form of Taylor's theorem, as described in §6.4.[14.18] The product of two finite group elements x and y is then obtained from the expression $\mathrm{e}^{\boldsymbol{\xi}}\mathrm{e}^{\boldsymbol{\eta}}$. This differs from $\mathrm{e}^{\boldsymbol{\xi}+\boldsymbol{\eta}}$ (compare §5.3) by an expression that is constructed entirely from Lie algebra expression[6] in $\boldsymbol{\xi}$ and $\boldsymbol{\eta}$.

It may be noted that a version of this exponentiation operation $\mathrm{e}^{\boldsymbol{\xi}}$ also applies to a vector field $\boldsymbol{\xi}$ in a general manifold $\mathcal{M}$ (where $\mathcal{M}$ and $\boldsymbol{\xi}$ are assumed analytic—i.e. C^{ω}-smooth, see §6.4). Recall from §12.3 (and Fig. 10.6) that, with ε chosen small, $\varepsilon\boldsymbol{\xi}(\Phi)$ measures the $O(\varepsilon)$ increase of a scalar function Φ from the tail to the head of the 'arrow' that represents $\varepsilon\boldsymbol{\xi}$. More exactly, the quantity $\mathrm{e}^{t\boldsymbol{\xi}}(\Phi)$ measures the *total* value Φ that is reached as we follow along the '$\boldsymbol{\xi}$-arrows' from a starting point O, to a

---

[14.17] Try to explain why there is torsion but no curvature.

[14.18] Explain (at a formal level) why $\mathrm{e}^{a\,\mathrm{d}/\mathrm{d}y}f(y)=f(y+a)$ when a is a constant.

final point given by the parameter value $u = t$, where the parameter u is scaled so that $\boldsymbol{\xi}(u) = 1$ (cf. §14.5 and Fig. 14.8). All the derivatives (i.e. the rth derivative, in the case of $\boldsymbol{\xi}^r(\Phi)$) in the power series expression for $e^{t\boldsymbol{\xi}}(\Phi)$ are to be evaluated at O (convergence being assumed). 'Following along the arrows' would mean following along what is called an 'integral curve' of $\boldsymbol{\xi}$, that is, a curve whose tangent vectors are $\boldsymbol{\xi}$-vectors. See Fig. 14.16.[7]

What, then, is the definition of Lie *derivative*? First, we simply rewrite the Lie bracket as an operation $\underset{\xi}{\pounds}$ (depending on $\boldsymbol{\xi}$) which acts upon the vector field $\boldsymbol{\eta}$:

$$\underset{\xi}{\pounds}\boldsymbol{\eta} = [\boldsymbol{\xi}, \boldsymbol{\eta}].$$

This is to be the definition of the Lie derivative $\underset{\xi}{\pounds}$ (with respect to $\boldsymbol{\xi}$) of a $[{}^1_0]$-tensor (i.e. a vector) $\boldsymbol{\eta}$. We wish to write this in terms of some given torsion-free connection $\boldsymbol{\nabla}$. The required expression (see Fig. 14.17a, for the diagrammatic form)

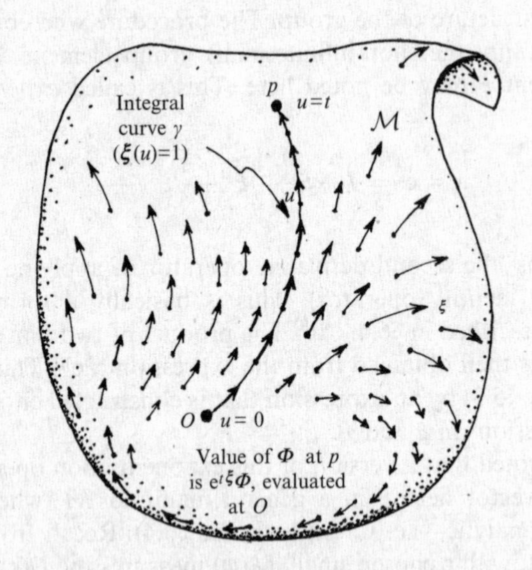

Fig. 14.16 An integral curve of a vector field $\boldsymbol{\xi}$ in $\mathcal{M}$ is a curve γ that 'follows the $\boldsymbol{\xi}$-arrows', i.e. whose tangent vectors are $\boldsymbol{\xi}$-vectors, with associated parameter u, in the sense $\boldsymbol{\xi}(u) = 1$ (cf. §14.5 and Fig. 14.8). Assume that $\mathcal{M}$ and $\boldsymbol{\xi}$ are analytic (i.e. C^ω), as is the scalar field Φ, and that γ stretches from some base point $O\,(u = 0)$ to another point $p\,(u = t)$. Then (assuming convergence) the value of Φ at p is given by the quantity $e^{t\boldsymbol{\xi}}(\Phi)$ evaluated at O, where $e^{t\boldsymbol{\xi}} = 1 + t\boldsymbol{\xi} + \frac{1}{2}t^2\boldsymbol{\xi}^2 + \frac{1}{6}t^3\boldsymbol{\xi}^3 + \ldots$ and where $\boldsymbol{\xi}^r$ stands for the rth derivative d^r/du^r at O along γ.

$$\pounds_{\boldsymbol{\xi}}\boldsymbol{\eta} = \nabla_{\boldsymbol{\xi}}\boldsymbol{\eta} - \nabla_{\boldsymbol{\eta}}\boldsymbol{\xi}, \quad \text{i.e. } (\pounds_{\boldsymbol{\xi}}\boldsymbol{\eta})^a = \xi^a\nabla_a\eta^b - \eta^a\nabla_a\xi^b,$$

can be directly obtained using $\boldsymbol{\xi}(\Phi) = \boldsymbol{\xi}^a\nabla_a\Phi$, etc.[14.19],[14.20] To obtain the Lie derivative of a general tensor, we employ the rule that (except for the absence of linearity in $\boldsymbol{\xi}$) $\pounds_{\boldsymbol{\xi}}$ satisfies rules similar to that of a connection $\nabla_{\boldsymbol{\xi}}$. These are: $\pounds_{\boldsymbol{\xi}}\Phi = \boldsymbol{\xi}(\Phi)$ for a scalar Φ; $\pounds_{\boldsymbol{\xi}}(\boldsymbol{T}+\boldsymbol{U}) = \pounds_{\boldsymbol{\xi}}\boldsymbol{T} + \pounds_{\boldsymbol{\xi}}\boldsymbol{U}$ for tensors $\boldsymbol{T}$ and $\boldsymbol{U}$ of the same valence; $\pounds_{\boldsymbol{\xi}}(\boldsymbol{T}\bullet\boldsymbol{U}) = (\pounds_{\boldsymbol{\xi}}\boldsymbol{T})\bullet\boldsymbol{U} + \boldsymbol{T}\bullet\pounds_{\boldsymbol{\xi}}\boldsymbol{U}$ with the arrangement of contractions being the same in each term. From these, and $\pounds_{\boldsymbol{\xi}}\boldsymbol{\eta} = [\boldsymbol{\xi}, \boldsymbol{\eta}]$, the action of $\pounds_{\boldsymbol{\xi}}$ on any tensor follows uniquely.[8] In particular, for a covector α (valence $\left[{0 \atop 1}\right]$),

$$\pounds_{\boldsymbol{\xi}}\boldsymbol{\alpha} = \nabla_{\boldsymbol{\xi}}\boldsymbol{\alpha} + \boldsymbol{\alpha}\bullet(\nabla\boldsymbol{\xi}), \quad \text{i.e. } (\pounds_{\boldsymbol{\xi}}\alpha)_a = \xi^b\nabla_b\alpha_a + \alpha_b\nabla_a\xi^b$$

(∇ being torsion-free); see Fig. 14.17b. For a tensor $\boldsymbol{Q}$ of valence $\left[{1 \atop 2}\right]$, say, we then have (Fig. 14.17c)[14.21]

$$\pounds_{\boldsymbol{\xi}} Q^c_{ab} = \xi^u\nabla_u Q^c_{ab} + Q^c_{ub}\nabla_a\xi^u + Q^c_{au}\nabla_b\xi^u - Q^u_{ab}\nabla_u\xi^c.$$

We note that the Lie derivative, considered as a function both of $\boldsymbol{\xi}$ and of the quantity $\boldsymbol{Q}$ (tensor field) upon which it acts is *independent* of the connection, i.e. it is the same whichever torsion-free operator ∇_a we choose. (This follows because $\pounds_{\boldsymbol{\xi}}$ is uniquely defined from the gradient 'd' operator.) In particular, we could use the coordinate derivative

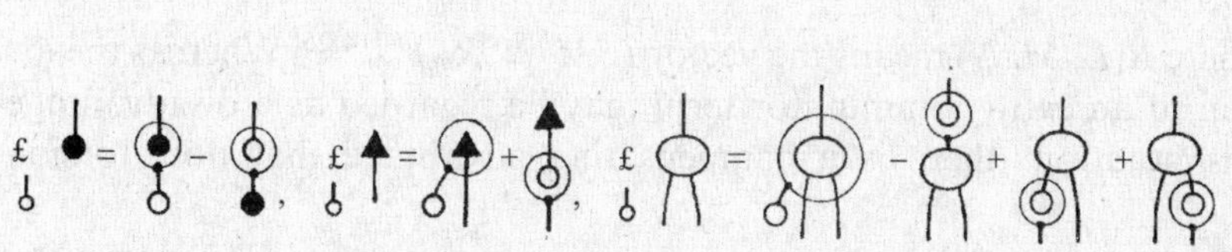

Fig. 14.17 Diagrams for Lie derivative (a) of a vector $\boldsymbol{\eta}$: $(\pounds_{\boldsymbol{\xi}}\boldsymbol{\eta})^a = \xi^a\nabla_a\eta^b - \eta^a\nabla_a\xi^b$; (b) of a covector α: $(\pounds_{\boldsymbol{\xi}}\boldsymbol{\alpha})_a = \xi^b\nabla_b\alpha_a + \alpha_b\nabla_a\xi^b$; and (c) of a ($\left[{1 \atop 2}\right]$-valent) tensor $\boldsymbol{Q}$: $\pounds_{\boldsymbol{\xi}} Q^c_{ab} = \xi^u\nabla_u Q^c_{ab} + Q^c_{ub}\nabla_a\xi^u + Q^c_{au}\nabla_b\xi^u - Q^u_{ab}\nabla_u\xi^c$.

[14.19] Derive this formula for $\pounds_{\boldsymbol{\xi}}\boldsymbol{\eta}$.

[14.20] How does torsion modify the formula of Exercise [14.9]?

[14.21] Establish uniqueness, verifying above covector formula, and give explicitly the Lie derivative of a general tensor.

operator $\partial/\partial x^a$ (in any local coordinate system we choose) in place of ∇_a, and the answer comes out the same. Even if we have a connection with torsion, we could still use it, by expressing it in terms of a second connection, uniquely defined by the given one, which *is* torsion-free, obtained by 'subtracting off' the given connection's torsion.[14.22]

The Lie derivative shares with the exterior derivative (see §12.6) this connection-independent property, whereby for any p-form $\boldsymbol{\alpha}$, with index expression $\alpha_{b...d}$,

$$(\mathrm{d}\boldsymbol{\alpha})_{ab...d} = \nabla_{[a}\alpha_{b...d]},$$

where $\boldsymbol{\nabla}$ is any torsion-free connection; see Fig. 14.18. This is the same expression as in §12.6, except that there the *coordinate connection* $\partial/\partial x^a$ was explicitly used. It is readily seen that the above expression is actually independent of the choice of torsion-free connection.[14.23] Moreover, the key property $\mathrm{d}^2\alpha = 0$ follows immediately from this expression.[14.24] There are also certain other special expressions that are connection-independent in this sense.[9]

Returning, finally, to the question of curvature, on our manifold $\mathcal{M}$, with connection $\boldsymbol{\nabla}$, we find that we need the Lie bracket for the definition of the curvature tensor in the mathematician's notation:

$$\Big(\underset{L}{\boldsymbol{\nabla}}\underset{M}{\boldsymbol{\nabla}} - \underset{M}{\boldsymbol{\nabla}}\underset{L}{\boldsymbol{\nabla}} - \underset{[L,\,M]}{\boldsymbol{\nabla}}\Big)\boldsymbol{N} = \boldsymbol{R}(\boldsymbol{L}, \boldsymbol{M}, \boldsymbol{N}),$$

where $\boldsymbol{R}(\boldsymbol{L}, \boldsymbol{M}, \boldsymbol{N})$ means the vector $L^a M^b N^c R_{abc}{}^d$.[14.25] Whereas the inclusion of an extra commutator term may be regarded as a disadvantage of this notation, there is a compensating advantage that now torsion is

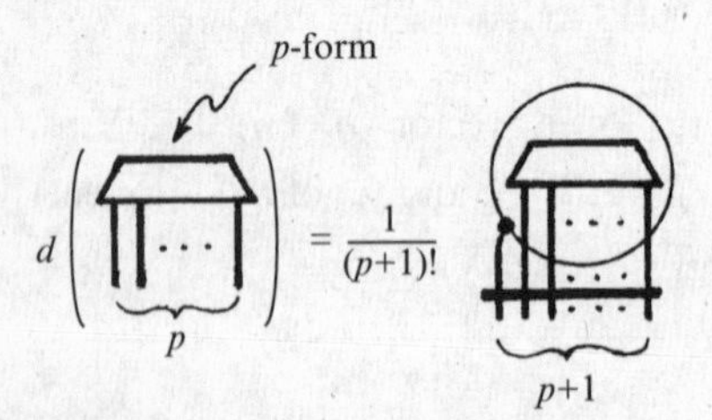

Fig. 14.18 Diagram for exterior derivative of a p-form: $(\mathrm{d}\boldsymbol{\alpha})_{ab...d} = \nabla_{[a}\alpha_{b...d]}$.

[14.22] Show how to find this second connection, taking the '$\boldsymbol{T}$' for the difference between the connections to be antisymmetric in its lower two indices. (See Exercise [14.5].)

[14.23] Establish this and show how the presence of a torsion tensor $\boldsymbol{\tau}$ modifies the expression.

[14.24] Show this.

[14.25] Demonstrate equivalence (if torsion vanishes) to the previous physicist's expression.

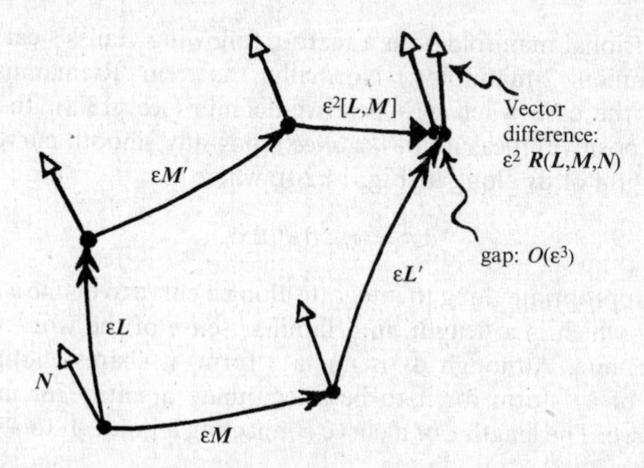

Fig. 14.19 Curvature, in the 'mathematician's notation' $(\underset{L}{\nabla} M - \underset{M}{\nabla} L - \underset{[M,L]}{\nabla})N = \boldsymbol{R}(\boldsymbol{L},\boldsymbol{M},\boldsymbol{N})$, from the $O(\varepsilon^2)$ discrepancy in parallel transport of a vector $\boldsymbol{N}$ around the (incomplete) 'quadrilateral' with sides $\varepsilon\boldsymbol{L}$, $\varepsilon\boldsymbol{M}$, $\varepsilon\boldsymbol{L}'$, $\varepsilon\boldsymbol{M}'$. The Lie bracket contribution $\varepsilon^2[\boldsymbol{L},\boldsymbol{M}]$ fills an $O(\varepsilon^2)$ gap, to order $O(\varepsilon^3)$. (The index form of the vector $\boldsymbol{R}(\boldsymbol{L},\boldsymbol{M},\boldsymbol{N})$ is $L^a M^b N^c R_{abc}{}^d$.)

automatically allowed for (in contrast with torsion needing an extra term in the physicist's notation). Recall the geometrical significance of the commutator term (Fig. 14.14). It allows for an $O(\varepsilon^2)$ 'gap' in the $O(\varepsilon)$ quadrilateral built from the vector fields $\boldsymbol{L}$ and $\boldsymbol{M}$. In fact, there is now the additional advantage that the loop around which we carry our vector $\boldsymbol{N}$ need not be thought of as a 'parallelogram' (to the order previously required), but just as a (curvilinear) quadrilateral. See Fig. 14.19. If $[\boldsymbol{L}, \boldsymbol{M}] = 0$, then this quadrilateral closes (to order $O(\varepsilon^2)$).

14.7 What a metric can do for you

Up to this point, we have been considering that the connection ∇ has simply been *assigned* to our manifold $\mathcal{M}$. This provides $\mathcal{M}$ with a certain type of structure. It is quite usual, however, to think of a connection more as a secondary structure arising from a *metric* defined on $\mathcal{M}$. Recall from §13.8 that a metric (or pseudometric) is a non-singular symmetric $[{}^0_2]$-valent tensor $\boldsymbol{g}$. We require that $\boldsymbol{g}$ be a smooth tensor *field*, so that $\boldsymbol{g}$ applies to the tangent spaces at the various points of $\mathcal{M}$. A manifold with a metric assigned to it in this way is called *Riemannian*, or perhaps *pseudo-Riemannian*.[10] (We have already encountered the great mathematician Bernhard Riemann in Chapters 7 and 8. He originated this concept of an n-dimensional manifold with a metric, following Gauss's earlier study of 'Riemannian' 2-manifolds.) Normally, the term 'Riemannian' is reserved for the case when **g** is positive-definite (see §13.8). In this case there is a (positive) measure of *distance* along any smooth curve, defined by the integral of ds along it (Fig. 14.20), where

$$\mathrm{d}s^2 = g_{ab}\,\mathrm{d}x^a\,\mathrm{d}x^b.$$

This is an appropriate thing to integrate along a curve to define a *length* for the curve—which is a 'length' in a familiar sense of the word when ***g*** is positive definite. Although ds is not a 1-form, it shares enough of the properties of a 1-form for it to be a legitimate quantity for integration along a curve. The length ℓ of a curve connecting a point A, to a point B is thus expressed as[11]

$$\ell = \int_A^B \mathrm{d}s, \qquad \text{where } \mathrm{d}s = (g_{ab}\mathrm{d}x^a\mathrm{d}x^b)^{\frac{1}{2}}.$$

It may be noted that, in the case of Euclidean space, this is precisely the ordinary definition of length of a curve, seen most easily in a Cartesian coordinate system, where the components g_{ab} take the standard 'Kronecker delta' form of §13.3 (i.e. 1 if $a = b$, and 0 if $a \neq b$). The expression for ds is basically a reflection of the Pythagorean theorem (§2.1) as noted in §13.3 (see Exercise [13.11]), but operating at the infinitesimal level. In a general Riemannian manifold, however, the measure of length of a curve, according to the above formula, provides us with a geometry which differs from that of Euclid. This reflects the failure of the Pythagorean theorem for finite (as opposed to infinitesimal) intervals. It is nevertheless remarkable how this ancient theorem still plays its fundamental part—now at the infinitesimal level. (Recall the final paragraph of §2.7.)

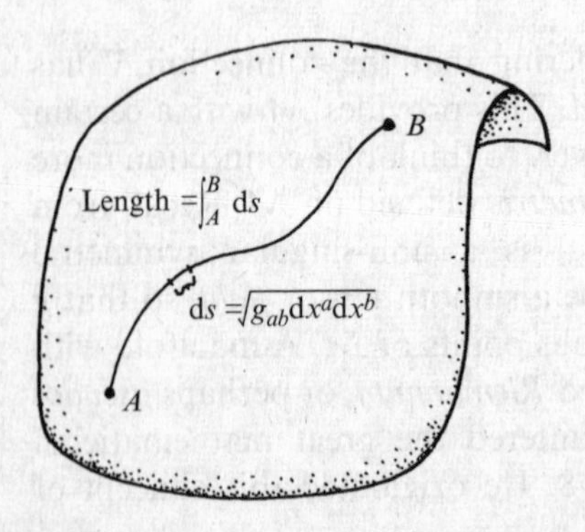

Fig. 14.20 The length of a smooth curve is $\int \mathrm{d}s$, where $\mathrm{d}s^2 = g_{ab}\mathrm{d}x^a\mathrm{d}x^b$.

We shall be seeing in §17.7 that the case of signature $+---$ has particular importance in relativity, where the (pseudo)metric now directly measures time as registered by an ideal clock. Also, any vector $\boldsymbol{v}$ has a *length* $|\boldsymbol{v}|$, defined by

$$|\boldsymbol{v}|^2 = g_{ab}v^a v^b,$$

which, for a positive-definite $\boldsymbol{g}$, is positive whenever $\boldsymbol{v}$ does not vanish. In relativity theory, however, we need a *Lorentzian* metric instead (see §13.8), and $|\boldsymbol{v}|^2$ can be of either sign. We shall see the significance of this later on (§17.9, §18.3).

How does a non-singular (pseudo)metric $\boldsymbol{g}$ uniquely determine a torsion-free connection ∇? One way of expressing the requirement on ∇ is simply to say that the parallel transport of a vector must always preserve its length (a property that I asserted, in §14.2, for parallel transport on the sphere S^2). Equivalently, we can express this requirement as

$$\nabla \boldsymbol{g} = 0.$$

This condition (together with the vanishing of torsion) suffices to fix ∇ completely.[14.26] This connection ∇ is variously termed the *Riemannian*, *Christoffel*, or *Levi-Civita* connection (after Bernhard Riemann (1826–66), Elwin Christoffel (1829–1900), and Tullio Levi-Civita (1873–1941), all of whom contributed important ideas in relation to this notion).[14.27]

There is another way of understanding the fact that a (let us say positive-definite) metric $\boldsymbol{g}$ determines a connection. The notion of a *geodesic* can be obtained directly from the metric. A curve on $\mathcal{M}$ that minimizes its length $\int ds$ (the quantity illustrated in Fig. 14.20) between two fixed points is actually a geodesic for the metric $\boldsymbol{g}$. Knowing the geodesic loci is most of what is needed for knowing the connection ∇. The remaining information needed to fix ∇ completely is a knowledge of the *affine parameters* along the geodesics. These turn out to be the parameters that measure arc length along the curves, and the constant multiples of such parameters, and this is again fixed by $\boldsymbol{g}$.[14.28] When $\boldsymbol{g}$ is not positive definite, the argument is basically the same, but now the

---

[14.26] Derive the explicit component expression $\Gamma^a_{bc} = \frac{1}{2}g^{ad}(\partial g_{bd}/\partial x^c + \partial g_{cd}/\partial x^b - \partial g_{cb}/\partial x^d)$ for the connection quantities Γ^a_{bc} (Christoffel symbols). (See Exercise [14.6]).

[14.27] Derive the classical expression $R_{abc}{}^d = \partial\Gamma^d_{cb}/\partial x^a - \partial\Gamma^d_{ca}/\partial x^b + \Gamma^u_{cb}\Gamma^d_{ua} - \Gamma^u_{ca}\Gamma^d_{ub}$ for the curvature tensor in terms of Christoffel symbols. *Hint*: Use the definition in §14.4 of the curvature tensor, where ξ^d is each of the coordinate vectors $\delta^a_1, \ldots, \delta^a_n$, in turn. (As in Exercise [14.6], the quantities δ^a_1, δ^a_2, etc. are to be thought of as actual individual vectors, where the upper index a may be viewed as an abstract index, in accordance with §12.8).

[14.28] Supply details for this entire argument.

geodesics do not minimize $\int ds$, the integral being what is called 'stationary' for a geodesic. (This issue will be addressed again later; see. §17.9 and §20.1.)

In (pseudo)Riemannian geometry, the metric g_{ab} and its inverse g^{ab} (defined by $g^{ab}g_{bc} = \delta^a_c$) can be used to raise or lower the indices of a tensor. In particular, vectors can be converted to covectors and covectors to vectors (and back again), as in §13.9:

$$v_a = g_{ab}\, v^b \text{ and } \alpha^a = g^{ab}\alpha_b.$$

It is usual to stick to the same kernel symbol (here v and α) and to use the index positioning to distinguish the geometrical character of the quantity. Applying this procedure to lower the upper index of the curvature tensor, we define the *Riemann* or *Riemann–Christoffel tensor*

$$R_{abcd} = R_{abc}{}^{e}\, g_{ed},$$

which has valence $\left[{0 \atop 4}\right]$. It possesses some remarkable symmetries in addition to the two relations (antisymmetry in *ab* and Bianchi symmetry, i.e. vanishing of antisymmetric part in *abc*) that we had before. We also have[14.29] antisymmetry in *cd* and symmetry under interchange of *ab* with *cd*:

$$R_{abcd} = -R_{abdc} = R_{cdab}.$$

See Fig. 14.21 for the diagrammatic representation of these things. A general $\left[{0 \atop 4}\right]$-valent tensor in an n-manifold has n^4 components; but for a Riemann tensor, because of these symmetries, only $\frac{1}{12}n^2(n^2-1)$ of these components are independent.[14.30]

At this point, it is appropriate to bring to the attention of the reader the notion of a *Killing vector* on a (pseudo-)Riemannian manifold $\mathcal{M}$. This is a vector field $\boldsymbol{\kappa}$ which has the property that Lie differentiation with respect to it annihilates the metric:

$$\underset{\boldsymbol{\kappa}}{\pounds}\boldsymbol{g} = 0.$$

This equation can be rewritten in the index notation (with parentheses denoting symmetrization, as in §12.7; see also Fig. 14.21) as

$$\nabla_a\kappa_b + \nabla_b\kappa_a = 0, \text{ i.e. } \nabla_{(a}\kappa_{b)} = 0,$$

[14.29] Establish these relations, first deriving the antisymmetry in *cd* from $\nabla_{[a}\nabla_{b]}g_{cd} = 0$ and then using the two antisymmetries and Bianchi symmetry to obtain the interchange symmetry.

[14.30] Verify that the symmetries allow only 20 independent components when $n = 4$.

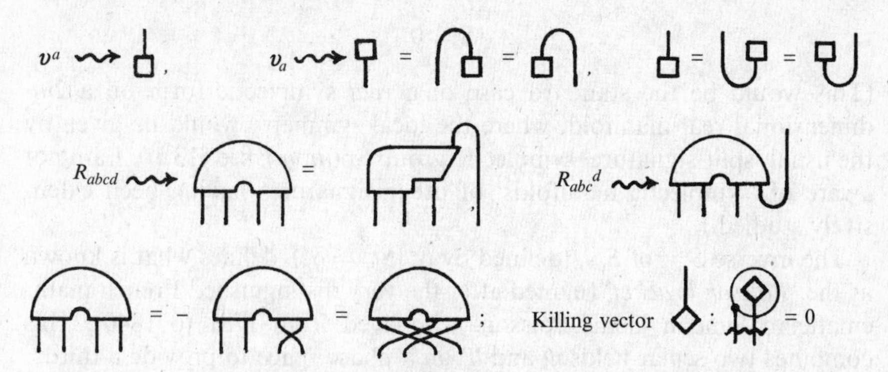

Fig. 14.21 Raising and lowering indices in the 'hoop' notation: $v_a = g_{ab}v^b = v^b g_{ba}$, $v^a = g^{ab}v_b = v_b g^{ba}$, $R_{abcd} = R_{abc}{}^e g_{ed}$, $R_{abc}{}^d = R_{abce}g^{ed}$, $R_{abcd} = -R_{abdc} = R_{cdab}$; κ^a is a Killing vector if $\nabla_{(a}\kappa_{b)} = 0$.

where ∇ is the standard Levi-Civita connection.[14.31] A Killing vector on a (pseudo-)Riemannian manifold $\mathcal{M}$ is the generator of a continuous *symmetry* of $\mathcal{M}$ (which may only be a local[12] symmetry, if $\mathcal{M}$ is non-compact). If $\mathcal{M}$ contains more than one independent Killing vector, then the commutator of the two is a further Killing vector.[14.32] Killing vectors have particular importance in relativity theory, as we shall be seeing in §19.5 and §§30.4,6,7.

14.8 Symplectic manifolds

It should be remarked that there are not many local tensor structures that define a unique connection, so we are fortunate that metrics (or pseudometrics) are often things that are given to us physically. An important family of examples for which this uniqueness is *not* the case, however, is obtained when we have a structure given by a (non-singular) *anti*symmetric tensor field $\boldsymbol{S}$, given by its components S_{ab}. Such a structure is present in the *phase spaces* of classical mechanics (§20.1). I shall have more to say about these remarkable spaces later, in §§20.2,4, §27.3. They are examples of what are known as *symplectic manifolds*. Apart from being antisymmetric and non-singular, the *symplectic structure* $\boldsymbol{S}$ must satisfy[14.33]

[14.31] Derive this equation.

[14.32] Verify this 'geometrically obvious' fact by direct calculation—and why is it 'obvious'?

[14.33] Explain why this can be written $\nabla_a S_{bc} + \nabla_b S_{ca} + \nabla_c S_{ab} = 0$, using any torsion-free connection ∇.

$$\mathrm{d}\boldsymbol{S} = 0.$$

(This would be the standard case of a *real* symplectic form on a $2m$-dimensional real manifold, where the local symmetry would be given by the usual 'split-signature' symplectic group $\mathrm{Sp}(m, m)$; see §13.10. I am not aware of 'symplectic manifolds' of other signatures having been extensively studied.)

The inverse S^{ab}, of S_{ab}, (defined by $S^{ab}S_{bc} = \delta^a_c$), defines what is known as the '*Poisson bracket*' (named after the very distinguished French mathematician Siméon Denis Poisson, who lived from 1781 to 1840). This combines two scalar fields Φ and Ψ on a phase space to provide a third:

$$\{\Phi, \Psi\} = -\tfrac{1}{2}S^{ab}\nabla_a\Phi\nabla_b\Psi$$

(where the factor $-\frac{1}{2}$ is inserted merely for consistency with the conventional coordinate expressions). This is an important quantity in classical mechanics. We shall be seeing later (in §20.4) how it encodes *Hamilton's equations*, these equations providing a fundamental general procedure that encompasses the dynamics of classical physics and supplies the link to quantum mechanics. The antisymmetry of $\boldsymbol{S}$ and the condition $\mathrm{d}\boldsymbol{S} = 0$ provide us with the elegant relations[14.34]

$$\{\Phi, \Psi\} = -\{\Psi, \Phi\}, \quad \{\Theta, \{\Phi, \Psi\}\} + \{\Phi, \{\Psi, \Theta\}\} + \{\Psi, \{\Theta, \Phi\}\} = 0.$$

This may be compared with the corresponding commutator (Lie bracket) identities of §14.6. (Recall the Jacobi identity.) We shall return to the remarkably rich geometry of symplectic manifolds when we consider the geometrical description of classical mechanics in §20.4.

The local structure of a symplectic manifold is an example of what might be called a 'floppy' structure. There is, for example, no notion of curvature for a symplectic manifold, which might serve to distinguish one symplectic manifold from another, locally. If we have two real symplectic manifolds of the same dimension (and the same 'signature', cf. §13.10), then they are locally completely identical (in the sense that for any point p in one manifold and any point q in the other, there are open sets of p and q that are identical[13]). This is in stark contrast with the case of (pseudo-) Riemannian manifolds, or manifolds in which merely a connection is specified. In those cases, the curvature tensor (and, for example, its various covariant derivatives) defines some distinguishing local structure which is likely to be different for different such manifolds.

There are other examples of such 'floppy' structures, among them being the complex structure defined in §12.9 which enables a $2m$-dimensional real manifold to be re-interpreted as an m-dimensional complex manifold.

[14.34] Demonstrate these relations, first establishing that $S^{a[b}\nabla_a S^{cd]} = 0$.

In this case the floppiness is evident, because there is clearly no feature, apart from the complex dimension m, which locally distinguishes one complex manifold from another (or from $\mathbb{C}^m$). It would still remain floppy if a complex (holomorphic) symplectic structure were assigned to it[14.35] (and now we do not even have to worry about a notion of 'signature' for the complex S_{ab}; see §13.10).

Many other examples of floppy structures can be specified. One such would be a real manifold with a nowhere vanishing vector field on it. On the other hand, a real manifold with *two* general vector fields on it would not be floppy.[14.36] The issue of floppiness has some importance for *twistor theory*, as we shall be seeing in §33.11.

Notes

Section 14.2

14.1. In fact there is a topological reason that there can be no way whatever of assigning a 'parallel' to $\boldsymbol{v}$ at all points of S^2 in a continuous way (the problem of 'combing the hair of a spherical dog'!). The analogous statement for S^3 is not true, however, as the construction of Clifford parallels (given in §15.4) shows.

Section 14.3

14.2. In much of the physics literature and older mathematics literature, the coordinate derivative $\partial/\partial x^a$ is indicated by appending a lower index a, preceded by a comma, to the right-hand end of the list of indices attached to the quantity being differentiated. In the case of ∇_a, a semicolon is frequently used in place of the comma.

The '∇_a' notation works well with the abstract–index notation (§12.8) and the subsequent equations in the main text of this book can (should) be read in this way. Coordinate expressions can also be powerfully treated in this notation, but two distinguishable types of index are needed, component and abstract (see Penrose 1968a; Penrose and Rindler 1984).

Section 14.4

14.3. The index staggering is needed for when a metric is introduced (§14.7) since spaces are needed for the raising and lowering of indices.

Section 14.5

14.4. Strictly, $\boldsymbol{\nabla}$ acts on fields defined on $\mathcal{M}$, not just along curves lying within $\mathcal{M}$. But this equation makes sense because the operator differentiates only in the direction along the curve. If we like, we may think of the region of definition of $\boldsymbol{t}$ as being extended smoothly outwards away from γ into $\mathcal{M}$ in some arbitrary way. The precise way in which this is done is irrelevant, since it is only along γ that we are asking for the equation on $\boldsymbol{t}$ to hold.

14.5. See, for example, Nayfeh (1993); Simmonds and Mann (1998).

[14.35] Explain why.

[14.36] Explain why, in each case. *Hint*: Construct a coordinate system with $\boldsymbol{\xi} = \partial/\partial x^1$; then take repeated Lie derivatives to construct a frame, etc.

Section 14.6

14.6. We see the explicit role of the Lie algebra of commutators in the *Baker–Campbell–Hausdorff* formula, the first few terms of which are given explicitly in $e^{\boldsymbol{\xi}}e^{\boldsymbol{\eta}} = e^{\boldsymbol{\xi}+\boldsymbol{\eta}+\frac{1}{2}[\boldsymbol{\xi},\boldsymbol{\eta}]+\frac{1}{12}([\boldsymbol{\xi},[\boldsymbol{\xi},\boldsymbol{\eta}]]+[[\boldsymbol{\xi},\boldsymbol{\eta}],\boldsymbol{\eta}])+\cdots}$, where the continuation dots stand for a further expression in multiple commutators of $\boldsymbol{\xi}$ and $\boldsymbol{\eta}$, i.e. an element of the Lie algebra generated by $\boldsymbol{\xi}$ and $\boldsymbol{\eta}$.

14.7. Somewhat more precisely, we can choose coordinates $x^2, x^3, \ldots, x^n$ constant along this curve, with $x^1 = t$; then $\boldsymbol{\xi} = \partial/\partial t$, along the curve. It is simply Taylor's theorem (§6.4) that tells us that the above prescription gives $e^{t\boldsymbol{\xi}}(\Phi)$.

14.8. Analogous to the exponentiation $e^{t\boldsymbol{\xi}}$ of $\boldsymbol{\xi}$, which obtains the value of a scalar quantity Φ a finite distance away, there is a corresponding expression with $\pounds_{\boldsymbol{\xi}}$ in place of $\boldsymbol{\xi}$, to obtain a tensor $\boldsymbol{Q}$ a finite distance away, as measured against a 'dragged' reference frame.

14.9. See Schouten (1954); Penrose and Rindler (1984), p. 202.

Section 14.7

14.10. In some mathematical books the term 'semi-Riemannian' has been used for the indefinite case (see O'Neill 1983), but it seems to me that 'pseudo-Riemannian' is a more appropriate terminology.

14.11. A common way to give meaning to this expression is to introduce a parameter, say u, along the curve and to write $ds = (ds/du)du$. The quantity ds/du is an ordinary function of u, expressed in terms of dx^a/du.

14.12. This 'locality' can be understood in the following sense. For each point p of $\mathcal{M}$, there is an exponentiation (§14.6) of some small constant non-zero multiple of $\boldsymbol{\kappa}$ that takes some open set containing p into some other open set in $\mathcal{M}$ with an identical metric structure.

Section 14.8

14.13. Here, 'identical' refers to the fact that each can be mapped to the other in such a way that the symplectic structures correspond.

15
Fibre bundles and gauge connections

15.1 Some physical motivations for fibre bundles

THE machinery introduced in Chapters 12 and 14 is sufficient for the treatment of Einstein's general relativity and for the phase spaces of classical mechanics. However, a good deal of the modern theory of particle interactions depends upon a generalization of the specific notion of 'connection' (or covariant derivative) that was introduced in §14.3, this generalization being referred to as a *gauge connection*. Basically, our original notion of covariant derivative was based upon what we mean by the parallel transport of a vector along some curve in our manifold $\mathcal{M}$ (§14.2). Knowing parallel transport for vectors, we can uniquely extend this to the transport of any tensor quantity (§14.3). Now, vectors and tensors are quantities that refer to the tangent spaces at points of $\mathcal{M}$ (see §12.3, §14.1, and Fig. 12.6). But a gauge connection refers to 'parallel transport' of certain quantities of particular physical interest that are best thought of as referring to some kind of 'space' other than the tangent space at a point p in $\mathcal{M}$, but still to be thought of as being, in a sense, 'located at the point p'.

To clarify, a little, what is needed here, we recall from §§12.3,8 that once we have a vector space—here the space of tangent vectors at a point—we can construct its dual (space of covectors) and all the various spaces of $[{}^{p}_{q}]$-valent tensors. Thus, in a clear sense, the spaces of $[{}^{p}_{q}]$-tensors (including the cotangent spaces, covectors being $[{}^{0}_{1}]$-tensors) are 'not anything new', once we have the tangent spaces T_p at points p. (An almost similar remark would apply—at least according to my own way of viewing things—to the spaces of *spinors* at p; see §11.3. Some others might try to take a different attitude to spinors; but these alternative perspectives on the matter will not be of concern for us here.) The spaces that we need for the gauge theories of particle interactions (other than gravity), are different from these (and so they *are* something new), and it is best to think of them as referring to a kind of 'spatial' dimension that is additional to those of ordinary space and time. These extra 'spatial' dimensions are frequently referred to as *internal* dimensions, so that moving along in such an 'internal direction' does not actually carry us away from the spacetime point at which we are situated.

To make geometrical sense of this idea, we need the notion of a *bundle*. This is a perfectly precise mathematical notion, and we shall be coming to it properly in §15.2. It had been found to be useful in pure mathematics[1] long before physicists realized that some of the important notions that they had been previously using were actually to be understood in bundle terms. In subsequent years, theoretical physicists have become very familiar with the required mathematical concepts and have incorporated them into their theories. However, in some modern theories, these notions are presented in a modified form, in relation to which spacetime itself is thought of as acquiring extra dimensions.

Indeed, in many (or most?) of the current attempts at finding a deeper framework for fundamental physics (e.g. supergravity or string theory), the very notion of 'spacetime' is extended to higher dimensionality. The 'internal dimensions' then come about through the agency of these extra spatial dimensions, where these extra spatial dimensions are put on an essentially equal footing with those of ordinary space and time. The resulting 'spacetime' thus acquires more dimensions than the standard four. Ideas of this nature go back to about 1919, when Theodor Kaluza and Oskar Klein provided an extension of Einstein's general relativity in which the number of spacetime dimensions is increased from 4 to 5. The extra dimension, enables Maxwell's superb theory of electromagnetism (see §§19.2,4) to be incorporated, in a certain sense, into a 'spacetime geometrical description'. However, this '5th dimension' has to be thought of as being 'curled up into a tiny loop' so that we are not directly aware of it as an ordinary spatial dimension.

The analogy is often presented of a hosepipe (see Fig. 15.1), which is to represent a Kaluza–Klein-type modification of a 1-dimensional universe. When looked at on a large scale, the hosepipe indeed looks 1-dimensional: the dimension of its length. But when examined more closely, we find that the hosepipe surface is actually 2-dimensional, with the extra dimension looping tightly around on a much smaller scale than the length of the hosepipe. This is to be taken as the direct analogy of how we would perceive only a 4-dimensional *physical* spacetime in a 5-dimensional Kaluza–Klein *total* 'spacetime'. The Kaluza–Klein 5-space is to be the direct analogue of the hosepipe 2-surface, where the 4-spacetime that we actually perceive is the direct analogue of the basically 1-dimensional appearance of the hosepipe.

In many ways, this is an appealing idea, and it is certainly an ingenious one. The proponents of the modern speculative physical theories (such as supergravity and string theory that we shall encounter in Chapter 31) actually find themselves driven to consider yet higher-dimensional versions

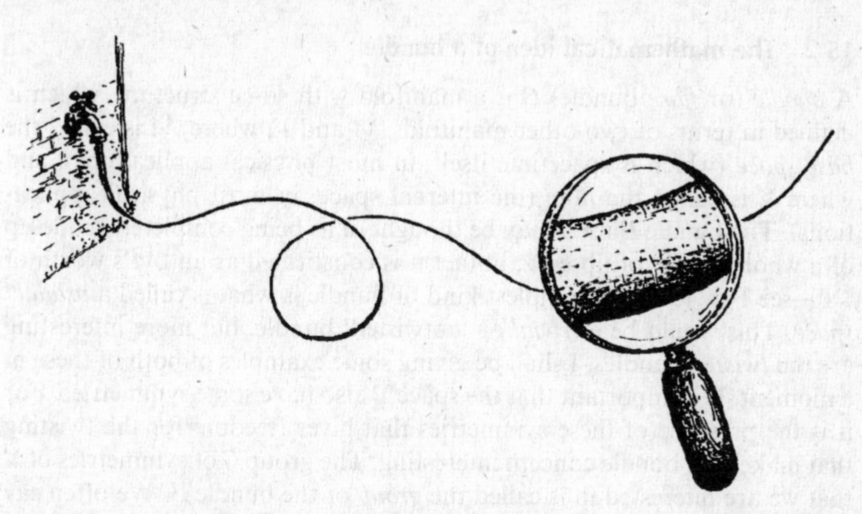

Fig. 15.1 The analogy of a hosepipe. Viewed on a large scale, it appears 1-dimensional, but when examined more minutely it is seen to be a 2-dimensional surface. Likewise, according to the Kaluza–Klein idea, there could be 'small' extra spatial dimensions unobserved on an ordinary scale.

of the Kaluza–Klein idea (a total dimensionality of 26, 11, and 10 having been among the most popular). In such theories, it is perceived that interactions other than electromagnetism can be included by use of the gauge-connection idea that we shall be coming to shortly.

However, it must be emphasized that the Kaluza–Klein idea is still a speculative one. The 'internal dimensions' that the conventional current gauge theories of particle interactions depend upon are not to be thought of as being on a par with ordinary spacetime dimensions, and therefore do not arise from a Kaluza–Klein-type scheme. It is a matter of interesting speculation whether it is sensible to regard the internal dimensions of current gauge theories as ultimately arising from this kind of (Kaluza–Klein-type) 'extended spacetime', in any significant sense.[2] I shall return to this matter later (§31.4).

Instead of regarding these internal dimensions as being part of a higher-dimensional spacetime, it will be more appropriate to think of them as providing us with what is called a *fibre bundle* (or simply a *bundle*) over spacetime. This is an important notion that is central to the modern gauge theories of particle interactions. We imagine that 'above' each point of spacetime is another space, called a *fibre*. The fibre consists of all the internal dimensions, according to the physical picture referred to above. But the bundle concept has much broader applications than this, so it will be best if we do not necessarily tie ourselves to this kind of physical interpretation, at least for the time being.

15.2 The mathematical idea of a bundle

A *bundle* (or *fibre* bundle) $\mathcal{B}$ is a manifold with some structure, which is defined in terms of two other manifolds $\mathcal{M}$ and $\mathcal{V}$, where $\mathcal{M}$ is called the *base space* (which is spacetime itself, in most physical applications), and where $\mathcal{V}$ is called the *fibre* (the internal space, in most physical applications). The bundle $\mathcal{B}$ itself may be thought of as being completely made up of a whole family of fibres $\mathcal{V}$; in fact it is constituted as an '$\mathcal{M}$'s worth of $\mathcal{V}$s'—see Fig. 15.2. The simplest kind of bundle is what is called a *product space*. This would be a *trivial* or 'untwisted' bundle, but more interesting are the *twisted* bundles. I shall be giving some examples of both of these in a moment. It is important that the space $\mathcal{V}$ also have some symmetries. For it is the presence of these symmetries that gives freedom for the twisting that makes the bundle concept interesting. The group $\mathcal{G}$ of symmetries of $\mathcal{V}$ that we are interested in is called the *group* of the bundle $\mathcal{B}$. We often say that $\mathcal{B}$ is a $\mathcal{G}$ *bundle over* $\mathcal{M}$. In many situations, $\mathcal{V}$ is taken to be a vector space, in which case we call the bundle a *vector bundle*. Then the group $\mathcal{G}$ is the general linear group of the relevant dimension, or a subgroup of it (see §§13.3,6–10).

We are not to think of $\mathcal{M}$ as being a part of $\mathcal{B}$ (i.e. $\mathcal{M}$ is not inside $\mathcal{B}$); instead, $\mathcal{B}$ is to be viewed as a separate space from $\mathcal{M}$, which we tend to regard as standing, in some sense, *above* the base space $\mathcal{M}$. There are many copies of the fibre $\mathcal{V}$ in the bundle $\mathcal{B}$, one entire copy of $\mathcal{V}$ standing above each point of $\mathcal{M}$. The copies of the fibres are all disjoint (i.e. no two intersect), and together they make up the entire bundle $\mathcal{B}$. The way to think of $\mathcal{M}$ in relation to $\mathcal{B}$ is as a *factor space* of the bundle $\mathcal{B}$ by the family of fibres $\mathcal{V}$. That is to say, each point of $\mathcal{M}$ corresponds precisely to a separate individual copy of $\mathcal{V}$. There is a continuous map from $\mathcal{B}$ down

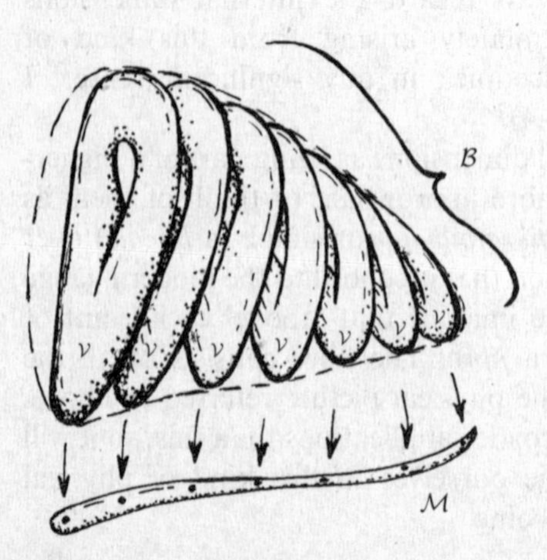

Fig. 15.2 A bundle $\mathcal{B}$, with base space $\mathcal{M}$ and fibre $\mathcal{V}$ may be thought of as constituted as an '$\mathcal{M}$'s worth of $\mathcal{V}$s'. The canonical projection from $\mathcal{B}$ down to $\mathcal{M}$ may be viewed as the collapsing of each fibre $\mathcal{V}$ down to a single point.

to $\mathcal{M}$, called the *canonical projection* from $\mathcal{B}$ to $\mathcal{M}$, which collapses each entire fibre $\mathcal{V}$ down to that particular point of $\mathcal{M}$ which it stands above. (See Fig. 15.2.)

The *product space* of $\mathcal{M}$ with $\mathcal{V}$ (trivial bundle of $\mathcal{V}$ over $\mathcal{M}$) is written $\mathcal{M}\times\mathcal{V}$. The points of $\mathcal{M}\times\mathcal{V}$ are the *pairs* of elements (a, b), where a belongs to $\mathcal{M}$ and b belongs to $\mathcal{V}$; see Fig. 15.3a. (We already saw the same idea applied to groups in §13.2.)[3] A more general 'twisted' bundle $\mathcal{B}$, over $\mathcal{M}$, resembles $\mathcal{M}\times\mathcal{V}$ *locally*, in the sense that the part of $\mathcal{B}$ that lies over any sufficiently small open region of $\mathcal{M}$, is identical in structure with that part of $\mathcal{M}\times\mathcal{V}$ lying over that same open region of $\mathcal{M}$. See Fig. 15.3b. But, as we move around in $\mathcal{M}$, the fibres above may twist around so that, as a whole, $\mathcal{B}$ is different (often topologically different) from $\mathcal{M}\times\mathcal{V}$. The dimension of $\mathcal{B}$ is always the sum of the dimensions of $\mathcal{M}$ and $\mathcal{V}$, irrespective of the twisting.[15.1]

All this may well be confusing, so to get a better feeling for what a bundle is like, let me give an example. First, take our space $\mathcal{M}$ to be a circle S^1, and the fibre $\mathcal{V}$ to be a 1-dimensional vector space (which we can picture topologically as a copy of the real line $\mathbb{R}$, with the origin 0 marked). Such a bundle is called a (real) *line bundle* over S^1. Now $\mathcal{M}\times\mathcal{V}$ is a 2-dimensional cylinder; see Fig. 15.4a. How can we construct a *twisted* bundle $\mathcal{B}$, over $\mathcal{M}$,

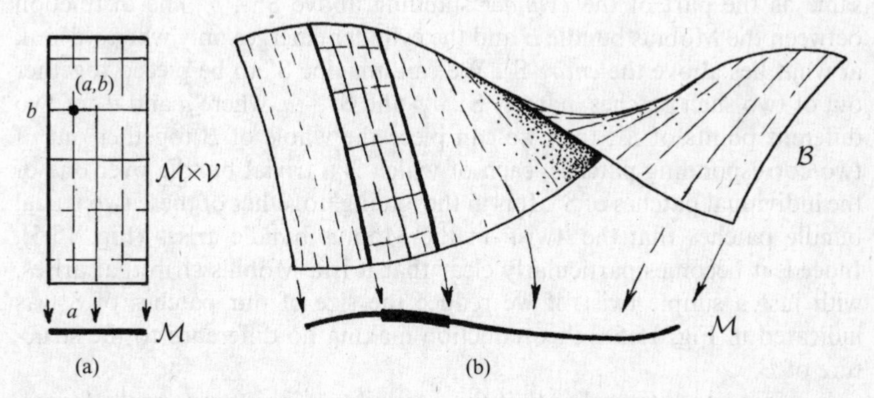

Fig. 15.3 (a) The particular case of a 'trivial' bundle, which is the product space $\mathcal{M}\times\mathcal{V}$ of $\mathcal{M}$ with $\mathcal{V}$. The points of $\mathcal{M}\times\mathcal{V}$ can be interpreted as pairs of elements (a,b), with a in $\mathcal{M}$ and b in $\mathcal{V}$. (b) The general 'twisted' bundle $\mathcal{B}$, over $\mathcal{M}$, with fibre $\mathcal{V}$, resembles $\mathcal{M}\times\mathcal{V}$ locally—i.e. the part of $\mathcal{B}$ over any sufficiently small open region of $\mathcal{M}$ is identical to that part of $\mathcal{M}\times\mathcal{V}$ over the same region of $\mathcal{M}$. But the fibres twist around, so that $\mathcal{B}$ is globally not the same as $\mathcal{M}\times\mathcal{V}$.

[15.1] Explain why the dimension of $\mathcal{M}\times\mathcal{V}$ is the sum of the dimensions of $\mathcal{M}$ and of $\mathcal{V}$.

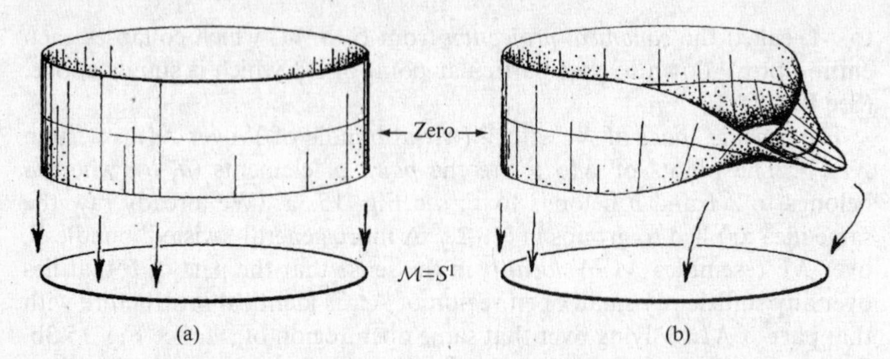

Fig. 15.4 To understand how this twisting can occur, consider the case when $\mathcal{M}$ is a circle S^1 and the fibre $\mathcal{V}$ is a 1-dimensional vector space (i.e. a space modelled on $\mathbb{R}$, but where only the origin 0 is marked, but no other value (such as the identity element 1). (a) The trivial case $\mathcal{M}\times\mathcal{V}$, which is here an ordinary 2-dimensional cylinder. (b) In the twisted case, we get a Möbius strip (as in Fig. 12.15).

with fibre $\mathcal{V}$? We can take a *Möbius strip*; see Fig. 15.4b (and Fig. 12.15). Let us see why this is a bundle—'locally' the same as the cylinder. We can produce an adequately 'local' region of the base space S^1 by removing a point p from S^1. This breaks the base circle into a simply-connected[4] segment[5] $S^1 - p$, and the part of $\mathcal{B}$ lying above such a segment is just the same as the part of the *cylinder* standing above $S^1 - p$. The distinction between the Möbius bundle $\mathcal{B}$ and the cylinder emerges only when we look at what lies above the *entire* S^1. We can imagine S^1 to be pieced together out of two such patches, namely $S^1 - p$ and $S^1 - q$, where p and q are two different points of S^1; then we can piece the whole of $\mathcal{B}$ together out of two corresponding patches, each of which is a trivial bundle over one of the individual patches of S^1. It is in the 'gluing' together of these two trivial bundle patches that the 'twist' in the Möbius bundle arises (Fig. 15.5). Indeed, it becomes particularly clear that it is a Möbius strip that arises, with just a simple twist, if we reduce the size of our patches of S^1, as indicated in Fig. 15.5b, this reduction making no difference to the structure of $\mathcal{B}$.

It is important to realize that the possibility of this twist results from a particular *symmetry* that the fibre $\mathcal{V}$ possess, namely the one which reverses the sign of the elements of the 1-dimensional vector space $\mathcal{V}$. (This is $\boldsymbol{v} \mapsto -\boldsymbol{v}$, for each $\boldsymbol{v}$ in $\mathcal{V}$.) This operation preserves the structure of $\mathcal{V}$ as a vector space. We should note that this operation is not actually a symmetry of the real-number system $\mathbb{R}$. In fact, $\mathbb{R}$ itself possesses no symmetries at all. (The number 1 is certainly different from -1, for example, and $x \mapsto -x$ is *not* a symmetry of $\mathbb{R}$, not preserving the

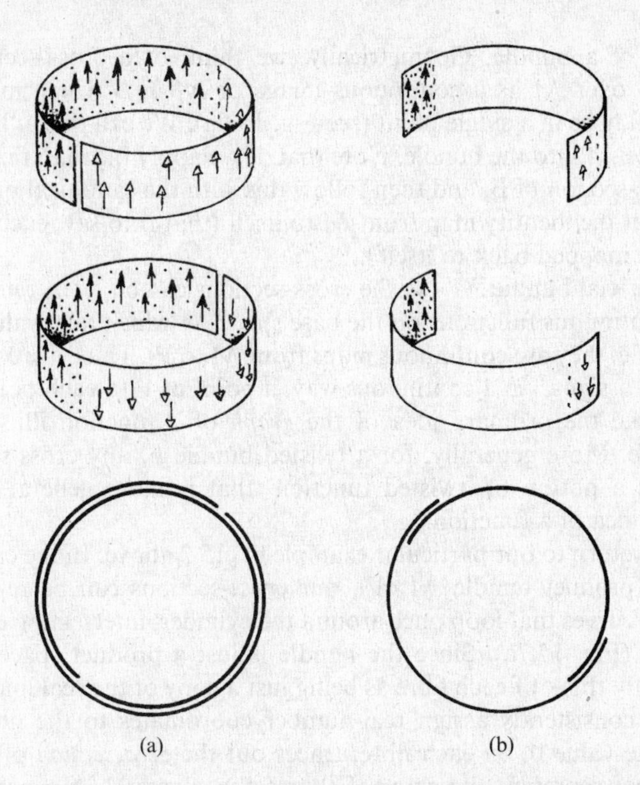

Fig. 15.5 (a) We can produce an adequately 'local' (simply-connected) region of the base S^1 by removing a point p from it, the part of the bundle above $S^1 - p$ being just a product. The same applies to the part of $\mathcal{B}$ above $S^1 - q$ where q is a different point of S^1. We get a cylinder if we can match the two parts of $\mathcal{B}$ directly, but we get the Möbius bundle, as illustrated above, if we apply an up/down reflection (a symmetry of $\mathcal{V}$) to one of the two matched portions. (b) The resulting Möbius strip is little more obvious if we reduce the size of the two parts of S^1 so that there are only small regions of overlap.

multiplicative structure of $\mathbb{R}$.[15.2]) It is for this reason that $\mathcal{V}$ is taken as a 1-dimensional real vector space rather than just *as* the real line $\mathbb{R}$ itself. We sometimes say that $\mathcal{V}$ is *modelled* on the real line. We shall be seeing shortly how other fibre symmetries provide opportunities for other kinds of twist.

15.3 Cross-sections of bundles

One way that we can characterize the difference between the cylinder and the Möbius bundle is in terms of what are called *cross-sections* (or simply

[15.2] Explain this.

sections) of a bundle. Geometrically, we think of a cross-section of a bundle $\mathcal{B}$ over $\mathcal{M}$ as a continuous image of $\mathcal{M}$ in $\mathcal{B}$ which meets each individual fibre in a single point (see Fig. 15.6a). We call this a 'lift' of the base space $\mathcal{M}$ into the bundle. Note that, if we apply the map that lifts $\mathcal{M}$ to a cross-section of $\mathcal{B}$, and then follow this with the canonical projection, we just get the identity map from $\mathcal{M}$ to itself (that is to say, each point of $\mathcal{M}$ is just mapped back to itself).

For a trivial bundle $\mathcal{M}\times\mathcal{V}$, the cross-sections can be interpreted simply as the continuous functions on the base space $\mathcal{M}$ which take values in the space $\mathcal{V}$ (i.e. they are continuous maps from $\mathcal{M}$ to $\mathcal{V}$). Thus, a cross-section of $\mathcal{M}\times\mathcal{V}$ assigns,[6] in a continuous way, a point of $\mathcal{V}$ to each point of $\mathcal{M}$. This is like the ordinary idea of the *graph* of a function illustrated in Fig. 15.6b. More generally, for a twisted bundle $\mathcal{B}$, any cross-section of $\mathcal{B}$ defines a notion of 'twisted function' that is more general than the ordinary idea of a function.

Let us return to our particular example in §15.2 above. In the case of the cylinder (product bundle $\mathcal{M}\times\mathcal{V}$), our cross-sections can be represented simply as curves that loop once around the cylinder, intersecting each fibre just once (Fig. 15.7a). Since the bundle is just a product space, we can consistently think of each fibre as being just a copy of the real line, and we can thus consistently assign real-number coordinates to the fibres. The coordinate value 0, on each fibre, traces out the *zero section* of 'marked points' that represent the *zeros* of the vector spaces $\mathcal{V}$. A general cross-section provides a continuous real-valued function on the circle (the 'height' above the zero section being the value of the function at each point of the circle). Clearly there are many cross-sections that do not

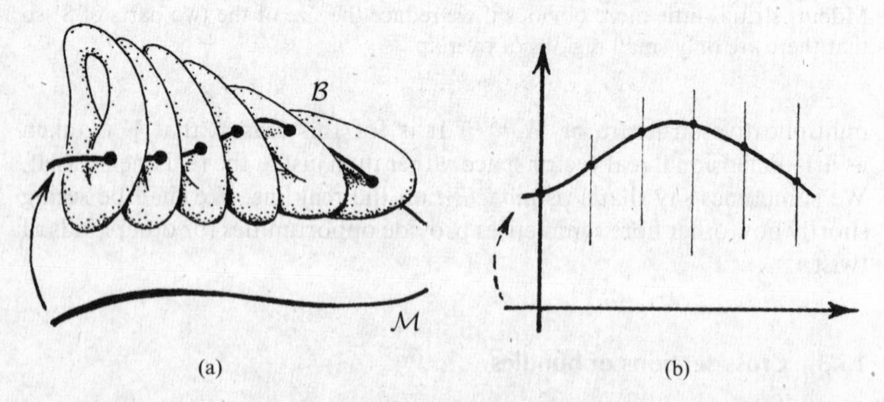

Fig. 15.6 (a) A cross-section (or section) of a bundle $\mathcal{B}$ is a continuous image of $\mathcal{M}$ in $\mathcal{B}$ which meets each individual fibre in single point. (b) This generalizes the ordinary idea of the graph of a function.

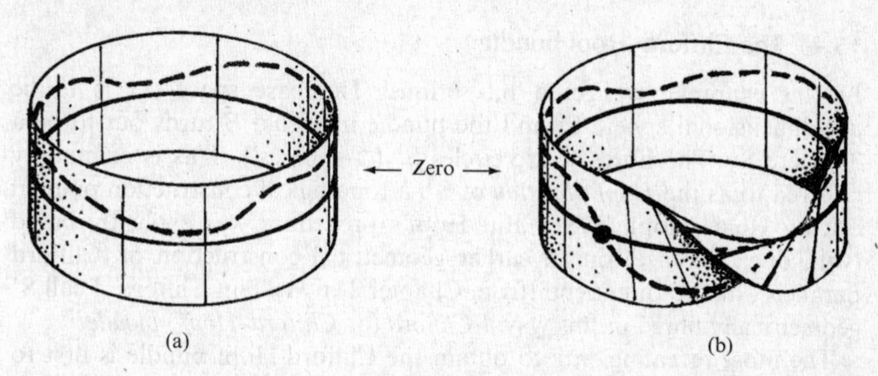

Fig. 15.7 A (cross-)section of a line bundle over S^1 is a loop that goes once around, intersecting each fibre just once. (a) Cylinder: there are sections that nowhere intersect the zero section. (b) Möbius bundle: every section intersects the zero section.

intersect the zero section (non-vanishing functions on S^1). For example, we can choose a section of the cylinder that is parallel to the zero section but not coincident with it. This represents a *constant* non-zero function on the circle.

However, when we consider the Möbius bundle $\mathcal{B}$, we find that things are very different. The reader should not find it hard to accept that now every cross-section of $\mathcal{B}$ must intersect the zero section (Fig. 15.7b). (The notion of zero section still applies, since $\mathcal{V}$ is a vector space, with its zero 'marked'.) This qualitative difference from the previous case makes it clear that $\mathcal{B}$ must be topologically distinct from $\mathcal{M} \times \mathcal{V}$. To be a bit more specific, we can begin to assign real-number coordinates to the various fibres $\mathcal{V}$, just as before, but we need to adopt a convention that, at some point of the circle, the sign has to be 'flipped' ($x \mapsto -x$), so that a cross-section of $\mathcal{B}$ corresponds to a real-valued function on the circle that would be continuous except that it changes sign when the circle is circumnavigated. Any such cross-section must take the value zero somewhere.[15.3]

In this example, the nature of the family of cross-sections is sufficient to distinguish the Möbius bundle from the cylinder. An examination of the family of cross-sections often leads to a useful way of distinguishing various different bundles over the same base space $\mathcal{M}$. The distinction between the Möbius bundle and the product space (cylinder) is a little less extreme than in the case of certain other examples of bundles, however. Sometimes a bundle has no cross-sections at all! Let us consider a particularly important and famous such example next.

[15.3] Spell this argument out, using the construction of B from two patches, as indicated above.

15.4 The Clifford–Hopf bundle

In this example, we get a bit serious! The base space $\mathcal{M}$ is to be a 2-dimensional sphere S^2 and the bundle manifold $\mathcal{B}$ turns out to be a 3-sphere S^3. The fibres $\mathcal{V}$ are circles S^1 ('1-spheres'). This is commonly referred to as the *Hopf fibration* of S^3, a topological construction pointed out by Heinz Hopf (1931). But Hopf's procedure was explicitly based (with due reference) on an earlier geometrical construction of 'Clifford parallels', due to our friend (from Chapter 11) William Clifford. I call S^3 geometrically fibred in this way a *Clifford* (or *Clifford–Hopf*) *bundle.*

The most revealing way to obtain the Clifford-Hopf bundle is first to consider the space $\mathbb{C}^2$ of pairs of complex numbers (w, z). (The relevant structure of $\mathbb{C}^2$, here, is simply that it is a 2-dimensional complex vector space; see §12.9.) Our bundle space B ($= S^3$) is to be thought of as the unit 3-sphere S^3 sitting in $\mathbb{C}^2$, as defined by the equation (see the end of §10.1)

$$|w|^2 + |z|^2 = 1.$$

This stands for the real equation $u^2 + v^2 + x^2 + y^2 = 1$, the equation of a 3-sphere, where $w = u + iv$ and $z = x + iy$ are the respective expressions of w and z in terms of their real and imaginary parts. (This is in direct analogy with the equation of an ordinary 2-sphere $x^2 + y^2 + z^2 = 1$ in Euclidean 3-space with real Cartesian coordinates x, y, z.)

To obtain the fibration, we are going to consider the family of *complex straight lines* through the origin (i.e. complex 1-dimensional vector subspaces of $\mathbb{C}^2$). Each such line is given by an equation of the form

$$Aw + Bz = 0,$$

where A and B are complex numbers (not both zero). Being a 1-complex-dimensional vector space, this line is a copy of the complex plane, and it meets S^3 in a circle S^1, which we can think of as the unit circle in that plane (Fig. 15.8). These circles are to be our fibres $\mathcal{V} = S^1$. The different lines can meet only at the origin, so no two distinct S^1s can have a point in common. Thus, this family of S^1s indeed constitute fibres giving S^3 a bundle structure.

What is the base space $\mathcal{M}$? Clearly, we get the same line $Aw+Bz=0$ if we multiply both A and B by the same non-zero complex number, so it is really the ratio $A : B$ that distinguishes the lines from one another. Either of A or B can be zero, but not both. The space of such ratios is the *Riemann sphere* as described at some length in §8.3. We are thus to identify the base space $\mathcal{M}$ of our bundle as this Riemann sphere S^2. Accordingly, we can

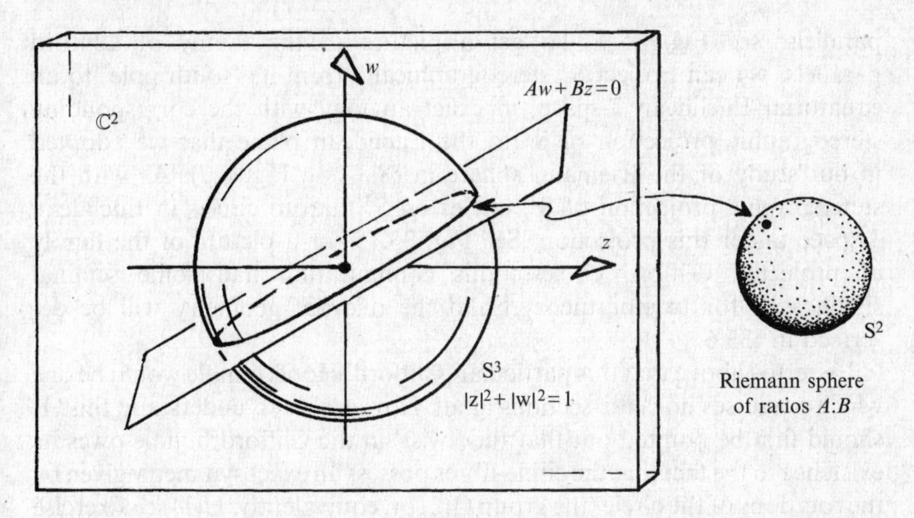

Fig. 15.8 The Clifford–Hopf bundle. Take $\mathbb{C}^2$ with coordinates (w,z), containing the 3-sphere $\mathcal{B} = S^3$ given by $|w|^2 + |z|^2 = 1$. Each fibre $\mathcal{V} = S^1$ is the unit circle in a complex straight line through the origin $Aw+Bz=0$ (complex 1-dimensional vector subspace of $\mathbb{C}^2$), and is determined by the ratio $A{:}B$. The Riemann sphere S^2 of such ratios is the base space $\mathcal{M}$.

see that S^3 may be regarded as an S^1 bundle over S^2. (We must not expect such a relation as this for other dimensions, if we require bundle, base space, and fibre all to be spheres. However, it actually turns out that S^7 may be viewed as an S^3 bundle over S^4, as can be obtained (with care) by replacing the complex numbers w and z in the above argument by quaternions;[15.4] also, S^{15} can be regarded as an S^7 bundle over S^8, where w and z are now replaced by octonions (see §11.2 and §16.2); but this does not work for any other higher-dimensional sphere.[7]

This family of circles in S^3, called *Clifford parallels*, is a particularly interesting one. The circles, which are great circles, twist around each other, remaining the same distance apart all along (which is why they are referred to as 'parallels'). Any two of the circles are linked, so they are *skew* (not co-spherical). In Euclidean 3-space, straight lines that are skew (not coplanar) have the property that they get farther apart from one another as they move out towards infinity. The 3-sphere, however, has positive curvature, so that the Clifford circles, which are geodesics in S^3, have a compensating tendency to bend towards each other in accordance with the geodesic deviation effect considered in §14.5 (see Fig. 14.12). These two effects exactly compensate one another in the case of Clifford

[15.4] Carry out this argument. Can you see how to do the S^{15} case?

parallels; see Fig. 15.9. To get a picture of the family of Clifford parallels, we can project S^3 stereographically from its 'south pole' to an equatorial Euclidean 3-space, in exact analogy with the corresponding stereographic projection of S^2 to the Euclidean plane that we adopted in our study of the Riemann sphere in §8.3 (see Fig. 8.7). As with the stereographic projection of S^2, circles on S^3 map to circles in Euclidean 3-space under this projection. See Fig. 33.15 for a picture of the family of projected Clifford circles. This configuration had some seminal significance for twistor theory,[8] and the relevant geometry will be described in §33.6.

I asserted above that this particular (Clifford–Hopf) bundle would be one which possesses no cross-sections at all. How are we to understand this? It should first be pointed out that the 'twist' in the Clifford bundle owes its existence to the fact that the circle-fibres possess an exact symmetry given by the rotations of the circle (the group O(2) or, equivalently, U(1) see Exercise [13.59]). We cannot identify each of these fibres with some specifically given circle, such as the unit circle in the complex plane $\mathbb{C}$. If we could, then we could consistently choose some specific point on the circle (e.g. the point 1 on the unit circle in $\mathbb{C}$) and thereby obtain a cross-section of the Clifford bundle. The non-existence of cross-sections can occur because the Clifford circles are only modelled on the unit circle in $\mathbb{C}$, not identified with it.

Of course, this in itself does not tell us why the Clifford bundle has no continuous cross-sections. To understand this it will be helpful to look at the Clifford bundle in another way. In fact, it turns out that each point of our sphere S^3 can be interpreted as a unit-length 'spinorial' tangent vector to S^2 at one of its points.[15.5] Recall from §11.3 that a spinorial object is a

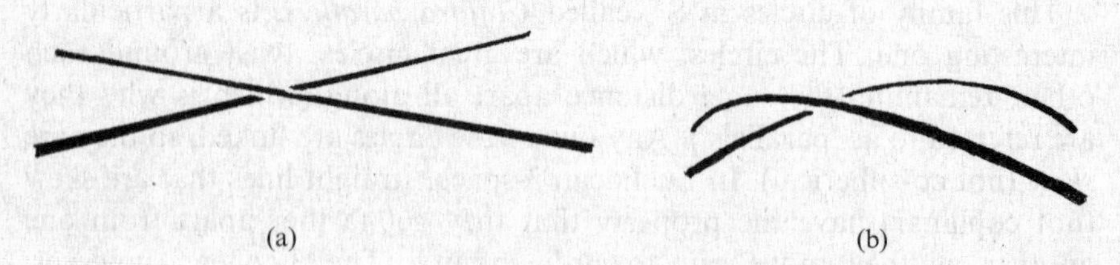

Fig. 15.9 (a) In Euclidean 3-space, skew straight lines get increasingly distant from each other as they go off. (b) In S^3, the positive curvature provides a compensating tendency to bend geodesics (great circles) towards each other (by geodesic deviation; see Fig. 14.12). For Clifford parallels the compensation is exact.

[15.5] Show this. *Hint*: Take the tangent vector to be $u\partial/\partial v - v\partial/\partial u + x\partial/\partial y - y\partial/\partial x$.

quantity which, when completely rotated through 2π, becomes the negative of what it was originally. According to the above statement, a cross-section of our bundle $\mathcal{B}$ (= S^3) would represent a continuous field of such spinorial unit vectors on $\mathcal{M}$ (= S^2). Now, it is a well-known topological fact that there is no global continuous field of ordinary unit tangent vectors on S^2. (This is the problem of combing the hair of a 'spherical dog'! It is impossible for the hairs to lie flat in a continuous way, all over the sphere.) Making these directions 'spinorial' clearly does not help, so no global continuous field of unit spinorial tangent vectors can exist either. Hence our bundle $\mathcal{B}$ (= S^3) has no cross-sections.

This deserves some further discussion, for there is a good deal more to be gained from this example. In the first place, we can obtain the actual bundle $\mathcal{B}'$ of unit tangent vectors to S^2 by slightly modifying the Clifford bundle described above. Since any ordinary unit tangent vector has just two manifestations as a spinorial object (one being the 'negative' of the other), we must identify these two if we wish to pass from the spinorial vector to the ordinary vector. What this means, in terms of the Clifford bundle $\mathcal{B}$ (= S^3), is that two points of S^3 must be identified in order to give a single point[9] of the bundle $\mathcal{B}'$ of unit vectors to S^2. The pairs of points of S^3 that must be identified are the antipodal points on this 3-sphere. See Fig. 15.10. The fibres of $\mathcal{B}'$ are still circles. It is just that each circle-fibre of $\mathcal{B}$ (= S^3) 'wraps around twice' each circle-fibre of $\mathcal{B}'$. Each point of $\mathcal{B}'$ now represents a point of S^2 with a unit tangent vector at that point. In fact, the space $\mathcal{B}'$ is topologically identical with the space $\mathcal{R}$ that we encountered in §12.1, and which represents the different spatial orientations of an

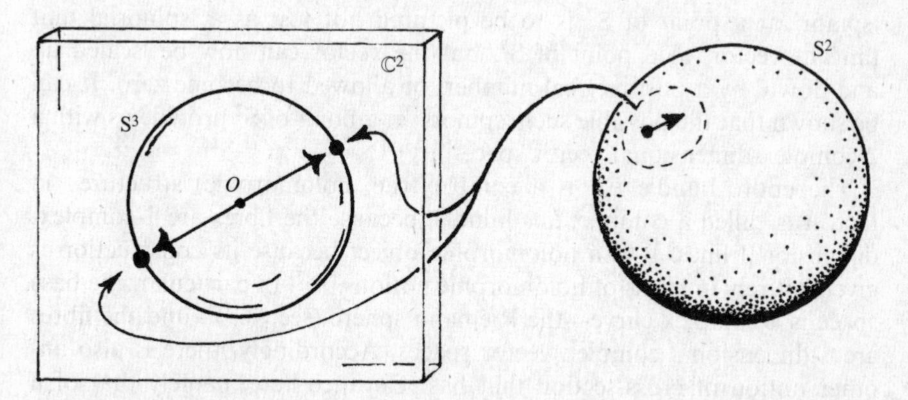

Fig. 15.10 The bundle $\mathcal{B}'$ of unit tangent vectors to S^2 is a slight modification of the Clifford bundle, where antipodal points of S^3 are identified. Without this identification, we obtain S^3 as the (Clifford–Hopf) bundle $\mathcal{B}$ of *spinorial* tangent vectors to S^2. The fibres of $\mathcal{B}'$ are still circles, but each circle-fibre of $\mathcal{B}$ wraps twice around each circle-fibre of $\mathcal{B}'$.

object (such as the book, considered in §11.3) in Euclidean 3-space. This is made evident if we think of our 'object' to be the sphere S^2 with an arrow (unit tangent vector) marked on it at one of its points. This marked arrow will completely fix the spatial orientation of the sphere.

15.5 Complex vector bundles, (co)tangent bundles

A slight extension of the idea behind the Clifford bundle (and also of $\mathcal{B}'$) gives us a good example of a *complex vector bundle*, in this case, a bundle that I shall call $\mathcal{B}^{\mathbb{C}}$ (or correspondingly $\mathcal{B}'^{\mathbb{C}}$). Each of the lines $Aw+Bz=0$ is itself a 1-dimensional complex vector space. (The entire line consists of the family of multiples of a single vector (w, z) by complex numbers λ, where (w, z) multiplies to $(\lambda w, \lambda z)$.) We now think of this complex vector 1-space as our fibre $\mathcal{V}$. The Riemann sphere S^2 is our base space $\mathcal{M}$, just as before.

There is one further thing that we need to do in order to get the correct complex vector bundle $\mathcal{B}^{\mathbb{C}}$, however. In $\mathbb{C}^2$, the different fibres are not disjoint, all having the origin (0, 0) in common. Thus, to get $\mathcal{B}^{\mathbb{C}}$, we must modify $\mathbb{C}^2$ by replacing the origin by a copy of the entire Riemann sphere ($\mathbb{CP}^1$; see §15.6), so that instead of having just one zero, we have a whole Riemann sphere's worth of zeros, one for each fibre, giving the *zero section* of the bundle (see Fig. 15.11). This procedure is known as *blowing up* the origin of $\mathbb{C}^2$ (an important idea for algebraic geometry, complex-manifold theory, string theory, twistor theory, and many other areas). Since we are now allowed zero on the fibres, there do exist continuous cross-sections of $\mathcal{B}$. It turns out that these cross-sections represent the *spinor fields* on S^2. A 'spinor' at a point of S^2 is to be pictured not just as a 'spinorial unit tangent vector' at a point of S^2, but the vector can now be 'scaled up and down' by a positive real number, or allowed to become zero. It can be shown that the possible such 'spinors' at a point of S^2 provide us with a 2-complex-dimensional vector space.[10,[15.6]]

The entire bundle $\mathcal{B}^{\mathbb{C}}$ is a complex (i.e. holomorphic) structure—in fact, it is called a complex *line* bundle, because the fibres are 1-complex-dimensional lines. It is a holomorphic object because its construction is given entirely in terms of holomorphic notions.[[15.7]] In particular, the base space is a complex curve—the Riemann sphere (see §8.3)—and the fibres are 1-dimensional complex vector spaces. Accordingly, there is also another notion of cross-section that has relevance here, namely that of a *holomorphic* cross-section. A holomorphic cross-section is a cross-section of a complex bundle that is itself a complex submanifold of the bundle

[15.6] Why does every such spinor field take the value zero at at least one point of S^2?

[15.7] Explain this in detail.

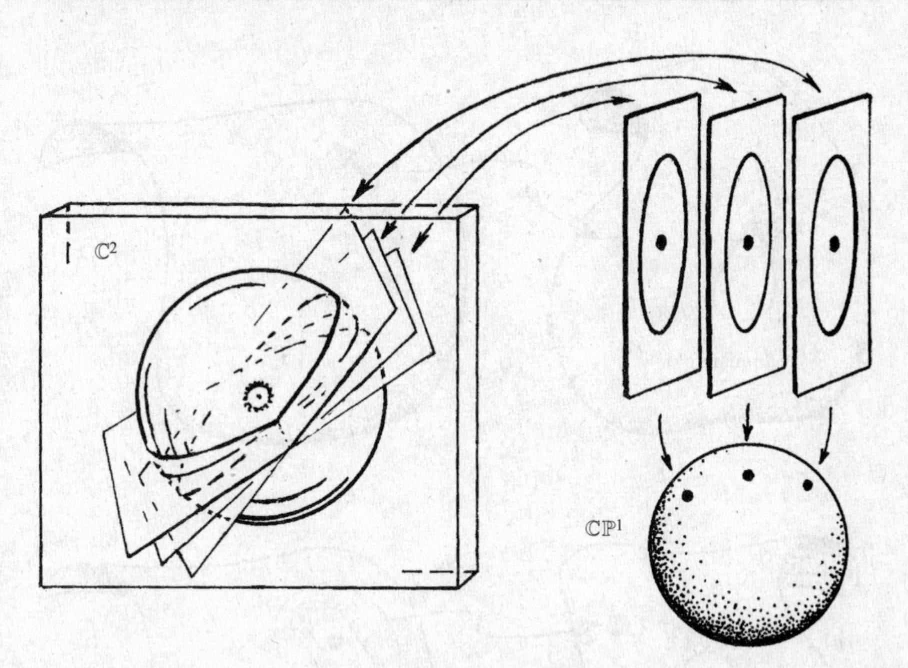

Fig. 15.11 By taking the entire line $Aw + Bz = 0$ (a complex plane), rather than just its unit circle, we get an example of a complex line bundle $\mathcal{B}^{\mathbb{C}}$, the fibre $\mathcal{V}$ being now a complex 1-dimensional vector space. The Riemann sphere $S^2 = \mathbb{CP}^1$ (also a complex manifold, see §8.3, §15.6) is still the base space $\mathcal{M}$. But to make the different fibres disjoint, we must 'blow up' the origin (0,0), replacing it with an entire Riemann sphere, giving us a Riemann sphere's worth of zeros.

(which just means that it is given locally by holomorphic equations). Sometimes, in the case of a complex line bundle, such a cross-section is referred to as a *twisted* holomorphic function on the base space. Such things have considerable importance in many areas of pure mathematics and mathematical physics.[11] They also play a particular role in twistor theory (see §33.8). Holomorphic sections constitute a tightly controlled but important family. In the case of $\mathcal{B}^{\mathbb{C}}$, it turns out that there are no (global) holomorphic sections other than the *zero* section (i.e. zero everywhere).

In a minor modification of this construction (corresponding to the passage from $\mathcal{B}$ to $\mathcal{B}'$) we obtain vector fields, rather than spinor fields, on S^2. The appopriate bundle $B'^{\mathbb{C}}$ can again be interpreted as a complex vector bundle—in fact it is what is called the *square* of the vector bundle $\mathcal{B}^{\mathbb{C}}$. It is constructed in just the same way as $\mathcal{B}^{\mathbb{C}}$, except that we now *identify* each point (w, z) with its 'antipodal' point $(-w, -z)$, multiplication of (w, z) by the complex number λ now being given by $(\lambda^{1/2}w, \lambda^{1/2}z)$ (rather than by $(\lambda w, \lambda z)$).

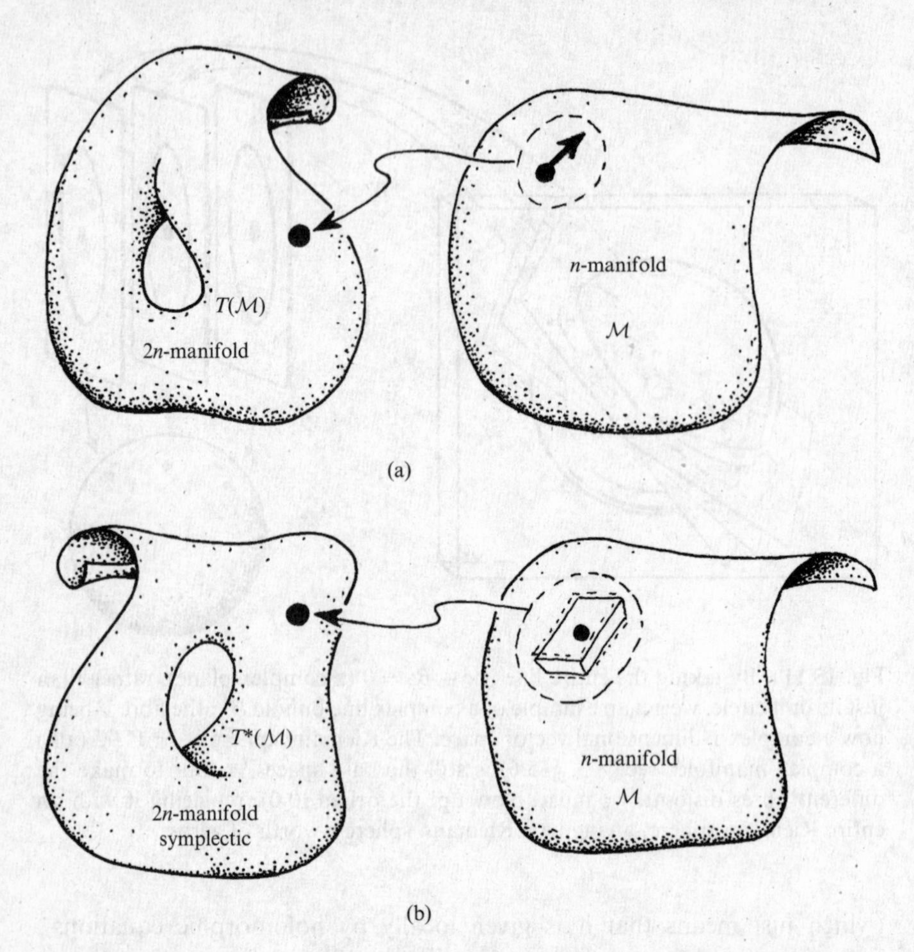

Fig. 15.12 (a) For a general manifold $\mathcal{M}$, each point of its tangent bundle $T(\mathcal{M})$ represents a point of $\mathcal{M}$ together with a tangent vector to $\mathcal{M}$ there. A cross-section of $T(\mathcal{M})$ represents a vector field on $\mathcal{M}$. (b) The cotangent bundle $T^*(\mathcal{M})$ is similar, but with covectors instead of vectors. Cotangent bundles are always symplectic manifolds.

To end this section, I should point out that the bundle $\mathcal{B}'^{\mathbb{C}}$ can be loosely re-interpreted, in real terms, as what is called the *tangent bundle* $T(S^2)$ of S^2. The tangent bundle $T(\mathcal{M})$ of a general manifold $\mathcal{M}$ is that space each of whose points represents a point of $\mathcal{M}$ together with a tangent vector to $\mathcal{M}$ at that point. See Fig. 15.12a.[15.8] A cross-section of $T(\mathcal{M})$ represents a vector field on $\mathcal{M}$. A notion of perhaps even greater physical importance is that of the *cotangent bundle* $T^*(\mathcal{M})$ of a manifold $\mathcal{M}$, each of whose points represents a point of $\mathcal{M}$, together with a *covector* at that point (Fig. 15.12b). In

[15.8] Show that $\mathcal{B}'^{\mathbb{C}}$, interpreted as a real bundle over S^2 is indeed the same as $T(S^2)$. *Hint*: Re-examine Exercise [15.5].

Chapter 20, we shall be glimpsing something of the importance of these ideas. Cross-sections of $T^*(\mathcal{M})$ represent covector fields on $\mathcal{M}$. It turns out that the cotangent bundles are always symplectic manifolds (see §14.8, §§20.2,4), a fact of considerable importance for classical mechanics. We can also correspondingly define various kinds of tensor bundles. A tensor field may be interpreted as a cross-section of such a bundle.

15.6 Projective spaces

Another important notion, associated with a general vector space, is that of a *projective space*. The vector space itself is 'almost' a bundle over the projective space. If we remove the origin of the vector space, then we do get a bundle over the projective space, the fibre being a line with the origin removed; alternatively, as with the particular example of $\mathcal{B}^{\mathbb{C}}$ given above, in §15.5, we can 'blow up' the origin of the vector space. (I shall come back to this in a moment.) Projective spaces have a considerable importance in mathematics and have a particular role to play in the geometry of quantum mechanics (see §21.9 and §22.9)—and also in twistor theory (§33.5). It is appropriate, therefore, that I comment on these spaces briefly here.

The idea of a projective space appears to have come originally from the study of perspective in drawing and painting, this being taken within the context of Euclidean geometry. Recall that, in the Euclidean plane, two distinct lines always intersect unless they are parallel. However, if we draw a picture, on a vertical piece of paper, of a pair of parallel lines receding into the distance on a horizontal plane (say of the boundaries of a straight road), then we find that in the drawing, the lines appear to intersect at a 'vanishing point' on the horizon (see Fig. 15.13). Projective geometry takes these vanishing points seriously, by adjoining 'points at infinity' to the Euclidean plane which enable parallel lines to intersect at these additional points.

There are many theorems about lines in ordinary Euclidean 3-space which are awkward to state because of exceptions having to be made for parallel lines. In Fig. 15.14, I depict two remarkable examples, namely the theorems of Pappos[12] (found in the late 3rd century AD) and of Desargues (found in 1636). In each case, the theorem (which I am stating in 'converse' form) asserts that if all the straight lines indicated in the diagram (9 lines for Pappos and 10 for Desargues) intersect in triples at all but one of the points marked with black spots (there being 9 black spots in all for Pappos and 10 in all for Desargues), then the triple of lines indicated as intersecting at the remaining black spot do in fact have a point in common. However, stated in this way, these theorems are true only if we consider

Fig. 15.13 Projective geometry adjoins 'points at infinity' to the Euclidean plane enabling parallel lines to intersect there. In the artist's picture, painted on a vertical canvas, a pair of horizontal parallel lines receding into the distance—the boundaries of a straight horizontal road—appear to intersect at a 'vanishing point' on the horizon.

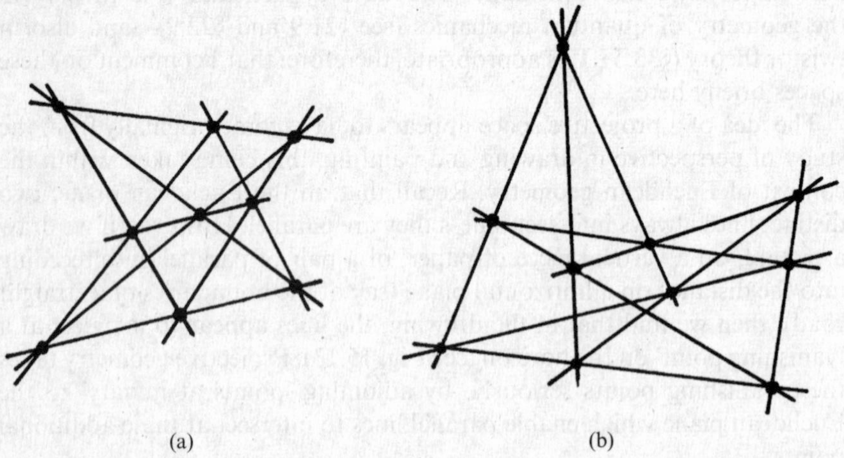

Fig. 15.14 Configurations of two famous theorems of plane projective geometry: (a) that of Pappos, with 9 lines and 9 marked points, and (b) of Desargues, with 10 lines and 10 marked points. In each case, the assertion is that if each but one of the marked points is the intersection of a triple of the lines, then the remaining marked point occurs in this way also.

that a triple of mutually *parallel* lines are counted as having a point in common, namely a 'point at infinity'. With this interpretation, the theorems remain true when the lines are parallel. They also remain true even if one of the lines lies entirely at infinity. Thus, the theorems of Pappos and Desargues are more properly theorems in projective geometry than in Euclidean geometry.

How do we construct an n-dimensional projective space $\mathbb{P}^n$? The most immediate way is to take an $(n+1)$-dimensional vector space $\mathbf{V}^{n+1}$, and regard our space $\mathbb{P}^n$ as the space of the 1-dimensional vector subspaces of $\mathbf{V}^{n+1}$. (These 1-dimensional vector subspaces are the lines through the origin of $\mathbf{V}^{n+1}$.) A straight line in $\mathbb{P}^n$ (which is itself an example of a $\mathbb{P}^1$) is given by a 2-dimensional subspace of $\mathbf{V}^{n+1}$ (a plane through the origin), the collinear points of $\mathbb{P}^n$ arising as lines lying in such a plane (Fig. 15.15). There are also higher-dimensional flat subspaces of $\mathbb{P}^n$, these being projective spaces $\mathbb{P}^r$ contained in $\mathbb{P}^n$ $(r < n)$. Each $\mathbb{P}^r$ corresponds to an $(r+1)$-dimensional vector subspace of $\mathbf{V}^{n+1}$.

This construction (in the case $n=2$) formalizes the procedures of perspective in pictorial representation; for we can consider the artist's eye to be situated at the origin O of the vector space $\mathbf{V}^3$, this space representing the artist's ambient Euclidean 3-space. A light ray through O (artist's eye) is viewed by the artist as a single point. Thus, the artist's 'field of vision', taken as the totality of such light rays, can be thought of as a projective plane $\mathbb{P}^2$. (See Fig. 15.15 again.) Any straight line in space (not through O), that the artist perceives, corresponds to the plane joining that line to O, in accordance with the definition of a 'straight line' in $\mathbb{P}^2$, as given above.

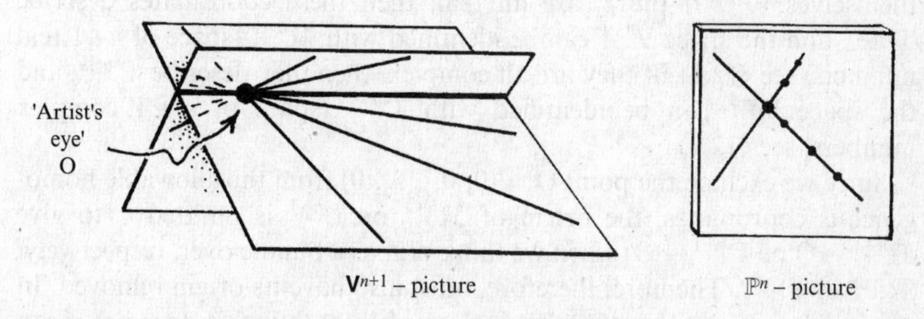

Fig. 15.15 To construct n-dimensional projective space $\mathbb{P}^n$, take an $(n+1)$-dimensional vector space $\mathbf{V}^{n+1}$, and regard $\mathbb{P}^n$ as the space of the 1-dimensional vector subspaces of $\mathbf{V}^{n+1}$ (lines through the origin of $\mathbf{V}^{n+1}$). A straight line in $\mathbb{P}^n$ is given by a 2-dimensional subspace of $\mathbf{V}^{n+1}$ (plane through origin), collinear points of $\mathbb{P}^n$ arising as lines through O in such a plane. This applies both to the real case ($\mathbb{RP}^n$) and the complex case ($\mathbb{CP}^n$). The geometry of $\mathbb{RP}^2$ formalizes the procedures of perspective in pictorial representation: consider the artist's eye to be at the origin O of $\mathbf{V}^3$, taking $\mathbf{V}^3$ as the artist's ambient Euclidean 3-space. A light ray through O is viewed by the artist as single point. What the artist depicts as a 'straight line' ($\mathbb{RP}^1$ in $\mathbb{RP}^2$) (on any particular choice of artist's canvas) indeed corresponds to the plane ($\mathbf{V}^2$) joining that line to O. Pairs of planes through O always intersect, even when joining parallel lines in $\mathbf{V}^3$ to O. (For example, the two bottom boundary lines in the left-hand picture play the role of the road boundaries of Fig. 15.13.)

Imagine that the artist paints an accurate picture of the perceived scene on some canvas that coincides with some particular flat plane (not through O). Any such plane will capture only part of the entire $\mathbb{P}^2$. It will certainly not intersect those light rays that are parallel to it. But several such planes will provide an adequate 'patchwork' covering the whole of $\mathbb{P}^2$ (three will suffice[13,[15.9]]). Parallel lines in one such plane, will be depicted as lines with a common *vanishing point* in another.

We can consider either real projective spaces, $\mathbb{P}^n = \mathbb{RP}^n$, or complex ones, $\mathbb{P}^n = \mathbb{CP}^n$. We have already considered one example of a complex projective space, namely the *Riemann sphere*, which is $\mathbb{CP}^1$. Recall that the Riemann sphere arises as the space of ratios of pairs of complex numbers (w, z), not both zero, which is the space of complex lines through the origin in $\mathbb{C}^2$. (See Fig. 15.8.) More generally, any projective space can be assigned what are called *homogeneous* coordinates. These are the coordinates $z^0, z^1, z^2, \ldots, z^n$ for the $(n+1)$-dimensional vector space $\mathbf{V}^{n+1}$ from which $\mathbb{P}^n$ arises, but the 'homogeneous coordinates' for $\mathbb{P}^n$ are the n independent *ratios*

$$z^0 : z^1 : z^2 : \ldots : z^n$$

(where the zs are not all zero), rather than the values of the individul zs themselves.[15.10] If the z^r are all real, then these coordinates describe $\mathbb{RP}^n$, and the space $\mathbf{V}^{n+1}$ can be identified with $\mathbb{R}^{n+1}$ (space of $n+1$ real numbers; see §12.2). If they are all complex, then they describe $\mathbb{CP}^n$, and the space $\mathbf{V}^{n+1}$ can be identified with $\mathbb{C}^{n+1}$ (space of $n+1$ complex numbers; see §12.9).

Since we exclude the point $O = (0; 0, \ldots, 0)$ from the allowable homogeneous coordinates, the origin of $\mathbb{R}^{n+1}$ or $\mathbb{C}^{n+1}$ is omitted[14] (to give $\mathbb{R}^{n+1} - O$ or $\mathbb{C}^{n+1} - O$) when we think of it as a bundle over, respectively, $\mathbb{RP}^n$ or $\mathbb{CP}^n$. The fibre, therefore, must also have its origin removed. In the real case, this splits the fibre into two pieces (but this does not mean that the bundle splits into two pieces; in fact, $\mathbb{R}^{n+1} - O$ is connected, when $n > 0$).[15.11] In the complex case, the fibre is $\mathbb{C} - O$ (often written $\mathbb{C}^*$), which is connected. In either case, we may prefer to reinstate the origin in the fibre, so that we get a vector bundle. But if we do this, then this amounts to more than simply putting the origin back into $\mathbb{R}^{n+1}$ or $\mathbb{C}^{n+1}$. As with the particular case of $\mathbb{C}^2$, considered above, we must put

---

[15.9] Explain how to do this. *Hint*: Think of Cartesian coordinates (x, y, z). Take two at a time, with the canvas given by the third set to unity.

[15.10] Explain why there are n independent ratios. Find $n+1$ sets of n ordinary coordinates (constructed from the zs), for $n+1$ different coordinate patches, which together cover $\mathbb{P}^n$.

[15.11] Explain this geometry, showing that the bundle $\mathbb{R}^{n+1} - O$ over $\mathbb{RP}^n$ can be understood as the composition of the bundle $\mathbb{R}^{n+1} - O$ over S^n (the fibre, $\mathbb{R}^+$, being the positive reals) and of S^n as a twofold cover of $\mathbb{RP}^n$.

back the origin in each fibre separately, so that the origin is 'blown up'. The bundle space becomes $\mathbb{R}^{n+1}$ with an $\mathbb{RP}^n$ inserted in place of O, or $\mathbb{C}^{n+1}$ with a $\mathbb{CP}^n$ in place of O.

In the complex case, we can also consider the unit $(2n+1)$-sphere S^{2n+1} in $\mathbb{C}^{n+1}$, just as we did in the particular case $n = 1$ when constructing the Clifford bundle. Each fibre intersects S^{2n+1} in a circle S^1, so now we obtain S^{2n+1} as an S^1 bundle over $\mathbb{CP}^n$. This structure underlies the geometry of quantum mechanics—although this beautiful geometrical fact impinges only infrequently on the thinking of quantum physicists—where we shall find that the space of physically distinct quantum states, for an $(n+1)$-state system, is a $\mathbb{CP}^n$. In addition, there is a quantity known as the *phase*, which is normally thought of as being a complex number of unit modulus ($e^{i\theta}$, with θ real; see §5.3), whereas it is really a *twisted* unit-modulus complex number.[15] These matters will be returned to at the end of this chapter, and when we consider quantum mechanics in earnest in Chapters 21 and 22 (see §21.9, §22.9).

15.7 Non-triviality in a bundle connection

I have just taken the reader on a whirlwind tour of some important fibre-bundle and bundle-related concepts! Some of the geometry and topology involved is rather intricate, so the reader should not be disconcerted if it all seems a little bewildering. Let us now return to something much simpler—in the sense that we do not need so many dimensions (at first, at least!) in order to get the idea across. Although my next example of a bundle is indeed a very simple one, it expresses an important subtlety involved in the bundle notion that we have not encountered before. In all the bundles considered above, the non-triviality of the bundle was revealed in some topological feature of the geometry, the 'twist' being of a topological character. However, it is perfectly possible for a bundle to be non-trivial in an important sense, despite being topologically trivial.

Let us return to our original example, where the base space $\mathcal{M}$ is an ordinary circle S^1 and the fibre $\mathcal{V}$ is a 1-dimensional real vector space. We shall now construct our bundle $\mathcal{B}$ in a somewhat different way from the simple 'flipping over' of the fibre $\mathcal{V}$, when we circumnavigate $\mathcal{M}$, that gave us the Möbius bundle. Instead, let us give it a *stretch* by a factor of 2. This is depicted in Fig. 15.16. This exploits a different symmetry of a 1-dimensional real vector space from the 'flip' symmetry $\boldsymbol{v} \mapsto -\boldsymbol{v}$ used in the Möbius bundle. The 'stretch' transformation $\boldsymbol{v} \mapsto 2\boldsymbol{v}$ preserves the vector-space structure of $\mathcal{V}$ just as well. Now, the topology of the bundle is not the issue. Topologically, we simply have a cylinder $S^1 \times \mathbb{R}$, just as in our first example of Fig. 15.4a, but now there is a different kind of

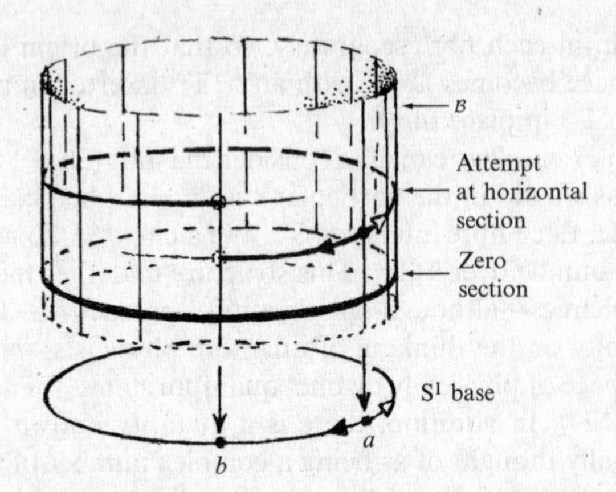

Fig. 15.16 A 'strained' line bundle $\mathcal{B}$ over $\mathcal{M}=\mathrm{S}^1$, using a different symmetry of the fibre $\mathcal{V}$ from that of Figs. 15.4, 15.5, and 15.7 (where $\mathcal{V}$ is still a 1-dimensional real vector space $\mathbf{V}^1$), namely a stretch by a positive factor (here 2). The topology is just that of the cylinder $\mathrm{S}^1\times\mathbb{R}$, but there is a 'strain' that can be recognized in terms of a *connection* on $\mathcal{B}$. This connection defines a local notion of 'horizontal', for curves in $\mathcal{B}$. But consider two paths from a to b in the base, the direct path (black arrow) and the indirect one (white arrow). When we arrive at b we find a discrepancy (by a factor of 2), indicating that the notion of 'horizontal' here is path dependent.

'strain' in the bundle, which we can recognize is terms of an appropriate kind of *connection* on it.

Our previous type of connection, as discussed in Chapter 14, was concerned with a notion of 'parallelism' for tangent vectors carried along curves in the manifold $\mathcal{M}$. The way to view this, in the present context, is to think in terms of the tangent bundle $T(\mathcal{M})$ of $\mathcal{M}$. Since a point of $T(\mathcal{M})$ represents a tangent vector $\boldsymbol{v}$ to $\mathcal{M}$ at a point a of $\mathcal{M}$, the transport of $\boldsymbol{v}$ along some curve γ in $\mathcal{M}$ will be represented just by a curve $\gamma_{\boldsymbol{v}}$ in $T(\mathcal{M})$. See Fig. 15.17a. Having a notion of what 'parallel' means for the transport of $\boldsymbol{v}$ is equivalent to having a notion of 'horizontal' for the curve $\gamma_{\boldsymbol{v}}$ in the bundle (since keeping $\gamma_{\boldsymbol{v}}$ 'horizontal' in the bundle amounts to keeping $\boldsymbol{v}$ 'constant' along γ in the base). The idea here is to generalize this notion so that it applies to bundles other than the tangent bundle; see Fig. 15.17b. We have already seen, in Chapter 14, the beginnings of such a generalization, because we extended the notion of connection so that it applies to entities other than tangent vectors, namely to covectors and to $[^p_q]$-tensors generally. However, as noted in §15.1, this is a very limited kind of generalization, because the

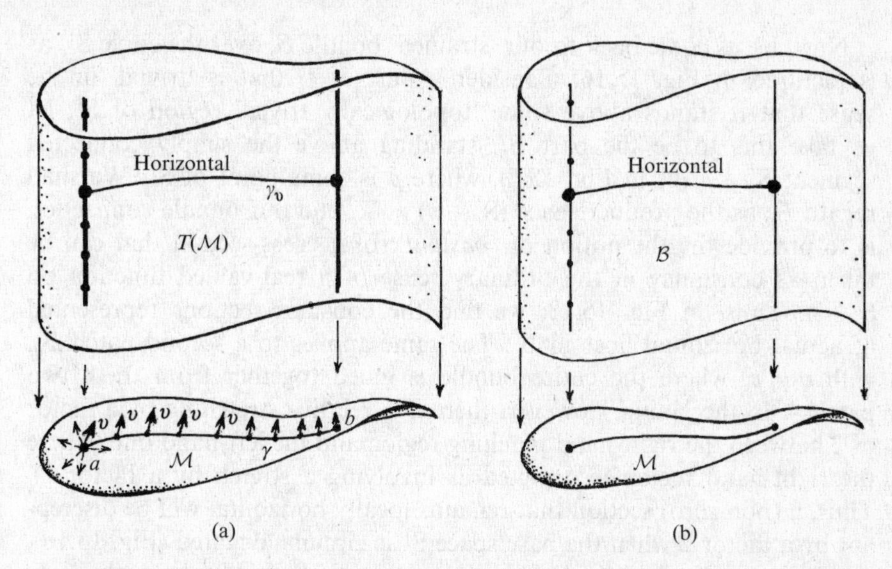

Fig. 15.17 Types of connection on a general manifold $\mathcal{M}$ compared. (a) The original notion (§14.3), defining a notion of 'parallel' for tangent vectors transported along curves in $\mathcal{M}$, is described in terms of the tangent bundle $T(\mathcal{M})$ of $\mathcal{M}$ (Fig. 15.12a). A particular tangent vector $\boldsymbol{v}$ at a point a of $\mathcal{M}$ is represented in $T(\mathcal{M})$ by a particular point of the fibre above a. A 'horizontal' curve $\gamma_{\boldsymbol{v}}$ in $T(\mathcal{M})$ from this point represents the parallel transport of $\boldsymbol{v}$ along a curve γ in $\mathcal{M}$. (b) The same idea applies to a bundle $\mathcal{B}$ over $\mathcal{M}$, other than $T(\mathcal{M})$, where 'constant transport' in $\mathcal{M}$ is defined from a notion of 'horizontal' in $\mathcal{B}$.

extension of the connection from vectors to these different kinds of entity is uniquely prescribed, with no additional freedom left (essentially because cotangent bundle and the tensor bundles are completely determined by the tangent bundle). For a general bundle over $\mathcal{M}$, there need be no association with the tangent bundle, so that the way that the connection acts on such a bundle can be specified independently of the way that it acts on tangent vectors. For a bundle over $\mathcal{M}$ which is unassociated with $T(\mathcal{M})$, it is not so appropriate to speak in terms of a 'parallelism', because the (local) notion of 'parallel' is something that refers to *directions*, which basically means directions of tangent vectors. Accordingly, it is more usual to refer to a local 'constancy' for the quantity that is described by the bundle, rather than to the 'parallelism' that refers to the tangent vectors described by $T(\mathcal{M})$. Such a local notion of 'constancy'—i.e. of 'horizontality' in the bundle—provides the structure known as a *bundle connection*.

Now, let us come back to our 'strained' bundle $\mathcal{B}$, over the circle S^1, as is pictured in Fig. 15.16. Consider a part of $\mathcal{B}$ that is 'trivial' in the sense that it stands above some 'topologically trivial' region of S^1; let us take this to be the part $\mathcal{B}_p$, standing above the simply connected segment $S^1 - p$ (as in Fig. 15.5), where p is some point of S^1. We shall regard $\mathcal{B}_p$ as the product space $(S^1 - p) \times \mathbb{R}$, and our bundle connection is to provide the the notion of *constancy* of a cross-section that can be taken as constancy in the ordinary sense of a real-valued function on $S^1 - p$. Thus, in Fig. 15.18, we find the constant sections represented as actual horizontal lines in $\mathcal{B}_p$. The same applies to a second patch $\mathcal{B}_q$, with $q \neq p$, where the entire bundle is glued together from these two patches. In the gluing, however, there is a relative stretching by a factor of 2 between the right-hand patching region and the left-hand one (where the right-hand region is depicted as involving a stretch by a factor 2). Thus, a (non-zero) section that remains locally horizontal will be discrepant by a factor 2 when the base space S^1 is circumnavigated (Fig. 15.16). Accordingly, the bundle $\mathcal{B}$ has no cross-sections (apart from the zero section) that are locally horizontal according to our specified bundle connection.

We can look at this situation slightly differently. We imagine a curve in the base space S^1 which starts at a point a and ends at b, and we envisage the 'constant transport', of a fibre-valued function on S^1, from a to b. That is to say, we look for a curve on $\mathcal{B}$ that is locally a horizontal cross-section above this curve. See Fig. 15.16. Now, there is more than one curve from a to b on the base space; if we go one way around, then we get a different

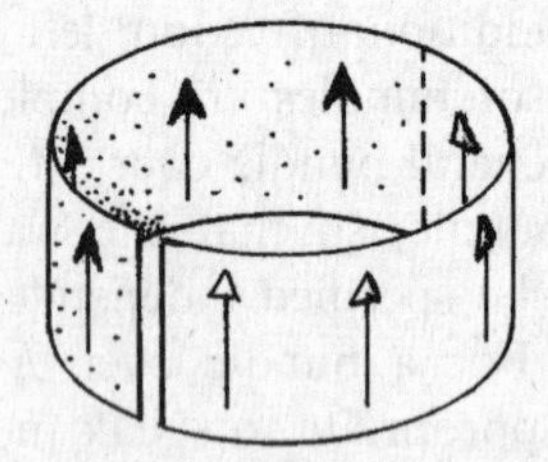

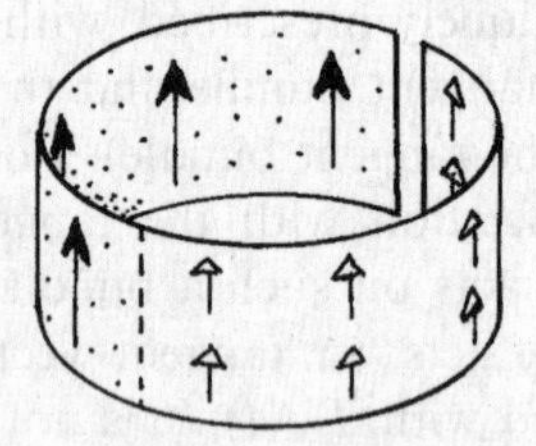

Fig. 15.18 Consider a part $\mathcal{B}_p$, of $\mathcal{B}$ (of Fig. 15.16) that stands above a 'trivial' region $S^1 - p$ of S^1, and similarly for $\mathcal{B}_q$, just as in Fig. 15.5a. Take 'horizontal' in each patch to mean horizontal in the ordinary sense. In the gluing, however, there is a relative stretching by a factor of 2 between one region of gluing and the other (illustrated in the right-hand patching). This provides the connection illustrated in Fig. 15.16.

answer for the final value at b from the answer that we obtain when we go the other way around. The notion of constant transport that we have defined is *path-dependent*.

This is not quite the same as the path dependence that we encountered for our tangent-bundle connection ∇, which we studied in Chapter 13. For, in that case, there was a local path dependence that occurred even for infinitesimal loops, and was manifested in the curvature of the connection. In the case of our 'strained' bundle $\mathcal{B}$, the path dependence is of a global character instead. Of course, there is no possibility of a *local* path dependence in this example, since the base space is 1-dimensional. But this example incidentally shows that it is possible to have path dependence globally even when none is present locally.

15.8 Bundle curvature

We can, however, modify our example so as to obtain a bundle over a *2-dimensional* space, within which we choose a particular circle to represent our original S^1. For convenience, let us take our S^1 to be the unit circle in the complex plane, so we shall take the base space $\mathcal{M}^{\mathbb{C}}$ of our new bundle $\mathcal{B}^{\mathbb{C}}$, to be given by $\mathcal{M}^{\mathbb{C}} = \mathbb{C}$. See Fig. 15.19. The fibres are to remain copies of the real line $\mathbb{R}$. Let us see how we can extend our bundle connection to this space.

If there were to be no 'strain' in our new bundle $\mathcal{B}^{\mathbb{C}}$, then we could take this connection to be given by straightforward differentiation with respect to the standard coordinates $(z, \bar{z})$ for the complex plane $\mathcal{M}^{\mathbb{C}}$. Then 'constancy' of a cross-section Φ (a real-valued function of z and $\bar{z}$) could be thought of simply as constancy in the ordinary sense, namely $\partial\Phi/\partial z = 0$ (whence also $\partial\Phi/\partial\bar{z} = 0$, since Φ is real). When we introduce 'strain' into the bundle connection, we can do this by modifying the operator $\partial/\partial z$ to become a new operator ∇ where

$$\nabla = \frac{\partial}{\partial z} - A,$$

the quantity A being a complex (not necessarily holomorphic) smooth function of z, which 'operates' simply by (scalar) multiplication. The operator ∇ acts on quantities like Φ. Topologically, our bundle $\mathcal{B}^{\mathbb{C}}$ is to be just the trivial bundle $\mathbb{C}\times\mathbb{R}$, so we can use global coordinates (z, Φ) for $\mathcal{B}^{\mathbb{C}}$, with z complex and Φ real.

A cross-section of $\mathcal{B}^{\mathbb{C}}$ is determined by Φ being given as a function of z:

$$\Phi = \Phi(z, \bar{z}),$$

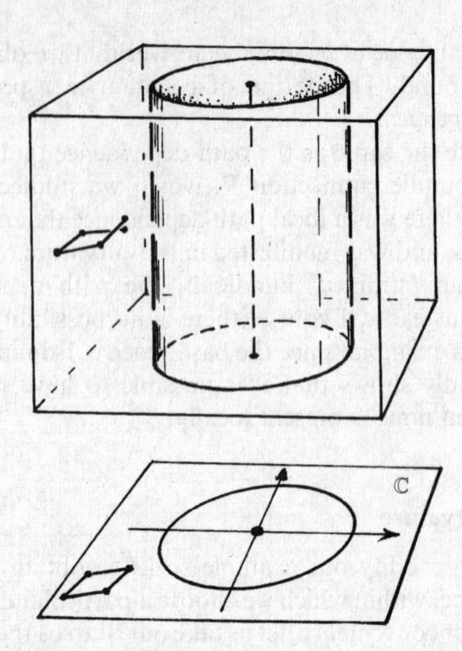

Fig. 15.19 To obtain a local path dependence (with curvature), in our bundle (now $\mathcal{B}^{\mathbb{C}}$), we need at least 2 dimensions in the base $\mathcal{M}^{\mathbb{C}}$, now taken as the complex plane $\mathbb{C}$, where the S^1 of Fig. 15.16 is its unit circle. The fibres are to remain $\mathbf{V}^1$ (i.e. modelled on the real line $\mathbb{R}$). Using z as a complex coordinate for $\mathbb{C} = \mathcal{M}^{\mathbb{C}}$, we use the explicit connection $\nabla = \partial/\partial z - A$, where A is a complex smooth function of z. When A is holomorphic the bundle curvature vanishes, but if $A = \mathrm{i}k\bar{z}$ (with suitable k), we get the strained bundle of Fig. 15.16 for the part over the unit circle. The bundle curvature is manifested in the failure to close of a horizontal polygon above a small parallelogram in $\mathcal{M}^{\mathbb{C}}$.

(the appearance of $\bar{z}$ indicating lack of holomorphicity; see §10.5). For the cross-section to be *constant* (i.e. horizontal), we require $\nabla\Phi = 0$ (whence $\overline{\nabla}\Phi = 0$ also, because Φ is real), i.e.

$$\frac{\partial \Phi}{\partial z} = A\Phi.$$

If A is holomorphic, then there is no problem about solving this equation, because an expression of the form $\Phi = \mathrm{e}^{(B+\bar{B})}$ will fit the bill, where $B = \int A\mathrm{d}z$.[15.12] However, in the general case, with a non-holomorphic A, we do not tend to get non-zero solutions, because of the commutator relation

[15.12] Check this.

$$\boldsymbol{\nabla}\overline{\boldsymbol{\nabla}} - \overline{\boldsymbol{\nabla}}\boldsymbol{\nabla} = \frac{\partial A}{\partial \bar{z}} - \frac{\partial \bar{A}}{\partial z}$$

acting on Φ.[15.13] (The right-hand side gives a number multiplying Φ that does not generally vanish, although the left-hand side annihilates any real solution of the equation $\partial\Phi/\partial z = A\Phi$.) This commutator serves to define a *curvature* for $\boldsymbol{\nabla}$, given by the imaginary part of $\partial A/\partial\bar{z}$, this curvature measuring the local degree of 'strain' in the bundle.

By making a specific choice of A, for which this commutator takes a constant non-zero value, such as $A = \mathrm{i}k\bar{z}$ for a suitable real constant k, we can get a 'stretching factor', when we travel around a closed loop in $\mathcal{M}^{\mathbb{C}}$, that is simply proportional to the area of the loop. This applies, in particular, to the unit circle S^1, so that we can reproduce our original 'strained' bundle $\mathcal{B}$ over S^1 by taking just that part of the bundle that lies above this S^1. We get the required 'stretching by a factor of 2' over the unit circle by taking an appropriate value of k.[15.14]

This commutator is the direct analogue of the commutator of operators ∇_a that we considered in §14.4, and which give rise to torsion and curvature. We may as well assume that the torsion is zero. (Torsion has to do with the action of the connection on tangent vectors, and is not of any concern for us in relation to bundles, like the one under consideration here, that are not associated with the tangent bundle.) For an n-dimensional base space $\mathcal{M}$, we have quantities just like the ∇_a and $\underset{x}{\boldsymbol{\nabla}}$ of Chapter 14, except that they now act on bundle quantities.[16] When we form their commutators appropriately, we extract the curvature of the bundle connection. When this curvature vanishes, then we have many locally constant sections of the bundle; otherwise, we run into obstructions to finding such sections, i.e. we find a local path dependence of the connection. The curvature describes this path dependence at the infinitesimal level. This is illustrated in Fig. 15.19.

In terms of indices, the connection is usually expressed, in some coordinate system, as an operator of the general form

$$\nabla_a = \frac{\partial}{\partial x^a} - A_a,$$

where the quantity A_a may be considered to have some suppressed 'bundle indices'. We can use Greek letters for these[17] (assuming that we are concerned

---

[15.13] Verify this formula.

[15.14] Confirm the assertions in this paragraph, finding the explicit value of k that gives this required factor 2.

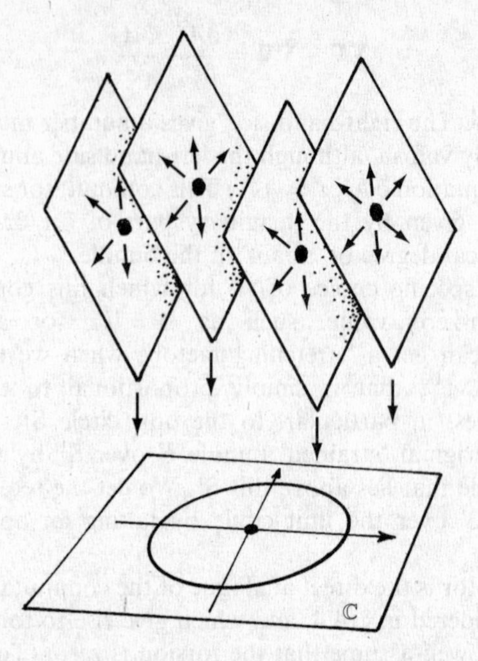

Fig. 15.20 We can also make the fibre into a complex 1-dimensional vector space, the 'stretch' corresponding to multiplication by a real number.

with a vector bundle, so that tensor ideas will apply), and then the quantity A_a looks like $A_a{}^\mu{}_\lambda$. (For the full index expression, there would be a δ^μ_λ multiplying the other two terms.) The bundle *curvature* would be a quantity

$$F_{ab}{}^\mu{}_\lambda,$$

where the antisymmetric pair of indices *ab* refers to tangent 2-plane directions in $\mathcal{M}$, in just the same way as for the curvature tensor that we had before, but now the indices λ and μ refer to the directions in the fibre (and are normally suppressed in most treatments). There is also a direct analogue of the (second) Bianchi identity (see §14.4). (The use of complex coordinates in the specific example of $\mathcal{B}^{\mathbb{C}}$ was a convenience only, and an index notation could have been used, just as in the *n*-dimensional case.)

It should be pointed out that, in many cases of fibre bundles, the relevant symmetry involved in the bundle's construction need not completely coincide with the symmetry of the fibre. For example, in the example of the 'strained' bundle $\mathcal{B}$ over S^1, or $\mathcal{B}^{\mathbb{C}}$ over $\mathbb{C}$, we could think of the 1-dimensional fibre as being broadened out into a 2-dimensional real vector space, where the 'stretch' of the fibre is represented as a uniform expansion of the vector 2-space. We could also provide this real

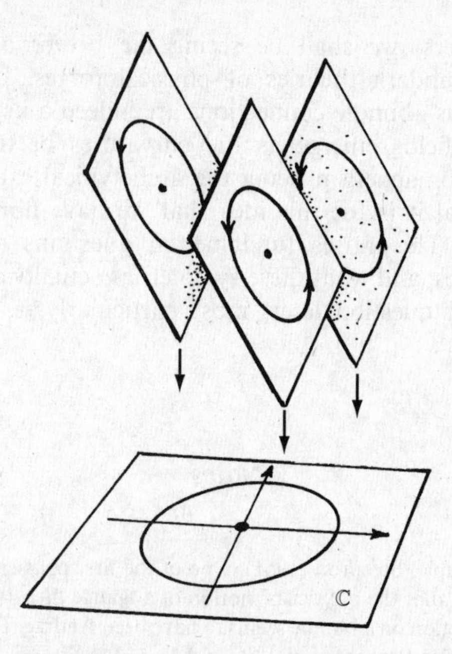

Fig. 15.21 Alternatively, we can impose a 'complex stretch' instead, such as multiplication by a complex phase ($e^{i\theta}$, with θ real), so the group of the bundle is now U(1), the multiplicative group of these complex numbers.

vector 2-space with the additional structure that makes it a 1-dimensional *complex* vector space, the 'stretch' corresponding to multiplication by a real number (Fig. 15.20). This leads us to consider what happens when we impose a 'complex stretch' instead. A particular case of this would be multiplication by a complex number of unit modulus ($\times e^{i\theta}$, with θ real), which would provide a rotation, rather than an actual stretch (Fig. 15.21) (which is the sort of thing that is involved in the Clifford–Hopf bundle, considered above). In this case, the group involved is U(1), the multiplicative group of unimodular complex numbers (see §13.9). Bundle connections with this U(1) symmetry group are of particular importance in physics, because they describe electromagnetic interactions, as we shall be seeing in §19.4. The essence of such a bundle is captured if the fibre is taken to be modelled on just the unit circle S^1, rather than on the whole complex plane $\mathbb{C}$. This is in a certain sense, more 'economical' since the rest of the plane is simply 'carried along' with the circle, and it provides no extra information. Nevertheless some advantage could be obtained from using the complex plane as fibre, because the bundle then becomes a (complex) vector bundle.[18]

In later chapters, we shall be seeing the power of these ideas in relation to the modern theories of physical forces. In their guise as 'gauge connections', bundle connections are indeed a key ingredient, and certain physical fields emerge as the curvatures of these connections (Maxwell's electromagnetism being the archetypical example). We have seen how essential it is for this idea that we have fibres possessing an *exact symmetry*. This raises fundamental questions as to the origin of such symmetries, and what these symmetries actually are. I shall return to this important question later, most particularly in Chapters 28, 31 and 34.

Notes

Section 15.1

15.1. See, for example, Steenrod (1951). One of the first physicists to appreciate, in around 1967, that the physicists' notion of a 'gauge theory' is really concerned with a connection on a bundle seems to have been Andrzej Trautman; see Trautman (1970) (also Penrose *et al.* 1997, p.A4).

15.2. In fact, the extra spacetime dimensions (Calabi–Yau spaces; see §31.14) of string theory are not to be thought of directly as the 'fibres' of a fibre bundle. Those fibres would be spaces of certain spinor fields in the Calabi–Yau spaces.

Section 15.2

15.3. Further information is required for a complete definition of product space, so that the notions of topology and smoothness are correctly defined for $\mathcal{M}\times\mathcal{V}$. When a volume measure can be assigned to each of $\mathcal{M}$ and $\mathcal{V}$, then the volume of $\mathcal{M}\times\mathcal{V}$ is the product of the volumes of $\mathcal{M}$ and $\mathcal{V}$. It would be distracting for me to go into these matters properly here, even though, technically speaking, they are necessary. For an appropriate reference, see Kelley (1965); Lefshetz (1949); or Munkres (1954).

15.4. See §12.1 for the general meaning of 'simply-connected'.

15.5. For notational simplicity, I am adopting a (mild) abuse of notation by writing 'S^1-p' for the space which consists of S^1 but with the point p removed. Purists would write '$S^1-\{p\}$', or more probably '$S^1\backslash\{p\}$' (see Note 9.13). The 'difference' expressed in these notations is between two *sets*, and '$\{p\}$' denotes the set whose only element is the point p.

Section 15.3

15.6. Normally pure mathematicians are relatively respectful of grammar, but many of them have adopted the habit of using the dreadful phrase 'associated to' when they seem to feel that 'associated with' has not a sufficiently specific flavour. I am at a loss to understand why they do not use the perfectly grammatical 'assigned to' instead. In my view, 'associated to' is rather worse than another common mathematician's abuse of language namely 'according as' (which I must confess to having used myself on various occasions) since the phrase 'according to whether', which it stands in for, is a bit of a mouthful.

Section 15.4

15.7. See Adams and Atiyah (1966).

15.8. See Penrose (1987a); Penrose and Rindler (1986).

15.9. We say that $\mathcal{B}$ is a *covering space* of $\mathcal{B}'$. In fact $\mathcal{B}$ is what is called the *universal* covering space of $\mathcal{B}'$. Being simply connected, it cannot be covered further.

Section 15.5

15.10. This geometrical description of 2-spinors is discussed in some detail in Penrose and Rindler (1984), Chap. 1.

15.11. For example, in §9.5, the splitting of functions (of a real variable) into positive- and negative-frequency parts (crucial for quantum field theory) was analysed in terms of extensions to holomorphic functions; but the reader may recall a certain awkwardness in relation to the constant functions. This issue is greatly clarified when we allow these to be twisted holomorphic functions and has relevance to twistor theory in §§33.8,10.

Section 15.6

15.12. I use the Greek spelling here, although the Latinized version 'Pappus' is somewhat more usual.

15.13. It would not be unreasonable to take the position that the artist's field of vision is more properly thought of as a *sphere* S^2, rather than $\mathbb{P}^2$, where we take the directed light rays through O as the artist's field of vision, rather that the undirected ones that I have been (implicitly) using in the text. The sphere is just a twofold cover of the projective plane, and the only trouble with it as providing a 'geometry', in this context, is that pairs of 'lines' (namely great circles) intersect in pairs of points rather than single points. The artist would need four canvases, rather than three, to cover the sphere S^2.

15.14. See Note 15.5.

15.15. This fact has relevance to an intriguing and important quantum-mechanical notion known as the 'Berry phase' (see Berry 1984, 1985; Simon 1983; Aharonov and Anandan 1987; Shankar (1994); also, Woodhouse 1991, pp. 225–49), which takes account of the fact that we do not know where '1' is on the unit circle—i.e. such a 'number' is an element of an S^1-fibre for an S^1-bundle, in this case, S^{2n+1} over $\mathbb{CP}^n$.

Section 15.8

15.16. In the case of ∇_a, we also need it to act on (co)tangent vectors so that ∇_a can operate on quantities with spacetime indices, in order that the commutator $\nabla_{[a}\nabla_{b]}$ can be given meaning. In the case of $\underset{X}{\boldsymbol{\nabla}}$, we can use the commutator expression $\underset{L}{\boldsymbol{\nabla}}\underset{M}{\boldsymbol{\nabla}} - \underset{M}{\boldsymbol{\nabla}}\underset{L}{\boldsymbol{\nabla}} - \underset{[L,\,M]}{\boldsymbol{\nabla}}$, which does not require this.

15.17. This type of index notation for bundle indices is developed explicitly in Penrose and Rindler (1984), Chap. 5.

15.18. On the other hand, when the fibre is the unit circle, the bundle becomes an example of a *principal bundle* which has advantages in other contexts. A principal bundle is one in which the fibre $\mathcal{V}$ is actually modelled on the group $\mathcal{G}$ of its own symmetries. Roughly speaking, $\mathcal{G}$ and $\mathcal{V}$ are the 'same' for a principal bundle, but where, more correctly, $\mathcal{V}$ is $\mathcal{G}$ but where one 'forgets' which is $\mathcal{G}$'s identity element; accordingly $\mathcal{V}$ is a (not necessarily Abelian) *affine* space, in accordance with §14.1 and Exercises [14.1], [14.2].

16
The ladder of infinity

16.1 Finite fields

It appears to be a universal feature of the mathematics normally believed to underlie the workings of our physical universe that it has a fundamental dependence on the *infinite*. In the times of the ancient Greeks, even before they found themselves to be forced into considerations of the real-number system, they had already become accustomed, in effect, to the use of rational numbers (see §3.1). Not only is the system of rationals infinite in that it has the potential to allow quantities to be indefinitely large (a property shared with the natural numbers themselves), but it also allows for an unending degree of refinement on an indefinitely small scale. There are some who are troubled with both of these aspects of the infinite. They might prefer a universe that is, on the one hand, finite in extent and, on the other, only finitely divisible, so that a fundamental discreteness might begin to emerge at the tiniest levels.

Although such a standpoint must be regarded as distinctly unconventional, it is not inherently inconsistent. Indeed, there has been a school of thought that the apparently basic physical role for the real-number system $\mathbb{R}$ is some kind of approximation to a 'true' physical number system which has only a finite number of elements. (This kind of approach has been pursued, particularly, by Y. Ahmavaara (1965) and some coworkers; see §33.1.) How can we make sense of such a finite number system? The simplest examples are those constructed from the integers, by 'reducing them *modulo p*', where p is some prime number. (Recall that the prime numbers are the natural numbers 2, 3, 5, 7, 11, 13, 17, ... which have no factors other than themselves and 1, and where 1 is itself *not* regarded as a prime.) To reduce the integers modulo p, we regard two integers as *equivalent* if their difference is a multiple of p; that is to say,

$$a \equiv b \pmod{p}$$

if and only if

$$a - b = kp \quad \text{(for some integer } k\text{).}$$

The integers fall into exactly p 'equivalence classes' (see the Preface, for the notion of equivalence class), according to this prescription (so a and b belong to the same class whenever $a \equiv b$). These classes are regarded as the elements of the *finite field* $\mathbb{F}_p$ and there are exactly p such elements. (Here, I am adopting the algebraists' use of the term 'field'. This should not be confused with the 'fields' on a manifold, such as vector or tensor fields, nor a physical field such as electromagnetism. An algebraist's field is just a *commutative division ring*; see §11.1.) Ordinary rules of addition, subtraction, (commutative) multiplication and division hold for the elements of $\mathbb{F}_p$.[16.1] However, we have the additional curious property that if we add p identical elements together, we always get *zero* (and, of course, the prime number p itself has to count as 'zero').

Note that, as $\mathbb{F}_p$ has been just described, its elements are themselves defined as 'infinite sets of integers'—since the 'equivalence classes' are themselves infinite sets, such as the particular equivalence class $\{\ldots, -7, -2, 3, 8, 13, \ldots\}$ which defines the element of $\mathbb{F}_5$ ($p = 5$) that we would denote by '3'. Thus, we have appealed to the infinite in order to define the quantities that constitute our finite number system! This is an example of the way in which mathematicians often provide a rigorous prescription for a mathematical entity by defining it in terms of infinite sets. It is the same 'equivalence class' procedure that is involved in the definition of fractions, as referred to in the Preface, in relation to the 'cancelling' that my mother's friend found so confusing! I imagine that to someone convinced that the number system $\mathbb{F}_p$ (for some suitable p), is 'really' directly rooted in nature, the 'equivalence class' procedure would be merely a mathematician's convenience, aimed at providing some kind of a rigorous prescription in terms of the more (historically) familiar infinite procedures. In fact we do not need to appeal to infinite sets of integers here; it is just that this is the most systematic procedure. In any given case, we could, alternatively, simply list all the operations, since these are finite in number.

Let us look at the case $p = 5$ in more detail, just as an example. We can label the elements of $\mathbb{F}_5$ by the standard symbols 0, 1, 2, 3, 4, and we have the addition and multiplication tables

[16.1] Show how these rules work, explaining why p has to be prime.

+	0	1	2	3	4
0	0	1	2	3	4
1	1	2	3	4	0
2	2	3	4	0	1
3	3	4	0	1	2
4	4	0	1	2	3

×	0	1	2	3	4
0	0	0	0	0	0
1	0	1	2	3	4
2	0	2	4	1	3
3	0	3	1	4	2
4	0	4	3	2	1

and we note that each non-zero element has a multiplicative inverse:

$$1^{-1} = 1, \quad 2^{-1} = 3, \quad 3^{-1} = 2, \quad 4^{-1} = 4,$$

in the sense that $2 \times 3 \equiv 1 \pmod 5$, etc. (From here on, I use '=' rather than '≡', when working with the elements of a particular finite number system.)

There are also other finite fields $\mathbb{F}_q$, constructed in a somewhat more elaborate way, where the total number of elements is some *power* of a prime: $q = p^m$. Let me just give the simplest example, namely the case $q = 4 = 2^2$. Here we can label the different elements as 0, 1, ω, ω^2, where $\omega^3 = 1$ and where each element x is subject to $x + x = 0$. This slightly extends the multiplicative group of complex numbers 1, ω, ω^2 that are cube roots of unity (described in §5.4 and mentioned in §5.5 as describing the 'quarkiness' of strongly interacting particles). To get $\mathbb{F}_4$, we just adjoin a zero '0' and supply an 'addition' operation for which $x + x = 0$.[16.2] In the general case $\mathbb{F}_{p^m}$, we would have $x + x + \cdots + x = 0$, where the number of xs in the sum is p.

16.2 A finite or infinite geometry for physics?

It is unclear whether such things really have a significant role to play in physics, although the idea has been revived from time to time. If $\mathbb{F}_q$ were to take the place of the real-number system, in any significant sense, then p would have to be very large indeed (so that the '$x + x + \cdots + x = 0$' would not show up as a serious discrepancy in observed behaviour). To my mind, a physical theory which depends fundamentally upon some absurdly enormous prime number would be a far more complicated (and improbable) theory than one that is able to depend upon a simple notion of infinity. Nevertheless, it is of some interest to pursue these matters. Much of geometry survives, in fact, when coordinates are given as elements of some $\mathbb{F}_q$. The ideas of calculus need more care; nevertheless, many of these also survive.

[16.2] Make complete addition and multiplication tables for $\mathbb{F}_4$ and check that the laws of algebra work (where we assume that $1 + \omega + \omega^2 = 0$).

It is instructive (and entertaining) to see how *projective geometry* with a finite total number of points works, and we can, accordingly, explore the projective n-spaces $\mathbb{P}^n(\mathbb{F}_q)$ over the field $\mathbb{F}_q$. We find that $\mathbb{P}^n(\mathbb{F}_q)$ has exactly $1 + q + q^2 + \cdots + q^n = (q^{n+1} - 1)/(q - 1)$ different points.[16.3] The projective *planes* $\mathbb{P}^2(\mathbb{F}_q)$ are particularly fascinating because a very elegant construction for them can be given. This can be described as follows. Take a circular disc made from some suitable material such as cardboard, and place a drawing pin through its centre, pinning it to a fixed piece of background card so that it can rotate freely. Mark $1 + q + q^2$ points equally spaced around the circumference on the background card, labelling them, in an anticlockwise direction, by the numbers $0, 1, 2, \ldots, q(1 + q)$. On the rotating disc, mark $1 + q$ special points in certain carefully chosen positions. These positions are to be such that, for any selection of two of the marked points on the background, there is exactly one position of the disc for which the two selected points coincide with two of these special points on the disc. Another way of saying this is as follows: if $a_0, a_1, \ldots, a_q$ are the successive distances around the circumference between these special points, taken cyclically (where the distance around the circumference between successive marked points on the background circle is taken as the *unit* distance) then every distance $1, 2, 3, \ldots, q + 1$ can be uniquely represented as a sum of a cyclically successive collection of the as. I call such a disc a *magic disc*. In Fig. 16.1, I have depicted magic discs for $q = 2$, 3, 4, and 5, for which $a^0, \ldots, a^q$ can be taken as 1, 2, 4; 1, 2, 6, 4; 1, 3, 10, 2, 5; 1, 2, 7, 4, 12, 5, respectively.[16.4] In the cases $q = 7, 8, 9, 11, 13$, and 16, we can make magic discs defined by 1, 2, 10, 19, 4, 7, 9, 5; 1, 2, 4, 8, 16, 5, 18, 9, 10; 1, 2, 6, 18, 22, 7, 5, 16, 4, 10; 1, 2, 13, 7, 5, 14, 34, 6, 4, 33, 18, 17, 21, 8; 1, 2, 4, 8, 16, 32, 27, 26, 11, 9, 45, 13, 10, 29, 5, 17, 18, respectively. It is a mathematical theorem that magic discs exist for every $\mathbb{P}^2(\mathbb{F}_q)$ (with q a power of a prime).[1] The reader may find it amusing to check various instances of the theorems of Pappos and Desargues (see §15.6, Fig. 15.14).[2] (Take $q > 2$, so as to have enough points for a non-degenerate configuration!) Two examples (Desargues for $q = 3$, and Pappos for $q = 5$, using the discs of Fig. 16.1) are illustrated in Fig. 16.2.

The simplest case $q = 2$ has particular interest from other directions.[16.5] This plane, with 7 points, is called the *Fano plane*, and it is depicted in Fig. 16.3, the circle being counted as a 'straight line'. Although

---

[16.3] Show this.

[16.4] Show how to construct *new* magic discs, in the cases $q = 3$, 5 by starting at a particular marked point on one of the discs that I have given and then multiplying each of the angular distances from the other marked points by some fixed integer. Why does this work?

[16.5] The finite field $\mathbb{F}_8$ has elements $0, 1, \varepsilon, \varepsilon^2, \varepsilon^3, \varepsilon^4, \varepsilon^5, \varepsilon^6$, where $\varepsilon^7 = 1$ and $1 + 1 = 0$. show that either (1) there is an identity of the form $\varepsilon^a + \varepsilon^b + \varepsilon^c = 0$ whenever a, b, and c are numbers on the background circle of Fig. 16.1a which can line up with the three spots on the disc, or else (2) the same holds, but with ε^3 in place of ε (i.e. $\varepsilon^{3a} + \varepsilon^{3b} + \varepsilon^{3c} = 0$).

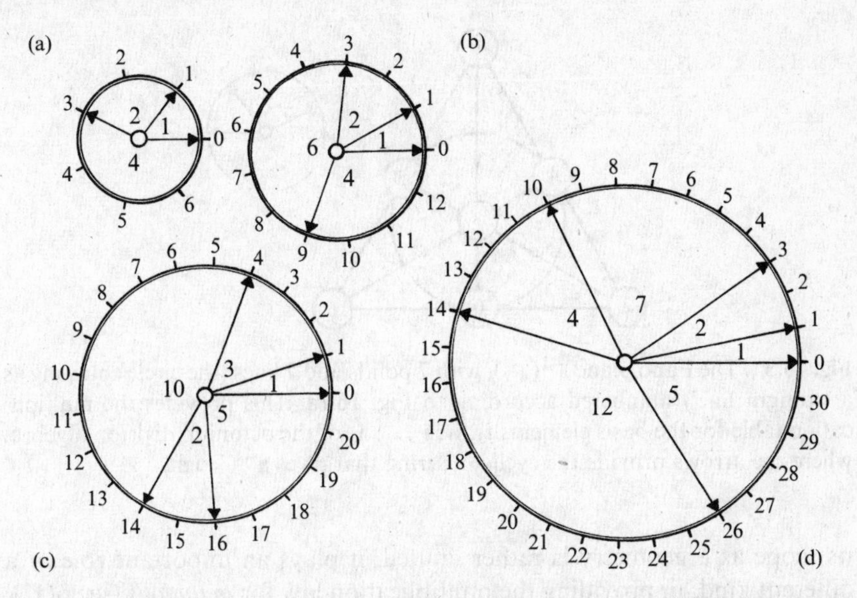

Fig. 16.1 'Magic discs' for finite projective planes $\mathbb{P}^2(\mathbb{F}_q)$ (q being a power of a prime). The $1+q+q^2$ points are represented as successive numerals 0, 1, 2, $\ldots$, $q(1+q)$ placed equidistantly around a background circle. A freely rotating circular disc is attached, with arrows labelling $1+q$ particular places: the points of a line in $\mathbb{P}^2(\mathbb{F}_q)$. These are such that for each pair of distinct numerals, there is exactly one disc setting so that arrows point at them. Magic discs are shown for (a) $q=2$; (b) $q=3$; (c) $q=4=2^2$; and (d) $q=5$.

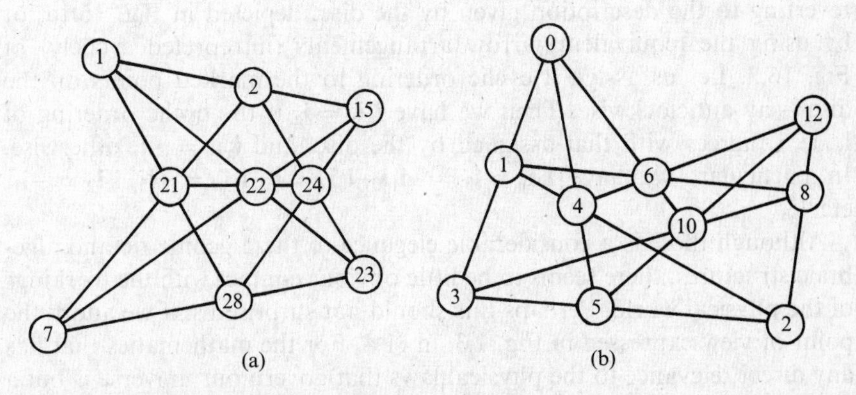

Fig. 16.2 Finite-geometry versions of the theorems of Fig. 5.14. (a) Pappos (with $q=5$) and (b) Desargues (with $q=3$), illustrated by respective use of the discs shown in Fig. 16.1d and 16.1b.

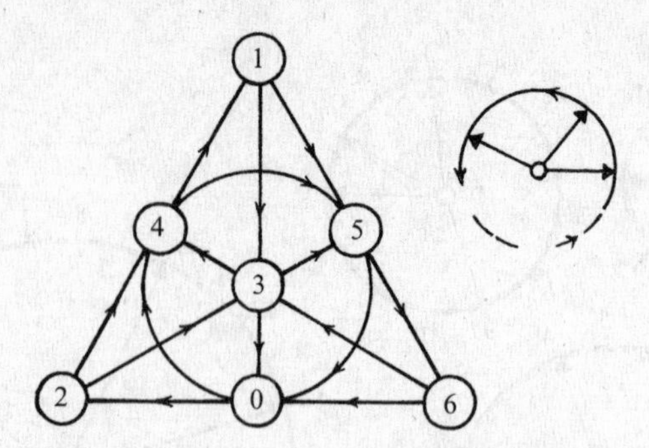

Fig. 16.3 The Fano plane $\mathbb{P}^2(\mathbb{F}_2)$, with 7 points and 7 lines (the circle counting as a 'straight line') numbered according to Fig. 16.1a. This provides the multiplication table for the basis elements $\mathbf{i}_0, \mathbf{i}_1, \mathbf{i}_2, \ldots, \mathbf{i}_6$ of the octonion division algebra, where the arrows provide the cyclic ordering that gives a '+' sign.

its scope as a geometry is rather limited, it plays an important role of a different kind, in providing the multiplication law for *octonions* (see §11.2, §15.4). The Fano plane has 7 points in it, and each point is to be associated with one of the generating elements $\mathbf{i}_0, \mathbf{i}_1, \mathbf{i}_2, \ldots, \mathbf{i}_6$ of the octonion algebra. Each of these is to satisfy $\mathbf{i}_r^2 = -1$. To find the product of two *distinct* generating elements, we just find the line in the Fano plane which joins the points representing them, and then the remaining point on the line is the point representing the product (up to a sign) of these other two. For this, the simple picture of the Fano plane is not quite enough, because the *sign* of the product needs to be determined also. We can find this sign by reverting to the description given by the disc, depicted in Fig. 16.1a, or by using the (equivalent) arrow arrangements (intrepreted cyclicly) of Fig. 16.3. Let us assign a cyclic ordering to the marked points on the disc—say anticlockwise. Then we have $\mathbf{i}_x\mathbf{i}_y = \mathbf{i}_z$ if the cyclic ordering of $\mathbf{i}_x, \mathbf{i}_y, \mathbf{i}_z$ agrees with that assigned by the disc, and $\mathbf{i}_x\mathbf{i}_y = -\mathbf{i}_z$ otherwise. In particular, we have $\mathbf{i}_0\mathbf{i}_1 = \mathbf{i}_3 = -\mathbf{i}_1\mathbf{i}_0$, $\mathbf{i}_0\mathbf{i}_2 = \mathbf{i}_6$, $\mathbf{i}_1\mathbf{i}_6 = -\mathbf{i}_5$, $\mathbf{i}_4\mathbf{i}_2 = -\mathbf{i}_1$, etc.[16.6]

Although there is a considerable elegance to these geometric and algebraic structures, there seems to be little obvious contact with the workings of the physical world. Perhaps this should not surprise us, if we adopt the point of view expressed in Fig. 1.3, in §1.4. For the mathematics that has any direct relevance to the physical laws that govern our universe is but a tiny part of the Platonic mathematical world as a whole—or so it would seem, as far as our present understanding has taken us. It is possible that,

[16.6] Show that the 'associator' $a(bc) - (ab)c$ is antisymmetrical in a, b, c when these are generating elements, and deduce that this (whence also $a(ab) = a^2b$) holds for *all* elements. *Hint*: Make use of Fig. 16.3 and the full symmetry of the Fano plane.

as our knowledge deepens in the future, important roles will be found for such elegant structures as finite geometries or for the algebra of octonions. But as things stand, the case has yet to be convincingly made, in my opinion.[3] It seems that mathematical elegance alone is far from enough (see also §34.9). This should teach us caution in our search for the underlying principles of the laws of the universe!

Let us drag ourselves back from such flirtations with these appealing finite structures and return to the awesome mathematical richness that is inherent in the infinite. As a preliminary, it should be pointed out that infinite structures (such as the totality of natural numbers $\mathbb{N}$) might be part of some mathematical formalism aimed at a description of reality, whereas it is not intended that these infinite structures have direct physical interpretation as infinite (or infinitesimal) physical entities. For example, some attempts have been made to develop a scheme in which discreteness (and indeed finiteness) appears at the smallest level, while there is still the potential for describing indefinitely (or even infinitely) large structures. This applies, in particular, to some old ideas of my own for building up space in a finite way, using the theory of spin networks which I shall describe briefly in §32.6, and which depends upon the fact that, according to standard quantum mechanics, the measure of spin of an object is given by a natural number multiple of a certain fixed quantity ($\frac{1}{2}\hbar$). Indeed, as I mentioned in §3.3, in the early days of quantum mechanics, there was a great hope, not realized by future developments, that quantum theory was leading physics to a picture of the world in which there is actually discreteness at the tiniest levels. In the successful theories of our present day, as things have turned out, we take spacetime as a continuum even when quantum concepts are involved, and ideas that involve small-scale spacetime discreteness must be regarded as 'unconventional' (§33.1). The continuum still features in an essential way even in those theories which attempt to apply the ideas of quantum mechanics to the very structure of space and time. This applies, in particular, to the Ashtekar–Rovelli–Smolin–Jacobson theory of loop variables, in which discrete (combinatorial) ideas, such as those of knot and link theory, actually play key roles, and where spin networks also enter into the basic structure. (We shall be seeing something of this remarkable scheme in Chapter 32 and, in §33.1, we shall briefly encounter some other ideas relating to 'discrete spacetime'.)

Thus it appears, for the time being at least, that we need to take the use of the infinite seriously, particularly in its role in the mathematical description of the physical continuum. But what kind of infinity is it that we are requiring here? In §3.2 I briefly described the 'Dedekind cut' method of constructing the real-number system in terms of infinite sets of rational numbers. In fact, this is an enormous step, involving a notion of infinity that greatly surpasses that which is involved with the rational numbers themselves. It will have some significance for us to address this issue here. In fact, as the great Danish/Russian/German mathematician Georg Cantor showed, in 1874, as part of a theory that he continued to develop until 1895, there are *different sizes* of infinity! The infinitude of natural numbers is actually the smallest of these, and different infinities continue unendingly to larger and larger scales. Let us try to catch a glimpse of Cantor's ground-breaking and fundamental ideas.

16.3 Different sizes of infinity

The first key ingredient in Cantor's revolution is the idea of a 1–1 (i.e. a 'one-to-one') correspondence.[4] We say that two sets have the same *cardinality* (which means, in ordinary language, that they have the 'same number of elements') if it is possible to set up a correspondence between the elements of one set and the elements of the other set, one to one, so that there are no elements of either set that fail to take part in the correspondence. It is clear that this procedure gives the right answer ('same number of elements') for finite sets (i.e. sets with a finite number 1, 2, 3, 4, ... of members, or even 0 elements, where in that case we require the correspondence to be vacuous). But in the case of infinite sets, there is a novel feature (already noticed, by 1638, by the great physicist and astronomer Galileo Galilei)[5] that an infinite set has the same cardinality as some of its proper subsets (where 'proper' means other than the whole set).

Let us see this in the case of the set $\mathbb{N}$ of natural numbers:

$$\mathbb{N} = \{0, 1, 2, 3, 4, 5, \ldots\}.$$

If we remove 0 from this set,[6] we find a new set $\mathbb{N} - 0$ which clearly has the same cardinality as $\mathbb{N}$, because we can set up the 1–1 correspondence in which the element r in $\mathbb{N}$ is made to correspond with the element $r + 1$ in $\mathbb{N} - 0$. Alternatively, we can take Galileo's example, and see that the set of square numbers $\{0, 1, 4, 9, 16, 25, \ldots\}$ must also have the same cardinality as $\mathbb{N}$, despite the fact that, in a well-defined sense, the square numbers constitute a vanishingly small proportion of the natural numbers as a whole. We can also see that the cardinality of the set $\mathbb{Z}$ of all the integers is again of this same cardinality. This can be seen if we consider the ordering of $\mathbb{Z}$ given by

$$\{0, 1, -1, 2, -2, 3, -3, 4, -4, \ldots\},$$

which we can simply pair off with the elements $\{0, 1, 2, 3, 4, 5, 6, 7, 8, \ldots\}$ of the set $\mathbb{N}$. More striking is the fact that the cardinality of the *rational* numbers is again the same as the cardinality of $\mathbb{N}$. There are many ways of seeing this directly,[16.7],[16.8] but rather than demonstrating this in detail here, let us see how this particular example falls into the general framework of Cantor's wonderful theory of infinite cardinal numbers.

First, what is a *cardinal number*? Basically, it is the 'number' of elements in some set, where we regard two sets as having the 'same number of elements' if and only if they can be put into 1–1 correspondence with each other. We could try to be more precise by using the 'equivalence class' idea (employed in §16.1 above to define $\mathbb{F}_p$ for a prime p; see also the Preface) and say that the cardinal number α of some set A *is* the equivalence class of all sets with the same cardinality as A. In fact the logician Gottlob Frege tried to do just this in 1884, but it turns out that there are fundamental difficulties with open-ended concepts like 'all sets', since serious contradictions can arise with them (as we shall be seeing in §16.5). In order to avoid such contradictions, it seems to be necessary to put some restriction on the size of the 'universe of possible sets'. I shall have some remarks to make about this disturbing issue shortly. For the moment, let us evade it by taking refuge in a position that I have been taking before (as referred to in the Preface, in relation to the 'equivalence class' definition of the rational numbers). We take the cardinals as simply being mathematical entities (inhabitants of Plato's world!) which can be abstracted from the notion of 1–1 equivalence between sets. We allow ourselves to say that the set A 'has cardinality α', or that it 'has α elements', provided that we are consistent and say that the set B also 'has cardinality α', or that *it* 'has α elements', if and only if A and B can be put into 1–1 correspondence. Notice that the natural numbers can all be thought of as cardinal numbers in this sense—and this is a good deal closer to the intuitive notion of what a natural number 'is' than the 'ordinal' definition ($0 = \{\}, 1 = \{0\}, 2 = \{0, \{0\}\}, 3 = \{0, \{0\}, \{0, \{0\}\}\}, \ldots$) given in §3.4! The natural numbers are in fact the *finite* cardinals (in the sense that the *infinite* cardinals are the cardinalities of those sets, like $\mathbb{N}$ above, which contain proper subsets of the same cardinality as themselves).

Next, we can set up relationships between cardinal numbers. We say that the cardinal α is *less than or equal to* the cardinal β, and write

$$\alpha \leq \beta$$

(or equivalently $\beta \geq \alpha$), if the elements of a set A with cardinality α can be put into 1–1 correspondence with the elements of some subset (not necessarily a proper subset) of the elements of some set B, with cardinality β. It

---

[16.7] See if you can provide such an explicit procedure, by finding some sort of systematic way of ordering all the fractions. You may find the result of Exercise [16.8] helpful.

[16.8] Show that the function $\frac{1}{2}((a+b)^2 + 3a + b)$ explicitly provides a 1–1 correspondence between the natural numbers and the pairs (a, b) of natural numbers.

should be clear that, if $\alpha \leq \beta$ and $\beta \leq \gamma$, then $\alpha \leq \gamma$.[16.9] One of the beautiful results of the theory of cardinal numbers is that, if

$$\alpha \leq \beta \text{ and } \beta \leq \alpha,$$

then

$$\alpha = \beta,$$

meaning that there is a 1–1 correspondence between A and B.[16.10] We may ask whether there are pairs of cardinals α and β for which neither of the relations $\alpha \leq \beta$ and $\beta \leq \alpha$ holds. Such cardinals would be *non-comparable*. In fact, it follows from the assumption known as the *axiom of choice* (referred to briefly in §1.3) that non-comparable cardinals do not exist.

The axiom of choice asserts that if we have a set A, all of whose members are non-empty sets, then there exists a set B which contains exactly one element from each of the sets belonging to A. It would appear, at first, that the axiom of choice is merely asserting something absolutely obvious! (See Fig. 16.4.) However, it is not altogether uncontroversial that the axiom of choice should be accepted as something that is universally valid. My own position is to be cautious about it. The trouble with this axiom is that it is a pure 'existence' assertion, without any hint of a rule whereby the set B might be specified. In fact, it has a number of alarming consequences. One of these is the Banach–Tarski theorem,[7] one version of which says that the ordinary unit sphere in Euclidean 3-space can be cut into five pieces with the property that, simply by Euclidean motions

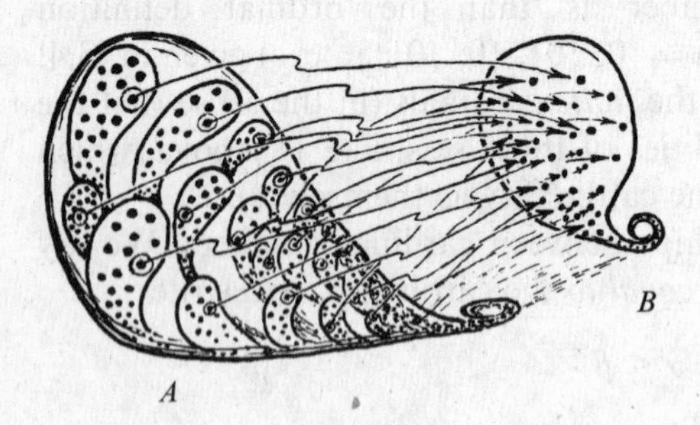

Fig. 16.4 The axiom of choice asserts that for any set A, all of whose members are non-empty sets, there exists a set B which contains exactly one element from each of the sets belonging to A.

[16.9] Spell this out in detail.

[16.10] Prove this. Outline: there is a 1–1 map b taking A to some subset bA $(= b(A))$ of B, and a 1–1 map a taking B to some subset aB of A; consider the map of A to B which uses b to map $A-aB$ to $bA-baB$ and $abA-abaB$ to $babA-babaB$, etc. and which uses a^{-1} to map $aB-abA$ to $B-bA$ and $abaB-ababA$ to $baB-babA$, etc., and sort out what to do with the rest of A and B.

(i.e. translations and rotations), these pieces can be reassembled to make *two* complete unit spheres! The 'pieces', of course, are not solid bodies, but intricate assemblages of points, and are defined in a very non-constructive way, being asserted to 'exist' only by use of the axiom of choice.

Let me now list, without proof, a few very basic properties of cardinal numbers. First, the symbol $\leq$ gives the normal meaning (see Note 3.1) when applied to the natural numbers (the *finite* cardinals). Moreover, any natural number is less than or equal to ($\leq$) any infinite cardinal number—and, of course, it is *strictly* smaller, i.e. less than ($<$) and not equal to it. Now suppose that $\beta \leq \alpha$, with α infinite, then (in stark contrast with what we are familiar with for finite numbers) the cardinality of the *union* $A \cup B$ is simply the greater of the two, namely α, and the cardinality of the *product* $A \times B$ is also α. (We have seen examples of the product before, e.g. §13.2, §15.2. The set $A \times B$ is consists of all *pairs* (a, b), where a is taken from A and b from B. For finite sets, the cardinality of their product, as sets, is the ordinary numerical product of their cardinalities, which for finite sets with more than one member is always larger than the cardinality of either individually.) This does not seem to get us very far if we want to find infinities that are bigger than the ones that we have already. We seem to have got 'stuck' at α.

We shall be seeing how to get 'unstuck' in the next section. For the moment, however, we can see that what we have done above is at least enough to show us that the number of rational numbers is the same as the number of natural numbers. Following Cantor, let us use the symbol $\aleph_0$ ('aleph nought' or, in the US, 'aleph null') for the cardinality of the natural numbers $\mathbb{N}$ which, as we have seen above, is the same as the cardinality of the integers $\mathbb{Z}$. In fact, the infinite number $\aleph_0$ is the smallest of the infinite cardinals. Now, what is the cardinality ρ of the rationals? Any rational number can be written (in many ways) in the form a/b, where a and b are integers. Choosing one of these ways (say, 'lowest terms') for each rational, we have found a 1–1 correspondence between the set of rationals with a *subset* of the set $\mathbb{N} \times \mathbb{N}$. Therefore ρ is less than or equal to the cardinality of $\mathbb{N} \times \mathbb{N}$. But by the above (or, by direct application of Exercise [16.8], the cardinality of $\mathbb{N} \times \mathbb{N}$ is equal to the cardinality of $\mathbb{N}$, namely $\aleph_0$. Thus, $\rho \leq \aleph_0$. But the integers are contained in the rationals, so $\aleph_0 \leq \rho$. Hence, $\rho = \aleph_0$.

16.4 Cantor's diagonal slash

Now we come to Cantor's astounding early achievement, namely his demonstration that there are indeed infinities strictly greater than $\aleph_0$, and that the cardinality of the set $\mathbb{R}$ of real numbers is such an infinity.

I shall give this result here as a particular instance of Cantor's more general

$$\alpha < 2^{\alpha},$$

where $\alpha < \beta$ means $\alpha \leq \beta$ and $\alpha \neq \beta$ (and, of course, we can also write $\alpha < \beta$ as $\beta > \alpha$). Cantor's remarkable proof of this result (and the result itself) constitutes one of the most original and influential achievements in the whole of mathematics. Yet it is simple enough that I can give it in its entirety here.

First I should explain the notation. If we have two sets A and B, then the set B^A is the *set of all mappings from A to B*. What is the rationale for this use of notation? We think of the set A spread out before us, each element of A being represented as a 'point'. Then, to picture an element of B^A, we place one of the elements of B at each of these points. This is a mapping from A to B because it provides an assignment of an element of B to each element of A (see Fig. 16.5). The reason for the 'exponential notation' B^A is that when we apply this procedure to finite sets, say to a set A, with a elements, and a set B, with b elements, then the total number of ways of assigning an element of B to each element of A is indeed b^a. (There are b ways for the first member of A; there are b ways for the second; there are b ways for the third; and so on, for each of the a members of A. The total number in all is therefore $b \times b \times b \times \ldots \times b$, the number of bs in the product being a, so this is just b^a.) Cantor's notation is

$$\beta^{\alpha}$$

for the cardinality of B^A, where β and α are the respective cardinalities of B and A.

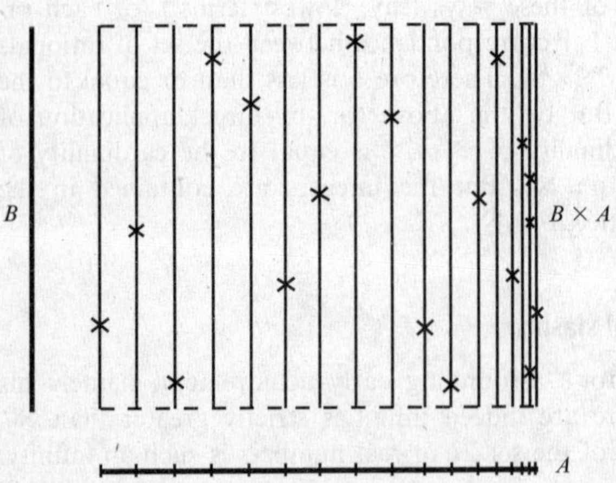

Fig. 16.5 For general sets A, B, the set of all mappings from A to B is denoted B^A (see also Fig. 6.1). Each element of A is assigned a particular element of B. This provides a cross-section of $B \times A$, regarded as a bundle over A (as in Fig. 15.6a), except that there is no notion of continuity involved.

This takes on a particular significance when $\beta = 2$. Here we can take B to be a set with two elements that we shall think of being the labels 'in' and 'out'. Each element of B^A is thus an assignment of either 'in' or 'out' to every element of A. Such an assignment amounts simply to choosing a subset of A (namely the subset of 'in' elements). Thus, B^A is, in this case, just the set of subsets of A (and we frequently denote this set of subsets of A by 2^A). Accordingly:

2^α is the total number of subsets of any set with α elements.

Now for Cantor's astonishing proof. This proceeds in accordance with the classic ancient Greek tradition of 'proof by contradiction' (§2.6, §3.1). First, let us try to suppose that $\alpha = 2^\alpha$, so that there is some 1–1 correspondence between some set A and its set of subsets 2^A. Then each element a of A will be associated with a particular subset $S(a)$ of A, under this correspondence. We may expect that sometimes the set $S(a)$ will contain a itself as a member and sometimes it will not. Let us consider the collection of all the elements a for which $S(a)$ does not contain a. This collection will be some particular subset Q of A (which we allow to be either the empty set or the whole of A, if need be). Under the supposed 1–1 correspondence, we must have $Q = S(q)$, for some q in S. We now ask the question: 'Is q in Q or is it not?' First suppose that it is not. Then q must belong to the collection of elements of A that we have just singled out as the subset Q, so q must belong to Q after all: a contradiction. This leaves us with the alternative supposition, namely that q *is* in Q. But then q cannot belong to the collection that we have called Q, so q does not belong to Q after all: again a contradiction. We therefore conclude that our supposed 1–1 correspondence between A and 2^A cannot exist.

Finally, we need to show that $\alpha \leq 2^\alpha$, i.e. that there is a 1–1 correspondence between A and some subset of 2^A. This is achieved by simply using the 1–1 correspondence which assigns each element a of A to the particular subset of A that contains just the element a and no other. Thus, we have established $\alpha < 2^\alpha$, as required, having shown $\alpha \leq 2^\alpha$ but $\alpha \neq 2^\alpha$.

Though this argument may be a little confusing (and any confused reader may care to study it all over again), it is extremely 'elementary' in the sense that it does not appeal to mathematical ideas requiring any expert knowledge. In view of this, it is very remarkable that its implications are extraordinarily far-reaching. Not only does it enable us to see that there are fundamentally more real numbers than there are natural numbers, but it also shows that there is no end to the hugeness of the possible infinite numbers. Moreover, in a slightly modified form, the argument shows that there is no computational way of deciding whether a general computation will ever come to an end (Turing), and a related consequence is Gödel's famous *incompleteness theorem* which shows that no set of pre-assigned trustworthy mathematical rules can encapsulate all the procedures whereby mathematical truths are ascertained. I shall try to give the flavour of how such results are obtained in the next section.

To end this section, however, let us see why the above result actually establishes Cantor's first remarkable breakthrough concerning the infinite, namely that there are actually far more real numbers than there are natural numbers—despite the fact that there are exactly as many fractions as natural numbers. (This breakthrough established that there is, indeed, a non-trivial theory of the infinite!) This will follow if we can see that the cardinality of the reals, usually denoted by C, is actually equal to $2^{\aleph_0}$:

$$\mathsf{C} = 2^{\aleph_0}.$$

Then, by the above argument, $\mathsf{C} > \aleph_0$ as required.

There are many ways to see that $\mathsf{C} = 2^{\aleph_0}$. To show that $2^{\aleph_0} \leq \mathsf{C}$ (which is actually all that we now need for $\mathsf{C} > \aleph_0$), it is sufficient to establish that there is a 1–1 correspondence between $2^{\mathbb{N}}$ and some *subset* of $\mathbb{R}$. We can think of each element of $2^{\mathbb{N}}$ as an assignment of either 0 or 1 ('out' or 'in') to each natural number, i.e. such an element can be thought of as an infinite sequence, such as

$$100110001011101\ldots.$$

(This particular element of $2^{\mathbb{N}}$ assigns 1 to natural number 0, it assigns 0 to the natural number 1, it assigns 0 to the natural number 2, it assigns 1 to the natural number 3, it assigns 1 to the natural number 4, etc., so our subset is $\{0,3,4,8,\ldots\}$.) Now, we could try to read off this entire sequence of digits as the binary expansion of a some real number, where we think of a decimal point situated at the far left. Unfortunately, this does not quite work, because of the irritating fact that there is an ambiguity in certain such representations, namely with those that end in an infinite sequence consisting entirely of 0s or else consisting entirely of 1s.[16.11] We can get around this awkwardness by any number of stupid devices. One of these would be to interleave the binary digits with, say, the digit 3, to obtain

$$.313030313130303031303131313031\ldots,$$

and then read this number off as the ordinary decimal expression of some real number. Accordingly, we have indeed set up a 1–1 correspondence between $2^{\mathbb{N}}$ and a certain subset of $\mathbb{R}$ (namely the subset whose decimal expansions have this odd-looking interleaved form). Hence $2^{\aleph_0} \leq \mathsf{C}$ (and we now obtain Cantor's $\mathsf{C} > \aleph_0$), as required.

[16.11] Explain this.

To deduce that $\mathsf{C} = 2^{\aleph_0}$, we have to be able to show that $\mathsf{C} \leq 2^{\aleph_0}$. Now, every real number strictly between 0 and 1 has a binary expansion (as considered above), albeit sometimes redundantly; thus that particular set of reals certainly has cardinality $\leq 2^{\aleph_0}$. There are many simple functions that take this interval to the whole of $\mathbb{R}$,[16.12] establishing that $\mathsf{C} \leq 2^{\aleph_0}$, and hence $\mathsf{C} = 2^{\aleph_0}$, as required.

Cantor's original version of the argument was given somewhat differently from the one presented above, although the essentials are the same. His original version was also a proof by contradiction, but more direct. A hypothetical 1–1 correspondence between $\mathbb{N}$ and the real numbers strictly between 0 and 1 was envisaged, and presented as a vertical listing of all real numbers, each written out in decimal expansion. A contradiction with the assumption that the list is complete was obtained by a 'diagonal argument' whereby a new real number, not in the list, is constructed by going down the main diagonal of the array, starting at the top left corner and differing in the nth place from the nth real number in the list. (There are many popular accounts of this; see, for example, the version of it given in Chapter 3 of my book *The Emperor's New Mind*).[16.13] This general type of argument (including that which we used at the beginning of this section to demonstrate $\alpha < 2^{\alpha}$), is sometimes referred to as Cantor's 'diagonal slash'.

16.5 Puzzles in the foundations of mathematics

As remarked above, the cardinality, $2^{\aleph_0}$, of the continuum (i.e. of $\mathbb{R}$) is often denoted by the letter C. Cantor would have preferred to be able to label it '$\aleph_1$', by which he meant the 'next smallest' cardinal after $\aleph_0$. He tried, but failed, to prove $2^{\aleph_0} = \aleph_1$; in fact the contention '$2^{\aleph_0} = \aleph_1$', known as the *continuum hypothesis*, became a famous unresolved issue for many years after Cantor proposed it. It is still unresolved, in an 'absolute' sense. Kurt Gödel and Paul Cohen were able to show that the continuum hypothesis (and also the axiom of choice) is not decidable by the means of standard set theory. However, because of Gödel's incompleteness theorem, which I shall be coming to in a moment, and various related matters, this does not in itself resolve the issue of the *truth* of the continuum hypothesis. It is still possible that more powerful methods of proof than those of standard set theory might be able to decide the truth or otherwise of the continuum hypothesis; on the other hand, it could be the case that its truth or falsehood is a subjective issue depending

[16.12] Exhibit one. *Hint*: Look at Fig. 9.8, for example.

[16.13] Explain why this is essentially the same argument as the one I have given here, in the case $\alpha = \aleph_0$ for showing $\alpha < 2^{\alpha}$.

upon what mathematical standpoint one adheres to.[8] This issue was referred to in §1.3, but in relation to the axiom of choice, rather than the continuum hypothesis.

We see that the relation $\alpha < 2^\alpha$ tells us that there cannot be any greatest infinity; for if some cardinal number Ω were proposed as being the greatest, then the cardinal number 2^Ω is seen to be even greater. This fact (and Cantor's argument establishing this fact) has had momentous implications for the foundations of mathematics. In particular, the philosopher Bertrand Russell, being previously of the opinion that there must be a largest cardinal number (namely that of the class of all classes) had been suspicious of Cantor's conclusion, but changed his mind, by around 1902, after studying it in detail. In effect, he appplied Cantor's argument to the 'set of all sets', leading him at once to the now famous 'Russell paradox'!

This paradox proceeds as follows. Consider the set $\mathcal{R}$, consisting of 'all sets that are not members of themselves'. (For the moment, it does not matter whether you are prepared to believe that a set can be a member of itself. If no set belongs to itself, then $\mathcal{R}$ is the set of *all* sets.) We ask the question, what about $\mathcal{R}$ itself? Is $\mathcal{R}$ a member of itself? Suppose that it is. Then, since it then belongs to the set $\mathcal{R}$ of sets which are *not* members of themselves, it does not belong to itself after all—a contradiction! The alternative supposition is that it does not belong to itself. But then it must be a member of the entire family of sets that are not members of themselves, namely the set $\mathcal{R}$. Thus, $\mathcal{R}$ belongs to $\mathcal{R}$, which contradicts the assumption that it does not belong to itself. This is a clear contradiction!

It may be noticed that this is simply what happens to the Cantor proof $\alpha < 2^\alpha$, if it is applied in the case when α is taken to be the 'set of all sets'.[16.14] Indeed this is how Russell came across his paradox.[9] What this argument is actually showing is that there is no such thing as the 'set of all sets'. (In fact Cantor was already aware of this, and knew about the 'Russell paradox' some years before Russell himself.[10] It might seem odd that something so straightforward as the 'set of all sets' is a forbidden concept. One might imagine that any proposal for a set ought to be perfectly acceptable if there is a well-defined rule for telling us when something belongs to it and when something does not. Here is seems that there certainly is such a rule, namely that *every* set is in it! The catch seems to be that we are allowing the same status to this stupendous collection as we are to each of its members, namely calling both kinds of collection simply a 'set'. The whole argument depends upon our having a clear idea about what a *set* actually is. And once we have such an idea,

[16.14] Show that this is what happens.

the question arises: is the collection of all these things itself actually to count as a set? What Cantor and Russell have told us is that the answer to this question has to be no!

In fact, the way that mathematicians have come to terms with this apparently paradoxical situation is to imagine that some kind of distinction has been made between 'sets' and 'classes'. (Think of the classes as sometimes being large unruly things that are not supposed to join clubs, whereas sets are always regarded as respectable enough to do so.) Roughly speaking, any collection of sets whatever could be allowed to be considered as a whole, and such a collection would be called a *class*. Some classes are respectable enough to be considered as sets themselves, but other classes would be considered to be 'too big' or 'too untidy' to be counted as sets. We are not necessarily allowed to collect *classes* together, on the other hand, to form larger entities. Thus, the 'set of all sets' is not allowed (nor is the 'class of all classes' allowed), but the 'class of all sets' is considered to be legitimate. Cantor denoted this 'supreme' class by Ω, and he attributed an almost deistic significance to it. We are not allowed to form bigger classes than Ω. The trouble with '2^{Ω}' would be that it involves 'collecting together' all the different 'subclasses' of Ω, most of which are not themselves sets, so this is disallowed.

There is something that appears rather unsatisfactory about all this. I have to confess to being decidedly dissatisfied with it myself. This procedure might be reasonable if there were a clear-cut criterion telling us when a class actually qualifies as being a set. However, the 'distinction' appears often to be made in a very circular way. A class is deemed to be a set if and only if it can itself be a member of some other class—which, to me, seems like begging the question! The trouble is that there is no obvious place to draw the line. Once a line has been drawn, it begins to appear, after a while, that the line has actually been drawn too narrowly. There seems to be no reason not to include some larger (or more unruly) classes into our club of sets. Of course, one must avoid an out-and-out contradiction. But it turns out that the more liberal are the rules for membership of the club of sets, the more powerful are the methods of mathematical proof that the set concept now provides. But open the door to this club just a crack too wide and disaster strikes—CONTRADICTION!—and the whole edifice falls to the ground! The drawing of such a line is one of the most delicate and difficult procedures in mathematics.[11]

Many mathematicians might prefer to pull back from such extreme liberalism, even taking a rigidly conservative 'constructivist' approach, according to which a set is permitted only if there is a direct construction for enabling us to tell when an element belongs to the set and when it does not. Certainly 'sets' that are defined solely by use of the axiom of choice would be a disallowed membership criterion under such strict rules! But it turns out that these extreme conservatives are no more immune from Cantor's diagonal slash than are the extreme liberals. Let us try to see, in the next section, what the trouble is.

16.6 Turing machines and Gödel's theorem

First, we need a notion of what it means to 'construct' something in mathematics. It is best that we restrict attention to subsets of the set $\mathbb{N}$ of natural numbers, at least for our primitive considerations here. We may ask which such subsets are defined 'constructively'? It is fortunate that we have at our disposal a wonderful notion, introduced by various logicians[12] of the first third of the 20th century and put on a clear footing by Alan Turing in 1936. This is the notion of *computability*; and since electronic computers have become so familiar to us now, it will probably suffice for me to refer to the actions of these physical devices rather than give the relevant ideas in terms of some precise mathematical formulation. Roughly speaking, a *computation* (or *algorithm*) is what an idealized computer would perform, where 'idealized' means that it can go on for an indefinite length of time without 'wearing out', that it never makes mistakes, and that it has an unlimited storage space. Mathematically, such an entity is effectively what is called a *Turing machine*.[13]

Any particular Turing machine $\boldsymbol{T}$ corresponds to some specific computation that can be performed on natural numbers. The action of $\boldsymbol{T}$ on the particular natural number n is written $\boldsymbol{T}(n)$, and we normally take this action to yield some (other) natural number m:

$$\boldsymbol{T}(n) = m.$$

Now, a Turing machine might have the property that it gets 'stuck' (or 'goes into a loop') because the computation that it is performing never terminates. I shall say that a Turing machine is *faulty* if it fails to terminate when applied to some natural number n. I call it *effective* if, on the other hand, it always does terminate, whatever number it is presented with.

An example of a non-terminating (faulty) Turing machine $\boldsymbol{T}$ would be the one that, when presented with n, tries to find the smallest natural number that is not the sum of n square numbers ($0^2 = 0$ included). We find $\boldsymbol{T}(0) = 1$, $\boldsymbol{T}(1) = 2$, $\boldsymbol{T}(2) = 3$, $\boldsymbol{T}(3) = 7$ (the meaning of these equations being exemplified by the last one: '7 is the smallest number that is not the sum of 3 squares'),[16.15] but when $\boldsymbol{T}$ is applied to 4, it goes on computing forever, trying to find a number that is not the sum of four squares. The cause of this particular machine's hang-up is a famous

---

[16.15] Give a rough description of how our algorithm might be performed and explain these particular values.

theorem due to the great 18th century French–Italian mathematician Joseph C. Lagrange, who was able to prove that in fact every natural number is the sum of four square numbers. (Lagrange will have a very considerable importance for us in a different context later, most particularly in Chapters 20 and 26, as we shall see!)

Each separate Turing machine (whether faulty or effective) has a certain 'table of instructions' that characterizes the particular algorithm that this particular Turing machine performs. Such a table of instructions can be completely specified by some 'code', which we can write out as a sequence of digits. We can then re-interpret this sequence as a natural number t; thus t codifies the 'program' that enables the machine to carry out its particular algorithm. The Turing machine that is thereby encoded by the natural number t will be denoted by $\boldsymbol{T}_t$. The coding may not work for all natural numbers t, but if it does not, for some reason, then we can refer to $\boldsymbol{T}_t$ as being 'faulty', in addition to those cases just considered where the machine fails to stop when applied to some n. The only effective Turing machines $\boldsymbol{T}_t$ are those which provide an answer, after a finite time, when applied to any individual n.

One of Turing's fundamental achievements was to realize that it is possible to specify a single Turing machine, called a *universal* Turing machine $\boldsymbol{U}$, which can imitate the action of any Turing machine whatever. All that is needed is for $\boldsymbol{U}$ to act first on the natural number t, specifying the particular Turing machine $\boldsymbol{T}_t$ that is to be mimicked, after which $\boldsymbol{U}$ acts upon the number n, so that it can proceed to evaluate $\boldsymbol{T}_t(n)$. (Modern general-purpose computers are, in essence, just universal Turing machines.) I shall write this combined action $\boldsymbol{U}(t, n)$, so that

$$\boldsymbol{U}(t, n) = \boldsymbol{T}_t(n).$$

We should bear in mind, however, that Turing machines, as defined here, are supposed to act only on a single natural number, rather than a pair, such as (t, n). But it is not hard to encode a pair of natural numbers as a single natural number, as we have seen earlier (e.g. in Exercise [16.8]). The machine $\boldsymbol{U}$ will itself be defined by some natural number, say u, so we have

$$\boldsymbol{U} = \boldsymbol{T}_u.$$

How can we tell whether a Turing machine is effective or faulty? Can we find some algorithm for making this decision? It was one of Turing's important achievements to show that the answer to this question is in fact 'no'! The proof is an application of Cantor's diagonal slash. We shall consider the set $\mathbb{N}$, as before, but now instead of considering *all* subsets of $\mathbb{N}$, we consider just those subsets for which it is a *computational* matter to decide whether or not an element is in the set. (These cannot be all the subsets of ℕ because the number of different computations is only $\aleph_0$, whereas the number of all subsets of ℕ is C.) Such computationally defined sets are called *recursive*. In fact any recursive subset of ℕ is defined by the output of an effective Turing machine ***T***, of the particular kind that it only outputs 0 or 1. If $\boldsymbol{T}(n) = 1$, then n is a member of the recursive set defined by ***T*** ('in'), whereas if $\boldsymbol{T}(n) = 0$, then n is not a member ('out'). We now apply the Cantor argument just as before, but now just to *recursive* subsets of ℕ. The argument immediately tells us that the set of natural numbers t for which $\boldsymbol{T}_t$ is effective cannot be recursive. There is no algorithm, applicable to any given Turing machine ***T***, for telling us whether or not ***T*** is faulty!

It is worth while looking at this reasoning a little more closely. What the Turing/Cantor argument really shows is that the set of t for which $\boldsymbol{T}_t$ is effective is not even *recursively enumerable*. What is a recursively enumerable subset of ℕ? It is a set of natural numbers for which there is an effective Turing machine ***T*** which eventually generates each member (possibly more than once) of this set when applied to 0, 1, 2, 3, 4, ... successively. (That is, m is a member of the set if and only if $m = \boldsymbol{T}(n)$ for some natural number n.) A subset S of ℕ is recursive if and only if it is recursively enumerable and its *complement* ℕ $- S$ is also recursively enumerable.[16.16] The supposed 1–1 correspondence with which the Turing/Cantor argument derives a contradiction is a recursive enumeration of the effective Turing machines. A little consideration tells us that what we have learnt is that there is no general algorithm for telling us when a Turing machine action $\boldsymbol{T}_t(n)$ will *fail* to stop.

What this ultimately tells us is that despite the hopes that one might have had for a position of 'extreme conservatism', in which the only acceptable sets would be ones—the recursive sets—whose membership is determined by clear-cut computational rules, this viewpoint immediately drives us into having to consider sets that are non-recursive. The viewpoint even encounters the fundamental difficulty that there is no computational way of *generally* deciding whether or not two recursive sets are the same or different sets, if they are defined by two different effective Turing machines $\boldsymbol{T}_t$ and $\boldsymbol{T}_s$![16.17] Moreover, this kind of problem is encountered again and again at different levels, when we try to restrict our notion of 'set' by too conservative a point of view. We are always driven to consider classes that do not belong to our previously allowed family of sets.

[16.16] Show this.

[16.17] Can you see why this is so? *Hint*: For an arbitrary Turing machine action of ***T*** applied to n, we can consider an effective Turing machine ***Q*** which has the property that $\boldsymbol{Q}(r) = 0$ if ***T*** applied to n has not stopped after r computational steps, and $\boldsymbol{Q}(r) = 1$ if it has. Take the modulo 2 sum of $\boldsymbol{Q}(n)$ with $\boldsymbol{T}_t(n)$ to get $\boldsymbol{T}_s(n)$.

These issues are closely related to the famous theorem of Kurt Gödel. He was concerned with the question of the methods of proof that are available to mathematicians. At around the turn of the 20th century, and for a good many years afterwards, mathematicians had attempted to avoid the paradoxes (such as the Russell paradox) that arose from an excessively liberal use of the theory of sets, by introducing the idea of a mathematical *formal system*, according to which there was to be laid down a collection of absolutely clear-cut rules as to what lines of reasoning are to count as a mathematical proof. What Gödel showed was that this programme will not work. In effect, he demonstrated that, if we are prepared to accept that the rules of some such formal system $\boldsymbol{F}$ are to be trusted as giving us only mathematically correct conclusions, then we must also accept, as correct, a certain clear-cut mathematical statement $\mathrm{G}(\boldsymbol{F})$, while concluding that $\mathrm{G}(\boldsymbol{F})$ is not provable by the methods of $\boldsymbol{F}$ alone. Thus, Gödel shows us how to transcend any $\boldsymbol{F}$ that we are prepared to trust.

There is a common misconception that Gödel's theorem tells us that there are 'unprovable mathematical propositions', and that this implies that there are regions of the 'Platonic world' of mathematical truths (see §1.4) that are in principle inaccessible to us. This is very far from the conclusion that we should be drawing from Gödel's theorem. What Gödel actually tells us is that whatever rules of proof we have laid down beforehand, if we already accept that those rules are trustworthy (i.e. that they do not allow us to derive falsehoods) and are not too limited, then we are provided with a new means of access to certain mathematical truths that those particular rules are not powerful enough to derive.

Gödel's result follows directly from Turing's (although historically things were the other way around). How does this work? The point about a formal system is that no further mathematical judgements are needed in order to check whether the rules of $\boldsymbol{F}$ have been correctly applied. It has to be an entirely computational matter to decide the correctness of a mathematical proof according to $\boldsymbol{F}$. We find that, for any $\boldsymbol{F}$, the set of mathematical theorems that can be proved using its rules is necessarily recursively enumerable.

Now, some well-known mathematical statements can be phrased in the form 'such-and-such Turing machine action does not terminate'. We have already seen one example, namely Lagrange's theorem that every natural number is the sum of four squares. Another even more famous example is 'Fermat's last theorem', proved at the end of the 20th century by Andrew Wiles (§1.3).[14] Yet another (but unresolved) is the well-known 'Goldbach conjecture' that every even number greater than 2 is the sum of two primes. Statements of this nature are known to mathematical logicians as *Π_1-sentences*. Now it follows immediately from Turing's argument above that the family of true Π_1-sentences constitutes a non-recursively enumerable set (i.e. one that is not recursively enumerable). Hence there are true Π_1-sentences that cannot be obtained from the rules of $\boldsymbol{F}$ (where we assume that $\boldsymbol{F}$ is trustworthy) This is the basic form of Gödel's theorem. In fact, by examining the details of this a little more closely, we can refine the argument so as to obtain the version of it stated above, and obtain a specific Π_1-sentence $\mathrm{G}(\boldsymbol{F})$ which, if we believe $\boldsymbol{F}$ to yield only true Π_1-sentences, must escape the net cast by $\boldsymbol{F}$ despite the remarkable fact that we must conclude that $\mathrm{G}(\boldsymbol{F})$ is also a true Π_1-sentence![16.18]

16.7 Sizes of infinity in physics

Finally, let us see how these issues of infinity and constructibility lie, in relation to the mathematics of our previous chapters and to our current understanding of physics. It is perhaps remarkable, in view of the close relationship between mathematics and physics, that issues of such basic importance in mathematics as transfinite set theory and computability have as yet had a very limited impact on our description of the physical world. It is my own personal opinion that we shall find that computability issues will eventually be found to have a deep relevance to future physical theory,[15] but only very little use of these ideas has so far been made in mathematical physics.[16]

With regard to the *size* of the infinities that have found value, it is rather striking that almost none of physical theory seems to need our going beyond $\mathsf{C}(=2^{\aleph_0})$, the cardinality of the real-number system $\mathbb{R}$. The cardinality of the complex field $\mathbb{C}$ is the same as that of $\mathbb{R}$ (namely C), since $\mathbb{C}$ is just $\mathbb{R}\times\mathbb{R}$ (pairs of real numbers) with certain addition and multiplication laws defined on it. Likewise, the vector spaces and manifolds that we have been considering are built from families of points that can be assigned coordinates from some $\mathbb{R}\times\mathbb{R}\times\ldots\times\mathbb{R}$ (or $\mathbb{C}\times\mathbb{C}\times\ldots\times\mathbb{C}$) or from finite (or countably many, i.e. $\aleph_0$'s worth of) such coordinate patches, and again the cardinality is C.

What about the families of functions on such spaces? If we consider, say, the family of all real-number-valued functions on some space with C points, then we find, from the above considerations, that the family has C^{C} members (being mappings from a C-element space to a C-element space). This is certainly larger than C. In fact $\mathsf{C}^{\mathsf{C}}=2^{\mathsf{C}}$. (This follows because each element of $\mathbb{R}^{\mathbb{R}}$ can be re-interpreted as a particular element of $2^{\mathbb{R}\times\mathbb{R}}$, namely as a (usually far from continuous) cross-section of the bundle $\mathbb{R}\times\mathbb{R}$, and the cardinality of $\mathbb{R}\times\mathbb{R}$ is C.) However, the *continuous* real (or complex) functions (or tensor fields, or connections) on a manifold are only C in number, because a continuous function is

[16.18] See if you can establish this.

determined once its values on the set of points with rational coordinates are known. The number of these is just $C^{\aleph_0}$, since the number of points with rational coordinates is just $\aleph_0$. But $C^{\aleph_0} = (2^{\aleph_0})^{\aleph_0} = 2^{\aleph_0 \times \aleph_0} = 2^{\aleph_0} = C$.[16.19] In §§6.4,6, we considered certain generalizations of continuous functions, leading to the very great generalization known as *hyperfunctions* (§9.7). However the number of these is again no greater than C, as they are defined by pairs of holomorphic functions (each C in number).

In §22.3, we shall be seeing that quantum theory requires the use of certain spaces, known as *Hilbert spaces*, that may have infinitely many dimensions. However, although these particular infinite-dimensional spaces differ significantly from finite-dimensional spaces, there are not more continuous functions on them than in the finite-dimensional case, and again we get C as the total number. The best bet for going higher than this is in relation to the path-integral formulation of quantum field theory (as will be discussed in §26.6), when a space of wild-looking curves (or of wild-looking physical field configurations) in spacetime are considered. However, we still seem just to get C for the total number, because despite their wildness, there is a sufficient remnant of continuity in these structures.

The notion of cardinality does not seem to be sufficiently refined to capture the appropriate concept of *size* for the spaces that are encountered in physics. Almost all the spaces of significance simply have C points in them. However, there is a vast difference in the 'sizes' of these spaces, where in the first instance we think of this 'size' simply as the dimension of the vector space or manifold $\mathcal{M}$ under consideration. This dimension of $\mathcal{M}$ may be a natural number (e.g. 4, in the case of ordinary spacetime, or 6×10^{19}, in the case of the phase space considered in §12.1), or it could be infinity, such as with (most of) the Hilbert state-spaces that arise in quantum mechanics. Mathematically, the simplest infinite-dimensional Hilbert space is the space of sequences $(z_1, z_2, z_3, \ldots)$ of complex numbers for which the infinite sum $|z_1|^2 + |z_2|^2 + |z_3|^2 + \ldots$ converges. In the case of an infinite-dimensional Hilbert space, it is most appropriate to think of this dimensionality as being $\aleph_0$. (There are various subtleties about this, but it is best not to get involved with these here.) For an n-real-dimensional space, I shall say that it has '∞^n' points (which expresses that this continuum of points is organized in an n-dimensional array). In the infinite-dimensional case, I shall refer to this as '∞^∞' points.

We are also interested in the spaces of various kinds of field defined on $\mathcal{M}$. These are normally taken to be smooth, but sometimes they are more general (e.g. distributions), coming within the compass of hyperfunction theory (see §9.7). They may be subject to (partial) differential equations,

[16.19] Explain why $(A^B)^C$ may be identified with $A^{B \times C}$, for sets A, B, C.

which restrict their freedom. If they are not so restricted, then they count as 'functions of n variables', for an n-dimensional $\mathcal{M}$ (where $n=4$ for standard spacetime). At each point, the field may have k independent components. Then I shall say that the freedom in the field is $\infty^{k\infty^n}$. The explanation for this notation[17] is that the fields may be thought (crudely and locally) to be maps from a space with ∞^n points to a space with ∞^k points, and we take advantage of the (formal) notational relation

$$(\infty^k)^{\infty^n} = \infty^{k\infty^n}.$$

When the fields are restricted by appropriate partial differential equations, then it may be that they will be completely determined by the *initial data* for the fields (see §27.1 particularly), that is, by some subsidiary field data specified on some lower-dimensional space $\mathcal{S}$ of, say, q dimensions. If the data can be expressed *freely* on $\mathcal{S}$ (which means, basically, not subject to *constraints*, these being differential or algebraic equations that the data would have to satisfy on $\mathcal{S}$), and if these data consist of r independent components at each point of $\mathcal{S}$, then I shall say that the freedom in the field is $\infty^{r\infty^q}$. In many cases, it is not an altogether easy matter to find r and q, but the important thing is that they are *invariant* quantities, independent of how the fields may be re-expressed in terms of other equivalent quantities.[18] These matters will have considerable importance for us later (see §23.2, §§31.10–12, 15–17).

Notes

Section 16.2

16.1. See Howie (1989), pp. 269–71; Hirschfeld (1998), p. 098; magic discs are equivalent to what are called *perfect difference sets*.

16.2. It is apparently unknown whether magic discs exist (necessarily *not* arising from a $\mathbb{P}^2(\mathbb{F}_q)$) for which the theorem of Desargues (or, equivalently, of Pappos) ever *fails*—or whether *n-point* projective planes (necessarily non-Desargueian and non-Pappian) exist for $n \neq q^2 + q + 1$ with q a prime power.

16.3. A physical role for octonions has nevertheless been argued for, from time to time (see, for example, Gürsey and Tze 1996; Dixon 1994; Manogue and Dray 1999; Dray and Manogue 1999); but there are fundamental difficulties for the construction of a general 'octonionic quantum mechanics' (Adler 1995), the situation with regard to a 'quaternionic quantum mechanics' being just a little more positive. Another number system, suggested on occasion as a candidate for a significant physical role, is that of 'p-adic numbers'. These constitute number systems to which the rules of *calculus* apply, and they can be expressed like ordinary decimally expanded real numbers, except that the digits represent

0, 1, 2, 3, ..., $p-1$ (where p is the chosen prime number) and they are allowed to be infinite *the opposite way around* from what is the case with ordinary decimals (and we do not need minus signs). For example,

$$......24033200411.3104$$

represents a particular 5-adic number. The rules for adding and multiplying are just the same as they would be for 'ordinary' p-ary arithmetic (in which the symbol '10' stands for the prime p, etc.). See Mahler (1981); Gouvea (1993); Brekke and Frend (1993); Vladimirov and Volovich (1989); Pitkäenen (1995).

Section 16.3

16.4. The modern mathematical terminology is to call this a set isomorphism. There are other words such as 'endomorphism', 'epimorphism', and 'monomorphism' (or just 'morphism') that mathematicians tend to use in a general context for characterizing mappings between one set or structure to another. I prefer to avoid this kind of terminology in this particular book, as I think it takes rather more effort to get accustomed to it than is worth while for our needs.

16.5. For some even earlier deliberations of this nature, see Moore (1990), Chap. 3.

16.6. Recall from Note 15.5 that I have been prepared to adopt an abuse of notation whereby $\mathbb{N} - 0$ indeed stands for the set of non-zero natural numbers. There is the irony here that if one were to adopt the seemingly 'more correct' $\mathbb{N} - \{0\}$, while also adopting the procedures of §3.4 whereby $\{0\} = 1$, we should be landed with the even more confusing '$\mathbb{N} - 1$' for the set under consideration!

16.7. See Wagon (1985); see Runde (2002) for a popular account.

Section 16.5

16.8. Similar remarks apply to Cantor's *generalized* continuum hypothesis: $2^{\aleph_\alpha} = \aleph_{\alpha+1}$ (where α is now an 'ordinal number', whose definition I have not discussed here), and these remarks also apply to the axiom of choice.

16.9. See Russell (1903), p. 362, second footnote [in 1937 edn].

16.10. See Van Heijenoort (1967), p. 114.

16.11. See Woodin (2001) for a novel approach to these matters. For general references on the foundations of mathematics, see Abian (1965) and Wilder (1965).

Section 16.6

16.12. These precursors of Turing were, in the main, Alonzo Church, Haskell B. Curry, Stephen Kleene, Kurt Gödel, and Emil Post; see Gandy (1988).

16.13. For a detailed description of a Turing machine, see Penrose (1989), Chap. 2, for example, Davis (1978), or the original reference: Turing (1937).

16.14. See Singh (1997); Wiles (1995).

Section 16.7

16.15. See Penrose (1989, 1994, 1997a,b).

16.16. See Komar (1964); Geroch and Hartle (1986), §34.7.

16.17. I owe this useful notation to John A. Wheeler, see Wheeler (1960), p. 67.

16.18. See Cartan (1945) especially §§68,69 on pp. 75, 76 (original edition). Some care needs to be taken in order to ensure that the quantity r in $\infty^{r\infty^q}$ is correctly counted. Two systems may be equivalent, but having r values that nevertheless appear at first sight to differ. However, there can be no ambiguity in the determination of the value of q. The rigorous modern treatment of these issues makes things clearer; it is given in terms of the theory of *jet bundles* (see Bryant *et al.* 1991). It may be mentioned that there is a refinement of Wheeler's notation (see Penrose 2003) where, for example, $\infty^{2\infty^2+3\infty^1+5}$ stands for 'the fields depend on 2 functions of 2 variables, 3 functions of 1 variable, and 5 constants'. We are thus led to consider expressions like $\infty^{p(\infty)}$, where p denotes a polynomial with non-negative integer coefficients.

17
Spacetime

17.1 The spacetime of Aristotelian physics

FROM now on, in this book, our attention will be turned from the largely mathematical considerations that have occupied us in earlier chapters, to the actual pictures of the physical world that theory and observation have led us into. Let us begin by trying to understand that arena within which all the phenomena of the physical universe appear to take place: *spacetime*. We shall find that this notion plays a vital role in most of the rest of this book!

We must first ask why 'spacetime'?[1] What is wrong with thinking of space and time separately, rather than attempting to unify these two seemingly very different notions together into one? Despite what appears to be the common perception on this matter, and despite Einstein's quite superb use of this idea in his framing of the general theory of relativity, spacetime was not Einstein's original idea nor, it appears, was he particularly enthusiastic about it when he first heard of it. Moreover, if we look back with hindsight to the magnificent older relativistic insights of Galileo and Newton, we find that they, too, could in principle have gained great benefit from the spacetime perspective.

In order to understand this, let us go much farther back in history and try to see what kind of spacetime structure would have been appropriate for the dynamical framework of Aristotle and his contemporaries. In Aristotelian physics, there is a notion of Euclidean 3-space $\mathbb{E}^3$ to represent physical space, and the points of this space retain their identity from one moment to the next. This is because the state of rest is dynamically preferred, in the Aristotelian scheme, from all other states of motion. We take the attitude that a particular spatial point, at one moment of time, is the *same* spatial point, at a later moment of time, if a particle situated at that point remains at rest from one moment to the next. Our picture of reality is like the screen in a cinema theatre, where a particular point on the screen retains its identity no matter what kinds of vigorous movement might be projected upon it. See Fig. 17.1.

Fig. 17.1 Is physical motion like that perceived on a cinema screen? A particular point on the screen (here marked '×') retains its identity no matter what movement is projected upon it.

Time, also, is represented as a Euclidean space, but as a rather trivial one, namely the 1-dimensional space $\mathbb{E}^1$. Thus, we think of time, as well as physical space, as being a 'Euclidean geometry', rather than as being just a copy of the real line $\mathbb{R}$. This is because $\mathbb{R}$ has a preferred element 0, which would represent the 'zero' of time, whereas in our 'Aristotelian' dynamical view, there is to be no preferred origin. (In this, I am taking an idealized view of what might be called 'Aristotelian dynamics', or 'Aristotelian physics', and I take no viewpoint with regard to what the *actual* Aristotle might have thought!)[2] Had there been a preferred 'origin of time', the dynamical laws could be envisaged as changing when time proceeds away from that preferred origin. With no preferred origin, the laws must remain the same for all time, because there is no preferred *time parameter* which these laws can depend upon.

Likewise, I am taking the view that there is to be no preferred spatial origin, and that space continues indefinitely in all directions, with complete uniformity in the dynamical laws (again, irrespective of what the actual Aristotle might have believed!). In Euclidean geometry, whether 1-dimensional or 3-dimensional, there is a notion of *distance*. In the 3-dimensional spatial case, this is to be ordinary Euclidean distance (measured in metres, or feet, say); in the 1-dimensional case, this distance is the ordinary time interval (measured, say, in seconds).

In Aristotelian physics—and, indeed, in the later dynamical scheme(s) of Galileo and Newton—there is an absolute notion of temporal *simultaneity*. Thus, it has absolute meaning to say, according to such dynamical schemes, that the time here, *at this very moment*, as I sit typing this in my office at home in Oxford, is 'the same time' as some event taking place on the Andromeda galaxy (say the explosion of some supernova star). To return to our analogy of the cinema screen, we can ask whether two projected images, occurring at two widely separated places on the screen, are taking place simultaneously or not. The answer here is clear. The events are to be taken as simultaneous if and only if they occur in the same projected frame. Thus, not only do we have a clear notion of whether or not two (temporally separated) events occur at the same spatial location on the screen, but we also have a clear notion of whether or not two (spatially separated) events occur at the same time. Moreover, if the spatial locations of the two events are different, we have a clear notion of the *distance* between them, whether or not they occur at the same time (i.e. the distance measured along the screen); also, if the times of the two events are different, we have a clear notion of the *time interval* between them, whether or not they occur at the same place.

What this tells us is that, in our Aristotelian scheme, it is appropriate to think of *spacetime* as simply the product

$$\mathcal{A} = \mathbb{E}^1 \times \mathbb{E}^3,$$

which I shall call *Aristotelian spacetime*. This is simply the space of pairs $(t, \mathbf{x})$, where t is an element of $\mathbb{E}^1$, a 'time', and $\mathbf{x}$ is an element of $\mathbb{E}^3$, a 'point in space'. (See Fig. 17.2.) For two different points of $\mathbb{E}^1 \times \mathbb{E}^3$, say $(t, \mathbf{x})$ and $(t', \mathbf{x}')$—i.e. two different *events*—we have a well-defined notion of their spatial separation, namely the distance between the points $\mathbf{x}$ and $\mathbf{x}'$ of $\mathbb{E}^3$, and we also have a well-defined notion of their time difference, namely the separation between t and t' as measured in $\mathbb{E}^1$. In particular, we know whether or not two events occur at the same place (vanishing of spatial displacement) and whether or not they take place at the same time (vanishing of time difference).

17.2 Spacetime for Galilean relativity

Now let us see what notion of spacetime is appropriate for the dynamical scheme introduced by Galileo in 1638. We wish to incorporate the *principle of Galilean relativity* into our spacetime picture. Let us try to

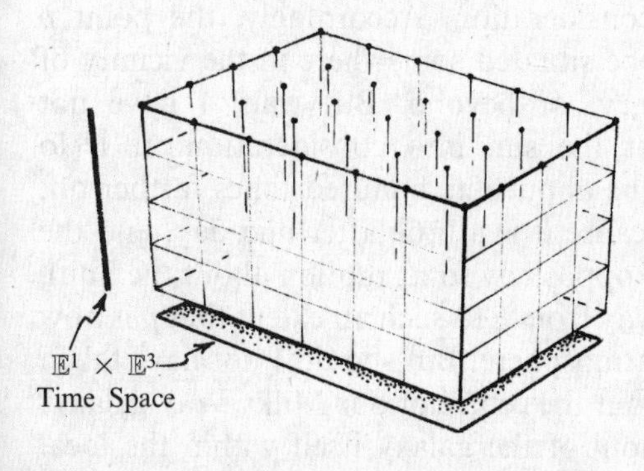

Fig. 17.2 Aristotelian spacetime $\mathcal{A} = \mathbb{E}^1 \times \mathbb{E}^3$ is the space of pairs $(t, \mathbf{x})$, where t ('time') ranges over a Euclidean 1-space $\mathbb{E}^1$, and $\mathbf{x}$ ('point in space') ranges over a Euclidean 3-space $\mathbb{E}^3$.

recall what this principle asserts. It is hard to do better than quote Galileo himself (in a translation due to Stillman Drake[3] which I give here in abbreviated form only; and I strongly recommend an examination of the quote as a whole, for those who have access to it):

> Shut yourself up with some friend in the main cabin below decks on some large ship, and have with you some flies, butterflies, and other small flying animals ... hang up a bottle that empties drop by drop into a wide vessel beneath it ... have the ship proceed with any speed you like, so long as the motion is uniform and not fluctuating this way and that. ... The droplets will fall ... into the vessel beneath without dropping toward the stern, although while the drops are in the air the ship runs many spans ... the butterflies and flies will continue their flights indifferently toward every side, nor will it ever happen that they are concentrated toward the stern, as if tired out from keeping up with the course of the ship. ...

What Galileo teaches us is that the dynamical laws are precisely the same when referred to any uniformly moving frame. (This was an essential ingredient of his wholehearted acceptance of the Copernican scheme, whereby the Earth is allowed to be in motion without our directly noticing this motion, as opposed to its necessarily stationary status according to the earlier Aristotelian framework.) There is nothing to distinguish the physics of the state of rest from that of uniform motion. In terms of what has been said above, what this tells us is that there is no dynamical meaning to saying that a particular point in space is, or is not, the same point as some chosen point in space at a later time. In other words, our cinema-screen analogy is inappropriate! There is no background space—a 'screen'—which remains fixed as time evolves. We cannot meaningfully say that a particular point p in space (say, the point of the exclamation mark on the keyboard of my laptop) is, or is not, the *same* point in space as it was a minute ago. To address this issue more forcefully, consider the rotation of the Earth. According to this motion, a point fixed to the Earth's surface (at the latitude of Oxford, say) will have moved by some 10 miles in the minute under consideration. Accordingly, the point p that I had just selected will now be situated somewhere in the vicinity of the neighbouring town of Witney, or beyond. But wait! I have not taken the Earth's motion about the sun into consideration. If I do that, then I find that p will now be about one hundred times farther off, but in the opposite direction (because it is a little after mid-day, and the Earth's surface, here, now moves oppositely to its motion about the Sun), and the Earth will have moved away from p to such an extent that p is now beyond the reach of the Earth's atmosphere! But should I not have taken into account the sun's motion about the centre of our Milky Way galaxy? Or what about the 'proper motion' of the galaxy itself within the local group? Or the motion of the local group about the centre of the Virgo cluster of which it is a tiny part, or of the Virgo cluster in relation to the vast Coma supercluster, or perhaps the Coma cluster towards 'the Great Attractor' (§27.11)?

Clearly we should take Galileo seriously. There is no meaning to be attached to the notion that any particular point in space a minute from now is to be judged as the *same* point in space as the one that I have chosen. In Galilean dynamics, we do not have just one Euclidean 3-space $\mathbb{E}^3$, as an arena for the actions of the physical world evolving with time, we have a *different* $\mathbb{E}^3$ for each moment in time, with no natural identification between these various $\mathbb{E}^3$s.

It may seem alarming that our very notion of physical space seems to be of something that evaporates completely as one moment passes, and reappears as a completely different space as the next moment arrives! But here the mathematics of Chapter 15 comes to our rescue, for this situation is just the kind of thing that we studied there. *Galilean spacetime* $\mathcal{G}$ is not a product space $\mathbb{E}^1 \times \mathbb{E}^3$, it is a *fibre bundle*[4] with base space $\mathbb{E}^1$ and fibre $\mathbb{E}^3$! In a fibre bundle, there is no pointwise identification between one fibre and the next; nevertheless the fibres fit together to form a connected whole. Each spacetime event is naturally assigned a *time*, as a particular element of one specific 'clock space' $\mathbb{E}^1$, but there is no natural assignment of a spatial location in one specific 'location space' $\mathbb{E}^3$. In the bundle language of §15.2, this natural assignment of a time is achieved by the *canonical projection* from $\mathcal{G}$ to $\mathbb{E}^1$. (See Fig. 17.3; compare also Fig. 15.2.)

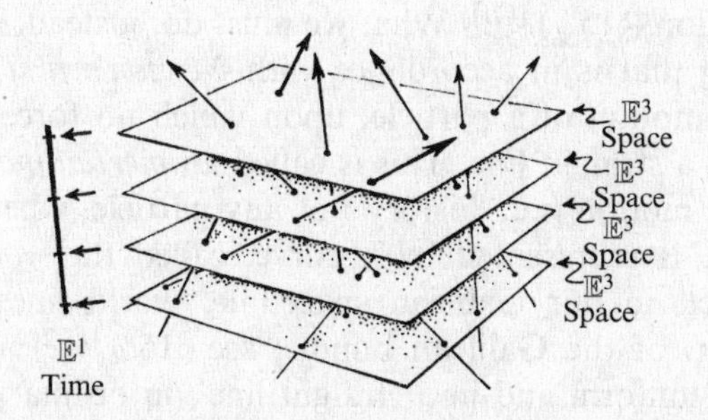

Fig. 17.3 Galilean spacetime $\mathcal{G}$ is a fibre bundle with base space $\mathbb{E}^1$ and fibre $\mathbb{E}^3$, so there is no given pointwise identification between different $\mathbb{E}^3$ fibres (no absolute space), whereas each spacetime event is assigned a time via the canonical projection (absolute time). (Compare Fig. 15.2, but the canonical projection to the base is here depicted horizontally.) Particle histories (world lines) are cross-sections of the bundle (compare Fig. 15.6a), the inertial particle motions being depicted here as what $\mathcal{G}$'s structure specifies, that is: 'straight' world lines.

17.3 Newtonian dynamics in spacetime terms

This 'bundle' picture of spacetime is all very well, but how are we to express the *dynamics* of Galileo–Newton in terms of it? It is not surprising that Newton, when he came to formulate his laws of dynamics, found himself driven to a description in which he appeared to favour a notion of 'absolute space'. In fact, Newton was, at least initially, as much of a Galilean relativist as was Galileo himself. This is made clear from the fact that in his original formulation of his laws of motion, he explicitly stated the Galilean principle of relativity as a fundamental law (this being the principle that physical action should be blind to a change from one uniformly moving reference frame to another, the notion of *time* being absolute, as is manifested in the picture above of Galilean spacetime $\mathcal{G}$).

He had originally proposed five (or six) laws, law 4 of which was indeed the Galilean principle,[5] but later he simplified them, in his published *Principia*, to the three 'Newton's laws' that we are now familiar with. For he had realized that these were sufficient for deriving all the others. In order to make the framework for his laws precise, he needed to adopt an 'absolute space' with respect to which his motions were to be described. Had the notion of a 'fibre bundle' been available at the time (admittedly a far-fetched possibility), then it would have been conceivable for Newton to formulate his laws in a way that is completely 'Galilean-invariant'. But without such a notion, it is hard to see how Newton could have proceeded without introducing some concept of 'absolute space', which indeed he did.

What kind of structure must we assign to our 'Galilean spacetime' $\mathcal{G}$? It would certainly be far too strong to endow our fibre bundle $\mathcal{G}$ with a bundle connection (§15.7).[17.1] What we must do, instead, is to provide it with something that is in accordance with *Newton's first law*. This law states that the motion of a particle, upon which no forces act, must be uniform and in a straight line. This is called an *inertial motion*. In spacetime terms, the motion (i.e. 'history') of any particle, whether in inertial motion or not, is represented by a curve, called the *world line* of the particle. In fact, in our Galilean spacetime, world lines must always be *cross-sections* of the Galilean bundle; see §15.3.[17.2] and Fig. 17.3.) The notion of 'uniform and in a straight line', in ordinary spatial terms (an inertial motion), is interpreted simply as 'straight', in spacetime terms. Thus, the Galilean bundle $\mathcal{G}$ must have a structure that encodes the notion of 'straightness' of world lines. One way of saying this is to assert that $\mathcal{G}$ is an *affine* space (§14.1) in which the affine structure, when restricted to individual $\mathbb{E}^3$ fibres, agrees with the Euclidean affine structure of each $\mathbb{E}^3$.

[17.1] Why?

[17.2] Explain the reason for this.

Another way is simply to specify the ∞^6 family of straight lines that naturally resides in $\mathbb{E}^1 \times \mathbb{E}^3$ (the 'Aristotelian' uniform motions) and to take these over to provide the 'straight-line' structure of the Galilean bundle, while 'forgetting' the actual product structure of the Aristotelian spacetime $\mathcal{A}$. (Recall that ∞^6 means a 6-dimensional family; see §16.7.) Yet another way is to assert that the Galilean spacetime, considered as a manifold, possesses a connection which has both vanishing curvature and vanishing torsion (which is quite different from it possessing a bundle connection, when considered as a bundle over $\mathbb{E}^1$).[17.3]

In fact, this third point of view is the most satisfactory, as it allows for the generalizations that we shall be needing in §§17.5,9 in order to describe gravitation in accordance with Einstein's ideas. Having a connection defined on $\mathcal{G}$, we are provided with a notion of *geodesic* (§14.5), and these geodesics (apart from those which are simply straight lines in individual $\mathbb{E}^3$s) define Newton's *inertial motions*. We can also consider world lines that are not geodesics. In ordinary spatial terms, these represent particle motions that accelerate. The actual magnitude of this acceleration is measured, in spacetime terms, as a curvature of the world-line.[17.4] According to *Newton's second law*, this acceleration is equal to the total force on the particle, divided by its mass. (This is Newton's $f = ma$, in the form $a = f \div m$, where a is the particle's acceleration, m is its mass, and f is the total force acting upon it.) Thus, the curvature of a world line, for a particle of given mass, provides a direct measure of the total force acting on that particle.

In standard Newtonian mechanics, the total force on a particle is the (vector) *sum* of contributions from all the other particles (Fig. 17.4a). In any particular $\mathbb{E}^3$ (that is, at any one time), the contribution to the force on one particle, from some other particle, acts in the line joining the two that lies in that particular $\mathbb{E}^3$. That is to say, it acts *simultaneously* between the two particles. (See Fig. 17.4b.) *Newton's third law* asserts that the force on one of these particles, as exerted by the other, is always equal in magnitude and opposite in direction to the force on the other as exerted by the one. In addition, for each different variety of force, there is a *force law*, informing us what function of the spatial distance between the particles the magnitude of that force should be, and what parameters should be used for each type of particle, describing the overall scale for that force. In the particular case of gravity, this function is taken to be the inverse square of the distance, and the overall scale is a certain constant, called Newton's gravitational constant G, multiplied by the product of the two masses

---

[17.3] Explain these three ways more thoroughly, showing why they all give the same structure.

[17.4] Try to write down an expression for this curvature, in terms of the connection ∇. What normalization condition on the tangent vectors is needed (if any)?

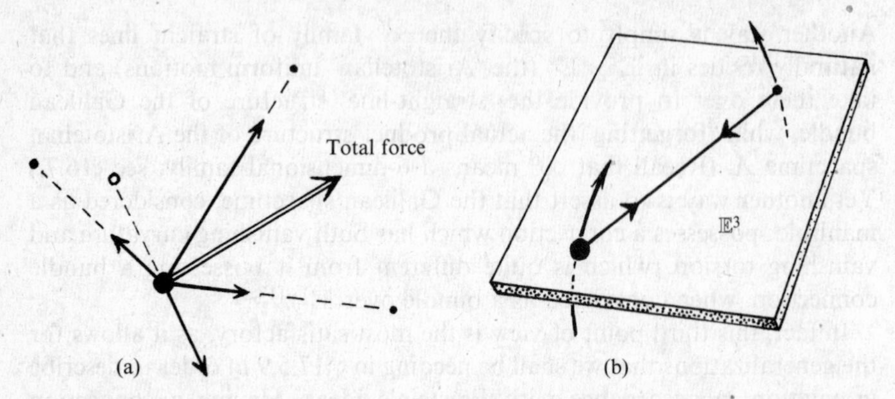

Fig. 17.4 (a) Newtonian force: at any one time, the total force on a particle (double shafted arrow) is the vector sum of contributions (attractive or repulsive) from all other particles. (b) Two particle world lines and the force between them, acting 'instantaneously', in a line joining the two particles, at any one moment, within the particular $\mathbb{E}^3$ that the moment defines. Newton's Third Law asserts that force on one, as exerted by the other, is equal in magnitude and opposite in direction to the force on the other as exerted by the one.

involved. In terms of symbols, we get Newton's well-known formula for the attractive force on a particle of mass m, as exerted by another particle of mass M, a distance r away from it, namely

$$\frac{GmM}{r^2}.$$

It is remarkable that, from just these simple ingredients, a theory of extraordinary power and versatility arises, which can be used with great accuracy to describe the behaviour of macroscopic bodies (and, for most basic considerations, submicroscopic particles also), so long as their speeds are significantly less than that of light. In the case of gravity, the accordance between theory and observation is especially clear, because of the very detailed observations of the planetary motions in our solar system. Newton's theory is now found to be accurate to something like one part in 10^7, which is an extremely impressive achievement, particularly since the accuracy of data that Newton had to go on was only about one ten-thousandth of this (one part in 10^3).

17.4 The principle of equivalence

Despite this extraordinary precision, and despite the fact that Newton's great theory remained virtually unchallenged for nearly two and one half centuries, we now know that this theory is not absolutely precise; moreover, in order to improve upon Newton's scheme, Einstein's deeper and very revolutionary perspective with regard to the nature of gravitation was required. Yet, this particular perspective does not, in itself, change Newton's theory at all, with regard to any observational consequences. The changes come about only when Einstein's perspective is combined with other considerations that relate to the finiteness of the speed of light and the ideas of special relativity, which will be described in §§17.6–8. The full combination, yielding Einstein's *general* relativity, will be given in qualitative terms in §17.9 and in fuller detail in §§19.6–8.

What, then, is Einstein's deeper perspective? It is the realization of the fundamental importance of the *principle of equivalence*. What is the principle of equivalence? The essential idea goes back (again!) to the great Galileo himself (at the end of the 16th century—although there were precursors even before him, namely Simon Stevin in 1586, and others even earlier, such as Ioannes Philiponos in the 5th or 6th century). Recall Galileo's (alleged) experiment, which consisted of dropping two rocks, one large and one small, from the top of the Leaning Tower of Pisa (Fig. 17.5a). Galileo's great insight was that each of the two would fall at the same rate, assuming that the effects of air resistance can be neglected. Whether or not he actually dropped rocks from the Leaning Tower, he certainly performed other experiments which convinced him of this conclusion.

Fig. 17.5 (a) Galileo's (alleged) experiment. Two rocks, one large and one small, are dropped from the top of Leaning Tower of Pisa. Galileo's insight was that if the effects of air resistance can be ignored, each would fall at the same rate. (b) Oppositely charged pith balls (of equal small mass), in an electric field, directed towards the ground. One charge would 'fall' downwards, but the other would rise upwards.

Now the first point to make here is that this is a particular property of the *gravitational* field, and it is not to be expected for any other force acting on bodies. The property of gravity that Galileo's insight depends upon is the fact that the strength of the gravitational force on a body, exerted by some given gravitational field, is proportional to the *mass* of that body, whereas the resistance to motion (the quantity m appearing in Newton's second law) is also the mass. It is useful to distinguish these two mass notions and call the first the *gravitational mass* and the second, the *inertial mass*. (One might also choose to distinguish the *passive* from the *active* gravitational mass. The passive mass is the contribution m in Newton's inverse square formula GmM/r^2, when we consider the gravitational force on the m particle due to the M particle. When we consider the force on the M particle due to the m particle, then the mass m appears in its *active* role. But Newton's third law decrees that passive and active masses be equal, so I am not going to distinguish between these two here.[6]) Thus, Galileo's insight depends upon the equality (or, more correctly, the proportionality) of the gravitational and inertial mass.

From the perspective of Newton's overall dynamical scheme, it would appear to be a fluke of Nature that the inertial and gravitational masses are the same. If the field were not gravitational but, say, an electric field, then the result would be completely different. The electric analogue of passive gravitational mass is *electric charge*, while the role of inertial mass (i.e. resistance to acceleration) is precisely the same as in the gravitational case (i.e. still the m of Newton's second law $f = ma$). The difference is made particularly obvious if the analogue of Galileo's pair of rocks is taken to be a pair of pith balls of equal small mass but of opposite charge. In a background electric field directed towards the ground, one charge would 'fall' downwards, but the other would rise upwards—an acceleration in completely the opposite direction! (See Fig. 17.5b.) This can occur because the electric charge on a body has no relation to its inertial mass, even to the extent that its sign can be different. Galileo's insight does not apply to electric forces; it is a particular feature of gravity alone.

Why is this feature of gravity called 'the principle of equivalence'? The 'equivalence' refers to the fact that a uniform gravitational field is equivalent to an acceleration. The effect is a very familiar one in air travel, where it is possible to get a completely wrong idea of where 'down' is from inside an aeroplane that is performing an accelerated motion (which might just be a change of its direction). The effects of acceleration and of the Earth's gravitational field cannot be distinguished simply by how it 'feels' inside the plane, and the two effects can add up in two different directions to provide you with some feeling of where down 'ought to be' which (perhaps to your surprise upon looking out of the window) may be distinctly different from the actual downward direction.

To see why this equivalence between acceleration and the effects of gravity is really just Galileo's insight described above, consider again his falling rocks, as they descend together from the top of the Leaning Tower. Imagine an insect clinging to one of the rocks and looking at the other. To the insect, the other rock appears simply to hover without motion, as though there were no gravitational field at all. (See Fig. 17.6a.) The acceleration that the insect partakes of, when falling with the rocks, cancels out the gravitational field, and it is as though gravity were completely absent—until rocks and insect all hit the ground, and the 'gravity-free' experience[7] comes abruptly to an end.

We are familiar with astronauts also having 'gravity-free' experiences—but they avoid our insect's awkward abrupt end to these experiences by being in orbit around the Earth (Fig. 17.6b) (or in an aeroplane that comes out of its dive in the nick of time!). Again they are just falling freely, like the insect, but with a more judiciously chosen path. The fact that gravity can be cancelled by acceleration in this way (by use of the principle of equivalence) is a direct consequence of the fact that (passive) gravitational mass is the same as (or is proportional to) inertial mass, the very fact underlying Galileo's great insight.

If we are to take seriously this equivalence principle, then we must take a different view from the one that we adopted in §17.3, with regard to what should count as an 'inertial motion'. Previously, an inertial motion was distinguished as the kind of motion that occurs when a particle is subject to a zero total external force. But with gravity we have a difficulty. Because of the principle of equivalence, there is no local way of telling whether a

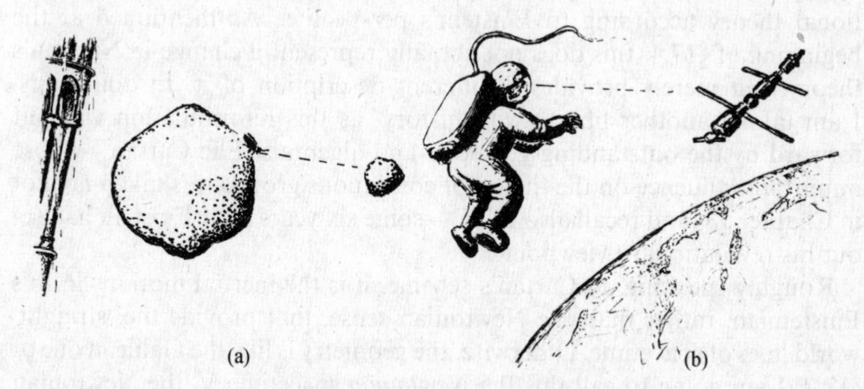

Fig. 17.6 (a) To an insect clinging to one rock of Fig. 17.5a, the other rock appears simply to hover without motion, as though gravitational field is absent. (b) Similarly, a freely orbiting astronaut has gravity-free experience, and the space station appears to hover without motion, despite the obvious presence of the Earth.

gravitational force is acting or whether what 'feels' like a gravitational force may just be the effect of an acceleration. Moreover, as with our insect on Galileo's rock or our astronaut in orbit, the gravitational force can be eliminated by simply falling freely with it. And since we can eliminate the gravitational force this way, we must take a different attitude to it. This was Einstein's profoundly novel view: regard the *inertial motions* as being those motions that particles take when the total of *non*-gravitational forces acting upon them is zero, so they must be falling freely with the gravitational field (so the *effective* gravitational force is also reduced to zero). Thus, our insect's falling trajectory and our astronauts' motion in orbit about the Earth must both count as inertial motions. On the other hand, someone just standing on the ground is *not* executing an inertial motion, in the Einsteinian scheme, because standing still in a gravitational field is not a free-fall motion. To Newton, that would have counted as inertial, because 'the state of rest' must always count as 'inertial' in the Newtonian scheme. The gravitational force acting on the person is compensated by the upward force exerted by the ground, but they are not separately zero as Einstein requires. On the other hand, the Einstein-inertial motions of the insect or astronaut are, according to Newton, *not* inertial.

17.5 Cartan's 'Newtonian spacetime'

How do we incorporate Einstein's notion of an 'inertial' motion into the structure of spacetime? As a step in the direction of the full Einstein theory, it will be helpful to consider a reformulation of Newton's gravitational theory according to Einstein's perspective. As mentioned at the beginning of §17.4, this does not actually represent a change in Newton's theory, but merely provides a different description of it. In doing this, I am taking another liberty with history, as this reformulation was put forward by the outstanding geometer and algebraist Élie Cartan—whose important influence on the theory of continuous groups was taken note of in Chapter 13 (and recall also §12.5)—some six years after Einstein had set out his revolutionary viewpoint.

Roughly speaking, in Cartan's scheme, it is the inertial motions in this Einsteinian, rather than the Newtonian sense, that provide the 'straight' world lines of spacetime. Otherwise, the geometry is like the Galilean one of §17.2. I am going to call this the *Newtonian* spacetime $\mathcal{N}$, the Newtonian gravitational field being completely encoded into its structure. (Perhaps I should have called it 'Cartannian', but that is an awkward word. In any case, Aristotle didn't know about product spaces, nor Galileo about fibre bundles!)

The spacetime $\mathcal{N}$ is to be a bundle with base space $\mathbb{E}^1$ and fibre $\mathbb{E}^3$, just as was the case for our previous Galilean spacetime $\mathcal{G}$. But now there is to be some kind of structure on $\mathcal{N}$ different from that of $\mathcal{G}$, because the family of 'straight' world lines that represents inertial motions is different; see Fig. 17.7a. At least it is essentially different in all cases except those in which the gravitational field can be eliminated completely by some choice of freely falling global reference frame. One such exception would be a Newtonian gravitational field that is completely constant (both in magnitude and in direction) over the whole of space, but perhaps varying in time. To an observer who falls freely in such a field, it would appear that there is

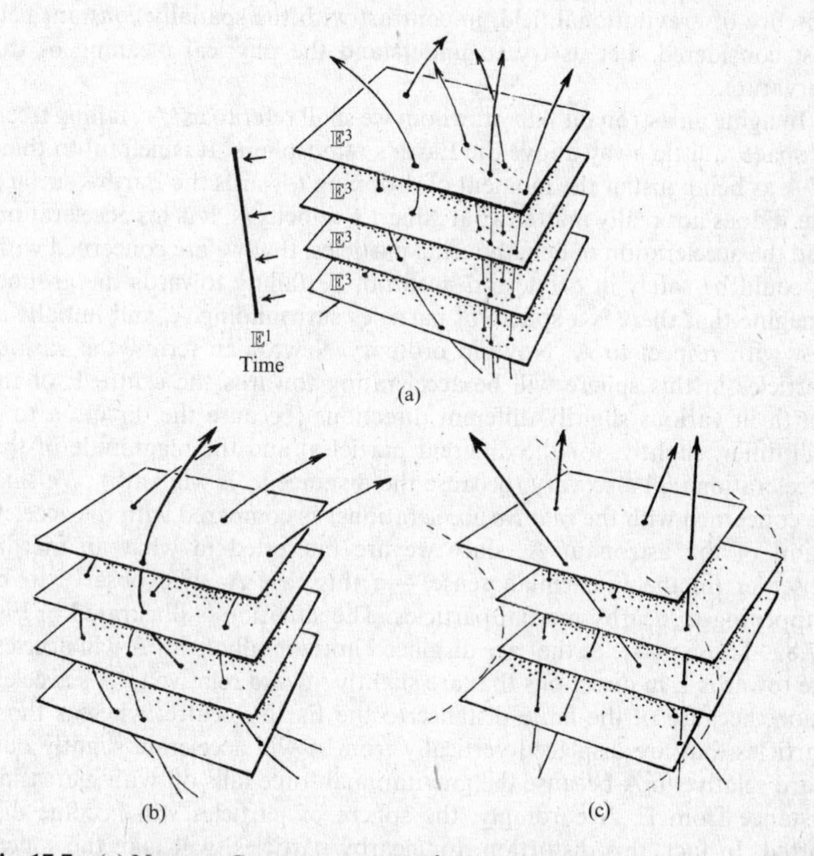

Fig. 17.7 (a) Newton–Cartan spacetime $\mathcal{N}$, like the particular Galilean case $\mathcal{G}$, is a bundle with base-space $\mathbb{E}^1$ and fibre $\mathbb{E}^3$. Its structure is provided by the family of motions, 'inertial' in Einstein's sense, of free fall under gravity. (b) The special case of a Newtonian gravitational field *constant* over all space. (c) Its structure is completely equivalent to that of $\mathcal{G}$, as can be seen by 'sliding' the $\mathbb{E}^3$ fibres horizontally until the world lines of free fall are all straight.

no field at all![17.5] In such a case, the structure of $\mathcal{N}$ would be the same as that of $\mathcal{G}$ (Fig. 17.7b,c). But most gravitational fields count as 'essentially different' from the absence of a gravitational field. Can we see why? Can we recognize when the structure of $\mathcal{N}$ is different from that of $\mathcal{G}$? We shall come to this in a moment.

The idea is that the manifold $\mathcal{N}$ is to possess a connection, just as was the case for the particular case $\mathcal{G}$. The geodesics of this connection, ∇ (see §14.5), are to be the 'straight' world lines that represent inertial motions in the Einsteinian sense. This connection will be torsion-free (§14.4), but it will generally possess curvature (§14.4). It is the presence of this curvature that makes some gravitational fields 'essentially different' from the absence of gravitational field, in contrast with the spatially constant field just considered. Let us try to understand the physical meaning of this curvature.

Imagine an astronaut Albert, whom we shall refer to as 'A', falling freely in space, a little away above the Earth's atmosphere. It is helpful to think of A as being just at the moment of dropping towards the Earth's surface, but it does not really matter what Albert's velocity is; it is his acceleration, and the acceleration of neighbouring particles, that we are concerned with. A could be safely in orbit, and need not be falling towards the ground. Imagine that there is a sphere of particles surrounding A, and initially at rest with respect to A. Now, in ordinary Newtonian terms, the various particles in this sphere will be accelerating towards the centre E of the Earth in various slightly different directions (because the direction to E will differ, slightly, for the different particles) and the magnitude of this acceleration will also vary (because the distance to E will vary). We shall be concerned with the *relative* accelerations, as compared with the acceleration of the astronaut A, since we are interested in what an inertial observer (in the Einsteinian sense)—in this case A—will observe to be happening to nearby inertial particles. The situation is illustrated in Fig. 17.8a. Those particles that are displaced horizontally from A will accelerate towards E in directions that are slightly inward relative to A's acceleration, because of the finite distance to the Earth's centre, whereas those particles that are displaced vertically from A will accelerate slightly outward relative to A because the gravitational force falls off with increasing distance from E. Accordingly, the sphere of particles will become distorted. In fact, this distortion, for nearby particles, will take the sphere into an *ellipsoid of revolution*, a (prolate) ellipsoid, having its major axis (the symmetry axis) in the direction of the line AE. Moreover, the initial distortion of the sphere will be into an ellipsoid whose volume is equal to

[17.5] Find an explicit transformation of $\mathbf{x}$, as a function of t, that does this, for a given Newtonian gravitational field $\mathbf{F}(t)$ that is spatially constant at any one time, but temporally varying both in magnitude and direction.

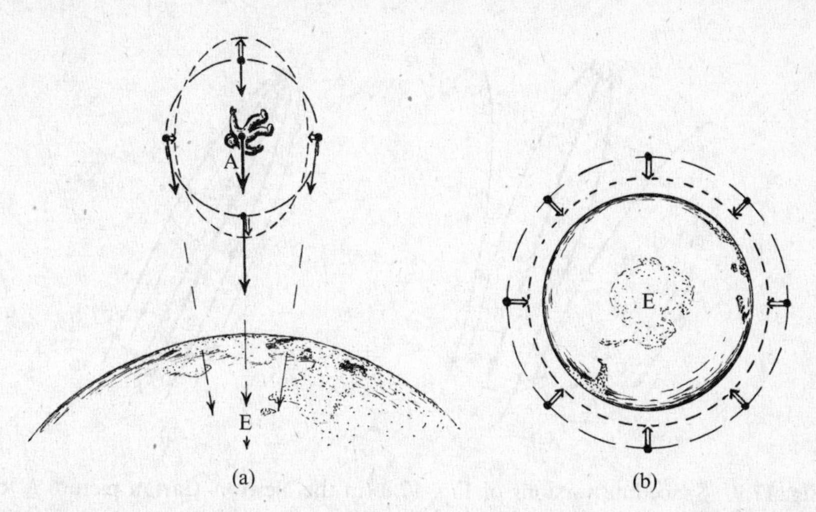

Fig. 17.8 (a) Tidal effect. The astronaut A (Albert) surrounded by a sphere of nearby particles initially at rest with respect to A. In Newtonian terms, they have an acceleration towards the Earth's centre E, varying slightly in direction and magnitude (single-shafted arrows). By subtracting A's acceleration from each, we obtain the accelerations relative to A (double-shafted arrows); this relative acceleration is slightly inward for those particles displaced horizontally from A, but slightly outward for those displaced vertically from A. Accordingly, the sphere becomes distorted into a (prolate) ellipsoid of revolution, with symmetry axis in the direction AE. The initial distortion preserves volume. (b) Now move A to the Earth's centre E and the sphere of particles to surround E just above the atmosphere. The acceleration (relative to A = E) is inward all around the sphere, with an initial volume reduction acceleration $4\pi GM$, where M is the total mass surrounded.

that of the sphere.[17.6] This last property is a characteristic property of the inverse square law of Newtonian gravity, a remarkable fact that will have significance for us when we come to Einstein's general relativity proper. It should be noted that this volume-preserving effect only applies initially, when the particles start at rest relative to A; nevertheless, with this proviso, it is a general feature of Newtonian gravitational fields, when A is in a vacuum region. (The rotational symmetry of the ellipsoid, on the other hand, is an accident of the symmetry of the particular geometry considered here.)

Now, how are we to think of all this in terms of our spacetime picture $\mathcal{N}$? In Fig. 17.9a, I have tried to indicate how this situation would look for the world lines of A and the surrounding particles. (Of

[17.6] Derive these various properties, making clear by use of the $O(\ )$ notation, at what order these statements are intended to hold.

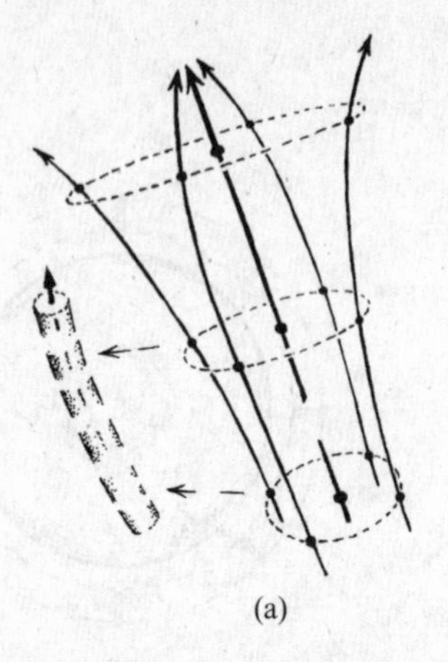

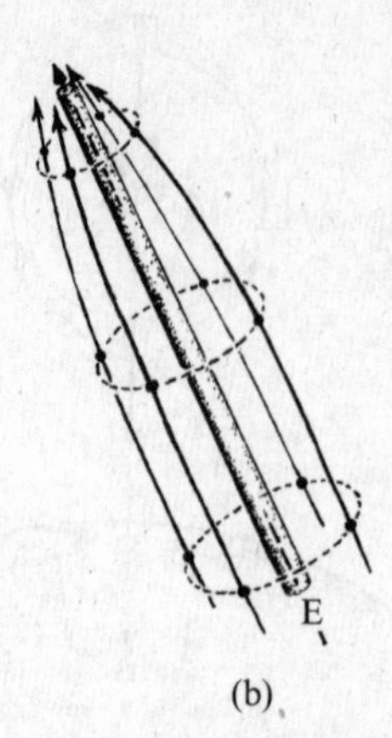

(a) (b)

Fig. 17.9 Spacetime versions of Fig. 17.8 (in the Newton–Cartan picture $\mathcal{N}$ of Fig. 17.7), in terms of the relative distortion of neighbouring geodesics. (a) Geodesic deviation in empty space (basically Weyl curvature of §19.7) as seen in the world lines of A and surrounding particles (one spatial dimension suppressed), as might be induced from the gravitational field of a nearby body E. (b) The corresponding inward acceleration (basically Ricci curvature) due to the mass density within the bundle of geodesics.

course, I have had to discard a spatial dimension, because it is hard to depict a genuinely 4-dimensional geometry! Fortunately, two space dimensions are adequate here for conveying the essential idea.) Note that the distortion of the sphere of particles (depicted here as a circle of particles) arises because of the geodesic deviation of the geodesics that are neighbouring to the geodesic world line of A. In §14.5, I indicated why this geodesic deviation is in fact a measure of the *curvature* $\boldsymbol{R}$ of the connection $\boldsymbol{\nabla}$.

In Newtonian physical terms, the distortion effect that I have just described is what is called the *tidal* effect of gravity. The reason for this terminology is made evident if we let E swap roles with A, so we now think of A as being the Earth's centre, but with the Moon (or perhaps the Sun) located at E. Think of the sphere of particles as being the surface of the Earth's oceans, so we see that there is a distortion effect due to the Moon's (or Sun's) non-uniform gravitational field.[17.7] This distortion is the cause

[17.7] Show that this tidal distortion is proportional to mr^{-3} where m is the mass of the gravitating body (regarded as a point) and r is its distance. The Sun and Moon display discs, at the Earth, of closely equal angular size, yet the Moon's tidal distortion on the Earth's oceans is about five times that due to the Sun. What does that tell us about their relative densities?

of the ocean tides, so the terminology 'tidal effect', for this direct physical manifestation of spacetime curvature, is indeed apposite.

In fact, in the situation just considered, the effect of the Moon (or Sun) on the relative accelerations of particles at the Earth's surface is only a small correction to the major gravitational effect on those particles, namely the gravitational pull of the Earth itself. Of course, this is inwards, namely in the direction of the Earth's centre (now the point A, in our spatial description; see Fig. 17.8b) as measured from each particle's individual location. If the sphere of particles is now taken to surround the Earth, just above the Earth's atmosphere (so that we can ignore air resistance), then there will be free fall (Einsteinian inertial motion) inwards all around the sphere. Rather than distortion of the spherical shape into that of an ellipse of initially equal volume, we now have a *volume reduction*. In general, there could be both effects present. In empty space, there is only distortion and no initial volume reduction; when the sphere surrounds matter, there is an initial volume reduction that is proportional to the total mass surrounded. If this mass is M, then the initial 'rate' (as a measure of inward acceleration) of volume reduction is in fact

$$4\pi GM$$

where G is Newton's gravitational constant.[17.8],[17.9]

In fact, as Cartan showed, it is possible to reformulate Newton's gravitational theory completely in terms of mathematical conditions on the connection ∇, these being basically equations on the curvature $\boldsymbol{R}$ which provide a precise mathematical expression of the requirements outlined above, and which relate the matter density ρ (mass per unit spatial volume) to the 'volume-reducing' part of $\boldsymbol{R}$. I shall not give Cartan's description for this in detail here, because it is not necessary for our later considerations, the full Einstein theory being, in a sense, simpler. However, the idea itself is an important one for us here, not only for leading us gently into Einstein's theory, but also because it has a role to play in our later considerations of Chapter 30 (§30.11), concerning the profound puzzles that the quantum theory presents us with, and their possible resolution.

17.6 The fixed finite speed of light

In our discussions above, we have been considering two fundamental aspects of Einstein's general relativity, namely the principle of *relativity*,

---

[17.8] Establish this result, assuming that all the mass is concentrated at the centre of the sphere.

[17.9] Show that this result is still true quite generally, no matter how large or what shape the surrounding shell of stationary particles is, and whatever the distribution of mass.

which tells us that the laws of physics are blind to the distinction between stationarity and uniform motion, and the principle of *equivalence* which tells us how these ideas must be subtly modified in order to encompass the gravitational field. We must now turn to the third fundamental ingredient of Einstein's theory, which has to do with the finiteness of the speed of light. It is a remarkable fact that all three of these basic ingredients can be traced back to Galileo; for Galileo also seems to have been the first person to have such a clear expectation that light ought to travel with finite speed that he actually took steps to measure that speed. The method he proposed (in 1638), involving the synchronizing of lantern flashes between distant hills, was, as we now know, far too crude. But he had no way to anticipate the extraordinary swiftness with which light actually travels.

It appears that both Galileo and Newton[8] seem to have had powerful suspicions concerning a possibly deep role connecting the nature of light with the forces that bind matter together. But the proper realization of these insights had to wait until the twentieth century, when the true nature of chemical forces and of the forces that hold individual atoms together were revealed. We now know that these forces are fundamentally *electromagnetic* in origin (concerning the involvement of electromagnetic field with charged particles) and that the theory of electromagnetism is also the theory of light. To understand atoms and chemistry, further ingredients from the quantum theory are needed, but the basic equations that describe both electromagnetism and light were those put forward in 1865 by the great Scottish physicist James Clark Maxwell, who had been inspired by the magnificent experimental findings of Michael Faraday, over 30 years earlier. We shall be coming to Maxwell's theory later (§19.2), but its immediate importance for us now is that it requires that the speed of light has a definite fixed value, which is usually referred to as c, and which in ordinary units is about 3×10^8 metres per second.

This, however, provides us with a conundrum, if we wish to preserve the relativity principle. Common sense would seem to tell us that if the speed of light is measured to take the particular value c in one observer's rest frame, then a second observer, who moves with a very high speed with respect to the first one, will measure light to travel at a different speed, reduced or increased, according to the second observer's motion. But the relativity principle would demand that the second observer's physical laws—these defining, in particular, the speed of light that the second observer perceives—should be identical with those of the first observer. This apparent contradiction between the constancy of the speed of light and the relativity principle led Einstein—as it had, in effect, previously led the Dutch physicst Hendrick Antoon Lorentz and, more completely, the French mathematician Henri Poincaré—to a remarkable viewpoint whereby the contradiction is completely removed.

How does this work? It would be natural for us to believe that there is an irresolvable conflict between the requirements of (i) a theory, such as that of Maxwell, in which there is an absolute speed of light, and (ii) a relativity principle, according to which physical laws appear the same no matter what speed of reference frame is used for their description. For could not the reference frame be made to move with a speed approaching, or even exceeding that of light? And according to such a frame, surely the apparent light speed could not possibly remain what it had been before? This undoubted conundrum does not arise with a theory, such as that originally favoured by Newton (and, I would guess, by Galileo also), in which light behaves like *particles* whose velocity is thereby dependent upon the velocity of the source. Accordingly Galileo and Newton could still live happily with a relativity principle. But such a picture of the nature of light had encountered increasing conflict with observation over the years, such as with observations of distant double stars which showed light's speed to be *independent* of that of its source.[9] On the other hand, Maxwell's theory had gained in strength, not only because of the powerful support it obtained from observation (most notably the 1888 experiments of Heinrich Hertz), but also because of the compelling and unifying nature of the theory itself, whereby the laws governing electric fields, magnetic fields, and light are all subsumed into a mathematical scheme of remarkable elegance and essential simplicity. In Maxwell's theory, light takes the form of waves, not particles; and we must face up to the fact that, in this theory, there is indeed a fixed speed according to which the waves of light must travel.

17.7 Light cones

The spacetime-geometric viewpoint provides us with a particularly clear route to the solution of the conundrum presented by the conflict between Maxwell's theory and the principle of relativity. As I remarked earlier, this spacetime viewpoint was not the one that Einstein originally adopted (nor was it Lorentz's viewpoint nor, apparently, even Poincaré's). But with hindsight, we can see the power of this approach. For the moment, let us ignore gravity, and the attendant subtleties and complications provided by the principle of equivalence. We shall start with a blank slate—or, rather, with a featureless real 4-manifold. We wish to see what it might mean to say that there is a fundamental speed, which is to be the speed of light. At any point (i.e. 'event') p in spacetime, we can envisage the family of all different rays of light that pass through p, in all the different spatial directions. The spacetime description is a family of world lines through p. See Fig. 17.10a,b.

It will be convenient to refer to these world lines as 'photon histories' through p, although Maxwell's theory takes light to be a wave effect. This

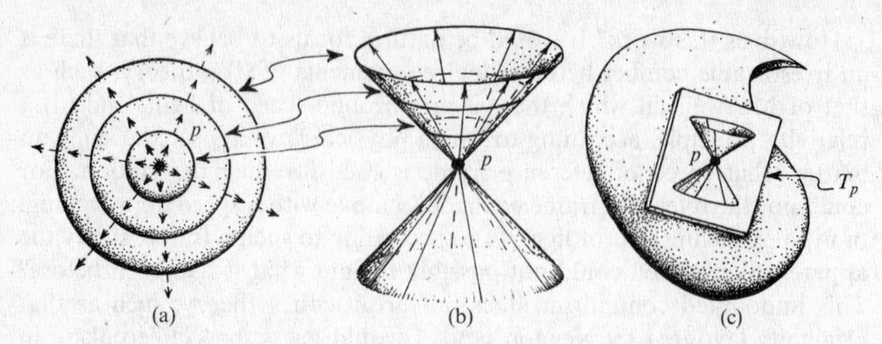

Fig. 17.10 The light cone specifies the fundamental speed of light. Photon histories through a spacetime point (event) p. (a) In purely spatial terms, the (future) light cone is a sphere expanding outwards from p (wavefronts). (b) In spacetime, the photon histories encountering p sweep out the light cone at p. (c) Since we shall later be considering curved spacetimes, it is better to think of the cone—frequently called the *null* cone at p—as a local structure in spacetime, i.e. in the tangent space T_p at p.

is not really an important conflict, for various reasons. One can consider a 'photon', in Maxwell's theory, as a tiny bundle of electromagnetic disturbance of very high frequency, and this will behave, quite adequately for our purposes, as a little particle travelling with the speed of light. (Alternatively, we might think in terms of 'wave fronts' or of what the mathematicians call 'bi-characteristics', or we may prefer to appeal to the quantum theory, according to which light can also be considered to consist of 'particles', which are, indeed, referred to as 'photons'.)

In the neighbourhood of p, the family of photon histories through p, as depicted in Fig. 17.10b, describes a cone in spacetime, referred to as the *light cone* at p. To take the light speed as fundamental is, in spacetime terms, to take the light cones as fundamental. In fact, from the point of view that is appropriate for the geometry of manifolds (see Chapters 12, 14), it is often better to think of the 'light cone' as a structure in the tangent space T_p at p (see Fig. 17.10c). (We are, after all, concerned with velocities at p, and a velocity is something that is defined in the tangent space.) Frequently, the term *null cone* is used for this tangent-space structure—and this is actually my own preference—the term 'light cone' being reserved for the actual locus in spacetime that is swept out by the light rays passing through a point p. Notice that the light cone (or null cone) has two parts to it, the *past* cone and the *future* cone. We can think of the past cone as representing the history of a flash of light that is imploding on p, so that all the light converges simultaneously at the one event p; correspondingly, the future cone represents the history of a flash of light of an explosion taking place at the event p; see Fig. 17.11.

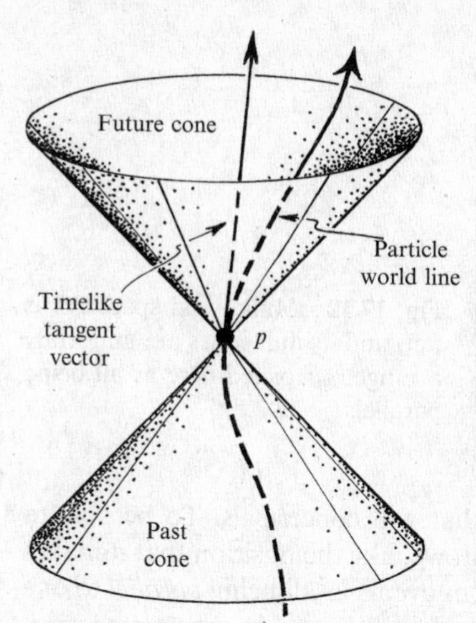

Fig. 17.11 The past cone and the future cone. The past null cone (of past-null vectors) refers to light imploding on p in the same way that the future cone (of future null vectors) refers to light originating at p. The world line of any massive particle at p has a tangent vector that is (future-)timelike, and so lies within the (future) null cone at p.

How are we to provide a mathematical description of the null cone at p? Chapters 13 and 14 have given us the background. We require the speed of light to be the same in all directions at p, so that an instant after a light flash the spatial configuration surrounding the point appears as a sphere rather than some other ovoid shape.[10] By referring to 'an instant', I really mean that these considerations are to apply to the infinitesimal temporal (as well as spatial) neighbourhood of p, so it is legitimate to think of this as indeed referring to structures in the tangent space at p. To say that the null cone appears 'spherical' is really only to say that the cone is given by an equation in the tangent space that is *quadratic*. This means that this equation takes the form

$$g_{ab}v^a v^b = 0,$$

where g_{ab} is the index form of some non-singular symmetric $[{}^0_2]$-tensor $\boldsymbol{g}$ of Lorentzian signature (§13.8).[17.10] The term 'null' in 'null cone' refers to the fact that the vector $\boldsymbol{v}$ has a *zero length* ($|\boldsymbol{v}|^2 = 0$) with respect to the (pseudo)metric $\boldsymbol{g}$.

At this stage, we are concerned with $\boldsymbol{g}$ only in its role in defining the null cones, according to the above equation. If we multiply $\boldsymbol{g}$ by any non-zero real number, we get precisely the same null cone as we did before (see also §27.12 and §33.3). Shortly, we shall require $\boldsymbol{g}$ to play the further physical role of providing the spacetime metric, and for this we shall require the appropriate scaling factor; but for the moment, it is just the family of null

[17.10] Explain why.

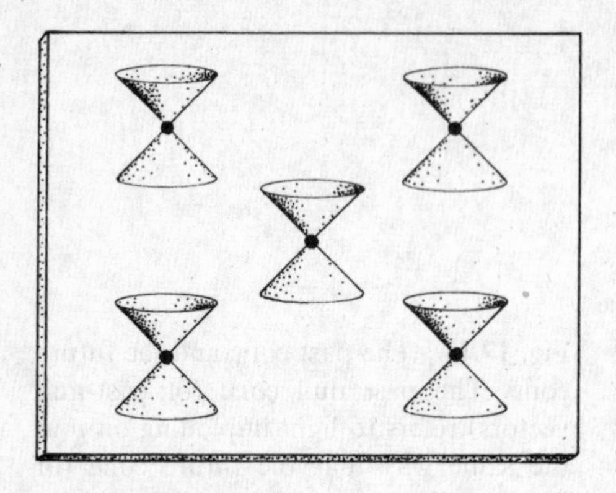

Fig. 17.12 Minkowski space $\mathbb{M}$ is flat, and its null cones are uniformly arranged, depicted here as all being parallel.

cones, one at each spacetime point, that will concern us. To be able to assert that the speed of light is constant, we take the position that it makes sense to regard the null cones at different events as all being *parallel* to one another, since 'speed' in spatial terms, refers to 'slope' in spacetime terms. This leads us to the picture of spacetime depicted in Fig. 17.12.

17.8 The abandonment of absolute time

We may now ask whether the bundle structure of Galilean spacetime $\mathcal{G}$ would be appropriate to impose in addition. In other words, can we include a notion of *absolute time* into our picture? This would lead us to a picture like that of Fig. 17.13. The $\mathbb{E}^3$ slices through the spacetime would give us a *3-plane element* in each tangent space T_p, in addition to the null cone, as depicted in Fig. 17.13. But, as I shall explain more fully in the next chapter, $\boldsymbol{g}$ determines a notion of *orthogonality* which means that there is now a preferred *time-direction* at each event p (the orthogonal complement, with respect to $\boldsymbol{g}$, of this 3-plane element), and this preferred time-direction gives us a preferred *state of rest* at each event. We have lost the relativity principle!

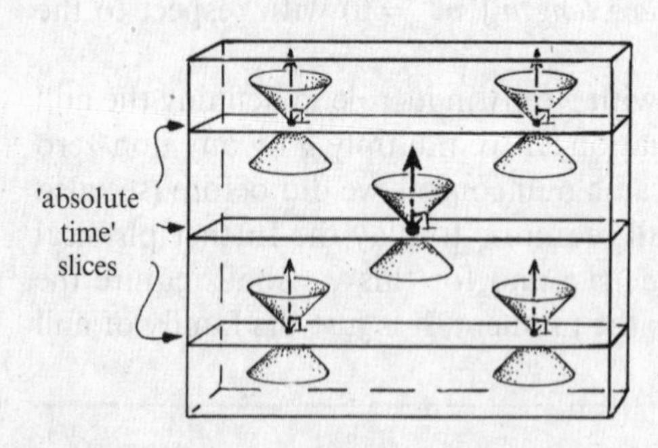

Fig. 17.13 A notion of absolute time introduced into $\mathbb{M}$ would specify a family of $\mathbb{E}^3$-slices cutting through $\mathbb{M}$ and hence a local 3-plane-element at each event. But each null cone defines a (pseudo) metric $\boldsymbol{g}$, up to proportionality, whose notion of orthogonality thereby determines a state of rest.

In more prosaic terms, this argument is simply expressing the 'common-sense' notion that if there is an absolute light speed, then there is a preferred 'state of rest' with respect to which this speed appears to be the same in all directions. What is less obvious is that this conflict arises *only* if we try to retain the notion of an absolute time (or, at least, a preferred 3-space in each T_p). It should now be clear how we must proceed. The notion of an absolute time (and therefore of the bundle structure of $\mathcal{G}$ and $\mathcal{N}$) must be abandoned. At the stage of sophistication that we have arrived at by now, this should not shock us particularly. We have already seen that absolute space has to be abandoned as soon as even a Galilean relativity principle is seriously adopted (although this perception is not recognized nearly as widely as it should be). So, by now, the acceptance of the fact that time is not an absolute concept, as well as space not being an absolute concept, should not seem to be such a revolution as we might have thought.

Thus we must indeed bid farewell to the $\mathbb{E}^3$ slices through spacetime, and accept that the only reason for having an absolute time so firmly ingrained in our thinking is that the speed of light is so extraordinarily large by the standards of the speeds familiar to us. In Fig. 17.14, I have redrawn part of Fig. 17.13., with a horizontal/vertical scale ratio that is a little closer to that which would be appropriate for the normal units that we tend to use in every-day life. But it is only a very little closer, since we must bear in mind that in ordinary units, say seconds for time and metres for distance, we find that the speed of light c is given by

$$c = 299\,792\,458 \text{ metres/second}$$

where this value is actually exact![11] Since our spacetime diagrams (and our formulae) look so awkward in conventional units, it is a common practice, in relativity theory work, to use units for which $c = 1$. All that this means is that if we choose a *second* as our unit of time, then we must use a *light-second* (i.e. 299792458 metres) for our unit of distance; if we use the *year* as our unit of time, then we use the *light-year* (about 9.46×10^{15} metres) as the unit of distance; if we wish to use a metre as our distance measure, then we must use for our time measure something like $3\frac{1}{3}$ nanoseconds, etc.

Fig. 17.14 The null cone redrawn so that the space and time scales are just slightly closer to those of normal experience.

The spacetime picture of Fig. 17.12. was first introduced by Hermann Minkowski (1864–1909), who was an extremely fine and original mathematician. Coincidentally, he was also one of Einstein's teachers at ETH, The Federal Institute of Technology in Zurich, in the late 1890s. In fact, the very idea of spacetime itself came from Minkowski who wrote, in 1908,[12] 'Henceforth space by itself, and time by itself, are doomed to fade away into mere shadows, and only a kind of union of the two will preserve an independent reality.' In my opinion, the theory of special relativity was not yet complete, despite the wonderful physical insights of Einstein and the profound contributions of Lorentz and Poincaré, until Minkowski provided his fundamental and revolutionary viewpoint: *spacetime*.

To complete Minkowski's viewpoint with regard to the geometry underlying special relativity, and thereby define *Minkowskian spacetime* $\mathbb{M}$, we must fix the *scaling* of $\boldsymbol{g}$, so that it provides a measure of 'length' along world lines. This applies to curves in $\mathbb{M}$ that we refer to as *timelike* which means that their tangents always lie within the null cones (Fig. 17.15a and see also Fig. 17.11) and, according to the theory, are

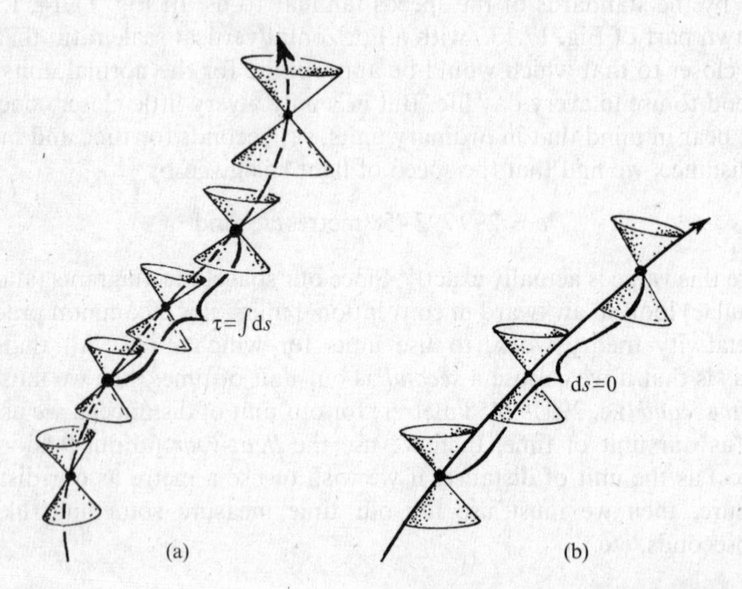

Fig. 17.15 (a) The world line of a massive particle is a timelike curve, so its tangents are always within the local null cones, giving $ds^2 = g_{ab}dx^a dx^b$ positive. The quantity $ds = \left(g_{ab}dx^a dx^b\right)^{1/2}$ measures the infinitesimal time-interval along the curve, so the 'length' $\tau = \int ds$, is the time measured by an ideal clock carried by the particle between two events on the curve. (b) In the case of a massless particle (e.g. a photon) the world lines have tangents on the null cones (null world line), so the time-interval $\tau = \int ds$ always vanishes.

possible world lines for ordinary massive particles. This 'length' is actually a *time* and it measures the actual time τ that an (ideal) clock would register, between two points A and B on the curve, according to the formula (see §14.7, §13.8)

$$\tau = \int_A^B \mathrm{d}s, \quad \text{where } \mathrm{d}s = (g_{ab}\mathrm{d}x^a\mathrm{d}x^b)^{\frac{1}{2}}.$$

For this, we require the choice of spacetime metric $\boldsymbol{g}$ to have signature $+ - - -$ (which is my own preferred choice, rather than $+ + + -$, which some other people prefer, for different reasons). Photons have world lines that are called *null* (or *lightlike*), having tangents that are *on* the null cones (Fig. 17.15b). Accordingly the 'time' that a photon experiences (if a photon could actually have experiences) has to be zero!

In my discussion above, I have chosen to emphasize the null-cone structure of spacetime, even more than its metric. In certain respects, the null cones are indeed more fundamental than the metric. In particular, they determine the causality properties of the spacetime. As we have just seen, material particles are to have their world lines constrained to lie within the cones, and light rays have world lines along the cones. No physical particle is permitted to have a *spacelike* world line, i.e. one outside its associated light cones.[13] If we think of actual signals as being transmitted by material particles or photons, then we find that no such signal can pass outside the constraints imposed by the null cones. If we consider some point p in $\mathbb{M}$, then we find that the region that lies on or within its future light cone consists of all the events that can, in principle, receive a signal from p. Likewise, the points of $\mathbb{M}$ lying on or within p's past light cone are precisely those events that can, in principle send a signal *to* the point p; see Fig. 17.16. The situation is similar when we consider propagating fields and even quantum-mechanical effects (although some strangely puzzling situations can arise with what is called *quantum entanglement*—or '*quanglement*'—as we shall be seeing in §23.10). The null cones indeed define the *causality* structure of $\mathcal{M}$: no material body or signal is permitted to travel faster than light; it is necessarily constrained to be within (or on) the light cones.

What about the relativity principle? We shall be seeing in §18.2 that Minkowski's remarkable geometry has just as big a symmetry group as has the spacetime $\mathcal{G}$ of Galilean physics. Not only is every point of $\mathbb{M}$ on an equal footing, but all possible velocities (timelike future-pointing directions) are also on an equal footing with each other. This will all be explained more fully in §18.2. The relativity principle holds just as well for $\mathbb{M}$ as it does for $\mathcal{G}$!

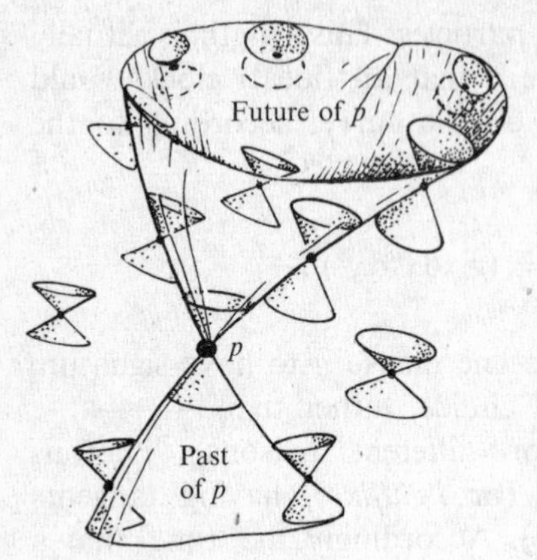

Fig. 17.16 The future of p is the region that can be reached by future-timelike curves from p. A curved-spacetime case is indicated (see Fig. 17.17). The boundary of this region (wherever smooth) is tangential to the light cones. Signals, whether carried by massive particles or massless photons, reach points within this region or on its boundary. The past of p is defined similarly.

17.9 The spacetime of Einstein's general relativity

Finally, we come to the *Einsteinian* spacetime $\mathcal{E}$ of general relativity. Basically, we apply the same generalization to Minkowski's $\mathbb{M}$, as we previously did to Galileo's $\mathcal{G}$, when we obtained the Newton(–Cartan) spacetime $\mathcal{N}$. Rather than having the uniform arrangement of null cones depicted in Fig. 17.12, we now have a more irregular-looking arrangement like that of Fig. 17.17. Again, we have a Lorentzian $(+ - - -)$ metric $\boldsymbol{g}$ whose physical interpretation is to define the time measured by an ideal clock, according to precisely the same formula as for $\mathbb{M}$, although now $\boldsymbol{g}$ is a more general metric without the unifomity that is the characteristic of the metric of $\mathbb{M}$.

The null-cone structure defined by this $\boldsymbol{g}$ specifies $\mathcal{E}$'s causality structure, just as was the case for Minkowski space $\mathbb{M}$. Locally, the differences are slight, but things can get decidedly more elaborate when we examine the global causality structure of a complicated Einsteinian spacetime $\mathcal{E}$. An

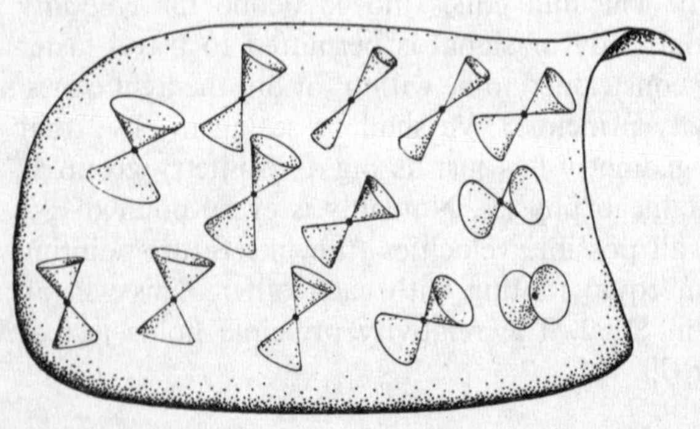

Fig. 17.17 Einsteinian spacetime $\mathcal{E}$ of general relativity. This generalization of Minkowski's $\mathbb{M}$ is similar to the passage from $\mathcal{G}$ to $\mathcal{N}$ (Figs. 17.12, 17.3, 17.7a, respectively). As with $\mathbb{M}$, the Lorentzian $(+ - - -)$ pseudo-metric $\boldsymbol{g}$ defines the physical measure of time.

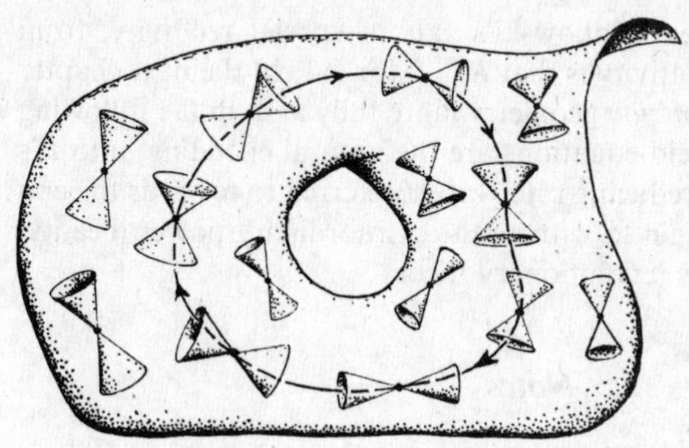

Fig. 17.18 The causality structure of $\mathcal{E}$ is determined by $\boldsymbol{g}$ (as with $\mathbb{M}$, see Fig. 17.16), so extreme unphysical situations with 'closed timelike curves' might hypothetically arise, allowing future-directed signals to return from the past.

extreme situation arises when we have what is referred to as *causality violation* in which 'closed timelike curves' can occur, and it becomes possible for a signal to be sent from some event into the past of that same event! See Fig. 17.18. Such situations are normally ruled out as 'unphysical', and my own position would certainly be to rule them out, for a classically acceptable spacetime. Yet some physicists take a considerably more relaxed view of the matter[14] being prepared to admit the possibility of the *time travel* that such closed timelike curves would allow. (See §30.6 for a discussion of these issues.) On the other hand, less extreme—though certainly somewhat exotic—causality structures can arise in some interesting spacetimes of great relevance to modern astrophysics, namely those which represent black holes. These will be considered in §27.8.

In §14.7, we encountered the fact that a (pseudo)metric $\boldsymbol{g}$ determines a unique torsion-free connection $\boldsymbol{\nabla}$ for which $\boldsymbol{\nabla g} = 0$, so this will apply here. This is a remarkable fact. It tells us that Einstein's concept of inertial motion is completely determined by the spacetime metric. This is quite different from the situation with Cartan's Newtonian spacetime, where the '$\boldsymbol{\nabla}$' had to be specified in addition to the metric notions. The advantage here is that the metric $\boldsymbol{g}$ is now non-degenerate, so that $\boldsymbol{\nabla}$ is completely determined by it. In fact, the timelike *geodesics* of $\boldsymbol{\nabla}$ (inertial motions) are fixed by the property that they are (locally) the curves that *maximize* what is called the *proper time*. This proper time is simply the length, as measured along the world line, and it is what is measured by an ideal clock having that world line. (This is a curious 'opposite' to the 'stretched-string' notion of a geodesic on an ordinary Riemannian surface with a positive-definite metric; see §14.7. We shall see, in §18.3, that this maximization of proper time for the unaccelerated world line is basically an expression of the 'clock paradox' of relativity theory.)

The connection $\boldsymbol{\nabla}$ has a curvature tensor $\boldsymbol{R}$, whose physical interpretation is basically just the same as has been given above in the case of $\mathcal{N}$.

What locally distinguishes Minkowski's $\mathbb{M}$, of special relativity, from Einstein's $\mathcal{E}$ of general relativity is that $\boldsymbol{R} = 0$ for $\mathbb{M}$. In the next chapter we shall explore this *Lorentzian* geometry more fully and, in the following one, see how Einstein's field equations are the natural encoding, into $\mathcal{E}$'s structure, of the 'volume-reducing rate' $4\pi GM$ referred to towards the end of §17.5. We shall also begin to witness the extraordinary power, beauty, and accuracy of Einstein's revolutionary theory.

Notes

Section 17.1

17.1. Although in the past I have been a proponent of the hyphenated 'space-time', I have found that there are places in this book where that would cause complications in phraseology. Accordingly I am adopting 'spacetime' consistently here.

17.2. It appears that Aristotle may well have had difficulties with the notion of an infinite physical space, as is required if Euclidean geometry $\mathbb{E}^3$ is to provide an accurate description of spatial geometry, but his views with regard to time may have been more in accord with the '$\mathbb{E}^1$' of the $\mathbb{E}^1 \times \mathbb{E}^3$ picture. See Moore (1990), Chap. 2.

Section 17.2

17.3. See Drake (1953), pp. 186–87.

17.4. See Trautman (1965); Arnol'd (1978); Penrose (1968a), p. 126.

Section 17.3

17.5. This was in his manuscript fragment *De motu corporum in mediis regulariter cedentibus*—a precursor of *Principia*, written in 1684. See also Penrose (1987c), p. 49.

Section 17.4

17.6. But see Bondi (1957).

17.7. Now there are 'tourist opportunities', in Russia, for such experiences for humans, in aeroplanes in parabolic flight!

Section 17.6

17.8. See Drake (1957), p. 278, concerning a remark Galileo made in the *Assayer*; see also Newton (1730), Query 30; Penrose (1987c), p. 23.

17.9. See de Sitter (1913).

Section 17.7

17.10. There is a knotty issue of how one actually tells a 'sphere' from an 'ellipsoid', because distances can be recalibrated in different directions, so as to make any ellipsoid appear 'spherical'. However, what recalibrations cannot do is to make a non-ellipsoidal ovoid look spherical, at least with 'smooth' recalibrations. Such ovoids would give rise to a *Finsler space*, which does not have the pleasant local symmetry of the (pseudo-)Riemannian structures of relativity theory.

Section 17.8

17.11. The reader might well be puzzled that the speed of light comes out as an exact integer when measured in metres per second. This is no accident, but merely a reflection of the fact that very accurate distance measurements are now much harder to ascertain than very accurate time measurements. Accordingly, the most accurate standard for the metre is conveniently *defined* so that there are exactly 299792458 of them to the distance travelled by light in a standard second, giving a value for the metre that very accurately matches the now inadequately precise standard metre rule in Paris.

17.12. See Minkowski (1952). This is a translation of the Address Minkowski delivered at the 80th Assembly of German Natural Scientists and Physicians, Cologne, 21 September, 1908.

17.13. Some physicists have toyed with the idea of hypothetical 'particles' known as *tachyons* that would have spacelike world lines (so they travel faster than light). See Bilaniuk and Sudarshan (1969); for a more technical reference, see Sudarshan and Dhar (1968). It is difficult to develop anything like a consistent theory in which tachyons are present, and it is normally considered that such entities do not exist.

Section 17.9

17.14. See, for example, Novikov (2001); Davies (2003).

18
Minkowskian geometry

18.1 Euclidean and Minkowskian 4-space

THE geometries of Euclidean 2-space and 3-space are very familiar to us. Moreover, the generalization to a 4-dimensional Euclidean geometry $\mathbb{E}^4$ is not difficult to make in principle, although it is not something for which 'visual intuition' can be readily appealed to. It is clear, however, that there are many beautiful 4-dimensional configurations—or they surely would be beautiful, if only we could actually see them! One of the simpler (!) such configurations is the pattern of Clifford parallels on the 3-sphere, where we think of this sphere as sitting in $\mathbb{E}^4$. Of course we can do a little better here, with regard to visualization, because S^3 is only 3-dimensional, and its stereographic projection, as presented in Fig. 33.15, gives us some idea of the actual Clifford configuration. (If we could really 'see' this configuration as part of $\mathbb{E}^4$, we ought to be able to gain some feeling for what the complex vector 2-space structure of $\mathbb{C}^2$ actually 'looks like';[1] see §15.4, Fig. 15.8.) Minkowski space $\mathbb{M}$ is in many respects very similar to $\mathbb{E}^4$, but there are some important differences that we shall be coming to.

Algebraically, the treatment of $\mathbb{E}^4$ is very close to the coordinate treatment of 'ordinary' 3-space $\mathbb{E}^3$. All that is needed is one more Cartesian coordinate w, in addition to the standard x, y, and z. The $\mathbb{E}^4$ distance s between the points (w, x, y, z) and (w', x', y', z') is given by the Pythagorean relation

$$s^2 = (w - w')^2 + (x - x')^2 + (y - y')^2 + (z - z)^2.$$

If we think of (w, x, y, z) and (w', x', y', z') as only 'infinitesimally' displaced from one another, and formally write $(\mathrm{d}w, \mathrm{d}x, \mathrm{d}y, \mathrm{d}z)$ for the difference $(w', x', y', z') - (w, x, y, z)$, i.e.[2]

$$w' = w + \mathrm{d}w,\ x' = x + \mathrm{d}x,\ y' = y + \mathrm{d}y,\ z' = z + \mathrm{d}z,$$

then we find

$$\mathrm{d}s^2 = \mathrm{d}w^2 + \mathrm{d}x^2 + \mathrm{d}y^2 + \mathrm{d}z^2.$$

The length of a curve in $\mathbb{E}^4$ is given by the same formula as in $\mathbb{E}^3$, namely $\int ds$ (taking the positive sign for ds).

Now the geometry of *Minkowski* spacetime $\mathbb{M}$ is very close to this, the only difference being signs. Many workers in the field prefer to concentrate on the $(+++-)$-signature pseudometric

$$d\ell^2 = -dt^2 + dx^2 + dy^2 + dz^2,$$

since this is convenient when considering spatial geometry, the quantity represented above by '$d\ell^2$' being positive for *spacelike* displacements (i.e. displacements that are neither on nor within the future or past null cones; see Fig. 18.1). But the quantity 'ds^2' defined by the $(+---)$-signature quantity

$$ds^2 = dt^2 - dx^2 - dy^2 - dz^2$$

is more directly physical, because it is positive along the *timelike* curves that are the allowable worldlines of massive particles, the integral $\int ds$ (with $ds > 0$) being directly interpretable as the actual *physical time* measured by an ideal clock with this as its world line. I shall use *this* signature $(+---)$ for my choice of (pseudo)metric tensor $\boldsymbol{g}$, with index form g_{ab}, so that the above expression can be written in index form (see §13.8)

$$ds^2 = g_{ab}dx^a\, dx^b.$$

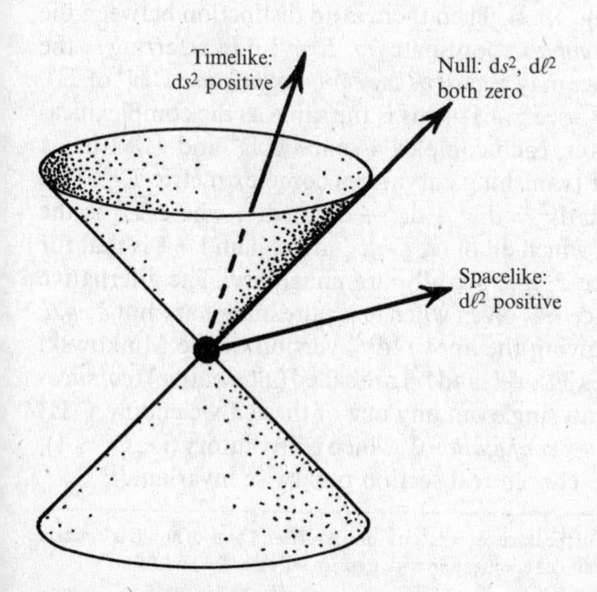

Fig. 18.1 In Minkowski space $\mathbb{M}$, the $d\ell^2$ metric provides a measure of spatial (distance)^2 for spacelike displacements (neither on nor within future or past null cones). For timelike displacements (within the null cone), ds^2 provides a measure temporal (interval)^2, where $\int ds$ is physical time as measured by an ideal clock. For a null displacement (along the null cone) both $d\ell^2$ and ds^2 give zero.

We should, however, recall from §17.8 that, unlike the case for a massive particle, $\int ds$ is *zero* for a world line of a photon (so non-coincident points on the world-line can be 'zero distance' apart). This would also be true for any other particle that travels with the speed of light. The time 'experienced' by such a particle would always be zero, no matter how far it travels! This is allowed because of the non-positive-definite (Lorentzian) nature of g_{ab}.

In the early days of relativity theory, there was a tendency to emphasize the closeness of $\mathbb{M}$'s geometry to that of $\mathbb{E}^4$ by simply taking the time coordinate t to be purely imaginary:

$$t = \mathrm{i}w,$$

which makes the '$d\ell^2$' form of the Minkowskian metric look just the same as the ds^2 of $\mathbb{E}^4$. Of course, appearances are somewhat illusory, because of the unnatural-looking hidden 'reality' condition that time is measured in purely imaginary units whereas the space coordinates use ordinary real units. Moreover, in a moving frame, the reality conditions get complicated because the real and imaginary coordinates are thoroughly mixed up. In fact, there is a modern tendency to do something very similar to this, in various different guises, in the name of what is called 'Euclidean quantum field theory'. Later, in §28.9, I shall come to my reasons for being considerably less than happy with this type of procedure (at least if it is regarded as a key ingredient in an approach to a new fundamental physical theory, as it sometimes is; the device is also used as a 'trick' for obtaining solutions to questions in quantum field theory, and for this it can indeed play an honest and valuable role).

Rather than adopting such a procedure that (to me, at least) looks as unnatural as this, let us try to 'go the whole hog' and allow *all* our coordinates to be complex (see Fig. 18.2). Then there is no distinction between the different signatures, our *complex* coordinates ω, ξ, η, ζ now referring to the complex space $\mathbb{C}^4$, which we may regard as the *complexification* $\mathbb{CE}^4$ of $\mathbb{E}^4$. As a complex affine space—see §14.1—this is the same as the complexification $\mathbb{CM}$ of $\mathbb{M}$. Moreover, each complex 4-space $\mathbb{CE}^4$ and $\mathbb{CM}$ has a completely equivalent flat (vanishing curvature) complex metric $\mathbb{C}g$. This metric can be taken to be $ds^2 = d\omega^2 + d\xi^2 + d\eta^2 + d\zeta^2$, where $\mathbb{E}^4$ is the real subspace of $\mathbb{CM}$ for which all of ω, ξ, η, ζ are real and $\mathbb{M}$ is that for which ω is real, but where ξ, η, ζ are all pure imaginary. The alternative Minkowskian real subspace $\tilde{\mathbb{M}}$, given when ω is pure imaginary but ξ, η, ζ are all real, has its 'ds^2' giving the above '$d\ell^2$' version of the Minkowski metric. The three subspaces $\mathbb{E}^4$, $\mathbb{M}$, and $\tilde{\mathbb{M}}$ are called (alternative) *real slices* or *sections* of $\mathbb{CE}^4$. We can single out any one of these if we endow $\mathbb{CE}^4$ with an operation of *complex conjugation* C, which is involutory (i.e. $\mathrm{C}^2 = 1$), and which leaves only the chosen real section pointwise invariant.[18.1]

[18.1] Find C explicitly for each of the three cases $\mathbb{E}^4$, $\mathbb{M}$, and $\tilde{\mathbb{M}}$. *Hint*: Think of how C is to act on ω, ξ, η, and ζ. Modify standard complex conjugation with *signs*, in the cases $\mathbb{M}$ and $\tilde{\mathbb{M}}$.

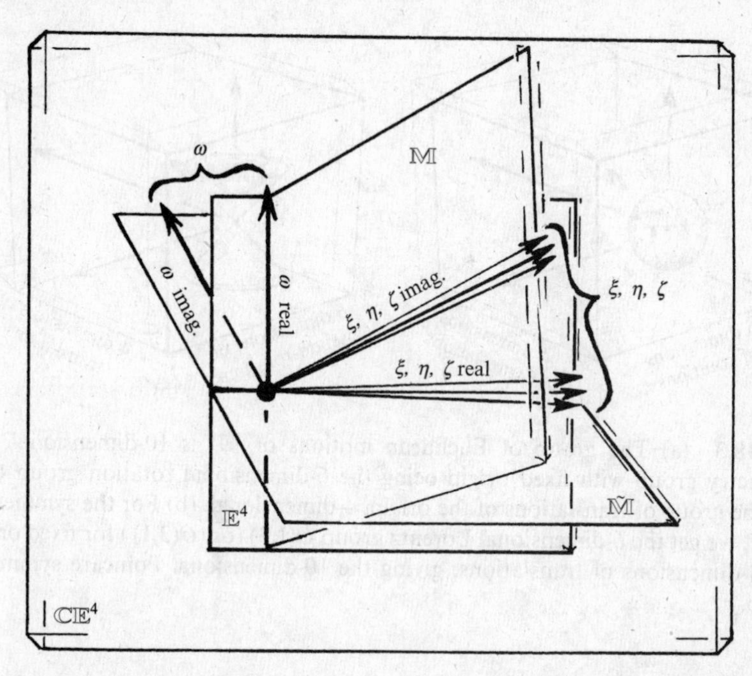

Fig. 18.2 Complex Euclidean space $\mathbb{CE}^4$ has a complex (holomorphic) metric $ds^2 = d\omega^2 + d\xi^2 + d\eta^2 + d\zeta^2$ in complex Cartesian coordinates $(\omega, \xi, \eta, \zeta)$. Euclidean 4-space $\mathbb{E}^4$ is the 'real section' for which ω, ξ, η, ζ are all real. Minkowski spacetime $\mathbb{M}$, with the $+---$ ds^2 metric, is a different real section, ω being real and ξ, η, ζ pure imaginary. We get another Lorentzian real section $\tilde{\mathbb{M}}$ by taking ω to be pure imaginary and ξ, η, ζ real, where the induced ds^2 now gives the $+++-$ '$d\ell^2$' version of the Minkowski metric.

18.2 The symmetry groups of Minkowski space

The group of symmetries of $\mathbb{E}^4$ (i.e. its group of Euclidean motions) is 10-dimensional, since (i) the symmetry group for which the origin is fixed is the 6-dimensional rotation group O(4) (because $n(n-1)/2 = 6$ when $n = 4$; see §13.8), and (ii) there is a 4-dimensional symmetry group of translations of the origin see Fig. 18.3a. When we complexify $\mathbb{E}^4$ to $\mathbb{CE}^4$, we get a 10-*complex*-dimensional group—clearly, because if we write out any of the real Euclidean motions of $\mathbb{E}^4$ as an algebraic formula in terms of the coordinates, all we have to do is allow all the quantities appearing in the formula (coordinates and coefficients) to become complex rather than real, and we get a corresponding complex motion of $\mathbb{CE}^4$. Since the first preserves the metric, so will the second. Moreover, all continous motions

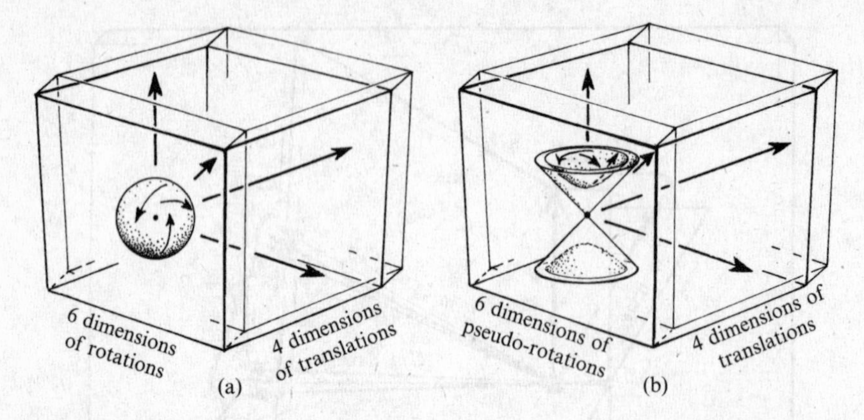

Fig. 18.3 (a) The group of Euclidean motions of $\mathbb{E}^4$ is 10-dimensional, the symmetry group with fixed origin being the 6-dimensional rotation group $O(4)$ and the group of translations of the origin, 4-dimensional. (b) For the symmetries of $\mathbb{M}$, we get the 6-dimensional Lorentz group $O(1,3)$ (or $(O(3,1))$ for fixed origin and 4-dimensions of translations, giving the 10-dimensional Poincaré symmetry group.

of $\mathbb{CE}^4$ to itself which preserve the complexified metric $\mathbb{C}\boldsymbol{g}$ are of this nature.[18.2]

Now it is very plausible, but not completely obvious at this stage, that the group would have the same dimension, namely 10 (but now *real* dimensional), if we specialize to a different 'real section' of $\mathbb{CE}^4$, such as the one for which the coordinates $(\omega, \xi, \eta, \zeta)$ have the reality condition that ω is pure imaginary and ξ, η, ζ are real (signature $+++-$) or else for which ω is real and ξ, η, ζ are pure imaginary (signature $+---$); see Fig. 18.3. The translational part is obviously still 4-dimensional. In fact, this part tells us that the group is *transitive* on $\mathbb{M}$, which means that any specified point of $\mathbb{M}$ can be sent to any other specified point of $\mathbb{M}$ by some element of the group, just as was the case for $\mathbb{E}^4$. But what about the Lorentz group (O(3, 1) or O(1, 3))? How can we see that this is 'just as 6-dimensional' as is O(4)? In fact the Lorentz group *is* 6-dimensional (see Fig. 18.3b). The most general way of seeing such a thing is to examine the Lie algebra—see §13.6—and check that this still works with the required minor sign changes.[18.3] We shall be seeing a rather remarkable alternative way of looking at O(1,3) shortly (§18.5), and checking its 6-dimensionality, by relating it to the symmetry group of the Riemann sphere.

[18.2] Can you see why?

[18.3] Confirm it in this case examining the 4×4 Lie algebra matrices explicitly.

The full 10-dimensional symmetry group of Minkowski space $\mathbb{M}$ is called the *Poincaré group*, in recognition of the achievement of the outstanding French mathematician Henri Poincaré (1854–1912), in building up the essential mathematical structure of special relativity in the years between 1898 and 1905, independently of Einstein's fundamental input of 1905.[3] The Poincaré group is important in relativistic physics, particularly in particle physics and quantum field theory (Chapters 25 and 26). It turns out that, according to the rules of quantum mechanics, individual particles correspond to *representations* (§§13.6,7) of the Poincaré group, where the values for their mass and spin determine the particular representations (§22.12).

It is, in essence, the extensiveness of this group that allows us to assert that the relativity principle still holds for $\mathbb{M}$, even though we have a fixed speed of light (§§17.6,8). In the first place, we see that every point of the spacetime $\mathbb{M}$ is on an equal footing with every other, because of the transitive nature of the translation subgroup. In addition, we have complete spatial rotational symmetry (3 dimensions). This leaves 3 more dimensions to express the fact that there is complete freedom to move from one velocity ($<c$) to any another, and the whole structure remains the same—which is basically $\mathbb{M}$'s relativity principle! A little more formally, what the relativity principle asserts is that the Poincaré group acts transitively on the *bundle of future-timelike directions* of $\mathbb{M}$.[4] These are the directions that point into the interiors of the future null cones, such directions being the possible tangent directions to observers' world lines.[18.4] It may be noted, however, that this only works because we have given up the family of 'simultaneity slices' through the the Galilean or Newtonian spacetime. Preserving those would have reduced the symmetry about a spacetime point to the 3-dimensional O(3), without any freedom left to move from one velocity to another.

18.3 Lorentzian orthogonality; the 'clock paradox'

This point of view regards $\mathbb{M}$ as just a 'real section' or 'slice' of the complex space $\mathbb{CE}^4$ (or $\mathbb{C}^4$), but a section with a different character from $\mathbb{E}^4$ itself. This is very convenient viewpoint, so long as we can adopt the correct attitude of mind. For example, in the Euclidean $\mathbb{E}^4$, we have a notion of 'orthogonal' (which means 'at right angles'). This carries over directly to $\mathbb{CE}^4$ by the process of 'complexification'.[5] However, there are certain types of property that we must expect to be a little different after we apply this procedure. For example, we find that, in $\mathbb{CE}^4$, a direction can now be orthogonal to itself, which is something that certainly cannot happen in $\mathbb{E}^4$. This feature persists, however, when we

[18.4] Explain this action of the Poincaré group a little more fully.

pass back to our new real section, the Lorentzian $\mathbb{M}$. Thus, we retain a notion of orthogonality in $\mathbb{M}$—but we find that now there are real directions that are orthogonal to themselves, these being the *null* directions that point along photon world-lines (see below).

We can carry this orthogonality notion further and consider the *orthogonal complement* $\boldsymbol{\eta}^{\perp}$ of an r-plane element $\boldsymbol{\eta}$ at a point p. This is the $(4-r)$-plane element $\boldsymbol{\eta}^{\perp}$ of all directions at p that are orthogonal to all the directions in $\boldsymbol{\eta}$ at p. Thus the orthogonal complement of a line element is a 3-plane element, the orthogonal complement of a 2-plane element is another 2-plane element, and the orthogonal complement of a 3-plane element is a line element. In each case, taking the orthogonal complement again would return to us the element that we started with; in other words $(\boldsymbol{\eta}^{\perp})^{\perp} = \boldsymbol{\eta}$. Recall that in §13.9 and §14.7 we considered the operations of lowering and raising indices, on a vector or tensor quantity, with g_{ab} or g^{ab}. When applied to the simple r-vector or simple $(4-r)$-form that represents an r-surface element, in accordance §§12.4,7 (e.g. $\boldsymbol{\eta}_{ab} \mapsto \boldsymbol{\eta}^{ab} = \boldsymbol{\eta}_{cd} g^{ac} g^{bd}$; $\boldsymbol{\eta}^{ab} \mapsto \boldsymbol{\eta}_{ab} = \boldsymbol{\eta}^{cd} g_{ac} g_{bd}$), this raising/lowering operation corresponds to passing to the orthogonal complement; see also §19.2.

In $\mathbb{E}^4$, the orthogonal complement of a 3-plane element $\boldsymbol{\eta}$, for example, is a line element $\boldsymbol{\eta}^{\perp}$ (normal to $\boldsymbol{\eta}$) which is never contained in $\boldsymbol{\eta}$; see Fig. 18.4. But as in Fig. 18.2, we can pass to the complexification $\mathbb{C}\mathbb{E}^4$ and thence to the different real section $\mathbb{M}$. In effect, we were

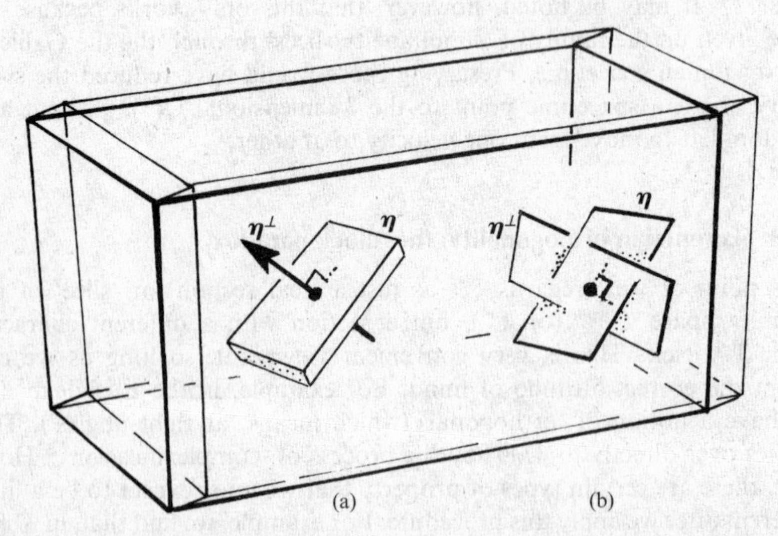

Fig. 18.4 In $\mathbb{E}^4$, an r-plane element $\boldsymbol{\eta}$ at a point p has an orthogonal complement $\boldsymbol{\eta}^{\perp}$ which is a $(4-r)$-plane element, where $\boldsymbol{\eta}$ and $\boldsymbol{\eta}^{\perp}$ never have a direction in common. (a) In particular, if $\boldsymbol{\eta}$ is a 3-plane element, then $\boldsymbol{\eta}^{\perp}$ is the normal direction to it. (b) If $\boldsymbol{\eta}$ is a 2-plane element, then $\boldsymbol{\eta}^{\perp}$ is another 2-plane element.

appealing to this procedure in the previous chapter (§17.8) when we asked for the orthogonal complement of a time slice (spacelike 3-plane element) at a point p to find a timelike direction ('state of rest'), which showed us that a relativity principle cannot be maintained if we wish to have both a finite speed of light and an absolute time (see Fig. 17.13).[18.5] However, now let us read this in the opposite direction. Consider an inertial observer at a particular event p in $\mathbb{M}$. Suppose that the observer's world line has some (timelike) direction $\boldsymbol{\tau}$ at p. Then the 3-space $\boldsymbol{\tau}^{\perp}$ represents the family of 'purely spatial' directions at p for that observer, i.e. those neighbouring events that are deemed by the observer to be simultaneous with p.

It is not my purpose here to develop the full details of the special theory of relativity nor to see why, in particular, this is a reasonable notion of 'simultaneous'. For this kind of thing, the reader may be referred to several excellent texts.[6] The point should be made, however, that this notion of simultaneity actually depends upon the observer's velocity. In Euclidean geometry, the orthogonal complement of a direction in space will change when that direction changes (Fig. 18.5a). Correspondingly, in Lorentzian geometry, the orthogonal complement will also change when the direction (i.e. observer's velocity) changes. The only distinction is that the change tilts the orthogonal complement the opposite way from what happens in the Euclidean case (see Fig. 18.5b) and, accordingly, it is possible for the orthogonal complement of a direction to *contain* that direction (see Fig. 18.5c), as remarked upon above, this being what happens for a *null* direction (i.e. along the light cone).

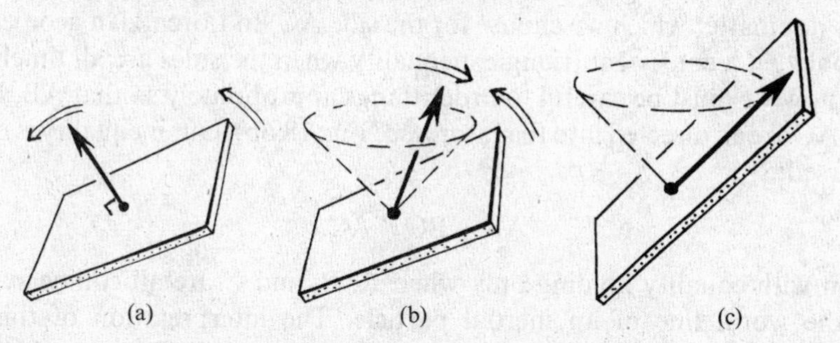

Fig. 18.5 (a) In Euclidean 4-geometry, if a direction rotates, so also does its orthogonal complement 3-plane element. (b) This is true also in Lorentzian 4-geometry, but for a timelike direction the slope of the orthogonal complement 3-plane (spatial directions of 'simultaneity') moves in the reverse sense; (c) accordingly, if the direction becomes null, the orthogonal complement actually contains that direction.

[18.5] (i) Under what circumstances is it possible for a 3-plane element $\boldsymbol{\eta}$ to contain its normal $\boldsymbol{\eta}^{\perp}$, in $\mathbb{M}$? (ii) Show that there are two distinct families of 2-planes that are the orthogonal complements of themselves in $\mathbb{CE}^4$, but neither of these families survives in $\mathbb{M}$. (These so-called 'self-dual' and 'anti-self-dual' complex 2-planes will have considerable importance later; see §32.2 and §33.11.)

In passing from $\mathbb{E}^4$ to $\mathbb{M}$, there are also changes that relate to inequalities. The most dramatic of these contains the essence of the so-called 'clock paradox' (or 'twin paradox') of special relativity. Some readers may be familiar with this 'paradox'; it refers to a space traveller who takes a rocket ship to a distant planet, travelling at close to the speed of light, and then returns to find that time on the Earth had moved forward many centuries, while the traveller might be only a few years older. As Bondi (1964, 1967) has emphasized, if we accept that the passage of time, as registered by a moving clock, is really a kind of 'arc length' measured along a world line, then the phenomenon is not more puzzling than the fact that the distance between two points in Euclidean space depens upon the path along which this distance is measured. Both are measured by the same formula, namely $\int ds$, but in the Euclidean case, the straight path represents the minimizing of the measured distance between two fixed endpoints, whereas in the Minkowski case, it turns out that the straight, i.e. *inertial*, path represents the *maximizing* of the measured time between two fixed end events (see also §17.9).

The basic inequality, from which all this springs, is what is called the *triangle inequality* of ordinary Euclidean geometry. If ABC is any Euclidean triangle, then the side lengths satisfy

$$AB + BC \geq AC,$$

with equality holding only in the degenerate case when A, B, and C are all collinear (see Fig. 18.6a). Of course, things are symmetrical, and it does not matter which we choose for the side AC. In Lorentzian geometry, we only get a consistent triangle inequality when the sides are all timelike, and now we must be careful to order things appropriately so that AB, BC, and AC are all directed into the future (see Fig. 18.6b). Our inequality is now reversed:

$$AB + BC \leq AC,$$

again with equality holding only when A, B, and C are all collinear, i.e. on the world line of an inertial particle. The interpretation of this is precisely the so-called 'clock paradox'. The space traveller's world line is the broken path ABC, whereas the inhabitants of Earth have the world line AC. We see that, according to the inequality, the space traveller's clock indeed registers a shorter total elapsed time than those on Earth.

Some people worry that the acceleration of the rocket ship is not properly accounted for in this description, and indeed I have idealized things so that the astronaut appears to be subjected to an impulsive (i.e.

infinite) acceleration at the event B (which ought to be fatal!). However, this issue is easily dealt with by simply smoothing over the corners of the triangle, as is indicated in Fig. 18.6d. The time difference is not greatly affected, as is obvious in the corresponding situation for the Euclidean

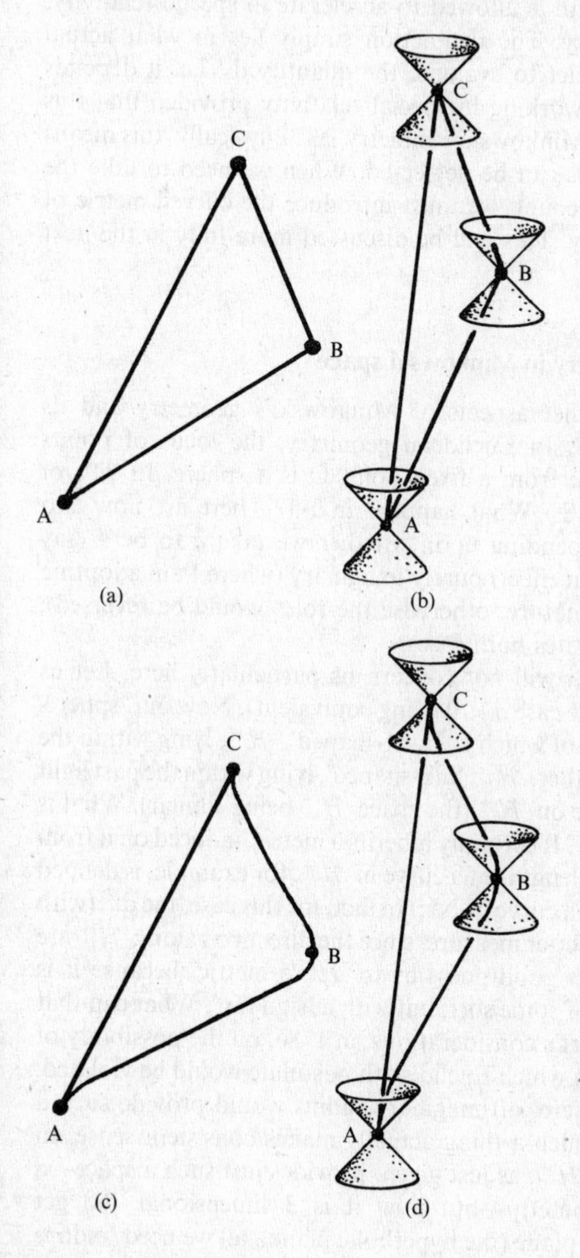

Fig. 18.6 (a) The Euclidean triangle inequality AB + BC ≥ AC, with equality holding only in the degenerate case when A, B, C are collinear. (b) In Lorentzian geometry, with AB, BC, AC all future-timelike, the inequality is reversed: AB + BC ≤ AC, with equality holding only when A, B, C are all on the world-line of an inertial particle. This illustrates the 'clock paradox' of special relativity whereby a space traveller with world-line ABC experiences a shorter time interval than the Earth's inhabitants AC. (c) 'Smoothing' the corners of a Euclidean triangle makes little difference to the edge lengths, and the straight path is still the shortest. (d) Similarly, making accelerations finite (by 'smoothing' corners) makes little difference to the times, and the straight (inertial) path is still the longest.

'smoothed-off' triangle depicted in Fig. 18.6c. It used to be frequently argued that it would be necessary to pass to Einstein's *general* relativity in order to handle acceleration, but this is completely wrong. The answer for the clock times is obtained using the formula $\int ds$ (with $ds>0$) in both theories. The astronaut is allowed to accelerate in special relativity, just as in general relativity. The distinction simply lies in what actual metric is being used in order to evaluate the quantity ds; i.e. it depends on the actual g_{ij}. We are working in special relativity provided that this metric is the flat metric of Minkowski geometry $\mathbb{M}$. Physically, this means that the gravitational fields can be neglected. When we need to take the gravitational fields into account, we must introduce the curved metric of Einstein's general relativity. This will be discussed more fully in the next chapter.

18.4 Hyperbolic geometry in Minkowski space

Let us look at some further aspects of Minkowski's geometry and its relation to that of Euclid. In Euclidean geometry, the locus of points that are a fixed distance a from a fixed point O is a sphere. In $\mathbb{E}^4$, of course, this is a 3-sphere S^3. What happens in $\mathbb{M}$? There are now two situations to consider, depending upon whether we take a to be a (say positive) real number or (in effect) purely imaginary (where I am adopting my preferred $+ - - -$ signature; otherwise the roles would be reversed); see Fig. 18.7, which illustrates both cases.

The case of imaginary a will not concern us particularly here. Let us therefore assume $a > 0$ (the case $a < 0$ being equivalent). Now our 'sphere' consists of two pieces, one of which is 'bowl-shaped', $\mathcal{H}^+$, lying within the future light cone, and the other, $\mathcal{H}^-$, 'hill-shaped', lying within the past light cone. We shall concentrate on $\mathcal{H}^+$ (the space $\mathcal{H}^-$ being similar). What is the intrinsic metric on $\mathcal{H}^+$? It certainly inherits a metric, induced on it from its embedding in $\mathbb{M}$. (The lengths of a curve in $\mathcal{H}^+$, for example, is defined simply by considering it as a curve in $\mathbb{M}$.) In fact, for this case, the $d\ell^2$ (with signature $+ + + -$) is the better measure, since the directions along $\mathcal{H}^+$ are spacelike. We can make a good guess as to $\mathcal{H}^+$'s metric, because it is essentially just a 'sphere' of some sort, but with a 'sign flip'. What can that be? Recall Johann Lambert's considerations, in 1786, on the possibility of constructing a geometry in which Euclid's 5th postulate would be violated. He considered that a 'sphere' of imaginary radius would provide such a geometry, provided that such a thing actually makes consistent sense. In fact, our construction of $\mathcal{H}^+$, as just given, provides just such a space—a model of hyperbolic geometry—but now it is 3-dimensional. To get Lambert's non-Euclidean plane (the hyperbolic plane), all we need to do is

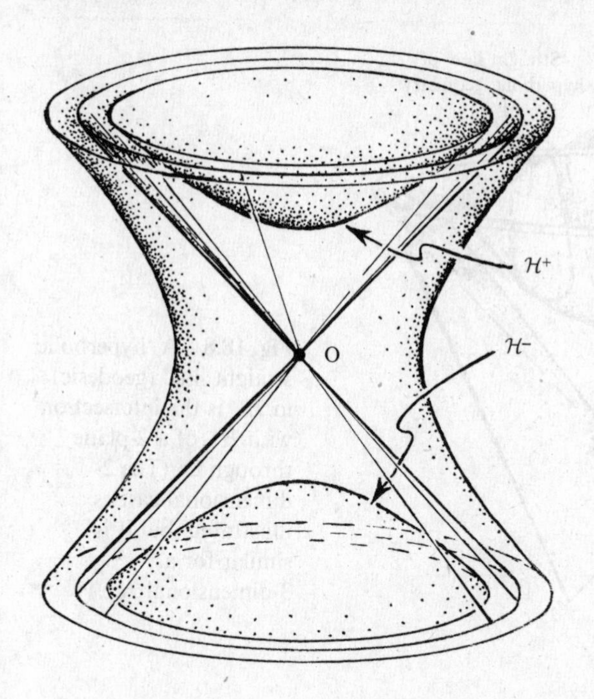

Fig. 18.7 'Spheres' in $\mathbb{M}$, as the loci of points a fixed Minkowski distance a from a fixed point O. If $a > 0$ (with the $+---$ ds^2 signature) we get two 'hyperbolic' pieces, the 'bowl-shaped' $\mathcal{H}^+$ (within the future light cone) and the 'hill-shaped' $\mathcal{H}^-$, (within the past light cone). For imaginary a (or with real a and the $+++-$ $d\ell^2$ signature) we get a one-sheeted hyperboloid, spacelike-separated from O.

dispense with one of the spatial dimensions in what has been described above. In each case the 'hyperbolic straight lines' (geodesics) are simply intersections of $\mathcal{H}^+$ with 2-planes through O (Fig. 18.8).

Of course, it is somewhat fanciful to imagine that Lambert might have had something like this construction hidden at the back of his mind. Nevertheless, it illustrates something of the inner consistency of ideas of this general kind, in which signatures can be 'flipped' and real quantities made imaginary and imaginary quantities made real. This is something about which Lambert could easily have had very creditable instincts. It is perhaps instructive to examine Fig. 18.9. Here I have drawn a light cone $t^2 - x^2 - y^2 - z^2 = 0$ (y suppressed), for Minkowski 4-space $\mathbb{M}$, with coordinates (t, x, y, z), and I have taken a family of sections of the cone by the planes

$$z + t + \lambda(t - z) = 2,$$

for various values of λ, all taken through a particular plane $t = 1 = z$. This intersection is 2-dimensional (the cone itself being 3-dimensional), and it turns out that, for each positive value of λ, the metric of this 2-surface is exactly that of a sphere, of radius $\lambda^{-1/2} = 1/\sqrt{\lambda}$ (with respect to the $d\ell^2$ metric). When $\lambda = 0$, we get the metric of an ordinary Euclidean

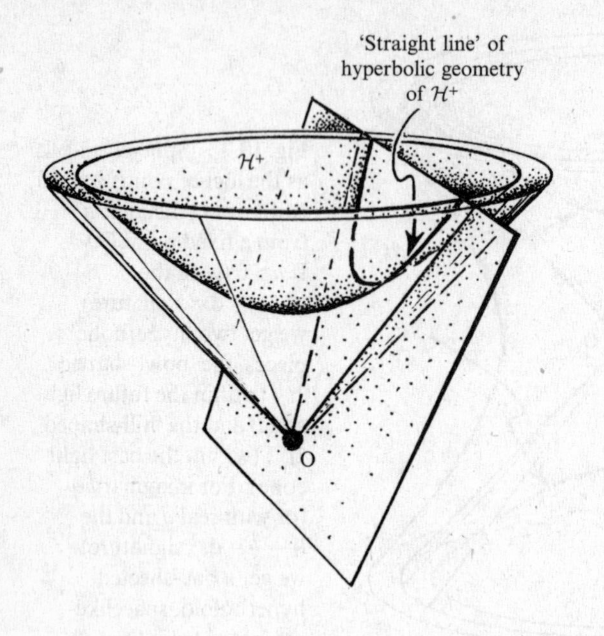

Fig. 18.8 A 'hyperbolic straight line' (geodesic) in $\mathcal{H}^+$ is the intersection with $\mathcal{H}^+$ of a 2-plane through O. (The 2-dimensional case is illustrated, but it is similar for a 3-dimensional $\mathcal{H}^+$.)

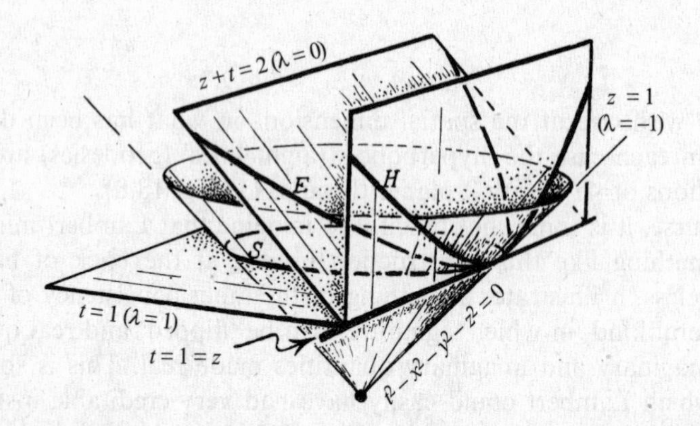

Fig. 18.9 Sections of the light cone $t^2 - x^2 - y^2 - z^2 = 0$, by 3-planes $(z + t)+ \lambda(t - z) = 2$, through the 2-plane $t = 1 = z$. The coordinate y is suppressed, so dimensions appear reduced by 1. When $\lambda > 0$ the section S has a 2-sphere $d\ell^2$ metric, illustrated by the horizontal case $\lambda = 1$. When $\lambda = 0$ we get the flat Euclidean $d\ell^2$ metric of the paraboloidal section E. When $\lambda < 0$ we get a hyperbolic $d\ell^2$ metric, illustrated by the vertical hyperbolic section H, in the case $\lambda = -1$.

plane. (This intersection does not look 'flat', but 'paraboloidal' instead; nevertheless its intrinsic metric is indeed flat.)[18.6] When λ becomes nega-

[18.6] Show all this. *Hint*: It is handy to make use of coordinates x, y, and w, where $w = (t - z - 1/\lambda)\sqrt{\lambda} = (1 - t - z)/\sqrt{\lambda}$.

tive, the intersection is Lambert's sphere of imaginary radius ($=1/\sqrt{\lambda}$). It indeed has an intrinsic metric (from $d\ell^2$) of hyperbolic geometry. In this way, we see that Lambert's tentative insight that imaginary-radius spheres might make sense was perfectly justified, albeit centuries ahead of its time.

The construction for hyperbolic geometry as the 'pseudosphere' $\mathcal{H}^+$ can be directly related to Beltrami's conformal and projective representations that were described (in the 2-dimensional case) in §§2.4,5. In Fig. 18.10, I have illustrated the way that both of these can be obtained directly from $\mathcal{H}^+$, explicitly depicting the 2-dimensional case of pseudospheres in Minkowski 3-space $\mathbb{M}^3$ (with coordinates t, x, y). Taking $\mathcal{H}^+$ to have equation $t^2 - x^2 - y^2 = 1$, we obtain Beltrami's 'Klein' (i.e. projective) representation by projecting it from the origin (0, 0, 0) to the plane $t = 1$, and we obtain Beltrami's 'Poincaré' (i.e. conformal) representation by projecting from the 'south pole' (-1, 0, 0) to the 'equatorial plane' $t = 0$ (i.e. 'stereographic projection'; see §8.3, Fig. 8.7).[18.7]

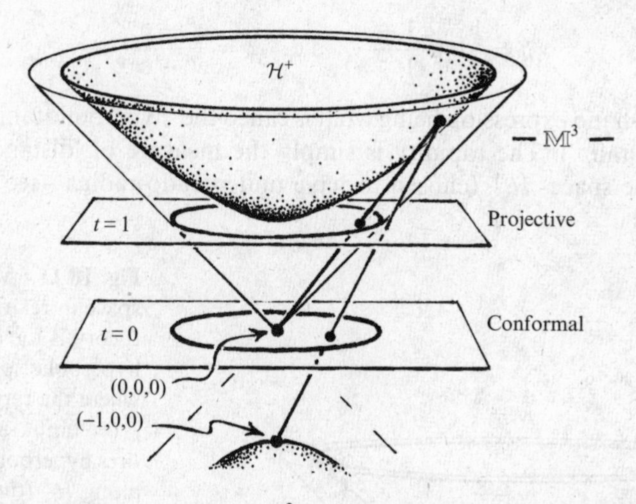

Fig. 18.10 In Minkowski 3-space $\mathbb{M}^3$, the hyperbolic 2-geometry of $\mathcal{H}^+$ (given by $t^2 - x^2 - y^2 = 1$) directly relates to Beltrami's conformal and projective representations (illustrated in Figs. 2.11 and 2.16 respectively—M.C.Escher's print and in its distorted version). Beltrami's projective ('Klein') model is obtained by projecting $\mathcal{H}^+$ from the origin (0,0,0) to the interior of the unit circle in the plane $t = 1$. Beltrami's conformal ('Poincaré') model is obtained by projecting $\mathcal{H}^+$ from (−1,0,0) to the interior of the unit circle in $t = 0$. (See also Beltrami's geometry of Fig. 2.17.) The analogous construction works also for hyperbolic 3-geometry in $\mathbb{M}$.

[18.7] Show why the hyperbolic straight lines are represented as straight in the 'Klein' case and by circles meeting the boundary orthogonally in the 'Poincaré' case, indicating, by use of a 'signature flip', why this second case is indeed conformal.

Notice that the future-timelike *directions* are represented by the points of $\mathcal{H}^+$ (where, for definiteness, I take $a = 1$). These are simply the possible velocities of a massive particle. Thus, $\mathcal{H}^+$ can be thought of as *velocity space* in relativity theory. (Recall that this issue was raised at the end of §2.7.) It is one of the aspects of relativity that people often find most disturbing that one cannot simply add up velocities in the normal way. Thus, in particular, if a rocket ship were to travel in some direction at $\frac{3}{4}c$, relative to the Earth, and it were to eject a missile in the same spatial direction at $\frac{3}{4}c$, relative to the ship, then the missile travels at only $\frac{24}{25}c$, relative to the Earth, not the superluminal $(\frac{3}{4}+\frac{3}{4})c = \frac{3}{2}c$. (Here c is light speed, re-introduced for clarity only; units are chosen so that $c = 1$.) This is understood here as an effect of adding *lengths* in the hyperbolic geometry (see Fig. 18.11).[18.8]

To appreciate this, we need to understand the physical interpretation of this hyperbolic 'length'. In fact, it is a quantity, known as the *rapidity*, for which I shall use the Greek letter ρ, defined in terms of the speed v by the formulae (graphed in Fig. 18.12)

$$\rho = \frac{1}{2}\log\frac{1+v}{1-v}, \quad \text{i.e.} \quad v = \frac{e^{\rho} - e^{-\rho}}{e^{\rho} + e^{-\rho}},$$

(the right-hand expression being what is called the 'hyperbolic tangent' of ρ, written '$\tanh\rho$'). The rapidity is simply the measure of 'distance' in the hyperbolic space $\mathcal{H}^+$ (chosen to have unit pseudo-radius—see §§2.4,6—

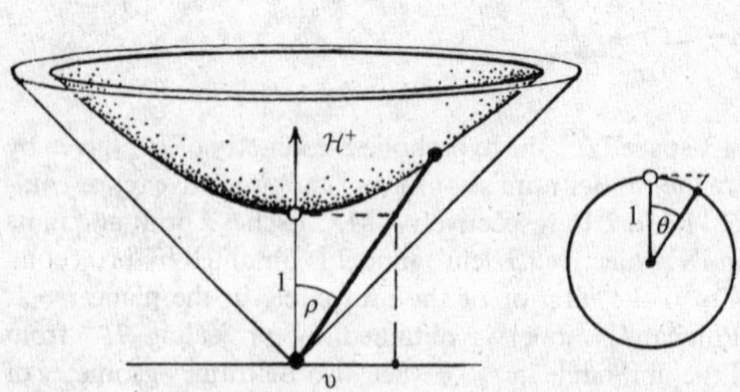

Fig. 18.11 Velocity space in relativity theory is the (unit) hyperbolic space $\mathcal{H}^+$, where the rapidity ρ (= $\tanh^{-1}v$) measures hyperbolic distance along $\mathcal{H}^+$ (the speed of light $c = 1$ corresponding to infinite ρ). This is analogous (by 'signature flip') to distance along a unit circle being the angle θ subtended at its centre.

[18.8] Use a 'signature-flip' argument, to see why adding lengths in hyperbolic geometry should give rise to the addition formula being used here, namely $(u + v)c/(1 + uv)$, for 'adding' the velocities uc and vc in the same spatial direction. Consider adding arc lenghts around a circle or sphere, the 'velocity' corresponding to each arc length being the tangent of the angle it subtends at the centre.

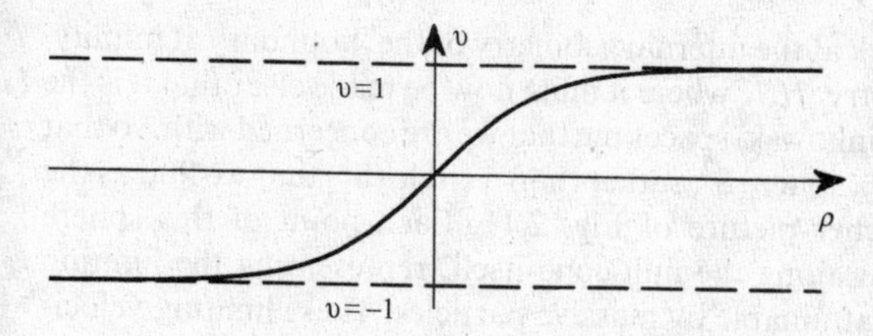

Fig. 18.12 The graph of velocity v (with $c = 1$) in terms of rapidity ρ defined by $\rho = \frac{1}{2}\log\{(1+v)/(1-v)\}$, i.e. $v = (e^{\rho} - e^{-\rho}) \div (e^{\rho} + e^{-\rho}) = \tanh\rho$.

since $a = 1$). For speeds v that are small compared with that of light, the rapidity is the same as v.[18.9] Note that the boundary, in the Escher picture shown in Fig. 2.11, which describes infinity for hyperbolic geometry ($\rho = \infty$), represents the unattainable limiting velocity c $(= 1)$.

Composing velocities in the same direction is described simply by adding their rapidities (i.e. adding hyperbolic lengths); see Fig. 18.13a. We can compose velocities in different directions simply by using the procedure given for ordinary rotations in §11.4, as illustrated in Fig. 11.4 (appropriately 'signature-flipped'). Here we use a hyperbolic *triangle law*, applied to the two velocities to be composed, where each is represented by a hyperbolic segment whose hyperbolic length is exactly one half of the rapidity that it represents (corresponding to the fact that the arc lengths in Fig. 11.4 are exactly one-half of the angle that is being rotated through); see Fig. 18.13b.

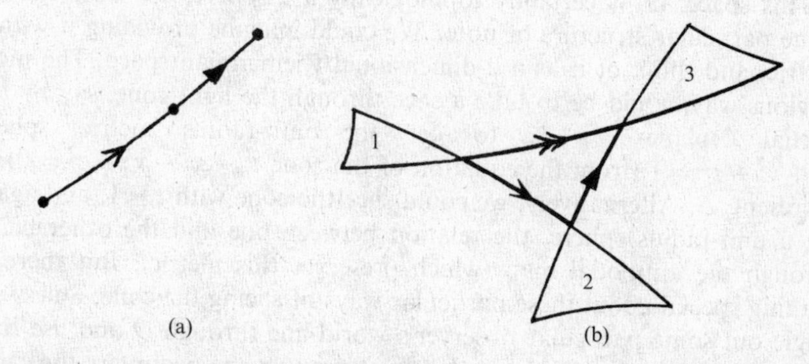

Fig. 18.13 Composing relativistic velocities in hyperbolic velocity space $\mathcal{H}^+$. (a) For velocities in the same direction, we simply add the rapidities. (b) For velocities in different directions, we use a triangle law to compose them, where the hyperbolic side-lengths are one-half of their respective rapidities. (Compare Fig. 11.4b, describing the composition of ordinary rotations in 3-space, the proof being the same.)

[18.9] Justify this assertion; prove the equivalence of the above two displayed formulae.

18.5 The celestial sphere as a Riemann sphere

Let us next have a look at the internal geometry of the 'boundary at infinity' for hyperbolic geometry $\mathcal{H}^+$, where it must now be made clear that it is the full 4-dimensional Minkowski spacetime that we are concerned with, so that this boundary is now a sphere S^2, rather than a circle (S^1) that we find as the boundary of the Escher picture of Fig. 2.11. Each point of this sphere represents a direction along the null cone itself, representing the limiting light speed that is unattainable by massive particles. These limiting velocities are attainable for *massless* particles however; in fact, these are the only velocities available to massless particles in free flight. Fortunately, photons are such massless particles, and you can see photons. If you look up at the sky on a clear cloudless night, you appear to see a hemispherical dome above you, punctuated by myriads of stars. In fact, you are realistically picturing the family of light rays that constitute the light cone centred at the event O that is occupied by your eye at the moment that you perceive the celestial scene. Actually, you are only perceiving about half of the rays of the light cone, but if you imagine that you are out in space, with a full view of the celestial sphere surrounding you, then you will have a better picture of the sphere of rays that make up the entire light cone of O. Perhaps it is easier to picture this sphere as representing O's *past* cone, because our concern is with the light coming into your eye, not coming out of it. But light rays, in the sense of null straight lines extend both ways, from past to future, so the celestial sphere may also be thought of as simply representing this family $\mathcal{S}$ of *entire* light rays through O. (See also §33.2.)

This space $\mathcal{S}$ is certainly topologically a 2-sphere, but does it have some particular structure of note? We could imagine providing it with a metric, and think of it as a 2-dimensional Riemannian space. The most obvious way would be to take a slice through the light cone, say by the spatial 3-plane $t = -1$, to get the unit-radius metric sphere $x^2 + y^2 + z^2 = 1$ (from the equation of the cone $t^2 - x^2 - y^2 - z^2 = 0$) to represent $\mathcal{S}$. Alternatively, we could slice the cone with $t = 1$, and again get a unit-radius sphere, the relation between one and the other being through the antipodal map (which preserves this metric). But there is nothing special about these particular ways of slicing the cone, unless we single out some particular observer's world-line through O and use that observer's 't coordinate'. For another observer who encounters the same event O, but who might be travelling at some high speed with respect the first, there may be some distortion between the map of the celestial sphere that one observer makes and the map that the other makes.

Indeed, there *is* some kind of distortion, because of the effect known as *stellar abberation*, which was observed by James Bradley in 1725. According to this effect, the apparent position of a star on the celestial sphere is seasonally slightly displaced, owing to the fact that the velocity of the Earth changes when it is at different places in its orbit about the Sun. This effect is akin to that commonly observed by motorists when travelling at speed in the rain. To those who are in the car, it appears that the rain is coming almost directly from the front, whereas from the perspective of an observer standing on the ground, the rain may be falling essentially vertically downwards. This effect comes about from the fact that the finite velocity of the rain must be composed, appropriately, with the velocity of the car in order that the observed relative effect can be ascertained. In fact, in this situation, the car's speed is being taken to be much greater than that of the rain, so that the main apparent effect comes from the car's motion. In the case of the star, on the other hand, the variation in the Earth's orbital velocity is much smaller than the speed that of star's light, as it travels towards us. Accordingly, the seasonal variation in the star's apparent position on the celestial sphere is very small (about 20 seconds of arc, in fact). Nevertheless, the effect is present, and it represents a velocity-dependent distortion of the celestial sphere, telling us that we cannot regard this sphere as having a natural metric structure, independent of the velocity of the observer.

The question that I am posing here is whether there is some nice mathematical structure on $\mathcal{S}$, weaker than a metric structure, which is preserved when we pass from the celestial map that one observer makes to the map that another makes, when both pass by each other at the event O, at high relative speed. In fact there *is* such a structure; and, remarkably, it is just that structure that we studied earlier in §§8.2,3, when we considered the Riemann sphere. Recall that the Riemann sphere possesses a conformal structure: thus, although it does not have a particular metric assigned to it, so that there is no notion of distance defined between nearby points, or lengths assigned to curves, there is an absolute notion of *angle* defined between curves on the sphere. Any allowable, i.e. *conformal*, transformation of the Riemann sphere to itself must preserve this notion of angle. Consequently, (infinitesimally) small shapes are preserved under such transformations, although their sizes may change. Moreover, circles of any size on the sphere are transformed again to circles. This is indeed the very structure that is possessed by the celestial sphere $\mathcal{S}$. Accordingly, any circular pattern of stars, as perceived by one observer, must also be perceived as circular by any other.[18.10] This suggests that a convenient

[18.10] Try to fill in the details of an ingenious argument for this, due to the highly original and influential Irish relativity theorist John L. Synge, which requires no calculation! The argument proceeds roughly as follows. Consider the geometrical configuration consisting of the past light cone C of an event O and a (timelike) 3-plane P through O. Let Σ be the intersection of C and P. Describe the 'history', as time progresses, of the respective spatial descriptions of C, P, and Σ, according to some particular Minkowskian reference frame. Explain why any observer at O sees Σ as a circle and, moreover, that this geometrical construction characterizes, in a frame-independent way, those bundles of rays that appear to an observer as a circle.

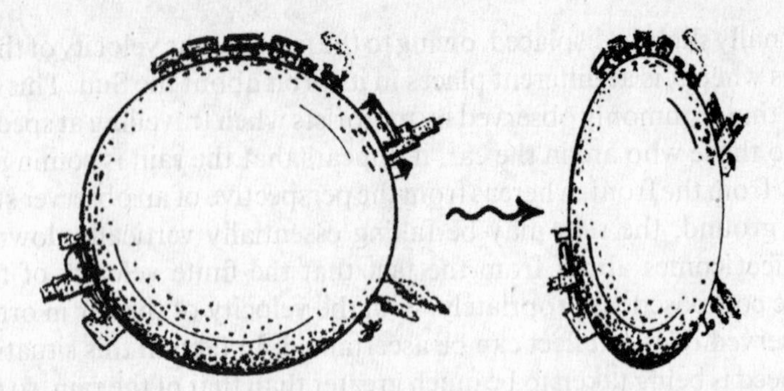

Fig. 18.14 FitzGerald–Lorentz 'flattening effect'. A spherical planet moves to the right at a speed v (close to that of light) with respect to a fixed reference system. In that system it would be described as being flattened by a factor $(1 - v^2/c^2)^{1/2}$ in its direction of motion.

labelling of the stars in the sky might be to assign a complex number to each (allowing also ∞)! I am not aware that such a proposal has been taken up in astronomy, but the use of such a complex parameter, called a 'stereographic coordinate', related to standard spherical polar angles (§22.11, Fig. 22.16) by the formula $\zeta = e^{i\varphi}\cot\frac{1}{2}\theta$,[18.11] is common in general relativity theory.[7]

This property may seem surprising, especially to those familiar with the FitzGerald–Lorentz contraction, whereby a sphere, moving rapidly with speed v, is regarded as being flattened in its direction of motion, by a factor $\gamma = \surd(1 - v^2/c^2)$, see Fig. 18.14. (I have not explicitly discussed this flattening effect here. It arises when we consider the spatial description of a moving object, and it can be found in most standard accounts of relativity theory).[8,[18.12]] Imagine that the sphere passes horizontally overhead at a speed approaching that of light. It is easy to imagine that this flattening ought surely to be perceivable to an observer standing at rest on the ground. By the relativity principle, the effect should be identical with what the observer perceives if it is the observer who moves with speed v in the opposite direction and the sphere remains at rest. But to an observer at rest viewing a sphere at rest, the sphere is certainly perceived as something with a circular outline. This would seem to contradict the 'perceived circles go to perceived circles' assertion of the preceding paragraph. In fact, there is no contradiction, because this FitzGerald–Lorentz 'flattening effect' is, in fact, not directly observable. This follows by detailed consideration of the path lengths of the light that appears to be coming to an observer, with respect to whom the sphere is in motion. See Fig. 18.15. The

[18.11] Derive this formula.

[18.12] Try to derive this formula using the spacetime geometry ideas above.

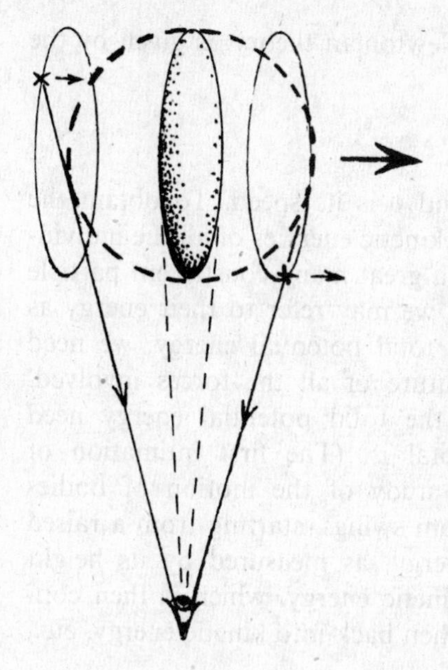

Fig. 18.15 The FitzGerald–Lorentz flattening is not directly visible because what appears to an observer to be the rear of the sphere involves a longer path length than what appears to be the front of the sphere (the rear part moving out of the way of the light and the front part moving into it). Accordingly, the apparent rear edge refers to an earlier position of the sphere than does the front edge, whereby the image is compensatingly stretched in the direction of motion.

light which appears to come from the rear of the sphere reaches the observer from a more distant point than that which appears to be coming from the sphere's front.[9,[18.13]]

18.6 Newtonian energy and (angular) momentum

There is one final aspect of Minkowskian geometry that I wish to discuss in this chapter. This concerns the important issues of *energy*, *momentum*, and *angular momentum* in relativity theory. We shall come to this shortly in §18.7, but I should first make some remarks about these essential concepts in Newtonian theory, as I have not introduced them before in this book. The vital importance of these quantities is that they are things with a well-defined meaning in Newtonian theory which are *conserved*—for a system not acted upon by external forces—in the sense that the total energy, momentum, and angular momentum are constant in time.

The energy of a system may be considered to be composed of two parts, namely the *kinetic* energy (i.e. energy of motion) and the *potential* energy (the energy stored in the forces between particles). The kinetic

---

[18.13] Develop this argument in detail, to show why the FitzGerald–Lorentz flattening exactly compensates for the effect arising from the path-length difference. Show that for small angular diameter, the apparent effect is a rotation of the sphere, rather than a flattening.

energy of a (structureless) particle, in Newtonian theory, is given by the expression

$$\tfrac{1}{2}mv^2,$$

where m is the mass of the particle and v is its speed. To obtain the entire kinetic energy, we simply add the kinetic energies of all the individual particles (although, when there are a great many constituent particle components moving around randomly, we may refer to their energy as *heat* energy; see §27.3). To obtain the total potential energy, we need to know something of the detailed nature of all the forces involved. Neither the total kinetic energy nor the total potential energy need be individually conserved, but the total is. (The first intimation of this can be traced back to Galileo's study of the motion of bodies under gravity. As the bob of a pendulum swings, starting from a raised position, its gravitational potential energy, as measured by its height above the ground, is converted into kinetic energy, which is then converted back into potential energy, and then back into kinetic energy, etc., etc.)

The *momentum* $\mathbf{p}$ of our particle is a vector quantity, given by the expression

$$\mathbf{p} = m\mathbf{v}$$

where $\mathbf{v}$ is the vector describing its velocity. To get the entire momentum, we take the vector sum of all the individual momenta. This total quantity is also conserved in time.[18.14]

Now, we recall from §§17.2,3 that a relativity principle holds for Newtonian theory (Galilean relativity). How do our conservation laws manage to survive when neither the energy nor the momentum is left unchanged as we move from one inertial frame to another? If the second frame moves uniformly, with respect to the first, with a velocity given by the vector $\mathbf{u}$, then a particle whose velocity is $\mathbf{v}$, in the first frame, has its velocity described as $\mathbf{v} - \mathbf{u}$ in the second. It turns out that conservation of energy and momentum in the first frame goes over to conservation of energy and momentum in the second frame provided we take into account that mass is also conserved (and we must also make use of Newton's third law; see Fig. 17.4b, §17.3).[18.15]

It should be mentioned that in Newtonian mechanics there are also other conserved quantities, the most important of which is *angular*

[18.14] Use conservation of energy and momentum to show that if a stationary billiard ball is hit by another of the same mass, then they emerge at right angles (assuming an elastic collision, so there is no conversion of kinetic energy to heat).

[18.15] Show all this.

momentum (or *moment* of momentum), taken about some origin point O. Suppose that the *position vector* relative to O of some particle is

$$\mathbf{x} = (x^1, x^2, x^3),$$

x^1, x^2, x^3 being its Cartesian coordinates and $\mathbf{p}$ is its momentum; then the angular momentum is given by the quantity

$$\mathbf{M} = 2\mathbf{x} \wedge \mathbf{p}$$

(see §11.6, for the meaning of $\wedge$).[10] To, get the angular momentum of the entire system, we simply add the quantities $\mathbf{M}$ for all the individual particles.[18.16]

There is also another quantity that is conserved in time in the absence of external forces, in Newtonian theory, which is less often discussed than angular momentum. For a single particle, this is

$$\mathbf{N} = t\mathbf{p} - m\mathbf{x},$$

where t is the time, and we get the total value of $\mathbf{N}$ by adding the individual values for each particle. This total has the same form as $\mathbf{N}$ given above, but where $\mathbf{x}$ is now the position vector of the mass centre and $\mathbf{p}$ the total momentum. The constancy of this total $\mathbf{N}$ expresses the fact that the mass centre moves uniformly in a straight line; see Fig. 18.16.[18.17]

We shall need to ask the question: how is all this affected by the upheavals of special relativity? Do we still have concepts of conserved energy, momentum, angular momentum, and mass-centre motion? What about conservation of mass? The answer to the first four questions is 'yes', although we have to be careful to define these quantities correctly. As regards *mass* conservation, something very curious happens. The two

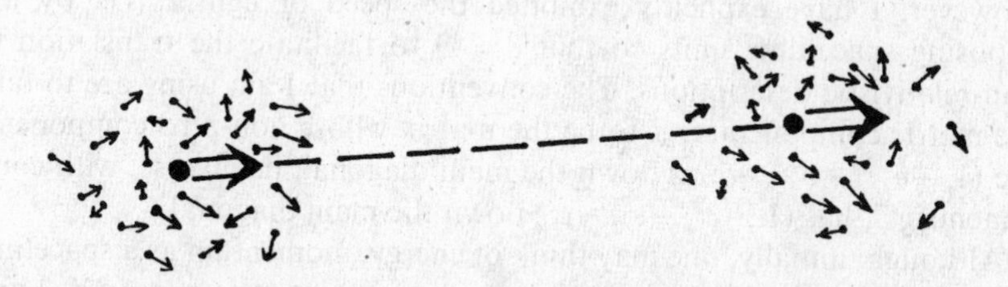

Fig. 18.16 Uniform motion of mass centre. The quantity $\mathbf{N} = t\mathbf{p} - m\mathbf{x}$, where t is the time and $\mathbf{x}$ is the position vector of mass centre, is conserved. This expresses fact that mass centre moves uniformly in a straight line, with velocity $\mathbf{p}/m$.

[18.16] Why do spinning skaters pull in their arms to increase their rate of rotation?

[18.17] Show this. (N.B. The position vector of the mass centre is the sum of the quantities $m\mathbf{x}$ divided by the sum of the masses m.)

separate Newtonian conservation laws for energy and mass become subsumed into one. In a clear sense, mass and energy become completely equivalent to one another, according to Einstein's most famous equation

$$E = mc^2,$$

where E is the total energy of the system and m is its total mass, c being the speed of light, as before. In the final section of this chapter, we shall see how this all works.

18.7 Relativistic energy and (angular) momentum

Recall the way that space and time become united in relativity theory to become the single entity 'spacetime', the time coordinate t being adjoined to the 3-space position vector $\mathbf{x} = (x^1, x^2, x^3)$ to give the 4-vector

$$(x^0, x^1, x^2, x^3) = (t, \mathbf{x}).$$

We shall find that momentum and energy become similarly united. Any finite system in special relativity will have a total energy E and a total momentum 3-vector $\mathbf{p}$. These unite into what is called the *energy–momentum* 4-vector, whose spatial components are

$$(p^1, p^2, p^3) = c^2\mathbf{p},$$

and whose time-component p^0 measures not only the total energy but also, equivalently, the total *mass* m of the system according to

$$p^0 = E = mc^2,$$

which incorporates Einstein's famous mass–energy relation.

With more natural units with $c = 1$, energy and mass are simply equal. However, I have explicitly exhibited the speed of light c (i.e. by not choosing space/time units so that $c = 1$) to facilitate the translation to non-relativistic descriptions. The conventions that I am using are to take the metric components g_{ab} to be the matrix whose non-zero components are $(1, -c^{-2}, -c^{-2}, -c^{-2})$ down the main diagonal; its inverse, with components g^{ab}, has $(1, -c^2, -c^2, -c^2)$ down the main diagonal.

Although, initially, one may think of energy–momentum as a spacetime vector in this way, it turns out that it is more appropriate (see §20.2 and §21.2) to regard it as a *covector*, described by the index-lowered quantity p_a with components

$$(p_0, p_1, p_2, p_3) = (E, \ -\mathbf{p}).$$

This has an irritating minus sign (although the c has now gone). Whichever version is used (p_a or p^a), the 4-momentum satisfies a *conservation law*.

Thus, in an encounter between two or more particles (or systems), or in the decay of a single particle (or system) into two or more, or the capture of a particle by another, the sum of all the 4-momenta before the encounter is equal to the sum of all the 4-momenta afterwards. Thus, the law of energy conservation, of momentum conservation, and also of mass conservation, are all subsumed into this one law. The reason for collecting them together in this way is that, under change of reference frame, these quantities get transformed among themselves in the correct way for relativity theory, as demanded by the index notation (see §12.8).

We note that the total mass of a system is not a scalar quantity in relativity theory, so that its value depends on the reference frame with respect to which it is measured. For example, a particle whose mass is m, as measured in its own rest frame, appears to have a larger mass when measured in a second frame with respect to which it is moving. For this to be a significant effect, however, the relative velocity between the two frames would need to be comparable with the velocity of light.[18.18]

However, these comments apply only to the kind of mass which is *conserved* in the additive sense just described (for a system not acted upon by external forces). There is another concept of mass in relativity, namely the *rest mass* μ (≥ 0), which does not depend on the reference frame. It is equal to the mass measured in the system's own rest frame—i.e. in the frame for which the momentum is zero. The rest mass μ is c^{-2} times the *rest energy* $(p_a p^a)^{1/2}$, so that

$$(c^2\mu)^2 = p_a p^a = E^2 - c^2\mathbf{p}^2;$$

and we have $\mu = c^{-2}(E^2 - c^2\mathbf{p}^2)^{1/2}$. Here, I am adopting the 3-space vector notation whereby, for an arbitrary 3-vector $\mathbf{a}$, we define $\mathbf{a}^2 = \mathbf{a}\bullet\mathbf{a} = a_1^2 + a_2^2 + a_3^2$. The 'dot' defines 'scalar product' (similarly to the notation of §12.3):

$$\mathbf{a}\bullet\mathbf{b} = a_1b_1 + a_2b_2 + a_3b_3,$$

with $\mathbf{a} = (a_1, a_2, a_3)$ and $\mathbf{b} = (b_1, b_2, b_3)$. (This notation will be handy later.)

For a single particle which is *massive* in the sense that $\mu > 0$, we can take the 4-momentum to be the *4-velocity* scaled up by the rest mass μ. The 4-velocity v^a is the (future-)timelike vector tangent to the particle's world-line, having a (Minkowskian) length of c (i.e. a *unit* vector if $c = 1$):

$$p^a = \mu v^a, \text{ where } v_a v^a = c^2;$$

[18.18] Show that the formula for the increased mass is $m(1 - v^2/c^2)^{-1/2}$, where v is the velocity of the particle in the second frame; see below.

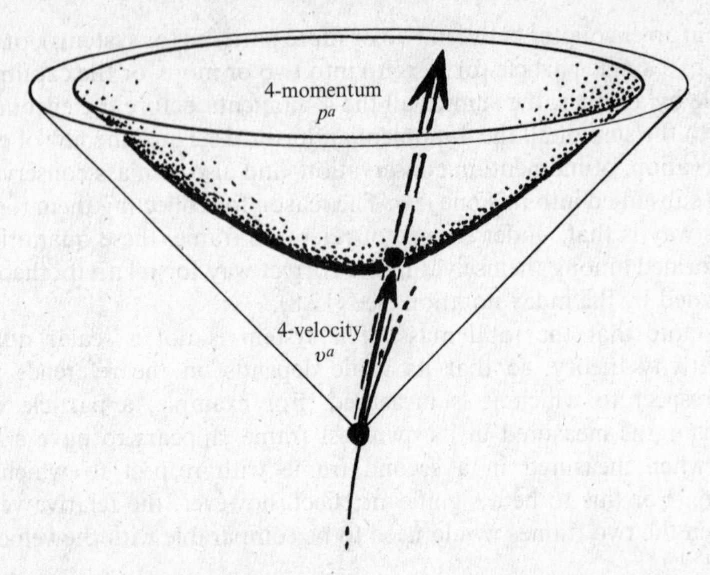

Fig. 18.17 For a massive particle, the 4-momentum p^a is the 4-velocity v^a scaled up by the rest mass μ (> 0), where v^a is a (future-timelike) unit 4-vector tangent to the particle's world-line (taking $c = 1$).

see Fig. 18.17. As remarked above, the rest mass of a massive particle is the mass (mass–energy) of that particle as measured in its own rest frame. Taking the particle's ordinary 3-velocity to be $\mathbf{v}$, so that $\mathbf{v} = (\mathrm{d}x^1/\mathrm{d}t, \mathrm{d}x^2/\mathrm{d}t, \mathrm{d}x^3/\mathrm{d}t)$, where $t = x^0$, we get[18.19],[18.20]

$$\mathbf{p} = m\mathbf{v}, \quad m = \gamma\mu, \quad v^a = \gamma(c^2, \mathbf{v}),$$

where

$$\gamma = (1 - \mathbf{v}^2/c^2)^{-1/2}.$$

Particles can also be *massless* (i.e. with *zero rest mass*, $\mu = 0$), the photon being the prime example. Then the 4-momentum is a *null vector*. Since rest mass is not conserved, there is nothing against a massive particle decaying into massless ones, or massless particles coming together to produce massive ones. In fact, a massive particle known as the 'neutral pion' (denoted by π^0) will normally decay into two photons in about 10^{-16} seconds.

[18.19] Why?

[18.20] Use the Taylor series of §6.4, to derive $(1 + x)^{1/2} = 1 + \frac{1}{2}x - \frac{1}{8}x^2 + \frac{1}{16}x^3 - \ldots$. Hence, obtain a power series expansion for the energy $E = [(c^2\mu)^2 + c^2\mathbf{p}^2]^{1/2}$ of a particle of rest-mass μ and 3-momentum $\mathbf{p}$. Show that the leading term is just Einstein's $E = mc^2$ applied to the rest energy μ, and that the next term is the Newtonian expression for kinetic energy. Write down the next two terms, so as to give better approximations to the full relativistic energy.

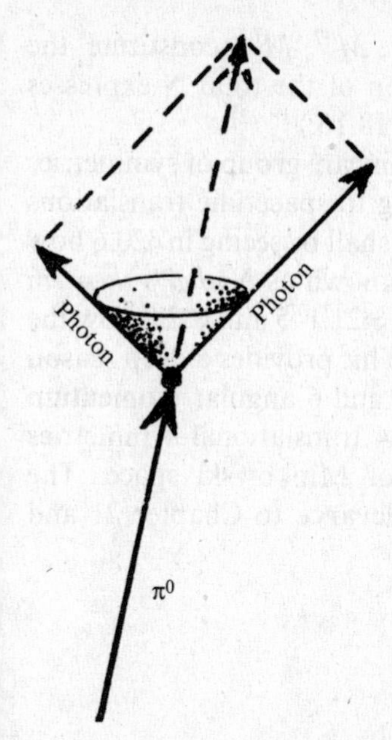

Fig. 18.18 The decay of a massive 'neutral pion' π^0 to 2 massless photons. The mass/energy 4-vector is additively conserved (although rest-mass is not).

In any particular frame, the total mass–energy (*not* the rest mass) is additively conserved, the mass–energy of each individual photon being non-zero. The way that the 4-momenta add up is illustrated in Fig. 18.18.

Finally, let us see how *angular* momentum needs to be treated in special relativity. It is described by a tensor quantity M^{ab}, antisymmetrical in its two indices:

$$M^{ab} = -M^{ba}.$$

(See §22.12 for the relevance of M^{ab} to quantum mechanics.) For a single structureless point particle, we have[11]

$$M^{ab} = x^a p^b - x^b p^a,$$

where x^a is the position 4-vector (in index form) of the point on the particle's world line at the time that its angular momentum is being considered. If the particle is in inertial motion, then M^{ab} is the same for *all* points on its world line.[18.21] To obtain the total relativistic angular momentum, we simply add the angular momentum tensors for each particle separately. For an individual (non-spinning) particle, the three independent purely spatial components M^{23}, M^{31}, M^{12} are the components ($\times\, c^2$) of ordinary angular momentum $\mathbf{M} = 2\mathbf{x} \wedge \mathbf{p}$ considered in §18.6 above, and

[18.21] Why?

the remaining independent components M^{01}, M^{02}, M^{03} constitute the quantity $\mathbf{N} = t\mathbf{p} - m\mathbf{x}$ ($\times\, c^2$). (The conservation of the total **N** expresses the uniform motion of the mass centre; see Fig.18.16.)[18.22]

Recall from §18.2 that the 10-dimensional Poincaré group of symmetries of Minkowski space has 4 dimensions referring to spacetime translations and the remaining 6 to (Lorentz) rotations. We shall be seeing in §20.6 how an important principle of classical mechanics known as *Nöther's theorem* relates symmetries to conservation laws, and in §§21.1–5 and §22.8 how the same kind of thing occurs in quantum theory. This provides a deep reason for the conservation laws for 4-momentum p_a and 6-angular momentum M^{ab}, since these arise, respectively, from the 4 translational symmetries and the 6 (Lorentz) rotational symmetries of Minkowski space. The conservation of p_a and M^{ab} has important relevance to Chapter 21 and §§22.8,12,13.

Notes

Section 18.1

18.1. Tom Banchoff, of Brown University, has for many years been developing interactive computer systems aimed at developing 4-dimensional intuition, and in particular complex function visualization in terms of Riemann surfaces in $\mathbb{C}^2$. See Banchoff (1990, 1996).

18.2. The quantities '$\mathrm{d}s$', in this expression should simply be read as 'infinitesimal quantities' (like the ε of §13.6). Compare Note 12.8.

Section 18.2

18.3. For a particular detailed discussion of the roles of Lorentz, Poincaré, and Einstein in the development of special relativity, see Stachel (1995), pp. 249–356. In my own view, even Einstein did not completely have special relativity in 1905, and it took Minkowski's 4-dimensional perspective of 1908 to complete the picture; see §17.8.

18.4. There are also *time-reversing* elements of the Poincaré group, which send future-timelike directions into past-timelike directions.

Section 18.3

18.5. I should emphasize, particularly to those readers already familiar with quantum mechanics, that the complex notion of 'orthogonality' that I am using here is necessarily the holomorphic one (this being what 'complexification' is all about), and not the Hermitian notion of §13.9 that brings in complex conjugation, and which is used in many other areas of mathematics and physics.

18.6. See, for example, Rindler (1982, 2001); Synge (1956); Taylor and Wheeler (1963); Hartle (2003). For an axiomatic geometric approach, see Schutz (1997).

Section 18.5

18.7. See, in particular, Newman and Penrose (1966); Penrose and Rindler (1984, §§ 1.2–4, §4.15; 1986, §9.8).

[18.22] Explain, in detail, in the relativistic case.

18.8. See, for example, Rindler (1982, 2001).
18.9. See, for example, Terrell (1959); Penrose (1959).

Section 18.6

18.10. Some readers may be confused by the presence of a '2' in this expression, but they should re-examine the definition of '$\wedge$' that I have given in §11.6. The components of $\mathbf{x} \wedge \mathbf{p}$ are $x^{[i}p^{j]} = \frac{1}{2}(x^i p^j - x^j p^i)$. Hence, $\mathbf{M}$ has components $x^i p^j - x^j p^i$.

Section 18.7

18.11. We shall be seeing in §22.8 that most (quantum) particles also possess an *intrinsic spin* which provides a (constant) 'spin' contribution to M^{ab} (see §22.12) added to the 'orbital M^{ab}' that is given here.

19
The classical fields of Maxwell and Einstein

19.1 Evolution away from Newtonian dynamics

In the period between the introduction of Newton's superb dynamical scheme, which we can best date as the publication of his *Principia* in 1687, and the appearance of special relativity theory, which could reasonably be dated at Einstein's first publication on the subject, in 1905, many important developments in our pictures of fundamental physics took place. The biggest shift that occurred in this period was the realization, mainly through the 19th century work of Faraday and Maxwell, that some notion of *physical field*, permeating space, must coexist with the previously held 'Newtonian reality' of individual particles interacting via instantaneous forces.[1] Later, this 'field' notion also became a crucial ingredient of Einstein's 1915 curved-spacetime theory of gravity. What are now called the *classical* fields are, indeed, the electromagnetic field of Maxwell and the gravitational field of Einstein.

But we now know that there is much more to the nature of the physical world than just classical physics. Already in 1900, Max Planck had revealed the first hints of the need for a 'quantum theory', although more than another quarter century was required before a well formulated and comprehensive theory could be provided. It should also be made clear that, in addition to all these profound changes to the 'Newtonian' foundations of physics that have taken place, there had been other important developments, both prior to these changes and coexistent with some of them in the form of powerful mathematical advances, within Newtonian theory itself. These mathematical advances will be the subject of Chapter 20. They have important interrelations with the theory of classical fields and, even more significantly, they form an essential prerequisite to the proper understanding of quantum mechanics, as will be described in subsequent chapters. As a further important area of advance, the subject of *thermodynamics* (and its refinement, referred to as *statistical mechanics*) should certainly be considered. This concerns the behaviour of systems of large numbers of bodies, where the details of the motions are not regarded as important, the behaviour of the system being described in terms of *averages* of appropriate quantities. This was an achievement initiated in the mid-19th to early 20th centuries, and the names of Carnot, Clausius, Maxwell, Boltzmann, Gibbs, and Einstein feature most strongly. I shall address some of the most fundamental and puzzling issues raised by thermodynamics later, in Chapter 27.

In this chapter, I shall describe the physical *field* theories of Maxwell and Einstein: the 'classical physics' of electromagnetism and gravitation. The theory of electromagnetism also plays an important part in quantum theory, providing the archetypical 'field' for the further development of *quantum field theory*, which we shall encounter in Chapter 26. On the other hand, the appropriate quantum approach to the gravitational field remains enigmatic and controversial. Addressing these quantum/gravitational issues will be an important part of the later chapters in this book (Chapter 28 onwards). For the physics that we shall be examining next, however, we shall confine our investigation to physical fields in their *classical* guise.

I referred, at the beginning of this chapter, to the fact that a profound shift in Newtonian foundations had already begun in the 19th century, before the revolutions of relativity and quantum theory in the 20th. The first hint that such a change might be needed came from the wonderful experimental findings of Michael Faraday in about 1833, and from the pictures of reality that he found himself needing in order to accommodate these. Basically, the fundamental change was to consider that the 'Newtonian particles' and the 'forces' that act between them are not the only inhabitants of our universe. Instead, the idea of a 'field', with a disembodied existence of its own was now having to be taken seriously. It was the great Scottish physicist James Clark Maxwell who, in 1864, formulated the equations that this 'disembodied field' must satisfy, and he showed that these fields can carry energy from one place to another. These equations unified the behaviour of electric fields, magnetic fields, and even light, and they are now known simply as *Maxwell's equations*, the first of the relativistic field equations.

From the vantage point of the 20th century, when profound advances in mathematical technique have been made (and here I refer particularly to the calculus on manifolds that we have seen in Chapters 12–15), Maxwell's equations seem to have a compelling naturalness and simplicity that almost make us wonder how the electric/magnetic fields could ever have been considered to obey any other laws. But such a perspective on things ignores the fact that it was the Maxwell equations themselves that led to a very great many of these mathematical developments. It was the form of these equations that led Lorentz, Poincaré, and Einstein to the spacetime transformations of special relativity which, in turn, led to Minkowski's conception of *spacetime*. In the spacetime framework, these equations found a form that developed naturally into Cartan's theory of differential forms (§12.6); and the charge and magnetic flux conservation laws of Maxwell's theory led to the body of integral expressions that are now encapsulated so beautifully by that marvellous formula referred to, in §§12.5,6, as the *fundamental theorem of exterior calculus*.

Perhaps, in seeming to attribute all these advances to the influence of Maxwell's equations, I have taken a somewhat too extreme position with these comments. Indeed, while Maxwell's equations undoubtedly had a key significance in this regard, many of the precursors of these equations, such as those of Laplace, D'Alembert, Gauss, Green, Ostrogradski, Coulomb, Ampère, and others have also had important influences. Yet it was still the need to understand electric and magnetic fields that largely supplied the driving force behind these developments—these, and the gravitational field also. The remainder of this chapter is devoted to understanding the electromagnetic and the gravitational fields and how they fit in with the modern mathematical framework.

19.2 Maxwell's electromagnetic theory

What, then, are the Maxwell equations? They are partial differential equations (see §10.2) which describe the time-evolutions of the three components E_1, E_2, E_3 of the electric field and of the three components B_1, B_2, B_3 of the magnetic field, where the electric charge density ρ and the three components of the electric current density j_1, j_2, j_3 are considered as given quantities. Certain other field quantities having to do with an ambient material within which the fields may be considered to be propagating can also be incorporated. In discussions of *fundamental* physics, as is our concern here, it is usual to ignore those aspects of Maxwell's equations that relate to such an ambient medium, since the medium itself would, in reality, consist of many tiny constituents, each of which could in principle be treated at the more fundamental level. It will be convenient, also, to choose what are called 'Gaussian' units, and use *standard Minkowski coordinates* (of §18.1), namely $x^0 = t$, $x^1 = x$, $x^2 = y$, $x^3 = z$ ($+---$ signature) with spacetime units so that the velocity of light c is taken to be unity ($c = 1$).

The electromagnetic field and the charge-current density are, respectively, collected together (according to a prescription originally due, in effect, to Minkowski) into a spacetime 2-form $\boldsymbol{F}$, called the *Maxwell field tensor*, and a spacetime vector $\boldsymbol{J}$, called the *charge-current vector*, with components displayed in matrix form as

$$\begin{pmatrix} F_{00} & F_{01} & F_{02} & F_{03} \\ F_{10} & F_{11} & F_{12} & F_{13} \\ F_{20} & F_{21} & F_{22} & F_{23} \\ F_{30} & F_{31} & F_{32} & F_{33} \end{pmatrix} = \begin{pmatrix} 0 & E_1 & E_2 & E_3 \\ -E_1 & 0 & -B_3 & B_2 \\ -E_2 & B_3 & 0 & -B_1 \\ -E_3 & -B_2 & B_1 & 0 \end{pmatrix},$$

$$\begin{pmatrix} J^0 \\ J^1 \\ J^2 \\ J^3 \end{pmatrix} = \begin{pmatrix} \rho \\ j_1 \\ j_2 \\ j_3 \end{pmatrix}.$$

Note that the antisymmetry $F_{ba} = -F_{ab}$ holds, as is required for a 2-form. I shall also make use of what are referred to as the *Hodge duals* of $\boldsymbol{F}$ and $\boldsymbol{J}$, these being, respectively, the 2-form ${}^*\boldsymbol{F}$ and the 3-form ${}^*\boldsymbol{J}$, defined by

$$\begin{pmatrix} {}^*F_{00} & {}^*F_{01} & {}^*F_{02} & {}^*F_{03} \\ {}^*F_{10} & {}^*F_{11} & {}^*F_{12} & {}^*F_{13} \\ {}^*F_{20} & {}^*F_{21} & {}^*F_{22} & {}^*F_{23} \\ {}^*F_{30} & {}^*F_{31} & {}^*F_{32} & {}^*F_{33} \end{pmatrix} = \begin{pmatrix} 0 & -B_1 & -B_2 & -B_3 \\ B_1 & 0 & -E_3 & E_2 \\ B_2 & E_3 & 0 & -E_1 \\ B_3 & -E_2 & E_1 & 0 \end{pmatrix}$$

$$\begin{pmatrix} {}^*J_{123} \\ {}^*J_{023} \\ {}^*J_{013} \\ {}^*J_{012} \end{pmatrix} = \begin{pmatrix} -\rho \\ j_1 \\ -j_2 \\ j_3 \end{pmatrix}.$$

Where the required antisymmetry properties ${}^*F_{ab} = {}^*F_{[ab]}$ and ${}^*J_{abc} = {}^*J_{[abc]}$ hold. In terms of the *Levi-Civita tensor* $\boldsymbol{\varepsilon}$ (§12.7), with totally antisymmetric components $\varepsilon_{abcd} (= \varepsilon_{[abcd]})$ and normalized so that $\varepsilon_{0123} = 1$, the duals can be written as

$${}^*F_{ab} = \tfrac{1}{2}\varepsilon_{abcd}F^{cd} \text{ and } {}^*J_{abc} = \varepsilon_{abcd}J^d,$$

where the raised version F^{ab} of F_{ab} is simply $g^{ac}g^{bd}F_{cd}$, in accordance with §14.7. Note that the 'raised' version $\varepsilon^{abcd} = g^{ap}g^{bq}g^{cr}g^{ds}\varepsilon_{pqrs}$ satisfies $\varepsilon^{0123} = -1$, whence the $\boldsymbol{\epsilon}$ of §12.7 is given by[19.1] $\epsilon^{abcd} = -\varepsilon^{abcd}$. See Fig. 19.1 for the diagrammatic form of these 'dualizing' operations (and also of the Maxwell equations themselves). We shall find that the notion of a 'dual' in this sense (and other related senses) will have importance for us later, in various different contexts.

A remark should be made about the geometrical significance of the Hodge dual. We recall from §12.7 that the operation of passing from a *bivector* $\boldsymbol{H}$, as described by the antisymmetric quantity H^{ab}, to its 'dual' 2-form $\boldsymbol{H}^\#$, as given by $\tfrac{1}{2}\varepsilon_{abcd}\boldsymbol{H}^{cd}$, does not make much difference to

[19.1] Check both these statements.

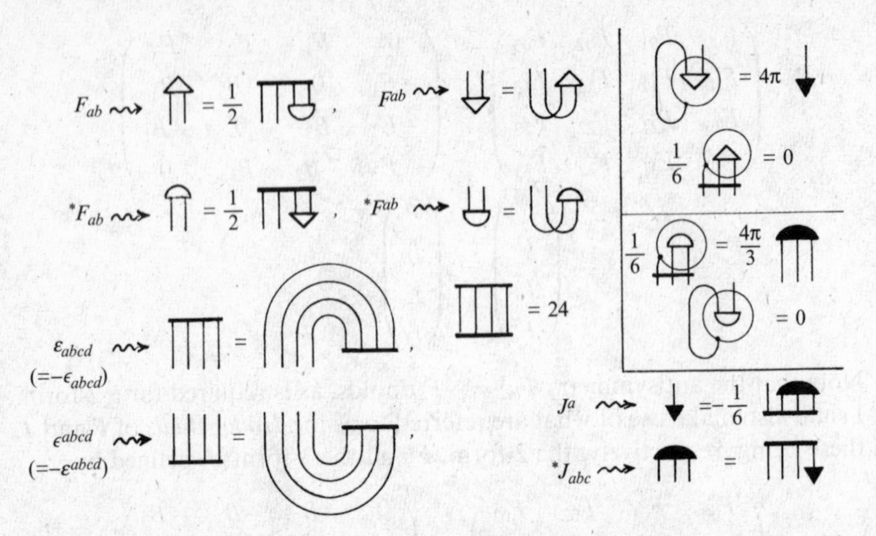

Fig. 19.1 Diagrams for Hodge duals and Maxwell equations. The quantities $\varepsilon_{abcd}\left(=\varepsilon_{[abcd]}\right)$ and $\epsilon^{abcd}\left(=\epsilon^{[abcd]}\right)$, normalized so that $\epsilon_{0123} = \epsilon^{0123} = 1$ in a standard Minkowski frame, are related to their raised/lowered versions (via g^{ab} and g_{ab}) by $\varepsilon_{abcd} = -\epsilon_{abcd}$ and $\epsilon^{abcd} = -\varepsilon^{abcd}$. In the diagrams (left middle, lower two lines) this sign change is absorbed by an effective index reversal. Boxed off at the top right are the Maxwell equations, first using the field tensor $\boldsymbol{F}$ (with its raised form $F^{ab} = g^{ac}g^{bd}F_{cd}$; cf. Fig. 14.21) so the equations are $\nabla_a F^{ab} = 4\pi \boldsymbol{J}^b$, $\nabla_{[a}F_{bc]} = 0$, and beneath that, correspondingly using the dual $^*\boldsymbol{F}$ (where $^*F_{ab} = \frac{1}{2}\varepsilon_{abcd}F^{cd}$, $^*\boldsymbol{J}_{abc} = \varepsilon_{abcd}\boldsymbol{J}^d$) so the equations are $\nabla^*_{[a}F_{bc]} = \frac{4\pi}{3}\,{}^*\boldsymbol{J}_{abc}$, $\nabla^*_a F^{ab} = 0$.

its geometrical interpretation. If $\boldsymbol{H}$ were a *simple* bivector, for example, so that the 2-form $\boldsymbol{H}^\#$ would also be simple (see the end of §12.7), then the 2-plane element determined by $\boldsymbol{H}^\#$ would be precisely the same as the 2-plane element determined by $\boldsymbol{H}$ (the only difference being that, strictly, $\boldsymbol{H}^\#$ has the quality of a density, as pointed out in §12.7). On the other hand, the index-raising that takes us from a 2-form H_{ab} to a bivector $H^{ab}\,(= H_{cd}g^{ca}g^{db})$, has a more significant geometrical effect. In the case of a simple bivector, the 2-plane element determined by H_{ab} is the *orthogonal complement* of the 2-plane element determined by H^{ab} (see §18.3). The Hodge dual, as applied to the 2-form H_{ab}, taking us to $\frac{1}{2}\varepsilon_{abcd}H^{cd}$ (i.e. to $\boldsymbol{H}^\#$), employs the index raising $H_{ab} \mapsto H^{ab}$ and therefore involves passing to the orthogonal complement. See Fig. 19.2. Accordingly, the Hodge dual taking us from $\boldsymbol{F}$ to $^*\boldsymbol{F}$ also involves an orthogonal complement.

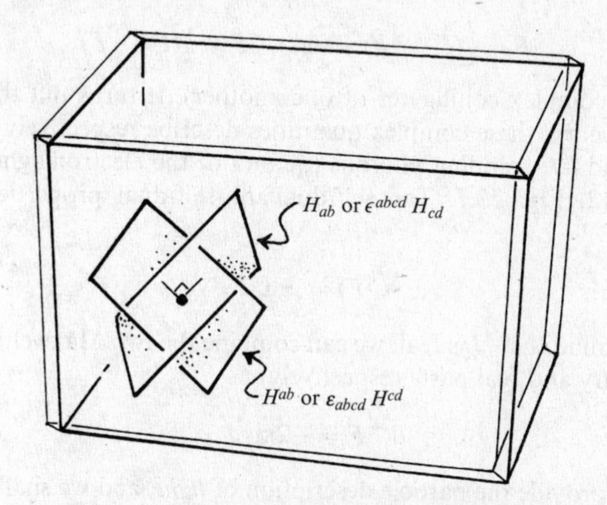

Fig. 19.2 In 4-space, a simple bivector $\boldsymbol{H}$ (H^{ab}) represents the same 2-plane element as its 'dual' 2-form $\boldsymbol{H}^{\#}\left(\frac{1}{2}\varepsilon_{abcd}H^{cd}\right)$. But the index-lowered version of $\boldsymbol{H}$, the simple 2-form H_{ab}, which is equivalent to its 'dual' bivector $\frac{1}{2}\epsilon^{abcd}H_{cd}$, represents the orthogonal complement 2-plane element (see Fig. 18.4). Hence it is the index raising/lowering in the Hodge dual that leads to the passage to the orthogonal complement.

Having set up this notation, we can now write *Maxwell's equations* very simply as[19.2]

$$\mathrm{d}\boldsymbol{F} = 0, \quad \mathrm{d}^{*}\boldsymbol{F} = 4\pi^{*}\boldsymbol{J}.$$

We can also write the Maxwell equations entirely in index form as[19.3]

$$\nabla_{[a}F_{bc]} = 0, \quad \nabla_{a}F^{ab} = 4\pi J^{b}.$$

Note that, if we apply the exterior derivative operator d to both sides of the second Maxwell equation $\mathrm{d}^{*}\boldsymbol{F} = 4\pi^{*}\boldsymbol{J}$, and use the fact that $\mathrm{d}^2 = 0$ (§12.6), we deduce that the charge-current vector $\boldsymbol{J}$, satisfies the 'vanishing divergence' equation[19.4]

$$\mathrm{d}^{*}\boldsymbol{J} = 0 \qquad \text{or equivalently} \qquad \nabla_{a}J^{a} = 0.$$

At this point, as a slight digression which will have considerable importance for us later (§32.2 and §§33.6,8,11—see §18.3, exercise [18.5](ii)), it is worth while to point out the *self-dual* and *anti-self-dual* parts of the Maxwell tensor, given respectively by

[19.2] Write these out fully, in terms of the electric and magnetic field components, showing how these equations provide a time-evolution of the electric and magnetic fields, in terms of the operator $\partial/\partial t$.

[19.3] Show the equivalence to the previous pair of equations.

[19.4] Show that the two versions of this vanishing divergence are equivalent.

$$^{+}\boldsymbol{F} = \tfrac{1}{2}(\boldsymbol{F} - \mathrm{i}\,{}^{*}\boldsymbol{F}) \quad \text{and} \quad {}^{-}\boldsymbol{F} = \tfrac{1}{2}(\boldsymbol{F} + \mathrm{i}\,{}^{*}\boldsymbol{F})$$

(which are complex conjugates of one another). It turns out that, in the quantum theory, these complex quantities describe respectively the *right*-spinning and *left*-spinning photons (quanta of the electromagnetic field); see §§22.7,12, Fig. 22.7. The self-dual/anti-self-dual properties are expressed in[19.5]

$${}^{*}({}^{\pm}\boldsymbol{F}) = \pm\,\mathrm{i}\,{}^{\pm}\boldsymbol{F}.$$

Bearing in mind that ${}^{*}\boldsymbol{J}$ is real, we can combine the two Maxwell equations (as imaginary and real parts respectively) as

$$\mathrm{d}\,{}^{+}\boldsymbol{F} = -2\pi\mathrm{i}\,{}^{*}\boldsymbol{J}.$$

Photons provide the particle description of *light*, and we shall be seeing in Chapter 21 how quantum theory allows a particle and wave description of light to coexist. It was one of Maxwell's supreme achievements to show, by means of his equations, that there are electromagnetic waves which travel with the speed of light, and have all the known polarization properties that light has (and which we shall be examining in §22.7). In accordance with these remarkable facts, Maxwell proposed that light is indeed an electromagnetic phenomenon. In 1888, almost a quarter century after Maxwell published his equations, Heinrich Hertz experimentally confirmed Maxwell's marvellous theoretical prediction.

In the explicit descriptions above, I have assumed that the background spacetime is flat Minkowski space $\mathbb{M}$, and the discussions to follow, in §§19.3,4, and the first part of §19.5 can all be taken on this basis, also. However, this is not really necessary, and all the conclusions still apply if spacetime curvature is present. For this, the components given above must be regarded as being taken with respect to some local Minkowskian frame, and the index notation will take care of the rest.[19.6]

19.3 Conservation and flux laws in Maxwell theory

The vanishing divergence of the charge-current vector provides us with the equation of *conservation of electric charge*. The reason that it is

---

[19.5] Show this, first demonstrating that dualizing twice yields minus the original quantity. Does this sign relate to the Lorentzian signature of spacetime? Explain.

[19.6] Can you spell this out? What happens to the components of $\boldsymbol{F}$ and ${}^{*}\boldsymbol{F}$ in a general curvilinear coordinate system? Why are the Maxwell equations unaffected if expressed correctly?

referred to as a 'conservation equation' comes from the fact that, by the fundamental theorem of exterior calculus (see §12.6), we have $\int_{\mathcal{R}} \mathrm{d}\,{}^*\boldsymbol{J} = \int_{\partial\mathcal{R}} {}^*\boldsymbol{J}$, so that

$$\int_{\mathcal{Q}} {}^*\boldsymbol{J} = 0,$$

integrated over any closed 3-surface $\mathcal{Q}$ in Minkowski space $\mathbb{M}$. (Any closed 3-surface in $\mathbb{M}$ is the boundary $\partial\mathcal{R}$ of some compact 4-dimensional region $\mathcal{R}$ in $\mathbb{M}$.) See Fig. 19.3. The quantity ${}^*\boldsymbol{J}$ can be interpreted as the 'flux of charge' (or 'flow' of charge) across $\mathcal{Q} = \partial\mathcal{R}$. Thus, what the above equation tells us is that the net flux of electric charge across this

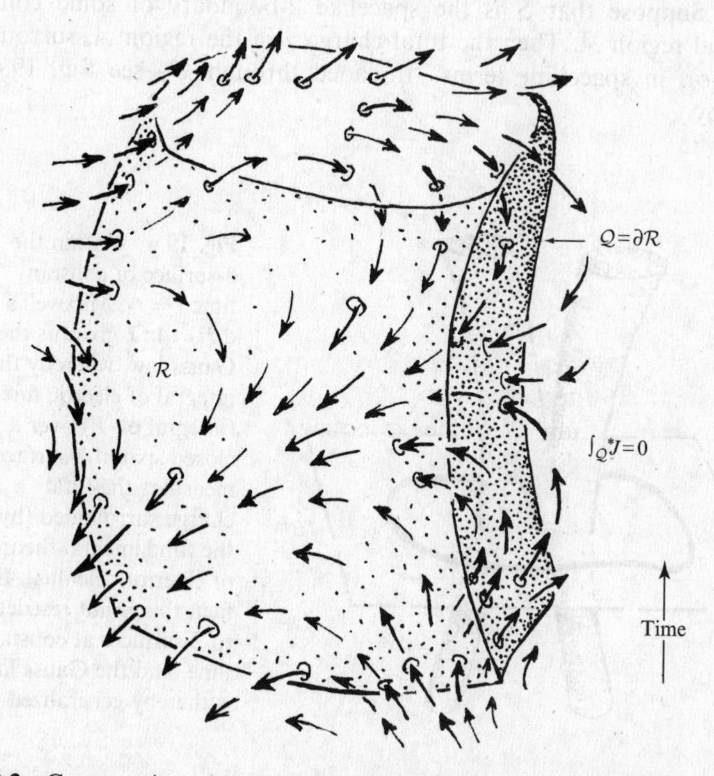

Fig. 19.3 Conservation of electric charge in spacetime. The closed 3-surface $\mathcal{Q}$ is the boundary $\mathcal{Q} = \partial\mathcal{R}$ of a compact 4-volume $\mathcal{R}$, in Minkowski spacetime $\mathbb{M}$, so the fundamental theorem of exterior calculus tells us $\int_{\mathcal{Q}} {}^*\boldsymbol{J} = \int_{\mathcal{R}} \mathrm{d}{}^*\boldsymbol{J} = 0$, since $\mathrm{d}{}^*\boldsymbol{J} = 0$. The quantity ${}^*\boldsymbol{J}$ describes the 'flux' (or 'flow') of charge across $\mathcal{Q}$, so the total charge flowing in across $\mathcal{Q}$ is equal to that flowing out, expressing charge conservation.

boundary has to be zero; i.e. the total coming into $\mathcal{R}$ has to be exactly equal to the total going out of $\mathcal{R}$: *electric charge is conserved*.[19.7]

We can also use the second Maxwell equation $\mathrm{d}\,{}^*\boldsymbol{F} = 4\pi\,{}^*\boldsymbol{J}$ to derive what is called a 'Gauss law'. This particular law applies at one given time $t = t_0$, so we are now using the three-dimensional version of the fundamental theorem of exterior calculus. This tells us the value of the total charge lying within some closed 2-surface $\mathcal{S}$ at time t_0 (see Fig. 19.4), by expressing this charge as an integral over $\mathcal{S}$ of the dual of the Maxwell tensor ${}^*\boldsymbol{F}$—which amounts to saying that we can obtain the total charge surrounded by $\mathcal{S}$ if we integrate the total flux of electric field $\boldsymbol{E}$ across $\mathcal{S}$.[19.8]

More generally, this applies even if $\mathcal{S}$ does not lie in some fixed time $t = t_0$. Suppose that $\mathcal{S}$ is the spacelike 2-boundary of some compact 3-spatial region $\mathcal{A}$. Then the total charge χ in the region $\mathcal{A}$, surrounded by $\mathcal{S}$ (or, in spacetime terms, 'threaded through' $\mathcal{S}$—see Fig. 19.4), is given by

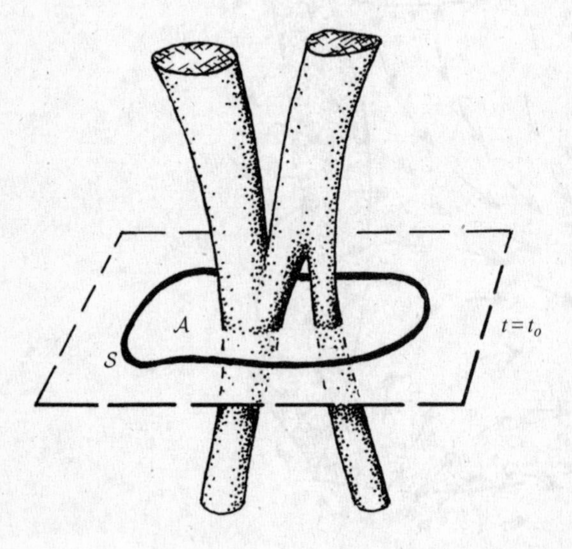

Fig. 19.4 Within the 3-surface of constant time $t = t_0$, Maxwell's $\mathrm{d}{}^*\boldsymbol{F} = 4\pi{}^*\boldsymbol{J}$ gives us the Gauss law, whereby the integral of electric flux (integral of ${}^*\boldsymbol{F}$) over a closed spatial 2-surface measures the total charge surrounded (by the fundamental theorem of exterior calculus). In fact, this is not restricted to 2-surfaces at constant time, and the Gauss law is thereby generalized.

[19.7] Although correct, this argument has been given somewhat glibly. Spell out the details more fully, in the case when $\mathcal{R}$ is a spacetime 'cylinder' consisting of some bounded spatial region that is constant in time, for a fixed finite interval of the time coordinate t. Explain the different notions of 'flux of charge' involved, contrasting this for the spacelike 'base' and 'top' of the cylinder with that for the timelike 'sides'.

[19.8] Spell out why this is just the electric flux.

$$\int_S {}^*\boldsymbol{F} = 4\pi\chi, \quad \text{where } \chi = \int_A {}^*\boldsymbol{J}.$$

We can also obtain a related kind of conservation law from the first Maxwell equation $\mathrm{d}\boldsymbol{F} = 0$. This has just the same form as the second Maxwell equation, except that $\boldsymbol{F}$ replaces ${}^*\boldsymbol{F}$ and the source corresponding to ${}^*\boldsymbol{J}$ is now zero. Thus, for any closed 2-surface in Minkowski space,[2] we always have the flux law

$$\int_S \boldsymbol{F} = 0.$$

Note that in passing from ${}^*\boldsymbol{F}$ to $\boldsymbol{F}$ (or from $\boldsymbol{F}$ to ${}^*\boldsymbol{F}$) we simply interchange the electric and magnetic field vectors (with a change of sign for one of them). The absence of a source for $\boldsymbol{F}$ is an expression of the fact that (as far as is known) there are *no magnetic monopoles* in Nature. A magnetic monopole would be a magnetic north pole or a magnetic south pole on its own—rather than north and south poles always appearing in pairs, which is what happens in an ordinary magnet. (These poles are not independent physical entities, but arise from the circulation of *electric* charges.) It appears that in Nature there is never a net 'magnetic charge' (non-zero 'pole strength') on a physical object. From the point of view of the Maxwell equations alone, there does not seem to be any good reason for the absence of magnetic monopoles, since we could simply supply a right-hand side to the first Maxwell equation $\mathrm{d}\boldsymbol{F} = 0$ without any loss of consistency. In fact, from time to time, physicists have contemplated the possibility that magnetic monopoles might actually exist and have tried to look for them. Their existence would have important implications for particle physics (see §28.2) but there is no indication, as of now, that there are any such monopoles in the actual universe.

19.4 The Maxwell field as gauge curvature

The first Maxwell equation $\mathrm{d}\boldsymbol{F} = 0$ also has the implication that

$$\boldsymbol{F} = 2\mathrm{d}\boldsymbol{A},$$

for some 1-form $\boldsymbol{A}$. (This is taking advantage of the 'Poincaré lemma', which states that, if the r-form $\boldsymbol{\alpha}$ satisfies $\mathrm{d}\boldsymbol{\alpha} = 0$, then *locally* there is always an $(r-1)$-form β for which $\boldsymbol{\alpha} = \mathrm{d}\boldsymbol{\beta}$; see §12.6.) Moreover, in a region with Euclidean topology, this local result extends to a global one.[3] The quantity $\boldsymbol{A}$ is called the *electromagnetic potential*. It is not uniquely determined by the field $\boldsymbol{F}$, but is fixed to within the addition of a quantity $\mathrm{d}\Theta$,[19.9] where Θ is some real scalar field:

$$\boldsymbol{A} \mapsto \boldsymbol{A} + \mathrm{d}\Theta.$$

In index form, these relations are

$$F_{ab} = \nabla_a A_b - \nabla_b A_a$$

with freedom

$$A_a \mapsto A_a + \nabla_a \Theta.$$

This 'gauge freedom' in the electromagnetic potential tells us that $\boldsymbol{A}$ is not a locally measurable quantity. There can be no experiment to measure 'the value of $\boldsymbol{A}$' at some point because $\boldsymbol{A} + \mathrm{d}\Theta$ serves exactly the same physical purpose as does $\boldsymbol{A}$. However, the potential provides the mathematical key to the procedure whereby the Maxwell field interacts with some other physical entity Ψ. How does this work? The specific role of A_a is that it provides us with a *gauge connection* (or bundle connection; see §15.8)

$$\nabla_a = \partial/\partial x^a - \mathrm{i}eA_a,$$

where e is a particular real number that quantifies the *electric charge* of the entity described by Ψ. In fact, this 'entity' will generally be some charged quantum particle, such as an electron or proton, and Ψ would then be its quantum-mechanical wavefunction. The full meaning of these terms will have to await the discussion in Chapter 21, when the notion of a wavefunction will be explained. All that we shall need to know about it now is that Ψ is to be thought of as a cross-section of a bundle (§15.3), a bundle describing charged fields, and it is this bundle on which $\boldsymbol{\nabla}$ acts as a connection.

The electromagnetic field quantities $\boldsymbol{F}$ and $\boldsymbol{A}$ are uncharged ($e = 0$ for them), so that all our Maxwell equations, etc., are undisturbed by having this new definition for ∇_a; i.e. we still have $\nabla_a = \partial/\partial x^a$ in those equations, in flat Minkowski coordinates—or the appropriate generalization (see §14.3) if we are considering curved spacetime. What is the geometrical nature of the bundle that this connection acts upon? One possible viewpoint is to think of this bundle as having fibres that are circles (S^1s), over the spacetime $\mathbb{M}$, where this circle describes a phase multiplier $e^{i\theta}$ for Ψ. (This is the kind of thing that happens in the 'Kaluza–Klein' picture referred to in §15.1 but where in that case the entire bundle is thought of as 'spacetime'.) More appropriate is to think of the bundle as the vector bundle of the possible Ψ values at each point, where the freedom of phase multiplications make the bundle a U(1) bundle over the spacetime $\mathbb{M}$. (This kind of issue was considered at the end of §15.8.) For this to make sense, Ψ must be a

[19.9] Why can we add such a quantity?

complex field whose physical interpretation is, in some appropriate sense, insensitive to the replacement $\Psi \mapsto e^{i\theta}\Psi$ (where θ is some real-valued field on the manifold $\mathcal{M}$). This replacement is referred to as an electromagnetic *gauge transformation*, and the fact the physical interpretation is insensitive to this replacement is called *gauge invariance*. The *curvature* of our bundle connection then turns out to be the Maxwell field tensor F_{ab}.[19.10]

Before exploring with these ideas further, it is appropriate to make some brief historical comments. Shortly after Einstein introduced his general theory of relativity in 1915, Weyl suggested, in 1918, a generalization in which the very notion of *length* becomes path-dependent. (Hermann Weyl, 1885–1955, was an important 20th-century mathematical figure. Indeed, among the work of those mathematicians who wrote entirely in the 20th century, his was, to my mind, the most influential—and he was important not only as a pure mathematician but also as a physicist.) In Weyl's theory, the null cones retain the fundamental role that they have in Einstein's theory (e.g. to define the limiting velocities for massive particles and to provide us with the local 'Lorentz group' that is to act in the neighbourhood of each point), so a Lorentzian (say $+---$) metric $\boldsymbol{g}$ still is locally required for the purpose of defining these cones. However, there is no absolute scaling for time or space measures, in Weyl's scheme, so the metric is given only *up to proportionality*. Thus, transformations of the form

$$\boldsymbol{g} \mapsto \lambda \boldsymbol{g},$$

for some (say positive) scalar function λ on the spacetime $\mathcal{M}$, are allowed, these not affecting the null cones of $\mathcal{M}$. (Such transformations are referred to as *conformal rescalings* of the metric $\boldsymbol{g}$; in Weyl's theory, each choice of $\boldsymbol{g}$ provides us with a possible *gauge* in terms of which distances and times can be measured.) Although Weyl may have had spatial separations more in mind, it will be appropriate for us to think in terms of time measurements (in accordance with the viewpoint of Chapter 17). Thus, in Weyl's geometry, there are no absolute 'ideal clocks'. The rate at which any clock measures time would depend upon its history.

The situation is 'worse' than in the standard 'clock paradox' that I described in §18.3 (Fig. 18.6d). In Weyl's geometry, we can envisage a space traveller who journeys to a distant star and then returns to Earth to find not just that those on the Earth had aged much more, but also that the clocks on Earth are now found to run at a different *rate* from those on the rocket ship! See Fig. 19.5a. Using this very striking idea, Weyl was able to incorporate the equations of Maxwell's electromagnetic theory into the spacetime geometry.

[19.10] Show this. *Hint*: Have a look at §15.8.

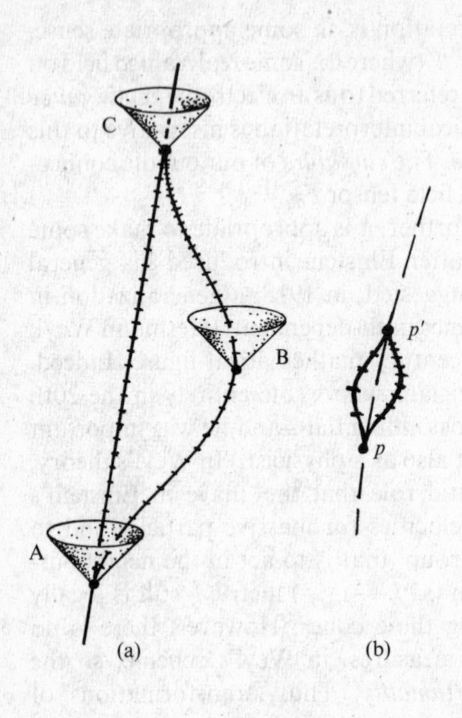

Fig. 19.5 In Weyl's original gauge theory of electromagnetism, the notion of time interval (or space interval) is not absolute but depends on the path taken. (a) A comparison with the 'clock paradox' illustrated in Fig. 18.6: in Weyl's theory we find that the space traveller arrives home (world-line ABC) to find not only differing clock readings between those on Earth (direct route AC) and those on the rocket ship, but also differing clock *rates*! (b) Weyl's gauge curvature (giving the Maxwell field $\boldsymbol{F}$) comes about from this (conformal) time scale change as we go around an infinitesimal loop (difference between two routes from p to neighbouring point p').

The essential way that he did this was to encode the electromagnetic potential into a bundle connection, just as I have done above, but without the imaginary unit 'i' in the expression for ∇_a. We can think of the relevant bundle over $\mathcal{M}$ as being given by the Lorentzian metrics $\boldsymbol{g}$ that share the same null cones. Thus, the fibre above some point x in $\mathcal{M}$ consists of a family of *proportional* metrics (where we can, if desired, choose the proportionality factors to be positive). These factors are the possible 'λs' in $\boldsymbol{g} \mapsto \lambda\boldsymbol{g}$ above. For any particular choice of metric, we have a *gauge* whereby distances or times along curves are defined. But there is to be no absolute choice of gauge, and so no preferred choice of metric $\boldsymbol{g}$ from the equivalence class of proportional ones. There is some structure additional to that of the null cones (i.e. to the *conformal* structure), however, namely a bundle connection—or *gauge connection*—which Weyl introduced, in order to have Maxwell's $\boldsymbol{F}$ (i.e. F_{ab}) as its *curvature*. This curvature measures the discrepancy in the clock rates as illustrated in Fig. 19.5a when the world-lines differ only by an infinitesimal part; see Fig. 19.5b. (This may be compared with the 'strained bundle' $\mathcal{B}^{\mathbb{C}}$, over $\mathbb{C}$, considered in §15.8, Figs. 15.16 and 15.19; the basic bundle concept is very similar.)

When Einstein heard about this theory, he informed Weyl that he had a fundamental *physical* objection to it, despite the mathematical elegance of Weyl's ideas. Spectral frequencies, for example, appear to be completely unaffected by an atom's history, whereas Weyl's theory would predict otherwise. More fundamentally, although not all the relevant quantum-mechanical rules had been fully formulated at the time (and we shall be coming to these later, in §21.4, §§23.7,8) Weyl's theory is in conflict with the necessarily exact identity between different particles of the same type (see §23.7). In particular, there is a direct relation between clock rates and particle masses. As we shall see later, a particle of rest-mass m has a natural frequency mc^2h^{-1}, where h is Planck's constant and c the speed of light. Thus, in Weyl's geometry, not just clock rates but also a particle's *mass* will depend upon its history. Accordingly, two protons, if they had different histories, would almost certainly have different masses, according to Weyl's theory, thereby violating the quantum-mechanical principle that particles of the same kind have to be *exactly* identical (see §§23.7,8).

Although this was a damning observation, with regard to the original version of Weyl's theory, it was later realized[4] that the same idea would work if his 'gauge' referred not to the *real* scaling (by λ), but to a scaling by a *complex number of unit modulus* ($e^{i\theta}$). This may seem like a strange idea, but as we shall see in Chapter 21 and onwards (see §§21.6,9 most particularly), the rules of *quantum mechanics* force upon us the use of complex numbers in the description of the state of a system. There is, in particular, a unit-modulus complex number $e^{i\theta}$ which can multiply this 'quantum state'—the state often being referred to as ψ—without observable consequences, locally. This 'non-observable' replacement $\psi \mapsto e^{i\theta}\psi$ is still referred to today as a 'gauge transformation' even though there is now no change in length scale involved, the change being a rotation in the complex plane (a complex plane with no direct connection with either space or time dimensions). In this strangely twisted form, Weyl's idea provided the appropriate physical setting for a $U(1)$ connection, of the kind that I illustrated at the end of Chapter 15, and it now forms the basis of the modern picture of how the electromagnetic field actually interacts. The operator $\boldsymbol{\nabla}$ that is defined above from the electromagnetic potential (i.e. $\nabla_a = \partial/\partial x^a - ieA_a$) provides a $U(1)$-bundle connection on the bundle of charged quantum wavefunctions ψ (See §21.9).

It is interesting that the path dependence of the connection (which we may compare with the path dependence illustrated in Fig. 19.5) shows up in a striking way in certain types of experimental situation, illustrating what is known as the *Aharonov–Bohm effect*.[5] Since our connection $\boldsymbol{\nabla}$ operates only at the level of quantum phenomena, we do not see this path dependence in classical experiments; instead, the Aharonov–Bohm effect depends upon *quantum interference* (see §21.4 and Fig. 21.4). In the best-known version, electrons are aimed so as to pass through two regions that are free of electromagnetic field ($\boldsymbol{F} = 0$), but which are separated from each other by a long cylindrical solenoid (which contains magnetic lines of force), arriving at a detector screen behind (see Fig. 19.6a). At no stage do the electrons encounter any non-zero field $\boldsymbol{F}$. However, the relevant field-free region $\mathcal{R}$ (starting at the source, bifurcating so that they pass on either side of the solenoid, and reuniting at the screen) is not simply-connected, and the field $\boldsymbol{F}$ outside $\mathcal{R}$ is such that there is no gauge choice for which the potential $\boldsymbol{A}$ vanishes everywhere within $\mathcal{R}$. The presence of this non-zero potential in the non-simply-connected $\mathcal{R}$—or more correctly, the path dependence of ∇ in $\mathcal{R}$—leads to a displacement in the interference fringes at the screen.

In fact, the fringe-shifting effect does not depend upon any particular local values that $\boldsymbol{A}$ might have (which it cannot, because $\boldsymbol{A}$ is not locally observable, as mentioned above) but upon a certain non-local integral of $\boldsymbol{A}$. This is the quantity $\oint \boldsymbol{A}$, taken around a topologically non-trivial loop within $\mathcal{R}$. See Fig. 19.6b. Since $\mathrm{d}\boldsymbol{A}$ vanishes within $\mathcal{R}$ (because $\boldsymbol{F} = 0$ in $\mathcal{R}$), the integral $\oint \boldsymbol{A}$ is unaffected if we continuously move our closed loop around within $\mathcal{R}$.[19.11] From this it is clear that the non-vanishing of $\oint \boldsymbol{A}$, within a field-free region, and thence the Aharonov–Bohm effect itself, depends upon this field-free region being topologically non-trivial.

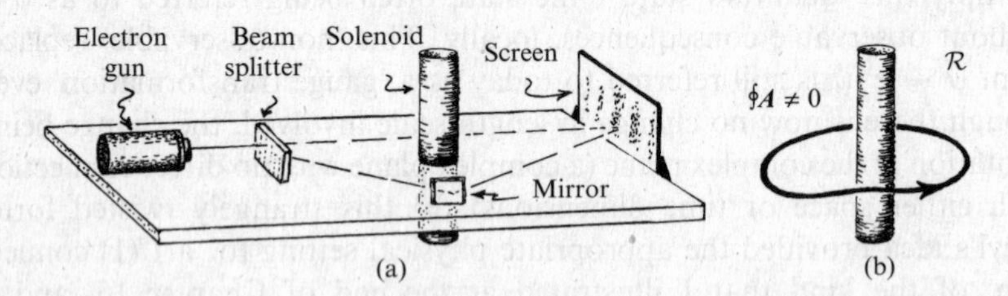

Fig. 19.6 Aharonov–Bohm effect. (a) A beam of electrons is split into two paths that go to either side of a collection of lines of magnetic flux (achieved by means of a long solenoid). The beams are brought together at a screen, and the resulting quantum interference pattern (compare Fig. 21.4) depends upon the magnetic flux strength—despite the fact that the electrons only encounter a zero field strength ($\boldsymbol{F} = 0$). (b) The effect depends on the value of $\oint \boldsymbol{A}$, which can be non-zero over the relevant topologically non-trivial closed path despite $\boldsymbol{F}$ vanishing over this path. The quantity $\oint \boldsymbol{A}$ is unchanged for continuous deformations of the path within the field-free region.

[19.11] Explain this.

Because of its historical origins in Weyl's remarkable idea (which originally did play a role as a path-dependent 'gauging'), we call this electromagnetic connection ∇ a *gauge* connection—and this name is also adopted for the generalizations of electromagnetism, referred to as 'Yang-Mills' theory, that are used in the description of both weak and strong interactions in modern particle physics. We note that the 'gauge-connection' idea really does depend, strictly speaking, on the existence of a symmetry (which for electromagnetism is the $\psi \mapsto e^{i\theta}\psi$ symmetry) that is supposed to be *exact* and not directly observable. We recall Einstein's objection to Weyl's original gauge idea which, in effect, was that the mass of a particle (and therefore its natural frequency) is directly measurable, and so cannot be used as a 'gauge field' in the sense required. We shall be finding later that this issue becomes distinctly muddied in some modern uses of the 'gauge' idea.

19.5 The energy–momentum tensor

As a prerequisite to turning our attentions to that other fundamental classical field with its 'gauge theory' aspects, namely the gravitational field, it will be important first to consider the question of the *energy density* of a field, this density being the source of gravity. For Einstein's famous equation $E = mc^2$ tells us that mass and energy are basically the same thing (see §18.6) and, as Newton had already informed us, it is mass that is the source of gravitation. Thus, we need to understand how to describe the energy density of a field, such as Maxwell's, and how this can act as a source of gravity. What Einstein tells us is that it does so via a tensor quantity known as the *energy–momentum tensor*. This is a symmetric $[{}^0_2]$-valent tensor $\boldsymbol{T}$ (index form $T_{ab} = T_{ba}$) which satisfies a 'conservation equation'

$$\nabla^a T_{ab} = 0.$$

(For the rest of this chapter, we use the spacetime covariant derivative operator ∇_a in place of $\partial/\partial x^a$. Since our fields here are all uncharged, our earlier expressions will carry over unchanged; see also the final paragraph of §19.2, Note 19.2, and Exercise [19.6].) We may compare this expression with the conservation equation $\nabla^a J_a = 0$ for electric charge. The reason for the extra index on T_{ab} is that what is conserved, namely energy–momentum, is a 4-(co)vector quantity (the energy–momentum 4-(co)vector p_a, considered in §18.7) as opposed to the *scalar* electric charge. To describe the physical content of T_{ab} a little more fully, it is convenient to pass to the equivalent quantity $T^a{}_b = g^{ac}T_{cb}$, where one index has been raised by use of the metric tensor g^{ab}.[19.12] The quantity $T^a{}_b$ collects together all the

[19.12] How do the individual components $T^a{}_b$ relate to T_{ab}, in a local Minkowskian frame, where the components g_{ab} have the diagonal form $(1, -1, -1, -1)$?

different densities and fluxes of the energy and momentum in the fields and particles. More specifically, in a standard Minkowski coordinate system, the covector $T^0{}_b$ defines the density of 4-momentum, and the three covectors $T^1{}_b$, $T^2{}_b$, $T^3{}_b$, provide the flux of 4-momentum in the three independent spatial directions. This is directly analogous to the case of J^a, since J^0 is the density of charge and the three quantities J^1, J^2, J^3, provide the flux of charge (i.e. the current) in the three independent spatial directions. It is the extra index b that tells us that our conservation law now refers to a (co-)vector quantity. It turns out that the quantity T_{00} measures the energy density, and T_{11}, T_{22}, T_{33}, measure the *pressure*, in the three directions of the spatial coordinate axes.

Recall that, as Maxwell taught us, electromagnetic fields themselves carry energy. In the index notation, the energy–momentum tensor of the electromagnetic field turns out to be[19.13]

$$\frac{1}{8\pi}(F_{ac}F^c{}_b + {}^*F_{ac}{}^*F^c{}_b).$$

Other physical fields also have their energy–momentum tensors, and various different such contributions would have to be added together in order to yield the full energy–momentum tensor $\boldsymbol{T}$, satisfying the conservation equation $\nabla^a T_{ab} = 0$.

However, something very different happens with the energy–momentum of gravity itself, as we shall be seeing shortly. When gravity is absent, spacetime is flat (i.e. Minkowski space), and we can use flat (Minkowskian) coordinates. Then each of the four vectors $T^a{}_0$, $T^a{}_1$, $T^a{}_2$, and $T^a{}_3$ individually satisfies exactly the same conservation equation as does the vector J^a (namely $\nabla_a T^a{}_0 = 0$, etc., analogous to $\nabla_a J^a = 0$), with the implication that there is an *integral* conservation law exactly analogous to that of charge (i.e. analogous to $\int_Q {}^*\boldsymbol{J} = 0$), for each of the 4 component of energy–momentum separately. Thus, total mass is conserved, and so are the three components of total momentum. But recall the discussion given in Chapter 17 of Einstein's equivalence principle, and of why this leads us to a curved spacetime. Thus, when gravity is present, we must take into account the fact that '∇_a' is no longer simply '$\partial/\partial x^a$', but (in accordance with §14.3) there are extra $\Gamma^b{}_{ac}$ terms that confuse the very meaning of '$\nabla_a T^a{}_0$' and which certainly prevent us from deriving an integral conservation law for energy and momentum just from our 'conservation equation' $\nabla^a T_{ab} = 0$. The problem can be phrased as the fact that the extra index b in T_{ab} prevents it from being the dual of a 3-form, and we cannot write a coordinate-independent formulation of a 'conservation equation' (like the vanishing exterior derivative of the 3-form

[19.13] Show that this satisfies the conservation equation $\nabla^a T_{ab} = 0$ if $\boldsymbol{J} = 0$. Obtain the 00 component of this tensor, and recover Maxwell's original expression $(E^2 + B^2)/8\pi$ for the energy density of an electromagnetic field in terms of (E_1, E_2, E_3) and (B_1, B_2, B_3).

$^*\!J$ in '$\mathrm{d}\,^*\!J = 0$'). We seem to have lost those most crucial conservation laws of physics, the laws of conservation of energy and momentum!

In fact, there is a more satisfactory perspective on energy/momentum conservation, which refers also to certain curved spacetimes $\mathcal{M}$ as well as to Minkowski space, and it applies also to angular-momentum conservation (see §18.6 and §§22.8,11). For this perspective, suppose that we have a *Killing vector* $\boldsymbol{\kappa}$ for $\mathcal{M}$ (this satisfies $\nabla_{(a}\kappa_{b)} = 0$; see §14.7), which describes some *continuous symmetry* of $\mathcal{M}$. In Minkowski space, there are 10 independent such symmetries, referring to the 4 independent *translational* symmetries (3 space and 1 time) and 6 independent spacetime *rotations* (the non-reflective part of the Lorentz group O(1,3)). See Fig. 18.3b. Thus, Minkowski space has 10 independent Killing vectors. As we shall be encountering in the next chapter, the Lagrangian formalism (Nöther's theorem) allows us to derive a conservation law from each continuous symmetry that the laws of the system possess. Time-translational symmetry provides energy conservation, whereas space-translational symmetry provides 3-momentum conservation. Rotational symmetry gives angular momentum. (Ordinary spatial rotations give us the 3 components of ordinary angular momentum, but there are also 3 components coming from the Lorentzian 'boosts', that take us from one velocity to another. These give us the conservation of mass-centre movement; see §§18.6,7, Fig. 18.16.) To obtain the appropriate conservation law from any particular Killing vector $\boldsymbol{\kappa}$, we construct the *flux* quantity

$$L_a = T_{ab}\kappa^b,$$

which satisfies the conservation law $\nabla_a L^a = 0$ whenever the symmetric T_{ab} satisfies $\nabla^a T_{ab} = 0$.[19.14] Hence, as in §19.3, there is an integral conservation law $\int_{\mathcal{Q}} {}^*\!\boldsymbol{L} = 0$.

These conservation laws hold only in a spacetime for which there *is* the appropriate symmetry, given by the Killing vector $\boldsymbol{\kappa}$. Physically, the reason for this is that the degrees of freedom in the spacetime geometry—i.e. gravity—are decoupled from the fields. The spacetime geometry serves merely as a background, so it is undisturbed by the fields within it; moreover, the fields are unable to pick up the quantity in question from the background (or lose it to the background) because of the symmetry. These considerations will have importance for us later, particularly in Chapter 30 (§§30.6,7). Nevertheless, they do not really help us in understanding what the fate of the conservation laws will be when gravity itself becomes an active player. We still have not regained our missing

[19.14] Why? Why does this procedure specialize to the above $\nabla_a T^a{}_0 = 0$, etc.? Can you find an analogue of the continuous-field conservation law $\nabla^a(T_{ab}\kappa^b) = 0$, for a discrete system of particles where 4-momentum is conserved in collisions? *Hint*: Find a quantity, given the Killing vector κ^a, that is constant for each particle between collisions.

conservation laws of energy and momentum, when gravity enters the picture.

This awkward-seeming fact has, since the early days of general relativity, evoked some of the strongest objections to that theory, and reasons for unease with it, as expressed by numerous physicists over the years.[6] We shall be seeing later, in §19.8, that in fact Einstein's theory takes account of energy–momentum conservation in a rather sophisticated way—at least in those circumstances where such a conservation law is most needed. For the moment, we take note of the fact that in Einstein's theory, the symmetric $[^0_2]$-valent tensor $\boldsymbol{T}$ that appears in his field equation is to include the energy–momentum of all *non*-gravitational fields (and particles). Whatever energy there is in the gravitational field itself is to be excluded from having any representation within $\boldsymbol{T}$.

This point of view is made somewhat plausible if we think again of the principle of equivalence. Imagine an observer in free orbit, say within some spaceship without windows, so that it appears, at least to a first approximation, that there is no gravitational field. That observer would expect that energy is conserved within the spaceship, and would therefore expect that equation $\nabla^a T_{ab} = 0$ holds without there being any contribution from the gravitational field. This 'conservation' is, however, only an approximation, which is expected to need correction as soon as the relative acceleration (tidal) effects due to the non-uniformity of the gravitational field (as studied in §17.5; see Figs. 17.8a, 17.8b and 17.9) begin to play a role. Now this is a slightly delicate issue, and it becomes necessary to examine the 'orders' at which different kinds of effect begin to play a role. The upshot of it all is that the quantity $\boldsymbol{T}$ and its equation $\nabla^a T_{ab} = 0$ should remain undisturbed by the non-uniformity of the gravitational field—i.e. unaffected by the curvature $\boldsymbol{R}$ of the spacetime connection $\boldsymbol{\nabla}$—and that the contributions of gravity to energy–momentum conservation should somehow enter *non-locally* as corrections to the calculation of total energy–momentum. (The only real exception to this comment might occur if one needs to contemplate spacetime-curvature corrections to those mathematical expressions that tell us how physical fields contribute to $\boldsymbol{T}$. Normally there are no such corrections, and it is not an important issue for our considerations here.) From this perspective, gravitational contributions to energy–momentum, in a sense, 'slip in through the cracks' that separate the *local* equation $\nabla^a T_{ab} = 0$ from an integral conservation law of *total* energy–momentum.

19.6 Einstein's field equation

I shall return to this issue in §19.8, but for the present we shall need to know the actual form of Einstein's field equation. This equation is expressed in terms of the *tensor* formalism that, by now, the reader will (I hope!) find not too uncongenial. Part of the reason that tensors are needed is that spacetime curvature, in 4-dimensions, is a complicated thing. Recall Albert, our astronaut A of §17.5, orbiting freely in the gravitational field of the Earth. In various directions out from A there are inward accelerations, and in other directions there are outward accelerations. These represent the *tidal* forces experienced by A. Tidal forces are manifestations of spacetime curvature. In order to collect together these complicated effects, a tensor quantity with components R_{abcd} is used, which has 10 independent components in empty space, and a total of 20, when there is also matter density around. In fact, R_{abcd} is simply the index form of the Riemann(–Christoffel) tensor $\boldsymbol{R}$ that we have previously encountered in §14.7.

But there is another reason, apart from just organizing complication, that the tensor calculus plays such a fundamental role in Einstein's theory. This goes back to the foundational *principle of equivalence* which started Einstein's whole line of thinking. Gravitation is not to be regarded as a force; for, to an observer who is falling freely (such as our astronaut A), there is no gravitational force to be felt. Instead, gravitation manifests itself in the form of spacetime curvature. Now it is important, if this idea is to work, that there be no 'preferred coordinates' in the theory.[7] For, if a certain limited class of coordinate systems were taken to be Nature's preferred choices, then these would define 'natural observer systems' with respect to which the notion of a 'gravitational force' could be reintroduced, and the central role of the principle of equivalence would be lost. The point is, in fact, a rather delicate one, and many physicists have, from time to time and in one way or another, departed from it. To my way of thinking, it is essential for the spirit of Einstein's theory that this notion of coordinate independence be maintained. This is what is referred to as *the principle of general covariance*. It tells us not only that there are to be no preferred coordinates, but also that, if we have two different spacetimes, representing two physically distinct gravitational fields, then there is to be no naturally preferred pointwise identification between the two—so we cannot say which particular spacetime point of one is to be regarded as the *same* point as some particular spacetime point of the other! This philosophical issue will concern us later (§30.11), regarding how Einstein's theory relates to the principles of quantum mechanics. For the moment, the importance of the principle of general covariance to us is that it forces us into a coordinate-free description of gravitational physics. It is for this reason, most particularly, that the tensor formalism is central to Einstein's theory.

Let us now see what Einstein's equation actually is. The form of this equation is driven, basically, by the two further requirements: (i) that the (local) source of gravity should, in effect, be the energy–momentum tensor $\boldsymbol{T}$, subject to $\nabla^a T_{ab} = 0$, and (ii) that, in the appropriate Newtonian limit (small velocities, as compared with that of light, and weak gravitational fields), standard Newtonian gravitational theory should be recovered. We must return to the discussion of §17.5, where we found that, in Newtonian theory, there is a volume-reducing effect for geodesics that are neighbouring to, and initially parallel to, an observer's geodesic world line γ. These neighbouring geodesics accelerate relative to γ in such a way that the (infinitesimal) spacelike *3-volume* δV that they enclose has an overall acceleration that is equal to $-4\pi G\,\delta M$, where δM is the active gravitational mass within the (infinitesimal) volume enclosed by the geodesics. The minus sign comes from the fact that it is a volume *reduction* that is involved; see Fig. 17.8b. This is a full expression of Newton's theory, with regard to the active gravitational effect of a distribution of mass.

How are we to translate this into an equation relating the spacetime curvature $\boldsymbol{R}$ to the energy–momentum tensor $\boldsymbol{T}$? The key geometrical fact is that this inward acceleration of volume that occurs in this situation, is measured by a $\left[{0 \atop 2}\right]$-valent symmetric tensor, called the *Ricci tensor*, defined by

$$R_{ab} = R_{acb}{}^{c},$$

R_{abcd} being the Riemann tensor.[19.15] (See Fig. 19.7 for the diagrammatic notation for this.) Again, there are innumerable different conventions with regard to signs, index orderings, signatures, etc. As before, I am imposing upon the reader my own preferences; see §14.4.) More specifically, the acceleration of volume (starting from rest) is given by[19.16]

$$\mathbf{D}^2(\delta V) = R_{ab}t^a t^b \delta V.$$

Here, $\mathbf{D}$ represents the rate of change with respect to the observer's proper *time* (see §17.9), along the observer's world line γ, so $\mathbf{D}^2$ indeed denotes acceleration. We have

R_{ab} ⇝ [diagram] = [diagram]

Fig. 19.7 Diagrammatic notation for Ricci-tensor definition $R_{ab} = R_{acb}{}^{c}$ (see Fig. 14.21).

[19.15] Why is R_{ab} symmetric?

[19.16] See if you can prove this using the Ricci identity and the properties of Lie derivative.

$$\mathbf{D} = t^a\nabla_a = \underset{t}{\boldsymbol{\nabla}},$$

where t^a is the future-timelike unit vector tangent to γ (so $t^a t_a = 1$).

The *mass density* (which is the same as the *energy* density, by '$E = mc^2$' with $c = 1$; see §18.6), as measured by the observer, is the '00 component' of T_{ab} in the observer's local frame. This is just the quantity $T_{ab}t^a t^b$, so the mass δM within the volume δV enclosed by the neighbouring geodesics is

$$\delta M = T_{ab}t^a t^b \delta V.$$

Thus, the 'Newtonian expectation' $-4\pi G\,\delta M$ (§17.5) for the volume acceleration due to matter density is

$$-4\pi G T_{ab} t^a t^b \delta V.$$

But we have just seen that the volume-acceleration effect due to spacetime curvature is $R_{ab}t^a t^b\ \delta V$, so we come up with the expectation

$$R_{ab}t^a t^b \delta V = \ -4\pi G T_{ab} t^a t^b\ \delta V.$$

Dividing through by δV and realizing that this applies to *all* observers through the same event, so we can remove $t^a t^b$,[19.17] we arrive at the suggested field equation

$$R_{ab} = -4\pi G T_{ab}$$

which, indeed, was Einstein's initial proposal. This, however, is not satisfactory, because the 'conservation equation' $\nabla^a T_{ab} = 0$ then leads to $\nabla^a R_{ab} = 0$ which, in turn, leads to trouble!

What is this trouble? Recall, from §14.4, the *Bianchi identity* equation $\nabla_{[a}R_{bc]d}{}^e = 0$. By taking a contraction of this equation, we get[19.18]

$$\nabla^a(R_{ab} - \tfrac{1}{2}R\, g_{ab}) = 0,$$

where the *Ricci scalar* (or *scalar curvature*—although '$-R$' might fit in better with most mathematical conventions for the positive-definite case) is defined by

$$R = R_a{}^a$$

(where R is not to be confused with the bold-face $\boldsymbol{R}$ that stands for the *entire* curvature tensor). The 'trouble' with the above proposed equation $R_{ab} = -4\pi G T_{ab}$ is that, when combined with the contracted Bianchi iden-

---

[19.17] Show fully why we can 'lop off' all the t^as, explaining the role of the symmetry of the tensors.

[19.18] Show this, using the diagrammatic notation, if you like.

tity, it leads to the conclusion that the *trace* T of the energy–momentum tensor, defined by

$$T = T^a_a,$$

has to be *constant* throughout spacetime.[19.19] This is blatantly inconsistent with ordinary (non-gravitational) physics. Accordingly Einstein eventually concluded (in 1915) that, for consistency, the two tensors satisfying the 'conservation equation' $\nabla^a(\dots) = 0$ should be *equated* (to within a constant factor), and he came up with what we now know as *Einstein's field equation*:[8,][19.20]

$$R_{ab} - \tfrac{1}{2}R\, g_{ab} = -8\pi G T_{ab}.$$

In the particular situation when there is no matter present (including electromagnetic field), we have $T_{ab} = 0$. This is referred to as *vacuum*. Einstein's equation—the *vacuum equation*—becomes $R_{ab} - \frac{1}{2}Rg_{ab} = 0$, which can be rewritten as[19.21]

$$R_{ab} = 0.$$

A space with vanishing Ricci tensor is sometimes referred to as *Ricci-flat*.

19.7 Further issues: cosmological constant; Weyl tensor

At this point, we should consider the additional term that Einstein suggested in 1917, called the *cosmological constant*. This is an exceedingly tiny constant quantity Λ, whose actual presence is strongly suggested by modern cosmological observations, but which cannot differ from 0 by more than the very tiny amount of about $10^{-55}\ \mathrm{cm}^{-2}$. It has no direct observational relevance until cosmological scales are reached. The quantity $R_{ab} - \frac{1}{2}Rg_{ab}$, in the above expression, is accordingly replaced by $R_{ab} - \frac{1}{2}Rg_{ab} + \Lambda g_{ab}$. This still satisfies the 'conservation equation' since Λ is constant (and $\nabla \mathbf{g} = 0$). The Einstein equation now reads

$$R_{ab} - \tfrac{1}{2}R\, g_{ab} + \Lambda g_{ab} = -8\pi G\, T_{ab}.$$

Einstein originally introduced this extra term, in order to have the possibility of a *static* spatially closed universe on the cosmological scale.[9] But when it became clear, from Edwin Hubble's observations in 1929, that the universe is expanding, and therefore *not* static, Einstein withdrew his support for the cosmological constant, asserting that it had been 'his

[19.19] Why?

[19.20] Explain the coefficient $-8\pi G$, as compared with $-4\pi G$.

[19.21] Why?

greatest mistake' (perhaps because he might otherwise have predicted the expansion of the universe!). Nevertheless, ideas once put forward do not necessarily go away easily. The cosmological constant has hovered in the background of cosmological theory ever since Einstein first put it forward, causing worry to some and solace to others. Very recently, observations of distant supernovae have led most theorists to re-introduce Λ, or something similar, referred to as 'dark energy', as a way of making these observations consistent with other perceived requirements.[10] I shall return to the issue of the cosmological constant later (see §28.10, particularly). For my own part, in common with most relativity theorists, although normally allowing for the possibility of a non-zero Λ in the equations, I had myself been rather reluctant to accept that Nature would be likely to make use of this term. However, as we shall be seeing in §28.10, much recent cosmological evidence does seem to be pointing in this direction.

We can also write Einstein's field equation (including the cosmological constant) the opposite way around:[19.22] $R_{ab} = -8\pi G(T_{ab} - \frac{1}{2}Tg_{ab}) + \Lambda g_{ab}$. Using a local coordinate frame with the time axis given by t^a, so that contracting this with $t^a t^b$ gives the 00-component, we find that the inward acceleration of the volume is given by $8\pi G(T_{00} - \frac{1}{2}Tg_{00}) - \Lambda$, which is $4\pi G(\rho + P_1 + P_2 + P_3) - \Lambda$, where P_1, P_2, and P_3 are the values of the *pressure* of the matter along three (orthogonal) spatial axes. Let us now make the comparison with the $4\pi G\,\delta M$ that Newton's theory gives, finding the density ρ_G of *active gravitational mass*, in Einstein's general relativity to be

$$\rho_G = \rho + P_1 + P_2 + P_3 - \frac{\Lambda}{4\pi G},$$

rather than $\rho_G = \rho$, the latter being what we might have expected simply from '$E = mc^2$'. (Units have been chosen so that $c = 1$.) The Λ contribution is extremely tiny and, indeed, the extra pressure terms are also normally very tiny by comparison with the energy, roughly speaking because the little particles that make up the material in question are moving around relatively slowly, compared with the speed of light. However, the pressure contributions to active gravitational mass do play significant roles under certain extreme conditions. When a very massive star is getting close to a situation in which it is in danger of collapsing under its own inward gravitational pull, we find that an increased pressure in the star, which we might expect to help keep the star supported, actually *increases* the tendency to collapse because of the extra gravitational mass that it produces!

As pointed out above (§19.5), the energy–momentum tensor T_{ab} is analogous, in Einstein's theory, to the charge-current vector J_a of Maxwell theory. The quantitity T_{ab} may be regarded as describing the *source* of

[19.22] Why?

gravitation, in the same way as J_a is the source of electromagnetism. We may ask what might be the appropriate analogue of the Maxwell field tensor F_{ab} describing the *gravitational degrees of freedom*? The answer is *not* the metric tensor **g**, which is more analogous to the electromagnetic potential **A**. Some people might regard the full Riemann curvature tensor R_{abcd} as the analogue of $\boldsymbol{F}$, but it is more appropriate to choose what is called the *Weyl tensor* (or *conformal tensor*) C_{abcd}, which is like the full Riemann tensor, but has the Ricci tensor part 'removed'. This is reasonable, because the Ricci tensor can be closely identified with the source T_{ab}, so we need to remove these 'source degrees of freedom' if we wish to identify those degrees of freedom that directly describe the gravitational field. In free space, where there is no matter (and for simplicity take the cosmological constant Λ to be zero), the Weyl tensor is *equal* to the Riemann curvature tensor, but generally the Weyl tensor is defined by the somewhat complicated-looking formula which removes the Ricci tensor part from the full curvature (where I have raised two indices in order to make full use of the square-bracket notation of §11.6):[19.23]

$$C_{ab}{}^{cd} = R_{ab}{}^{cd} - 2R_{[a}{}^{[c}g_{b]}{}^{d]} + \tfrac{1}{3}Rg_{[a}{}^{c}g_{b]}{}^{d}.$$

We shall be seeing a key physical role for the Weyl tensor in §28.8. The vanishing of this quantity is the condition for conformal flatness of the spacetime.

19.8 Gravitational field energy

Let us return to the question of the mass/energy in the gravitational field itself. Although there is no room for such a thing in the energy–momentum tensor $\boldsymbol{T}$, it is clear that there are situations where a 'disembodied' gravitational energy is actually playing a physical role. Imagine two massive bodies (planets, say). If they are close together (and we can suppose that they are instantaneously at rest relative to each other), then there will be a (negative) gravitational potential energy contribution which makes the total energy, and therefore the total mass, smaller than it would be if they are far apart (see Fig. 19.8). Ignoring much tinier energy effects, such as distortions of each body's shape due to the gravitational tidal field of the other, we see that the total contributions from the actual energy–momentum tensor $\boldsymbol{T}$ will be the same whether the two bodies are close together or far apart. Yet, the total mass/energy will differ in the two cases,

[19.23] Show that all the 'traces' of $\boldsymbol{C}$ vanish (e.g. $C_{abc}{}^{a} = 0$, etc.). Do this calculation in diagrammatic form, if you wish.

Fig. 19.8 Non-locality of gravitational potential energy. Imagine two planets (which for simplicity we may suppose to be instantaneously relatively at rest). If (a) they are far apart, then the (Newtonian) negative potential energy contribution is not so great as (b) when they are close together. Thus the total energy (and hence the total mass of the whole system) is larger in case (a) than in case (b) despite the total integrated energy densities, as measured by the energy–momentum tensors, being virtually the same in the two cases.

and this difference would be attributed to the energy in the gravitational field itself (in fact a negative contribution, that is more sizeable when the bodies are close than when they are far apart).

Now let us consider that the bodies are in motion, in orbit about one another. It is a consequence of Einstein's field equation that *gravitational waves*—ripples in the fabric of spacetime—will emanate from the system and carry (positive) energy away from it. In normal circumstances, this energy loss will be very small. For example, the largest such effect in our own solar system arises from the Jupiter–Sun system, and the rate of energy loss is only about that emitted by a 40-watt light bulb! But for more massive and violent systems, such as the final coalescence of two black holes that have been spiralling into each other, it is expected that the energy loss would be so large that detectors presently being constructed here on Earth might be able to register the presence of such gravitational waves at a distance of 15 megaparsecs or about 4.6×10^{23} metres.

Intermediate between these two extremes are the gravitational waves emitted by the remarkable double-neutron-star system known as PSR 1913 + 16, studied by the Nobel prize-winning team of Joseph Taylor and Russell Hulse; see Fig. 19.9. (A neutron star is an extremely compact star, composed mainly of neutrons, so tightly squeezed together that the star's overall density is comparable with that of an atomic nucleus. A tennis ball filled with such material would have a total mass comparable with that of Mars's moon Deimos!) This system has now been observed over a period of some 25 years, and its detailed motion has been tracked to great precision (which is possible because one of the stars is a *pulsar* which emits very precisely timed electromagnetic 'blips' some 17 times a second).

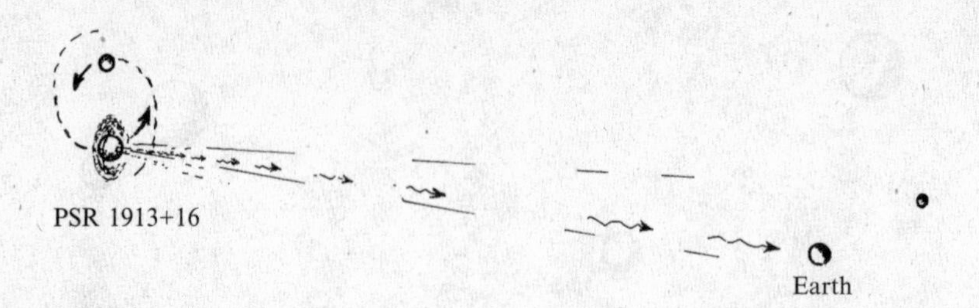

Fig. 19.9 The Hulse–Taylor double neutron star system PSR 1913 + 16. One member is a pulsar which sends out precisely timed electromagnetic signals that are received at Earth, enabling the orbits to be determined with extraordinary accuracy. It is observed that the system loses energy in exact accord with Einstein's prediction of energy-carrying gravitational waves emitted by such a system. These waves are ripples in the spacetime vacuum, where the energy–momentum tensor vanishes. (Not to scale.)

The timing of these signals is so precise, and the system itself so 'clean', that comparison between observation and theoretical expectation provides a confirmation of Einstein's general relativity to about one part in 10^{14}, an accuracy unprecedented in the scientific comparison between the observation of a particular system and theory. This figure refers to the overall timing precision over a period of more than 20 years.[11]

With observations of this nature—and also the impressive gravitational lensing effects that will be addressed in §28.8—observational general relativity has come a long way from the early days of the subject. However, in the years 1915–1969 (where 1969 marks the year that radio observations of distant quasars initiated a new family of tests of general relativity[12]) there were only the famous, but comparatively unimpressive, 'three tests' to give support to the theory. The most significant of these was Einstein's explanation of the 'anomolous perihelion advance' of the planet Mercury. This was a very slight deviation from the predictions of Newtonian gravitational theory (of just 43 seconds of arc per century, or about one orbital rotation in 3 million years!) that had been observed for over half a century[13]. (See also §30.1 and §34.9.) A second observational effect was the tiny bending of distant starlight by the Sun, seen by Arthur Eddington's expedition to the island of Principe (off the coast of West Africa), during the solar eclipse of 1919. This is an instance of the same 'gravitational lensing' phenomenon, referred to above, that is now impressively used out to near cosmological distances to obtain important information about the mass distribution in the universe. (See §28.8.) Finally, there is the slowing of clock rates in a gravitational potential, predicted by Einstein's theory. This was tentatively (and questionably) observed by W.S. Adams in 1925, for a white-dwarf star known as the companion of Sirius (a star some thousands of times denser than the sun). But a far more convincing confirmation was later obtained, in a delicate experiment performed by Pound and Rebka in 1960, for the Earth's own gravitational field. (However, this effect is to be expected simply from general energy and basic quantum considerations and is a rather weak test of Einstein's theory.) There is also a different kind of 'time-delay' effect, for light signals reaching the Earth from objects almost directly behind the Sun, which was first proposed (in 1964) by Irwin Shapiro, and later confirmed by him in 1968–1971, for observations of Mercury and Venus, and more precisely (to 0.1 per cent, in 1971) by Reasenberg and Shapiro using transponders on the Viking spacecraft in orbit around Mars, in comparison with one on the ground on Mars.

It is clear that Einstein's theory is now very well supported observationally. The existence of gravitational waves seems to be clearly confirmed by the Hulse–Taylor observations, even though this is not a direct detection of such waves. There are now several projects for the direct detection of gravitational waves which constitute a worldwide concerted effort to use such waves to probe violent activity (such as black hole collisions) in distant parts of the universe. In effect, these combined projects[14] will provide a gravitational-wave telescope, so that Einstein's theory has the potential to give us yet another powerful way of exploring the distant universe.

We see that, despite some people's worries about energy conservation, general relativity has some very remarkable observational confirmations. Let us, therefore, return to the question of gravitational energy. It is an essential point of consistency, both in theory and observation, that the ripples of empty space that constitute the gravitational waves emitted by PSR 1913 + 16 and other such systems indeed carry actual energy away. The energy–momentum tensor in empty space is zero, so the gravitational wave energy has to be measured in some other way that is not locally attributable to an energy 'density'. Gravitational energy is a genuinely non-local quantity. This does not imply that there is no mathematical description of gravitational energy, however. Although I believe that it is fair to say that we do not yet have a complete understanding of gravitational mass/energy, there is an important class of situations in which a very complete answer can be given. These situations are those referred to as *asymptotically flat*, and they refer to gravitating systems that may be regarded as being isolated from the rest of the universe, essentially because of their very large distance from everything else. It might be, say, a double-star system, like the Hulse–Taylor binary pulsar, in which one is concerned with the energy that is lost through gravitational radiation. The work of Hermann Bondi and his collaborators as generalized by Rayner Sachs[15] (to remove Bondi's simplifying asssumption of axial symmetry), provided

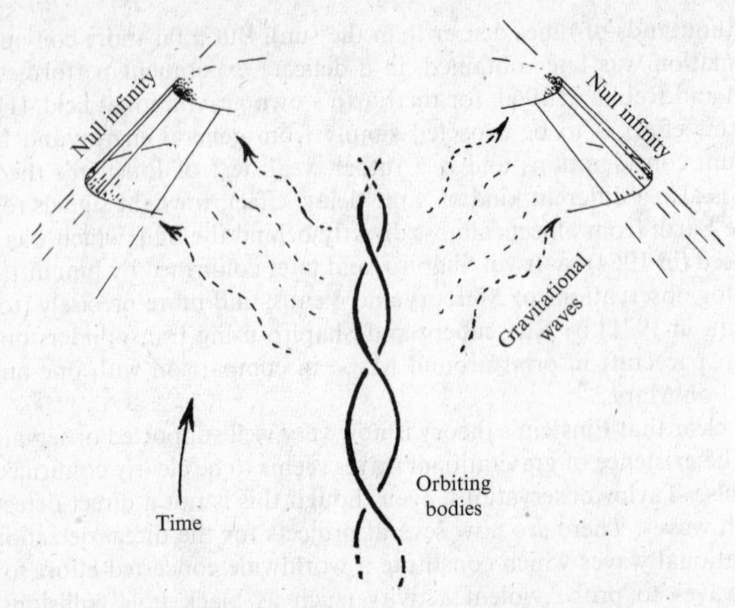

Fig. 19.10 For an isolated system emitting gravitational waves, where it may be assumed that the spacetime is asymptotically flat, there is a precise measure of total mass/energy–momentum and of its loss through gravitational radiation, referred to as the Bondi–Sachs mass/energy conservation law. The relevant mathematical quantities are non-local and defined at 'null infinity' (a geometrical notion which will be discussed in §27.12).

a clear-cut mathematical accounting of the mass/energy carried away from such a system in the form of gravitational waves, and a conservation law for energy–momentum was accordingly achieved;[16] see Fig. 19.10. This conservation law does not have the local character of that for non-gravitational fields, as manifested in the 'conservation equation' $\nabla^a T_{ab} = 0$, and it only applies in an exact way in the limit when the system becomes completely spatially isolated from everything else. Yet, there is something a little 'miraculous' about how things all fit together, including certain 'positivity' theorems that were later proved, which tell us that the *total mass* of a system (including the 'negative gravitational potential energy contributions' discussed above) cannot be negative.[17]

There are general prescriptions for obtaining conservation laws for systems of interacting fields. These come from the Lagrangian approach, which will be introduced in the next chapter. The Lagrangian approach is very powerful, general, and beautiful, despite the fact that it does not (or, at least, not directly) seem to give us everything that we need in the case of gravitation. It, and the closely related Hamiltonian approach, both form central parts of modern physics, and it is important to know something about them. Let us venture into this fabulous territory next.

Notes

Section 19.1

19.1. It seems doubtful that Newton himself would have held so dogmatically to such a particle-based picture (see Newton's *Queries* in his *Opticks* 1730). This 'Newtonian' view was, however, argued for forcefully in the 18th century by R.G. Boscovich; see Barbour (1989).

Section 19.3

19.2. The result would also apply in curved topologically trivial spacetime, so that (more specifically) a closed 2-surface always spans a compact 3-volume).

Section 19.4

19.3. See, for example, Flanders (1963).

19.4. See Weyl (1928), pp.87–8 (transl., pp. 100–1); also Weyl (1929). This observation was also made independently by W. Gordon, and by Pauli and Heisenberg; see Pais (1986), p.345.

19.5. See Aharonov and Bohm (1959). In fact this effect had already been noted 10 years earlier by Ehrenberg and Siday (1949). It was experimentally verified by Chambers, and then more convincingly established by Tonomura *et al.* (1982, 1986).

Section 19.5

19.6. See Pais (1982).

Section 19.6

19.7. The requirement, in the text, of 'no preferred coordinates' is not only rather vague, but also something that might be regarded as somewhat too strong. In flat space, for example, it could reasonably be said that the choices of 'Cartesian coordinates' (here the Minkowski coordinates (t, x, y, z) of §18.1, for which the metric takes the particularly simple form $ds^2 = dt^2 - dx^2 - dy^2 - dz^2$) are 'preferred' over all other coordinate systems, and cosmological models also have special coordinate systems in which the metric form looks particularly simple (see §27.11, Exercise [27.18]). The point is, rather, the more subtle one that such special coordinates should not have a physical role to play, and that the equations of the theory should be such that their most natural expression does not depend on any particular choice of coordinates.

19.8. See Stachel (1995), p. 353–64. Among the many excellent texts on general relativity are Synge (1960) Weinberg (1972); Misner, Thorne and Wheeler (1973); Wald (1984); Ludvigsen (1999); Rindler (1977, 2001); Schutz (2003); and Hartle (2003).

Section 19.7

19.9. Einstein's model was the space $\mathcal{E}$, with topology $S^3 \times E^1$, that we shall encounter in §31.16.

19.10. Einstein's introduction of a cosmological term was one of a number of modifications of the original theory of general relativity that have been introduced over the years. In addition to Weyl's theory discussed in §19.4 and the higher-dimensional

Kaluza–Klein ideas referred to in §31.4 (nowadays usually combined with supersymmetry; see §§31.2,3), there is the Brans–Dicke modification in which there is an additional scalar field, and Einstein's own numerous attempts at a 'unified field theory' put forward in the period between 1925 and 1955. See Einstein (1925); Einstein (1945); Einstein and Straus (1946); Einstein (1948); Einstein and Kaufman (1955); Schrödinger (1950); for a more recent reference, see Antoci (2001). Most of these proposals were intended to incorporate electromagnetism, and perhaps other fields, into the overall framework of general relativity. Noteworthy also is the scheme referred to as the Einstein–Cartan–Sciama–Kibble theory, in which a torsion is introduced (§14.4) and considered to describe a direct gravitational effect of a density of spin (see §22.8); see Kibble (1961), Sciama (1962), and the accounts by Trautman (1972, 1973)—for which the references Cartan (1923, 1924, 1925) are relevant.

Section 19.8

19.11. Here, Einstein's theory is taken to include Newton's, and it should be emphasized that the '10^{14}' figure does not represent an increase of accuracy over Newton's scheme. Moreover, it should be borne in mind that some of the timing accuracy goes to determine the unknown parameters, such as the masses, orbit inclination, eccentricity, etc., that are needed to compute the details of the system. The '10^{14}' is really a measure of the overall consistency of the picture.

19.12. The 1991 results of D.S. Robinson and collaborators, using 'Very Long Baseline Interferometry', now confirm the light-bending effects of general relativity to an accuracy of 10^{-4}.

19.13. For a detailed account of Mercury's perihelion anomaly, see Roseveare (1982).

19.14. These gravitational wave searches go by such colourful acronyms as LIGO, LISA, and GEO. See Shawhan (2001); Abbott (2004); Grishchuk *et al.* (2001); Thorne (1995b); as well as John Baez's very useful web-commentary http://math.ucr.edu/home/baez/week143.html

19.15. Bondi (1960); Bondi *et al.* (1962); Sachs (1961, 1962a). This work was partially anticipated by Trautman (1958).

19.16. See also Newman and Unti (1962); Penrose (1963, 1964); Sachs (1962b); Bonnor and Rotenberg (1966); Penrose and Rindler (1986), pp. 423–7.

19.17. Schoen and Yau (1979, 1982); Witten (1981); Nester (1981); Parker and Taubes (1982); Ludvigsen and Vickers (1982); Horowitz and Perry (1982); Reula and Tod (1984); see also Penrose and Rindler (1986) and §32.3, particularly Note 32.11.

20
Lagrangians and Hamiltonians

20.1 The magical Lagrangian formalism

In the centuries following Newton's introduction of his dynamical laws, an extremely impressive body of theoretical work was built up from these Newtonian foundations. Euler, Laplace, Lagrange, Legendre, Gauss, Liouville, Ostrogradski, Poisson, Jacobi, Hamilton, and others came forth with reformulating ideas that led to a profound unifying overview. I shall give a brief introduction here to this dynamical overview, although I am afraid that my account of it will provide only a very inadequate impression of the magnitude of the achievement. It should also be remarked that just the existence of such a mathematically elegant unifying picture appears to be telling us something deep about the mathematical underpinnings of our physical universe, even at the level of the laws that were revealed in 17th century Newtonian mechanics. Not many suggested laws for a physical universe could lead to mathematical structures of such imposing splendour.

What elegant unifying picture is it that resulted from Newton's mechanics? It occurs basically in two different but closely related forms, each having its characteristic virtues. Let us refer to the first as the *Lagrangian* picture and the second as the *Hamiltonian* one. (There is the usual difficulty with names here. Apparently, both pictures were known to Lagrange, significantly before Hamilton, and the Lagrangian one was at least partially anticipated by Euler.) Let us consider that we have a Newtonian system consisting of a (finite) number of individual particles and perhaps some rigid bodies each considered as an indivisible entity. There will be a *configuration space* $\mathcal{C}$ of some large number N of dimensions, each of whose points represents a single spatial arrangement of all these particles and bodies (see §12.1). As time evolves, the *single point* of $\mathcal{C}$ that represents the entire system will move about in $\mathcal{C}$ according to some law which encapsulates the Newtonian behaviour of the system; see Fig. 20.1. It is a remarkable (and computationally very valuable) fact that this law can be obtained by a direct mathematical procedure from a *single function*. In the Lagrangian picture (at least in its

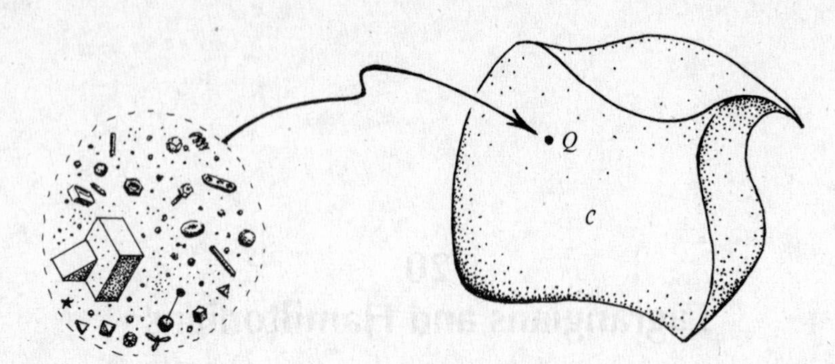

Fig. 20.1 Configuration space. Each point Q of the N-dimensional manifold $\mathcal{C}$ represents an entire possible configuration of (say) a family of Newtonian point particles and rigid bodies. As the system evolves in time, Q describes some curve in $\mathcal{C}$.

simplest and most usual form[1]), this function—called the *Lagrangian* function—is defined on the *tangent bundle* $\mathrm{T}(\mathcal{C})$ of the configuration space $\mathcal{C}$ (Fig. 20.2a); see §15.5. In the Hamiltonian picture, the function—called the *Hamiltonian* function—is defined on the *cotangent bundle* $\mathrm{T}^*(\mathcal{C})$ (see §15.5), called the *phase space* (Fig. 20.2b). We note that $\mathrm{T}(\mathcal{C})$ (each of whose points stands for a point Q of $\mathcal{C}$, together with a tangent vector at Q) and $\mathrm{T}^*(\mathcal{C})$ (each of whose points stands for a point Q of $\mathcal{C}$, together with a cotangent vector at Q) are both $2N$-dimensional manifolds.

In this section, we investigate the Lagrangian picture, leaving the Hamiltonian one to the next. Coordinates for Lagrange's $\mathrm{T}(\mathcal{C})$ would serve to determine the positions of all the Newtonian bodies (including appropriate angles to specify the spatial orientations of the rigid bodies, etc.) and also their velocities (including corresponding angular velocities of rigid bodies, etc.). The position coordinates $q^1, \ldots, q^N$, usually termed 'generalized coordinates', label the different points q of the configuration space $\mathcal{C}$ (perhaps just given 'patchwise', see §12.2). Any (adequate) system of coordinates will do. They need not be 'Cartesian' or of any other standard kind. This is the beauty of the Lagrangian (and also Hamiltonian) approach. The choice of coordinates is governed merely by convenience. This is just the same role for coordinates used in Chapters 8, 10, 12, 14, and 15, etc., when general manifolds of various kinds were considered. Corresponding to the chosen set of generalized coordinates are the 'generalized velocities' $\dot{q}^1, \ldots, \dot{q}^N$, where the 'dot' means the rate of change '$\mathrm{d}/\mathrm{d}t$' with respect to time:

$$\dot{q}^1 = \frac{\mathrm{d}q^1}{\mathrm{d}t}, \ldots, \dot{q}^N = \frac{\mathrm{d}q^N}{\mathrm{d}t}.$$

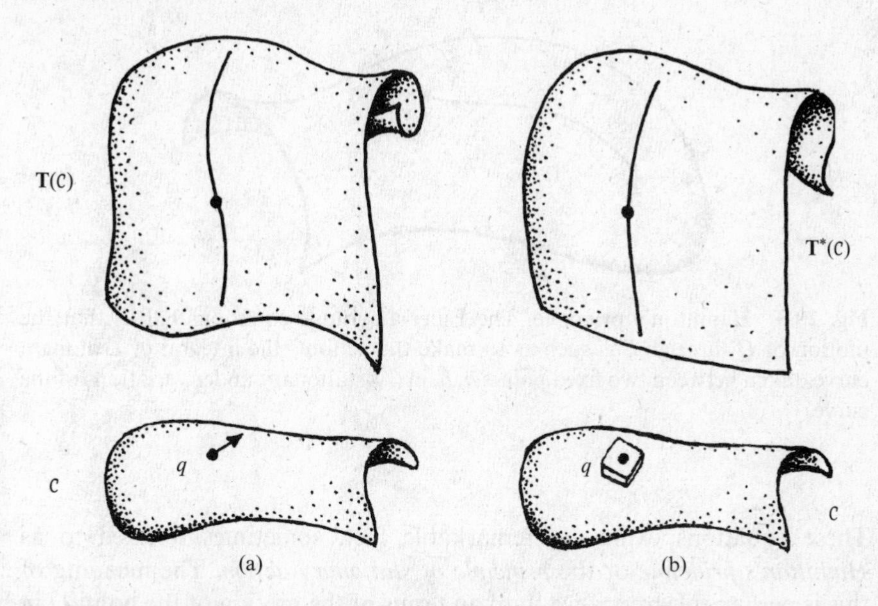

Fig. 20.2 (a) In the standard Lagrangian picture, the Lagrangian $\mathcal{L}$ is a smooth function on the tangent bundle $T(C)$ of configuration space C. (b) In the Hamiltonian picture, the Hamiltonian $\mathcal{H}$ is a smooth function on the cotangent bundle $T^*(C)$ called phase space.

The Lagrangian $\mathcal{L}$ would be written as a function of *all* of these.[2]

$$\mathcal{L} = \mathcal{L}(q^1, \ldots, q^N; \dot{q}^1, \ldots, \dot{q}^N).$$

Each $\dot{q}^r$ has to be treated as an *independent* variable (independent of q^r, in particular) in this expression. This is one of the initially baffling features of Lagrangians—but it works![3]

The normal physical interpretation of the actual value of the function $\mathcal{L}$ would be the difference $\mathcal{L} = K - V$ between the kinetic energy K of the system and the potential energy V due to the external or internal forces, expressed in these coordinates (see §18.6). The equations of motion of the system—encoding its *entire* Newtonian behaviour—are given by what are called the *Euler–Lagrange* equations, which are astonishing in their extraordinary scope and essential simplicity:

$$\frac{\mathrm{d}}{\mathrm{d}t}\frac{\partial \mathcal{L}}{\partial \dot{q}^r} = \frac{\partial \mathcal{L}}{\partial q^r} \quad (r = 1, \ldots, N).$$

Remember that each $\dot{q}^r$ is to be treated as an independent variable, so that the expression '$\partial\mathcal{L}/\partial\dot{q}^r$' (which means 'differentiate $\mathcal{L}$ *formally* with respect to $\dot{q}^r$, holding all the other variables fixed') actually makes sense!

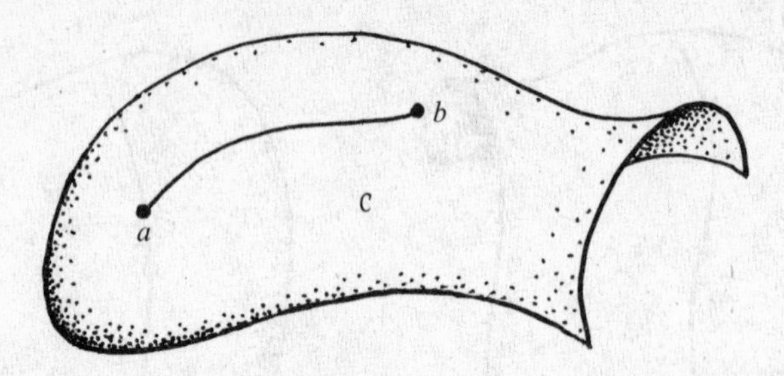

Fig. 20.3 Hamilton's principle. The Euler–Lagrange equations tell us that the motion of Q through $\mathcal{C}$ is such as to make the action—the integral of $\mathcal{L}$ along a curve, taken between two fixed points a, b, in $\mathcal{C}$—stationary under variations of the curve.

These equations express a remarkable fact, sometimes referred to as *Hamilton's principle* or the *principle of stationary action*. The meaning of this is perhaps clearest if we think in terms of the motion of the point Q in $\mathcal{C}$, where we recall that $\mathcal{C}$ represents the space of possible spatial configurations of the entire system (i.e. all the locations of all its parts). The point Q, whose position at any time is labelled by the q^r, moves along some curve in $\mathcal{C}$ at a certain rate, this rate together with the tangent direction to the curve being determined by the values of the $\dot{q}^r$. The Euler–Lagrange equations basically tell us that the motion of Q through $\mathcal{C}$ is such as to minimize the *action*, this 'action' being the integral of $\mathcal{L}$ along the curve, taken between two fixed end points, a and b, in the configuration space $\mathcal{C}$; see Fig. 20.3.

More correctly, this may not be actually a 'minimum', but the term 'stationary' would be appropriate. The situation is basically similar to that which happens in ordinary calculus (see §6.2), where the occurrence of a *minimum* of a smooth real-valued function $f(x)$ requires $\mathrm{d}f/\mathrm{d}x = 0$, but where sometimes $\mathrm{d}f/\mathrm{d}x = 0$ occurs when the function f is not a minimum: it might be a maximum or possibly a point of inflexion or, in higher dimensions, what is called a *saddle point* (Fig. 20.4b). All places where $\mathrm{d}f(x)/\mathrm{d}x = 0$ are called *stationary*. See Figs. 6.4 and 20.4. Recall the basically similar characterization of a *geodesic* in (pseudo)Riemannian space, given in §14.7, §17.9, and §18.3 as a 'minimum-length path' in the positive-definite case (locally), and sometimes as a 'maximum length timelike path' in the Lorentzian case, although merely of 'stationary length' in the general case. Thus, Q's trajectory can be thought of as some sort of 'geodesic' in the space $\mathcal{C}$.

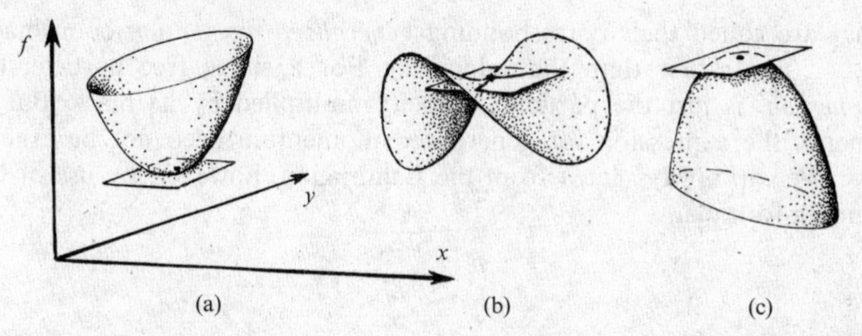

Fig. 20.4 Stationary values of a smooth real-valued function f of several variables. Illustrated is the case of a function $f(x,y)$ of two variables. This is stationary where its graph (a 2-dimensional surface) is horizontal ($\partial f/\partial x = 0 = \partial f/\partial y$). This occurs (a) where F has a minimum, but also in other situations such as (b) at a saddle point and (c) at a maximum. In the case of Hamilton's principle (Fig. 20.3)—or a geodesic connecting two points a, b—the Lagrangian $\mathcal{L}$ takes the place of f, but the specification of a path requires infinitely many parameters, rather than just x and y. Again, $\mathcal{L}$ may not be a minimum, though a stationary point of some kind.

It is helpful to consider a simple example of a Lagrangian, such as that for a single Newtonian particle of mass m, moving in some fixed external field given by a potential V which depends on position: $V = V(x, y, z; t)$. The meaning of V is that it defines the *potential energy* of the particle due to this external field. For the case of the gravitational field of the Earth (near the Earth's surface), thought of as a constant downward pull, we can take $V = mgz$, where z is the height above the ground and g is the downward acceleration due to gravity. The three components of velocity are $\dot{x}$, $\dot{y}$, $\dot{z}$, so using the expression $\frac{1}{2}mv^2$ for kinetic energy (see §18.6), we find the Lagrangian

$$\mathcal{L} = \tfrac{1}{2}m(\dot{x}^2 + \dot{y}^2 + \dot{z}^2) - mgz.$$

The Euler–Lagrange equation for z now gives us $d(m\dot{z})/\mathrm{d}t = -mg$, from which Galileo's constancy of acceleration, in the direction of the Earth, follows.[20.1]

20.2 The more symmetrical Hamiltonian picture

In the Hamiltonian picture, we still use generalized coordinates, but now the generalized position coordinates $q^1, \ldots, q^N$ are taken together with

[20.1] Fill in the full details, completing the argument to obtain Galileo's parabolic motion for free fall under gravity.

what are called their corresponding *generalized momentum* coordinates $p_1, \ldots, p_N$ (rather than the velocities). For a single free particle, the *momentum* is just the particle's velocity multiplied by its mass. But in general, the expression for generalized momentum need not be exactly this. We can always get it from the Lagrangian, however, by use of the defining formula

$$p_r = \frac{\partial \mathcal{L}}{\partial \dot{q}^r}.$$

In any case, these parameters p_r serve to provide coordinates for the *cotangent spaces* to $\mathcal{C}$, so that a covector can be written as

$$p_a \mathrm{d}q^a$$

(where we recall the summation convention of §12.7 which we adopt here, although it is legitimate to read this also as an abstract–index expression, as in §12.8). This, of course, is a 1-form, and its exterior derivative (§12.6)

$$\boldsymbol{S} = \mathrm{d}p_a \wedge \mathrm{d}q^a$$

is a 2-form (satisfying $\mathrm{d}\boldsymbol{S}=0$)[20.2] which assigns a natural *symplectic structure* to the phase space $\mathrm{T}^*(\mathcal{C})$ (see §14.8). Much of the strength of the Hamiltonian picture lies in the fact that phase spaces are *symplectic manifolds*, and this symplectic structure is independent of the particular Hamiltonian that is chosen to provide the dynamics. Classical physics is thereby intimately connected with the beautiful and surprising geometry of symplectic manifolds that we shall be coming to in §20.4.

As a preliminary to understanding the role that this geometry plays, let us see the form of Hamilton's dynamical equations. These describe the time-evolution of a system as a trajectory, within the phase space $\mathrm{T}^*(\mathcal{C})$, of a point P representing the entire Newtonian system. This evolution is completely governed by the *Hamiltonian function*

$$\mathcal{H} = \mathcal{H}(p_1, \ldots, p_N;\ q^1, \ldots, q^N).$$

which (in the case of the time-independent Lagrangians and Hamiltonians that we are concerned with here) describes the *total energy* of the system, in terms of the (generalized) momenta and positions. We can actually obtain it from the Lagrangian by means of the expression (summation convention or abstract indices)

$$\mathcal{H} = \dot{q}^r \frac{\partial \mathcal{L}}{\partial \dot{q}^r} - \mathcal{L},$$

[20.2] Why?

which then has to be rewritten by eliminating all the generalized velocities in favour of the generalized momenta (not an easy task, in general!). In terms of these momentum and position coordinates, Hamilton's evolution equations are beautifully symmetrical:

$$\frac{\mathrm{d}p_r}{\mathrm{d}t} = -\frac{\partial \mathcal{H}}{\partial q^r}, \quad \frac{\mathrm{d}q^r}{\mathrm{d}t} = \frac{\partial \mathcal{H}}{\partial p_r}.$$

These equations describe the velocity of a point P in $\mathrm{T}^*(\mathcal{C})$. This velocity is defined for every P, so we have a vector field on $\mathrm{T}^*(\mathcal{C})$, defined by the Hamiltonian $\mathcal{H}$. In terms of the 'partial differentiation operator' notation for a vector field given in §12.3 this is[20.3]

$$\frac{\partial \mathcal{H}}{\partial p_r}\frac{\partial}{\partial q^r} - \frac{\partial \mathcal{H}}{\partial q^r}\frac{\partial}{\partial p_r},$$

written as $\{\mathcal{H}, \quad\}$ in §20.4. This provides a 'flow' on $\mathrm{T}^*(\mathcal{C})$ which describes the Newtonian behaviour of the system (Fig. 20.5).

In the particular example of a particle falling in a constant gravitational field, as given above (§20.1) in Lagrangian form, the Hamiltonian is

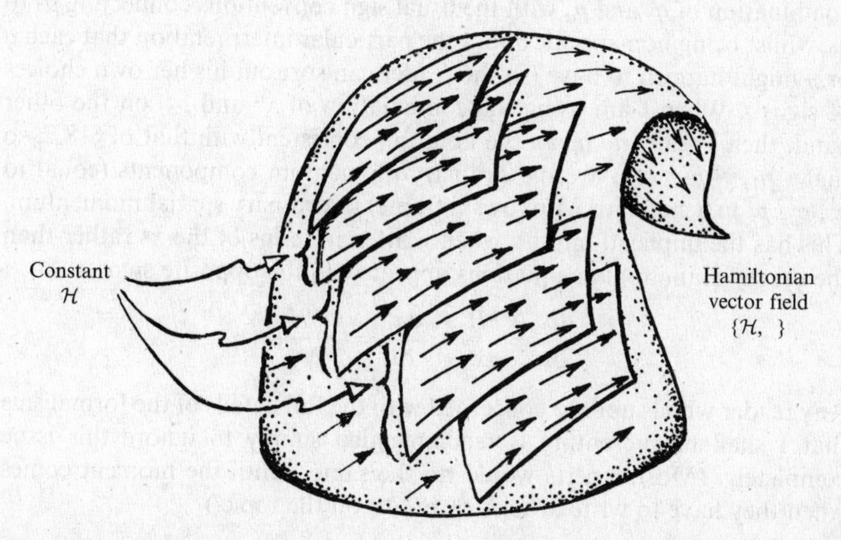

Fig. 20.5 The Hamiltonian flow $\{\mathcal{H}, \quad\}$, representing the Newtonian time-evolution of the system (see §20.4), is a vector field on phase space $\mathrm{T}^*(\mathcal{C})$. For the hypersurfaces of fixed $\mathcal{H}$-values (fixed energy, taking $\mathcal{H}$ to be time independent), the trajectories remain within the fixed-$\mathcal{H}$ hypersurface, in accordance with energy conservation. (See Note 27.36 for the term 'hypersurface'.)

[20.3] Explain this.

$$\mathcal{H} = \frac{p_x^2 + p_y^2 + p_z^2}{2m} + mgz$$
$$= \frac{p^2}{2m} + mgz,$$

where p_x, p_y, and p_z are the ordinary spatial momentum components in the directions of the Cartesian x, y, and z axes, respectively. This can be written down directly from knowledge of what the total energy of the particle ought to be when expressed in terms of position and momentum components, or else we can obtain it from the Lagrangian, as given by the above procedures.[20.4]

At this point I should confess to a notational awkwardness that I see no way around, so I had better come clean! We saw in §18.7, that the spatial momentum components p_1, p_2, p_3, in standard Minkowski coordinates for flat spacetime, with my preferred $(+ - - -)$ signature, are the *negatives* of the normal momentum components. Thus we have, in the above example, $p_x = -p_1$, $p_y = -p_2$, and $p_z = -p_3$. In the *general* discussion of Hamiltonians it is natural to use the 'downstairs' versions of the momenta p_a, yet this is inconsistent with the 'p_a' (i.e. p_1, p_2, p_3) that are natural in relativity with $(+ - - -)$ signature. The way that I am dealing with this notational problem, in this book, is simply to give the general formalism using the combination of q^a and p_a with the usual sign conventions connecting ps to qs, whilst being non-specific about the particular interpretation that each q or p might happen to have (so the reader can sort out his/her own choices of signs!). When I am using the combination of x^a and p_a, on the other hand, then I really *do* mean the notation consistent with that of §18.7, so that $-p_1$, $-p_2$, $-p_3$ are the ordinary momentum components (equal to p^1, p^2, p^3 in a standard Minkowski frame) of ordinary spatial momentum. This has the implication that, when written in terms of the xs rather than the qs, my Hamiltonian equations appear with the opposite sign

$$\frac{\mathrm{d}p_r}{\mathrm{d}t} = \frac{\partial \mathcal{H}}{\partial x^r}, \quad \frac{\mathrm{d}x^r}{\mathrm{d}t} = -\frac{\partial \mathcal{H}}{\partial p_r}.$$

Any reader who is not too concerned with the full details of the formalisms that I shall be presenting is recommended simply to ignore this issue completely. (Most experts would do the same—until the moment comes when they have to write articles or books on the topic!)

20.3 Small oscillations

Before moving on, in the next section, to the remarkable geometry that the Hamiltonian description of things leads us to, it will be illuminating, first,

---

[20.4] Do this explicitly. Use Hamilton's equations to obtain the Newtonian equations of motion for a particle falling in a constant gravitational field.

to consider the important topic of vibrations of a physical system about a state of equilibrium. The topic has considerable relevance in a number of different areas, and it has particular significance for us later in the context of quantum mechanics (§22.13). The theory of vibrations can be conveniently described either in the Lagrangian or the Hamiltonian formalism, each of which is very well suited to its treatment. I shall give my descriptions explicitly here in the Hamiltonian formalism primarily because this more directly leads us into the quantum-mechanical version of vibrations, which we shall catch a good glimpse of in §22.13. The Lagrangian theory of vibrations, which is very similar to the Hamiltonian one, is left to the reader (see Exercise [20.10]).

A simple example of a vibrating system occurs with an ordinary pendulum, swinging under gravity. When the oscillations are small, then the motion of the bob, backwards and forwards, describes a sine wave, as a function of time (see Fig. 20.6). (This is the kind of behaviour encountered with the individual 'Fourier components' studied in §9.1.) The period of vibration, for such small oscillations, is actually independent of the amplitude of the oscillation (i.e. of the distance through which the bob swings)—a famous early observation of Galileo's, in 1583. This type of motion is referred to as *simple harmonic motion*.

We shall be seeing, in this section, how ubiquitous this motion is. A general physical structure (supposing that frictional effects can be disregarded) can 'wobble' about its equilibrium state only in very specific ways. We shall find that every small-scale wobble can be broken down into particular modes of vibration—called *normal modes*—in which the whole

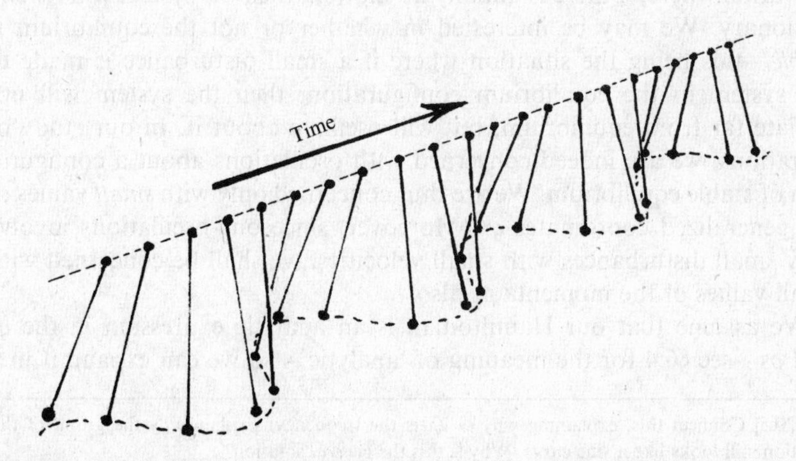

Fig. 20.6 A pendulum, swinging under gravity. For small oscillations, the motion of the bob approximates simple harmonic motion, the displacement of the bob (mapped out as a function of time) giving a 'sine wave'.

structure partakes of a simple harmonic motion with a very specific frequency, called a *normal frequency*.

Let us first see how simple harmonic motion is described analytically. Let q denote the horizontal distance of our pendulum bob out from the lowest point—or else the outward displacement from equilibrium of whatever other vibrating quantity we might be considering. Then the equation of motion, for small displacements q, is

$$\frac{d^2q}{dt^2} = -\omega^2 q,$$

where the positive constant quantity $\omega/2\pi$ is the *frequency* of the oscillation. This tells us that the inward acceleration d^2q/dt^2 is proportional (with factor ω^2) to the outward displacement. We see from §6.5 that $q = \cos \omega t$ and $q = \sin \omega t$ both satisfy this equation and so also does the general linear combination

$$q = a \cos \omega t + b \sin \omega t,$$

where a and b are constants.[20.5] For a pendulum of length h swinging under gravity (in one plane), we find an equation of motion that closely approximates the one given above, when q is small, with $\omega^2 = g/h$; but for larger values of q, deviations from this equation arise.[20.6]

Suppose that we have a general Hamiltonian system, which is in equilibrium when the qs take some particular values $q^a = q^a_0$. It will be convenient to choose the *origin* of our generalized coordinates to represent our equilibrium state, i.e. we choose $q^a_0 = 0$. 'Equilibrium' refers to a configuration where, if there is initially no motion, then the system will *remain* stationary. We may be interested in whether or not the equilibrium is *stable*—this being the situation where if a small disturbance is made to the system in the equilibrium configuration, then the system will not deviate far from equilibrium, but will oscillate about it. In our study of vibrations, we are indeed concerned with oscillations about a configuration of stable equilibrium. We are thus concerned only with *small* values of the generalized coordinates q^a. Moreover, since our oscillations involve only small disturbances with small velocities, we shall be concerned with small values of the momenta p_a also.

We assume that our Hamiltonian is an analytic expression in the qs and ps—see §6.4 for the meaning of 'analytic'—so we can expand it in a

---

[20.5] Confirm this, explaining why $\omega/2\pi$ is the frequency. Explain why the graph of this function still looks like a sine curve. Why is this the *general* solution?

[20.6] Show this, finding the *full* equation, (a) using the Lagrangian method, (b) using the Hamiltonian method, and (c) directly from Newton's laws. *Hint*: Show that $\mathcal{L} = \frac{1}{2} mh^2\dot{q}^2(h^2 - q^2)^{-1} + mg(h^2 - q^2)^{1/2}$. (Note that the Lagrangian and Hamiltonian methods do not gain us anything in this simple case; their power resides in treating more general situations.)

power series in the qs and ps. For a stable equilibrium configuation, $q^a = 0$ must represent a (local) minimum of potential energy.[20.7] Moreover, when motion is introduced, this can only increase the energy (the kinetic energy); the kinetic energy is minimum when $p_a = 0$. The *total* energy—which is the value of the Hamiltonian $\mathcal{H}$—is therefore locally a minimum at $q^a = 0 = p_a$. It follows that our power series expansion must start off (terms *linear* in the qs, ps, or both, being absent) as

$$\mathcal{H} = \text{constant} + \tfrac{1}{2}Q_{ab}q^a q^b + \tfrac{1}{2}P^{ab}p_a p_b$$
$$+ \text{ terms of order 3 or more in } q\text{s and } p\text{s},$$

where Q_{ab} and P^{ab} are the components of *positive definite* constant symmetric matrices (so $Q_{ab}q^a q^b > 0$ if $q^a \neq 0$ and $P^{ab}p_a p_b > 0$ if $p_a \neq 0$; see §13.8). The factors $\frac{1}{2}$ are put in for convenience.[20.8]

Let us ignore the higher-order terms, so as to find the nature of the small oscillations. Hamilton's equations then give

$$\frac{\mathrm{d}q^a}{\mathrm{d}t} = \frac{\partial \mathcal{H}}{\partial p_a} = P^{ab}p_b;$$

whence, differentiating once more with respect to t,

$$\frac{\mathrm{d}^2 q^a}{\mathrm{d}t^2} = \frac{\mathrm{d}}{\mathrm{d}t}P^{ab}p_b = P^{ab}\frac{\mathrm{d}p_b}{\mathrm{d}t}$$
$$= -P^{ab}\frac{\partial \mathcal{H}}{\partial q^b} = -P^{ab}Q_{bc}q^c = -W^a{}_c\, q^c.$$

where $W^a{}_c = P^{ab}Q_{bc}$ is the matrix product of Q_{ab} with P^{ab} (see §13.3), which we can write in the form

$$\mathbf{W} = \mathbf{PQ},$$

so the conclusion of our previous displayed equation can now be rewritten

$$\frac{\mathrm{d}^2\mathbf{q}}{\mathrm{d}t^2} = -\mathbf{Wq}.$$

We are interested in the *eigenvectors* of the matrix $\mathbf{W}$ (see §13.5) which are the vectors $\mathbf{q}$ satisfying

$$\mathbf{Wq} = \omega^2\mathbf{q},$$

where ω^2 is the *eigenvalue* of $\mathbf{W}$ corresponding to $\mathbf{q}$. In fact, this eigenvalue must be positive, because the matrices $\mathbf{P}$ and $\mathbf{Q}$ are both positive-

[20.7] Why?

[20.8] Can you explain all this more fully? Can we have the linear terms if the equilibrium is unstable? Explain.

definite[20.9] so we can write it as the square of the positive quantity ω. We see that any such eigenvector $\mathbf{q}$ must satisfy the equation

$$\frac{d^2\mathbf{q}}{dt^2} = -\omega^2\,\mathbf{q},$$

representing simple harmonic motion with frequency $\omega/2\pi$.[20.10]

Each such eigenvector $\mathbf{q}$ is some combination of generalized coordinates q^a, so the oscillation corresponding to $\mathbf{q}$ would require these coordinates all to vibrate together, all at the same frequency. This is referred to as a *normal mode* of oscillation, and the corresponding $\omega/2\pi$ is called the *normal frequency* corresponding to this mode. In the most general case, these frequencies are all distinct, but in special 'degenerate' cases, some of these normal frequencies may coincide.[20.11] Degenerate eigenvalues must count with their appropriate *multiplicities*; whence the total number of normal modes is still equal to the number N of generalized coordinates $q_1, \ldots, q_N$. It may be noted that any two normal modes $\mathbf{q}$ and $\mathbf{r}$, corresponding to different frequencies, are 'orthogonal' to each other with respect to the 'metric' defined by $\mathbf{Q}$, in the sense $\mathbf{r}^{\mathrm{T}}\mathbf{Q}\mathbf{q} = 0$.[20.12]

What have we learnt from all this? We have formed a very general and remarkable conclusion about how a classical system, with N degrees of freedom, can vibrate about a configuration of stable equilibrium. Any such vibration is composed of normal modes—which can be treated as independent of one another—each mode having its own characteristic frequency, and where there are N modes altogether. In this description, we ignore the effects of *dissipation*, according to which, in practice, a vibration in a macroscopic system would eventually die away, its energy being transferred to the random motions of constituent particles. When all the constituent parts are taken into account (as with a molecule, for example), then dissipation does not occur.

Up to this point, I have been considering the ordinary situation in which the number of degrees of freedom N in the system is finite, but the foregoing theory also applies to systems that are—at least in idealization—infinite-dimensional. We are familiar with this idea when we are concerned with the sounds that a musical instrument might make. A drum, for example, or a musical triangle, will oscillate in accordance with various frequencies when it is struck, these frequencies determining its particular

[20.9] See if you can prove this deduction. *Hint:* Show that the inverse of a positive-definite matrix is positive-definite.

[20.10] See if you can carry out the foregoing analysis in the Lagrangian, rather than Hamiltonian, form.

[20.11] Describe the system of eigenvectors in such degenerate cases.

[20.12] Prove this. (Recall from §13.7 that 'T' stands for 'transposed'.)

timbre. The sound of a wind instrument is similar, coming from the oscillation of the column of air within it. Similar also is the vibration of a string of a stringed instrument, etc.

Fourier analysis, which we studied in Chapter 9, enables us to express the vibrations of finite-length string. We could take this string to be fixed at its endpoints, or perhaps bent into a circle. Fourier analysis expresses the general vibrations as a linear combination of modes, these being the pure-tone sine waves or cosine waves—and are infinite in number. In this case, the frequencies are all integral multiples of that of the primary mode. This is the kind of thing that one strives for in the construction of a sonorous musical instrument! But generally (as with a drum or bell), the normal frequencies are not so simply related.

In such situations, the Hamiltonian or Lagrangian formalisms can readily be extended, so as to cover the case $N = \infty$, but some care is needed. In a sense, we are taken naturally to the Lagrangian (or Hamiltonian) theory of *fields*, which we shall have a look at in §20.5. This has many applications in modern physics. In particular, the approach to a fundamental theory of Nature—referred to as *string theory*—where point particles are replaced by little loops (or else open-ended 'strings') requires this formalism. Here, the various fields or particles of Nature are taken to arise from normal modes of vibration of the 'strings' (see §§31.5,7,14).

One final point should be made here. The discussion of this section has been concerned only with oscillation about *stable* equilibrium, but it applies also to movements from *unstable equilibrium*. The basic difference is that our real symmetric matrix $\mathbf{Q}$ is not now positive-definite (or even non-negative-definite) so that $\mathbf{W} = \mathbf{PQ}$ can have *negative* eigenvalues. The corresponding small disturbances then diverge exponentially away from equilibrium.[20.13]

20.4 Hamiltonian dynamics as symplectic geometry

Let us step back to see how Hamilton's equations, in a finite number of dimensions, tie in with symplectic geometry. As described in §14.8, any symplectic manifold possesses an operation, that can be performed on pairs of scalar fields Φ and Ψ on the manifold to produce another scalar field Θ, called their *Poisson bracket*[20.14]

$$\Theta = \{\Phi, \Psi\} = \frac{\partial \Phi}{\partial p_a}\frac{\partial \Psi}{\partial q^a} - \frac{\partial \Phi}{\partial q^a}\frac{\partial \Psi}{\partial p_a}.$$

[20.13] Describe this behaviour.

[20.14] Confirm that this expression for $\{\Phi, \Psi\}$ agrees with that of §14.8.

If the 'Ψ slot' is left blank, then we get a differential operator $\{\Phi, \quad\}$, a *vector field* (see §12.3) whose action on Ψ gives $\{\Phi, \Psi\}$. Let us substitute $\mathcal{H}$ for Φ. We find that the vector field $\{\mathcal{H}, \}$ 'points along' the trajectories on $T^*(\mathcal{C})$ that represent *time-evolution*; in fact $\{\mathcal{H}, \}$ is just this evolution according to Hamilton's equations (see §20.2). One of the remarkable features of symplectic geometry is that the dynamical evolution of a system can thus be geometrically encapsulated in a single scalar function (namely the Hamiltonian).

Symplectic geometry does many other things for us. For example, there is a famous result due to Liouville which asserts that phase-space volume is always preserved by the dynamics; see Fig. 20.7. The *volume* element in phase space is taken to be the $2N$-form

$$\boldsymbol{\Sigma} = \boldsymbol{S} \wedge \boldsymbol{S} \wedge \ldots \wedge \boldsymbol{S},$$

there being N of the $\boldsymbol{S}$s wedged together; here, as we recall, the symplectic 2-form $\boldsymbol{S}$ is given by $\boldsymbol{S} = \mathrm{d}p_a \wedge \mathrm{d}q^a$. Now, it is not hard to check that $\boldsymbol{S}$ itself is preserved by the Hamiltonian evolution (i.e. that the Lie derivative of $\boldsymbol{S}$, with respect to the vector field $\{\mathcal{H}, \}$, vanishes).[20.15] It immediately follows that the full volume form $\boldsymbol{\Sigma}$ is preserved by this evolution, also. This is Liouville's theorem.

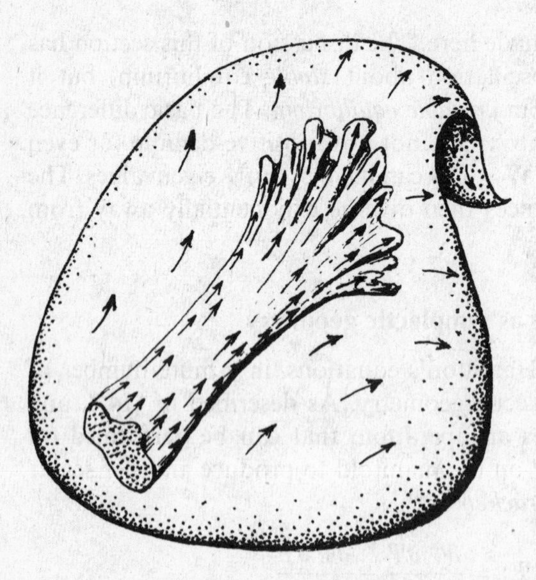

Fig. 20.7 Liouville's theorem. The Hamiltonian flow preserves the volume of the initial phase-space region (representing a range of possible initial states), even though the shape of this region may become grossly distorted in the time-evolution.

[20.15] Show this.

Since $\{\mathcal{H}, \mathcal{H}\} = 0$,[20.16] it follows that the Hamiltonian itself is preserved, i.e. it is constant along the trajectories, which is a reflection of the fact that the *total energy of a closed system is constant*. Thus, each trajectory lies on a $(2N - 1)$-dimensional surface given by $\mathcal{H}$ = constant; see Fig. 20.5. Now, we can think of the entire history of the system as being represented by its trajectory on $T^*(\mathcal{C})$, The space of these trajectories, for a fixed value of $\mathcal{H}$, is $(2N - 2)$-dimensional, see Fig. 20.8. (We lose one dimension because we hold $\mathcal{H}$ fixed, and lose another because we 'factor out' by the 1-dimensional trajectories.) It is a striking and important fact that the resulting $(2N - 2)$-manifold is again symplectic. This procedure (not just when Φ is chosen to be $\mathcal{H}$) has many elegant applications in classical mechanics and symplectic geometry.

There is undoubted beauty in this wonderfully comprehensive picture of Newtonian dynamics. Nevertheless, as we shall be seeing also in relation to later physical theories, it is important not to allow ourselves to be carried away by the beauty and seeming finality of such apparently tightly knit mathematical schemes. Nature has had a habit, in the past, of first tempting us to a euphoric complacency by the power and elegance of the

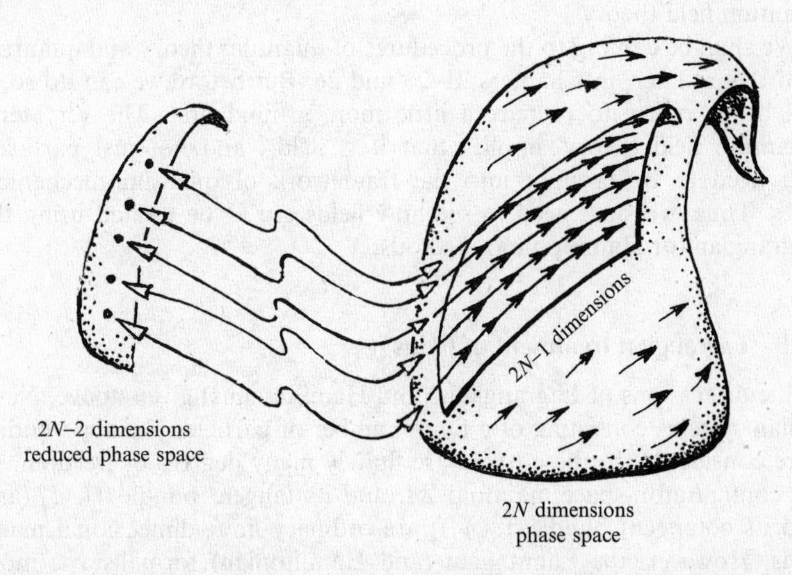

Fig. 20.8 Phase space $T^*(\mathcal{C})$ is a $2N$-dimensional symplectic manifold, for an N-dimensional $\mathcal{C}$. For a given energy value (constant $\mathcal{H}$, as in Fig. 20.5), we have a $(2N–1)$-dimensional region containing a $(2N–2)$-dimensional family of the Hamiltonian flow trajectories. The reduced phase space, whose points represent these trajectories, is itself a $2(N–1)$-dimensional symplectic manifold.

[20.16] Why?

mathematical structures that she appears to force us to accept as guiding her world, but then jolting us, from time to time, out of our conceptual torpor by showing us that our picture could not have been correct, after all! Yet the shift has always been a subtle one which leaves the previous edifice still standing proud, despite the fact that the foundations on which it had stood have now been completely replaced.

The Hamiltonian view provides a marvellous example. Although the classical mechanics that it embodies is contradicted by some harsh facts of the quantum world, the Hamiltonian framework provides us with an important lead into the actual theory of quantum mechanics. Moreover, the quantum versions of Hamiltonians provide essential ingredients for the standard quantum formalism. This, I should say, is for the standard non-relativistic quantum theory, in which there is no serious attempt to combine time and space together in accordance with the principles of relativity. In the case of relativistic quantum theory, however, it is the Lagrangian framework that has generally been found to provide the more natural leaping-off point. But where do we leap to? It is the need for an appropriate combination of the principles of special relativity with those of quantum mechanics that entices us to plunge into the deep quagmire of quantum field theory!

We shall be coming to the procedures of quantum theory and quantum field theory later, in Chapters 21–23 and 26. But before we can do so, it will be necessary to prepare a little more ground first. The very term 'quantum field theory' implies that it is fields, and not just particles, that need to be brought into the framework of quantum-mechanical rules. Thus, we shall need to see how fields are to be treated using the Lagrangian (or Hamiltonian) methods.

20.5 Lagrangian treatment of fields

In the discussions of Lagrangians (and Hamiltonians) given above, Newtonian systems consisting of a finite number of particles and rigid bodies were considered. In these, there are finitely many degrees of freedom, so the configuraton-space manifold $\mathcal{M}$, and its tangent bundle $T(\mathcal{M})$ (and also its cotangent bundle $T^*(\mathcal{M})$) are ordinary finite-dimensional manifolds. However, the Lagrangian (and Hamiltonian) formalism is more general than this, and it can indeed be applied also to physical fields. A field varies continuously from place to place, and it cannot be specified by a finite number of parameters. The configuration space of Maxwellian free fields in some region, for example, will be infinite-dimensional.

It is still possible to use the Lagrangian (or Hamiltonian) formalism in the case of an infinite-dimensional configuration space; indeed, this is the standard procedure in both classical and quantum field theory. The main novelty in the required formal mathematical procedures is the concept of *functional differentiation*. The Lagrangian, rather than being a function of a finite number of generalized coordinates $q^1, \ldots, q^N$ and a finite number of generalized velocities $\dot{q}^1, \ldots, \dot{q}^N$, is taken to be a function of a number of *fields* $\Phi, \ldots, \Psi$ (each of which is itself a function on spacetime, and perhaps possessing indices to indicate its tensorial or spinorial character) and the *derivatives* of these fields $\nabla_a\Phi, \ldots, \nabla_a\Psi$ (where usually only first derivatives appear, but higher-order derivatives are also allowed). Note that there is now no special role for the time derivatives (as was indicated by the 'dot' in the arguments of our original Lagrangians), and we are now being more even-handed by using the ∇_a operator instead. Accordingly, the formalism is being brought into line with the requirements of relativity.

Particularly in this kind of situation, Lagrangians are often called *functionals* because we are concerned with the functional form that they have, rather than just their actual values for specific values of their arguments. The Euler–Lagrange equations now involve 'derivatives with respect to the fields' and 'with respect to the gradients of the fields'. The formal carrying out of these operations mirrors very closely the operations of ordinary calculus, as described in Chapter 6. There are often mathematical subtleties involved, if one wants to be sure that the results are rigorously true, but it is customary for physicists not to be too worried about these, and the main concern is the correct following of the formal rules.

It is not my purpose, here, to go into these issues in detail, but it is worth writing down the Euler–Lagrange equations for this 'functional derivative' case (where functional derivative is denoted by using 'δ' in place of '∂'):

$$\nabla_a \frac{\delta\mathcal{L}}{\delta\nabla_a\Phi} = \frac{\delta\mathcal{L}}{\delta\Phi}, \ldots, \nabla_a \frac{\delta\mathcal{L}}{\delta\nabla_a\Psi} = \frac{\delta\mathcal{L}}{\delta\Psi}.$$

As mentioned above, the fields $\Phi, \ldots, \Psi$ may also possess indices. Carrying out a functional derivative in practice is essentially just applying the same rules as for ordinary calculus, and using a fair amount of 'mathematical common sense' (e.g. if $\mathcal{L} = \Phi^a\Phi^b\nabla_a\Psi_b$, then $\delta\mathcal{L}/\delta\Phi^c = \Phi^b\nabla_c\Psi_b + \Phi^a\nabla_a\Psi_c$, $\delta\mathcal{L}/\delta\nabla_c\Phi^d = 0$, $\delta\mathcal{L}/\delta\Psi_c = 0$, $\delta\mathcal{L}/\delta\nabla_c\Psi_d = \Phi^c\Phi^d$).

There is an analogue of Hamilton's principle for such Lagrangians, where we recall that this principle expresses the Euler–Lagrange equations as the stationarity of the *action*, the action being the integral of the Lagrangian along a curve joining two fixed points a and b of the configuration space (recall Fig. 20.3). In our more general situations under consideration here, the fixed end points a and b in $\mathcal{C}$ are replaced by field configurations in some 3-dimensional region(s) of spacetime. Often these are taken to be two 3-space regions $\mathcal{A}$ and $\mathcal{B}$, in spacetime, spanning the same 2-space $\mathcal{S}$ (where $\mathcal{S}$ is perhaps being taken at infinity)—see

Fig. 20.9—and this picture is also of importance in the path-integral formulation of quantum field theory that we shall be coming to later (§26.6). If desired, we can take $\mathcal{A}$ and $\mathcal{B}$ together (reversing the orientation of one of them) to constitute the *boundary* $\partial\mathcal{D}$ of a (possibly compact—see §12.6) spacetime 4-volume $\mathcal{D}$; see Fig. 20.10. In any case, the Hamilton principle expresses the stationarity of the spacetime integral of the Lagrangian over the region $\mathcal{D}$. Thus, the Lagrangian $\mathcal{L}$ is to be thought of as a spacetime *density* which, strictly speaking, means that the invariant entity is the 4-form $\mathcal{L}\boldsymbol{\varepsilon}$, where the natural 4-form $\boldsymbol{\varepsilon}$ is the quantity[4] which is commonly expressed as $\boldsymbol{\varepsilon} = \mathrm{d}x^0 \wedge \mathrm{d}x^1 \wedge \mathrm{d}x^2 \wedge \mathrm{d}x^3 \sqrt{(-\det \mathbf{g})}$. The *action* integral is then

$$S = \int_{\mathcal{D}} \mathcal{L}\boldsymbol{\varepsilon}.$$

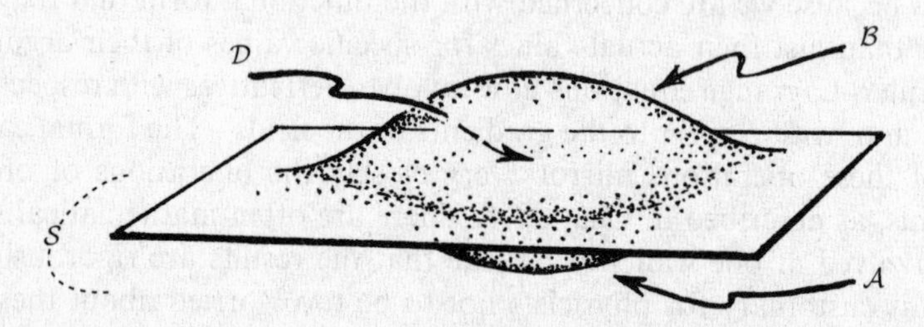

Fig. 20.9 Hamilton's principle for *field* Lagrangians. The two fixed end-points a, b, in $\mathcal{C}$, of Fig. 20.3, represent field configurations in two 3-dimensional spacetime regions $\mathcal{A}$, $\mathcal{B}$, respectively—forming a 'blister' which encloses the 4-region $\mathcal{D}$. We may take $\mathcal{A}$ and $\mathcal{B}$ to come together, terminating at a finite 2-surface $\mathcal{S}$ (not drawn in the figure), or we may regard '$\mathcal{S}$' to be out at infinity, perhaps out along a spacelike hypersurface throughout which $\mathcal{A}$ and $\mathcal{B}$ coincide beyond the region $\mathcal{D}$ (the case illustrated).

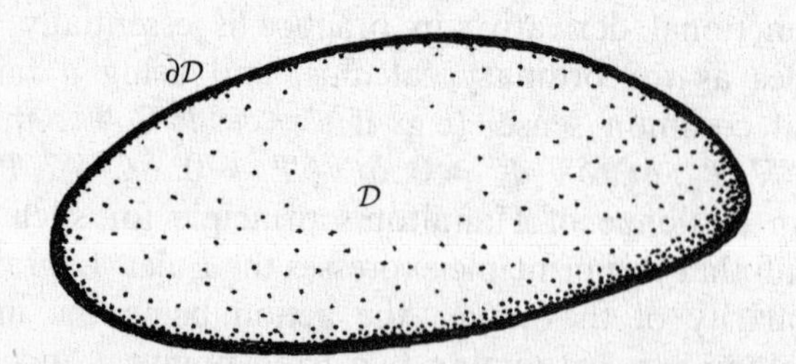

Fig. 20.10 If desired, we may regard the $\mathcal{A}$ and $\mathcal{B}$ of Fig. 20.9 to be joined together—but taken with opposite orientations (see Fig. 12.16)—so as to constitute the boundary $\partial\mathcal{D}$ of a (compact) spacetime 4-volume $\mathcal{D}$. Hamilton's principle (Fig. 20.3) expresses the stationarity of $\int_{\mathcal{D}} \mathcal{L}\boldsymbol{\varepsilon}$ for a given field configuration on the boundary $\partial\mathcal{D}$.

The field equations then arise from the assertion that the quantity S is stationary with respect to variations of all the variables (so it gives the analogue of a geodesic; see Fig. 20.3), which means that the variational derivative of $\mathcal{L}$ with respect to all the constituent fields and their derivatives has to vanish. This condition is written

$$\delta S = 0.$$

The quantity S is central to the path-integral approach to quantum field theory, which we shall come to in §26.6.

20.6 How Lagrangians drive modern theory

Lagrangian theory (as well as Hamiltonian theory) has a highly influential role in modern physics, there being many remarkable uses to which it can be put. For example, there is an important theorem, known as *Nöther's theorem*, which tells us that, if an ordinary Lagrangian possesses some continuous (smooth) symmetry, then there will be a conservation law associated with that symmetry. In particular, if there is invariance of the Lagrangian under *time translation* (i.e. independent of time), then there is a conserved *energy*; if it is invariant under some *spatial translation*, then a *momentum* is conserved. Furthermore, if there is invariance under *angular rotation* about some axis, then there is a conserved *angular momentum* about the same axis. For an isolated system in flat spacetime, these symmetries are to be expected. If we choose coordinates so that our given symmetry of the Lagrangian $\mathcal{L}$ is expressed in the fact that $\mathcal{L}$ is independent of some generalized 'position' coordinate q_r, then the conserved quantity will be the 'conjugate momentum' p_r to this coordinate q_r, as given by the above prescription (in §20.2): $p_r = \partial\mathcal{L}/\partial\dot{q}^r$. It is immediate from the Euler–Lagrange equations that this p_r is indeed constant in time.[20.17]

This procedure can be generalized to Lagrangian functionals of fields. For example, if there is a 'gauge invariance', then we expect to find a corresponding 'conserved charge' (e.g. the electric charge, in the case of the electromagnetic gauge invariance under $\Psi \mapsto e^{i\theta}\Psi$). Complicating issues begin to arise in such situations, however. For example, it is not at all a clear-cut matter to apply these ideas to obtain energy–momentum conservation in general relativity, and strictly speaking, the method does not work in this case. The apparent gravitational analogue of the gauge symmetry $\Psi \mapsto e^{i\theta}\Psi$ is 'invariance under general coordinate transformations' (which general relativity takes care of by having its equations

[20.17] Explain why.

written in terms of tensor operations), but the Nöther theorem does not work in this situation, giving something of the nature '0 = 0'. It appears that some quite different kinds of insight are needed for general relativity, despite the powerful understanding that can be gained from the use of Nöther's theorem in other circumstances, some of it carrying over in impressive ways to quantum theory, as will be indicated in §21.1. To exemplify the limitations of Nöther's theorem in the case of gravitational theory, it should be pointed out that a significant question mark still hangs over the issue of angular momentum in general relativity, even in the case of an asymptotically flat spacetimes.[5]

Einstein's theory can certainly be derived from a Lagrangian approach, as was first shown by the profound and versatile mathematician David Hilbert (1915). Hilbert's gravitational Lagrangian is basically the scalar curvature R, divided by the constant $-16\pi G$, but this needs to be made into a density (or a 4-form) by multiplying it by the natural 4-form $\boldsymbol{\varepsilon}$ of §20.5. This needs to be added to the matter Lagrangian $\mathcal{L}$, and we get, for the total action

$$S = \int_{\mathcal{D}} \left(\mathcal{L} - \frac{1}{16\pi G} R \right) \boldsymbol{\varepsilon}.$$

When Hilbert suggested this action, he was taken with a theory of matter which was very popular at the time, namely *Mie's theory*, and he stated his gravitational action principle only for the case when the matter Lagrangian was, in fact, the one that is appropriate to Mie's theory. He appears to have believed that his total Lagrangian gives us what we would now refer to as a 'theory of everything'. That was in 1915. Who remembers Mie's theory today?

Although Mie's theory involved a departure from Maxwell's theory, an appropriate Lagrangian for the *standard* Maxwell theory of the electromagnetic field had actually been known many years previously[6] namely

$$\mathcal{L}_{\mathrm{EM}} = -\tfrac{1}{4} F_{ab} F^{ab}.$$

But in order to make this work, we need to make sure that this Lagrangian is written out in terms of the *electromagnetic potential* A_a. When there are charged fields also, then additional terms expressing this interaction are needed, these also involving A_a. As an important point, the whole thing needs to be checked for gauge invariance. When gravity is incorporated as well, then there needs to be the 'gauge invariance' appropriate to gravity, namely coordinate invariance. This is usually handled by writing things appropriately in tensor form (or else according to other invariant prescriptions using basis frames or appropriate spinor formalisms).

In modern attempts at fundamental physics, when some suggested new theory is put forward, it is almost invariably given in the form of some Lagrangian functional. This has many advantages, such as the fact that there is a greater chance (but not an absolute certainty) of the resulting theory having required consistency and invariance properties, and that some form of 'Newton's third law' is implicit (in the sense that if two fields interact then the interaction is mutual: if one acts upon the other then the other acts equally back on the one). Moreover Lagrangians have the pleasant property that, if a new field is introduced, then its contributions can usually simply be added to the Lagrangian that one had before, with any required interaction terms added also. More importantly, perhaps, there is a direct route to the formation of a quantum theory, via the path-integral approach that I have alluded to above, that we shall come to in §26.6.

However, I must confess my unease with this as a *fundamental* approach. I have difficulties in formulating my unease, but it has something to do with the generality of the Lagrangian approach, so that little guidance may be provided towards finding the correct theories. Also the choice of Lagrangian is often not unique, and sometimes rather contrived—even to the extent of undisguised complication. There tends to be a remoteness from actual physical 'hands-on' understanding, particularly in the case of Lagrangians for fields. Even the Lagrangian for free Maxwell theory, $\frac{1}{4}F_{ab}F^{ab}$, has no obvious physical significance (this quantity being $\frac{1}{8}$ of the difference between the squared lengths of the electric and magnetic field vectors, in 3-dimensional terms.[20.18] Moreover, the 'Maxwell Lagrangian' does not work as a Lagrangian unless it is expressed in terms of a potential, although the actual value of the potential A_a is not a directly observable quantity. In the case of gravity (unlike the case of electromagnetism), the Lagrangian for free Einstein theory *vanishes identically* when the field equation is satisfied (since $R_{ab} - \frac{1}{2}Rg_{ab} = 0$ implies $R = 0$). Again, R does not work as a Lagrangian unless it is expressed in terms of quantities (normally the metric components in some coordinate system) that are again not invariantly meaningful. In most situations, the Lagrangian density does not itself seem to have clear physical meaning; moreover, there tend to be many different Lagrangians leading to the same field equations.

Lagrangians for fields are undoubtedly extremely useful as mathematical devices, and they enable us to write down large numbers of suggestions for physical theories. But I remain uneasy about relying upon them too strongly in our searches for improved fundamental physical theories. This unease has a relevance also to the issues of quantum field theory that we shall come to in §26.6, but this is enough for now.

[20.18] Show this.

Notes

Section 20.1

20.1. More general types of Lagrangian (for non-Newtonian systems) can involve higher derivatives and are defined on what are called 'jet bundles' of C, but we need not concern ourselves with these here.

20.2. I am simplifying the general discussion of Lagrangians, in my account here, by assuming that our system is what is called *holonomic*. With a non-holonomic system, there are not enough velocity coordinates available, in relation to the generalized positions. A good example of a non-holonomic system is the rolling of a hoop on a horizontal plane, where the hoop is constrained not to slip, so that its contact point can move only in the direction of the hoop's tangent, by a rolling motion. Two coordinates are needed for the location of this contact point, but only one is available for its velocity.

One may consider that for systems treated at the *fundamental* level of description, such non-holonomicity does not occur. In the case of our hoop, the rolling constraint is an idealization in which the possibility of slipping is denied. Once a small amount of slipping is allowed for, the system becomes a holonomic one.

20.3. I am taking the case of a 'time-independent Lagrangian' here, for simplicity of description. But we can easily bring in a time-dependence of external forces, simply by including another 'generalized coordinate' $q^0 = t$ and a formal quantity $\dot{q}^0$ which ultimately takes the value 1.

Section 20.5

20.4. Another way of specifying $\boldsymbol{\varepsilon}$ is to say that the component ε_{0123} of $\boldsymbol{\varepsilon}$, in a local right-handed orthonormal frame, satisfies $\varepsilon_{0123}=1$ (§19.2). The $[^0_4]$-tensor $\boldsymbol{\varepsilon}$ is fixed up to sign by the metric by $\varepsilon_{abcd}\varepsilon_{pqrs}g^{ap}g^{bq}g^{cr}g^{ds} = -24$, the choice of sign for $\boldsymbol{\varepsilon}$ sign providing the *orientation* of the spacetime volume.[20.19]

Section 20.6

20.5. See Penrose (1982); Penrose and Rindler (1986); Winicour (1980); Rizzi (1998).

20.6. See Pais (1986), p. 342, and references 46, 47, 48 on p. 357.

[20.19] Show that this prescription is equivalent to that given in the main text.

21
The quantum particle

21.1 Non-commuting variables

It is probable that most physicists would regard the changes in our picture of the world that quantum mechanics has wrought as being far more revolutionary even than the extraordinary curved spacetime of Einstein's general relativity. In fact, what quantum theory actually tells us to believe about 'reality' at the submicroscopic levels of atoms or of fundamental particles is, as we shall be seeing in this chapter and in the next two, so greatly removed from our ordinary classical pictures that we may choose simply to give up on quantum-level 'pictures' altogether. Indeed many physicists appear even to doubt the very existence of a true 'reality' at quantum scales and, instead, rely merely upon the quantum-mechanical mathematical formalism to obtain answers. (In Chapter 29, I shall return more fully to the controversial issue of 'quantum reality'.)

Yet, despite all this, it is very remarkable how much of the Lagrangian/Hamiltonian collection of procedures of Chapter 20—that comprehensive but entirely classical scheme that grew out of 17th century Newtonian mechanics—provides the essential background to quantum-mechanical theory. Of course, there had to be changes in the mathematical formalism. Otherwise the new theory would be just a copy of the old. But it is as though the formalism that grew out of Newton's scheme was already waiting for quantum mechanics to come, with parts of its machinery of just the right shape and size, so that the new quantum ingredients could simply be inserted in their place.

The key mathematical property that allows this to happen is an apparent 'curiosity' that had already been noted towards the end of the 19th century by the highly original electrical engineer and mathematical physicist Oliver Heaviside (1850–1925), whom we remember from §6.1. Heaviside's observation was that differential operators can often be treated in just the same way as can ordinary numbers, a fact that is often useful in solving certain types of differential equation. Let us look at an example. Consider the differential equation[1]

$$y + \frac{d^2y}{dx^2} = x^5,$$

for example (and see §6.3 for the meaning of the symbols). We wish to find some particular function $y = y(x)$ which satisfies this relation. Heaviside's method was to treat d/dx as though this operator were an ordinary number. To make this look more 'plausible', let us denote the operator by the single letter D:

$$D = \frac{d}{dx}.$$

The entity represented by 'D^2' is then the repeated differentiation $d^2/dx^2 = (d/dx)^2$, which is a *second* derivative operator; that represented by 'D^3' is the third derivative d^3/dx^3, etc. Then our equation becomes $y + D^2y = x^5$, which we can express as

$$(1 + D^2)y = x^5.$$

We can 'solve' it by formally 'dividing through by $1 + D^2$', writing the answer as $y = (1+D^2)^{-1}x^5$. Expanding $(1 + D^2)^{-1}$ out as a 'power series in D', we find

$$y = (1 - D^2 + D^4 - D^6 + \cdots)x^5.$$

(Recall that we already considered this series in §4.3, with x in place of D.) Noting (§6.5) that $Dx^5 = 5x^4$, $D^2x^5 = 20x^3$, $D^3x^5 = 60x^2$, $D^4x^5 = 120x$, $D^5x^5 = 120$, $D^6x^5 = 0$, etc., we find the (correct!) particular solution[21.1], [21.2], [21.3]

$$y = x^5 - 20x^3 + 120x.$$

With careful attention to the appropriate rules, this kind of formal procedure can be made perfectly rigorous—although Heaviside encountered a great deal of opposition to the use of it at first!

Although the quantity $D(= d/dx)$ can be treated (if with due care) in an algebraic way like an ordinary number, we must be cautious when we have Ds and xs mixed up together, because they do not commute. We have to think of 'x' and 'D' as acting on some invisible function on the right, say $\Psi(x)$. The operator x simply multiplies what lies to the right of it by x, whereas D differentiates what lies to the right of it with respect to x. Then we find that we have the *commutation* relation

---

[21.1] Show that $(1 + D^2)\cos x = 0$ and $(1 + D^2)\sin x = 0$ (referring to formulae in §6.5, if you need them).

[21.2] Taking note of Exercise [21.1], find the *general* solution of $(1 + D^2)y = x^5$, providing a proof that your solution is, in fact, the most general.

[21.3] See if you can explain why the procedure given in the text misses most of the solutions given in Excerise [21. 2]. Can you suggest a modified general procedure which finds them all? *Hint*: To what extent does '$1 - D^2 + D^4 - D^6 + \ldots$' really satisfy the requirements for an inverse to $1 + D^2$? Try acting on $(1 + D^2)\cos x$ with this infinite expression.

$$Dx - xD = 1.$$

Why is this? Recall the 'Leibniz law' property of §6.5, which tells us that $D(x\psi) = (D(x))\psi + xD(\psi)$, i.e. $D(x\psi) - xD(\psi) = (D(x))\psi$. This is simply the relation $(Dx - xD)\psi = 1\psi$, where we bear in mind that $D(x) = 1$ (i.e. D applied directly to x is 1), which is the above displayed relation applied to an arbitrary $\psi = \psi(x)$ on the right.

Let us now extend this to many variables $x^1, \ldots, x^N$, and to the corresponding operators $D_1 = \partial/\partial x^1, \ldots, D_N = \partial/\partial x^N$ (now partial derivative operators—and remember that x^N is just the Nth coordinate, not N copies of x multiplied together), where the 'invisible' function on the right-hand side is now some function of all these variables: $\psi = \psi(x^1, \ldots, x^N)$. We obtain the commutation relations

$$D_b x^a - x^a D_b = \delta_b^a.$$

(Recall the Kronecker delta δ_b^a of §13.3; the above expression contains both the previous commutator, when $a = b$, and the fact that the x and D commute[21.4] when $a \neq b$.) We could suppose that the coordinates x^a are ordinary spatial or spacetime coordinates in flat space, but we may imagine that they could also be something more general, such as the generalized coordinates q^a of the Lagrangian or Hamiltonian formalisms. There are some profound difficulties with taking this generality too far, however. Thus, for the moment at least, it will be better to imagine that we are dealing with some *flat* N-space $\mathbb{E}^N$ (of not necessarily just 3 or 4 dimensions). The operators $D_1, \ldots, D_N$ then describe infinitesimal translations of $\mathbb{E}^N$ in the directions of each of the axes (Fig. 21.1), each of these expressing an independent symmetry of the affine space $\mathbb{E}^N$.

We recall, from Nöther's theorem (in §20.6), that there is a close association between such symmetries of the space and *momentum conservation*: if a Lagrangian is unchanged by spatial translation in some direction,

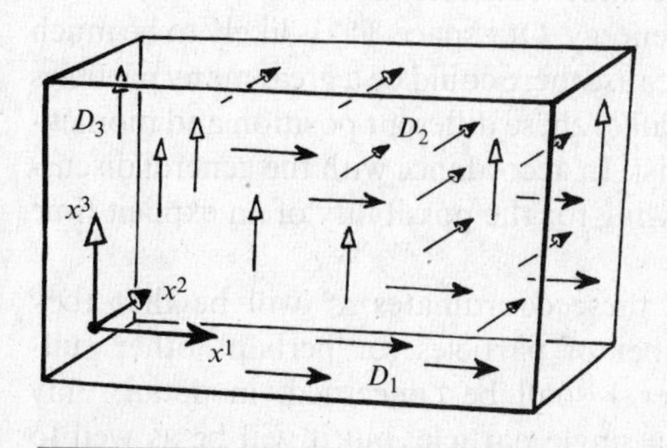

Fig. 21.1 In (affine) Euclidean N-space $\mathbb{E}^N$, there are N independent translational symmetries generated by the operators (vector fields) $D_1 = \partial/\partial x^1$, $D_2 = \partial/\partial x^2$, ..., $D_N = \partial/\partial x^N$, satisfying commutation relations $D_b x^a - x^a D_b = \delta_b^a$ with the respective Cartesian coordinates $x^1, x^2, \ldots, x^N$. (The case $N = 3$ is illustrated.)

[21.4] Why?

then the momentum in that direction that it provides is conserved. This is an elegant and important fact, and it is mathematically perfectly comprehensible. Quantum mechanics does something that looks quite a bit like this, but it is not nearly so mathematically comprehensible. In fact, it seems to be mathematically completely crazy! Yet, there is an undoubted mathematical elegance in this strange quantum-mechanical procedure. For in quantum mechanics, not only is there a conserved momentum associated with any such symmetry, but the momentum itself is actually *identified* with the differential operator that generates that particular symmetry!

21.2 Quantum Hamiltonians

How can a momentum actually be identified with a differential operator? This indeed sounds crazy! To be more correct, there is a factor of $\hbar$ (Dirac's version of Planck's constant, namely $h/2\pi$, where h is the original Planck's constant; see below), and also of the imaginary unit i, to be incorporated. Thus, we make the absurd-looking definition $p_a = \mathrm{i}\hbar D_a$, that is

$$p_a = \mathrm{i}\hbar \frac{\partial}{\partial x^a},$$

for the momentum associated with x^a. Going along with this, we are led to a commutation law called a *canonical commutation rule* relating position and momentum

$$p_b x^a - x^a p_b = \mathrm{i}\hbar\ \delta_b^a.$$

What are we to do with this crazy-looking operator/momentum? The role of this 'quantum-mechanical momentum', $\mathrm{i}\hbar\partial/\partial x^a$, is that it is to be slotted into the classical Hamiltonian function $\mathcal{H}(p_1, \ldots, p_N; x^a, \ldots, x^N)$, just where the old classical momentum p_a used to be. This is the key to the procedure known as (canonical) *quantization*. We are not worrying about relativity just yet, so the 'momenta' under consideration above are indeed to be the spatial momenta,[2] and not energy. Our space $\mathbb{E}^N$ is likely to be much larger than just 3-dimensional, because there could be a great many particles or other structures involved, and all of these different position and momentum components are to be in the list. In accordance with the general discussion of Chapter 20, I am not allowing for the possibility of an explicit time dependence in the Hamiltonian.[3]

The normal interpretation of these coordinates x^a will be that they provide the positions of a number of particles (or perhaps other suitable parameters). In this chapter I shall be concerned, in detail, only with the quantum mechanics of a single particle, but it will be as well to have the general formalism ready for when more complicated many-particle systems are considered in Chapter 23. Particularly for a single particle, there turns out to be some evident relativistic symmetry between its time component x^0 and its three spatial components x^1, x^2, x^3. We shall be seeing in a moment how this plays an important role in defining the actual time evolution of quantum mechanics. Nevertheless, as the procedures of (canonical) 'quantization' stand, particularly when many particles are involved, they provide a non-relativistic procedure, where the treatments of the spatial and temporal aspects of physics are very different.

Let us have a look at a simple example of a quantum Hamiltonian, so as to see how this crazy idea is to proceed. We can consider the case of a single Newtonian particle of mass m, moving in some external field given by a potential energy function V which can depend on position: $V = V(x, y, z)$. We have seen the *classical* Hamiltonian already, in §20.2, and we recall that this is $\mathcal{H} = (p_x^2 + p_y^2 + p_z^2)/2m + V(x, y, z)$, where p_x, p_y, and p_z are the spatial momenta in the directions of the Cartesian x, y, and z axes. The quantum (canonically quantized) Hamiltonian is therefore

$$\mathcal{H} = \frac{p_x^2 + p_y^2 + p_z^2}{2m} + V(x, y, z) = -\frac{\hbar^2}{2m}\nabla^2 + V(x, y, z),$$

where $\nabla^2 = (\partial/\partial x)^2 + (\partial/\partial y)^2 + (\partial/\partial_z)^2$ (which means $\partial^2/\partial x^2 + \partial^2/\partial y^2 + \partial^2/\partial z^2$) is the Laplacian (as considered earlier, in §10.5, but now in the 3-dimensional case).

In this example, everything has moved along smoothly (but to *where*—we shall need to wait for the next section!). In general, however, the replacing of classical momenta by quantum ones in the Hamiltonian may not be an unambiguous procedure, mainly because of non-commutation between the quantum-mechanical p and its corresponding x. For example, if a *product* term of the form px appears in the classical Hamiltonian, it is not clear whether in the corresponding quantum Hamiltonian it should appear as px, or as xp, or perhaps as $\frac{1}{2}(px + xp)$ or as any one of an infinite number of other possibilities. This kind of ambiguity is referred to as the *factor-ordering problem*. In many practical circumstances, this ambiguity may not be very serious, as there often turns out to be some 'obvious' choice. The choice may be governed by some overriding guiding principle, such as a symmetry or invariance requirement, or perhaps by some compelling physical or mathematical instinctive or aesthetic demands. Or it may sometimes be that different alternatives nevertheless result in equivalent quantum theories. Yet, the fact that such ambiguities do exist, in general, tells us that the process of 'quantizing' some given classical theory may well sometimes involve serious matters of choice.

There is a related issue which concerns the 'generality' of the choice of coordinates $x^1, \ldots, x^N$. Recall that in §§20.1,2 we were allowed complete freedom in our choice of generalized coordinates $q^1, \ldots, q^N$ on the configuration space $\mathcal{C}$. We may ask: is this complete freedom still allowed when we pass to the quantum theory? In fact, the answer is 'no', if we are expecting that the classical conjugate momentum p_a, of each q^a, is to be 'quantized' simply as $-i\hbar\partial/\partial q^a$. The issue is a very delicate one, and it takes us into the fascinating area known as *geometric quantization*.[4] It has a particular importance in relation to general relativity, whether one is proposing to 'quantize the gravitational field' or merely to discuss quantum fields in a curved spacetime background. (I shall return to the matter of quantum theory in curved backgrounds in §30.4.) There are, however many standard situations in which we can get away with coordinates that are more general than just flat ones, so long as we are appropriately careful. In particular, angular coordinates are useful to use, and the conjugate momenta are then angular momenta. We shall be concerned with angular momentum later (§22.8; and §22.12 for the relativistic case).

21.3 Schrödinger's equation

Let us ignore these issues of factor ordering and generalized coordinates etc., at least for the moment, and suppose that we have a quantum-mechanical Hamiltonian that we are satisfied with. What use are we to put it to? The answer is that it plays a crucial role in that equation, fundamental to our understanding of how a quantum system evolves with time, known as the *Schrödinger equation*. In fact, the form of this equation is effectively already determined by the rules set up above. How does this work? In the first place, we must bring into view the 'invisible' function ψ that has been hiding unseen at the extreme right-hand side of all our commutator relations. The Hamiltonian is now an operator, after all, because of all those $\partial/\partial x$s, and it needs something (potentially, at least) to operate on, at its extreme right-hand end. Since Schrödinger's equation, being a time-evolution equation, is now going to make ψ vary with time, we need to write ψ as a function of t, as well as of all our spatial x^as:

$$\psi = \psi(x^1, \ldots, x^N; t).$$

But it cannot depend on the p_as, because these quantities are not now 'independent variables', but are to be interpreted as differentiations with respect to the x^as. Such a function ψ is called a *wavefunction*. It provides the *quantum state* of the system. We shall be looking at the physical interpretation of wavefunctions in due course.

How does differention with respect to t fit into this? Here is where the remarkable time evolution of Schrödinger comes in. Recall from §20.2 that the (classical time-independent) Hamiltonian represents the total energy of the system. We also take note of the fact—as hinted above (in §21.2)—that if our quantum theory is ever to fit in with the requirements of relativity, the quantum rule $p_a = i\hbar\partial/\partial x^a$ (for a single particle) ought to extend to the $a = 0$ component as well as to the three spatial components (see §18.7). Accordingly, in the 'quantization' procedure, energy should get replaced by differentiation with respect to time ($E = i\hbar\partial/\partial t$). What the Schrödinger equation expresses is precisely this 'quantum role' of the total-energy interpretation of the Hamiltonian:

$$i\hbar\frac{\partial\psi}{\partial t} = \mathcal{H}\psi,$$

where

$$\mathcal{H} = \mathcal{H}\left(i\hbar\frac{\partial}{\partial x^1}, \ldots, i\hbar\frac{\partial}{\partial x^N}; x^1, \ldots, x^N\right)$$

As a simple example, using the particular case of a quantum Hamiltonian given in §21.2 above, we can now write down the Schrödinger equation for a single particle of mass m, moving in an external field whose energy contribution is $V = V(x, y, z)$:[21.5],[21.6]

$$i\hbar\frac{\partial\psi}{\partial t} = -\frac{\hbar^2}{2m}\nabla^2\psi + V\psi.$$

Of course, all this replacing of momentum and energy by differential operators looks like so much mathematical mumbo-jumbo, and we may well ask what such amusements have to do with the momentum imparted by a boxer's fist or by a golfer's swing. We may well ask! But according to quantum mechanics, it has everything to do with it. The key to momentum is that it is conserved, and the effect of a blow on its recipient is simply a result of the inevitability of that conservation. The momentum must go somewhere; it cannot just disappear, because it is conserved. The same applies to energy.

[21.5] Solve this Schrödinger equation explicitly in the case of a particle of mass m in a constant Newtonian gravitational field: $V = mgz$. (Here z is the height above the Earth's surface and g is the downward gravitational acceleration.)

[21.6] By transforming to the freely-falling frame with coordinates $X = x, Y = y, Z = z - \frac{1}{2}t^2g, T = t$, show that the Schrödinger equation of Exercise [21.5] transforms to one without a gravitational field, with wavefunction $\Psi = e^{i(\frac{1}{6}mt^3g^2+mtzg)}\psi$. What does this tell us about Einstein's principle of equivalence (see §17.4), as applied to quantum systems? (Take note of §21.9.)

But surely we already have momentum conservation and energy conservation in *classical* Hamiltonian theory, the skeptical reader may well complain, so why the need for the curious identification between a physical quantity and an effectively disembodied differential operator, whatever good *that* does?[21.7] To attempt to answer this, and to make the proposal more plausible, one needs to appeal to experiment. (Such a ridiculous-looking proposal could hardly carry much weight otherwise!) Detailed experimental issues are not things that I can go into here, but the essential point that emerges from a vast amount of experimental evidence is that there is a direct association between frequency and energy, and a corresponding association between wave number (the reciprocal of wavelength) and momentum; moreover, these associations appear to be universal, in all phenomena. We shall be seeing the relevance of this to '$p_a = i\hbar\partial/\partial x^a$' in §21.5. In the meantime, let us have a look at some of the experimental reasons for believing that energy and momentum indeed have this kind of 'wavy' association.

21.4 Quantum theory's experimental background

Perhaps one of the most direct manifestations of this sort of association occurs with crystaline materials. In such structures we have a spatial periodicity in the crystal's atomic arrangements. As was first shown in a famous experiment by C. J. Davisson and L. H. Germer (1927), if electrons, having a suitable choice of initial 3-momentum, are fired at such a material, then they are deflected (or reflected) from it through certain very special angles. These directions and angles are found to depend specifically on the incoming and outgoing 3-momenta in relation to the periodic nature of the crystalline lattice. A manifest implication of these experimental results is that there is a precise relationship of inverse proportionality between the electrons' 3-momenta and a distance of periodic displacement; see Fig. 21.2. The same holds for other types of particle. The upshot is that a particle of momentum p seems to be a periodic thing, like a wave, where there is a universal relationship between the wavelength λ and the magnitude p of its momentum, according to the reciprocal formula (involving Planck's constant $h = 2\pi\hbar$)

$$\lambda = hp^{-1} = \frac{2\pi\hbar}{p}.$$

[21.7] Show that, if the quantum Hamiltonian $\mathcal{H}$ has a translation invariance, say being independent of the position variable x^3, then the corresponding momentum p_3 is conserved in the sense that the operator p_3 commutes with the time evolution $\partial/\partial t$. Explain, in the light of the interpretations given later, why this commutation implies conservation.

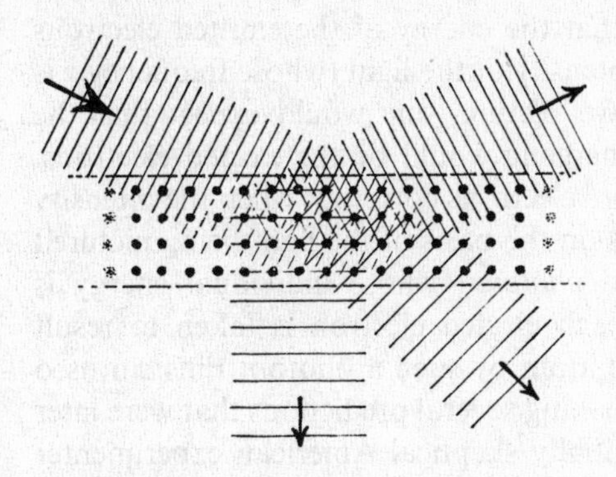

Fig. 21.2 The Davisson–Germer experiment. A beam of electrons of 3-momentum **p** encounters a material of periodic crystalline structure. Scattering or reflection occurs when the atomic pattern matches with that of the electrons, these being regarded as waves with a wavelength λ related to the magnitude of the momentum p according to $\lambda = h/p$, where h is Planck's constant.

The wavelength λ, associated with a particle of momentum p, is called its *de Broglie* wavelength, after the highly insightful French aristocrat and physicist Prince Louis de Broglie, who first suggested, in 1923, that all material particles have a wavelike nature with a wavelength given by the above formula. Moreover, in accordance with the requirements of relativity (see §18.7), the particle should also have a frequency ν, given in terms of its energy E by the Planck formula.

$$E = h\nu = 2\pi\hbar\nu$$

that we shall come to shortly.[21.8] In its own rest-frame the particle's energy is Einstein's $E = \mu c^2$, where μ is its rest-mass, so it is intrinsically associated with the frequency $\mu c^2/2\pi\hbar$, that is, $\mu c^2/h$.

These kinds of consideration led to the conclusion that an ordinary particle displays wavelike behaviour, this having a universal relationship to the particle's rest-mass as determined by the Planck and de Broglie formulae. But, in the previous two decades, a converse to this had already been established, demonstrating that entities previously thought of as purely wavelike—basically Maxwell's oscillating electric and magnetic fields as the constituents of light (recall §19.2)—had also to be viewed as having a *particulate* nature, again consistent with the Planck and de Broglie formulae. The most convincing evidence for this was in the *photoelectric effect*, first observed by Heinrich Hertz in 1887 and whose most puzzling aspects, as demonstrated by Philipp Lenard in 1902, were magnificently explained by Einstein in 1905 using a particle picture of light. (This is what earned Einstein the 1921 Nobel Prize, not relativity theory!) The photoelectric effect occurs when light of suitably high frequency ν shines on an appropriate metallic material, causing electrons to be emitted.

[21.8] See if you can see why the requirements of special relativity enable Planck's $E = h\nu$ to be deduced from de Broglie's $p = h\lambda^{-1}$. (*Hint*: You may assume that the hyperplanes in $\mathbb{M}$ along which the wave takes a constant value are Lorentz-orthogonal to the particle's 4-velocity.)

The puzzle arises from the fact that the energy of the emitted electrons does not at all depend upon the intensity of the light (whose frequency ν is taken to be constant). On a wave picture, one would expect that the greater the intensity, the more energetic would be the ejected electrons. This does not happen (though *more* electrons come out when the intensity is greater). Einstein explained this on the basis of the light being pictured as incident particles—now called *photons*—whose individual energy is given by Planck's $E = h\nu$, and each ejected electron is taken to result from the impact on an individual atom by such a photon. Einstein used Planck's formula to great effect, making several predictions that were later confirmed, particularly by the initially skeptical American experimenter Robert Millikan in the years up to 1916.

In fact, this quantum-mechanical particulate nature of light had already begun to reveal itself somewhat earlier. This was in 1900, when Max Planck launched the quantum revolution. He did this by providing a very remarkable analysis of *black-body radiation*, which concerned electromagnetic radiation in equilibrium with its 'black'[5] material surroundings, all kept at some specific temperature T (see Fig. 21.3a). He obtained the (correct) formula, plotted in Fig. 21.3b,

$$\frac{2h\nu^3}{(e^{h\nu/kT} - 1)^{-1}}$$

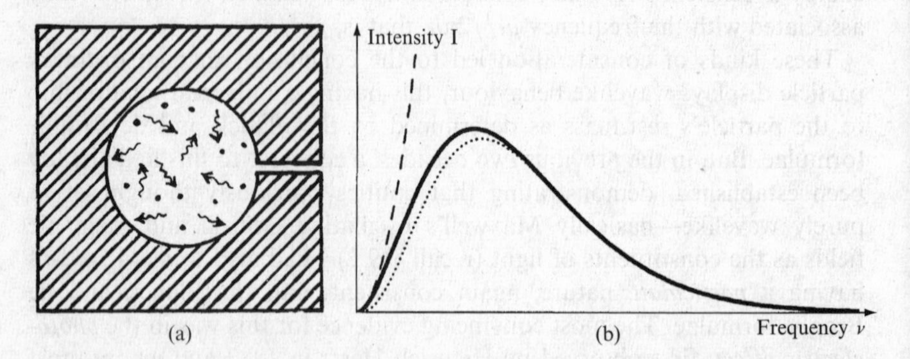

Fig. 21.3 Blackbody radiation. (a) The 'black' cavity ensures that the contained radiation is in thermal equilibrium, at temperature T, with its heated surroundings. (b) For a given T, the intensity I at each frequency ν is found to be a specific function of ν. The continuous curve is the observed one, given by Planck's famous formula $I = 2h\nu^3/(e^{h\nu/kT} - 1)^{-1}$ (where h and k are Planck's and Boltzmann's constants, respectively). The broken curve is that of Rayleigh–Jeans, $I = 2kT\nu^2$, in which the radiation is treated as classical waves, approximating Planck's formula for small ν, but diverging for large ν. The dotted curve depicts Wien's law $I = 2h\nu^3 e^{-h\nu/kT}$ in which the radiation is treated as classical particles.

for the specific *intensity* I, as a function of the frequency ν, where k is Boltzmann's constant (related to the units in which temperature is measured, cf. §27.3).

Planck's formula turns out to fit the observations perfectly. Prior to Planck, the nature of the black-body spectrum had been a mystery. The entirely wave picture of electromagnetic radiation had led to the paradoxical Rayleigh–Jeans formula $I = 2kT\nu^2$, accurate for small ν, but for which the intensity would diverge to infinity for large ν. An apparent improvement was Wien's proposal $I = 2h\nu^3 e^{-h\nu/kT}$, accurate for large ν, which could be given a 'justification' on the basis of treating the radiation as though it were a bath of classical particles. The quantity h, as it appeared in Planck's own formula, was postulated, by him to be a *new* fundamental constant of Nature (now called *Planck's constant*), whose extremely tiny value is found to be about 6.62×10^{-34} Joule-seconds. To obtain his formula, Planck found himself forced to the view that electromagnetic oscillations could be absorbed or emitted only in bundles of a specific energy E, directly related to the frequency ν of oscillation according to the above relation

$$E = h\nu,$$

where he also adopted a 'crazy' statistical counting that amazingly foreshadowed the (quantum-mechanically correct) Bose–Einstein statistics that we shall come to in §23.7.

Here the physical puzzle was the other way around from that of the electrons encountering a crystal, since electromagnetic effects were thought of, at that time, as *being* just waves, yet now they seemed to be having properties of particles! Using Dirac's form of Planck's constant, we find $E = 2\pi\hbar\nu$, so the time-period of the oscillation ν^{-1} satisfies the corresponding formula to our earlier one ($\lambda = 2\pi\hbar/p$) relating wavelength to momentum, namely $\nu^{-1} = 2\pi\hbar/E$. Nowadays, (following the further insights of Einstein, Bose, and others) we understand Planck's relation to refer not just to 'oscillations of electromagnetic field', but to *actual* 'particles'—the quanta of Maxwell's electromagnetism that we call photons—although it took many years for Einstein's original lonely insights to become accepted. Among the further confirmations, following these successes with the photoelectric effect, was the crucial experiment of Arthur Compton (1923), which demonstrated that photons, in their encounters with charged particles, indeed behaved just like massless particles, in accordance of the relativistic dynamics of §18.7 (see §25.4, Fig. 25.10). Accordingly, the energy and momentum are both reciprocally related to a period (of time for energy and of space for momentum), the period being always scaled in terms of $2\pi\hbar$.

One of the most convincing (and best known) reasons for our having to face up to the fact that particles can behave as waves and waves as particles is the *two-slit experiment*.[6] Here we have a source of particles and a detector screen, where there is a barrier with a pair of narrowly separated parallel slits in it, situated between source and screen; see Fig. 21.4a. We suppose that one particle at a time is emitted, aimed at the screen. If we start with one slit open and the other closed, then a haphazard pattern of dots will appear at the screen, forming one at a time as individual particles from the source hit it. The intensity of the pattern (in the sense of the greatest density of dots) is most extreme in a central strip close to the plane connecting source to slit, as is to be expected, and it falls off uniformly in both directions from this central strip (Fig. 21.4b). This pattern is effectively the same if the experiment is repeated with the other slit being the open one (Fig. 21.4c). No puzzle here. But if the experiment is run once more when both slits are now open, then something extraordinary happens; see Fig. 21.4d. The particles still make dots on the screen one at a time, but now there is a wavy *interference* pattern of parallel bands of intensity, where we even find that there are regions on the screen that are never reached by particles from the source, despite the fact that when just one or the other of the slits was open, then particles could reach those regions perfectly happily! Although the spots appear at the screen one at a time at localized positions, and although each occurrence of a particle meeting the screen can be identified with a particular particle-emission event at the source, the behaviour of the particle *between* source and screen, including its ambiguous encounter with the two slits in the barrier, is like a wave, where the wave/particle feels out both slits during this encounter. Moreover—and this is the matter of particular relevance

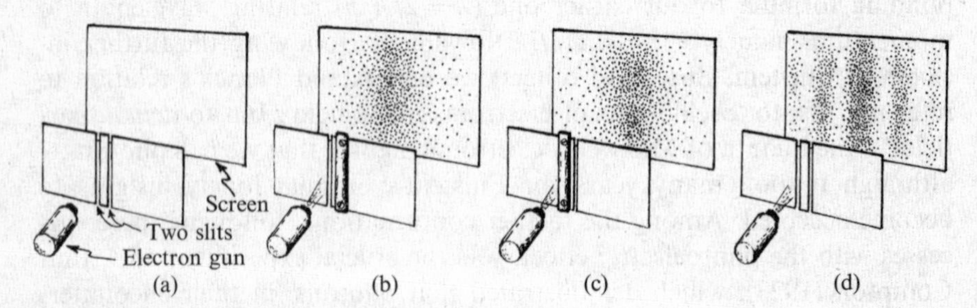

Fig. 21.4 (a) Arrangement for the two-slit experiment. One electron is emitted at a time, aimed at the screen through the pair of slits. (b) Pattern on the screen when the right-hand slit is covered. (c) The same, when the left-hand slit is covered. (d) Interference occurs when both slits are open. Some regions on the screen cannot now be reached despite the fact that they can be with just one or the other slit open.

for our most immediate purposes—the spacing of the bands on the screen tells us what the wavelength of our wave/particle must be, and this wavelength λ is indeed given in terms of the particles' momentum p by just the same formula as before, namely $\lambda = 2\pi\hbar/p$.

21.5 Understanding wave–particle duality

All this is as may be, so the hard-headed skeptic might well interject, but it still does not force us into making this absurd-looking identification between energy–momentum and an operator! Indeed not, but we should not turn down a *miracle* when it is presented to us! What miracle is this? The miracle is the fact that these seemingly gross absurdities of experimental fact—that waves are particles and that particles are waves—can all be accommodated within a beautiful mathematical formalism, a formalism in which momentum is indeed identified with 'differentiation with respect to position', and energy with 'differentiation with respect to time'.

How does this formalism help us to understand this mysterious wave–particle duality? To describe our wave/particle, we shall require a mathematical entity that can provide us with a clearly defined particle's 4-momentum P_a, while possessing a spatial and temporal wavelike periodicity of the amount prescribed above. (I am now using a *capital* letter P, because in this particular case I am referring to a specific 'classical' value of 4-momentum that our particle might turn out to have. We are still thinking of the 'quantum 4-momentum' as being described by a differential operator.) A natural such wavelike mathematical entity would be a wavefunction with the particular spatio-temporal dependence of the form (see §5.3)

$$\psi(x^a) = \mathrm{e}^{-\mathrm{i}P_a x^a/\hbar}$$

(a plane wave). This quantity becomes itself again if we increase $P_a x^a$ by $2\pi\hbar$ (since that adds $-2\pi\mathrm{i}$ to the exponent, so the expression is multiplied by $\mathrm{e}^{-2\pi\mathrm{i}} = 1$). It therefore has a temporal periodicity of $2\pi\hbar/P_0$ and a spatial periodicity of $2\pi\hbar/P_1$ in the x_1-direction, and similarly for the other three spatial directions. This agrees exactly with the requirements set out above.

Now, what is special about this particular entity? It is what is called an *eigenfunction* of our quantum momentum operator

$$p_a = \mathrm{i}\hbar\frac{\partial}{\partial x^a}.$$

This means that if we apply this operator to the above $\psi(x^a)$ we simply get a constant multiple of $\psi(x^a)$ again (§6.5):

$$\mathrm{i}\hbar\frac{\partial}{\partial x^a}\psi(x^b) = \mathrm{i}\hbar\frac{\partial}{\partial x^a}\mathrm{e}^{-\mathrm{i}P_b x^b/\hbar} = P_a \mathrm{e}^{-\mathrm{i}P_b x^b/\hbar} = P_a\psi(x^b).$$

Notice that this constant multiplier is in fact just the (classical) 4-momentum P_a that we require our entity to have. Thus, when $\psi(x^a)$ has the right form, namely that given above, our mysterious quantum momentum $p_a = \mathrm{i}\hbar\partial/\partial x^a$ converts itself into a straightforward classical momentum P_a, when applied to this ψ:

$$p_a\psi = P_a\psi,$$

but it does not do this for other kinds of state. We say that the above ψ has a *definite value* for the 4-momentum, and we call it a *momentum state*. We are to think of a particle in free flight, which happens to have a definite classically identifiable momentum P_a, as being mathematically described by this particular wavefunction ψ, which is the eigenfunction of the quantum operator p_a, having eigenvalue P_a. The only wavefunctions that have a definite classical momentum value are, indeed, those which are eigenfunctions of the quantum momentum operator.

Recall that, in §13.5, the notion of an *eigenvector* of a linear operator $\boldsymbol{T}$ was introduced, this being a vector $\boldsymbol{v}$ for which $\boldsymbol{T}\boldsymbol{v} = \lambda\boldsymbol{v}$, for some scalar quantity λ called its *eigenvalue*. This is exactly the situation that we have here, with $\mathrm{i}\hbar\partial/\partial x^a$ standing for the operator $\boldsymbol{T}$, and with P_a standing for λ (taking each particular value for a in turn), except that in §13.5 we were basically referring to *finite*-dimensional vector spaces and their linear transformations. Here, we are concerned with the vector space $\mathbf{W}$ of possible $\psi(x^a)$s, and that will be an *infinite*-dimensional vector space. (It is a vector space because we can add together functions of x^a and we can multiply functions of x^a by numbers, and in each case we just get new functions of x^a. It is infinite-dimensional because, for example, all the functions of the particular kind that we have just been considering above, for infinitely many different choices of P_a, are linearly independent.[21.9]

Eigenfunctions (or *eigenstates*, as they are frequently called in quantum mechanics), play a key role in the quantum formalism. In the language of quantum mechanics, various operators (like the $p_a = \mathrm{i}\hbar\partial/\partial x^a$ which we have just been considering, and others, such as position or angular momentum, which we consider later) are called *dynamical variables*. Our wavefunction ψ, that initially simply played the role of the 'invisible function' that we imagined as sitting in the shadows, over on the right-hand side of all our operators, is now definitely beginning to play an active role. We think of it as the *state* of the physical system, as mentioned above. Sometimes it is called a *state vector* (although this is really a more

[21.9] Why? Here linear dependence can involve continuous sums, namely integrals.

general term, where the particular descriptions in terms of space and time coordinates that I have been using for ψ need not apply). As with the case of 4-momentum, considered above, the *eigenstates* of some dynamical variable are those states for which that particular dynamical variable has what is called 'a definite value', and the eigenvalue is the actual 'value' that the dynamical variable has for that state.

A point should be made about the fact that, up until this point, I have been treating our momentum eigenstate in a completely 4-dimensional spacetime way, consistently with the requirements of special relativity. This is economical, in that the expression[21.10]

$$\mathrm{e}^{-\mathrm{i}P_a x^a/\hbar} = \mathrm{e}^{-\mathrm{i}Et/\hbar}\,\mathrm{e}^{\mathrm{i}\mathbf{P}\cdot\mathbf{x}/\hbar}$$

(with $P_a = (E, -\mathbf{P})$ and $x^a = (t, \mathbf{x})$, as in §18.7) contains both the *spatial* dependence that makes it an eigenstate of the ordinary spatial 3-momentum

$$\mathbf{p} = (-p_1, -p_2, -p_3) = -\mathrm{i}\hbar\left(\frac{\partial}{\partial x^1}, \frac{\partial}{\partial x^2}, \frac{\partial}{\partial x^3}\right)$$

with eigenvalue $\mathbf{P}$ and the *temporal* dependence that makes it a solution of the Schrödinger equation with energy eigenvalue E. However, the Schrödinger formalism as a whole is not a relativistic scheme, in that it treats time differently from the spatial variables, so in the discussions that follow, in this chapter, it is better that I revert to non-relativistic descriptions.

21.6 What is quantum 'reality'?

Let us step back from these detailed matters for a moment, and ask what all this is trying to tell us about 'reality'. Are the dynamical variables 'real things'? Are the states 'real'? Or should we say that we have achieved reality only when we have arrived at the seemingly 'classical' quantities that arise as eigenvalues of the dynamical variables (or of other operators)? In fact, quantum physicists tend not to be very clear about this issue. Most of them are distinctly uncomfortable about addressing the issue of 'reality' at all. They may claim to take what they would call a 'positivist' stand, and refuse to consider what 'reality' is supposed to mean, regarding such an inquiry as 'unscientific'. All that we should ask of our formalism, they might claim, would be that it give answers to appropriate questions that we may pose of a system, and that those answers agree with observational fact.

[21.10] Why can I split it this way?

If we are to believe that any one thing in the quantum formalism is 'actually' real, for a quantum system, then I think that it has to be the wavefunction (or state vector) that describes quantum reality. (I shall be addressing some other possibilities later, in Chapter 29; see also the end of §22.4.) My own viewpoint is that the question of 'reality' *must* be addressed in quantum mechanics—especially if one takes the view (as many physicists appear to) that the quantum formalism applies universally to the whole of physics—for then, if there is no quantum reality, there can be no reality at any level (all levels being quantum levels, on this view). To me, it makes no sense to deny reality altogether in this way. We need a notion of physical reality, even if only a provisional or approximate one, for without it our objective universe, and thence the whole of science, simply evaporates before our contemplative gaze!

All right then, what about the state vector? What is the difficulty about taking *it* as representing reality? Why is it that physicists often express extreme reluctance about taking this philosophical stance? To understand the difficulties, we must look more carefully into the nature of wavefunctions and their physical interpretations.

Let us first examine our momentum state $\psi = e^{i\mathbf{P}\cdot\mathbf{x}/\hbar}$ more closely (where I have taken it at time $t = 0$, for convenience). We note that it is in no way localized like an ordinary particle. It is spread out evenly over the whole of the universe. Its 'magnitude', as measured by its modulus $|e^{i\mathbf{P}\cdot\mathbf{x}/\hbar}|$, has the same value 1 everywhere in space (see §5.1). The reader will be excused for thinking that this is a strange picture to have to entertain, for a single particle, with just a well-defined momentum in some spatial direction. What has happened to our ordinary picture of a particle, as something (at least approximately) localized at a single point? Well, we might say that a momentum state is only an idealization. We can still get away with having a very well-defined (if not perfectly precisely defined) momentum if we pass to somewhat similar states referred to as 'wave packets'. These are given by wavefunctions that peak sharply in magnitude at some position and are 'almost' eigenfunctions of momentum, in an appropriate sense. In one dimension such wave packets can be presented explicitly, by taking the product of a momentum state with a Gaussian e^{-x^2} or, more appropriately, with a general Gaussian

$$Ae^{-B^2(x-C)^2},$$

(where A, B and C are real constants). This is the well-known 'bell-shaped' curve of statistics (for an illustration of which, see Fig. 27.5 in §27.4) where, in the above expression, its 'peak' is centred at the point

$x = C$. It is of some interest (and calculational advantage) that the wave packet obtained by taking such a product can be succinctly expressed by allowing C to be a *complex number* in above expression.[21.11] Wave packets in the full three dimensions of space can also be similarly constructed, such as by using a Gaussian $Ae^{-B^2(x^2+y^2+z^2)}$, with its peak displaced in complex direction. In each case, B^{-1} provides a measure of its spread. In fact, there is a theorem, underlying what is called 'Heisenberg's uncertainty principle', that tells us that there is an absolute limit to how small its spread can be in relation to how closely 'almost' a momentum state it is. We shall see this a little more explicitly in §21.11.

For now, let us try to get a better picture of what momentum states and wave packets are actually like. Now we must bear in mind that a wavefunction is a complex-valued wave, and its 'wavy' character is not necessarily to be seen as an oscillation in its magnitude (or intensity). In the case of a momentum state, it is the wavefunction's *argument* (§5.1), namely $-P_a x^a/\hbar$, taken as measured round a circle—i.e. $e^{-iP_a x^a/\hbar}$ taken on the unit circle in the complex plane—that has a 'wavy' character. In quantum theory, we tend to refer to the argument of the value of the wavefunction as its *phase*. We find that the phase is not so much 'wavy' as 'twisting round and round'. In Fig. 21.5a, I have tried to indicate this behaviour of the wavefunction in some particular direction, by plotting that direction at an angle sloping off to the right (x axis of the picture), and taking a plane perpendicular to that direction (remaining u and v axes in the picture) to represent the complex plane of values that the

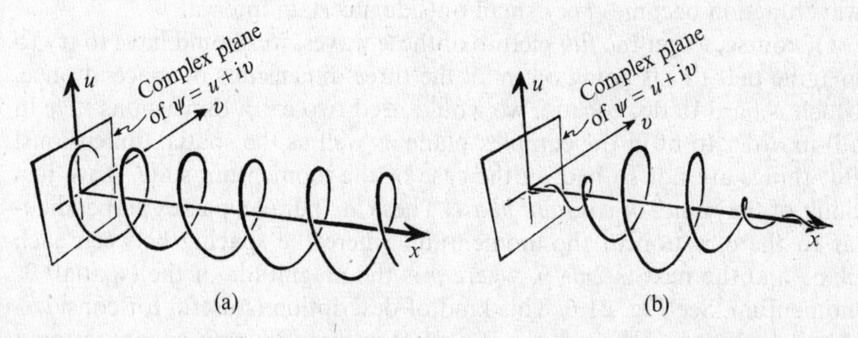

Fig. 21.5 A particle wavefunction: ψ as a complex function of position x. (a) Momentum state $e^{-iPx/\hbar}$, depicted as a corkscrew (eigenfunction of momentum p). (b) A wave packet $e^{-A^2x^2}e^{-iPx/\hbar}$.

[21.11] Replacing the real number C in the above displayed expression by the complex number $C + iD$ (where C and D are real), find the frequency of the wave packet and the location of its peak.

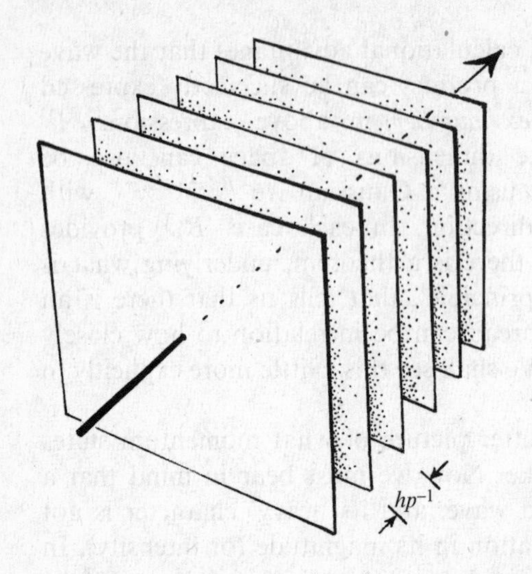

Fig. 21.6 Planes of a given phase, for a momentum eigenstate with spacing hp^{-1}, where p is the magnitude of the spatial 3-momentum. (Compare Fig. 21.2.)

wavefunction ψ can take (so the picture plots $\psi = u + iv$ on that plane). Thus, the x direction of my picture corresponds to some actual direction in ordinary space, but the u and v directions are not ordinary spatial directions; they are put in to represent the complex plane of possible values of the wavefunction. We notice that, for our momentum state, the wavefunction is a corkscrew (which is right-handed for positive momentum in the spatial direction represented by our picture's x direction). In Fig. 21.5b, I have drawn the corresponding picture for a wave packet. It is like a corkscrew for a stretch (so it has only a moderately well-defined momentum), but then this 'corkscrewness' tails off in both directions, and the wavefunction becomes very small outside a certain interval.

Of course, to get the *full* picture of these waves, we should have to try to imagine that this is going on in all the three dimensions of space at once, which is hard to do, because we would need two extra dimensions (five in all) in order to fit in the complex plane as well as the spatial dimensions! But things are not so bad, in the case of the momentum state, if we just think of the *planes of constant phase*. These are parallel planes perpendicular to the direction of the momentum, where the spacing between each plane and the next is $2\pi\hbar/p$, where p is the magnitude of the (spatial) 3-momentum. See Fig. 21.6. This kind of description is useful for considering such things as the picture of a photon wavefunction encountering a crystal, as was done in Fig. 21.2. We can also use this description in the case of the two-slit experiment, if we take the slits as being a long way from the screen; then we can think of the wavefunction of each particle, as it approaches some localized region of the screen, as being composed as the sum of two parts, each of which is closely a momentum state (being

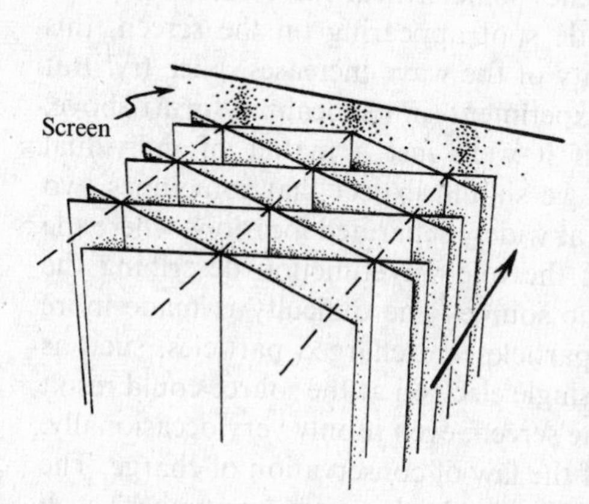

Fig. 21.7 An electron's wavefunction, approaching the screen of Fig. 21.4 in the 2-slit experiment, may be regarded as a superposition of two of the plane waves of Fig. 21.6, mutually tilted at a slight angle. Where the phases agree (along the broken lines) the two reinforce each other, giving rise to the greatest probability of arrival at the screen. Half way between these maxima, the phases are opposite and the waves cancel, giving zero probability of the electron reaching the screen.

essentially a single-frequency plane wave—owing to the great distance of the slits from the screen), but where the direction of each of the two component parts is slightly different. At some places on the screen, the two waves will reinforce each other, while at other places they will cancel out, giving the bands of greater and lesser intensity that I described above (in Fig. 21.4d). We can see this geometry in Fig. 21.7, in which the planes represent regions in space where each component wave has a constant value for its phase. In the full wavefunction, these two component parts must be added together. So, if we assume that each part individually has the same intensity, they will cancel out at places where they are out of phase and reinforce where they are in phase. This provides us with the bands of intensity that are actually observed in the two-slit experiment.

Yes, yes, the impatient reader might well interject, but this is just how *waves* behave. I have not faced up to the fact that our wave/particles are wave/*particles*! Apart from the apparently very minor embellishment that my waves are complex waves, all that I have been doing is to describe wave interference of a kind that would occur with ordinary waves in the sea, or sound waves, or Maxwell's waves built from classical electromagnetic field (radio waves, visible light, X-rays, etc). But the whole point of the two-slit experiment—so I was supposed to have been insisting—is that the experiment shows up a conflict between the wave picture and a particle picture. Indeed so; the most obvious manifestation of particle nature, in this experiment, occurs when these little fellows make their tiny marks on the screen: one at a time . . . !

21.7 The 'holistic' nature of a wavefunction

There is something that should be emphasized here. One could imagine that a little spot on the screen comes about from time to time, when the local intensity of the wave reaches some critical value or, rather, that there is some probability of a little spot appearing on the screen, this probability increases as the intensity of the wave increases. Nice try! But as I have formulated the two-slit experiment (in its idealized form) above, this simply will not work. For if it were just a matter of individual probabilities at individual places, we should expect that sometimes two spots would appear on the screen, at widely separated locations where the intensity is appreciable, with just the one wavefunction describing the emission of a single particle at the source. The difficulty is made more manifest if we imagine that our particles are charged particles, such as electrons. For if the emission of a single electron at the source could result in a pair of electrons arriving at the screen, even if only very occasionally, then we should have a violation of the law of conservation of charge. The same would apply to any other conserved particle 'quantum number', such as baryon conservation (§25.6), for example, if we were to use neutrons.[21.12] Such non-conservation behaviour would be in gross contradiction with an enormous amount of experimental evidence. Yet, electrons and neutrons *do* exhibit the kind of self-interference that results in a two-slit-experiment behaviour as I have just described!

So we have just got exactly *nowhere* in understanding wave/particles—some irate reader will surely object with increasingly justified impatience! But hold on please, we are not through with interpreting our wavefunctions. We have to think of the entire wave as describing (or 'being') just a single particle. Although it does, in a definite sense, determine the probability that a spot will occur at the various places on the screen, this probability refers to just the one particle. This interpretation will not work if we think of the wavefunction in a local way, as independently providing a probability of spot formation at each separate place on the screen. We must think of a wavefunction as one entire thing. If it causes a spot to appear at one place, then it has done its job, and this apparent act of creation forbids it from causing a spot to appear somewhere else as well. Wavefunctions are quite unlike the waves of classical physics in this important respect. The different parts of the wave cannot be thought of as local disturbances, each carrying on independently of what is happening in a remote region. Wavefunctions have a strongly non-local character; in this sense they are completely holistic entities.

This point can be made even more forcefully in a somewhat different experimental situation. This has the additional advantage of making quite clear to us that the wave-packet picture of a wave/particle is, by itself,

---

[21.12] Show that the probability of such double-spot appearances, according to such a picture, must be quite appreciable, whatever the law of probability of spot appearance in terms of wavefunction intensity might be. *Hint*: Divide the screen into two parts, with equal probability of spot appearance in each.

quite inadequate for explaning particle-like quantum behaviour. Let us imagine that there is a particle source, just as before, and we are going to suppose that it only emits a single particle. Instead of using a barrier with a pair of slits, we are going to suppose that there is what is called a *beam-splitter* in the particle's path. It will help our imaginations if we think of our particle as a photon, and we can imagine that the beam-splitter is a kind of 'half-silvered mirror'[7] which is to split our photon wave packet into two widely seperated parts. For clarity of our conceptions, let us envisage our 'experiment' being carried out in interstellar space (and the reader should be warned that I am not proposing anything remotely practical here—our example will serve merely to exhibit some very basic predictions of quantum mechanics under extreme circumstances). If we choose, we can imagine the photon's wavefunction to start out from the source in the form of a neat little wave packet, but, after encountering the beam-splitter, it will divide itself in two, with one wave-packet part reflected from the beam-splitter and the other wave-packet part transmitted through it, say in perpendicular directions (Fig. 21.8). The entire wavefunction is the sum of these two parts. We could wait for a year, if we like, before choosing to intercept the photon's wavefunction with a photographic plate or other kind of detector. The two parts will be a very long way away from each other by now, but we can imagine that I have two colleagues (in two different space laboratories), more than 1.4 light years separated from one another. Each of my colleagues has a separate detector, and although each of the two wave-packet parts may individually have dispersed considerably by now, each colleague has a large paraboidal reflecting mirror which collects the dispersed wave packet, focusing it on that particular colleague's detector. What does quantum mechanics say will happen? It says that one or other of my colleagues will indeed detect the photon, but that they cannot *both* detect the photon. This is not the

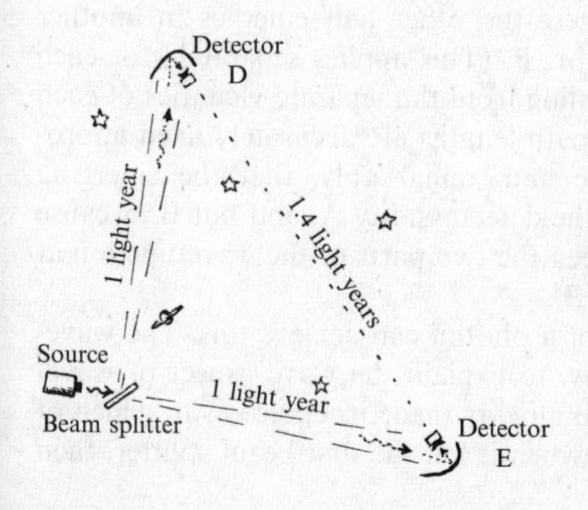

Fig. 21.8 A hypothetical space experiment illustrating the non-local nature of a wavefunction under measurement. The photon's wavefunction starts from the source as a neat little wave packet, but divides into two after encountering the beam-splitter, to arrive, after one year, at light-year distant detectors D and E. But just one of D or E can register the photon.

kind of thing that a classical wave does. Remember that my two colleagues are over 1.4 light years apart. Relativity insists that no signal can pass between them in less than 1.4 years (§17.8); yet the fact that one wave-packet part yields up a photon prevents the other one, 1.4 light years away, from doing so, and vice versa. In only a year's time, I learn from each of them what has happened, and I find that only one of them has received a photon. The part of the wavefunction that each colleague has access to seems to 'know' what the other part of the wavefunction is up to! Every time I perform this experiment, I find that one or other of them receives the photon, but not both. No classical type of wave effect could achieve this apparently 'instantaneous communication' between the two parts of the wavefunction. Quantum wavefunctions are just *different* from classical waves.

Yet, the sceptical reader may still not be convinced: for no such communication would be needed if the photon simply made its choice to go one way or the other at the *beam splitter*. Quite true. What the above experimental set-up is exhibiting is the particle-like aspect of a photon. If the photon were to remain localized and particle-like, then its decision as to which way to go would have to be made at the beam-splitter. (A localized particle can't stretch simultaneously over light-year spans!) If experiments like this were all that photons had to contend with, then wavefunctions would not be needed. But there are other experiments that might be performed on the photon after it emerges from the beam splitter. How can our poor little photon know, when it is about to emerge, that my colleagues do not plan a different type of fate for it? Suppose that, instead of each individually trying to detect the photon, they had concocted the following plan. They would separately reflect their parts of the wavefunction to a fourth location, where the two reflected parts would, say after a further year, simultaneously encounter a second beam splitter (Fig. 21.9). There, each arriving wave packet part would be individually split in two, so that one half emerges from this beam splitter in one direction to encounter a detector A, and where the other half emerges in another direction to go to another detector, B. (This applies separately for each of the two wave-packet parts, coming from the separate vicinities of each of my two colleagues.) If all the path lengths are accurately fixed appropriately (say all equal), then we find, remarkably, that the emerging photon can only activate *one* of the detectors, say A, and not B, because of constructive interference between the two parts of the wavefunction at A and destructive interference at B.

No purely particulate picture of a photon can achieve this. The wavefunction is definitely needed, now, to explain the wave aspect of wave/particle duality. If the photon had already made its choice as to which of my colleagues to travel towards, when it left the first beam splitter, then

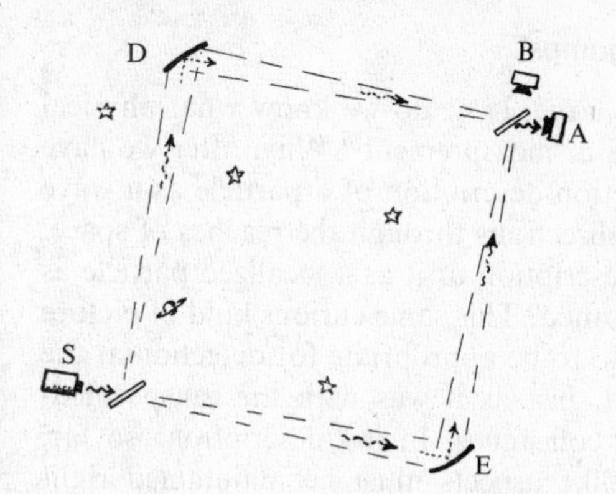

Fig. 21.9 Mach–Zehnder interferometer on an interstellar scale. How can the photon know, upon emergence from the first beam splitter, that instead of the arrangement of Fig. 21.8, mirrors at D and E reflect the wavefunction portions to a second beam-splitter? Following this encounter, only detector A is able to receive the photon.

the other route would become irrelevant. In that case, when the photon finally reaches the second beam splitter it comes from only one direction, and it could go either way, to reach *either* A *or* B. There is now no possibility of the needed destructive interference that prevents it from reaching the detector at B. Since A is always the detector that registers, it cannot just be the case that the photon has simply made its choice when it leaves the first beam splitter. It is necessary that both of the alternative routes that the photon might take are simultaneously felt out by the photon in its passage from the first to the second beam splitter.[8]

Of course, I have been grossly over-fanciful in my specific astronomical-scale situations described above. It is clear that no quantum experiment with anything like such a baseline has actually been performed! On the other hand, ground-based versions of this kind of experiment (the latter being what is called a Mach–Zehnder interferometer) have frequently been carried out, with arm lengths of perhaps metres rather than light years, and the expectations of quantum mechanics have never been contradicted. The key puzzle is that somehow a photon (or other quantum particle) seems to have to 'know' what kind of experiment is going to be performed upon it well in advance of the actual performing of that experiment. How can it have the foresight to know whether to put itself into 'particle mode' or 'wave mode' as it leaves the (first) beam splitter?

The way that quantum theory works is not to give the particle any such 'foresight' but simply to accept the non-local holistic character of a wavefunction. In both of the above experiments, we take the wavefunction to be split into two parts at the initial beam splitter, and the particle-like aspect of the wave/particle only shows up at the detector, when the measurement is finally performed. The measurement makes the holistic character of the wavefunction manifest, in the sense that the particle always shows up in just one place, its appearance at one location forbidding its simultaneous appearance anywhere else.

21.8 The mysterious 'quantum jumps'

But now another question looms large. How do we know what physical circumstance it is that constitutes a 'measurement'? Why, after we have been happily using this wavefunction description of a particle as a wave spread out in two quite different directions through the reaches of space, should we suddenly revert to a description of it as a localized particle as soon as the detection of it is performed? This same curious kind of picture of a quantum particle appears also to be appropriate for detection at the screen in our two-slit experiment, just as it was with the (unspecified) 'detectors' used by my far-flung colleagues. In my descriptions so far, it certainly seems that the wavelike aspects must be maintained right up until we choose to 'perform a measurement' to detect the particle, but then we suddenly revert to a particle-like description, where there is an awkward discontinuous (and non-local) change of the state—a *quantum jump*—as we pass from the wavefunction picture to the 'reality' presented by the measurement. Why? What is it about the detection process that demands that a different (and highly non-local) mathematical procedure should be adopted, in the event of a 'measurement', from the standard quantum-evolution procedure provided by Schrödinger's equation?

I shall try to address this puzzling issue in some depth later, in Chapters 23, 29, and 30. But even if we accept that, at least at the level of formal mathematical description, we must adopt this curious 'jumping' procedure, there is the question of what this tells us about the 'reality' of the wavefunction. This 'jumping' of the quantum state—a process that does not seem to be covered by any continuous evolution in accordance with the Schrödinger equation—is what leads a great many physicists to doubt that the evolution of the state vector can possibly be taken seriously as an adequate description of physical reality. Schrödinger himself was extremely uncomfortable with 'quantum jumps', and he once remarked in a conversation with Niels Bohr:[9]

> If all this damned quantum jumping were really here to stay then I should be sorry I ever got involved with quantum theory.

For the moment, let us accept this curious description, at least as a mathematical model of the quantum world, whereby the quantum state evolves for a while in the form of a wavefunction, usually spreading out through space (but possibly being focused in again to a more localized region); but then, when a measurement is performed, the state collapses down to something localized and specific. This instant localization happens no matter how spread out the wavefunction may have been before the measurement, whereafter the state again evolves as a Schrödinger-guided wave, starting from this specific localized configuration, usually spreading out again until the next measurement is performed. From the above experimental (and 'thought-experimental') situations, the impression could be gained that the particle-like aspects of a wave/particle are what show up in a measurement, whereas it is the wavelike ones that show up between measurements.

This is not so far from the truth of what quantum mechanics tells us, but the two wave/particle aspects are by no means so simply delineated as this. Whereas some physicists have indeed taken the view that all measurements are ultimately measurements of position,[10] I would myself regard such a perspective as being much too narrow. Indeed, the way that the quantum formalism is normally presented certainly does not require measurements to be only of position. For example, the measurement of a particle's momentum (or, say, of its angular momentum about some axis) would constitute just as good a measurement as one of position. I shall discuss the relationship between measurements of position and momentum in §21.11, but the general question of how the quantum formalism treats measurements will be left to the next chapter. The mathematical description of the physical measurement of a quantum system will be found to be something very different from a (Schrödinger) quantum evolution. The controversial issues arising from this curious fact will be discussed later, and most completely in Chapter 29.

21.9 Probability distribution in a wavefunction

Let us here address the more limited question of what the wavefunction ψ is supposed to be telling us about the particle's position. The rules of quantum theory tell us that ψ's *squared modulus* $|\psi|^2 (= \bar{\psi}\psi$; see §10.1) is to be interpreted as the probability distribution, giving the likelihood of a position measurement finding the particle at the various possible spatial locations. Thus, wherever the wavefunction is largest in absolute value, the particle is most likely to be found. Wherever it is zero, the particle will not be found. Now, the total probability of finding the particle somewhere in space has to be 1; therefore the integral of $|\psi|^2$ over the whole of space,[11] i.e.

$$\|\psi\| = \int_{E^3} |\psi(\mathbf{x})|^2 \mathrm{d}x^1 \wedge \mathrm{d}x^2 \wedge \mathrm{d}x^3,$$

is equal to 1:

$$\|\psi\| = 1.$$

We say that the wavefunction ψ is *normalized* if this condition holds.

This normalization requirement has the irritating implication that it rules out the 'momentum-state' wavefunctions $\psi = \mathrm{e}^{\mathrm{i}\mathbf{P}\cdot\mathbf{x}/\hbar}$ that we started with, because $|\psi|^2 = 1$ over the whole of infinite space, so the above integral (being equal to the total volume of space) *diverges*. We thus have to regard the momentum states as unrealizable idealizations. We can make life a little easier for momentum states, on the other hand, if we adopt a somewhat more relaxed attitude with regard to wavefunctions. We can still call ψ a 'wavefunction' even if it does not satisfy this normalization condition, but we call it a *normalized wavefunction* if it does.

A wavefunction ψ will be *normalizable* if the integral defining $\|\psi\|$ converges. In this case we an divide ψ by the square root of $\|\psi\|$ to obtain the normalized wavefunction: $\psi\|\psi\|^{-1/2}$. Only the normalizable wavefunctions have a chance of being physically realized. The others (such as momentum states) indeed represent physical idealizations. The complex vector space of (not necessarily normalized) wavefunctions is our state space $\mathbf{W}$. I shall also have to allow that some of our wavefunctions might actually be hyperfunctional (§9.7), the reason for which will become apparent shortly.

With regard to physical interpretation (allowing for this more relaxed attitude), we consider that if ψ is multiplied by a non-zero constant complex number, then it represents the same physical situation that it did before. It is, in any case, standard in quantum theory to regard ψ and $e^{i\theta}\psi$ as being physically equivalent, where θ is a real constant. In other words, multiplying the wavefunction by a constant phase makes no difference to the physical state. (Clearly this does not affect the value of $|\psi(\boldsymbol{x})|^2$.) It is not unreasonable to carry this a little further and allow multiplication by any non-zero complex constant κ, still regarding the wavefunctions as equivalent:

$$\psi \equiv \kappa\psi.$$

(The Schrödinger equation is clearly also unaffected by this replacement.) Factoring out by this equivalence amounts to passing from the complex vector space $\mathbf{W}$ of wavefunctions to its projective space $\mathbb{P}\mathbf{W}$ of idealized 'physical states'. (See §15.6, for the notion of a projective space.[12]) Of course, the general constant scaling $\psi \mapsto \kappa\psi$ does not preserve $|\psi|^2$, so we need to re-interpret the probability density for the particle's location, so that it applies when ψ is not normalized. This we do by providing the revised rule that the probability density is to be obtained by taking $|\psi|^2$ divided by the integral of $|\psi|^2$ over the whole of space:

$$\frac{|\psi(\boldsymbol{x})|^2}{\|\psi\|}.$$

For some states, such as momentum states, $\|\psi\|$ diverges, so we do not get a sensible probability distribution in this way (the probability *density* being zero everywhere, which is reasonable for a single particle in an infinite universe).

In accordance with this probability interpretation, it is not uncommon for the wavefunction to be called a 'probability wave'. However, I think that this is a very unsatisfactory description. In the first place, $\psi(\boldsymbol{x})$ itself is complex, and so it certainly cannot be a probability. Moreover, the phase of ψ (up to an overall constant multiplying factor) is an essential ingredient for the Schrödinger evolution. Even regarding $|\psi|^2$ (or $|\psi|^2/\|\psi\|$) as a 'probability wave' does not seem very sensible to me. Recall that for a momentum state, the modulus $|\psi|$ of ψ is actually *constant* throughout the whole of spacetime. There is no information in $|\psi|$ telling us even the direction of motion of the wave! It is the phase, alone, that gives this wave its 'wavelike' character.

Moreover, probabilities are never negative, let alone complex. If the wavefunction were just a wave of probablities, then there would never be any of the cancellations of destructive interference. This cancellation is a characteristic feature of quantum mechanics, so vividly portrayed (Fig. 21.4d) in the two-slit experiment!

At this point, it is appropriate to widen the discussion slightly and make contact with our considerations, in §19.4, of the electromagnetic field and the gauge connection $\boldsymbol{\nabla}$ that is associated with it. If our wavefunction describes a charged particle, then we are now allowed to make *gauge* transformations of the form $\psi \mapsto e^{i\theta}\psi$, where θ $(= \theta(\mathbf{x}))$ is an arbitrary real-number function of position, providing the necessary 'gauge symmetry' which enables electromagnetism to act as a gauge connection. But have I not just asserted that Schrödinger time-evolution depends *essentially* upon the knowledge of how the phases of the wavefunction vary from place to place? The application of a gauge transformation $\psi \mapsto e^{i\theta}\psi$ would allow us to change the way the phases vary to anything we please! Does that not contradict what I have just been claiming about the crucial physical importance of how the phases vary?

Not at all: whereas these non-constant phase changes *are* allowed, this is only if they are accompanied by a compensating change in the $\partial/\partial x^a$ operators (i.e. in the momentum). This change $(\partial/\partial x^a \mapsto \partial/\partial x^a - ieA_a$, where $A_a = \nabla_a\theta$ and $e = 1)$ is precisely such as to leave the action of the bundle connection $\boldsymbol{\nabla}$ unaltered. The 'phase information' is still there, but it is now mixed up with the definition of $\boldsymbol{\nabla}$. One cannot simply apply $\psi \mapsto e^{i\theta}\psi$ alone, with an arbitrarily varying θ, and hope to leave the physical situation unaltered. The details of the spatial variation of θ (in relation to $\boldsymbol{\nabla}$) are essential to the dynamical evolution of the state, and I would argue that ψ is clearly much more than a probability wave. In any case, if ψ describes an *un*charged wave/particle $(e = 0)$, then the situation is exactly as it was before.

21.10 Position states

It seems to me to be clear that the wavefunction must be something a good deal more 'real' than would be the case for merely 'a probability wave'. The Schrödinger equation provides us with a precise evolution in time for this entity (whether it is charged or not), an evolution that depends critically upon how the phase indeed varies from place to place. But if we ask of a wavefunction 'where is the particle?', by performing upon it a position measurement, we must be prepared to lose this phase-distribution information. In fact, after the measurement, we have to start all over again with a new wavefunction. If the result of the measurement asserts 'the particle is here', then our new wavefunction has to be very strongly peaked at the position 'here', but then it rapidly disperses again, in accordance with Schrödinger evolution. If our position measurement were *absolutely* precise, then the new state would be 'infinitely peaked' at that location; in fact it would have to be described by a Dirac delta function, a quantity that we encountered briefly in §6.6, and in §9.7 in the guise of a hyperfunction.

Let us see how the formalism deals with this. For simplicity, consider the measurement of just one component of a particle's position, say the coordinate x^1. The *result* of our measurement ought to be a state with a 'definite value for x^1'; so, in accordance with what was said in the case of momentum, we require ψ to be an eigenstate of the operator x^1 (i.e. of multiplication by x^1), the eigenvalue being the particular value X^1 of the coordinate x^1 that the particle is found to have. In order for the action of x^1, namely

$$\psi \mapsto x^1\psi,$$

to have the definite x^1-coordinate value X^1 (a real number), we require the eigenvalue equation

$$x^1\psi = X^1\psi$$

(where we recall that x^1 is a linear operator and X^1 is a number). This is satisfied by

$$\psi = \delta(x^1 - X^1),$$

where $\delta(x)$ is Dirac's 'delta function', which was defined (as a hyperfunction) in §9.7. For it has the property[21.13] that $x\delta(x) = 0$, whence $(x^1 - X^1)\delta(x^1 - X^1) = 0$, i.e. $x^1\delta(x^1 - X^1) = X^1\delta(x^1 - X^1)$, as required. This 'wavefunction' is not a function in the ordinary sense, but it is an

[21.13] Check this from the hyperfunctional definition given in §9.7.

idealized function (a hyperfunction or distribution), being infinitely peaked at the eigenvalue $x^1 = X^1$, as mentioned above.

This particular measurement says nothing about the remaining spatial coordinates, and the wavefunction can still have arbitrary variation in these coordinates, providing us with a scaling for the delta function that is an arbitrary function of the remaining coordinates x^2 and x^3, so we get

$$\psi = \phi(x^2, x^3)\, \delta(x^1 - X^1)$$

for the general eigenstate of the operator x^1. We can proceed further and ask for a state that is simultaneously an eigenstate of all three of the spatial coordinates. (This is a legitimate request because x^1, x^2, x^3 all commute. There is, indeed, a general property of quantum-mechanical observables that if we have a collection of them, all of which commute among themselves, then common eigenstate(s) exist for all of them together; see §22.13.[13] The answer is that, for the resulting value (*eigenvalue*) $\mathbf{X} = (X^1, X^2, X^3)$ for the (triple) spatial measurement, we require (up to an overall scale factor)

$$\begin{aligned}\psi &= \delta(x^1 - X^1)\, \delta(x^2 - X^2)\, \delta(x^3 - X^3) \\ &= \delta(\mathbf{x} - \mathbf{X}),\end{aligned}$$

the final line being defined by the one above it.[14] This is what a *position state* is like.

Such 'position states' are idealized wavefunctions in the opposite sense from the momentum states. Whereas the momentum states are infinitely spread out, the position states are infinitely concentrated. Neither is normalizable (the trouble with $\psi = \delta(\boldsymbol{x} - \boldsymbol{X})$ being that delta functions cannot be squared, cf. §9.7). I shall end this chapter by pointing out that there is an important duality between position and momentum which elucidates this issue.

21.11 Momentum-space description

Up until this point, I have been representing quantum states entirely as functions of position: wavefunctions. What this means, in effect, is that each *state*—element of **W**—is thought of as a linear combination of eigenstates of the position operator $\mathbf{x}$, i.e. of *position states* (states $\delta(\mathbf{x} - \mathbf{X})$). Expressing a wavefunction ψ as a function of position means, in effect, that it is regarded as a linear combination of such delta functions. We achieve this by the formula $\psi(\mathbf{x}) = \int \psi(\mathbf{X})\delta(\mathbf{x} - \mathbf{X})\mathrm{d}^3\mathbf{X}$, expressing $\psi(\mathbf{x})$ as a *continuous* combination of them, where $\mathrm{d}^3\mathbf{X} = \mathrm{d}X^1 \wedge \mathrm{d}X^2 \wedge \mathrm{d}X^3$. In this formula, the 'coefficients' in this linear combination are the complex numbers $\psi(\mathbf{X})$.

But there are many other ways of representing a quantum state ψ. We can, alternatively, represent it as a linear combination of *momentum* states $e^{i\mathbf{P}\cdot\mathbf{x}/\hbar}$. Now the 'coefficients' are different complex numbers, which we take to be $(2\pi)^{-3/2}$ times the quantities $\tilde{\psi}(\mathbf{P})$, so we arrive at the formula:

$$\psi(\mathbf{x}) = (2\pi)^{-3/2} \int_{\mathbb{E}^3} \tilde{\psi}(\mathbf{P})\, e^{i\mathbf{P}\cdot\mathbf{x}/\hbar}\, d^3\mathbf{P}.$$

(The reason for the $(2\pi)^{-3/2}$ will be explained very shortly.) This formula expresses $\psi(\mathbf{x})$ as a Fourier transform of some function $\tilde{\psi}(\mathbf{P})$ just as was done in §9.4, except that here we have a 3-dimensional Fourier transform—which amounts to applying the formula of §9.4 three times over.

This suggests that $\tilde{\psi}$ (as a function of $\mathbf{P}$, but we can now write it as a function of $\mathbf{p}$), provides just as good a representation of the particle's quantum states as does the original function $\psi(\mathbf{x})$. There is, indeed, a very precise symmetry between the position and momentum variables. We can now consider regarding the *momentum* variables $\mathbf{p}$ as the primary ones, and represent the position variables $\boldsymbol{x}$ as 'differentiation with respect to $\mathbf{p}$', so we can make the *reverse* interpretation (noting the sign change):

$$x^a = -i\hbar \frac{\partial}{\partial p_a}$$

(at least for the spatial variables x^1, x^2, x^3).[21.14] Indeed, commutation relations are satisfied that are identical to those that we had before:

$$p_b x^a - x^a p_b = i\hbar\, \delta^a_b.$$

The 'invisible' function at the extreme right is now to be a function of momentum p_a, rather than position x^a. It is the momentum states that are now represented by delta functions $\delta(\mathbf{p} - \mathbf{P})$, and the position states are represented as the plane waves $e^{-i\mathbf{p}\cdot\mathbf{X}/\hbar}$. The representation of momentum 'wavefunctions' in terms of the position eigenstates $e^{-i\mathbf{p}\cdot\mathbf{X}/\hbar}$ is given by the virtually identical (inverse) Fourier transform:

$$\tilde{\psi}(\boldsymbol{p}) = (2\pi)^{-3/2} \int_{\mathbb{E}^3} \psi(\mathbf{X})e^{-i\mathbf{p}\cdot\mathbf{X}/\hbar}\, d^3\mathbf{X},$$

with only the very minor sign change in the exponent. (We now see the reason for the $(2\pi)^{-3/2}$; it is to balance things so that the inverse Fourier transform is virtually the same as the original one.)

[21.14] Show that replacing ψ by $x^1\psi$ or by $i\hbar\partial\psi/\partial x^1$ corresponds, respectively, to replacing $\tilde{\psi}$ by $-i\hbar\partial\tilde{\psi}/\partial p_1$ or by $p_1\tilde{\psi}$. Show that replacing $\psi(x^a)$ by $\psi(x^a + C^a)$ corresponds to replacing ψ by $e^{-iC^a p_a/\hbar}\,\tilde{\psi}$ (where a ranges over 1, 2, 3).

Wave packets can be described just as well in the momentum-space representation as in the position representation.[21.15] One can introduce a precise notion of the 'spread' (or lack of localization) of a wave packet in either the position description or the momentum description. Let us denote these spread measures, respectively, by Δx and by Δp. *Heisenberg's uncertainty relation* tells us that the product of these spreads cannot be smaller than the order of Planck's constant, and we have[15]

$$\Delta p \; \Delta x \gtrsim \tfrac{1}{2}\hbar.$$

Position states, momentum states, and wave packets are illustrated in Fig. 21.10, in the both position and momentum representations. We note that, in the extreme case of a pure momentum state, the spread in the momentum is zero, so $\Delta p = 0$ (i.e. a delta function in momentum space). From the Heisenberg relation, Δx is now infinite, in accordance with the picture described above (in §21.6), where the wavefunction becomes spread uniformly over the whole of position space. The situation is just the opposite with a position state, where now $\Delta x = 0$, the position being defined with complete precision, but where the spread Δp in the momentum now becomes infinite.

It is interesting to see that here we have examples that clearly illustrate the incompatibility of non-commuting measurements in quantum

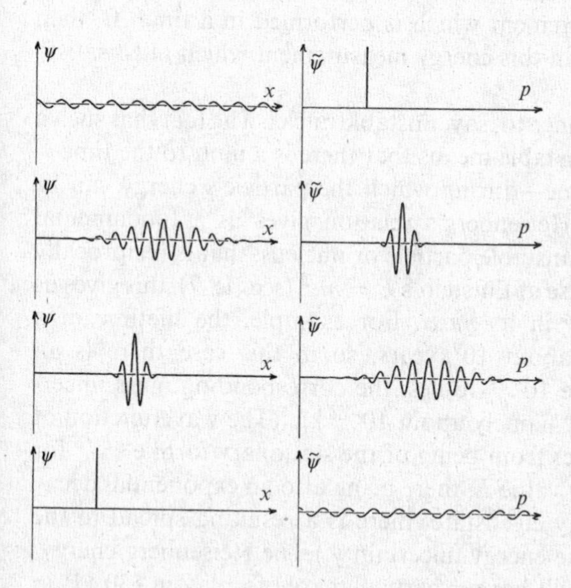

Fig. 21.10 Position-space pictures of wavefunctions ψ are on the left, with corresponding momentum-space pictures of $\tilde{\psi}$ on the right. The top pair depict a momentum state and the bottom pair, a position state. The two between them depict wave packets. The Heisenberg uncertainty relation is illustrated by the greater spread in position being accompanied by a smaller spread in momentum and *vice versa*.

[21.15] Use the results of Exercises [21.11], [21.13], and [21.14] to show that the Fourier transform of the wave packet $\psi = A\mathrm{e}^{-B^2(x-C)^2}\mathrm{e}^{\mathrm{i}\omega x}$ is $\tilde{\psi} = \left(A\mathrm{e}^{\mathrm{i}\omega C}/B\sqrt{2}\right)\mathrm{e}^{-(p-\omega)^2/4B^2}\mathrm{e}^{-\mathrm{i}Cp}$ (putting $\hbar = 1$, for convenience).

mechanics (which is a general phenomenon that we shall encounter frequently in our later considerations). A measurement of a particle's momentum would put it into a momentum state, corresponding to some classical value P, and any subsequent measurement of the momentum in this state would yield the same result P. However, if the state were instead subjected to a subsequent *position* measurement following an initial measurement of momentum, the result would be completely uncertain, and any one result for the position would be as likely as any other. This measurement makes the state a delta function in position. In momentum space, this state is a plane wave, spread out uniformly in all possible values for the momentum. A subsequent *momentum* measurement would then be completely uncertain. Thus, the very act of intermediate position measurement has completely ruined the purity of the original momentum state.

It should also be mentioned that, consistently with relativity (§18.7), there is a similar Heisenberg uncertainty relation between energy and time:

$$\Delta E \; \Delta t \gtrsim \tfrac{1}{2}\hbar.$$

This is normally considered to have a somewhat different physical status from the more familiar momentum/position uncertainty relation, since time is just treated as an external parameter in standard quantum mechanics, rather than as a dynamical variable. The usual interpretation of the energy/time uncertainty is that if the energy of a quantum system is ascertained in some measurement which is performed in a time Δt, then there is an uncertainty ΔE in this energy measurement which must satisfy the above relation.

This has particular relevance to, say, unstable nuclei. The fact that such a nucleus (say uranium) is unstable means that there is a limit to the time—namely the particle's lifetime—during which the particle's energy can be ascertained. Accordingly, Heisenberg's relation gives us a fundamental energy uncertainty, for an unstable particle or nucleus, that is reciprocally related to its lifetime. Because of Einstein's $E = mc^2$ (see §18.7), this gives us a fundamental uncertainty in its *mass*. For example, the lifetime of a uranium U_{238} nucleus is about 10^9 years, so in this case there is an energy uncertainty of some 10^{-51} Joules; the corresponding mass uncertainty being utterly minute, namely about 10^{-68} kg. (The wavefunction of an unstable particle deviates from being of the stationary form $e^{-iEt/\hbar}$ for some definite real[16] energy value E, there being also an exponential decay factor. Being not an energy eigenstate, there is a resulting spread in the measured energy, giving the energy uncertainty.) The Heisenberg energy/time uncertainty relation will have a particular role to play in §30.11, in connection with a particular approach to the resolution of the enigma of quantum measurement!

Notes

Section 21.1

21.1. This is an example of what is called an *ordinary* differential equation, or ODE, since it is an equation that involves only ordinary differential operators such as d/dx, d/dy, etc. or its powers, e.g. d^3/dx^3. A *partial* differential equation, or PDE, would be an equation involving the partial differential operators $\partial/\partial x$, $\partial^2/\partial x^2$, $\partial^2/\partial x\partial y$, etc., such as in the Maxwell or Einstein equations of Chapter 19.

Section 21.2

21.2. However, for consistency sake (see also §21.3), I am sticking to the notation appropriate to relativity (§18.7), so the spatial components of momenta in 'p_a' are the *negatives* of the usual momentum components (which are c^{-2} times the spatial components p^a). This choice is compatible with my comments of §20.2 because I am now using x (rather than the q of the general Lagrange/Hamilton formalism)

21.3. This time independence ensures that the interpretation of $\mathcal{H}$ as a conserved *total energy* can be maintained. The reader might be disturbed by the fact that since a dependence on space coordinates has been allowed for, the requirements of a fundamental-level relativistic invariance might demand that we allow for time dependence also (and see Note 20.3). But at a *fundamental* level, *both* time and space independence would be a normal requirement.

21.4. See Woodhouse (1991).

Section 21.4

21.5. The term 'black' refers here to the (as nearly as possible) completely absorbent nature of the body surrounding the radiation. In these early experiments an almost completely spherical dark cavity was used for containing the radiation, with a very narrow opening connecting the internal volume to the outside. The surrounding body might well be glowing, however, owing to the temperature and so might not actually look black.

21.6. In my descriptions of this experiment, I am idealizing the situation, leaving out all the practical difficulties, in order to get the essential point across.

Section 21.7

21.7. In accurate experiments, such a thing would be unlikely to involve actual silvering, but would use interference effects between the reflected waves from the two sides of a thin transparent material.

21.8. In the description of quantum mechanics referred to as the de Broglie–Bohm theory (Bohm and Hiley 1994), *both* the wave and the particle aspects are, in effect, simultaneously retained. Here the particle does indeed make its choice at the beam-splitter, but the wave carries on, exploring both routes simultaneously. When the final beam splitter is reached, it is the wave that instructs the particle to reach the detector at A, forbidding it to reach B. I shall try to assess this interesting but 'unconventional' (and still non-local) viewpoint in §§29.2,9.

Section 21.8

21.9. As reported by Heisenberg (1971), p. 73.

21.10. See Goldstein (1987); Bell (1987).

Section 21.9

21.11. Many authors might define the 'norm' as the *square root* of what I mean by $\|\psi\|$ here, that is, their $\|\psi\|^2$ is my $\|\psi\|$.

21.12. Various authors have developed the quantum-mechanical formalism in an elegant way, completely within the projective framework. See particularly Brody and Hughston (2001); Hughston (1995); Ashtekar and Schilling (1998).

Section 21.10

21.13. This property of commuting observables is discussed in any text on quantum mechanics; see, for example, Shankar (1994).

21.14. It is legitimate to multiply delta functions if they refer to *different* variables. See Arfken and Weber (2000) for properties of delta functions.

Section 21.11

21.15. See Shankar (1994); Hannabuss (1997).

21.16. It is common practice, in particle physics, to use the same $e^{-iEt/\hbar}$ time dependence, but with a *complex E*, whose real part is the mean value of the energy and whose imaginary part is $-\frac{1}{2}\hbar \log 2$ times the reciprocal of the half-life. (See for example, Das and Ferbel 2004)[21.16]

[21.16] Can you see how to justify the factor $-\frac{1}{2}\hbar$ log 2? (The half-life is the time at which the probability of decay has reached one half.)

22
Quantum algebra, geometry, and spin

22.1 The quantum procedures U and R

THE non-intuitive nature of quantum mechanics—or, rather, of Nature herself at the level of quantum-mechanical activity—leads many people to despair of finding any kind of trustworthy picture of quantum-level phenomena. Yet, there is much beautiful geometry associated with quantum mechanics in addition to its elegant algebraic structure, and it would be a pity to feel that one must necessarily rely merely upon a pictureless, unvisualizable formalism in order to make headway with the description of quantum actions. Although we have seen that even a single featureless 'point particle' appears to be a mysterious spread-out wavy thing in the quantum formalism, it is a 'thing' that can be pictured, having a fascinating mathematical structure in which many of the aspects of complex-number magic start to show themselves.

This picture enables us to begin to come to terms with the quantum description of a single point particle, and having understood what a single quantum particle is like, we might suppose that we shall be able to sit back and relax a little, since this surely provides us, in principle, with an understanding of complicated systems involving many different kinds of particle. Unhappily, this expectation is premature, and we shall need a broader perspective if we are to arrive at a comprehensive quantum picture of the world. We shall be seeing, in Chapter 23, how much more confusing our picture of things becomes when several particles have to be considered together in a system. Instead of each particle individually having its separate 'state vector', we find that the entire quantum system requires a thoroughly self-entangled *single* state vector.

But even individual 'point particles' tend to have more structure to them than that encompassed by the descriptions that I have so far provided. For they often possess what is called *spin*, which leads to extra complication. Fortunately, as we shall see later in this chapter, spin is itself a phenomenon with a mathematical description of particular richness and elegance, where other aspects of geometry and complex-number magic come to the fore.

Let us review the descriptions in the previous chapter, where we had to become accustomed to the (non-relativistic) quantum particle as being something described by what we have called a *state vector* (or wavefunction) whose evolution is, in a very precise way, provided by the Schrödinger equation—until some measurement is performed on the system. As we shall be seeing more explicitly, in Chapter 23, the same will apply to the state vectors that describe entire complicated quantum systems. The measurement itself is described mathematically in a completely different way from Schrödinger evolution. We have seen indications of this in §§21.4,7,8. In §§21.10,11 we considered position measurements, in the course of which, a particle's state would *jump* to a (generally different) state, now localized in some particular location—i.e. to a state which is an eigenvector of the position operator **x** (this eigenvector being a delta function in the position coordinates). We also considered the results of momentum measurements (in §§21.5,6,11), whereby a particle's state has been made to jump into an eigenstate of the momentum operator **p**, so the particle's state is now spread out in a wavelike form (in principle over all space). More generally, a measurement would correspond to an operator $\boldsymbol{Q}$ of some sort (usually a Hermitian operator; see §22.5), and the effect of the measurement on the state would be to make it jump into some eigenstate of $\boldsymbol{Q}$. Which eigenstate of $\boldsymbol{Q}$ is it to jump to? This is a matter of pure chance, according to quantum theory, but there are precise rules for calculating the probabilities (see §22.5).

The jumping of the quantum state[1] to one of the eigenstates of $\boldsymbol{Q}$ is the process referred to as *state-vector reduction* or *collapse of the wavefunction*. It is one of quantum theory's most puzzling features, and we shall be coming back to this issue many times in this book. I believe that most quantum physicists would not regard state-vector reduction as a *real* action of the physical world, but that it reflects the fact that we should not regard the state vector as describing an 'actual' quantum-level physical reality. We shall come to this contentious issue in more detail in Chapter 29. Nevertheless, irrespective of whatever attitude we might happen to have about the physical reality of the phenomenon, the way in which quantum mechanics is used in practice is to take the state indeed to jump in this curious way whenever a measurement is deemed to take place. Immediately after the measurement, Schrödinger evolution takes over again—until another measurement is performed on the system, and so on.

I denote Schrödinger evolution by **U** and state reduction by **R**. This alternation between these two completely different-looking procedures would appear to be a distinctly odd type of way for a universe to behave! See Fig. 22.1. Indeed, we might imagine that, in actuality, this is an approximation to something else, as yet unknown. Perhaps there *is*

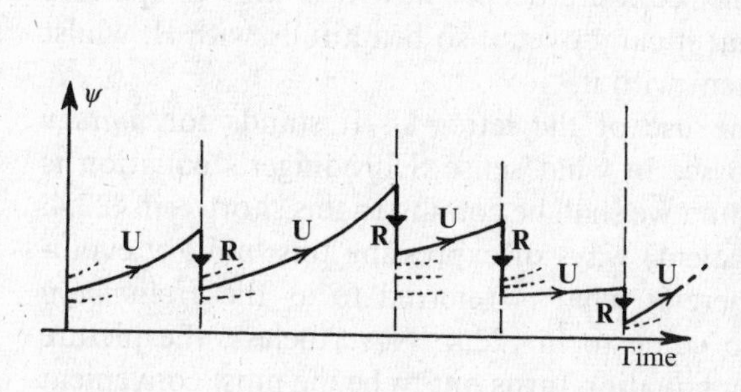

Fig. 22.1 The time-evolution of the state ψ for a physical system, according to accepted tenets of quantum mechanics, alternates between two completely different procedures: unitary (Schrödinger) evolution **U** (continuous, deterministic) and state reduction **R** (discontinuous, probabilistic).

a more general mathematical equation, or evolution principle of some-coherent mathematical kind, which has both **U** and **R** as limiting approximations? My personal opinion is that this kind of change to quantum theory is very likely to be correct—as a part of a new 21st century physics, perhaps—and in Chapter 30 I shall be making some specific suggestions directed towards this possibility. However most physicists appear not to believe that this kind of route is a fruitful one to follow.

Their reason for preferring not to contemplate altering the basic framework of quantum mechanics is (in addition to the great mathematical elegance of its **U** formalism) the tremendously impressive and precise agreement between quantum theory and experimental fact, where nothing is known that tells against quantum theory (in its present hybrid form) and many varied results confirm it to great accuracy. Accordingly, most quantum physicists would adopt a philosophical standpoint (or, rather, one of the various different alternative philosophical standpoints to be described in §29.1) which try to come to terms with the apparent contradiction between the **U** and **R** procedures, while not attempting to change the present-day quantum formalism in any significant way. One of my purposes in this particular chapter, and in the next, is to begin to examine this formalism, but without deviating from what is now conventional in quantum theory. I shall come back to the **U**/**R** issue later, particularly in §§29.1,2,7–9, and also in §§30.10–13 where I shall give my own perspectives on the matter more fully.

I think that it would be fair to say that a common thread in much of what might be called 'conventional' attitudes to quantum mechanics is that the **U** process is to be taken as an 'underlying truth' and that one must come to terms with **R**, in one way or another, as being some type of approximation, illusion, or convenience, and there are many accounts in the literature which pursue this kind of approach.[2] Even those (myself included) who are of the opinion that some change in the quantum formalism is needed at some stage, would argue that the present-day scheme is at least a marvellous approximation, so it is necessary to understand it thoroughly if there is to be any hope of moving beyond it. Accordingly, we must try to see more deeply how it is that **U** operates and, moreover, how it is that it can dovetail so beautifully with **R**, whilst nevertheless being inconsistent with it!

I should also explain the use of the letter **U**. It stands for *unitary evolution*. We shall need to see in what sense Schrödinger's equation is indeed 'unitary' (see §13.9), and we shall be coming to this shortly, in §22.4. There are also other (equivalent) ways of expressing this 'unitary evolution'; most particularly, there is what is referred to as the *Heisenberg picture* which we shall also come to in §22.4. Nevertheless, the picture provided by the Schrödinger equation turns out to be the most convenient for our descriptions here.

22.2 The linearity of U and its problems for R

Before addressing the full issue of unitarity, let us examine the more primitive question of the *linearity* of **U**. We shall see that this aspect alone, of **U**, presents a serious incompatibility with **R**. Let us, therefore, again examine Schrödinger's equation $i\hbar\partial\psi/\partial t = \mathcal{H}\psi$. We shall imagine the Hamiltonian $\mathcal{H}$ to be known (being specified by the nature of the particles that it describes and the forces between them, and by any external conservative—i.e. energy conserving—forces that might have an influence on the system). There are certain consequences that are immediate from the general form of the equation, and are quite independent of the detailed nature of the Hamiltonian.

One thing that we note is that it is a *deterministic* equation (the time-evolution being completely fixed once the state is known at any one time). This may come as a surprise to some people, who may well have heard of 'quantum uncertainty', and of the fact that quantum systems behave in non-deterministic ways. This lack of determinism comes about in the application of the **R**-process only. It is not to be found in the (**U**) time-evolution of the quantum state, as described by the Schrödinger equation. Another thing that we immediately see in the Schrödinger equation is that it is a *complex* equation, owing to the manifest appearance of i on the left (and there are many more possibilities for occurrence of i in the Hamiltonian).

Finally, we see that the Schrödinger equation is indeed linear, in the sense that if ψ and ϕ are solutions (with the same $\mathcal{H}$) of

$$i\hbar\frac{\partial\psi}{\partial t} = \mathcal{H}\psi, \quad i\hbar\frac{\partial\phi}{\partial t} = \mathcal{H}\phi,$$

then so also is any *linear combination* $w\psi + z\phi$, where w and z are complex constants. For, by adding w times the first of the above equations to z times the second we get (§6.5):

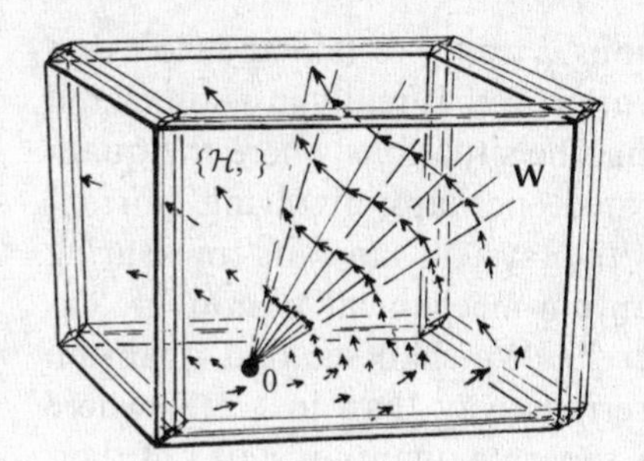

Fig. 22.2 The Hamiltonian flow $\{\mathcal{H}, \}$ (a vector field) defines an infinitesimal linear transformation of state space **W**, giving the change in the state after an infinitesimal time. To get the (unitary) change after a finite time, we must 'exponentiate' this infinitesimal Hamiltonian action.

$$i\hbar\frac{\partial}{\partial t}(w\psi + z\phi) = \mathcal{H}(w\psi + z\phi).$$

From this, we see that the Schrödinger evolution preserves the *complex-vector-space* structure of the state space **W** (which is usually an infinite-dimensional space).

The Hamiltonian $\mathcal{H}$ defines the infinitesimal linear transformation of **W** that describes the change in a state that takes place after it has evolved for an infinitesimal time. This Hamiltonian action is therefore described by a vector field on **W** (see Fig. 22.2). After a finite time, the states would have changed according to a finite linear transformation, obtained by what is called 'exponentiating' the infinitesimal Hamiltonian action. This is very similar to the 'exponentiation' that we encountered earlier (§14.6), describing the process whereby a Lie group element is obtained from the exponentiation of an element of the corresponding Lie algebra. However, the exponentiation in Hamiltonian evolution can be a good deal more difficult to carry out. (Also, further difficulties arise because of the infinite-dimensional nature of **W**.)

But difficult or not, the essential point here is that, after any finite time T, the transformation of the space **W** of quantum states will always be linear. This amounts to the following assertion (where I shall use the symbol $\rightsquigarrow$ to indicate how a state will have evolved after the specified time-period T):

If

$$\psi \rightsquigarrow \psi' \quad \text{and} \quad \phi \rightsquigarrow \phi',$$

then

$$w\psi + z\phi \rightsquigarrow w\psi' + z\phi'.$$

Here, ψ and ϕ are two arbitrarily chosen states (wavefunctions) and w and z are arbitrary complex constants.[22.1]

[22.1] Make it clear why the action of any Schrödinger evolution is linear, despite the fact that $\mathcal{H}$ may be a highly non-linear function of the ps and xs.

This has certain very curious implications if we try to take the view that **U** is the whole story, and the measurement process is really just some kind of 'convenience' that one calls upon to handle situations where the quantum state gets unmanageably complicated, perhaps involving horrendously many 'entangled' particles in the system and its measuring apparatus. (We shall come to the quantum-mechanical notion of 'entanglement' more specifically in Chapter 23. We shall see that quantum states are 'holistic' entities in a more serious way than in §21.7, where different parts of the system do not have separate quantum states of their own, but are parts of one entangled 'whole'. None of this affects the present discussion, however.) According to such a 'convenience' view of **R**, one imagines that **R** would emerge as some kind of approximation to a 'true' underlying **U** evolution. But this viewpoint leads to serious paradoxes.

For example, let us recall the thought experiment of §21.7, where my two colleagues in space had individual detectors, and try to imagine that the response of each detector is simply the result of a Schrödinger evolution starting from its interaction with the wave-packet part that it receives. The quantum state before detection is actually a sum of the two individual wave-packet parts, one reaching one detector and the other part reaching the other detector; therefore, by linearity, the subsequent Schrödinger-evolved response of each detector must coexist in superposition with a response in the other. The Schrödinger evolution leads to one detector response *plus* the other detector response ('plus' in the sense of quantum superposition of the two detector responses), not one detector response *or* the other detector response (the 'or' being what actually always happens in practice). It seems to me untenable to maintain that **U** tells the whole story (and the 'conventional' quantum mechanics of Niels Bohr's 'Copenhagen interpretation' certainly does not try to do this; for it treats the detectors themselves as 'classical entities').

As far as I can see, the only way to insist that **U** holds for all processes, including measurement, would be to pass to a 'many-worlds' type of view (see §29.1) in which the two detector responses do actually coexist, but in what are referred to as 'different worlds'.[3] But even then, **U** cannot be 'the whole story', because we would need a theory to explain that aspect of our conscious perceptions which allows only individual detector responses to be consciously perceived, whereas superpositions of responses with non-responses are never consciously perceived! (These issues will be returned to in §§29.1,8.) I should register, at this point, that I do not myself believe that 'many worlds' is the right way to go; I am merely arguing that it seems to be where one is led if one insists on '**U** at all levels'.

We shall be coming back to these issues later, in Chapters 29 and 30, where the question of whether **U** and **R** must be treated as approximations to some more comprehensive future theory will be addressed. For the moment, let us follow the prescriptions of the conventional formalism. If an improved theory is needed, it will in any case have to accord with the prescriptions of this conventional theory to a very high degree of accuracy. Any reader with aspirations towards the finding of a new theory (and I hope that there are some!) will be well advised to come fully to terms with what the conventional theory has to say.

22.3 Unitary structure, Hilbert space, Dirac notation

I have not yet properly addressed the 'unitary' aspect of Schrödinger evolution. This has to do with the 'normalization' property of wavefunctions referred to in the previous chapter. Recall that, for the wavefunction ψ of a single (spinless) particle, the 'norm' referred to the quantity $\|\psi\|$, defined as the integral of $|\psi(\boldsymbol{x})|^2$ over the whole of space. The normalization condition on ψ is that $\|\psi\| = 1$ (and when this condition is imposed, then $|\psi(\boldsymbol{x})|^2$ is the probability density for a position measurement finding the particle at the point $\boldsymbol{x}$). In a general quantum-mechanical situation, where there may be many interacting spinning particles (or perhaps entities of more general type, such as strings, etc.), we always demand some corresponding notion of *norm* $\|\psi\|$, which is to be a positive real number[4] for any properly acceptable quantum state ψ. Although this norm is something pertinent to the **U** part of the quantum formalism, it plays a crucial role also in the **R** part, determining, in effect, all the *probabilities* that arise.

We can think of the norm, mathematically, as providing a notion of *squared length*, which ought to be finite for 'acceptable' vectors belonging to the state space **W**. The adjective 'unitary', as applied to temporal evolution, tells us that this norm is preserved throughout the evolution. We shall be seeing shortly (in §22.4) why this indeed applies in the case of Schrödinger evolution.

First, it will be helpful to set up some notation and to investigate some of the properties of normalizable quantum states. It will be valuable to think of the norm as a particular case of a Hermitian scalar product (§13.9) between states. For states ϕ and ψ, this is usually written $\langle\phi|\psi\rangle$, in the quantum-mechanical literature, and the norm of ψ is the special case of this when $\phi = \psi$:

$$\|\psi\| = \langle\psi|\psi\rangle.$$

In the case of a single (spinless) particle, the scalar product is:

$$\langle\phi|\psi\rangle = \int_{\mathbb{E}^3} \bar{\phi}\psi \mathrm{d}x^1 \wedge \mathrm{d}x^2 \wedge \mathrm{d}x^3,$$

generalizing the particular expression for $\|\psi\|$ given in §21.9. This gives us a positive-definite Hermitian scalar product defined between any two normalizable 1-particle wavefunctions ϕ and ψ.[22.2]

In fact, the normalizable wavefunctions constitute a complex vector space **H** (a subspace of **W**), and it is a vector space of a particular kind known as a *Hilbert space*.[22.3] The definition of a Hilbert space is that it is a complex vector space possessing a scalar product operation $\langle\,|\,\rangle$, whose value is a complex number, which satisfies the algebraic properties

$$\langle\phi|\psi+\chi\rangle = \langle\phi|\psi\rangle + \langle\phi|\chi\rangle,$$
$$\langle\phi|a\psi\rangle = a\langle\phi|\psi\rangle,$$
$$\langle\phi|\psi\rangle = \overline{\langle\psi|\phi\rangle}$$
$$\psi \neq 0 \text{ implies } \langle\psi|\psi\rangle > 0$$

(all of which are immediate, in the case of the 1-particle integral given above).[22.4] These equations also imply $\langle\phi+\chi|\psi\rangle = \langle\phi|\psi\rangle + \langle\chi|\psi\rangle$ and $\langle a\phi|\psi\rangle = \bar{a}\langle\phi|\psi\rangle$.[22.5] Moreover, once the norm is known, the scalar product can be defined in terms of it,[22.6] so linear transformations that preserve the norm must also preserve the scalar product. In addition, a Hilbert space should satisfy certain very basic continuity properties.[5]

The above notation forms part of the valuable and widely used notational framework for quantum mechanics introduced by the great 20th-century physicist Paul Dirac. As an ingredient of this general scheme, it proves to be useful to regard expressions like

$$|\psi\rangle,\ |\uparrow\rangle,\ |\rightarrow\rangle,\ |\leftrightarrow\rangle,\ |0\rangle,\ |7\rangle,\ |+\rangle,\ |\boldsymbol{X}\rangle,\ |\text{DEAD}\rangle, \text{ or } |\text{OFF}\rangle$$

as representing various state vectors belonging to the Hilbert space **H**, where the symbol within the $|\ldots\rangle$ is some appropriate (and perhaps memorable) label indicating the state in question. These are sometimes

---

[22.2] See if you can explain why the $\langle\phi|\psi\rangle$ integral converges whenever both $\langle\phi|\phi\rangle$ and $\langle\psi|\psi\rangle$ converge. *Hint*: Consider what is implied by the integral of $|\phi - \lambda\psi|^2$ being non-negative over any *finite* region of $\mathbb{E}^3$, deriving an inequality connecting the squared modulus of the integral of $\bar{\phi}\psi$ with the product of the integral of $\bar{\phi}\phi$ with the integral of $\bar{\psi}\psi$. As an intermediate step, find conditions on complex numbers a, b, c, d that imply $a + \lambda b + \bar{\lambda}c + \bar{\lambda}\lambda d \geq 0$ for all λ.

[22.3] Following on from Exercise [22.2], show that the normalizable wavefunctions indeed constitute a vector space.

[22.4] Verify this, stating carefully which properties of integration are being used.

[22.5] Show why.

[22.6] Show how $\langle\phi|\psi\rangle$ can be defined from the norm. *Hint*: Work out the norms of $\phi + \psi$ and $\phi + \mathrm{i}\psi$.

called 'ket' vectors. For each such ket, there will be a particular member of the dual space $\mathbf{H}^*$ (§12.3), called the corresponding 'bra' vector, which is the *Hermitian conjugate* of that state (in the sense of §13.9), respectively written

$$\langle\psi|,\ \langle\uparrow|,\ \langle\rightarrow|,\ \langle\leftrightarrow|,\ \langle 0|,\ \langle 7|,\ \langle +|,\ \langle \boldsymbol{X}|,\ \langle \text{DEAD}|, \text{ or } \langle \text{OFF}|.$$

Since the bra vectors are dual to the ket vectors, they have a scalar product in the same sense as the dot product of §12.3. This scalar product—or 'bracket'—of a bra vector $\langle\psi|$ with a ket vector $|\phi\rangle$ is precisely the Hermitian scalar product written $\langle\psi|\phi\rangle$ above. This is consistent with the complex number $\langle\psi|\phi\rangle$ being the complex conjugate of $\langle\phi|\psi\rangle$. The two states $|\phi\rangle$ and $|\psi\rangle$ are said to be *orthogonal* if $\langle\phi|\psi\rangle = 0$, i.e. if $\langle\psi|\phi\rangle = 0$.

The action of some linear operator $\boldsymbol{L}$ on $|\psi\rangle$ is written $\boldsymbol{L}|\psi\rangle$, and the scalar product of the bra $\langle\phi|$ with $\boldsymbol{L}|\psi\rangle$ is written

$$\langle\phi|\boldsymbol{L}|\psi\rangle.$$

This is also the scalar product of a certain bra '$\langle\phi|\boldsymbol{L}$' with $|\psi\rangle$. What is the bra '$\langle\phi|\boldsymbol{L}$'? It is the complex conjugate of a certain ket $\boldsymbol{L}^*|\phi\rangle$, where $\boldsymbol{L}^*$ is the *adjoint*[6] of $\boldsymbol{L}$. This 'adjoint' operation, applied to a linear operator $\boldsymbol{L}$, is just the Hermitian conjugate operation $*$ that was considered in §13.9, in the finite-dimensional case. The complex conjugate of the complex number $\langle\phi|\boldsymbol{L}|\psi\rangle$ is the complex number $\langle\psi|\boldsymbol{L}^*|\phi\rangle$.

22.4 Unitary evolution: Schrödinger and Heisenberg

We are now in a good position to look at the 'unitary' nature of Schrödinger evolution. We have already seen, in §22.3, that this evolution is linear, so all we need to establish is that it preserves the scalar product $\langle\phi|\psi\rangle$ between two elements $|\phi\rangle, |\psi\rangle$, of $\mathbf{H}$. That is to say, $\langle\phi|\psi\rangle$ is *constant* in time: $\mathrm{d}\langle\phi|\psi\rangle/\mathrm{d}t = 0$. (From what has been said above, preserving the norm and preserving the scalar product are equivalent requirements.) Basically, what we need from our quantum Hamiltonian $\mathcal{H}$ is (i) that it keeps us in the Hilbert space, and (ii) it is Hermitian. These are very minimal requirements, and will be satisfied by any reasonable suggestion for a Hamiltonian. Its Hermitian nature, for example, is a natural requirement to ensure that its eigenvalues—the possible values for the energy of the system—should be real numbers. It is usual also to demand that $\mathcal{H}$ be *positive-definite*, which means that $\langle\psi|\mathcal{H}|\psi\rangle > 0$ for all non-zero $|\psi\rangle$, whence all the eigenvalues of $\mathcal{H}$ (energy values) are positive—although this is not needed for the unitary nature of the evolution. We quickly obtain (using the Leibniz property for the derivative of a product; see §6.5 and properties above)

$$\begin{aligned}\frac{\mathrm{d}}{\mathrm{d}t}\langle\phi|\psi\rangle &= \left\langle\frac{\mathrm{d}}{\mathrm{d}t}\phi\middle|\psi\right\rangle + \left\langle\phi\middle|\frac{\mathrm{d}}{\mathrm{d}t}\psi\right\rangle\\ &= \langle -\mathrm{i}\hbar^{-1}\mathcal{H}\phi|\psi\rangle + \langle\phi| - \mathrm{i}\hbar^{-1}\mathcal{H}\psi\rangle\\ &= \mathrm{i}\hbar^{-1}\langle\phi|\mathcal{H}|\psi\rangle - \mathrm{i}\hbar^{-1}\langle\phi|\mathcal{H}|\psi\rangle = 0,\end{aligned}$$

showing that scalar products are indeed preserved, i.e. Schrödinger evolution is unitary.[22.7] The same argument applies to other Hermitian operators, such as the generators of spatial translations or of rotations, showing that these also correspond to unitary transformations of **H**.

The above equation shows that the *rate* of change of a scalar product $\langle\phi|\psi\rangle$ is zero. From this it follows that $\langle\phi|\psi\rangle$ remains unchanged for *all* time, where $|\phi\rangle$ and $|\psi\rangle$ individually undergo Schrödinger evolution according to the same $\mathcal{H}$. Suppose we have quantum states $|\phi\rangle$ and $|\psi\rangle$ at time $t = 0$, and take them to evolve by Schrödinger's prescription until a later time T, when the states become respectively $|\phi_T\rangle$ and $|\psi_T\rangle$:

$$|\phi\rangle \rightsquigarrow |\phi_T\rangle \quad \text{and} \quad |\psi\rangle \rightsquigarrow |\psi_T\rangle$$

(using the notation of §22.2). Then

$$\langle\phi|\psi\rangle = \langle\phi_T|\psi_T\rangle.$$

This tells us that the linear action of the Schrödinger evolution, on the Hilbert space **H**, taken from $t = 0$ until some definite time $t = T$, is *unitary*, in the sense that there is an operator $\boldsymbol{U}_T$ effecting this transformation, so that

$$|\phi_T\rangle = \boldsymbol{U}_T|\phi\rangle, \quad |\psi_T\rangle = \boldsymbol{U}_T|\psi\rangle, \quad \text{etc.},$$

where this operator $\boldsymbol{U}_T$ is *unitary* in the sense of §13.9, namely that its inverse is equal to its adjoint:

$$\boldsymbol{U}_T^{-1} = \boldsymbol{U}_T^*, \quad \text{i.e.} \quad \boldsymbol{U}_T\boldsymbol{U}_T^* = \boldsymbol{U}_T^*\boldsymbol{U}_T = \boldsymbol{I}.$$

Here $\boldsymbol{I}$ is the identity operator on **H**. (See §13.9 for the demonstration of this property of $\boldsymbol{U}_T$.)

As mentioned in §22.1, there are other ways of representing the evolution of a quantum system, and what is called the *Heisenberg picture* is the most familiar alternative. In the Heisenberg picture, the 'state' of the system is considered to be *constant* in time, the time-evolution being taken up by the dynamical variables instead. The reader might well question how the quantum state could be regarded as being 'unchanging' even though some actual physical change might well be taking place in the

[22.7] Spell this argument out a little more fully. Can you explain why we should expect the Leibniz property to hold for a Hilbert-space scalar product?

quantum system! Indeed; but passing from the Schrödinger picture to the Heisenberg picture is really just a matter of redefining our symbols.

Consider, first, the ordinary Schrödinger picture that we have been adopting up until now. We have some quantum state $|\psi\rangle$ at time $t = 0$, which we take to be evolving according to Schrödinger's prescription, as defined by a given quantum Hamiltonian $\mathcal{H}$, so that at some later time T the state is $|\psi_T\rangle$:

$$|\psi\rangle \rightsquigarrow |\psi_T\rangle = \boldsymbol{U}_T|\psi\rangle.$$

Recall that the action of $\boldsymbol{U}_T$ really applies linearly to the entire Hilbert space $\mathbf{H}$, so that any other state $|\phi\rangle$ would undergo a corresponding evolution $|\phi\rangle \rightsquigarrow |\phi_T\rangle = \boldsymbol{U}_T|\phi\rangle$, with the same $\boldsymbol{U}_T$ as was used for $|\psi\rangle$. In the Heisenberg picture, we simply think of the 'state', at time T, as

$$|\psi\rangle_{\mathrm{H}} = \boldsymbol{U}_T^{-1}|\psi\rangle = \boldsymbol{U}_T^{*}|\psi\rangle.$$

It is clear that this 'Heisenberg state' $|\psi\rangle_{\mathrm{H}}$ does not change (basically by definition!) as time passes. On the other hand, in order that all the algebraic procedures follow through as before, so that the eigenvalues (measured physical parameters) are the same as in the Schrödinger picture, we require the dynamical variables also to evolve compensatingly. Thus, any linear operator $\boldsymbol{Q}$ (on $\mathbf{H}$), must be replaced by its Heisenberg version

$$\boldsymbol{Q}_{\mathrm{H}} = \boldsymbol{U}_t^{-1}\boldsymbol{Q}\boldsymbol{U}_t = \boldsymbol{U}_t^{*}\boldsymbol{Q}\boldsymbol{U}_t.$$

It follows directly that the Heisenberg version of any eigenvalue or of any scalar product is the same as the Schrödinger version.[22.8] The Heisenberg evolution now applies to the operators $\boldsymbol{Q}$ (assumed constant in Schrödinger picture) and in particular to the dynamical variables. We find that[22.9]

$$i\hbar\frac{\mathrm{d}}{dt}\boldsymbol{Q}_{\mathrm{H}} = \mathcal{H}\boldsymbol{Q}_{\mathrm{H}} - \boldsymbol{Q}_{\mathrm{H}}\mathcal{H},$$

which are *Heisenberg's equations of motion*. (Note that an obvious consequence of this is conservation of energy, given when $\boldsymbol{Q}_{\mathrm{H}} = \mathcal{H}$.)

The reader may ask what we have gained by phrasing things in this way. In some contexts, there are technical advantages in the Heisenberg picture, but the Heisenberg picture does not help things with regard to the interpretative puzzles of quantum mechanics. The problem of the 'quantum jumps' has not gone away, but we may have the choice as to whether we put the blame on the state, allowing the $|\psi\rangle_{\mathrm{H}}$ to 'jump' to something else at

---

[22.8] Explain all this in detail.

[22.9] See if you can confirm this.

the operation of **R** or, instead, taking the Heisenberg dynamical variables to do the 'jumping'! For my own part, I find that these 'jumping' matters are simply made more obscure in the Heisenberg picture, without anything being resolved.

At least in the Schrödinger picture we have an evolving state vector which has a chance of giving us some glimpse of what 'quantum reality' might be like! The Heisenberg picture does not seem to have much chance of doing this, since its state vector just sits there unmoving even though physical action is taking place. Moreover, the evolution of the dynamical variables cannot represent the change in any specific physical system, because they do not describe specific systems at all, but rather questions that can be asked of a system, such as 'what is your position?', etc.

The reason for having these two different pictures is, to a large extent historical. Heisenberg was first, producing his scheme in July 1925, and Schrödinger put forward his proposal half a year later, in January 1926, realizing the equivalence between the two schemes shortly afterwards. It was Max Born who first recognized the probability interpretation for the squared modulus $|\psi|^2$ of Schrödinger's wavefunction (§21.9), in June 1926. Schrödinger himself had tried to hold to a more 'classical-field' picture of ψ. The general operator framework of quantum mechanics arose out of the work of Heisenberg, Born, and Pascual Jordan, and was fully formulated by Dirac, and described in detail in his highly influential book, *The Principles of Quantum Mechanics*, first published in 1930.[7]

Of course it may be that when a change is eventually introduced into quantum theory, then there may be good reasons for preferring one formalism over another, and the equivalence between the two may be broken. This is mildly the case even with quantum field theory (see Chapter 26), which tries to bring quantum theory and (special) relativity theory together in one consistent scheme. Dirac has made some arguments for preferring[8] the Heisenberg picture in this case. Neither the Heisenberg picture nor the Schrödinger picture is relativistically invariant, however, and sometimes the hybrid 'interaction picture' is preferred in this context.[9]

22.5 Quantum 'observables'

Let us now consider how a measurement of a quantum system is to be represented in the formalism. As noted in §22.1, the examples of position and momentum measurements given in Chapter 21 are illustrative of what happens in the general case of a quantum measurement. Some 'measurable' quality of a quantum system would be represented by a certain kind of operator $\boldsymbol{Q}$, called an *observable*, and this operator could be applied to the quantum state. The dynamical variables (say position or momentum) would be examples of observables.[10] The theory demands that an observable $\boldsymbol{Q}$ be represented as a linear operator (as with the examples of position or momentum operators), so that its action on the space H would be to effect a linear transformation of H—although possibly a singular one (§13.3). We say that the state ψ has a *definite value* for the observable $\boldsymbol{Q}$ if ψ is an eigenstate of $\boldsymbol{Q}$, and the corresponding eigenvalue q would be that definite value.[11] This is just the same terminology that we already encountered in §§21.5,10,11 for position and momentum.

In conventional quantum mechanics, one normally demands that all the eigenvalues have to be real numbers. One can ensure this (assuming normalizability of eigenvectors) by requiring that $\boldsymbol{Q}$ be Hermitian in the sense that $\boldsymbol{Q}$ is equal to its adjoint $\boldsymbol{Q}^*$:[22.10]

$$\boldsymbol{Q}^* = \boldsymbol{Q}.$$

In my opinion, this Hermitian requirement on an observable $\boldsymbol{Q}$ is an unreasonably strong requirement, since complex numbers are frequently used in classical physics, such as for the Riemann sphere representation of the celestial sphere (§18.5), and in many standard discussions of the harmonic oscillator (§22.13), etc.[12] An essential requirement of an observable is that its eigenvectors, corresponding to distinct eigenvalues, are orthogonal to one another. This is a characteristic property of operators that are referred to as 'normal'. A *normal* operator $\boldsymbol{Q}$ is one that commutes with its adjoint:

$$\boldsymbol{Q}^*\boldsymbol{Q} = \boldsymbol{Q}\,\boldsymbol{Q}^*,$$

and any pair of (normalizable) eigenvectors of such a normal $\boldsymbol{Q}$, corresponding to distinct eigenvalues, must indeed be orthogonal.[22.11] Since I am happy for the results of measurements (eigenvalues) to be complex numbers, while insisting on the standard requirement of orthogonality between the alternative states that can result from a measurement, I shall demand only that my quantum 'observables' be normal linear operators, rather than the stronger conventional requirement that they be Hermitian.

I should comment, here, on a further requirement of quantum observables that their eigenvectors span the entire Hilbert space H (so that any element of H can be expressed linearly in terms of these eigenvectors). In the finite-dimensional case, this property is a mathematical consequence of the Hermitian (or normal) nature of $\boldsymbol{Q}$. But for an infinite-dimensional H we need this as a separate assumption for any $\boldsymbol{Q}$ that is to play a role as a

---

[22.10] Show that any eigenvalue of a Hermitian operator $\boldsymbol{Q}$ is indeed a *real* number.

[22.11] See if you can prove this. *Hint*: By considering the expression $\langle\psi|(\boldsymbol{Q}^* - \bar{\lambda}\boldsymbol{I})(\boldsymbol{Q} - \lambda\boldsymbol{I})|\psi\rangle$, show first that if $\boldsymbol{Q}|\psi\rangle = \lambda|\psi\rangle$, then $\boldsymbol{Q}^*|\psi\rangle = \bar{\lambda}|\psi\rangle$.

quantum observable. A Hermitian Q with this property is called *self-adjoint*.

The orthogonality requirement for a quantum observable is important for the quantum measurement process. According to the rules of quantum mechanics, the result of a measurement, corresponding to some operator Q, will always be one of its eigenstates: this is the 'jumping' of the quantum state that occurs with the **R** process (see §22.1). Whatever state the system is in before measurement, it jumps to one of the eigenstates of Q just as the state is measured, in accordance with **R**. After the measurement, the state acquires a definite value for the observable Q, namely the corresponding eigenvalue q. Thus, for each of the different possible results of the measurement of the observable Q—that is, for each different eigenvalue $q_1, q_2, q_3, \ldots$—we get one of a set of alternative resulting states, all of which are mutually orthogonal.

Why is this important? We shall be seeing in a moment what the quantum rules are for calculating the probabilities of each of these alternative outcomes. One implication of these rules will be that the probability is always zero for a state to jump, as a result of a measurement, to an orthogonal state. Accordingly, if the measurement defined by an observable Q is repeated, then the second measurement will give the same eigenvalue—i.e. the same result for the measurement—as did the first measurement. To give a different result would involve it in effecting a jump from one state to an orthogonal one, which the probability rules do not allow. But this happy conclusion depends upon the orthogonality of the eigenstates of Q for different eigenvalues, which is why we are requiring Q to be a normal operator.

Let us now turn to the assignment of probabilities to the different alternative eigenstates of the observable Q, when it is presented with the state $|\psi\rangle$ that is being 'observed'. It is remarkable feature of the quantum **R** process that the quantum-mechanical probability depends only upon the quantum states before and after measurement, and not upon any other aspect of the observable Q (such as the value of the measured eigenvalue, for example). The rule is that the probability of the state jumping from $|\psi\rangle$ to the eigenstate $|\phi\rangle$, of Q is given by

$$|\langle\psi|\phi\rangle|^2,$$

assuming that $|\psi\rangle$ and $|\phi\rangle$ are normalized ($\|\psi\| = 1 = \|\phi\|$). Otherwise, we need to divide the above by $\|\psi\|$ and $\|\phi\|$ before the probability is obtained. We may prefer to write this probability, for non-normalized states, in the elegant form

$$\frac{\langle\phi|\psi\rangle\langle\psi|\phi\rangle}{\langle\psi|\psi\rangle\langle\phi|\phi\rangle}.$$

This is always a real number between 0 and 1, taking the value 1 only if the states are proportional.[22.12] Recall, from the above discussion, that Schrödinger evolution preserves the scalar products $\langle\phi|\psi\rangle$. This is an important consistency relation between the **U** and **R** processes, and it expresses the fact that, despite being inconsistent with each other, **U** and **R** indeed neatly 'dovetail' with each other. We see that, indeed, a state never jumps directly to an orthogonal state, in a measurement, because $\langle\phi|\psi\rangle = 0$ implies that the probability for this would be zero.

In a quantum superposition between orthogonal normalized states ψ and ϕ, say $w\psi + z\phi$, the complex-number weighting factors, w and z, are sometimes called *amplitudes*—or 'probability amplitudes'. In this case, an experiment set up to distinguish ψ from ϕ in the state $w\psi + z\phi$ would get ψ with probability $\bar{w}w = |w|^2$ and ϕ with probability $\bar{z}z = |z|^2$, i.e. we take the squared moduli of the amplitudes to get the probabilities. A similar comment applies to superpositions of more than two such states.

A useful property of a normal operator $\boldsymbol{Q}$ (assuming that its eigenvectors span the whole of **H**) is that it always possesses a family of eigenstates that form an orthonormal basis for the Hilbert space. An *orthonormal basis* (compare §13.9) is a set of elements $\boldsymbol{e}_1, \boldsymbol{e}_2, \boldsymbol{e}_3, \ldots$ of **H** such that

$$\langle \mathbf{e}_i \,|\, \mathbf{e}_j \rangle = \delta_{ij}$$

(δ_{ij} being the Kronecker delta) and where every element ψ of **H** can be expressed as

$$\psi = z_1\boldsymbol{e}_1 + z_2\boldsymbol{e}_2 + z_3\boldsymbol{e}_3 + \cdots,$$

($z_1, z_2, z_3, \ldots$ being complex 'cartesian coordinates' for ψ). This is similar to the expression of a general wavefunction, for a single structureless particle, as a continuous linear combination of momentum states (as is achieved using a Fourier transform) or of position states (using $\psi(\boldsymbol{x}) = \int \psi(\boldsymbol{X})\delta(\boldsymbol{x} - \boldsymbol{X})\mathrm{d}^3\boldsymbol{X}$) (§21.11), since the momentum and position states are the eigenstates of the momentum and position operators **p** and **x** respectively. Passing from the position representation to the momentum representation amounts to a change of basis in the Hilbert space **H** (see Fig. 22.3). However, neither momentum states nor position states actually form a basis, in the technical sense, because they are not normalizable and certainly do not themselves actually belong to **H**! (Quantum mechanics is full of irritating issues of this kind. As the state of the art stands, one can either be decidedly sloppy about such mathematical niceties and even pretend that position states and momentum states are actually states, or else spend the whole time insisting on getting the mathematics right, in which case there is a contrasting danger of getting trapped in a 'rigour

[22.12] Show this, from the algebraic properties of $\langle \;|\; \rangle$ by methods used in Exercise [22.2].

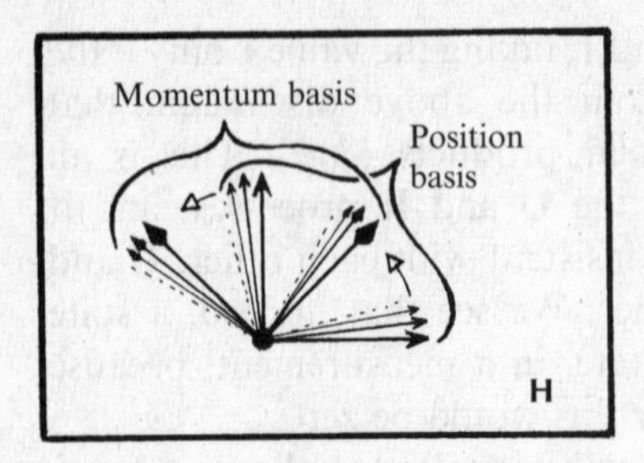

Fig. 22.3 Passing from a position to a momentum representation is just changing the basis in Hilbert space **H** (although, technically, neither momentum nor position states, being non-normalizable, actually belong to **H**).

mortis'. I am doing my best to steer a middle path, but I am not at all sure what the correct answer is for making progress in the subject!)

22.6 YES/NO measurements; projectors

In the case of operators such as position or momentum, where the eigenstates are not normalizable, we get zero for the probability of finding a particle in such a state. This is actually the 'correct' answer, because the probability of the position or momentum having any particular value would indeed be zero (position and momentum being continuous parameters). This is not very helpful to us, so we might prefer to use other kinds of observable, such as that which poses a question: 'is the position within such-and-such a range of values', and a similar question might be posed for momentum (or for any other continuous observable). Yes/no questions such as this can be incorporated into the quantum formalism by, say, assigning the eigenvalue 1 to the YES answer and the eigenvalue 0 to NO. An observable of this kind is described by what is called a *projector*.

A projector $\boldsymbol{E}$ has the property that it is self-adjoint and squares to itself[22.13]

$$\boldsymbol{E}^2 = \boldsymbol{E} = \boldsymbol{E}^*.$$

Such things provide the most primitive kind of measurement, and for many purposes the issues raised by 'measurement' in quantum mechanics are best carried out in terms of such operators. There is, however, a particular issue that becomes especially prominent when such a YES/NO measurement is performed, because (in more than 2 dimensions) these operators are (thoroughly) *degenerate*.

We say that $\boldsymbol{Q}$ is degenerate, with respect to some eigenvalue q, if the space of eigenvectors corresponding to q is more than 1-dimensional, i.e. if there are non-proportional eigenvectors of $\boldsymbol{Q}$ corresponding to the same

[22.13] Show that if an observable $\boldsymbol{Q}$ satisfies some polynomial equation, then every one of its eigenvalues satisfies the same equation.

eigenvalue q (§13.5). The entire linear subspace of **H** consisting of all eigenvectors corresponding to the same eigenvalue q is referred to as the *eigenspace* of Q corresponding to q. In such cases, the obtaining of the 'result' of the measurement (i.e. the determining of the eigenvalue) does not, in itself, tell us which state the state vector is supposed to 'jump' to. The issue is resolved by the so-called *projection postulate* which asserts that the state $|\psi\rangle$ being subjected to the measurement is orthogonally projected to the eigenspace[13] of Q corresponding to q. In fact, the term 'projection postulate' is often used simply for the standard quantum-mechanical procedure of §22.1 (as made explicit by von Neumann[14]) that, as a result of the measurement of an observable Q, the state jumps to an eigenstate of Q, corresponding to the eigenvalue that the measurement provides. In this section and the next, I am stressing the importance of the projection aspect of this postulate in the case of degenerate eigenvalues.[15]

One of the best ways of expressing this projection is by use of an appropriate projector E, namely the one whose eigenspace corresponding to its YES eigenvalue 1 is identical to Q's eigenspace corresponding to q. (This can always be done; E is simply asking a more basic question than that posed by Q, namely: 'is q the result of the Q measurement?') Then what the projection postulate asserts is that the result of the measurement (either Q with the result q, or E with the result 1) is that

$$|\psi\rangle \quad \text{jumps to} \quad E|\psi\rangle.$$

In this, I have not bothered about normalizations (and there is no need to bother if we do not wish to). If we ask that the resulting state be normalized, we can take $|\psi\rangle$ to jump to the more messy-looking $E|\psi\rangle\langle\psi|E|\psi\rangle^{-1/2}$. In my descriptions here, however, I shall find it more convenient not to have to normalize my states. This makes many of the formulae look simpler than otherwise.

In Fig. 22.4, I have indicated the geometrical nature, within the Hilbert space **H**, of the projection postulate. Notice that if we replace the projector E by $I–E$ (also a projector), then we find that the YES and NO eigenspaces are simply interchanged. (Here I is the identity operator on **H**.) Thus if the measurement obtains 0 for the E measurement, then $|\psi\rangle$ jumps to $(I–E)|\psi\rangle(=|\psi\rangle–E|\psi\rangle)$ instead. Note that $|\psi\rangle$ is the sum of the two states $E|\psi\rangle$ and $(I–E)|\psi\rangle$, which are orthogonal to each other[22.14] and the measurement E decides between the two, YES for the first and 'NO' for the second:

$$|\psi\rangle = E|\psi\rangle + (I–E)|\psi\rangle.$$

[22.14] Show this.

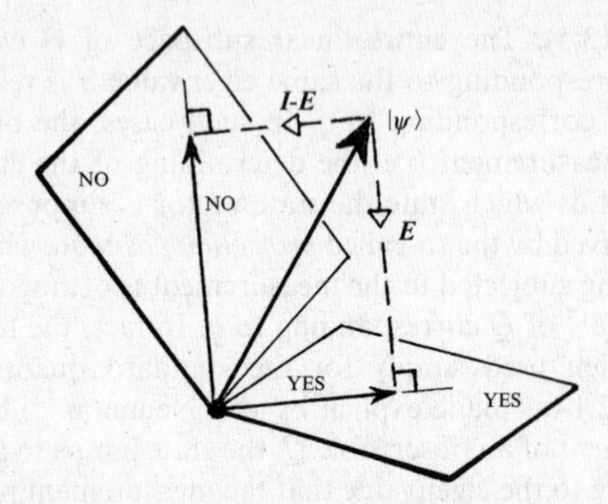

Fig. 22.4 The geometric nature, within **H**, of the projection postulate. The eigenspaces of the projector $\boldsymbol{E}$ are indicated, the horizontal plane representing the eigenvalue 1 (YES) and the vertical plane, the eigenvalue 0 (NO). The picture illustrates the decomposition $|\psi\rangle = \boldsymbol{E}|\psi\rangle + (\boldsymbol{I} - \boldsymbol{E})|\psi\rangle$ of $|\psi\rangle$ into two orthogonal parts, where $\boldsymbol{E}|\psi\rangle$ is the projection of $|\psi\rangle$ to within the YES space (the result of the measurement yielding YES) and $(\boldsymbol{I}-\boldsymbol{E})|\psi\rangle$, the projection to within the NO space (from the result NO). The probability in each case is given by the exact proportionality factor whereby the squared (Hermitian) length $\|\psi\|$ of $|\psi\rangle$ is reduced in the projection (state vectors not normalized).

There is a direct geometrical way expressing the probabilities of these two alternatives, namely the factor whereby the 'norm' (squared length) of the state gets reduced in each respective projection.[22.15] This simple geometrical fact is obscured if we insist on normalizing our states!

22.7 Null measurements; helicity

Some physicists have expressed doubts about the projection postulate (or that it is 'unnecessary' or 'unobservable'), the difficulty being that we may have no means of determining what the state has actually become after measurement, perhaps because the measurement process itself has caused the observed entity to become entangled with the measuring apparatus, so that the state of the entity being observed cannot be considered on its own. Indeed, that might sometimes be a complicating issue, but there are certainly circumstances where the projection postulate manifestly describes a (degenerate if necessary) measurement. The clearest case of this occurs with what is called a *null* (or *interaction-free*) measurement. This fascinating type of situation is of interest in its own right, and it illustrates one of the strangest aspects of quantum-mechanical behaviour. Accordingly, it will be worth our while having a look at an example or two.

[22.15] Why?

Let us consider a situation of the kind that was discussed in §21.7, where a single photon is aimed at a beam splitter, and its state is partially reflected and partially transmitted. After the encounter, the state is thus a sum of these two orthogonal parts, the transmitted part $|\tau\rangle$ and the reflected part $|\rho\rangle$ (where, to make it a nice straightforward sum, we absorb any relative phase factor into the definitions of $|\tau\rangle$ and $|\rho\rangle$ and we do not insist upon normalization):

$$|\psi\rangle = |\tau\rangle + |\rho\rangle$$

(see Fig. 22.5). Suppose that a detector is placed in the transmitted beam where, for the purposes of argument, we assume that the detector has 100% detection efficiency. Moreover, the photon source is to be such that each photon emission event is recorded (at the source) with 100% efficiency. (These are clearly idealizations; in an actual experiment it might be hard to come very close to such efficiencies. Nevertheless, these are reasonable idealizations to make, to illustrate how quantum mechanics works.) If we find that, on some occasions, the source has emitted a photon but the detector has not received it, then we can be sure that on these occasions the photon has 'gone the other way', and its state is therefore the reflected one: $|\rho\rangle$. The remarkable thing is that the measurement of *non*-detection of the photon has caused the photon's state to undergo a quantum jump (from the superposition $|\psi\rangle$ to the reflected state $|\rho\rangle$), despite the fact that the photon has not interacted with the measuring apparatus at all! This is an example of a *null* measurement.

An impressive use of this kind of thing has been suggested by Avshalom Elitzur and Lev Vaidman.[16] Let us think of our beam-splitter as being part of a Mach–Zehnder type of interferometer (recall the final part of my astronomical thought experiment described in §21.7; see Fig. 21.9), but

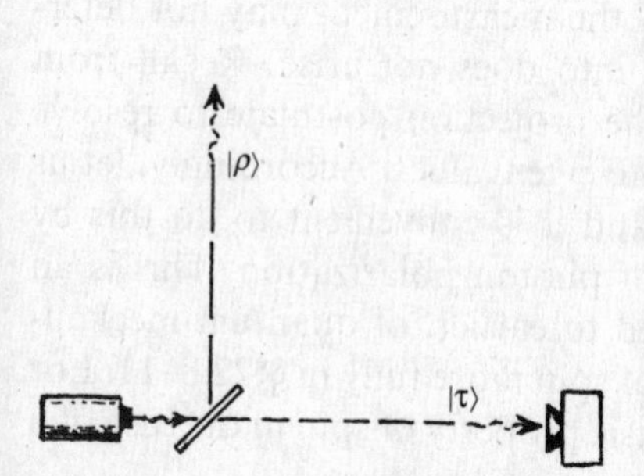

Fig. 22.5 Null measurement, requiring the projection postulate. A single photon is aimed at a beam-splitter. The resulting state $|\psi\rangle$, being partially reflected and partially transmitted, is the sum $|\psi\rangle = |\tau\rangle + |\rho\rangle$ of the transmitted part $|\tau\rangle$ and the reflected part $|\rho\rangle$ (absorbing any relative phase factor into the definitions and not insisting on normalization). If it is found that the source has emitted a photon but the detector has not received it, then we know that photon is in state $|\rho\rangle$ even though it has not interacted with the detector at all.

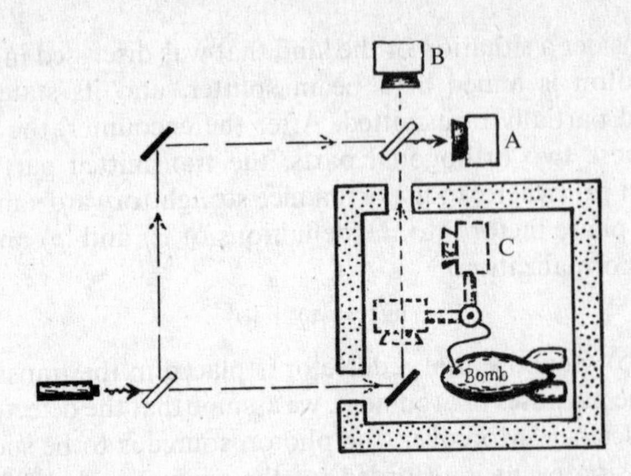

Fig. 22.6 Elitzur–Vaidman bomb test. A detector C, attached to a bomb, may or may not be inserted into a Mach–Zehnder type of interferometer (see Fig. 21.9). (The white thin rectangles specify beam splitters; the black ones, mirrors). Arm lengths within the interferometer are equal, so that a photon emitted by the source must reach detector A whenever C is not inserted. In the event that detector B receives the photon (without the bomb exploding), we know that C is in place in the beam, even though it has not encountered the photon.

where we do not know whether a detector C has, or has not, been placed in the transmitted beam of the first beam splitter. Let us suppose that the detector C triggers a bomb, so that the bomb would explode if C were to receive the photon. There are two final detectors A and B, and we know (from §21.7) that only A and not B can register receipt of the photon if C is absent. See Fig. 22.6. We wish to ascertain the presence of C (and the bomb) in some circumstance where we do not actually lose it in an explosion. This is achieved when detector B actually does register the photon; for that can occur only if detector C makes the measurement that it does *not* receive the photon! For then the photon has actually taken the other route, so that now A and B each has probability $\frac{1}{2}$ of receiving the photon (because there is now no interference between the two beams), whereas in the absence of C, only A can ever receive the photon.[17]

In the examples just given, there is no degeneracy, so the issue that was addressed above that the mere result of the measurement may not determine the state that the system 'jumps' into does not arise. Recall from §22.6 that we need the proper use of the projection postulate to resolve these ambiguities arising from degenerate eigenvalues. Accordingly, let us introduce another degree of freedom, and it is convenient to do this by taking into account the phenomenon of photon polarization. This is an example of the physical quality, referred to earlier, of quantum-mechanical spin. I shall be coming to the ideas of spin more fully in §§22.8–11. For the moment we shall only need a very basic property of spin in the case of a

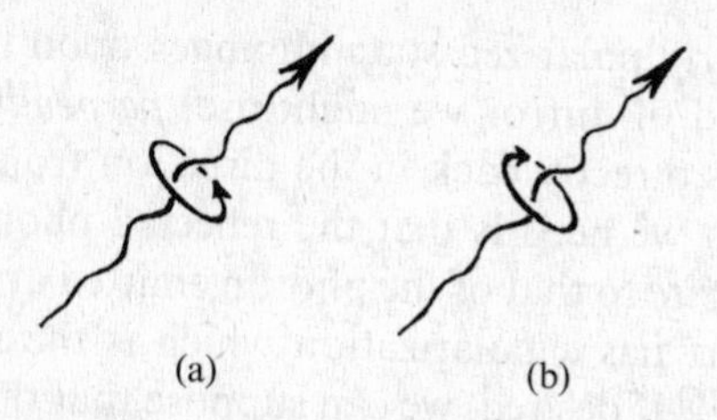

Fig. 22.7 A massless particle, such as a photon, can spin only about its direction of motion. The magnitude $|s|$ of this spin is always the same, for a given type of massless particle, but if the helicity s is non-zero (as is the case for a photon), then the spin can be either (a) right-handed ($s > 0$: positive helicity) or (b) left-handed ($s < 0$: negative helicity). For a photon, we have $|s| = 1$ (in units of $\hbar$), giving the two cases $s = 1$, for right-handed circular polarization, and $s = -1$, for left-handed circular polarization. By the quantum superposition principle, we can form complex linear combinations of these, yielding the other possible states of photon polarization, as shown in Figs. 22.12 and 22.13.

massless particle. Photons are indeed particles that possess spin but, being massless, their spin behaves in a way that is a little different from the more usual spin of a massive particle (e.g. an electron or proton) that we shall come to in §§22.8–10. We must think of a photon (or other massless particle) as necessarily spinning about its direction of motion; see Fig. 22.7.

The amount $|s|$ of this spin is always the same, for a given type of massless particle, but the spin can be either right-handed ($s > 0$) or left-handed ($s < 0$), about the direction of motion. In addition, in accordance with the general principles of quantum mechanics, the spin state can be any (quantum) linear combination of the two. The quantity s itself is called the *helicity* of the massless particle (§22.12), and its value always has to be an integer or half integer (or, bringing the appropriate units in, we ought to say that the helicity is an integral multiple of $\frac{1}{2}\hbar$). A massless particle is said to have *spin* j if $|s| = j$ (or, with units brought in, $|s| = j\hbar$). A photon has spin 1 (so its helicity is ± 1); a graviton has spin 2 (helicity ± 2). Neutrinos have spin $\frac{1}{2}$, and if there are massless neutrinos,[18] such a neutrino would have helicity $-\frac{1}{2}$, its corresponding antineutrino having helicity $\frac{1}{2}$.

In the case of a photon, the helicity states (states of definite helicity) are the states of *circular polarization*, right-handed for $s = 1$ and left-handed for $s = -1$, respectively. There are other possible states of polarization of a photon, such as plane polarization, but these are simply linear combinations of the right- and left-handed states. I shall come to the geometry of all this shortly, at the end of §22.9, but for the moment this will not be required. All that we need for now is one particular fact about how circular polarization behaves upon reflection. I am supposing that a photon in a circularly polarized state impinges upon the beamsplitter (or whatever other kind of mirror we might use) *perpendicularly*, so that the reflected beam goes directly back in the direction from which the photon came. The fact that we need is that the reflected photon's state of polarization is then *opposite* to that of the photon emitted at the source, whereas the transmitted part has a polarization which is the *same* as that of the emitted photon.[22.16] If desired, we can suppose that there is a very tiny tilt in the initial beam direction, so that the reflected photon beam does not simply re-enter the source. This will not significantly affect our considerations.

Let us return to our original 'null-measurement' experiment of Fig. 22.5, but with the photon now impinging perpendicularly, as in Fig. 22.8. Suppose that our source can be tuned so that it emits its photons in either a right-handed or a left-handed circularly polarized state. On a particular occasion, it emits a right-handed photon (and takes note of this fact). After the photon has encountered the beam splitter, the photon's state is now a linear combination (a *sum* with appropriate conventions about phase factors, as before):

$$|\psi+\rangle = |\tau+\rangle + |\rho-\rangle,$$

where the + or – inside the ket refers to the sign of the helicity. Let us place our detector in the transmitted beam, as before (and suppose that it is insensitive to polarization). Then if, as before, the source registers that it has emitted the right-handed photon but the detector fails to register, so that it has not received the photon, then it must be concluded that the state has jumped (upon 'non-detection' by the source) to the reflected left-handed state $|\rho-\rangle$. The point that I am making here, is that the full projection postulate is

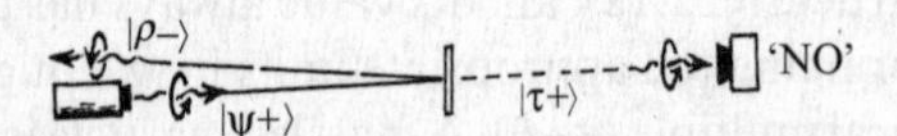

Fig. 22.8 A return to the experiment of Fig. 22.5, but now the photon impinges almost perpendicularly. The source emits a right-handed photon. After the photon has encountered the beam-splitter, its state is now $|\psi+\rangle = |\tau+\rangle + |\rho-\rangle$ where the '+' or '–' inside the ket refers to the sign of helicity. If the (polarization-insensitive) detector records not receiving the photon, we conclude that state has jumped (upon non-detection) to the reflected left-handed state $|\rho-\rangle$. This requires the full projection postulate (to the Lüders point; see Fig. 22.9), because there is a degeneracy both in the case NO (the 2-space spanned by $|\rho+\rangle$ and $|\rho-\rangle$) and in the case YES (that spanned by $|\tau+\rangle$ and $|\tau-\rangle$). The actual starting state $|\tau+\rangle + |\rho-\rangle$ is needed to determine where the state jumps to upon measurement (here a non-detection).

[22.16] Can you suggest a simple reason for this?

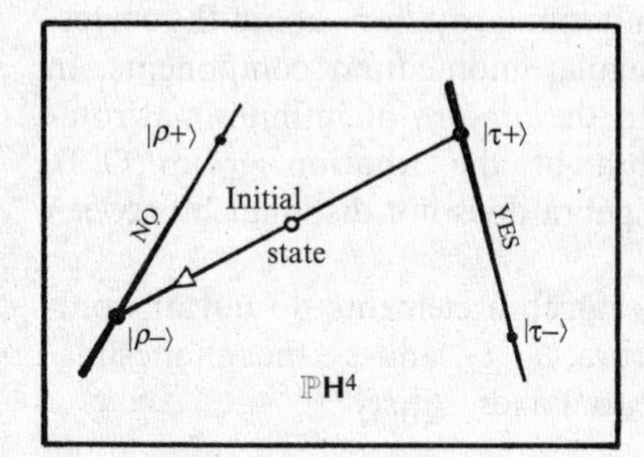

Fig. 22.9 A description in projective Hilbert space $\mathbb{P}\mathbf{H}^4$ (see Fig. 15.15) of the projection postulate of Fig. 22.4, for the photon polarization states in Fig. 22.8. The initial state is $|\tau+\rangle + |\rho-\rangle$ indicated within $\mathbb{P}\mathbf{H}^2$, the full space being spanned by $|\tau+\rangle$, $|\tau-\rangle$, $|\rho+\rangle$, and $|\rho-\rangle$. The white triangle arrow shows the projection to (the Lüders point) $|\rho-\rangle$, this being along the line which is the unique transversal to the YES and NO lines from the initial point $(|\tau+\rangle + |\rho-\rangle)$. The (non-)detection itself would merely tell us that the resulting state lies on the NO line, but the choice of the initial state breaks this degeneracy, according to the full projection postulate.

required to ascertain the nature of this resulting state; see Fig. 22.9. The measurement is of a purely YES/NO character, because the result is either 'non-detection' (NO) or 'detection' (YES). There is a degeneracy for both these alternatives, because the eigenspace of the NO answer is the 2-space spanned by $|\rho+\rangle$ and $|\rho-\rangle$, and the eigenspace of the YES answer is that spanned by $|\tau+\rangle$ and $|\tau-\rangle$. Since, in this case, the initial state is $|\tau+\rangle + |\rho-\rangle$, the projection postulate[19] correctly carries us to $|\rho-\rangle$, in the case of NO, rather than to $|\rho+\rangle$ or to $|\rho+\rangle + |\rho-\rangle$ (or to any other linear combination of $|\rho+\rangle$ and $|\rho-\rangle$) for the result of non-detection.[20,[22.17]]

22.8 Spin and spinors

This is hardly a very exciting experiment, but it illustrates a point. We shall be seeing some much more remarkable things in Chapter 23. But in preparation for this, it will be appropriate to say a little more about spin. What this refers to, in the case of a massive particle, is the angular momentum about its centre of mass.[21] In §§21.1–5, we encountered the significance of mass–energy conservation and momentum conservation, as features, respectively, of time-translation symmetry and space-translation symmetry of our quantum laws. Rotational symmetry, in a similar way, gives rise to *angular* momentum conservation (see also §18.7 and §20.6).

[22.17] Explain more fully why the correct answer is given by 'projection'.

For a massive particle, we can imagine that we are in the particle's rest frame, and then the relevant rotations are those which constitute the rotation group O(3) taken about the particle's location in that frame.

Corresponding to the way that a component of momentum, in quantum mechanics, would be represented as $i\hbar$ times the operator generating infinitesimal translations in the direction of the corresponding position coordinate (§§21.1,2) so also is a component of angular momentum represented as $i\hbar$ times the generator of infinitesimal rotations about the corresponding (Cartesian spatial) axis. Angular momentum components, in quantum mechanics, therefore refer to the *algebra* of infinitesimal rotations (§§13.6–8), i.e. the Lie algebra of the rotation group O(3), or equivalently SO(3), since the Lie algebra does not distinguish between the two.

Since SO(3) is non-Abelian, the Lie algebra elements do not all commute; in fact the generators of this algebra, $\boldsymbol{\ell}_1$, $\boldsymbol{\ell}_2$, and $\boldsymbol{\ell}_3$, the infinitesimal rotations about the three Cartesian spatial axes, satisfy[22.18]

$$\boldsymbol{\ell}_1\boldsymbol{\ell}_2 - \boldsymbol{\ell}_2\boldsymbol{\ell}_1 = \boldsymbol{\ell}_3, \qquad \boldsymbol{\ell}_2\boldsymbol{\ell}_3 - \boldsymbol{\ell}_3\boldsymbol{\ell}_2 = \boldsymbol{\ell}_1, \qquad \boldsymbol{\ell}_3\boldsymbol{\ell}_1 - \boldsymbol{\ell}_1\boldsymbol{\ell}_3 = \boldsymbol{\ell}_2.$$

These are related, according to the rules of quantum mechanics, to the components $\boldsymbol{L}_1$, $\boldsymbol{L}_2$, $\boldsymbol{L}_3$ of angular momentum about the three axes, according to:

$$\boldsymbol{L}_1 = i\hbar\boldsymbol{\ell}_1, \quad \boldsymbol{L}_2 = i\hbar\boldsymbol{\ell}_2, \quad \boldsymbol{L}_3 = i\hbar\boldsymbol{\ell}_3.$$

So our angular momentum commutation rules are[22]

$$\boldsymbol{L}_1\boldsymbol{L}_2 - \boldsymbol{L}_2\boldsymbol{L}_1 = i\hbar\boldsymbol{L}_3, \quad \boldsymbol{L}_2\boldsymbol{L}_3 - \boldsymbol{L}_3\boldsymbol{L}_2 = i\hbar\boldsymbol{L}_1, \quad \boldsymbol{L}_3\boldsymbol{L}_1 - \boldsymbol{L}_1\boldsymbol{L}_3 = i\hbar\boldsymbol{L}_2.$$

As with practically everything else in quantum mechanics, the angular momentum components $\boldsymbol{L}_1$, $\boldsymbol{L}_2$, $\boldsymbol{L}_3$ must act as a linear operators on the Hilbert space $\mathbf{H}$. Thus, quantum systems possessing angular momentum provide a representation of the Lie algebra of SO(3) in terms of linear transformations of $\mathbf{H}$ (see §§13.6–8,10, §14.6).

This leads to one of the most elegant and revealing aspects of quantum mechanics, and it is a subject which amply repays much detailed study. This is not the place for full detail, however, and I shall try to provide only a few points of particular significance. In the first place, we take note of the fact that the explicit matrices

$$\boldsymbol{L}_1 = \frac{\hbar}{2}\begin{pmatrix} 0 & 1 \\ 1 & 0 \end{pmatrix}, \boldsymbol{L}_2 = \frac{\hbar}{2}\begin{pmatrix} 0 & -i \\ i & 0 \end{pmatrix}, \boldsymbol{L}_3 = \frac{\hbar}{2}\begin{pmatrix} 1 & 0 \\ 0 & -1 \end{pmatrix},$$

[22.18] Use quaternions to check this.

called (without the $\hbar/2$) the *Pauli matrices*, satisfy the required commutation relations.[22.19] They provide the simplest (non-trivial) representation of angular momentum, and we imagine that these 2×2 matrices act on a wavefunction with two components $\{\psi_0(\boldsymbol{x}), \psi_1(\boldsymbol{x})\}$ (thought of as a column vector). When we start to rotate this state, the components $\psi_0(\boldsymbol{x})$ and $\psi_1(\boldsymbol{x})$ get churned around, in accordance with the matrix multiplication rules that the Pauli matrices generate.

We can label this 2-component wavefunction ψ_A, using a lower index A (which takes the values 0 and 1, or else we can think of this as an abstract index according to the 'abstract-index notation' referred to in §12.8). The quantity described by ψ_A is called a *spinor*, and its index A is referred to as a *2-spinor* index. It turns out that ψ_A is indeed a spinorial object in the sense described in §11.3 (a continuous 2π rotation takes it to its negative). Indeed, if we continuously 'exponentiate' (see §14.6) one of the Pauli matrices until we get an entire rotation through 2π, we find that we get the operator $-\boldsymbol{I}$, which sends ψ_A to $-\psi_A$.[22.20]

This notation is part of a powerful formalism that can be developed to supplement (or even replace[23]) the formalism of tensor calculus, by using 'tensor-like' quantities built up from things like 'ψ_A'. Although not fully needed here, its real power arises when we take advantage of the relativistic version of this formalism. For this, we also need 'primed' indices $A', B', C', \ldots$, in addition to the 'unprimed' ones $A, B, C, \ldots$, the primed and unprimed indices being, in an appropriate sense, *complex conjugates* of one another; see §13.9. The notation has great value in quantum field theory (a fact perhaps less well appreciated than it ought to be; [24] see §25.2 and §34.3) and in general relativity[25] (and it plays a basic role in twistor theory; see §33.6). It is not appropriate for me to enter into this at the present stage (although we shall come back to it in §25.2), but it will be helpful to borrow a little from this 2-spinor formalism. All that we need of it here and in §§22.9–11 is to represent general spin states in a neat way. We shall not need the primed indices (until §§25.2,3 and §§33.6,8) since we are only doing non-relativistic physics here.

Before entering into this, I wish to make a notational simplification. For the remainder of this section, and up until the end of §22.11, I shall adopt the convenient assumption that units have been chosen so that $\hbar = 1$. In fact, this is always possible—and we shall be seeing in §27.10 (and §31.1) that we could go much further than this and describe things in terms of what are called 'Planck units', where the speed of light and the gravitational constant are also both set equal to unity. There is no need to go this far here, and in any case it is not hard to reinstate $\hbar$, if required, from simple

[22.19] Check this. Explain how their multiplication rules relate to those of quaternions.

[22.20] Do this explicitly.

considerations of physical dimensions. (For example, to reinstate $\hbar$ in any physical formula for which $\hbar$ has been set to unity, we replace any quantity scaling as the qth power of mass, ignoring length and time, by $\hbar^{-q}$ times that quantity. In particular, mass, energy, momentum, and angular momentum would simply be divided by $\hbar$.)

Now, returning to the 2-spinor formalism, we recall that a univalent spinor quantity ψ_A can be used to describe a particle of spin $\frac{1}{2}$. The same kind of notation can be adopted for higher values of the spin, corresponding to other representations of the Lie algebra of SO(3). The value of the spin is always a non-negative integer multiple of $\frac{1}{2}$:

$$0, \tfrac{1}{2}, 1, \tfrac{3}{2}, 2, \tfrac{5}{2}, \ldots$$

(or, reinstating $\hbar$ we would say that the spin/$\hbar$ takes these values) and the wavefunction can be described by an object $\psi_{AB\ldots F}$ (a 'spin-tensor') which is completely symmetric in its n indices in the case of spin $\frac{n}{2}$

$$\psi_{AB\ldots F} = \psi_{(AB\ldots F)}$$

(where the round brackets denote symmetrization over all n indices; see §12.7). In fact, all representations of SO(3)—where we include the 2-valued spinorial ones—can be built up as *direct sums* of these particular ones, the *irreducible* representations (see §13.7). This amounts to saying that the general representation can be expressed as a (possibly infinite) collection of wavefunctions

$$\{\psi_{AB\ldots F}, \phi_{GH\ldots K}, \chi_{LM\ldots R}, \ldots\},$$

each of which is totally symmetric in its spinor indices.

For an individual particle, there would be only one such symmetric field, e.g. $\psi_{AB\ldots F}$, for its wavefunction. (It would be an understandable mistake to think that for two particles there would be two of them separately, for three particles, three of them, etc. We shall be seeing how systems of more than one particle are actually described in the next chapter. It is something distinctly more subtle than this.) For a spin 0 particle, called a *scalar* particle (such as a π meson), the wavefunction has 0 indices, and this was the situation treated in Chapter 21. The most familiar particles, electrons, muons, neutrinos, protons, neutrons, and also their constituent quarks, all have spin $\frac{1}{2}$ (just 1 index). The deuteron (nucleus of heavy hydrogen) and the W-boson (see §25.4) have spin 1 (2 symmetric spinor indices). Many heavier nuclei, or even whole atoms, can be treated like single particles with much higher spin. For spin $\frac{1}{2}n$, the n-index object $\psi_{\mathrm{AB\ldots F}}$ has $n+1$ independent[26] complex components.[22.21] Although the spin-tensor $\psi_{AB\ldots F}$ is frequently referred to as an n-index *spinor*, it is a

[22.21] See if you can work this out, from the information given.

spinorial object (§11.3) only when n is odd, these being the cases where the spin is half-odd integral, not integral. It should be remarked, also, that the spin value itself, $j = \frac{1}{2}n(\geq 0)$, determines (and is determined by) the eigenvalue $j(j+1)$ of the 'total spin' operator[27]

$$\mathbf{J}^2 = L_1^2 + L_2^2 + L_3^2;$$

this being the 'squared length' of the 3-vector operator $\mathbf{J} = (L_1, L_2, L_3)$.

The total spin $\mathbf{J}^2$ *commutes*[22.22],[22.23] with each component L_1, L_2, L_3 of angular momentum (despite the fact that these components do not commute among themselves). This property characterizes $\mathbf{J}^2$ as a *Casimir operator* for SO(3); see §22.12. To delineate quantum states completely, we usually form a *complete set of commuting operators* (§22.12), and look for states that are simultaneously eigenstates of all the operators of the set. For angular momentum, this is normally done by taking the operator L_3 of angular momentum about the *upward* ('z') direction, to accompany $\mathbf{J}^2$. The two 'quantum numbers' j and m are then taken to label the state, where $j(j+1)$ is the eigenvalue of $\mathbf{J}^2$, and m is the eigenvalue of L_3. We take $j \geq 0$ and $-j \leq m \leq j$, where j and m are both half odd integers (spinorial case) or both integers. The $2j+1$ $(=n+1)$ different possible m values correspond to the different components of $\psi_{AB...F}$.

The choice of the upward direction is, of course arbitrary (and it corresponds to choosing an up/down basis (the $|\uparrow\rangle, |\downarrow\rangle$ of §22.9) for the spinor components. Any other spatial direction could equally well be chosen in place of 'up'. Accordingly, I shall occasionally refer to the 'm value' in some *other* given direction (as with the Majorana description of §22.10).

22.9 The Riemann sphere of two-state systems

Let us consider the remarkably concise—even magical—quantum geometry of the individual spin states for spin $\frac{1}{2}$ (e.g. electron, proton, neutron, quark). This is also illuminating for the understanding of 2-state quantum systems generally. Such a system is described by a complex 2-dimensional Hilbert space $\mathbf{H}^2$, and the case of spin $\frac{1}{2}$ nicely represents its geometry.

For our spin $\frac{1}{2}$ particle, we shall be concerned with the *spin degree of freedom* only, in the particle's rest frame. To make this explicit, we can

---

[22.22] Check this commutation directly from the angular-momentum commutation rules.

[22.23] Consider the operators $L^+ = L_1 + iL_2$ and $L^- = L_1 - iL_2$ and work out their commutators with L_3. Work out $\mathbf{J}^2$ in terms of $L^\pm$ and L_3. Show that if $|\psi\rangle$ is an eigenstate of L_3, then so also is each of $L^\pm|\psi\rangle$, whenever it is non-zero, and find its eigenvalue in terms of that of $|\psi\rangle$. Show that if $|\psi\rangle$ belongs to a finite-dimensional irreducible representation space spanned by such eigenstates, then the dimension is an integer $2j$, where $j(j+1)$ is the eigenvalue of $\mathbf{J}^2$ for all states in the space.

imagine that the particle is 'at rest' in the sense that it is in the eigenstate of zero momentum, so its state has to be constant[22.24] in the space variables $\mathbf{x}$. Then ψ_0 and ψ_1 are just complex numbers, say $\psi_0 = w$ and $\psi_1 = z$, and we write the state as $\{w, z\}$. We can arrange 'spin up' $|\Uparrow\rangle$ (right-handed about the upward vertical) to be spin state $\{1, 0\}$; correspondingly, 'spin down' $|\Downarrow\rangle$ (right-handed about the downward vertical) is to be $\{0, 1\}$. These two basis states are orthogonal:

$$\langle\Uparrow|\Downarrow\rangle = 0.$$

We also normalize:

$$\langle\Uparrow|\Uparrow\rangle = 1 = \langle\Downarrow|\Downarrow\rangle.$$

The general spin-$\frac{1}{2}$ state $\psi_A = \{w, z\}$ (general element of $\mathbf{H}^2$), is the linear combination

$$\{w, z\} = w|\Uparrow\rangle + z|\Downarrow\rangle$$

of these two basis states. The scalar product of another general state $\{a, b\}$ (i.e. $a|\Uparrow\rangle + b|\Downarrow\rangle$) with $\{w, z\}$ is given by[22.25]

$$\langle\{a, b\}|\{w, z\}\rangle = \bar{a}w + \bar{b}z.$$

It now turns out that every spin-$\frac{1}{2}$ state must actually be a pure state of spin that is right-handed about *some* direction in space, so we can write (say)

$$w|\Uparrow\rangle + z|\Downarrow\rangle = |\nearrow\rangle,$$

where '$\nearrow$' is some actual direction in space![22.26] This gives us a remarkable identification between the projective space $\mathbb{P}\mathbf{H}^2$ (§15.6) and the geometry of directions in space, these directions being thought of as spin directions. The physically distinct spin $\frac{1}{2}$ states are indeed provided by this projective space (see §21.9), the different points of $\mathbb{P}\mathbf{H}^2$ being labelled by the distinct ratios

$$z : w.$$

[22.24] Why?

[22.25] Obtain this expression from what has been said above.

[22.26] See if you can derive this fact in two different ways: (i) finding the direction explicitly in some suitable Cartesian frame, where the state $\{a, b\}$ defines b/a as a point on the complex plane of Fig. 8.7a; (ii) without direct calculation, using the fact that because $\mathbf{H}^2$ is a representation space of SO(3), every direction of spin is included, yet $\mathbb{P}\mathbf{H}^2$ is not 'big enough' to contain any more states than this.

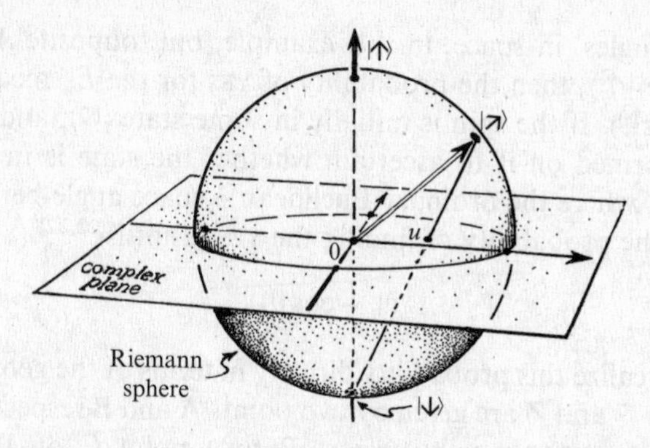

Fig. 22.10 The projective space $\mathbb{P}\mathbf{H}^2$ for 2-state system is a Riemann sphere (see Fig. 8.7). For the spin states of a massive particle of spin $\frac{1}{2}$, we can use the north pole to represent the spin state $|\uparrow\rangle$ (spin 'up') and the south pole, the state $|\downarrow\rangle$ (spin 'down'). A general spin state $|\nearrow\rangle$ is represented (with appropriate phases for $|\uparrow\rangle$ and $|\downarrow\rangle$) by the point on the sphere whose direction out from the centre is that of $|\nearrow\rangle$ (i.e. which gives the result 'YES' with *certainty* for a spin measurement $\boldsymbol{E}_{\nearrow}$ in that direction), as illustrated by the double-shafted arrow. We can express the state $|\nearrow\rangle$ as a linear combination $|\nearrow\rangle = w|\uparrow\rangle + z|\downarrow\rangle$ (where we can regard the complex numbers z, w as the components $w = \psi_0$, $z = \psi_1$ of a 2-spinor ψ_A). The points on the sphere correspond to the distinct ratios z:w. Each of these can be represented by a complex number $u = z/w$ (allowing ∞) in the complex plane, this plane being taken to be the equatorial plane of the sphere. The point u projects stereographically from the south pole to the point on the sphere representing $|\nearrow\rangle$.

In other words, $\mathbb{P}\mathbf{H}^2$ is just a copy of our old friend the *Riemann sphere* that we first became acquainted with in §8.3. Each point of this Riemann sphere labels a distinct spin-$\frac{1}{2}$ state, this being the '$m = \frac{1}{2}$' eigenstate of the particular spin measurement that is taken in the direction out to this point, from the centre of the sphere (Fig. 22.10).

We see this geometrical relationship more explicitly if we use the stereographic projection of the sphere from its south pole to its equatorial plane described in §8.3 (Fig. 8.7a). This plane is to be regarded as the complex plane of the *ratio* $u = z/w$ (rather than of the 'z' of §8.3) of quantum-mechanical amplitudes z and w. This relates the particular point on the sphere, corresponding to the spatial direction $\nearrow$, directly to the ratio z/w.

Let us use the projector $\boldsymbol{E}_{\nearrow}$ to denote the measurement that asks the question 'is the spin in the direction $\nearrow$?', so the eigenvalue is 1 (YES) if the spin state is found to be (or is projected to) $|\nearrow\rangle$ and it is 0 (NO) if the spin is thereby projected to the orthogonal spin state $|\swarrow\rangle$ in the *opposite* spatial direction (corresponding to the antipodal point on the Riemann sphere). (Note that 'orthogonal' in the Hilbert space does not correspond to 'at right angles' in space, in this example, but 'opposite'.) If we start with the state $|\uparrow\rangle$, then the probability of YES for the $\boldsymbol{E}_{\nearrow}$ measurement is $|w|^2/(|w|^2+|z|^2)$. If the spin is initially in some state $|\nwarrow\rangle$, and a measurement is performed on it to ascertain whether the state is in some other direction $|\nearrow\rangle$, where the ordinary Euclidean 3-space angle between $\nwarrow$ and $\nearrow$ is θ, then the probability of finding the YES result is[22.27]

$$\tfrac{1}{2}(1+\cos\theta).$$

We can also realize this probability directly in terms of the geometry of the sphere, where $\nwarrow$ and $\nearrow$ are given by two points A and B respectively on the sphere, and we orthogonally project B to a point C on the diameter through A (Fig. 22.11). If A′ is the point antipodal to A, then the probability of YES is the length A′C divided by the sphere's diameter AA′.[22.28]

Note that the 'Riemann sphere' used here has more structure than that of §8.3 and the celestial sphere of §18.5, in that now the notion of 'antipodal point' is part of the sphere's structure (in order that we can tell which states are 'orthogonal' in the Hilbert-space sense). The sphere is now a 'metric sphere' rather than a 'conformal sphere', so that its symmetries are given by rotations in the ordinary sense, and we lose the conformal motions that were exhibited in aberration effects on the celestial sphere. Nevertheless, our present use of the Riemann sphere clearly exhibits an explicit connection between the complex-number ratios that arise in quantum mechanics and ordinary directions in space. We see that the complex numbers that appear in the quantum-state formalism are not completely abstract things; they are intimately related to geometrical and dynamical behaviour. (Recall, also, the role of the complex phases in determining the dynamics of a momentum state, as described in §21.6.)

It should be pointed out that the geometry of Fig. 22.11, expressing the probabilities that arise in a quantum measurement in relation to $\mathbb{P}\mathbf{H}^2$, is not restricted to the case of spin, but is quite general for a 2-state system. What is special for the case of spin $\frac{1}{2}$ is the immediate association between oridinary *spatial directions* with the points of the Riemann sphere $\mathbb{P}\mathbf{H}^2$. The Riemann sphere is always there, in a 2-state system, providing the 'quantum spread' of a pair of classical alternatives. In many physical situations, however, the geometrical role of this sphere, and of the underlying quantum-mechanical complex numbers (amplitudes), is not very direct, and there is a tendency for physicists to regard them as entirely 'formal' quantities. This attitude arises partly from the fact that the overall

[22.27] Show this.

[22.28] Confirm this.

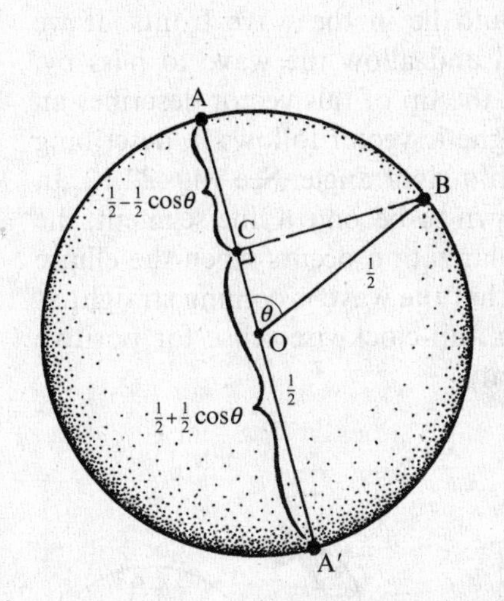

Fig. 22.11 Suppose that the initial state of a 2-state system (like that of Fig. 22.10) is represented by the point B on the Riemann sphere and we wish to perform a YES/NO measurement corresponding to some other point on the sphere, where YES would find the state at A and NO would find it at the point A′, antipodal to A. Taking the sphere to have radius $\frac{1}{2}$, and projecting B orthogonally to C on the axis A′A, we find that the probability of YES is the length A′C, which is $\frac{1}{2}(1+\cos\theta)$, and the probability of NO is the length CA, which is $\frac{1}{2}(1-\cos\theta)$, where θ is the angle between OB and OA, the sphere's centre being O.

phase of the state vector for an entire physical system is taken to be unobservable, so people often ignore the potential geometrical richness of the internal complex cofficients. The *relative* phases between one part and another certainly play an observable role. One way of expressing this is in the fact that the complex geometry of the entire projective Hilbert space $\mathbb{P}\mathbf{H}$ for a system *is* physically meaningful. Although the overall phase is completely taken out in the definition of $\mathbb{P}\mathbf{H}$, all relative phases feature in its geometry. Indeed, there are elegant approaches to quantum mechanics that exploit the complex projective geometry of $\mathbb{P}\mathbf{H}$.[28]

There are also other situations in which the Riemann sphere's geometry relates the complex numbers of quantum mechanics directly to spatial properties of spin. Most significantly, this applies to the general spin states of a higher-spin massive particle, as will be described shortly (in §22.11). But to end this section, let us return to the photon polarization that we briefly encountered in §22.7. Recall that the general polarization state of a photon is a complex linear combination of the states of positive helicity $|+\rangle$ and negative helicity $|-\rangle$:

$$|\phi\rangle = w|+\rangle + z|-\rangle.$$

The physical interpretation of such a state is in terms of what is called *elliptical polarization* which generalizes the particular cases of plane polarization and circular polarization. It is not my purpose to describe this in full detail here, but a good enough picture is obtained if we think in terms of a classical electromagnetic plane wave. The 'planes' are the wave fronts, which are perpendicular to the direction of motion. At each point in space, there will be an electric vector **E** and a magnetic vector **B**, and for a plane wave these are always perpendicular and lie in the wave fronts. If we imagine keeping a point in space fixed and allow the wave to pass by, the electric vector swings round so that the tip of this vector describes an ellipse in the wave-front plane. The magnetic vector follows it, describing an identical ellipse, but rotated through a right angle. See Fig. 22.12. In particular cases, the ellipse squashes down to become a line segment: the cases of *plane* polarization. *Circular* polarization occurs when the ellipse becomes a circle. If we orient things so that the wave is coming straight at us, then the vectors swing round in an anti-clockwise sense for positive helicity and clockwise for negative helicity.

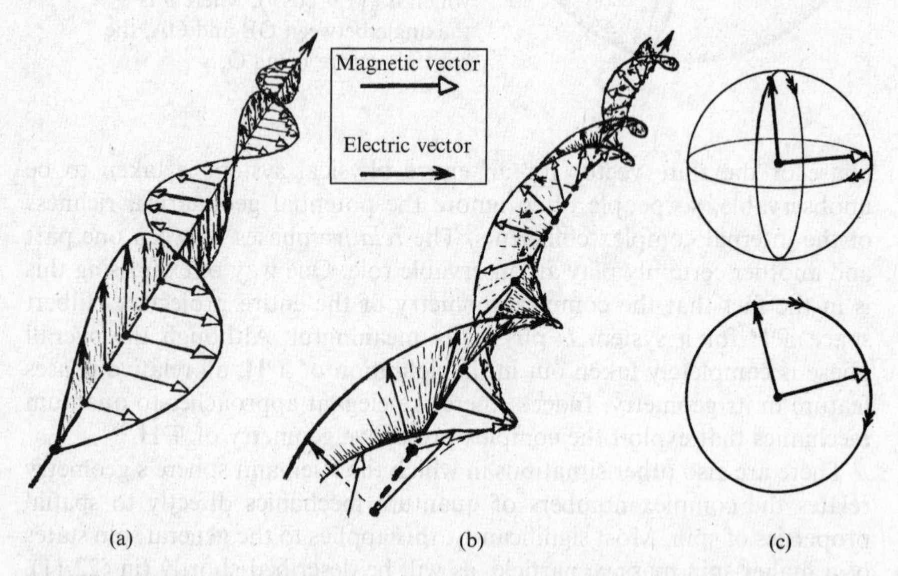

Fig. 22.12 Photon polarization (see Fig. 21.7) as a feature of electromagnetic plane waves. (a) A plane-polarized wave receding from the viewer. The electric vectors (black-headed arrows) and magnetic vectors (white-headed arrows) each oscillate back and forth, in two fixed perpendicular planes. (b) In a circularly polarized plane wave, the electric and magnetic vectors rotate about the direction of motion, always remaining perpendicular and of equal constant length. (c) Viewed from behind, the diagrams show how the electric and magnetic vectors rotate along the wave (positive helicity case), the lower figure showing the situation for circular polarization, and the upper one, for the general case of elliptical polarization, the heads of the two arrows tracing out congruent ellipses with perpendicular major axes. A single photon's wavefunction would exhibit behaviour of this kind.

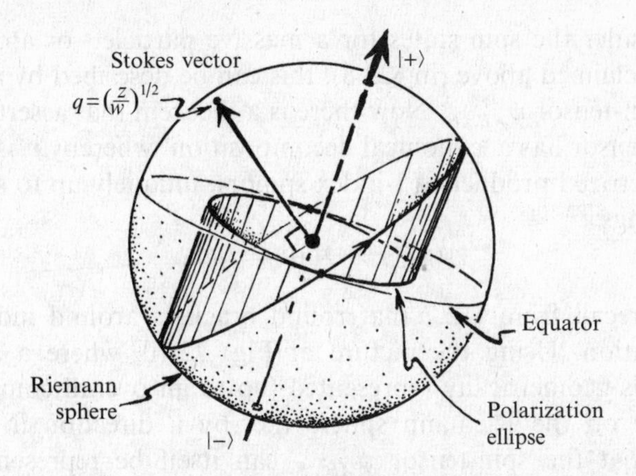

Fig. 22.13 Photon polarization states represented on the Riemann sphere. Take the north pole to represent the positive helicity state $|+\rangle$ and south pole, the negative helicity state $|-\rangle$, where we think of the photon's momentum to be in the direction of north. The general polarization state $w|+\rangle + z|-\rangle$ is represented by the point $q = (z/w)^{1/2}$ on the Riemann sphere. Consider the semi-diameter of the sphere out to q, called the 'Stokes vector', and draw the great circle lying in the diametral plane perpendicular to it. Orient this circle right-handed about the Stokes vector. Then project this circle orthogonally down to the sphere's equatorial plane. This gives us the required polarization ellipse and the correct orientation.

Let us see how the Riemann sphere fits in with all this. Let us take the north pole to represent the positive helicity state $|+\rangle$, and the south pole to represent the negative helicity state $|-\rangle$. We assume that the photon is travelling upwards, in the direction of $|+\rangle$. Now, instead of marking z/w on the Riemann sphere, we are going to look at its square root $q = (z/w)^{1/2}$ (it doesn't matter which one), so $q = 0, \infty$ give $|+\rangle, |-\rangle$, respectively. Let us consider the radius of the sphere out through q ('Stokes vector') and draw the great circle on the sphere lying in the diametral plane perpendicular to this line. Orient this circle in a right-handed sense about the vector pointing out to q. Then project this circle orthogonally to the sphere's equatorial plane. We get the required polarization ellipse, together with the correct orientation. See Fig. 22.13.[22.29]

22.10 Higher spin: Majorana picture

As a further example, illustrating the close relation between the apparently abstract complex numbers in quantum mechanics and the geometry of

[22.29] Verify all this. Why do I not worry about the *sign* of q?

space, consider the spin states for a massive particle—or atom—of spin $j = \frac{1}{2}n$. As claimed above (in §22.8), this can be described by a symmetric n-index spin-tensor $\psi_{AB...F}$. Now there is a theorem that asserts that every such spin-tensor has a 'canonical decomposition' whereby it is expressible as a symmetrized product of 1-index spinors, uniquely up to scale factors and orderings:[22.30]

$$\psi_{AB...F} = \alpha_{(A}\beta_B \cdots \varphi_{F)},$$

where we recall from §12.7 that round brackets around indices denote symmetrization. Using the picture of Fig. 22.10, where a single-index spinor ψ_A is geometrically represented (up to an overall complex factor) by a point on the Riemann sphere (i.e. by a direction in space), we conclude that the spin-tensor $\psi_{AB...F}$ can itself be represented on the Riemann sphere, up to an overall scale factor, as an unordered set of n points on the sphere (i.e. n unordered directions in space); see Fig. 22.14. This representation of a general spin n state is called the *Majorana* description. It was found originally in 1932 (but by a different kind of procedure[29] which I shall indicate briefly in §22.11) by the brilliant Italian physicist Ettore Majorana. (At the young age of 31, he disappeared mysteriously from a ship in the Bay of Naples, perhaps by suicide.)

There is a standard basis of states for spin $j = \frac{1}{2}n$. In the Majorana description, these are realized as the states for which the points in the Majorana description are all at either the north or south pole:

$$|\Uparrow\Uparrow\Uparrow\ldots\Uparrow\rangle,\ |\Downarrow\Uparrow\Uparrow\ldots\Uparrow\rangle,\ |\Downarrow\Downarrow\Uparrow\ldots\Uparrow\rangle,\ \ldots,\ |\Downarrow\Downarrow\Downarrow\ldots\Downarrow\rangle.$$

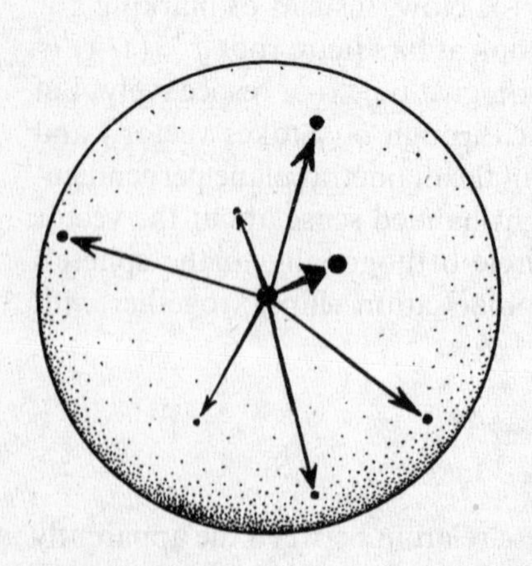

Fig. 22.14 Majorana's description of the general (projective) spin state for a massive particle of spin $\frac{n}{2}$, given by n unordered points on the Riemann sphere. We can think of the vectors out from the centre to each of these points to be contributing spin $\frac{1}{2}$, in accordance with the prescription of Fig. 22.10. The symmetrical product of these spins gives the total. (In 2-spinor notation, the complete spin state is the symmetric n-valent spinor, which factorizes $\psi_{AB...F} = \alpha_{(A}\beta_B \cdots \varphi_{F)}$, where $\alpha_A, \beta_A, \ldots, \varphi_A$, determine the n points, as in Fig. 22.10.)

[22.30] See if you can prove this, using the 'fundamental theorem of algebra' stated in Note 4.2. *Hint*: Consider the polynomial $\psi_{AB...F}\zeta^A\zeta^B \cdots \zeta^F$, where the components of ζ^A are $\{1, z\}$.

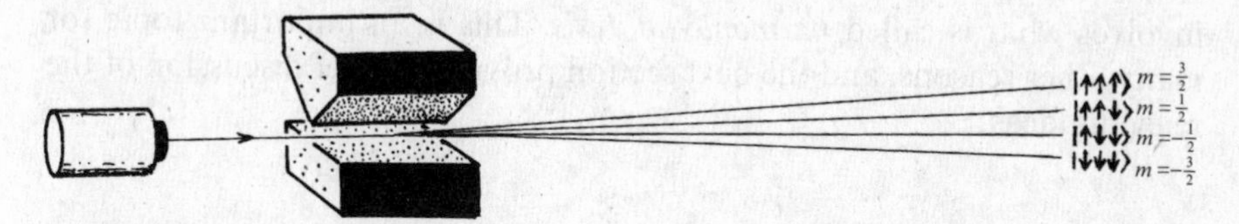

Fig. 22.15 Stern–Gerlach apparatus, used to measure the 'm-value' of an atom's magnetic moment (coupled to its spin). The atoms pass through a strongly inhomogeneous magnetic field, deflecting their paths slightly differently for each m-value.

These $n+1$ states are the eigenstates of the observable $\boldsymbol{L}_3$ (the x^3 axis being the 'up' direction) and are therefore all orthogonal to one another. They can be distinguished by the $n+1$ different spin eigenvalues, called *m values* (§22.8), respectively: $j, j-1, j-2, \ldots, -j$. We shall see a little more about this in §22.11.

There is a standard measuring device, known as a Stern–Gerlach apparatus, which can often be used to measure this 'm value' for an atom. For this to work, we require that the atom possess a magnetic moment (so it is a tiny magnet), where the magnetic-moment vector is a certain multiple of the spin vector. The atoms are passed through a strongly inhomogeneous magnetic field. This deflects their paths slightly differently for each m value, since m determines how each atom's magnetic moment vector is oriented in relation to the inhomogeneous magnetic field; see Fig. 22.15.

Although the states for each different m value are all orthogonal to one another, the orthogonality conditions for *general* Majorana descriptions are complicated.[30] It may be remarked, however, that a Majorana state for which some direction $\nearrow$ features, is necessarily orthogonal to the state $|\swarrow\swarrow\swarrow\ldots\swarrow\rangle$, where $\swarrow$ is diametrically opposite to $\nearrow$. Moreover, if $\nearrow$ features in the Majorana description with multiplicity r, then the state is orthogonal to any other spin $\frac{1}{2}n$ state whose Majorana description involves the opposite direction $\swarrow$ to multiplicity at least $n-r+1$.[22.31]

These results enable us to interpret the Majorana directions physically. The Majorana directions are precisely those for which a Stern–Gerlach measurement in that direction has probability zero to find the spin to be entirely in the opposite direction. For a Majorana direction of multiplicity r, the probability is zero for the m value in that direction to be anything from $-j$ to $-j+r-1$.[31]

It should be pointed out that the procedure, outlined above, for representing the general spin state for a massive particle is not very familiar to most physicists. Instead, they would adopt a different procedure which

[22.31] See if you can show all this using the geometry of §22.9. Apply this result to the orthogonality of the various eigenstates of L_3.

involves what is called *harmonic analysis*. This is an important topic for many other reasons, and the next section provides a brief discussion of the relevant ideas.

22.11 Spherical harmonics

In §20.3, we encountered the classical theory of vibrations (of small-amplitude and without dissipation). Our main discussion was concerned with systems having a finite number of degrees of freedom. But also (briefly)—as with the vibrations of a drum or a column of air—we considered systems for which the number of degrees of freedom is treated as being infinite. These vibrations (in either case) are composed of normal modes, each having its own frequency of vibration called a normal frequency. If the vibrating object is *compact* (see §12.6, Figs. 12–14, for the meaning of this term), its modes will constitute a discrete family, providing a discrete spectrum of different normal frequencies. In the particular case of a sphere S^2, the different modes of vibration (which we can visualize as the vibrational modes of a soap bubble, say, or a spherical balloon) correspond to what are called *spherical harmonics*. What has this to do with the quantum mechanics of angular momentum? We shall be seeing shortly.

To classify these harmonics, we look for eigenstates of the Laplacian operator ∇^2 defined on S^2. We have encountered the ordinary 2-dimensional Laplacian in §10.5, defined on the Euclidean plane by $\nabla^2 = \partial^2/\partial x^2 + \partial^2/\partial y^2$. On the unit sphere S^2, this expression must be modified to take into account the curved metric. This metric form is

$$ds^2 = g_{ab}dx^a dx^b = d\theta^2 + \sin^2\theta\, d\phi^2$$

in the usual *spherical polar* coordinates (θ, ϕ)—which is a labelling of points on S^2 for which Cartesian coordinates (in ordinary 3-space) of a point are $x = \sin\theta\cos\phi$, $y = \sin\theta\sin\phi$, $z = \cos\theta$ (Fig. 22.16). Thus, ϕ is essentially the *longitude* and $\frac{1}{2}\pi - \theta$ the *latitude* (all in radians). The Laplacian (with covariant derivative ∇_a; see §14.3) is[22.32]

$$\nabla^2 = g^{ab}\nabla_a\nabla_b$$
$$= \frac{\partial^2}{\partial\theta^2} + \frac{\cos\theta}{\sin\theta}\frac{\partial}{\partial\theta} + \frac{1}{\sin^2\theta}\frac{\partial^2}{\partial\phi^2}.$$

The possible eigenvalues of ∇^2 turn out to be the numbers $-j(j+1)$ (for $j = 0, 1, 2, 3, \ldots$), so

$$\nabla^2\Phi = -j(j+1)\Phi,$$

[22.32] Can you derive this spherical polar expression?

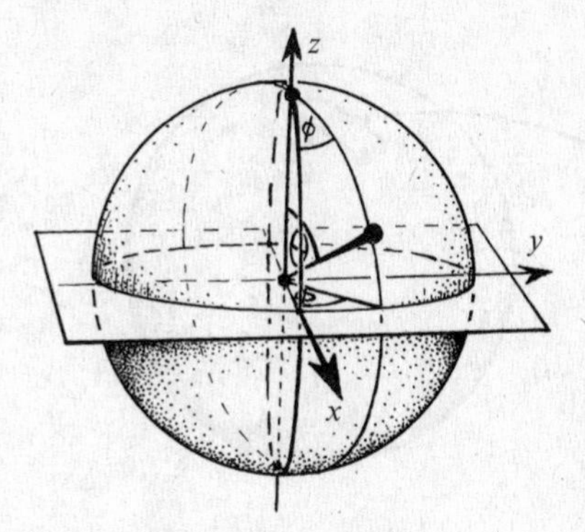

Fig. 22.16 Standard spherical polar coordinates θ and ϕ on the sphere are related to Cartesian coordinates by $x = \sin\theta\cos\phi$, $y = \sin\theta\sin\phi$, $z = \cos\theta$. Thus ϕ is basically a measure of longitude (marked here both at the north pole and on the equator) and $\pi/2 - \theta$ is the latitude.

where Φ is the corresponding eigenfunction.[32] These eigenfunctions are the *spherical harmonics*, and it is usual to demand that the harmonics be simultaneously also eigenfunctions of the operator $\partial/\partial\phi$ (which commutes with ∇^2). The possible eigenvalues of $\partial/\partial\phi$ are $\mathrm{i}m$, where the integer m lies in the range $-j \le m \le j$:

$$\frac{\partial\Phi}{\partial\phi} = \mathrm{i}m\Phi.$$

Examples of such eigenfunctions are $\Phi = 1$(for $j = m = 0$), $\Phi = \cos\theta$ (for $j = 1$ with $m = 0$), $\Phi = \mathrm{e}^{\pm\mathrm{i}\phi}\sin\theta$ (for $j = 1$, $m = \pm1$), $\Phi = 3\cos^2\theta - 1$ (for $j = 2$, $m = 0$) etc.[33]

The remarkable similarity with the eigenvalues $j(j+1)$ for the total angular momentum operator $\mathbf{J}^2 = \boldsymbol{L}_1^2 + \boldsymbol{L}_2^2 + \boldsymbol{L}_3^2$, and with m for the component $\boldsymbol{L}_3$, as referred to at the end of §22.8 and in §22.10 respectively, should not be lost on the reader. Indeed, the angular dependence of the wavefunction, for a particle with integral spin j, is necessarily a j-spherical harmonic. Moreover, the eigenstates of $\boldsymbol{L}_3$ correspond to harmonics that are eigenfunctions of $\partial\Phi/\partial\phi$. In fact, we can 'identify'

$$\mathbf{J}^2 = -\nabla^2 \quad \text{and} \quad \boldsymbol{L}_3 = -\mathrm{i}\frac{\partial}{\partial\phi}$$

for the angular behaviour of such wavefunctions.[34]

This does not give us the 'spinorial' cases for which j is a half-odd integer (as is m, accordingly). For this, we can generalize to what are referred to as 'spin-weighted spherical harmonics'.[35] These are not simply functions on the sphere S^2, but they also have a dependence on a unit (spinorial) tangent vector at each point of S^2 (Fig. 22.17). (They may be thought of as functions on the S^3 that represents the bundle of 'spinorial' unit tangent vectors to S^2 that the Clifford bundle provides us with, as described in §15.4.[36] Here is not the place to go into the details of this, and the reader is referred to the literature.

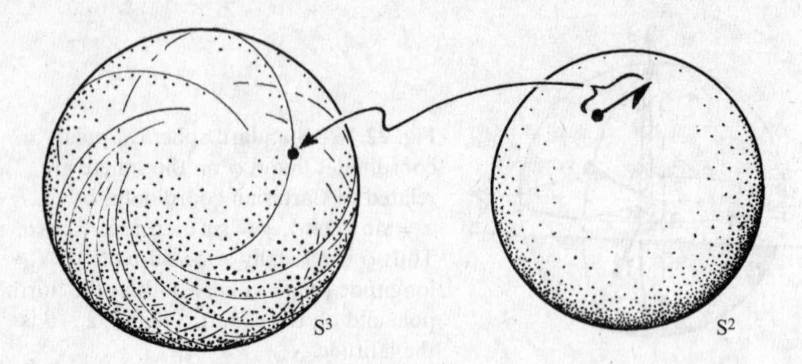

Fig. 22.17 Spin-weighted spherical harmonics. Spin-weighted functions on the sphere S^2 (drawn on the right) are not simply functions on S^2, but also have a dependence on a unit (spinorial) tangent vector to S^2 at the point in question (represented here as a 'half arrow' to signal its spinorial nature). Such functions are more properly on the S^3 depicted on the left, which is the Clifford spin-vector bundle of Fig. 15.10. (A function of 'spin-weight s' has an $e^{is\chi}$ dependence on the angle χ through which the spin-vector rotates within its tangent plane to S^2. As χ increases, the corresponding point on S^3 describes a Clifford circle.)

In fact, the 2-spinor description of spin states, as introduced in §22.8 and used for the Majorana representation of §22.10, is closely related to the theory of spherical harmonics and spin-weighted spherical harmonics. Any n-index symmetric spin-tensor $\psi_{AB\ldots F}$ corresponds explicitly to a collection of (spin-weighted) spherical harmonics for $j = \frac{1}{2}n$. To find them, we take two 2-spinors ξ^A and η^A, with components

$$\{\xi^0, \xi^1\} = e^{i\phi/2}\cos\tfrac{\theta}{2},\ e^{-i\phi/2}\sin\tfrac{\theta}{2},$$

$$\{\eta^0, \eta^1\} = -e^{i\phi/2}\sin\tfrac{\theta}{2},\ e^{-i\phi/2}\cos\tfrac{\theta}{2},$$

so that ξ^A and η^A represent diametrically opposite points on S^2.[22.33] To write down each (spin-weighted) harmonic, we take the components of $\psi_{AB\ldots F}$ with respect to ξ^A and η^A (regarded as a variable spinor reference frame). These 'components' are the quantities

$$\psi_{A\ldots CD\ldots F}\xi^A \ldots \xi^C\ \eta^D \ldots \eta^F.$$

If the number of ξs is equal to the number of ηs ($= j$), in this expression, then we obtain ordinary (rather than spin-weighted) harmonics. (Generally, the number of ξs and ηs is $j + s$ and $j - s$ respectively, where s is the 'spin-weight'.) We get (multiples[37] of) the *standard* spherical harmonics, which are eigenstates of $\partial/\partial\phi$, if we take $\psi_{AB\ldots F}$ to be, in turn, each of the *standard*

[22.33] Explain why the points are antipodal.

basis states referred to in §22.10, namely $|\downarrow \ldots \downarrow\uparrow \ldots \uparrow\rangle$. (For these, just one of the $n+1$ independent components of $\psi_{AB\ldots F}$ is non-zero.) We must bear in mind that these basis states are *symmetrized*. For example, $|\downarrow\uparrow\uparrow\rangle$ is a multiple of $|\downarrow\rangle|\uparrow\rangle|\uparrow\rangle + |\uparrow\rangle|\downarrow\rangle|\uparrow\rangle + |\uparrow\rangle|\uparrow\rangle|\downarrow\rangle$. In this particular case, all components of ψ_{ABC} vanish except for the single independent component $\psi_{011} = \psi_{101} = \psi_{110}$. Although my description of these matters is rather inadequate owing to its brevity,[38] it conveys an outline of what is involved, and the reader may begin to appreciate that spinors provide what is actually a remarkably effective (though unconventional) route to spherical harmonics.[22.34] Recall from §22.8 (cf. §13.7) that the spin-tensor quantities $\psi_{AB\ldots F}$ provide an $(n+1)$-dimensional irreducible representation space for the rotation group SO(3), so the same applies to the space of $j = 2n$ (spin-weighted) spherical harmonics.

The Majorana description can be readily obtained in this way, the spinors $\alpha_A, \beta_A, \ldots, \varphi_A$ in the decomposition $\psi_{AB\ldots F} = \alpha_{(A}\beta_B \cdots \varphi_{F)}$ corresponding to the zeros of the (spin-weighted) spherical harmonics arising in the above description in which only ξs appear, and no ηs. In fact, it was through considerations corresponding to this that Majorana first found his description. It is possible to obtain some valuable insights into spherical harmonics by using the 2-spinor formalism. In many respects the spinor approach is simpler to use, but it is not very familiar.

Spherical harmonics are important in many other areas, such as in classical physics, and in most applications there is no particular connection with angular momentum. (In such situations, it is usual to use the letter ℓ in place of j, as the latter seems to have connotations of angular momentum.) Small oscillations of a soap bubble would be an example. Another would be in the analysis of the temperature distribution, over the celestial sphere, of the microwave (2.7 K) radiation coming from the depths of space, where one is interested particularly in high values of ℓ, of 200 and more. This analysis has great significance for cosmology, as we shall be seeing in §§27.7,10,11, §28.4, and especially §28.10.

The contrast between quantum and classical manifestations of spherical harmonics is striking and non-intuitive. In a quantum system in which the θ and ϕ coordinates have the standard angular spatial interpretation, the j (or ℓ) value always has the interpretation as an angular momentum, but this is far from so for a classical system. In particular, a system of zero angular momentum in quantum mechanics must be spherically symmetrical because a wavefunction with $j = 0$ is composed solely of the spherical harmonic constant on the sphere; but in classical physics, zero angular momentum (i.e. 'non-spinning') certainly does not imply spherical symmetry!

[22.34] Calculate the ordinary spherical harmonics explicitly this way (up to an overall factor) for $j = 1,2,3$. Check that they are indeed eigenstates of ∇^2 and $\partial/\partial\phi$.

In the opposite direction, we see that a randomly chosen quantum system with a *large* angular momentum (large j value) has a state defined by a Majorana description consisting of $2j$ points more-or-less randomly peppered about the sphere S^2. This bears no resemblance to the classical angular momentum state of a system of large angular momentum, despite the common impression that a quantum system with large values for its quantum numbers[39] should approximate a classical system! For a classical-like quantum state, we require that the Majorana points mainly cluster about a particular direction out from the centre of S^2, namely the direction that is the (positive) axis of classical spin. Why is there such a discrepancy between these two pictures? The answer is that almost all 'large' quantum states do *not* resemble classical ones. The most famous such example is Schrödinger's hypothetical cat, which is in a quantum superposition of being alive and dead (see §29.7). Why do we not actually see things like this at a classical level? This is an aspect of the *measurement paradox* which will be discussed in Chapters 29 and 30.

Harmonic analysis for spaces more general than S^2 forms an important part of many areas of scientific research. It is extremely valuable when small perturbations or oscillations of a system are considered. A word of warning may be appropriate, however. In a space that is non-compact, the situation can get far more complicted than in the situation of S^2, considered above. We saw something of this in Chapter 9, when we moved from Fourier analysis (on the compact circle) to the Fourier transform (on the non-compact open line). There is sometimes a tendency for people to believe that one can take over the analysis from a compact to a non-compact form—say from the sphere to hyperbolic space—with just a few changes of sign (and with trigonometric functions being replaced by their hyperbolic analogues, in accordance with the 'signature flip' ideas of §18.4). Unfortunately the truth can be a lot more complicated than this. Such an incomplete 'harmonic analysis' captures only a vanishingly small proportion of the relevant functions on hyperbolic space, owing to the extreme non-completeness of the system of harmonics.

22.12 Relativistic quantum angular momentum

Let us now address the issue of relativistic angular momentum. Recall the classical expressions, described in §18.7. Analogous to mass/energy and momentum combining into a 4-vector p_a, there is an antisymmetric 6-tensor quantity M^{ab} describing an object's *angular* momentum and mass-centre movement. How are we to deal with these quantum-mechanically?[40]

We have seen in §§21.1–3 how the quantum notions of energy and momentum mysteriously represent—or (essentially) *are*—the generators of time- and space-translational motions of spacetime. Similarly, the components of 6-angular momentum M^{ab} represent—are—the generators of the (Lorentz) rotational motions of Minkowski space $\mathbb{M}$. Together with the translational motions p_a, these rotational motions give rise to the entire (non-reflective) Poincaré group (§18.2)—the Minkowskian analogue of the rigid motions of Euclidean geometry.

More explicitly, the generators of translational Poincaré motions are the components p_0, p_1, p_2, p_3 of the 4-momentum vector p_a, where the energy $\mathrm{E} = p_0 = \mathrm{i}\hbar\partial/\partial x^0$ generates time translation and the remaining three components (i.e. momentum) similarly generate spatial displacements: $p_1 = \mathrm{i}\hbar\partial/\partial x^1$, $p_2 = \mathrm{i}\hbar\partial/\partial x^2$, $p_3 = \mathrm{i}\hbar\partial/\partial x^3$—where we bear in mind that $(-p_1, -p_2, -p_3)$ are the components of the 3-momentum $\mathbf{p}$; see §18.7. The 3-space rotational Poincaré motions are generated by the components $c^{-2}M^{23} = \boldsymbol{L}_1 = \mathrm{i}\hbar\boldsymbol{\ell}_1$, $c^{-2}M^{31} = \boldsymbol{L}_2 = \mathrm{i}\hbar\boldsymbol{\ell}_2$, $c^{-2}M^{12} = \boldsymbol{L}_3 = \mathrm{i}\hbar\boldsymbol{\ell}_3$, which we already considered in §22.8, defining the quantum notion of ordinary angular momentum. These are the entirely spatial components of the 6-angular momentum M^{ab}, and the remaining 3 independent[41] components $c^{-2}M^{01}$, $c^{-2}M^{02}$, $c^{-2}M^{03}$, which generate the Lorentz velocity transformations, refer to the uniform motion of the mass centre in accordance with §18.7 (see Fig. 18.16). (In this section, $c = 1$ is not assumed.)

Since the Poincaré group is non-Abelian, its generators do not all commute. Their commutation laws tell us the commutation laws for our quantum operators p_a and M^{ab}:

$$[p_a, p_b] = 0,$$

$$[p_a, M^{bc}] = \mathrm{i}\hbar(g_a{}^b p^c - g_a{}^c p^b),$$

$$[M^{ab}, M^{cd}] = \mathrm{i}\hbar(g^{bc}M^{ad} - g^{bd}M^{ac} + g^{ad}M^{bc} - g^{ac}M^{bd}).$$

These may look somewhat complicated, but they have a fundamental significance in relativistic physics, as they define the Lie algebra (§14.6) of the Poincaré group. They look a little simpler in the diagrammatic notation, as depicted in Fig. 22.18.[22.35]

Recall that, for non-relativistic angular momentum, we were able to describe a basis for states in terms of the eigenvalues $j(j+1)$ and m of the two commuting observables $\mathbf{J}^2$ and $\boldsymbol{L}_3$; see §§22.8,11. These operators provide a *complete* commuting set (in the sense that any other operator constructed from the generators $\boldsymbol{L}_1$, $\boldsymbol{L}_2$, and $\boldsymbol{L}_3$, and which commutes with $\mathbf{J}^2$ and $\boldsymbol{L}_3$, gives nothing new, since it must itself be a function of these two). It is an important part of quantum mechanics, generally, to find such a complete commuting set, for a given system under consideration. Most particularly, we should like to be able to do this for operators constructed

---

[22.35] Show that the commutators given in §22.8 for 3-dimensional angular momentum are contained in these.

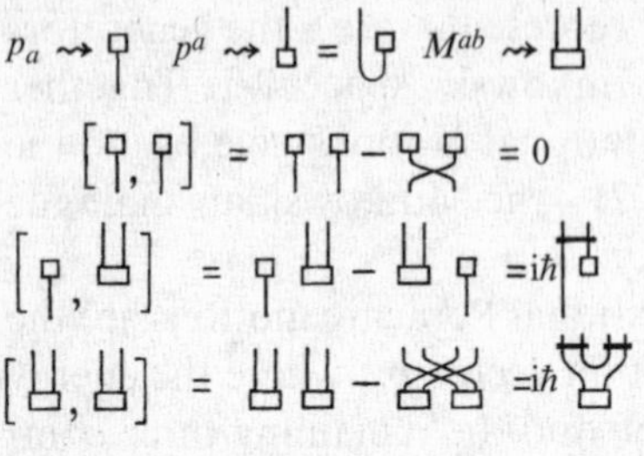

Fig. 22.18 Diagrammatic form of the relativistic 4-momentum and 6-angular momentum quantum commutators $[p_a, p_b] = 0$, $[p_a, M^{bc}] = i\hbar(g_a{}^b p^c - g_a{}^c p^b)$, $[M^{ab}, M^{cd}] = i\hbar(g^{bc}M^{ad} - g^{bd}M^{ac} + g^{ad}M^{bc} - g^{ac}M^{bd})$.

from the components of p_a and M^{ab}, and use their eigenvalues to classify relativistic particles, or relativistic systems.

Why are we interested in commuting observables? The reason is that if $\boldsymbol{A}$ and $\boldsymbol{B}$ are two such—so $\boldsymbol{AB} = \boldsymbol{BA}$—then we can find states $|\psi_{rs}\rangle$ that are simultaneously eigenstates of both, and the pair of corresponding eigenvalues (a_r, b_s) can be used to label these states.[42] If we have a complete set of commuting observables $\boldsymbol{A}, \boldsymbol{B}, \boldsymbol{C}, \boldsymbol{D}, \ldots$ (whose eigenstates span the space under consideration), then we have a family of basis states $|\psi_{rstu\ldots}\rangle$ where the corresponding family of eigenvalues $(a_r, b_s, c_t, d_u, \ldots)$ can be used to label these states.[22.36]

In obtaining a complete commuting set, it is usual to start by finding the *Casimir* operators, which are (scalar) operators that commute with *all* the operators of the system under consideration. In the case of ordinary 3-dimensional angular momentum (recall §22.8), there is just one (independent[43]) Casimir operator, namely $\mathbf{J}^2 = \boldsymbol{L}_1^2 + \boldsymbol{L}_2^2 + \boldsymbol{L}_3^2$. An important question is: what are the Casimir operators for the system generated by p_a and M^{ab}, satisfying the above commutation laws?

Now, the spin about the mass centre is defined by the quantity

$$S_a = \tfrac{1}{2}\varepsilon_{abcd}M^{bc}p^d,$$

called the *Pauli–Lubanski spin vector*, where Levi-Civita's antisymmetrical ε_{abcd} was defined in §19.2, but here we have $\varepsilon_{0123} = c^{-3}$ since $c = 1$ is not assumed. (In the 'mathematician's notation' we might write $\boldsymbol{S} = {*}(\boldsymbol{M} \wedge \boldsymbol{p})$, now using $\boldsymbol{p}$ to represent the 4-momentum, rather than the previous 3-momentum; cf. §11.6, §12.7, §19.2.) We have seen that a single classical structureless particle has $M^{ab} = x^a p^b - x^b p^a$, where x^a is the position vector of a point on the particle's world line (see the end of §18.7). We take the same expression in the quantum case—from which it follows that $S^a = 0$ for such a particle. But S^a need not vanish for a total system of two

[22.36] Work out the details of these claims—where you may assume, for convenience, that the eigenvalues form a discrete rather than a continuous system. Assume first that there are no degenerate eigenvalues, and then show how the argument carries through when there *are* degeneracies. *Hint*: Express each eigenvector of $\boldsymbol{A}$ in terms of eigenvectors of $\boldsymbol{B}$, and so on.

or more such particles. Moreover, for a single particle with spin, the angular momentum M^{ab} does not have this simple form, there being an additional *spin* term $\mu^{-2}\varepsilon^{abcd}S_c p_d$, assuming $\mu \neq 0$ (see Note 18.11). We find that S^a is always orthogonal to p_a ($p_a S^a = 0$) and that it commutes with p_a (i.e. $[S^a, p_b] = 0$), so that S^a, like p_a, is origin-independent.[22.37]

There are two independent Casimir operators (Casimirs for the Poincaré group), namely

$$p_a p^a = c^4\mu^2 \quad \text{and} \quad S_a S^a = -\mu^2 \mathbf{J}^2,$$

μ being the rest-mass of the entire system.[22.38] We find that the '$\mathbf{J}^2$' defined in the second of the above equations is indeed $\mathbf{J}^2 = \boldsymbol{L}_1^2 + \boldsymbol{L}_2^2 + \boldsymbol{L}_3^2$, where $\boldsymbol{L}_1$, $\boldsymbol{L}_2$, and $\boldsymbol{L}_3$ are the components of the angular momentum about the mass centre in its rest frame. To complete the set of commuting operators we may choose p_1, p_2, p_3, and a component, say S_3, of the spin vector which—together with $p_a p^a$ and $S_a S^a$—give us six in all. (Although a great many other choices are possible, in detail, the total number of independent operators is always[44] six.). This has considerable relevance to the discussions of §22.13 and §31.10.

The situation is therefore very similar to the non-relativistic case where, in order to include translations in time and space, we could choose the energy E as a 'Casimir operator' to supplement the quantity $\mathbf{J}^2$, and the three components of momentum in addition to $\boldsymbol{L}_3$. It should be noted that, in the relativistic case, we do not directly get $\mathbf{J}^2$, but rather

$$\mathbf{J}^2 = -c^4(p_a p^a)^{-1}\, S_a S^a,$$

which gives us something basically equivalent, provided that $p_a p^a \neq 0$. Indeed, in the above discussion, we assumed that the rest mass μ does not vanish. If $\mu = 0$, we cannot express the magnitude of spin in this way.

How do we deal with the massless case $\mu = 0$? We retrieve, instead, the *helicity* s—a quantity that we already encountered, in the case of a photon, in §§22.7,9. This is defined by a physical requirement that the Pauli–Lubanski vector S^a be proportional to the 4-momentum p_a:[22.39]

$$S_a = s\, p_a.$$

Right-handed helicity is given by $s > 0$, and left-handed by $s < 0$, while $s = 0$ is also allowed. Now, we have four independent commuting observables, which we can take to be s, p_1, p_2, p_3. In fact, it turns out that by far the neatest way to handle the massless case is to appeal to *twistor theory*.

[22.37] Establish the properties claimed in these four sentences.

[22.38] Provide a simple reason why these two displayed operators must commute with p_a and M^{ab}. *Hint*: Have a look at §22.13.

[22.39] How can S_a and p_a be both orthogonal and proportional?

We shall come to this in §33.6 (where we shall see that the 'twistor variables' Z^0, Z^1, Z^2, Z^3 can also be used as four independent commuting operators).

22.13 The general isolated quantum object

How does quantum mechanics describe an isolated object generally, such as an atom or a molecule? I am assuming that there are no external forces acting on the object and that it remains localized, but there could be internal forces acting within it. As an important feature of the description of such an object, we separate this description into (i) the external characterization of the object as a whole and (ii) its internal detailed workings and geometrical structure.

This external characterization (i) refers to its overall mass/energy, its momentum, the location and movement of its mass centre, and its angular momentum. Let us take these quantities in the relativistic sense, and use the p_a and M^{ab} of §22.12 to describe the external parameters. The internal workings (ii) refer to the constituent particles, their particular nature, the nature of the forces between them, and their geometrical relations. These relations will be taken to be given by some generalized coordinates q_r (§20.1) of an entirely relative[45] nature (e.g. the distance of some part out from the mass centre, or the angle that various parts make with one another, or their distances from one another). Thus, they are not changed if the whole object is displaced by a spatial or temporal translational motion, or rotated through some definite angle, or moved in some direction with a uniform velocity.

Because of their relative nature, all the internal coordinates are unchanged by any symmetry of the Poincaré group. It follows that they must commute with p_a and M^{ab}. Why? Suppose that some symmetry operator $\boldsymbol{S}$ acts on a quantum system according to

$$|\psi\rangle \mapsto \boldsymbol{S}|\psi\rangle,$$

and $\boldsymbol{Q}$ is some quantum operator, then the action of the symmetry operator on $\boldsymbol{Q}$ is[22.40]

$$\boldsymbol{Q} \mapsto \boldsymbol{SQS}^{-1}.$$

If $\boldsymbol{Q}$ is unchanged by $\boldsymbol{S}$, then $\boldsymbol{SQS}^{-1} = \boldsymbol{Q}$, whence

$$\boldsymbol{SQ} = \boldsymbol{QS}.$$

Thus, taking $\boldsymbol{S}$ to be each of the components of p_a and M^{ab} in turn, we see that each internal parameter must indeed commute with p_a and M^{ab}.

[22.40] Explain why. *Hint*: A glance at §22.4 may help.

In the present context, this means that we can *separate* the part of the wavefunction that refers to internal degrees of freedom from that which refers to the external parameters of 4-momentum and 6-angular momentum. In the usual treatments, we suppose that the system is in an eigenstate of the appropriate complete system of external observables. In particular, the energy and momentum would be taken to be given by definite eigenvalues, and it would be usual to refer things to the reference frame in which the 3-momentum is zero ($\mathbf{P}=0$, in the notation of §21.5). Then the angular momentum can be treated according to the non-relativistic discussions of §§22.8–11, so we can ask for the system to be in an eigenstate of total angular momentum $\mathbf{J}^2$, and also of L_3 if desired.

The internal parameters will, of course, depend upon the details of the particular system under consideration. In some circumstances, it may be a good approximation to consider the internal degrees of freedom to be well decribed by small oscillations about equilibrium. Then the classical analysis of §20.3 has relevance. We recall from §20.3 that, if we take a Hamiltonian in the form

$$\mathcal{H} = \tfrac{1}{2} Q_{ab} q^a q^b + \tfrac{1}{2} P^{ab} p_a p_b,$$

where Q_{ab} and P^{ab} are symmetric, positive-definite, and constant in time, then, in the classical case, each normal frequency $\omega/2\pi$ arises from an eigenvalue ω^2 of the matrix $\mathbf{W}=\mathbf{PQ}$ (i.e. of $W^a{}_c = P^{ab} Q_{bc}$).

But what about quantum mechanics? Recalling Planck's relation $E=h\nu=2\pi\hbar\nu$, where ν is the frequency, we might expect an energy $E=\hbar\omega$ for an oscillation in that particular normal mode. Perhaps we might anticipate higher values for the energy also, since classically the amplitude of the oscillation could become as large as we like (so long as the 'small oscillation' nature of the approximation is not disturbed), and with greater amplitude we get greater energy. If we suppose that 'higher harmonics' are involved—where we recall from §9.1 that these occur with frequencies that are integer multiples of the basic frequency $\omega/2\pi$—then we might imagine that the allowed quantum energy eigenstates would be:

$$0,\ \hbar\omega,\ 2\hbar\omega,\ 3\hbar\omega,\ 4\hbar\omega,\ \ldots\ .$$

In fact, this is not far off the correct quantum-mechanical answer, but it turns out that there is an additional contribution $\tfrac{1}{2}\hbar\omega$ to the energy, called the *zero-point energy*[46]. The allowed energy eigenstates are then:

$$\tfrac{1}{2}\hbar\omega,\quad \tfrac{3}{2}\hbar\omega,\quad \tfrac{5}{2}\hbar\omega,\quad \tfrac{7}{2}\hbar\omega,\quad \tfrac{9}{2}\hbar\omega,\quad \ldots\ .$$

This comes from the standard quantum discussion of the 1-dimensional harmonic oscillator,[47] for which the Hamiltonian is $\mathcal{H} = (m^2\omega^2 q^2 + p^2)/2m$. For each mode separately, there is an energy contribution from one of these values, for each eigenvalue ω of the matrix **W**.

For a general quantum system, these values would only be approximations, because higher-order terms could start to become important. However, various systems can be very well approximated in this way. Moreover, rather remarkably, it turns out that the quantum field theory of photons—or of any other particle of the kind referred to as a 'boson' (see §23.7 and §26.2)—can be treated as though the entire system of bosons were a collection of oscillators. These oscillators are exactly of the simple harmonic type, as considered above (where there are no higher-order terms in the Hamiltonian) when the bosons are in a stationary state with no interactions between them.[48] Accordingly, this 'harmonic oscillator' picture provides a quite broadly applicable scheme. Nevertheless, to proceed more thoroughly, a detailed knowledge of the interactions is needed.

For example, a hydrogen atom consists of an electron in orbit around its proton nucleus (usually taken to be fixed, as a good approximation, since the proton moves little, being so much more massive than the electron—by a factor of approximately 1836). But the rules of quantum mechanics tell us that the quantum-mechanical orbit will not involve just a single classical trajectory about the nucleus, but is basically a quantum superposition of many such. These superposed 'quantum orbits' will be stationary solutions of the Schrödinger equation, with a Hamiltonian that is basically the same as in the classical case, but 'canonically quantized' in accordance with the rules of §§21.2,3 (and of §23.8, where needed). To be an eigenstate of angular momentum, we find wavefunctions that are spherical harmonics in their angular dependence (§22.11). Generally, we could use the energy eigenvalue E and the angular momentum eigenvalue j (together with m, if appropriate) as quantum numbers labelling the various states. In the case of the hydrogen atom (if we ignore the spins of the electron and proton and take a non-relativistic form of the Hamiltonian), we find that the energy eigenvalue E happens to be determined by the total angular momentum eigenvalue j but, fortuitously, j is not determined by E. In a more accurate theory of the hydrogen atom (and for more complicated atoms) we find, generally, that E does determine j, so all the different states are actually characterized by the energy eigenvalue alone.

In the original Bohr theory of the atom, put forward in 1913, more than a decade before the much more precise full quantum mechanics of Heisenberg, Schrödinger, and Dirac, the allowed angular momentum and energy values of hydrogen were calculated as though the orbits were the classical

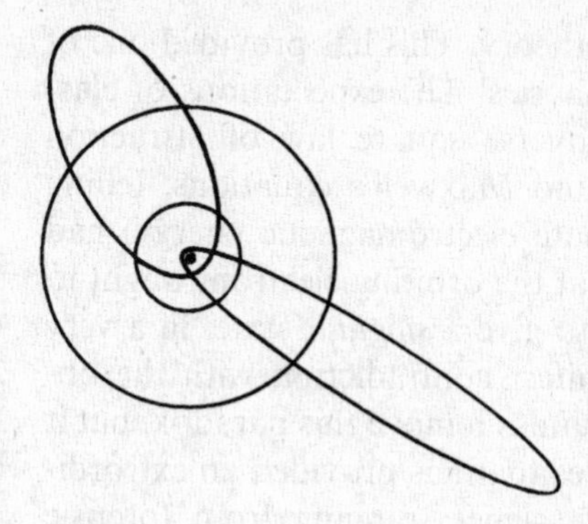

Fig. 22.19 'Bohr atom', where the orbiting electrons are primarily viewed as having classical Kepler–Newton elliptical orbits, according to the inverse square law of electrostatic attraction, but where their energies and angular momenta are constrained by the 'quantum condition' that the orbital angular momenta must be integer multiples of $\hbar$. The idea applied most successfully to circular orbits for the single electron of hydrogen.

Kepler–Newton elliptical orbits—given by the inverse square law of electrostatic attraction between nucleus and orbiting electron—but with the 'quantum condition' that the electron's orbital angular momentum should be an integral multiple of $\hbar$. Such 'quantized orbits' are sometimes referred to as *orbitals*; see Fig. 22.19. This procedure worked remarkably well,[49] but was unsupported by the theoretical underpinnings that the subsequent quantum mechanics provided, these leading to results of far greater generality and accuracy. More complicated atoms, simple molecules, relativistic effects, the presence of electron spin and nuclear spin, etc. can all be treated by the quantum formalism using the ideas outlined above, although approximation techniques and numerical computation are to be expected, rather than exact mathematical treatments.

The above use of electrostatics is also an approximation, and one must allow for transitions from one stationary state to another by the emission/absorption of photons. These require the Maxwell theory, but in its quantized form which, strictly speaking, needs the formalism of *quantum field theory* (to be outlined in Chapter 26). Dirac's relativistic electron, of Chapter 24, would also be needed for full accuracy. An atom in an eigenstate of lowest energy, called the *ground state*, will remain in that state (assuming that it is fully isolated from environmental disturbance), but if it is in a more energetic state—referred to as an *excited state*—then there there is likely[50] to be a finite probability for it to drop into the ground state, with the emission of one or more photons. For this reason, one expects to find free atoms or molecules in their ground states, or near to their ground states. The *frequency* ν of a single photon emitted when an atom or molecule drops from one state to another, is fixed, via Planck's $E = 2\pi\hbar\nu$ (see §21.4) and energy conservation, by the energy *difference* E between the two states.

Such frequencies have long been obesrved, in *spectral lines*, the explanation of which had been a long-standing scientific puzzle. The extraordinary richness of information in these observed spectral-line patterns is explained in the above way by quantum theory. This has provided one of the supreme triumphs of 20th-century physics! The expectations of classical physics (involving the Coulomb inverse square law of attraction between positive and negative charges and Maxwell's equations, telling us how accelerating electrons must radiate electromagnetic energy) had previously yielded the clear prediction that the orbiting electrons ought to spiral into the nucleus catastrophically, to give a *singular* state, in a very short time. This conclusion was in manifest contradiction with the observed facts. Not only did quantum mechanics remove this paradox, but it provided a detailed theory of spectral lines that has provided an extrordinarily powerful tool in many areas of science, ranging from forensic science to nuclear physics and cosmology.

As a final general comment, of considerable importance, it may be remarked that the existence of *discrete* quantum numbers, such as the j and m of angular momentum, or of the energy eigenstates for the harmonic oscillator or the hydrogen atom, etc. arise ultimately from the *compactness* of some space.[51] In the case of angular momentum, this comes from the compactness of the *sphere of spatial directions*, which is the S^2 to which the harmonic analysis of §22.11 applies. Without something like compactness (or periodicity), we would just have solutions of equations like $\nabla^2\Phi = -k\Phi$, in which the eigenvalue k is unrestricted. It is ironic that in the *absence* of such compactness, the general formalism of quantum mechanics would not have provided the striking *discreteness* that started the subject off, and from which the very name 'quantum' originally arose!

Notes

Section 22.1

22.1. These jumps seem to be what stimulated the now colloquial expression 'quantum leap'. To a physicist, this is an extremely odd choice of words, as the quantum jumps that occur in quantum state reduction tend to be extraordinarily tiny, barely detectable, and possibly unreal events!

22.2. For a general discussion of different viewpoints on quantum mechanics, see Rae (1994); Polkinghorne (2002); Home (1997); or DeWitt and Graham (1973).

Section 22.2

22.3. See Chapter 29, and Everett (1957); Wheeler (1957); DeWitt and Graham (1973); Geroch (1984), Deutch (2000).

Section 22.3

22.4. As with the case of a single particle (§21.9), some authors might use '$\|\psi\|^2$' for my $\|\psi\|$.

22.5. For a lovely introduction to the study of such spaces, see Chen (2002); Reed and Simon (1972).

22.6. In the quantum-mechanical literature, the notation $\boldsymbol{Q}^\dagger$ is frequently used, rather than the $\boldsymbol{Q}^*$ of most relevant mathematical literature; see §13.9.

Section 22.4

22.7. See Dirac (1982) for the most recent reprinting. See Shankar (1994) for a more recent treatment.

22.8. See Dirac (1966) for an argument that the Schrödinger picture does not exist in relativistic quantum field theory.

22.9. The interaction picture, for example, is often used in 'time-dependent perturbation theory' calculations, where the Hamiltonian is time-dependent. See Shankar (1994), Chap. 18; Dirac (1966).

Section 22.5

22.10. I am here ignoring the fact that the eigenstates are not normalizable, which might disqualify position or momentum from being *true* 'observables' in some formulations.

22.11. More generally, whether or not it is an eigenvalue, we say that q is the *expectation value* of $\boldsymbol{Q}$ for the normalized state $|\psi\rangle$ if $q = \langle\psi|\boldsymbol{Q}|\psi\rangle$.

22.12. See Dirac (1982a). Complex parameters are used in many fields, e.g. Fortney (1997).

Section 22.6

22.13. An elegant treatment of projection measurements may be found in Kraus (1983); Nielsen and Chuang (2000).

22.14. See von Neumann (1955).

22.15. See Lüders (1951), see also Penrose (1994).

Section 22.7

22.16. See Elitzur and Vaidman (1993).

22.17. The original idea of interaction-free measurements appears to be due to Robert Dicke (cf. 1981). It has some very striking applications, such as in gravitational wave detection; see Braginsky (1977). The extraordinary Elitzur–Vaidman 'bomb-testing' thought experiment described here (see also Penrose 1994) might lead to other kinds of application.

22.18. There is now good evidence that at least most of the various types of neutrino are *massive*, and perhaps *all* are. Even so, the assumption that they are 'massless' provides a good approximation to their behaviour. I shall come back to this issue in §25.3.

22.19. This refinement of the projection postulate appears to be due to Lüders (1951) and, in this case the point $|\rho-\rangle$ in $\mathbb{P}\mathbf{H}^4$ would be referred to as the 'Lüders point'.

22.20. For a more economical (and more interesting) example, we could consider a slightly different situation in which the surface of a refracting medium is used for the beam splitter, where rather than being polarized, the incoming beam is aimed at the *Brewster angle* for the medium. The reflected beam then has a specific linear polarization, and the transmitted beam, the opposite linear polarization. The analysis is basically the same as the above (with linear rather than circular polarization), but now we do not need to ensure that the incoming photon is polarized, the mere fact that it comes from outside the medium (at the appropriate angle), rather than from inside, being sufficient to ensure that the null measurement produces the required resulting polarized state. For a good general, reference on electromagnetism, see Becker (1982); Jackson (1998).

Section 22.8

22.21. A 'structureless particle' would have no angular momentum about its mass centre, since the expression $\mathbf{M} = 2\mathbf{x} \wedge \mathbf{p}$ of §18.6 vanishes when $\mathbf{x} = 0$. But, as noted in Note 18.11, a quantity describing 'intrinsic spin' needs to be added to the angular momentum when there is the 'structure' defined by the particle's spin. We shall see this more explicitly in §22.12.

22.22. The attentive reader may wonder whether there are any subtleties of sign involved here, of the sort that we encountered in §21.5, arising from the signature of the metric. For a detailed development of angular momentum theory in quantum mechanics, from the Lie algebra of SO(3), the reader should see the lucid expositions of Jones (2002); and Elliot and Dawber (1984). An alternative, and somewhat more 'physical' (albeit more complicated) derivation of the angular momentum algebra is given in Shankar (1994).

22.23. See Penrose and Rindler (1984).

22.24. See Geroch (University of Chicago Lectures, unpublished).

22.25. Witten (1959); Penrose (1960, 1968a); Geroch (1968, 1970); Penrose and Rindler (1984, 1986), O'Donnell (2003).

22.26. The word 'independent' is used in the sense that all components of $\psi_{AB...F}$ can be obtained algebraically from this independent set, but not from a smaller set. Here this arises simply from the symmetry, so the total of 2^n components reduce to $n+1$ independent ones (e.g. ψ_{001}, ψ_{010}, and ψ_{100} are not independent components trivially, since $\psi_{001} = \psi_{010} = \psi_{100}$).

22.27. See Shankar (1994).

Section 22.9

22.28. See Note 21.12; Nielsen and Chuang (2000) also discuss some aspects of quantum information science from a similar standpoint.

Section 22.10

22.29. See Majorana (1932).

22.30. See Biedenharn and Louck (1981) for a general review. For an interesting, modern application, see Swain (2004). See also Note 22.31.

22.31. See Penrose (1994, 2000b); Zimba and Penrose (1993).

Section 22.11

22.32. In the context of spherical harmonics, the letter ℓ is frequently used, rather than the j that I am employing here.

22.33. For more details see any text on quantum mechanics, e.g. Shankar (1994) or Arfken and Weber (2000).

22.34. Shankar (1994).

22.35. See Newman and Penrose (1966); Penrose and Rindler (1984).

22.36. See Goldberg, *et al.* (1967).

22.37. There are also orthogonality and (to fix the overall scale) normalization properties of spherical harmonics that are important for using and calculating with them. These matters would take us too far afield, however, and the reader is referred to the following expositions of the theory of spherical harmonics: Groemer (1996); Byerly (2003).

22.38. The reader who wishes to follow through the spinor algebra and geometry a little more thoroughly should take note of the fact that spinor indices can be 'raised' or 'lowered' according to the scheme: $\xi_1 = \xi^0$, $\xi_0 = -\xi^1$. See Penrose and Rindler (1984); Zee (2003), Appendix.

22.39. The term 'quantum number' usually refers to the possible discrete eigenvalues of some significant quantum observable, such as angular momentum, charge, baryon number, etc. which is used to classify a particle or simple quantum system. See §3.5.

Section 22.12

22.40. We shall be seeing in Chapters 24–26 that a properly (special-)relativistic quantum mechanics requires a good deal more than the basic considerations of this section, but this will not affect the present discussion.

22.41. Recall Note 22.6. Here the independence takes into account the *anti*symmetry of M^{ab}.

22.42. There is a connection between this and the phenomenon of 'separation of variables', which happens when a general function $f(\theta, \phi)$, say, can be written as a sum $f(\theta, \phi) = \sum \lambda_{ij} g_i(\theta) h_j(\phi)$, where $g_i(\theta)$ and $h_j(\phi)$ are respective eigenfunctions of appropriate (commuting) operators $\boldsymbol{A}$ and $\boldsymbol{B}$. Spherical harmonics have this property. See Groemer (1996); Byerley (2003).

22.43. 'Independent' refers, here, to functional independence (compare Note 22.26). Thus, whereas $2\mathbf{J}^2$, $(\mathbf{J}^2)^3$, and $\cos \mathbf{J}^2$ are not the same Casimir operator as $\mathbf{J}^2$, they are not independent of $\mathbf{J}^2$.

22.44. Some caution is required concerning the invariance of the 'number of independent commuting operators'. Strictly, this refers to the dimension of a space which applies to the *local* solutions of partial differential equations. In quantum-mechanical problems there are likely to be compactness requirements on the solution space (e.g. the S^2 of §22.11) which severely restrict the allowed eigenvalues and confuse the counting of degrees of freedom.

Section 22.13

22.45. Issues of general relativity are being ignored here, so 'relative' is being taken in the sense of *special* relativity.

22.46. There is, however, the freedom to add a constant to the Hamiltonian, as allowed for in §20.3, which simply redefines the zero of energy (cf. Exercise [24.2] of §24.3), so this addition of $\frac{1}{2}\hbar\omega$ is sometimes regarded as being of no direct physical relevance.

22.47. See, for example, Dirac's classic treatment in *The Principles of Quantum Mechanics*, Dirac (1982a).

22.48. The quantities $\eta = (2m\hbar\omega)^{-1/2}(p + imq)$, in the Heisenberg picture §22.4, play the role of the creation operators of §26.2.

22.49. In particular, it yielded the previously incomprehensible Balmer formula for spectral-line frequencies of hydrogen: $\nu = R(N^{-2} - M^{-2})$, where R is a constant (known as the Rydberg–Ritz constant) and where $M > N > 0$ are integers.

22.50. There may be 'selection rules', arising from conservation laws, forbidding some of these transitions.

22.51. Compare Note 22.44.

23
The entangled quantum world

23.1 Quantum mechanics of many-particle systems

We have seen, in the previous two chapters, how mysterious is the behaviour of individual quantum particles, with or without spin, and how a strange and wonderful mathematical formalism has been evolved in order to cope with this behaviour. It would not be unreasonable to expect that, since our formalism has described for us the quantum behaviour of individual particles or other isolated entities, so also should it have told us how to describe systems containing several separate particles, perhaps interacting with one another in various ways. In a sense this is true—up to a point—since the general formalism of §21.2 is broad enough for this, but some distinctly new features arise, when more than just one particle is present in a system. The underlying quality that is new is the phenomenon of *quantum entanglement*, whereby a system of more than one particle must nevertheless be treated as a single holistic unit, and different manifestations of this phenomenon present us with yet more mystery in quantum behaviour than we have encountered already. Moreover, particles that are identical to each other are always *automatically* entangled with one another, although we shall find that this can happen in two quite distinct ways, depending upon the nature of the particle.

Let us return to what has been set out in the preceding two chapters, for the mathematics of a quantum system. The quantum-Hamiltonian approach, which provides us with the Schrödinger equation for the evolution of the quantum state vector, still applies when there are many particles, possibly interacting, possibly spinning, just as well as it did with a single particle without spin. All we need is a suitable Hamiltonian to incorporate all these features. We do not have a separate wavefunction for each particle; instead, we have *one* state vector, which describes the entire system. In a position-space representation, this single state vector can still be thought of as a wavefunction Ψ, but it would be a function of all the position coordinates of *all* the particles—so it is really a function on the *configuration* space of the system of particles (see §12.1), and it could also depend upon some discrete parameters to label the spin states (e.g. if we use a 2-spinor description $\Psi_{AB...F}$ to describe a spinning particle, as in §22.8, then the 'discrete parameters' would be labelling the different individual components). The Schrödinger equation will tell us how Ψ evolves in time, so Ψ will need to depend upon the time variable t also.

A noteworthy feature of standard quantum theory is that, for a system of many particles, there is only one time coordinate, whereas each of the independent particles involved in the quantum system has its own independent set of position coordinates. This is a curious feature of non-relativistic quantum mechanics if we like to think of it as some kind of limiting approximation to a 'more complete' *relativistic* theory. For, in a relativistic scheme, the way that we treat space is essentially the way that we should also treat time. Since each particle has its own space coordinates, it should also have its own time coordinate. But this is not how ordinary quantum mechanics works. There is only one time for all the particles.

When we think about physics in an ordinary 'non-relativistic' way, this may indeed seem sensible, since in non-relativistic physics, time is external and absolute, and it simply 'ticks away' in the background, independently of the particular contents of the universe at any one moment. But, since the introduction of relativity, we know that such a picture can only be an approximation. What is the 'time' for one observer is a mixture of space and time for another, and vice versa. Ordinary quantum theory demands that each particle individually must carry its own space coordinate. Accordingly, in a properly relativistic quantum theory, it should also individually carry its own time coordinate. Indeed, this viewpoint has been adopted from time to time by various authors,[1] going back to the late 1920s, but it does not seem to have been developed into a full-blown relativistic theory. A basic difficulty with allowing each particle its own separate time is that then each particle seems to go on its merry way off into a separate time dimension, so further ingredients would be needed to get us back to reality.

In §26.6, I shall introduce the 'path-integral' approach to relativistic quantum theory, which is based on a relativistic Lagrangian rather than on a Hamiltonian formalism, and the 'one-time/many-spaces' problem is circumvented; however, as we shall see later, serious new problems come in, as they always seem to, no matter what (known) procedure is used. Moreover, we shall be seeing shortly that the ordinary Schrödinger equation itself is not immune from the difficulties of 'getting back to reality'. In my opinion, this simple spacetime asymmetry of the Schrödinger approach hides something deep that is still missing from our quantum picture of things; but this should not concern us at the moment. For now, I shall ignore these issues, and present things merely from the standpoint of non-relativistic quantum theory, where the notion of a universal external time can be considered to apply. But the issue of relativity will not go away, and we shall need to return to it at the end of this chapter, in §23.10.

How, then, are we to treat many-particle systems according to the standard non-relativistic Schrödinger picture? As described in §21.2, we shall have a single Hamiltonian, in which *all* momentum variables must appear for all the particles in the system. Each of these momenta gets replaced, in the quantization prescription of the position-space (Schrödinger) representation, by a partial differentiation operator with respect to the relevant position coordinate of that particular particle. All these operators have to act on something and, for consistency of their interpretation, they must all act on the same thing. This is the wavefunction. As stated above, we must indeed have *one* wavefunction Ψ for the entire system, and this wavefunction must indeed be a function of the different position coordinates of *all* the separate particles.

23.2 Hugeness of many-particle state space

This sounds harmless enough, but is it? Let us pause to digest the enormity of this apparently simple last requirement. If it were the case that each particle had its own separate wavefunction, then for n scalar (i.e. non-spinning) particles, we should have n different complex functions of position. Although this is a bit of a stretch of our visual imaginations, for n little particles, it is something that we can perhaps just about cope with. (I am ignoring the time, in these considerations; just take everything at just one instant.) For visualization purposes, we could have a picture not so unlike that of a *field* in space with n different components, where each component could itself be thought to describe a separate 'field'. (Each separate such field might represent an individual particle's wavefunction.) Perhaps we should consider this as $2n$ components, if we are talking about real components, because wavefunctions are complex. An electromagnetic field has 6 real components, after all—that is 6 functions of 3 variables (analogous to three complex scalar wavefunctions)—and a field of electric and magnetic vectors is not such a terrible strain on the imagination!

How are we to count the 'freedom' in a complex scalar field, such as the wavefunction for a scalar particle in 3-space? What is the 'number' of different possible such fields? Recall that according to the notation of §16.7, the expression $\infty^{a\infty^b}$ denotes the freedom available to a freely chosen (smooth) field with a real components in a space of b real dimensions. Thus, for a complex scalar field, $a = 2$ (because a complex number counts as two real numbers), so that the freedom would be $\infty^{2\infty^3}$. This is taking the field at just at one time—i.e. t is constant—so we are considering ordinary 3-space, giving $b = 3$ (rather than the spacetime value $b = 4$). We could also consider spacetime, but in that case we have field equations restricting the freedom. In the case of the wavefunction, that restriction is the Schrödinger equation, which reduces the freedom down to what can be specified freely, as initial data, on an initial 3-space, so we still get $\infty^{2\infty^3}$ as the freedom in the field.

As an incidental consideration, we may examine the case of a free Maxwell field with no sources (charges) to worry about. Here we have 6 real components in ordinary 3-space, so if we take the field just at some fixed t and ignore the Maxwell equations, we get the freedom $\infty^{6\infty^3}$. But the Maxwell equations imply that two constraints must hold on any initial data 3-space: namely the vanishing of the divergence of the electric and magnetic field vectors.[23.1] This reduces the effective number of free components on the initial data 3-surface by 2, so the freedom is actually $\infty^{4\infty^3}$.

Let us now consider the quantum-mechanical description of n scalar particles. If the description were just n different wavefunctions, then the freedom would be $\infty^{2n\infty^3}$, since that is the freedom in choosing n complex numbers per point in 3-space. But in the case of an *actual* quantum wavefunction describing n scalar particles, we have one complex function of $3n$ real variables. This is like a complex scalar field in a space of $3n$ dimensions, so the freedom is $\infty^{2\infty^{3n}}$ instead, which is *stupendously larger*!

It is probably not so easy to appreciate the enormity of this increase when it is hidden in all those ∞s. So let me consider a 'toy' universe which just has 10 points in it. We can label these points **0**, **1**, **2**, **3**, **4**, **5**, **6**, **7**, **8**, **9**. The wavefunction of a scalar particle in this universe would consist of a complex number at each one of these 10 points, i.e. 10 complex numbers $z_0, z_1, z_2, \ldots, z_9$. The space of all these wavefunctions would be the 10-complex-dimensional (20-real-dimensional) Hilbert space $\mathbf{H}^{10}$. If we normalize the wavefunction so that the sum of the squares of the moduli of these zs is unity, then $|z_6|^2$ would represent the probability of a position measurement finding the particle at **6**, and so on.

This discrete model is not really such an absurdity. In actual physical situations, one might have what is, in effect, a sequence of 10 boxes, with an electron that might be in one of the boxes. See Fig. 23.1. Experimenters can construct things of this nature called *quantum dots*, and they have relevance to the theoretical possibility of constructing *quantum computers*, which would make use of the vastness of the sizes of the kinds of space of wavefunctions that I am about to consider.

Suppose that there are now *two* particles in our universe. It is better that they are not both the same kind of particle, for a reason that I shall be coming to later. So let us call them an A particle and a B particle. Each of the two particles could be in 10 different alternative places, so there are 100

[23.1] Can you explain this vanishing? Recall the 4-dimensional notion of 'divergence' described in §19.3; here we need the 3-space version. *Hint*: See Exercise [19.2].

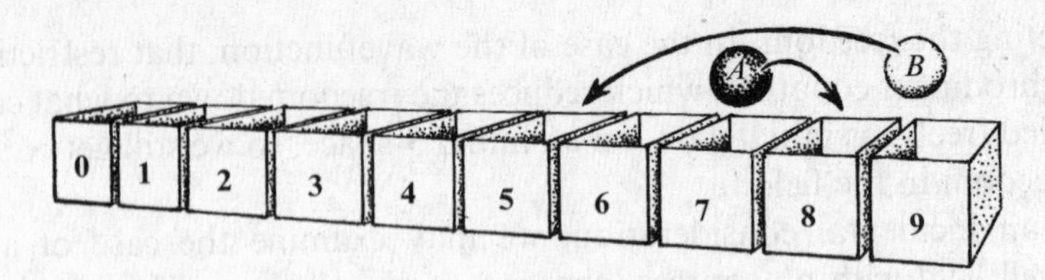

Fig. 23.1 We imagine a 'toy universe' with just 10 possible locations for particles, here illustrated by 10 boxes. Two distinguishable particles A and B are shown, each of which can occupy any one of the boxes, independently of the other.

different possible placings for the pair of them (allowing them to be both in the same box). We now need 100 different complex numbers, say $z_{00}, z_{01}, \ldots, z_{09}, z_{10}, z_{11}, \ldots, z_{19}, z_{20}, \ldots, z_{99}$ to define the wavefunction, one complex number being assigned to each pair of placings. If we normalize so that the sum of the squares of the moduli of all these zs is unity, then $|z_{38}|^2$, for example, would represent the probability of finding that the A particle is at **3** and the B particle is at **8**. We are now dealing with $\mathbf{H}^{100}$. If we had *three* different particles—an A particle, a B particle, and a C particle—then the wavefunction would consist of 1000 complex numbers $z_{000}, z_{001}, \ldots, z_{999}$, and our state-space is $\mathbf{H}^{1000}$. Had the rules been to have merely three individual wavefunctions, than the state space would have been only $\mathbf{H}^{30}$. For *four* different particles, we have $\mathbf{H}^{10000}$, whereas for four individual wavefunctions, merely $\mathbf{H}^{40}$, and so on.

Reverting to the '$\infty^{a\infty^{3n}}$' notation that I used above, we take note of the fact that the upper '∞^3' refers to the 'number of points' in Euclidean 3-space $\mathbb{E}^3$. That number is now replaced by 10, the actual number of points in our *toy* universe, so that $\infty^{a\infty^{3n}}$ becomes ∞^{a10^n} (which denotes the 'number of points' in an $(a \times 10^n)$-real-dimensional space). Thus, instead of $\infty^{2\infty^{3n}}$, for the freedom in an n-particle scalar wavefunction in $\mathbb{E}^3$, we now have $\infty^{2\times 10^n}$ for the freedom in an n-particle wavefunction for our toy universe. The complex Hilbert space is now $\mathbf{H}^{10^n}$, for our toy universe's n-particle wavefunction, as compared with $\mathbf{H}^{10n}$ for n separate 1-particle complex wavefunctions. Thus, our n-particle wavefunction is defined on a 2×10^n-dimensional space (this 10^n-complex-dimensional Hilbert space), rather than a mere $20n$-dimensional space for n separate wavefunctions. For just 8 particles, for example, this is 200 000 000 dimensions instead of a mere 160.

23.3 Quantum entanglement; Bell inequalities

What is all this extra information doing? It is expressing what are known as the 'entanglement' relations between the particles. How are we to understand these? Entanglements between particles, a notion first made explicit by Schrödinger (1935), are what lead to the extremely puzzling but actually observed phenomena known as *Einstein–Podolsky–Rosen* (EPR) effects.[2] They are, however, rather subtle features of the quantum world which are quite hard to demonstrate experimentally in a convincing way. It is remarkable that we seem to have to turn to something so esoteric and hidden from view when, for many-particle systems, almost the *entire* 'information' in the wavefunction is concerned with such matters! This is a puzzle that I shall be coming back to shortly (§23.6). In my opinion, this puzzle is trying to tell us something about what kind of new directions our present-day quantum formalism ought to be moving in. But, be that as it may, it is certainly telling us something of the potential power of *quantum computing*[3]—a subject of very active current research which aims to exploit the enormous 'information' resources that lie hidden in these entanglement relations.

So what is quantum entanglement? What are EPR effects? It will be clearest if we consider just a finite-dimensional situation, which we can do if we just concentrate on states of spin. The simplest EPR situation is that considered by David Bohm (1951). In this, we envisage a pair of spin $\frac{1}{2}$ particles, let us say, particle P_L and particle P_R, which start together in a combined spin 0 state, and then travel away from each other to the left and right to respective detectors L and R at a great distance apart (see Fig. 23.2). Let us suppose that each of the detectors is capable of measuring the spin of the approaching particle in some direction that is only decided upon when the two particles are well separated from each other. The problem is to see whether it is possible to reproduce the expectations of quantum mechanics using some model in which the particles are regarded as unconnected independent classical-like entities, each one being unable to communicate with the other after they have separated.

It turns out, because of a remarkable theorem due to the Northern Irish physicist John S. Bell, that it is not possible to reproduce the predictions of quantum theory in this way. Bell derived inequalities[4] relating the joint probabilities of the results of two physically separated measurements that

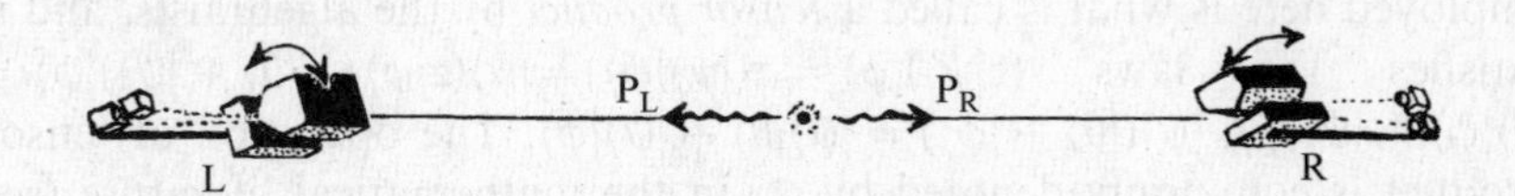

Fig. 23.2 The EPR–Bohm thought experiment. A pair of spin-$\frac{1}{2}$ particles $\mathbf{P}_L$ and $\mathbf{P}_R$ originate in a combined spin 0 state, and then travel out in opposite directions, left and right, to respective widely separated detectors L and R. Each detector is set up to measure the spin of the approaching particle, but in some direction which is decided upon only after the particles are in full flight. Bell's theorem tells us that there is no way of reproducing the expectations of quantum mechanics with a model in which the two can act as classical-like independent objects that cannot communicate after they have become separated.

are violated by the expectations of quantum mechanics, yet which are necessarily satisfied by any model in which the two particles behave as independent entities after they have become physically separated. Thus, Bell-inequality violation demonstrates the presence of essentially quantum-theoretic effects—these being effects of quantum *entanglements* between physically separated particles—which cannot be explained by any model according to which the particles are treated as unconnected and independent actual things.

There are many striking particular examples of this kind of Bell-inequality violation in the literature.[5] Some of these, referred to as 'Bell inequalities without probabilities'[6] are particularly remarkable in that they just involve yes/no issues, and we do not need to worry about probabilities—or, rather, we worry only about the extreme *definite* cases of probabilities 0 ('never') and 1 ('always'). Here I shall give just two explicit versions of the Bell-inequality type of contradiction between quantum particles and individual particles. Both of these involve a pair of spin $\frac{1}{2}$ particles going off separately to a detector L on the left and another detector R on the right. The first, which follows an argument due to Henry Stapp (1971), (1979), is a direct example of the original Bohm version of EPR, as referred to above, and in which we need to examine actual probability values. The second, due to Lucien Hardy (1992), (1993), is 'almost' a version without probabilities, but it has a slight extra twist to it.

Before giving these in detail, I shall need one more bit of (Dirac) notation. Suppose that we have a quantum system that consists of two parts $|\psi\rangle$ and $|\phi\rangle$, which can be taken to be independent of each other. Then if we wish to consider the quantum state that consists of both of them together, we write this

$$|\psi\rangle|\phi\rangle.$$

This is still a single state, and it would be legitimate to write an equation such as $|\chi\rangle = |\psi\rangle|\phi\rangle$, which expresses this fact. The type of product that is being employed here is what is called a *tensor product* by the algebraists, and it satisfies the laws $(z|\psi\rangle)|\phi\rangle = z(|\psi\rangle|\phi\rangle) = |\psi\rangle(z|\phi\rangle)$, $(|\theta\rangle + |\psi\rangle)|\phi\rangle = |\theta\rangle|\phi\rangle + |\psi\rangle|\phi\rangle$, $|\psi\rangle(|\theta\rangle + |\phi\rangle) = |\psi\rangle|\theta\rangle + |\psi\rangle|\phi\rangle$. The operation of tensor product is commonly denoted by $\otimes$, in the mathematical literature (see also §13.7), and the product $|\psi\rangle|\phi\rangle$ might then be denoted $|\psi\rangle\otimes|\phi\rangle$.

In any case, it is handy to use the $\otimes$ symbol in connection with the (Hilbert) spaces to which such products belong. Thus, if $|\psi\rangle$ belongs to $\mathbf{H}^p$ and $|\phi\rangle$ belongs to $\mathbf{H}^q$ then $|\psi\rangle|\phi\rangle$ belongs to $\mathbf{H}^p\otimes\mathbf{H}^q$. The dimension of $\mathbf{H}^p\otimes\mathbf{H}^q$ is the product of the dimensions of its two factors, so we could legitimately write $\mathbf{H}^p\otimes\mathbf{H}^q = \mathbf{H}^{pq}$. I am allowing either or both of p and q to be ∞, in which case we take the product to be also ∞. Only a very small part of $\mathbf{H}^p\otimes\mathbf{H}^q$ consists of elements of the form $|\psi\rangle|\phi\rangle$ (assuming

$p, q > 1$), where $|\psi\rangle$ belongs to $\mathbf{H}^p$ and $|\phi\rangle$ belongs to $\mathbf{H}^q$. These are the *unentangled* states. A *general* element of $\mathbf{H}^p \otimes \mathbf{H}^q$ would be a linear combination of these unentangled states (possibly involving an infinite sum or integral, if both of p and q are infinite).[7] We should bear in mind, however, that the very notion of entanglement depends upon the *particular* splitting of our entire Hilbert space $\mathbf{H}^{pq}$ into something of the form $\mathbf{H}^p \otimes \mathbf{H}^q$. (No one such splitting of a general Hilbert space $\mathbf{H}^{pq}$ is to be preferred over any other. Algebraically, there will always be many ways of expressing $\mathbf{H}^n$ as a tensor product, whenever n is a composite number.) In situations where one is interested in the 'entanglement' notion, the particular splitting of physical interest is something reasonably obvious, most notably when there are 'individual' particles separated by a large distance, which is what EPR is all about.

It is sometimes useful to use an abstract-index formulation of operations such as this (see §12.8). The ket vector $|\psi\rangle$ could be written ψ^α, with an upper abstract index, and its corresponding (complex conjugate) bra vector $\langle\psi|$ by $\bar{\psi}_\alpha$, with a lower abstract index. The full bracket $\langle\psi|\phi\rangle$ would be $\bar{\psi}_\alpha\phi^\alpha$ and an expression $\langle\psi|\boldsymbol{Q}|\phi\rangle$ would be $\bar{\psi}_\alpha Q^\alpha{}_\beta\phi^\beta$. The tensor product $|\psi\rangle|\phi\rangle$ of ψ^α with ϕ^β could then be written $\psi^\alpha\phi^\beta$. Unentangled states always split in this way. But a general (probably entangled) state would simply be an entity of the form $\phi^{\alpha\beta}$. We shall be seeing a particular use for this kind of notation later in this chapter.

23.4 Bohm-type EPR experiments

Let us now return to the Bohm version of EPR. Consider the initial state, just before a measurement is performed on it. The two separated spin $\frac{1}{2}$ particles, taken together, must constitute a state of spin 0. This is because angular momentum is conserved, and the particles start out in a combined state of spin 0. We therefore need a combination of basis states for the spin of each particle for which the total spin is zero. This is achieved by the state $|\Omega\rangle$, of spin 0, given by

$$|\Omega\rangle = |\uparrow\rangle|\downarrow\rangle - |\downarrow\rangle|\uparrow\rangle$$

(where I still do not worry about normalizing my states).[23.2],[23.3] In the literature, one often sees this kind of thing written in some way such as $|\uparrow\mathrm{L}\rangle|\downarrow\mathrm{R}\rangle - |\downarrow\mathrm{L}\rangle|\uparrow\mathrm{R}\rangle$, where it is made explicit in the notation which state

---

[23.2] If $|\uparrow\rangle$ and $|\downarrow\rangle$ are normalized, what factor does $|\Omega\rangle$ need to make it normalized? (You may assume that $\||\alpha\rangle|\beta\rangle\| = \|\alpha\|\,\|\beta\|$.)

[23.3] Can you see quickly why this has spin 0? *Hint*: One way is to use the index notation to show that any such anti-symmetrical combination must essentially be a scalar, bearing in mind that the spin space is 2-dimensional.

refers to the left-hand particle and which refers to the right-hand one. In my opinion, this is not necessary because (i) the notation is only singling out the spin part of the wavefunction in any case, and not the particle's position or momentum or whatever, so the direction of spin fixes what it is we are concerned with, and (ii) since tensor products do not commute, we can unambiguously tell which 'side' of the product is which. My convention is that the left-hand term of the product refers to the left-hand particle and the right-hand term to the right-hand particle. Readers who find this confusing can reinstate L and R into the kets throughout the discussion, if they wish.

This a clear example of an entangled state, since it cannot be rewritten in the form $|\alpha\rangle|\beta\rangle$ with $|\alpha\rangle$ localized at L and $|\beta\rangle$ localized at R.[23.4] Let us try to see what implications this entangled state has. Now, I am going to imagine that I am sitting at the left, at L, and I am going to perform a measurement of the spin of the left-hand particle P_L in the 'up' direction $\Uparrow$ (YES if $\Uparrow$; NO if $\Downarrow$). This would project the entire state $|\Omega\rangle$ into $|\Uparrow\rangle|\Downarrow\rangle$ if I get the answer YES, and it would project it into $(-)|\Downarrow\rangle|\Uparrow\rangle$ if I get NO. The result would now be unentangled—except that standard **U**-evolution would tell us that P_L is now likely to be hopelessly entangled with my own measuring apparatus L. What can be clearly stated is that if I get the answer YES, then my colleague, situated at the right-hand detector R, will be presented with P_R having spin state $|\Downarrow\rangle$, whereas if I measure NO, then my colleague will be presented with $|\Uparrow\rangle$. Upon subsequently performing an 'up' measurement on P_R, my colleague will necessarily obtain the opposite result to my own.

There is nothing special about the up/down choice in all this; for whatever direction I choose to measure, say $\swarrow$, then if my colleague chooses the same direction $\swarrow$, the result will be opposite. This should be clear from the rotational invariance of the spin 0, but it is instructive to perform a direct algebraic calculation to verify (where $\propto$ means 'equals, up to a non-zero overall factor', see §12.7) that

$$|\Omega\rangle \propto |\swarrow\rangle|\nearrow\rangle - |\nearrow\rangle|\swarrow\rangle,$$

where $\nearrow$ is the direction opposite to $\swarrow$. (Note: if $|\swarrow\rangle = a|\Uparrow\rangle + b|\Downarrow\rangle$, then $|\nearrow\rangle \propto \bar{b}|\Uparrow\rangle - \bar{a}|\Downarrow\rangle$.)[23.5]

We also conclude from all this what would be the joint probabilities for YY, YN, NY, NN (abbreviating YES to Y and NO to N) if I and my colleague choose *different* directions in which to measure the spin. Suppose I choose $\nwarrow$ and my colleague chooses $\nearrow$, where the angle between $\nwarrow$ and $\nearrow$ is θ. Then, using the probability value given in §22.9 (see Fig. 22.11), we find the joint probabilities

[23.4] Why not? Find a way of doing this, however, if $|\alpha\rangle$ and $|\beta\rangle$ are not so localized.

[23.5] Confirm this parenthetic comment, and give a direct calculational verification of the above expression for $|\Omega\rangle$. *Hint*: See Exercise [22.26].

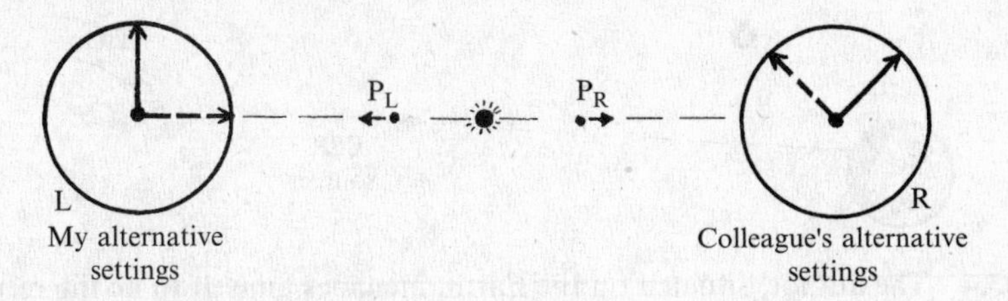

Fig. 23.3 Polarization arrangements for Stapp's version of EPR–Bohm; an example of Bell inequalities. Initially, we take the spin measurements at either side to be in the directions given by the solid-shafted arrows, but by a change of mind, either or both might be rotated to be in the direction of the broken-shafted arrows. The quantum joint probabilities cannot be modelled by any classical-like scheme with particle pairs behaving as non-communicating independent entities without foreknowledge of the directions of proposed spin measurements.

$$\text{agree: } \tfrac{1}{2}(1-\cos\theta), \quad \text{disagree: } \tfrac{1}{2}(1+\cos\theta)$$

(where 'agree' means YY or NN and 'disagree' means YN or NY).

Now, let us consider Stapp's example. Things are arranged so that my own apparatus can be oriented to measure spin either in the direction ↑, taken to be vertically upwards, or in the direction →, which is a horizontal direction (perpendicular to ↑). My colleague's apparatus is oriented to measure spin either in the direction ↗, which lies in the plane of the directions ↑ and →, at 45° to each of them, or in the direction ↖, which lies in the same plane, but is at 45° to ↑ and at 135° to → (Fig. 23.3). There are three possibilities where my measurement direction is at 45° to that of my colleague, and there is just one where the angle between them is 135°. In the 45° cases, we get a probability of agreement of a little under 15%, whereas for 135° we get just over 85%.

Let us allow that the decision as to which of the two possible measurements that I might perform need not actually be made until the particles are in full flight, and the same applies to my colleague. OK, let's put my colleague on Titan (one of Saturn's moons) and have the source of the particles somewhere between the two of us, so that even at the speed of light we would have something like three-quarters of an hour to make up our minds! See Fig. 23.4. The particles have no way of 'knowing' which way my colleague and I are (independently) going to orient our measuring devices.

Let us suppose that I have chosen ↑ and my colleague has chosen ↗ when we each receive a stream of seemingly randomly oriented particles. They come one at a time, each being a member of an EPR–Bohm pair, sent from the mid-way source, one to me and one to my colleague. When we compare notes (perhaps some years later, when my colleague returns) we would find that there is only a little less than 15% agreement between our corresponding results, in accordance with the above.

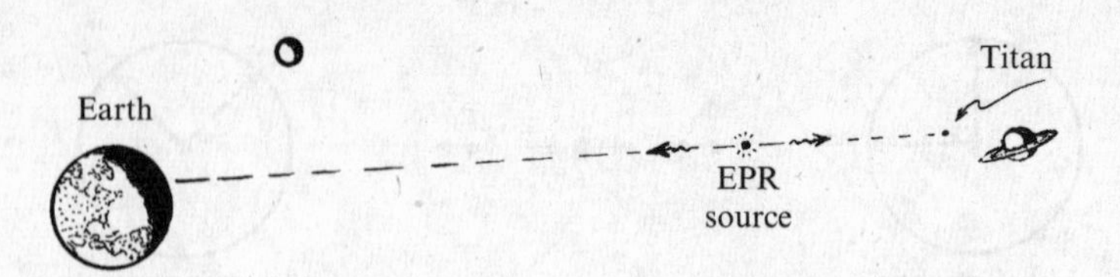

Fig. 23.4 The author, situated on the Earth, imagines himself to be the receiver of one component of a succession of EPR particle pairs, the other being a colleague on Titan, where the source of the pairs is roughly equidistant between the two receivers. Even for particles travelling at the speed of light, there would be about 45 minutes to decide on the detector orientations.

Now if the particles have no prior knowledge about how we are going to orient our measuring devices, and behave as separate individual non-communicating (classical-like) entities, it should make no difference to my colleague's actual measurements had I suddenly changed my mind at the last minute and measured the → direction instead. Were I to do so, then—because the angle between the directions is still 45°—there would still have been only 15% agreement between the measurements that I would now obtain and the original ones of my colleague. On the other hand, suppose instead that it was my *colleague* who had a last minute change of heart and measured ↖ instead of ↗, but I had not changed my mind. My colleague's change should likewise not have affected my own original ↑ measurements. Again, we would find that my colleague's new ↖ measurements would have to be in just under 15% agreement with my own original ↑ measurements.

But suppose that *both* of us had decided to change orientations at the last minute so that my own measurement would be → and my colleague's ↖. Now the angle between them has come around to 135°, so the expectations of quantum mechanics tell us that the agreement ought then to be more than 85%. Is that consistent with the pairs of particles providing the correct joint probabilities for each of the possible pairs of detector orientations just considered? Well, let's see. The particle pairs have to be prepared to encounter any of the four possible combinations of detector settings, and to give the correct quantum-mechanical probabilities in each case. Let us recall what these are. The results of my altered apparatus setting → would be expected to have no more than 15% agreement with my colleague's original ↗ setting. This, in turn, should have no more than 15% agreement with my own original ↑ setting, and this should have no more than 15% agreement with my colleague's altered ↖ setting. If a particular particle pair is going to give agreement in the case →, ↖, then it cannot be in *dis*agreement in all of the cases →, ↗ and ↑, ↗ and ↑, ↖. (Three disagreements must give disagreement, not agreement.) So in at least one of these three possible pairs of settings there must be an agreement. But this happens in less than 15% of the cases, for each possible pair of setting. There are only three of these, so this allows not more than 15% + 15% + 15% = 45% agreement when we get around to the case →, ↖. (In fact the agreement percentage comes out as a bit less than this, because I have effectively counted the case where there is agreement in all three pairs of settings three times.) But 45% is nowhere near 85%, so we have a blatant contradiction with our 'classical-like' assumptions for the particle pairs!

Some might worry that this argument seems to have been phrased in terms of hypothetical measurements that 'might have happened but didn't' (the philosopher's 'counterfactuals'). But this is not important. The key issue is that the particles have been assumed to behave independently of each other after they have left the source, and to give the correct joint quantum probabilities whatever combination of detector settings confronts them. The point is that the particles have to mimic the expectations of quantum mechanics. We have found that these cannot be split into separate expectations for the two particles individually. The only way that the particles can consistently provide the correct quantum-mechanical answers is by being, in some way, 'connected' to each other, right up until one or the other of them is actually measured. This mysterious 'connection' between them is quantum entanglement.

Of course, no experiment of this nature has been performed over such distances. But many EPR-type experiments of an essentially similar kind have actually been performed (usually using photon polarization, not the spin directions of particles of spin $\frac{1}{2}$, but the distinctions are not important). The expectations of quantum mechanics (rather than of common sense) have been consistently vindicated! Although direct quantum entanglements of this nature have certainly not yet been observed over Earth–Saturn distances, some recent experiments have confirmed Bell-inequality violations over distances of more than 15 kilometres.[8]

23.5 Hardy's EPR example: almost probability-free

Let us now come to Lucien Hardy's beautiful example.[9] Again, my colleague and I are poised to make spin measurements, where I select between the ↑ and → measurements (vertically up and the horizontal rightward direction), as before, but now my colleague also selects between ↑ and →, quite independently of my own choice. The crucial new feature is that the source of particle pairs does not now emit them in a combined state of spin 0, but in a particular state of spin 1. I am taking this initial state to be the one with the Majorana description $|{\leftarrow}{\nearrow}\rangle$ (§22.10, Fig. 22.14), where the direction of ↗ lies in the quarter plane spanned by the perpendicular directions $\uparrow$ and $\rightarrow$, and has an upward slope of $\frac{4}{3}$ (so the angle θ between $\rightarrow$ and $\nearrow$ satisfies $\cos\theta = \frac{3}{5}$), and where $\leftarrow$ is opposite to $\rightarrow$; see Fig. 23.5. We can express this state as[23.6]

$$|\leftarrow\nearrow\rangle = |\leftarrow\rangle|\nearrow\rangle + |\nearrow\rangle|\leftarrow\rangle$$

ignoring an overall factor, and it has the important feature that, while it is not orthogonal to

$$|\downarrow\rangle|\downarrow\rangle$$

(with $\downarrow$ opposite to $\uparrow$), it *is* orthogonal to each of [23.7]

$$|\downarrow\rangle|\leftarrow\rangle,\ |\leftarrow\rangle|\downarrow\rangle,\ |\rightarrow\rangle|\rightarrow\rangle.$$

These orthogonality relations are respectively responsible for the following key 'YES/NO' properties (0), (1), (2), and (3):

(0) sometimes I obtain NO for a $\uparrow$ measurement when my colleague obtains NO for a $\uparrow$ measurement;
(1) if I obtain NO for a $\uparrow$ measurement, then my colleague must obtain YES for a $\rightarrow$ measurement;
(2) if my colleague obtains NO for a $\uparrow$ measurement, then I must obtain YES for a $\rightarrow$ measurement;
(3) I never obtain YES for a $\rightarrow$ measurement when my colleague obtains YES for a $\rightarrow$ measurement.

It may be mentioned, in connection with (0), that the actual quantum-mechanical probability for both of us to obtain a NO answer, given that we both choose to perform the $\uparrow$ measurement, is exactly $\frac{1}{12}$, in this

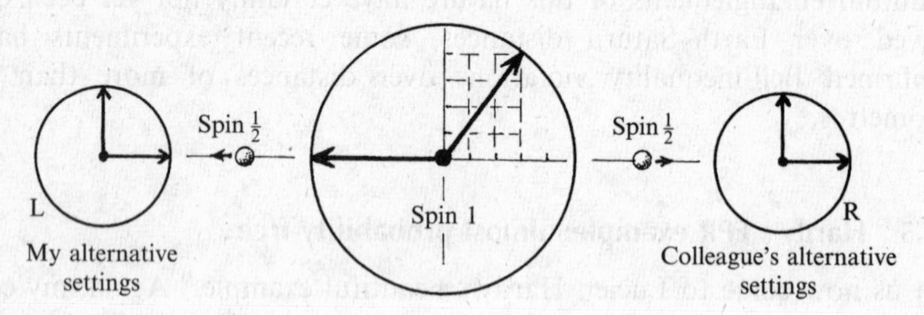

Fig. 23.5 Hardy's version of EPR 'almost' without probabilities. The initial state, of spin 1, is $|\leftarrow\nearrow\rangle = |\leftarrow\rangle|\nearrow\rangle + |\nearrow\rangle|\leftarrow\rangle$, where the direction $\nearrow$ lies in quarter plane spanned by the vertical $\uparrow$ and horizontal $\rightarrow$, at an upward slope of $\frac{4}{3}$. Each detector measures the spin of the approaching particle either vertically or horizontally.

[23.6] Why?

[23.7] See if you can prove these. *Hint*: Use the coordinate and/or geometric descriptions of §22.9.

experiment.[23.8] Note that $\frac{1}{12} = 8.33\%$, whereas Hardy's optimal value, with some slight adjustments, is very close[10] to 9.017%

I should make it quite clear why there is no way of contriving the results (0), ... ,(3), with the two particles being separate non-communicating entities and without foreknowledge of the experiments to be performed upon them. Because of (0), the two particles (now assumed not in communication, and without foreknowledge) must each be prepared jointly to provide a NO answer, from time to time ($\frac{1}{12}$ of the time, in fact), to the eventuality that both I and my colleague might simultaneously perform ↑ measurements. Moreover, the preparation of the particles must have been careful to have pre-arranged (when they were together) that on those occasions when they might do this (i.e. give simultaneous NO answers to our simultaneous ↑ measurements) they must also *definitely* give the YES answer, on those occasions, to a → measurement if just one or the other of us switches to this, so as not to violate (1) or (2). Yet that very decision places them in dire jeopardy of (3), because I and my colleague might *both* happen to perform → measurements, on some such occasion, and thereby obtain the forbidden result YES, YES.

23.6 Two mysteries of quantum entanglement

It seems to me that there are two quite distinct mysteries presented by quantum entanglement, and I believe that the answer to each of them is something of a completely different (although interrelated) character. The first mystery is the phenomenon itself. How are we to come to terms with quantum entanglement and to make sense of it in terms of ideas that we can comprehend, so that we can manage to accept it as something that forms an important part of the workings of our actual universe? The second mystery is somewhat complementary to the first. Since, according to quantum mechanics, entanglement is such a ubiquitous phenomenon—and we recall that the stupendous majority of quantum states are actually entangled ones—why is it something that we barely notice in our direct experience of the world? Why do these ubiquitous effects of entanglement not confront us at every turn? I do not believe that this second mystery has received nearly the attention that it deserves, people's puzzlement having been almost entirely concentrated on the first.

Let me begin by addressing this second mystery. I shall be returning to the first in due course. A puzzle that must be faced is the fact that entanglements tend to spread. It would seem that eventually every particle in the universe must become entangled with every other. Or are they already all entangled with each other? Why do we not just experience an

[23.8] Show this.

entangled mess, with no resemblance whatsoever to the (almost) classical world that we actually perceive? The Schrödinger evolution of a system does not help with this. It tends to make things worse and worse, with more and more parts of the universe becoming entangled with any system that we start with, as time goes on. In terms of Hilbert space **H**, I think that it is generally accepted that the Schrödinger equation (**U** process) will not, by itself, get us out of our difficulties. If we start safely in a relatively unentangled part of **H**, then the Schrödinger evolution will (usually) almost immediately plunge us into the depths of entanglement and will not, in itself, provide us with any route, or even guidance, back out of this enormous seaweed-strewn ocean of entangled states (see Fig. 23.6).

Yet, we seem to get along pretty well, in everyday life, without even noticing these entanglements. Why is this? If we are to get no help from quantum theory's **U** process, then we must turn to its other essential ingredient: the **R** process. In fact, we have already seen something of the way that this might help in our considerations of EPR effects. Recall that I envisaged performing a measurement on an EPR pair, the other member of which was approaching my colleague on Titan. If I make my measurement first, then upon my performing this measurement, this very act would cut my colleague's particle free of its entanglement with mine, and from then on (until it became measured by my colleague) it would possess a state vector of its own, unencumbered by any further responsibility to its partner, no matter what I might subsequently do to it. Thus, it seems, it is *measurements* that slash through these entanglements. Can this be true? Is **R** the general answer to the second puzzle presented by the very phenomenon of quantum entanglement?

I think that this must be true, at least if we are thinking in terms of the way that quantum mechanics is used in practice. This has relevance to how we set up any quantum experiment, such at the (thought) experiments that we have just been considering. Recall that, in our EPR considerations, we required a number of pairs of particles that were arranged to be in

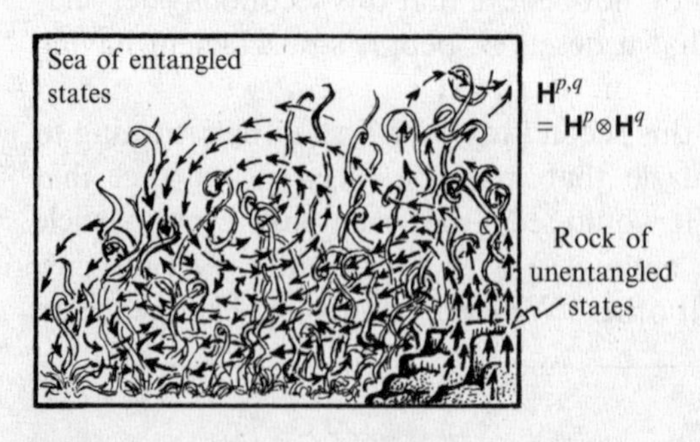

Fig. 23.6 Schrödinger evolution, away from an initial unentangled state (illustrated by the rock at the lower right) almost always leads to increasing entanglements (illustrated by the seaweed-strewn sea). So why do the ordinary objects of experience appear as separated independent things?

a particular quantum state: of spin 0 in the Stapp example and spin 1 in the Hardy example. How, using only **U**-process effects, could we ensure that our particles were not already horrendously entangled with everything else around. It seems to me that something of the nature of a 'measurement' is always an essential part of the *setting up* of a quantum experiment, to ensure that the state is uncontaminated by swarms of these unwanted entanglements. In saying this, I do not mean to imply that the experimenter deliberately sets up a 'measurement' to achieve this. It is my own view that Nature herself is continually enacting R-process effects, without any deliberate intentions on the part of an experimenter or any intervention by a 'conscious observer'.

I am entering controversial waters here, and my own position on these issues will need to be returned to later (in §§30.9–13). But how is the matter dealt with in 'conventional' quantum mechanics? It seems that 'in practice' physicists always assume that these supposed entanglements with the outside world can be ignored. Otherwise neither classical mechanics nor conventional quantum mechanics could ever be trusted. The view seems to be that all the entanglements will somehow 'average out' so that they do not need to be considered in practice, in any actual situation. Yet I am unaware of any remotely convincing demonstration that this is likely to be the case. Rather than averaging out, it would appear to be the case that everything just gets less and less like the universe we know, with individual objects not even having approximately defined locations that are not conditional on vastly many other occurrences elsewhere in the universe. I do not see any way out of this conundrum, if we are to see it as a problem in isolation from the **U**/**R** paradox that lies at the centre of the interpretation of quantum mechanics.

However we look at the issue of this pervasive entanglement with the rest of the universe, we cannot divorce it from the broader issue of why it is that, on the one hand, the **U** procedures work so supremely well for simple enough systems, whereas on the other, we have to give up on **U** and abruptly, yet stealthily, interpose the **R** process from time to time. Why, but also when and how? This is the *measurement problem* or (I think, more accurately) the measurement *paradox*, in the words of the Nobel Laureate Tony Leggett. I shall be returning to the matter in Chapter 29.

I have not yet finished with the other puzzles presented to us by entanglement. Some of these have to do with the way that the measurement of an entangled system sits extremely uncomfortably with the requirements of relativity, since a measurement of one part of an entangled pair would seem to have to affect the other *simultaneously* which, as we have seen in Chapter 17, is not a notion that we ought to countenance if we are to remain true to relativistic principles. Before attempting to face this problem, I should address one other aspect of entanglement. It is an aspect that is even more ubiquitous than those that we have been addressing in the preceding few paragraphs. It is so ubiquitous that even measurements do not cut through it, in whatever way we may choose to look at the measurement paradox. Moreover, it is a characteristic feature of quantum mechanics that appears to be independent of the others that we have been addressing so far. I refer to the remarkable way in which quantum mechanics treats systems of *identical particles*.

23.7 Bosons and fermions

Recall (§23.2) our entertaining of a 'toy universe' in which there were just 10 distinct locations open to a particle, labelled **0, 1, 2,** ..., **9**. When I considered that this universe might be inhabited by more than one particle, I was careful to require that the particles were not to be thought of as 'the same kind of particle', and I referred to them as an 'A particle' and a 'B particle', etc., rather than, say, 'two electrons', or some such. The reason for this is that quantum mechanics treats Nature's actual particles by procedures that are characteristically different from those of our earlier discussion. In fact, we must make a distinction, at this point, between two quite *different* such procedures! One of these procedures applies to particles known as *bosons* and the other to those known as *fermions*. The bosons turn out to be particles with *integral spin* (i.e. where the spin, in units of $\hbar$, takes one of the values 0, 1, 2, 3, ...) and the fermions, particles with half-odd-integer spin: values $\frac{1}{2}, \frac{3}{2}, \frac{5}{2}, \frac{7}{2}, \ldots$. (This association follows from a famous mathematical theorem, in the context of quantum field theory, known as the *spin-statistics theorem*; see §26.2.) Composite particles, such as nuclei or whole atoms or, indeed, individual hadrons such as protons or neutrons (taken as composed of quarks) can also be treated to an appropriate degree of approximation, as individual bosons or fermions. Thus, photons are bosons, and so are mesons (pions, kaons, etc.) and the particles responsible for weak interactions (W and Z particles) and strong interactions (gluons). The clearly composite α particles (2 protons, 2 neutrons), deuterons (1 proton, 1 neutron), etc. also behave closely as bosons. On the other hand, electrons, protons, neutrons, their constituent quarks, neutrinos, muons, and many other particles are fermions. We may take note of the fact that the wavefunctions of fermions are *spinorial objects*, in the terminology of §11.3 (compare §22.8), whereas those of the bosons are not.

To see what really distinguishes the bosons from the fermions, let us return to our toy universe with just 10 points in it, labelled **0**, **1**, ..., **9**. Recall that the appropriate analogue of a wavefunction would simply be a collection of complex numbers $z_0, z_1, \ldots, z_9$, for a single particle,

$z_{00}, z_{01}, \ldots, z_{99}$, for a pair of distinguishable particles, $z_{000}, z_{001}, \ldots, z_{999}$ for three such particles, etc. For a pair of bosons, however the requirement is that the collection of complex numbers z_{ij} should be *symmetric* in its indices:

$$z_{ij} = z_{ji},$$

so that $z_{38} = z_{83}$, for example. Thus, it makes no difference, as far as this 'wavefunction' is concerned, which of the particles is the one at **3** and which is the one at **8**. There is just a *particle pair*, occupying the two points **3** and **8**. Note that the pair of bosons can perfectly well have both of them in the same place; for example z_{33} is the complex weighting factor for both bosons simultaneously to occupy the point **3**. We see that there are only $\frac{1}{2}(10 \times 11) = 55$ distinguishable ways of putting the (unordered) pair of particles down on the 10 points, and only *this* number of complex numbers is needed (i.e. $\mathbf{H}^{55}$ rather than $\mathbf{H}^{100}$). With three identical bosons, we have symmetry in all three arguments:

$$z_{ijk} = z_{jik} = z_{jki} = z_{kji} = z_{kij} = z_{ikj},$$

so that we now have $\frac{1}{6}(10 \times 11 \times 12) = 220$ complex numbers to define the state: an element of $\mathbf{H}^{220}$ instead of $\mathbf{H}^{1000}$. For n identical bosons, the number is $(9+n)!/9!n!$ where this is the number of independent complex numbers $z_{ij\ldots m}$, required to be totally *symmetric* in the indices (see §§12.4,7 and §14.7 for the notation):

$$\mathrm{z}_{ij\ldots m} = \mathrm{z}_{(ij\ldots m)}.$$

Now let us consider fermions. The difference from bosons is that, for fermions, the wavefunction is required to be *anti*symmetric in its arguments,

$$z_{ij} = -\,z_{ji},$$

$$z_{ijk} = -z_{jik} = z_{jki} = -\,z_{kji} = z_{kij} = -z_{ikj},$$

$$\mathrm{z}_{ij\ldots m} = \mathrm{z}_{[ij\ldots m]},$$

so that we have $\frac{1}{2}(10 \times 9) = 45$ complex numbers for two identical fermions, $\frac{1}{6}(10 \times 9 \times 8) = 120$ complex numbers for three identical fermions, and $10!/n!(10-n)!$ for n identical fermions.[23.9] The difference in the counting arises from the fact that now we are not allowed to have two fermions at the same point, because the antisymmetry implies that the complex weightings $z\ldots$ must vanish when that occurs: $z_{33} = 0$, $z_{474} = 0$, etc.

[23.9] Explain all these numbers in both the boson and fermion cases.

Note that when we have more than 5 identical fermions in our toy universe, the numbers start coming down again. When we get to 10 fermions, there is only one possible state, and we cannot have *more* than 10 identical fermions altogether in our toy universe. We see, in this model, a manifestation of an important principle in quantum physics, called the *Pauli exclusion principle*. This tells us that two identical fermions cannot be in the same state (which is simply a feature of the antisymmetry of the fermionic wavefunction). The fact that solid materials do not collapse in on themselves ultimately depends upon this principle. Ordinary solid matter is basically composed of fermions: electrons, protons, neutrons. These have to 'keep out of each other's way', because of the Pauli principle.

In the case of bosons, things are the other way around. There is a slight tendency for bosons to 'prefer' to be in the same state. (This comes about as a purely statistical effect, when we compare the counting of different boson states with the corresponding different classical states.) When the temperature is very low, this effect can become important, and a phenomenon known as *Bose–Einstein condensation* may take place, when most of the relevant particles collect together into the same state. Superfluids are examples of this sort of thing, and even lasers are taking advantage of it. In a superconductor, electrons have a way of 'pairing up' and these *Cooper pairs* have the capability of behaving as though they were individual bosons. Some of the most impressive and counter-intuitive practical uses of quantum mechanics come about from this type of 'collective' phenomenon.

23.8 The quantum states of bosons and fermions

Although I have stated the symmetry and antisymmetry requirements of bosons and fermions only in reference to our 'toy universe', the boson/fermion symmetry requirements for a collection of actual bosons or fermions in ordinary space are basically the same. The wavefunction will be a function of a number of points in space, labelled $\mathbf{u}, \mathbf{v}, \ldots, \mathbf{y}$, as well as of various discrete parameters labelled by $u, v, \ldots, y$, respectively, to encompass each particle's group of (spinor or tensor) indices. We ask, first, how a wavefunction ψ for a *pair* of identical bosons would look. The requirement is that the function $\psi = \psi(\mathbf{u}, u; \mathbf{v}, v)$ should be *symmetric* under interchange of the particles:

$$\psi(\mathbf{u}, u; \mathbf{v}, v) = \psi(\mathbf{v}, v; \mathbf{u}, u).$$

For three identical bosons, our wavefunction should be symmetric under permutations of all three particles:

$$\psi(\mathbf{u}, u; \mathbf{v}, v; \mathbf{w}, w) = \psi(\mathbf{v}, v; \mathbf{u}, u; \mathbf{w}, w) = \psi(\mathbf{v}, v; \mathbf{w}, w; \mathbf{u}, u) = \ldots,$$

and so on.

For the case of fermions, these relations are replaced by *anti*-symmetry under interchange of the particles:

$$\psi(\mathbf{u}, u; \mathbf{v}, v) = -\psi(\mathbf{v}, v; \mathbf{u}, u).$$

$$\psi(\mathbf{u}, u; \mathbf{v}, v; \mathbf{w}, w) = -\psi(\mathbf{v}, v; \mathbf{u}, u; \mathbf{w}, w) = \psi(\mathbf{v}, v; \mathbf{w}, w; \mathbf{u}, u) = \ldots,$$

and so on. Note that, in each case, the spin state (as characterized by the discrete variables $u, v, \ldots$) must be carried with the particle in these interchanges. This has the implication that, when applying the Pauli exclusion principle, we regard the states as identical only if the spin states are also identical, in addition to their locations being identical. This is important in chemistry, for example, where two electrons can share the same orbital provided that their spins are opposite (see §24.8, Fig. 24.2).

Here is a place where the (abstract) index notation for states, referred to in §23.3 above, is handy (and a diagrammatic notation, as described in §12.8 can also be used—illustrated in Fig. 26.1). Accordingly, we could use the notation ψ^α for the wavefunction of a particular particle to which the label α has been assigned, and ϕ^β for the wavefunction of a second particle to which the label β has been assigned, and so on. If the particles are not identical, then the wavefunction for the pair of them would be the (tensor-) product state

$$\psi^\alpha \phi^\beta,$$

whereas if they are identical bosons, the state (not worrying about normalizing factors) is

$$\psi^\alpha \phi^\beta + \phi^\alpha \psi^\beta.$$

(A point about the abstract-index formulation: we have commutative multiplication, e.g. $\phi^\alpha \psi^\beta = \psi^\beta \phi^\alpha$. The non-commuting of tensor product is dealt with by the index ordering, so that $|\phi\rangle|\psi\rangle \neq |\psi\rangle|\phi\rangle$ is expressed as $\phi^\alpha \psi^\beta \neq \psi^\alpha \phi^\beta$.) We can write this symmetrized state (ignoring a factor 2) as

$$\psi^{(\alpha} \phi^{\beta)},$$

using our round-bracket notation for symmetrization (§12.7, §22.8). This has the advantage that we can immediately write down the quantum state for n identical bosons, whose individual states would be ψ^α, ϕ^β ,..., χ^κ, as the symmetrized product

$$\psi^{(\alpha} \phi^\beta \ldots \chi^{\kappa)}.$$

We can do just the same thing for fermions, where for individual states ψ^α, ϕ^β , . . . , χ^κ, the collection of all n identical fermions, would have the antisymmetrized state (§12.4)

$$\psi^{[\alpha}\phi^{\beta}\ldots\chi^{\kappa]}.$$

Notice that these many-particle states are technically all *entangled* (as we see, in particular, for the description of a pair of identical fermions being the combination $\psi^\alpha\phi^\beta - \phi^\alpha\psi^\beta$). It is a mild type of entanglement, however, because the superposition is between states that are 'physically indistinguishable', being applied only to identical particles. The states $\psi^{(\alpha}\phi^{\beta}\ldots\chi^{\kappa)}$ and $\psi^{[\alpha}\phi^{\beta}\ldots\chi^{\kappa]}$, for bosons and fermions respectively, are the nearest that we can get to 'unentangled' states, and we could take the alternative position to call such states 'unentangled'. (The general n-particle boson state, in this notation, would be some $\Psi^{\alpha\beta\ldots\kappa} = \Psi^{(\alpha\beta\ldots\kappa)}$ which does not split in this way. Similarly, a general n-particle fermion state is a 'non-splitting' $\Phi^{\alpha\beta\ldots\kappa} = \Phi^{[\alpha\beta\ldots\kappa]}$.) In terms of the ket notation, we could envisage a 'wedge product' notation $|\psi\rangle \wedge |\phi\rangle \wedge \ldots \wedge |\chi\rangle$ to deal with these symmetry and antisymmetry requirements,[11] where we bear in mind that terms commute or anticommute depending on the 'grades' of the individual factors (see §11.6).

Although the type of 'entanglement' that occurs with identical bosons or fermions is relatively 'harmless' (and, in fact, serves to reduce rather than increase the large number of alternatives that are open to a quantum system), it has at least one significant implication for an effect stretching over large physical distances. The bosonic 'entangled' nature of photons, arriving at the Earth from opposite sides of a relatively nearby star, has been be used to measure such stars' diameters, using a method due to Hanbury Brown and Twiss (1954, 1956). When their method was first proposed, it met with great opposition from many (even distinguished) quantum physicists, who argued that 'photons can only interfere with themselves, not with other photons'; but they had overlooked the fact that the 'other photons' were part of a boson-entangled whole.

23.9 Quantum teleportation

To end this chapter, we return to the puzzles that are inherent in the interpretation of EPR effects. Most particularly, we recall the seeming conflict with special relativity: that the 'communication' between EPR pairs seems to pay no respect to Einstein's own requirements that signalling faster than light should not be allowed. In order to illuminate these issues I shall give one further rather mysterious implication of quantum entanglement known as *quantum teleportation*. In my opinion, this implication leads us in a direction which may well be the one that we have to explore, if we are to come properly to terms with EPR effects generally. Yet, we shall find this direction leading into a territory that many people would, no doubt, be most reluctant to enter—and with reason, as we shall see!

What is meant by the term 'teleportation'? It evokes 'Star-Trek', with images of Captain Kirk and some of his crew being beamed down to an unexplored planet's surface, and where the view is taken that, for a person's 'identity' to be successfully beamed, it is necessary for an actual *quantum state* to be faithfully projected to the planet's surface, not just some classical listing of particle locations, etc. Such a perspective has the philosophical advantage that the teleportation procedure could not be used to *duplicate* an individual—which might present delicate conundrums concerning which of the two represents the continuation of the individual's 'stream of consciousness'.[12] Why is it not possible to copy an unknown quantum state? The question has been convincingly addressed in the literature,[13] but we can see from basic considerations that such a possibility would lead to a contradiction with the principles of standard **U**/**R** quantum mechanics. Unless one is prepared to destroy the original, then one cannot make an exact copy, and one certainly cannot make two exact copies of an unknown quantum state.

Why not? If one could, then repeating the process, one could have 4 copies, then 8, then 16, etc. Suppose that the state is just a simple spin state $|\nearrow\rangle$ of a massive particle of spin $\frac{1}{2}$. Then, after copying it many times, we should have $|\nearrow\rangle|\nearrow\rangle\dots|\nearrow\rangle = |\nearrow\nearrow\dots\nearrow\rangle$, which for large enough angular momentum could be measured in a classical way, and the spatial direction $\nearrow$ thereby obtained. By this means, we should have obtained a measurement of what the state actually is (up to a proportionality factor). But the standard **U**/**R** procedures of quantum mechanics do not allow us to do this. The only measurements on a state $|\nearrow\rangle$ that **R** permits us to perform are given by some Hermitian (or normal) operator, and these simply provide questions: 'Is the spin in some direction $\searrow$? Answer YES or NO.' After measurement, the state will be in either the asked direction $\searrow$ (YES) or in its opposite $\nwarrow$ (NO). There are actually other measurements that we can perform if we regard the spin state as entangled with other things (and we shall be seeing the value of this sort of thing shortly). But if the state being examined is to be regarded as unentangled with the outside world, then we cannot do better than perform a direct measurement on it. All that we can get from the state is a single YES/NO answer, i.e. just one *bit* (binary digit) of information. We can rotate the measuring apparatus to whatever angle we choose, but the system will not tell us the direction $\nearrow$ that the state actually points in. True, that direction is distinguished by the fact that it is the only one for which the YES answer comes with certainty (probability 1), but we cannot know beforehand which direction this is. (If someone who had set up the quantum state told us that the direction is ↗, then we could copy that; but this is not how the problem is posed: it is a previously *unknown* quantum state that we are examining and which we propose to copy.)

What quantum teleportation aims to achieve is the *sending* of a quantum state from one place to another, say from Kirk's spaceship Enterprise to the unexplored planet's surface. Quantum mechanics certainly provides no bar to achieving this; indeed, we could just transport the quantum object bodily from one place to another in an ordinary way. But we suppose that, in the particular situation envisaged, the conditions have become far too 'noisy' for the trustworthy transportation of a quantum object or of a quantum signal of any kind. The transmitting of ordinary classical information is to be all that the conditions allow. However, it is not possible to transmit a quantum state using only classical signalling. The reason for this should be clear, because classical signals, by their very nature, can be copied. If they could be used to transmit a quantum state, then quantum states could also be copied, and we have seen above that this should not be possible. What we need to do is to 'prepare the ground' first. Since the 'unexplored planet' image is a little inappropriate for this, let me enlist the help of my colleague on Titan instead, and it is this colleague to whom I intend to transmit an unknown quantum state of spin $\frac{1}{2}$.

The 'ground-preparation' that is required is that each of us must be already in possession of one member of an EPR pair of particles of spin $\frac{1}{2}$. We can suppose that the particles started together in a state of spin 0, just as in the original Bohm version of EPR. Our supposition was that it is now unreliable to send quantum states across the reaches of space between Earth and Titan. But let us say that, five years ago, before my colleague left for Titan, we each took with us our respective particle of the aforesaid entangled pair, and each particle was kept perfectly isolated from external disturbance. If our particles still remain undisturbed by the time my colleague returns, then on bringing the two of them together, the state of spin 0 would again be retrieved.

Now, we suppose, some friend presents me with another particle of spin $\frac{1}{2}$, again kept isolated from external disturbance. I am asked to transmit the state of spin of that particle, intact, to my colleague on Titan straight away. Bearing in mind that conditions are now not supposed to allow that a quantum state can be trusted to the reaches of space between here and Titan, I am permitted to send only a classical radio signal. But before doing so, I take my friend's particle to where I have my EPR particle stored and bring these two particles together. Each particle has spin $\frac{1}{2}$, so together their states would constitute a 4-dimensional system (just $\mathbf{H}^4$, if it were not for the entanglement between my particle and my colleague's particle on Titan). I now perform a measurement on the pair of them (my friend's and mine) together, which distinguishes the four orthogonal states (called *Bell states*)

$$\begin{aligned}&(0)\ \ |\uparrow\rangle|\downarrow\rangle - |\downarrow\rangle|\uparrow\rangle,\\&(1)\ \ |\uparrow\rangle|\uparrow\rangle - |\downarrow\rangle|\downarrow\rangle,\\&(2)\ \ |\uparrow\rangle|\uparrow\rangle + |\downarrow\rangle|\downarrow\rangle,\\&(3)\ \ |\uparrow\rangle|\downarrow\rangle + |\downarrow\rangle|\uparrow\rangle.\end{aligned}$$

I convey the result of this measurement, to my colleague on Titan by an ordinary classical signal, coded (say) by the numbers 0, 1, 2, 3, corresponding respectively to whichever of the above four states my measurement reveals. On receiving my message, my colleague takes out the other member of the EPR pair—carefully shielded, up until that point, from external disturbance—and performs the following rotation on it:

(0) leave alone,

(1) 180° about x axis,

(2) 180° about y axis,

(3) 180° about z axis.

It may be directly checked that this achieves the successful 'teleporting' of my friend's quantum state to my colleague on Titan.[23.10]

What is particularly striking about quantum teleportation is that, by sending my colleague merely 2 bits of classical information (one of the numbers 0, 1, 2, 3, which could have been coded as 00, 01, 10, 11 respectively), I have conveyed the 'information' of a point on the entire Riemann sphere. (Recall Fig. 22.10.) In ordinary classical terms, this would have needed the information contained in the unrestricted choice of a point in a continuum: strictly $\aleph_0$ bits, for perfect accuracy! (See §§16.3,4.) How have we achieved this feat? At this point I should mention that real experiments have been performed which confirm the expectations of quantum-mechanical teleportation (over distances of the order of metres, not Earth–Saturn spans, of course)[14] so we must take these things seriously. Not only that, but the blossoming subject of quantum cryptography depends upon things of this general nature; so do many of the ideas of quantum computation.

Have a look at Fig. 23.7. This is a spacetime diagram in which worldlines of me, my friend, and my colleague are indicated and, more importantly, of all the particles of relevance to the story, together with the classical signal that I send to my colleague on Titan. Somehow, the 'information' of

[23.10] Confirm this, with appropriate conventions concerning coordinate axes etc.

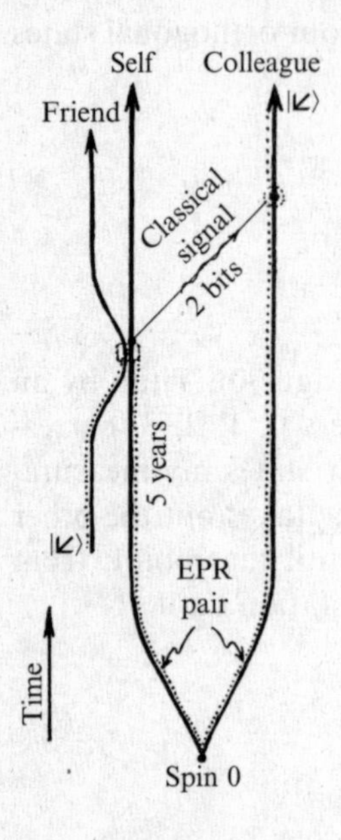

Fig. 23.7 'Quantum teleportation', demonstrating the acausal propagation of quanglement. A space-time diagram illustrates the process whereby the unknown quantum state (|↙⟩) of spin $\frac{1}{2}$, given to me by my friend, can be conveyed to my colleague on Titan by the mere transmission of 2 bits of classical information, provided that my colleague and I already share an EPR pair. The acausal quanglement link is depicted as a dotted path.

the spin direction of my friend's particle (indicated by |↙⟩) has conveyed itself to Titan despite the fact that only 2 bits of information have been actually transmitted. How have all the other $\aleph_0$ bits got across to my colleague?

Some people might take refuge in the viewpoint that quantum states are 'not real things' being 'not measurable', or something. I find it hard, myself, to come to terms with this particular kind of perspective on the world. For the direction of a spin $\frac{1}{2}$ state is saying something very definite about the world. It is saying that if someone (on Titan, in this example) chooses to measure the spin in that particular direction and *only* in this particular direction, then the answer YES will be obtained with certainty. Moreover, my friend, or a friend of my friend, could well have *prepared* the original particle to have a spin in some preassigned direction, and would know the result of a measurement to be performed on Titan in that same (or opposite) direction with certainty. That sounds real to me. (And don't be put off by the fact that my examples are a little outlandish; the principle's the point!)

So let us look again at Fig. 23.7. Something real has conveyed itself from my friend to my colleague, but the classical channel (of just 2 bits) is far too narrow to provide scope for the remaining $\aleph_0$ bits to get across. Yet there *is* a connecting link. It consists of a small stretch from my friend to me, a long one—back in time—from me to the origin of our EPR pair, and another long stretch from that point of origin to my colleague on Titan. This, indeed, is the *only* connecting link between us that is capable of supporting the amount of 'information' that is required. The trouble with it, of course, is that it contains a stretch that extends 5 years into the past!

23.10 Quanglement

I must make it very clear that I am not trying to give support to the idea that ordinary information can be propagated backwards in time (nor can EPR effects be used to send classical information faster than light; see later). That kind of thing would lead to all sorts of paradoxes that we should have absolutely no truck with (I shall return to this kind of issue in §30.6). Information, in the ordinary sense, cannot travel backwards in time. I am talking about something quite different that is sometimes referred to as *quantum information*. Now there is a difficulty about this term, namely the appearance of the word 'information'. In my view, the prefix 'quantum' does not do enough to soften the association with ordinary information, so I am proposing that we adopt a new[15] term for it:

QUANGLEMENT

At least in this book, I shall refer to what is commonly called 'quantum information' as *quanglement*. The term suggests 'quantum mechanics' and it suggests 'entanglement'. This is very appropriate. This is what quanglement is all about. Quanglement also does have something very much to do with information, but it is not information. There is no way to send an ordinary signal by means of quanglement alone. This much is made clear from the fact that past-directed channels of quanglement can be used just as well as future-directed channels. If quanglement were transmittable information, then it would be possible to send messages into the past, which it isn't. But quanglement can be used in conjunction with ordinary information channels, to enable these to achieve things that ordinary signalling alone cannot achieve. It is a very subtle thing. In a sense, quantum computing and quantum cryptography, and certainly quantum teleportation, depend crucially on the properties of quanglement and its interrelation with ordinary information.

As far as I can make out, quanglement links are always constrained by the light cones, just as are ordinary information links, but quanglement links have the novel feature that they can zig-zag backwards and forwards in time,[16] so as to achieve an effective 'spacelike propagation'. Since quanglement is not information, this does not allow actual signals to be sent faster than light. There is also an association between quanglement and ordinary spatial geometry (via the connections between the Riemann sphere and spin, as pictured in Figs. 22.10, 22.13, 22.14), this association

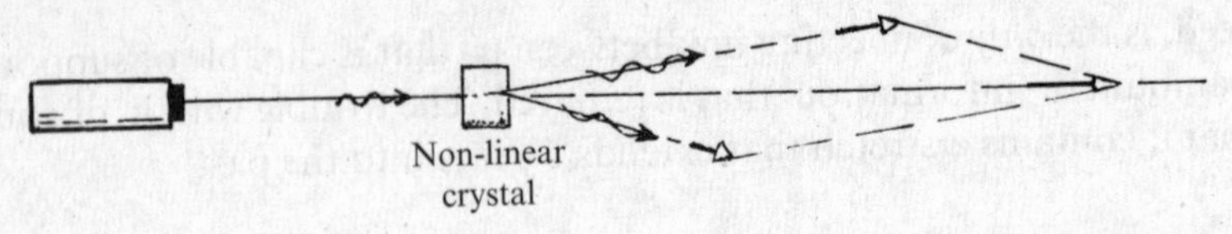

Fig. 23.8 Parametric down-conversion. A photon, emerging from a laser, impinging upon a suitable 'non-linear crystal', produces a pair of entangled photons. This entanglement manifests itself in the EPR nature of the correlated polarization states of the secondary photons, but also in the fact that their 3-momentum states must sum to that of the incident photon.

being spatially reflected at a reversal of time direction, with interesting implications.[17] It would take us too far afield to explore these in detail.

One of the most direct uses of the idea of quanglement is in certain experiments where a pair of entangled photons is produced according to the process referred to as *parametric down-conversion* (see Fig. 23.8). This occurs when a photon, produced by a laser, enters a particular type of ('non-linear') crystal which converts it into a pair of photons. These emitted photons are entangled in various ways. Their momenta must add up to the momentum of the incident photon, and their polarizations are also related to one another in an EPR way, like the examples given earlier, above.

In one particularly striking experiment, one of the photons (photon A) passes through hole of a particular shape as it speeds towards its detector D_A. The other photon (photon B) passes through a lens that is positioned so as to focus it, appropriately, at its detector D_B. The position of detector D_B is moved around slightly as each photon pair is emitted. The situation is illustrated schematically in Fig. 23.9a. Whenever D_A registers reception of photon A and D_B also registers reception of B, the position of D_B is noted. This is repeated many times, and gradually an image is built up by the detector D_B, where only the positions of B are counted when simultaneously D_A registers. The shape of the hole that A encounters is gradually built up at D_B, even though photon B never directly encounters the hole at all! It is as though D_B 'sees' the shape of the hole by looking backwards in time to the emission point C at the crystal, and then forwards in time in the guise of photon A. It can do this because the 'seeing' process in this situation is achieved by quanglement. This flitting back and forth in time is precisely the kind of thing that quanglement is allowed to do. Even the strength and positioning of the lens can be understood in terms of quanglement. To obtain the lens location, think of a mirror placed at the emission point C. The lens (a positive lens) is placed so that the image of the hole, as reflected in this mirror at C, is focused at the detector D_B. Of course there is no actual mirror at C, but the quanglement links act as though reflected at a mirror, but they are reflected in time as well as space.[23.11]

[23.11] See if you can give a fuller explanation of this, using quanglement ideas or otherwise.

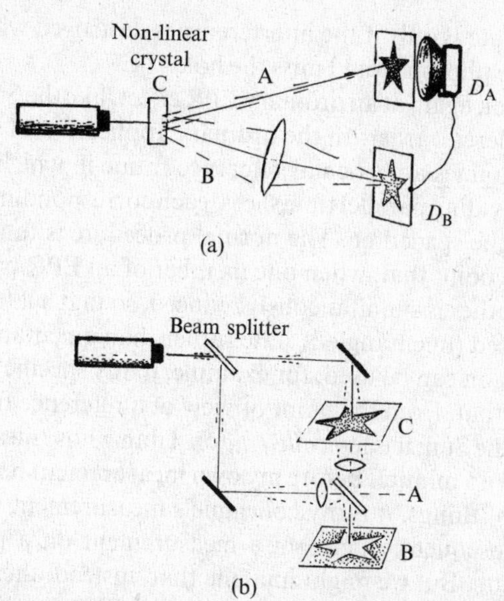

Fig. 23.9 Transmission of an image via quantum effects. (a) Entangled photons A, B are produced by parametric down-conversion at C. Photon A has to pass through a hole of some special shape to reach detector D_A, while B passes through a lens, positioned so as to focus it at detector D_B. Detector positions are gradually moved, appropriately in conjunction, and when they both register, the position of D_B is noted. Repeated many times, an image of the hole shape is gradually built up by D_B, where only those positions of B are counted when D_A also registers. (This is schematically illustrated here by having, instead, D_B as a fixed photographic plate that is only activated when D_A registers.) Quanglement is illustrated by the lens positioning being determined as though C were a 'mirror' that reflects the photon backwards in time as well as in direction. (b) An alternative scheme using an adaption of the Elitzur–Vaidman bomb test of Fig. 22.6 (which is to be reflected in a horizontal line). The photographic plate at B receives the photon only when the photon 'would have been stopped' by the template at C, but actually took the lower route!

In case the reader finds this experiment far-fetched, I should make clear that this is a real effect. It has been successfully confirmed in experiments[18] performed at the University of Baltimore, Maryland. Various other related experiments involving parametric down-conversion, which can be best understood in terms of quanglement, have also been carried out.[19]

On the other hand, the general type of situation illustrated in Fig. 23.9a might be regarded as not being 'essentially quantum mechanical'. For one could envisage a device at C which simply ejects classical particles pairwise in the appropriate directions and, apart from the lensing, similar results could be obtained. We can remedy this by using a modification of the Elitzur–Vaidman set-up illustrated in Fig. 22.6 (reflected horizontally); see Fig. 23.9b. Now there is only one photon at a time. It can register at the photographic plate B only if the interference is destroyed when the alternative route for the photon would miss the hole at C.

Now, let us look again at an ordinary EPR effect, like the Stapp and Hardy examples considered earlier. In the ordinary application of the quantum **R** process, one imagines a particular reference frame in which there is a time coordinate t providing parallel time slices, each corresponding to a constant t value, through the spacetime. The normal procedure is to adopt the (non-relativistic) viewpoint that, when one member of an EPR pair is measured, the state of the other is simultaneously reduced, so that a later measurement looks at a reduced (unentangled) state rather than an entangled state. This kind of description can be used, for example, in my specific EPR examples. Let us suppose that, from the point of view of a reference frame stationary with respect to the Sun, it is my *colleague* on Titan whose measurement takes place first, some 15 minutes before my own measurement here on Earth. So, in this picture of things, it is my colleague's measurement that reduces the state, and I subsequently perform a measurement on a particle with an unentangled state. But we might imagine that, instead, the whole situation is described from the perspective of some observer O passing by at great speed (say $\frac{2}{3}c$) in the general direction from my colleague on Titan to me. From O's viewpoint, I was the one who first made the measurement on the EPR pair, thereby reducing the state, and it was my colleague who measured the reduced unentangled state (Fig. 23.10) (see §18.3, Fig. 18.5b). The joint probabilities come out the same either way, but O has a different picture of 'reality' from the one that I and my colleague had before. If we think of **R** as a *real* process, then we seem to be in conflict with the principle of special relativity, because there are two incompatible views as to which of us

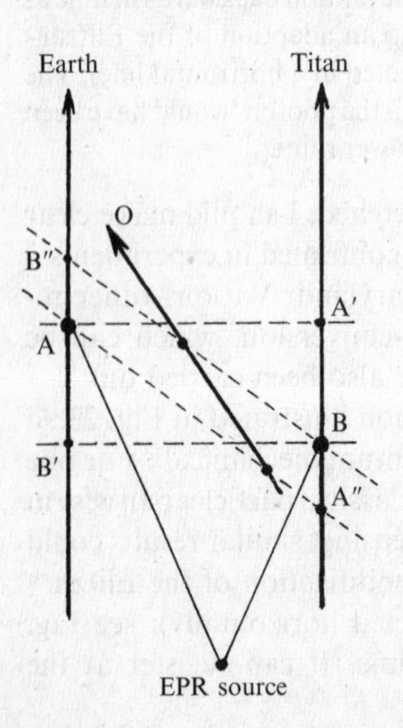

Fig. 23.10 Conflict between relativity and the objectivity of state reduction? Spacetime diagram of an EPR situation, with detectors on Earth and Titan and source closer to Titan than Earth. From the perspective of an inertial frame, stationary with respect to the Sun, the detector on Titan registers first (at B) and this reduces the state simultaneously (at B′) on Earth. Only later does detection on Earth take place (at A) of a state now unentangled (simultaneous with A′ on Titan). However, to an observer O, travelling towards Earth from Titan with very great speed, detection takes place first on Earth (at A, simultaneous with A″ on Titan, according to the 'sloping' simultaneity lines of O) and Titan receives the reduced unentangled state (at B, simultaneous with B″ on Earth).

effected the reduction of the state and which of us observed the reduced state after reduction.

We may deduce from this that EPR effects, despite their seemingly acausal nature, cannot be directly used to transmit ordinary information acausally, which one might imagine could influence the behaviour of a receiver at spacelike separation from the transmitter. A reference frame can always be chosen in which it is the 'reception event' which occurs first, and the 'transmitter' then has only the reduced state to examine. It is 'too late', by then, for the entanglement to be used for a signal because it has already been destroyed by the state reduction.

What is the quanglement perspective on these matters?[20] See §30.3. On this picture, it is not correct to think of either measurement (mine or my colleague's) as effecting the reduction and the other (my colleague's or mine) as measuring the reduced state. The two measurement events are on an equal footing with one another, and we think of the quanglement as providing a connection between these events which correlates the two. It makes no difference which event is viewed as being to the past of the other, for quanglement can equally be thought of as propagating into the past as propagating into the future. Not being capable directly of carrying information, quanglement does not respect the normal restrictions of relativistic causality. It merely effects constraints on the joint probabilities of the results of different measurements.

Although quanglement is a useful idea in 'making sense' of this kind of puzzling quantum experiment, I am not sure how far these ideas can be carried, nor how precisely the effects of quanglement can be delineated. The idea of quanglement certainly does not resolve the issue of quantum measurement, telling us little, if anything, about the circumstances under which **R** takes over from **U**. That issue will be addressed more fully in Chapters 29 and 30, especially in §30.12, but the precise role of quanglement in this in this is not yet very clear, to my mind. A more promising connection is with some of the ideas of twistor theory, and these will be examined briefly in §33.2.

Notes

Section 23.1

23.1. See Eddington(1929b); Mott(1929); Dirac (1932).

Section 23.3

23.2. See Einstein *et al.* (1935); Schrödinger (1935); also Afriat (1999).

23.3. In detail what it tells us about quantum computing is a subtle issue, however; see Jozsa and Linden (2002).

23.4. See Bell (1987). Perhaps the neatest and most widely quoted version of this inequality is that due to Clauser *et al.* (1969). It takes the form

$|E(A,B) - E(A,D)| + |E(C,B) + E(C,D)| \leq 2$, where $E(x,y)$ is the expectation value of agreement ($E = 1$ for complete agreement and $E = -1$ for complete disagreement) between the results of alternative measurements A, C for one component of the EPR pair and B, D for the other. Highly relevant, also, are Gleason (1957) and Kochen and Specker (1967). See Redhead (1987).

23.5. See Bohm (19); Redhead (1987); Afriat (1999). For some recent extrordinary experimental confirmation of EPR effects, see Tittel *et al.* (1998).

23.6. For various specific examples relevant to this (Heywood and Redhead 1983; Stairs 1983), see Kochen and Specker (1967); Peres (1991, 1995); Conway and Kochen (2002); Penrose (1994), Section 5.3; Penrose (2000b); Zimba and Penrose (1993).

23.7. See, for example, Hannabuss (1997). See Nielsen and Chuang (2000) for a general discussion of such issues in entanglement.

Section 23.4

23.8. See Note 23.5.

Section 23.5

23.9. See Hardy (1992, 1993).

23.10. See Hardy (1993).

Section 23.8

23.11. A scheme of this nature was effectively adopted in my (1994) book *Shadows of the Mind*, §5.15, but without explicitly using 'wedges'.

Section 23.9

23.12. See Penrose (1989).

23.13. Wooters and Zurek (1982).

23.14. See Jennewein *et al.* (2002).

Section 23.10

23.15. See Penrose (2002).

23.16. See Jozsa (1998); Peres (2000).

23.17. See Penrose (1998a).

23.18. See Shih *et al.* (1995).

23.19. See Gisin *et al.* (2003), for example, for a taste of this important area.

23.20. Compare with Aharonov and Vaidman (2001); Cramer (1988); Costa de Beauregard (1995); and Werbos and Dolmatova (2000).

24
Dirac's electron and antiparticles

24.1 Tension between quantum theory and relativity

THE considerations of §23.10 only just begin to touch upon some of the profound issues about the relation between the principles of quantum mechanics and those of relativity. Indeed, in presenting the detailed way in which quantum theory operates, in the preceding three chapters, I have taken a very non-relativistic standpoint, appearing to ignore those important lessons that Einstein and Minkowski have taught us (as described in Chapter 17) about the interdependence of time and space. In fact, this is quite usual in quantum theory. The standard approach adopts a 'picture of reality' in which time is treated differently from space. As remarked early in Chapter 22, there is a single external time coordinate; but there are many spatial ones, each particle requiring its own set. This asymmetry is usually regarded as a 'temporary' feature of non-relativistic quantum theory, which would be merely an approximation to some more complete fully relativistic scheme. In this chapter, and in the following two, we shall begin to witness the profound issues that arise when we try seriously to bring the principles of quantum theory together with those of special relativity. (The more ambitious union with Einstein's *general* relativity—where gravitation and spacetime curvature are also brought into the picture—requires something considerably more, and there is as yet no consensus on the most promising lines of pursuit. I shall address some of these lines in Chapters 28 and 30–33.)

It is a particular feature of combining quantum theory with special relativity that the resulting theory becomes not just a theory of quantum particles, but a theory of quantum *fields*. The reason for this can be boiled down to the fact that the bringing in of relativity implies that individual particles are no longer conserved, but can be created and destroyed in conjunction with their *antiparticles*. This comment needs some explanation. Why is there this need for 'antiparticles' in a relativistic quantum theory? Why does the presence of antiparticles lead us from a quantum theory of particles to a quantum theory of fields? This chapter is largely aimed at the answers to these two questions, but particularly the first, and with particular reference to Dirac's wonderful insights into the mathematical description of electrons.

Quantum field theory itself will be discussed in Chapter 26, and we shall be glimpsing some of the pervasive tension between special relativity and quantum theory that has guided the subject of particle physics into more and more elaborate mathematical schemes. We shall find ourselves enticed into a long and fascinating journey. When the tension can be resolved appropriately, as with the standard model of particle physics, discussed in Chapter 25, the resulting theory is found to have a very remarkable agreement with observational fact.

Yet, in many respects, this tension has remained, and has never been fully resolved. Strictly speaking, quantum field theory (at least in most of the fully relevant non-trivial instances of this theory that we know) is *mathematically inconsistent*, and various 'tricks' are needed to provide meaningful calculational operations. It is a very delicate matter of judgement to know whether these tricks are merely stop-gap procedures that enable us to edge forward within a mathematical framework that may perhaps be fundamentally flawed at a deep level, or whether these tricks reflect profound truths that actually have a genuine significance to Nature herself. Most of the recent attempts to move forward in fundamental physics indeed take many of these 'tricks' to be fundamental. We shall be seeing several examples of such ingenious schemes in this and later chapters. Some of these appear to be genuinely unravelling some of Nature's secrets. On the other hand, it might well turn out that Nature is a good deal less in sympathy with some of the others!

24.2 Why do antiparticles imply quantum fields?

The theoretical anticipation of antiparticles, in a relativistic quantum theory, appears to have unravelled one of Nature's true secrets, now well supported by observation. We shall be seeing something of the theoretical reasons for antiparticles later in this chapter, and most specifically in §24.8. For the moment, instead of addressing that issue, let us restrict attention to the second of the two questions raised above, namely: why does the presence of antiparticles lead us away from a quantum theory of particles and into a quantum theory of fields? Let us, for now, just accept that there is an antiparticle to each type of particle, and try to come to terms with the consequences of this remarkable fact.

The key property of an antiparticle (at least, the antiparticle of a massive particle) is that the particle and antiparticle can come together and annihilate one another, their combined mass being converted into energy, in accordance with Einstein's $E = mc^2$; conversely, if sufficient energy is introduced into a system, localized in a suitably small region, then there arises the strong possibility that this energy might serve to create some particle together with its antiparticle. Thus, with this potential for the production of antiparticles, there is always the possibility of more and more particles coming into the picture, each particle appearing together with its antiparticle. Thus, our relativistic theory certainly cannot just be a theory of single particles, nor of any fixed number of particles whatever. (In quantum theory, as we shall be seeing in Chapters 25 and 26 particularly, if there is the potential for something to happen—e.g. the production of numerous particle/antiparticle pairs—then this potential possibility actually makes its contribution to the quantum state.) In attempting to come up with a theory of relativistic particles, therefore, one is driven to provide a theory in which there is a potential for the creation of an unlimited number of particles.

This takes us outside the framework of Chapters 21–23; but we shall see in Chapter 26 how the quantum theory of fields enables us to accommodate such behaviour. Indeed, according to a common viewpoint, the primary entities in such a theory are taken to be the quantum fields, the particles themselves arising merely as 'field excitations'. Yet, we shall find that this is not the only way to look at quantum field theory. In the Feynman-graph approach, which we shall address in Chapters 25 and 26, there is a strong 'particle-like' perspective on the basic processes that go to make up the quantum field theory where, indeed, an unlimited number of particles can be created or destroyed.

It is instructive to elaborate, a little more, on the reasons underlying particle creation, as a feature of a sensible relativistic quantum theory. I am still assuming, for the moment, that antiparticles exist. Essentially, the reason to expect particle creation comes down to Einstein's famous $E = mc^2$. Energy is basically interchangeable with mass (c^2 being merely a 'conversion constant' between the units of energy and mass that are being used). When enough energy is available, then a particle's mass can be created out of that energy.

However, having the means to produce the particle's mass is not, in itself, sufficient for the conjuring up the particle itself. There are likely to be various *conserved* (additive) quantum numbers, such as electric charge (or other things, e.g. baryon number) which are not supposed to be able to change in a physical process. Simply to conjure a charged particle out of pure energy, for example, would represent a violation of charge conservation (and the same would apply to other conserved quantities, such as baryon number, etc.). However, with the assumption that for every kind of particle there is a corresponding *antiparticle*, for which every additive quantum number is reversed in sign, a particle together with its antiparticle can be created out of pure

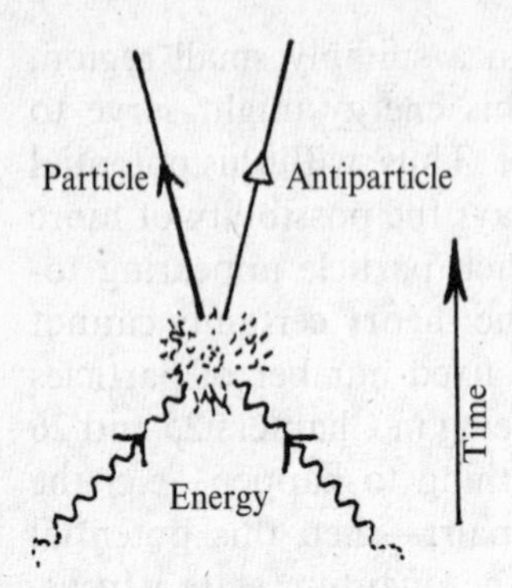

Fig. 24.1 A particle and its antiparticle can be created out of energy. All the particle's conserved additive quantum numbers are reversed in sign for the antiparticle, to ensure conservation of these quantities in the creation process.

energy (see Fig. 24.1). All the additive quantum numbers will be conserved in this process.

The rest mass of the antiparticle (rest mass being non-additive) is, on the other hand, the same as that of the original particle. We need sufficient energy—at least twice that of the rest mass/energy of the particle itself—to create both the particle and its antiparticle in this process. Conversely, if a particle of a given type encounters another particle which is of its antiparticle's type, then it is possible for them to annihilate one another with the production of energy. Again, the energy has to be at least twice the rest mass/energy of the individual particle. In either the creation or the annihilation process, the energy can be more than this value, because the particle and antiparticle are likely to be in relative motion, and there will be an energy residing in this motion—the kinetic energy—that adds in to the total. In any event, we see that the presence of antiparticles indeed forces us away from the quantum theory of individual particles, as described in Chapters 21–23.

24.3 Energy positivity in quantum mechanics

Let us now return to the road that ultimately leads us to the requirement of antiparticles in a relativistic quantum theory. We shall need to examine the framework of quantum theory from a somewhat deeper perspective than before. First, let us recall the basic form of the Schrödinger equation

$$i\hbar\frac{\partial\psi}{\partial t} = \mathcal{H}\psi.$$

Suppose that we require our quantum system to have a definite value E for its energy, so that ψ is an *eigenstate of energy*, with eigenvalue E; that is (since $\mathcal{H}$ is the operator defining the total energy of the system), we require

$$\mathcal{H}\psi = E\psi.$$

According to the quantum-mechanical **R** process (§§22.1,5), such a state ψ would result from our having performed a measurement on a system asking it the question 'what is your energy', where we have received the specific answer 'E'. Schrödinger's equation then tells us

$$i\hbar\frac{\partial\psi}{\partial t} = E\psi.$$

The solutions of this equation have the form[24.1]

$$\psi = C\,e^{-iEt/\hbar},$$

where C is *independent* of t (i.e. a complex function of spatial variables only).

Now, it is important that the energy value E be a positive number. Negative-energy states are 'bad news' in quantum mechanics, for various reasons (their presence leading to catastrophic instabilities[1]).[24.2] When the energy E is indeed positive, the coefficient $-iE/\hbar$ of t in the exponent (in $e^{-iEt/\hbar}$) is a *negative* multiple of i. Recall from §9.5 (and see Note 9.3) that functions $\psi(t)$ of this nature, or linear combinations of such functions, are said (a little confusingly) to be of *positive frequency*.

Recall, also, that in §9.3 we addressed the splitting of a function $f(x)$ (of a real variable x) into its positive- and negative-frequency parts in an apparently completely different way, namely in terms of the geometry of the Riemann sphere.[2] There we treated this as just an elegant piece of pure mathematics. The real line could be thought as wrapped once around the equator of the Riemann sphere, and the positive-frequency part of the function f was understood as that part which extended—holomorphically (see §7.1)—into the southern hemisphere, the negative-frequency part extending, likewise, into the northern hemisphere. But now we have come to a remarkable *physical* reason for the great importance of this notion. Any self-respecting wavefunction, though it need not itself be an eigenstate of energy, ought to be expressible as a linear combination of eigenstates of energy, and each energy eigenvalue ought to be positive. Thus the time dependence of any decent wavefunction ought indeed to have this crucial positive-frequency property. It seems to me that this remarkable relation between an essential physical requirement, on the one hand, and an elegant mathematical property, on the other, is a

---

[24.1] Check that this is indeed a solution.

[24.2] Explain why adding a constant K to the Hamiltonian simply has the effect that all solutions of the Schrödinger equation are multiplied by the same factor. Find this factor. Does this substantially affect the quantum dynamics? Suppose we are concerned with the gravitational effect of a quantum system. Why can we not simply 'renormalize' the energy in this way under these circumstances?

wonderful instance of the deep, subtle, and indeed mysterious relationship between sophisticated mathematical ideas and the inner workings of our actual universe.

In non-relativistic quantum mechanics, this requirement of positive frequency tends to come about automatically, as a natural feature of the theory, provided that the Hamiltonian comes from a reasonable physical problem where classical energies are positive. For example, in the case of a single free non-relativistic (spinless) particle of (positive) mass μ, we have the Hamiltonian $\mathcal{H} = p^2/2\mu$ (recall §20.2, §21.2). The expression p^2, and hence the Hamiltonian $\mathcal{H}$ itself, is what is called 'positive-definite'[3] (§§13.8,9). Classically, this comes about because p^2 is a sum of squares, and this cannot be negative: $p^2 = \mathbf{p} \bullet \mathbf{p} = (p_1)^2 + (p_2)^2 + (p_3)^2$. Quantum-mechanically, we must make the replacement of $\mathbf{p}$ by $-i\hbar\boldsymbol{\nabla}$, where $\boldsymbol{\nabla} = (\partial/\partial x^1, \partial/\partial x^2, \partial/\partial x^3)$, and now the 'positive-definite' assertion refers to the *eigenvalues* of the operator $-\nabla^2$ (for normalizable states, i.e. elements of an appropriate Hilbert space **H**), and again these cannot be negative, esentially for the same reason as in the classical case.[24.3]

24.4 Difficulties with the relativistic energy formula

Now, let us consider a *relativistic* quantum particle. In this case, the Hamiltonian is obtained from the relativistic expression for the energy, which is not $p^2/2\mu$ but

$$[(c^2\mu)^2 + c^2p^2]^{\frac{1}{2}}.$$

This expression comes directly from the equation $(c^2\mu)^2 = E^2 - c^2p^2$ of §18.7, where μ is now the rest mass of the particle. The reader who worries that this expression does not look much like $p^2/2\mu$ should refer back to Exercise [18.20]. That told us, from a power series expansion of $[(c^2\mu)^2 + c^2p^2]^{\frac{1}{2}}$, that our relativistic expression incorporates Einstein's famous $E = mc^2$ as a first term. This term is the energy contribution coming from the particle's rest mass, and it is additional to the kinetic energy of the particle's motion. The second term indeed gives us the Newtonian (kinetic energy) Hamiltonian $p^2/2\mu$.

The reader may thereby be reassured about our choice of relativistic Hamiltonian! Nevertheless, it would be decidedly awkward (and not very

---

[24.3] Schrödinger's equation, here, is $\partial\psi/\partial t = (i\hbar/2\mu)\nabla^2\psi$. Confirming, first, that for an energy eigenstate with energy E we have $-\nabla^2\psi = A\psi$, where $A = 2\mu\hbar^{-2}E$, use *Green's theorem* $\int\bar{\psi}\nabla^2\psi\, d^3x = -\int\boldsymbol{\nabla}\bar{\psi} \bullet \boldsymbol{\nabla}\psi d^3x$ to show that A must be positive for a normalizable state. (Conversely, it is in fact true that, for positive A, there are many solutions of $-\nabla^2\psi = A\psi$, which tail off suitably towards infinity so that the norm $\|\psi\|$ remains finite[4] and we can normalize to $\|\psi\| = 1$, if we wish.) Show how to derive Green's theorem, from the fundamental theorem of exterior calculus.

illuminating) to attempt to use this actual power series expression for our Hamiltonian, particularly because the classical series does not even converge when $p^2 > \mu^2$. Yet, we shall find that the square root (half power) in the exact expression $[(c^2\mu)^2 + c^2p^2]^{\frac{1}{2}}$ carries its own profound difficulties, in relation to preserving the positive-frequency requirement. Let us try to understand something of the importance of this.

To avoid cluttering our expressions unnecessarily, I shall return to units for which the speed of light is unity

$$c = 1,$$

so that our relativistic Hamiltonian (including the rest energy) is now

$$\mathcal{H} = (\mu^2 + p^2)^{\frac{1}{2}}.$$

We must bear in mind that the p^2, in quantum mechanics, is really the second-order partial differential operator $-\hbar^2\nabla^2$, so we shall need some considerable mathematical sophistication, if we are actually to assign a consistent meaning to the expression $(\mu^2 - \hbar^2\nabla^2)^{\frac{1}{2}}$, which is the square root of a partial differential operator! (To appreciate the difficulty, think of trying to assign a meaning to a thing like $\sqrt{(1 - \mathrm{d}^2/\mathrm{d}x^2)}$, for example.[24.4])

There is a more serious difficulty with this square-root expression, because it contains an implicit *sign ambiguity*. In classical physics, such things might not worry us, because the quantities under consideration are ordinary real-valued functions, and we can imagine that we could keep the positive values separate from the negative ones. However, in quantum mechanics, this is not so easy. Part of the reason for this is that quantum wavefunctions are complex, and the two square roots of a complex-number expression do not tend to separate neatly into 'positive' and 'negative' in a globally consistent way (§5.4). This should be considered in relation to the fact that quantum mechanics deals with operators acting on complex functions, and things like square roots can lead to essential ambiguities that are not simply resolved by just saying 'take the positive root'.

There is another way of expressing this difficulty. In quantum mechanics, one has to consider that the various possible things that 'might' happen, in a physical situation, can all contribute to the quantum state, and therefore all these alternatives have an influence on whatever it is that does happen. When there is something like a square root involved, each of the two roots has to be considered as a 'possibility', so even an 'unphysical negative energy' has to be considered as a 'physical possibility'. As soon as there is the potential for such a negative-energy state, then there is opened

[24.4] Make some suggestions, either using Fourier transforms (§9.4), or a power series, or contour integrals, or otherwise.

up the likelihood of a spontaneous transition from positive to negative energy, which can lead to a catastrophic instability. In the case of a non-relativistic free particle, we do not have this problem of the possibility of negative energy, because the positive-definite quantity $p^2/2\mu$ does not have this awkward square root. However, the relativistic expression $(\mu^2 + p^2)^{\frac{1}{2}}$ is more problematic in that we do not normally have a clear-cut procedure for ruling out negative square roots.

It turns out that in the case of a single free particle (or a system of such non-interacting particles), this does not actually cause a real difficulty, because we can restrict attention to superpositions of positive-energy plane-wave solutions of the free Schrödinger equation, which are just those considered in §21.5, and there are no transitions to negative energy states. However, when *interactions* are present, this is no longer the case. Even for just a single relativistic charged particle in a fixed electromagnetic background field the wavefunction cannot, in general, maintain the condition that it be of positive frequency. In this, we begin to perceive the tension between the principles of quantum mechanics and those of relativity.

As we shall be seeing in §24.8, the great physicist Paul Dirac found a way to resolve this particular tension. But as a first step, he put forward an ingenious and deeply insightful proposal—his now famous equation for the electron—which got rid of the troublesome square root in a marvellous and unexpected way. This subsequently led to a highly original point of view in which negative energies are eliminated, their effects being taken over by what was then a startling prediction: the existence of *antiparticles*. In order to understand all this, let us return to that essential feature of relativity theory from which the square root originates.

24.5 The non-invariance of $\partial/\partial t$

Let us recall the reason underlying our apparent need to adopt the Hamiltonian $(\mu^2 + p^2)^{\frac{1}{2}}$ in the relativistic case. This ultimately comes down the fact that Schrödinger's equation makes use of the operator $\partial/\partial t$ (i.e. 'rate of change with respect to time') whereas, in relativity, $\partial/\partial t$ is not an invariant thing because time and space cannot be considered separately but are just particular aspects of a combined 'spacetime'. Thus, it is not 'relativistically invariant' to regard $\partial/\partial t$ as a fundamental thing. Now, as we saw in §21.3, the $\partial/\partial t$ in Schrödinger's equation comes from the general 'quantization trick' whereby the standard spacetime 4-momenta p_a (i.e. the energy E, and negative 3-momentum $-\mathbf{p}$) are replaced by the differential operators $i\hbar\partial/\partial x^a$ (i.e. the energy E by $i\hbar\partial/\partial t$ and $-\mathbf{p}$ by $i\hbar\nabla$). The 'relativistic non-invariance' of $\partial/\partial t$ is thus closely related to the non-invariance of the energy. In the same way that time and space get mixed up in relativity theory, energy and momentum also get mixed up (as we saw in §18.7).

Recall, also, that Einstein's $E = mc^2$ (with the convention that $c = 1$) tells us that energy is mass and mass is energy, so mass, also, is 'non-invariant'. This, however, refers to the additive 'mass' concept m (time component of the energy–momentum 4-vector) which is not intrinsic to a particle itself, but which is the mass measured in some reference frame that need not share that particle's velocity. The larger the particle's velocity, the larger will be this 'perceived' mass (which, indeed, is the reason that m is not an invariant quantity). The rest mass μ, for a particle, is invariant, but the trouble with rest mass is that it is not additive and it is not conserved in particle transformations, so it makes a poor choice for something to be equated to a Hamiltonian. Moreover, μ is given as a square root of an expression in the energy and momentum, namely (taking $c = 1$)

$$\mu^2 = p_a p^a = m^2 - p^2, \quad \text{i.e.} \quad \mu = (m^2 - p^2)^{\frac{1}{2}},$$

which re-expresses the square-root expression for the mass/energy $m = E(= \mathcal{H})$ that we had earlier, namely $m = (\mu^2 + p^2)^{\frac{1}{2}}$.

Nevertheless, we might toy with the idea of using this invariant rest energy μ or its square μ^2 in a Schrödinger-type equation, instead of the non-invariant energy component m. The quantization trick (i.e. m replaced by $\mathrm{i}\hbar\partial/\partial t$ and $\mathbf{p}$ by $-\mathrm{i}\hbar\nabla$) applied to the squared rest energy, namely to $\mu^2 = m^2 - p^2$, provides us with $(\mathrm{i}\hbar)^2$ times the operator[5]

$$\begin{aligned}\Box &= \left(\frac{\partial}{\partial t}\right)^2 - \nabla^2 \\ &= \left(\frac{\partial}{\partial t}\right)^2 - \left(\frac{\partial}{\partial x}\right)^2 - \left(\frac{\partial}{\partial y}\right)^2 - \left(\frac{\partial}{\partial z}\right)^2\end{aligned}$$

in Minkowskian coordinates (t, x, y, z). This is called the *wave operator* or *D'Alembertian*, and it indeed has an invariant meaning. (Recall that $(\partial/\partial x)^2$ means the second derivative operator $\partial^2/\partial x^2$, etc.) Although the conventional Schrödinger equation does not allow us to employ this operator directly (for reasons indicated above, the Schrödinger equation requiring the first-order '$\partial/\partial t$' and not the second-order $(\partial/\partial t)^2$), we can nevertheless anticipate that the second-order equation $(\mathrm{i}\hbar)^2\Box\psi = \mu^2\psi$ (where $(\mathrm{i}\hbar)^2\Box$ is obtained from μ^2 by the quantization trick, and the μ in the equation actually *is* the rest mass) should have significance as a wave equation for a relativistic particle. This equation can be re-written as

$$(\Box + M^2)\psi = 0,$$

where $M = \mu/\hbar$ and, indeed, it does have significance in relativistic quantum theory. This equation is now frequently referred to as the 'Klein–Gordon equation', although Schrödinger himself appears to have been about the first to put forward this relativistically invariant equation, which he did even before he settled on his now more famous 'Schrödinger equation' (as described in §21.3).[6]

In the context of modern quantum field theory, the Klein–Gordon equation can be used, if interpreted in the appropriate way, to describe massive spinless particles, notably those particles referred to as *mesons* (intermediate mass particles such as pions or kaons). But this interpretation requires the full framework of quantum field theory, which was only in an embryonic form when Dirac first suggested his quite different-looking equation for the electron in 1928. Dirac had argued in favour an equation in which the time-derivative $\partial/\partial t$ appears in a first-order form (as it appears in Schrödinger's equation) rather than in the second-order form $(\partial/\partial t)^2$ that it has in the wave operator $\Box$. His reasons were related to those indicated above, but more specifically he reasoned from a requirement that the wavefunction of a particle ought to provide an expression for a probability density for finding the particle at any chosen place, qualitatively similar to the $\bar{\Psi}\Psi$ of standard non-relativistic quantum mechanics (§21.9), which should be positive-definite so that this probability can never become negative. This is not quite the same as the requirement that the energy be positive-definite, but it is a complementary requirement of essentially equal importance.[7]

24.6 Clifford–Dirac square root of wave operator

By an ingenious and magnificently insightful resolution of the seemingly irresolvable conflict between the demands of relativity and his perceived need for a first-order $\partial/\partial t$, Dirac managed to find an equation that *is* of the first order in $\partial/\partial t$ by explicitly taking the square root of the wave operator $\Box$ in a way that is subtly relativistically invariant. He did this by allowing the introduction of certain additional non-commuting quantities. Such quantities are legitimate in quantum mechanics because they are to be treated as linear operators acting on the wavefunction, in the manner of the non-commuting position and momentum operators that we originally encountered in §21.2. As we shall shortly see, what is remarkable is that these non-commuting operators, which Dirac found himself driven to introduce, describe the physical *spin* degrees of freedom of the most fundamental fermions (see §23.7) of Nature, namely the electrons and protons that were known in Dirac's day, and the neutrons, muons, quarks, and many other spin $\frac{1}{2}$ particles that are known today.

In fact, in finding his non-commuting 'spin' quantities, Dirac rediscovered (an instance of) the *Clifford algebras* that we encountered in §11.5. He appears not to have been aware of William Kingdon Clifford's earlier work, nor of the fact that Clifford (1877), and even Hamilton before him, had already noticed that elements of these algebras can be used to 'take the square root' of Laplacians—the wave operator $\square$ being a particular kind of generalized Laplacian, where the dimension is 4 and the signature $+ - - -$. In fact, as Clifford himself knew, William Rowan Hamilton had already shown around 1840 that a square root of the ordinary 3-dimensional Laplacian can be obtained by the use of *quaternions*:[8]

$$\left(\mathbf{i}\frac{\partial}{\partial x}+\mathbf{j}\frac{\partial}{\partial y}+\mathbf{k}\frac{\partial}{\partial z}\right)^2=-\left(\frac{\partial}{\partial x}\right)^2-\left(\frac{\partial}{\partial y}\right)^2-\left(\frac{\partial}{\partial z}\right)^2=-\nabla^2$$

(see §11.1). Clifford's procedure generalized this to higher dimension.[9] It is perhaps not surprising that Dirac was unaware of Clifford's discoveries of over half a century earlier, because this work was not at all well known in the 1920s, even to many specialists in algebra. Even if Dirac had known of Clifford algebras before, this would not have dimmed the brilliance of the realization that such entities are of importance for the quantum mechanics of a spinning electron—this constituting a major and unexpected advance in physical understanding.

In Dirac's case, it is indeed the wave operator that needs to have its square root taken, this being the 4-dimensional (Lorentzian) Laplacian of relevance to Minkowskian geometry:

$$\square=\left(\frac{\partial}{\partial t}\right)^2-\nabla^2.$$

We thus use 'Lorentzian' Clifford algebra elements $\gamma_0, \ldots, \gamma_3$, satisfying

$$\boldsymbol{\gamma}_0^2=1,\ \boldsymbol{\gamma}_1^2=-1,\ \boldsymbol{\gamma}_2^2=-1,\ \boldsymbol{\gamma}_3^2=-1.$$

In a standard ($+ + \ldots +$ signature) Clifford algebra, each of these squares would be -1. Here, I follow what appears to be the standard physicist's convention, with regard to signs, whereby the spatial $\boldsymbol{\gamma}$s retain the original Clifford negative squares.[10] The temporal $\boldsymbol{\gamma}_0$ has a positive square, however. It is in this sense that Dirac's Clifford algebra is 'Lorentzian'. For different $\boldsymbol{\gamma}$s, Clifford's anti-commutation still holds (§11.5):

$$\boldsymbol{\gamma}_i\boldsymbol{\gamma}_j=-\boldsymbol{\gamma}_j\boldsymbol{\gamma}_i\ (i\neq j).$$

The key fact that Dirac made use of is that the wave operator is the square of a first-order operator defined with the help of these Clifford elements[24.5]

$$\square = (\boldsymbol{\gamma}_0\partial/\partial t - \boldsymbol{\gamma}_1\partial/\partial x - \boldsymbol{\gamma}_2\partial/\partial y - \boldsymbol{\gamma}_3\partial/\partial z)^2.$$

We can write this more concisely in vector notation, with $\boldsymbol{\gamma}=(\boldsymbol{\gamma}_1, \boldsymbol{\gamma}_2, \boldsymbol{\gamma}_3)$, as

$$\square = (\boldsymbol{\gamma}_0\partial/\partial t - \boldsymbol{\gamma}\cdot\boldsymbol{\nabla})^2,$$

or more concisely still, as

$$\square = \not{\partial}^2,$$

where the quantity

$$\begin{aligned}\not{\partial} &= \boldsymbol{\gamma}_0\partial/\partial t - \boldsymbol{\gamma}\cdot\nabla\\ &= \boldsymbol{\gamma}^a\partial/\partial x_a\end{aligned}$$

(with $\boldsymbol{\gamma}^a = g^{ab}\boldsymbol{\gamma}_b$) is called the *Dirac operator*. This handy 'slash' notation was introduced by Richard Feynman, where more generally, a vector A^a could be represented, by the Clifford–Dirac algebra element

$$\not{A} = \boldsymbol{\gamma}_a A^a.$$

24.7 The Dirac equation

Let us now return to our 'wave equation' $(\square + M^2)\psi = 0$. Using the Dirac operator $\not{\partial}$, we can now factorize the quantity $\square + M^2$ appearing in this equation:

$$\begin{aligned}\square + M^2 &= \not{\partial}^2 + M^2,\\ &= (\not{\partial} - \mathrm{i}M)(\not{\partial} + \mathrm{i}M),\end{aligned}$$

where $M = \mu/\hbar$. The *Dirac equation* for the electron is then $(\not{\partial} + \mathrm{i}M)\psi = 0$, i.e.

$$\not{\partial}\psi = -\mathrm{i}M\psi,$$

or, reinstating $\hbar$ by writing it in terms of the rest-mass μ

$$\hbar\not{\partial}\psi = -\mathrm{i}\mu\psi.$$

It is clear from the above factorization that, whenever this equation holds, the wave equation $(\square + M^2)\psi = 0$ must hold also. (This would also apply to the 'anti-Dirac equation' $(\not{\partial} - \mathrm{i}M)\psi = 0$, but with standard

[24.5] Check.

conventions, that would refer to a particle with negative mass $-\hbar M$.) Thus, wavefunctions that satisfy the above Dirac equation also must satisfy the 'wave equation' that governs relativistic particles of rest mass $\hbar M$.

The Dirac equation has the advantage over the wave equation that it is first-order in $\partial/\partial t$. Indeed, Dirac's equation can be rewritten in the form of a Schrödinger equation[24.6]

$$\mathrm{i}\hbar\,\frac{\partial\psi}{\partial t} = (\mathrm{i}\hbar\boldsymbol{\gamma}_0\boldsymbol{\gamma}\cdot\boldsymbol{\nabla} + \boldsymbol{\gamma}_0\mu)\psi,$$

where $\mathrm{i}\hbar\boldsymbol{\gamma}_0\boldsymbol{\gamma}\cdot\boldsymbol{\nabla} + \boldsymbol{\gamma}_0\mu$ plays the role of a Hamiltonian operator. The singling out of the operator $\partial/\partial t$ is, of course, not relativistically invariant, but the entire Dirac equation $\not{\partial}\psi = -\mathrm{i}M\psi$ *is* relativistically invariant. (To see this, one needs a careful examination of the interplay between the Clifford algebra elements and Lorentz transformations.[24.7]) It came as a considerable shock to the physicists of the day to learn from this that there are relativistically invariant entities that lie outside the standard framework of the vector/tensor calculus (Chapters 12 and 14). What Dirac had effectively initiated was a powerful new formalism, now known as *spinor calculus*,[11] a calculus that goes beyond what had been the conventional vector/tensor calculus of the day.

The 'price' that we seem to have pay for this remarkable elimination of the awkward square root, while retaining relativistic invariance, is the appearance of these strange non-commuting Clifford algebra elements γ_a. What do they mean? Well, we have to think of these things as operators acting on the wavefunction. Since these particular operators are new things, not directly arising from the (non-commuting) position and momentum quantum variables for a particle that we had been considering before, they must refer to (and act upon) some new degrees of freedom for our particle. We must ask what physical purpose these new degrees of freedom can serve. With the hindsight that our present-day terminology provides us with, we see the answer in the very name 'spinor'—the new degrees of freedom describe the *spin* of the electron.[12] Let us recall what was said in §11.5: 'A spinor may be thought of as an object upon which the elements of the Clifford algebra act as operators'. In the Dirac equation, the Clifford elements act on the wavefunction ψ. Thus, ψ itself must be a spinor. It has extra degrees of freedom (of a nature that we shall be examining shortly) over and above the mere dependence on position and time of an ordinary scalar wavefunction, and these extra degrees of freedom indeed describe the electron's spin!

[24.6] Show this.

[24.7] Explain this. *Hint*: The equation given in Exercise [26.3] may inspire!

We now begin to see that the price that we have had to pay for being able to factorize the wave operator by use of Clifford elements has bought us an almost unbelievable bargain! Not only does it give us a theory which precisely describes the electron's spin, but when we add the standard term to the Hamiltonian that gives interaction with a background electromagnetic field—a term that introduces electrodynamics precisely in accordance with the 'gauge prescriptions'[13] of §19.4 and §21.9—then we find that Dirac's electron correctly responds to the electromagnetic field as the charged electron should, including some subtle terms arising from the electron's relativistic motion.

But, not only is the electron's charged-particle behaviour correctly described; in addition Dirac's electron responds in accordance with its possessing a *magnetic moment* of a very specific amount, namely

$$\hbar^2 e/4\mu c,$$

where $-e$ is the electron's charge and μ is its mass. This is to say that Dirac's electron is not only electrically charged, but it also behaves as a little magnet, whose strength is given by the above value. Remarkably, Dirac's clear-cut value for the magnetic moment of the electron is a very close approximation to the actual observed value, to about one part in a thousand. The best modern determination of the electron's magnetic moment differs from Dirac's original value, as given above, by the multiplicative factor

$$1.001\,159\,652\,118\,8\ldots.$$

Even this small discrepancy is now explained, to the above *explicit precision*, from correction effects coming from quantum electrodynamics, which incorporates the Dirac equation as one of its fundamental ingredients. The accord with Nature that is revealed in Dirac's subtle little equation $\partial\!\!\!/\psi = -\mathrm{i}M\psi$ is indeed extraordinary!

24.8 Dirac's route to the positron

But we are by no means finished with this story; I have described merely its barest beginnings. Let us continue by next observing a seeming anomaly in the mathematics of Dirac's equation, with regard to the electron's spin. This apparent anomaly has to do with the number of independent components that there are to be found in the Dirac spinor ψ. It turns out that the Dirac's ψ has *four* independent components, whereas superficially we might expect *two* because a spin-$\frac{1}{2}$ particle has just two independent spin states (see §22.8). Let us try to understand this problem a little more fully.

In 1925, less than three years before Dirac published his equation (in 1928), George Uhlenbeck and Samuel Goudsmit had come to the conclusion that the electron must possess a quantum-mechanical spin, built from two basic spin states. In 1927 Wolfgang Pauli showed how to represent the transformations of these spin states under rotation by use of what we now call the 'Pauli matrices' (see §22.8, and also the Riemann sphere picture of the states for spin $\frac{1}{2}$ described in Fig. 22.10). Pauli matrices (which are basically quaternions, with a factor i) are also Clifford algebra elements, but for the 3-dimensional rotation group.[24.8]

In fact, there is a strong physical need for the electron's two spin states. Indeed, the very subject of chemistry, as we know it, depends upon this. In an atom, the electrons surrounding the nucleus are constrained to orbit the nucleus in particular states known as 'orbitals' (see §22.13). By Pauli's exclusion principle, it would seem that each electron orbital can be occupied by no more than *one* electron, yet we find that a second electron is always allowed in each of the orbitals. The pair of them can coexist and still satisfy the exclusion principle because their states are *not* identical but have opposite spins. There can be no more than two electrons in any one orbital, however, because there are only two independent spin states for the electron. The chemical notion of 'covalent bond' depends upon the same phenomenon, the two shared electrons seeming to coexist in the same state because their spins are opposite; see Fig. 24.2.

Pauli's description of the electron is a two-component entity $\psi_A = (\psi_0, \psi_1)$, corresponding to the fact that the Pauli matrices are 2×2. But we find that Dirac's Clifford elements ($\boldsymbol{\gamma}_0$, $\boldsymbol{\gamma}_1$, $\boldsymbol{\gamma}_2$, $\boldsymbol{\gamma}_3$) require 4×4 matrices in order to represent the Clifford multiplication laws.[24.9] Thus, the Dirac electron is a 4-component entity, rather than just having the 2 components of a 'Pauli spinor', describing the 2 independent states of spin that a non-relativistic particle of spin $\frac{1}{2}$ possesses, as described in §22.8.

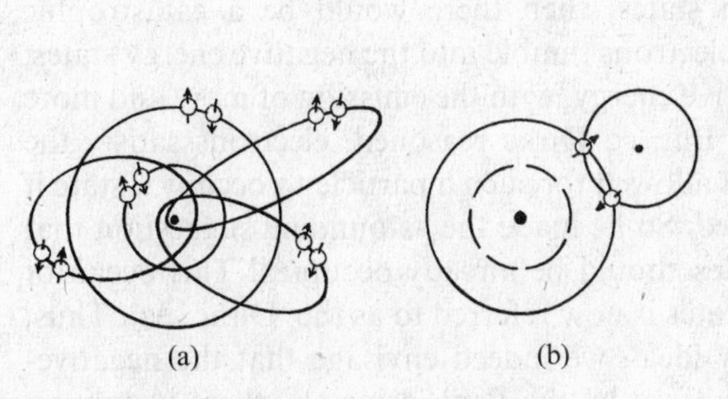

Fig. 24.2 Evidence of electron's spin $\frac{1}{2}$. (a) In an atom, two electrons, but not more, can occupy the same orbital. This is achieved by their spin states being opposite, so the Pauli exclusion principle is not violated. (b) Chemistry's 'covalent bond' involves a pair of electrons of opposite spin sharing orbitals of two separate atoms.

[24.8] Explain this comment in relation to the connection between quaternions and Clifford elements explained in §11.5.

[24.9] Show why 2×2 matrices connot satisfy all the conditions; find a set of 4×4 matrices that do.

In fact, there are only 2 components of spin for a particle described by Dirac's equation, despite there being 4 components for the wave-function. *Mathematically*, the reason for this is closely related to the fact that the Dirac equation $\nabla\!\!\!/\psi = \mathrm{i}M\psi$ is a first-order equation, and its space of solutions is spanned by only half as many solutions as in the case of the second-order wave equation $(\Box + M^2)\psi = 0$. (That equation is also satisfied by the 'anti-Dirac' equation $\nabla\!\!\!/\psi = +\mathrm{i}M\psi$, which is the Dirac equation for the negative rest mass $-M$.) *Physically*, this 'counting'[14] of solutions of the Dirac equation must take into account the fact that the degrees of freedom of the electron's antiparticle, namely the positron, are also hiding in the solutions of the Dirac equation. It would be misleading to think of two of the components of the Dirac equation as referring to the electron and the other two to the positron, however (and compare §25.2). Things are very much more subtle than this, as we shall see.

Recall that one of our main tasks, leading up to our consideration of the Dirac equation, had been to see what do about the unwanted negative-frequency (i.e. negative-energy) solutions to Schrödinger's equation. But it turns out that the solutions of the Dirac equation are *not* restricted to being of positive frequency, despite all our (or, rather, Dirac's) cleverness and hard work to eliminate the square root in the Hamiltonian. As with the previous attempts, described earlier, the presence of interactions, such as a background electromagnetic field, will cause an initially positive-frequency wave to pick up negative-frequency parts.

But Dirac's ingenuity was not going to be turned back at this stage. When he finally became convinced that the negative-frequency solutions could not be mathematically eliminated, he argued basically as follows. What, after all, is the danger in the negative-frequency solutions? The problem would be that if negative-energy states exist, then an electron could fall into such a state with the emission of energy, and if there is an unlimited number of such states, then there would be a catastrophic instability, in which all the electrons tumble into the negative energy states, of greater and greater negative energy, with the emission of more and more energy, without any limit. But, so Dirac reasoned, electrons satisfy the Pauli principle, and it is not allowed for such a particle to occupy a state if that state is already occupied. So he made the astounding suggestion that all the negative-energy states should be already occupied! This ocean of occupied negative-energy states is now referred to as the 'Dirac sea'. Thus, according to Dirac's 'crazy idea', we indeed envisage that the negative-energy states are already full up; by the Pauli principle, there is now no room for an electron to fall into such a state.

But, as Dirac further reasoned, occasionally there might be a few negative-energy states that are unoccupied. What would happen then?

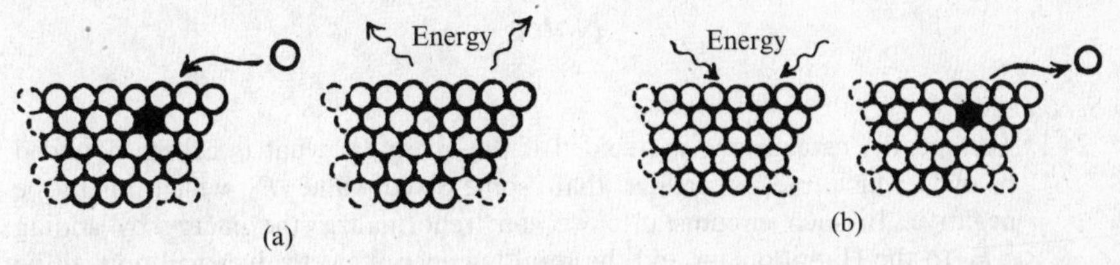

Fig. 24.3 Positrons as 'holes' in Dirac's 'sea' of negative-energy electron states. Dirac proposed that almost all negative energy states of the electron are filled, the Pauli principle preventing an electron from falling into such a state. The occasional unoccupied such state—a 'hole' in this negative-energy sea—would appear as an antielectron (positron), thereby having positive energy. (a) An electron falling into such a hole would be interpreted as the annihilation of the electron and positron, with the release of energy—the sum of the positive contributions from the electron and positron. (b) In reverse, the supplying of sufficient energy to the Dirac sea could produce an electron-positron pair. (The pictures are schematic only, the lattice structure depicted having no actual relevance to Dirac's sea.)

Such a 'hole' in the Dirac sea of negative-energy states would appear just like a positive-energy particle (and hence a positive-mass particle), whose electric charge would be the opposite of the charge on the electron. Such an empty negative-energy state could now be occupied by an ordinary electron; so the electron might 'fall into' that state with the emission of energy (normally in the form of electromagnetic radiation, i.e. photons). This would result in the 'hole' and the electron annihilating one another in the manner that we now understand as a particle and its anti-particle undergoing mutual annihilation (Fig. 24.3a). Conversely, if a hole were not present initially, but a sufficient amount of energy (say in the form of photons) enters the system, then an electron can be kicked out of one of the negative-energy states to leave a hole (Fig. 24.3b). Dirac's 'hole' is indeed the electron's antiparticle, now referred to as the *positron*.

At first Dirac was cautious about making the claim that his theory actually predicted the existence of antiparticles to electrons, initially thinking (in 1929) that the 'holes' could be protons, which were the only massive particles known at the time having a positive charge. But it was not long before it became clear[15] that the mass of each hole had to be equal to the mass of the electron, rather than the mass of a proton, which is about 1836 times larger. In the year 1931, Dirac came to the conclusion that the holes must be 'antielectrons'—previously unknown particles that we now call positrons. In the next year after Dirac's theoretical prediction, Carl Anderson announced the discovery of a particle which indeed had the properties that Dirac had predicted: the first antiparticle had been found!

Notes

Section 24.3

24.1. Technically, catastrophe is avoided if the energy is what is called 'bounded below', which means greater than some fixed value E_0 which might be negative. In such circumstance we can 'renormalize' the energy by adding $-E_0$ to the Hamiltonian, and the resulting energy eigenvalues will now all be positive.

24.2. There is a subtlety about the treatment of the point ∞, as f is likely to be singular there. The hyperfunctional treatment of §9.7 is appropriate; see Bailey *et al.* (1982).

24.3. Strictly, we should say positive semi-definite, since the (continuous) eigenvalue spectrum does go down to, and include, zero.

24.4. See Shankar (1994) for the applications to quantum mechanics, Arfken and Weber (2000) for a general discussion.

Section 24.5

24.5. Some people define this operator with the opposite sign, usually because they adopt the $+++-$ signature rather than the $+---$ that I am using here.

24.6. See Pais (1986); Miller (2003); Dirac (1983).

24.7. These two requirements combine together as essential ingredients of the proof of the CPT theorem that we shall encounter in §25.4.

Section 24.6

24.8. See Trautman (1997) for a reference on these 'square root' ideas.

24.9. In Clifford (1882), pp. 778–815; see also Lounesto (2001) for a general treatment.

24.10. This convention seems to be at variance with the usual mathematician's convention (see Harvey 1990; Budinich and Trautman 1988; Lounesto 2001; Lawson and Michelson 1990) and also with my own (see Penrose and Rindler 1986, Appendix). If the $+---$ signature for spacetime is adopted, as here. The defining equation for a general Clifford algebra is $\gamma_i\gamma_j - \gamma_j\gamma_i = -2g_{ij}$.

Section 24.7

24.11. See Clifford (1878); Cartan (1966); van der Waerden (1929); Infeld and van der Waerden (1933); Laporte and Uhlenbeck (1931); Penrose (1960); Penrose and Rindler (1984, 1986); O'Donnell (2003). In the 2-spinor notation of §22.8, this leads us to the 'zig-zag' form of the Dirac equation to be given in §25.2.

24.12. The term 'spinor' was apparently introduced by Paul Ehrenfest, in a letter to Bartel van der Waerden.

24.13. The added term is $\mathrm{i}e\not{A}$, where $\not{A} = g^{ab}A_a\gamma_b$ and A_a is the electromagnetic potential, this amounting to replacing the $\not{\partial}$ operator by $\not{\partial} - \mathrm{i}\,e\not{A}$.[24.10]

Section 24.8

24.14. The counting of solutions, for relativistic equations, is most easily carried out by the method of 'exact sets' in the calculus of 2-spinors (see Penrose and Rindler 1984, pp. 389, 90).

24.15. This work was done by Igor Tamm, Hermann Weyl, and J. Robert Oppenheimer; see Oppenheimer (1930) for an example of the reasoning behind it. There are some subtleties in the route to the positron that would take us rather far afield; see Zee (2003) for a complete, rigorous, and lovely treatment of all this.

[24.10] Explain why this is the standard 'gauge prescription'.

25
The standard model of particle physics

25.1 The origins of modern particle physics

THE Dirac equation for the electron provided a turning point for physics in many ways. In 1928, when Dirac put his equation forward, the only particles known to science were electrons, protons, and photons. The free Maxwell equations describe the photon, as was effectively foreseen by Einstein in 1905, this early work being gradually developed by Einstein, Bose, and others, until in 1927 Jordan and Pauli provided an overall mathematical scheme for describing free photons according to a quantized free-field Maxwell theory. Moreover, the proton, as well as the electron, seemed to be well-enough described by the Dirac equation. The electromagnetic interaction, describing how electrons and protons are influenced by photons, was excellently handled by Dirac's prescription, namely by the *gauge* idea (as basically introduced by Weyl in 1918; see §19.4), and a start to the formulation of a full theory of interacting electrons (or protons) with photons (i.e. quantum electrodynamics) had already been made by Dirac himself in 1927.[1] Thus, the basic tools seemed to be more or less to hand, for the description of all the known particles of Nature, together with their most manifest interactions.

Yet most physicists of the day were not so foolish as to think that this could shortly lead to a 'theory of everything'. For they were aware that neither the forces needed to hold the nucleus together—which we now call *strong* forces—nor the mechanisms responsible for radioactive decay—now called *weak* forces—could be accommodated without further major advances. If Dirac-style protons and electrons, interacting merely electromagnetically, were the only ingredients of atoms, including their nuclei, then all ordinary nuclei (except the single proton that constitutes the nucleus of hydrogen) would instantly disintegrate, owing to the electrostatic repulsion of the preponderence of positive charges. There must indeed be an unknown something else, amounting to a very strong attractive influence in operation within the nucleus! In 1932, Chadwick discovered the neutron, and it was eventually realized that the proton/electron model for the nucleus that had been popular earlier, must be replaced by one in which both protons and neutrons would be present, and where a strong proton–neutron interaction holds the nucleus together. But even this strong force was not all that was missing from the understanding of the day. Radioactivity in uranium had been known since the observations of Henri Becquerel in 1896, and this proved to be the result of yet another interaction—the *weak* force—different from the strong and from the electromagnetic. Even a neutron itself, if left on its own, would indulge in radioactive disintegration, in a period of about fifteen minutes. One of the mysterious products of radioactivity was the elusive neutrino, put forward as a tentative hypothesis by Pauli in about 1929, but not observed directly until 1956. It was the study of radioactivity that would eventually lead the physicists to an unaccustomed notoriety and influence, towards the end of the Second World War, and in its aftermath . . . .

Things have moved a great deal from these beginnings of an understanding of particle physics, as it stood in the first third of the 20th century. As we embark on the 21st century, a much more complete picture is to hand, known as the *standard model* of particle physics. This model appears to accommodate almost all of observed behaviour concerning the vast array of particles that are now known. The photon, electron, proton, positron, neutron, and neutrino have been joined by various other neutrinos, the muon, pions (effectively predicted by Yukawa in 1934), kaons, lambda and sigma particles, and the famously predicted omega-minus particle. The antiproton was directly observed in 1955 and the antineutron, in 1956. There are new kinds of entity known as quarks, gluons, and W and Z bosons; there are vast hordes of particles whose existence is so fleeting that they are never directly observed, tending to be referred to merely as 'resonances'. The formalism of modern theory also demands transient entities called 'virtual' particles, and also quantities known as 'ghosts' that are even further removed from direct observability. There are bewildering numbers of proposed particles—as yet unobserved—that are predicted by certain theoretical models but are by no means implications of the general framework of accepted particle physics, namely 'X-bosons', 'axions', 'photinos', 'squarks', 'gluinos', 'magnetic monopoles', 'dilatons', etc. There is also the shadowy Higgs particle—still unobserved at the time of writing—whose existence, in some form or other (perhaps not as a single particle), is essential to present-day particle physics, where the related Higgs field is held responsible for the mass of every particle.

25.2 The zigzag picture of the electron

In this chapter, I shall provide a brief guide to the standard model of the particle physics of today—though my particular approach to it could be judged a mite 'non-standard' in places. Let us indeed start in a slightly non-standard way, by re-examining the Dirac equation in terms of the '2-spinor notation', briefly introduced in §22.8. As remarked in §24.8 above, a 'Pauli spinor' description for a particle of spin $\frac{1}{2}$ is a 2-component quantity ψ_A. (The components are ψ_0 and ψ_1.) When considering relativity, according to §22.8, we also need quantities with *primed* indices A', B', C', ..., the primed indices resulting from complex conjugation applied to unprimed indices. It turns out[2] that the Dirac spinor ψ, as described above, with its 4 complex components, can be represented, as a pair of 2-spinors[3] α_A and $\beta_{A'}$, one with an unprimed index and one with a primed index:

$$\psi = (\alpha_A, \beta_{A'}).$$

The Dirac equation can then be written as an equation coupling these two 2-spinors, each acting as a kind of 'source' for the other, with a 'coupling constant' $2^{-1/2}M$ describing the strength of the 'interaction' between the two:

$$\nabla^A_{B'}\alpha_A = 2^{-1/2}M\beta_{B'}, \ \nabla^{B'}_A\beta_{B'} = 2^{-1/2}M\alpha_{A'}.$$

The operators $\nabla^A_{B'}$ and $\nabla^{B'}_A$ are just 2-spinor translations of the ordinary gradient operator $\boldsymbol{\nabla}$. Do not worry about all those indices, the $2^{-1/2}$s, and the exact form of these equations. I present them here just to indicate how the Dirac equation can be brought into the general framework of the 2-spinor caculus, and that, when this is done, some new insights as to the nature of Dirac's equation are revealed.[4]

From the form of these equations, we see that the Dirac electron can be thought of as being composed of two ingredients α_A and $\beta_{B'}$. It is possible to obtain a kind of physical interpretation of these ingredients. We form a picture in which there are two 'particles', one described by α_A and the other by $\beta_{A'}$, each of which is massless,[25.1] and where each one is continually converting itself into the other one. Let us call these the 'zig' particle and the 'zag' particle, where α_A describes the zig and $\beta_{A'}$ describes the zag. Being massless, each of these should be travelling with the speed of light, but we can think of them, rather, as 'jiggling' backwards and forwards where the forward motion of the zig is continually being converted to the backward motion of the zag and vice versa. In fact, this is a realization of the phenomenon referred to as 'zitterbewegung', according to which, the electron's instantaneous motion is always measured to be the speed of light, owing to the electron's jiggling motion, even though the overall averaged motion of the electron is less than light speed.[5] Each ingredient has a *spin* about its direction of motion, of magnitude $\frac{1}{2}\hbar$, where the spin is

[25.1] By referring to Weyl's neutrino equation, given in §25.3, explain why it is reasonable to take the view that α_A and $\beta_{A'}$ each describe massless particles, coupled by an interaction converting each into the other.

left-handed in the case of the zig and right-handed for the zag. (This has to do with the fact that the zig's α_A has an unprimed index, which is associated with negative helicity, whereas the zag's $\beta_{B'}$ has a primed index, which indicates positive helicity. All this has relevance to the discussion of §§33.6–8, but it is not appropriate to go into details at this point.) We note that, although the velocity keeps reversing, the spin direction remains constant in the electron's rest-frame (Fig. 25.1). In this interpretation, the zig particle acts as the source for the zag particle and the zag particle as a source for the zig particle, the coupling strength being determined by M.

In Fig. 25.2, I have given a diagrammatic illustration of how this process is to contribute to the full 'Feynmann propagator' (see §26.7), in the manner of the Feynman graphs[6] that we shall be coming to in more detail in the next chapter. Each constituent zigzag process is of finite length, but the totality of these, involving zigzags of ever-increasing length, contributes to the entire propagation of the electron, according to the 2×2 matrix depicted in Fig. 25.2. Typically a zig particle becomes a zag, and the zag then becomes a zig, this zig becoming a zag again, and so on for some finite stretch. In the total process, we find that the average rate at which this happens is (reciprocally) related to the mass coupling parameter M; in fact, this rate is essentially the *de Broglie frequency* of the electron (see §21.4).

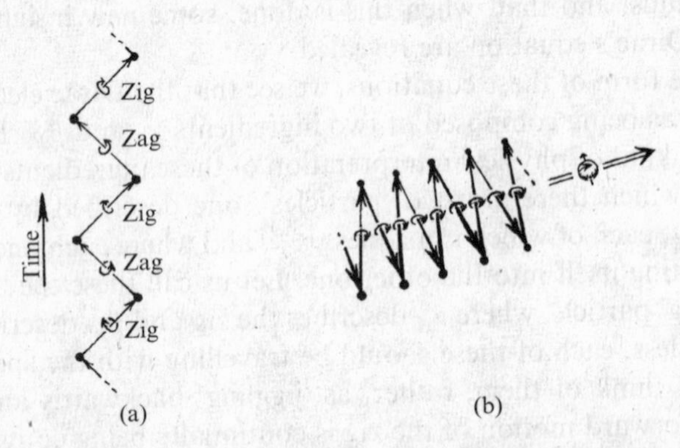

Fig. 25.1 Zigzag picture of the electron. (a) The electron (or other massive particle of spin $\frac{1}{2}$) can be viewed, in spacetime, as oscillating between a left-handed massless zig particle (helicity $-\frac{1}{2}$, as described by the unprimed 2-spinor α_A or, in the more usual physicist's notation, by the part projected out by $\frac{1}{2}(1-\gamma_5)$) and a right-handed massless zag particle (helicity $+\frac{1}{2}$, as described by the primed 2-spinor $\beta_{B'}$, the part projected out by $\frac{1}{2}(1+\gamma_5)$). Each is the source for the other, with the rest-mass as coupling constant. (b) From a 3-space perspective, in the 'rest-frame' of the electron, there is a continual reversal of the velocity (always the speed of light), but the direction of spin remains constant. (For reasons of clarity, the figure is drawn not quite in the electron's rest-frame, the electron drifting slowly off to the right.)

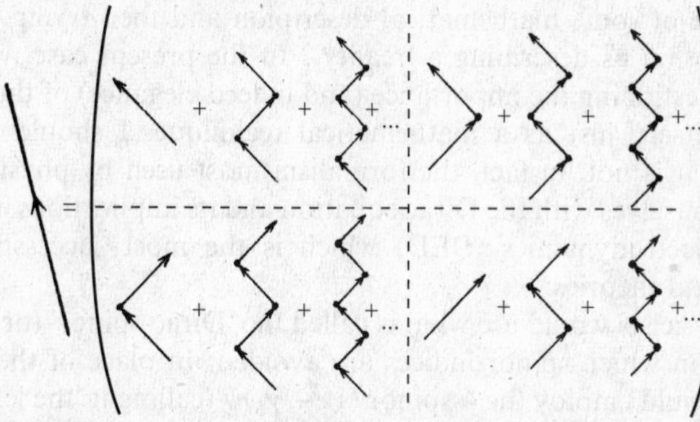

Fig. 25.2 Each zigzag process separately contributes, as part of an infinite quantum superposition, to the total 'propagator' in the manner of a Feynman graph. The conventional single-line Feynman propagator is drawn at the left, and it stands for the entire matrix of infinite sums of finite zigzags, drawn on the right.

There is a word of warning that I must give, however, about how we are to interpret Feyman-graph diagrams. We can legitimately think of the process that is being depicted as a spacetime description of what is going on; but at the quantum level of things, we must take the view that, even for a single particle, there are a great many such processes going on simultaneously. Each individual one of these processes is to be viewed as taking part in some enormous quantum superposition of vast numbers of different processes. The actual quantum state of the system consists of the entire superposition. An individual Feynman graph represents merely one component of it.

Accordingly, my above description of the electron's motion as consisting of this jiggling back and forth, where a zig is continually being converted into a zag and back again, must be taken appropriately in this spirit. The actual motion is composed of a vast number of such individual processes (in fact infinitely many of them) all superposed, and we may think of the electron's perceived motion as being some sort of 'average' (though strictly a quantum superposition) of these. Even this describes merely the free electron. An actual electron will be continually undergoing interactions with other particles (such as photons, the quanta of the electromagnetic field). All such interaction processes should also be included in the overall superposition.

Bearing this in mind, let us raise the question as to whether these zig and zag particles are 'real'. Or are they perhaps artefacts of the particular mathematical formalism that I have been adopting here for the description of the Dirac equation for the electron? This raises a more general question: what is the physical justification in allowing oneself to be carried along by the elegance of some mathematical descripion and then trying to regard that description as describing a 'reality'? In the present case, we should begin by questioning the importance (and indeed elegance) of the 2-spinor formalism itself, just as a mathematical technique. I should warn the reader that it is not, in fact, the formalism most used by physicists who concern themselves with the Dirac equation and its implications, such as in quantum electrodynamics (QED) which is the most successful of the quantum field theories.[7]

Most physicists would use what is called the 'Dirac-spinor' (or 4-spinor) formalism, in which spinor indices are avoided. In place of the 2-spinor 'α_A', they would employ the 4-spinor $(1 - \boldsymbol{\gamma}_5)\psi$ (calling it 'the left-handed helicity part of the Dirac electron', or something similar, rather than using my 'zig particle').[8] Here, the quantity $\boldsymbol{\gamma}_5$ is the product

$$\boldsymbol{\gamma}_5 = -\mathrm{i}\boldsymbol{\gamma}_0\boldsymbol{\gamma}_1\boldsymbol{\gamma}_2\boldsymbol{\gamma}_3,$$

and it has the property that it anticommutes with every element of the Clifford algebra, and $(\gamma_5)^2 = 1$.[25.2] Similarly, they would use $(1 + \boldsymbol{\gamma}_5)\psi$ instead of $\beta_{A'}$ (the right-handed helicity part). One could say that this is merely a notational matter, and indeed it is possible to translate back and forth between the 2-spinor and 4-spinor formalisms. The 'zigzag' picture that I have presented here is certainly a valid (but not altogether usual) description in either formalism, but it is more directly suggested by 2-spinors than by 4-spinors.

So are these zigs and zags real? For my own part, I would say so; they are as real as the 'Dirac electron' is itself real—as a highly appropriate idealized mathematical description of one of the most fundamental ingredients of the universe. But is this *real* 'reality'? In §§1.3,4, I touched upon this general question of mathematical and physical reality, and the relation between the two. At the end of the book, in §34.6, I shall take up this question again.

25.3 Electroweak interactions; reflection asymmetry

Each of the zig and zag particles has the same electric charge—which must be the case since charge is conserved, and each particle continually converts itself into the other. In the Feynman-graph picture, the interaction with the electromagnetic field that charged particles indulge in is represented by the attachment of a line representing a photon. This is depicted in Fig. 25.3a according to conventional procedures, whereby the electron's trajectory is represented as a single Dirac 4-spinor line in

[25.2] Show both of these things.

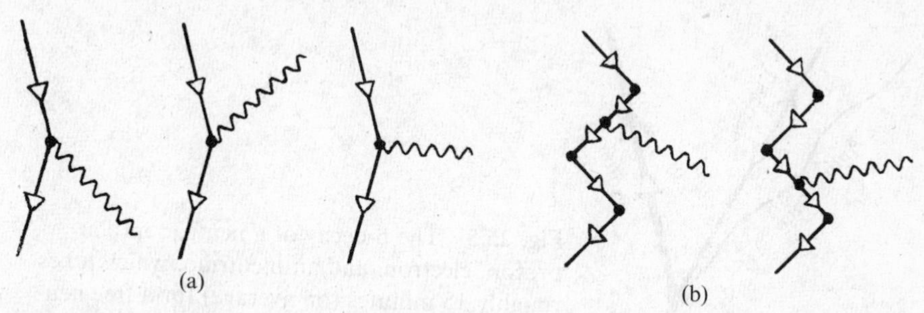

Fig. 25.3 (a) Feynman graph (in conventional form, without zigzags) of an electron in interaction with a quantum of electromagnetic field, or photon. Whereas in the left-hand figure we might view the process as the absorption of a photon, the middle figure as the emission of one, and the right-hand figure as an electrostatic influence, these processes are to be thought of as all being the same, and are referred to as an interaction with a 'virtual' (off-shell) photon. (b) The same, with zigzags represented. The (virtual) photon interacts equally with the zag and the zig. In all these figures, the propagation of electric charge is represented by the white triangle-like arrow. The ones illustrated all point into the past because we are considering electrons, which are negatively charged.

the usual way, and in Fig. 25.3b using this (somewhat unconventional) 'zigzag' description.

Note that both the left-handed (zig) and right-handed (zag) indulge equally in electromagnetic interaction. It turns out, however, that there is another physical interaction—the *weak* interaction—according to which these things are completely *un*equal, in the sense that only the electron's zig takes part in these weak interactions, and the zag, not at all (see Fig. 25.4). Weak interactions are mediated by photon analogues, called W and Z bosons. As remarked earlier, these interactions are responsible for radioactive decay, whereby, for example, a uranium U^{238} nucleus will in about 5×10^9 years, on the average, spontaneously disintegrate into a thorium and a helium nucleus (α particle), or whereby a free neutron will decay into a proton, an electron, and an antineutrino in an average of about 15 minutes; see Fig. 25.5. These processes are referred to as 'β decay', the electron being referred to, in this context, as a 'β particle' (for historical reasons).

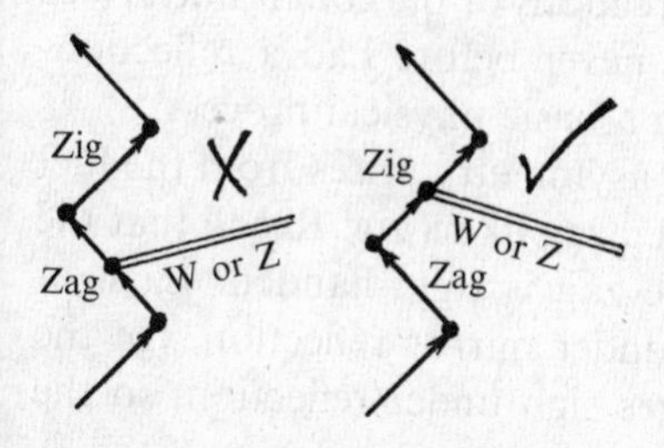

Fig. 25.4 In the case of weak interactions, on the other hand, only the zig of a weakly interacting particle interacts with the W or Z boson. (But for what is classified as an 'antiparticle', it would be the zag which interacts weakly.)

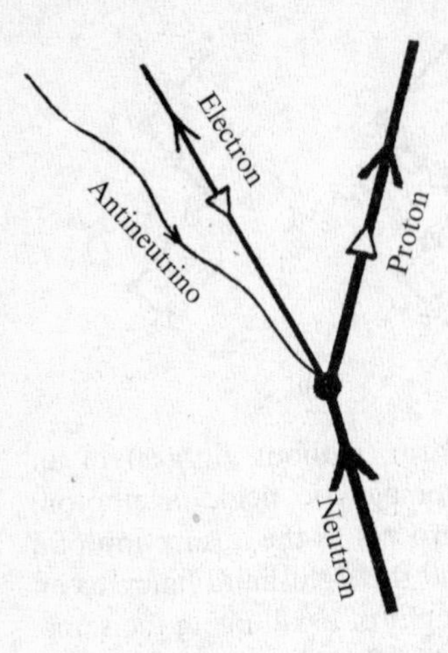

Fig. 25.5 The β-decay of a neutron into a proton, electron, and antineutrino, which takes roughly 15 minutes (on average) for a free neutron. The reverse arrow on the antineutrino indicates that it is an 'antiparticle' in the lepton classification scheme. As in Fig. 25.4, the white arrow on the electron and proton lines indicates electric charge.

For many years, weak interactions had been treated as single point processes—illustrated as the single point of decay in Fig. 25.5—in accordance with a scheme dating back to 1933, that had been put forward by the outstanding Italian physicist Enrico Fermi. But later this began to run into theoretical difficulties which were finally resolved in the electroweak theory of Weinberg, Salam, Ward, and Glashow, which we shall glimpse something of in §25.5. As part of these newer ideas it was realized that in place of the pointlike Fermi interaction, there would be an intermediate 'gauge boson'—the W or Z particles just referred to—mediating the weak interaction process, according to which the β-decay of Fig. 25.5 is now interpreted as shown in Fig. 25.6. What is the significance of the zig/zag asymmetry? In 1956 there was a great shock to physicists when Tsung Dao Lee and Chen Ning Yang made an astonishing proposal,[9] concerning β decay—and concerning weak interactions generally—that they should *not* be reflection-invariant, this proposal being startlingly confirmed experimentally by Chien-Shiung Wu and her associates shortly afterwards, in January 1957. According to this, the *mirror reflection* of a weak interaction process surprisingly would *not* generally be an allowed weak interaction process, so weak interactions exhibit *chirality*. In particular, Wu's experiment examined the emission pattern of electrons from radioactive cobalt 60, finding a clearly mirror-asymmetric relation between the distribution of emitted electrons and the spin directions of the cobalt nuclei (see Fig. 25.7). This was astounding, because never before had a reflection-asymmetric phenomenon been observed in a basic physical process!

In terms of our zigs and zags, the chiral asymmetry arises from the fact that, in a mirror, a zig looks like a zag and a zag like a zig. Recall that the zig has a left-handed helicity whereas the zag is right-handed. Each of these is indeed converted into the other under mirror reflection. (In the more conventional terminology, γ_5 changes sign under reflection, so the

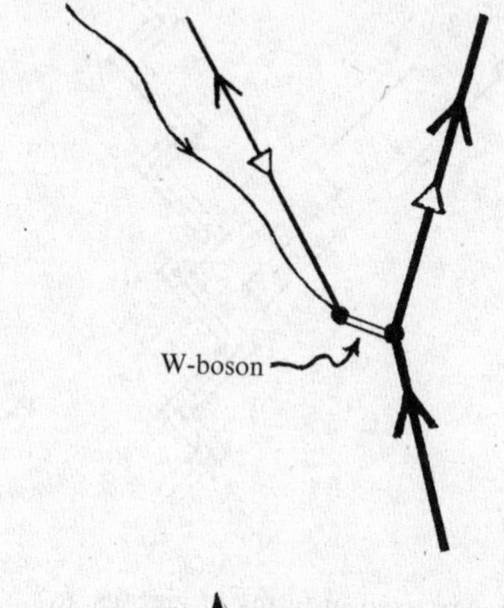

Fig. 25.6 Weak interactions are not 'pointlike', as would be suggested by Fig. 25.5 (original Fermi theory), but occur through the intermediary of a 'vector boson' ($W^{\pm}$ or Z^0)—here a W particle.

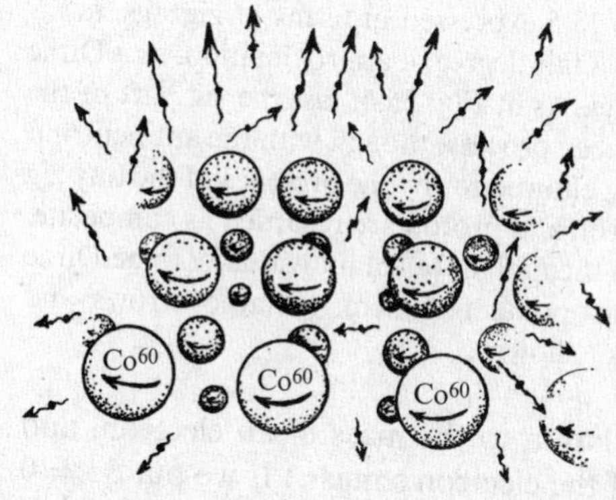

Fig. 25.7 Wu's experiment examined the emission pattern of electrons from radioactive cobalt 60, finding a clearly mirror-asymmetric relation between the distribution of emitted electrons and the spin directions of the cobalt nuclei. Here more electrons come out at the top of the picture than at the bottom.

roles of the left- and right-helicity parts of the electron's wavefunction, $(1-\gamma_5)\psi$ and $(1+\gamma_5)\psi$, are interchanged.) Thus, the non-invariance of weak interactions under reflection symmetry is realized in the fact that only the zig part of the electron indulges in weak interactions. The same can be said of the neutron as it undergoes spontaneous β decay, and also of the resulting proton. A neutron and a proton can, to a fair degree of approximation, also be described by the Dirac equation, whence the zigzag description becomes appropriate for each. Again, it is just the zig part of the neutron and of the proton that engage in the weak decay process, and this is illustrated in Fig. 25.8a. More appropriate, in accordance with the modern picture, is to regard both the neutron and proton as composite particles, where each is made up of three quarks. The quarks themselves are taken to be individually described by Dirac's equation, so the zigzag picture becomes appropriate for each of them also, and in Fig. 25.8b, the neutron's β decay is represented in these terms.

The neutrino, also, attracts a special interest in this respect. At least to a very good aproximation, it can be treated as a *massless* particle. (Its mass

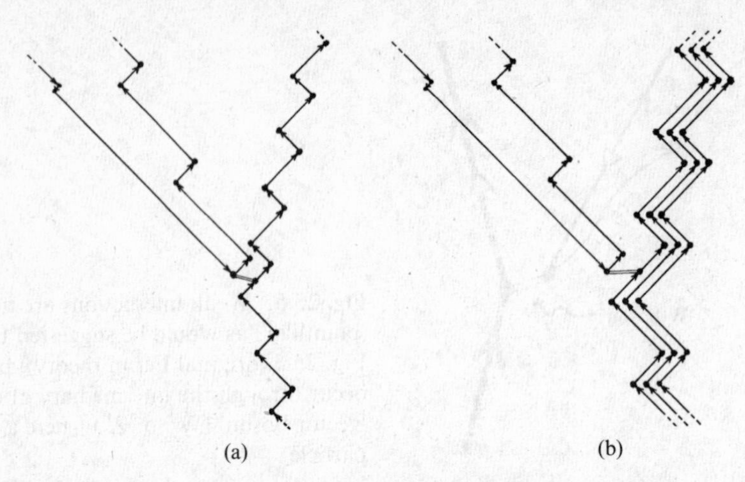

Fig. 25.8 The β-decay process of Fig. 25.5 expressed in terms of zigzags. (a) A neutron and proton can be described, to a fair degree of approximation, as a Dirac particle, so zigzags are not inappropriate. As in Fig. 25.4, just the zig part of the neutron and proton engage in a weak decay process, though with the antineutrino it is zag (left handed), a small mass being allowed for by the presence of the tiny zig at the upper left. (b) However, the neutron and proton are regarded as composite, each made up of 3 quarks, these quarks themselves taken individually to be Dirac particles, so the zigzag picture is appropriate for them. (Charge arrows and connecting gluons for the quarks are not shown).

is, in any case, extremely tiny in relation to the mass of an electron, and certainly not more than 6×10^{-6} of the electron's mass.) If we put $M = 0$ in the 2-spinor version of the Dirac equation, the equations *decouple*, to become

$$\nabla^A_{B'}\alpha_A = 0, \quad \nabla^{B'}_A \beta_{B'} = 0.$$

Either one could exist in the absence of the other (and either one of these, by itself, is referred to as the 'Weyl equation'[10] for the neutrino). But only the zig version (given by the *unprimed* α_A, subject to $\nabla^A_{B'}\alpha_A = 0$) indulges in weak interactions, or could be created in a weak interaction process. Thus, neutrinos are particles with a left-handed helicity.

Do neutrinos actually possess mass? There now appears to be good experimental evidence that at least two out of the three neutrino types must indeed be massive. These three types are the 'electron neutrino' ν_e (which is the one involved in ordinary β decay, its antiparticle $\bar{\nu}_e$ being what is emitted in the decay of the neutron; see Fig. 25.5), the 'muon neutrino' ν_μ, and the 'tau neutrino' ν_τ. Observations at the Japanese detector Superkamiokande clearly indicate that the differences in the mass of these three neutrino types, though very small (around 10^{-7} of an electron's mass, in all) cannot be zero, owing to the fact that they have a tendency to flip into one another ('neutrino oscillations'), which cannot happen with zero mass. I gather that it is still possible that ν_e' (or conceivably some appropriate quantum 'linear combination' of the three of them) could have zero mass, but definitive evidence on these matters is still missing. A massless neutrino could be entirely zig, but with a small mass, the picture would be more like that depicted in Fig. 25.9a, where the zig very occasionally flips momentarily into a zag and back again. However, as viewed with respect to a rest frame moving with the neutrino, the zig and zag aspects would appear to contribute equally to its overall motion (Fig. 25.9b).

A few words of clarification are necessary here. When I have commented, above, that it is the zig (i.e. left-handed) particles that indulge in weak interactions and not the zag particles, I have presupposed that we know how to distinguish a 'particle' from an 'antiparticle'. With the antiparticle, things are the other way around. In the case of the electron's antiparticle, the *positron*, we can again present a 'zigzag' description in which the zig is left-handed and the zag right-handed, but the positron's zig is the antiparticle of the electron's zag, and vice versa. Thus, in the case of a positron, it is the right-handed *zag* (the antiparticle to the electron's zig) that indulges in weak interactions, rather than the zig. A similar remark would apply to the antiproton and the antineutron and, indeed, to the antiquark. It would also apply to the antineutrino which, if it were massless, would be entirely zag.

Now, this could cause some confusion because I have given no criterion for deciding whether a (spin $\frac{1}{2}$) particle-like entity is to be thought of as a 'particle' or as an 'antiparticle' in order that we can know whether it is its zig or its zag that is to indulge in weak interactions. Although, in the previous chapter, I have given the notion of an antiparticle only in terms of Dirac's original concept of a 'hole' in the 'sea of negative energy states', an antiparticle should not really be thought of as a totally separate kind of entity from a particle. In the context of modern quantum field theory, it is not necessary to present things in Dirac's original (seemingly asymmet-

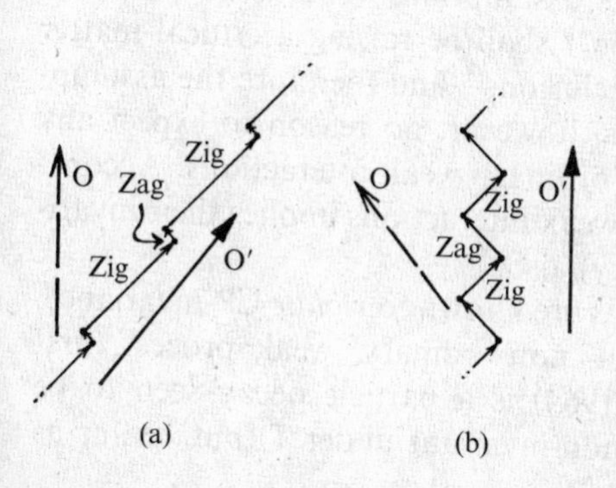

Fig. 25.9 (a) A massless neutrino could be entirely zig; but with a small mass we must envisage the occasional momentary 'flip' to a zag and back. The picture is displayed from the perspective of the laboratory's rest-frame O. (b) As viewed with respect to a second rest-frame O′, moving with the neutrino, the zig and zag aspects appear to contribute equally to the overall motion.

rical) way. Antiparticles are just as much 'particles' as are the particles that they are the 'antis' of. Moreover, the notion of antiparticle refers to bosons (particles of integer spin) as well as it does to fermions, whereas only fermions are subject to the Pauli principle (see §§23.7,8), so the 'Dirac sea' perspective on antiparticles cannot be applied to bosons. The positively charged pion (the π^+ meson), for example, which is a boson, has an antiparticle which is the negatively charged pion (π^- meson). In fact, several bosons are their own antiparticles. A photon is an example of this; so also is the neutral pion (π^0 meson). As far as is known (and certainly according to standard theory), every particle in nature has an antiparticle.

25.4 Charge conjugation, parity, and time reversal

The operation that replaces every particle by its antiparticle is referred to as C (which stands for *charge conjugation*). A physical interaction that is invariant under the replacement of particles by their antiparticles (and vice-versa) is called C-invariant. The operation of spatial reflection (reflection in a mirror) is referred to as P (which stands for *parity*). In accordance with the above discussion, in §25.3, ordinary weak interactions are not invariant under either P or C separately, but it turns out that they are invariant under the combined operation CP (= PC). We may regard CP as the operation performed by an unusual mirror, in which each particle is reflected as its antiparticle. We see that CP sends a particle's zig into its antiparticle's zag, and vice versa. There is one further operation that is normally discussed in relation to these, which is that of *time reversal*, referred to as T. An interaction is invariant under T if it unaltered if we view it from the prespective of a time direction that is the reverse of normal. There is a famous theorem in quantum field theory, referred to as the *CPT theorem* which asserts every physical interaction is invariant if all three of the operations C, P, and T are applied to it at once. Of course, a theorem is 'just a piece of mathematics', so its physical validity is dependent upon the physical validity of its assumptions. This issue will have importance for us later (§30.2), when I shall be raising a critical matter that may lead us to question the conclusions—and therefore the assumptions—of the CPT theorem. There is, however, no reason to expect any difficulty of this sort, in relation to ordinary weak interactions. Accordingly, the CP invariance of ordinary weak interactions implies their invariance under T (time-reversal symmetry) also.

A very few observed physical effects are known to *violate* CP-invariance. The most long-standing example (a 'non-ordinary' weak process, first observed by Fitch and Cronin in 1964) is a particle decay seen to be non-invariant under CP. It is also non-invariant under T (but, as far as can be told, invariant under CPT, in accordance with the CPT theorem). This is the decay of the K^0 meson (which can be into 2 pions or 3 pions, where there is a sophisticated issue concerning K^0 flipping into its antiparticle $\bar{K}^0$, with an oscillation between these two taking place).

The CPT theorem provides us with an alternative perspective on antiparticles which is different from using Dirac's 'sea', and it is more satisfactory since it can also be applied to bosons. Assuming CPT, we can regard C—the interchange of particles with their antiparticles—as equivalent to PT, so we can regard the antiparticle of some particle as being the 'space–time reflection' (PT) of that particle. Ignoring the space-reflection aspect of this, we obtain the interpretation of an antiparticle as being the particle travelling *backwards in time*. This, indeed, is the way that Richard Feynman liked to interpret anti-particles. It provides a very convenient and consistent way of treating antiparticles within the context of Feynman graphs. (The idea had been suggested to Feynman by John A. Wheeler and, it had been earlier proposed, independently, by Stückelberg (1942)). In its different way, it is just as 'crazy' an idea as was Dirac's sea!

In a Feynman graph, particles that are not their own antiparticles have to have lines in the diagrams that are directed in some way, such as by attaching an appropriate kind of arrow on each line. We might think of this arrow as pointing into the future—when the line depicts the particle itself—but, in this case, when it points into the past, we get that particle's antiparticle. This perspective on antiparticles has the great advantage that many very different-looking particle processes are revealed as being basically the same process, but viewed from different 'angles' in spacetime. As an example, in Fig. 25.10 (but without bothering with the zigzags), I have depicted electron–positron annihilation into a pair of photons, showing that this is 'essentially the same' (i.e. spacetime re-organized) process as the Compton scattering of an electron by a photon. (We shall be seeing shortly that we also need to allow that the particles' lines can point in

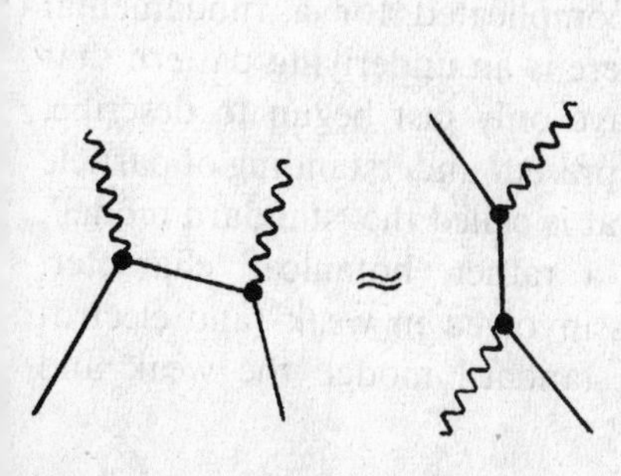

Fig. 25.10 Crossing symmetry. Processes which differ only with respect to the time-orderings in various places, but without the topology of the diagram being affected, are basically mathematically equivalent (through analytic continuation, §7.4). This is illustrated by such an equivalence between the particle–antiparticle pair annihilation into two photons, as illustrated on the left, and the Compton scattering process on the right (drawn without zigzags).

spacelike directions, describing what is referred to as a 'virtual particle', but this is enough confusion for the moment!)

Let us now return to the problem of deciding, for an entity of spin $\frac{1}{2}$, whether it is its zig or the zag that takes part in weak interactions. We need a clear rule for telling us whether it is to count as a 'particle' or as an 'antiparticle'. The rule that is used decrees that those particle referred to as 'leptons' (electrons, their heavier sister particles, the muons and tauons, and their corresponding neutrinos ν_e, ν_μ, and ν_τ), and also the quarks that compose protons and neutrons (and other hadrons), are to count as 'particles'. These have zigs that undergo weak interactions. The 'antis' of all of these count as antiparticles, and in those cases it is the zags that undergo weak interactions. The situation is complicated by the fact that there are also (massive) entities of spin 1 involved in weak interactions,[11] namely the W and Z bosons. These are the mediators of weak interactions, playing roles similar to the photons that mediate electromagnetic interactions (photons being the quanta of the electromagnetic field). Such particles are sometimes called 'gauge quanta', for reasons that we shall be coming to. There are *two* different W bosons, labelled W^+ and W^- (antiparticles of each other), having respective electric charges 1 and -1 (in units given by the charge on the positron), whereas there is only the uncharged Z^0 (its own antiparticle). Each of these takes part in weak interactions, having a Feynman graph line that attaches itself at either end to a zig part of a lepton or quark or to a zag part of an antilepton or antiquark (see Fig. 25.11). Throughout each weak-interaction process, electric charge is conserved, and so is lepton number. In fact, there are three different kinds of lepton number, each of which is separately conserved (electron, muon, and tauon number) in the standard model of weak interactions, the lepton numbers of the Ws and Z^0 counting as zero. To check these four conservation laws, in a Feynman graph, all we need to do is make sure that each of the four kinds of arrow on the lines follow through the diagram as a continuous, consistently oriented, path.

25.5 The electroweak symmetry group

All this no doubt sounds somewhat complicated for a fundamental theory. Yes, it *is* complicated, though there is an underlying pattern that I have not yet explained; however, I have only just begun to describe, and in very qualitative terms, what is our present understanding of particle physics according to less than half of what is called the 'standard model'. Moreover, my remarks have been of a rather 'botanical' character, so far, concerning the different particles involved in weak (and electromagnetic) interactions. In fact, in the standard model, the weak and

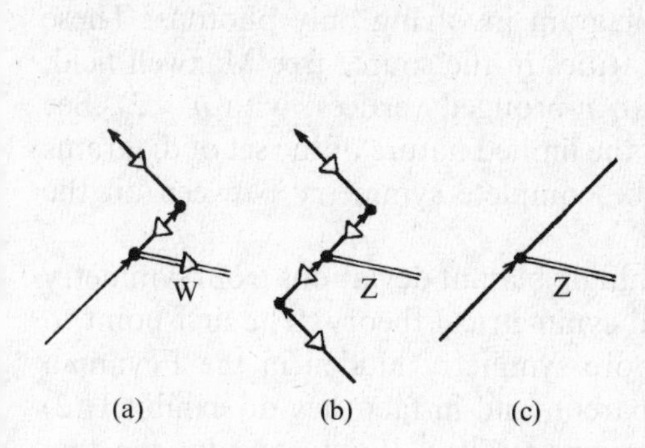

Fig. 25.11 Illustrations of interactions between a zig particle and a weak-interaction gauge boson. (a) The charged W^+ and W^- (antiparticles of each other) induce a change in the zig's electric charge (to ensure electric charge conservation), whereas (b) the uncharged Z^0 does not (and Z^0 is its own antiparticle). (c) The neutrino's zig can interact with the uncharged Z^0.

electromagnetic interactions are unified in what is called *electroweak* theory, where there is a special symmetry relating W^+, W^-, Z^0, and the *photon* γ, according to the group $SU(2) \times U(1)$ or, more correctly,[12] U(2). (See §13.9, if you need reminding what these groups are.) It is this (hidden) symmetry that supplies the underlying pattern referred to above.

I shall explain the role of this symmetry a little more completely later in this chapter. This symmetry also interrelates the zig parts of various leptons and quarks. The idea has the consequence that, from a more primitive perspective, all of W^+, W^-, Z^0, and γ can, in a certain sense, be continuously 'rotated into one another', so that various sets of (quantum) *linear combinations* of these particles are on an equal footing with the individual particles themselves!

As I have described things above, this 'symmetry' appears to be very strange and subtle, particularly because pure electromagnetism is reflection-invariant, both zig and zag parts of the sources being equally involved, whereas the weak interactions are about as *non*-invariant under reflection as they can be, involving only the zig parts of the particles. Moreover, the photon appears to be clearly singled out, among all the bosons in the theory, by being a *massless* particle. Indeed, the mass of the photon, if non-zero, would certainly have to be less than 10^{-20} of an electron's mass for good observational reasons, and so it is less than about 5×10^{-26} of the measured mass of the W and Z bosons. Furthermore, the W bosons are electrically charged, whereas the photon does not, conversely, carry a weak charge.

In Fig. 25.12, I have listed all the possible 3-pronged Feynman vertices involving only gauge bosons (i.e. W^+, W^-, Z^0, or γ). There are just two of them. But the fact that there are any at all is an expression of a *non-linearity* in the free gauge field, which arises from the gauge group being non-Abelian—and this indeed applies to U(2). (Pure electrodynamics comes from the *Abelian* gauge group U(1) and, consequently, there is no analogous 3-pronged Feynman diagram involving only photons. These would have given rise to non-linearities in the source-free Maxwell field. An analogous statement applies to n-pronged vertices, with $n > 2$). See §15.8, §19.2. It would appear from the limited nature of the set of diagrams in Fig. 25.12 that there cannot be complete symmetry between all the gauge bosons.

How do we reconcile these seemingly blatant deviations from symmetry with the required goal of a unified symmetrical theory? The first point to realise is that there is actually more symmetry hidden in the Feynman diagrams than is immediately apparent, and in fact they do exhibit U(2) symmetry if looked at in the appropriate way. First consider the two diagrams in Fig. 25.12. To get a better idea of the underlying symmetry here, think of a 2×2 Hermitian matrix (see §13.9). We may imagine that its two *real* diagonal elements are analogous to Z^0 and γ, and that its two remaining off-diagonal elements—*complex conjugates* of each other—are analogous to W^+ and W^-. The real-number nature of the diagonal elements corresponds to Z^0 and γ being the same as their respective antiparticles (lines without arrows in Fig. 25.12), whereas the complex-conjugate nature of the off-diagonal elements corresponds to the fact that W^+ and W^- are antiparticles of each other (reversal of arrow direction in passing from one to the other). A general U(2) transformation of this Hermitian matrix (which we must bear in mind involves both pre-multiplication by the U(2) matrix and post multiplication by the inverse of that matrix) does 'churn around' the elements of this Hermitian matrix, in very specific ways, but its Hermitian character is always preserved. In fact this analogy is very close to the way in which U(1) indeed acts in electroweak theory (the only complication being that we must allow for a linear combination of the diagonal elements with the trace, in this identification, related to the 'Weinberg angle' that we shall be coming to in §25.7). The *a*symmetry that we seem to see in the actual world, with respect to these particles, comes about in electroweak theory merely because Nature chooses certain particular combinations—i.e. particular quantum superpositions of these elements—to be realised as actual free particles.

But what about the other seemingly most blatant asymmetry in our Feynman diagrams, that the Z^0 and $W^{\pm}$ can attach only to the zig lines of

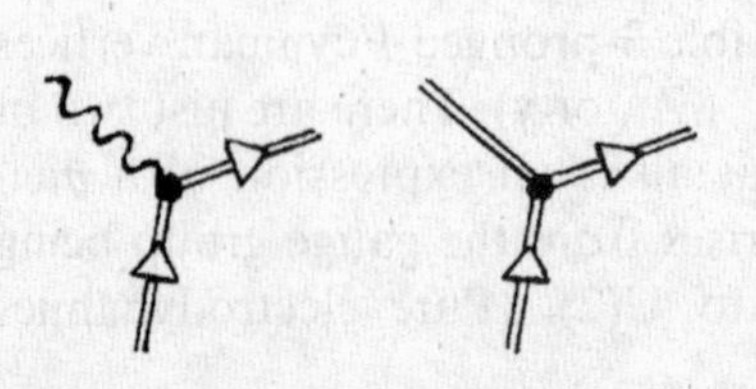

Fig. 25.12 Electroweak 3-particle gauge-boson vertices that can theoretically occur, owing to the non-Abelian nature of the gauge group. (Photon lines are wiggly; Z^0 lines are double without arrow; $W^{\pm}$ lines are double with white arrow.

particles whereas γ attaches indiscriminately to zig or to zag? Here again it is a matter of which superpositions Nature allows us to find as free particles. For example, there would be some particular superposition of Z^0 and γ—let us call it Y—which sees only the *zag* part of a particle. (Roughly: 'subtract' Z^0 from γ so as to kill off the zig interaction, leaving only the zag.) We could retrieve our original γ from Z^0 and Y, but there would be many other possible such superpositions which could equally have played the role of the photon, had Nature chosen things in some other way.

A key issue, therefore, is: what criteria does Nature adopt, in allowing us to find certain particular superpositions as free particles and not others? The basic answer is that, for a free particle, we need it to be an eigenstate of *mass*, so we need to know what it is that determines the mass of particles generally. Here, we cannot expect full symmetry under U(2); in other words, mass involves some kind of *symmetry breaking*. How is this done in the standard model? The idea, at least as it is normally presented, is that the asymmetry that we actually observe today, in particle interactions, is the result of a *spontaneous* symmetry breaking that is taken to have occurred in the early stages of the universe. Before that period, conditions were very different from those holding today, and standard electroweak theory asserts that in the extremely high temperatures in the early universe the U(2) symmetry held *exactly*, so that W^+, W^-, Z^0, and γ would be completely equivalent to many other sets of quantum superpositions of these particles, and where the photon γ is on an equal footing with all sorts of other combinations that could arise in this way. But, as this idea goes, when the temperature in the universe cooled (to below about 10^{16}K, at about 10^{-12} seconds after the Big Bang; see §§28.1–3), the particular W^+, W^-, Z^0, and γ that we observe today were 'frozen out' by this process of spontaneous symmetry breaking. Thus, in this process, four actual particles get resolved out of the completely symmetrical manifold of initial possibilities. Just three of them acquire mass, and are referred to as the Ws and the Z^0; the other one remains massless and is called the photon. In the initial 'pure' unbroken version of the theory, when there was complete U(2) symmetry, the Ws, Z^0, and γ would all have to be effectively massless. As a fundamental aspect of this symmetry-breaking proposal, another particle/field needs to come in, known as the *Higgs* (particle). The Higgs (field) is regarded as being responsible for assigning mass to all these particles (including the Higgs particle itself) and also to the quarks that compose other particles in the universe.

How does it do this? The full details of this remarkable and ingenious body of ideas must, unfortunately, remain outside the scope of this book, but I shall be giving some of its ingredients later, in §26.11 and §28.1. For the moment, I think that I can best describe the role of the Higgs field (very incompletely) by referring back to the 'zigzag' description of the Dirac electron as displayed in Fig. 25.2. Recall that the electron was viewed in terms of a continual oscillation between a left-handed zig (α_A) and a right-handed zag ($\beta_{B'}$), each of which is, on its own, massless. There was taken to be a 'coupling constant' $2^{-\frac{1}{2}}M$ governing the rate of 'flipping' between the α_A and the $\beta_{B'}$ parts of the Dirac spinor. In effect, the 'Higgs' point of view is to regard $2^{-\frac{1}{2}}M$ as a *field*—essentially the Higgs field—that enters as an interaction where we previously had the coupling constant $2^{-\frac{1}{2}}M$. (See Fig. 25.13.) One of the effects of the act of spontaneous symmetry breaking in the very early universe is taken to be that the Higgs field settles down to have a constant value everywhere. This value would fix an overall scale for the determination of the masses of all particles, the differing values of these masses being scaled by some numerical factor that depends upon the details of each particular particle.

I shall postpone my assessment of this extraordinary collection of ideas until §§28.1–3, where it fits in more appropriately. But whatever we may think of these ideas, the resulting unified theory of weak and electromagnetic forces—electroweak theory[13]—has been remarkably successful. Among its predictions was the very existence of the Z^0 (and also $W^{\pm}$, but the existence of $W^{\pm}$ had already been inferred on the basis of earlier ideas) and some rather specific values for the masses of $W^{\pm}$ and Z^0 (about 80 and 90 GeV, respectively).[14] The $W^{\pm}$ and Z^0 were observed in experiments at CERN (in Geneva, Switzerland) in 1983 and the predicted mass values were confirmed to a rather good precision, the modern observed values being about 81.4 and 91.2 GeV, respectively. Numerous other kinds of prediction have also been confirmed, and the electroweak theory stands in excellent shape, observationally, at the time of writing.

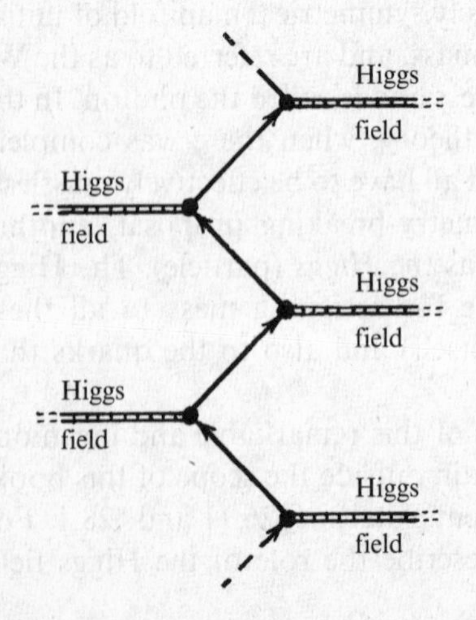

Fig. 25.13 In the zigzag picture of a Dirac particle, the vertices may be viewed as interactions with the (constant) Higgs field.

25.6 Strongly interacting particles

Now, what of *strong* interactions? The modern theory that describes them constitutes the other 'half' of the standard model, and it is referred to as *quantum chromodynamics* or QCD. This may seem an odd name, since the Greek *khroma*, from which the name comes, means 'colour', and we may ask what place 'colour' has in a theory of the strong interactions that govern nuclear forces. The answer is that the notion of 'colour' referred to here is entirely whimsical and has nothing to do with the ordinary concept of colour, which is concerned with the frequency of visible light.[15] In order to explain what the notion of 'colour' in (nuclear) particle physics might be, it will be appropriate to backtrack a little and consider the mystifying array of particles known as *hadrons*, of which neutrons and protons are particular examples.

The name 'hadron' is from the Greek *hadros* meaning 'bulky'. Hadrons are the more massive of the basic particles of Nature, and they take part in strong interactions (the strength of these interactions providing a large energy contribution to this mass). The family of hadrons includes those fermions known as 'baryons' and also those bosons referred to as 'mesons'. All hadrons are taken to be composed of *quarks*, in conventional theory, about which more will be said shortly. In particular, those hadrons known as *baryons* are the ordinary 'nucleons' (neutrons or protons) and their heavier cousins, called 'hyperons' (discovered in cosmic ray showers and in particle accelerators). The original *mesons* were a remarkable theoretical prediction by the Japanese physicist Hideki Yukawa in 1934, on the basis of his analysis of nuclear forces, these being the *pions* (π mesons) that were eventually found by C.F. Powell, in 1947, in cosmic ray tracks. Now many other meson cousins to the pion are also known.

The term 'baryon' comes from the Greek *barys* meaning 'heavy', in contrast with 'lepton', from *leptos* meaning 'small'. The *leptons* are the electron and its sister particles, the muon, and the tauon, together with their corresponding neutrinos; the anti-particles of these are referred to as *antileptons*. Both leptons and baryons are spin $\frac{1}{2}$ fermions, but leptons are distinguished from baryons by the fact that they do not directly indulge in strong interactions—which is perhaps the main 'reason' that leptons tend to be much less massive than baryons (though the tauon is an exception, being almost twice as massive as the proton or the neutron).

Since the late 1940s, vast numbers of hadrons have been discovered, in cosmic rays and in accelerators: Λ^0, $\Sigma^\pm$, Σ^0, Ξ^-, Ξ^0, Δ^{++}, $\Delta^\pm$, Δ^0, Ω^-, ρ^0, $\rho^\pm$, ω^0, η^0, $K^\pm$, K^0, and numerous heavier versions of many of these particles having higher spin (indicated, here, by the attachment of asterisks to the symbols, e.g. Ξ^{*-}) referred to as 'Regge recurrences' (see Fig. 31.6). This would have been totally bewildering had it not been for the fact that they were observed to fall into certain families, called *multiplets*. A good understanding of the nature of these multiplets was obtained (by Murray Gell-Mann and Yuval Ne'eman, in 1961) on the basis that these multiplets provide representations of the group SU(3) or, more correctly, $SU(3)/\mathbb{Z}_3$ (see §13.6 for the notion of 'representation' and §13.2 for the interpretation of the notion of 'factor group' that is involved in the use of the 'division' /; here $\mathbb{Z}_3$ stands for the cyclic group with 3 elements, which arises naturally as a normal subgroup of SU(3);[25.3] see also §§5.4,5).

The best way of understanding what is involved in these representations is to make the hypothesis (as was made explicit by Zweig and by Gell-Mann in 1963) that each hadron is constructed out of certain basic entities of spin $\frac{1}{2}$, that Gell-Mann christened 'quarks' (three types) and 'anti-quarks' (three types). Each baryon is taken to consist of just three of these quarks, and each meson, of one quark and one antiquark, where the three types of quark—referred to as three *flavours*—are called (rather unimaginatively) 'up', 'down', and 'strange'. A mysterious feature of quarks was that they have to possess fractional electric charge (in proton-charge units), the up, down, and strange quarks having respective charge values $\frac{2}{3}$, $-\frac{1}{3}$, and $-\frac{1}{3}$.

Perhaps largely because of these implausible-seeming values for the quarks' electric charges—and the related fact that quarks are never observed on their own (observed particles always having integral values for their charges; see §5.5)—quarks were not originally thought of as real particles, but were taken simply to provided a convenient 'bookkeeping' for the different representations of $SU(3)/\mathbb{Z}_3$. The bookkeeping only worked, however, if the quarks were treated as entities that satisfied the 'wrong statistics' for particles of spin $\frac{1}{2}$. That is to say, one had to pretend that quarks are 'bosons' for the multiplets to come out right, not the fermions that the spin-statistics theorem (see §23.7 and §26.2) would seem to demand.

In order to understand this last point, let us consider two examples. The most clear-cut of the two is that provided by the decuplet of 10 spin $\frac{3}{2}$ particles which led to the prediction of the Ω^- particle, by Gell-Mann and Ne'eman in 1962 (where all the other particles in the multiplet were already known), this prediction being confirmed in 1964:[16]

$$\begin{matrix} \Delta^{++}\ \Delta^{+}\ \Delta^{0}\ \Delta^{-} \\ \Sigma^{*+}\ \Sigma^{*0}\ \Sigma^{*-} \\ \Xi^{*0}\ \Xi^{*-} \\ \Omega^{-} \end{matrix}$$

[25.3] Find this normal subgroup. *Hint*: Think of the determinant of a 3 × 3 matrix.

This array can be understood if we regard each particle to be made up of three quarks, of various flavours, where d stands for down, u for up, and s for strange:[25.4]

uuu uud udd ddd
uus uds dds
uss dss
sss

Now, this works only because the three quarks are in a symmetrical state. For example, uud is not distinguished from udu. Moreover, states with two of the same kind of quark, such as uuu and uud, do not vanish identically, which they would in the case of an antisymmetrical state for which the Pauli principle holds. The fact that the spin is $\frac{3}{2}$ means that all the three quark spins (each being of value $\frac{1}{2}$) are aligned, so there is complete symmetry with regard to the spin aspect of the state. If the quarks behaved as fermions, then we would get antisymmetry under interchange of the quarks, not symmetry, which is inconsistent with this picture.[25.5]

A similar (but more involved) comment applies to the more complicated situation that arises for the octet of 8 spin $\frac{1}{2}$ particles to which the ordinary proton (N^+) and neutron (N^0) belong:17

$$\begin{matrix} & N^+ & & N^0 & \\ \Sigma^+ & & \Sigma^0 \ \Lambda^0 & & \Sigma^- \\ & \Xi^0 & & \Xi^-. & \end{matrix}$$

Here we must think of Σ^0 and Λ^0 as occupying basically the 'same slot' at the centre of a hexagonal array. This arrangement comes about when we consider that the spin is now $\frac{1}{2}$, so we can think of two of the quark spins as parallel and one of them antiparallel. It turns out that there are just two linearly independent ways of arranging this for the quark composition uds that is represented at the centre (corresponding to the pair Σ^0 and Λ^0); there is none at all for uuu, ddd, and sss, which explains the hexagonal rather than a triangular array; and there is just one for each of the rest.[25.6]

[25.4] Check that the charge values, indicated by the superfixes in the first table, come out right.

[25.5] Explain this more completely, using the 2-spinor index description for the quark spins, as described in §22.8, and using a new 3-dimensional 'SU(3) index' which takes 3 values u, d, s.

[25.6] See if you can explain all this in some appropriate detail. Care is needed for the treatment of the 2-spinor spin indices, if you wish to use them. An antisymmetry in a pair of them allows that pair to be removed (as when representing a spin 0 state in terms of a pair of spin $\frac{1}{2}$ particles, as in §23.4). Yet there is a (hidden) symmetry also, because there are only two independent spin states for each quark.

25.7 'Coloured quarks'

How can we treat quarks as real particles, if they have the wrong 'spin-statistics' relation (see §23.7 and §26.2). The way that this problem is dealt with,[18] in the standard model, is to demand that each flavour of quark also comes in three (so-called) 'colours', and that any actual particle, composed of quarks, must be completely antisymmetrical in the colour degree of freedom. This antisymmetry passes over to the quark states themselves, so that antisymmetry between individual (fermionic) quarks gets effectively converted into symmetry, in a three-quark particle.[25.7] The colours are never to manifest themselves in free particles, so colour is, in an essential way 'unobservable'. Any free particle has to be 'colour-neutral'. We do not, for example, have three different versions of the Δ^+ particle, depending upon which colour the d-quark is in 'uud'. The antisymmetry in the colour degree of freedom, for actual free particles, ensures this.[25.8]

The 'colours' are sometimes referred to as 'red', 'white', and 'blue', which strikes me as being both confusing (since I do not think of white as a colour) and revealing of some sort of misplaced patriotism. Sometimes they are called 'red', 'green', and 'blue', which is better; but since the association between 'quark colour' and the colour receptors in the eye has in any case no scientific justification, I shall use 'red' (R), 'yellow' (Y), and 'blue' (B) instead. This choice of terminology has the advantage that I can 'mix' my colours more easily, and we note that 'orange', 'green', and 'purple' (these being regarded as quantum superpositions of the original R, Y, and B) would do just as well as the original set. There is indeed a symmetry here that goes well beyond merely permuting the colours. There is to be a full 8-real-dimensional SU(3) of colour symmetry, in which R, Y, and B provide merely one set of basis elements for the vector space on which the SU(3) matrices act (see §13.9).

At this stage, the introduction of these apparently unobservable 'colour' degrees of freedom would appear to be rather contrived, since we now have nine basic quarks (together with their various antiparticles and quantum superpositions):

$$\mathrm{d_R}\,,\ \mathrm{d_Y}\,,\ \mathrm{d_B}\,;\quad \mathrm{u_R}\,,\ \mathrm{u_Y}\,,\ \mathrm{u_B}\,;\quad \mathrm{s_R}\,,\ \mathrm{s_Y}\,,\ \mathrm{s_B}\,;$$

none of which can be directly observed. In fact the situation, in the standard model, is actually 'twice as bad' as this, because three more flavours of quark have had to be introduced, called (equally unimaginatively) 'charm' (c), 'bottom' (b), and 'top' (t), so we also have:

$$\mathrm{c_R}\,,\ \mathrm{c_Y}\,,\ \mathrm{c_B};\quad \mathrm{b_R}\,,\ \mathrm{b_Y}\,,\ \mathrm{b_B};\quad \mathrm{t_R}\,,\ \mathrm{t_Y}\,,\ \mathrm{t_B};$$

[25.7] Use indices to explain this comment, where there is a new 3-dimensional SU(3) *colour* index, in addition to a 3-dimensional flavour index of Exercise [25.5].

[25.8] Explain.

giving eighteen independent quarks in all, every one of which is not directly observable.

If satisfying the spin-statistics relation were to be the only benefit of this proliferation of hypothesized unobservable particles, then the scheme would look decidedly artificial. But the complete unobservability of 'free' quark colour actually brings a bountiful reward! For this unobservability, and the (closely related) totally unbroken nature of the colour SU(3) symmetry, provides us with the potential to use this symmetry directly as a basis for the wonderful idea of a gauge connection, as described in §§15.1,8. Recall that this is how the electromagnetic interaction is described, the gauge group being U(1) in that case (see §19.4, §21.9, and §24.7). Indeed, the U(1) gauge symmetry of electromagnetism is taken to be exact and unbroken.[19] Recall, also, that at the very basis of the fibre-bundle idea, as described in Chapter 15, is the presence of an exact symmetry group acting on the fibres. The hadronic SU(3) 'colour group' of strong interactions provides just such an exact symmetry, and the analogy with the U(1) electromagnetic gauge group is very close. The generalization of electromagnetism, which is based on a gauge connection for the abelian group U(1), to a corresponding theory based on a gauge connection for a non-abelian groups such as SU(2) or SU(3) is called *Yang–Mills* theory.[20]

This, indeed, is the basis of QCD (quantum chromodynamics). As with electromagnetism, we can use a quantity like the electromagnetic potential A_a to modify the derivative $\partial/\partial x^a$, when acting on quark fields, to an appropriate notion of 'covariant derivative operator' (like the $\partial/\partial x^a - \mathrm{i}eA_a$ of electromagnetism) that provides us with a bundle connection (see §15.8 and §19.4). Because the colour space is 3-dimensional, we have something more complicated than that which we had for the 1-dimensional electromagnetic case, and it is convenient to introduce indices to cope with these extra degrees of freedom. A crucial difference between the electrodynamic case and that of strong interactions is that whereas U(1) is Abelian (i.e. commutative; see §13.1), the colour group SU(3) is non-Abelian, and the theory is accordingly referred to as a *non-Abelian gauge theory*. This leads to special complicated and interesting features. For full details of what is involved here, I refer the reader to the literature,[21] but the essential idea of how the strong interactions manifest themselves is basically as I have just described it.

The 'gauge bosons' of QCD (the SU(3) analogues of photons) are quantities referred to as *gluons*. In the Feynman-graph descriptions, the gluon lines attach themselves to quark lines in the same way that photon lines attach themselves to charged particle lines (Fig. 25.14a). The non-Abelian nature of SU(3) manifests itself in the fact that the gluon lines themselves possess 'colour charge', so that three-pronged (or more)

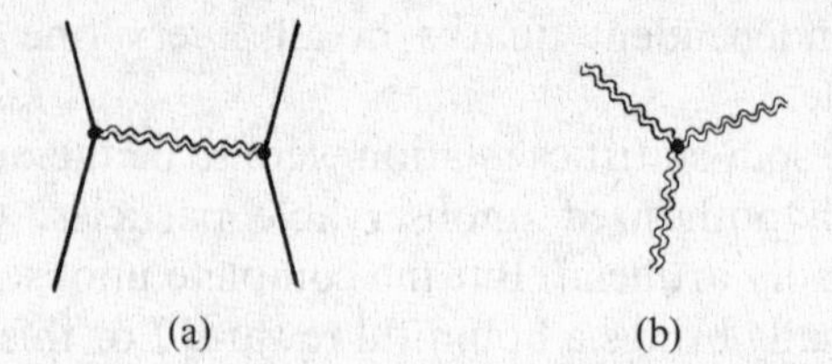

Fig. 25.14 Gluons are the 'gauge bosons' of QCD. (a) Gluon exchange between quarks (here drawn without zigzags) underlies nuclear forces and quark confinement. (b) The gauge theory being non-Abelian, gluon lines themselves possess 'colour charge', so three-pronged gluon Feynman graphs can occur (as in Fig. 25.12).

gluon Feynman graphs can occur (Fig. 25.14b) which is something that does not happen for the Abelian electromagnetic case.

Thus, the main role of the group SU(3), in the standard model, has moved from the 'flavour symmetry' that it had in the 1960s and 1970s to the 'colour symmetry' of the present-day standard model. In fact, in this standard model, the three flavours d, u, and s are not now fundamentally grouped together at all. Instead, the groupings provide three *generations* of doublets (d, u), (s, c), (b, t). The notion of having three generations applies also to the leptons, the generations being that of the electron, the muon, and the tauon (and their corresponding neutrinos).

In the standard model as a whole, there are complicated interrelations between the strong and electroweak interactions. In particular, there are certain 'rotation angles' between the basic entities that are recognized by strong interactions and those recognized by weak interactions. An example occurs with the K^0 meson, which can be produced in high-energy proton–proton collisions. We say that K^0 is an *eigenstate of strong interactions*. When the K^0 itself decays, however, it decays weakly, and for that it has to be considered as a quantum linear combination of the two eigenstates K_L (K-long) and K_S (K-short) of weak interactions. (The K_L usually decays to three pions in about 5×10^{-8} seconds, whereas the K_S normally goes to two pions in the much shorter timescale of 10^{-10} seconds.) Each of K_L and K_S is a linear combination of K^0 and its antiparticle $\bar{K}^0$, into which K^0 can 'flip' by means of weak but not strong interactions. The 'rotation' between the strong-interaction basis states (K^0, $\bar{K}^0$) and the weak-interaction basis states (K_L, K_S) takes place through an (abstract) angle referred to as the *Cabibbo angle* (which is about 0.26 radians). This same angle features in the interrelations between strong and weak interactions generally.

In a somewhat similar way, there is an angle referred to as the *Weinberg angle* or the *weak mixing angle* (§25.5) which features in the interrelation between weak and electromagnetic interactions, and forms an integral part of electroweak theory. Indeed, some of the most impressive confirmations of electroweak theory come from various (seemingly independent) types of observational determination of this angle giving answers that closely agree with one another. However, as theory stands, there is a difference in the roles between the Cabibbo and Weinberg angles; for the weak and electromagnetic interactions are considered to be unified, and the view may be taken, concerning the Weinberg angle, that it was something that got 'frozen in' when the U(2) symmetry of electroweak theory was 'broken' at about 10^{-12} seconds after the Big Bang (§28.1), but the Cabibbo angle does not have such a status in the standard model, since that model makes no assertion as to how the electroweak and strong interactions might be unified. The basic symmetry group[22] of the entire standard model is taken to be $SU(3) \times SU(2) \times U(1)/Z_6$.

25.8 Beyond the standard model?

On the other hand, one *could* adopt a perspective on the Cabibbo angle corresponding to that taken for the Weinberg angle, but this would require something beyond the present-day standard model of particle physics. We would need a model in which both strong and weak interactions are united under some larger symmetry group that includes SU(3) and U(2) together. Such a theory is referred to as a *grand unified theory* or GUT. There is no commonly accepted GUT, but there have been many attempts (the main ones being based on SU(5), or SO(10), or the exceptional group E_8; see §13.2). We shall be seeing in §31.14 that string theory has something to say about these matters also. Some remarkable implications of certain GUT models will be considered in §28.2.

In any case, the standard model is clearly not the 'ultimate answer', with regard to particle physics, because it contains many unexplained features and 'ragged edges', despite its undoubted success. It involves about 17 unexplained parameters that simply need to be taken from observation (such as the Cabibbo and Weinberg angles, the masses of the quarks and leptons, and a number of other features). Also there is the rather strange asymmetry between the roles of SU(3) and U(2)—in that SU(3) is taken to be exact, whereas U(2) is severely broken. Indeed, in my view, there does appear to be something strange about the particular way that U(2) is taken as a 'gauge group', which would seem to require an exact unbroken symmetry (see Chapter 15, particularly the final paragraph of §15.8).

At this point, it is pertinent to refer to another development, distinct from the GUT idea, that addresses this particular question in a novel way. It has a special appeal for me personally, for reasons that will become apparent in §33.13. This is a proposal due to the Chinese–British husband-and-wife team Chan Hong-Mo and Tsou Sheung Tsun (2002). In their scheme, each (non-Abelian) particle symmetry group has a corresponding *dual* group, which is the same abstract group as the original one but which plays a kind of opposite role. Recall the dual $^*\boldsymbol{F}$ of the Maxwell tensor $\boldsymbol{F}$, that was introduced in §19.2. We could imagine a 'dual' U(1) gauge connection that has $^*\boldsymbol{F}$ as its bundle curvature (see §15.8), rather than $\boldsymbol{F}$. The idea is to do something similar for the remaining symmetry groups of the standard model SU(2) and SU(3). But, since these groups are non-Abelian, it is not possible simply to regard the corresponding dual curvatures directly as bundle curvatures again[25.9] and something more sophisticated is required (where 'path-dependent' quantities need to be considered).

One of the attractive aspects to this scheme is that the group and dual group play qualitatively different roles, one of the two being exact, like the SU(3) of QCD (or the U(1) of electromagmetism), and the other being broken like the SU(2) of electroweak theory, and where 'confinement' (which is what prevents the 'colour-charged' quarks escaping into the wide world, in the case of SU(3)) is expected for the exact group. (This property relates to earlier work by 't Hooft and Weinberg.[23]) In the Chan–Tsou scheme, there would be a new exact SU(2) (dual to the broken one that currently features in electroweak theory) which would refer to a hitherto undiscovered symmetry, relating analogues of quarks that would be confined '2-coloured' lepton constituents. (These sub-particles would be very heavy, which is why they have not yet been detected, and why leptons appear as point particles at present-day energies.) Correspondingly, there must also be a broken SU(3) (dual to the colour SU(3)), and this is taken simply to be the 'SU(3)' of the 3 generations of quarks and of leptons that seems so puzzling in the standard model as it is conventionally understood. The Chan–Tsou scheme also has clear predictions concerning the 17 (or so) free parameters of the standard model, calculating 14 of them from 3 adjustable parameters. This strikes me as a definite step forwards, provided that the predictions of the scheme are borne out. As things stand, the outlook seems promising.

It is less clear to me how, in the conventional attitude to the standard model, the group SU(2) can be taken actually as a gauge group, while being so severely broken. Some might take this SU(2) as reflecting some kind of 'hidden symmetry' which is really exact, and which acts only 'potentially' as a gauge group, and the SU(2) of electroweak theory is some kind of external manifestation of this. (Perhaps this is not so far from the Chan–Tsou idea, but not so explicit.) The conventional perspective on electroweak theory's SU(2) seems to be that it really is (or, rather, was) exact and has become broken in extreme processes that took place in the early universe. We shall be having a look at some of the unpleasant implications of this in Chapter 28. In the meantime, as part of the discus-

---

[25.9] Can you see what the difficulty is? *Hint*: Work out expressions for gauge curvature, Bianchi identities, etc.

sion in the next chapter, we shall be seeing something of the exotic but essential mathematical ideas that lie behind the way that symmetry breaking is presently handled in the standard model.

Notes

Section 25.1

25.1. See Pais (1986), pp. 334 and 356, refs. 25,26.

Section 25.2

25.2. I have not entered into the details of how the Dirac equation as described in §24.7 can be transcribed into the 2-spinor form given here. The interested reader is referred to Zee (2003), Appendix. Weyl introduced 2-spinors in Weyl (1929). See van der Waerden (1929); Infeld and van der Waerden (1933); Penrose and Rindler (1984), pp. 221–23.

25.3. These are the *reduced spinors* (or half-spinors) referred to in §11.5.

25.4. See Infeld and van der Waerden (1933); Laporte and Uhlenbeck (1931); Penrose and Rindler (1984).

25.5. See Schrödinger (1930); Dirac (1982a); Huang (1952) for an alternative picture; or, for an interesting modern perspective, Hestenes (1990).

25.6. Those readers who already have some familiarity with Feynman graphs may find my vertical temporal ordering confusing. It is more usual, in the QFT community, to depict increasing time as off to the right. My own preference for increasing time to be depicted upwards is in accord with the notation of much of the relativity community, since this is consistent with most spacetime diagrams (see Chapter 17, most particularly).

25.7. In fact, such physicists' lives could have been made significantly simpler by use of the 2-spinor formalism in QED! See Geroch (University of Chicago lecture notes, unpublished), and also §34.3.

25.8. My own conventions would have been to write $(1 \pm i\gamma_5)\psi$ here, rather than $(1 \pm \gamma_5)\psi$ (see Penrose and Rindler 1984, 1986, Appendix), as would some other authors. Here, I am accommodating myself to what appears to be standard in the physics community.

Section 25.3

25.9. This may have been partly influenced by a suggestion made by Martin Block (and conveyed by Richard Feynman); see the fascinating account by Martin Gardner in *The New Ambidextrous Universe* (W.H. Freeman 1990), Chap. 22.

25.10. This equation was proposed by Weyl in 1929, and it had been considered also by Dirac before he came across his 'Dirac equation for the electron'; Dirac (1928); Dirac (1982b). Pauli had vehemently objected to Weyl's equation on account of its lack of invariance under spatial reflection. Unfortunately, Weyl died in the year before the non-invariance of reflection was proposed in weak interactions, vindicating his proposal. Zee (2003) discusses both equations.

Section 25.4

25.11. Massive particles of spin 1 can be described as having *three* ingredients, a left-handed zig (helicity 1), a right-handed zag (helicity -1), and a non-spinning 'zog' (helicity 0), let us say. (The zig 2-spinor and the zag 2-spinor has two unprimed indices and two primed indices, respectively, while the zog 2-spinor has one of each.) We may take the view that it is just the zog particle that mediates the weak interactions.

Section 25.5

25.12. The group might be expressed as $SU(2) \times U(1)/Z_2$, where the '$/Z_2$' means 'factor out by a Z_2 subgroup'. However, there is more than one such subgroup, so this notation is not fully explicit. The notation 'U(2)' automatically picks out the correct one. (I am grateful to Florence Tsou for this observation.) It seems that the reason that the electroweak symmetry group is not conventionally referred to as 'U(2)' is that this does not easily extend to the symmetry of the full standard model, which also incorporates the strong symmetry group SU(3), the full group being a version of $SU(3) \times SU(2) \times U(1)/Z_6$; see §25.7.

25.13. The electroweak theory was sorted out by Stephen Weinberg, Sheldon Glashow, and Abdus Salam in the late 1960s—work earning all three of them a Nobel Prize. See Weinberg (1967); Salam and Ward (1959); Glashow (1959); for a general reference on the electroweak theory, see Zee (2003) or Halzen and Martin (1984); Kaku (1993).

25.14. GeV are giga electron-volts. Giga is a Greek prefix, indicating multiplication by 10^9; and an electron-volt a measure of energy, specifically how much energy a single unbound electron will gain when falling through a potential difference of 1 volt. It is around 1.6×10^{-19} J.

Section 25.6

25.15. Visible light falls between wavelengths $\lambda = 400$–700 nanometres, where one may convert between wavelength and frequency ν according to the relation $\nu = \frac{c}{\lambda}$.

25.16. See Gell-Mann and Ne'eman (2000) for the theory; V. E. Barnes's paper on the observation of the Ω^-, originally published in 1964, is in the same work, on pp. 88–92.

25.17. In the terminology of modern particle physics, 'N^+' and 'N^0' seem to have replaced 'p' and 'n', to denote the proton and neutron, respectively. This is consistent with the notation for other particles in that (N^+, N^0) constitutes a *doublet*, in the SU(3) classification scheme, like (Ξ^0, Ξ^-), etc., and it allows us to refer to a *nucleon* generically as 'N'.

Section 25.7

25.18. See Han and Nambu (1965).

25.19. See Weinberg (1992).

25.20. C.N. Yang and R. L. Mills found this theory in 1954, although the basic idea had been discovered earlier (and rejected because the gauge particles had to be massless) by Wolfgang Pauli, in the years after World War II, and Ronald Shaw in 1955. See Abdus Salam (1980) for an exhaustive history of these matters, presented in his Nobel lecture. The trick which is now used in order to circumvent the 'masslessness' problem is the symmetry-breaking 'Higgs mechanism' that was alluded to in §25.5 and will be discussed further in §26.11.

25.21. See Aitchison and Hey (2004), Vol. 2, or Zee (2003) for the technical details. See Chan and Tsou (1993) for an overview of gauge theory concepts.

25.22. See Note 25.12. An overview of the Standard Model can be found in any good Quantum Field Theory textbook, for example, Zee (2003).

Section 25.8

25.23. The Chan–Tsou idea is laid out in Chan and Tsou (2002); it is based on a property developed in t'Hooft (1978).

26
Quantum field theory

26.1 Fundamental status of QFT in modern theory

We have made our brief acquaintance, in the previous chapter, with the 20th century's standard model for particle physics. It is a mathematical model in remarkable accord with observational facts over a broad range of phenomena, and it involves some ingenious mathematical ingredients that seem to find a deep harmony with Nature's ways. Yet, as I have presented things, the mathematical structure of this model would appear to be somewhat complicated and arbitrary. Of course much of this structure has been motivated by brute facts of particle physics, and physicists have had to come to terms with these facts, as Nature has presented them. This is as it should be, for any serious scientific theory. But there are also powerful *theoretical* reasons underlying the particular choices of structure that are found in the standard model. The predictive power of the theory indeed depends crucially upon the mathematical consistency of such theoretical underpinnings.

The theoretical driving force is a continuation of the story that we began in Chapter 24: how do we find a quantum theory for particle physics which is consistent with the requirements of Einstein's special theory of relativity? We saw, in that chapter, the importance of Dirac's introduction of antiparticles in a relativistic quantum theory, and that we were forced into the framework of a quantum theory of fields. In fact the standard model is a particular instance of a quantum theory of interacting fields, and has been driven largely by certain powerful consistency requirements, hard to satisfy in such theories. In order to appreciate something of the force behind these consistency requirements (which continue to drive the more modern speculative theories of today, such as string theory), we shall need to look at something of the structure of quantum field theory (QFT). This will also help us to appreciate the meaning of the Feynman graphs that we encountered in the previous section. In addition, we shall gain yet another perspective on anti-particles, which is somewhat broader than those that we have encountered in Chapters 24 and 25.

Quantum field theory constitutes the essential background underlying the standard model, as well as practically all other physical theories that attempt to probe the foundations of physical reality. It is therefore necessary for us to catch a glimpse of that magnificent and imposing scheme of things, a scheme that arose, in good measure, from those remarkable insights of Paul Dirac with which we made some acquaintance in Chapter 24. It should be pointed out that Dirac was himself the main initiator of QFT, although important initial contributions came also from Jordan, Heisenberg, and Pauli. However, as it stood, and for most problems of interest, that theory was not able to provide *finite* answers—rather than the '∞' that practically always seemed to arise. It took powerful later developments from Bethe, Tomonaga, Dyson, Schwinger, and particularly Feynman to make the theory workable for suitable QFTs referred to as 'renormalizable'. More recent input from Ward, Weinberg, Salam, Wilson, Veltman, and 't Hooft, among others, has led us to a very appropriate class of renormalizable theories, giving a vital input towards what is now the standard model of particle physics (Chapter 25), from which consistent answers can indeed be obtained.[1] (The theoretical requirements appear to be so tight that it might seem almost incidental that these answers are actually in excellent agreement with experiment!) The basic problem has always been to circumvent the infinities in some appropriate way, and it has been this drive, together with important input from observation, which has taken the theory in its appropriate and fruitful directions.

In fact, QFT appears to underlie virtually all of the physical theories that attempt, in a serious way, to provide a picture of the workings of the universe at its deepest levels. Many (and perhaps even most) physicists would take the view that the framework of QFT is 'here to stay', and that the blame for any inconsistencies (these being usually infinities coming from divergent integrals, or from divergent sums, or both) lies in the particular scheme to which QFT is being applied, rather than in the framework of QFT itself. Such schemes are normally specified by a Lagrangian, subject to certain symmetry principles. We shall be seeing in §§26.6,10 the general way in which Lagrangian ideas are applied in QFT.

Many modern attempts to remove the infinities in QFT look to *gravity* to alter spacetime behaviour profoundly at extremely tiny scales, and thereby supply the 'cut-offs' that could render the presently still divergent expressions finite (see §31.1, in particular). Yet there remains a question as to whether QFT itself might need modification when the (gravitational) principles of Einstein's *general* relativity are brought in (see Chapter 30). However, as judged by the activities of the vast majority of current researchers in this kind of area, QFT in its present form is not normally questioned, and it will be important for us to come to terms with its sophisticated ideas, as best we can. I shall certainly not be able to go into great detail in my description of this magnificent, profound, difficult, sometimes phenomenonally accurate, and yet often tantalizingly inconsistent scheme of things. But I shall try, briefly, to convey something of QFT's flavour, albeit very incompletely, before finally returning to those features that supply the theoretical driving force underlying the standard model.

26.2 Creation and annihilation operators

One of the earliest ideas for QFT was the procedure which goes under the rather misleading name of 'second quantization'. According to this procedure, we try to pretend that the wavefunction ψ of some particle itself becomes an 'operator', acting on some shadowy state vector which may be denoted by $|0\rangle$, hiding over on the far right (compare §21.3 and the Heisenberg picture of §22.4). I shall denote this 'wavefunction operator' by the boldface *capital* Greek letter $\boldsymbol{\Psi}$, corresponding to the Greek letter ψ that denotes our one-particle wavefunction. As in ordinary quantum particle mechanics, $\boldsymbol{\Psi}$ can be thought of as a function of the particle's 3-space position $\mathbf{x}$, i.e. $\boldsymbol{\Psi} = \boldsymbol{\Psi}(\mathbf{x})$, or else of its 3-momentum $\mathbf{p}$, if a momentum representation is preferred, i.e. $\tilde{\boldsymbol{\Psi}} = \tilde{\boldsymbol{\Psi}}(\mathbf{p})$.

How are we to interpret this strange 'wavefunction operator' $\boldsymbol{\Psi}$ (or $\tilde{\boldsymbol{\Psi}}$)? It does not now represent the actual quantum state, but it describes the operation that 'creates' a new particle having this given wavefunction[2] ψ, introducing it into the state that is there previously—this 'previous' state being represented by the expression that follows immediately to the right of the operator symbol $\boldsymbol{\Psi}$ (or $\tilde{\boldsymbol{\Psi}}$). Such an operator is referred to as a *creation operator*.

The shadowy state-vector $|0\rangle$ over on the extreme right is normally taken to be the 'vacuum state', representing the complete absence of particles of any kind. A succession of these creation operators then creates a succession of particles, added one by one into the vacuum, so that

$$\boldsymbol{\Psi}\boldsymbol{\Phi}\ldots\boldsymbol{\Theta}|0\rangle$$

is the state that results from introducing particles successively with wavefunctions

$$\theta, \ldots, \phi, \psi.$$

Since any particular type of particle will be either a fermion or a boson, this fact needs to be taken into account. In particular, the Pauli principle has to be incorporated, which prevents us introducing two fermions into the same state, one after the other. The Pauli principle is expressed, in this formalism by the property $\boldsymbol{\Psi}^2 = 0$ (i.e. $\boldsymbol{\Psi\Psi} = 0$) for any fermion wavefunction ψ, which tells us that, if we try to introduce this particular fermion wavefunction into the state twice, we get zero, which is not an allowable state vector. This 'Pauli-principle' rule is just a particular instance of the anticommutation property

$$\boldsymbol{\Psi\Phi} = -\boldsymbol{\Phi\Psi},$$

where $\boldsymbol{\Psi}$ and $\boldsymbol{\Phi}$ are creation operators describing the same type of fermion. For creation operators $\boldsymbol{\Theta}$ and $\boldsymbol{\Xi}$ describing the same type of boson we have the commutation property[26.1]

$$\boldsymbol{\Theta\Xi} = \boldsymbol{\Xi\Theta}.$$

Thus, we see that creation operators satisfy the rules of a (graded) Grassmann algebra, as described in §11.6, where the fermion creation operators are considered to be of odd grade and the boson creation operators, of even grade.

In accordance with the discussion of §24.3, the wavefunctions that are introduced into a state, for the creation of a particle, must be of positive frequency. Negative-frequency quantities also play a role in the formalism, namely as annihilation operators. The complex conjugate $\overline{\psi}$ of the positive-frequency wavefunction ψ is a quantity of negative frequency. It is associated with the annihilation operator $\boldsymbol{\Psi}^*$, which is the Hermitian conjugate[3] of the creation operator $\boldsymbol{\Psi}$ (see §13.9). The interpretation of $\boldsymbol{\Psi}^*$ is that it represents the removal of a particle from the total state (that total state being the one which, as before, is described by whatever lies to the right of $\boldsymbol{\Psi}^*$ in the expression. Since our shadowy vacuum state $|0\rangle$, way over on the right, contains no particles whatever, the action of any annihilation operator directly on it must give zero:

$$\boldsymbol{\Psi}^*|0\rangle = 0.$$

Of course, this does not mean that annihilation operators always yield zero, because we could put some particles in first. An expression like $\boldsymbol{\Psi}^*\boldsymbol{\Phi\Theta}|0\rangle$ need not be zero, for example. This holds even if neither of the states $\boldsymbol{\Phi}$ and $\boldsymbol{\Theta}$ is the same as the $\boldsymbol{\Psi}$ that we are removing. For we should not try to think of the operator $\boldsymbol{\Psi}^*$ as simply removing the particle's specific wavefunction ψ from that state.[4] In general, the specific wavefunction ψ is unlikely to feature exactly as part of the state. Instead, what $\boldsymbol{\Psi}^*$ does is, in effect, to form a scalar product with the part of the state that refers to the type of particle that is being removed. (In Fig. 26.1—mainly for the

[26.1] Explain why this gives states with the correct symmetries for bosons and fermions, as described in §23.8.

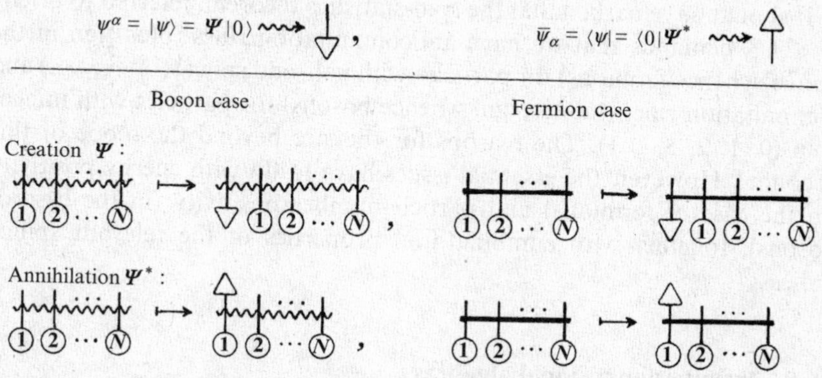

$\psi^{\alpha} = |\psi\rangle = \boldsymbol{\Psi}|0\rangle \rightsquigarrow$, $\quad \bar{\psi}_{\alpha} = \langle\psi| = \langle 0|\boldsymbol{\Psi}^{*} \rightsquigarrow$

Boson case — Fermion case

Creation $\boldsymbol{\Psi}$:

Annihilation $\boldsymbol{\Psi}^{*}$:

Fig. 26.1 Diagrammatic form of the action of a creation operator $\boldsymbol{\Psi}$ in the boson case $\phi_1^{(\beta}\phi_2^{\gamma} \cdots \phi_N{}^{\nu)} \mapsto \psi^{(\alpha}\phi_1^{\beta}\phi_2^{\gamma} \cdots \phi_N{}^{\nu)}$ and in the fermion case $\phi_1^{[\beta}\phi_2^{\gamma}\cdots\phi_N{}^{\nu]} \mapsto \psi^{[\alpha}\phi_1^{\beta}\phi_2^{\gamma} \cdots \phi_N{}^{\nu]}$; and of an annihilation operator $\boldsymbol{\Psi}^{*}$ in the boson case $\phi_1^{(\alpha}\phi_2^{\beta} \cdots \phi_N{}^{\mu]} \mapsto \bar{\psi}_{\alpha}\phi_1^{(\alpha}\phi_2^{\beta} \cdots \phi_N{}^{\mu)}$ and in the fermion case $\phi_1^{[\alpha}\phi_2^{\beta}\cdots\phi_N{}^{\mu]} \mapsto \bar{\psi}_{\alpha}\phi_1^{[\bar{\alpha}}\phi_2^{\beta} \cdots \phi_N{}^{\mu]}$.

amusement of the experts—I have indicated the diagrammatic form of what I mean by this, both in the fermion and the boson case, where I have also given diagrams representing the creation as well as the annihilation process.)[26.2] In accordance with this, it turns out that the creation and annihilation operators (for the same type of particle) must satisfy (anti)-commutation rules

$$\boldsymbol{\Psi}^{*}\boldsymbol{\Phi} \pm \boldsymbol{\Phi}\boldsymbol{\Psi}^{*} = \mathrm{i}^{k}\langle\psi|\phi\rangle\boldsymbol{I},$$

where the 'plus' sign refers to fermions and the 'minus' sign to bosons, where $\boldsymbol{I}$ represents the *identity* operator, with $\langle\ |\ \rangle$ standing for the ordinary Hilbert-space scalar product for individual particles (the spinless case having been considered in §22.3, there being an appropriate generalization for particles with spin[5]), and where i^{k} stands for one of $1, \mathrm{i}, -1, -\mathrm{i}$, depending on the spin (and I shan't worry you about which). We also have the following (anti)commutation rules for two creation operators (also given above), and for two annihilation operators (plus sign for fermions, minus for bosons):

$$\boldsymbol{\Psi}\boldsymbol{\Phi} \pm \boldsymbol{\Phi}\boldsymbol{\Psi} = 0, \qquad \boldsymbol{\Psi}^{*}\boldsymbol{\Phi}^{*} \pm \boldsymbol{\Phi}^{*}\boldsymbol{\Psi}^{*} = 0.$$

[26.2] Make sense of all this (and verify this commutation law for creation and annihilation of particles of a given type) by referring to the index notation of §23.8 or the diagrammatic notation of Fig. 12.17, or both, using expressions like $\bar{\psi}_{\alpha}\psi^{[\alpha}\phi^{\beta}\cdots\chi^{\kappa]}$. Sort out all the factorial factors which preserve normalization of the state, both in the fermion and the boson case.

It should be remarked that the spin-statistics theorem, referred to briefly in §23.7, demands that we have anticommutation rules (plus sign in the above, whence fermions) for particles with half-odd spin ($\frac{1}{2}$, $\frac{3}{2}$, $\frac{5}{2}$, ...) and commutation rules (minus sign, whence bosons) for particles with integer spin (0, 1, 2, 3, ...). The reasons for this are beyond the scope of this account.[6] However, the essential issues have to do with energy positivity (in the case of fermions) and particle-number positivity (in the case of bosons), together with combinatorial properties of the relevant spinor indices.[7]

26.3 Infinite-dimensional algebras

It is a remarkable fact that, in the case of fermions, these anticommutation rules are precisely of the same algebraic form as those defining a Clifford algebra, as described in §11.5.[26.3] The only essential difference is that ordinary Clifford algebras are finite-dimensional, whereas the space of creation and annihilation operators, for a fermion field, is infinite-dimensional—the space of one-particle wavefunctions being infinite-dimensional. The reader should be warned, however, that infinite-dimensional spaces, though often analogous to finite-dimensional ones, can have some very different properties, and are frequently much harder to work with.

It is of interest that the formalism of QFT also involves infinite-dimensional versions of some of the other types of finite-dimensional algebraic structure that we have considered earlier in this book. The scalar product $\langle\ |\ \rangle$, for example, is really an infinite-dimensional version of the Hermitian scalar product considered in §13.9 (cf. §22.3). In fact, in QFT, it turns out that not only are we concerned with 'Hermitianness' (unitarity), but we find that symmetric forms ('pseudo-orthogonality'), antisymmetric (symplectic) forms, and complex structures also play their roles.[8] The ordinary finite-dimensional versions of pseudo-orthogonal and symplectic forms were considered §§13.8,10; ordinary finite-dimensional complex structures featured in §12.9.

There is a particular significance, for QFT, in how an (infinite-dimensional) complex structure arises here. We have already seen that complex numbers, holomorphic functions, and complex vector spaces have fundamental roles in quantum theory (and therefore in the basic structure of our world). But the particular infinite-dimensional complex structure that comes in at this point, in the study of QFT, seems to have a

---

[26.3] Explain this Clifford-algebra structure, spelling out the role of the scalar product more explicitly. (Take the defining laws for a Clifford algebra in the form $\gamma_p\gamma_q + \gamma_q\gamma_p = -2g_{pq}\boldsymbol{I}$.) *Hint*: g_{pq} need not be diagonal.

somewhat different (though interrelated) significance from these earlier instances of complex-number magic. It goes beyond the mere statement that the Hilbert spaces of quantum theory are complex (i.e. that quantum superposition takes place with complex coefficients). Let us try to understand what this is about.

Let us recall how the notion of *complex structure* was introduced in §12.9. A complex vector space of n dimensions can be thought of as a real vector space of $2n$ dimensions where there is an operation $\boldsymbol{J}$, satisfying $\boldsymbol{J}^2 = -1$, whose action on the real $2n$-space amounts to 'multiplication by i' on the complex n-space. The infinite-dimensional version of this relevant to QFT has to do with the passage from a classical field to a quantum field. Up to this point, I have been phrasing things in a particle/wavefunction language. But we also need to know how to go straight from a classical field to a quantum field, since with classical fields there is no classical particle picture to hand which we can 'quantize' according to the procedures of Chapters 21–23.

It is useful to keep the electromagnetic field particularly in mind, as a model. Here, the linearity of Maxwell's equations (§19.2) makes things easier. The space $\mathcal{F}$ of solutions of the free Maxwell equations (with suitable fall-off conditions at infinity to make the relevant integrals converge) is an infinite-dimensional real vector space.[26.4] Using procedures related to those described in §§9.2,3,5, we can express each solution $\boldsymbol{F}$ of Maxwell's equations[9] as a *sum* of a positive-frequency solution $\boldsymbol{F}^+$ and a negative-frequency solution $\boldsymbol{F}^-$:

$$\boldsymbol{F} = \boldsymbol{F}^+ + \boldsymbol{F}^- .$$

This splitting into positive and negative frequencies is crucial for the construction of the appropriate QFT (recall the comments on this issue in §24.3 and §26.2). The operation $\boldsymbol{J}$, as applied to this infinite-dimensional real vector space $\mathcal{F}$, transforms it into a complex (infinite-dimensional) vector space and, in doing so, provides a way of encapsulating this positive/negative frequency splitting. $\boldsymbol{J}$ does this by acting on each free Maxwell field $\boldsymbol{F}$ in the following way:

$$\boldsymbol{J}(\boldsymbol{F}) = \mathrm{i}\boldsymbol{F}^+ - \mathrm{i}\boldsymbol{F}^- .$$

The eigenstates of $\boldsymbol{J}$ with eigenvalue i are the positive-frequency (complex) fields and those with eigenvalue −i are the negative-frequency fields.[26.5] The positive-frequency fields supply the single-photon wavefunctions that the creation operators are to introduce. There is also an

---

[26.4] Explain what addition, and multiplication by a scalar constant, mean in this space.

[26.5] Show this. (Don't worry about subleties like 'fall-off conditions'?)

explicit scalar-product expression that can be used to normalize states when required (involving an integral over any spacelike 3-surface of an expression involving Maxwell field components multiplied by Maxwell potential components;[10] compare §21.9 and §22.3 for the scalar case). Other classical fields can be treated in a similar way, but when the 'free field equations' are not linear (such as with general relativity) profound difficulties can arise. We may refer to non-linear fields as 'self-interacting' fields, and we can attribute such difficulties to the problems associated with quantization in the presence of interactions, which we shall be coming to very shortly.

26.4 Antiparticles in QFT

Before doing so, let us return to the issue of antiparticles. In Chapter 24, and again in §26.1, I stressed the importance of the antiparticle concept for QFT. How do antiparticles feature in the present QFT formalism? As remarked in §25.3, some particles are their own antiparticles, whereas most particles are not. Mathematically, the issue is whether or not the operation of complex conjugation, as applied directly to the classical field quantities (or to the 1-particle wavefunction), yields a quantity of the same kind as it was before, or not. In the case of a scalar field, this is usually (but not quite adequately) expressed as the issue of whether the classical field is a real field, or not. The complex fields are taken to be charged fields, where the complex phase angle (the $e^{i\theta}$ factor) is to be treated according to the 'gauge-field' prescriptions for electromagnetic interactions, as described in §15.8 and §19.4. The complex conjugate of such a field has the opposite charge, and so it is not a 'quantity of the same kind' (so, for example, we could not meaningfully add the field to its complex conjugate). In such situations, the particle and its antiparticle are certainly different. However, the complex nature—or 'charged' nature—of the classical field is not by any means the whole story. The uncharged K^0 meson, for example, differs from its antiparticle, whereas the uncharged π^0 meson is the same as its antiparticle. In both cases, the classical field would be a real scalar.

What about spinor (or fermion) fields? With Dirac's electron, its charge is sufficient to characterize its complex conjugate as having a different character from the original field. But in the case of neutrinos, if they are massive, there is more than one possibility. For example, in the situation of what is referred to as a *Majorana* spinor field, the (massive) neutrino would be its own antiparticle. (In the descriptions given in Chapter 25, the neutrino's zig would be the antineutrino's zag, and vice-versa.) According to current understanding,[11] neutrinos all differ from their antiparticles.

So, how does the QFT formalism deal with antiparticles? Consider the case when the operation of complex conjugation on the classical field—or on the 1-particle wavefunction ψ—yields a quantity of a different character, so the quantum particle differs from its antiparticle. (The situation where particle and antiparticle are the same is handled by our earlier discussion, in §26.2) Now an ordinary wavefunction ψ should have positive frequency, but we can also consider some quantity ϕ, of the same kind as ψ, but which is of negative frequency. Then the complex conjugate $\overline{\phi}$, of ϕ, would be a wavefunction of a different type from ψ, although both $\overline{\phi}$ and ψ are now of positive frequency. The quantity $\overline{\phi}$ would provide a wavefunction for a 1-antiparticle state. The corresponding creation operator for the antiparticle in that state would be $\overline{\boldsymbol{\Phi}}$ and for the annihilation operator, $\overline{\boldsymbol{\Phi}}^*$.

Let us try to make contact with Dirac's original 'sea' (described in §24.8) by thinking of the 'shadowy' state, over on the far right of all the operators, in Dirac's case, as being different from the usual vacuum state $|0\rangle$, where we recall that $|0\rangle$ is to be viewed as totally devoid of particles or antiparticles. Instead, we take this new 'vacuum' state to be Dirac's 'sea' itself, denoted by $|\Sigma\rangle$, which is to be completely full of all negative-energy electron states, but nothing else. Let us now consider the situation of a single positron which, in Dirac's original picture, is described by a single 'hole' in the negative-energy electron states. All the other negative-energy states are to be filled except for this particular missing one, which is given by some negative-frequency ϕ. The quantum-field-theoretic description, using this $|\Sigma\rangle$ vacuum, would be result of the annihilation operator $\boldsymbol{\Phi}^*$ acting on $|\Sigma\rangle$, since this operator removes the negative energy state ϕ from this vacuum, giving the total state $\boldsymbol{\Phi}^*|\Sigma\rangle$.[26.6]

If we were to use the description that employs the more usual vacuum $|0\rangle$, then we would think that, instead of removing the negative energy electron state ϕ, we are inserting the positron state with wavefunction $\overline{\phi}$. This is achieved by applying the creation operator $\overline{\boldsymbol{\Phi}}$ to $|0\rangle$, giving the total state $\overline{\boldsymbol{\Phi}}|0\rangle$. This does not *look* the same as the $\boldsymbol{\Phi}^*|\Sigma\rangle$ that we had with the 'Dirac sea' description, but there is a certain sense in which the states $\overline{\boldsymbol{\Phi}}|0\rangle$ and $\boldsymbol{\Phi}^*|\Sigma\rangle$ are basically equivalent. The operators $\overline{\boldsymbol{\Phi}}$ and $\boldsymbol{\Phi}^*$ both involve introducing the same algebraic quantity into the total state, namely that particular vector defined by $\overline{\phi}$ in the Hilbert space of 1-particle wavefunctions. The difference between the actual operators $\overline{\boldsymbol{\Phi}}$ and $\boldsymbol{\Phi}^*$ lies merely in the algebraic[26.7] way in which this Hilbert-space vector is deemed to act on the total state. Since we can always use anticommutation laws to move $\overline{\boldsymbol{\Phi}}$ or $\boldsymbol{\Phi}^*$ over to the far right, the way in which the $\overline{\phi}$ is deemed to act on the total state boils down to its action on the chosen vacuum state, $|0\rangle$ or

[26.6] Explain why we can remove a specific state in this way, despite my earlier qualifications about what an annihilation operator actually does. (*Hint*: See Exercise [26.2].)

[26.7] By referring back to Exercise [26.2] and Fig. 12.18, exhibit this algebraic difference in the abstract–index or diagrammatic notation.

$|\Sigma\rangle$ as the case may be, over on the far right. This information can be considered to be part of the specification of what this vacuum state actually means.

26.5 Alternative vacua

Some important comments should be made here about this situation in which there appear to be alternative choices for our 'vacuum state'. We shall find that the issue of 'alternative vacua' has some considerable importance in modern QFT. Let us consider the algebra $\mathcal{A}$, consisting of all operators $\boldsymbol{A}$ that can be constructed from algebraic expressions, or convergent power series expressions, in the creation and annihilation operators. Two proposals for a 'vacuum state', say $|0\rangle$ and $|\Sigma\rangle$, may have the property that there is no element of $\mathcal{A}$ that can be applied to either one of $|0\rangle$ or $|\Sigma\rangle$ in order to get the other. In such cases, the states $|0\rangle$ and $|\Sigma\rangle$ have to be regarded as belonging to different Hilbert spaces, and we are then likely to find that there are expressions of the form

$$\langle\Sigma|\boldsymbol{A}|0\rangle \text{ or } \langle 0|\boldsymbol{A}|\Sigma\rangle,$$

where $\boldsymbol{A}$ belongs to $\mathcal{A}$, which give us infinite answers, or no meaningful answer at all. This forbids us from constructing a consistent quantum theory in which both the states $|0\rangle$ and $|\Sigma\rangle$ appear. (Recall from the discussion of §22.5 that quantities like $\langle\Sigma|A|0\rangle$ are expressions of the general kind that we would need to use in order to compute probabilities; see §22.3 for the notation.)

The issue coming up here is a profound one in QFT, and it plays a vital role in modern approaches to particle physics. The 'choice of vacuum state' is a matter of importance comparable with (and complementary to) the choice of the algebra $\mathcal{A}$ generated by creation and annihilation operators, the latter defining, in a sense, the *dynamics* of the QFT. In the case of free electrons, the two vacuua that we have been considering, namely $|0\rangle$ (containing no particles and no antiparticles) and $|\Sigma\rangle$ (in which all the negative-energy particle states are filled) can be considered as being, in a sense, effectively equivalent despite the fact that $|0\rangle$ and $|\Sigma\rangle$ give us different Hilbert spaces. We can regard the difference between the $|\Sigma\rangle$ vacuum and the $|0\rangle$ vacuum as being just a matter of where we draw a line defining the 'zero of charge'.

Indeed, we might think that Dirac's sea is physically different from a proper vacuum, because the sea of negative-energy electrons should provide us with an enormous—indeed infinite—electric charge. For the state $|\Sigma\rangle$ to make physical sense, we have to 'renormalize' the charge so that the infinite total charge value of the 'sea' (in fact negatively infinite, the electron's charge being negative) counts as zero. A similar situation would come about if we were to consider the *mass* of Dirac's sea, where we might now be concerned with its (active) gravitational influence. The infinite total negative energy of Dirac's sea would (by $E = mc^2$) provide an infinite negative mass, which is physically nonsensical, just as in the case of infinite electric charge. Again, if we are to take Dirac's sea seriously, we must also renormalize the mass of the vacuum, by 'adding an infinite mass density' to that of the sea, so that the total provides us with the zero value that we require for the mass density of the observed vacuum.

The reader may well have formed the impression that this question of 'alternative vacua' and of this apparent need to 'renormalize' such things as charge and mass, adding in an infinite constant in order to get sensible physical answers, is merely an artefact of the strange 'sea' idea that Dirac originally found the need to introduce. However, we shall be finding that these two features are by no means specific to Dirac's extraordinary 'sea'. They appear to be ubiquitous in all serious approaches to a realistic theory of particle physics—at least as those approaches stand today. The standard model makes fundamental use both of renormalization and of alternative vacua. Far from being an anomaly of history, Dirac's sea serves as a kind of model that we do well to keep in mind when we try to move forward, at least within the scheme of things that is available to us today. The twin criteria of agreement with observation and of mathematical consistency, although incompletely fulfilled, have taken us on a route which, so far, has been dependent upon the ideas of renormalization and non-unique vacua.

26.6 Interactions: Lagrangians and path integrals

The difficulties that have led us in these directions arise from the problems that come about when we attempt to treat *interactions* within the framework of QFT. Indeed, my discussion so far has been basically concerned only with the case of free fields, and although I have not supplied all the details, I hope that the reader will trust me when I say that things proceed in an essentially trouble-free way when interactions are absent. States can be constructed in which there are superpositions of different numbers of particles and antiparticles, and even unlimited numbers of such particles. These states are obtained by acting on $|0\rangle$ with an arbitrary element of $\mathcal{A}$, i.e. an expression in creation and annihilation operators (polynomial or power series, paying due attention to convergence issues in the latter case). The space of such states is referred to as *Fock space* (after the Russian physicist V.A. Fock, who was one of the first to study such things), and it can be thought of as what is called a *direct sum*[12] (see §13.7) of Hilbert spaces with increasing numbers of particles. The number of particles in a state may be unlimited, such as with the *coherent states* which are, in a certain well-defined sense,[13] the most 'classical-like' of the quantum field states. These are states of the form

$$e^{\boldsymbol{\Xi}}|0\rangle,$$

where $\boldsymbol{\Xi}$ is the *field operator* associated with the particular field configuration $\boldsymbol{F}$ (which let us take to be a free real Maxwell field with suitable fall-off properties at infinity to ensure that a finite norm exists). We define $\boldsymbol{\Xi}$ to be the sum of the creation and annihilation operators (not normalized) corresponding to the positive- and negative-frequency parts of **F**, respectively.

We recall from §26.2 that the creation and annihilation operators satisfy certain commutation relations. It follows that the various components of the field operators do not generally commute with each other. For example, in the case of the electromagnetic field, the components of the operator defining the magnetic field **B** and those defining the electromagnetic potential **A** (see §§19.2,4) satisfy *canonical* commutation rules (like those between position and momentum of a particle, see §21.2).[14] It follows that Heisenberg uncertainly relations (see §21.11) must hold between these quantities, providing a limit to the accuracy that they can be simultaneously measured.

How, then do we deal with interactions? Crucial to the general framework of modern QFT is the *Lagrangian* (see §20.1), which is in many respects more appropriate than a Hamiltonian when we are concerned with a relativistic theory. As we recall from §21.2, §23.1, and Chapter 24, the standard Schrödinger/Hamiltonian quantization procedures lie uncomfortably with the spacetime symmetry of relativity. However, unlike the Hamiltonian, which is associated with a choice of time coordinate, the Lagrangian can be taken to be a completely relativistically invariant entity (see §20.5). How do we construct a QFT starting from a Lagrangian? The basic idea, like so many of the ideas underlying the formalism of quantum theory, is one that goes back to Dirac,[15] although the person who carried it through as a basis for relativistic quantum theory was the brilliant American physicist Richard Feynman.[16] Accordingly, it is commonly referred to as the formulation in terms of *Feynman path integrals* or Feynman *sum over histories*. It is also the basis of the Feynman graphs that we considered in Chapter 25.

The basic idea is a different perspective on the fundamental quantum-mechanical principle of complex linear superposition that we encountered earlier, and was made particularly explicit in §22.2. Here, we think of that principle as applied, not just to specific quantum states, but to entire

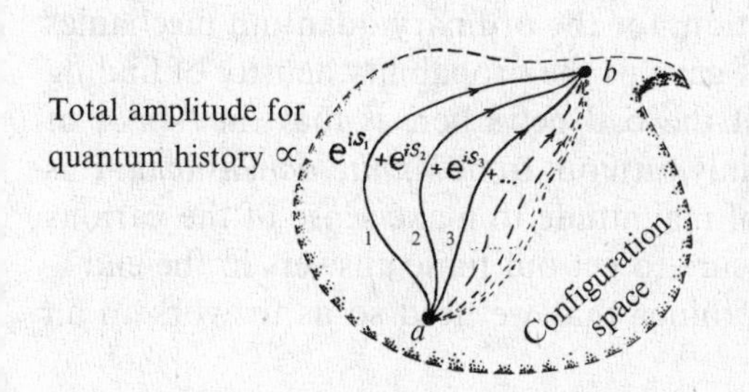

Fig. 26.2 In the path-integral approach to quantum theory and QFT, we consider quantum superpositions of alternative classical histories, a history being a path in configuration space, here taken between fixed points a and b. The amplitude assigned to such a path is $e^{iS/\hbar}$ (times a fixed constant), where the action S is the integral of the Lagrangian along the path, as in §20.1 (Fig. 20.3). The total amplitude to get from a to b is the sum of these.

spacetime histories. We tend to think of these histories as 'possible alternative classical trajectories' (in configuration space). The idea is that in the quantum world, instead of there being just one classical 'reality', represented by one such trajectory (one history), there is a great *complex superposition* of all these 'alternative realities' (superposed alternative histories). Accordingly, each history is to be assigned a complex weighting factor, which we refer to as an *amplitude* (§22.5) if the total is normalized to modulus unity, so the squared modulus of an amplitude gives us a probability. We are usually interested in amplitudes for getting from a point a to a point b in configuration space.

The magic role of the Lagrangian is that it tells us what amplitude is to be assigned to each such history; see Fig. 26.2. If we know the Lagrangian $\mathcal{L}$, then we can obtain the *action* S, for that history (the action being just the integral of $\mathcal{L}$ for that classical history, according to the prescription given in §20.5; see Fig. 20.3.). The complex amplitude to be assigned to that particular history is then given by the deceptively simple formula

$$\text{amplitude} \propto e^{iS/\hbar}.$$

Part of the deception, in the simplicity of this formula, lies in the fact that the 'amplitude' is not really a (complex) number, here (which, as written, would have to have unit modulus), but some kind of density. If we had just a discrete family of alternative classical histories, numbered 1, 2, 3, 4, ..., say, then we could imagine that the nth history could be assigned a genuine complex number α_n as its amplitude, whose squared modulus $|\alpha_n|^2$ could be interpreted as the probability of that history, in accordance with the rules of quantum measurement (§22.5), and we should normalize, to have $\sum |\alpha_n|^2 = 1$, summed over all the classical alternatives, to give total probability 1. But here we have a *continuous* infinity of classical alternatives. Our above 'amplitude' thus has to be thought of as an 'amplitude density', and we need something like $\int |\alpha(X)|^2 \mathrm{d}X = 1$, instead, where we have to integrate over the space of classical states to achieve our requirement that the total probability comes out as 1. This would not be particularly troublesome—we did this sort of thing before, in §21.9, in the case of a wavefunction, for the ordinary quantum mechanics of a point particle (where $|\psi(\boldsymbol{x})|^2$ gave us the probability density of finding the particle at the point $\boldsymbol{x}$). But the bad news here is that the 'space of classical paths' will almost certainly turn out to be *infinite-dimensional*. It is a problem of a different order of magnitude to make sense of the various quantities involved—and to be sure to get out finite answers in the end—when we have to define all the things that we need so as to work in an infinite-dimensional space.

The most accessible illustration of a path integral is the case of a single point particle moving in some field of force (so the configuration space is now space itself). Here, we consider all the various histories, starting at some spacetime point a and finishing at some other spacetime point b as in Fig. 26.3a. These histories are taken to be continuous spacetime paths winding their way from a to b. We do not require that the path be a 'legal' one, according to the rules of special relativity (i.e. that it be constrained to lie within the light cones, as required by classical relativity; see §17.8), nor do we even require that the path proceed entirely into the future. The 'history' can wiggle up and down in time if it wants to (Fig. 26.3b)![26.8] Let

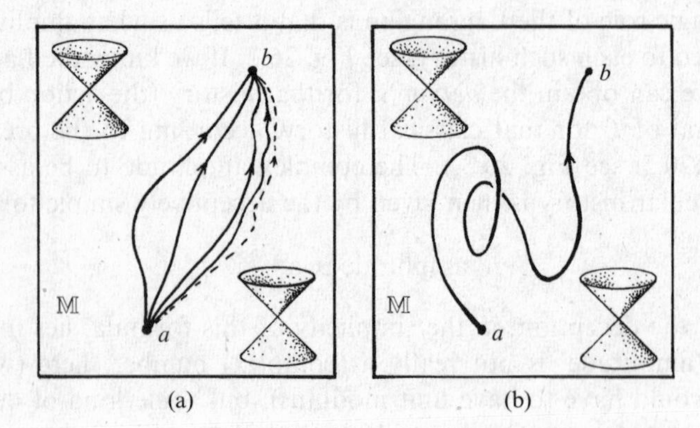

Fig. 26.3 (a) For a single structureless particle, a classical history is a curve in spacetime (here Minkowski space $\mathbb{M}$), taken between fixed events a, b. (b) The curve need not be a classically allowable smooth world-line, with tangent always future-timelike; it can even wander backwards and forwards in time.

[26.8] Give a 'physical interpretation' of the history of Fig. 26.3b, in terms of particle creation and annihilation.

us suppose that we have some Lagrangian $\mathcal{L}$, describing (in accordance with §20.1) the particle's kinetic energy minus the potential energy due to the field of force. For each history there will be some action S, where S is the integral of the Lagrangian along the path (recall Fig. 20.3). In classical mechanics, our friend Joseph L. Lagrange would have told us to search for a particular history for which the action integral is stationary (Hamilton's principle; see §20.1), this being the actual particle motion consistent with the classical motion under the given force. In the path-integral approach to quantum mechanics, we are to take a different view. *All* the histories are supposed to 'coexist' in quantum superposition, and each history is assigned an amplitude $e^{iS/\hbar}$. How are we to make contact with Lagrange's requirement, perhaps just in some approximate sense, that there should be a particular history singled out for which the action is indeed stationary?

The idea is that those histories within our superposition that are far away from a 'stationary-action' history will basically have their contributions cancel out with the contributions from neighbouring histories (Fig. 26.4a). This is because the changes in S that come about when the history is varied will produce phase angles $e^{iS/\hbar}$ that vary all around the clock, and so will cancel out on the average. (This applies, in particular, to the very 'non-physical' contributions coming from the wildly acausal histories of Fig. 26.3b.) Only if the history is very close to one for which the action is large and stationary (so the argument runs), will its contribution begin to be reinforced by those of its neighbours, rather than cancelled by them (Fig. 26.4b), because in this case there will be a large bunching of phase angles in the same direction.[26.9]

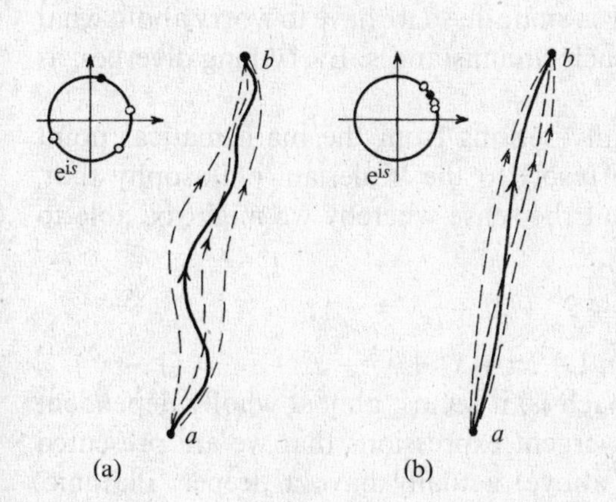

Fig. 26.4 The quantum 'Hamilton's principle'. (a) A history for which S is not stationary (and is large compared with $\hbar$). The values of $e^{iS/\hbar}$ for histories close by tend to vary greatly around the unit circle, and consequently there is much cancellation in the sum. (b) A history for which S is stationary (and large). For nearby histories, the values of $e^{iS/\hbar}$ do not change much, so there is much less cancellation.

[26.9] Try to make these statements more precise by referring to first-order changes in the path, using 'O' symbols (as in §14.5), and relating this to the discussion given in §20.1, concerning the meaning of 'stationary action'. (Assume that S is large in units of $\hbar$.)

This is indeed a very beautiful idea. In accordance with the 'path-integral' philosophy, not only should we obtain the classical history as the major contributor to the total amplitude—and therefore to the total probability—but also the smaller *quantum corrections* to this classical behaviour, arising from the histories that are not quite classical and give contributions that do not quite cancel out, which may often be experimentally observable. Although my descriptions above have been phrased in terms of a point particle moving in a field of force, the ideas apply extremely generally, and can be applied to the dynamics of fields as well as to the motion of particles. Again, the 'field-histories' that represent classical solutions of the field equations should emerge as providing the main contributions, and there will also be quantum corrections arising from the near-classical histories.

26.7 Divergent path integrals: Feynman's response

At least, that is what is *supposed* to happen. But does it? Are the crude descriptions that I have presented above mathematically justified? Even if not, and we barge through, ignoring mathematical niceties, do we get good physical answers that are in agreement with experiment?

I can only give very mixed answers to these questions. The issue of mathematical justification is particularly troublesome, and the fairest answer to give, on this point, would be: 'No; not as things stand today.' Even the case of the single point particle, as described above, is decidedly problematic. The space of paths is certainly infinite-dimensional,[26.10] and an appropriate 'measure' (the infinite-dimensional version of a volume) is required to handle this. It turns out that this measure is heavily weighted in favour of histories that are not even smooth, so we have to worry about what the Lagrangian even means in such circumstances. Everything diverges, as the definition stands.

These divergences are certainly serious from the mathematical point of view, and we may prefer to resort to the 'Eulerian' philosophy that, in §4.3, led us to speculate about the sense whereby we might be able to trust the 'nonsense' summation

$$1 + 2^2 + 2^4 + 2^6 + 2^8 + \ldots = -\tfrac{1}{3},$$

obtained by substituting $x = 2$ in $1 + x^2 + x^4 + x^6 + x^8 + \cdots = (1 - x^2)^{-1}$. Indeed, the path-integral approach is, it seems, almost wholly dependent upon a faith that the wildly divergent expressions that we are presented with (like the divergent series above) actually have a deeper 'Platonic' meaning that we may not yet properly perceive. We appear to be forced to admit that something of this nature must be the case because, on the

[26.10] Why?

physical side, we are not infrequently presented with answers of uncanny physical accuracy when (if I may be permitted to conjure up an improbable-sounding metaphor) we bulldoze our way through the mathematics with great sensitivity and precision! For example, these calculational procedures yield the correction factor 1.001 159 652 188, referred to in §24.7, to Dirac's original value for the electron's magnetic moment, providing an agreement between theory and observation[17] which is discrepant by less than 10^{-11}.

It is indeed remarkable how far the application of mathematical/physical sensitivity can lead us to excellent answers in many cases. A valuable first step towards making sense of such path integrals,[18] in the case of individual free quantum particles, is the replacement of the wild collection of histories by what is called the *Feynman propagator*.[19] This gives us the mathematical interpretation of one of the lines in a Feynman graph (such as those encountered in Chapter 25).

More specifically, let us consider a sum over histories for which some free particle is to start at a point p and to end at some other point q, in spacetime. In principle, we are to form the sum (integral) of all the $e^{iS/\hbar}$ for paths originating at p and terminating at q, but this is certainly wildly divergent, as it stands. On the other hand, we can suppose that the sum $K(p, q)$ has some kind of mathematical ('Eulerian/Platonic') existence, and we ask what formal algebraic and differential properties this sum ought to have, if it existed. These properties (including an appropriate 'positive-frequency' condition; see §24.3) fix the form of $K(p, q)$ uniquely (if we are reasonably fortunate in the example that we have chosen), and this gives us the Feynman propagator that we seek. In fact, it is more usual (though by no means essential[20]) to describe these things in momentum space rather than position space, the momentum-space descriptions looking significantly simpler.

In the case of a Dirac particle (e.g. an electron), the momentum-space propagator turns out to take the form $i(\not{P} - M + i\varepsilon)^{-1}$, where $\not{P} = \gamma^a P_a$ (see §§24.6,7), the quantity P_a being the 4-momentum that the particle happens to have, for the chosen path under consideration. The quantity 'ε' is taken to be a very small positive real number, which is a device geared to ensure the positive/negative-frequency requirements of the Feynman propagator. In the limit $\varepsilon \to 0$, it turns out that we get a singularity in the propagator—an infinite value—when the 'rest mass' $(P_a P^a)^{1/2}$ that the particle happens to have for that chosen path, takes the value M that is the particle's actual rest mass.[26.11] For a classical particle, we would require that this 'rest mass' does take this value, i.e. that $P_a P^a = M^2$, but with the

[26.11] Explain how this singularity arises, by first rewriting $(\not{P} - M + i\varepsilon)^{-1}$ as a quotient for which the denominator is $P_a P^a - M^2 - \varepsilon^2$.

quantum-mechanical sum over histories, we must allow that the particle feels out values of the momentum for which the rest mass comes out 'wrong'. Because of the singularity just referred to, however, we find that the amplitude gets very large when P_aP^a gets very close to the value M^2, so the classical value for the mass gives the dominant contribution. This is a feature that is not specific to a Dirac particle, and it applies quite generally.

26.8 Constructing Feynman graphs; the S-matrix

What has been described in the previous section is the first step towards obtaining a Feynman graph. This needs some further explanation. What we have found is a single line (segment) of such a graph. A particular such line in a Feynman graph would normally be just part of a complicated expression, involving other particle lines and various vertices where the lines come together. The contributions to the total amplitude coming from the *vertices* are normally[21] just simple factors, involving a scalar coupling constant (such as the electric charge) governing the stength of the interaction, perhaps a term such as γ_a that is needed to 'get the indices to match up', and a 'delta-function' term (§9.7) to ensure that the only non-zero contributions to the total amplitude arise when conservation of 4-momentum takes place at each vertex.[22] There will be various kinds of term arising from the different varieties of line in the graph (depending upon the spin and rest-mass value for the particle that the line represents). The *infinities* in the expression (apart from those of the delta-function terms, which are normally just regarded as providing constraints ensuring 4-momentum conservation) arise when the momenta P_a acquire the particular values expected for classical paths (basically $P_aP^a = M^2$). This makes sense because we expect classical behaviour to dominate the path integral. The presence of these singularities (the infinities apart from those in the delta functions) is therefore intimately related to the requirement that classical behaviour provides, in some rough sense, the leading contribution to the quantum-mechanical amplitude. Yet, there is danger lurking in these singularities, as we shall shortly see.

Just to emphasize the necessity of these singular expressions, I should point out that we cannot regard the condition $P_aP^a = M^2$ as a constraint (like the momentum conservation at vertices), because of the existence of basic processes such as that of Fig. 26.5, in which two electrons 'exchange a photon' (the photon being indicated by a wiggly line, as in §§25.3–5). This is the basic quantum-mechanical manifestation of the electrostatic (Coulomb) repulsion between the two negatively charged particles (Møller scattering). The two incoming lines (at the bottom of the graph) represent

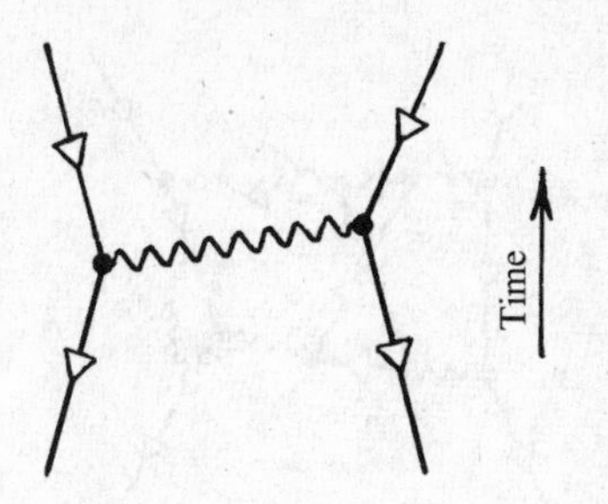

Fig. 26.5 Møller scattering of electrons: the most primitive quantum manifestation of the electrostatic (Coulomb) force between two charged particles. The electrostatic force comes about, here, from the 'exchange' of a single photon (wiggly line) between two electrons. The photon is necessarily 'off shell' and therefore virtual, as follows from 4-momentum conservation at each vertex.

the two electrons in their initial state and the two outgoing lines (at the top[23] of the graph) represent the electrons in their final state. These are taken as 'given things'—providing the *external* momenta—and are not to be 'integrated over' in calculating the final amplitude.

For these external states (and for these only) we take the momenta to satisfy the classical relation $P_a P^a = M^2$. We say that a particle's mass is *on shell* when this relation holds, this being the 'mass shell', which is the momentum-space version of the bowl-shaped hyperboloid depicted in Fig. 18.7. See Fig. 26.6. Real particles (the ones that are actually observed as free particles) are always on shell. However, with regard to the *internal* lines of a Feynman graph, we do not expect this on-shell requirement to hold. In particular, the exchanged photon, in the Feynman graph of Fig. 26.5 cannot be on shell (i.e. its 4-momentum does not satisfy $P_a P^a = 0$) whenever there is a non-trivial interaction.[26.12] Such off-shell particles are referred to as *virtual* particles, and they can occur only in the interior of a

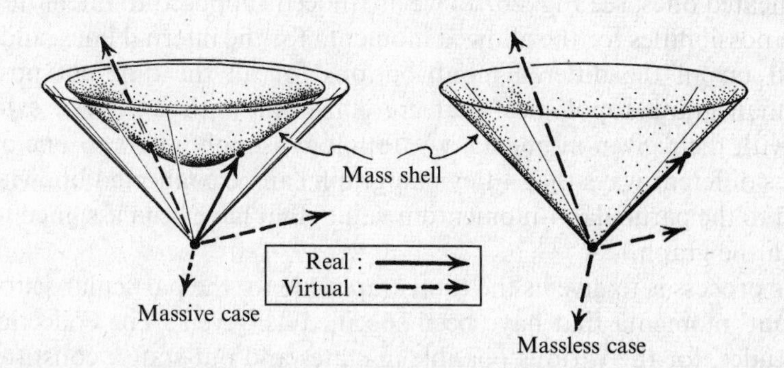

Fig. 26.6 The mass shell in momentum space. (Compare Figs. 18.7, 18.17.) For real (free) particles of rest-mass M, the 4-momentum p^a lies on the mass shell (so p^a is future timelike or future null with $p_a p^a = M^2$), but virtual particles, in the interior of a Feynman graph, can be 'off shell'.

[26.12] Why not? Explain how 4-momentum conservation at each vertex determines the 4-momentum of the virtual photon. *Hint*: All electrons have the same rest mass!

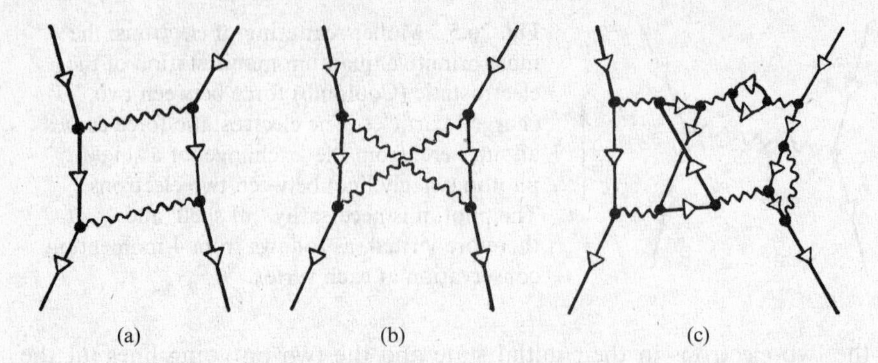

Fig. 26.7 Higher-order processes giving corrections to Møller scattering. (a) and (b) depict 2-photon exchanges between two electrons, while (c) is a much higher-order process involving internal pair creations and annihilations. Each such Feynman graph represents an integral, and the contributions from all of them must be added up.

Feynman graph. The exchanged photon of Fig. 26.5 is virtual, and it cannot 'escape' to be observed at large distances.

The process of Fig. 26.5 is rather special in that the state of the internal line (virtual photon) is completely fixed by the external lines. In this case, the 'integration over internal states' that is generally required is completely trivial, consisting of just a single term. However, in a more complicated process, such as those depicted in Fig. 26.7a,b, where *two* photons are exchanged, there is some freedom in the 4-momenta of the internal lines.[26.13] The idea is that in such cases (and in myriads of ever more complicated ones; see Fig. 26.7c) we are indeed supposed to integrate over all the possibilities for the allowed momenta for the internal lines, and also to add up all the different contributions for all the different possible 'Feynman graph topologies' that are consistent with the given external lines with their given momenta. (A 'topology' simply refers to one of the various different ways that a Feyman graph can be connected up, without regard to the particular 4-momentum values that have been assigned to the lines in the graph.)

This process is to give us the total amplitude for the particular set of 'in' and 'out' momenta that have been specified as 'given'. The collection of amplitudes, for the various possible in-states and out-states, constitutes a kind of matrix (though an infinite-dimensional one) whose 'rows' and 'columns' correspond to a basis for the out-states and in-states, respectively. This is referred to as the *scattering matrix* or, more usually, simply as the *S-matrix*. The calculation of the S-matrix is considered to be a major objective of QFT.[24]

[26.13] What is this freedom?

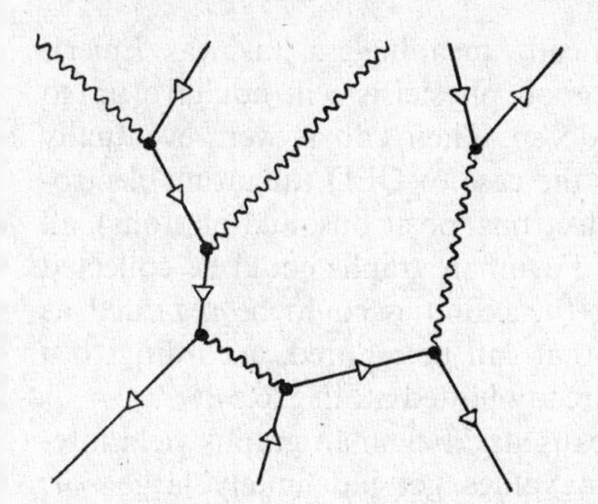

Fig. 26.8 A tree graph contains no loops. The internal momenta are consequently fixed by the external momenta, so no integration is involved. Tree graphs reproduce the classical theory.

The above procedure is a vast improvement, in calculational terms, over the original 'sum over histories', because in effect we have already effectively done the infinite-dimensional (and seemingly hopelessly divergent) path integrals that correspond to each separate line in the graph. Each choice of Feynman graph topology represents an ordinary finite-dimensional integral (like those considered in §12.6), and this is a considerable advance from the wildly divergent infinite-dimensional integrals that the direct interpretation of a path integral would lead us into. Moreover, these finite-dimensional integrals can be treated by the powerful methods of complex contour integration (as discussed in §7.2). Feynman's parameter ε, appearing in the propagator (see the last paragraph of §26.7), is really just a prescription for guiding the contour of integration to the appropriate side of the singularities that appear in the expressions.

Yet, we are very far from being 'out of the woods', because the merely finite-dimensional integral that we have been left with, for each Feynman graph topology, is itself going to be divergent, whenever there are *closed loops* in the Feynman graph. This would appear to be 'very bad news'; for it is only with closed loops that we begin to come to terms with the performing of any integration at all. In all other cases (i.e. the so-called 'tree graphs', which have no closed loops; see Fig. 26.8), the internal momenta are simply fixed by the external values. Tree graphs merely reproduce the *classical* theory!

26.9 Renormalization

So it seems that for all our (or, rather, Feynman's) efforts, we are still stuck with a divergent expression for the total amplitude of any genuinely quantum process. The weary reader may be legitimately wondering, at this stage, what good all this has done us. Indeed, from a strict mathematical standpoint, we have got officially 'nowhere', in the sense that all our expressions are still 'mathematically meaningless' (as was Euler's $1 + 2^2 + 2^4 + 2^6 + \cdots = -\frac{1}{3}$). Yet good physicists will not give up so easily. And they were right not to do so. Their efforts were eventually rewarded[25] when it emerged that, in the case of QED (quantum electrodynamics: the theory of interacting electrons, positrons, and photons), all the divergent parts of the individual Feynman graphs could be collected together in various 'parcels' so that the infinities could be regarded as merely providing 'rescaling' factors that can be ignored, according to a process known as *renormalization* (already hinted at, in §26.5).

These particular infinities arise because the Feynman graphs yield integrals that diverge when momentum values get indefinitely large—or, equivalently, when distances get indefinitely small. (Recall Heisenberg's momentum–position uncertainty relation $\Delta p\, \Delta x \geq \frac{1}{2}\hbar$; see §21.11.) The infinities are referred to as *ultraviolet* divergences. Although not the only divergences in QFT, they are considered to be the most serious ones. There are also *infrared* divergences that we can regard as coming about from indefinitely large distances (i.e. from indefinitely small momenta). These are usually regarded as 'curable' by various means, often by restricting the type of question that is regarded as being physically sensible to ask of a system.

In order to get some feeling for what is involved in the ultraviolet divergences, let us examine the physical meaning of the clearest instance of renormalization. This occurs with the value of the electric charge possessed by the electron. Imagine an electron to be a point charge, situated at some point E in space. There is an effect known as *vacuum polarization* which can be understood in the following way. We envisage that, at some point close to E, there might be the creation of a (virtual) pair of particles: an electron and positron, which after a very short period of time annihilate each other. (We regard this period of time as being short enough that the energy required to produce the pair falls within the uncertainties of Heisenberg's energy–time relation $\Delta E\, \Delta t \geq \frac{1}{2}\hbar$ (§21.11.) The Feynman graph for this process is indicated in Fig. 26.9a. The presence of the (virtual) photon line at the beginning (and also at the end) of this process is to indicate that the creation (and subsequent annihilation) occurs in the ambient electric field of the electron at E. (We could also contemplate completely disconnected Feynman 'loops', see Fig. 26.9b, in which the creation and annihilation processes take place without the presence of the ambient field of the electron at E; but such 'totally disconnected' processes are considered to have no physically observable effects.) The effect of this ambient field is that the created electron is slightly repelled by the electron at E, whereas the created positron is slightly attracted by it, so there is a slight physical separation between these charges during their momentary existence. This is happening all the

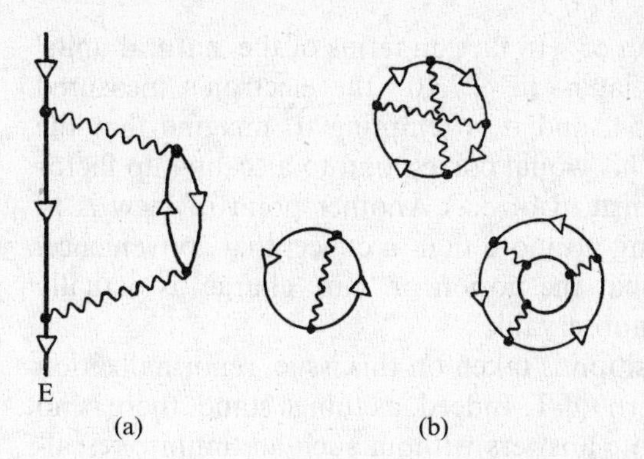

Fig. 26.9 (a) A Feynman graph involved in charge renormalization. This represents a positron–electron pair creation and subsequent annihilation in the field of a background electron (see Fig. 26.10). (b) Completely disconnected Feynman graphs. These are considered to have no directly observable effects.

time all around the electron at E, and the net effect, referred to as 'vacuum polarization', is to reduce[26] the apparent value of that electron's charge, as measured by its effect on other charges; Fig. 26.10.[26.14] The vacuum serves to 'shield' the electron's charge, and make it appear to have a smaller value—called the *dressed* value of the charge—than the 'actual' *bare* charge value of the electron. It is the dressed value that would be the one measured directly in physical experiments.

This seems all very reasonable. But the trouble is that the calculated numerical factor whereby the bare value must be scaled up, in relation to the dressed value, tuns out to be infinity! This infinity can be clearly identified as one of the infinities in the quantum electrodynamical calculation (basically diagrams like that of Fig. 26.9a and elaborations of it). One may take the view that according to some future theory, the divergent integrals should be replaced by something finite, perhaps because there is a 'cut-off' coming in at very small distances, i.e. to very large momenta (§21.11), and the correct renormalization factor should be some rather

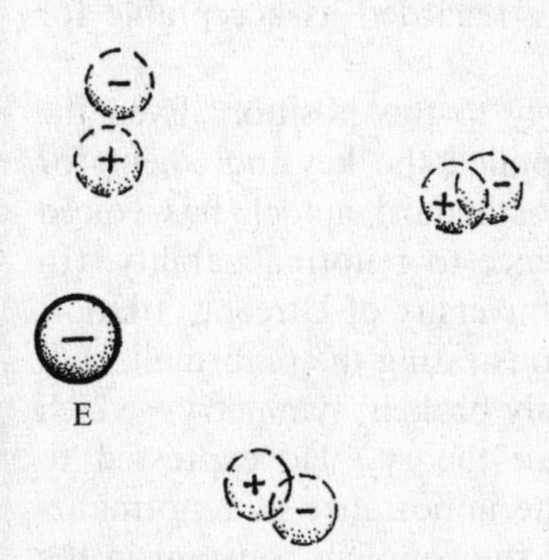

Fig. 26.10 Vacuum polarization: the physical basis of charge renormalization. The electron E induces a slight charge separation in virtual electron–positron pairs momentarily created out of the vacuum. This somewhat reduces E's effective charge from its bare value—unfortunately by an infinite factor, according to direct calculation.

[26.14] Can you see why this should be?

large finite number, rather than ∞. (In fact, in terms of the 'natural units' that we shall be coming to later—in §27.10—the electron's measured dressed charge is about 0.0854, and it is tempting to imagine that the bare value should be 1, say. This would correspond to a scaling up factor of 11.7062, or about $\sqrt{137}$, instead of ∞.) Another point of view is to regard the *bare* charge as being no more than a conceptual convenience, and to take the standpoint that the notion of 'bare charge' is actually 'meaningless', because it is 'unobservable'.

Whatever philosophical position is taken on this issue, renormalization is an essential feature of modern QFT. Indeed, as things stand, there is no accepted way of obtaining finite answers without such an 'infinite rescaling' procedure applied not necessarily only to charge, or to mass, but to other quantities also. Theories in which this kind of procedure works are called *renormalizable*. In a renormalizable QFT, it is possible to collect together all the divergent parts of the Feynman graphs into a finite number of 'parcels'[27] which can be 'scaled away' by renormalization, any remaining divergent expressions being deemed to cancel out with each other in accordance with certain overall principles (such as the symmetry principles that play important roles in the standard model). QED is a renormalizable theory, and so is the standard model as a whole. Most QFTs, on the other hand, are non-renormalizable. It is a common standpoint, among particle physicists, to take renormalizability as a selection principle for proposed theories. Accordingly any non-renormalizable theory would be automatically rejected as inappropriate to Nature. Indeed, this principle has provided a powerful guide towards the particular choice of theory that became the 20th century's standard model of particle physics which we encountered in Chapter 25. Thus, on this viewpoint, the prevalence of infinities in QFTs is not a 'bad' thing at all, but is a feature that can be turned powerfully to our advantage.[28] Very few theories pass the test of renormalizability, and only those that do pass have a chance of being regarded as acceptable for physics.

Yet, not all physicists subscribe rigorously to this position. Even the Nobel Laureate Gerard 't Hooft, who supplied the key ingredient for demonstrating the renormalizability of the standard model, has voiced certain reservations about the strict adherence to renormalizability. (In 1971, while still a graduate student at the University of Utrecht, 't Hooft had shaken the physics community by demonstrating the renormalizability of theories where there is a 'spontaneously broken' symmetry—which became an essential feature of electroweak theory.) He expressed to me, on one occasion, his viewpoint that the importance of renormalizabiliy to a theory depends upon the size of the coupling constant in the interaction under consideration. He referred specifically to gravity, which is extraordinarily weak compared with the forces of particle physics, yet its quantum theory turns out to be non-renormalizable according to standard approaches to the quantization of Einstein's equation for general relativity; see §19.6 and §31.1. (The gravitational attraction between the electron and proton in a hydrogen atom is weaker than the electric force by a factor of some 10^{-40}, which makes gravity almost unimaginably weaker than the 'weak interactions' of radioactive decay.) His remarks expressed what might be called a *pragmatic* view of QFT. Even renormalizable theories are not free from infinities, an issue that I shall elaborate upon in a moment. His point was to question whether infinities, potentially present in a theory, are actually physically relevant at energies even remotely accessible to experiment. In the case of a 'quantized gravity', the relevant extreme energies are ridiculously beyond what is feasible, and many other uncertainties in physical theory would enter the picture long before gravity's non-renormalizability would begin to make its mark.

On the other end of the scale, he argued, we have the *strong* interactions, with a coupling constant so large that it is doubtful that a description solely in terms of Feynman graphs is fully useful, because the series of increasing terms would diverge too wildly. Renormalizability alone is insufficient to ensure that quantum chromodynamics can supply finite answers. In this case, one takes advantage of what is called the *asymptotic freedom* of the strong force (work gaining the 2004 Nobel Prize for Gross, Politzer, and Wilczek). For very large momenta—which, in quantum theory, amounts to very tiny distances—the strong force has the remarkable property that it effectively disappears. This is in complete contrast with the familiar electric or gravitational force between particles, where the inverse square law tells us that the force increases when the distance gets smaller. The strong force is more like an elastic band, where the strength of the force increases in proportion to the distance of stretch, and it drops to zero when the distance becomes zero.[29] This force law is held responsible for the fact—referred to as *confinement* (see §§25.7,8)—that quarks cannot be individually pulled out of a hadron. Unlike an ordinary elastic band, the strong force cannot 'snap', although if you pull hard enough, other entities such as anti-quarks or quark pairs can be dragged out of the vacuum as well—which is the kind of thing that happens with the 'jets' that can occur in particle accelerators. This remarkable property of asymptotic freedom is what saves the theory of strong interactions from being calculationally useless, despite its renormalizability. For the record, the strong coupling constant is about 10, which may be contrasted with the electromagnetic coupling constant—the so-called *fine-structure constant*—which is about $\frac{1}{137}$, and the weak force, though not directly numerically comparable, is vastly weaker (see also §31.1).

26.10 Feynman graphs from Lagrangians

In my descriptions of Feynman graphs, renormalization, etc., I have rather leaped ahead, and have not explained how these diagrams are obtained, for any particular field theory. Nor have I related the Feynman-graph description to the general formalism of QFT with which this chapter began. Let me now go a little way towards rectifying this omission and make clearer the status of Feynman graphs within the general framework of QFT.

The starting point would be a Lagrangian, appropriate to the theory under consideration. The Feynman graphs then represent a *perturbation expansion* of the quantum theory associated with that Lagrangian. A perturbation expansion is basically just a power series expansion, in terms of some parameter (or family of parameters) that we normally think of as small. This type of expansion is the same sort of thing that was discussed in §4.3, where a function $f(x)$ is expanded as a power series in x. The analogue of x for the Feynman graphs would normally be some coupling constant. In the case of QED, for example, this parameter would be electric charge e. For each vertex of a Feynman graph there would be a factor of e, so the terms of the series would be graphs with increasing numbers of vertices, where the graphs with n vertices would, together, provide the coefficient of e^n. For theories with more than one coupling constant, we would get a more complicated power series, in more than one variable. An example would be a version of QED in which the electron lines of the standard approach are replaced by zigzags, in accordance with Figs. 25.2 and 25.3b. The two 'coupling constants' would then be the electric charge and the mass M of the electron.

I have remarked that renormalizable theories are not necessarily finite. Even that archetypal renormalizable theory, QED, is not actually a finite theory, even after renormalization. How can this be? Renormalization refers to the removal of infinities from finite collections of Feynman graphs. It does not tell us that the *summation* of all these resulting finite quantities is actually convergent. What QED gives us is a power series like $f_0 + f_1 e + f_2 e^2 + f_3 e^3 + \cdots$, where each of the coefficients $f_0, f_1, f_2, f_3, \ldots$ is a finite quantity, obtained from the accepted 'renormalizing' procedures of working out Feynman graph integrals, at each successive order, 0, 1, 2, 3, ... (In fact, only even or only odd powers would appear, in any particular case.[26.15]) Renormalizability does not tell us that the sum of the entire series is finite. In fact it is not finite but has a 'logarithmic divergence' (like the series $1 + \frac{1}{2} + \frac{1}{3} + \frac{1}{4} + \cdots$ for $-\log(1 - x)$ at $x = 1$) which, for QED, does not begin to show up until

[26.15] Can you see why?

we reach terms of order 137, or so, which is far beyond what is normally considered to be relevant.

For a general quantum field theory, to work out exactly what graphs occur at each order, we need to appeal to the original path-integral expression, even though this expression represents something that would be very badly divergent, if we tried to sum it directly. The procedure is to treat the path integral as an entirely *formal* quantity, to which straightforward but formal functional derivative procedures are applied (§20.5). Feynman graphs with successively more vertices are obtained when higher and higher order functional derivatives are performed. I do not propose to go into this matter in any more detail here, except to say that the generation of the Feynman graphs is unambiguous,[30] according to this formal procedure. The Lagrangian will, of course, be a *field* Lagrangian, of the general type discussed in §20.5. The 'path' that is involved, for such a Lagrangian, would not be an ordinary 1-dimensional curve, say in some infinite-dimensional configuration space. For a fully relativistically invariant picture, the 'history' must be an entire 4-dimensional field configuration in a specified spacetime region. The integral of the Lagrangian density over that region would be the action S, and $e^{iS/\hbar}$ then provides the amplitude (density) that is to be assigned to that particular configuration.

26.11 Feynman graphs and the choice of vacuum

For a theory with a symmetry under some group, like the U(2) symmetry of electroweak theory or the SU(3) symmetry of quantum chromodynamics or both, this symmetry would normally be taken to be a manifest symmetry of the Lagangian. The presence of such a symmetry would be important for the renormalizability of the QFT. Roughly speaking, the symmetry is used in order to ensure that certain divergent terms cancel each other out, the cancellation occurring (or being deemed to occur) because if there were to be a surviving diverging expression, such an expression could not share the postulated symmetry of the theory.

At least, that is the general idea. In the case of electroweak theory, however, there is another subtlety, because the resulting theory does not, after all, possess the originally postulated U(2) symmetry.[31] The lack of U(2) symmetry is taken to be the result of symmetry breaking (§25.5), but to understand how this is to be accommodated, we need to return to our general quantum-field-theoretic formalism. The basic idea is that the breaking of the symmetry is accounted for by a U(2)-*a*symmetric choice of *vacuum state*. Accordingly, the shadowy '$|0\rangle$' state that is imagined to be over to the far right of all the creation and annihilation operators, and which has been largely ignored in our considerations of Feynman graphs up to this point, must begin to emerge from its shadows.

First, we shall need to see, very roughly, how to relate the elements of the QFT algebra $\mathcal{A}$ to Feynman graphs. A key point of importance is that the *Feynman propagators*, which the lines of a Feynman graph represent, are basically the values of the *commutators* or *anticommutators* that we came across in §26.2 (i.e. the '$\langle\psi|\phi\rangle$' in these expressions). In practice, these are normally expressed in terms of momentum space—although there are some subtleties in defining the precise Feynman propagators, arising from the positive/negative-frequency issues (which may perhaps be best understood from the hyperfunctional perspective of §9.7). Let us not concern ourselves with these subtleties here.

Now, suppose we are concerned with a situation which starts off with a certain collection of incoming particles, and where some collection of outgoing particles finally emerges. We start with the vacuum state $|0\rangle$, and then apply the various creation operators that are needed to produce the required state for the incoming particles. This procedure yields the initial state $|\psi_{\text{in}}\rangle$. Similarly, we can adopt the same procedure, but now using the creation operators for the outgoing particles, again acting on $|0\rangle$, so as to produce the final state $|\psi_{\text{out}}\rangle$. The amplitude $\langle\psi_{\text{out}}|\psi_{\text{in}}\rangle$ is what we wish to calculate, from which we can obtain the probability of getting from 'in' to 'out' by simply using the standard formula, given in §22.5, which is just $|\langle\psi_{\text{out}}|\psi_{\text{in}}\rangle|^2$, if the states are normalized.

Now, the expression $\langle\psi_{\text{out}}|\psi_{\text{in}}\rangle$ involves annihilation operators on the left (because the Hermitian conjugation involved in passing from $|\psi_{\text{out}}\rangle$ to $\langle\psi_{\text{out}}|$ changes all the creation operators into annihilation operators). These all lie to the left of the creation operators in $|\psi_{\text{in}}\rangle$, so we can envisage 'pushing' all these annihilation operators through the creation operators on their right, until they hit the $|0\rangle$ at the far right. Whenever this happens, the $|0\rangle$ gets 'killed' (see §26.2), so the expression obtained is zero. But each time we push an annihilation operator through a creation operator we must take into account the commutator (and positive/negative frequency requirements) referred to above, giving us one line of a Feynman graph, as I have indicated. Each time we do this, another such line appears. Finally, all we get is $\langle 0|0\rangle$ multiplied by the collection of Feynman propagators representing the lines of a Feynman graph—and $\langle 0|0\rangle = 1$, for a normalized vacuum state, so we just get the Feynman graph itself.

So far, our Feynman graph is completely trivial, having no vertices at all—but this is because I have not included any interactions in the algebra of operators $\mathcal{A}$. To do so, we would need to examine the specific Lagrangian that is relevant to our particular problem and use it to generate the correct $\mathcal{A}$. Basically, these procedures would just mirror those referred to in §26.10 for generating Feynman graphs, with their appropriate vertex terms.

So far, we may have gained little, but an advantage of bringing our Feynman graphs into the general framework of QFT is that now we can replace the vacuum state $|0\rangle$ by an alternative vacuum $|\Theta\rangle$, which can be inequivalent to it (as with the Dirac sea state $|\Sigma\rangle$ that we considered in §26.4). The virtue of this, with regard to electroweak theory, and other theories which depend crucially on a fundamentally broken symmetry, is that whereas the Lagrangian—and consequently the Feynman graphs of the theory—are subject to an exact symmetry (the group U(2), in the case of electroweak theory), the actual states of the system are subject to only a lower symmetry (the gauge group U(1) of electromagnetism, in the case of electroweak theory), because the vacuum state $|\Theta\rangle$ possesses only this lower symmetry. By this means, the renormalizability of the theory that the full unbroken symmetry confers is undisturbed, despite the fact that the theory as a whole exhibits only a smaller 'broken' symmetry group.

This is clearly a marvellous device for producing physical theories which can reap the benefit of an exact symmetry while the observational situation is one in which the symmetry is far from satisfied. It is the kind of thing that has provided a great temptation to physicists in their further gropings for better and deeper schemes. Indeed, all the modern ideas for going beyond the standard model try to take advantage of this type of 'symmetry breaking'. Yet, all such attempts, no matter how popular—such as those that I shall be addressing in §§28.1–5—must still be regarded as very speculative. We shall need to keep a critical and skeptical eye on proposals of this nature, lest we get carried away too easily.

As a prelude to addressing some of these proposals, we shall need to gain some acquaintance with the Big Bang in the next chapter. Then in Chapter 28, we shall try to come to terms with some of the alarming issues that can accompany the idea of spontaneous symmetry breaking in the particular context of the early universe. Finally, we shall need to brace ourselves even harder, for the needed uses of this ubiquitous idea, when, in Chapter 31, we come to examine supersymmetry, the original ideas of string theory, and then some of their extraordinary descendents.

Notes

Section 26.1

26.1. See Aitchison and Hey (2004); or Zee (2003).

Section 26.2

26.2. I am going to be a little 'non-standard' in my descriptions here, by allowing the 'wavefunction' ψ to be just a *general*, not necessarily normalized, positive-frequency field. The lack of normalization correspondingly also applies to the creation operator $\boldsymbol{\Psi}$ (and to the annihilation operator $\boldsymbol{\Psi}^*$). In many conventional descriptions, ψ would be taken to be some momentum state.

26.3. In much standard literature, the symbol $\boldsymbol{a}$ is used for an annihilation operator, where $\boldsymbol{a}^\dagger$ (the Hermitian conjugate of $\boldsymbol{a}$) is used for the corresponding creation operator, and a momentum-space description is normally adopted; see Shankar (1994) and Zee (2003).

26.4. Some readers, familiar with the standard literature, may be confused by this because it is frequently the case that the creation and annihilation operators that are used are restricted to be those for the various different momentum states, which form an *orthogonal basis*. In that case, the annihilation operators *do* remove specific states.

26.5. See Zee (2003); Peskin and Schröder (1995).

26.6. Zee (2003) for an incisive demonstration of this requirement.

26.7. There are also some intriguing topological issues that interconnect particle exchange with 2π rotation, but the full implication of these with regard to QFT remains unclear. See Finkelstein and Rubinstein (1968); Feynman (1987); Berry and Robbins (1997).

Section 26.3

26.8. See Landsman (1998) for a rather challenging technical reference; also Ashtekar and Magnon (1980).

26.9. Perhaps written in terms of a potential.

26.10. General issues of state normalisation are treated, for example, in Ryder (1996). Quantization of the electromagnetic field is given a somewhat more traditional treatment in Shankar (1994).

Section 26.4

26.11. See Shrock (2003) for some of the latest news on neutrinos—currently a very 'hot' area in physics!

Section 26.6

26.12. The Fock space for the simple case of a boson field, where the particle is its own antiparticle, can be written $\mathbb{C}\oplus\mathcal{H}\oplus\{\mathcal{H}\odot\mathcal{H}\}\oplus\{\mathcal{H}\odot\mathcal{H}\odot\mathcal{H}\}\oplus\{\mathcal{H}\odot\mathcal{H}\odot\mathcal{H}\odot\mathcal{H}\}$ $\oplus\ldots$, where the direct sum operation is denoted by $\oplus$ and where the symbol $\odot$ denotes *symmetrized* tensor product. More complicated cases where there is spin and charge, etc., can be treated correspondingly. See Shankar (1994) for the general idea; Davydov (1976) may also be useful.

26.13. See Hannabuss (1997); Shankar (1994) for a discussion of coherent states—which can come in many varieties (fermionic, spin, etc.).

26.14. See Wald (1994); Birrell and Davies (1984).

26.15. See Dirac (1933); Schwinger (1958).

26.16. See Feynman (1948, 1949). Feynman and Hibbs (1965) is an excellent overview of the idea. Schwinger's competing approach to quantum electrodynamics (see, for example, Schwinger 1951) was in many ways more rigorous, but most workers use the more intuitive picture of path integrals and Feynman diagrams today.

Section 26.7

26.17. As has been pointed out by Feynman, this degree of precision would determine the distance between Los Angeles and New York to less than the thickness of a human hair!

26.18. Some other noteworthy ideas, such as so-called 'Euclideanization' will be discussed in §28.9.

26.19. This is an example of what is called a *Green's function* (after the very remarkable miller's son and self-taught English mathematician George Green; 1793–1841). The Feynman propagator is a particular Green's function $K(p, q)$, defined by the positive-frequency requirements of quantum theory referred to in §24.3.

26.20. See, for example, the classic Bjorken and Drell (1965).

Section 26.8

26.21. There are quantities referred to as 'running coupling constants' which have functional dependence on the rest-energy of the total system of incoming particles at a Feynman vertex. These have a significance in many modern theories of particle physics.

26.22. Thus, if $P_a^{(1)}, P_a^{(2)}, \ldots$ are the ingoing momenta and $Q_a^{(1)}, Q_a^{(2)}, \ldots$ the outgoing momenta, at a vertex, then include the term $\delta(P_a^{(1)} + P_a^{(2)} + \cdots - Q_a^{(1)} - Q_a^{(2)} - \cdots)$.

26.23. See Note 25.6.

26.24. The important concept of an 'S-matrix' (due, basically, to Heisenberg and the highly original American physicist J.A. Wheeler) is not tied to the notion of a Feynman graph, and it may be evaluated by some other means. See Eden *et al.* (2002).

Section 26.9

26.25. See Zee (2003); or Ryder (1996) for more (gory) detail.

26.26. Since the electron's charge is negative, 'reduce', here, means 'make the modulus smaller'.

26.27. There are certain elegant mathematical procedures, geared to the systematizing of this method, that take advantage of the notion of a 'co-product', related to the ideas of non-commutative geometry dicussed briefly in §33.1; see Connes and Kreimer (1998).

26.28. One important body of techniques is that supplied by the notion of the 'renormalization group'. Zee (2003) and Ryder (1996) as well as Peskin and Schröder (1995) treat these ideas; their bearing on statistical mechanics is elucidated in the encyclopaedic Zinn-Justin (1996).

26.29. Gravitational forces are still noticeable (even to beyond galactic scales), despite their falling off with distance, according to the inverse square law. The reader might worry why the strong force, on the other hand, is hardly noticeable at all at greater than nuclear distances even though it actually *increases* with distance. The reason is that, whereas gravitation accumulates, being always attractive, the strong force is a composite of attractive and repulsive components which necessarily cancel one another completely between separated nuclei (individual nucleus being necessarily 'colour singlets').

Section 26.10

26.30. Zee (2003) and Zinn-Justin (1996) teach the algorithm; for a rather amusing and intuitive take, Mattuck (1976) is also recommended.

Section 26.11

26.31. See Note 25.12.

27
The Big Bang and its thermodynamic legacy

27.1 Time symmetry in dynamical evolution

WHAT sorts of laws shape the universe with all its contents? The answer provided by practically all successful physical theories, from the time of Galileo onwards, would be given in the form of a *dynamics*—that is, a specification of how a physical system will develop with time, given the physical state of the system at one particular time. These theories do not tell us what the world is like; they say, instead: 'if the world was like such-and-such at one time, then it will be like so-and-so at some later time'. Such a theory will not tell us how the world *is* shaped unless we tell it how the world *was* shaped.

There have been important exceptions to this form of things, such as Kepler's wonderful conclusion, in 1609, that the orbits of the planets about the Sun have certain geometrical shapes—ellipses with the Sun at one focus—described with speeds satisfying specific rules. That was an assertion about how the universe *is*, rather than how its state might develop from moment to moment, in accordance with some dynamical law. But our present perspective on Kepler's geometrical motions is that they are mere consequences of 17th century gravitational dynamics, as first shown by Newton and published in his great *Principia* of 1687, and Kepler's laws are not to be thought of as directly fundamental to the ways of Nature. Indeed, it could be argued that Kepler—and science as a whole—was immensely fortunate that the nature of the specific law of force governing Newton's gravity, the inverse square law (§17.3), has the property that all the orbits of small bodies about a centre of force are actually simple and elegant mathematical shapes (and, indeed, shapes that had been intensively studied by the ancient Greeks, some eighteen centuries earlier). For this is a very exceptional property, shared by hardly any other simple central force law. In general, our modern perspective holds that it is the *dynamical laws* that we expect to have an elegant mathematical form, and it is a matter of good fortune for us if we happen to find simple mathematical shapes as consequences of these laws.

The usual way of thinking about how these dynamical laws act is that it is the choice of initial conditions that determines which particular realization of the dynamics happens to occur. Normally, one thinks in terms of systems evolving into the future, from data specified in the past, where the particular evolution that takes place is determined by differential equations. (These would be partial differential equations—field equations—when there are dynamically evolving fields or wavefunctions; see §10.2, §§19.2,6, §21.3, Exercise [19.2], and Note 21.1) One does not, on the other hand, tend to think of evolving these same equations into the *past*, despite the fact that the dynamical equations of classical and quantum mechanics are symmetrical under a reversal of the direction of time! As far as the mathematics is concerned, one can just as well specify final conditions, at some remote future time, and evolve backwards in time. Mathematically, final conditions are just as good as initial ones for determining the evolution of a system.

Some comments are called for, concerning this time-symmetrical dynamical determinism. First, the reader may be reassured that it is not substantially invalidated by the framework of either special or general relativity. Data defining the state of the system are specified at some initial 'time', which is some initial spacelike 3-surface, and these data evolve according to the dynamical equations to determine the physical state of the system to the future, and also to the past, of that 3-surface. There are, however, some new issues that are raised by *general* relativity, because the very structure of the spacetime into which the evolution flows is part of the physical state to be determined. (This has particular implications in the context of black holes, that we shall need to confront later; see §§27.8,9, §28.8, §§30.4,9.)

In the case of quantum mechanics, the determinism refers to the **U** part of that theory only, the quantum state being taken to be governed by Schrödinger's equation (or equivalent). Under time reversal—the T referred to in §25.4—the time-derivative operator $i\hbar\partial/\partial t$ of Schrödinger's equation (§21.3) must be replaced by $-i\hbar\partial/\partial t$ (since $t \mapsto -t$). Provided that the Hamiltonian is an ordinary one, which goes to itself under the action of T, we see that Schrödinger evolution also goes to itself, so long as we accompany the time reversal $t \mapsto -t$ by a reversal of the sign of the imaginary unit: $\mathrm{i} \mapsto -\mathrm{i}$. Indeed, this is how we think of the action of T in quantum mechanics. (We may note that a positive-frequency function $f(t)$ is converted back to a positive-frequency function under the combined replacements $t \mapsto -t$ and $\mathrm{i} \mapsto -\mathrm{i}$ so all is well, in this respect.[27.1]) The behaviour of quantum state reduction **R** under the action of T is another matter, however, and it will provide an important issue for our deliberations in Chapter 30 (§30.3).

[27.1] Why? Also, explain why spatial momentum is handled consistently by this replacement.

27.2 Submicroscopic ingredients

There are, however, other questions that might worry the knowing reader, even just with regard to classical dynamics. Time-reversal symmetry is certainly true of the submicroscopic dynamics of individual particles and their accompanying fields, in classical mechanics. But in practice, one has little knowledge of the behaviour of the individual ingredients of a system. A knowledge of the detailed position and momentum of every particle is normally deemed to be both unobtainable and unnecessary, the overall behaviour of the system being well enough described in terms merely of some appropriate averages of the physical parameters of individual particles. These would be things like the distribution of mass, momentum, and energy, the location and velocity of the centre of mass, the temperature and pressure at different places, the elasticity properties, the moments of inertia, the detailed overall shape and its orientation in space, etc. An important issue, therefore, is whether or not a good initial knowledge of such averaged 'overall' parameters will, in practice, suffice for determining the dynamical behaviour of the system to an adequate degree.

This is certainly not always the case. Systems known as *chaotic* have the property that the final behaviour depends critically on exactly how they are started off. As a familiar example, there is an 'executive toy' in which a magnetic pendulum swings just above a collection of magnets placed in some arrangement on the base. See Fig. 27.1. The dynamical behaviour is well-enough governed, in a deterministic way, by Newton's laws and the laws of magnetostatics, together with the slowing down, due to frictional resistance of the air. Yet the final resting place of the pendulum depends so critically upon the initial state that it is effectively unpredictable, although a fully detailed knowledge of this initial state, with all the constituent particles and fields would certainly fix this evolution uniquely.[1] Many other examples of such 'chaotic systems' are known. A good measure of the vagary of weather prediction is commonly attributed to the chaotic nature of the dynamical systems involved. Even the highly ordered (and very predictable) Newtonian gravitational motion of bodies in the solar

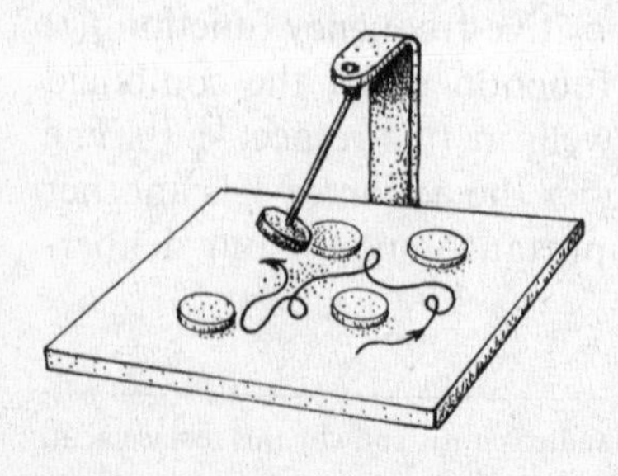

Fig. 27.1 Chaotic motion. An 'executive toy' consisting of a magnetic pendulum swinging just above a collection of fixed magnets. The actual path taken by the pendulum depends extremely sensitively on its initial position and velocity.

system (probably) constitutes, technically, a chaotic system, although the timescales that are relevant to such 'chaos' are vastly longer than those of astronomical observation.

What about evolution into the past, rather than the future? It would be a fair comment that such 'chaotic unpredictability' is normally much worse for the 'retrodiction' that is involved in past-directed evolution than for the 'prediction' of the normal future-directed evolution. This has to do with the *Second Law of thermodynamics*, which in its simplest form basically asserts:

Heat flows from a hotter to a colder body.

In accordance with this law, if we connect a hot body to a cold one using some heat-conducting material, then the hot body will become cooler and the cold body warmer until they settle down to the same temperature.[2] This is the expectation of *prediction*, and this evolution has a deterministic character. If, on the other hand, we view this process in the reverse-time direction, then we find the two bodies at effectively the same temperature spontaneously evolving to bodies of unequal temperature, and it would be a practical impossibility to decide which body will get hotter and which colder, how much, and when. This procedure of dynamical *retrodiction*, for this system, is clearly a hopeless prospect in practice.

In fact, this difficulty would apply to the retrodiction of almost any macroscopic system, with large numbers of constituent particles, behaving in accordance with the Second Law. For this kind of reason, physics is normally concerned with prediction, rather than retrodiction.[3] As another aspect of this, the Second Law is considered an essential ingredient to the predictive power of physics, as it removes those problems that we just encountered with retrodiction.

Nevertheless, many physicists would take the view that this law is not 'fundamental' in the same sense that, say, the law of conservation of energy, the principle of linear superposition in quantum mechanics, and perhaps the standard model of particle physics are fundamental. They would argue that the Second Law is an almost 'obvious' necessary ingredient to any sensible physical theory. Many would take the view that it is something vague and imprecise, and that it in no way can compare with the extraordinary precision that we find in the dynamical laws that control fundamental physics. I wish to argue very differently, and to demonstrate the almost 'mind-blowing' precision that lies behind that seemingly vague statistical principle that we usually simply refer to as the 'Second Law'.

27.3 Entropy

Let us examine, somewhat more exactly, what the Second Law actually states. As a preliminary, I should inform the reader of the First Law of thermodynamics. The First Law is simply the statement that the total energy is conserved in any isolated system. The reader might well complain that this is hardly something new (§18.6, §20.4, §21.3). But when this law was put forward (initially by Sadi Carnot in the early 1820s, although not published by him[4]), it had not been clear, previously, that heat is just a form of energy—nor was the ordinary macroscopic notion of energy itself completely clear. The first law makes it explicit that the total energy is not lost when, say, a body loses its kinetic energy (§18.6) as it slows down because of air resistance. For this energy is simply taken up in heating the air and the body. This *heat* energy is understood as (primarily) kinetic energy in the motions of air molecules and vibrations of particles composing the body. Moreover, *temperature* is simply a measure of energy per degree of freedom, so the thermodynamic notions of heat and temperature are basically the same as previously understood dynamical notions, but applied at the level of the individual constituents of materials and treated in a statistical way. The First Law has the kind of precision that we are familiar with: the value of something, namely the total energy, remains constant despite the fact that all kinds of complicated processes may be taking place. The total energy after the process is equal to the total energy before the process.

Whereas the First Law is an equality, the Second Law is an inequality. It tells us that a different quantity, known as the *entropy* has a larger (or, at least, not smaller) value after some process takes place than it had before. Entropy is, very roughly speaking, a measure of the 'randomness' in the system. Our body moving through the air starts with its energy in an organized form (its kinetic energy of motion) but when it slows down from air resistance, this energy gets distributed in the random motions of air particles and individual particles in the body. The 'randomness has increased'; more specifically, the entropy has increased.

The notion of entropy was introduced by Clausius in 1865, but it was the outstanding Austrian physicist Ludwig Boltzmann who, in 1877 made the definition of entropy clear (or, at least, as clear as it seems possible to make it). To understand Boltzmann's idea (for a classical system), we need the notion of the *phase space* (§12.1, §§14.1,8, §§20.1,2,4) which, we recall, for a classical system of n (featureless) particles, is a space $\mathcal{P}$ of $6n$ dimensions, each of whose points represents the entire family of positions and momenta of all n particles. In order to make the notion of entropy precise, we require a concept of what is called *coarse graining*.[5] We can think of this as a division of the phase space $\mathcal{P}$ into a number of subregions which I shall

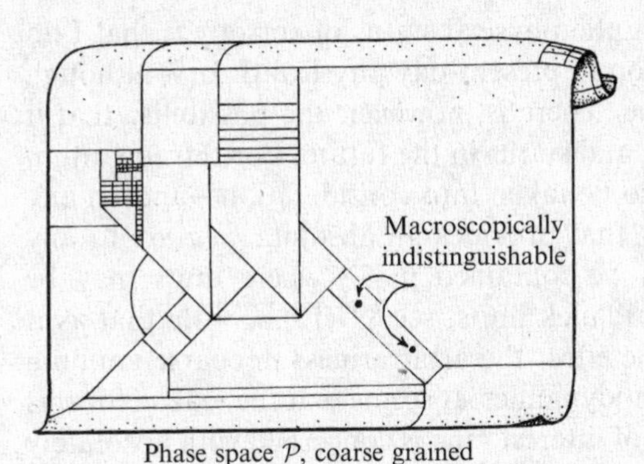

Fig. 27.2 Boltzmann entropy. This involves the division of phase space $\mathcal{P}$ into sub-regions ('boxes')—called a 'coarse graining' of $\mathcal{P}$—where the points of a given box represent physical states that are macroscopically indistinguishable. Boltzmann's definition of the entropy of a state x, in a box $\mathcal{V}$ of volume V, is $S = k \log V$, where k is Boltzmann's constant.

refer to as 'boxes'. See Fig. 27.2. The idea is that collections of points of $\mathcal{P}$ that represent states of the system that are indistinguishable from one another with regard to macroscopic observations, are considered to be grouped together in the same box, but points of $\mathcal{P}$ belonging to different boxes are deemed to be macroscopically distinguishable. The Boltzmann entropy S, for the state of the system represented by some point x of $\mathcal{P}$ is

$$S = k \log V,$$

where V is the volume of the box $\mathcal{V}$ that contains x (this being a *natural* logarithm; see §5.3), and where k is *Boltzmann's constant*,[6] having the value

$$k = 1.38 \times 10^{-23} \text{J K}^{-1}$$

(where J means joules and K^{-1} means 'per degree kelvin').

I have said that Boltzmann's definition makes the notion of entropy 'clear'. But for the above formula for S to represent something physically precise, it would be necessary to have a clear-cut prescription for the coarse graining that our family of 'boxes' is supposed to represent. There is undoubtedly something 'arbitrary' in the particular division into boxes that one might happen to select. The definition seems to depend upon how closely one chooses to examine a system. Two states that are 'macroscopically indistinguishable' to one experimenter might be distinguishable to another. Moreover, exactly where the boundary between two boxes happens to be drawn is again very arbitrary, since two neighbouring points of $\mathcal{P}$, with one on each side of the boundary, might be assigned quite different entropies, despite their being virtually identical. There is still something very subjective about this definition of S, despite its being a distinct advance on earlier notions of more limited applicability, and an undoubted improvement on the idea of just a measure of the 'randomness' in a system.

My own position concerning the physical status of entropy is that I do not see it as an 'absolute' notion in present-day physical theory, although it is certainly a very useful one. There is, however, the possibility that it might acquire a more fundamental status in the future. For this, quantum physics would certainly need to be taken into consideration—and, in any case, it is quantum mechanics that provides an absolute *measure* to any particular phase-space region $\mathcal{V}$, contained in $\mathcal{P}$, where units may be chosen so that $\hbar = 1$ (as with Planck units, see §27.10).[27.2] Be that as it may, it is remarkable how little effect the arbitrariness in coarse graining has in the calculations of thermodynamics. It seems that the reason for this is that, in most considerations of interest, one is concerned with absolutely enormous ratios between the sizes of the relevant phase-space box volumes and it makes little difference where the boundaries are drawn, provided that the coarse graining 'reasonably' reflects the intuitive idea of when systems are to be considered to be macroscopically distinguishable. Since the entropy is defined as a logarithm of the box volume, it would indeed need a 'stupendous' redrawing of boundaries to get any significant change in S.[27.3] In my view, entropy has the status of a 'convenience', in present-day theory, rather than being 'fundamental'—though there are indications that, in a deeper context where quantum-gravitational considerations become important (especially in relation to black-hole entropy), there may be a more fundamental status for this kind of notion. We shall be coming to this issue later in this chapter (§27.10) and in §§30.4–8, §31.15, and §32.6.

27.4 The robustness of the entropy concept

A simple illustration may make the role of Boltzmann's entropy formula a little clearer. Consider a closed container in which a region $\mathcal{R}$ is marked off as *special*, say a bulb-shaped protuberence of one tenth of the container's volume having access to the rest of the container through a small opening; see Fig. 27.3. Suppose that there is a gas in this container consisting of m molecules. We are going to ask for the entropy S to be assigned to the situation in which the entire gas finds itself in $\mathcal{R}$, in comparison with the entropy to be assigned to the situation in which the gas is randomly distributed throughout the container.

We shall have $S = k \log V_{\mathcal{R}}$, by Boltzmann's formula, where $V_{\mathcal{R}}$ is the volume of the phase-space region $\mathcal{V}_{\mathcal{R}}$ representing all molecules being in $\mathcal{R}$.

[27.2] Show how to assign an absolute measure to a phase-space volume if units are chosen so that $\hbar = 1$.

[27.3] How does the logarithm in Boltzmann's formula relate to the 'stupendous' discrepancies in box volumes?

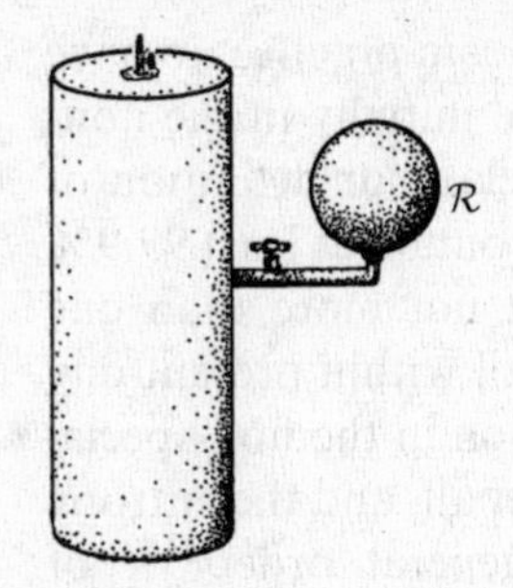

Fig. 27.3 A closed container, part of which is a bulb-shaped region $\mathcal{R}$ whose volume is $\frac{1}{10}$ of that of the entire container. How much does the entropy increase when a gas, initially all contained in $\mathcal{R}$, is allowed to flow into the entire volume?

For simplicity, we assume what is called *Boltzmann statistics*, as opposed to the 'Bose–Einstein statistics' of bosons and the 'Fermi–Dirac statistics' of fermions, as described in §23.7; that is, we suppose that all the gas molecules are *distinguishable* from each other (at least in principle).[27.4] Taking the gas to be ordinary air at atmospheric pressure, with a container of a litre in volume, we have $m = 10^{22}$ molecules, approximately. The ratio of the volume of the phase-space region $\mathcal{V}_{\mathcal{R}}$ to the volume of the entire phase space $\mathcal{P}$ for the gas in the container is $10^{-m}(=\frac{1}{10^m})$,[27.5] which is

$$10^{-10000000000000000000000},$$

so we see something of the 'stupendous' volume ratios that come up with considerations of this kind. The figure above represents the ridiculously tiny probability of finding—just by pure chance—that all the gas molecules are indeed in $\mathcal{R}$. The entropy of that exceedingly improbable situation is much smaller than the entropy of the situation where the gas is randomly distributed, the difference being about

$$\begin{aligned} -k\log(10^{-10000000000000000000000}) &= 2.3\times 10^{22}k \\ &= 0.32\ \mathrm{J\,K^{-1}} \end{aligned}$$

by Boltzmann's formula,[27.6] where I have used the fact that the natural logarithm of 10 is about 2.3. Thus, if we suppose that the gas is initially held in $\mathcal{R}$, say by a valve closing it off from the rest of the container, and if we then open the valve to release the gas into the remainder of the container, we find an appreciable entropy increase of some $2.3\times 10^{22}k$, which in ordinary units is about one-third of a joule per kelvin.

[27.4] Explain how the counting would differ, in the three cases.

[27.5] Why?

[27.6] Show why this result is not significantly affected if we bring in fermion/boson considerations or if we take into account the probable momentum decrease of the gas molecules when they cease to be constrained to $\mathcal{R}$.

The reader might worry that it would not be feasible, in practice, to have a container in which absolutely no gas molecules are initially in the non-special part of the container. Let us, accordingly, relax our definition of the region $\mathcal{V}_\mathcal{R}$ a little, defining it so that $\mathcal{R}$ need only contain at least 99.9% of the gas molecules. Thus, $\mathcal{V}_\mathcal{R}$ now demands that not more than one thousandth of these molecules lie outside $\mathcal{R}$. It is well within present-day technology to have a vacuum of this relative perfection in the non-special region. It turns out that the result is hardly affected at all, and the entropy increase, on opening the valve, is *still* of the general order[27.7] of $2.3 \times 10^{22}k$. This is a striking illustration of the fact that although there is subjectivity in drawing the coarse-graining boxes (say $\mathcal{V}_\mathcal{R}$), this causes no serious problems so long as the boxes are drawn 'reasonably'.

The logarithm in Boltzmann's formula has an important purpose, in addition to making the enormously large numbers look manageable. This is that the resulting definition of entropy, for independent systems, is *additive*. Thus, if the entropies assigned to two independent systems are S_1 and S_2, then the entropy assigned to the total system, consisting of the two taken together, will be $S_1 + S_2$. I am assuming that the phase-space for the total system is $\mathcal{P} = \mathcal{P}_1 \times \mathcal{P}_2$, where $\mathcal{P}_1$ and $\mathcal{P}_2$ are the phase-spaces for the two individual systems, and that the coarse-graining boxes of the total system are the products of the coarse-graining boxes of $\mathcal{P}_1$ and $\mathcal{P}_2$, this being a very natural assumption,[27.8] for independent systems S_1, S_2. (See §15.2, Exercise [15.1], and Fig. 15.3a for the definition of $\times$, applied to spaces.) Since the box volumes are multiplied, the corresponding entropies are added (by the standard property of the logarithm; see §5.2).

In normal examples of physical systems—and certainly the much-studied case of an ordinary gas in an ordinary container—there is a particular box $\mathcal{E}$ of the coarse graining whose volume E far exceeds the volume of any of the other boxes. This represents the state of *thermal equilibrium*. Indeed, E will normally be practically equal to the volume P of the entire phase space, and so E will easily exceed the volumes of all the other boxes put together. See Fig. 27.4. For an ordinary gas, which we consider to be made up of identical spherically symmetrical balls in thermal equilibrium, the distribution of velocities takes up a particular form known as the *Maxwell distribution* (found by the same James Clark Maxwell that we have encountered before, in connection with electromagnetism). It has a probability density of the form

---

[27.7] See if you can see why the entropy increase has dropped just a very little, from $2.30 \times 10^{22}k$ to $2.29 \times 10^{22}k$, approximately, by making various rough estimates for the mathematical quantities involved. (Use Stirling's formula $n! \approx (n/e)^n(2\pi n)^{1/2}$, if you wish.).

[27.8] Why is this course-graining assumption natural?

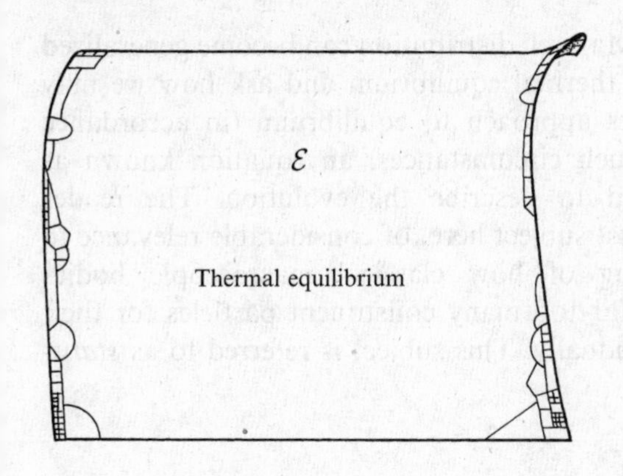

Fig. 27.4 The particular box $\mathcal{E}$, representing thermal equilibrium, has a volume E, that is normally practically equal to the volume P of the entire phase space $\mathcal{P}$, and therefore far exceeds the volumes of all the other boxes together.

$$A\,\mathrm{e}^{-\beta v^2},$$

(which those in the know will recognize as that of a *Gaussian* distribution, sometimes called the 'bell curve') where v is the magnitude of the 3-velocity of the gas particle in question, β is a constant related to the temperature, and A is a constant such that the probability integral over the space of all possible velocities is 1; see Fig. 27.5. Thermal equilibrium, having easily the largest possible entropy for the system, is the state that one would expect a system to settle into if left for a sufficiently long time, in accordance with the second law.

The Maxwell distribution, as just described, refers to a gas made from identical classical bodies, with no internal degrees of freedom. Things can get considerably more complicated when there are many kinds of constituents of different sizes and various internal degrees of freedom (such as spin or vibrations between its constituent parts). There is a general principle, for such systems in thermal equilibrium, known as *equipartition of energy* according to which the energy of the system is distributed equally (with a statistical spread) between all the different degrees of freedom of the system.

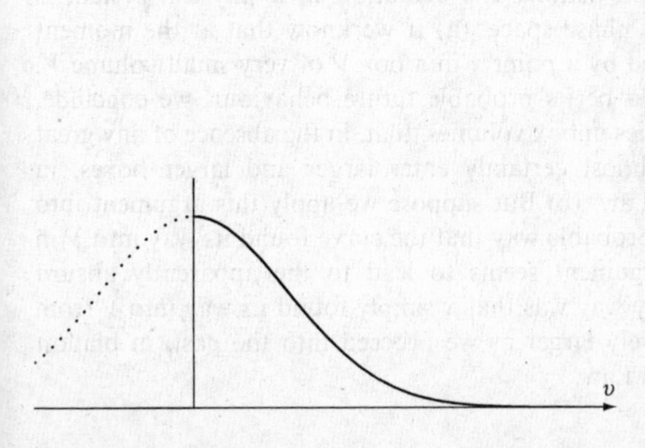

Fig. 27.5 The Maxwellian distribution of velocities, for a gas in equilibrium, has the form $A\mathrm{e}^{-\beta v^2}$, where A and β are constants, with β related to the temperature of the gas, v being the particle's speed. The dotted part of the curve extends this to negative values of v, revealing the familiar 'bell-curve' form of the Gaussian distribution of statistics.

Another way in which the Maxwell distribution can become generalized is to move away from exact thermal equilibrium and ask how we may expect a gas will move in its approach to equilibrium (in accordance with the Second Law). In such circumstances, an equation known as *Boltzmann's equation* is used to describe the evolution. The reader will perceive that there is a vast subject here, of considerable relevance to the theoretical understanding of how classical macroscopic bodies will behave, when there are far too many constituent particles for their dynamics to be tracked individually. This subject is referred to as *statistical mechanics*.

27.5 Derivation of the second law—or not?

Let us now try to understand what lies behind the second law. Imagine that we have some physical system represented by a point x in some suitably coarse-grained phase-space $\mathcal{P}$. Suppose (Fig. 27.6a) that x starts NOW in some small coarse-graining box $\mathcal{V}$ of volume V. The point x will move around in $\mathcal{P}$ in some way, in accordance with the dynamical equations

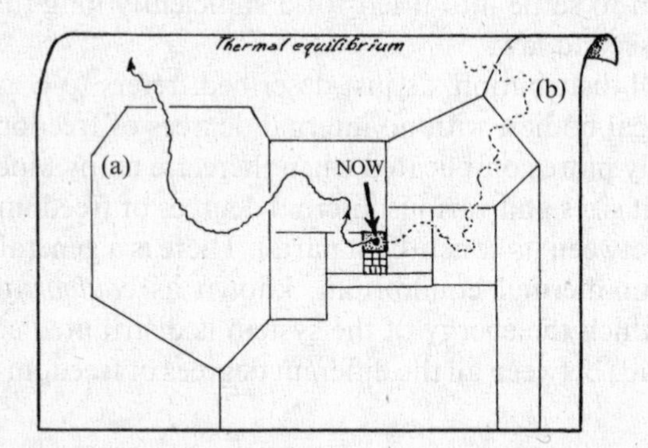

Fig. 27.6 The Second Law in action. The evolution of a physical system is represented by some curve in phase space. (a) If we know that at the moment NOW, our system is represented by a point x in a box $\mathcal{V}$ of very small volume V, and we try to see what would be its probable future behaviour, we conclude, because of the vast discrepancies in box volumes, that, in the absence of any great bias in its motion, it will almost certainly enter larger and larger boxes, in accordance with the Second Law. (b) But suppose we apply this argument into the *past*, asking for the most probable way that the curve found its way into $\mathcal{V}$ in the first place. The same argument seems to lead to the apparently absurd conclusion that the most likely way was that x simply found its way into $\mathcal{V}$ from boxes that become progressively larger as we proceed into the past, in blatant contradiction with the Second Law.

appropriate to the physical situation under consideration. Bearing in mind the enormous discrepancy in the sizes of the different coarse-graining boxes, and anticipating no particular bias with regard to x's dynamical motion in relation to the box locations, we expect that, in overwhelmingly many cases, x will wander into boxes of larger and larger volume. In other words, the entropy of the system will indeed get larger and larger, as time marches forward. Once x finds its way into a box with a certain entropy measure, it becomes overwhelmingly unlikely that, in any sensible period of time, it can find itself back in a box of significantly smaller entropy than that. To reach a significantly smaller entropy would mean finding an absurdly tinier volume, and the odds are immensely against it. Think of the example that we have just been considering, and the absolutely stupendous reduction in phase-space volume that would accompany a very modest reduction in entropy, owing to the logarithm in Boltzmann's formula and the small size of Boltzmann's constant. Once the gas has found its way out of $\mathcal{R}$, it is ridiculously unlikely that it will find its way back again into $\mathcal{R}$ (at least not within any time-scale that is not 'utterly ridiculously long').[27.9]

This argument contains the essential reason for expecting the second law to hold. We note that the argument does not appear to depend on the particulars of the dynamics at all, except that we require no bias having the effect that the point x deliberately seeks out smaller boxes. Is this really all there is to the second law? It all seems too easy—and the apparently universal nature of this kind of argument is perhaps the reason that many physicists take the view that there is nothing fundamentally puzzling about the second law, and that any reasonable physical theory must satisfy it. A wonderful quote from the outstanding astrophysicist Sir Arthur Eddington is pertinent here:

> If someone points out to you that your pet theory of the universe is in disagreement with Maxwell's equations—then so much the worse for Maxwell's equations. If it is found to be contradicted by observation—well, these experimentalists do bungle things sometimes. But if your theory is found to be against the second law of thermodynamics I can offer you no hope; there is nothing for it but to collapse in deepest humiliation.[7]

But a few moments' reflection tells us that there is something odd about the conclusion of the argument that I have just presented, or perhaps something significantly missing from these elementary considerations. What we seem to have deduced is a time-*asymmetrical* law when the underlying physics may be taken to be *symmetrical* in time. How has this

---

[27.9] Try to estimate how long this would have to be, both in the case where *all* the gas returns to $\mathcal{R}$ and in the case where 99.9% of it does. Do you really need to know how fast the gas molecules are moving?

come about? We could imagine trying to apply the very same argument in the *past* time direction (Fig. 27.6b). We seem to conclude that if we place our phase-space point x in the same small box that we chose before, at time NOW, and then examine the past-directed evolution prior to NOW, then we conclude that it is overwhelmingly likely that x entered this box from boxes that, as we proceed farther and farther into the past, become larger and larger! But this would tell us that the *reverse* of the second law held in the past, with entropy increasing in *past* directions, despite our expectations, on the basis of this argument, that the familiar version of the second law should indeed apply into the future. This conclusion is grossly at variance with observations of the way in which our universe actually behaved in the past. See Fig. 27.7.

What has gone wrong? To try to understand this, let us apply these arguments to the behaviour of our gas in its container, where we start (at time t_0) with the gas entirely in $\mathcal{R}$, so x lies in $\mathcal{V}_{\mathcal{R}}$. We seem to get the correct future behaviour for the gas, where after the valve is opened, the gas indeed floods from $\mathcal{R}$ into the entire container, the entropy going up very significantly as x fairly rapidly finds itself in the region $\mathcal{E}$, representing thermal equilibrium. But what about the past behaviour? We have to ask the question what happened just before time t_0. What, indeed, is the most likely way for the gas actually to have found its way into $\mathcal{R}$? If we imagine that the valve were open just before t_0, then the 'most probable evolution' that we seem to have obtained is that the gas started spread out throughout the container, effectively in thermal equilibrium, at sometime earlier than t_0, and then it spontaneously concentrated itself more and more in the region $\mathcal{R}$, finding itself entirely in $\mathcal{R}$ at the time t_0.

Absurd as it may seem, this is actually the correct answer to the problem posed in this particular way, with no interference from outside. In practice, we would just never find the gas entirely in $\mathcal{R}$. The argument merely tells us how a randomly moving gas would have to behave *if* we were to find the

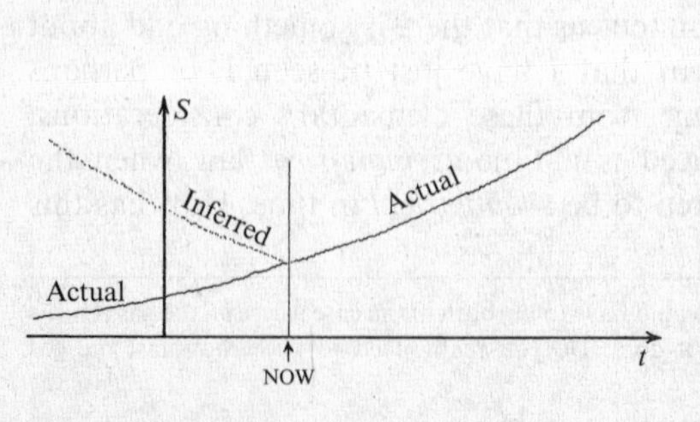

Fig. 27.7 The conclusions of Fig. 27.6, expressed in a plot of the entropy S against time t. The reasoning correctly leads us to expect a Second Law for behaviour to the future of us NOW, but it gives us the seemingly absurd answer that the reverse of the Second Law held in the past, in gross contradiction with actual experience.

gas *spontaneously* all in $\mathcal{R}$, which we won't. There is no paradox here. But this evades the problem that I wish the reader to consider, which is how could such a situation as the gas being entirely in $\mathcal{R}$ arise in practice? There is no problem with this happening in our actual universe (where we take the relaxed definition of the region $\mathcal{V}_{\mathcal{R}}$ which permits one thousandth of the gas to lie outside $\mathcal{R}$). We could imagine that some experimenter initially pumped ten times the required amount of gas into the entire container, next closed the valve, and finally applied a vacuum pump to the main part of the container, to remove practically all of that 90% of the gas. Throughout this entire process, the entropy would have been going up all the time, in accordance with the second law. Of course, to discuss this in terms of phase space, we need a bigger phase space in which the experimenter is also incorporated—probably along with a good deal more of the universe too, perhaps extending out to the Sun or beyond. The entropy in the experimenter's body is kept very low through the acts of eating and breathing. For simplicity, let us suppose that the pump is hand-operated—otherwise we need to worry about origin of the low entropy in the fuel source (which raises issues inessential to our present purposes). Part of the experimenter's lowness of entropy is transferred to the container with its gas, and is employed in getting the gas into $\mathcal{R}$. The low entropy in the experimenter's meals and in the air ultimately comes from the external Sun. I shall be returning to this specific role of the Sun shortly.

Thus, we have been able to obtain the required situation, in which virtually all the gas in the container is in the region $\mathcal{R}$, without violating the Second Law, which has its physically appropriate form: 'entropy increases with time'. What has happened to our difficulty with the apparent deduction of a time-reversed second law, for behaviour into the past? Has it been resolved? No, it certainly has not! The experimenter's body ought to (and indeed does) act in accordance with the actual second law, as does the Sun and everything described by our enlarged phase space. But if we try to apply our phase-space argument—now to this enlarged phase space—then we still seem to obtain the physical absurdity that the entropy must have gone up again into the past, prior to any time when we examine our entire system.

27.6 Is the whole universe an 'isolated system'?

Some theoreticians try to make a distinction between 'isolated' and 'open' systems, arguing that while entropy increases (until equilibrium is reached) in an isolated system, there is always the possibility of an input from the outside world that can serve to reduce entropy from time to time—such as intervention by an experimenter or a low-entropy input from the Sun, etc. It seems to me that any attempted explanation for the time-asymmetry in the second law along these lines can have only a provisional status, because these outside influences can, in their totality be incorporated into the system. This implies that the 'system' under consideration really has to be the universe as a whole. Sometimes people object to this, but I see no justification for such an objection. It may well be that the universe is indeed infinite in extent, but that is no bar to its being considered as a whole (see Chapter 16). In any case, the universe might be spatially finite (a distinct possibility that we shall come to shortly) and it would seem strange to rely on an argument for the second law whose validity is dependent upon the universe being actually spatially infinite. As we shall see later, the infinite/finite distinction, is only very mildly relevant to the question of the origin of the second law. The discussion of entropy can indeed be applied to the entire universe $\mathcal{U}$, where a phase-space $\mathcal{P}_{\mathcal{U}}$ (whose volume might be infinite) describes a broad totality of possible universes, incorporating all their evolutions according to the dynamical equations of (the appropriate) classical dynamics.

There are, however, some awkward issues to be faced. To treat the universe as a whole, we need to enter the realm of cosmology, which cannot be done adequately without bringing in general relativity. In order to have a discussion that is completely in accord with the principle of general covariance of general relativity (§19.6), it would be necessary to employ a description in which there is no special choice of time coordinate with respect to which the universe is supposed to 'evolve'. A temporally evolving picture is explicitly the way that we have been regarding a physical system, when we represent it as a point x moving in a phase-space $\mathcal{P}$. Each location of x represents a spatial description (momentum included) of the system at one time. But to adopt a more relativistic view would complicate this description unnecessarily, and I do not think that, for the points that I wish to make here, it is helpful to try to take a strictly relativistic viewpoint. In fact, as we shall shortly see, the standard cosmological models possess a naturally defined time coordinate, and this gives a good approximation to a 'time parameter' t that the whole universe can be described as evolving with respect to. Each point of $\mathcal{P}_{\mathcal{U}}$ will be taken to describe not only the material contents of the universe at time t, but also the distribution (and momentum) of continuous fields. The gravitational field is one such field, so the universe's *spatial geometry* (together with its rate of change—given by appropriate initial data for the gravitational field)[8] will also be encoded in the location of x within $\mathcal{P}_{\mathcal{U}}$.

In fact, $\mathcal{P}_{\mathcal{U}}$ will be infinite-dimensional, but this happens whether or not the universe $\mathcal{U}$ is infinite in extent, and it is a feature that occurs with all other fields also, such as the electromagnetic field. This causes some technical problems for the definition of entropy, since each required phase-space region $\mathcal{V}$ will have infinite volume. It is usual to deal with this problem by borrowing ideas from quantum (field) theory, which enables a finite answer to be obtained for the phase-space volumes which refer to systems that are appropriately bounded in energy and spatial dimension. The details of this are not important for us. Although there is no fully satisfactory way of dealing with these issues in the case of gravity—owing to a lack of a satisfactory theory of quantum gravity—I am going to regard these as technicalities that do not affect the general discussion of the issues raised by the second law.

At this point, I should mention a misconception that frequently causes great confusion with regard to the second law in a cosmological setting. There is a common view that the entropy increase in the second law is somehow just a necessary consequence of the expansion of the universe. (We shall be coming to this expansion in §27.11). This opinion seems to be based on the misunderstanding that there are comparatively few degrees of freedom available to the universe when it is 'small', providing some kind of low 'ceiling' to possible entropy values, and more available degrees of freedom when the universe gets larger, giving a higher 'ceiling', thereby allowing higher entropies. As the universe expands, this allowable maximum would increase, so the actual entropy of the universe could increase also.

There are many ways to see that this viewpoint cannot be correct. It implies for example that, in those universe models where there is a collapsing phase, the entropy necessarily starts to decrease, in violation of the second law. Some might not be unhappy about this,[9] but this viewpoint encounters fundamental difficulties, particularly in the presence of black holes.[10]

We shall be considering black holes shortly (in §27.8), but we really do not need to know about them to see why the aforementioned viewpoint—demanding a 'ceiling' on entropy values depending upon the universe's size—is misconceived. This cannot be the correct explanation for the entropy increase; for the degrees of freedom that are available to the universe are described by the total phase space $\mathcal{P}_{\mathcal{U}}$. The dynamics of general relativity (which includes the degree of freedom defining the universe's size) is just as much described by the motion of our point x in the phase space $\mathcal{P}_{\mathcal{U}}$ as are all the other physical processes involved. This phase space is just 'there', and it does not in any sense 'grow with time', time not being part of $\mathcal{P}_{\mathcal{U}}$. There is no such 'ceiling', because all states that are dynamically accessible to the universe (or family of universes) under consideration must be represented in $\mathcal{P}_{\mathcal{U}}$. It may take some while for x to reach some large coarse-graining box from some given smaller one, but the notion of an 'entropy ceiling' is inappropriate. (See also §27.13.)

Let us return to the argument given above for demonstrating the second law. We shall use the phase space $\mathcal{P}_{\mathcal{U}}$ appropriate to the entire universe, so the evolution of the universe as a whole is described by the point x moving along a curve ξ in $\mathcal{P}_\mathcal{U}$. The curve ξ is parametrized by the time coordinate t, and we can expect that, from the second law, ξ enters immensely larger and larger coarse-graining boxes as t increases. We suppose that some 'reasonable' coarse graining has been applied to $\mathcal{P}_\mathcal{U}$, but if we wish to obtain finite values for the entropies that x encounters, we would want the volumes of these boxes to be finite. It would appear that for this to be achieved, for a physically appropriate coarse graining, the universe has to be taken to be finite, with a finite bound on its available energy. In fact, as we shall shortly see, one of the three standard cosmological models is indeed of this nature, so we can imagine that the argument is being applied in such a situation. But there is no definite requirement for this, if we do not mind an actual infinite numerical value for the entropy, at any one time. (We can still make mathematical sense of the notion that some boxes are 'immensely larger' than other boxes, even if the actual volumes of some of them, and therefore their entropies, are infinite.)

27.7 The role of the Big Bang

How are we to envisage that our parametrized curve ξ, representing a possible universe history, is to be placed in the phase-space $\mathcal{P}_\mathcal{U}$? If ξ were simply to be thrown randomly into $\mathcal{P}_\mathcal{U}$, then we would expect that, with overwhelming probability, it would lie entirely (or almost entirely) in the most enormous thermal-equilibrium box $\mathcal{E}$, and there would be no consistently discernible measure of 'entropy increase' throughout its length; see Fig. 27.8a. Such a situation is completely at variance with the universe as we actually know it, in which a second law holds sway. So also is the situation indicated in Fig. 27.8b,c, where for some particular time $t_0(>0)$ that represents NOW, the point x on ξ is constrained to be in some reasonable-sized, but not particularly large, region $\mathcal{V}$ (representing a universe of an entropy value that we happen to observe now) but where the curve ξ is otherwise chosen randomly. This corresponds to a universe whose entropy goes up in the future from NOW, but also goes up into the past from NOW—in violation of the Second Law! What we actually find is something like Fig. 27.8b,d, for a universe with our familiar Second Law, where ξ has one end—the past end (say $t = 0$)—in an exceedingly tiny region $\mathcal{B}$ in $\mathcal{P}_\mathcal{U}$ (which therefore has a exceedingly small entropy), but from there on it flaps around as it will (following the dynamical laws), finding volumes of immensely greater and greater size as t increases, and where for our particular t value t_0, representing NOW, we happen to find x in the still rather small volume $\mathcal{V}$ corresponding to the universe we observe. This is simply what the Second Law asserts, and we get (d),(b), as opposed to (c), (b).

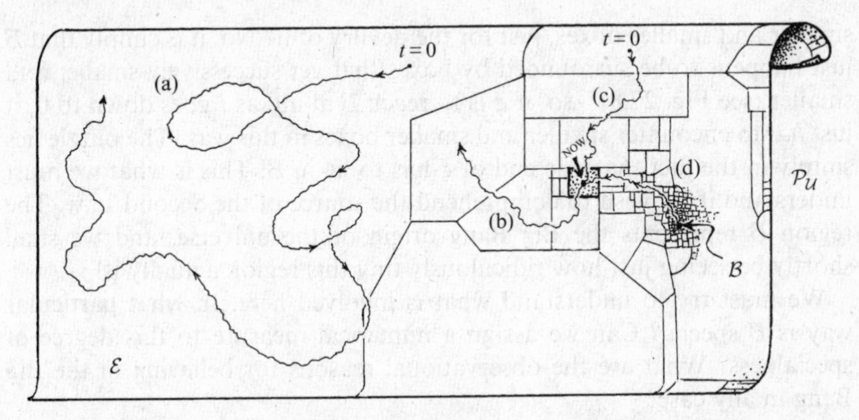

Fig. 27.8 Different possible universe evolutions, described by a parameterized curve ξ in the phase space $\mathcal{P}_{\mathcal{U}}$, of possible universe states (of fixed overall mass, say, or of any other conserved quantity). (a) If the curve ξ is thrown randomly into $\mathcal{P}_{\mathcal{U}}$, it spends almost all of its life in $\mathcal{E}$ and, apart from minor fluctuations, the universe hardly differs from 'thermal equilibrium' (and could resemble Fig. 27.20d if it is closed). (b) If we merely specify that the curve starts NOW in a very small box $\mathcal{V}$ (shown shaded), taking the points of $\mathcal{V}$ to resemble the universe we now live in, but where ξ is otherwise thrown randomly, then we find a future evolution consistent with what we continue to see, with the increasing entropy of the Second Law. (c) If we apply the same consideration to the situation where the entire curve ξ is constrained merely to pass through $\mathcal{V}$ at some particular time $t_N > 0$, (NOW), then we find a reasonable future for the universe, but, as with Fig. 27.6b, we find gross violation of the Second Law in the past. (d) This is remedied if we further specify that the initial end ($t = 0$) of ξ lies in the absurdly tiny region $\mathcal{B}$, at which the universe starts with the extraordinarily special Big Bang that apparently occurred in our actual universe.

Let me try to rephrase what has been said above. Suppose we look at things from the vantage point of some particular time t_0 (> 0) that we call 'NOW', finding x, at time t_0, in some reasonable-sized region $\mathcal{V}$. Then looking at where ξ wanders for larger values of t, we indeed see it enter larger and larger box sizes as t increases. This is consistent with the Second Law and with the above assumption that x exhibits 'no particular bias' in relation to the box locations. But as viewed from a reversed-time perspective, 'starting' at t, with x in $\mathcal{V}$, the point x gives the impression of being purposefully guided, back in time, towards the absurdly tiny region of phase-space that I have labelled $\mathcal{B}$.

With respect to this reverse-time direction, the behaviour of x indeed seems unbelievably 'biased', seeking out boxes that become successively smaller and smaller to an extraordinary degree, as time proceeds into the past. Are we to understand this as a perverse 'deliberate' seeking of smaller and smaller boxes, just for the devilry of it? No, it is simply that $\mathcal{B}$ just happens to be surrounded by boxes that get successively smaller and smaller (see Fig. 27.8)—so, if ξ is to reach $\mathcal{B}$ at all, as t goes down to 0, it just *has* to encounter smaller and smaller boxes in this way. The puzzle lies simply in the fact that one end of ξ has to lie in $\mathcal{B}$! This is what we must understand if we wish to comprehend the source of the Second Law. The region $\mathcal{B}$ represents the Big Bang origin of the universe, and we shall shortly be seeing just how ridiculously tiny this region actually is!

We must try to understand what is involved here. In what particular way is $\mathcal{B}$ special? Can we assign a numerical measure to this degree of specialness? What are the observational reasons for believing in the Big Bang in any case?

The reasons for believing in an explosive origin to the universe came initially from a theoretical study of Einstein's equation in a cosmological context, made by Alexandr Friedmann in 1922 (see §27.11 below). Then, in 1929, Edwin Hubble made the remarkable discovery that the distant galaxies are indeed receding from us[11] in a way that seemed to be implying that the matter in the universe was the result of a stupendous explosion. On modern reckoning, the explosion—now called the Big Bang—took place some 1.4×10^{10} years ago. Hubble's conclusions were based on the fact that the light from rapidly receding objects is *redshifted* (so that spectral lines are displaced to the 'red end of the spectrum', i.e. to longer wavelengths) owing to the Doppler effect.[27.10] He found that this redshift was systematically greater the more distant the galaxy appeared to be, indicating a velocity of recession that is proportional to the distance from us, consistently with the 'explosion' picture.

But the most impressive direct piece of observational support for the Big Bang is the universal presence of radiation permeating space, having the temperature of about 2.7 K (i.e. 2.7 °C above absolute zero).[12] Although this may seem to be an extraordinarily low temperature for such a violent event, this radiation is believed to be the 'flash' of the Big Bang itself, enormously attenuated ('redshifted') and cooled, owing to the vast expansion of the universe. The 2.7 K radiation plays an exceedingly important role in modern cosmology. It is commonly referred to as the '(cosmic)microwave background', or sometimes as the 'background black-body radiation', or the 'cosmic relic' radiation. It is exceedingly uniform (to something like one part in 10^5), indicating that the early universe was itself extremely uniform just after the Big Bang, and very well described by the cosmological models that we shall be considering in §27.11.

Now, let us try to gain some physical insights as to the *nature* of the enormously low-entropy constraint on the Big Bang that restricts $\mathcal{B}$ to

---

[27.10] Derive the special-relativistic Doppler frequency shift for a source receding with speed v (a) using a wave picture of light, and (b) by using scalar products of 4-vectors and $E = h\nu$.

have such a tiny volume.[13] We shall find that what was extraordinarily special about the Big Bang was actually its great uniformity, as just referred to. We must try to understand why this corresponds to a very low entropy, and how it provides us with a Second Law that is relevant to us here on Earth in the familiar form that we know.

First, consider, again, the Sun's role as a low-entropy source. There is a common misconception that the *energy* supplied by the Sun is what our survival depends upon. This is misleading. For that energy to be of any use to us at all, it must be provided in a low-entropy form. Had the entire sky been uniformly illuminated, for example, with some uniform temperature—whether that of the Sun or anything else—then there would be no way of making use of this energy (whatever kind of creature we might imagine having evolved to try to cope with it). An energy supply in thermal equilibrium is useless. We, however, are fortunate that the Sun is a *hot spot* in an otherwise *cold background*. During the day, energy reaches the Earth from the Sun, but during the course of the day and night it all goes back again into space. The net balance of energy is (on the average) simply that we send back all the energy that we receive.[14]

However, what we get from the Sun is in the form of individual photons of *high* energy (basically yellow high-frequency photons because of the Sun's high temperature), whereas this energy mostly goes back into space in the form of photons of *low* energy (infrared, low frequency). (This photon energy relation comes from Planck's formula $E = h\nu$ and his insights into black-body radiation; see §21.4). Because of their higher energy (higher temperature) there are many fewer photons from the Sun than there are photons going back into space, because the *total* energy carried by them is the same coming in as going out. The Sun's smaller number of photons means fewer degrees of freedom, and therefore a smaller phase-space region and hence a smaller entropy, than in the photons returned to space. The plants make use of this low-entropy energy in their photosynthesis, thereby reducing their own entropy. Then we take advantage of the plants to reduce ours, by eating them, or eating something that eats them, and by breathing the oxygen that the plants release; see Fig. 27.9.

But why is the Sun a hot spot in a cold sky? Although the detailed story is complicated, it ultimately comes down to the fact that the Sun—and all other stars—have condensed gravitationally from a previously uniform gas (of mainly hydrogen). Whatever other influences are present (primarily nuclear forces), the Sun could not even exist without gravity! The 'lowness' in the Sun's entropy (considerable remoteness from thermal equilibrium) comes from a huge reservoir of low entropy that is potentially available in the *uniformity* of the gas from which the Sun has gravitationally condensed.

Fig. 27.9 The Earth gives back the same amount of energy that it receives from the Sun, but what it receives from the Sun is in a much lower entropy form, owing to the fact that the Sun's yellow light has higher frequency than the infrared that the Earth returns. Accordingly, by Planck's $E = h\nu$, the Sun's photons carry more energy per photon than do those that Earth returns, so the energy from the Sun is carried by fewer photons than that returned by the Earth. Fewer photons means fewer degrees of freedom and therefore a smaller phase-space region and thus lower entropy than in the photons returned to space. Plants make use of this low entropy energy in photosynthesis, thereby reducing their own entropy, and we take advantage of the plants to reduce ours, by eating them, or eating something that eats them, and by breathing the oxygen that the plants release. This ultimately comes from the temperature imbalance in the sky that resulted from the gravitational clumping that produced the Sun.

Gravitation is somewhat confusing, in relation to entropy, because of its universally attractive nature. We are used to thinking about entropy in terms of an ordinary gas, where having the gas concentrated in small regions represents *low* entropy (as with our container in Fig. 27.3), and where in the *high*-entropy state of thermal equilibrium, the gas is spread uniformly. But with gravity, things tend to be the other way about. A uniformly spread system of gravitating bodies would represent relatively *low* entropy (unless the velocities of the bodies are enormously high and/or the bodies are very small and/or greatly spread out, so that the gravitational contributions become insignificant), whereas *high* entropy is achieved when the gravitating bodies clump together (Fig. 27.10).

What about the maximum-entropy state? Whereas with a gas, the maximum entropy of thermal equilibrium has the gas uniformly spread throughout the region in question, with large *gravitating* bodies, maximum entropy is achieved when all the mass is concentrated in

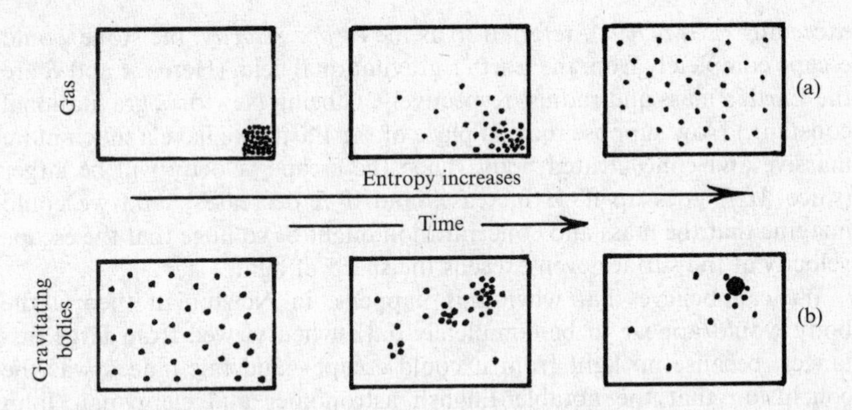

Fig. 27.10 Increasing entropy, with increasing time, left to right. (a) For gas in a box, initially all tucked in one corner, entropy increases as the gas starts to spread itself throughout the box, finally reaching the uniform state of thermal equilibrium. (b) With gravity, things tend to be the other way about. An initial uniformly spread system of gravitating bodies represents a relatively low entropy, and clumping tends to occur as the entropy increases. Finally, there is a vast increase in entropy as a black hole forms, swallowing most of the material.

one place—in the form of an entity known as a *black hole*. We shall need to understand something of these strange and wonderful objects in order to proceed further, and thereby obtain a remarkably good estimate of the entropy that is potentially available in the universe as a whole. This will then enable us to estimate the required volumes of $\mathcal{B}$ and $\mathcal{P}_{\mathcal{U}}$.

27.8 Black holes

What is a black hole? Roughly speaking, it is a region of spacetime that has resulted from the inward gravitational collapse of material, where the gravitational attraction has become so strong that even light cannot escape. To get an intuitive picture of why such a situation might come about, think of the Newtonian notion of *escape velocity*. If a stone is hurled upwards from the ground at a certain speed v, then it will fall back to the ground after it has reached a certain height, this height being that for which the kinetic energy of the stone has been entirely used up in overcoming the gravitational potential energy from ground level (§17.3, §18.6). The height from the ground is entirely dependent upon the speed of projection, ignoring the effects of air resistance.[27.11] However, for a speed

[27.11] Show that this height is $v^2R(2gR - v^2)^{-1}$, where R is the Earth's radius and g the acceleration due to gravity at the Earth's surface.

exceeding $(2GM/R)^{1/2}$, referred to as the *escape velocity*, the stone would escape completely from the Earth's gravitational field. (Here, M and R are the Earth's mass and radius, respectively, G being Newton's gravitational constant.) Now suppose that, in place of the Earth, we have a much more massive and concentrated body. Then the escape velocity will be larger (since M/R goes up if M increases and if R decreases), and we could imagine that the mass and concentration might be so huge that the escape velocity at the surface even exceeds the speed of light.

We can believe that when this happens, in Newtonian theory, the body would appear to be completely dark when viewed from large distances, because no light from it could escape—and this indeed was the conclusion that the notable English astronomer and clergyman John Michell came to in 1784. Later, in 1799, the great French mathematical physicist Pierre Simon Laplace came to the same conclusion.[15] However, the situation does not seem to me to be that clear, because the speed of light has no absolute status in Newtonian theory, and one can argue a good case that for such a body, the speed of light at its surface ought to be considerably greater than that measured in free space, and that light could still escape to infinity, no matter how massive and concentrated that body might be.[16,[27.12]] Thus, Michell's 'dark star', though a prescient precursor of the black-hole concept, does not, to my mind, provide a persuasive case for 'invisible' gravitating objects in Newtonian theory.

This issue is much more pertinent in the context of relativity theory, since there the speed of light is fundamental and indeed represents the limiting speed for all signalling (§17.8). Since we are concerned with a gravitational phenomenon, however, we require a *general*-relativistic spacetime, rather than just Minkowski space. In general relativity the expectations are indeed that situations will occur in which the escape velocity exceeds the speed of light, resulting in what we now call a *black hole*.

A black hole is to be expected when a large massive body reaches a stage where internal pressure forces are insufficient to hold the body apart against the relentless inward pull of its own gravitational influence. Indeed, such gravitational collapse is to be expected when a large star, of a total mass several times that of the Sun—let us say $10M_\odot$ (where $1M_\odot$ is a solar mass, the Sun's mass)—uses up all its available internal sources of energy, so that it cools and cannot keep up sufficient pressure to avoid collapse. When this happens, the collapse may become unstoppable, as the gravitational effects mount relentlessly.

[27.12] Can you see why? *Hint*: Think in terms of a particle theory of light, and of external light falling to the surface of the body. What happens if the light falls on a horizontal mirror at the body's surface?

The detailed picture can become very complicated, especially since under conditions of great pressure, sophisticated issues concerning the behaviour of matter become important. Of particular relevance is electron or neutron *degeneracy pressure*. This has to do with the Pauli principle which, as we recall from §23.7, prevents two or more identical fermions from being in the same quantum state. A *white dwarf*, which could have something like a solar mass concentrated into roughly the size of the Earth, is held apart by electron degeneracy pressure; a *neutron star* of the same mass would be a body of merely some 10 km across, held apart mainly by neutron degeneracy pressure. (A tennis ball filled with neutron star material would weigh as much as Mars's moon Deimos!) However, because of the requirements of relativity, it turns out that degeneracy pressure alone cannot hold such a star apart if the mass is more than about $2M_{\odot}$. The key result was obtained by Subrahmanyan Chandrasekhar, in 1931, when he established such a limit of about $1.4M_{\odot}$ for white dwarfs. Later refinements obtained a slightly larger limit for neutron stars.[17] The upshot of all this is that there is no resting configuration for a cold object of more than roughly $2M_{\odot}$ (and probably not more than $1.6M_{\odot}$). Such an object would collapse inwards, and would continue to collapse right down to the kind of dimension at which Michell's considerations begin to become relevant. What then happens?

Let us return to our large star, of, say, $10M_{\odot}$, assumed to be initially at a high enough temperature that thermal pressures can support it. As the star cools, however, at a certain stage its compressed core will exceed the Chandrasekhar limit, and will collapse. The infall of the outer parts could trigger a violent explosion, known as a *supernova*. Such exploding stars have frequently been observed, mainly in other galaxies, and for a few days the supernova can outshine the entire galaxy in which it resides. But if sufficient material is not shed in such an explosion—and for a star initially of $10M_{\odot}$, it is unlikely that it would lose that much—the expectations are, indeed, that the star would collapse unstoppably until it reaches the scale at which Michell's considerations apply. Let us examine Fig. 27.11, which is a spacetime diagram depicting collapse down to a black hole. (Of course, one of the spatial dimensions has had to be suppressed.) We see that the matter continues to collapse inwards, through that surface—called an (absolute) *event horizon*—where the escape velocity indeed becomes the speed of light. Thereafter, no further information from the star itself can reach any outside observer, and a black hole is formed.

The picture in Fig. 27.11 is based on the famous Schwarzschild solution of the Einstein equation, discovered by Karl Schwarzschild 1916,[18] shortly after the publication of Einstein's theory, and only a few months before Schwarzschild died of a rare disease contracted on the eastern front during

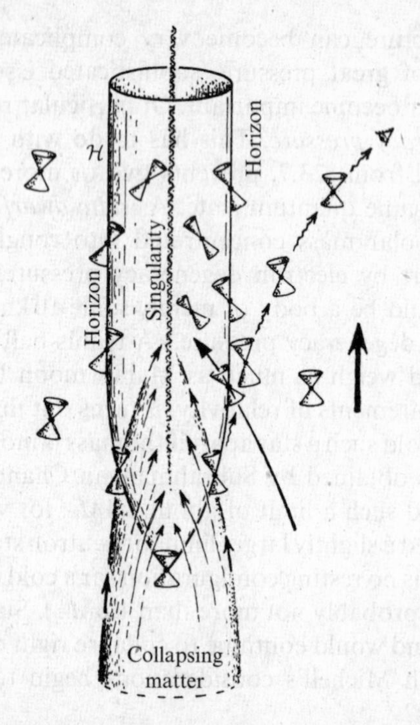

Fig. 27.11 Spacetime diagram of collapse to a black hole. (One spatial dimension is suppressed.) Matter collapses inwards, through the 3-surface that becomes the (absolute) event horizon. No matter or information can escape the hole once it has been formed. The null cones are tangent to the horizon and allow matter or signals to pass inwards but not outwards. An external observer cannot see inside the hole, but only the matter—vastly dimmed and red shifted—just before it enters the hole.

the first world war. This solution describes the static gravitational field surrounding a spherically symmetrical body, whether or not the body is contracting. The horizon occurs at the radial distance $r = 2MG/c^2$ (exactly Michell's critical value).[27.13]

[27.13] Schwarzschild's original metric form was $ds^2 = (1 - 2M/r)dt^2 - (1 - 2M/r)^{-1}dr^2 - r^2(d\theta^2 + \sin^2\theta\, d\phi^2)$, where units are chosen so that $G = c = 1$ and where θ and ϕ are standard spherical polar coordinates (§22.11). Explain how the radial coordinate r is fixed by a requirement on the area of the spheres of constants r and t. This metric form does not extend smoothly to the $r \leq 2M$ region; for this, the Eddington–Finkelstein form of the metric $ds^2 = (1 - 2M/r)dv^2 - 2\, dv\, dr - r^2(d\theta^2 + \sin^2\theta\, d\phi^2)$ can be used. Find a coordinate change explicitly relating the two. Explain why the null curves in each (v, r) plane must be the radial null geodesics, and use this fact to obtain their equations and to plot them. (Draw the constant-r lines as vertical and the constant-v lines as sloping inwards from the right at 45°.) Identify the event horizon and the singularity (§27.9).

The event horizon is not made of any material substance. It is merely a particular (hyper)surface in spacetime, separating those places from which signals can escape to external infinity from those places from which all signals would inevitably be trapped by the black hole. A hapless observer who falls through the event horizon, from the outside to the inside, would not notice anything locally peculiar just as the horizon is crossed. Moreover, the black hole itself is not a ponderable body; we think of it merely as a gravitating region of spacetime from within which no signal can escape. And what of the fate of the poor star itself? We shall be coming to this conundrum shortly, in §27.9.

First, let us consider the observational situation. Is there evidence for the existence of black holes? Indeed there is. In the 1970s a number of examples of curious 'double-star' systems were known, where only one member of the pair would be luminous, in visible light. The existence, the mass, and the motion of the other was inferred from the fine details of the motion of its visible partner. Moreover, from the emission of X-ray signals coming from its vicinity, the invisible partner was deduced to be a compact object, with a mass too great for the object to be of either of the two types—a white dwarf or neutron star—that accepted physical principles could allow a compact star to be. The X-ray emission was consistent with the invisible object being a black hole surrounded by what is known as an 'accretion disc' of gas and dust, spiralling gradually closer and closer in towards the hole at ever greater velocities, becoming enormously heated as it gets nearer to the centre. Eventually, X-rays would be emitted before the material actually enters the hole (see Fig. 27.12a). The best known (and at that time observationally most persuasive) of these black-hole candidates was the X-ray source Cygnus X-1, the compact and dark member of the pair having a mass of about $7M_{\odot}$, which certainly bars it from being a white dwarf or neutron star, according to accepted theory.

This kind of evidence was always rather indirect and less than totally satisfactory, because it relied upon *theory* to tell us that such massive compact objects cannot exist as extended bodies. Now, however, there is some rather more directly impressive evidence for black holes. Accretion discs are not the only configurations taken up by material falling into black holes. In some cases the material simply falls 'straight in', and this type of behaviour appears now to have been observed (Fig. 27.12b). If the attracting compact object were to have a material surface of any kind, then the infalling material would heat that surface up, and its glow would become visible after a time. But no such glow is seen. There is thus now some direct evidence that such a compact entity has no surface at all, and it may be rather convincingly inferred that the entity is indeed a black hole.[19]

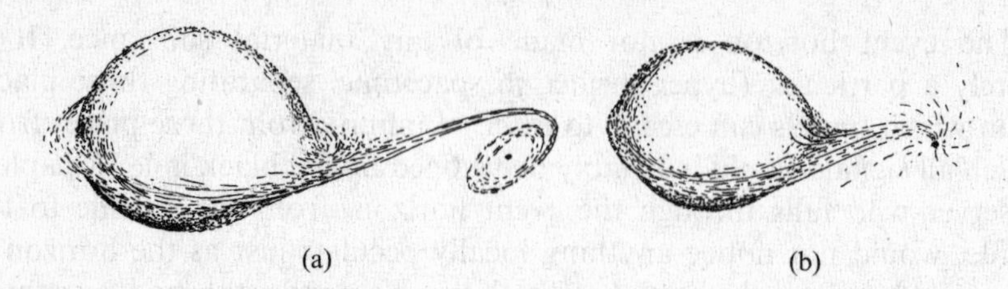

Fig. 27.12 Double star systems, one member of which is a (tiny) black hole. (a) Matter dragged from the larger star by the black hole forms an accretion disc around it, gradually spiralling in and becoming heated until X-rays are emitted before the material actually enters the hole. (b) In some cases there is no accretion disc and the material simply falls 'straight in'. If the attracting compact object were to have a ponderable surface, in-falling material would heat it up, but no glow is seen, confirming the presence of a black hole.

All this refers to 'stellar' black holes, whose mass would be just a few times that of the Sun. There is also some impressive evidence for very much larger black holes. It seems that most—and perhaps all—galaxies have very sizeable black holes at their centres. In particular, there appears to be a massive $3 \times 10^6 \, M_{\odot}$, black hole at the centre of our own Milky Way galaxy, and the actual motions of stars orbiting about it have been tracked in detail, being fully consistent with this black-hole picture.

27.9 Event horizons and spacetime singularities

I have drawn in some of the null cones in Fig. 27.11, so that the causality properties of the spacetime should become reasonably apparent. The most essential feature is the existence of the black hole's event horizon which, in the spacetime, is a 3-surface $\mathcal{H}$. As stated in §27.8, this has the property that no signal originating in the region inside $\mathcal{H}$ can escape to the outside region. This can be seen to be an effect of the tilting of the cones inwards, so that they find themselves to be tangent to $\mathcal{H}$. Any world line that crosses from the inside to the outside of $\mathcal{H}$ would have to violate the causality that the cones define (§17.7). I have depicted the case where the gravitational collapse is completely spherically symmetrical, which was the original situation studied by J. Robert Oppenheimer and Hartland Snyder (1939), and which employs Schwarzschild's geometry to describe the region external to the infalling matter.

Although the horizon $\mathcal{H}$ has strange properties, the local geometry there is not significantly different from elsewhere. As noted above, an observer in a space ship would notice nothing particular happening as the horizon is crossed from the outside to the inside. Yet, as soon as that perilous journey has been undertaken, there is no return. The tipping of the null cones is such that there is no escape, and the observer would encounter rapidly increasing tidal effects (spacetime curvature; see §17.5 and §19.6) that diverge to infinity at the spacetime singularity at the centre ($r = 0$). These features are not specific to the case of spherical symmetry, but are quite general. There are, indeed, very comprehensive theorems which tell us that singularities cannot be avoided in any gravitational collapse that passes a certain 'point of no return'.[20] Some of the relevant issues will be discussed in a little greater detail in §28.8.

For a black hole of a few solar masses, the tidal forces would be easily enough to kill a person long before the horizon is even reached, let alone crossed, but for the large black holes of $10^6 M_\odot$, or more, that are believed to inhabit galactic centres, there would be no particular problem from tidal effects as the horizon is crossed (the horizon being some millions of kilometres across). In fact, for our own galaxy, the curvature at the horizon of its central black hole is only about twenty times the spacetime curvature here at the surface of the Earth—which we don't even notice! Yet, the relentless dragging of the observer inwards to the singularity at the centre would subsequently cause tidal effects to mount very rapidly to infinity, totally destroying the observer in less than a minute! Destruction by rapidly mounting tidal forces is, indeed, what awaits any physical material as it plunges inwards towards the centre of a black hole. Recall our concern about the fate of the material of our $10M_\odot$ collapsing star. Even the individual particles of which it is composed will, in short order, encounter tidal forces so strong that they will be torn apart—to what, no-one knows!

At least, what we *do* know is that, so long as Einstein's picture of a classical spacetime can be maintained, acting in accordance with Einstein's equation (with non-negative energy densities and some other mild and 'reasonable' assumptions), then a spacetime singularity will be encountered within the hole.[21] The expectation is that Einstein's equation will tell us that this singularity cannot be avoided by any of the matter in the hole and that the 'tidal forces' (i.e. Weyl curvature; see §19.7) will diverge to infinity—very possibly in a wildly quasi-oscillatory fashion, in the general case.[22] In fact, it seems unavoidable that the realm of *quantum gravity* (or whatever is the appropriate term) will be entered, so that these expectations of the classical theory will have to be modified in accordance with this. We do not yet know what the correct 'quantum-gravitational' theory must be, but these black-hole considerations supply us with an important input; and this input should be guiding us in the appropriate directions in our search for the correct 'quantum gravity'. These issues will be important for us in later chapters, particularly Chapters 30, 31, and 32.

It is generally believed that the spacetime singularities of gravitational collapse will necessarily always lie within an event horizon, so that whatever happen to be the extraordinary physical effects at such a singularity, these will be hidden from view of any external observer. This is not a mathematically established property of general relativity, however. The assumption that the singularities will always be so hidden is referred to as *cosmic censorship*,[23] and it will be discussed more fully in §28.8.

On the other hand, we do not have to go so far as the singularity in order to find extraordinary effects resulting from gravitational collapse. There are some very violent processes visible in the universe. For example, the exceptionally luminous *quasars* are believed to be powered by rotating black holes at galactic centres, where the rotation of the black hole is the powerhouse, although the actual material ejected (apparently along the axis of rotation) comes from outside the hole (see §30.7). The energy emitted by some quasars, though coming from a tiny region (about the size of the Solar System), can outshine an entire galaxy by a factor of 10^2 or 10^3 or more! They can be seen at enormous distances, and are important observational tools for cosmology. There are also sources of powerful γ rays (extremely energetic photons) that are also believed to involve black holes, perhaps pairs of black holes in collision.[24]

27.10 Black-hole entropy

Let us return to consideration of the 'safer' external regions of isolated stationary ('dead') black holes. We shall see how extraordinarily large is the entropy that is to be assigned to such an object. First, we should take note of the fact that there are mathematical theorems[25] providing compelling evidence that general black holes, which may initially possess complicated irregularities due to an asymmetrical collapse—in a possibly wildly spiralling and irreversible catastrophe—will nevertheless rapidly settle down (as far as their external spacetime geometries are concerned) to a remarkably simple and elegant geometrical form. This is described by the *Kerr metric*,[26] and it is characterized by just two physical/geometrical parameters (real numbers) labelled m and a.[27] Here, m describes the total mass of the black hole, and $a \times m$ is the total angular momentum (in units where $G = c = 1$). As the Nobel Laureate Subrahmanyan Chandrasekhar (whose famous 1931 result, as we recall from §27.8, set astrophysics on the route to the black hole) has written:

> The black holes of nature are the most perfect macroscopic objects there are in the universe: the only elements in their construction are our concepts of

space and time. And since the general theory of relativity provides only a single unique family of solutions for their descriptions, they are the simplest objects as well.[28]

The relentless nature of a black hole, as it sweeps up all kinds of material—which could have an immense amount of detailed structure—converting it into a single configuration describable by only ten parameters (these being a, m, the direction of the spin axis, the position of the mass centre, and its 3-velocity) is a powerful manifestation of the Second Law. These ten parameters are all that are needed for an adequate macroscopic characterization of the final state.[29] Although a black hole does not look like ordinary matter in thermal equilibrium, it shares with it the key property that huge numbers of microscopically distinct states lead to something which can be described by very few parameters. For this reason, the corresponding phase-space coarse-graining box is indeed stupendous, and black holes, consequently, have enormous entropies.

In fact, black-hole entropy has a remarkable geometrical interpretation: it is proportional to the area of the hole's horizon! According to the famous *Bekenstein–Hawking formula*, a well-defined entropy can indeed be attributed to a black hole, which is

$$S_{\mathrm{BH}} = \frac{kc^3 A}{4G\hbar}$$

where A is the surface area of the black hole's horizon—and where you can take BH to stand for Bekenstein–Hawking or black hole, as you wish! Note the appearance of Planck's constant, as well as the gravitational constant, indicating that this is this entropy is a 'quantum-gravitational' effect. Indeed, this is the first place where we have encountered both the fundamental constant of quantum mechanics (Planck's constant, written in Dirac's form $\hbar$) and that of general relativity (Newton's gravitational constant G) appearing together in the same formula.

For issues of fundamental physics, in which quantum mechanics and general relativity are both involved, it is often convenient to adopt units in which both these constants are taken to be *unity*. We have already seen, in §17.8 and §§19.2,6,7 (and elsewhere such as in Chapter 24), that it is often extremely convenient to adopt units in which the speed of light c is taken to be unity. Without loss of consistency, we can extend this convention so that $\hbar$ and G are also unity. This has the remarkable implication that all the units of time, space, mass, and electric charge are now completely *fixed*, providing what are known as *Planck* units (or *natural* units or *absolute* units). Moreover, we can take Boltzmann's constant k to be unity also (see §27.3)

$$G = c = \hbar = k = 1,$$

and then the unit of temperature becomes an absolute thing as well.

These are far from practical units for everyday use, as can be seen when we try to put our conventional units in terms of Planck units:

$$\text{gram} = 4.7 \times 10^{4},$$
$$\text{metre} = 6.3 \times 10^{34},$$
$$\text{second} = 1.9 \times 10^{43},$$
$$\text{degree Kelvin} = 4 \times 10^{-33}.$$

In these units, the charge on the proton (or *minus* that on the electron) comes out as roughly $e = \frac{1}{\sqrt{137}}$, and more precisely as[30]

$$e = 0.0854245\ldots$$

We can also express these relations the other way around, finding that

$$\text{Planck mass} = 2.1 \times 10^{-5}\ \text{g},$$
$$\text{Planck length} = 1.6 \times 10^{-35}\ \text{m},$$
$$\text{Planck time} = 5.3 \times 10^{-44}\ \text{s},$$
$$\text{Planck temperature} = 2.5 \times 10^{32}\ \text{K}$$
$$\text{Planck charge} = 11.7\ \text{proton charges}$$

We shall be seeing more about Planck units in §31.1

Returning to the Bekenstein–Hawking formula for black-hole entropy, we now find that, in Planck units, the entropy S_{BH} of a black hole of surface area A is simply

$$S_{BH} = \tfrac{1}{4}A$$

In the case of the Kerr solution, we find explicitly

$$A = \frac{8\pi G^2}{c^4} m(m + \sqrt{m^2 - a^2})$$

$$S_{BH} = \frac{2\pi Gk}{c\hbar} m(m + \sqrt{m^2 - a^2})$$

(in general units). We shall be coming to some of the reasons lying behind the remarkable Bekenstein–Hawking formula in §30.4.

To get some idea of the extraordinary entropy values that can be achieved in a black hole, let us first consider what had been thought, in the 1960s, to provide the greatest of all contributions to the entropy of the universe, namely the entropy in the 2.7K microwave radiation: the remnants of the 'flash' of the Big Bang. This entropy is about 10^8 or 10^9 per baryon, in natural units. (Roughly speaking, this is the number of photons per baryon left over from the Big Bang.) Let us compare this admittedly enormous figure with the entropy due to black holes in the universe. Astronomers do not have a fully clear idea about how many black holes there are, nor what size they all might be, but there is very good evidence for a black hole at the centre of our own Milky Way galaxy of about $3 \times 10^6 M_\odot$, which may well be reasonably typical. Some galaxies have much larger black holes, and these ought easily to compensate for large numbers of other galaxies that might have smaller ones, since it is the large black holes that easily dominate the entropy values over all.[27.14] As a rough (probably very conservative) estimate, take our galaxy as typical, and we find an entropy per baryon of about 10^{21}, which completely dwarfs the 10^8 or 10^9 figure for the microwave background. Moreover, whatever the figure is now, it will relentlessly and stupendously grow in the future.

27.11 Cosmology

Before trying to find an estimate for the admittedly colossal entropy figure that is potentially accessible to our universe—so that we can get a feeling for how 'special' our universe actually is now, and of how 'particularly special' our universe must have been at the time of the Big Bang—we shall need to know something about cosmology. We shall try to use cosmological evidence to estimate the size of the box $\mathcal{B}$ of phase space that represents the Big Bang, and compare it with the size of the entire phase space $\mathcal{P}_\mathcal{U}$, and also with the phase-space volume of the coarse-graining box $\mathcal{N}$ that represents the universe as it is now.

Let me start by briefly describing what are known as the *standard models* of cosmology, of which there are (in essence) three. As we recall from §27.7 the discussion goes back to the Russian Aleksandr Friedmann, who in 1922 first found the appropriate cosmological solutions of the Einstein equation, with a material source that can be used to approximate a completely uniform distribution of galaxies on a large scale (sometimes called a 'pressureless fluid' or 'dust'). The cosmological models of the general class that Friedmann studied (sometimes with a different type of matter source from Friedmann's 'dust') are now commonly referred to as the Friedmann–Lemaître–Robertson–Walker (FLRW) models, owing to later contributions, clarifications, and generalizations from these others.

[27.14] Can you see why?

Basically, an FLRW model is characterized by the fact that it is completely *spatially homogeneous* and *isotropic*. Roughly speaking, 'isotropic' means that the universe looks the same in all directions, so it has an O(3) rotational symmetry group. Also, 'spatially homogeneous' means that the universe looks the same at each point of space, at any one time; accordingly, there is group of symmetries that is *transitive* (§18.2) on each member of a family of spacelike 3-surfaces, these being the 3-surfaces $\mathcal{T}_t$ of 'space' at constant 'time' t (giving a 6-dimensional symmetry group in all [27.15]). This pair of assumptions is in good accord with observations of the matter distribution on a very large scale, and with the nature of the microwave background. Spatial isotropy is found directly to be a very good approximation (from observations of very distant sources, and primarily from the 2.7 K radiation). Moreover, if the universe were *not* homogeneous, it could appear to be isotropic only from very particular places,[27.16] so we would have to be in a very privileged location for the universe to *appear* to us to be isotropic unless it were also homogeneous. Of course, the observational isotropy is not exact, since we see individual galaxies, clusters of galaxies, and superclusters of galaxies only in certain directions. There are uneven distributions of material, not always visible, on mind-boggling scales, such as that referred to as the 'Great Attractor' which seems to be pulling on not only our own galaxy but several neighbouring galaxy clusters. But it appears to be the case that the deviations from particular spatial uniformity get proportionally smaller, the farther away we look. The best information that we have for the most distant regions of the universe that are accessible to us comes from the 2.7 K black-body background radiation. The COBE, BOOMERANG, and WMAP data etc. tell us that, although there are very slight temperature deviations, at the tiny level of only a few parts in 10^5, isotropy is well supported.[31]

It indeed seems that the homogeneous and isotropic cosmologies—the FLRW models—are excellent approximations to the structure of the actual universe, at least out to the limits of the observable universe which extends to a distance that includes around 10^{11} galaxies, containing some 10^{80} baryons. (We shall be seeing what this notion of 'observable universe' means shortly.) Spatial isotropy and homogeneity implies[32] that 3-dimensional 'constant-time' spatial sections $\mathcal{T}_t$ fill up the whole spacetime $\mathcal{M}$ (without intersecting each other), each 3-geometry $\mathcal{T}_t$ sharing $\mathcal{M}$'s homogeneity/isotropy symmetry group; see Fig. 27.13. The (essentially) three different possibilities for the 3-geometry depend upon whether the (constant) spatial curvature is positive ($K>0$), zero ($K=0$), or negative ($K<0$). It is usual, in the cosmological literature, to normalize the

[27.15] Why 6-dimensional?

[27.16] Give a general argument to show why a connected (3-)space cannot be isotropic about two distinct points without being homogeneous.

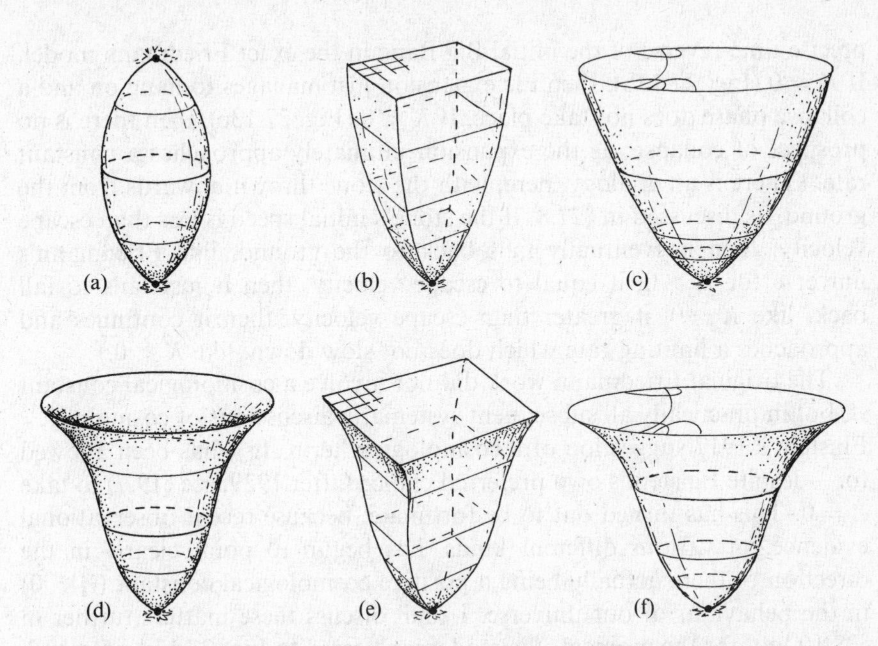

Fig. 27.13 Friedmann–Lemaître–Robertson–Walker (FLRW) spatially homogeneous and isotropic cosmological models. Time is depicted upwards and each model starts with a Big Bang. Each is filled with a 1-parameter family of non-intersecting homogeneous spacelike 3-surfaces $\mathcal{T}_t$, giving 'space' at time t. In Friedman's models, matter is treated as a pressureless fluid ('dust'). The three cases are illustrated: (a) $K > 0$, where the $\mathcal{T}_t$ are 3-spheres S^3 (indicated in the diagrams as a bounding circle S^1), where the model ultimately collapses in a Big Crunch; or (b) the $\mathcal{T}_t$ being Euclidean 3-spaces $\mathbb{E}^3$, depicted as the 2-plane at the top; or (c) the $\mathcal{T}_t$ being hyperbolic 3-spaces (indicated by a conformal representation at the top). In (d), (e), and (f), a positive cosmological constant Λ is incorporated into (a), (b), (c), respectively, with ultimate exponential expansion, where in case (d) it is assumed that Λ is large enough to prevent the collapsing phase.

radius of curvature, in the $K \neq 0$ cases, referring to $K > 0$ and $K < 0$ as simply $K = 1$ and $K = -1$, respectively. I shall not do this here, however, for clarity in later discussions, preferring the respective descriptions $K > 0$ and $K < 0$.

In Fig. 27.13a,b,c, I have tried to depict the time-evolution of the universe, according to Friedmann's original analysis of the Einstein equation, for the different alternative choices of spatial curvature. In each case, the universe starts from a *singularity*—the so-called Big Bang—where spacetime curvatures become infinite and then it expands rapidly outwards. The ultimate behaviour depends critically on the value of K. If $K > 0$ (Fig. 27.13a), the expansion eventually reverses, and the universe returns to a singularity, often referred to as the *Big Crunch*, which is a precise time-reverse of the initial Big Bang in the exact Friedmann model. If $K = 0$ (Fig. 27.13b), then the expansion just manages to hang on and a collapse phase does not take place. If $K < 0$ (Fig. 27.13c), then there is no prospect of collapse, as the expansion ultimately approaches a constant rate. (There is an analogy, here, with the stone thrown upwards from the ground, as discussed in §27.8. If the stone's initial speed is less than escape velocity, then it eventually falls back to the ground, like Friedmann's universe for $K > 0$; if equal to escape velocity, then it just fails to fall back, like $K = 0$; if greater than escape velocity, then it continues and approaches a limiting rate which does not slow down, like $K < 0$.)

The original Friedmann work did not involve a cosmological constant Λ, but in practically all subsequent systematic discussions of cosmology,[33] Einstein's 1917 suggestion of a cosmological term Λg_{ab} has been allowed for—despite Einstein's own preferred choice (after 1929, see §19.7) to take $\Lambda = 0$. This has turned out to be fortunate, because recent observational evidence, of various different kinds, has begun to point clearly in the direction of there actually being a positive cosmological constant ($\Lambda > 0$) in the behaviour of our universe. I shall discuss these matters further in §28.10 but, for the moment, the reader is referred to Fig. 27.13d,e,f, which provide the analogues of Fig. 27.13a,b,c in which a (sufficiently large) positive Λ is incorporated into Friedmann's equations. According to the present-day balance of observation and opinion among cosmologists, one of these models would seem to be a fair description of the history of our actual universe, at least from the time of *decoupling*, when the universe was a mere $\sim 3 \times 10^5$ years old, which is about 1/50000 of its present age of roughly 1.5×10^{10} years, decoupling being the time that we effectively 'look back to' when we observe the microwave background.

Prior to decoupling, the universe would have been basically 'radiation dominated' and after decoupling, 'matter dominated'. We would not expect Friedmann's 'dust' model to be appropriate in the radiation-dominated phase, and more appropriate might be the radiation-filled model of Tolman (1934). This does not make a great deal of difference in our pictures. It shortens the universe's lifetime from Big Bang to decoupling by a factor of about $\frac{3}{4}$ from what would have been the 'Friedmannian' prediction[27.17] as I have indicated in Fig. 27.14. The proponents of inflationary cosmology suggest a much greater change in the evolution, namely an exponential expansion, increasing the scale of the universe by a factor of perhaps 10^{60}. But this would have come to an end by the time the universe was only about 10^{-32} seconds old, so it does not make any

[27.17] See if you can derive this $\frac{3}{4}$ factor, assuming that the behaviour of a Friedmannian 'dust' model is of the form $t = AR^{3/2}$ for small values of the time t, and that of Tolman's 'radiation' is $t = BR^2$, where $R = R(t)$ is a measure of 'radius' of the universe, and A and B are constants. *Hint*: Must the tangents to the curves match?

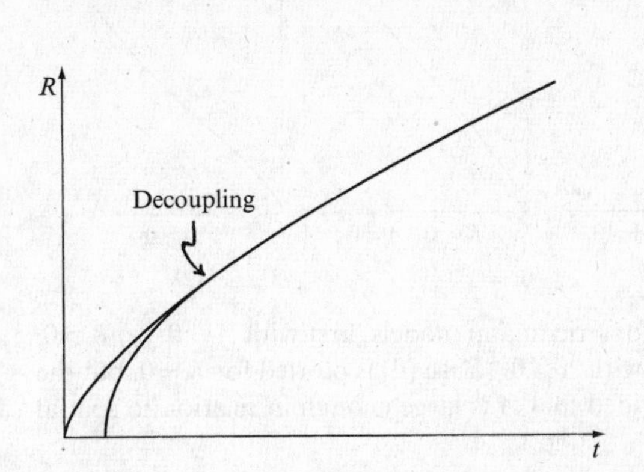

Fig. 27.14 Before 'decoupling', which occurred when the universe was about 300 000 years old (only about 1/50 000 of its present age)—the epoch we 'look back' to with the microwave background—the universe was 'radiation dominated' and Friedmann's 'dust' approximation did not hold. Instead, we have the somewhat more rapid Tolman expansion, indicated by the inner curve.

difference to the appearance of Fig. 27.13 or 27.14! Yet the implications in other respects could be enormous, if the inflationary picture is correct. I shall consider inflationary cosmology in §§28.4,5. In any case, I think it is reasonable not to include inflation in what is to be called 'the standard model of cosmology', and I shall not do so here.[34]

But which particular one of the three models of Fig. 27.13d,e,f is likely to be the appropriate for the actual universe? I shall discuss this issue in §28.10. For the moment, let us consider that any one of them *might* be basically correct. Let us examine each of these different spatial geometries a little further.

The case $K > 0$ is normally represented as the 3-sphere. It should be mentioned, however, there is also the *projective space* $\mathbb{RP}^3$ obtained by identifying antipodal points of S^3 (see §2.7, §§15.4–6); it is hard to imagine that the two would be observationally distinguishable, in practice. There are other identifications between separated points of S^3, giving what are called *lens spaces*, but none of these is globally isotropic.[35] The (isotropic) case $K = 0$ is ordinary Euclidean 3-space, and $K < 0$ likewise gives hyperbolic 3-geometry, which we studied in §§2.4–7 and §18.4. See Fig. 2.22a,b and c, respectively, for M.C. Escher's elegant and ingenious representations of (the 2-dimensional versions of) the respective spatial geometries for $K > 0$, $K = 0$, and $K < 0$. The usual $K > 0$ case is called a *closed* universe, which means *spatially closed* (i.e. contains a compact spacelike hypersurface[36]). Frequently cosmologists refer to $K < 0$ as the 'open' case, whereas technically the $K = 0$ case is also spatially open. Accordingly, I shall not use this somewhat confusing terminology here. If we abandon global isotropy then, as with the $K > 0$ lens spaces referred to above, there are (non-isotropic) closed-universe models also for $K = 0$ and $K < 0$.[37]

The full 4-space $\mathcal{M}$ is described in terms of a time-evolution for the spatial 3-geometry, as we have seen, where there is an overall scale that changes with time. In the standard picture, the universe initially expands

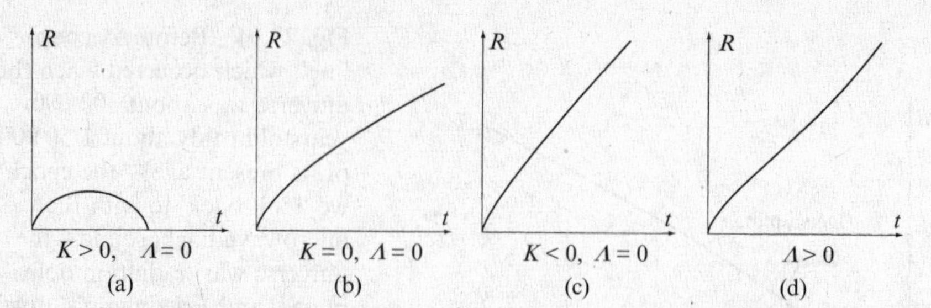

Fig. 27.15 Graphs of $R = R(t)$ for Friedmann models, first with $\Lambda = 0$: (a) $K > 0$, (b) $K = 0$, (c) $K < 0$, and then (d) with $\Lambda > 0$. (Case (d) is plotted for $K = 0$, but the other cases are very similar, provided that Λ is large enough in relation to spatial curvature.)

very rapidly away from a Big Bang, but it is an incorrect picture to think of a 'central point' at which the explosion occurred, and from which everything recedes. A more appropriate image, in the case of two spatial dimensions, is the surface of a balloon as it is being blown up. Each point on the surface gradually recedes from each other point, as time passes, and there is no 'central point' in the universe-model. In this analogy, the surface is to represent the entire universe. Thus, the centre of the balloon does not count as part of the expanding universe; nor does any other point that does not lie on the surface.

Let us use the notation $\mathrm{d}\Sigma^2$ to denote the metric form of one of these three 3-geometries, where in the cases $K \neq 0$ we now normalize the metric to be that of the *unit* 3-sphere or *unit* hyperbolic space (i.e. we take $K = 1$ or $K = -1$ respectively).[27.18] The 4-metric of the entire spacetime can then be expressed in the form

$$\mathrm{d}s^2 = \mathrm{d}t^2 - R^2\mathrm{d}\Sigma^2,$$

where t is a 'cosmic-time' parameter, whose constant values determine the individual $\mathcal{T}_t$, and where

$$R = R(t)$$

is some function of the time-parameter t giving the 'size' of the spatial universe 'at time t'. Thus, the metric for each $\mathcal{T}_t$ is given by $R^2\mathrm{d}\Sigma^2$. In Fig. 27.15a,b,c, I have plotted the graph of $R = R(t)$ for $K = 1, 0, -1$,

[27.18] See if you can show that $\mathrm{d}\Sigma^2 = \mathrm{d}r^2 + \sin^2\varphi\,(\mathrm{d}\varphi^2 + \sin^2\theta\,\mathrm{d}\theta^2)$ describes the metric of a unit 3-sphere, and deduce that $\mathrm{d}\Sigma^2 = \mathrm{d}r^2 + \sinh^2\chi\,(\mathrm{d}\chi^2 + \sin^2\theta\,\mathrm{d}\theta^2)$ describes unit hyperbolic space, using the procedures of §18.1. *Hint*: Write down the metric for a 3-sphere of arbitrary radius first.

respectively, in Friedmann's original case of 'dust' (pressureless fluid)[27.19] with $\Lambda = 0$, and in Fig. 27.15d, I have shown what happens with a positive Λ, the curves for all three values of K being very similar (provided that, in the case $K > 0$, Λ is large enough to overcome subsequent collapse—as observations indeed suggest). The ultimate expansion rate is then exponential.

27.12 Conformal diagrams

To understand what is meant by the term 'observable universe', it is helpful to employ what is known as a *conformal diagram*[38] in which a (frequently 2-dimensional) representation of the entire spacetime is presented so that the null directions are drawn at 45° to the vertical, and where infinity is also represented, as (part of) the boundary of the diagram. The script letter $\mathscr{I}$ is commonly used—and pronounced 'scri'—for this notion of 'infinity', where $\mathscr{I}^+$ is used for *future* (or future null) infinity, ultimately 'reached' by outgoing light rays, and $\mathscr{I}^-$ for *past* infinity, for incoming light rays. They normally turn out to be *null* 3-surfaces in standard Einstein theory with $\Lambda = 0$, and *spacelike* 3-surfaces if $\Lambda > 0$.[39]

Conformal diagrams depict the *causality* structure of the spacetime, where it is the family of null cones, rather than the full spacetime metric, that we are interested in. This is the Lorentzian version of the conformal geometry that we encountered in §2.4, §8.2, and §§18.4,5 (defined by an equivalence class of metrics, $\boldsymbol{g}$ being equivalent to $\Omega^2\boldsymbol{g}$ where Ω is a positive scalar function on the spacetime, so Ω modifies the distance scale from place to place). In §2.2, we saw how the entire hyperbolic plane can be represented conformally in a finite region of the Euclidean plane (Figs. 2.11, 2.12, 2.13). The idea of a *conformal spacetime diagram* is basically the same, but now it is the Lorentzian (non-positive-definite) metric of spacetime that is being conformally represented. The key new feature is that, in Lorentzian geometry, the null cones themselves define the conformal geometry.

In two dimensions, the null cone consists of a pair of null directions, and this determines the 2-metric up to a local conformal factor. A circumstance where such a 2-dimensional representation is particularly valuable is where there is spherical symmetry in the entire 4-space. Then we can think of this 4-spacetime as being a 2-spacetime that is 'rotated around', so that each point of the 2-space represents an entire S^2 in the 4-space. For such spacetimes, the conformal diagrams can be made quite precise, and

---

[27.19] Friedmann's 'dust' solution for $K > 0$, $\Lambda = 0$ can be expressed in the form $R = C(1 - \cos\xi)$, $t = C(\xi - \sin\xi)$, where C is a constant and ξ is a convenient parameter. Show that this is the equation of a *cycloid*—the curve traced out by a point on the circumference of a circle rolling on a horizontal straight line. Can you see how to get from the $K > 0$ to the $K < 0$ case using a 'trick' similar to that used in §18.1 and Exercise [27.16], and to $K = 0$ by taking a suitable limit (involving a coordinate rescaling)?

for these I shall make use of the notion of a *strict* conformal diagram. Conformal diagrams that are not strict will be called *schematic*. The *points* of a strict conformal diagram indeed represent entire (metric) spheres S^2. (In the case of an n-dimensional Lorentzian 'spacetime' of the kind that one might consider in string theory etc.—see §§31.4,7—these would be $(n-2)$-spheres S^{n-2}.) Exceptional places, where points of the diagram represent single spacetime points, occur on those parts of the boundary of the diagram that describe a *symmetry axis*. These are indicated by dashed lines, so you have to think of the diagram as being rotated about such a dashed line.[27.20] The parts of the boundary that represents infinity are indicated by solid lines, and those parts that represent singularities are indicated by jagged lines. See Fig. 27.16a. There are also certain corners where different boundary lines of a conformal diagram meet. Those that are indicated by little open circles ∘ are to be thought of as representing entire 2-spheres (like the boundary of hyperbolic 3-space; see §2.4 and §18.4), while those indicated by filled-in points • are best thought of as representing points (spheres of zero radius). Figure 27.16b is the strict conformal diagram for Minkowski space and Fig. 27.16c, for gravitational collapse to a Schwarzschild black hole (the spherically symmetric collapse described in Fig. 27.11). In Fig. 27.17, I have depicted the respective cosmological models of Fig. 27.13.[27.21]

Conformal diagrams are useful because they make the causality properties of the spacetimes particularly manifest. Note, for example, that in the spherically symmetrical collapse to a black hole depicted in Fig. 27.16c, the black hole's horizon lies at 45°. Any material particle's world line cannot tilt at more that 45° to the vertical, so it cannot escape from the interior region behind the horizon once it has crossed into it. Moreover, once inside that region, it is forced into the singularity (Fig. 27.18a). The singularity appears to be a spacelike future boundary to the interior part of the spacetime, a fact that is somewhat counter-intuitive from the more conventional perspective of Fig. 27.11. The situation exhibited by the Big Bang plays a role like the temporal reverse of this, acting as a spacelike past boundary to the spacetime (Fig. 27.18b). This is again somewhat counter-intuitive, since we tend to think of the Big Bang as a (singular) *point*.[40]

The spacelike nature of this initial boundary leads us to the notion of a particle horizon, which is an important aspect of the Big Bang. Consider

---

[27.20] See if you can obtain Minkowski 4-space of Fig. 27.16b explicitly, by taking the right-hand half of the entire Minkowski 2-space (metric $ds^2 = dt^2 - dr^2$, with $r \geq 0$), and rotating about the vertical axis in this way. Express the 4-space metric, using suitable functions of t, r, and spherical polar angles θ and ϕ (see Exercise [27.18]). (For visualization purposes, try obtaining Minkowski 3-space first, where the rotation is now of a more familiar type.

[27.21] See if you can see how the diagrams of Figs. 27.11 and 27.16b match up. Find appropriate conformal factors multiplying the metrics for each of the examples of Figs. 27.16,27.17.

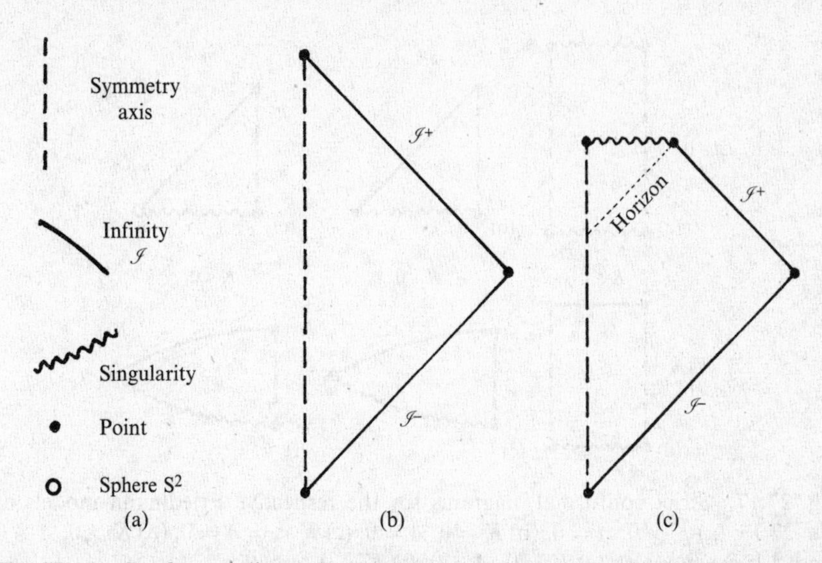

Fig. 27.16 Conformal diagrams are plane representations of spacetimes, usually drawn so that the spacetime null lines which lie in the plane itself are oriented at 45° to the vertical, and where 'infinity' tends to be represented as a finite boundary to the picture—where the conformal factor from the physical metric to that of the diagram goes to *zero* on the boundary. (a) In a *strict* (as opposed to *schematic*) conformal diagram, each point in the interior of the diagram represents an exact 2-sphere; but on a symmetry axis (shown with a dashed line) this 2-sphere shrinks to a point, as it does at a corner marked with •; but at a corner marked with ∘, the boundary point remains conformally a 2-sphere. Infinity is indicated with a solid-line boundary (often denoted $\mathscr{I}$—and pronounced 'scri'); singularities are indicated as wiggly-line boundaries. (b) A strict conformal diagram for Minkowski space $\mathbb{M}$. (c) A strict conformal diagram for Fig. 27.11, depicting spherically symmetric collapse to a black hole.

Fig. 27.18b, where an observer is at a point p close to the Big Bang boundary. The region of the universe that can transmit information to the observer is that region on or within the past light cone of p, and we note that this intersects only a portion P of the Big Bang initial hypersurface.[41] Particles created in the Big Bang in the region outside P are not accessible to observation at p. These regions are beyond p's *particle horizon*. We say that they lie outside p's *observable universe*—this observable part of the universe being that lying on or within p's past light cone.[27.22]

[27.22] By referring to the conformal diagrams given here, show that for, $K = 0$ or $K < 0$, where $\Lambda = 0$, the observable universe of a particle originating at p increases to include the entire universe, in the particle's future limit of time, whereas this is not true for the case $K > 0$, or for any K (in the cases depicted in Fig. 27.17) if $\Lambda > 0$ (where a 'cosmological event horizon' occurs).

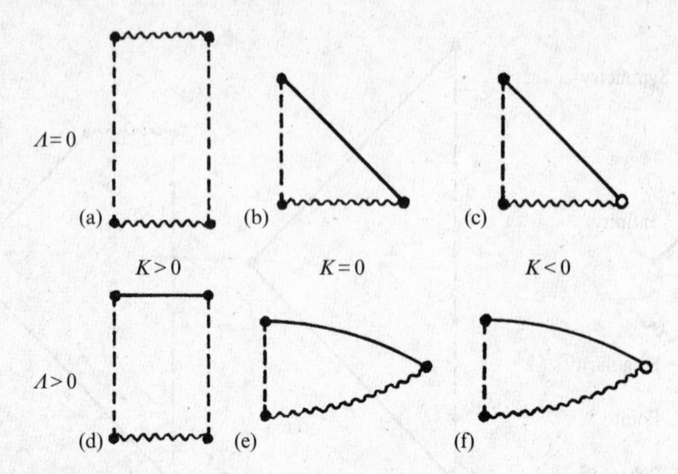

Fig. 27.17 Strict conformal diagrams for the respective Friedmann models of Fig. 27.13: (a) $K>0$, $\Lambda=0$; (b) $K=0$, $\Lambda=0$; (c) $K<0$, $\Lambda=0$; (d) $K>0$, $\Lambda>0$ (and Λ large enough); (e) $K=0$, $\Lambda>0$; (f) $K<0$, $\Lambda>0$.

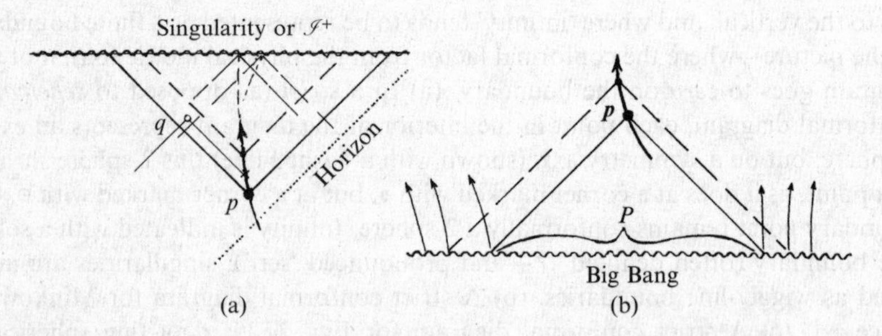

Fig. 27.18 Horizons. (a) Event horizons occur when a future boundary—either a singularity or infinity—is spacelike in a schematic conformal diagram. As an observer p approaches the boundary, there always remains some portion of the spacetime (whose boundary is defined as an event horizon) that p cannot see, although exactly which part depends upon how p moves. (For example, the event q is ultimately seen if p takes the left-hand route, but not the right-hand route.) In the case of a black hole the more familiar 'event horizon' of a more absolute nature (dotted, in the figure) being common to all external observers. (b) Particle horizons occur in all standard cosmologies, arising from the past singularity being spacelike. The observer at p sees only the limited portion P of the Big Bang (and of particles produced there), although this portion grows with time.

27.13 Our extraordinarily special Big Bang

Now let us return to the extraordinary 'specialness' of the Big Bang. The fact that it must have had an absurdly low entropy is already evident from the mere existence of the Second Law of thermodynamics. But low entropy

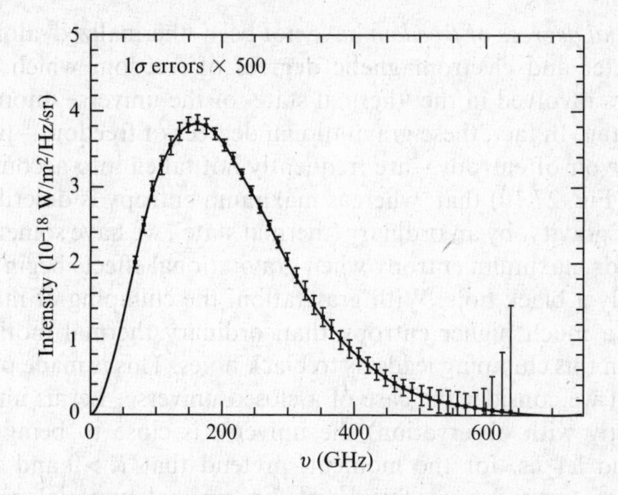

Fig. 27.19 The microwave background has an intensity, in terms of frequency that is extremely precisely in accord with Planck's black-body curve (Fig. 21.3b). (Note that the 'error bars' shown are exaggerated by a factor of 500.)

can take many different forms. We want to understand the *particular way* in which our universe was initially special.

One especially striking—and seemingly contradictory—property of the Big Bang comes from excellent observational evidence that the very early universe was in a *thermal state*. Part of this evidence is the exceptional closeness to the theoretical *Planck* 'black-body' curve (see §21.4, Fig. 21.3b) that is exhibited by the 2.7 K microwave background radiation that represents the actual 'flash' of the Big Bang, still in evidence today, though immensely cooled by the 'red-shift' caused by the expansion of the universe (Fig. 27.19). Other evidence comes from the remarkably detailed agreement between what theory and observation tell us about nuclear processes in the early universe. These theoretical calculations crucially depend on assuming the *thermal equilibrium* of the matter in the early universe—taken in conjunction with the universe's rapid expansion.

It seems to me that this apparent thermal equilibrium in the early universe has grossly misled some cosmologists into thinking that the Big Bang was somehow a high-entropy 'random' (i.e. thermal) state, despite the fact that, because of the second law, it must actually have been a very organized (i.e. low-entropy) state. A prevalent view seems to have been that the resolution of this paradox must lie in the fact that, soon after the Big Bang, the universe was 'small', so that comparatively few degrees of freedom were available to it, giving a low 'ceiling' to possible entropies. This point of view is fallacious, however, as pointed out in §27.6. The correct resolution of the apparent paradox lies in the fact that the *gravitational degrees of freedom* have not been 'thermalized' along with all those matter and electromagnetic degrees of freedom which define the parameters involved in the 'thermal state' of the universe, moments after the Big Bang. In fact, these gravitational degrees of freedom—providing a huge reservoir of entropy—are frequently not taken into account at all!

Recall (Fig. 27.10) that, whereas maximum entropy is described, in the *absence* of gravity, by an ordinary 'thermal state', we have something quite different for maximum entropy when gravitational effects begin to dominate, namely a black hole. With gravitation, the clumping of material can represent a much higher entropy than ordinary thermal motions, especially when this clumping leads us to black holes. This is made particularly manifest if we consider the case of a closed universe. Let us imagine that (consistently with observation) the universe is close to being a FLRW model, and let us, for the moment, pretend that $K>0$ and $\Lambda=0$. The presence of some irregularities[42] in the original material can lead to gravitational condensations, and let us suppose that these are sufficient to yield galaxies containing substantial black holes (say of $10^6 M_\odot$), providing us with an entropy per baryon of some 10^{21}. If we take our closed universe to have about 10^{80} baryons (about the baryon content of the observable universe), this gives a total entropy of 10^{101}, far larger than the 10^{88} that would be in the radiation and matter at the time of decoupling, about 300000 years after the Big Bang. The galactic black holes would gradually grow, but the major increase would occur during the universe's final collapse phase, when galaxies come back together and their black holes congeal. The final Big Crunch is not the tidy one depicted in Fig. 27.13a, which is the time-reversal of the neat symmetrical FLRW Big Bang; it is more like the dreadful mess of congealing black-hole singularities depicted in Fig. 27.20a. We can estimate the entropy of this mess of a Big Crunch by using the Bekenstein–Hawking entropy formula a little before its final state, when we can still think of the mess as being composed of actual black holes, approaching the ultimate black-hole agglomeration in which all 10^{80} baryons are involved. The value of S_{BH} for this number of baryons, which is about 10^{123}, should not be too far off the answer for the entropy to be assigned to this messy Big Crunch.

The reader may, of course, reasonably object, at this point, that even if it is true that $K>0$, present-day observations seem to be pointing strongly against the $\Lambda=0$ assumption that I have been making, where (together with observed limits on the spatial curvature) the observed positive value of Λ indeed seems to be easily large enough to prevent the occurrrence of the collapsing phase that I have been considering, but with the expectation of an ultimate exponential expansion instead. However, if phrased appropriately, the preceding discussion can still be applied, and the same measure of an entropy value ($\sim 10^{123}$) is found to be available to a closed

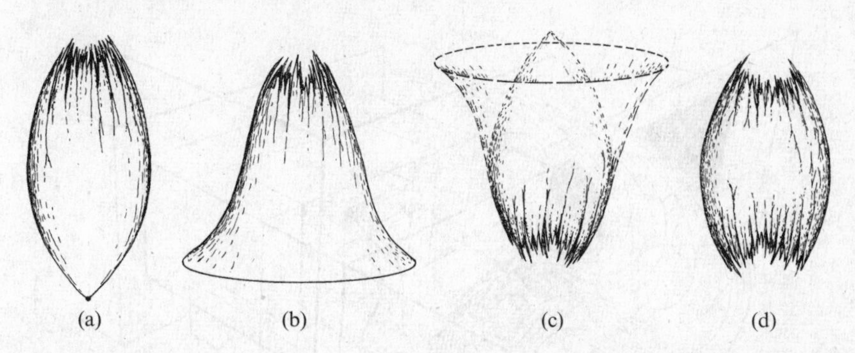

Fig. 27.20 (a) If, in the case $K > 0$, $\Lambda = 0$ of Fig. 27.13a, we allow for irregularities, of the type that we see in our actual universe, then instead of the 'clean' Big Crunch of the exact Friedmann model, we get a dreadful mess of congealing black-hole singularities of enormously higher entropy ($S \approx 10^{123}$). (b) This is not dependent on $\Lambda = 0$, as we could likewise consider corresponding perturbations of the time-reverse of Fig. 27.13d ($K > 0$, $\Lambda > 0$) and again we get a similar enormously high entropy mess ($S \approx 10^{123}$) of congealing black holes. (c) A generic Big Bang would look like that of the time-reverse of such a generic collapse (illustrated for $K > 0$ and either $\Lambda = 0$ or $\Lambda > 0$). (d) The most 'probable' situation (like the curve of Fig. 27.8a)—illustrated in the case $K > 0$, $\Lambda = 0$ for clarity—bears no similarity to the actual universe, in its early stages.

10^{80}-baryon universe irrespective of $\Lambda > 0$. For the time-reverse of the universe described by Fig. 27.13d is just as much a solution of the dynamical equations as is that described by Fig. 27.13d itself (as we are considering dynamical laws that should be reversible in time). If we consider perturbations of this universe, we can find models in which already-formed black holes come together and produce a similar kind of 'mess' of congealing black holes that we had before. (See Fig. 27.20b.) Again, we reach an entropy value which, by the same reasoning as before, is of the order of 10^{123}. (This type of reasoning will have relevance to us again, when we come to consider inflationary cosmology in §28.5.)

We are thus led to a reasonable estimate for the total volume of $\mathcal{P}_{\mathcal{U}}$ (which is essentially the same as the volume E of the maximum-entropy box $\mathcal{E}$ of Fig. 27.4), namely the exponential of this entropy value:

$$E = \mathrm{e}^{10^{123}} \approx 10^{10^{123}}$$

very closely.[27.23] (This comes from Boltzmann's $S = \log V$, in natural units.) Now, how does this compare with what we know of the volume N of the box $\mathcal{N}$ for entropy today, and with the volume B of the box $\mathcal{B}$ for

[27.23] Why are these figures—to within the precision expressed by the number '123'—virtually the same? Why does the actual value of B not appear in the conclusions below?

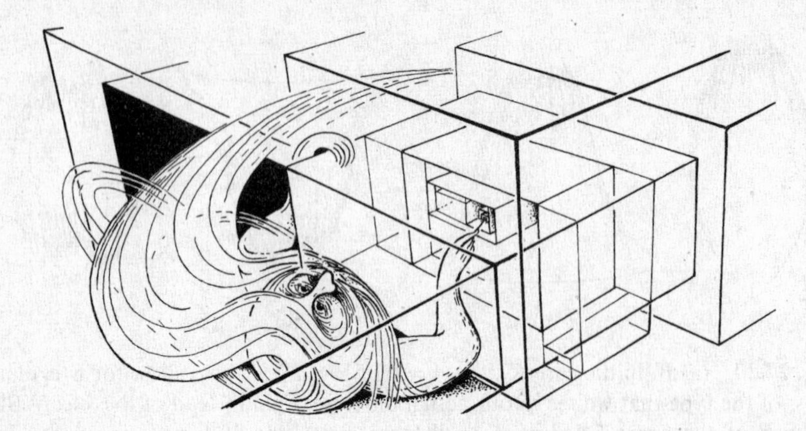

Fig. 27.21 Creation of the universe: a fanciful description! The Creator's pin has to find a tiny box, just 1 part in $10^{10^{123}}$ of the entire phase-space volume, in order to create a universe with as special a Big Bang as that we actually find.

the entropy in the Big Bang (assuming, for now, that we live in a 10^{80} baryon universe)? Taking the black-hole estimate, given above, for the entropy today, and the value of 10^8 for the entropy per baryon in the 2.7 K radiation, we find

$$B\colon N\colon E = 10^{10^{88}} : 10^{10^{101}} : 10^{10^{123}}.$$

It follows that each of B and N is only

$$\text{one part in } 10^{10^{123}}$$

of the total volume E. Moreover, the volume B is only

$$\text{one part in } 10^{10^{101}}$$

of the phase-space volume N of the universe today.

As a way of appreciating the problem posed by the absurdly tiny phase-space volume of $\mathcal{B}$, we can imagine the Creator trying to use a pin to locate this tiny spot in the space $\mathcal{P}_{\mathcal{U}}$, so as to start the universe off in a way that resembles what we know of it today. In Fig. 27.21, I have drawn a fanciful representation of this momentous event! If the Creator were to miss this spot by just the tiniest amount and plunge the pin effectively randomly into the maximum entropy region $\mathcal{E}$, then an uninhabitable universe like that of Fig. 27.20d in the case $\Lambda = 0$, $K > 0$, but otherwise rather like the ever-expanding case of 27.20c, would be the result, in which there is no Second Law to define a statistical time-directionality (like Fig. 27.8a). (Things are actually not much better explained if we imagine that our

Creator merely aims to construct a universe in which there are sentient beings, like ourselves. This raises the issue of the 'anthropic principle', and I shall discuss these matters in §§28.6,7 and in §34.7.)

On the other hand, it may well be that the universe is spatially *infinite*, like the FLRW models with $K = 0$ or $K < 0$. This does not invalidate the foregoing argument. We can imagine applying it to just the observable universe (at the present time) rather than to the entire universe. Taking the presently observable universe to contain about 10^{80} baryons, it is hard to see that the above considerations will be seriously affected. On the other hand, if we apply the arguments to the universe as a whole (still taking FLRW as a good approximation), we simply get a requirement of *infinite* precision on the part of the Creator, rather than merely absurdly large precision. I do not see how this in any way resolves the conundrum presented by the extraordinarily precise 'tuning' that was inherent in the Big Bang—an essential correlate of the Second Law.

What message do we carry from these considerations? We have learned not only that the Big Bang origin of the universe was extraordinarily special, but also something important about the nature of this specialness. As far as matter (including electromagnetism) was concerned, the description 'thermal equilibrium', in the context of an expanding universe, seems to have been highly appropriate. This is the successful 'Hot Big Bang' picture that is an important ingredient of the standard model of cosmology. After about 10^{-11}s, the universe seems to have had a temperature of about 10^{15} K, while after some 10^{2}s, the temperature dropped to 10^{9} K. This drop in temperature would have been in accordance with the Tolman–Friedmann expansion rate, and many observational details (e.g. hydrogen/deuterium/helium ratios) are consistent with the nuclear processes that would take place at those later temperatures.

Yet, for gravitation, things were completely different in that the gravitational degrees of freedom were not 'thermalized' at all. The very uniformity (i.e. FLRW nature) of the initial spacetime geometry was what was special about the Big Bang. The fact that an initial singular state for the universe 'need not have been so' is illustrated in Fig. 27.20c—or in the time-reversal of the physically appropriate Big Crunch of Fig. 27.20a. Gravity seems to have a very special status, different from that of any other field. Rather than sharing in the thermalization that, in the early universe, applies to all other fields, gravity remained aloof, its degrees of freedom lying in wait, so that the second law would come into play as these degrees of freedom begin to become taken up. Not only does this give us a Second Law, but it gives us one in the particular form that we observe in Nature. Gravity just seems to have been different!

But why was it different? We enter more speculative areas when we attempt answers to this kind of question. In Chapter 28, we shall see some of the ways that physicists have tried to come to terms with this puzzle and related ones, concerning the origin of the universe. In my opinion, none of these attempts comes at all close to dealing with the puzzle addressed in the preceding paragraph. In accordance with my own beliefs, we shall need to return to an examination of the very foundations of quantum mechanics, for it is my strong opinion that these issues are deeply connected. We shall do this in Chapter 29. Then, in Chapter 30, I shall try to present a good measure of my own perspective on these fundamental questions.

Notes

Section 27.2

27.1. This is taking the dynamics to be entirely classical. Technically, a 'chaotic system' is a classical system in which a tiny change in the initial state can result in a subsequent behavioural change that grows exponentially with time rather than, say, linearly. This 'unpredictability' is, of course, a matter of degree and not the matter of principle that it is sometimes assigned, with regard to determinism.

27.2. This assumes that the specific heats are positive, which is normally the case. But with black holes, this assumption is usually untrue; see §31.15.

27.3. There is the curious 'paradox', however, that in ordinary life, things are normally the other way around! One is frequently making accurate 'retrodictions' by simply remembering what happened in the past, whereas we have no corresponding access to the future. Moreover, archaeological investigations can extend such 'memories' to times far earlier than there were human beings. However, this retrodiction does not involve the evolution of dynamical equations, in any obvious sense, and its detailed connection with the second law still remains somewhat obscure to me. (See Penrose 1979a).

Section 27.3

27.4. See Pais (1986).

27.5. See Gibbs (1960); Ehrenfest and Ehrenfest (1959); Pais (1982).

27.6. Actually, Boltzmann himself never used this constant, since he did not concern himself with the actual units that might be used in practice; see Cercignani (1999). The formula $S = k \log V$, involving this constant, appears to have been first explicitly written by Planck; see Pais (1982).

Section 27.5

27.7. From Eddington (1929a).

Section 27.6

27.8. See Hawking and Ellis (1973); Misner *et al.* (1973); Wald (1984); Hartle (2002).

27.9. See Gold (1962); see Tipler (1997) for these ideas taken to a rather fanciful conclusion.

27.10. See Penrose (1979a).

Section 27.7

27.11. Some years before Hubble, in 1917, the American astronomer Vesto Slipher had already found some indication that the universe is expanding. See Slipher

(1917). Though he is rarely given credit for these observations, he also headed the team in which Clyde Tombaugh finally discovered Pluto!

27.12. This radiation was first theoretically predicted by George Gamow in 1946, on the basis of the Big Bang picture, and more explicitly by Alpher, Bethe, and Gamow in 1948; then again independently by Robert Dicke, in 1964. It was discovered observationally (accidentally), by Arno Penzias and Robert Wilson, in 1965, and immediately interpreted by Dicke and his colleagues; See Alpher *et al.* (1948); Dicke *et al.* (1965), and of course Penzias and Wilson (1965)—possibly the most modestly named scientific paper of all time!

27.13. For further discussion, see Penrose (1979a, 1989).

27.14. In fact, overall, the Earth sends back just slightly *more* energy than it receives. Ignoring the issue of human burning of fossil fuels, which finally returns some energy received from the Sun and stored in the Earth many millions of years ago (and, on the other side of the scales, ignoring the accompanying global warming that results from the 'greenhouse effect' whereby the Earth traps a little more of the Sun's energy than previously), there is the heating of the Earth's interior through radioactive decay, this energy being very gradually lost into space through the atmosphere. See §34.10.

Section 27.8

27.15. See Michell (1784); Tipler *et al.* (1980).

27.16. See Penrose (1978).

27.17. See van Kerkwijk (2000) for the state of the art on this matter.

27.18. See Schwarzschild (1916), or the modern presentation in Wald (1984).

27.19. See Narayan (2003) for recent evidence.

Section 27.9

27.20. The occurrence of what is known as a 'trapped surface' is one useful characterization of such a 'point of no return'. A trapped surface is a compact spacelike 2-surface $\mathcal{S}$ with the property that the two families of null normals to $\mathcal{S}$ both *converge* into the future. (In more 'colloquial' terms, this means that, if a flash of light originates at $\mathcal{S}$, then the areas of both the outgoing and ingoing parts of the flash will start to get smaller.) We expect to find trapped surfaces inside the horizon $\mathcal{H}$ of a black hole. The virtue of the trapped-surface criterion is that it does not depend upon any assumptions of symmetry, and it is 'stable' under small perturbations of the geometry. Once a trapped surface is formed, then singularities are inevitable (assuming certain very weak and reasonable conditions concerning causality and energy positivity in the Einstein theory). Similar results apply to the cosmological Big Bang singularity. See Penrose (1965b); Hawking and Penrose (1970).

27.21. See Penrose (1965b); Hawking and Penrose (1970). Wald (1984) reviews these theorems in a pedagogical setting.

27.22. See Penrose (1969a, 1998b); Belinskii *et al.* (1970).

27.23. See Penrose (1969a, 1998b).

27.24. See Reeves *et al.* (2002) for the most up-to-date view on these matters, and Cheng and Wang (1999); Hansen and Murali (1998) for the collision theory.

Section 27.10

27.25. See Israel (1967); Carter (1971); Hawking (1972); Robinson (1975).

27.26. See Kerr (1963); Newman *et al.* (1965) in the charged case. Wald (1984) has a pedagogical presentation.

27.27. Like Kepler's ellipses, as referred to at the beginning of this chapter, the Kerr metric supplies another of those exceptional situations where we have been blessed with the good fortune that relatively simple geometrical configurations actually arise from the dynamical laws.

27.28. See Chandrasekhar (1983), p. 1.

27.29. In fact (as we shall see in §31.15, and see Note 27.26), there is a further parameter which describes the total electric charge (this being a conserved quantity; see §19.3). But for realistic astrophysical black holes, it can be ignored in the black-hole geometry, being tiny in comparison with m and a, because of a strong tendency for a black hole to neutralize itself electrically.

27.30. One should, of course, be careful not to confuse this 'e' with the base of natural logarithms $\mathrm{e} = 2.7182818285\ldots$, see §5.3.

Section 27.11

27.31. See Smoot *et al.* (1991) for COBE evidence, Spergel *et al.* (2003) for WMAP.

27.32. Liddle (1999) is a superb introduction to cosmology. Wald (1984) covers the topic at a more sophisticated level.

27.33. See Bondi (1961); Rindler (2001); Dodelson (2003).

27.34. The term 'concordance model' has emerged to describe the situation for which $K = 0$ and $\Lambda > 0$ where inflation is also incorporated. See Blanchard *et al.* (2003); Bahcall *et al.* (1999). See §28.10 for my assesment of the present staus of this.

27.35. A rather peculiar possibility is that the ancient Greeks were right (Fig. 1.1) and the universe is actually a dodecahedron (or, rather a glued-up version of one). See Luminet (2003).

27.36. The term *hypersurface* refers to an $(n-1)$-dimensional submanifold of some n-manifold. Here, $\mathcal{T}_t$ is a spacelike 3-surface.

27.37. See Killing (1893); Wolf (1974).

Section 27.12

27.38. These are sometimes known as 'Penrose diagrams' or 'Carter–Penrose diagrams'. I first made use of them in my Warsaw lecture (1962); the systematic notion of a strict conformal diagram was introduced by Carter (1966a,b; see also Gibbons and Hawking (1977)).

27.39. See Penrose (1964, 1965a); Carter (1966b); Penrose and Rindler (1986), Chapter 9.

27.40. Certain hypothetical models in which the Big Bang (or, rather, the Big Crunch) is indeed conformally (i.e. causally) a point—referred to as 'the Ω point'—find favour with some theoreticians; see Tipler (1997). I am not aware of any discussion, compatible with the arguments of Chapter 27, which makes such models physically plausible, however.

27.41. See Note 27.36 for the term 'hypersurface'. In this case, we see that the Big Bang, in its conformal representation, is *3-dimensional*. (We may contrast this with certain other representations, see Rindler 2001).

Section 27.13

27.42. It is often considered that the phenomenon ultimately responsible for such irregularities is 'quantum fluctuations' in the initial matter density in the Big Bang. (This will be discussed in §30.14.)

28
Speculative theories of the early universe

28.1 Early-universe spontaneous symmetry breaking

Up to this point in the book, our considerations have been well within the scope of firmly established physical theory, where impressive observational data has supplied powerful support for the sometimes strange-looking theoretical ideas that have come into play. Some of my arguments have been presented in ways that may be a little different from those usually found in the literature, but I do not think that there is anything contentious about this. In this chapter, I shall begin to address some of the more speculative ideas that are concerned with issues raised by the special nature of the Big Bang.

In particular, I shall consider the ideas of inflationary cosmology, in addition to others that relate to spontaneous symmetry breaking in the early universe (see §25.5). Some readers, conversant with certain ideas in common use in cosmology, may find it puzzling that I am placing inflationary cosmology so firmly in the 'speculative' camp. Indeed, popular accounts often seem to take as an established fact that, in the very early stages of the universe, there was a period of exponential expansion in the course of which the universe inflated by a factor of about 10^{30}, or perhaps even 10^{60} or more. Other knowledgeable readers may be even more alarmed by the fact that I am regarding the general phenomenon of spontaneous symmetry breaking in the early universe as a speculative idea. Nevertheless, the notions that I wish to address in this chapter have, as yet, not a great deal (if any) of significant and unambiguous support from observation, and one may well raise the issue of whether or not these ideas have genuine relevance to Nature.

Let us start with the general idea of spontaneous symmetry breaking. Recall the power of this idea for producing renormalizable QFTs, in which the renormalizability takes advantage of a greater ('hidden') symmetry than is directly exhibited in observed behaviour. This lack of full symmetry in what is observed is attributed to the system's choice of a 'vacuum state' that does not share the full symmetry of the dynamical theory. In particular, this formed a key ingredient of the electroweak part of the standard model of particle physics. Moreover, this kind of idea, involving different possible 'vacua', is also an essential ingredient of inflation, and these notions of spontaneous symmetry breaking and 'false vacua' are also commonly invoked by theoreticians in search of ever-more unified schemes. However, I should make clear that spontaneous symmetry breaking itself is not a speculative idea. It has undoubted relevance to many genuine physical phenomena (superconductivity being an excellent example). It most certainly applies to a number of well-established phenomena, often in an elegant and satisfactory way. I am certainly not casting doubt on the idea in itself. My trouble with it is that its distinctive appeal may entice physicists sometimes to employ it too broadly, and sometimes in inappropriate circumstances.

The idea of spontaneous symmetry breaking is often graphically introduced by reference to the phenomenon of *ferromagnetism*. Imagine a spherical solid ball of iron. We may think of its atoms as little magnets for which, because of the forces involved, there is a tendency for them to line up parallel with their neighbours, with the same north/south orientation. When the temperature is high enough, namely above a critical value, which is about 770 °C (1043 K), the energetic thermal agitation of the atoms will override this tendency to magnetic alignment, and the material exhibits no propensity to become a magnet on a large scale, there being an effectively random arrangement of the orientations of the little atomic magnets. But at a lower temperature than 770 °C (the so-called 'Curie point'), it is energetically favourable for the atoms to line up, and in an ideal situation the iron would become magnetized.[1]

Now, imagine that our iron ball is initially heated to above 770 °C (but not to such a high temperature that it melts), so it is initially an unmagnetized spherical ball. Its environment is then gradually cooled to below the critical 770 °C. What happens? The natural tendency is for the ball to find a minimum-energy state, the energy in the internal vibrations of its atoms being conveyed out into the cooler environment. Because of the interactions between neighbouring atoms, the minimum energy occurs when all atoms are aligned, so that the ball becomes magnetized, with a definite direction for its north/south polarity. But none of these directions is favoured above any of the others. There is what is called a *degeneracy* in the states of minimum energy (compare §22.6). There being no favoured direction in its initial heated unmagnetized state, the final magnetization direction comes about *randomly*. This is an example of *spontaneous symmetry breaking*: the initial spherically symmetrical state settles down into a state with smaller symmetry, namely just the symmetry of rotation about the resulting magnetic north/south axis. An SO(3)-symmetrical state (namely the original hot unmagnetized ball) evolves to an SO(2)-symmetrical one (the cold magnetized ball; see §§13.1,2,3,8,10 for the meanings of these symbols).

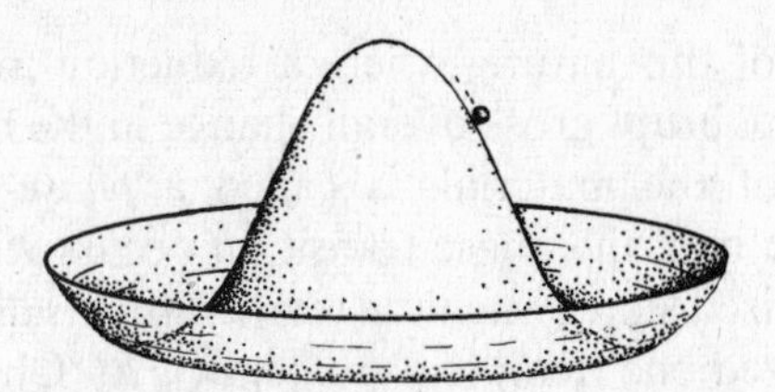

Fig. 28.1 Spontaneous symmetry breaking, with a 'Mexican hat' potential for the allowable states of a system, height measuring energy. The state of the system is represented as a marble, constrained to the surface of the hat. When the ambient temperature is high enough (Curie point), the equilibrium state of the system is represented by the marble resting at the peak, and has full rotational symmetry (SO(2), in this simplified picture). But when the temperature cools, the marble rolls down, finally reaching an arbitrary equilibrium point in the rim, breaking full rotational symmetry.

The picture used to describe this sort of situation is the 'Mexican hat' potential depicted in Fig. 28.1. The 'hat' represents the family of allowable states of the system (the ambient temperature having been cooled to zero), where 'height' represents the system's energy. We find that there is an *equilibrium state* (i.e. having a horizontal tangent plane) represented as the peak of the hat which possesses the complete symmetry of the original group—this group being represented, in the picture, as rotation about the vertical axis. (This SO(2) rotational symmetry is taken to be analogous to the full SO(3) symmetry of the iron ball, but we have had to lose one spatial dimension in order to make the picture visualizable. The peak of the hat represents complete lack of magnetization for the ball as a whole.) But this equilibrium—representing the unmagnetized state—is unstable, and it does not represent the minimum of the available energies. These minima are the states represented by the horizontal parts—a whole circle's worth—just inside the rim of the hat (different points in the rim representing different directions of total magnetization of the iron ball).

We may envisage the state being initially at the peak, as represented by a 'marble' initially perched at that point to represent the physical state, left there by the previous high-temperature state. But the lack of stability means that the marble will roll away from that point (assuming the existence of some random disturbing influences) and finally finds a resting point in the rim. Each point in the rim that the marble might settle in represents a different direction of magnetization that the ball might finally acquire. This marble location represents the final physical state. But because of the rotational degeneracy, there is no favoured place for the marble to come to rest. All of these equilibria in the rim are on an equal footing. The marble's choice is taken to be a random one, and when that choice is made, the symmetry has been broken—in some randomly chosen direction.

A phenomenon of this nature, where a reduction in the ambient temperature induces an abrupt gross overall change in the nature of the stable equilibrium state of the material, is called a *phase transition*. In our example of the iron ball, the phase transition occurs when the ball passes from the unmagnetized state (when the temperature is above 770 °C) to the uniformly magnetized one (temperature below 770 °C). More familiar are the phenomena of freezing (where the state passes from liquid to solid, as the temperature lowers) and, in a reverse process, boiling (where the state passes from liquid to gas, as the temperature rises). A phase transition, when the temperature is lowered, is often accompanied by a symmetry reduction, but this is not essential.

In QFT processes, a phase transition would frequently be described in terms of a new choice of vacuum state (like the $|\Theta\rangle$ of §26.11), where it is envisaged that the state 'tunnels'[2] from one vacuum into another. This description must be taken as an approximation, however, since there is, strictly speaking, no (unitary) quantum-mechanical process that can evolve a state from one *sector* to another (where a 'sector' refers to the states that can be built up from some particular choice of vacuum state $|\Theta\rangle$, the states in different sectors belonging to different Hilbert spaces; see §§26.5,11). The approximation, which involves taking a system as being infinite, when in practice it is finite, is evidently a good one, in practical situations. For example, the well-established phenomenon of superconductivity (where electrical resistance reduces to zero when the temperature is low enough) is treated in this way, superconductivity being a phase transition which accompanies the symmetry reduction that breaks the ordinary U(1) symmetry of electromagnetism.

In the specific example illustrated in Fig. 28.1, the symmetry is broken down from the group of axial rotations SO(2) to the trivial group ('SO(1)'), containing just one element (so all symmetry is finally lost, in this example, the marble's resting location completely breaking the symmetry).[3] But higher-dimensional versions of this 'hat' illustrate the spontaneous breaking of symmetry down from SO(p) to SO($p-1$), where $p > 2$.[28.1] (Our ball of iron illustrates the case $p = 3$.) We can also use the 'Mexican hat' picture to illustrate the breaking from U(2) down to U(1) that occurs in the standard model of particle physics,[28.2] whereby the electroweak U(2) symmetry (see §25.5) is taken to be broken to the U(1) symmetry of electromagnetism at a temperature of about 10^{16} K, which would have occurred a fleeting 10^{-12} s after the Big Bang. In the more general GUT theories (see §25.8), other groups are

---

[28.1] Show that the 'hat' with shape $E = (x_1^2 + \cdots + x_p^2 - 1)^2$ exhibits this symmetry breaking.

[28.2] Show this, with U(2) acting on $\mathbb{C}^2$, with complex coordinates (w, z), the 'hat' being given by $E = (|w|^2 + |z|^2 - 1)^2$. Can you see the geometry of this symmetry reduction in the configuration of Clifford parallels in S^3, as described in §15.4 and illustrated in Figs. 15.8 and 33.15?

involved, such as SU(5), and we can envisage different stages of symmetry breaking occurring at different temperatures. Thus, at some temperature much higher than 10^{16} K (i.e. at a significantly earlier time than 10^{-12} s, just after the Big Bang), SU(5) might first break down to something that appropriately[4] contains both the SU(3) for strong interactions and the $SU(2)\times U(1)/Z_2$ (i.e. U(2)) that is required for electroweak theory.

28.2 Cosmic topological defects

We should, however, bear in mind that this symmetry breaking is unlikely to take place 'all at once', and *domains* in which the symmetry is broken in different 'directions' may well occur. Consider our idealized iron ball once more; we may expect that the random initial choice of magnetization direction could be different at different places in the ball. We could imagine that if the cooling is slow enough then these non-uniformities might 'even themselves out', to give just one uniform magnet.[5] But, alternatively, with a more rapid cooling, we might find that we get a 'patchwork' of directions somewhat like that illustrated in Fig. 28.2. The size of the resulting cells, and the patterns that they present, might indeed depend upon the rate at which the cooling takes place, among other things. There is the issue of how readily 'communication' occurs between different regions, and of how readily the magnetization direction in one region of the ball might get 'turned around' under the influence of neighbouring regions.

More serious and interesting are the *topological defects* which cannot be removed at all by continuous wriggling around of magnetization directions in the interior of the ball. Such a defect is a 'Dirac magnetic monopole' (isolated north or south magnetic pole). However, such a monopole cannot be produced in ordinary space with any collection of magnets and currents.[28.3] However, an effective such monopole can be achieved if we

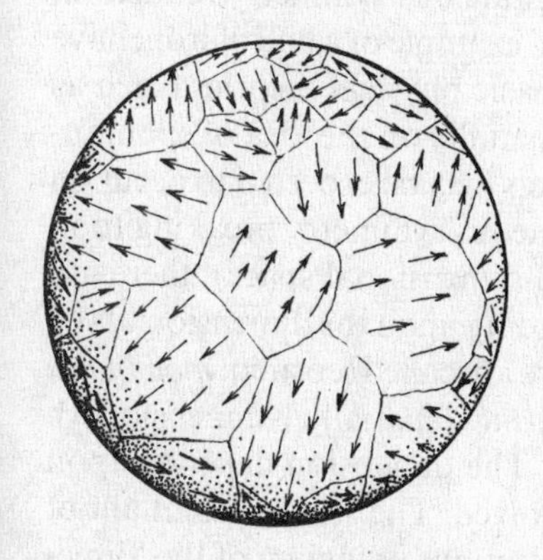

Fig. 28.2 Ideally, when a ferromagnet slowly cools from its Curie point, the directions of magnetization of its atom would all settle in the same (arbitrary) direction. But in practice (or with too rapid cooling), we get a 'patchwork' of such directions of magnetization.

[28.3] Show this, by appealing to the integral expressions of Chapter 19.

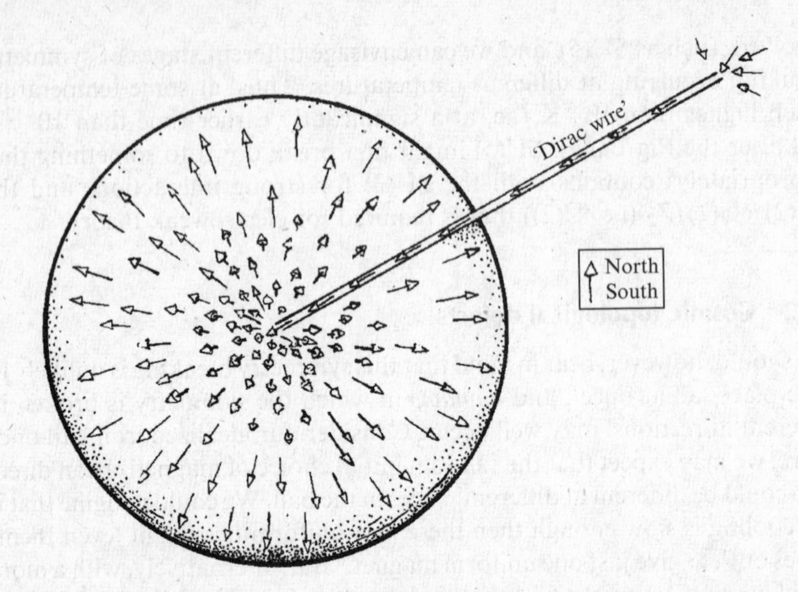

Fig. 28.3 A magnetic monopole could arise if we somehow 'pipe away' the excess 'south pole' at the sphere's centre along a 'magnetic wire'. With magnetic sources allowed in Maxwell's theory, such a pole could be inserted at the centre, and the (Dirac) 'wire' need occur only as a glitch in the potential ***A***. This glitch can be eliminated with the appropriate 'bundle' point of view (such monopoles also occurring in suitable non-Abelian gauge theories).

allow magnetic charge to be 'piped away' along a 'Dirac wire' as in Fig. 28.3. If magnetic charges are allowed for in Maxwell's theory (§19.2), then the 'wire' appears only in the potential ***A*** (§19.4), and can be eliminated altogether by the adoption of the appropriate 'bundle' point of view (§15.4). A similar kind of monopole will also exist in suitable non-Abelian gauge theories.

These complications in the picture of spontaneous symmetry reduction, partly illustrated above in the 'down-to-earth' example of a ball of iron, have relevance also at the more esoteric level of basic physical theories (such as electroweak or GUT) which depend fundamentally on the idea of spontaneously breaking symmetry. Topological defects may be expected to occur on a grand (cosmological) scale, if such spontaneous symmetry breaking took place in the early universe. In general (for 3-dimensional space), there are three basic kinds of topological defect, depending upon the dimension of the regions on which they essentially reside. These are called (cosmic) *monopoles* (which are spatially 0-dimensional), *cosmic strings* (spatially 1-dimensional), and *domain walls* (spatially 2-dimensional). The dimension depends upon topological issues to do with the groups involved. The serious point about topological defects is that no amount of continuous wriggling of the 'direction' of symmetry breaking can remove them (where we consider that, at the defect itself, there is no well-defined direction of symmetry breaking, whereas continuous variation of this direction takes place elsewhere). We must bear in mind that this notion of 'direction' does not refer to a direction in ordinary space, but to a more abstract notion of 'direction' that occurs within the physical model under consideration (e.g., in electroweak theory, that which tells us what degree of electron/neutrino mixture is being considered). Geometrically, we should be thinking in terms of a vector bundle over spacetime (see Chapter 15, if you wish to be reminded of this notion). Topological considerations still apply, and the topological defects would present serious issues that cannot just be 'laughed away' if the symmetry breaking is to be taken seriously as part of basic physical theory.

Indeed, cosmic strings on enormous (even longer than galactic) scales have been taken seriously as the essential agency responsible for inducing those inhomogeneities in the background gas that lead to galaxy formation.[6] We can think of the gravitational field of such a cosmic string as being constructed by a 'scissors-and-paste' procedure applied to Minkowski spacetime. In spatial terms (see Fig. 28.4), we picture a 'sector' removed from 3-space, this being bounded by a pair of half planes whose contained angle α is centred on the string itself. To construct the cosmic string geometry, the two plane surfaces are 'glued' together again. (In the suggested models, α is about 10^{-6}.)

The reader may, with some justification, feel that these are extreme measures for the production of such a 'commonplace' entity as an ordinary galaxy. Yet there are still some theoretical puzzles about galaxy formation, so such exotic ideas should not be dismissed out of hand, despite their seemingly outrageous nature. Indeed, one of the most plausible models of galaxy formation—which has some significant observational support—proposes that they are largely 'seeded' by the supermassive black holes that now reside at their centres.[7] But black holes must today be considered as conventional rather than exotic physics!

Most of these suggested topological defects refer to theories (such as the various GUTs) that do not have significant or unambiguous support from

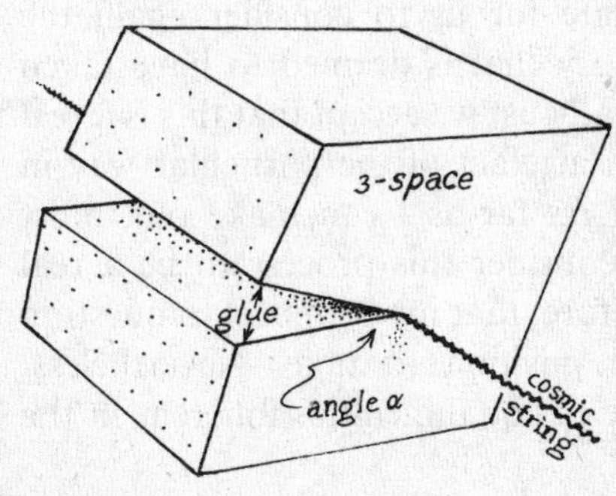

Fig. 28.4 The gravitational field of a cosmic string can be constructed by a 'scissors-and-paste' procedure applied to Minkowski 4-space. In 3-space, a sector is removed, bounded by two half-planes meeting at an angle α along the string. The half-plane surfaces are then 'glued'.

observation. Electroweak theory, on the other hand, is very well supported observationally, so we must pay heed to what this theory implies with regard to processes in the early universe. Cosmic monopoles, resulting from the symmetry breaking of electroweak theory, are a topological possibility, but they are not a necessity. They could arise in the spontaneous breaking from U(2) down to U(1), but only if what are called 'gauge monopoles' were already present in the unbroken U(2)-symmetric phase of the theory, which is taken to have occurred *before* 10^{-12} s. Such monopoles could arise from an earlier breaking from a larger GUT symmetry, but these ideas are by no means a necessary part of electroweak theory.[8]

Such *gauge monopoles* are the analogues, within some Yang–Mills (non-Abelian gauge) theory, of the 'magnetic monopoles' that Dirac once proposed (in 1931), in the context of the (Abelian gauge) theory of electromagnetism. By an ingenious argument, Dirac showed that, if even a single magnetic monopole (a separate magnetic north or south pole) were to exist in Nature, then all *electric* charges would have to have values that are integer multiples of some particular value, this value being reciprocally related to the magnetic strength of the monopole. In fact, present observations strongly suggest that electric charges *are* all integral multiples of a particular value (say that of the anti-d-quark charge, which is one third of that of the proton; see §3.5 and §25.6). Some would take this as circumstantial evidence for the actual existence of magnetic monopoles. Nevertheless, if such monopoles are not to be in gross conflict with observation, they would have to be exceptionally uncommon.[9] (Otherwise they would have the effect of 'short-circuiting' cosmic magnetic fields, whereas such fields are observed to exist throughout large reaches of the universe.) Similarly, Yang–Mills monopoles would cause severe observational conflicts if such monopoles were significantly present in the universe today. This issue has had important implications for the development of the subject of cosmology, as we shall soon be seeing!

28.3 Problems for early-universe symmetry breaking

Before coming to this, it is appropriate for us to consider again the symmetry breaking in electroweak theory that is deemed to have taken place at about 10^{-12} s after the Big Bang. Must we accept that this is a real phenomenon, or might it be merely an artefact of the particular way in which the theory is usually presented? As far as I can make out, most electroweak theorists would certainly consider this process to be a real one. The reader is hereby warned, therefore, that my proposal to question its reality here is an unconventional position to take. Nevertheless, let us press forward and consider some of the difficulties inherent in the symmetry-breaking idea.

Let us suppose that, contrary to my own (less than conventional) opinions on this matter, there was indeed a time in the early history of the universe—earlier than about 10^{-12} s after the Big Bang—when an exact U(2) symmetry held in which leptons and quarks were all massless, where 'zig' electrons and neutrinos were on an equal footing with each other, and where the W and Z bosons and the photon could be appropriately 'rotated' into combinations of each other according to a U(2) symmetry (see §25.5). Then, at time about 10^{-12} s, throughout the universe, the temperature dropped to just below the critical value. At this moment a particular choice of (W^-, W^+, Z^0, γ) was made, randomly, taken from the entire U(2)-symmetric manifold $\mathcal{G}$ of possible sets of gauge bosons. We do not expect this to happen exactly uniformly throughout space, simultaneously over the entire universe. We anticipate that, as with the domains of magnetization in the iron ball illustrated in Fig. 28.2, in some regions one particular choice will be made and in other places there will be different choices.

At this point, we should address the question of what is to be *meant* by the terms 'same' and 'different' in this context. The space $\mathcal{G}$ of possible gauge bosons is, at each spacetime point, completely U(2)-symmetric before the symmetry reduction takes place. As is inherent in the notion of a *bundle*, there is to be no particular way, favoured over any other, of making an identification between the $\mathcal{G}$ at one point and the $\mathcal{G}$ at another quite different point. Thus, we do not seem to be given an *a priori* rule for telling us which element of the $\mathcal{G}$ at one point is to be called 'the same' element as some element of the $\mathcal{G}$ at another point. This seems to give us the freedom of holding to the standpoint that we simply *define* the notion of 'the same' to be that provided by the particular choice that the spontaneous symmetry breaking provides. According to such a standpoint, the particular (W^-, W^+, Z^0, γ) that is 'frozen out' at one point would be identified with the corresponding (W^-, W^+, Z^0, γ) at any other point, so it seems accordingly that we would not witness the kind of 'inconsistency' between the symmetry breakings at different points that occurs with the domains of iron magnetization illustrated in Fig. 28.2.

However, such a standpoint flies in the face of the whole idea behind gauge theory, according to which not only are the $\mathcal{G}$ spaces the fibres of a fibre-bundle $\mathcal{B}_{\mathcal{G}}$, with base space the spacetime $\mathcal{M}$, but also the particular gauge theory—in this case unbroken electroweak theory—is defined in terms of a *connection* on this bundle (§§15.7,8). This connection defines the locally meaningful identification (parallelism) between the $\mathcal{G}$ spaces as we move along any given curve in $\mathcal{M}$.[10] In general this identification is not globally consistent as we go around closed loops (because of curvature in the connection, expressing the presence of non-trivial gauge field—see §15.8). In any case, the randomness involved in the symmetry breaking at different points will have the implication that local parallelism between the $\mathcal{G}$-spaces will generally not be consistent with the choices that are made in the spontaneous symmetry breaking, so the picture of Fig. 28.2 is not such an unreasonable analogy. We can imagine that, as with a sufficiently slowly cooled ball of iron, the inconsistencies will 'iron themselves out' if given sufficient time, where here it is assumed that there are no topological defects (as indicated in Figs. 28.3 and 28.4). The issue that I wish to raise is whether there can *ever* be 'enough time' in the case of electroweak spontaneous symmetry breaking.

The difficulty has to do with the particle horizons that we encountered in §27.12, Fig. 27.18b. Look at the schematic conformal diagram of Fig. 28.5. An observer situated at point p sees quasars (cf. §27.9) in two opposite directions, at respective spacetime points q and r. According to the standard FLRW models, if the red shift[11] (see §27.7) of the quasars is sufficiently great, then the past light cones of q and r will not intersect each other, so no kind of communication can have taken place between them. Being out of communication with each other, they will not have had time to 'iron out' their symmetry breaking to be consistent in the way indicated above. We shall shortly be considering the 'inflationary scenario' which pushes back the Big Bang line, in the conformal diagram, so as to bring q and r into 'communication' after all. But that will not help us

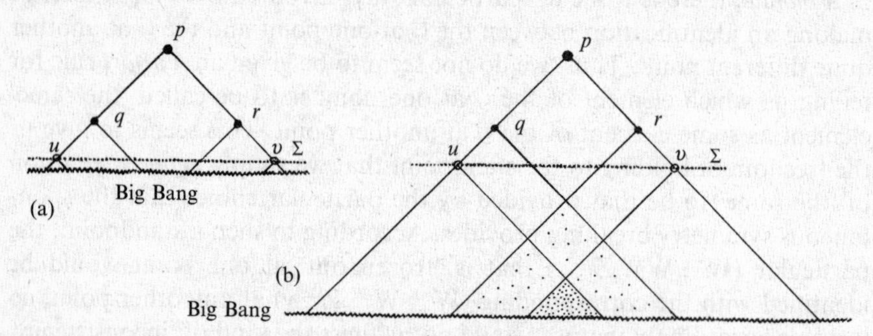

Fig. 28.5 Schematic conformal diagrams illustrating causal (in)dependence in the early universe. (a) Observer at p sees quasars in opposite directions, at q and r. If the dotted line represents the 3-surface Σ at time $\sim 10^{-12}$ s, along which exact prior U(2) electroweak symmetry (relating the photon γ to the W and Z bosons) is taken to be broken, then the particular 'frozen out' choice of γ at q almost certainly differs from that at r, the intersections of the pasts of q and r with Σ being disjoint; yet the respective γ choices cannot communicate their sameness/difference until p is reached. Similarly, if Σ now represents decoupling, at time 10^{13} s, temperatures at u and v cannot have equalized by thermalization, as their complete pasts are disjoint. (b) Inflation's 'resolution' of the latter 'horizon problem' is to push the Big Bang back so that the pasts of q and r now do intersect before reaching the Big Bang 3-surface. The former problem remains unresolved, however, since the intersections of their pasts occur prior to the 'freezing out' at $\sim 10^{-12}$ s.

here, because the 3-surface Σ, over which the electroweak symmetry breaking is to take place, effectively plays the role of the Big Bang in our present causality considerations, because the spontaneous symmetry breaking is taken as occurring randomly on the 3-surface Σ, with no effective common causal influence.

Now the lines qp and rp are null lines, so only the photon can travel from q to p or from r to p, not the W or Z bosons—the photon being the only massless member of the family of gauge bosons. Thus, all along these two null lines, we must have a consistent notion of what a photon is. The notion of 'photon' at q is highly likely to be inconsistent (in the sense indicated above) with the notion of 'photon' at r, because each was supposed to have been selected randomly without common causal influence, and without time for communication between them.[12] Can the 'different' kinds of photon 'iron themselves out' in time to save the observer at p from a baffling W–Z–γ confusion upon receiving them? I do not see how this can be possible without significant departures from the direct null (i.e. 'lightlike') connections from q to p and from r to p. This could lead to a gross conflict with the fact that distant objects are seen clearly through optical telescopes. It seems to me that here there is danger of being severely inconsistent with observation, although I have not seen it discussed in the literature.

But some readers will no doubt grumble (perhaps under their breaths) that I have seemed to ignore all the very impressive observational support for electroweak theory. Surely I am not going to abandon all that just because of some confusion that I may have concerning phenomena propagated from cosmological distances! Indeed not. I am in no way suggesting that we should abandon the essential beautiful insights of electroweak theory, but I prefer a slightly different attitude to the breaking of its U(2) symmetry from that which is usually put forward. As I see it, Nature's true scheme for particle physics has not yet come to light. Such a scheme should be mathematically consistent and will not have the nasty habit that our present-day QFTs have, of spitting out the answer '∞' to so many reasonably phrased physical questions. Why this (still unknown) 'correct' theory gives finite answers is not discernible to us today. Thus we have resorted to various 'tricks', which happen to have come our way through a combination of historical fortune and exceptional human ingenuity, that enable us to produce observation-matching finite answers. At our present stage of understanding, we certainly want a theory of weak and electromagnetic interactions that is renormalizable, and not only has the idea of broken non-Abelian gauge symmetry provided a route to such a renormalizable theory, but the constraints in doing so have guided us close to a family of deep truths about the way that these interactions fit together as part of a broader picture. But I do not see why a spontaneously broken symmetry need be Nature's true way, in particle physics. Indeed, there are other routes to seeing why the demands of renormalizability provide the needed relations between the parameters of electroweak theory.[13]

This raises an important issue (to which I shall return in §34.8): does the notion of *symmetry*, so prevalent in many ideas for probing Nature's secrets, really have the fundamental role that it is often assumed to have? I do not see why this need always be so. It does not necessarily strike me that basing particle physics on some large symmetry group (which is part of the GUT philosophy) is really a 'simple' picture, as far as a fundamental physical theory is concerned. To me, large geometrical symmetry groups are complicated rather than simple things. It might well be the case that there are fundamental asymmetries inherent in nature's laws, and that the symmetries that we see are often merely approximate features that do not persist right down to the deepest levels. I shall be coming back to this issue later (in §34.8).

28.4 Inflationary cosmology

Let us return to the question of the cosmic monopoles, whose proliferation is a feature of certain GUTs. The trouble with these monopoles is the lack of any indication of their actual existence. Worse than this, there are stringent observational limits on the cosmic abundance of such monopoles, far below the level predicted by the GUTs. However, in 1981, Alan Guth put forward the 'outrageous' proposal (also previously suggested independently, in essence, by Alexei Starobinski and Katsuoko Sato) that if the universe were to have expanded, by a factor of, say 10^{30} or perhaps even 10^{60} or more, at some period after the production of the monopoles (though before electroweak symmetry was broken, at the time of 10^{-12} s) then the unwanted monopoles would now be so sparse that they could easily escape detection, as was required from observation.

It was soon realized that this 'inflationary period' of extreme exponential expansion might serve other purposes as well, having to do with the uniformity of the universe. As emphasized in Chapter 27, the universe is indeed extremely uniform, and close to being spatially flat on a very large scale, and this presented a puzzle to cosmologists. For example, the observed temperature of the early universe is extremely closely the same in different directions (to at least about one part in 10^5). This might be taken as the result of a 'thermalization' in the very early universe, but only if the different parts of the universe in question were 'in communication' with each other. (Recall how the second law of thermodynamics serves to equalize the temperatures of a gas at different places, as part of the process of coming to thermal equilibrium; see §27.2.) Yet, an examination of Fig. 28.5a tells us that the equality of the temperatures at distant points u

and v, both observed from our present location p in spacetime, cannot be the result of thermalization in the conventional cosmological models because the points u and v (where we take Σ to be at the time of 'decoupling', when the cosmic black-body radiation was formed) were far too distant from each other to have ever been in causal communication, in the standard model.

This impossibility of the causal communication that would be required for thermalization, in the standard model, is referred to as the *horizon problem*. The effect of the period of inflation, in this respect, is depicted in the conformal diagram of Fig. 28.5b. The spacelike 3-surface that represents the Big Bang has now been displaced to a much 'earlier' location, so that the pasts of u and v now *do* intersect before reaching the 3-surface that describes the Big Bang, so that now thermalization does have the opportunity to take effect, and we may now imagine that the equality of temperatures at u and v can come about through this means.

Another perceived benefit of this proposed inflationary period was that it could provide an explanation of the remarkable uniformity of the matter distribution and spacetime geometry, this being referred to as the *smoothness problem*. The idea is that, with inflation, the initial state of the universe might have been very irregular in detail, but the enormous expansion of the universe during the inflationary stage would have served to 'iron out' these irregularities, and a closely FLRW universe is thereby anticipated. The inflationary viewpoint envisages that even a 'generic' initial state would look like a smooth manifold on a small scale, and we see this tiny smooth portion expanded out to cosmological scales—so as to appear to be spatially flat—during the course of the inflationary phase; see Fig. 28.6 (and compare Fig. 12.6). I shall be coming to my own assessment of this extraordinary idea shortly. For the moment, it is worth pointing out that, in this picture, not only is the universe uniform, but also it has zero spatial curvature ($K = 0$). This is an important factor for the historical development of the subject, as we shall be seeing. But whether or not the observable universe is, on the average, actually spatially flat, it is certainly remarkably close to being so, and this had presented a puzzle to many cosmologists—referred to as the *flatness problem*.

It is not immediate to the eye what the inflationary phase of expansion has to do with the moving back of the Big Bang 3-surface in the conformal diagrams, as in Fig. 28.5. It will be instructive, therefore, to examine the particular cosmological model on which this 'inflationary phase' is based. This is the 'steady-state' version of *de Sitter* space. The quickest way to describe de Sitter space mathematically is to say that it is a Lorentzian 4-sphere (signature $+---$) in Minkowski 5-space (signature $+----$). This description is in accordance with the geometrical 'signature flip' ideas of §18.4, but it is geometrically clearer if we picture de Sitter

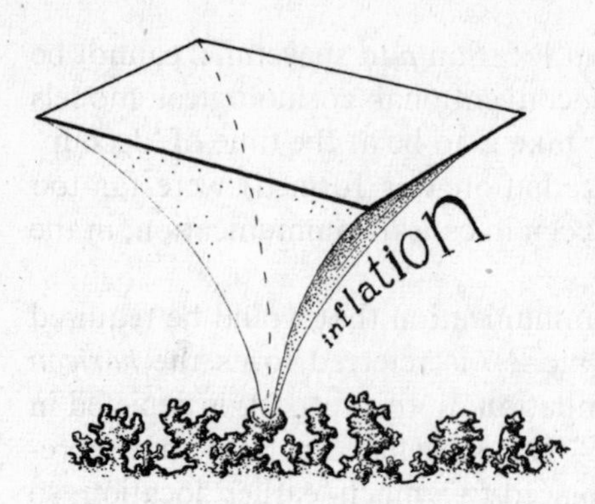

Fig. 28.6 One of the underlying motivations of inflation is that an exponential expansion scale of perhaps 10^{50} (say between times 10^{-35} s and 10^{-32} s) might serve to 'iron out' a generic initial state, so as to provide an essentially uniform, spatially flat, post-inflation universe.

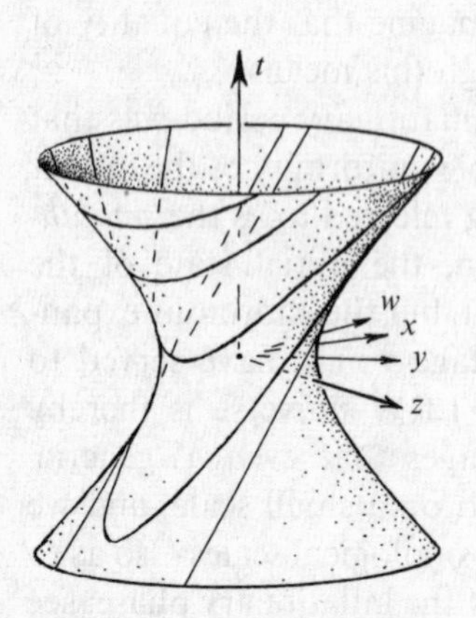

Fig. 28.7 The de Sitter spacetime (pictured as a hyperboloid, with two spatial dimensions suppressed) is a Lorentzian '4-sphere' (of imaginary radius, giving intrinsic metric signature $+ - - -$) in Minkowski 5-space $\mathbb{M}^5$ (whose metric is $ds^2 = dt^2 - dw^2 - dx^2 - dy^2 - dz^2$). To get the steady-state model, we 'cut' the hyperboloid into half, along $t = w$; constant time is given by constant positive $t - w$.

space as the *hyperboloid* of Fig. 28.7. At this point, it is worth mentioning another model, called *anti-de Sitter* space, which is a Lorentzian 4-sphere in a pseudo-Minkowskian 5-space of signature $+ + - - -$ (Fig. 28.8).[28.4] Note that anti-de Sitter space is not a very sensible spacetime, physically, because it possesses (causality violating) *closed timelike curves* (e.g. the circle in the plane spanning the t and w axes); see §17.9 and Fig. 17.18. Sometimes the term 'anti-de Sitter space' refers to an 'unwrapped' version in which each circle in a constant-(x, y, z) plane has been unwrapped into a line, and the whole space becomes simply-connected (§12.1). I have drawn a strict conformal diagram for de Sitter space in Fig. 28.9a, and also for the portion of it that represents the steady-state model (the *dotted*

[28.4] Write down explicitly the equations for the de Sitter and anti-de Sitter 4-spaces in the background 5-space, using the coordinates t, w, x, y, z indicated in Figs. 28.7 and 28.8. Find coordinates within 'half' of de Sitter space, so that its intrinsic metric takes the 'steady-state' form given later in this section.

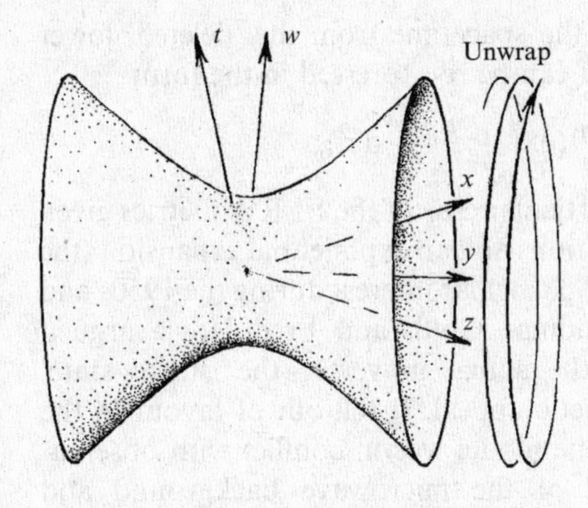

Fig. 28.8 Anti-de Sitter spacetime (pictured as a hyperboloid with two spatial dimensions suppressed) is a Lorentzian '4-sphere' (of positive radius, giving an intrinsic metric signature $+ - - -$), in pseudo-Minkowskian 5-space (with metric $ds^2 = dt^2 + dw^2 - dx^2 - dy^2 - dz^2$). As defined, we have closed timelike curves, but these can be removed by infinitely 'unwrapping' in the (t,w)-plane.

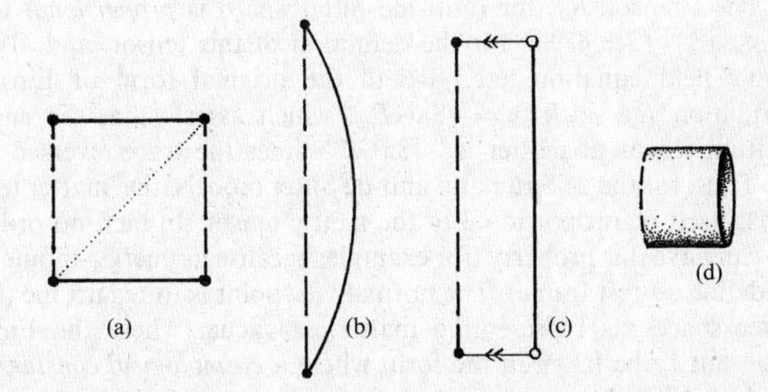

Fig. 28.9 Strict conformal diagrams (with the conventions of Fig. 27.16a) of: (a) de Sitter space, where the region above the internal dotted line gives the steady-state model; (b) anti-de Sitter space (fully unwrapped version, without causality violations); and (c) anti-de Sitter space in the original causality violating 'hyperboloid' form, where the top and bottom edges are to be identified. (d) The same as (c), but with the identification performed, so the diagram appears as a cylinder.

boundary line indicating the cut), for the causality-violating anti-de Sitter space in Fig. 28.9c (where the top and the bottom of the diagram must be identified) and in Fig. 28.9d, and for the unwrapped (causal) anti-de Sitter space in Fig. 28.9b.

To obtain the steady-state universe explicitly, we 'cut' de Sitter space in half, along the $t = w$ 4-plane of the Minkowski 5-space depicted in Fig. 28.7, retaining only the 'upper' half.[14] Curiously, although there is an 'incompleteness' in this model, owing to the cut (dotted line in Fig. 28.9a), this incompleteness is not usually considered as a defect, because no actual particle enters the spacetime from the 'deleted' lower half. The metric for the upper half can be re-expressed in the form

$$ds^2 = d\tau^2 - e^{A\tau}(dx^2 + dy^2 + dz^2),$$

(A being a constant), which is a particular case of the FLRW metrics given in §27.11, with flat $K = 0$ space sections and an exponential expansion (the factor $e^{A\tau}$).[28.5] (This metric was of particular interest during the 1950s and 1960s when Hermann Bondi, Thomas Gold, and Fred Hoyle argued strongly for it as a model for the actual universe—the 'steady-state' model of some considerable aesthetic appeal. It fell out of favour in the 1960s, after it became clear that the model was in conflict with observations, particularly measurements of the microwave background and counts of distant galaxies.)

The Ricci tensor R_{ab} for (anti-)de Sitter space is *proportional* to the metric g_{ab}.[28.6] (See §19.6, for the definition of this tensor, and also for Einstein's field equation, etc.) Recall the original form of Einstein's field equation $R_{ab} - \frac{1}{2}Rg_{ab} = -8\pi G T_{ab}$, which asserts that the energy–momentum tensor of matter is $-(8\pi G)^{-1}$ times the trace-reversed Ricci tensor. Thus, for the de Sitter and anti-de Sitter models, the 'matter tensor' T_{ab} must itself be proportional to the metric tensor. In fact, no ordinary matter can have this property (for example, because its energy–momentum would define no rest frame). The normal viewpoint is to regard the (anti-) de Sitter spaces as representing matterless vacua, where the Einstein equation has to be taken in the form where a *cosmological constant* Λ is included, so the field equations now give us

$$R_{ab} = \Lambda g_{ab}.$$

Here $\Lambda = A^2$, where A is the constant scaling the exponential-growth factor in the above steady-state metric. In inflationary cosmology, the inflationary 'material' is taken to be a 'false vacuum', about which I shall say some more in a moment.

To construct an inflationary-universe model, we take a portion of the steady-state universe, between two 3-surfaces of constant τ, and paste it on to two parts of a standard $K = 0$ FLRW model. This procedure is illustrated in Fig. 28.10. In Fig. 28.10a, the entire de Sitter space is cut to produce the steady-state model. In Fig. 28.10b, a greatly inflating portion of the steady-state model is selected. In Fig. 28.10c, a piece is cut from the

---

[28.5] Find metric forms for de Sitter and anti-de Sitter spaces of the FLRW type $ds^2 = dt^2 - (R(t))^2 d\Sigma^2$, where $d\Sigma^2$ gives the hyperbolic 3-metric according to the second expression in Exercise [27.18]. What portion of the full (anti-)de Sitter space does this cover?

[28.6] Can you see why this must be so, without doing any calculation?

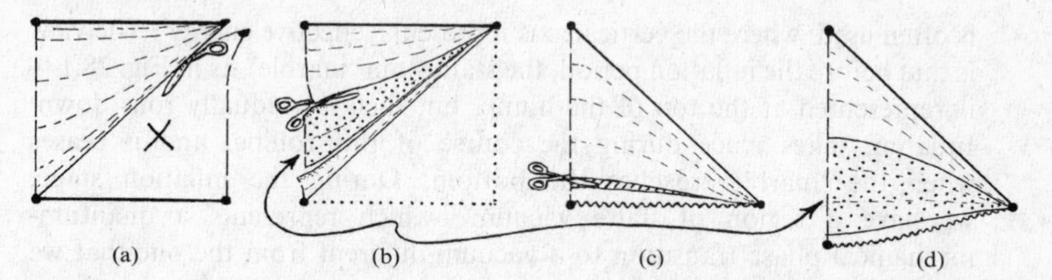

Fig. 28.10 Kit for constructing an inflationary-universe model. (a) de Sitter space cut to give steady-state model. (b) Greatly inflating portion, of steady-state model selected, between two constant-time lines. (c) Small constant-time interval removed from $K = 0$ FLRW model. (d) Portion from b inserted into c to obtain inflationary universe model. This moves the Big Bang back, as in Fig. 28.5b.

$K = 0$ FLRW model, so it can receive the inflationary portion to complete the model in Fig. 28.10d. The inserted steady-state portion in effect 'pushes back' the Big Bang (from the conformal, i.e. causal, point of view), so that the particle horizon is greatly expanded; see Fig. 28.5b.

In order to achieve this inflationary period, it is necessary to introduce a new scalar field φ into the menagerie of known (and conjectured) physical particle/fields. As far as I am aware, this field φ is not taken to be directly related to any of the other known fields of physics, but is introduced solely in order to obtain an inflationary phase in the early universe. It is sometimes referred to as a 'Higgs' field, but it does not seem to be the 'ordinary' one, related to electroweak theory (see §25.5). Some models require more than one separate inflationary phase, in which case there would have to be a different scalar field for each phase. The inflation process is described in terms of a picture bearing some relation to that of the 'Mexican hat', of Fig. 28.1, but without the initial symmetry. A diagram like that of Fig. 28.11

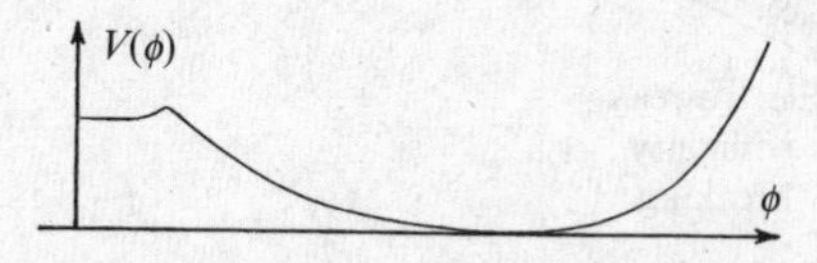

Fig. 28.11 The effective energy density of the very early universe, according to the inflationary model, would be dominated by the effective potential $V(\phi)$ for the scalar 'inflaton' quantum field ϕ. The graph shows one commonly assumed form of $V(\phi)$, where inflation is taken to occur as the state (the 'marble' of Fig. 28.1) 'rolls' down the hill on the left (where a 'false vacuum' is assumed to take place). Inflation ceases when the bottom is reached.

is often used, where the vertical axis represents 'effective energy'. The view is that before the inflation period, the state—our 'marble', as in Fig. 28.1—is represented at the top of the hump, but then it gradually rolls down. Inflation takes place during the course of this rolling, and it ceases when the 'marble' reaches the bottom. During the inflation stage, we have a region of 'false vacuum', which represents a quantum-mechanical phase transition to a vacuum different from the one that we are familiar with today.

As mentioned in §27.11, there is now some good evidence for a positive Λ in our present epoch, but this is extremely small in ordinary terms, corresponding to a density that is only about 10^{-30} of that of water. By contrast, the false vacuum of the inflationary phase would have had an effective Λ corresponding to a density exceeding that of water by about 10^{80}. This would completely dominate the energy-momentum tensor of any ordinary matter, and it is for this reason that the de Sitter model can be used for this phase.

In Fig. 28.12, I have indicated the kind of picture of the history of the very early universe that we are frequently presented with, and which has now become almost 'standard'. Note that the time and distance scales are 'logarithmic' ones (like the slide rule of Fig. 5.6) marked with different powers of 10, in units of one second (horizontal) or one centimeter (vertical). The 'radius' denotes the history of the '$R(t)$' of §27.11 (which must not be confused with the scalar curvature 'R' of §19.6). In my own opinion, this picture must be regarded as very speculative up to about $\frac{1}{10}$ (and certainly 10^{-30}) of a second, although it is often presented as virtually established fact!

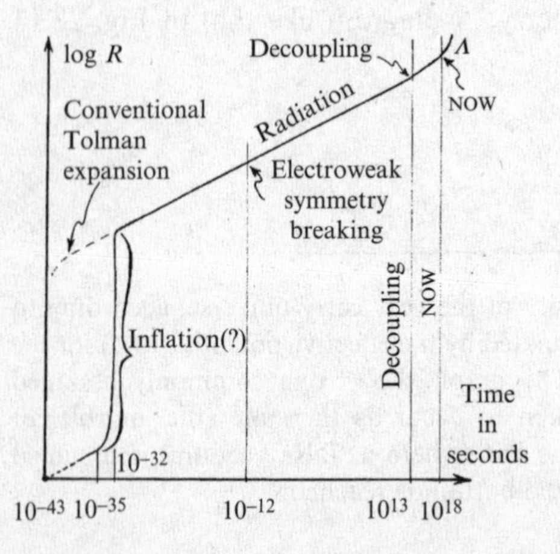

Fig. 28.12 A commonly described 'history of the universe', as a logarithmic plot, including an inflationary phase. Here $\log R(t)$ is plotted against $\log t$.

28.5 Are the motivations for inflation valid?

What reason is there to believe that such an inflationary picture of the universe is likely to be close to the truth? Despite its evident popularity, I wish to give my own reasons for casting considerable doubt on the entire idea! Again I must give my statutory warning to the reader. Inflationary cosmology has become a major part of the body of modern cosmological thinking. You will find that even among those who are not yet convinced of the necessity of inflation, there are few who will be as negative as I am going to be in the following critique. If you feel the need to 'balance' my account with one that finds more favour with the inflationary idea, see Alan Guth's very accessible book *The Inflationary Universe*.[15] For my own part, I must present things as I see them, and since I believe that there are powerful reasons for doubting the very basis of inflationary cosmology, I should not refrain from presenting these reasons to the reader.

But before making my critical assessment, I should make clear that my remarks do *not* tell us that inflationary cosmology is wrong. They merely provide strong reasons to doubt most of the initial motivations behind the inflationary idea. We may recall that many important scientific ideas of the past have, after all, been based (partly) on motivations that did not stand up in the light of subsequent understanding. One of the most important of these was Einstein's significant dependence on *Mach's principle* as a guide to his eventual discovery of general relativity. Mach's principle asserts that physics should be defined entirely in terms of the relation of one body to another, and that the very notion of a background space should be abandoned.[16] Later analysis of Einstein's theory showed that Mach's principle is not incorporated by general relativity,[17] however, irrespective of the motivational significance of Mach's idea.[18] Another example was Dirac's discovery of the electron's wave equation, which he based fundamentally on what he perceived to be the necessity for a first-order equation (see §§24.5,6). Later understanding of QFT showed that this requirement is not necessary (§ 26.6).

Similarly, if the observational predictions of inflationary cosmology are convincingly confirmed, then any inadequacy in the initial motivations would be less important, and the theory would be able to stand on its own without the original 'ladder' which led Guth and others to this particular scheme. In fact, inflationists have made some definite predictions which, in recent years, have measured up remarkably well against a number of impressive new observations.

I believe that particular caution is to be recommended in matters of cosmology, as opposed to most other sciences, especially in relation to the origin of the universe. People often have strong emotional responses to questions of the origin of the universe—and sometimes these are either implicitly or explicitly related to religious preferences. This is not unnatural; for the issue is indeed that of the creation of the entire world in which we live. As stressed in §27.13, because of the Second Law, there is an extraordinary degree of precision in the way that the universe started, in the Big Bang, and this presents what is undoubtedly a profound puzzle. We ask: is the solution of this puzzle of the Big Bang's precision something that may be answerable by a future scientific theory, even though it is still beyond our present-day scientific understanding? (This is essentially my own optimistic position; see §§30.10–14.) Or must we resign ourselves to it being some kind of 'act of God'? The view of the inflationists is different, namely that this puzzle is essentially 'solved' by their theory, and this belief provides a powerful driving force behind the inflationary position. However, I have never seen the profound puzzle raised by the Second Law seriously raised by inflationists!

Instead, three particular problems in the standard model of cosmology tend to be singled out by inflationists, these all being issues that are indeed related to the initial precision in the early universe. They were specifically addressed in §28.4, and are referred to as the horizon problem, the smoothness problem, and the flatness problem. In the standard model, these issues are handled by 'fine-tuning' of the initial Big Bang state, and this is regarded by inflationists as 'ugly'. The claim is that the need for such fine-tuning of the initial state is removed in the inflationary picture, and this is regarded as a more aesthetically pleasing physical picture. The conclusion of overall spatial flatness that comes about through inflation is also regarded as a positive feature, from the aesthetic point of view.[19]

It seems to me that great caution should be adopted in relation to such aesthetically based arguments. There are certainly some elements fundamental to the inflationary picture whose aesthetic status is somewhat questionable, such as the introduction of a scalar field (or perhaps several independent scalar fields, if more than one period of inflation is envisaged) unrelated to other known fields of physics and with very specific properties designed only for the purpose of making inflation work. Also, the aesthetic preference for $K = 0$ is very contentious. I know of many mathematicians (including myself) who regard the hyperbolic case ($K < 0$) as distinctly more beautiful! Yet others prefer the 'coziness' of a spatially finite (say $K > 0$) universe. The general issue of the role of beauty as a guide in basic theoretical physics will be discussed later in this book (see §34.9), as will further issues that relate specifically to inflation (§34.4) and to the role of scientific fashion (§34.3). Inflation is certainly extremely fashionable among present-day cosmologists, and it is important to try and see how far its fashionable status is justified.

As stated above, my basic objections to this idea of cosmic inflation have mainly to do with the underlying motivations behind it. Let us first consider the horizon problem, and how this is dealt with in inflationary cosmology where, for example, the almost equal background temperatures in different directions is perceived to be the result of *thermalization*. Inflation is brought in to remove the particle horizons that would otherwise preclude this thermalization.

There is, however, something fundamentally misconceived about trying to explain the uniformity of the early universe as resulting from a thermalization process (§28.4), whether this is a uniformity in the background temperature, the matter density, or in the spacetime geometry generally. Indeed, it is fundamentally misconceived to try to explain why the universe is special in *any* particular respect by appealing to a thermalization process. For, if the thermalization is actually doing anything (such as making temperatures in different regions more equal than they were before) then it represents a definite increasing of the entropy (§27.2). Thus, the universe would have been even more special before the thermalization than after. This only serves to increase whatever difficulty we might have had previously in trying to come to terms with the initial extraordinarily special nature of the universe (§27.13). There are certainly deep puzzles relating to the peculiarly constrained state of the early universe. But these constraints are fundamental to the very existence of the Second Law of thermodynamics, as was emphasized in Chapter 27. We cannot expect to be able to explain these constraints simply by appealing to *manifestations* of the Second Law (thermalization being one example)!

To elaborate upon this point, consider the issue of the equality of temperatures as seen in different directions from our particular vantage point in the universe. Suppose that the temperatures in two distant regions are indeed found to have been equal at some early cosmic time t_1, and suppose that we find this 'specialness' puzzling. Let us consider two possibilities. We might imagine (a) that in an even earlier era—time t_0—the temperatures were actually unequal and they became equal only after a thermalization process took place between the times t_0 and t_1. Alternatively, we might imagine (b) that at the earlier time t_0 the two temperatures were actually equal to each other, and no thermalization took place. In case (a), we find that there has been an entropy increase between t_0 and t_1, so we find an even greater degree of specialness at t_0 than there was at t_1, so we should be even more puzzled by the special nature of the universe at time t_0 than we were by its specialness at time t_1. The problem has got worse! In case (b), on the other hand, the problem of the specialness at t_0 is, at least, not any worse than that at t_1. In neither case have we explained the *puzzle of why the universe is special*, in this or any other particular respect, but we see that invoking arguments from thermalization, to address this particular problem, is worse than useless!

What about the uniformity (and flatness) of the universe? Here, the main inflationary argument is different. The claim is that the exponential expansion of the inflationary phase was what served to make the universe so uniform (and spatially flat). Again there is a fundamental misconception. The idea seems to be that if we start from a 'generic' initial state, then the 'stretching effect' of the exponential expansion of the inflationary phase will serve to iron out the irregularities of that initial state. Of course, in order to know whether such process has a chance, we need to have some idea of what a 'generic' initial geometry might be like. One important presumption is that such a state would have to be, on some small scale, smooth. But fractal sets, for example, never iron themselves out, no matter how much they are stretched. Recall the Mandelbrot set, portions of which are exhibited in Fig. 1.2. If anything, the Mandelbrot seems to get less smooth, the more that it is magnified.

But, I hear the reader muttering: surely that's just a quibble—OK, maybe there are some pathological situations in which stretching does not smooth things out, but surely in the general realistic case we should not expect such things. Unfortunately, this is by no means so clear; something fractal—or worse than fractal—is what we almost certainly have to be prepared for, in a generic starting state. Certainly, whatever this generic singular structure is, it is not something that we can expect to become ironed out simply because of a physics that allows inflationary processes. Why is this? The reasons have nothing to do with detailed technicalities, and are simply inherent in the misconceived nature of trying to assume that our actual universe might have started in a generic state[20]—which it cannot have done, because of the Second Law; see §27.7. If we want to get some idea of what such a 'generic' state might be like, consider the final stages of a collapsing closed universe, such as that schematically illustrated in Fig. 27.20a,b, and then reverse the flow of time, as in Fig. 27.20c (or Fig. 27.20d). The great mess of congealing black-hole singularities is the kind of thing that, in *time-reversed form*, we should expect for a generic Big Bang.

Of course, I am not asking that the reader have an instant understanding of the detailed complicated fractal-like geometry involved in a messy generic Big Crunch! I have little real conception of this myself, and I don't think that anyone else knows a very great deal about it.[21] But we do not need to know anything detailed about this geometry. To understand the essential issue, consider any collapsing-universe model, which we may construct starting from some highly irregular initial expanded state (compare Fig. 27.20b). It has to collapse to *something*; indeed, its collapse will result in some sort of generic *spacetime singularity*, as we can reasonably infer from precise mathematical theorems.[22] If we now reverse the direction of time in our model—assuming time-symmetrical dynamical laws—we obtain an evolution which starts from a general-looking singularity and then becomes whatever irregular type of universe we may care to choose. It might well be that there is no inflation in this evolution, although our time-reversed physical laws allow the possibility of inflation. The point is that whether or not we actually have inflation, the physical possibility of an inflationary period is of no use whatever in attempts to ensure that the evolution from a generic singularity will lead to a uniform (or spatially flat) universe.

Let us try to understand what the real problem is. This was discussed at length in Chapter 27. The universe *was* very special at the Big Bang. It had to be so for there to have been a Second Law of thermodynamics, extending right back to the beginning. All thermalization processes *depend upon* the Second Law; thus they explain neither why we have a Second Law nor why we had a very special universe at the beginning. Moreover, all spontaneous symmetry-breaking processes and all phase transitions (these being needed for inflation) take place only by the good grace of the Second Law. These processes do not explain the Second Law: they *use* it. Moreover, all the serious calculations in inflationary cosmology assume a spacetime geometry that is FLRW, or close to it, which gives no insight as to what would happen in the generic case. If we want to know why the universe was initially so very very special, in its extraordinary uniformity, we must appeal to completely different arguments from those upon which inflationary cosmology depends.

28.6 The anthropic principle

Before we come to these arguments, I need to address another issue that is frequently invoked as part of the inflationary standpoint. This is the *anthropic principle*, and this principle is also used in many other arguments to explain why the universe is as we find it. Roughly speaking, the anthropic argument takes as its starting point the fact that the universe we perceive about us must be of such a nature as will produce and accommodate beings who can perceive it. We could use this argument to explain why the planet upon which we live has such a congenial range of temperatures, atmosphere, abundance of water, etc. etc. If conditions were not so congenial on this particular planet, then we would not be here, but somewhere else![23]

One of the most impressive uses of the anthropic argument was that made by Robert Dicke in 1957 and Brandon Carter in 1973,[24] when they resolved a puzzle—pointed out by Dirac (1937)—concerning an apparently coincidental relation between the age of the universe, when measured in Planck units (§27.10) and the ratio between the strengths of electromagnetism and gravity.[25] If this coincidence were to reflect a fundamental relationship between the parameters of Nature, then it should be maintained as constant throughout the history of the universe. But since the age of the universe is something that increases with time (obviously!), then, accordingly, the strength of gravitational forces should reduce in comparison with electric ones. Indeed Dirac actually made that suggestion, but the evidence, now, is that such a variation in the gravitational constant is incompatible with the facts.[26] What Dicke and Carter showed was that there is another explanation for Dirac's coincidence. By examining the exact role that the constants of Nature play in determining the lifetime of an ordinary star—a star congenial to life as we know it—they were able to show that this timescale is of such an order that Dirac's coincidence would necessarily hold, pretty well, for beings who evolved on (and inhabit) a planet orbiting such a star. Thus, Dirac's coincidence has an anthropic explanation. It comes about because the parameters involved in the production of sentient life (in this case, those that determine a star's age) are related to those parameters that such sentient life will actually *see* in the outside world!

It should be evident to the reader that arguments from the anthropic principle are fraught with uncertainties, although they are not without genuine significance. We do not have much idea, for example, what conditions are actually necessary for the production of sentient life. Nevertheless, the situation is not so bad when used with examples, such as given above, where we are taking the laws of physics and the overall spacetime structure of the universe as given, and we ask merely questions like where or when in the universe are conditions likely to be so-and-so, in order to be conducive to sentient life. This version of the anthropic principle is referred to, by Carter, as the *weak* anthropic principle (Fig. 28.13a).

Much more problematic are versions of the *strong* anthropic principle, according to which we try to extend the anthropic argument to determine actual constants of nature (such as the ratio of the mass of the electron to that of the proton, or the value of the fine structure constant §26.9, §31.1). Some people might regard the strong anthropic principle as leading us to a belief in a 'Divine Purpose', whereby the Creator of the universe made sure that the fundamental physical constants were pre-ordained so as to have specific values that enable sentient life to be possible. On the other hand we may think of the strong principle as being an extension of the weak one where we broaden our questions of 'where' and 'when', so that they apply not just to a single spacetime, but to the whole ensemble of possible spacetimes (Fig. 28.13b).[27] Different members of the ensemble might be expected to possess different values for the basic physical constants. The where/when question now also involves a choice of universe within the ensemble, so again we must find ourselves in a universe which permits sentience to come about.

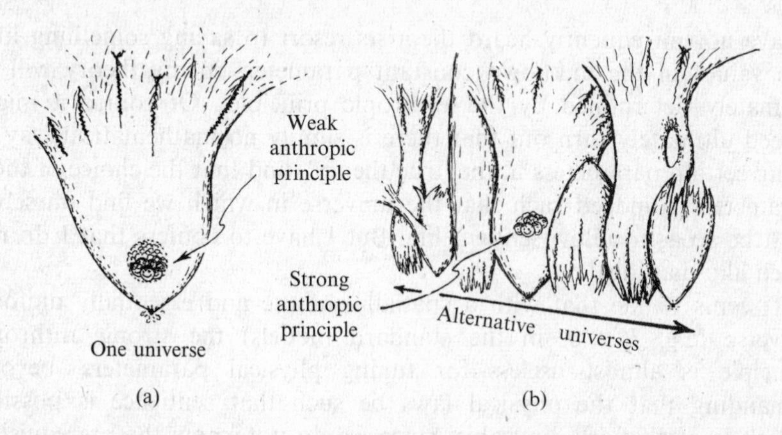

Fig. 28.13 Anthropic principle. (a) Weak form: sentient beings must find themselves in a spatio–temporal location in the universe, at which the conditions are suitable for sentient life. (b) Strong form: rather than considering just one universe we envisage an ensemble of possible universes—among which the fundamental constants of Nature may vary. Sentient beings must find themselves to be located in a universe where the constants of Nature (in addition to the spatio–temporal location) are congenial.

The first example of this kind of thing was, as far as I am aware, pointed out by Fred Hoyle, when he deduced that there must be a hitherto unobserved nuclear energy level of carbon, in order that it could be possible for stars to build up elements heavier than carbon in the process of stellar nucleosynthesis. This is the process whereby the heavier elements are produced (in stars—and finally spewed out in supernova explosions to provide the material for planet formation; see §27.8), and which our own bodies depend upon. We could not exist as living beings (of a kind that we know about) without it! On Hoyle's prompting, William Fowler and his associates[28] subsequently found Hoyle's energy level—confirming, in 1953, an impressive piece of prediction on the part of Hoyle. It is remarkable that the constants of Nature are so adjusted that such an energy level should be in just the right place, so life, as we know it, could come about. Another example of apparent cosmic good fortune is the fact that the neutron's mass is just slightly greater than that of the proton (1838 and 1836 electron masses, respectively). The existence of an appropriate family of stable nuclei, on which almost the whole of chemistry depends, rests upon this seemingly fortuitous fact.

My own position is to be extremely cautious about the use of the anthropic principle, most particularly the strong one. My impression is that the strong anthropic principle is often used as a kind of 'cop-out', when genuine theoretical considerations have seemed to reach their limit.

I have not infrequently heard theorists resort to saying something like: 'the values of the unknown constant parameters in my theory will be ultimately determined by the anthropic principle'. Of course it might indeed ultimately turn out that there is simply no mathematical way of fixing certain parameters in the 'true theory', and that the choice of these parameters is indeed such that the universe in which we find ourselves must be so as to allow sentient life. But I have to confess that I do not much like that idea!

It seems to me that with a spatially infinite and essentially uniform universe (e.g. $K \leq 0$, in the standard models) the strong anthropic principle is almost useless for tuning physical parameters, beyond demanding that the physical laws be such that sentience is possible (which is, itself fairly unusable, since we do not know the prerequisites for sentience). For if sentient life is possible at all, then we expect that, in a spatially infinite universe, it will occur. This will happen even if the conditions for sentience are extraordinarily unlikely to come about in any given finite region in the universe. In a spatially infinite universe, our expectation is that there should be *somewhere* in its infinite reaches where sentience does happen, if only by the mere chance coming together of all the necessary ingredients. This would indeed occur just by chance, even if extraordinarily infrequently.

Now, if we find that the fundamental physical constants happen to be such-and-such—perhaps fixed by mathematical criteria—then we can ask a better question: what are the *most probable circumstances* for intelligent life to come about, given these physical constant values? In the universe that we know, with the fundamental constant parameter values that we happen to have, the answer at least *seems* to be: 'on some planet rather like the Earth, near a star rather like the Sun, which has been around for perhaps 10^9 or 10^{10} years—time enough to allow appropriate Darwinian evolution to take place'. But for a universe with different constant parameter values, the answer might come out very differently.

To end this section, I should mention a related point of view with regard to the fundamental physical constants, originally put forward by John A. Wheeler in 1973. It has some connection with the anthropic principle. According to this viewpoint, the universe goes through cycles, where new 'big bangs' continually occur, each having been born out of a previous collapse phase.

Recall the Friedmann model in the case $K > 0$ $\Lambda = 0$. The universe expands from the initial Big Bang singularity and then contracts down to another singularity, the final Big Crunch. In the early days of cosmology, however, this was referred to as an 'oscillating' model, because the curve that plots $R(t)$ against t is a cycloid which indulges in an infinite number of cycles of expansion and contraction (see Fig. 27.15a, Exercise [27.19]). However, it is now better appreciated than it was in the early days that there is no way of 'smoothing out' the singularity that joins each 'crunch' to the following 'bang', within the confines of conventional classical general relativity.[29] If one ignores this fact, or presumes that some form of 'quantum gravity' will allow such a 'bounce' to take place, then one may speculate that the Friedmann cycloid is a plausible approximation to what could actually happen. Wheeler's idea was that the extreme quantum physics that takes place at the singular turn-around might entail a change in the fundamental constants of nature. Accordingly, the 'ensemble' of universes that is contemplated in connection with the strong anthropic principle is physically realized in Wheeler's proposal.

Lee Smolin, in his remarkable book *The Life of the Cosmos*,[30] suggests an intriguing modification of this idea. Instead of requiring a closed universe whose all-embracing Big Crunch converts itself into the Big Bang of the next universe phase, Smolin takes the singularities inside black holes to be the sources of new universe phases, where each black-hole singularity individually produces a different universe phase,[31] and where in each case there would be a slight readjustment to the fundamental physical constants. Smolin puts forward the ingenious idea that there could then be some form of 'natural selection' of universes, where the fundamental constants slowly evolve to obtain 'fitter' universe phases, and he takes the proliferation of black holes as a better indication of a universe's 'fitness' (because it produces many 'children') than any anthropic consideration. He argues that there is some indication that the fundamental physical constants that we actually find in our universe are indeed such as to favour a proliferation of black holes. However, it seems to me that the anthropic argument would also have a significant role in this discussion, since we could not find ourselves in a 'sentience-dead' universe phase, no matter how many of them there are!

The reader may well worry how the mass-energy of a single black hole could be converted into that for an entire universe, which might well be more than 10^{22} times more massive. Indeed, but since some unknown physics is needed in order to circumvent the singularity and alter the fundamental constants, 'all bets are off' with regard to the standard conservation laws of conventional physics. In any case, it may be argued that the law of conservation of mass-energy is problematic in the context of general relativity without the assumption of asymptotic flatness; see §19.8.

I have quite a lot of trouble with both the Wheeler and the Smolin proposals. In the first place, there is the extremely speculative nature of the key idea that some presently unknown physics can not only convert the spacetime singularity of collapse into a 'bounce', but also slightly readjust the fundamental physical constants when this happens. I know of no justification from known physics to suggest such an extrapolation. But, to my mind, it is even more geometrically implausible that the highly irregular singularities that result from collapse can magically convert themselves into (or glue themselves to) the extraordinarily smooth and uniform Big Bang that each new universe would need if it is to acquire a respectable Second Law of the kind that we are familiar with (see §27.13).

28.7 The Big Bang's special nature: an anthropic key?

Can the anthropic principle be invoked to explain the very special nature of the Big Bang? Can this principle be incorporated as part of the inflationary picture, so that an initially chaotic (maximum-entropy) state can nevertheless lead to a universe like the one we live in, in which the Second Law of thermodynamics holds sway? Basically the general argument is to say that the second law is essential to life as we know it; moreover the overall densities, temperatures, matter distributions and compositions, etc. must be so as to be conducive to life. In addition, the universe must have existed for long enough for evolution to operate, and so on. Sometimes this argument is used in conjunction with an inflationary argument. Accordingly, although a completely generic initial state might not inflate to give us a smoothed-out universe like the one we observe, we should ask merely for some small region of the initial spacetime 'manifold', just after the Big Bang, to be smooth enough for inflation to take over in that region, the entire observable universe today coming about as a result of an inflation of that tiny smooth region (see Fig. 28.14a). The argument would run roughly: 'for sentient life to exist, we need a large universe with timescales long enough for evolution to take place, in conducive conditions, etc.; this requires some inflation, originating from our tiny smooth initial region, and once it starts, the inflation goes on to provide us with the wonderfully enormous observable universe that we know'.

Although, it may seem that this picture is of such a marvelously romantic nature that it is completely immune from scientific attack, I do not believe that this is so. Let us return to the extraordinary degree of precision (or 'fine-tuning') that seems to be required for a Big Bang of the nature that we appear to observe. As was argued in §27.13, the required precision, in phase-space-volume terms, is one part in $10^{10^{123}}$ at least. The exponent '10^{123}' comes from the entropy of a black hole of mass equal to that in the observable universe.

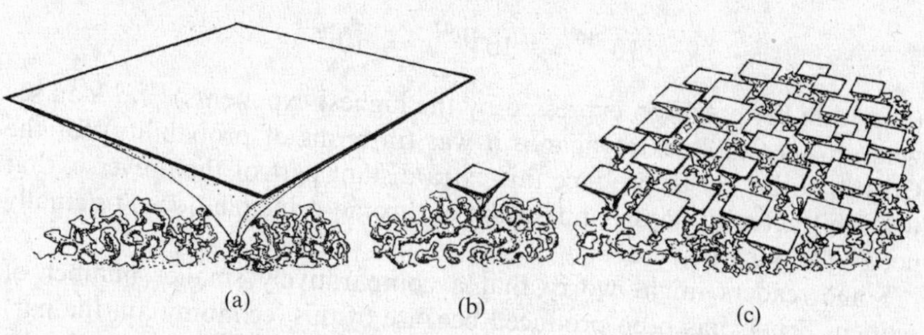

Fig. 28.14 (a) A completely general initial state for the universe does not inflate, but we can ask merely for a small initial region that is smooth enough to inflate to the universe that we observe (cost: $10^{10^{123}}$). (b) But how much of our vast universe is really needed for our sentient existence? Absurdly 'cheaper', for the creation of sentient life, is for the Creator to produce a universe of one tenth of the linear dimension (cost: merely $10^{10^{117}}$). (c) To create as many sentient beings as in (a), the Creator can far more cheaply simply produce 10^3 independent instances of the 'smaller' universes of (b) (at 'bargain' cost: $(10^{10^{117}})^{1000} = 10^{10^{120}}$). Hence the anthropic principle does not account for the apparent extravagence of inflation.

But do we really need the whole observable universe, in order that sentient life can come about? This seems unlikely. It is hard to imagine that even anything outside our galaxy would be needed. Yet, it might be that intelligent life is very rare, and it might be a bit more comfortable to have somewhat more space than that. Let us be very generous and ask that a region of radius one tenth of the distance out to the edge of the observable universe must resemble the universe that we know, but we do not care about what happens outside that radius. The phase-space volume can be calculated as before. We calculate the mass in that region to be 10^{-3} of what we had before, and that gives us a black-hole entropy of 10^{-6} of what we had before.[28.7] Thus, the precision needed, on the part of our 'Creator' (see Fig. 27.21), to construct this smaller region is now only about:

$$\text{one part in } 10^{10^{117}}.$$

Have a look at Fig. 28.14b. Our Creator now only requires a rather *smaller* 'tiny smooth region' of the initial 'manifold' than before. The Creator is much more likely to come across a smooth region of this smaller size than the somewhat larger one that we considered earlier. Assuming that the inflation acts in the same way on the small region as it would on the somewhat larger one, but producing a smaller inflated universe, in proportion, we can estimate how much more frequently the Creator comes across the smaller than the larger regions. The figure is no better than

[28.7] Why?

$$10^{-10^{117}} \div 10^{-10^{123}} = 10^{10^{123}}$$

(to within the precision expressed by the highest exponents).[28.8] You see what an incredible extravagance it was (in terms of probability) for the Creator to bother to produce this extra distant part of the universe, that we don't actually need—and so the anthropic principle doesn't actually need—for our existence!

Some readers might worry that a comparatively smaller number of sentient beings has been produced because of this 'economy' on the part of the Creator. Whether this is an issue or not, it is not the answer to why the 'extravagance' took place. It would be far far 'cheaper', in terms of probabilities (i.e. inverse box sizes in phase-space; see Fig. 27.2)—by a factor of about 1 to $10^{10^{123}}$—to have 10^3 of the smaller inflated universe regions (which gets us up to the same number of sentient beings as for a single larger one) than to have just 1 larger universe region (Fig. 28.14c).[28.9]

To see how impotent the anthropic argument is, in this context, consider the following facts. Life on Earth certainly does not directly need the microwave background radiation. In fact, we do not even need Darwinian evolution! It would have been far 'cheaper' in terms of 'probabilities' to have produced sentient life from the random coming together of gas and radiation. (One can estimate that the entire solar system, including its living inhabitants, could be created from the random collision of particles and radiation with a probability of one part in $10^{10^{60}}$ (or probably a good deal less than $10^{10^{60}}$). The figure $10^{10^{60}}$ is utter 'chicken feed' by comparison with the $10^{10^{123}}$ needed for the Big Bang of the observable universe.[32] We do not need a Big Bang to be in its observed uniform configuration. We do not need the Second Law at times earlier than life was around. It would be far 'cheaper' for the Creator not to bother with that. And inflation is of no help at all. The 'cheapskate' economy curve for the Creator to adopt, in Fig. 27.8, just in order to produce sentient life, would be much more like the curve (c)(b), rather than the observed (d)(b), inflation or no inflation!

All of this is simply reinforcing the argument that it is indeed misconceived to seek reasons of the above nature, where suitable universe conditions are supposed to have come about from some kind of random initial choice. There was indeed something very special about how the universe started off. It seems to me that there are two possible routes to addressing this question. The difference between the two is a matter of scientific attitude. We might take the position that the initial choice was an 'act of God' (rather like that fancifully illustrated in Fig. 27.21). or we might seek some scientific/mathematical theory to explain the extraordinarily special

[28.8] Explain these figures.

[28.9] Explain these figures carefully.

nature of the Big Bang. My own strong inclination is certainly to try to see how far we can get with the second possibility. We have become used to mathematical laws—laws of extraordinary precision—controlling the physical behaviour of the world. It appears that we again require something of exceptional precision, a law that determines the very nature of the Big Bang. But the Big Bang is a spacetime singularity, and our present-day theories are not able to handle this kind of thing. Our expectations, however, are that what is required is some appropriate form of *quantum gravity*,[33] where the rules of general relativity, of quantum mechanics, and perhaps also of some other unknown physical ingredients, must come together appropriately.

28.8 The Weyl curvature hypothesis

I shall postpone my main considerations of present-day activity in the field of quantum gravity until Chapters 30–33. For the moment, let us just concentrate on trying to understand what the geometrical constraints on the Big Bang appear to have been actually like. Afterwards, we examine the one proposal I know of, namely that of James Hartle and Stephen Hawking, that attempts to explain this sort of geometry on the basis of a serious quantum gravity theory.

Recall from §19.7 that the *gravitational* degrees of freedom are described by the Weyl conformal tensor C_{abcd}. Thus, in *empty space* (where a possible cosmological constant Λ, in any case small with regard to local physics, is for the moment being ignored) we find that the spacetime curvature is entirely Weyl curvature (the Ricci curvature vanishing). Weyl curvature is the kind of curvature whose effect on matter is of a distorting or tidal nature, rather than the volume-reducing one of material sources. The effect of Weyl curvature was illustrated in Fig. 17.9a (and the fact that this picture was originally a Newtonian spacetime picture in no way detracts from its validity). This picture is to be contrasted with Fig. 17.9b, in which we see the volume-reducing effect of matter, i.e. Ricci tensor. However, there are actually some complicating issues when we consider (as here) the effects of Weyl and Ricci curvature on *timelike* geodesics (freely moving massive particles), since the Ricci tensor can also sometimes have a distorting effect, in addition to its volume-reducing effect.

These complicating issues are eliminated if we think of the action of these kinds of curvature on *null* geodesics (light rays). Moreover, we can then reinstate a cosmological constant Λ, since a term of the form Λg_{ab} does not focus light rays.[28.10] We can think of the geodesics in Fig. 17.9 as being light rays belonging to some light cone (in the manner of Fig. 17.16). In fact, if we think of them as belonging to the past light cone of some observer, the

[28.10] Why not? *Hint*: Explain Note 28.34.

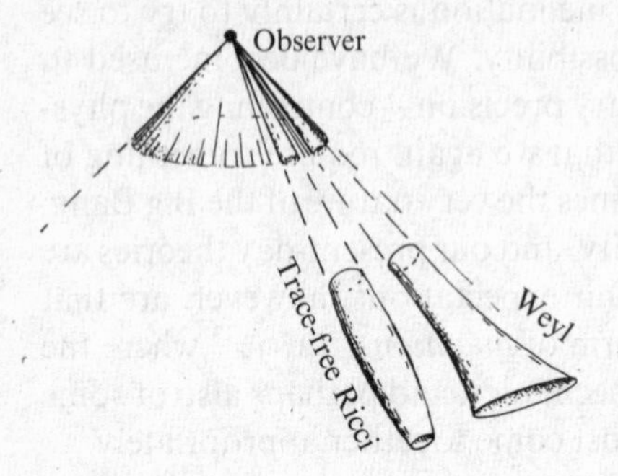

Fig. 28.15 The focusing effect of (trace-free) Ricci tensor (due to a matter distribution) is as a positively focusing lens, whereas that of Weyl tensor (due to free gravitational field) is as a purely astigmatic lens—with as much positive focussing in one plane as there is negative focussing in the perpendicular plane.

distortion effects can be understood very graphically in terms of lenses placed between a source of light and the observer. The effect of Ricci tensor,[34] due to a matter distribution, is as a positively focusing lens, whereas that of Weyl tensor, due to free gravitational field, is as a purely *astigmatic* lens—with as much positive focusing in one plane as there is negative focusing in a perpendicular plane (Fig. 28.15). We can get a very good impression of the (lowest-order) effects of these two different kinds of curvature if we imagine simply looking through a large transparent solid massive spherical body having the refractive index of the vacuum. (Perhaps we should think of 'looking' through the Sun with neutrinos—taken as *massless* particles—which pass straight through the Sun, paying attention only to its gravitational field!) To a reasonable approximation, we can consider the rays passing through the Sun to be mainly affected by Ricci curvature, so we get an apparent magnification (positive lens) of the star field behind the Sun. On the other hand, beyond the Sun's rim, we get, in effect, the purely astigmatic distorting effects of Weyl curvature, so that a small circular pattern in the background sky would appear to be elliptical to the observer. See Fig. 28.16.[28.11] This is essentially the way that the Sun's gravitational field distorts the background star pattern, as first seen in Eddington's 1919 expedition (see §19.8).

Let us now think of a universe evolving so that an initially uniform distribution of material (with some density fluctuations) gradually clumps gravitationally, so that eventually parts of it collapse into black holes. The initial uniformity corresponds to a mainly Ricci-curvature (matter) distribution, but as more and more material collects together gravitationally, we get increasing amounts of Weyl curvature, basically inhabiting the regions of spacetime distortion surrounding the clumped matter. The

[28.11] Show that areas are preserved, for an infinitesimal outward displacement, which varies inversely as the distance out.

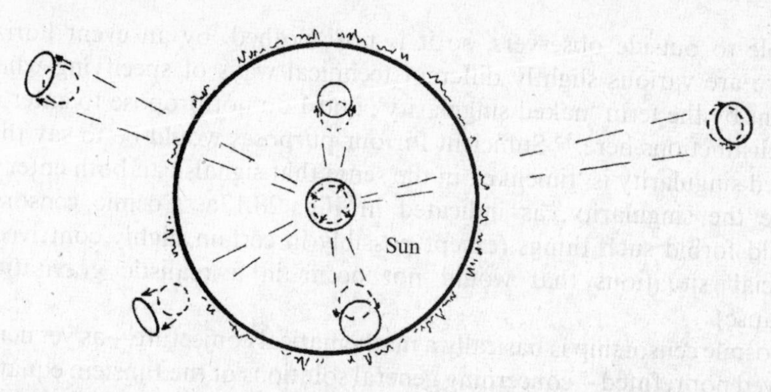

Fig. 28.16 We get a good impression of (lowest-order) effects of the two different kinds of spacetime curvature by 'looking' at the star field through a transparent non-refracting Sun (as though with massless neutrinos). To a reasonable approximation, rays passing through the Sun are focused just by Ricci curvature, resulting in magnification (as by a positive lens), whereas outside the Sun's rim, we get essentially purely astigmatic Weyl distortions, so a small circular pattern in the star field would appear elliptical.

Weyl curvature finally diverges to infinity as the black-hole singularities are reached. If we think of the material as having been originally spewed out from the Big Bang in an almost completely uniform way, then we start with a Weyl curvature that is, for all intents and purposes *zero*. Indeed, a characteristic feature of the FLRW models is that the Weyl curvature vanishes completely (these models being, accordingly, conformally flat, see §19.7). For a universe to start out *closely* FLRW, we expect the Weyl curvature to be extremely small, as compared with the Ricci curvature, the latter actually diverging at the Big Bang.

This picture strongly suggests what the geometrical difference is between the initial Big-Bang singularity—of exceedingly low entropy—and the generic black-hole singularities, of very high entropy. The Weyl curvature vanishes (or is, at least, very very small—e.g. merely finite—compared with what it might have been) at the initial singularity and is unconstrained, no doubt diverging wildly to infinity, at final singularities. It is this geometrical characterization that seems to distinguish Fig. 27.20a from Fig. 27.20d, for example even though it might be hard to recognize the distinction in terms of conformal diagrams.

This observation should be taken in conjunction with another conjectured feature of spacetime singularities, referred to as *cosmic censorship*. This is a (currently unproved) assertion that, roughly speaking, in unstoppable gravitational collapse, a black hole will be the result, rather than something worse, known as a *naked singularity*. A naked singularity would be a spacetime singularity, resulting from a gravitational collapse, which is visible to outside observers, so it is not 'clothed' by an event horizon. There are various slightly different technical ways of specifying what is meant by the term 'naked singularity', and I do not propose to enter into the distinctions here.[35] Sufficient for our purposes would be to say that a naked singularity is 'timelike', in the sense that signals can both enter and leave the singularity, as indicated in Fig. 28.17a. Cosmic censorship would forbid such things (except possibly in certain highly contrived or 'special' situations that would not occur in a realistic gravitational collapse).

Cosmic censorship is basically a mathematical conjecture—as yet neither proved nor refuted—concerning general solutions of the Einstein equation. If we assume this conjecture, then physical spacetime singularities have to be 'spacelike' (or perhaps 'null') but never 'timelike'. There are two kinds of spacelike (or null) singularities, namely 'initial' or 'final' ones, depending upon whether timelike curves can escape from the singularity into the future or enter it from the past; see Fig. 28.17b,c. The physical conjecture that I refer to as *the Weyl curvature hypothesis* asserts that (in some appropriate sense) the Weyl curvature is constrained to be zero (or at least very small) at *initial* singularities, in the actual physical universe. The creation of a universe in a way that satisfies the Weyl curvature hypothesis would represent an absolutely enormous constraint on the Creator's choice, in the process represented in Fig. 27.21. As a result, there would be a Second Law of thermodynamics, and it would actually take the form that we observe. There is now some good mathematical evidence that some form of 'Weyl curvature hypothesis' indeed adequately constrains the Big Bang in a way

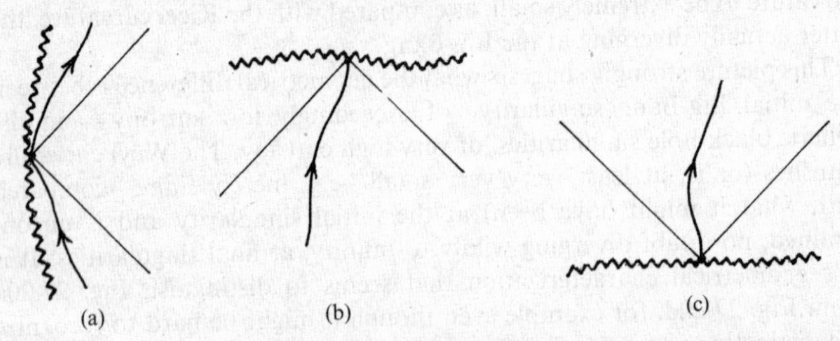

Fig. 28.17 (a) Causal signals can either enter or leave a 'naked singularity'. If these are excluded—by Cosmic Censorship—we are basically left with (b) 'future singularities' (resulting from gravitational collapse) which causal signals can enter but not leave, and (c) 'past singularities' (in the Big Bang, or perhaps more localized creation events) which causal signals can leave but not enter. The Weyl curvature hypothesis asserts that the Weyl curvature is (appropriately) constrained to zero (or to being very small) at the initial singularities (c) of the actual physical universe.

that the resulting universe model closely resembles an FLRW model in its early stages.[36]

28.9 The Hartle–Hawking 'no-boundary' proposal

Simply as an assertion, the Weyl curvature hypothesis is perhaps more like a claim for 'an act of God' than a physical theory. What is required is some theoretical justification for something of the nature of this hypothesis. What kind of theory will we have to appeal to? The usual point of view, with regard to spacetime singularities, is that this is the province of *quantum gravity*.

The difficulty here is that, despite over fifty years of determined efforts to bring general relativity and quantum mechanics together, there is still nothing that gets close to a consensus as to the correct approach to the subject. I shall be addressing some of the more currently popular schemes in Chapters 31 and 32, but even among those there is little serious attempt to come to terms with the particular nature of the Big Bang. There is, however, one notable exception, put forward by James Hartle and Stephen Hawking in 1983, and it is therefore appropriate that I make some comments in relation to their main idea.

One of the ingredients of the Hartle–Hawking proposal is what is commonly referred to as 'Euclideanization'. The underlying idea is closely related to that of a *Wick rotation* applied to Minkowski space, whereby the time coordinate t is 'rotated' into $\tau = it$. The (spatial) spacetime metric $d\ell^2$ then becomes $d\ell^2 = d\tau^2 + dx^2 + dy^2 + dz^2$ (see §18.1). The original (Gian Carlo Wick) idea[37] was that a (special-)relativistic quantum field theory can be constructed by first formulating it with Minkowski spacetime replaced by this Euclidean 4-space $\mathbb{E}^4$, where the theory is now taken to be invariant under the Euclidean group of symmetries of $\mathbb{E}^4$. Assuming that the quantities obtained in the Euclidean version of the theory are analytic in the coordinates, the Wick rotation can then be applied, with τ rotated continuously back into t, so that we now obtain a corresponding theory that is invariant under the Poincaré group of Minkowski 4-space. This procedure has two significant advantages. First, quantities that are liable to be divergent in Minkowski space may turn out to be *con*vergent in the Euclidean version of the theory. (The reason comes down to the Euclidean rotation group O(4) being compact, so of finite volume, whereas the relativistic Lorentz group O(3,1) is non-compact and of infinite volume.) In particular, path integrals (see §26.6) have a much better chance of a mathematically meaningful definition in the Euclidean rather than the Minkowskian version. The other advantage is that requirements of positive frequency (see §§9.3,5, §24.3) can be ensured by carefully applying the Wick rotation in the correct way.

In the Hartle–Hawking scheme, it is necessary to use Hawking's ingenious modification of the Wick idea, in which the 'rotation' is applied not to a space which is a *background* to the paths, in a path integral—which is the usual idea—but to the individual spacetimes which themselves constitute each path of the path integral.[38] These 'spacetimes' are, accordingly, allowed to have positive-definite Riemannian metrics, rather than the Lorentzian metrics that apply to a normal spacetime. (These Riemannian metrics are often confusingly referred to as 'Euclidean', despite the standard use of that name for a flat Euclidean space $\mathbb{E}^n$!) It should be made clear, however, that there is a 'leap of imagination' involved in the Hawking version of 'Euclideanization' going far beyond that of Wick's original idea. Whether or not this provides a fruitful route to the correct union of general relativity with quantum mechanics remains to be seen.[39]

Hartle and Hawking's striking proposal was that this path-integral approach of Hawking's could describe the relevant quantum theory for the Big Bang itself, and that in place of an actual singular spacetime there would be a quantum superposition (i.e. 'path integral') of 'spacetimes' which could have Riemannian in place of Lorentzian metrics. They referred to their idea as the 'no-boundary' proposal, because rather than having the singular boundary to the classical spacetime that the Big Bang represents, there would be a superposed family of *non*-singular spaces, dominated by Riemannian ones that simply 'close off' the bottom end in the manner indicated in Fig. 28.18, so the singular boundary disappears completely. Moments 'after' the Big Bang, there has to be a transition where the dominance of

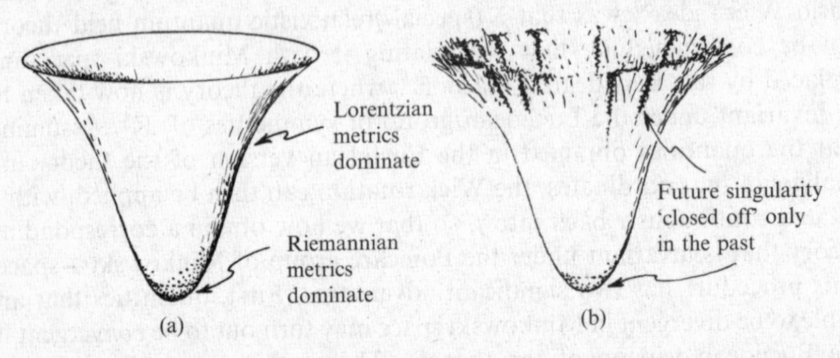

Fig. 28.18 The Hartle–Hawking 'no-boundary' proposal suggests that (a) the Big Bang can be treated according to a quantum-gravity procedure whereby Riemannian (rather than Lorentzian) geometries dominate the path integral near the classical singularity, and provide ways of closing off the spacetime in a non-singular way. (b) With regard to the singularities of collapse, the 'closing off' seems to be required only at the 'far end' of the spacetime, thereby allowing the high-entropy generic singularities that are expected to occur in gravitational collapse to black holes (or Big Crunch).

Riemannian geometry goes over to Lorentzian. (We can imagine this being achieved with the involvement of suitable *complex* metrics.) Even in the Lorentzian region there is still a superposition of 'spacetimes' (some of which are Riemannian), but away from the Big Bang, a classical Lorentzian spacetime is taken to dominate, whereas at the Big Bang region itself, the 'no-boundary' Riemannian metrics are taken to dominate. Not only does this scheme have genuine elegance, turning a seemingly intractable problem into one that appears to be vaguely manageable, but it also has the appearance of giving some direct support to a 'smooth early universe' that could be compatible with the Weyl curvature hypothesis.

So far, so good. But I also have some considerable difficulties with this proposal. First, the very idea of 'Euclideanization' is problematic in a number of respects, of relevance to the kind of context that finds its use here. Even in a flat-space context, it is normally out of the question to compute a path integral exactly, and many approximations need to be made. It would be usual to single out certain specific terms which may be regarded as dominating the integral, and to leave out the rest. This might be expected to give a reasonable approximation to the 'Euclidean' path integral, but recall that a process of analytic continuation needs then to be applied in order that the appropriate physical answer be obtained. This is a highly unreliable procedure, because something that approximates a holomorphic function in one region is likely to be wildly off in another region. To appreciate the essence of the difficulty, suppose that we have a real analytic function $f(x)$ which we know for real values of x, but only approximately, and we wish to infer its values for purely imaginary x. If we add a function of the form $\varepsilon \cos(Ax)$ to $f(x)$, where ε and A are real, with ε very small and A large, then $f(x)$ will not be much altered along the real axis of x; yet the behaviour along the imaginary axis will be changed completely, thereby illustrating the extreme instability of the analytic continuation process.[28.12] As far as I can see, the 'Euclideanization trick' can be very useful for producing exact model QFTs, but I have severe difficulties with it when used in conjunction with approximations, as here. (It is not clear to me how strongly the Hartle–Hawking proposal depends upon this analytic-continuation step, however.)

I also have some technical difficulties with the generality of Euclideanization. As far as I can see, it is a clever trick for producing consistent QFTs (and ensuring a positive-frequency condition §§9.3,5), but it would be grossly optimistic to expect that any particular QFT of interest can be obtained by means of it. Theories obtained via Euclideanization have, in effect, hidden structures, originating from their associated symmetry

[28.12] Explain this, using results from §5.3. (*Hint*: What is $e^{Aix} + e^{-Aix}$?)

group of the 'wrong signature'; see §13.8, and §18.2. I do not see why a 'correct' theory needs to have this special character.

28.10 Cosmological parameters: observational status?

There is also the question of agreement with observation. At least in its original form, the Hartle–Hawking no-boundary proposal would seem to point to a closed (in fact $K > 0$) universe, and for a number of years Hawking had given his support to such models. But in the face of mounting cosmological evidence, which had appeared to favour the hyperbolic ($K < 0$) case, Hawking, in collaboration with Turok, subsequently modified his arguments to enable the 'no-boundary' proposal to accommodate the hyperbolic case also.[40] There is an interesting parallel with the expectations of inflationary cosmology, which had for many years been argued to have the decisive implication that the observed universe must turn out to be spatially flat ($K = 0$). A number of inflationists, also, subsequently modified their arguments, in the face of this increasingly impressive cosmological data, to allow for the possibility[41] of $K < 0$.

What is the present observational position? Well, things have now shifted very significantly again, with the startling evidence (and from more than one source) that there seems to be a significant *positive cosmological constant* Λ. This has the implication that we could have $K = 0$ after all. And since if the observational evidence allows for $K = 0$, it cannot exclude a small positive spatial curvature (Hawking's preferred $K > 0$) or a small negative spatial curvature (my own preferred $K < 0$)—so all bets are off again!

What is this bearing of this discovery of $\Lambda > 0$ on the value of K? I should mention, first, the reasons behind the earlier belief that cosmological evidence favoured a negative value of K. The essential issue is the total *mass–energy content* of the universe, and that if this is too small, then it will not be able to close up the universe with positive curvature, or (in the Friedmann models) to haul it back in again following its initial expansion so as to produce a collapsing phase (see Fig. 27.15a,b,c). It had long been known that the density of ordinary visible 'baryonic' (see §25.6) material in galaxies is insufficient for this, being only about one thirtieth of the *critical* value that represents the division between the positive and negative K values, the critical density being that which gives us $K = 0$. The quantity Ω_b is commonly introduced to denote that fraction of the critical mass–energy density which is achieved by normal baryonic matter. Thus, if $\Omega_b = 1$, the baryonic matter would indeed supply the critical density, and any significant further (positive) mass–energy would lead to a $K > 0$ universe. However, as mentioned above, we seem to have something like $\Omega_b = 0.03$, instead, which had given a powerful indication of $K < 0$.

However, this fails to take into account the strong evidence that there is a good deal of more matter in the universe than the baryonic material that is directly observed in stars. For many years, it had become clear that the dynamics of stars within galaxies does not make sense, according to standard theory[42] unless there is a good deal of more material in the neighbourhood of a galaxy than is directly seen in stars. A similar comment applies to the dynamics of individual galaxies within clusters. Overall, there seems to be about 10 times more matter than is perceived in ordinary baryonic form. This is the mysterious *dark matter* whose actual nature is still not agreed upon by astronomers, and which may even be of some material different from any that is definitely known to particle physicists—though there is much speculation about this at the present time.[43] Since the dark matter seems to contribute about 10 times as much mass–energy as there is ordinary baryonic matter, the density supplied by the dark matter, as a fraction Ω_d of the critical density, is roughly given by $\Omega_d = 0.3$ (and the uncertainties are such that we can include the baryonic $\Omega_b = 0.03$ into this figure, if we choose). This still leaves us considerably short of the critical value. Moreover, various types of observation (including of gravitational lensing effects—which we recall from §19.8 provide a direct measure of the presence of mass) were beginning to show, fairly convincingly, that there can be no other significant concentrations of mass in the universe. So the conclusion $K < 0$ was now looking pretty firm, and consequently the inflationists and followers of Hartle–Hawking started looking for ways to incorporate $K < 0$ into their respective worldviews.

Then came the bombshell of the cosmological constant. We recall from §19.7 that Einstein had regarded the introduction of Λ as his 'greatest mistake' (perhaps mainly because it contributed to his failure to predict the expansion of the universe). Although since then it had always been taken as a possibility by cosmologists, rather few of them seem to have expected to find Λ to be non-zero in our actual universe. An additional issue was the fact that calculations of 'vacuum energy' by quantum field theorists (basically a renormalization effect like those of §26.9) had yielded an absurd answer that there should actually be an *effective* cosmological constant that is larger than what is seen by a factor of about 10^{120} (or at least 10^{60} if different assumptions are made)! This became known as the 'cosmological constant problem'. It might have been plausible that some unknown cancellation or general principle could give the value 0 for this vacuum energy, but to find a tiny residue that could have relevance to cosmology at the present epoch was in no way anticipated. (It should be mentioned that this 'vacuum energy' ought to be proportional to the metric g_{ab}, by local Lorentz invariance, so the form Λg_{ab} is anticipated, for a constant Λ, contributing to the Einstein equation precisely as Einstein had suggested in 1917. The only trouble is that the value of Λ comes out completely wrong!)

Nevertheless, when in 1998 two teams observing very distant supernovae (see §27.8)—one headed by Saul Perlmutter, in California, and the other, headed by Brian Schmidt in Australia and Robert Kirschner in eastern USA—came to the remarkable conclusion that the expansion of the universe had begun to accelerate, as is consistent with the upward turn in the graph of Fig. 27.15d, which is the hallmark of a positive cosmological constant! How big is this Λ that seems to be observed? There are still some uncertainties about this (and some theoreticians have argued that the case for a positive Λ has still not been convincingly made[44]), but the remarkable conclusion is that the effective mass–energy density Ω_Λ that Λ provides, as a fraction of the critical density, is given approximately by $\Omega_\Lambda = 0.7$, so we seem to have, for the total effective density, as a fraction of critical

$$\Omega \approx \Omega_d + \Omega_\Lambda \approx 0.3 + 0.7 = 1.$$

In other words, the observations seem to be now consistent with $K = 0$.

The inflationists (at least those who had the confidence not to shift their ground) are, of course, jubilant, and it can certainly be counted as a predictive success of their theory that against some seemingly powerful evidence to the contrary, the prediction of $K = 0$ seems to have won out. However, the uncertainties are still too great for this conclusion to be carried with conviction, and it is significant that other types of recent observation also have a powerful bearing on this question. As mentioned in §28.5, there have been several measurements of the detailed temperature variations in the microwave background, starting with the COBE satellite, launched in 1989, and the most recent survey (at the time of writing) being that made by the WMAP space explorer.

These temperature variations are normally analysed by decomposing the pattern over the sky into spherical harmonics, according to the procedures discussed in §22.11. We recall that the different spherical harmonics are labelled by a positive integer ℓ and an integer m in the range $-\ell$ to ℓ. (In the quantum mechanical situation, ℓ is normally called j, and both j and m could be half odd integers.) The m-quantity is less important, because it depends on an arbitrarily chosen direction in the sky, so the general intensity for each value of ℓ is regarded as being the quantity of most interest. In Fig. 28.19, I have shown the results of this analysis. Notice the indication that after the curve reaches its maximum, at around $\ell = 200$, it starts to oscillate. These local maxima are referred to as 'acoustic peaks' since they reflect a clear theoretical prediction that in an early stage of the universe, local concentrations of matter would start by falling inwards and then either

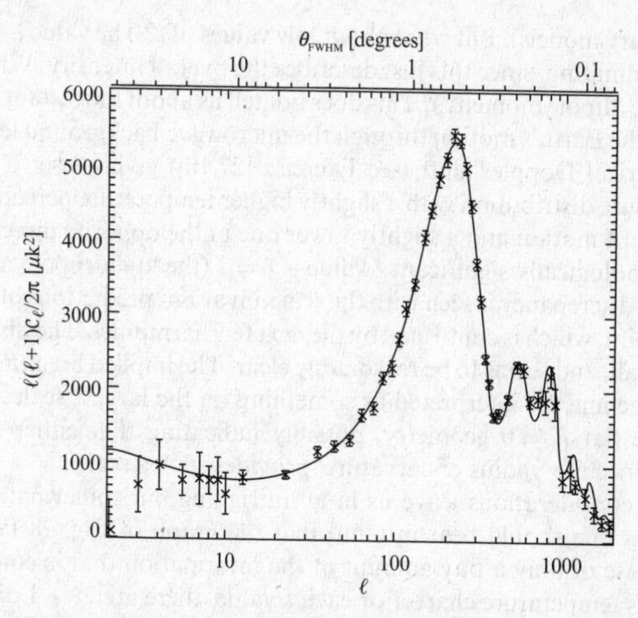

Fig. 28.19 The anticipated 'acoustic peaks' in the harmonic analysis of the cosmic microwave background (solid line), and the observed data points (crosses, with error bars). Be sure to notice the very significant discrepancy at the quadrupole ($\ell = 2$), almost hidden (accidentally?) by the vertical axis.

rebound or fall through themselves (which is what might be expected for dark matter), this resulting in a kind of sonic oscillation. The typical scale at which this oscillation would happen would be governed by a 'horizon scale' at decoupling (see Fig. 28.5a, and imagine points u and v moved around on the decoupling surface until their pasts just touch; this is the 'horizon scale').[45] It is at this scale that the main peak occurs.

There is, however, the question of what angular separation in the sky corresponds to what local distance separation in the universe at the time of decoupling, and it is here that the spatial curvature of the universe plays an important role, the acoustic peaks being shifted one way or the other, in ℓ, depending on the value of K (to smaller for positive K and larger for negative K). The issue is not quite straight forward, however, because the expansion rate of the universe also plays a role in this, so detailed calculations are necessary. The upshot is that this kind of analysis of the cosmic microwave background is, in a general way, consistent with $K = 0$, but there is still room for a positive or negative K value that would be of observational significance.

The results for high values of ℓ, thus seem to be consistent with the expectations of inflation (and there is also a scale-invariance in the observed temperature fluctuations that had also been a prediction of some inflationary models). But what about low values of ℓ? The value $\ell = 0$ is not very illuminating, since this just describes the overall intensity. What about $\ell = 1$ (the 'dipole moment')? This does not tell us about the distant universe, because the Earth's motion through the microwave background leads to an asymmetrical Doppler shift (see Exercise [27.10]) giving rise to an $\ell = 1$ temperature distribution with a slightly higher temperature perceived in the direction of motion and a slightly lower one in the opposite direction. The first cosmologically significant ℓ value is $\ell = 2$ (the 'quadrupole moment'). In fact, a discrepancy is seen with the scale-invariant predictions of inflation at this point, which is confirmed by the next few harmonics. The discrepancy is not small, and seems to be reasonably clear. The implied breaking of scale invariance may be interpreted as something on the largest scales differing from the flat $K = 0$ geometry, possibly indicating that either $K > 0$ or $K < 0$, since the 'radius of curvature' provides such a scale.

These considerations leave us in an intriguing but somewhat unsettled state. But one should bear in mind that the graph of Fig. 28.19 is really making use of only a tiny amount of the information that is contained in WMAP's temperature chart. For each ℓ value, there are $2\ell + 1$ different m values, and there is a real parameter for each of these. Most of that information is being ignored in this analysis, and there must be enormous amounts of hidden data telling us something of possibly great importance about the early universe.

Here, I mention only one alternative way of analysing this data, due mainly to Vahe Gurzadyan *et al.* (1992, 1994, 1997, 2002, 2003, 2004) which seems to have startling implications. In this approach, a harmonic analysis is not used; instead, one examines the distortions in the shape of distant regions of each particular temperature, due to intervening spatial curvature. If we imagine that the undistorted shape of such a region is actually circular, then curvature effects could cause this to become elliptical (recall Fig. 28.15). Of course, in practice, we do not know the shape of the region we are looking at, but there can be statistical effects, causing the regions of a specific temperature to become more (or less) stretched and spindly than they would be otherwise. This is clearly a delicate piece of statistical analysis, but the conclusion that Gurzadyan and his colleagues come to is that there is indeed a significant amount of ellipticity in the microwave maps (originally COBE, then BOOMERANG, and subsequently WMAP). What does this mean? The theoretical analysis of this situation tells us that only with $K < 0$ can we expect this degree of ellipticity—as a result of 'geodesic mixing'. These results are new, so one must wait to see if significant objections are levelled at this remarkable conclusion.

This analysis also provides independent evidence for a positive cosmological constant of about the size that is implied by the supernova data. Thus, the negative curvature is concluded to be small, in the sense that

$\Omega_d + \Omega_\Lambda$ cannot differ by much from unity, perhaps having a value of about 0.9. This emphasizes a puzzle that has worried many cosmologists. The quantities Ω_b, Ω_d, and Ω_Λ are not constant in time. In the early stages of the universe, Ω_b and Ω_d would have been far larger and Ω_Λ far smaller. In the very late stages of the universe, Ω_b and Ω_d would become negligible, with Ω_Λ dominating the effective mass–energy density. The seeming coincidence that Ω_Λ and Ω_d are, just at the present epoch, of the same general order of size presents a somewhat puzzling conundrum.

Curiously, the term 'cosmological constant' seems to have gone out of fashion almost as soon as Λ was observationally discovered, despite that being the standard terminology since Einstein's theoretical introduction of it in 1917. Instead, Λ is referred to as 'dark energy', or 'vacuum energy', or sometimes 'quintessence', perhaps because the cold term 'cosmological constant' does not carry with it a sufficient air of mystery, or perhaps, a little more rationally, because the presence of the word 'constant' rather implies that Λ cannot change with time! Many cosmologists seem to be happier with a varying Λ, possibly regarding the present 'Λ' as representing the onset of a 'new inflationary phase', where they point out the similarity to the supposed very early inflationary phase of the universe. We recall from §28.4 that this is taken to be dominated by a 'false vacuum' in which there is an effective cosmological constant that is so large that it completely dominates all the (already enormously dense) ordinary matter. If the universe was allowed to have an effective 'Λ' in those days, which was so extraordinarily different from the value that we find today—so the argument goes—then surely we should allow for a 'varying Λ' and the term 'cosmological constant' is, accordingly, inappropriate.

However, this idea, attractive as it may seem to some people, has its difficulties with the mathematics, as the term 'cosmological constant' was introduced with good reason. The constancy of Λ is a direct consequence of the energy conservation equation $\nabla^a T_{ab} = 0$, of §§19.5–7, since adding a multiple of g_{ab} to T_{ab} can leave that conservation equation undisturbed only if that multiple is a constant.[28.13] Thus, any non-constancy in 'Λ' would have to be accompanied by a compensating non-conservation of the mass–energy of the matter. It is certainly much more theoretically comfortable to have Λ constant—as is indeed consistent with observation.

Where does this leave us? Certainly in an interesting state. I do not see that inflationary cosmology is 'confirmed' by these observations, and even if it were, this would not resolve the cosmological problem that, in my opinion, overshadows all others, namely the extraordinarily 'special' Big Bang—to at least the degree of a part in $10^{10^{123}}$—which underlies the Second Law. Some cosmologists would regard the 'fine tuning' that is

---

[28.13] Why?

involved in this (see Fig. 27.21) as unacceptable, and they try to 'explain' it in terms of inflation or the anthropic principle (§§28.4,6), although, as we have seen, such procedures leave us very wide of the mark.

There is, indeed, a fundamental problem that I have with any proposal (e.g. inflation or the Hartle–Hawking proposal) that attempts to address the problem of spacetime singularities within an apparently *time-symmetrical* physics. There is no time-asymmetry in inflationary physics and, as far as I can make out, there is none in the Hartle–Hawking proposal either, so this proposal should be applied also to the final singularities of collapse (in black holes, or in the Big Crunch if there is one) as well as to the Big Bang. Hawking (1982) has argued that it *can* be, but in a decidedly exotic way, the space in the neighbourhood of a final singularity being 'closed off without boundary', by taking the universe all the way back to the Big Bang, the 'Euclideanization' being applied only there (Fig. 28.18)! His argument is that the no-boundary proposal merely asserts that there is *some* way of closing things off without boundary, and we *define* the 'beginning' (which determines the universe's time-sense) as the end at which the closing off occurs. I have to say that I have great difficulties with this argument—and indeed with *any* argument where there is no explicit time-asymmetry in the physical laws themselves. (In Hawking's 'exotic' argument, for example, it would appear that there is still a 'boundary', at the final singularity of collapse, even though there has been a smooth boundary-free closing off only 'on the other side' of the spacetime. It seems to me that only one-half of the boundary-removal problem has been attended to.)

Do we, then, have to address the possibility of an actually *time-asymmetrical* basic physics as I am claiming? In Chapter 30, I shall be confronting exactly this issue head on! And we shall find that it is related to something fundamentally puzzling that we have left hanging from our chapters on quantum mechanics. In the next chapter, therefore, I shall need to return to this important quantum-mechanical conundrum. Then in Chapter 30 I shall present my own ideas as to the correct route towards its resolution and hence, also, to the eventual resolution of the singularity time-asymmetry problem. Yet, I must again give the reader my statutory warning: many physicists may well be unhappy with the position that I shall be taking.

Notes

Section 28.1

28.1. See, for example, Weinberg (1992), p.195, where he also uses the example of ferromagnetism—as seems to be almost universal in popular expositions by experts. Yet, we must bear in mind that this is a considerable idealization, for a body of *actual* iron, in which the detailed effects of the forces can be very complicated. While for small enough regions within the iron this tendency to magnetization may be a good approximation, such magnetized regions tend to become randomly oriented in practice so that the iron as a whole does not tend to provide an effective magnet. Moreover, for the iron to become significantly magnetized, the cooling through the Curie point would need to be extremely slow, and the ideal situation is not easy to achieve. For the present theoretical discussion it is appropriate that we ignore such complications and accept the idealization that is being described.

28.2. Quantum-mechanical tunnelling occurs when a quantum system spontaneously undergoes a transition from one state to another of lower energy (with the emission of the excess energy) where there is an energy barrier preventing this from taking place classically.

28.3. Reflection symmetry has been excluded, in this example, because of the 'S' in SO(2).

28.4. This 'appropriate group' seems to be $SU(3) \times SU(2) \times U(1)/Z_6$.

Section 28.2

28.5. See Note 28.1.

28.6. See Vilenkin (2000); Gangui (2003); Sakellariadou (2002).

28.7. For a promising-looking theory of this kind, see Silk and Rees (1998). See Haehnelt (2003) for a review and further references.

28.8. See Chan and Tsou (1993).

28.9. The MACRO Collaboration has put stringent limits on the frequency of these particles. See MACRO (2002).

Section 28.3

28.10. This connection would be initially taken as a gauge connection ∇ on the smaller bundle $\mathcal{B}_{\mathcal{L}}$, over $\mathcal{M}$, whose fibres are the U(2)-symmetric spaces $\mathcal{L}$ of leptons at each point. But in just the same way that, as in §14.3, in ordinary tensor calculus, knowledge of how ∇ acts on vectors completely fixes how it acts on general tensors, the knowledge of ∇'s action on $\mathcal{B}_{\mathcal{L}}$ completely determines its action on the 'tensors' defined from $\mathcal{L}$. We can take $\mathcal{G}$ to be $\mathcal{L}^* \otimes \mathcal{L}$ (one 'index' down, one up).

28.11. The 'red shift' z is defined so that $1 + z$ measures the factor by which the wavelength is increased. Liddle (1999) is the most accessible text; Dodelson (2003) is a more advanced treatment.

28.12. One might contemplate a possible role for a *quanglement* connection (see §23.10) between q and r. This is certainly worth considering, but it goes beyond current ideas of 'spontaneous symmetry breaking'. My thinking on these issues has been influenced by conversations with George Sparling and Bikash Sinha.

28.13. See Llewellyn Smith (1973).

Section 28.4

28.14. See Schrödinger (1956).

Section 28.5

28.15. See Guth (1997). Dodelson (2003) or Liddle and Lyth (2000) are technical sources. For a careful and critical survey, Börner (2003) comes highly recommended.

28.16. See Barbour (1989, 2001a, 2001b); Barbour, Foster, and O'Murchadha (2002); Sciama (1959); Smolin (2002). An example of an entirely 'Machian' physical approach is that of *spin networks*, described briefly in §32.6.

28.17. See Ozsvath and Schücking (1962, 1969).

28.18. There are newer perspectives on these issues, however, which are argued to support the case that Einstein's theory *is* 'Machian' after all. See Barbour (2004); Barbour *et al.* (2002); Raine (1975).

28.19. These aesthetic desiderata are specifically argued for in Mario Livio's popular account: Livio (2000).

28.20. There was a precursor of this view in what was referred to a 'chaotic cosmology' put forward independently, in the 1960s by Charles W. Misner and by Yakov B. Zeldovich, whereby a random initial state was envisaged—despite the seeming fundamental conflict with the Second Law—thermal processes being invoked in an attempt to smooth the universe out. See Misner (1969).

28.21. The best proposal for a likely chaotic structure in this generic singularity comes from the 1970 work of Belinskii *et al.* (1970).

28.22. See Note 27.21, which provides the relevant references.

Section 28.6

28.23. I believe that I first heard of this 'weak' anthropic idea from Fred Hoyle's radio talks, given over BBC radio in the 1950s. I first became acquainted with the *stronger* form of the anthropic principle, which addresses the issue of 'anthropic' role of basic physical constants, from one of Hoyle's Cambridge lectures 'Religion as a Science', which referred to the building of heavy elements in stars requiring a specific nuclear energy level in carbon, to be described shortly.

28.24. See Dicke (1961) and Carter (1974).

28.25. Roughly: the cube root of the age of the universe in Planck units is remarkably closely the square root of the ratio of the electric to gravitational attraction between a proton and an electron.

28.26. See Dirac (1938); Buckley and Peat (1996); Guenther *et al.* (1998). A recent 'varying constant' idea is given an amusing account in Magueijo (2003).

28.27. My use of the term 'strong anthropic principle', here, follows Carter (1974). Barrow and Tipler (1988) break this down into several different categories.

28.28. For more information concerning this story, see Barrow and Tipler (1988); Smolin (1997), p. 111. For the profound implications for the contents of our universe, see Hoyle *et al.* (1956); Burbidge *et al.* (1957).

28.29. See Hawking and Penrose (1970).

28.30. See Smolin (1997).

28.31. In my Adams Prize essay of 1966 (see Penrose 1966, 1968), I put forward such an idea (but without the physical constant readjustments) in a non-serious way! Perhaps others had also done so earlier.

Section 28.7

28.32. See Penrose (1989).

28.33. An alternative standpoint has been stressed to me by Abhay Ashtekar that there might be something else, different from 'quantum gravity' that fixes the extraordinarily special nature of the Big Bang. Maybe so, but I cannot help being struck by the fact that it is gravity that was special at the Big Bang, and apparently gravity alone.

Section 28.8

28.34. In fact, only the trace-free part of the Ricci tensor $R_{ab} - \frac{1}{4}Rg_{ab}$ is relevant here, and the cosmological constant plays no role.

28.35. Penrose (1969a); see Penrose 1998b for a general overview of cosmic censorship.

28.36. See Newman (1993); Claudel and Newman (1998); Tod and Anguige (1999a, 1999b); Anguige (1999). A particularly appealing version of the Weyl curvature hypothesis is the one put forward by K.P. Tod, which simply asserts that at any initial singularity, there is a regular conformal geometry with boundary.

Section 28.9

28.37. See Wick (1956) for the first use of this technique, which is employed in Zinn-Justin (1996) to great and frequent effect.

28.38. See Hartle and Hawking (1983).

28.39. Recent work by Renate Loll and her collaborators suggests that there may be profound differences between the use of Riemannian metrics in the path integral, as with the Hawking proposal, and the more directly appropriate Lorentzian metrics. See Ambjorn *et al.* (1999).

Section 28.10

28.40. See Hawking and Turok (1998).

28.41. See Bucher *et al.* (1995) and Linde (1995).

28.42. Mordehai Milgrom (1994) has put forward the intriguing suggestion that there is no dark matter, but instead Newtonian gravitational dynamics needs alteration in a way different from Einstein's, where for very low accelerations the effect of gravity is increased in a certain specific way. Although this idea seems to fit the facts remarkably well, there is as yet no coherent theory of this which makes good overall theoretical sense. In my own opinion, such unconventional ideas should not just be dismissed, and it could be worth the effort to see whether this scheme can be made part of a broader consistent viewpoint. (I have not been able to see how to do this myself!)

28.43. See Krauss (2001) for an accessible discussion of dark matter (and also of 'dark energy'—i.e. a possibly varying Λ).

28.44. See Blanchard *et al.* (2003). For the more 'mainstream' interpretation, see Perlmutter *et al.* (1998); Bahcall *et al.* (1999).

28.45. Dodelson (2003) explains how to do this and related analysis of CMB data.

29
The measurement paradox

29.1 The conventional ontologies of quantum theory

THERE is no doubt that quantum mechanics has been one of the supreme achievements of the 20th century. It explains a great many phenomena that had been profoundly puzzling in the 19th, such as the existence of spectral lines, the stability of atoms, the nature of chemical bonds, the strengths and colours of materials, ferromagnetism, solid/liquid/gas phase transitions, and the colours of hot bodies in equilibrium with their hot surroundings (black-body radiation). Even some puzzling matters of biology, such as the extraordinary reliability of inheritance, are now seen to arise from quantum-mechanical principles. These phenomena—as well as many others which had become known in the 20th century, such as liquid crystals, superconductivity and superfluidity, the behaviour of lasers, Bose–Einstein condensates, the curious non-locality of EPR effects and of quantum teleportation—are now well understood on the basis of the mathematical formalism of quantum mechanics. This formalism has, indeed, provided us with a revolution in our picture of the real physical world that is far greater even than that of the curved spacetime of Einstein's general relativity.

Or has it? It is a common view among many of today's physicists that quantum mechanics provides us with *no* picture of 'reality' at all! The formalism of quantum mechanics, on this view, is to be taken as just that: a mathematical formalism. This formalism, as many quantum physicists would argue, tells us essentially nothing about an actual *quantum reality* of the world, but merely allows us to compute probabilities for alternative realities that might occur. Such quantum physicists' ontology—to the extent that they would be worried by matters of 'ontology' at all—would be the view (a): that there is simply no reality expressed in the quantum formalism. At the other extreme, there are many quantum physicists who take the (seemingly) diametrically opposite view (b): that the unitarily evolving quantum state completely describes actual reality, with the alarming implication that practically all quantum alternatives must always continue to coexist (in superposition). As already touched upon in §21.8, the basic difficulty that confronts quantum physicists, and that drives many of them to such views, is the conflict between the two quantum processes **U** and **R**, where (§22.1) **U** is the deterministic process of unitary evolution (as can be described by Schrödinger's equation) and **R** is the quantum state reduction which takes place when a 'measurement' is performed. The **U** process, when it was found, was something of the kind familiar to physicists: the clear-cut temporal evolution of a definite mathematical quantity, namely the state vector $|\psi\rangle$, controlled deterministically by a (partial) differential equation—the temporal evolution of the Schrödinger equation being not unlike that of the classical Maxwell equations (see §21.3 and Exercise [19.2]). On the other hand, the **R** process was something quite new to them: a discontinuous random jumping of this same $|\psi\rangle$, where only the probabilities of the different outcomes are determined. Had the physics of the observed world been described simply by a quantity $|\psi\rangle$, just acting according to **U** on its own, then physicists would have had no serious trouble with accepting **U** as providing a 'physically real' evolution process for a 'physically real' $|\psi\rangle$. But this is not how the observed world behaves. Instead, we seem to perceive a curious combination of **U** with the interjection of the very different process **R**, from time to time! (Recall Fig. 22.1.) This made it far harder for physicists to believe that $|\psi\rangle$ could actually be a description of physical reality after all. The puzzling issue of how **R** can somehow come about, when the state is supposed to be evolving in accordance with **U**-evolution, is the measurement problem—or, as I prefer it, *measurement paradox*—of quantum mechanics (discussed briefly in §23.6, and hinted at in §21.8 and §22.1).

The viewpoint (a) is basically the ontology of the *Copenhagen interpretation* as expressed specifically by Niels Bohr, who regarded $|\psi\rangle$ as not representing a quantum-level reality, but as something to be taken as merely describing the experimenter's 'knowledge' of a quantum system. The 'jumping', according to **R**, would then be understood as the experimenter's simply acquiring more knowledge about the system, so it is the *knowledge* that jumps, not the physics of the system. According to (a), one should not ask that any 'reality' be assigned to quantum-level phenomena, the only acknowledged reality being that of the classical world within which the experimenter's apparatus finds its home. As a variant of (a), one might take the view that this 'classical world' comes in not at the level of some piece of 'macroscopic machinery' that constitutes the observer's measuring apparatus, but at the level of the observer's own *consciousness*. I shall discuss these alternatives in more detail shortly.

The supporters of alternative (b), on the other hand, *do* take $|\psi\rangle$ to represent reality, but they deny that **R** happens at all. They would argue that when a measurement takes place, all the alternative outcomes actually *coexist* in reality, in a grand quantum linear superposition of alternative universes. This grand superposition is described by a wavefunction $|\psi\rangle$ for the entire universe. It is sometimes referred to as the 'multiverse',[1] but I believe that a more appropriate term is the *omnium*.[2] For although this viewpoint is commonly colloqually expressed as a belief in the parallel co-existence of different alternative worlds, this is misleading. The alternative worlds do not really 'exist' separately, in this view; only the vast *particular superposition* expessed by $|\psi\rangle$ is taken as real.

Why, according to (b), is the omnium not *perceived* as actual 'reality' by an experimenter? The idea is that the experimenter's states of mind also coexist in the quantum superposition, these different individual mind states being entangled with the different possible results of the measurement being performed. The view is that, accordingly, there is effectively a 'different world' for each different possible result of the measurement, there being a separate 'copy' of the experimenter in each of these different worlds, all these worlds co-existing in quantum superposition. Each copy of the experimenter experiences a different outcome for the experiment, but since these 'copies' inhabit different worlds, there is no communication between them, and each thinks that only one result has occurred. Proponents of (b) often maintain that it is the requirement that an experimenter have a consistent 'awareness state' that forces the impression that there is just 'one world' in which **R** *appears* to take place. Such a viewpoint was first explicitly put forward by Hugh Everett III in 1957[3] (although I suspect that many others had, not always with conviction, privately entertained this kind of view earlier—as I had myself in the mid-1950s—without daring to be open about it!).

Despite their diametrically opposing natures, the viewpoints (a) and (b) have some significant points in common, with regard to how $|\psi\rangle$ is taken to relate to our observed 'reality'—by which I mean to the seemingly real world that, on a macroscopic scale, we all experience. In this observed world, only one result of an experiment is taken to occur, and we may justly regard it as the job of physics to explain or to model the thing that we indeed normally refer to as 'reality'. Neither according to (a) nor according to (b) is the state vector $|\psi\rangle$ taken to describe that reality. And in each case, we must bring in the perceptions of some human experimenter to make sense of how the formalism relates to this observed real world. In case (a) it is the state vector $|\psi\rangle$ itself that is taken to be an artefact of that human experimenter's perceptions, whereas in case (b), it is 'ordinary reality' that is somehow delineated in terms of the perceptions of the experimenter, the state vector $|\psi\rangle$ now representing some kind of deeper overriding reality (the omnium) that is not directly perceived. In both cases the 'jumping' of **R** is taken to be not physically real, being, in a sense, 'all in the mind'!

I shall be explaining my own difficulties with both positions (a) and (b) in due course, but before doing so, I should mention a further possibility for interpreting conventional quantum mechanics. This, as far as I can make out, is the most prevalent of the quantum-mechanical standpoints—that of *environmental decoherence* (c)—although it is perhaps more of a pragmatic than an ontological stance. The idea of (c) is that in any measurement process, the quantum system under consideration cannot be taken in isolation from its surroundings. Thus, when a measurement is performed, each different outcome does not constitute a quantum state on its own, but must be considered as part of an entangled state (§23.3), where each alternative outcome is entangled with a different state of the environment. Now, the environment will consist of a great many particles, effectively in random motion, and the complete details of their locations and motions must be taken to be totally unobservable in practice.[4] There is a well-defined mathematical procedure for handling this kind of situation where knowledge is fundamentally lacking: one 'sums over' the unknown environmental states to obtain a mathematical object known as a *density matrix*, to describe the physical system under consideration. Density matrices are important for the general discussion of the measurement problem in quantum mechanics (and are important also in many other contexts), but their ontological status is hardly ever made clear. I shall explain what a density matrix is very shortly (in §29.3). However, we shall be seeing later why it is important for the position (c) that the ontology of the density matrix is *not* made completely clear! Holders of viewpoint (c) tend to regard themselves as 'positivists' who have no truck with 'wishy-washy' issues of ontology in any case, claiming to believe that they have no concern with what is 'real' and what is 'not real'. As Stephen Hawking has said:[5]

> I don't demand that a theory correspond to reality because I don't know what it is. Reality is not a quality you can test with litmus paper. All I'm concerned with is that the theory should predict the results of measurements.

My own position, on the other hand, is that the issue of ontology is crucial to quantum mechanics, though it raises some matters that are far from being resolved at the present time.

29.2 Unconventional ontologies for quantum theory

Before entering into the details of all this, let me consider three further general standpoints with regard to quantum mechanics. It should not be assumed that my list is in any way comprehensive, nor should it be taken that these new ones are completely independent of those that I have given in the previous section. The list (a), (b), (c), (d), (e), (f) that I shall be considering here represents the kind of spread of viewpoints that one most frequently finds in the current literature, but I make no claims as to the completeness, independence, or specificity of my list. The three additional ontologies that I consider here represent actual changes in the usual quantum formalism; but with two of them, (d) and (e), it is not anticipated that there will be experimental distinctions between the proposed formalism and standard quantum mechanics. The standpoint (d) is the 'consistent histories' approach due to Griffiths, Omnès, and Gell-Mann/Hartle, and (e) is the 'pilot-wave' ontology of de Broglie and Bohm/Hiley.[6] The final possibility (f) is that present-day quantum mechanics is merely an approximation to something better, and that—in this improved theory—both of **U** and **R** take place *objectively* as real processes; moreover, it is part of the perspective of (f) that future experiments should be able to distinguish such a theory from conventional quantum mechanics.

As soon as we have the necessary tools, I shall try to give my assessments of the various alternatives (a), ..., (f). However, in order that the reader can take a suitably objective attitude to these assessments, it is best that I 'come clean' with regard to my own position clearly at this stage. I am, in fact, a strong believer that some developments in line with (f) are necessary, in order that quantum mechanics can make fully consistent sense. In the next chapter, I shall actually be putting forward the particular version of (f) that seems to me to be most natural. With this warning to the reader, let us proceed, where I first list these alternatives, to aid the reader in keeping them explicitly in mind.

(a) 'Copenhagen'
(b) many worlds
(c) environmental decoherence
(d) consistent histories
(e) pilot wave
(f) new theory with objective **R**.

I shall need to make a few remarks about (d) and (e), since I have not really explained them. The 'consistent-histories' scheme (d) provides a generalization of the standard framework of quantum theory. Some proponents have provided (d) with an ontology that seems a bit like that of many worlds (b), although in one respect even more extravagant—but as far as I can see, such an extravagant ontology may well not be necessary. In both (b) and (d) we can take the position that we have, as basic ingredients, a Hilbert space **H**, a starting state $|\psi_0\rangle$ belonging to **H**, and a Hamiltonian $\mathcal{H}$.[7] In the many-worlds case (b), the ontological position is to regard reality (of the omnium, that is) as being described as a continuous 1-parameter family of states (elements of **H**, and with time parameter t), starting with $|\psi_0\rangle$ at $t = 0$ and completely governed, for $t > 0$ by the Schrödinger evolution determined by $\mathcal{H}$. There is no **R** here, only **U**. But the consistent-histories case (d) broadens this so as also to incorporate '**R**-type procedures' into its 'evolution'—even though these are not considered to be necessarily associated in any way with actual measurements.

To understand the mathematical nature of these procedures, we must first recall, from §§22.5,6, how a quantum-mechanical measurement is mathematically described (even though, for (d), we do not think of these procedures as measurements), in terms of the action of some Hermitian (or normal, see §22.5) operator $\boldsymbol{Q}$. If, just prior to measurement, the state of the system is $|\psi\rangle$, then immediately following the measurement it is taken to 'jump' to the eigenstate of $\boldsymbol{Q}$ corresponding to the eigenvalue of $\boldsymbol{Q}$ that the measurement yields. But as far as its effect on $|\psi\rangle$ is concerned, we may as well replace $\boldsymbol{Q}$ by a 'complete set of orthogonal projectors' $\boldsymbol{E}_1, \boldsymbol{E}_2, \boldsymbol{E}_3, \ldots, \boldsymbol{E}_r$ (supposing that $\boldsymbol{Q}$ has just r distinct eigenvalues, where for convenience we take our Hilbert space **H** to be finite dimensional). Then, if the measurement yields the eigenvalue q_j, we find that $|\psi\rangle$ jumps to a state proportional to $\boldsymbol{E}_j|\psi\rangle$ (projection postulate).

Let us look at this in a little more detail. We recall from §22.6 that a *projector* is an operator $\boldsymbol{E}$ that squares to itself and is Hermitian, i.e.

$$\boldsymbol{E}^2 = \boldsymbol{E} = \boldsymbol{E}^*.$$

The assertion that projectors $\boldsymbol{E}_1, \ldots, \boldsymbol{E}_r$ are *orthogonal* to each other is

$$\boldsymbol{E}_i\boldsymbol{E}_j = 0 \quad \text{whenever} \quad i \neq j$$

and their completeness is that they sum to the identity $\boldsymbol{I}$ on **H**:

$$\boldsymbol{E}_1 + \boldsymbol{E}_2 + \boldsymbol{E}_3 + \ldots + \boldsymbol{E}_r = \boldsymbol{I}.$$

Let us simply call a set of $\boldsymbol{E}$s satisfying all these conditions a *projector set*. The connection between $\boldsymbol{Q}$ and its corresponding projector set is that for each eigenvalue q_j of $\boldsymbol{Q}$, the corresponding eigenvector space consists of the vectors of the form $\boldsymbol{E}_j|\phi\rangle$. The role of the projector $\boldsymbol{E}_j$ is that it projects down to this eigenvector space, for the eigenvalue q_j.[29.1]

The *projection postulate* for the operation **R** (see §22.6), in the measurement represented by $\boldsymbol{Q}$, tells us that, if the result of the measurement is q_j,

---

[29.1] Explain why $\boldsymbol{E}_j|\psi\rangle$ is the result (normalization ignored) of a measurement given by $\boldsymbol{Q} = q_1\boldsymbol{E}_1 + q_2\boldsymbol{E}_2 + q_3\boldsymbol{E}_3 + \ldots + q_r\boldsymbol{E}_r$ applied to $|\psi\rangle$, where the eigenvalue is q_j, the quantities $q_1, q_2, q_3, \ldots, q_r$ being distinct real numbers. Can you prove that the general finite-dimensional Hermitian operator has this form? (You may assume that any finite-dimensional Hermitian matrix $\boldsymbol{Q}$ can be transformed to a diagonal matrix by a unitary transformation.) The $\boldsymbol{E}$s are called *principal idemponents* of $\boldsymbol{Q}$. What modifications are needed for a *normal* operator $\boldsymbol{Q}$?

then $|\psi\rangle$ jumps to (something proportional to) $\boldsymbol{E}_j|\psi\rangle$. This occurs with probability given by

$$\langle\psi|\boldsymbol{E}_j|\psi\rangle,$$

if we assume that $|\psi\rangle$ is *normalized*, i.e. $\langle\psi|\psi\rangle = 1$. Thus, to describe the effect on the quantum state, of the measurement corresponding to $\boldsymbol{Q}$ we need only consider the projector set defined by $\boldsymbol{Q}$.

Let us now return to the ontology of the consistent histories approach (d). The theory operates with entities called *coarse-grained histories*,[8] each of which largely resembles a Schrödinger evolving 'omnium' of the many-worlds approach (b), using the Hamiltonian $\mathcal{H}$. But with (d) we also allow projector sets to be inserted at various t-values during the course of the evolution.

The ontological status of the insertion of such a projector set is still not fully clear to me, but one is encouraged to adopt the attitude that the role of such a projector set is to provide some kind of 'refinement' of the history, rather than representing a fundamental change to what is happening in the world. The projectors certainly are not to be assigned the ontological status given by some objective measurement. A more appropriate analogy might be that the projector sets provide refinements for, or alterations to, coarse-graining 'boxes', as in classical phase space (see §27.3)—and this accounts for the term 'coarse-grained history' used here. In such a coarse-grained history, at the point at which a projector set is encountered (and similarly to the standard procedure adopted in quantum measurement), the current state $|\psi\rangle$ gets replaced by (something proportional to) $\boldsymbol{E}_j|\psi\rangle$, where $\boldsymbol{E}_j$ is some member of the projector set. This might be thought of as a loss of information, but there is no loss if we keep track of the *whole family* of $\boldsymbol{E}_j|\psi\rangle$, for all the $\boldsymbol{E}_j$ in the set, since $|\psi\rangle$ is simply the sum of all these.

In accordance with a desire for something to emerge which resembles the kind of classical world that we actually perceive, some particular families of coarse-grained histories are singled out and referred to as *consistent* (or, sometimes, 'decoherent') if a certain condition is satisfied—expressing the fact that the probabilities, calculated according to the standard quantum rules, satisfy the ordinary classical rules of probability.[9] A consistent set of coarse-grained histories is called *maximally refined* if one cannot insert another projector set (inequivalent to any that have been already incorporated) without destroying the consistency. A history from a maximally refined set seems to me to provide a strong candidate for what might be regarded as ontologically 'real', according to viewpoint (d).

Yet, I have not seen this viewpoint put forward explicitly, and something more akin to the *totality* of histories in a maximally refined set seems to be closer to the ontological viewpoint for 'consistent histories' that I have heard expressed.[10] This is perhaps more aligned with what we have seen in the many-worlds viewpoint (b), but the presence of many alternative possible consistent collections of projector sets seems to provide us with an even vaster ensemble of alternative 'worlds'. However, we recall that also in the many-worlds picture (b), there can arise something of an ontological confusion. The ontologically 'real' omnium (described by $|\psi\rangle\rangle$) is a *superposition* of numerous different worlds, and the *collection* of all these individual worlds (rather than just their particular superposition $|\psi\rangle\rangle$) is *not* to be taken as 'real'. Set against this kind of confusion is the advantage, in the consistent-histories viewpoint (d), that the correct quantum probabilities are provided by the theory, which does not seem to be the case with (b).

In the 'Bohmian' (pilot wave) case (e), the ontological position is, refreshingly, much more down to Earth, although even here there are some considerable subtleties—for there are, in a sense, *two* levels of reality, one of which is firmer than the other. It is simplest to put the case first for a system consisting of just a single spinless particle. Then this firmer level of reality is given by the particle's actual position. In a two-slit experiment (§21.4, Fig. 21.4), since the particle's location is ontologically real, it actually goes through one slit or it goes through the other, but its motion is 'guided', in effect, by ψ, so this provides a secondary, but nevertheless ontologically still 'real' status to the ψ also. It is fairly common, in this theory, to take somewhat different attitudes to the *modulus* and the *argument* of ψ (§5.1), where a quantity referred to as the 'quantum potential' is constructed from the former, and where the latter is employed to define what is called the 'pilot wave'. This kind of splitting is not necessary, however, and its significance seems to become less clear-cut with more complicated systems.

Generally, we can think of ψ as a complex function that is defined on configuration space $\mathcal{C}$, and it serves the function of 'guiding' the behaviour of a point P on $\mathcal{C}$. The firmer part of the system's reality is taken to be the classical configuration actually defined by P, but a kind of (weaker) reality is assigned also to the complex function ψ, by virtue of its role in guiding the behaviour of P. It is considered that all measurements can ultimately be reduced to 'position' measurements, which here means measurements of the system's configuration. The squared modulus $|\psi|^2$, at some point Q on $\mathcal{C}$, defines the probability density for finding the system in the configuration defined by Q, but the location of P on $\mathcal{C}$ determines what is considered to be the *actual* configuration of the system.

Now, this all seems almost 'too easy', but there are subtleties. Most particularly, the picture is a very non-local one, where ψ is a highly 'holistic' entity (as it must be, in order to accord with the holistic nature of wavefunctions that was stressed in §21.7). This, however, seems inevitable in quantum mechanics. Somewhat more serious is that there are important conditions that must be imposed on the probability distribution for the initial state $|\psi_0\rangle$, so that the $|\psi|^2$ quantum probability law holds true, and continues to hold true after sequential measurements. There is the further point that one might question the correctness of the assumption that all measurements can always ultimately be reduced to position measurements (particularly as strict position measurements are not fully legitimate in quantum mechanics, see §21.10), and whether the configuration-space picture is adequately unambiguous when non-classical parameters like spin are being considered. Nevertheless, the clarity of the ontological position of (e) is greatly to its credit (though, as we shall see in §29.9, there are further issues to be faced here also).[11]

Finally, there are many different proposals in accordance with (f). It is not appropriate for me to describe them all here in detail. But I can make some general comments about them. A good many of these proposals will accept (at least as a provisional stance) an ontologically real status for an evolving state vector $|\psi\rangle$. The time-evolution of $|\psi\rangle$, in such a theory, would be something closely approximating the alternation of **U** with **R**, that standard quantum mechanics tells us to adopt in practice; see Fig. 22.1. Despite the fact that theories in line with standpoint (f) are considered to be 'outside the mainstream' of quantum-mechanical thinking, it could be very reasonably said that (f) is actually the standpoint that is most *accepting* of the reality of the formalism of quantum mechanics as it is used in practice today, since *both* of the quantum-mechanical evolution processes **U** and **R** are taken seriously, ontologically, to describe the evolution of reality! The trouble is, however, that **U** and **R** are mathematically *inconsistent* with each other, which is why (f) demands that there must be changes from ordinary unitary evolution—and it is this that separates (f) from the mainstream!

Why is **R** mathematically inconsistent with **U**? Perhaps the most obvious reason is that **R** represents a discontinuous change in the state vector (except in the exceptional circumstance that the state prior to measurement is actually an eigenstate of the measurement operator), whereas **U** always acts continuously. But even if we imagine that the 'jump' induced by **R** is not absolutely instantaneous, there would be trouble with unitarity because of the lack of determinism in **R**. Different alternative outcomes can result from the same input, which is something that never happens with **U**. Moreover, a theory that makes **R** into a real process cannot ever be unitary when a (non-trivial) quantum jump—in accordance with **R**—actually takes place. Despite this, there is a remarkable accord, of a sort, between the two processes **U** and **R**, since the 'squared-modulus rule' that interrupts **U** to provide us with the probabilistic **R**, employs the very 'unitarity' of **U** to give us a law of conservation of probability for **R** (basically the fact that the scalar products $\langle\phi|\psi\rangle$, from which the quantum probabilities are computed, are preserved under unitary time-evolution, see §§22.4,5). This is an integral part of the wonders of quantum mechanics, and it provides one strong reason that people are reluctant to monkey with the principles of that theory in any way whatever—partly explaining why (f) is not particularly popular among today's quantum physicists.

Nevertheless, I believe that there are powerful reasons for expecting a change. Such a change would, in my view, represent a major revolution, and it cannot be achieved by just 'tinkering' with quantum mechanics. Yet, the necessary changes must themselves be thoroughly respectful of the central principles that lie at the heart of present-day physics. The very tightness of the quantum formalism, as indicated in the preceding paragraph, is a reason for both of these requirements. As a comparison, we recall the tightness of Newtonian physics. Relativity and quantum mechanics were not obtained from it by tinkering, but through the revolutionary changes of perspective that nevertheless paid due respect to Newtonian theory's highly organized Lagrangian/Hamiltonian/symplectic-geometric structure. Are the changes to quantum theory that have so far been suggested, by various people,[12] of such a respectful revolutionary character—or are they just tinkering? It must be said that, for the most part, these ideas have to be regarded as tinkering; yet some of these ideas could well provide pointers to the true road to an improved quantum theory.

29.3 The density matrix

But why is there a need for 'improving' quantum theory in any case? Most quantum physicists appear to believe that no such a theory is required, having made their peace with the seeming contradictions and obscure ontology of one or other of the standard pictures (or lack of such a picture). Before we can attempt to address any of the difficulties that there might be in any of the 'standard' pictures (a), (b), and (c), we should come to the notion of a density matrix, which is fundamental not only to standpoint (c), but which plays an important role in many other quantum-mechanical considerations. Moreover, it raises intriguing and profound issues concerning how reality should be represented in quantum mechanics.

Suppose that we have some quantum system whose state is not completely known to us. The state might be $|\psi\rangle$, or it might be $|\phi\rangle$, or it might be . . . , or it might be $|\chi\rangle$. The list could be infinite, but it will be sufficient for our purposes to consider a finite list of possibilities only. Each of these possibilities is to be assigned a probability, let us say $p, q, \ldots, s$, respectively. The possibilities are to be exhaustive, so the probabilities—real numbers between 0 and 1 (inclusive)—must sum to unity:

$$p+q+\cdots+s=1.$$

We suppose that each of $|\psi\rangle$, $|\phi\rangle$, ..., $|\chi\rangle$ is *normalized*:

$$\|\psi\|=1,\ \|\phi\|=1,\ldots,\ \|\chi\|=1.$$

(Recall, from §22.3, that $\|\psi\|=\langle\psi|\psi\rangle$, etc.) Then we define the *density matrix* to be the quantity

$$\boldsymbol{D}=p|\psi\rangle\langle\psi|+q|\phi\rangle\langle\phi|+\cdots+s|\chi\rangle\langle\chi|.$$

Recall from §22.3 that the bra vector $\langle\psi|$ is the Hermitian conjugate of the ket vector $|\psi\rangle$. The quantity $|\psi\rangle\langle\psi|$ is then the tensor product (or outer product) of $|\psi\rangle$ with $\langle\psi|$, etc. In the index notation, of §23.8, we can write $\langle\psi|$ as $\bar{\psi}_\alpha$, where ψ^α stands for $|\psi\rangle$. Then, $|\psi\rangle\langle\psi|$ could be written $\psi^\alpha\bar{\psi}_\beta$, etc. Accordingly, $\boldsymbol{D}$ itself would have the index structure $D^\alpha{}_\beta$. The density matrix has the algebraic properties that it is Hermitian, non-negative-definite (see §§13.8,9), and of trace unity:

$$\boldsymbol{D}^*=\boldsymbol{D},\qquad \langle\xi|\boldsymbol{D}|\xi\rangle\geq 0 \text{ for all } |\xi\rangle,\qquad \langle\boldsymbol{D}\rangle=1,$$

where $\langle\boldsymbol{D}\rangle=\text{trace}\,\boldsymbol{D}=D^\alpha{}_\alpha$ (see §13.4).[29.2]

The density matrix plays a role analogous to that which is frequently used in classical statistical mechanics, where we might not be particularly concerned with the precise (classical) state of a system, but we are content to consider some probability distribution of classical alternatives. This is most easily thought of in terms of the phase space $\mathcal{P}$ of the possible classical alternatives. Instead of the system being represented as a point P in $\mathcal{P}$, it would be thought of in terms of a probability distribution on $\mathcal{P}$. If we have just a finite number of alternatives[13] for the system, the various probabilities being p, q, ..., s, then we represent this as a finite set of points P, Q, ..., S, in $\mathcal{P}$, each of which is assigned its respective probability value p, q, ..., s; see Fig. 29.1. We could, indeed, envisage doing

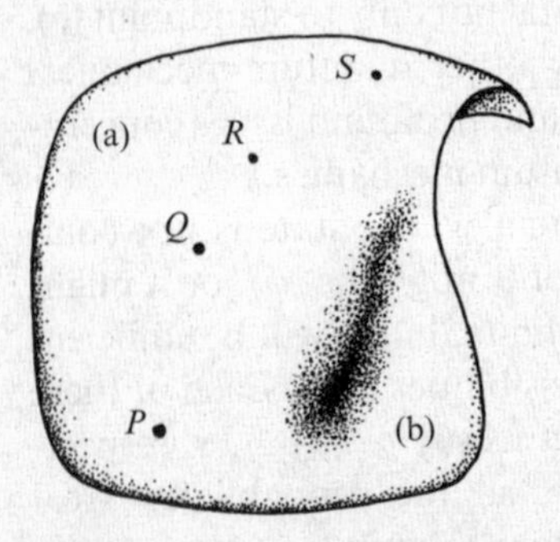

Fig. 29.1 Classical probability distributions represented in phase space $\mathcal{P}$. (a) For a finite set of points $P, Q, \ldots, S$, in $\mathcal{P}$, a probability value $p, q, \ldots, s$ (real numbers between 0 and 1) is assigned to each point, where $p+q+\ldots+s=1$. (b) A continuous distribution, with a probability measure (a non-negative real-number density), whose integral is 1, assigned in some region of $\mathcal{P}$.

[29.2] Derive these properties.

exactly the same thing, in quantum physics, with the Hilbert space **H** of the quantum system playing the role of the phase space $\mathcal{P}$; then we would have some probability distribution in **H**. In relation to the density matrix $\boldsymbol{D}$ that we have just been considering, this distribution would consist of just a finite number of points $P, Q, \ldots, S$, in **H**, each being assigned its respective probability value $p, q, \ldots, s$.

But this is *not* what is normally done in quantum mechanics; the density matrix is used instead.14 Why is this? The reason is that, in quantum mechanics, a *measurement*, having the form of question posed to a quantum system—and let us restrict attention to a YES/NO question—is phrased in terms of the action of some projector $\boldsymbol{E}$ applied to the (normalized) state vector $|\xi\rangle$. The probability of the answer YES is then given by[29.3]

$$\text{probability of YES} = \langle\xi|\boldsymbol{E}|\xi\rangle,$$

from which it follows that for the probability mixture of possible alternative states $|\psi\rangle, |\phi\rangle, \ldots, |\chi\rangle$, described above, with density matrix $\boldsymbol{D}$, we obtain the answer

$$\text{probability of YES} = \langle\boldsymbol{ED}\rangle.$$

The significance of this is that we do not need to know the complete information of the distribution of probabilities for the alternative states $|\psi\rangle, \ldots, |\chi\rangle$ in order to be able to calculate probabilities for a standard YES/NO question in quantum mechanics (or, indeed, for the expectation value of any other quantum-mechanical observable);[29.4] all the needed information is stored in the density matrix—and as we shall be seeing shortly that a given density matrix can be composed of many *different* probability distributions of states. There is a considerable economy and elegance in this remarkable mathematical entity (introduced by the outstanding Hungarian/American mathematician John von Neumann 1932). It combines together into one expression what would appear to be two quite distinct notions of probability. On the one hand, we have the numbers $p, q, \ldots, s$, which are the ordinary classical probabilities for the alternative states $|\psi\rangle, |\phi\rangle, \ldots, |\chi\rangle$, while on the other, we have the quantum probabilities obtained from the squared-modulus rule of §21.9. The density matrix combines the two and does not directly distinguish one kind from the other.

29.4 Density matrices for spin $\frac{1}{2}$: the Bloch sphere

Let me illustrate this point with a simple example. Suppose that we have a particle of spin $\frac{1}{2}$, whose state of spin we know to be either $|\uparrow\rangle$ or $|\downarrow\rangle$, with

[29.3] Explain why; also derive the expression $\langle\boldsymbol{ED}\rangle$, below.

[29.4] Can you see why this should be so?

probability $\frac{1}{2}$ for each alternative. If we choose to measure this spin in an up/down direction, we simply get 'up' if the state is $|\uparrow\rangle$, and 'down' if the state is $|\downarrow\rangle$. In each case the probability is $\frac{1}{2}$. These are just straightforward classical probabilities, and there is no quantum mystery here. But suppose that we measure the spin in the left/right direction instead. Then, if the state is $|\uparrow\rangle$, the quantum R rules tell us that we get a probability of $\frac{1}{2}$ that the spin is 'left' and a probability $\frac{1}{2}$ that the spin is 'right'. Exactly the same conclusion is obtained if the state is $|\downarrow\rangle$. Thus for the equal-probability mixture of $|\uparrow\rangle$ and $|\downarrow\rangle$, we still get probabilities of $\frac{1}{2}$ for each of 'left' and 'right'. Now, however, the probabilities are obtained entirely from the quantum-mechanical 'squared-modulus' law. We could also choose to measure the spin in any other direction. The probabilities would again turn out to be $\frac{1}{2}$ for each answer but this probability would, in general, be composed of a mixture of classical and quantum probabilities.[29.5]

We could, alternatively, imagine rotating the state mixture rather than the measuring apparatus. Thus, an equal-probability mixture of $|\leftarrow\rangle$ and $|\rightarrow\rangle$ would give just the same answers as the above equal-probability mixture of $|\uparrow\rangle$ and $|\downarrow\rangle$, and so also would an equal-probability mixture of $|\nwarrow\rangle$ and $|\searrow\rangle$ (where in each case we take these pairs of states to be orthogonal and normalized: $\langle\uparrow|\downarrow\rangle = \langle\leftarrow|\rightarrow\rangle = \langle\nwarrow|\searrow\rangle = 0$ and $\langle\uparrow|\uparrow\rangle = \langle\downarrow|\downarrow\rangle = \ldots = \langle\searrow|\searrow\rangle = 1$). We get, for the density matrix $\boldsymbol{D}$, in each case:

$$\boldsymbol{D} = \tfrac{1}{2}|\uparrow\rangle\langle\uparrow| + \tfrac{1}{2}|\downarrow\rangle\langle\downarrow|,$$
$$\boldsymbol{D} = \tfrac{1}{2}|\leftarrow\rangle\langle\leftarrow| + \tfrac{1}{2}|\rightarrow\rangle\langle\rightarrow|,$$
$$\boldsymbol{D} = \tfrac{1}{2}|\nwarrow\rangle\langle\nwarrow| + \tfrac{1}{2}|\searrow\rangle\langle\searrow|,$$

the remarkable property of the density matrix being that all these $\boldsymbol{D}$s are the *same*.[29.6] All the probabilities for spin measurements just referred to can be obtained by use of the above $\langle\boldsymbol{ED}\rangle$ formula; thus, since the $\boldsymbol{D}$s are the same, the respective probabilities must come out the same, as we have seen.

But how are we to regard the *ontology* of these probability mixtures of states? If we take the quantum state to have some kind of physical reality, then these three situations are definitely ontologically distinct. It is quite a different thing to say that the there is an equal probability that the state is one or other of the (physically real) alternatives $|\uparrow\rangle$, $|\downarrow\rangle$ than to say that there is an equal probability of it being $\langle\nwarrow|$ or $|\searrow\rangle$. However, this ontological issue is one that is extremely confused in much quantum-mechanical literature. Often, quantum physicists seem to be taking a quite different ontological position from that just described, regarding the density

[29.5] Work this out, for a general angle of slope θ of the measuring direction, using the expression $\frac{1}{2}(1+\cos\theta)$ of §22.9, for the probability.

[29.6] Show this by explicit calculation, using the results of §§22.8,9 and Exercise [22.25].

matrix itself as providing a better description of reality than the individual states. They might take the view that the three apparently distinct ontologies for $\boldsymbol{D}$ above (i.e. the three different probability-weighted collections of alternative quantum states) are physically indistinguishable. Accordingly, such physicists—often holders of the environmental-decoherence viewpoint (c)—might adopt the positivist or pragmatic position that it makes no sense to distinguish between these alternatives. Such people might adopt the view that it is the density matrix that best describes quantum reality.

Indeed, in many contexts, the word 'state' is often taken to refer to a density matrix rather than to the more primitive notion that, up until this point, I have been calling a 'quantum state'—namely a quantity describable by a ket such as $|\psi\rangle$. When the word 'state' is used in the sense of a density matrix, the term 'pure state' is then used for a density matrix of the special form $|\psi\rangle\langle\psi|$, and 'mixed state' for a more general density matrix that cannot be represented in this way. The 'pure states' in this sense refer to what I have been calling simply a 'state'. Personally, I find it very confusing to refer to a density matrix (pure or otherwise) as a 'state', and I shall refrain from using this terminology here. To me, a 'quantum state' is effectively a quantum state vector $|\psi\rangle$, not a density matrix. Yet, some might prefer to distinguish the terms 'quantum state' and 'quantum state vector', the latter being the ket $|\psi\rangle$ and the former being represented as the equivalence class of non-zero complex multiples of $|\psi\rangle$, i.e. the element of the *projective* Hilbert space $\mathbb{P}\mathbf{H}$ corresponding to the element $|\psi\rangle$ of $\mathbf{H}$ (see §15.6.) If we choose to normalize $|\psi\rangle$ by $\langle\psi|\psi\rangle = 1$, then the only freedom left in $|\psi\rangle$ (for a given point in $\mathbb{P}\mathbf{H}$) is the phase freedom $|\psi\rangle \mapsto e^{i\theta}|\psi\rangle$ (with θ real); see Fig. 29.2. The notion of a 'pure-state'

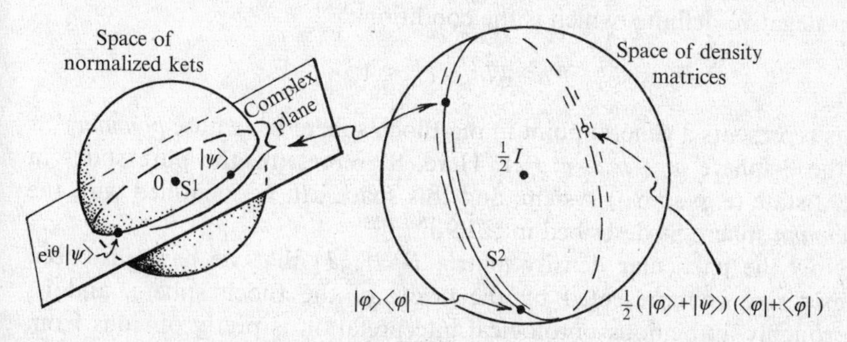

Fig. 29.2 How do we represent a pure quantum state? (a) Space of kets $|\psi\rangle$, normalized by $\langle\psi|\psi\rangle = 1$. (b) The density matrix $|\psi\rangle\langle\psi|$ is 'equivalent' to $|\psi\rangle$ up to the phase freedom $|\psi\rangle \mapsto e^{i\theta}|\psi\rangle$, and to the family of non-zero kets proportional to $|\psi\rangle$ (complex proportionality factors). Yet, basic quantum linearity is obscured in the density-matrix description.

density matrix is effectively equivalent to this 'projective' notion of a quantum state, since $|\psi\rangle\langle\psi|$ is invariant under this phase freedom. Thus we might reasonably take the position that a pure-state density matrix appropriately describes the physical quantum state.

Nevertheless, I feel uncomfortable about regarding such a 'pure-state density matrix' as the appropriate mathematical representation of a 'physical state'. The phase factor $e^{i\theta}$ is only 'unobservable' if the state under consideration represents the entire object of interest. When considering some state as part of a larger system, it is important to keep track of these phases. Moreover, the fundamental complex linearity of the basic structure of the Hilbert space of ket vectors becomes unnecessarily mathematically complicated if one has always to operate with the quantities $|\psi\rangle\langle\psi|$ instead of the mathematically simpler $|\psi\rangle$ (or $\langle\psi|$).[29.7] Partly for such reasons, my own position would be to take the density matrix not as 'reality', but as just a useful device. However, there are some intriguing aspects to the density matrix's confused ontological status, as we shall be seeing here and in §29.5.

Before coming to this, it will be helpful for us to become acquainted with the *Bloch sphere*, which represents the space of density matrices for a 2-state system. This is the closed solid sphere (or, in the mathematician's terminology, the *3-ball* or *3-disc*) B^3 residing in Euclidean 3-space. It represents the density matrices for spin $\frac{1}{2}$ (or for any other 2-state system); see §22.9. We can write the general Hermitian 2×2 matrix of trace unity as

$$\frac{1}{2}\begin{pmatrix} 1+a & b+ic \\ b-ic & 1-a \end{pmatrix},$$

where a, b, c are real numbers. For this to be a density matrix, it must be non-negative-definite, which is the condition[29.8]

$$a^2 + b^2 + c^2 \leq 1.$$

This represents a general point in the Bloch sphere B^3, whose *boundary* S^2 is the 2-sphere $a^2 + b^2 + c^2 = 1$. Here, S^2 represents the pure states in the 2-state (e.g. spin $\frac{1}{2}$) system, and this space can be identified with the *Riemann* sphere S^2 described in §22.9.[15]

Now the particular density matrix $\boldsymbol{D} = (\frac{1}{2}\boldsymbol{I})$ that we have just been considering is represented by the *origin* of the Bloch sphere, and its thoroughly ambiguous ontological interpretation is pretty obvious from the symmetry of the figure (Fig. 29.3). However *any* point (non-pure

[29.7] See if you can characterize the family of 'pure-state' density matrices that correspond to the linear combinations $w|\psi\rangle + z|\varphi\rangle$, for some arbitrary pair of fixed states $|\psi\rangle$ and $|\varphi\rangle$.

[29.8] Show this. *Hint*: What is the product of the eigenvalues, in terms of a, b, and c? What does it mean that this product be non-negative?

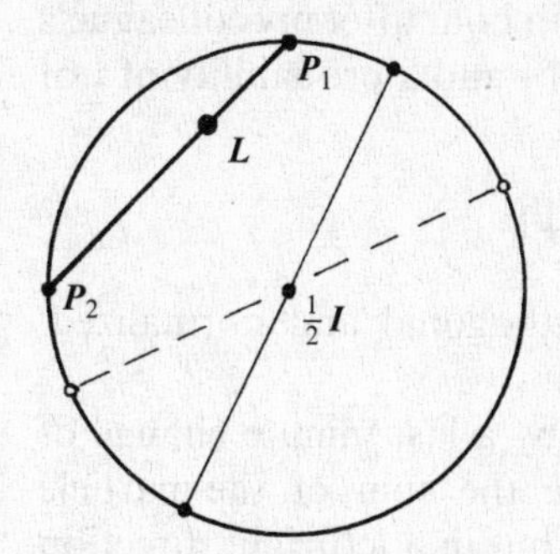

Fig. 29.3 Bloch sphere B^3 of density matrices for a 2-state system, centred at $\frac{1}{2}I$. Any (non-pure) density matrix $\boldsymbol{L}$ has an ambiguous ontological interpretation. An arbitrary chord through $\boldsymbol{L}$ meets the boundary S^2 in $\boldsymbol{P}_1$ and $\boldsymbol{P}_2$; then $\boldsymbol{L}$ has an interpretation as a probability mixture of the pure states $\boldsymbol{P}_1$ and $\boldsymbol{P}_2$.

density matrix) $\boldsymbol{L}$ in the interior of B^3 represents a density matrix with an equally ambiguous ontological interpretation. To see this, we simply draw an arbitrary straight line (chord) through $\boldsymbol{L}$ to meet the boundary S^2 in two points $\boldsymbol{P}_1$ and $\boldsymbol{P}_2$. These represent two pure states, and the density matrix $\boldsymbol{L}$ can then be interpreted as a probability mixture of these two.[29.9] The only thing that is particular about the origin $\boldsymbol{D}$ of the Bloch sphere is that all these pairs of pure states, in terms of which $\boldsymbol{D}$ can be represented, are *orthogonal* pairs. But there is nothing in the definition of a density matrix which requires that the probability mixture is between mutually orthogonal states. We shall be seeing, in §29.5, how non-orthogonal mixtures can certainly arise.

29.5 The density matrix in EPR situations

Let us examine a particularly clear-cut situation in which, in a natural way, a probability-weighted collection of possible state vectors arises. This comes about in the EPR–Bohm effect (§23.4). Suppose that, somewhere between Earth and Saturn's moon Titan—but let us say about twice as distant from Earth as from Titan—an EPR pair of spin $\frac{1}{2}$ particles is emitted in a combined state of spin 0. I am supposing that my colleague on Titan (our old acquaintance of §§23.4,5) measures the spin of the particle arriving there in an up/down direction and obtains some answer, roughly half an hour before I receive my particle here on Earth. Assume that when my particle does arrive, there has not been time for me to obtain any signal from my colleague about the result of that earlier measurement. (Titan is about three light hours from Earth.) As far as I am concerned, my particle has either spin $|\uparrow\rangle$ or spin $|\downarrow\rangle$. It will be $|\uparrow\rangle$ if my colleague happened to find the state $|\downarrow\rangle$ and my state will be $|\downarrow\rangle$ if my colleague actually found $|\uparrow\rangle$. Since I know that the chances of my colleague finding $|\downarrow\rangle$ or $|\uparrow\rangle$ are equal, I must take the

[29.9] Explain why this is so, showing that the two probabilities in the mixture have the same ratio as the two lengths into which $\boldsymbol{L}$ divides the chord.

view that the state of the particle that I receive (an hour after my colleague's measurement) has a probability of $\frac{1}{2}$ of being $|\uparrow\rangle$ and a probability of $\frac{1}{2}$ of being $|\downarrow\rangle$. I thus use the density matrix

$$\boldsymbol{D} = \tfrac{1}{2}|\uparrow\rangle\langle\uparrow| + \tfrac{1}{2}|\downarrow\rangle\langle\downarrow|$$

(the two states, $|\uparrow\rangle$ and $|\downarrow\rangle$ being taken orthogonal and normalized: $\langle\uparrow|\downarrow\rangle = 0$ and $\langle\uparrow|\uparrow\rangle = 1 = \langle\downarrow|\downarrow\rangle$).

However, it might have been the case that, by a last minute change of heart, my colleague had decided to measure the spin of the particle arriving on Titan not in an up/down direction but in a left/right direction instead. If my colleague obtained the result $|\leftarrow\rangle$, then I must find $|\rightarrow\rangle$ for my particle arriving here on Earth; if my colleague obtained $|\rightarrow\rangle$, then I must find $|\leftarrow\rangle$. Again the probability of my colleague's two alternatives would have been $\frac{1}{2}$ in each case; so, although I do not yet know which of these results my colleague obtained, I must conclude that my particle might be $|\rightarrow\rangle$ or it might be $|\leftarrow\rangle$, with probability $\frac{1}{2}$ in each case. I therefore assign the density matrix

$$\boldsymbol{D} = \tfrac{1}{2}|\rightarrow\rangle\langle\rightarrow| + \tfrac{1}{2}|\leftarrow\rangle\langle\leftarrow|$$

to my particle (where $\langle\leftarrow|\rightarrow\rangle = 0$ and $\langle\rightarrow|\rightarrow\rangle = 1 = \langle\leftarrow|\leftarrow\rangle$). Of course, as we have seen, this is just the same $\boldsymbol{D}$ as before. This is as it should be, because my colleague's decision as to which way to measure the particle on Titan should not affect the probabilities here on Earth (otherwise there would be a method of signalling from Titan to Earth faster than light[29.10]). Thus it would seem that, for the type of situation under consideration, the density matrix provides an excellent mathematical description of the physical situation. The spin state of the particle that I receive here on Earth, provided that I know nothing whatever of the proceedings on Titan—neither of the direction that my colleague chooses to measure the spin, nor of the result of that measurement—is very well described by the above density matrix $\boldsymbol{D}$.

Of course, this works well only if I receive no information from Titan. If I know the type of measurement that my colleague performs, this will affect my view of the ontology of the spin state that I receive, but it will not affect the expectations of the probabilities of the measurements that I might perform here on Earth.[16] I might take the view, if I know that my colleague's measurement is to be right/left, that the ontology of my particle's spin state is either right or left, but I do not know which—a viewpoint that I could not have taken had I not known of the direction of my colleague's measurement. But this ontological knowledge will not affect my estimations of the

---

[29.10] Explain how.

probabilities of the results of the spin measurements that I perform on Earth, so I might take the alternative position that the 'ontology' is of no importance and, perhaps, even scientifically meaningless, so the density matrix *is* all that is scientifically required. On the other hand, if I actually receive a message from Titan telling me the results of my colleague's measurement, then my probability estimates may well be affected. More than that, there will actually be consistency requirements constraining the results of our joint measurements (for example: I cannot obtain the result $\langle\leftarrow|$ if my colleague obtained $\langle\leftarrow|$). *Now* it is clear that the density matrix description is quite inadequate, and we must revert to a description in terms of an actual quantum state (vector) describing the entire entangled pair: $|\Omega\rangle = |\uparrow\rangle|\downarrow\rangle - |\downarrow\rangle|\uparrow\rangle$ $(= |\leftarrow\rangle|\rightarrow\rangle - |\rightarrow\rangle|\leftarrow\rangle$, etc.).

The particular density matrix that arises in the above example (already considered in §29.4) is very special. In terms of any orthonormal basis, it has the form

$$\boldsymbol{D} = \begin{pmatrix} \frac{1}{2} & 0 \\ 0 & \frac{1}{2} \end{pmatrix}.$$

What is special about it is that all its eigenvalues are equal (the two numbers $\frac{1}{2}$ down the diagonal). This has the implication that it has the same form whatever (orthonormal) basis is used—since it is just a multiple of the identity matrix. Thus there is nothing to distinguish the up/down basis from the left/right basis, etc.

It is important to point out that this is only a result of the particularly simple situation that we considered in this example. We have already seen, in §29.4, that there is nothing special about the particular (equal-eigenvalue) density matrix $\boldsymbol{D}$ with regard to its ontology confusion. With a very slight modification of the example, we can get any 2×2 density matrix we wish. Instead of the pair of spin $\frac{1}{2}$ EPR particles being produced in a spin 0 state, as in the case just considered, we take them to be initially in a state of spin 1. To see how this works in a particular case, we can consider Lucien Hardy's example, as studied in §23.5. Here, the initial state is $|\leftarrow\nearrow\rangle = |\leftarrow\rangle|\nearrow\rangle + |\nearrow\rangle|\leftarrow\rangle$ (in the Majorana description of §22.10, the tangent of the angle between $\rightarrow$ and $\nearrow$ being $\frac{4}{3}$), and I shall suppose that my colleague chooses to make a right/left measurement on the particle arriving on Titan. From the results of §23.5, we find that, if my colleague obtains $|\rightarrow\rangle$, then the state that I receive here on Earth is $|\leftarrow\rangle$, whereas if my colleague obtains $|\leftarrow\rangle$, then the state I receive is $|\uparrow\rangle$.[29.11] Thus, if I know that my colleague performed a right/left measurement (and I know that the initial state was $|\leftarrow\nearrow\rangle$), then I conclude that the spin state of the

[29.11] Why?

particle I receive on Earth is a probability mixture of $|\leftarrow\rangle$ and $|\uparrow\rangle$. Note that $|\leftarrow\rangle$ and $|\uparrow\rangle$ are *not orthogonal*. Orthogonality is not a requirement for the probability mixture of states composing a density matrix, and we see this explicitly in this example.

What is the density matrix that I would use for my particle? We can work this out if we know the probability values for the two alternative results, $|\rightarrow\rangle$ and $|\leftarrow\rangle$, that my colleague can obtain. In fact, these respective probabilities turn out to be $\frac{1}{3}$ and $\frac{2}{3}$, so I have a $\frac{1}{3}$ probability of receiving the state $|\leftarrow\rangle$ and a $\frac{2}{3}$ probalility of receiving $|\uparrow\rangle$. My density matrix is therefore now

$$\boldsymbol{L} = \tfrac{1}{3}|\leftarrow\rangle\langle\leftarrow| + \tfrac{2}{3}|\uparrow\rangle\langle\uparrow|.$$

In terms of an up/down basis frame, this matrix looks like

$$\boldsymbol{L} = \begin{pmatrix} \frac{5}{6} & -\frac{1}{6} \\ -\frac{1}{6} & \frac{1}{6} \end{pmatrix}$$

(taking $|\leftarrow\rangle = (|\uparrow\rangle - |\downarrow\rangle)/\sqrt{2}$). This certainly does not have equal eigenvalues, its eigenvalues in fact being $\frac{1}{2}+\frac{1}{6}\sqrt{5}$ and $\frac{1}{2}-\frac{1}{6}\sqrt{5}$.[29.12] The particular ontology '$|\leftarrow\rangle$ with probability $\frac{1}{3}$ and $|\uparrow\rangle$ with probability $\frac{2}{3}$' for this density matrix is, nevertheless, far from unique. For example, it is obvious from the symmetry between $\leftarrow$ and $\nearrow$, in the initial state $|\leftarrow\nearrow\rangle$, that if my colleague chose to measure in the direction of $\nearrow$, rather than left/right (direction of $\leftarrow$), then my own ontology for the density matrix $\boldsymbol{D}$ would be very much changed, involving $|\nearrow\rangle$ and another state perpendicular to it. Indeed, a different ontology would be obtained for each possible measuring direction that my colleague on Titan might happen to choose.[29.13]

Hosts of more complicated ontologies could be obtained, given any particular density matrix, if we allow the probability mixture to involve three or more different states. Such a situation would arise if the initial state had spin $\frac{1}{2}n$, for $n > 2$, which decays into a particle of spin $\frac{1}{2}$ aimed at Earth and one of spin $\frac{1}{2}n - \frac{1}{2}$ aimed at Titan, since my colleague's spin measurement would then allow n different outcomes, each with its own probability (§22.10); see Fig. 29.4. This clearly also generalizes to situations where the Hilbert space of states that I use for my particle as it arrives here on Earth is greater than 2-dimensional. All this serves to emphasize that there is no unique ontology of 'probability-weighted alternative states' whatever density matrix is used.[17] We shall see shortly that

---

[29.12] Derive this matrix form of $\boldsymbol{L}$, verify that these are its eigenvalues, and find the eigenvectors. The point $\boldsymbol{L}$ in the Bloch sphere of Fig. 29.3 is chosen to agree with this. How far out from the centre is it?

[29.13] Show that any preassigned 2×2 density matrix can be obtained by the above procedure, where the initial state for the EPR pair has spin 1. How do the eigenvector spin directions of the density matrix relate to the Majorana description of this initial state?

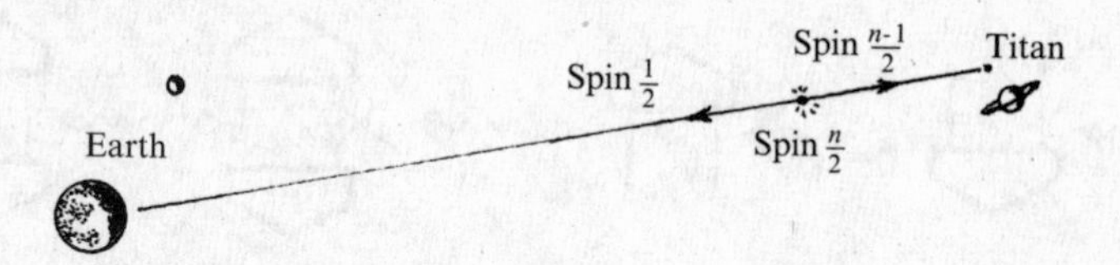

Fig. 29.4 A density matrix can represent a probability mixture of more states than the dimension of the space. In this example: at a point between Earth and Titan, but nearer to Titan, a known initial state of spin $\frac{n}{2}$ (for $n > 2$) splits into a spin $\frac{1}{2}$ particle aimed at Earth and a spin $\frac{1}{2}(n-1)$ particle aimed at Titan. A colleague on Titan measures the latter's spin m-value, and the probability for each of the n possible measurement results is a specific number that can be calculated (at Earth), knowing the initial state, so a specific 2×2 density matrix arises at Earth, composed as a probability mixture of n states. (This clearly also generalizes to a Hilbert space of more than 2 dimensions.)

this fact causes an awkwardness for the environmental-decoherence philosophy of viewpoint (c).

A remark should be made here, concerning the actual calculation of a density matrix—where, as above, part of the information in an entangled state is hidden (e.g. 'on Titan'). There is a very efficient method, referred to as 'summing over the unknown states'. This is most easily expressed in the index notation. Let us write our initial state (the normalized ket vector $|\psi\rangle$) as $\psi^{\alpha\rho}$, which is to be thought of as an entangled state, where α refers to *here* (say, the Earth) and ρ refers to *there* (say, Titan); see §§23.4,5. The complex conjugate of this state (the bra vector $\langle\psi|$) is $\bar{\psi}_{\alpha\rho}$. The normalization of the state is the condition

$$\bar{\psi}_{\alpha\rho}\psi^{\alpha\rho} = 1.$$

Then the density matrix that I would use here on Earth, in the absence of information from Titan, is the quantity

$$D_{\alpha}^{\beta} = \bar{\psi}_{\alpha\rho}\psi^{\beta\rho}$$

(with a contraction on the index ρ). Correspondingly, my colleague's density matrix would be $\bar{\psi}_{\alpha\rho}\psi^{\alpha\sigma}$.[29.14] See Fig. 29.5 for the diagrammatic version of this.

[29.14] Show why this works. (*Hint*: Choose a separate orthonormal basis for *here* and for *there* and work in terms of the joint probabilities for the various possible results of measurements in the two places.) Verify the above probabilities $\frac{1}{3}$ and $\frac{2}{3}$ for the case $|\psi\rangle = |\leftarrow\nearrow\rangle$, considered above.

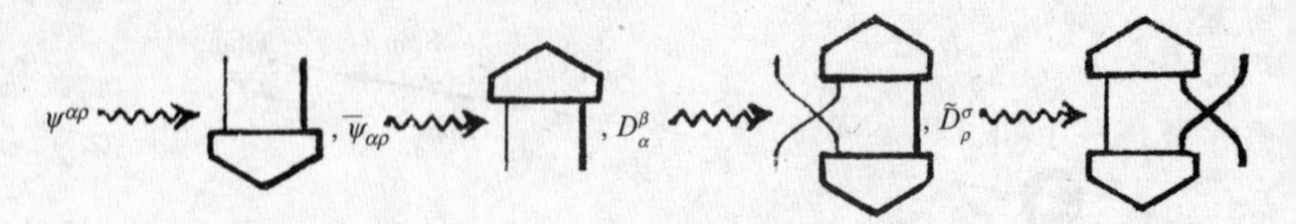

Fig. 29.5 Diagrammatic notation for density matrices constructed by 'summing over unknown states'. The normalized ket-vector $|\psi\rangle$ is expressed as $\psi^{\alpha\rho}$, where 'α' refers to 'here' (Earth) and 'ρ' refers to 'there' (Titan). The Hermitian conjugate (bra-vector $\langle\psi|$) is $\bar{\psi}_{\alpha\rho}$ and the normalization is $\bar{\psi}_{\alpha\rho}\psi^{\alpha\rho} = 1$. The density matrix used 'here' is $D^{\beta}_{\alpha} = \bar{\psi}_{\alpha\rho}\psi^{\beta\rho}$, while that used 'there' is $\tilde{D}^{\sigma}_{\rho} = \bar{\psi}_{\alpha\rho}\psi^{\alpha\sigma}$.

29.6 FAPP philosophy of environmental decoherence

The above considerations may be regarded as a 'prelude' to our investigation of the environmental-decoherence viewpoint (c), which maintains that state reduction **R** can be understood as coming about because the quantum system under consideration becomes inextricably entangled with its environment. To apply these ideas, we think of the system itself as the *here* part, and the environment as the *there* part. We take the environment to be extremely complicated and essentially 'random', so there is, in practice, no conceivable way of extracting the environment's *there* part of the information of the total quantum state. Accordingly, we 'sum over the unknown states' in the environment, to obtain a density matrix description for the *here* part of the state. Much work in this subject is geared to showing that if the environment is modelled in a 'reasonable' way, then in a very short period of time (for even a mildly 'noisy' environment) the density matrix becomes diagonal:

$$\boldsymbol{D} = \begin{pmatrix} p_1 & 0 & \cdots & 0 \\ 0 & p_2 & \cdots & 0 \\ \vdots & \vdots & \ddots & \vdots \\ 0 & 0 & \cdots & p_n \end{pmatrix}$$

to a high degree of approximation, when expressed in terms of some particular basis $|1\rangle, |2\rangle, \ldots, |n\rangle$, of special interest.[29.15] This is then interpreted as a probability mixture

$$\boldsymbol{D} = p_1|1\rangle\langle 1| + p_2|2\rangle\langle 2| + \ldots + p_n|n\rangle\langle n|,$$

[29.15] In fact, any density matrix is diagonal in *some* basis! Can you see why, in the particular case when all the eigenvalues are unequal?

of those particular basis states that correspond to the diagonal terms. This probability mixture is taken to reflect the alternatives that occur in the state-reduction process **R**, the probabilities for each outcome being the respective numbers $p_1, p_2, \ldots, p_n$.

Yet, as we have seen above, any density matrix has a host of ontological interpretations. We can never learn, merely from such an argument, that any one of these interpretations provides us with the 'real' state of affairs. Further, we cannot then even deduce that the state is *one* of $|1\rangle, |2\rangle, \ldots,$ or $|n\rangle$, with respective probabilities $p_1, p_2, \ldots, p_n$.

Under normal circumstances, moreover, one must regard the density matrix as some kind of *approximation* to the whole quantum truth. For there is no general principle providing an absolute bar to extracting detailed information from the environment. Maybe a future technology could provide means whereby quantum phase relations can be monitored in detail, under circumstances where present-day technology would simply 'give up'. It would seem that the resort to a density-matrix description is a technology-dependent prescription! With better technology, the state-vector description could be maintained for longer, and the resort to a density matrix put off until things get really *hopelessly* messy! It would seem to be a strange view of physical reality to regard it to be 'really' described by a density matrix. Accordingly, such descriptions are sometimes referred to as *FAPP*, an acronym suggested by John Bell (of Bell inequalities fame; see §23.3) denoting 'for all practical purposes'. The density-matrix description may be thus regarded as a pragmatic convenience: something FAPP, rather than providing a 'true' picture of fundamental physical reality.

There might, however, be a level at which the detailed phase relations indeed *actually* get lost, because of some deep overriding basic principle. Ideas aimed in this direction often appeal to *gravity* as possibly leading us to such a principle. Sometimes the idea of 'quantum fluctuations in the gravitational field' might be appealed to, according to which the very structure of spacetime would become 'foamlike', rather than resembling a smooth manifold (Fig. 29.6) at the 'Planck scale' of some 10^{-35}m.[18]

Fig. 29.6 What is the nature of spacetime at the Planck scale of 10^{-33} cm or 10^{-43} s? It has been argued that quantum fluctuations in the gravitational field may result in a seething mess of 'foam' with multiple topology changes, and where possibly detailed quantum phase relations may actually get lost at this level.

(I shall be referring to such ideas in §31.1 and §33.1.) One could imagine that the phase relations might indeed get inextricably 'lost in the foam' at such a scale. Another suggestion, due to Stephen Hawking, is that, in the presence of a black hole, information about the quantum state might get 'swallowed' by the hole, and become irretrievably lost *in principle*. In such circumstances, one might envisage that a quantum system—referring to some external physics that is entangled with a part that has fallen into the hole—should be actually described by a density matrix rather than by a 'pure state'.[19] I shall return to these ideas later, in §§30.4,7,8,14.

29.7 Schrödinger's cat with 'Copenhagen' ontology

Let us go back to the quantum-mechanical *measurement problem* of how **R** might—or might seem to—come about when it is supposed that the quantum state 'actually' evolves according to the deterministic **U** process (§21.8, §§22.1,2, §23.10). This problem is frequently presented, very graphically, in terms of the *paradox of Schrödinger's cat*. The version that I am presenting here differs, but only in inessential ways, from Schrödinger's original version. We suppose that there is a photon source *S* which emits a single photon in the direction of a beam splitter ('half-silvered' mirror), at which point the photon's state is split into two parts. In one of the two emerging beams, the photon encounters a detector that is coupled to some murderous device for killing the poor cat, while in the other, the photon escapes, and the cat remains alive. See Fig. 29.7. (Of course, this is only a 'thought experiment'. In an actual experiment—such as the one that we shall be coming to in §30.13—there is no need to involve a living creature. The cat is used only for dramatic effect!) Since these two alternatives for the photon must co-exist in quantum linear superposition, and since the linearity of Schrödinger's equation (i.e. of **U**) demands that the two subsequent time-evolutions must persist in constant complex-number-weighted superposition, as time passes (§22.2), the quantum

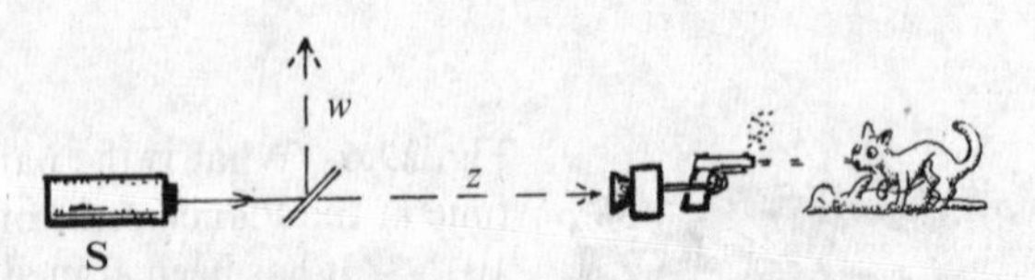

Fig. 29.7 Schrödinger's cat (modified from original). A photon source S emits a single photon aimed at a beam-splitter, whereupon the photon's state splits into a superposition of 2 parts. In one of these, the photon encounters a detector, triggering a murderous weapon that kills the cat; in the other, the photon escapes and the cat lives. **U**-evolution results in a superposition of a dead and a live cat.

state must ultimately involve such a complex-number superposition of a dead cat and a live cat: so the cat is both dead and alive at the same time!

Of course such a situation is an absurdity for the behaviour of a cat-sized object in the actual physical world as we experience it. How is this paradox dealt with according to the various 'standard' interpretations of quantum mechanics? Consider the Copenhagen viewpoint (a). As far as I can make out, this interpretation would simply regard the photon detector to be a 'classical measuring device', to which the rules of quantum superposition are not applied. The photon state between its emission and its detection (or non-detection) by the device is described by a wavefunction (state-vector), but no 'physical reality' is assigned to that. The wavefunction is used merely as a mathematical expression to be used for calculating probabilities. If the beam splitter is such that the photon amplitude is divided equally into two, then the calculation tells us that there is a 50% chance for the detector to register reception of the photon and a 50% chance that it will not. Therefore there is a 50% chance that the cat will be killed and a 50% chance that it will remain alive.

This is physically the correct answer, where 'physically' refers to the behaviour of the world that we actually experience. Yet this description provides us with a very unsatisfactory picture of things if we wish to pursue the physical events in greater detail. What actually goes on inside a detector? Why are we allowed to treat it as a 'classical device' when, after all, it is constructed from the same quantum ingredients (protons, electrons, neutrons, virtual photons, etc.) as any other piece of physical material, large or small? I can well appreciate that, in the early days of quantum mechanics, something of the nature of Niels Bohr's perspective on the subject was almost a necessity, so that the theory could actually be used, and progress in quantum physics could be made. Yet, as far as I can see, such a perspective can only be a temporary one, and it does not resolve the question of why, and at what stage, 'classical behaviour' might arise for large and complicated structures like 'detectors'. Since viewpoint (a) requires such 'classical structures' for its interpretation of quantum mechanics, it can only be a 'stop-gap' position, in which the deeper issues concerning what actually constitutes a measurement are not addressed at all.

Another variant of (a) would demand, in effect, that the 'classical measuring apparatus' is ultimately the observer's consciousness. Accordingly (if we discount the consciousness of the cat itself), it is only when a conscious experimenter examines the cat that classicality has been achieved. It seems to me that, once we have arrived at this level, we are driven to take a position that is more in line with (b) or with (f). If we take the view that the **U** rules of quantum linear superposition continue to hold right up to the level of a conscious being, then we are in the realm of the many-worlds perspective (b), but if we take the stand that **U** fails for conscious beings, then we are driven to a version of (f) according to which some new type of behaviour, outside the ordinary predictions of quantum mechanics, comes into play with beings who possess consciousness. A suggestion along this line was actually put forward by the distinguished quantum physicist Eugene Wigner in 1961.[20]

It seems to me, however, that any theory that demands the presence of a conscious observer, in order that **R** be effected, leads to a very lop-sided (and, I would argue, highly implausible) picture of the universe. Imagine some distant Earth-like planet without conscious life, and for which there is no consciousness for many many light years in all directions. What is the weather like on that planet? Weather patterns have the property that they are 'chaotic systems', in the sense that any particular pattern which develops will depend critically on the tiniest details of what happened before (see §27.2). Indeed, it is probable that, in a month, say, tiny quantum effects will become so magnified that the entire pattern of weather on the planet would depend upon them. The absence of consciousness, according to the particular version of (f) (or perhaps (a)) under discussion, would imply that **R** *never occurs* on such a planet, so that the weather is, in reality, just some quantum superposed mess that does not resemble an actual weather in the sense that we know it. Yet if a spacecraft containing conscious travellers, or a probe with the capacity to transmit a signal to a conscious being, is able to train its sensors on that planet, then immediately—and only at that point—its weather suddenly becomes an ordinary weather, just as though it had been ordinary weather all the time! There is no actual contradiction with experience here, but is this 'Wigner reality' a believable picture for the behaviour of an actual physical universe? It is not, to me; but I can (just about) understand others giving it more credence.

29.8 Can other conventional ontologies resolve the 'cat'?

What about the many-worlds standpoint (b), then? Here the 'reality' of the quantum superposition of a dead and a live cat is simply *accepted* (as would the quantum-superposed weather patterns of the previous paragraph); but this does not tell us what an observer, looking at the cat (or the weather), actually 'perceives'. The state of the observer's perception is considered to be entangled with the state of the cat. The perception state 'I perceive a live cat' accompanies the 'live-cat' state and the perception state 'I perceive a dead cat' accompanies the 'dead-cat' state. See Fig. 29.8. It is then assumed that a perceiving being always finds his/her perception state to be in one of these two; accordingly, the cat is, in the perceived

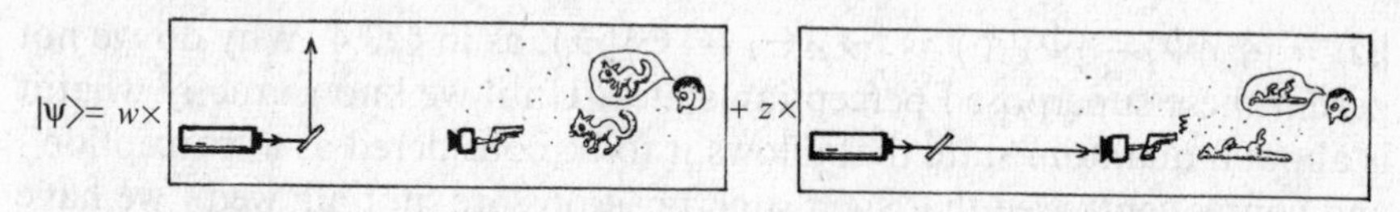

Fig. 29.8 The conclusion of Fig. 29.7 is unaffected by the presence of different environments entangled with the cat's states or by an observer's different responses. Thus the state takes the form

$$|\Psi\rangle = w \times |\text{live cat}\rangle|\text{live cat's environment}\rangle|\text{perceiving live cat}\rangle$$
$$+ z \times |\text{dead cat}\rangle|\text{dead cat's environment}\rangle|\text{perceiving dead cat}\rangle.$$

If **U**-evolution is to represent reality (many-worlds viewpoint (b)) then we must take the view that an observer's awareness can experience only one or the other alternative, and 'splits' into separate world-experiences at this stage.

world, either alive or dead. These two possibilities coexist in 'reality' in the entangled superposition:

$$|\Psi\rangle = w|\text{live cat}\rangle\,|\text{perceiving live cat}\rangle + z|\text{dead cat}\rangle\,|\text{perceiving dead cat}\rangle.$$

I wish to make clear that, as it stands, this is far from a resolution of the cat paradox. For there is nothing in the formalism of quantum mechanics that demands that a state of consciousness cannot involve the simultaneous perception of a live and a dead cat. In Fig. 29.9, I have illustrated this issue, where I have taken the simple situation in which the two amplitudes, z and w, for reflection and transmission at the beam splitter, are equal. As with the simple EPR–Bohm example with two particles of spin $\frac{1}{2}$ emitted in an initial state of spin 0, we can rewrite the resulting entangled state in many ways. In the example illustrated in Fig. 29.9, the state $|\text{live cat}\rangle + |\text{dead cat}\rangle$ is accompanied by $|\text{perceiving live cat}\rangle + |\text{perceiving dead cat}\rangle$ and the state $|\text{live cat}\rangle - |\text{dead cat}\rangle$ is accompanied by $|\text{perceiving live cat}\rangle - |\text{perceiving dead cat}\rangle$. This is exactly analogous to the rewriting the state

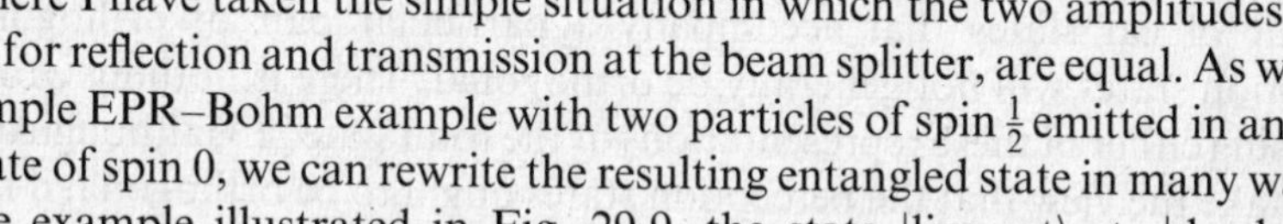

Fig. 29.9 Re-express Fig. 29.8 (in the case $z = w = \frac{1}{\sqrt{2}}$, and incorporating the environment state with that of the cat) as follows:

$$\sqrt{8}|\Psi\rangle = \{|\text{live cat}\rangle + |\text{dead cat}\rangle\}\{|\text{perceiving live cat}\rangle + |\text{perceiving dead cat}\rangle\}$$
$$+\{|\text{live cat}\rangle - |\text{dead cat}\rangle\}\{|\text{perceiving live cat}\rangle - |\text{perceiving dead cat}\rangle\}$$

$|\Omega\rangle = |\uparrow\rangle|\downarrow\rangle - |\downarrow\rangle|\uparrow\rangle$ as $|\rightarrow\rangle|\leftarrow\rangle - |\leftarrow\rangle|\rightarrow\rangle$, as in §23.4. Why do we not permit these superposed perception states? Until we know exactly what it is about a quantum state that allows it to be considered as a 'perception', and consequently see that such superpositions are 'not allowed', we have really got nowhere in explaining why the real world of our experiences cannot involve superpositions of live and dead cats.

Sometimes people object to this example on the grounds that the equality of the amplitudes for the two alternatives is a very special situation, and that in general there is not the freedom to re-express the entangled states in this way. When we look at this situation a little more deeply, however, we find that the 'equal-amplitude' aspect of this particular example is not really important at all. It is useful to bear in mind the example of an EPR pair of particles of spin $\frac{1}{2}$, considered above in §29.5. 'Equality of amplitudes' (actually 'equality of moduli of amplitudes' $|z| = |w|$) is what give rise to a density matrix with equal eigenvalues. We saw explicitly in §§29.4,5 that a 2×2 density matrix with unequal eigenvalues has many representations as a probability mixture of a pair of states, but the pair will in general be non-orthogonal. In fact, orthogonality occurs only when the two states are eigenvectors of the density matrix.[29.16] In the case of 'equal amplitudes' (strictly $|z| = |w|$), we may take the states |live cat⟩ and |dead cat⟩ to be orthogonal and, indeed, the accompanying states |perceiving live cat⟩ and |perceiving dead cat⟩ to be orthogonal (the 'eigenvectors'). But in the case $|z| \neq |w|$, the pair of perception states that accompany a particular orthogonal pair of superposed cat states will not generally be orthogonal, and the pair of cat states that accompany a particular pair of orthogonal perception states will not generally be orthogonal. There is nothing wrong with using either of these representations of the total state $|\Psi\rangle$, although one might take the view that the perception states ought to be orthogonal if it is those states that are to drive the appearance of reality in the many-worlds view. But since **R** does not actually take place at all, according to position (b), there is no special status for orthogonal alternatives (since nothing 'reduces' to them in any case).

In fact it turns out that, in the general case, there will be a unique pair of orthogonal perception states accompanying a pair of orthogonal cat states. This is something known as the *Schmidt decomposition* of an entangled state.[21] However, this is not of much use for resolving the measurement paradox (despite the popularity of the Schmidt decomposition in connection with quantum information theory[22]), because generally this 'mathematically preferred' pair of cat states (eigenstates of the cat's density matrix) would not be the desired |live cat⟩ and |dead cat⟩ at all, but some unwanted linear superpositions of these! We can see that these

---

[29.16] Show this.

density-matrix eigenstates that occur in a Schmidt decomposition need have nothing to do with one's expectations of what should be 'ontologically real', by looking again at Lucien Hardy's example considered in §29.5. We find (see Exercise [29.12]) that the eigenvectors of the density matrix (for the particle that I receive here on Earth) are quite different from the $|\leftarrow\rangle$ and $|\uparrow\rangle$ alternatives that are 'macroscopically distinguishable alternatives' according to the measurements of my colleague on Titan!

Since the mathematics alone will not single out the '$|$live cat$\rangle$' and '$|$dead cat$\rangle$' states as being in any way 'preferred', we still need a theory of perception before we can make sense of (b), and such a theory is lacking.[23] Moreover, the onus on such a theory would be not only to explain why superpositions of dead and live cats (or of anything else macroscopic) do not occur in the perceived world, but also why the wondrous and extraordinarily precise squared-modulus rule actually gives the right answers for probabilities in quantum mechanics! A theory of perception that could do this would itself need to be as precise as quantum theory. Supporters of (b) have come nowhere close to suggesting such a scheme.[24]

Now let us return to the attempts at a resolution of the cat paradox by environmental decoherence (c). Let us take the initial emission of the photon as ontologically real. (The source could be arranged to register this event in a macroscopic way.) Then, after the beam splitter is encountered, we have an ontologically real superposition of the photon in the two beams. The transmitted part of the photon's state evolves to a dead cat, together with its environment, and the reflected part to a live cat, together with a different environment. For the moment, the ontology is still the superposition of the two. The environmental alternatives, being 'unobservable' are next summed over, leaving us with a 2×2 density matrix. Now the ontological position stealthily shifts, and 'reality' becomes described by the density matrix itself. The environmental-decoherence argument now asserts that this matrix rapidly becomes extremely closely diagonal in the basis ($|$live cat$\rangle$, $|$dead cat$\rangle$), so there is another surreptitious shift in the ontology, and the state becomes a probability mixture of $|$live cat$\rangle$ and $|$dead cat$\rangle$. This is how we have been 'allowed' to get away with this ontology shift from the superposition

$$w|\text{live cat}\rangle|\text{live cat's environment}\rangle + z|\text{dead cat}\rangle|\text{dead cat's environment}\rangle$$

to the alternatives $|$live cat$\rangle$ or $|$dead cat$\rangle$! We recall that there is no uniqueness in the ontological interpretation of a density matrix as a probability mixture of states (whether or not the eigenvalues are equal). Indeed, to pass to the mixture of $|$live cat$\rangle$ and $|$dead cat$\rangle$ does represent a (double) ontology shift from the original superposition. Position (c) is indeed FAPP, and it gives us no consistent ontology for physical reality.

29.9 Which unconventional ontologies may help?

I should comment briefiy on (d) and (e). If the 'extravagant' ontology for the consistent-histories approach (d) is adopted, in which reality is represented as a totality of maximally refined consistent-history sets, then a criticism can be raised which is somewhat similar to that of the many-worlds case (b). As with (b), a detailed and precise theory of conscious perceivers seems to be needed in order that (d) can conjure up a picture that is consistent with the physical world that we know. Attempts, have been made in this direction (provided by the notion of an IGUS—'information gathering and using system') but, as yet, these seem to be rather far from sufficient.[25] Alternatively, one might prefer something like the more economical ontology hinted at in §29.2, in which a single maximally refined consistent history set might be considered as a plausible candidate for a 'real-world' ontology. However this (as well as the more extravagant ontology above) depends upon the criterion of 'consistent history' really achieving what it was designed for, namely to single out histories resembling the kind of world we actually live in. However, as was demonstrated by Dowker and Kent in 1996, this condition of 'consistency' alone is far from adequate. Some additional criteria seem to be required.

In my own view, a major drawback with (d) is that despite the introduction of **R**-like procedures (via the insertion of projector sets), it does not seem to get us any closer to an understanding of what a physical measurement actually is than do the more conventional ontologies of (a) or (b). Indeed, in (d), the **R**-like procedures are explicitly stated to be nothing directly to do with actual physical measurements. My difficulty with this is that by removing the association between these **R**-type replacements and physical measurements, we gain no insight as to what actually constitutes a physical measurement. Why, according to (d), do we not actually witness things like Schrödinger cats, in superposed limbo between life and death? The theory does not seem to give any improvement on the standard Copenhagen position (a) in explaining which systems (such as pieces of physical apparatus or cats) should behave classically, whereas neutrons or photons do not. The requirement of 'consistency' for (maximally refined) coarse-grained histories appears to be a long way from what is needed in order to provide a model[26] for observed physical reality.

Although it is a positive feature of (d) that it makes a serious attempt to incorporate **R**-like procedures at a fundamental level, the criteria that have so far been put forward do not do enough to narrow down the model's behaviour so that an unambiguous picture of something resembling the world we know can arise. This seems to be true both at the macroscopic 'classical-like' level (as I have been commenting earlier, in relation to the Dowker–Kent analysis of the 'consistent-history' criterion) and also at the 'quantum level' at which one would hope to see undisturbed unitary evolution. Since the measurement paradox is concerned with the seeming conflict between physical behaviour at these two different levels, it is hard to see how the consistent-history viewpoint (d) is yet in a position to shed much light on this paradox.

What about (e)? As remarked in §29.2, the de Broglie–Bohm 'pilot-wave' viewpoint (e) appears to have the clearest ontology among all those which do not actually alter the predictions of quantum theory. Yet, it does not, in my opinion, really address the measurement paradox in a clearly more satisfactory way than the others do. As I see it, (e) may indeed gain conceptual benefit from its two levels of reality—having a firmer 'particle' level of the reality of the configuration of the system, as well as a secondary 'wave' level of reality, defined by the wavefunction ψ, whose role is to guide the behaviour of the firmer level. But it is not clear to me how we can be sure, in any situation of actual experiment, which level we should be appealing to. My difficulty is that there is no parameter defining which systems are, in an appropriate sense, 'big', so that they accord with a more classical 'particle-like' or 'configuration-like' pictures, and which systems are 'small', so that the 'wavefunction-like' behaviour becomes important (and this criticism applies also to (d)). We know from §23.4, etc., that quantum behaviour can stretch over distances of tens of kilometres at least, so that it is not just physical distance that tells us when a system ceases to look quantum mechanical and begins to behave like a classical entity. But nevertheless there is a sense in which a large object (like a cat) seems not to accord with the small-scale unitary quantum laws. (In §30.11 I shall begin to explain my own particular views as to the type of 'scale-measure' that will be needed). But whether or not one believes that any particular such measure is appropriate, it seems to me that *some* measure of scale is indeed needed, for defining when classical-like behaviour begins to take over from small-scale quantum activity. In common with the other quantum ontologies in which no measurable deviations from standard quantum mechanics is expected, the point of view (e) does not possess such a scale measure, so I do not see that it can adequately address the paradox of Schrödinger's cat.

In relation to this issue, a general comment concerning attempts to 'derive' the apparent occurrence of **R** from the dynamics of (say) **U**, may be appropriate. We can see that ordinary (deterministic) dynamics alone can never achieve this—as is clear, if only for the reason that there are no probabilities in such a dynamical equation as the Schrödinger equation.

(I refer the reader to the discussion of §27.1.) Some probabilistic principle is, of necessity, also needed. **R** is, after all, a probabilistic law. Thus, as remarked upon in §29.2, it is indeed an essential ingredient of (e) that the appropriate successive probabilities of measurements are correctly encoded in the choice of (say) the initial state.

This leaves us with (f). The main difficulties with most of the many different (often heroic) proposals for an "objective **R**" lie in their unnatural appearance, their essentially non-relativistic character, their need for the introduction of arbitrary parameters unmotivated from known physics, their violations of the law of conservation of energy, and in some cases their direct conflict with observation. It would be inappropriate for me to discuss all of these proposals here, and it would be unfair of me to single some of them out at the expense of the others. In fact, I shall adopt the procedure of being uniformly unfair to all the proposals that others have put forward by imposing upon the reader, in Chapter 30, the one (in some ways minimilist) proposal that I myself believe to be the most likely to be correct (with apologies to many of my friends)! In fact, there has been a very significant stimulation and input from various proposals that others have put forward earlier, and I shall indeed refer to these (with appropriate gratitude), but only in relation to the specific ideas that I wish to argue for.

Notes

Section 29.1

29.1. See Deutsch (2000).

29.2. I owe this term to my classicist colleague Peter Derow. See Penrose (1987a).

29.3. See Everett (1957); Wheeler (1957); DeWitt and Graham (1973); Deutsch (2000).

29.4. Some physicists argue that there is 'no problem' about the quantum superposition of macroscopically different states—like Schrödinger's superposed dead and alive cat that we shall be coming to in §§29.7–9—because it would simply be 'far too expensive' (or a practical impossibility) to design an experiment to detect interference between the dead and alive states. This, again, is taking a 'pragmatic' stance that does not really address the ontological issues that are our concern here. I would place such physicists, generally, in category (c).

29.5. See Hawking and Penrose (1996), p. 121.

Section 29.2

29.6. For (d), see Gell-Mann and Hartle (1995); for (e), see Bohm and Hiley (1994). This list (a), (b), (c), (d) is representative only, and there are many different shades of standpoints within those that I have listed. For example, some have expressed the view (for example, Sorkin 1994) that 'quantum reality' is best understood in terms of the path integrals and/or Feynman graphs that we encountered in §§26.6–11. As far as I can make out, this particular family of ontologies would belong to the general class covered by (b) (although having some important elements in common with (d)), according to which a particular superposition that defines the 'quantum state' (or 'quantum history') would be assigned the status of 'reality'. I should mention also the 'transactional' ontologies of Aharonov and Vaidman (2001); Cramer (1988); Costa de Beauregard (1995); and Werbos and Dolmatova (2000), according to which a wavefunction Schrödinger-propagating into the future from the last measurement together with another wavefunction Schrödinger-propagating into the past from the next measurement are both enlisted in the description of reality (see §30.3). I do not see that, without further ingredients, the issue of the measurement paradox is better resolved in any of these alternative views than in (a), (b), (c), (d), or (e), however.

29.7. The formalism (d) also allows that the 'starting state' could be a density matrix (see §29.3).

29.8. Sometimes this is simply called a 'history', but this could cause confusion with the use of that term in the Feynman 'sum over histories' of §26.6.

29.9. This is a condition of the following type. Suppose we have a given succession of projector sets (and for the moment assume $\mathcal{H} = 0$); then we construct the expression $X = \langle\psi_0|\boldsymbol{E}'\boldsymbol{F}'\ldots\boldsymbol{K}'\boldsymbol{L}'\boldsymbol{D}_\infty\boldsymbol{L}\boldsymbol{K}\ldots\boldsymbol{F}\boldsymbol{E}|\psi_0\rangle$, where $|\psi_0\rangle$ is the initial state and where the 'final state' could be taken to be a density matrix $\boldsymbol{D}_\infty$ (see Note 29.7). The projectors in the successive pairs $(\boldsymbol{E},\boldsymbol{E}')$, $(\boldsymbol{F},\boldsymbol{F}')$, $\ldots$, $(\boldsymbol{K},\boldsymbol{K}')$, $(\boldsymbol{L},\boldsymbol{L}')$ belong, respectively, to the given succession of projector sets. The condition of consistency demands that the real part of X vanish whenever any of the pairs $(\boldsymbol{E},\boldsymbol{E}')$, $(\boldsymbol{F},\boldsymbol{F}')$, $\ldots$, $(\boldsymbol{K},\boldsymbol{K}')$, $(\boldsymbol{L},\boldsymbol{L}')$, is unequal. This is strictly the case only when the Schrödinger part of the evolution has been ignored (i.e. we take $\mathcal{H} = 0$), but a non-trivial Schrödinger evolution can be re-instated by introducing this evolution appropriately between the applications of the projectors. This 'consistency condition' on coarse-grained histories can be interpreted as the condition of 'no interference' between the histories being compared.

29.10. In fact, I have not located a clear statement of *any* actually intended '(d)-ontology' in the consistent-histories literature. What I am presenting here is merely my own attempt at coming to grips with this issue, based on extended discussions with Jim Hartle and, more particularly, some helpful correspondence with Fay Dowker. It is likely that, despite my efforts, I am still not adequately presenting an underlying intended ontology of the '(d)' community.

29.11. See Bohm and Hiley (1995); Valentini (2002). Antony Valentini also has a book on de Broglie-Bohm theory in the works, which we hope will see press soon!

29.12. See Károlyházy (1974); Frenkel (2000); Ghirardi *et al.* (1986); Ghirardi *et al.* (1990); Komar (1964); Pearle (1985); Pearle and Squires (1995); Kibble (1981); Weinberg (1989); Diósi (1984, 1989); Percival (1994, 1995); Gisin (1989, 1990); Penrose (1986a, 1989, 1996a, 2000a), Leggett (2002)—in no particular order.

Section 29.3

29.13. For a continuous probability distribution, we need a non-negative real-valued function f on $\mathcal{P}$ which 'integrates to 1'. The space $\mathcal{P}$ would have a natural volume form—the $2N$-form $\boldsymbol{\Sigma}$ of §20.4, which featured in Liouville's theorem—so that $f\boldsymbol{\Sigma}$ can legitimately be integrated over $\mathcal{P}$, our required condition being actually $\int f\boldsymbol{\Sigma} = 1$.

29.14. See Brody and Hughston (1998b). Nielsen and Chuang (2000) provide good conceptual coverage of the density matrix in action.

Section 29.4

29.15. For an n-state system, with $n > 2$, the picture is more complicated. Only part of the boundary of the $(n^2 - 1)$-dimensional space of density matrices is the space of pure states, this part being a complex projective $(n - 1)$-space $\mathbb{CP}^{n-1}$ (see §21.9 and §22.9).

Section 29.5

29.16. The reader may be wondering how the notion of *quanglement*, introduced in §23.10, might affect these ontological issues. This is an intriguing question, and it may well be that the whole issue of 'ontology', in a quantum context, will ultimately have to be viewed in a new light. But for the moment let us simply adopt a more 'common-sense' attitude to reality, in which the issues raised by relativity will not be entered into.

29.17. Nielsen and Chuang (2000) discuss this point; see also Hughston *et al.* (1993).

Section 29.6

29.18. The idea is originally due to Wheeler (like many things); see Ng (2004) for a modern perspective.

29.19. See Hawking (1976b); Preskill (1992); see, also, §30.14.

Section 29.7

29.20. I am not sure whether this viewpoint represented Wigner's actual position with regard to quantum measurement, which may, after all, have evolved during his life. I should also point out that my *own* position fundamentally differs from those, like the one referred to here, which assert that it is consciousness that reduces the state. (In this regard, my view has sometimes been misrepresented by other commentators.) See §§30.9–12.

Section 29.8

29.21. The Schmidt (or *polar*) decomposition of a general entangled state $|\Psi\rangle$ belonging to $\mathbf{H}^2 \times \mathbf{H}^2$, expresses it (essentially uniquely) as $|\Psi\rangle = \lambda|\alpha\rangle|\beta\rangle + \mu|\rho\rangle|\sigma\rangle$, where $|\alpha\rangle$ and $|\rho\rangle$, belonging to the first $\mathbf{H}^2$, are orthogonal (normalized eigenstates of its density matrix), and $|\beta\rangle$ and $|\sigma\rangle$ likewise correspond to the second $\mathbf{H}^2$. Here, $\bar{\lambda}\lambda$ and $\bar{\mu}\mu$ are density-matrix eigenvalues. A similar expression holds for $\mathbf{H}^n \times \mathbf{H}^n$, where the sum in $|\Psi\rangle$ has n terms. See Nielsen and Chuang (2000).

29.22. See Nielsen and Chuang (2000), which is after all about quantum information theory!

29.23. See Page (1995) for discussion of these issues.

29.24. See Gell-Mann (1994); Hartle (2004) for a ten-year slice of such thoughts in the context of 'consistent histories' (d).

Section 29.9

29.25. See Dowker and Kent (1996).

29.26. A striking example due to Adrian Kent shows clearly how far from sufficient the 'consistency' condition is for providing a physically plausible picture of 'reality'. In this example, a particle p can lie in one of three boxes A, B, C, as described by the respective normalized orthogonal states $|A\rangle$, $|B\rangle$, and $|C\rangle$. Take the Hamiltonian to be zero, giving a constant unitary evolution. The initial state is to be $|A\rangle + |B\rangle + |C\rangle$ and suppose that the final state is measured to be $|A\rangle + |B\rangle - |C\rangle$. (This is possible because $|A\rangle + |B\rangle + |C\rangle$ and $|A\rangle + |B\rangle - |C\rangle$ are not orthogonal.) The insertion of the projector set

$\{|A\rangle\langle A|, I - |A\rangle\langle A|\}$ between the two turns out to be 'consistent', and we seem to conclude that p must lie in the box A at this intermediate stage (basically because $|B\rangle + |C\rangle$ and $|B\rangle - |C\rangle$ *are* orthogonal). The same argument, with B in place of A, gives the equal conclusion that p must lie in B at the intermediate stage! This example appears to have been evolved from Yakir Aharonov's 'King problem', See Albert *et al.* (1985), p.5.

30
Gravity's role in quantum state reduction

30.1 Is today's quantum theory here to stay?

In this chapter, I shall put the case to the reader that there are powerful positive reasons, over and above the negative ones put forward in the preceding chapter, to believe that the laws of present-day quantum mechanics are in need of a fundamental (though presumably subtle) change. These reasons come from within accepted physical principles and from observed facts about the universe. Yet, I find it remarkable how few of today's quantum physicists are prepared to entertain seriously the idea of an actual change in the ground rules of their subject. Quantum mechanics, despite its extraordinary exception-free experimental support and strikingly confirmed predictions, is a comparatively young subject, being only about three-quarters of a century old (dating this from the establishment of the mathematical theory by Dirac and others, based on the schemes of Heisenberg and Schrödinger, in the years immediately following 1925). When I say 'comparatively', I am comparing the theory with that of Newton, which lasted for nearly three times as long before it needed serious modification in the form of special and then general relativity, and quantum mechanics. Even if we are to count Newton's theory as suffering its first modification with the introduction of Maxwellian fields, this still gave it an exception-free reign of over a century and three-quarters!

Moreover, Newton's theory did not have a *measurement paradox*. While the linearity of quantum theory's **U** process gives that theory a particular elegance, it is that very linearity (or unitarity) which leads us directly into the measurement paradox (§22.2). Is it so unreasonable to believe that this linearity might be an approximation to some more precise (but subtle) non-linearity?

We have a clear precedent. Newton's gravitational theory has the particular mathematical elegance that the gravitational forces always add up in a completely linear fashion; yet this is supplanted, in Einstein's more precise theory, by a distinctly subtle type of non-linearity in the way that gravitational effects of different bodies combine together. And Einstein's theory is certainly not short on elegance—of a quite different kind from that of Newton. We also see, in Einstein's theory, that the modifications to Newton's theory that were needed were nothing like the 'tinkering' that I referred to in §29.2. At various times, such tinkerings with Newton's theory had indeed been suggested, such as a replacement of the power 2 in Newton's inverse square formula GmM/r^2 (see §17.3) by 2.000 000 16, as suggested by Aspeth Hall in 1894, in order to accommodate those very slight deviations, as ascertained in 1843, from the Newtonian predictions of Mercury's motion about the Sun (and Hall's suggestion gets good fits also for the other planets, as shown by Simon Newcombe).[1] Einstein's theory subsequently explained these deviations, without fuss, but the new theory was by no means obtained just by tinkering with the old; it involved a completely radical change in perspective. This, it seems to me, is the general kind of change in the structure of quantum mechanics that we must look towards, if we are to obtain the (in my view) needed non-linear theory to replace the present-day conventional quantum theory.

Indeed, it is my own perspective that Einstein's general relativity will itself supply some necessary clues as to the modifications that are required. The 20th century gave us *two* fundamental revolutions in physical thought—and, to my way of thinking, general relativity has provided as impressive a revolution as has quantum theory (or quantum field theory). Yet, these two great schemes for the world are based upon principles that lie most uncomfortably with each other. The usual perspective, with regard to the proposed marriage between these theories, is that one of them, namely general relativity, must submit itself to the will of the other. There appears to be the common view that the rules of quantum field theory are immutable, and it is Einstein's theory that must bend itself appropriately to fit into the standard quantum mould. Few would suggest that the quantum rules must themselves admit to modification, in order to ensure an appropriately harmonious marriage. Indeed, the very name 'quantum gravity', that is normally assigned to the proposed union, carries the implicit connotation that it is a standard *quantum* (field) theory that is sought. Yet, I would claim that there is observational evidence that Nature's view of this union is very different from this! I contend that her design for this union must be what, in our eyes, would be a distinctly non-standard one, and that an objective state reduction must be one of its important features.

30.2 Clues from cosmological time asymmetry

What evidence is this? Let us first turn to those places where Nature's choice of quantum-gravity union most clearly reveals itself. I refer to the spacetime *singularities* of the Big Bang and of black holes (and also the Big Crunch, if such is to take place). In Chapter 27, the extraordinarily special nature of the Big Bang was presented, in stark contrast to the seemingly 'generic' nature of the singularities of collapse. Despite the brave suggestions made in accordance with the Hartle–Hawking proposal (as discussed in §28.9) I see no escape from a gross time-asymmetry being a necessary feature of Nature's quantum-gravity union.

Such a temporal asymmetry would seem to be completely at variance with the implications of any standard quantum field theory. Let us consider, for example, the CPT theorem, noted in §25.4. (Recall that 'T' stands for time reversal, whilst 'P' and 'C' stand respectively for space reflection reversal and for the replacement of particles by their antiparticles.) If we believe that the CPT theorem applies to our sought-for quantum-gravity union, then we are in trouble. If we apply CPT to any allowed 'generic' final singularity of gravitational collapse, then we get an initial-type singularity as a possibility for the Big Bang (or for part of the Big Bang). Recall the enormity of the available phase space, as described in §27.13 (and graphically illustrated in Fig. 27.21). Once such 'generic' initial singularities become *allowable*, then there is nothing to guide the Creator's pin into that absurdly (and, from the 'anthropic' perspective, §28.6, unnecessarily) tiny region $\mathcal{B}$, that seems to have been the actual starting point of our universe. It seems to me to be clear that the mystery of the extraordinarily special nature of the Big Bang cannot be resolved within the standard framework of quantum field theory.

At least this would be the case for any theory for which the word 'standard' entails the validity of the CPT theorem (§25.4). Strictly speaking, that theorem is not immediately applicable to a theory that fully respects the curved-spacetime basis of Einstein's general relativity. One of the premises of the CPT theorem is that the background spacetime is flat Minkowski space. Nevertheless, I suspect that most physicists would regard this as an unimportant 'technicality', taking the view that one can re-express Einstein's theory, if it is desired, in the form of a 'Poincaré-invariant field theory' by introducing a Minkowski background as a convenience. Personally, I have strong reservations about this type of procedure;[2] yet I would tend to agree that it seems unlikely that the completely time-symmetrical classical Einstein theory of general relativity should become so time-*a*symmetrical when submitted to the standard time-symmetrical procedures of quantum field theory.

On the other hand, we recall that, in §25.5, §§26.5,11, we encountered situations where a symmetry of the classical theory becomes broken when we pass to the quantum theory. Might it be that it is this that happens, when Einstein's theory becomes brought, appropriately, within the compass of standard QFT rules? I suppose that this is conceivable, but it is hard to see how this could be very much like the type of symmetry breaking that occurs in, say, electroweak theory, where the 'vacuum state' $|\Theta\rangle$ is taken not to share the symmetries of the quantum dynamics.

If this idea is to work, then $|\Theta\rangle$ has to be 'time-asymmetrical'. I am not sure how one could make sense of this kind of idea. It is true that the ket $|\Theta\rangle$ is what would be placed over on the right-hand side of all field operators, in the manner described in §26.11, and could be thought of as representing the initial state of the universe, which here means the very particular Big Bang state. But in standard QFT, the complex conjugate of $|\Theta\rangle$, namely the bra $\langle\Theta|$, would also feature in the formalism, being needed for the formulation of probabilities via expressions like $\langle\Theta|A|\Theta\rangle$, and it would play a completely symmetrical role to $|\Theta\rangle$, but with time reversed. Thus, $\langle\Theta|$ would have to represent the final state of the universe, and we have a final state of a similar structure to the initial one, in gross contradiction with the entire message of Chapter 27.

There are also other features that arise in the process of 'quantization', whereby the quantum theory might not share the symmetries of the classical theory, known as *anomalies*. These come about when the classical commutation rules, providing the classical symmetry (given by Poisson brackets—see §14.8) cannot be fully realized by quantum commutators, with only a subgroup of the whole classical symmetry group surviving in the quantum theory. Anomalies seem normally to be regarded as things to be avoided (and we shall see the contortions that theorists sometimes have to perform in order to eliminate such things when we come to consider string theory in the next chapter). Yet, one might imagine taking a different view, and regard an anomaly as being a 'good' thing, in those circumstances when the larger symmetry is something that one does not want to have. However, in our present case it is a discrete symmetry, namely CPT, in addition to T, CT, and PT—indeed anything with a 'T' in it—that one needs to violate, and it is hard to see the relevance of the usual anomaly idea, which usually (but not always) refers just to the continuous symmetries that can be realized in terms of Poisson brackets.

However one looks at it, it is hard to avoid the conclusion that, in those extreme circumstances where quantum effects and gravitational effects must both (have) come together—in the spacetime singularities at the Big Bang and in gravitational collapse—gravity just behaves *differently* from other fields. Recall the final conclusion in the penultimate paragraph of Chapter 27, concerning this point. For whatever reason, Nature has imposed a gross temporal asymmetry on the behaviour of gravity in such extreme circumstances.

30.3 Time-asymmetry in quantum state reduction

Does this relate to any other clues concerning the possible interrelation between gravity and quantum mechanics? I strongly believe so. Whereas we perceive no time-asymmetry in the **U** part of quantum theory (§27.1),

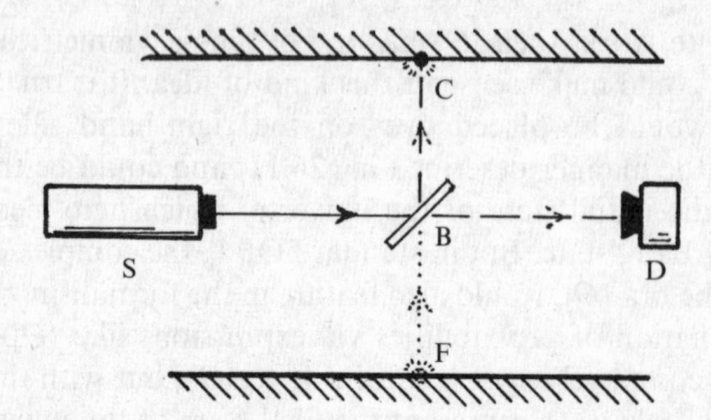

Fig. 30.1 A source S randomly emits single high-energy photons (each such event being recorded) aimed at a beam-splitter B, tilted at 45° to the beam. If transmitted through B, the photon activates a detector D (route SBD); if reflected, it is absorbed at the ceiling C (route SBC). The quantum squared-modulus rule correctly predicts probabilities $\frac{1}{2}$, $\frac{1}{2}$. On the other hand, given that D registers, the photon could have come from S (route SBD) or from the floor F (route FBD). Used in the reversed-time direction, the squared-modulus rule incorrectly retrodicts probabilities $\frac{1}{2}$, $\frac{1}{2}$, which should be 1, 0.

there *is* an essential time-asymmetry in **R**. We can see this very easily in a simple hypothetical quantum experiment. Suppose that there is a photon source S which emits single photons from time to time, and that, whenever it does so, this event is recorded.[3] I shall suppose that the photons have high energy, being possibly even X-ray photons. The photons are aimed at a beam splitter B ('half-silvered mirror') angled at 45° to the beam, so that if a photon is transmitted through, then it activates a detector D at the other side, while if it is reflected, then it gets absorbed into the ceiling C (see Fig. 30.1.) I am supposing equal amplitudes for these two alternatives, so that the detector will register reception of the photon in just one-half of the occasions that the source is registered as having emitted the photon.

This is just a straightforward application of the **R** procedure. There is an amplitude $\frac{1}{\sqrt{2}}$ (ignoring possible phase factors) for the photon history SBD and an amplitude $\frac{1}{\sqrt{2}}$ for the photon history SBC. Application of **R**'s squared-modulus rule then gives the (correct) answer that whenever there is an emission event at S, there is a 50% probability of a detection event at D and (by inference) a 50% probability of the photon reaching C. This is simply the correct answer.

But now let us imagine reading this particular experiment backwards in time. I am not proposing that we try to build a 'backward-time' source or detector. No, the physical processes are not to be altered in any way. It is just that I propose to phrase my *questions* about them in a reverse-time form. Rather than asking about the *final* probabilities, let us ask what the *initial* probabilities are, given that there is a detection event at D. The relevant amplitudes now refer to the two alternative histories SBD and FBD, where

F stands for a point on the floor with the property that if a photon were to be emitted from there it could be reflected at B to be received at D. Again the amplitudes are $\frac{1}{\sqrt{2}}$ for each of these two histories (ignoring phases). This must be so because the ratio of (the modulus of) the amplitudes for going one way or the other is just a property of the beam splitter. There is no time-asymmetry here. Now, if we were to apply the 'squared-modulus rule' to get the probabilities for these two alternatives, we would find a probability of 50% for emission at S and 50% (by inference) for the photon to come from the floor F, whenever there is a detection event at D.

This, of course, is an absurdity. There is virtually a zero chance that an X-ray photon will jump out of the floor, aimed at the beam splitter. The probabilities are more like 100% that there was an emission event at S and 0% that the photon came from the floor F, whenever there is a detection event at D. The squared-modulus rule, applied in the past direction, has simply given us completely the wrong answer![4]

Of course, this rule was not designed to be applied into the past, but it is instructive to see how completely wrong it would be to do so. Sometimes people have objected to this deduction, pointing out that I have failed to take into account all sorts of particular circumstances that pertain to my time-reversed description, such as the fact that the Second Law of thermodynamics only works one way in time, or the fact that the temperature of the floor is much lower than that of the source, etc. But the wonderful feature of the quantum-mechanical squared-modulus law is that we never have to worry about what the particular circumstances might be! The miracle is that the quantum probabilities for future predictions arising in the measurement process do not seem to depend at all on considerations of particular temperatures or geometries or anything.[5] If we know the amplitudes, then we can work out the future probabilities. All we need to know are the amplitudes. The situation is completely different for the probabilities for retrodiction. Then we do need to know all sorts of detailed things about the circumstances. The amplitudes alone are quite insufficient for computing past probabilities.

There are, however, situations in which the quantum probabilities can be computed in a way that is completely symmetrical in time, and it is perhaps instructive to have a look at these. These occur where the quantum state is measured to be something known both before and after some intermediate quantum measurement. To be more explicit, imagine a succession of three measurements, where the first one projects the state into $|\psi\rangle$ and the third one projects it into $|\phi\rangle$, there being a YES/NO measurement between these two, described by the projector $\boldsymbol{E}$ (§22.6). The probability of YES for the middle measurement is then given by[30.1]

[30.1] Why? Can you derive this formula?

$$|\langle \phi | \boldsymbol{E} | \psi \rangle|^2$$

(where we assume the normalizations $\langle \psi | \psi \rangle = 1 = \langle \phi | \phi \rangle$), which is certainly time-symmetrical. (To set up such a situation, one performs the succession of three measurements many many times over and picks out, for examination, only those cases for which the first measurement yields $|\psi\rangle$ and the third yields $|\phi\rangle$. The above probability then refers to the fraction of these cases for which the middle measurement yielded YES.)[6] This has led some people to conclude that there is, at root, no time-asymmetry in quantum measurement.[7]

However most quantum measurements are not of this kind. For the normal forward-time use of the squared-modulus rule, we do not specify a $|\phi\rangle$, and for the above backward-time attempted use, we do not specify $|\psi\rangle$. We see that we can perfectly well calculate the quantum probabilities while not specifying $|\phi\rangle$, but we cannot get away with not specifying $|\psi\rangle$. One might take the view that the reason that the quantum rules work well for the future probabilities has to do with $|\phi\rangle$ being, in some sense 'random', which has to do with the Second Law of thermodynamics. Perhaps there is something in this, but I find this requirement for $|\phi\rangle$ rather unclear. What does 'random' mean in this context? Nevertheless there would certainly seem to be some connection with the Second Law in the measurement question. We may take note of the fact that actual measuring devices tend to take advantage of this law in some part of their operation. That there is some connection between **R** and the Second Law is, indeed, part of my own perspective on the matter. And since we have seen that the Second Law is intimately tied up with the missing quantum/gravity union, we must expect an intimate relation between **R** and this anticipated union also.

Before coming to this issue more explicitly, it is worth pointing out that the other aspect of **R**, namely the 'jumping' of the quantum state—as opposed to the calculation of probabilities via the squared-modulus rule—can (apparently) be equally well be phrased according to a backward-time perspective as to a forward-time one. This is schematically illustrated in Fig. 30.2a,b, where in Fig. 30.2a I have depicted the 'normal' view of the ..., **U**, **R**, **U**, **R**, **U**,... alternation (see Fig. 22.1), where the state is an eigenstate of the measurement *after* it has taken place, and where in Fig. 30.2b. I have depicted the 'time-reversed' view, where the state is an eigenstate just *before* the measurement. The calculation of amplitudes comes out the same whichever viewpoint one adopts,[30.2] but the time-reversed one has a 'teleological' aspect to it that some may find disturbing. There is also a standpoint, the 'transactional' interpretation, (due, inde-

[30.2] Explain why this is basically an expression of the 'unitary' nature of **U**; see §22.4.

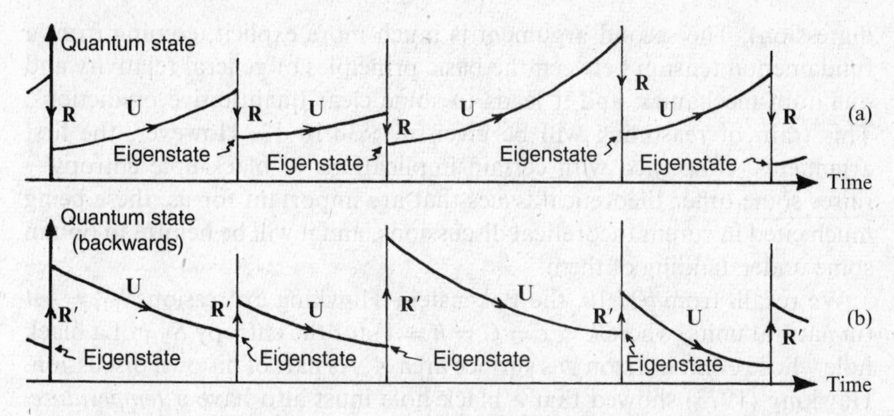

Fig. 30.2 Schematic illustration of the alternation ...,**U**, **R**, **U**, **R**, **U**,... of the two processes **U** and **R** as used in practice in quantum mechanics (compare Fig. 22.1), according to: (a) standard time-direction of evolution, where operator eigenstates occur at the past end of each stretch of **U**-evolution, and (b) time-reversed viewpoint of evolution, where operator eigenstates occur at the future end of each stretch of **U**-evolution. In the 'transactional' interpretation of quantum mechanics, there are two state vectors, one evolving according to (a) and the other according to (b).

pendently to various quantum theorists[8]) according to which both pictures are entertained simultaneously, and at any one time there are two simultaneously unitarily evolving state vectors describing the quantum system, one looking like Fig. 30.2a and the other, like Fig. 30.2b. This is held to have advantages with respect to interpreting the EPR phenomena of Chapter 23. To my own mind, this description is a little excessive, and it may be better to adopt a *quanglement* perspective in which the time direction of 'propagation' of the state is not important, quanglement simply providing connections between the states at different times (§23.10).

30.4 Hawking's black-hole temperature

Are there any ways of connecting **R** with the sought-for (time-asymmetric) quantum–gravity union, that are more direct than just the fact of temporal asymmetry in **R**? In my opinion there are, and I shall describe two such connections. The first of these follows on from the discussion of the previous section and has to do with the remarkable phenomenon of 'black-hole evaporation'. The argument is partly suggestive and certainly incomplete; moreover it is controversial in certain central respects. The ingredients of this discussion will be the subject of this section, and of following sections up until §30.9, (excluding §§30.5,6, which may be regarded as something of a digression). The second argument is much more explicit, coming from a fundamental tension between the basic principles of general relativity and quantum mechanics, and it leads to some clear quantitative predictions. This train of reasoning will be given in §§30.10–13. However, the first argument—concerned with certain implications of black-hole entropy—raises some other theoretical issues that are important for us, these being much cited in curent theoretical discussions, and it will be helpful to obtain some understanding of them.

We recall, from §27.10, the Bekenstein–Hawking expression $S_{\mathrm{BH}} = \frac{1}{4}A$ (in natural units, where $k = c = G = \hbar = 1$) for the entropy S_{BH} of a black hole whose event horizon has surface area A. As part of his own discussion, Hawking (1973) showed that a black hole must also have a *temperature*, which turns out to be proportional to what is called the 'surface gravity' of the hole. For a stationary rotating hole (Kerr geometry; see §27.10), we find

$$T_{\mathrm{BH}} = \frac{1}{4\pi m[1 + (1 - a^2/m^2)^{-\frac{1}{2}}]},$$

where, as in §27.10, m is the black hole's mass and am is its angular momentum. This temperature can be obtained from a standard formula of thermodynamics:

$$T\,\mathrm{d}S = \mathrm{d}E,$$

where, in varying the energy E, we hold the conserved angular momentum constant.[30.3] Accordingly, the black hole will emit photons, as though it were a physical object in thermal equilibrium, radiating energy with the characteristic 'black-body' (Planckian) spectrum, described in §21.4 (see Fig. 21.3b), for the temperature T_{BH}. It may be noted that although the Bekenstein–Hawking entropy of a black hole is enormous, and gives rise to the extraordinary figures discussed in §27.13, the Hawking temperature is absurdly tiny for black holes of a plausible size. For a black hole of a solar mass, for example, the Hawking temperature is only about 10^{-7} K, which is not much greater than the lowest man-made temperatures here on Earth (about 10^{-9} K).

Jacob Bekenstein (1972) had derived the black-hole entropy expression some years earlier, using a physical argument (based on applying the Second Law of thermodynamics to situations in which quantum particles are slowly lowered into the hole), but he had not obtained a clear value for the '$\frac{1}{4}$' that now appears in that expression, nor had he obtained a black-hole temperature. Stephen Hawking first supplied the temperature and the '$\frac{1}{4}$' in the entropy formula by employing techniques of QFT in a curved spacetime background. Here, the background describes a black hole that

[30.3] Obtain this formula for T_{BH}, assuming the expression for the area of a Kerr hole's horizon given in §27.10.

has come about from the collapse of some material (say a star) in the remote past. The situation is described by the conformal diagram of Fig. 27.16c (this being strict if the collapse is spherically symmetrical).

In my opinion, Hawking's remarkable calculation of the entropy and temperature of a black hole (together with the related 'Unruh effect'[9]) is the only reasonably reliable conclusion that has been obtained, to date, from any quantum-gravity theory. Even Hawking's conclusions were not strictly of quantum gravity but, rather, obtained from considerations of QFT in a curved background spacetime. In general, there are severe problems arising when one attempts to formulate quantum theory within a curved background, and it is striking that Hawking was nevertheless able to come to some firm conclusions.

One of the most crucial problems is to find an appropriate notion of 'positive frequency' in a curved background. As we saw in §24.3 and §26.2, this notion is a key ingredient of the standard view of quantum particles and QFT. The problem of formulating this issue in a general curved spacetime lies in the absence of a naturally defined 'time parameter' in terms of which the notion of 'positive frequency' can be formulated.

The alert reader might well point out that there is no naturally defined time parameter in flat Minkowski space either! However, a striking fact comes to our aid here, to tell us that, for solutions of relativistic wave equations (like those studied in Chapters 19, 24–26), positive frequency in one choice of Minkowski time parameter t is equivalent to positive frequency in any other such parameter—for which the temporal orientation is not reversed. For massless fields, one can even go further, and obtain the same positive-frequency condition by use of a 'time parameter' that is obtained from the standard Minkowski time parameter by a time-orientation-preserving conformal transformation (which is of relevance to twistor theory; see §§33.3,10).[10]

In a general spacetime, there is no natural analogue of such a parameter, and the positive-frequency notion would certainly come out differently for different choices of time parameter, in general. Apart from the case of the Hawking temperature, the most plausible results come from considerations of *stationary* spacetimes, for which there is a continuous family of time-displacements which preserve the spacetime geometry (see Fig. 30.3). Such spacetime motions are generated by a *timelike Killing vector* $\boldsymbol{\kappa}$ (see §14.7 and §30.6). The curves along which the vectors $\boldsymbol{\kappa}$ point (integral curves of $\boldsymbol{\kappa}$) are the curves along which a reasonably natural 'time parameter' t may be specified, so that

$$\boldsymbol{\kappa} = \frac{\partial}{\partial t},$$

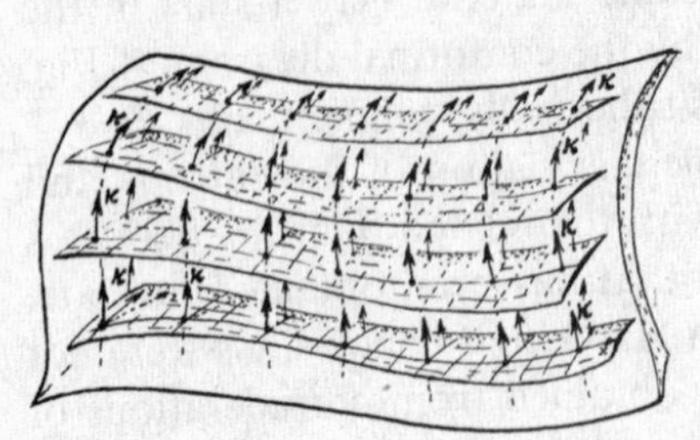

Fig. 30.3 Stationarity of a spacetime is expressed as the presence of a timelike Killing vector $\boldsymbol{\kappa}$. This generates a continuous family of time-displacements preserving the metric. If $\boldsymbol{\kappa} = \partial/\partial t$, where t is the 'time parameter' of a coordinate system (t,x,y,z), then x, y, and z must be constant along the integral curves of $\boldsymbol{\kappa}$. (See §14.7.)

where the three remaining coordinates x, y, z are taken to be constant along the curves. A notion of 'positive frequency' may then be defined with respect to this parameter.

A curious situation may arise when there is more than one timelike Killing vector, since then there may be more than one notion of 'positive frequency'. This multiplicity of timelike Killing vectors occurs with Minkowski space $\mathbb{M}$, of course, but from what has been said above, the notions of positive frequency agree when we pass from one Minkowskian inertial frame to another. This is not the case, however, when we pass from an inertial frame to an accelerating frame. Then we obtain a distinct notion of 'positive frequency', and the resulting QFT turns out to be in what is called a *thermal vacuum*, according to which an accelerating observer experiences a non-zero temperature—although absurdly low for any reasonable acceleration.

It should be made clear that, although this is a surprising effect, this 'acceleration temperature' is just the ordinary kind of temperature, as would be measured by an ordinary (though idealized) thermometer. In this case, the thermometer would be undergoing uniform acceleration, and it is taken to be in an ambient vacuum which would be measured to have zero temperature by an unaccelerated thermometer. (This notion of 'thermal vacuum' has connections with the QFT notions of 'alternative vacua' in §§26.5,11 and 'false vacuum' in §§28.1,4.) This is referred to as the 'Unruh effect', and it is consistent with Hawking's thermal state of a black hole. An observer being held stationary near a very large black hole would, according to the principle of equivalence (§17.4), experience an effective acceleration, and the Unruh temperature of this acceleration agrees with Hawking's temperature, as obtained by his own procedures.

The difficulty presented, in general, by the lack of a natural definition of 'positive frequency' is circumvented in some approaches, by abandoning the notion of 'particle' and concentrating on the algebra of quantum-mechanical operators.[11] At first sight, this may appear to be too great a sacrifice, and some ingenuity is indeed needed in the formulation of many questions of interest in such an approach. At the time of writing, I have not myself been able to assess the full merits of this intriguing and promising-looking type of

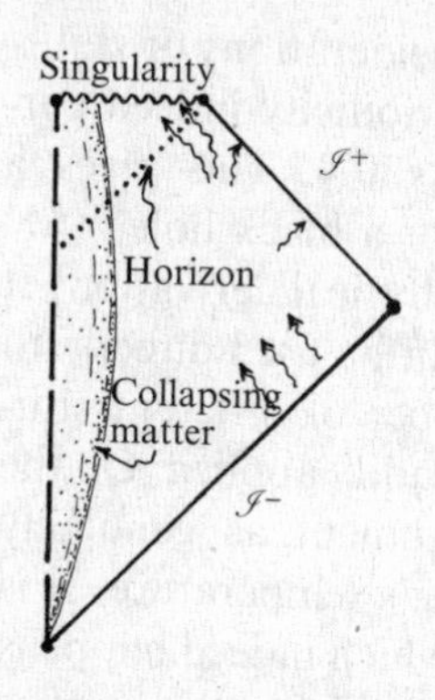

Fig. 30.4 Hawking's calculation of a black hole's temperature, involving the collapse of some matter to a black hole in the distant past, requires only the (standard) notion of positive/negative frequency splitting on $\mathscr{I}^+$ and on $\mathscr{I}^-$. The black hole's vacuum becomes a thermal state (a density matrix) because the initial information on $\mathscr{I}^-$ gets divided between that at $\mathscr{I}^+$ and that at the final singularity, the latter becoming lost.

theory. I suspect that it is superior to approaches based on specific timelike vector fields, however. In any case, I do not myself see why QFT in a fixed background should necessarily make complete physical sense. It is merely an approximation to a more exact scheme in which the degees of freedom in the gravitational field—i.e. in the geometry of the spacetime itself—must also take part in the quantum physics.

In Hawking's calculation of a black hole's temperature and entropy, he manages to avoid most of these issues by requiring a notion of positive/negative frequency splitting only at infinity. There is one such notion at $\mathscr{I}^-$ (past null infinity) and a different one at $\mathscr{I}^+$ (see §27.12). This difference leads to the production of Hawking's 'thermal state' by the black hole, which results in what is referred to as its *Hawking radiation*. It is noteworthy that this effect owes its existence to the fact that some of the information defined at $\mathscr{I}^-$ gets *lost* in the singularity, and does not all make it out to $\mathscr{I}^+$ (see Fig. 30.4). We shall be seeing the significance of this fact in §30.8.

30.5 Black-hole temperature from complex periodicity

At this point, it is instructive to consider an ingenious later derivation of the Hawking temperature, obtained by Gibbons and Perry in 1976, although it is a slight digression from the main lines of reasoning of this chapter. (These I shall resume in §30.8.) The Gibbons–Perry argument raises some interesting issues concerning the role of elegant mathematical ideas in the derivation of genuine physical phenomena. What they noticed was that, if the solution of the Einstein equation (namely the Schwarzschild or Kerr solution—see §§27.8,10) that represents the black hole in its final settled state is 'complexified' (i.e. extended from real to complex values of the coordinates—see §18.1), then a basic regularity condition on quantities defined on this complexified space implies that these quantities necessarily acquire a periodicity (see §9.1, Fig. 9.1a) in the complexified time, with purely imaginary period $2\pi i T_{BH}$. Considerations of statistical thermodynamics tell us that such a complex periodicity indeed corresponds to the exact temperature T_{BH}, given in §30.4. This gives a strikingly direct route to the Hawking temperature for a black hole.

But what are we to make of such a procedure as a physical derivation? It is certainly a remarkably elegant argument, and it can be used directly to obtain Hawking's black-hole temperature in a number of different situations to which his original discussion could not be readily applied. On the other hand, I have a difficulty in taking this argument as genuinely providing an actual physical justification of Hawking's temperature. It is a good example of a beautiful piece of mathematics, which indeed happens to give the correct answer (its 'correctness' being judged from its agreement with the answer obtained from more physically acceptable criteria—here, Hawking's original argument referred to above), despite the fact that some of the 'physical' assumptions going into the mathematics of this new argument may be judged as of dubious validity.

Let us look at the mathematical ingredients of this 'complexification' a little more closely. A good way to understand what is involved is to think first of the ordinary 2-dimensional Euclidean plane $\mathbb{E}^2$, and its standard complexification to $\mathbb{CE}^2 (= \mathbb{C}^2)$. The real space $\mathbb{E}^2$ is sometimes called a *real section* of $\mathbb{CE}^2$ (see Fig. 30.5a,b; see also Fig. 18.2). It is a Euclidean real section, because it possesses an ordinary Euclidean metric. But $\mathbb{CE}^2$ also has *Lorentzian* real sections (see Fig. 18.2, §18.2), and we can construct one of these, $\mathbb{M}^2$, by taking the coordinate y of a standard pair of Cartesian coordinates (x, y) for $\mathbb{E}^2$ (which are real-number parameters) and demanding that y take purely imaginary values rather than real ones. Then $t = iy$ serves as a time coordinate for $\mathbb{M}^2$. (This is just the 2-dimensional case of what we already did in §18.1.) Now let us consider *polar* coordinates (r, θ) for $\mathbb{E}^2$, rather than the Cartesian coordinates (x, y); see §5.1 and Fig. 30.5a. The non-negative real number r measures distance from the origin and the real-number angle θ gives the angle between the radius vector and the x-axis, measured in the anticlockwise sense. How do these coordinates extend to our Lorentzian section $\mathbb{M}^2$? Provided that we restrict attention to the right-hand quadrant $\mathbb{M}^R$, as indicated in Fig. 30.5b, the quantity r is still real and non-negative, but θ is now purely imaginary, so

$$\tau = i\theta$$

is real. The coordinate r now measures the Lorentzian spatial distance from the origin, and τ is the 'hyperbolic angle from the horizontal'.[30.4]

[30.4] Write these coordinates (r, τ) in terms of the Lorentzian Cartesian coordinates (x, t); see why the real part of θ indeed vanishes on $\mathbb{M}^2$.

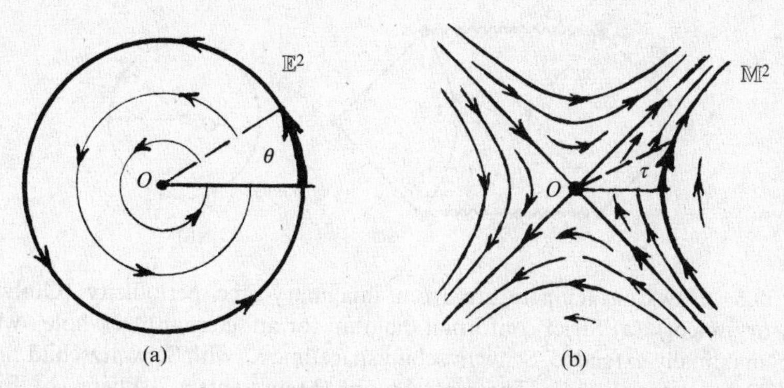

Fig. 30.5 Periodicity in imaginary time, illustrated by Minkowski 2-space $\mathbb{M}^2$, complexified to $\mathbb{CM}^2 = \mathbb{CE}^2$. (a) The Euclidean plane $\mathbb{E}^2$ is one real section of the complex space $\mathbb{CM}^2$. The Killing vector $\partial/\partial\theta$ generates rotations in $\mathbb{E}^2$, where we take polar coordinates (r, θ). Any function that is single-valued on $\mathbb{E}^2$ must be periodic in θ with period 2π. (b) In the Lorentzian real section $\mathbb{M}^2$ of $\mathbb{CM}^2$, the 'Rindler' (uniform-acceleration) time coordinate is $\tau = \mathrm{i}\theta$ (an analogue of the Schwarzschild time that is natural for a black hole), and a function that is analytic at the origin O must have an imaginary period in τ, with period $2\pi\mathrm{i}$. (Killing vector $\boldsymbol{\kappa} = \partial/\partial\tau = -\mathrm{i}\partial/\partial\theta$.)

We are going to interpret the coordinate τ (multiplied by a constant r_0) as measuring 'time' in the ordinary spacetime sense, for an observer (in this 2-dimensional flat spacetime geometry) who 'uniformly accelerates' away from the centre, and whose world line is given by $r = r_0$. ('Rindler coordinates'[12]). The spacetime itself is taken to be the entire $\mathbb{M}^2$, despite the fact that the observer's 'time' applies only to the quadrant $\mathbb{M}^R$. Let us suppose that the observer is concerned with analytic quantities defined on $\mathbb{M}^2$. Such quantities will have a holomorphic extension to the complexification of the spacetime (see §7.4 and §12.9), but this can only be guaranteed for some immediate neighbourhood of the 'real section', which, in this case, means the Lorentzian section. However, if such a quantity is actually analytic at the origin O of this section, then it must also be analytic at the origin of the Euclidean section (since this is exactly the same point O in each case). But a quantity that is smooth at the Euclidean origin must be periodic in θ, with period 2π, since if we increase θ by 2π (at some small radial value $r = \varepsilon$), we simply wind once around the origin to arrive at precisely the same point that we started at. Hence, the quantity τ, referred to the original Lorentzian spacetime coordinates, has an *imaginary period* of $2\pi\mathrm{i}$ in τ (complexified).

This is the basis of the Gibbons–Perry argument, where we now apply it to a full 4-space black-hole geometry rather than to our simplified 2-dimensional spacetime $\mathbb{M}^2$. The relevant geometry is now that of Schwarschild (for the non-spinning spherically symmetrical case, but we

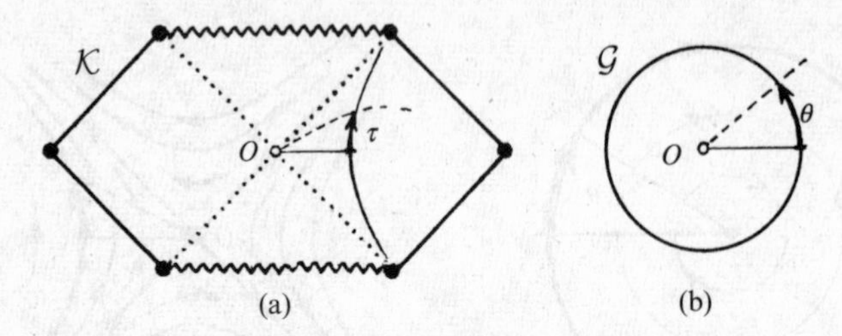

Fig. 30.6 Hawking temperature from imaginary-time periodicity (Gibbons–Perry argument). (a) Strict conformal diagram for an 'eternal black hole', which is the 'maximally extended' Schwarzschild spacetime $\mathcal{K}$, with Schwarzschild time τ and Killing vector $\boldsymbol{\kappa}=\partial/\partial\tau$. The central point O represents a 2-sphere. (b) 'Euclideanized' Schwarzschild space $\mathcal{G}$, near O, has a real angular coordinate θ, so the 'time' τ takes purely imaginary values $\tau=\mathrm{i}\beta\theta$, the constant real number β being the 'surface gravity' of the hole. Here, θ increases by 2π when we go once around O, so any function on $\mathcal{K}$, analytic at O (thereby extending to $\mathcal{G}$), has period 2π in θ, i.e. period $2\pi\mathrm{i}\beta$ in τ (near O), where β is to be interpreted as the black hole's Hawking temperature.

can also use that of Kerr for a rotating hole). For the argument to work, we need an analogue of the origin O in $\mathbb{M}^2$. We see this in Fig. 30.6a, where I have provided a strict conformal diagram for what is referred to as the 'maximally extended' Schwarzschild spacetime $\mathcal{K}$.[13] The spacetime $\mathcal{K}$ is sometimes called an 'eternal black hole' because it was not created from a gravitational collapse, but was 'always there'. The central point O of the diagram represents a *2-sphere*, in accordance with the conventions of strict conformal diagrams. $\mathcal{K}$ is the analogue of $\mathbb{M}^2$, but we also need an analogue of the Euclidean space $\mathbb{E}^2$. There *is* such a space, sometimes referred to as the 'Euclideanized' Schwarzschild space $\mathcal{G}$ (and often even called a 'Euclidean space', which strikes me as grossly confusing!), where the Schwarzschild 'time'[14] τ in $\mathcal{K}$ takes purely imaginary values $\tau = \mathrm{i}\beta\theta$ in $\mathcal{G}$, the quantity θ being an angular coordinate in $\mathcal{G}$ that is increased by 2π each time O is circumnavigated in a positive direction (Fig. 30.6b), and where β is a (constant, at constant r) real number called the 'surface gravity'. Any quantity that is regular (i.e. analytic; see §7.4) at O in $\mathcal{K}$ must also be regular at O in $\mathcal{G}$ (since this is exactly the same place 'O', in the complexified Schwarzschild space, for each of the two real sections $\mathcal{K}$ and $\mathcal{G}$). Being regular at O in $\mathcal{G}$, such a quantity must be periodic, with period $2\pi\mathrm{i}\beta$ in τ, because θ is an ordinary angular coordinate which when increased by 2π ($= 360°$) simply gets us back to the same place in the space(time) that we started from. This *imaginary periodicity* is characteristic of a 'thermal state of temperature β' according to the principles of statistical thermodynamics.

It is not my purpose here to discuss these thermodynamic principles. This would take us too far afield. The only point of concern, here, is whether we can trust the argument for this complex periodicity. It depends upon taking the region O seriously. Is this justified? This is by no means clear. For an actual physical black hole, this entire 'eternal' picture is certainly not appropriate. A physical hole must have been created from some gravitational collapse (say, of a supermassive star or cluster of material at a galactic centre), unless it was, in some sense a 'primordial' creation of the Big Bang itself. Even a primordial hole—a black hole rather than its time-reverse, namely a *white hole*—would still in some sense represent a 'collapse' and, whether black or white, is not well described by the full model of Fig. 30.6a. A certain *exterior* part of this model, however, is appropriate for the description of a collapse to a black hole, namely that part of Fig. 30.7 lying above and to the right of the indicated outer boundary line of the actual material indulging in the collapse. Below and to the left of this boundary, the spacetime metric would be that of the matter, and would be different from that of the eternal black hole. The complete collapse is sketched in Fig. 30.7, which amounts to a slight redrawing of Fig. 27.16c. Now, we note that O is always outside the region where the (extended) Schwarzschild metric applies. There appears to be no physical justification for assuming that physical quantities defined on the spacetime have a regularity at O, and it is hard to see why the argument provides a justification for the Hawking temperature, despite its mathematical elegance. (Any physically realistic model of a black hole would possess deviations from the exact Schwarzschild—or Kerr—metric, and these deviations can reasonably be expected to get larger and larger, finally diverging to infinity the closer we extend towards 'O'.)[30.5]

Yet the exact stationary black-hole model represents the ultimate limit of a realistic collapse, where all the irregularities are taken to iron themselves out as time progresses. It is the limiting spacetime that has this

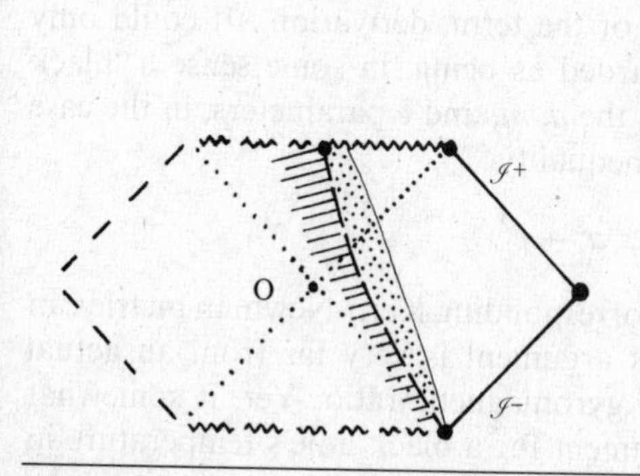

Fig. 30.7 A history of spherical gravitational collapse—a slight redrawing of Fig. 27.16c—represented in terms of the maximal Schwarzschild spacetime $\mathcal{K}$ of Fig. 30.6a. The region shaded with sloping lines is to be deleted and in that shaded with dots the metric differs from that of $\mathcal{K}$, because of the presence of matter. Note that O is always outside the region where $\mathcal{K}$'s metric applies.

[30.5] See if you can give an argument justifying this claim. *Hint*: Think of small linear perturbations. Do you expect exponential time-behaviour? (Consider eigenmodes of $\partial/\partial\tau$.)

regularity and therefore the required complex periodicity and hence the required temperature. Although I do not see how we can regard this argument as any actual physical *derivation* of the Hawking temperature (despite its commonly being taken as such), it does provide some sort of 'strong indication' of a hidden inner consistency of the whole idea of this 'black-hole temperature'.

At this juncture, I cannot resist making a comparison with another observation, due originally to Brandon Carter which, in a different context, has a significant similarity to the argument just given, although it has never been presented as a 'derivation' of anything. We recall that a stationary charge-free black hole is described by the two Kerr parameters m and a, where m is the hole's mass and am its angular momentum (and where for convenience I choose units for which $c = G = 1$, such as Planck units, as in §27.10). A generalization of the Kerr metric found by Ezra T. Newman[15] (usually referred to as the *Kerr–Newman metric*) represents an electrically *charged* rotating stationary black hole. We now have *three* parameters: m, a, and e. The mass and angular momentum are as before, but there is now a total electric charge e. There is also a magnetic moment $M = ae$, whose direction agrees with that of the angular momentum. Carter noticed that the *gyromagnetic ratio* (twice the mass times magnetic moment divided by the charge times angular momentum, which for a charged black hole is $2m \times ae/e \times am = 2$), being completely fixed for a black hole (i.e. independent of m, a, and e), actually takes precisely the value that Dirac originally predicted for the *electron*, namely 2 (where for the Dirac electron, the angular momentum is $\frac{1}{2}\hbar$ and the magnetic moment is $\frac{1}{2}\hbar e/mc$, again giving a gyromagnetic ratio of 2, taking $c = 1$). See §24.7. Newman (2001) has provided an interpretation of this 'coincidence' in terms of a displacement in a complex direction in space.

Can we regard this argument as providing a *derivation* of the electron's gyromagnetic ratio, independently of Dirac's original argument? Certainly it does *not*, in any ordinary sense of the term 'derivation'. It could only apply if an electron could be regarded as being, in some sense a 'black hole'. In fact, the actual values for the a, m, and e parameters, in the case of an electron, grossly violate an inequality

$$m^2 \geq a^2 + e^2$$

that is necessary in order that the corresponding Kerr–Newman metric can represent a black hole. Thus, this argument is very far from an actual derivation of the Dirac electron's gyromagnetic ratio. Yet, it somewhat resembles the Gibbons–Perry argument for a black hole's temperature in revealing a certain 'naturalness' of this value, via extensions into the complex.[16] The Gibbons–Perry argument does have the additional point in its favour that it is not constrained merely to consideration of the Schwarzschild/Kerr family of spacetime metrics. Nevertheless, in my opinion, this hardly justifies its common acceptance as an actual derivation.

30.6 Killing vectors, energy flow—and time travel!

The 'eternal black hole' has frequently caught people's attention for other reasons, despite the fact that it has some curious global properties which make it hard to take seriously as a model of a physically acceptable universe. Although some of these reasons would seem to have relevance more to the realms of science fiction rather than reality, the eternal black hole has some geometrical value for us, because it illustrates interesting mathematical features that will be important in §§30.7,10. We see that it has two distinct past null infinities ($\mathscr{I}^-$ and $\mathscr{I}^{-'}$) and two distinct future null infinities ($\mathscr{I}^+$ and $\mathscr{I}^{+'}$). This spacetime is often thought of as representing a time-evolution of two different universes that become connected by a 'wormhole', which subsequently 'pinches off' in a singularity; see Fig. 30.8.

With respect to each of the two 'external' regions, it would seem that each universe contains a black hole, but the black hole is an odd one in the sense that it is also a 'white hole' at the same time. Signals can escape to each external universe $\mathcal{E}$ and $\mathcal{E}'$ from the *past* internal region $\mathcal{B}^-$ ('white-hole' behaviour) as well as signals being able to propagate into the future internal region $\mathcal{B}^+$ from each external universe $\mathcal{E}$ and $\mathcal{E}'$ ('black-hole'

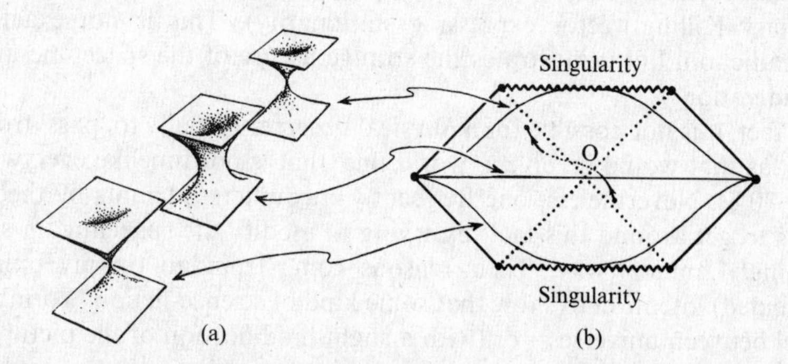

Fig. 30.8 (a) The spacetime $\mathcal{K}$ viewed globally as a 'time-evolving' 3-space, which represents a 'wormhole' connecting two asymptotically flat regions. The wormhole pinches off in a singular way, both into the future and into the past. (b) Any space traveller who proposes to travel through the wormhole from one region to the other cannot get through before it 'pinches off' as is manifest from the conformal diagram, since this would demand the traveller's worldline having a spacelike (superluminary) portion—shown dotted.

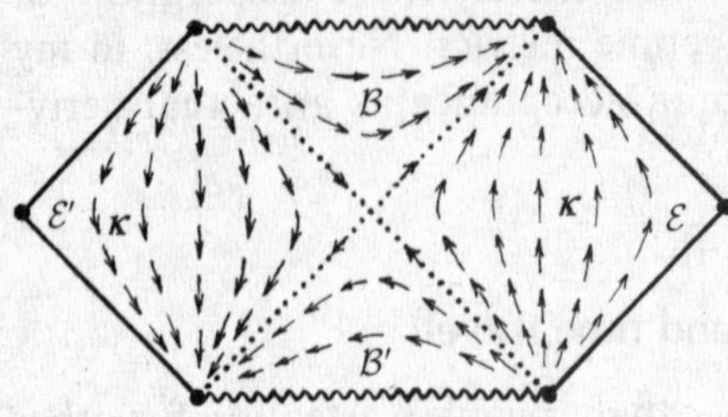

Fig. 30.9 The Killing vector $\boldsymbol{\kappa}$ is timelike in the two external regions $\mathcal{E}$ and $\mathcal{E}'$, but spacelike in the internal regions $\mathcal{B}$ and $\mathcal{B}'$. Comparing $\boldsymbol{\kappa}$ in $\mathcal{E}$ and in $\mathcal{E}'$, we find that it reverses time-orientation, so that the concept of conserved energy density $T_{ab}\kappa^a$ reverses its sign.

behaviour). The fact that spacetime is stationary is expressed in the existence of a Killing vector $\boldsymbol{\kappa}$ (see §14.7, §19.5, and §30.4). I have sketched this Killing vector in Fig. 30.9. We notice that the Killing vector is timelike in the two external regions $\mathcal{E}$ and $\mathcal{E}'$, but it is spacelike in the internal regions $\mathcal{B}^-$ and $\mathcal{B}^+$. The timelike nature of $\boldsymbol{\kappa}$ in the external regions means that $\boldsymbol{\kappa}$ expresses the stationarity of the black/white hole. A family of observers in $\mathcal{E}$, whose worldlines are tangent to the Killing vector field $\boldsymbol{\kappa}$, will perceive an unchanging universe. The same applies in $\mathcal{E}'$. However, the family of observers in $\mathcal{E}'$ with this property must apply these considerations to the Killing vector $-\boldsymbol{\kappa}$, rather than $\boldsymbol{\kappa}$, because we need to keep the future/past distinctions consistent for local observers throughout the spacetime. In a sense, the 'time direction' has reversed as we pass from $\mathcal{E}$ to $\mathcal{E}'$. The conserved energy-density quantity (§19.5) obtained from the contraction $T_{ab}\kappa^b$ of the energy–momentum tensor with the Killing vector $\boldsymbol{\kappa}$ provides a positive energy density (for normal matter) in $\mathcal{E}$, but a negative one for normal matter in $\mathcal{E}'$ (since $\boldsymbol{\kappa}$ is past-pointing in $\mathcal{E}'$, and $-\boldsymbol{\kappa}$ is now the ordinary Killing vector expressing stationarity). This is not exactly a contradiction, but it illustrates the strange nature of the spacetime under consideration.

In fact, it is not possible for a physical observer actually to 'pass' from $\mathcal{E}$ to $\mathcal{E}'$, as that would involve a 'world line' that is not timelike everywhere (Fig. 30.8). Nevertheless, one frequently encounters attempts by theoreticians to get around this fact, by trying to modify the spacetime in some seemingly 'minor' way. Their reasons come from an (in my opinion misguided) intention to show that some kind of science-fiction 'wormhole' travel between universes—or (with a slight modification of the picture, as in Fig. 30.10) from one region of spacetime to a distant one—could be achieved by some future technology. If such a proposal could succeed, then this would open up the potential possibility of having a form of space travel in which the normal limitations of relativity are transcended. In true 'Star Trek' traditions, a 'warp-drive' procedure is envisaged that allows the spaceship to travel through the wormhole to a distant region that might even be 'earlier' than when the spaceship entered the wormhole.

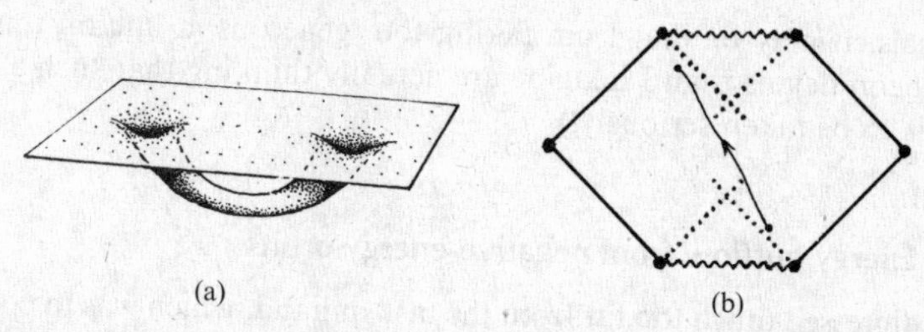

Fig. 30.10 A science-fiction suggestion for superluminary space travel, based on a modified wormhole spacetime. (a) By 'identifying' distant parts of the two external spatial regions of Fig. 30.8 one obtains a wormhole connecting distant regions of the same space, but again travel through the wormhole from one to the other cannot be achieved by a timelike curve. (b) For this to be possible, something of the nature of the 'stretched' version of $\mathcal{K}$, depicted here, would be needed (but such a model requires negative energy densities).

Strange as it may seem, even first-rate professionals in general relativity have entertained the possibility of such 'time travel'.[17] A serious reason for this is (presumably at least sometimes) not so much the possibility that time travel might actually be feasible within the confines of present-day (or imaginable) physics, but that we might learn something from the fact that, physically, it ought *not* to be.[30.6] In the 'spatial description' given in Fig. 30.8, the wormhole 'pinches off to zero size' before the space-traveller can get through. The idea is that one entertains the possibility that it might be allowable, within the confines of the theory, to 'hold the wormhole apart' for long enough for the traveller to get through to the other side, if negative energy densities are allowed. Such negative energy densities are normally considered to be forbidden in the classical theory, but might be allowed under special circumstances in the appropriate quantum field theory.

Do some relativity physicists really expect that such fanciful considerations might lead us to a notion of 'warp drive', in which travel to a distant part of the universe could be achievable through such a QFT-supported wormhole? Very few, I imagine.[18] A more serious issue seems to be that these considerations might provide a 'test' for one's ideas on quantum gravity. If those QFT ideas actually *do* allow such a 'holding apart', then this could be taken as a bad sign for those particular ideas about quantum gravity—so one needs to think again. In this way, some useful guidance could be obtained, as to the plausibility of the particular quantum gravity theory under consideration. (At least this is my own reading of such

[30.6] Explain why, in accordance with the tenets of special relativity, the possibility of travel between spacelike-separated events p and q, entails the possibility of travel from p to an event in the direct past of p, on a timelike world line through p.

proposals. It may be that I am taking too 'generous' a line on this, and more theoreticians than I imagine are actually thinking that such a 'warp drive' is to be taken seriously!)

30.7 Energy outflow from negative-energy orbits

I have digressed much too far from the task at hand, which was to consider the implications of the Hawking temperature of a black hole. Can we see from more physical reasons why, in the context of quantum mechanics, a black hole ought to emit radiation in accordance with it having a non-zero temperature? In fact, Hawking also provided an 'intuitive' derivation of the presence of this Hawking radiation. This is illustrated in Fig. 30.11. In the vicinity of the hole's horizon, virtual particle–antiparticle pairs are continually being produced out of the vacuum, only to annihilate each other in a very short period of time. (This is the process considered in §26.9, and illustrated in Figs. 26.9 and 26.10). However, the presence of a black hole modifies this activity because, from time to time, one of the particles of the pair falls into the hole, the other one escaping. This can only happen when the escaping particle becomes a *real* particle (i.e. 'on shell', as opposed to the virtual 'off shell' particle it started out as, see §26.8 and Fig. 26.6), and therefore the escaping particle must have positive energy, so that (from energy conservation) the particle falling into the hole has to become a real particle with negative energy (these energies being assessed from infinity). In fact, negative energies *can* occur for real

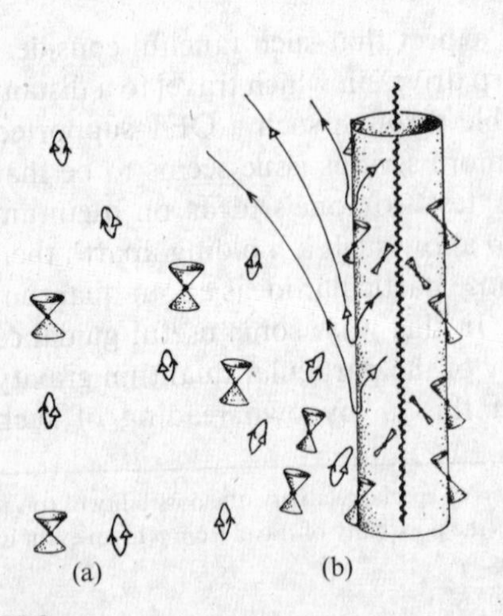

Fig. 30.11 Hawking's 'intuitive' derivation of Hawking radiation. (a) Far from the hole, virtual particle-anti-particle pairs are continually produced out of vacuum, but then annihilated in a very short time (see Fig. 26.9a). (b) Very close to the hole's horizon, we can envisage one of the pair falling into the hole, the other escaping to external infinity. For this, the virtual particles both become real, and energy conservation demands that the ingoing particles have negative energy. This it can do, because the Killing vector $\boldsymbol{\kappa}$ becomes spacelike inside the horizon. (If κ^a is spacelike, the conserved energy $p_a\kappa^a$ can be negative, where p_a is the particle's 4-momentum.)

particles inside the black hole. This possibility arises because the Killing vector κ^a becomes spacelike in the interior region $\mathcal{B}^+$ and a future-pointing timelike 4-momentum p_a can have a negative scalar product $p_a\kappa^a$, this being the (conserved) energy of the particle; see Fig. 30.11b.[30.7] The Hawking process comes about because a real (as opposed to virtual) particle can have negative energy if it is inside the hole's event horizon. The real partner of such a particle has to have positive energy, so positive energy can be carried away from the hole.

At this point it is worth remarking that a very similar thing happens in *classical* black-hole theory when the hole is *rotating*. And, in contrast with the case of Hawking radiation, for which the emission of energy is, for black holes of a plausible size, ridiculously small—and whose interest is purely theoretical—what happens with a classical rotating hole appears to have enormous astrophysical implications. Indeed, the most powerful sources of energy known in the universe (quasars and radio galaxies) appear to be fuelled by the rotational energy of a vast black holes.

The process has a similarity to that which leads to Hawking radiation, in that the energy comes from negative energy particles or fields being swallowed by the hole, which results in positive energy escaping from the hole to infinity. However, the important difference is that, with a rotating black hole, the part of spacetime within which the Killing vector $\boldsymbol{\kappa}$ becomes spacelike extends to a region *outside* the black hole's horizon. This region is referred to as the *ergosphere* (Fig. 30.12a). Thus, in the ergosphere, particles can have negative energy (as measured from infinity) while still being able to communicate with distant parts of the universe. It becomes possible, for example, for a particle to enter the ergosphere from the outside and then to split into two, where one of the resulting particles has negative energy so that the other escapes out again carrying more energy than the original particle brought in![19] The net result is to carry energy away from the hole, slightly reducing the energy stored in its rotational motion (Fig. 30.12b). A similar conclusion can be obtained if (electromagnetic) fields are involved, rather than particles.[20]

It should be emphasized that the 'negative-energy particle' falling into the hole is, when viewed locally, a perfectly ordinary particle (with an ordinary timelike 4-momentum of the kind described in §18.7). It is just that the quantity $p_a\kappa^a$, which measures the conserved energy, as viewed from infinity, happens to become negative, which can perfectly well happen when the particle lies within the ergosphere. This is a remarkable and very potent fact about black holes, but there is nothing mathematically inconsistent or physically unreasonable about it. However, it is this

---

[30.7] Explain how a spacelike κ^a can give a negative 'energy' value $p_a\kappa^a$.

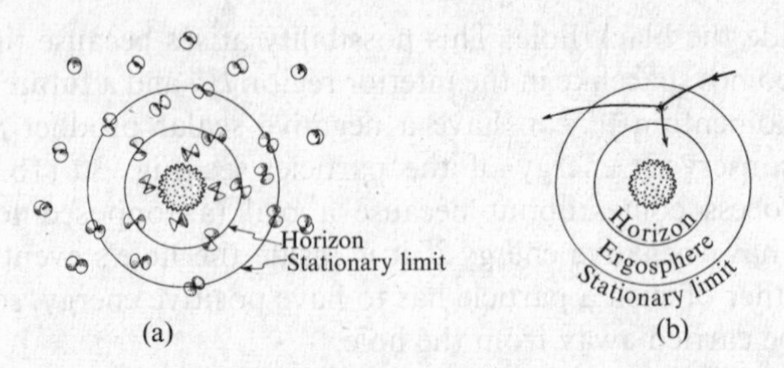

Fig. 30.12 Views 'down' along the time-axis of a rotating (Kerr) black hole. (a) For a Kerr black hole, there is a region—called the 'ergosphere'—within which the Killing vector $\boldsymbol{\kappa}$ of stationarity becomes spacelike *outside* the black hole's horizon. Within the ergosphere, particles can have negative conserved energy (as measured from infinity), and other particles they directly encounter are able to escape to infinity, carrying away the excess. (b) According to the so-called 'Penrose process', this fact can be harnessed, and the black hole's rotational energy extracted. In the simplest such process, a particle enters the ergosphere, splits into two particles, one of which enters the hole carrying negative energy, and the other escaping to infinity carrying away more energy than the original particle brought in.

fact that allows the often enormous rotational energy of a black hole to be hurled into the outside world.

In fact, the most plausible explanation of the vast energy output of a *quasar* (see §27.9) is that this energy comes from the rotation of a huge black hole. The hole's immense rotational energy is gradually extracted—and flung out into space—by what is essentially the above-mentioned process; see Fig. 30.13. It has commonly been proposed that the negative energy swallowed by the hole may be primarily in the form of an electromagnetic field (e.g. Blanford and Znajek 1977; Begelman *et al.* 1984) rather than to actual particles (e.g. Williams 1995, 2002, 2004). But the underlying reason is the same in each case.

30.8 Hawking explosions

Let us now return to the quantum-mechanical Hawking process. The temperature for a black hole of a solar mass 1 $M_\odot$ is extremely low, as noted above, in §30.4 (about 10^{-7} K). For larger black holes, this temperature would be even lower, in inverse proportion to the hole's mass (for a given $a:m$ ratio; see §27.10). There is no astrophysical evidence that there are any black holes of mass less than about 1 $M_\odot$, so black-hole temperatures are not believed to be of direct astrophysical interest.

Nevertheless, there is considerable *theoretical* interest in this temperature, as Hawking dramatically pointed out, in 1974.[21] For example, if the universe is of the ever-expanding kind (see §27.11 and §28.10), then there

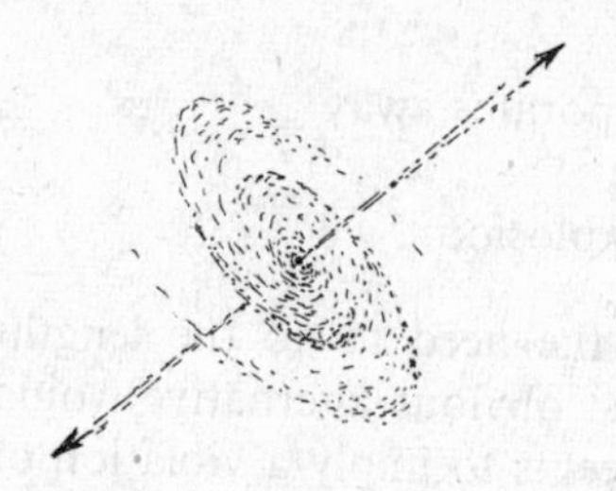

Fig. 30.13 The enormous energy output of a quasar appears to come from the rotational energy of a huge black hole at a galaxy's centre. This seems to be by a process of the general nature described in Fig. 30.12, but possibly mainly through the black hole swallowing electromagnetic fields of negative energy rather than particles.

will come a point when the ambient temperature will be lower than the value for any given black hole. (For a 1 $M_\odot$ black hole in a $K=0=\Lambda$ universe, this would take some 10^{16} years, which is about 10^6 times the present age of the universe.) After that, the black hole would start to lose energy by radiating away more energy than it absorbs from the background. As it loses energy, it loses mass, so its radius gets smaller, and accordingly it gets hotter. Let us imagine starting with a 1 $M_\odot$ black hole. It would continue to radiate at a very slow rate, gradually losing mass for some 10^{64} years, the temperature increasing slowly at first and then at an ever-accelerating rate, until it reaches about 10^9 K or 10^{10} K (the uncertainties lying in our lack of knowledge of particle physics at enormously high energies). At that point there is a runaway instability, and an explosion takes place with the remaining mass–energy in the hole being converted, almost instantaneously, completely into radiation! See Fig. 30.14.

At least, this appears to be the simplest and most natural-looking assumption, as originally put forward by Hawking. (Hawking had initially suggested that explosions of this nature might even be detectable now, if the Big Bang had been so kind as to furnish us with a significant number of 'mini-holes', say of the mass of a mountain and the diameter of a proton! However, from our present perspective, this seems unlikely, and no such explosions have yet been identified.) Other physicists[22] have argued that although such a final explosion would take place, the hole would not disappear completely, but would leave some 'remnant', or 'nugget'. The reason that they prefer this is that they are uncomfortable about the 'information' swallowed by the hole being lost to the system, and they prefer it to be 'stored' in this remaining nugget.[23] The problem is that it is hard to see how all the information concerning the details of the matter that collapsed into the hole—which might even have originally been a stellar-sized or even galactic black hole before the *thermal* (and therefore virtually 'information-free') radiation took away almost all the mass of the hole—could be stored in such a remnant. As an alternative, some researchers take the view that in the final explosion, all the information comes back out again 'at the last minute'.

These three alternatives are

LOSS: information is *lost* when the hole evaporates away
STORE: information is *stored* in final nugget
RETURN: information all *returned* in final explosion.

The reader might wonder why people feel the need to go the lengths required for STORE or RETURN, when the most obvious alternative would appear to be LOSS. The reason is that LOSS seems to imply a violation of unitarity, i.e. of the operation of **U**. If one's philosophy of quantum mechanics demands that unitarity is immutable, then one is in difficulty with LOSS. Hence we have the popularity, among many (and apparently most) particle physicists of the possibilities of STORE or RETURN, despite the seemingly contrived appearance of these alternatives.

My own view is that information LOSS is certainly the most probable. An examination of Fig. 30.14 conveys the clear picture that the collapsing physical material simply falls across the horizon, taking all its 'information' with it, to be finally destroyed at the singularity. Nothing particular, of local physical importance, should happen at the horizon. The matter does not even 'know' when it crosses the horizon. We should bear in mind

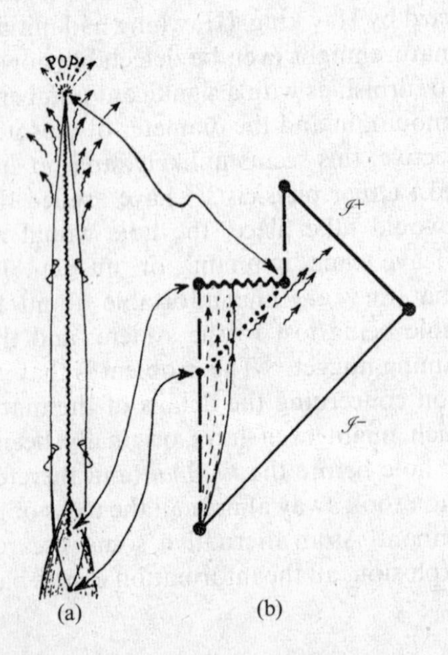

Fig. 30.14 Hawking black-hole evaporation. (a) A black hole forms through classical gravitational collapse. Then over an extremely long period it loses mass–energy at a very slow rate, through Hawking radiation, very gradually heating up as it loses mass. Finally, it appears to have an explosive disappearance (in an explosion that is small by astrophysical standards, and independent of the hole's original mass). (b) A strict conformal diagram of this process (spherically symmetrical case). This would seem to convey a clear picture in accordance with LOSS where collapsing material simply falls across the horizon, taking all its 'information' with it, to be destroyed at the singularity.

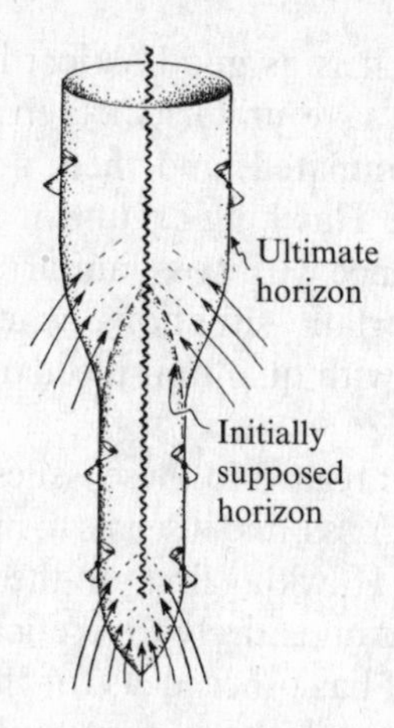

Fig. 30.15 The precise location of a black hole's horizon is determined 'teleologically', as it depends upon how much material ultimately falls into the hole.

that we could be considering an initially very large black hole, perhaps like the holes that are believed to inhabit galactic centres, which could be of a million solar masses or more. As the horizon is crossed, nothing particular happens. The spacetime curvature and density of material is not large: only of the kind that we find in our own solar system. Even the location of the horizon is not determined by local considerations, since that location depends upon how much material later falls into the hole. If more material falls in later, then the horizon would actually have been crossed earlier! See Fig. 30.15. I find it inconceivable that somehow 'at the moment just before the horizon is crossed' some sort of signal is emitted to the outside world conveying outwards the full details of all information contained in the collapsing material. In fact, simply a signal would by itself not be enough, since the material itself *is*, in a sense, really the 'information' that one is concerned with. Once it has fallen through the horizon, the material is trapped, and is inevitably destroyed in the singularity itself.

At least, that is the clear conclusion if we accept cosmic censorship (§28.8). I do not see that there is a great deal of leeway with this even if we do not. The picture is essentially that of Fig. 30.14. According to this picture, the material in the collapse is destroyed (and its 'information' is destroyed) only when it enters the singularity, not when it crosses the horizon. If one is to hold to the standpoint of RETURN[24]—that somehow the information of the collapsing material all comes out again at the event of the final explosion—as indicated by the word 'POP' in Fig. 30.14, then one must in some way explain how this information manages to sidle its way out, to get to this point from right across the singularity (which, according to a reasonable form of cosmic censorship, ought to be essentially spacelike; see §28.8). I do not find this at all plausible.

The situation with STORE is not much better, if at all. Even if the nugget *is* formed, it is not really of any use, since the information is 'locked inside' it forever, and it seems to me that it is as good as lost in any case. If the sole purpose of the nugget is to 'save unitarity', then a consistent QFT of nuggets would have to be formulated, and there are severe difficulties with doing this.[25] As I see it, the Hawking argument is presenting us with a powerful case that, in accordance with LOSS, unitarity must be expected simply to be violated in certain situations when general relativity enters the picture in conjunction with quantum-mechanical processes.

What is Stephen Hawking's own standpoint with regard to these issues? Right from the beginning, he has strongly argued for LOSS, and it seems to me that the case for this is as strong now as it was when Hawking first put these ideas forward. Of course, black-hole evaporation is an entirely theoretical notion, and it might be the case that Nature herself has other ideas for the remote future of black holes. It is hard to see that any such alternatives could occur, however, without some radical changes in the structure of either QFT or macroscopic general relativity (or both). Hawking's position—at least, as of 2003—was that unitarity should indeed be violated, but only in what I would regard as a rather mild sense. Hawking proposed that, in the presence of black holes, the quantum state of a system would actually evolve into a (non-pure-state) density matrix. In fact, this idea was briefly alluded to in §29.6, when I remarked on the fact that, if some part of an entangled quantum state could be genuinely lost—here by part of it falling into a black hole—as opposed to being lost only FAPP, then it might be reasonable to take the ontological position that quantum reality is actually to be described by a density matrix rather than a (pure) state. Hawking envisaged some kind of 'super-unitary' evolution that applies directly to density matrices and allows 'pure states' to evolve into 'mixed states'.[26,[30.8]]

30.9 A more radical perspective

My own standpoint is that, whereas I agree with Hawking that some form of LOSS, is indeed likely to be correct, I believe that something even more radical is needed. Hawking's proposal, as outlined in the preceding paragraph, does not incorporate any time-asymmetric features,[27] for example. But with time-symmetry, the 'white-hole' picture of Fig. 30.16a, which is the time-reverse of Fig. 30.4, would be allowed—as would the time-reverse of the evaporating black hole given in Fig. 30.14, as depicted in Fig. 30.16b. The 'general time-symmetrical situation', in which there is much destruction of information together with just as much creation of 'new information' is illustrated in Fig. 30.17. All of these violate the Weyl

[30.8] Use the index notation (e.g. ψ^α for $|\psi\rangle$) to indicate the kind of transformation that could achieve this. (*Hint*: Have a look at Fig. 29.5.)

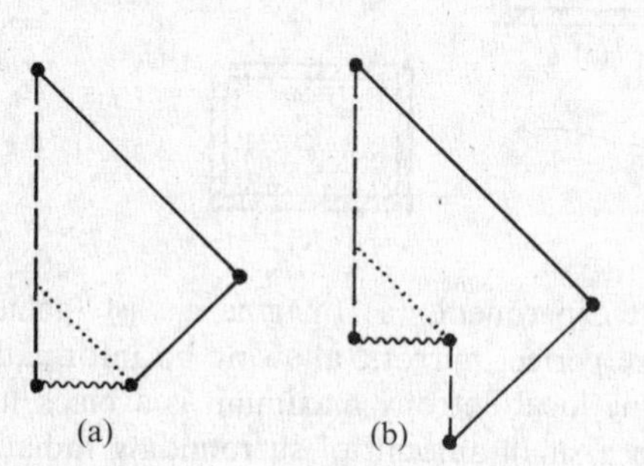

Fig. 30.16 White holes: time-reversed black holes. These violate the Weyl curvature hypothesis. (a) Conformal diagram of time-reverse of black-hole formation, as in Figs. 27.11, 27.16. (b) Conformal diagram of time-reverse of black-hole formation and subsequent disappearance through Hawking radiation, as in Fig. 30.14.

curvature hypothesis (§28.8). The 'symmetrical' case Fig. 30.17 involves the creation of a new white hole at the moment of final evaporation of the original black hole, the white hole growing until it reaches the size that the black hole had. I have never seen such an absurd-looking model seriously suggested! Once situations like those of Fig. 30.17 are permitted, then I fail to see why they do not proliferate in the Big Bang, leading to a gross inconsistency with the message of Chapter 27.

I do not propose to repeat all my arguments here,[28] but, roughly speaking, these rest upon the fact that Nature seems to be telling us that something closely resembling the Weyl curvature hypothesis holds true,[29] for the physical structure of those spacetime singularities that she actually allows in her universe. If we accept this, then there is indeed a net 'loss of information' in the singularities of black holes which is not regained. This is because, according to this hypothesis, the final singularities of collapse can contain—and therefore absorb—huge numbers of degrees of freedom (these residing in the Weyl curvature), whereas these degrees of freedom are forbidden for any initial singularity.

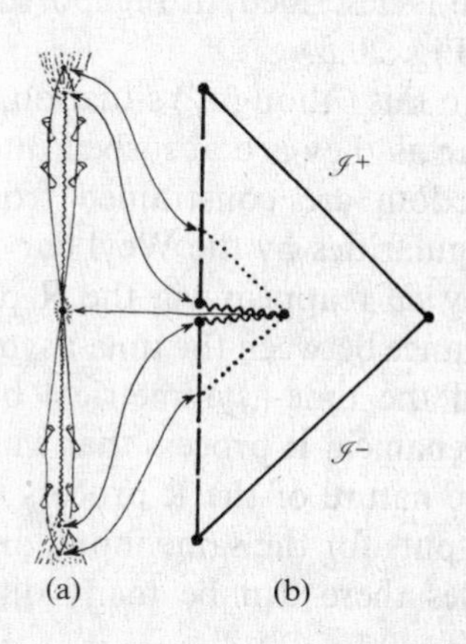

Fig. 30.17 (a) Time-symmetrical situation, where there is creation of a white hole at the moment of final evaporation of a black hole, which had been formed by gravitational collapse. The new white hole grows until reaching the previous black-hole size before disappearing with the ejection of a large amount of material. (b) conformal diagram of this.

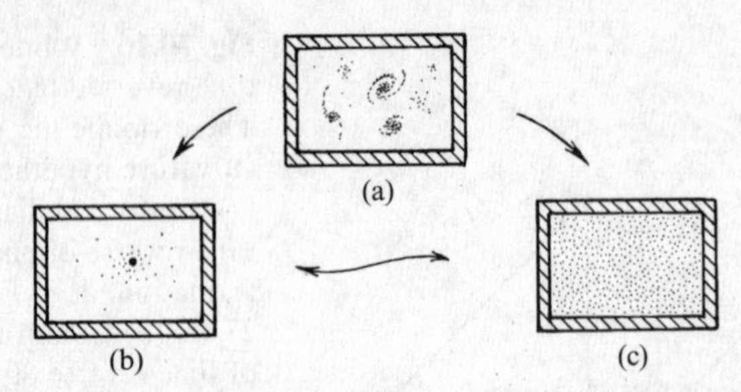

Fig. 30.18 Hawking's 'box' thought experiment. (a) Imagine a vast (galactic scale) 'box' of matter, whose walls are perfect mirrors, allowing no information or material to cross in or out. (b) One local entropy maximum is a black hole providing most of the mass, but with a small amount of surrounding radiation in thermal equilibrium with the hole. (c) Another local entropy maximum is just thermal radiation (and a few particles) but with no black hole.

Let us try to put this argument in terms of the *phase space* of a system, involving the formation and evaporation of black holes. Strictly, to make our phase-space argument work, we should be considering a closed system, containing a fixed finite quantity of energy. To assist our imaginations, we try to envisage a vast box, of greater than galactic dimensions, with walls that are to be considered as perfect mirrors, so that no information or material particles can cross either in or out; see Fig. 30.18. This, of course, is a practical absurdity—but I hasten to assure the reader that our system constitutes merely a 'thought experiment', not a real one! It is being contemplated[30] only to enable phase-space reasoning to be applied to a system involving the (apparent) loss of degrees of freedom in the process of Hawking radiation. The phase space $\mathcal{P}$ under consideration describes all possible physical states within our hypothetical box, with the given total energy. The dynamical evolution is described, in Fig. 30.19, by a family of arrows on $\mathcal{P}$, in the manner of Fig. 20.5.

In this ('thought') situation, as time proceeds, degrees of freedom disappear as they are absorbed into black-hole singularities. These degrees of freedom are constrained from reappearing in initial (white-hole-type) singularities by the Weyl curvature hypothesis, but my contention is that they *do* reappear via the **R** process. The idea is that there is an overall balance between the time-asymmetrical 'loss of information' in black holes and the time-asymmetrical behaviour of probabilities in the quantum-mechanical **R** process that was demonstrated in §30.3. The non-deterministic nature of the **R** process tells us that there can be several alternative outputs for the same input, and this is to balance the fact that with black holes there can be many different inputs giving the same output, the 'information' that distinguishes the various inputs being absorbed into the singularities. Recall that, in the hypothetical experiment of §30.3 (Fig. 30.1), we had two different outputs (photon reaching D and photon reaching C) for a single given input (photon emitted from S), whereas for a given output (photon reaching D) there was basically just one input (photon emitted at S). Thus, we get an effective spreading of the phase-space volume according to the **R** process, whereas the asymmetry in spacetime singularity structure causes an effective narrowing down of phase-space volume; see Fig. 30.19 again. The contention is that these two effects should, on average, balance each other out.

It should be made clear that this balancing is to be only an *overall* feature of the physical processes. It is, of course, not being claimed that there has to be the simultaneous presence of a black hole accompanying each instance of quantum state reduction. The idea is only that throughout the entire phase space there is a balance between these two effects. Accordingly, it is the *potential possibility* of the formation of black holes, with their ability to absorb information, that is to balance the future randomness in **R**.

We may note that both of these effects violate the theorem (Liouville's theorem; see §20.4, Fig. 20.7) that phase-space volume has to be preserved in dynamical evolution. But in each case we have something that goes

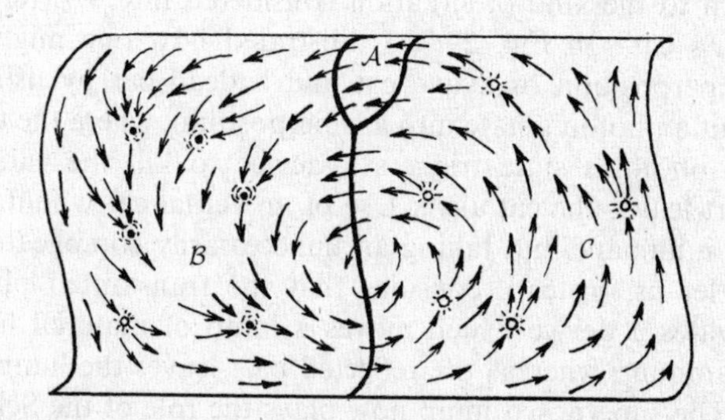

Fig. 30.19 Phase-space description of Hawking's box, with arrows describing (Hamiltonian) evolution (of the processes involved in Fig. 30.18). The regions $\mathcal{A}$,$\mathcal{B}$,$\mathcal{C}$, correspond, respectively, to (a), (b), (c) in Fig. 30.18. Accordingly, a black hole is present for region $\mathcal{B}$, but not for region $\mathcal{C}$. The presence of a black hole results in a confluence of flow lines (reduction of phase-space volume), according to LOSS, owing to information destruction at the black hole's (future) singularity. There is a compensating creation of flow lines (increase in phase-space volume) involved in the time-asymmetry of the **R** process (assumed to be objectively real); see Fig. 30.1. The proposal is that there should be an overall balance between these two Liouville-theorem violating processes, giving ultimate phase-space volume preservation in the flow.

beyond the ordinary classical dynamics which provides the scope for that theorem. Indeed, the very notion of a classical phase-space is not entirely appropriate here, when we are considering quantum and classical effects together. For a purely quantum system, we should be thinking entirely in terms of a *Hilbert space* instead. For those who believe that **U**-quantum evolution is the whole story, the Hilbert-space description is the correct one. But then the destruction of information (and therefore of unitarity) in black-hole evaporation presents a serious problem. My own viewpoint is that neither picture is entirely appropriate, and each should be regarded as an approximation to something else that we do not yet know how to describe.[31]

It has long been my intention to obtain a direct quantitive estimate of the rate of quantum state reduction by investigating the details of the balance between these two processes, as outlined in the above argument (and illustrated in Fig. 30.19), but I have so far been unable to carry this argument to completion. It is therefore fortunate that there is a quite different general line of reasoning that can be used to obtain an appropriate estimate. This is the subject of the remainder of this chapter.

30.10 Schrödinger's lump

Let us return to the kind of situation considered in §29.7, referred to as 'Schrödinger's cat'. In Fig. 29.7, I illustrated how one might set up a quantum superposition of a live cat and a dead cat by using a beam splitter to put a photon's state into a superposition, where the transmitted part of the photon's state triggers a device to kill the cat, while the reflected part leaves the cat alive. Use of an actual cat would, of course, be not only inhumane, but taking an unnecessarily complicated physical system. So let us, instead, consider that the transmitted photon state simply activates a device which moves a lump of material horizontally by a small amount, whereas the reflected part leaves the lump alone; see Fig. 30.20. The superposed lump now plays the role of the Schrödinger's cat—though not so dramatically as before!

The question that I now want to raise is the following: is the quantum superposition of the two lump locations a stationary state? In conventional quantum mechanics, this would certainly be the case if we consider that each lump location separately represents a stationary state and that the energy in each case is the same (so the resting place of the displaced lump is neither raised nor lowered in relation to its original location). This is just an elementary application of the rules that we learnt in Chapter 21 (see also §24.3). Representing the original lump location by the state $|\chi\rangle$ and the displaced one by $|\varphi\rangle$, we have the

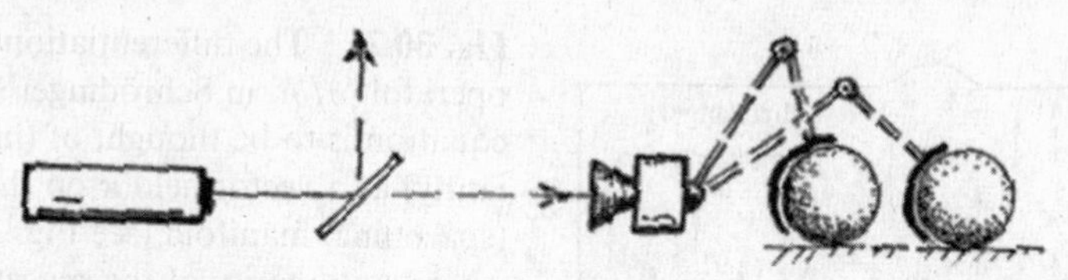

Fig. 30.20 Schrödinger's 'cat' of Fig. 29.7, but now the resulting quantum superposition is merely between two slightly differing locations of a lump of matter.

two Schrödinger equations describing stationarity for each of the two lump locations,

$$i\hbar\,\frac{\partial|\chi\rangle}{\partial t} = E|\chi\rangle, \quad i\hbar\,\frac{\partial|\varphi\rangle}{\partial t} = E|\varphi\rangle,$$

each giving us an eigenstate of energy, with energy eigenvalue E. If the superposition is represented as the state

$$|\Psi\rangle = w|\chi\rangle + z|\varphi\rangle,$$

then we directly obtain

$$i\hbar\frac{\partial|\Psi\rangle}{\partial t} = E|\Psi\rangle,$$

whatever values the (constant) amplitudes w and z might happen to have.[30.9] Thus, each quantum superposition $|\Psi\rangle$ is also a stationary state. If the states $|\chi\rangle$ and $|\varphi\rangle$ would each individually sit there forever, then so would every quantum superposition $|\Psi\rangle$ of them. This is just an expectation of standard quantum mechanics.

Now let us start to bring in the lessons that Einstein has taught us with his superb and now excellently confirmed general theory of relativity. In the first place, we might consider it important to bring in the gravitational field expressed in the background spacetime geometry. We can imagine that the experiment is being performed on the Earth, with the two instances of the lump sitting on a horizontal platform. The Earth's spacetime geometry is not quite flat, and we must consider what effect this spacetime curvature might have on the above considerations. Indeed, we must worry a little about the very meaning of the operator '$\partial/\partial t$' that appears in Schrödinger's equation. In general relativity we do not usually have a naturally presented coordinate system with respect to which the concept of '$\partial/\partial t$' would be defined. Recall, from §10.3 and §12.3 (see Fig. 10.5) that the 'invariant' way of thinking about a partial differentiation operator (like $\partial/\partial t$) is to consider it as a vector field on the (spacetime)

[30.9] Why? Explain what properties of a vector field $\boldsymbol{\kappa}$ are being used when we repeat this conclusion in the case of a stationary background spacetime, below.

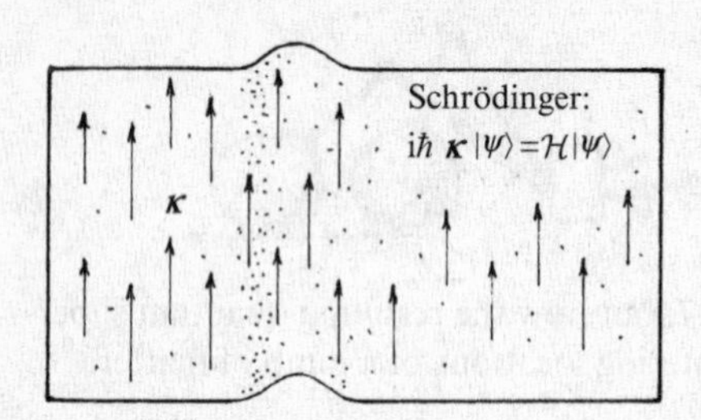

Fig. 30.21 The differentiation operator '$\partial/\partial t$' in Schrödinger's equation is to be thought of (invariantly) as a vector field $\boldsymbol{\kappa}$ on the (spacetime) manifold (see Fig. 30.3), where stationarity of the spacetime is expressed in $\boldsymbol{\kappa}(=\partial/\partial t)$ being a (timelike) Killing field ($\pounds_{\boldsymbol{\kappa}}\mathbf{g}=0$; see §14.7).

manifold—Fig. 30.21. Thus, we shall need a vector field on our spacetime in order to express our required notion of '$\partial/\partial t$'.

In the present situation we are not so badly off, because we are considering the issue of 'stationary states', so we must at least have a background spacetime that is itself stationary. Indeed, we consider the Earth's field to be stationary. As we saw above (§§30.4,6, Fig. 30.3), a stationary spacetime is characterized by the existence of a *timelike Killing vector* $\boldsymbol{\kappa}$. How does this particular vector field feature in the discussion? Our spacetime is stationary in the sense of being 'independent of t', telling us that we can simply make the replacement (Fig. 30.21) in the previous formulae,

$$\frac{\partial}{\partial t}\mapsto\boldsymbol{\kappa}.$$

There may be an issue of an overall constant scale factor, but this is not very important for us here. The usual way of fixing this overall factor is to demand that $\boldsymbol{\kappa}$ become an 'ordinary' time displacement at large distances, where the gravitational field is taken to tail off to zero. Locally, however, the magnitude of $\boldsymbol{\kappa}$ may change from place to place, in a way that takes into account the 'clock-slowing' effects of the Earth's gravitational field (§19.8).[30.10] Since $\boldsymbol{\kappa}$ is now taking over the role of $\partial/\partial t$, our individual Schrödinger equations, defining the stationarity of each of the separate states $|\chi\rangle$ and $|\varphi\rangle$, are

$$i\hbar\,\boldsymbol{\kappa}|\chi\rangle = E|\chi\rangle \quad\text{and}\quad i\hbar\,\boldsymbol{\kappa}|\varphi\rangle = E|\varphi\rangle,$$

and, just as before, we deduce that, for any superposition $|\Psi\rangle$, we still have

$$i\hbar\,\boldsymbol{\kappa}|\Psi\rangle = E|\Psi\rangle.$$

Thus, the presence of a stationary gravitational field as a background does not alter the fact that any quantum superposition of the two stationary states $|\chi\rangle$ and $|\varphi\rangle$ is itself stationary.

[30.10] See if you can give an account of this, using the conservation law provided by a Killing vector $\boldsymbol{\kappa}$, as described in §30.6 and taking note of the fact that the norm $\kappa_a\kappa^a$ may differ from unity in the vicinity of a gravitating body, even though it is normalized to unity at large distances from the body. How does this affect the measure of time?

But now let us see what happens when we take into account the lump's *own* gravitational field. If we consider each of the states $|\chi\rangle$ and $|\varphi\rangle$ individually, then we appear to have no real problem. Of course, each of $|\chi\rangle$ and $|\varphi\rangle$ is a quantum state and, in the absence of an accepted quantum gravity theory, we may not know how to treat its gravitational field. But it does not really matter that we do not know how to do this in detail. The conventional point of view would assert that the correct quantum gravity theory can accommodate things that appear like classical lumps of material with gravitational fields that are very accurately described according to the principles of Einstein's classical general relativity, even if not quite precisely. (To my mind, the validity of this 'conventional viewpoint' could well be questioned, but if we believe the standard twin assumptions—that the quantum formalism needs no change, and also that classical general relativity is to hold for macroscopic bodies—then we must accept it. The nature of the present argument, after all, is to explore the limits of the compatibility of these two presumptions.) Accordingly, there ought to be some quantum state $|\chi\rangle$ and some quantum state $|\varphi\rangle$ that very accurately describe the lump of material, sitting on the horizontal platform on the Earth, in each of its two separate locations, where each lump occurrence is accompanied by its nearly classical Einsteinian gravitational field.[32] Since each of these two lump location states is taken to be stationary in its accompanying spacetime, each will have its respective associated Killing vector,[33] $\boldsymbol{\kappa}_\chi$ and $\boldsymbol{\kappa}_\varphi$, and will satisfy its appropriate Schrödinger equation with eigenvalue E:

$$i\hbar\, \boldsymbol{\kappa}_\chi|\chi\rangle = E|\chi\rangle \quad \text{and} \quad i\hbar\, \boldsymbol{\kappa}_\varphi|\varphi\rangle = E|\varphi\rangle.$$

In the previous situation, when we ignored the gravitational fields of the lumps, we were able to write down the Schrödinger equation for any superposition $w|\chi\rangle + z|\varphi\rangle$ and ascertain that all of these are stationary. However, now there is trouble, because these two Killing vectors $\boldsymbol{\kappa}_\chi$ and $\boldsymbol{\kappa}_\varphi$ are different. What are we to do? It seems that we need an invariant notion of '$\partial/\partial t$' that applies to the superposed spacetimes, and neither $\boldsymbol{\kappa}_\chi$ nor $\boldsymbol{\kappa}_\varphi$ seems to fulfil this need. We shall be seeing in the next section that this problem is not a minor one, but it presents us with a fundamental difficulty, and it leads us to a direct clash between the foundational principles of quantum mechanics and those of general relativity.

30.11 Fundamental conflict with Einstein's principles

It is important to elaborate rather more deeply upon the fact that these two Killing vectors differ. When I say that the Killing vectors $\boldsymbol{\kappa}_\chi$ and $\boldsymbol{\kappa}_\varphi$ are different, I mean this in a profound sense. They are actually vector fields on different spacetimes! One might try to take the view that these two spacetimes only differ in that they have very slightly different metric structures. Accordingly, we could try to think of them as being really one and the same space, but with slightly different metric tensor fields specified, say $\boldsymbol{g}_\chi$ and $\boldsymbol{g}_\varphi$. But to take this position is to part company with one of the very basic principles of Einstein's theory, namely the *principle of general covariance* (see §19.6). To regard the sets of points of these two spacetimes as being, in some sense, the 'same' sets of points would, in effect, be to specify a pointwise identification between the two spacetimes. That would be like identifying a point in one spacetime with the point in the other that has the same coordinate labels. But the principle of general covariance denies any significance to particular coordinate systems. Indeed, it asserts that there should be no preferred pointwise identification between two different spacetimes.

Why does this lack of identification between the spacetimes of the two lump locations cause difficulties? We need to be able to write down the Schrödinger equation. But without a unique '$\boldsymbol{\kappa}$', how are we to do that? The most immediate suggestion might be to try to identify $\boldsymbol{\kappa}_\chi$ with $\boldsymbol{\kappa}_\varphi$, but that would certainly be committing a violation of a fundamental principle of Einstein's theory, since it would imply that we are thinking of these two Killing vectors as inhabiting the same space, which is cheating! It seems to me that, in this situation, we are indeed beginning to witness a clash between the fundamental principles of quantum mechanics and of general relativity.

Nevertheless, we should not just 'give up' at this point. Although, strictly speaking, we should need the appropriate new theory in order to know what to do next, it seems to me that we can make some genuine progress if we are prepared to accept this clash for the moment and merely ask for some measure of the *error* that is involved in our 'cheating'. Let us take the position that, in some sense, what Nature would be prepared to do would be to allow the identification of two spacetimes locally, provided that the notion of 'free-fall' is the same in each. This is some kind of reflection of the principle of equivalence; see §17.4. Our attempted identification would try to have the notion of a geodesic in one space coincide with the notion of a geodesic in the other. This cannot normally be arranged except in the immediate neighbourhood of a point; so instead we shall try to compute the error that is involved in this if we do identify the two spacetimes. This kind of thing is hard to do in full general relativity, but we can apply most of these ideas also in the limiting situation when the speed of light c is taken to be infinity, while nevertheless retaining much of the basic philosophy of Einstein's theory. This situation leads us to Cartan's formulation of Newtonian gravity, as was discussed in §17.5.[34]

We recall, from Chapter 17 that in the Newton/Cartan gravitational scheme, the spacetime is really a fibre bundle over the 1-dimensional Euclidean space $\mathbb{E}^1$ of different allowable 'times' t. The fibres are the different Euclidean 3-spaces $\mathbb{E}^3$, each of which refers to 'space' at some given time. Thus, we actually have an 'absolute time', described by the time coordinate t. The reader may be excused for perhaps thinking that, since we now have the same notion of time (as measured by t) for the spacetimes of both lump locations, our problems should have gone away. But the sad truth is that knowing t does not enable us to know $\partial/\partial t$. For the operator $\partial/\partial t$ requires knowing that the remaining coordinate variables (say x, y, z) are being held fixed. This is the issue of the 'second fundamental confusion of calculus', considered in §10.3 (see Fig. 10.7). We can see the issue clearly by referring to the geometry that is involved. Knowing t tells us where the $\mathbb{E}^3$ sections lie, but knowing $\partial/\partial t$ would tell us a Killing vector field, which defines a family of curves cutting across this family of 3-surfaces; see Fig. 30.22. In fact, this broad issue of not being able to specify Schrödinger's $\partial/\partial t$ is considered to be a profound one even in more 'conventional' approaches to quantum gravity. It relates to the so-called 'problem of time' in quantum cosmology.[35]

In the present context, I am not trying to be so ambitious as to resolve all these issues. All we need is some estimate of the error involved if we try to make an 'illegal' identification of the different vectors $\boldsymbol{\kappa}_\chi$ and $\boldsymbol{\kappa}_\varphi$. We do this by actually identifying the $\mathbb{E}^3$s but then taking the total error in the difference between the gravitational accelerations (differences between free falls, i.e. geodesics) in the two spaces. Suppose that the gravitational accelerations are given, respectively, by the 3-vectors $\boldsymbol{\Gamma}_\chi$ and $\boldsymbol{\Gamma}_\varphi$. Then we estimate our error by forming the squared length of their difference $(\boldsymbol{\Gamma}_\chi - \boldsymbol{\Gamma}_\varphi)^2$ and integrating this over the whole of $\mathbb{E}^3$. This integrated error is interpreted as a measure of the absolute uncertainty in the definition of the '$\partial/\partial t$' operator needed for Schrödinger's equation, at the time t that specifies that particular choice of $\mathbb{E}^3$. This uncertainty directly leads, via

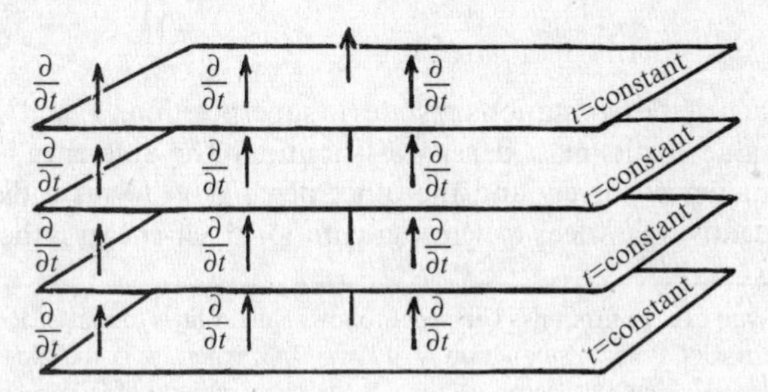

Fig. 30.22 Knowing t does not tell us $\partial/\partial t$ ('second fundamental confusion of calculus', see Fig. 10.7, §10.3); t tells us where the $\mathbb{E}^3$-sections lie, but $\partial/\partial t$ defines a family of curves cutting across this family of 3-surfaces.

Schrödinger's equation, to an absolute uncertainty E_G in the energy of the superposed states under consideration. The next step is to convert this expression for E_G into another (equivalent) mathematical form, which we can interpret[30.11] as:

E_G = gravitational self-energy of the difference between the two mass distributions in the states $|\chi\rangle$ and $|\varphi\rangle$.

The *gravitational self-energy* in a mass distribution is the energy that is gained in assembling that mass distribution out of point masses completely dispersed at infinity. The above difference could be thought of as the mass distribution in $|\chi\rangle$ taken positively, together with the mass distribution in $|\varphi\rangle$ taken negatively (see Fig. 30.23). (The reason that this does not just give us zero is that the energy is concerned with the effect of the gravitational field of each mass distribution on the other.)

This is a little difficult to appreciate, in ordinary terms, especially owing to the negative mass distributions involved. It is fortunate that, in the most usually considered situation, namely when the state $|\varphi\rangle$ is merely a rigid displacement of the state $|\chi\rangle$, then the quantity E_G can be interpreted more directly in another way. We consider the energy that it would cost to displace one instance of our lump, originally in location $|\chi\rangle$, but moved to location $|\varphi\rangle$, away from the gravitational field of the other, considered fixed in location $|\chi\rangle$. This energy turns out to be the same energy E_G as before, in the case of a rigid displacement,[30.12] but not always in other circumstances.

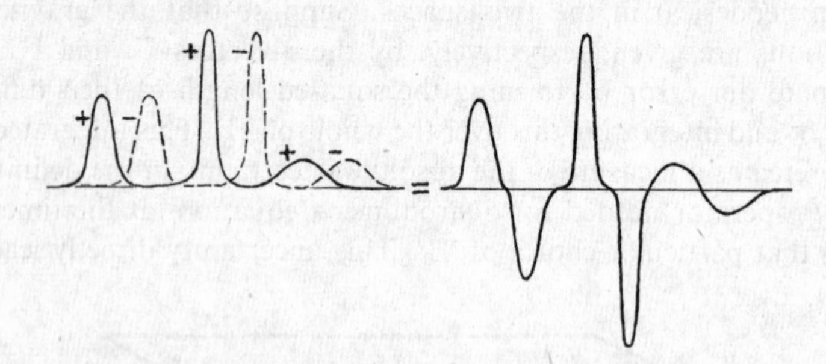

Fig. 30.23 Each of the two stationary states in superposition, $|\chi\rangle$ and $|\varphi\rangle$, defines an 'expectation value' for its mass density distribution. The difference between these two (i.e. one taken positively and the other negatively) forms a distribution of positive and negative mass density whose gravitational self-energy is the quantity E_G.

[30.11] See if you can confirm this. The proof follows similar lines to that of Exercise [24.3], in §24.3. We make use of Poisson's equation $\nabla^2\Phi = -4\pi\rho$, where Φ is the Newtonian (scalar) gravitational potential. Here our 'error' estimate is the space integral of $|\nabla\phi_1 - \nabla\phi_2|^2$.

[30.12] Can you see why this gives the same answer for E_G as before, in this particular situation? What would happen if the final location of the displaced lump is raised slightly with respect to its initial location? What happens if it is compressed?

In fact, one might consider adopting this second energy measure (namely the gravitational *interaction* energy) as an alternative definition of E_G. The first proposal, in terms of gravitational self-energy, seems to be better founded, as far as I can see, but one should not rule out other possibilities at the present stage of understanding. Diósi (1989) had considered both the above proposals, putting them to a purpose similar to the one that I am about to give, but proposing also a (stochastic) dynamics, which I am not doing here. These different suggestions (and some others) ought to be experimentally distinguishable, in experiments of the type that I shall come to shortly. However, it should be stressed that even the best founded of these proposals are somewhat incompletely motivated, and not totally free of controversy.[36]

So, what are we to do with our fundamental 'energy uncertainty' E_G? The next step is to invoke a form of Heisenberg's uncertainty principle (the time/energy uncertainty relation; see §21.11). It is a familiar fact, in the study of unstable particles or unstable nuclei (such as uranium U_{238}) that the average *lifetime* T having an inbuilt time uncertainty, is reciprocally related to an energy uncertainty, given by $\hbar/2T$. For example, as noted in §21.11, the lifetime of a U_{238} nucleus is about 10^9 years, so there is a fundamental energy uncertainty in each nucleus of about 10^{-51} Joules which translates, via Einstein's $E = mc^2$, to a mass uncertainty of about 10^{-44} of its total mass. Now, we are going to think of our superposed state $|\Psi\rangle = w|\chi\rangle + z|\varphi\rangle$ as being analogous to this, itself being unstable, with a lifetime T_G that is related, by Heisenberg's formula, to the fundamental energy uncertainty E_G discussed above. According to this picture,[37] any superposition like $|\Psi\rangle$ would therefore decay into one or the other constituent states, $|\chi\rangle$ or $|\varphi\rangle$, in an average timescale of

$$T_G \approx \hbar/E_G.$$

30.12 Preferred Schrödinger–Newton states?

The upshot of the above argument seems to be that a quantum superposition of two states ought indeed to decay into one or the other of its constituents in a time scale of the order $\hbar/E_G$. But the perceptive reader may well complain, at this point, that *any* quantum state $|\psi\rangle$ can be expressed as a linear superposition of a pair of other states (e.g. $|\psi\rangle = |\alpha\rangle + (|\psi\rangle - |\alpha\rangle)$, for any $|\alpha\rangle$). It would make no sense at all to regard all these states as decaying to such 'constituents', particularly if we choose $|\alpha\rangle$, for a given $|\psi\rangle$, so that the mass distributions in these alternatives differs sufficiently that the decay would have to be almost instantaneous!

An absurdity of this nature might be judged to be the conclusion of the above discussion even if we are to consider our superposition $|\Psi\rangle = w|\chi\rangle + z|\varphi\rangle$ to involve merely a single electron. For we could take $|\chi\rangle = |\alpha\rangle$, to represent the electron in an (almost) precise position. The mass distribution would be almost a delta function (§21.10), leading to an essentially infinite value for E_G, which would seem to imply an almost instantaneous reduction of the state $|\Psi\rangle$ to one or other of $|\chi\rangle$ or $|\varphi\rangle$. The same would hold for a system composed of point-like entities (e.g. quarks). Clearly this makes no sense; if such behaviour were true, there would be no quantum mechanics.

What we must do is be much more careful about what kind of states our $|\chi\rangle$ and $|\varphi\rangle$ are to be allowed to be. Recall that in the above argument, we considered $|\chi\rangle$ and $|\varphi\rangle$ to be stationary states. An electron in an (almost) fixed-position state is certainly not stationary. By Heisenberg's position/momentum uncertainty principle (§21.11), it would involve very large momenta and would instantly disperse. On the other hand, we seem to have a difficulty in applying the argument at all to individual particles if we require *exact* stationarity for both of $|\chi\rangle$ and $|\varphi\rangle$. For there are no stationary solutions of the ordinary Schrödinger equation, for a single free particle (of positive mass), that tail off towards spatial infinity.[30.13] The answer to this conundrum lies in the fact that we need to take into account the particle's gravitational field when writing down its Schrödinger equation. I am not asking for the gravitational field itself to be quantized, in this description, but merely that its effects be encapsulated in a Newtonian gravitational potential function Φ, whose source is to be what is called the 'expectation value' of the mass distribution in the wavefunction. It is perhaps not appropriate for this book for me to provide full description of what is involved here.[38] But this prescription does appear to give reasonable answers. The details of this are matters of active research. One concludes that for a single particle, this modified Schrödinger equation—I refer to this equation as the *Schrödinger–Newton* equation (on account of it incorporating a Newtonian gravitational field)—does indeed have well-behaved stationary solutions for a single particle that tail off appropriately towards infinity. (For a single electron, however, the spread in the wavefunction would exceed the extent of the observable universe; for a hydrogen atom, it would be a little less than the observable universe, the spread decreasing as the inverse cube of the mass of the particle.)

We now have what appears to be a plausible proposal for an objective state reduction which applies, at least, in situations when a quantum state is a superposition of two other states, each of which is stationary (in the aforementioned Schrödinger–Newton sense). According to this proposal,

---

[30.13] Why? (*Hint*: Have a look at Exercise [24.3] again).

such a superposed state will spontaneously reduce into one or the other of its stationary constituents in an average timescale of about $\hbar/E_G$, where E_G is the gravitational self-energy of the difference between the two mass distributions. I refer to this proposal as *gravitational* ***OR*** (where **OR** stands for the 'objective reduction' of the quantum state). For any pair of such constituent stationary states, the gravitational self-energy quantity E_G is indeed well defined. It refers to the difference between two mass distributions, each of these distributions being that same 'expectation value' expression used in defining the Schrödinger–Newton equation.

It is a feature of all other proposals for an **OR** scheme that they run into difficulties with *energy conservation*. In particular, the ingenious and ground-breaking proposal put forward by Giancarlo Ghirardi, Alberto Rimini and Tullio Weber, in 1986 ran into precisely this kind of trouble, as did various other proposals.[39] It has been a common attitude to 'live with' this problem, provided that the energy non-conservation can be reduced to an acceptably tiny level. My own perspective on this issue is to take it more seriously. There is the advantage with the gravitational **OR** scheme put forward above that the energy uncertainty in E_G would appear to cover such a potential non-conservation, leading to no actual violation of energy conservation. This is a matter that needs further study, however. It would seem that there is some kind of 'trade-off' between the apparent energy difficulties in the **OR** process and the decidedly non-local (and curiously 'slippery') nature of gravitational energy that was referred to in §19.8.

It is my own standpoint, with regard to quantum state reduction, that it is indeed an objective process, and that it is *always* a gravitational phenomenon. This would be the case even in situations where there has been substantial environmental decoherence leading to what might be considered as a FAPP state reduction, say in a system (such as a DNA molecule) that is much too small for gravitational **OR** to apply directly to it. In such situations, it would be the total displacement of mass in the environment that results in gravitational **OR**. In the particular situations that I have been considering, where the state in question is the superposition of two stationary states, I believe that this reduction process is indeed well approximated by the gravitational **OR** scheme that I have just described.

A full theory is certainly lacking, and I have provided no actual dynamics for the reduction of the state, according to this **OR** process, even in the case of the particular superpositions that I have been considering. In this respect, my proposal is a 'minimalist' one, and it does not aspire to a more complete dynamics, such as those inspirational proposals of Károlyházy; Károlyházy and Frenkel; Pearle; Kibble; Ghirardi, Rimini, and Weber; Ghirardi, Grassi, and Rimini; Diósi; Weinberg; Percival; Gisin; and others.[40] Nevertheless, my minimalist proposal seems to have clear experimental consequences, and I shall close this chapter by presenting the underlying idea for a class of actual experiments that have a definite potential for deciding whether or not such a gravitational **OR** scheme is really respected by Nature.

30.13 FELIX and related proposals

The basic scheme is to construct a 'Schrödinger's cat' that consists of a tiny mirror M, placed in a quantum superposition of two slightly differing locations, displaced from each other by about a nuclear diameter.[41] A reasonable size for this tiny mirror would be something comparable to a speck of dust, perhaps about one-tenth of the thickness of a human hair and containing something of the order of 10^{14} to 10^{16} nuclei (so its mass would be about 5×10^{-12}kg, and its diameter about 10^{-3} cm). Let us consider that this mirror M is placed in its superposition by the impact of a single X-ray photon which has been put in a superposition of two beams, one of which is aimed at M.

A possible experimental set-up is indicated in Fig. 30.24. The photon is produced by an X-ray laser L and directed at a beam-splitter B. The transmitted part (say) of the photon's resulting state is aimed at M, and its impact is such that it imparts a momentum to this tiny mirror when reflected from it. The mirror has to be of high quality, so it is of a 'rigid'

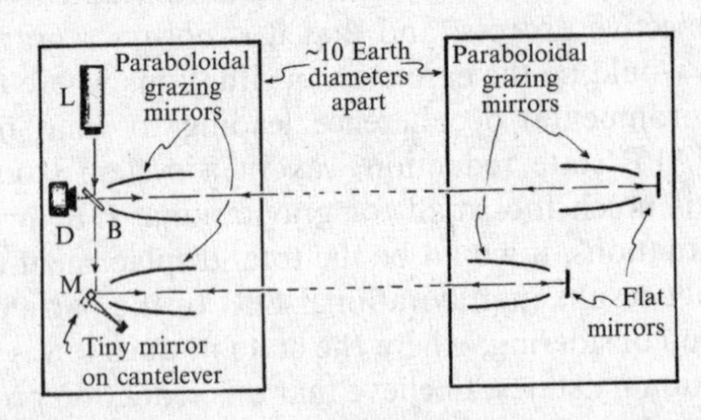

Fig. 30.24 FELIX (Free-orbit Experiment with Laser-Interferometry X-rays). A schematic set-up is indicated. A photon, produced by the X-ray laser L, is directed at the beam-splitter B. The transmitted part of the photon's resulting state is aimed at a tiny mirror M, roughly a 10-micron cube, the impact imparting a momentum to it when the photon is reflected. This puts M into a quantum superposition (Schrödinger's cat), which is to be held for, say, one second. In the meantime, the two parts of photon's wavefunction must be maintained coherently (here, by reflection between two space platforms) until this period has elapsed and whole process reversed. A perfect set-up (with equal path lengths) and conventional quantum mechanics would demand that the detector respond 0% of the time. Gravitational OR leads to an expectation of 50%.

nature, responding as a *whole* to the photon's impact without internal oscillations nor atoms being dislodged. The mirror is suspended in such a way that it would be restored to its original location in, say, one-tenth of a second (or even one second, as suggested in Fig. 30.24). In the meantime, the two parts of the photon's wavefunction have somehow to be coherently maintained, marking time until this period has elapsed, after which the entire process is to be reversed so that it can be ascertained whether phase coherence has been lost, as would indeed be the case if the quantum-superposed tiny mirror spontaneously reduces to one position or the other.

Keeping an X-ray photon coherent for one-tenth of a second is no mean task, however. (X-ray energies are unfortunately needed in order that a sufficient momentum can be imparted that an adequate movement in the tiny mirror occurs.) One suggestion for achieving coherence for this period of time is to perform the entire experiment in *space*, where the photon coherence is maintained by reflection between large mirrors on two space platforms of perhaps an Earth-diameter separation. It takes about one-tenth of a second for a photon to travel back and forth once over this distance. That part of the photon's wavefunction which had been reflected from M is then returned to M, whereas that part which had been reflected at the beam splitter B is returned to B. The timing is to be such that the entire physical process is precisely reversed. Thus, the part of the photon's wavefunction that was responsible for the M's motion encounters M again just as M returns to its original position, so the photon recovers the momentum that it had lost to M and reduces M to rest; moreover, the two parts of the photon's wavefunction are timed to recombine at the beam-splitter B. Provided that there has been no loss of phase coherence throughout this activity and path-lengths chosen appropriately, the photon's wavefunction will combine into the single beam aimed back into the laser L. Thus, a detector placed at the 'alternative' location D, that the photon might have arrived at when emerging from the beam-splitter B (see Fig. 30.24), will detect *nothing*. This has been termed the FELIX proposal (Free-orbit Experiment with Laser-Interferometry X-rays).

We note that for about one-tenth of a second, M's state will be a *superposition* of being displaced and not displaced, this being essentially the same situation as with the lump of material described above, illustrated in Fig. 30.20. According to the gravitational **OR** scheme, M's state should spontaneously reduce into having been displaced *or* not having been displaced in a timescale of the order of one-tenth of a second. The photon's state is entangled with that of M, so, as soon as M's state reduces, the photon's state also reduces with it. Then the photon is in either one beam or the other, so that when it finally returns to the beam-splitter B it will have equal probabilities of activating the detector D or of returning to the laser L. This procedure would then be repeated many

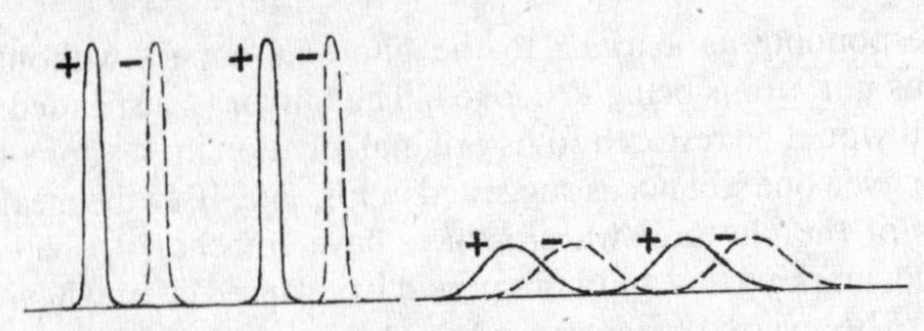

Fig. 30.25 A significant factor would be the amount of 'spread' in the mass distribution of the nuclei in the mirror. For a given total mass, more tightly localized mass distributions would give shorter reduction times.

times. The effect of **OR** would be that the detector responds in about 50% of the trials; whereas if, in accordance with standard quantum mechanics (for a perfect experiment), phase coherence is *not* lost, the detector does not respond at all.

Of course, in any practical situation, there could be many other ways that phase coherence might get lost. For this experiment to be successful, it would be necessary for these to be kept at a very low level, so that the particular signature of gravitational **OR** can be distinguished. The experiment would have to be repeated many times, using different tiny mirror sizes and materials, varying the timescale (perhaps using repeated reflections between space platforms). An important factor, in the particular **OR** scheme under consideration, would be the amount of 'spread' in the mass distribution of the nuclei in the tiny mirror. For a particular total mass, a more tightly localized mass distribution would give a shorter reduction time; see Fig. 30.25.

The above FELIX proposal is technically extremely difficult, for a number of reasons. A major problem would be the required precision of aiming of X-ray photons between space platforms of about 10 000 km apart. In any case, space experiments are inherently difficult and very expensive, and if there is a feasible ground-based alternative, this can have many advantages. Fortunately, it appears that such an alternative is indeed a practical possibility. Owing to an ingenious suggestion of William Marshall and some equally clever ideas for its implementation by Dik Bouwmeester and Christoph Simon, a feasible ground-based alternative indeed seems on the cards, and is now under active investigation. The proposal[42] is that instead of having a single X-ray photon's impact to produce the desired tiny mirror movement, a photon of considerably lower energy (such as a visible-light or even infra-red photon) could be used, reflected backwards and forwards (say) $\sim 10^6$ times so that there are now 10^6 impacts on the tiny mirror by the same photon, in place of the single impact of the X-ray photon as proposed earlier; see Fig. 30.26. At the time of writing, it seems that there is no fundamental obstruction to a preliminary experiment of this type being performed in a couple of years or so.

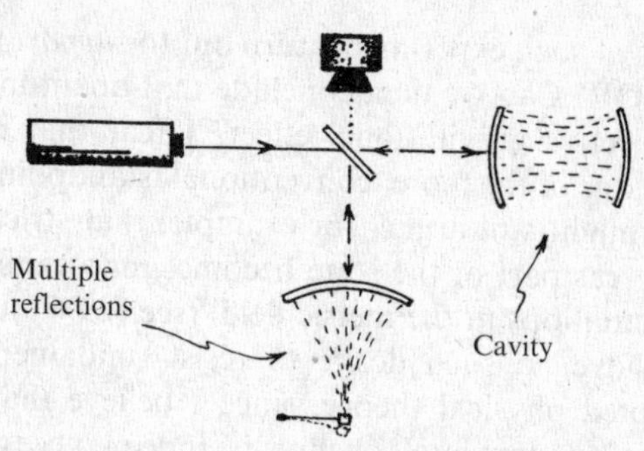

Fig. 30.26 A more practical version of 'FELIX' does not use X-rays, but requires some 10^6 impacts on the tiny mirror by a visible light photon, in place of the single impact by an X-ray photon.

If performed successfully, this preliminary experiment might still be some five or six orders of magnitude short of what is required for a definitive test of gravitational **OR**. Nevertheless, if quantum coherence for the superposition of the two tiny mirror locations can be maintained, this would represent an advance (in terms of mass) over the current 'record Schrödinger's cat' (C70 fullerene molecules[43]) by a factor of perhaps 10^{12}. It would seem likely that if this stage can be successfully reached, where the 'minimalist' gravitational **OR** scheme of §§30.9–12 predicts *agreement* with standard quantum mechanics, then the additional improvements needed to test the novel predictions of gravitational **OR** may well also be forthcoming within a few additional years.

It is perhaps remarkable that the extremely tiny gravitational energy uncertainty E_G that occurs in this class of experiment—say some 10^{-33} of a joule—is sufficient to give such a 'reasonable' collapse lifetime of one-tenth of a second or less. The smallness of gravitational effects, generally, has tended to lead many physicists to dismiss them altogether. Yet, we see that the effects of bringing gravitational considerations into our quantum picture could have profound observational consequences. It should be noted that the timescale $\hbar/E_G$ involves the *quotient* of the two small quantities $\hbar$ and G, and so need not be a small quantity in ordinary human terms. This is in stark contrast with the characteristic quantum-gravity quantities, the Planck length and Planck time (§27.10, §31.1), of sizes 10^{-33} cm and 10^{-43} s, which are absurdly small, and arise from the *product* of $\hbar$ and G.

Let us imagine that an experiment to test gravitational **OR** has been successfully performed. If phase coherence is *not* lost in the timescales predicted by the gravitational **OR** scheme outlined above, that particular scheme will have to be abandoned—or at least severely modified. But what if that the results of such experiments turn out to *support* these predictions of gravitational **OR**? Can we then conclude that quantum state-reduction is indeed an objective gravitational effect? I fear that many might still prefer to hold to one of the more 'conventional' standpoints with regard to this issue. They might still argue, for example, that strict unitarity **(U)** is maintained, whereas part of the state becomes inaccessible—perhaps lost in 'quantum fluctuations in the metric field' (see §29.6 and §30.14).

Personally I have no such desire to resist fundamental change in a previously accepted physical theory, since I believe that, in the case of quantum theory, fundamental change is indeed necessary—as I have argued at length above. But perhaps it is not too fanciful to make a comparison with those views of many highly esteemed physicists, such as Lorentz, who preferred to regard the effects of *special relativity* as merely 'corrections' to be applied within a 19th century world-view that accepts an absolute state of rest. No doubt there would be as many esteemed physicists who, likewise, might be as reluctant to relinquish their hard-won 20th century world-view of quantum mechanics, if it actually turns out that the predictions of gravitational **OR** *are* supported by a successfully performed experiment of the FELIX type. In my opinion, such a standpoint would be retrograde, and would relinquish the possibility of powerful new progress to be made, on the basis of a new quantum picture of the world that might actually make sense!

Of course, those of us who are expecting gravitational **OR** to bolster our less conventional standpoints must be prepared for the alternative eventuality that our views may be *contradicted* by such an experiment. My own reaction to this would be considerable bewilderment, despite the fact that many quantum physicists with whom I have discussed this issue have expressed the firm expectation that conventional quantum mechanics must again come through unscathed. My own bewilderment would arise primarily from a conviction that present-day quantum mechanics has no credible ontology, so that it *must* be seriously modified in order for the physics of the world to make sense. This does not in itself imply that it is *gravitational* **OR** that has to come to our rescue, nor is it imperative that the particular gravitational proposal outlined here must be the correct one.[44] Nevertheless, I feel that the steadfastness and resilience of modern quantum theory will not allow it to be shifted easily. In my own view, any such shift would require the agency of something equally formidable, and nothing else in known physics is of such stature save Einstein's general theory of relativity and its deep motivating principles. It is these that lead me to anticipate a gravitational **OR** scheme like the one I have been suggesting above. Whatever the final result of such deliberations, I anticipate many powerful and intriguing *new* quantum-mechanical issues to be raised and answered during the course of the 21st century!

30.14 Origin of fluctuations in the early universe

Before bringing this chapter to a close, I wish to raise just one of those numerous important issues that might indeed be profoundly affected by a change in the rules of quantum theory, in accordance with the deliberations of this chapter. In §27.13, I drew attention to the extraordinarily special state in which the universe appears to have started out. The main way in which this state was special, and which gave it its absurdly low entropy, was a very precise spatial isotropy and homogeneity, so that the universe's spacetime geometry is (still) in remarkably close accord with one of the standard cosmological FLRW models (§27.11). Of course, as is often argued, the universe cannot have been, absolutely and precisely, such a symmetrical model. If such high symmetry was *once* there, it must have remained there for all time; because the dynamics of Einstein's general relativity—and of the rest of classical physics—will preserve such symmetry precisely.

But what about *quantum* physics? Does not the 'randomness' inherent in the quantum evolution processes allow for deviations from this exact symmetry to arise? The notion of 'quantum fluctuations' is frequently invoked at this stage, as a means to providing the needed slight deviations from exact symmetry. The idea is that such 'fluctuations' might start out as tiny, but they would act as the seeds of irregularity in the mass distribution, which would be gradually increased through gravitational clumping, so that, eventually, stars, galaxies, and clusters of galaxies would be able to develop—in accordance with observation.

But what *are* quantum fluctuations? It is a feature of Heisenberg's uncertainty relations (§21.11), as applied to field quantities (see §§26.2,3, §26.9), that, if one tries to measure the value of a quantum field in some very small region to great accuracy, this will lead to a very large uncertainty in other (canonically) related field quantities, and hence to a very rapidly changing expected value of the quantity being measured. Thus, the very act of ascertaining the precise value of some field quantity will result in that quantity fluctuating wildly. This quantity could be some component of the spacetime metric, so we see that any attempt at measuring the metric precisely will result in enormous changes in that metric. It was considerations such as these that led John Wheeler, in the 1950s, to argue that the nature of spacetime at the Planck scale of 10^{-33} cm would be a wildly fluctuating 'foam' (see end of §29.6 and Fig. 29.6).

To clarify this picture, we must recall carefully what Heisenberg's uncertainty relations actually state. They do not tell us that there is something inherently 'fuzzy' or 'incoherent' in the way that nature behaves at the tiniest scales. Instead, Heisenberg uncertainty restricts the precision whereby two non-commuting measurements can be carried out. We recall that, for a single particle, both its position and momentum in some direction, being non-commuting, cannot be determined precisely at the same time, the product of their respective errors being not less than $\frac{1}{2}\hbar$ (§21.11). There is a perfectly well-defined quantum state, however, and if no actual measurement is performed, the state of the particle will evolve precisely, according to Schrödinger's equation (assuming that standard **U**-quantum mechanics holds).

Similarly, in standard quantum mechanics, all the variables defining a spacetime state cannot be determined together. The quantum description of spacetime should nevertheless be perfectly well defined. But the Heisenberg principle tells us that this description cannot resemble a classical (pseudo-)Riemannian manifold, as different spacetime geometric quantities do not commute with one another. Instead, according to Wheeler's picture, the state would consist of a vast superposition of different geometries, most of which would deviate wildly from flatness and so have the 'foamlike' character that he envisages.

Let us see how this applies to the state of the early universe. Can the deviations from exact symmetry indeed be attributed to 'quantum fluctuations', if the entire initial state possesses exact FLRW cosmological symmetry? The **U**-evolution of this state must continue to maintain this exact FLRW symmetry, irrespective of 'quantum fluctuations' or any other manifestation of Heisenberg uncertainty.[30.14] How is this consistent with the highly irregular 'foam-like' geometry that Wheeler envisages? There need not be any contradiction here because the entire state is a *superposition* of such irregular geometries, not an individual geometry. The superposition itself can possess a symmetry not possessed by the individual geometries of which it is composed. If one irregular geometry contributes, so do all the others obtained from that by the application of each FLRW symmetry.[45]

How then is this FLRW-symmetric vast quantum superposition of irregular geometries supposed to give rise to something resembling one specific 'almost FLRW-symmetric' universe which is perturbed only in some very minor way that is consistent with observations? It should be clear to the reader that there is no way that this can happen entirely within the **U**-evolution of standard quantum mechanics, since this must exactly preserve the symmetry. There must be something of the nature of an **R**-process taking place, which resolves this vast superposition of geometries into a single geometry or, rather, into some lesser superposition of geometries that more resembles a single geometry. The key is that irregularities arising from 'quantum fluctuations' cannot come about without some **R**-like action, whereby the single initial quantum state somehow resolves itself into a probability mixture of different states. This takes us back to the issues addressed in Chapter 29, where different attitudes to the 'reality' of **R** were discussed.

[30.14] Can you see why the maintaining of this symmetry follows merely from the deterministic *uniqueness* of **U**-evolution, together with a very weak general assumption about **U**-evolution?

We should bear in mind that we are here concerned with the very early universe, where the temperature would have been perhaps some 10^{32}K. There were no experimenters around at that time performing 'measurements', so it is hard to see how the standard 'Copenhagen' perspective ((a) of §29.1) can be applied. What about the many-worlds view ((b) of §29.1)? In that picture, there is no actual **R**, and the FLRW-symmetric state of the universe would be maintained until the present day, this state being representable as a grand superposition of many constituent spacetime geometries. Only when conscious observers try to make sense of the world, according to this view, would the resolution into alternative spacetime geometries be deemed appropriate—there now being a superposition of conscious observers, each one perceiving a single 'world'.[46] On the 'FAPP' view ((c) of §29.1), the presence of (sufficient) environmental decoherence is regarded as the signal, whereby our quantum superposition of different geometries is permitted to be regarded as a *probability mixture* of different geometries.

It is illuminating to make a comparison with an example in ordinary quantum mechanics.[47] Imagine a radioactive nucleus at rest, in a spherically symmetric state (i.e. spin 0, see §22.11) at some point O, centrally situated within a bubble chamber[48] (Fig. 30.27). Suppose that, by nuclear fission, it splits into two parts A and B, which are ejected in opposite directions from O. We may suppose that A and B are electrically charged, so that they leave tracks in the bubble chamber. In this example, we started with a state with spherical symmetry, centred at O. Yet, after decay, the spherical symmetry is broken by the axis along which the parts A and B have emerged. How are we to understand this in terms of the **U**-evolution

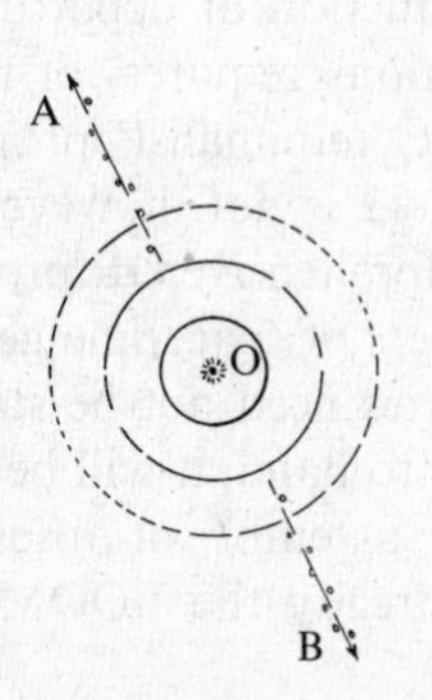

Fig. 30.27 Symmetry breaking by **OR**. A (spherically symmetrical) spin-0 nucleus splits into two parts, which are observed to occupy a specific pair of oppositely directed paths. The **U**-evolution of the initial state preserves spherical symmetry, but this consists of an (entangled) quantum superposition of pairs of opposite paths (Mott). **R** results in only one of these being perceived. This example is being taken as an illustrative model of what might be happening in the creation of density fluctuations in an initially highly symmetrical early-universe quantum state.

of the original state? Clearly, as stated above, spherical symmetry must be preserved, but the state achieves this by being composed as a linear superposition of all the possible situations given by different axis directions. The wavefunction has the form of a spherical wave centred at O—although we must bear in mind that the state is an entangled one involving both A and B, where each location of A is correlated with a location of B in the antipodal direction. As the influence of the charges on A and B begins to ionize the material in the bubble chamber, and bubbles are formed, the state becomes entangled with this material, so we find that the entire state consists of a superposition in which each component involves a pair of tracks of bubbles in opposite directions, one corresponding to the passage of A, and the other, the passage of B.

The situation just described is not essentially different from that in the early universe. Some version of **R** is needed in order that the symmetrical quantum superposition can be replaced by a probability mixture of less symmetrical alternatives. It seems that, in practice, theoreticians tend to adopt some form of FAPP interpretation ((c) of §29.1), where the size of the cosmological horizon is arbitrarily (and illogically) taken to supply some kind of 'cut-off' to quantum entanglements. The quantum superposition is then regarded as a probability mixture—although this actual position is hardly ever made clear. For example, in their graduate-level textbook 'The Early Universe' the prominent inflationary cosmologists Kolb and Turner (1994) assert, on p.286:

> As each mode crosses outside the horizon, it decouples from the microphysics and 'freezes in' as a classical fluctuation.

The 'mode' here refers to a component of a quantum superposition, so we see that the authors are attempting to use the horizon as somehow allowing a passage from a quantum amplitude to some probability of an actual classical alternative. This appears to be something along the lines of a FAPP proposal (see §29.6) and, as was argued in §§29.6,8, is strictly speaking, illogical.[49]

In my own view, it is clear that the introduction of departures from exact FLRW symmetry via quantum fluctuations requires, of necessity, some theory of objective state reduction. The 'minimalist' proposal for gravitational **OR** that was set forth in §§30.9–12 is not, however, strong enough as it stands. One needs some more comprehensive **OR** proposal, in which, quantum superpositions of large numbers of spacetime geometries can be handled, where the individual geometries need not be stationary, as they were in §30.10. When such a scheme is to hand, it will be immediately confronted with a mountingly impressive assembly of observational data, against which it must stand or fall. Already the BOOMERanG,

WMAP, and other observations have provided enormous quantities of data concerning density/temperature fluctuations in the early universe, and there will be a great deal more from other experiments now in the pipeline.

A final comment on this situation is appropriate here. We noted how the symmetrical state in our above example of nuclear fission was a highly entangled one. This would be true also, and to an even greater extent, for the state reductions that take us away from an initially FLRW-symmetrical state to a universe subject to 'quantum fluctuations'. Thus, in accordance with our discussion of EPR states given in §§23.3–6, we have 'Bell-inequality violations' that provide correlations between distant events that appear to violate classical causality. Such apparent causality violation need not be indicative of a mechanism such as inflation which would serve to bring such separated events into causal contact, but could arise as a result of any appropriate objective state reduction scheme (**OR**). However, we see from the discussion just given[50] that even within the standard FLRW cosmologies, such apparent 'causality violation' can occur, without the need for any inflation, if the initial fluctuations come about via some objective state-reduction scheme.

It is clear that we are far from a theory which can reliably address all these issues. But I hope, at least, that I have been able to persuade the reader of the fundamental importance of having a quantum mechanics with a viable ontology. The issues that are addressed in Chapters 29 and 30 of this book are not just matters of philosophical interest. The importance of having an ontologically coherent (improved) quantum mechanics cannot, in my view, be over-estimated. In this section, I have touched upon just one of the foundational issues that could be deeply affected by knowledge of such a theory. There are many more, including situations in biology (see §§34.7,10), where as with the early universe, the present-day 'Copenhagen' viewpoint cannot really be applied—there being no clear division into a quantum system and a classical measuring device.

Notes

Section 30.1

30.1. See Roseveare (1982).

Section 30.2

30.2. See Penrose (1980).

Section 30.3

30.3. There is no theoretical or technical bar to this, at least if we do not demand 100% accuracy. For example, one could arrange that the initial photon is always one of a pair (produced, say, by parametric down-conversion—see §23.10), with the other member of the pair triggering the registering device.

30.4. I find it remarkable how much difficulty people often have with this argument. The matter is perhaps clarified if we contemplate numerous occurrences of this experiment, taking place at various locations throughout spacetime. There are four alternative photon routes to be considered, SBD, SBC, FBD, and FBC. To see what the various probabilities are, we ask for the proportion of SBD, given S (forward-time situation), or for the proportion of SBD, given D (backward-time situation). The squared-modulus rule correctly gives the actual answer (50%) in the first case, but it does *not* give the actual answer (nearly 100%) in the second case.

30.5. They do, however, depend upon the initial state being what it is supposed to be and not part of some entangled state (§23.3) that might also involve something in the detector. We might raise the question as to whether the time reversal of such entanglements could be responsible for the time-reversed squared-modulus rule giving completely the wrong answers. But I am unable to see how to construct any plausible explanation along these lines. Perhaps some enterprising reader can do better.

30.6. See Aharonov and Vaidman (1990).

30.7. For a discussion of this issue, see Aharonov *et al.* (1964).

30.8. See Aharonov and Vaidman (2001); Cramer (1988); Costa de Beauregard (1995); and Werbos and Dolmatova (2000).

Section 30.4

30.9. See Unruh (1976); see also Wald (1994).

30.10. See Penrose (1968b, 1987b) and Bailey *et al.* (1982).

30.11. See Kay (2000); Kay and Wald (1991); Kay, Radzikowski, and Wald (1996); Hollands and Wald (2001); Haag (1992).

Section 30.5

30.12. See Wald (1984).

30.13. See Wald (1984); Synge (1950); Kruskal (1960); Szekeres (1960).

30.14. There is a slightly confusing discrepancy between the interpretation of 'τ' as *actual* time in the Schwarzschild case, considered here, whereas it is $r_0\tau$ that measures the accelerating observer's time in the flat (Rindler) case of Fig. 30.5a,b.

30.15. See Newman *et al.* (1965).

30.16. This gyromagnetic ratio refers to a 'pure Dirac particle', to which an electron is an excellent approximation, but an actual electron is subject to *radiative corrections* that come from quantum field theory, see end of §24.7. A proton or neutron is much further from being a Dirac particle, but that notion applies much more closely to their constituent quarks.

Section 30.6

30.17. See Novikov (2001); Thorne (1995a); Davies (2003).

30.18. Davies (2003) gives an amusing and readable discussion of such possibilities.

Section 30.7

30.19. See Penrose (1969a); Floyd and Penrose (1971).

30.20. See Blanford and Znajek (1977); Begelman *et al.* (1984). See also Williams (1995, 2002, 2004).

Section 30.8

30.21. See Hawking (1974, 1975, 1976a, 1976b); Kapusta (2001).

30.22. See Preskill (1992).

30.23. See Preskill (1992), or, for a different outlook Kay (1998a, 1998b, 2000).

30.24. See Preskill (1992); Susskind *et al.* (1993).
30.25. See Gottesman and Preskill (2003) for a critique of Horowitz and Maldacena (2003). See also Susskind (2003).
30.26. Hawking introduced a generalization of unitary evolution in which the S-matrix description of ordinary QFT (§26.8) is generalized to what he referred to as a 'super-scattering' operator (not connected with supersymmetry; see §31.2) denoted by a '$' sign. This operates between density-matrix states, rather than between the pure states treated by the S-matrix. See Hawking (1976b). It should be remarked that, as of 2004, Hawking has (regrettably, in my view) *retracted* his LOSS position, in favour of RETURN!

Section 30.9

30.27. The main disagreement between Stephen Hawking and me has, for about 20 years, centred on this temporal-asymmetry question. Throughout these arguments, he has steadfastly held to a time-symmetric physics and to either the immutability of **U**-quantum mechanics or the mild generalization referred to above (Note 30.26). As I shall be explaining, my own position is quite different on these matters.
30.28. See Penrose (1979).
30.29. The Weyl curvature hypothesis refers to *classical* geometry, so it says something about what happens just at the point where 'quantum geometry' crystalizes into a classical spacetime.
30.30. See Hawking (1976a, 1976b) and Gibbons and Perry (1978).
30.31. Perhaps some generalized notion of Hilbert space is required here, which could also take on some of the properties of a (curved) phase space, e.g. Mielnik (1974); Kibble (1979); Chernoff and Marsden (1974); Page (1987); and Brody and Hughston (2001).

Section 30.10

30.32. These might be *coherent states* as referred to in §26.6.
30.33. Bear in mind that the indices on $\boldsymbol{\kappa}_\chi$ and $\boldsymbol{\kappa}_\varphi$, are just labels and are not 'tensor indices' in the sense of §12.8. The same applies to $\boldsymbol{g}_\chi$ and $\boldsymbol{g}_\varphi$.

Section 30.11

30.34. See Christian (1995).
30.35. See Isham (1992); Kuchar (1992); Rovelli (1991); Smolin (1991); Barbour (1992).
30.36. See Note 29.12. for many of the objective state reduction theories. Those of Diósi, Percival, Kibble, Pearle, Squires, and myself involve gravitation crucially.
30.37. More recently, an idea has arisen for a more rigorous justification of this kind of proposal for gravitational **OR**. We recall from Exercise [21.6] that to make quantum theory consistent with the principle of equivalence, a phase factor involving a cubic term in the time t is needed, when passing from a freely falling frame to one fixed in a gravitational field. Accordingly, the two frames strictly describe different vacua (see §26.5), this being the remnant of the Unruh effect, mentioned in §30.4, that survives in the Galilean limit. Thus, if the principle of equivalence is to be fully respected, the superposition of two gravitational fields will involve the superposition of different vacua, and so should be unstable, even in the Galilean limit. Details of this argument will be published later.

Section 30.12

30.38. See §22.5 and Moroz *et al.* (1998). For 'expectation value', see Note 22.11.
30.39. See Note 29.12. for many of the seminal references in this field.

30.40. I have gained much from studying such proposals. Perhaps some of these may supply pointers towards a more complete gravitational OR theory. See Note 29.12, as well as Gisin (1989, 1990) for certain NO-GO theorems.

Section 30.13

30.41. The specifics of this proposal have had significant inputs from several colleagues. An important ingredient of the original idea (which involved the impact on a 'Mössbauer-type' crystal by a beam-split photon) came from Johannes Dapprich—some more specific ideas, including suggested parameters for the tiny mirror's size, photon energy, and many other things, arose in conversations with Anton Zeilinger and others in his experimental group in (at that time) Innsbruck. The idea of a space-based experiment (FELIX) arose from discussions with Anders Hansson. The ingenious ideas that appear to provide a more practical ground-based alternative are due to William Marshall, Dik Bouwmeester, and Christoph Simon. See Penrose (2000a), Marshall *et al.* (2003).

30.42. See Marshall *et al.* (2003).

30.43. See Arndt *et al.* (1999).

30.44. For example, both, the original gravitational **OR** scheme of Károlyházy (1974) and the more recent such proposal of Percival (1994), make very different predictions from those being put forward here.

Section 30.14

30.45. There is a subtlety here, however, because one might consider that the action of an abstract symmetry on a spacetime geometry simply yields the same geometry again (because of the principle of general covariance, see §19.6). There are different attitudes that one can take on this issue, but in any case, the general point raised in the text is not affected.

30.46. On Wheeler's own variant, the 'participatory universe', see Wheeler (1983), it would be the ultimate presence of conscious observers who somehow (teleologically) determine the particular selection of spacetime geometry that occurred in the early universe.

30.47. This bears some similarity to a discussion of the cloud-chamber tracks in α-particle emission, due to Neville Mott (1929).

30.48. This is a standard piece of apparatus, whereby the passage of a charged particle is indicated by a string of tiny bubbles; see Note 30.47, see Fernow (1989).

30.49. The real reason for a 'cut-off' at the Hubble radius (where recession reaches the speed of light), at that epoch, is not directly to do with the actual 'horizon size' (which is, in any case, far larger than the Hubble radius in the inflationary scheme; see Fig. 28.5) and is nothing to do with the passage from quantum to classical physics. It is a purely classical effect of the expansion of the universe on a field subject to the constraints of relativity.

30.50. Various people seem to have suggested that such 'acausal' EPR-type correlations could have been present in the early universe fluctuations, for such reasons. For example, a suggestion of this nature was put to me some years ago by Bikash Sinha.

31
Supersymmetry, supra-dimensionality, and strings

31.1 Unexplained parameters

Most physicists probably have ideas quite different from those outlined in the previous chapter, concerning what the physics of the 21st century may have in store for us. Very few of them appear to anticipate that there will be fundamental changes in the framework of quantum mechanics. Instead they argue for strange-sounding ideas like the need for extra dimensions to spacetime, or for point particles to be replaced by extended entities known as 'strings', or perhaps by higher-dimensional structures called 'membranes', or p-branes, or simply 'branes'—and where curious additional objects called 'D-branes' seem to play important roles. There are puzzling extensions of the idea of symmetry referred to as 'supersymmetry', or of 'quantum groups'. There are generalizations of the very notion of geometry described as 'non-commutative', and there are pictures of the world in which discreteness rather than continuity holds sway at the tiniest levels, or where the fabric of space itself consists of knots or links. There are suggestions that the very notion of spacetime will have to be abandoned, or reformulated in some other terms.

What are these various ideas and what are we to make of them? More importantly, what motivates so many physicists to describe a 'reality' bearing little resemblance to what we directly perceive at ordinary human scales. No doubt, part of the reason for contemplating such proposals lies in the success of quantum mechanics and, to a lesser extent, of general relativity. These 20th-century theories have shown us how our direct intuitions can mislead, and 'reality' may differ profoundly from those pictures provided by the physics of previous centuries. Yet, merely to be presented with a scheme for the world that is exotic or unusual does not give us grounds for believing it. We shall need to try to understand something of the underlying motivation of the research of modern theoreticians, as they attempt to probe more deeply into the inner workings of the universe.

We must pick up the threads of reasoning that we first encountered in Chapter 24 and continued in Chapters 25 and 26, where the combined requirements of special relativity and quantum theory forced us into the quagmire of the quantum theory of fields. This, in turn, led us into a minefield of infinities, and it needed great ingenuity to circumvent most of these, ultimately leading to the standard model of particle physics, which finds good accord with the measured workings of Nature. Yet the standard model itself is not free of infinities, being merely a 'renormalizable' rather that a finite theory. Renormalizability just allows certain calculations to be performed, giving finite answers to most questions of interest within the theory, but it does not provide us with any handle on certain of the most important parameters, such as the specific values of the mass or electric charge of particles described by the theory. These would have come out as 'infinity' (or perhaps 'zero'), were it not for the renormalization procedure itself, which evades these infinite scalings through a redefinition of terms, and allows finite answers for *other* quantities to be obtained. Basically, one 'gives up' on mass and charge, whose values are just inserted into the theory as unexplained parameters; indeed, there are some 17 or more such parameters, including coupling constants of various kinds in addition to the mass values of the basic quarks and leptons, the Higgs particle, etc. that need to be specified.

There are considerable mysteries surrounding the strange values that Nature's actual particles have for their mass and charge. For example, there is the unexplained 'fine structure constant' α, governing the strength of electromagnetic interactions, which is defined by the formula

$$\alpha = \frac{e^2}{\hbar c},$$

where $-e$ is the electron's charge. The reciprocal of the fine structure constant takes the value $\alpha^{-1} = 137$ rather closely, but more accurately

$$\alpha^{-1} = 137.0359\ldots.$$

For a number of years, some physicists thought that α^{-1} might actually take the *exact* value 137. In particular, Sir Arthur Eddington (1946) spent the latter part of his life trying to produce a 'fundamental theory', one consequence of which would indeed be '$\alpha^{-1} = 137$'. Many of today's physicsts might be less optimistic than their predecessors about finding a direct mathematical 'formula' for α, or for other 'constants of Nature'. Nowadays, physicists tend to regard these quantities as functions of the *energy* of the particles involved in an interaction, rather than simply as numbers, and refer to them as 'running coupling constants' (see Note 26.21). The observed scalar values that we refer to as 'constants of Nature' would then be 'low energy limits' of these 'running' values. Although one might still hope to find a purely mathematical reason for these specific limiting values, somehow such values may seem less 'fundamental' than they would if there were no dependence on energy.

It is often revealing to express quantities like charge and mass in terms of the *absolute* (Planck) units introduced in §27.10, for which Newton's gravitational constant G, the speed of light c, Dirac's form of Planck's constant $\hbar$, and Boltzmann's constant k are all put equal to unity:

$$G = c = \hbar = k = 1.$$

In these units, the charge on the proton (or minus that on the electron) comes out as roughly $e = 1/\sqrt{137}$, and more precisely as[1]

$$e = 0.085\,424\,6$$

and the basic quark charge (minus the charge of the down quark—see §25.6) has one third of this value. Absolute units are usually referred to as *Planck* units (or sometimes *Planck–Wheeler* units), because Max Planck (of quantum-mechanical fame—see §21.4) put an idea of this nature forward in a paper published in 1906. Ironically, in this paper, he used the electric charge as a basic unit, rather than his own 'Planck's constant' to fix things, and in that scheme we simply have $e = -1$. (The charge mystery has not gone away, of course, because, in his own scheme, $\hbar = 137.036$.) It was John Wheeler (e.g. 1973) who later emphasized the importance of these ideas in many of his writings (using $\hbar$, rather than Planck's choice of electric charge).

If that were all, then Planck units might more appropriately be called *Stoney* units, because the Irish physicist George Johnstone Stoney (who first measured the electron's charge) put forward the same idea as Planck did in 1906, but published way back in 1881. However, there is another paper by Planck, published in 1899, actually before his famous paper of 1900 which initiated quantum theory, in which 'Planck's constant' was used to define absolute units. Accordingly, I shall stick to the conventional terminology which refers to absolute units as 'Planck units'!

What about the mass values of particles? The problem of mass is a much thornier one than that of electric charge. It appears to be the case that all particles of Nature have charge values that are integral multiples of one basic charge. We can take this to be the charge on the proton if we are concerned only with particles that can exist freely on their own, or minus the down-quark charge, if we wish to include the internal constituents of hadrons. Although there is, as yet, no full understanding of this fact, and certainly no proper understanding of 137.036, this problem seems to be a good deal more manageable than the corresponding one for mass values. One of the mysterious aspects of the mass problem is the absurdly tiny size that the values that the masses of ordinary particles have when measured in absolute units. For example, the mass m_e of the electron, in absolute units, is about

$$m_e = 0.000\,000\,000\,000\,000\,000\,000\,043$$

and that of the proton is only about 1836 times this value. The mass of the electron neutrino ν_e is less than 10^{-5} of the above value. Another way of expressing the puzzle of these tiny mass values is to ask why the natural 'Planck mass', being a macroscopic 10^{-5}g (about the mass of a small midge), is so much larger than the masses of all the basic particles encountered in Nature. Yet another way of phrasing this puzzle is to ask why the Planck distance of 1.6163×10^{-35}m is some 20 orders of magnitude smaller than the tiniest scales normally encountered in particle physics. This distance is considered to be of profound relevance in quantum gravity theory, being a distance-scale below which the normal ideas of continuous spacetime seem to make no real sense.[2]

One way of viewing these mysteries would be to regard the small values of electric charge or mass as being the result of some renormalization process where the bare value (§26.9) might be some mathematically respectable number like 1 or 4π. Accordingly, the small observed values could result from some merely *large* rather than infinite renormalization factor. This could occur if the divergent sums and integrals of QFT could be replaced by something convergent. The divergences ('ultraviolet' divergences, that is; see §26.9) normally come about because they involve adding up larger and larger momenta, without limit, these referring to tinier and tinier distances, without limit. Accordingly, the infinities might be removed if there were a cut-off to the divergent integrals (or sums) at, say, the (gravitational) Planck scale[3] of 10^{-35}m. Indeed, this kind of idea was put forward by Oskar Klein in about 1935. All this suggests that, when gravitation is appropriately brought in to the QFT calculations, a finite theory, rather than a merely renormalizable one, might be the result, and that within such a finite theory one might find scope for understanding these unexplained numbers.

Whereas such hopes have been around for over half a century, now, the problems involved in bringing gravity directly into the picture have so far made things worse, rather than better. When standard techniques of quantization have been applied to Einstein's theory, a non-renormalizable theory has been the result, rather than a finite one. This has led many researchers to strive for something non-standard in their searches for a quantum theory of gravity. It has, of course, been one of the messages of earlier chapters of this book (most particularly Chapters 27–30) that we should indeed look for a non-standard union between quantum (field) theory and general relativity. But my contention that there should be some change on the quantum side of things is not one that has been seriously taken up. An unsatisfactory non-renormalizable[4] quantum gravity is indeed the result when Einstein's theory is submitted directly to the standard procedures of QFT, and whereas many have argued for a change in Einstein's theory, they have not argued for a change in QFT.

31.2 Supersymmetry

What kinds of changes have been suggested? In one of these, the ideas of *supersymmetry* have been adopted, and amalgamated with Einstein's theory (also with a torsion included; see §14.4 and Note 19.10) to produce a scheme known as *supergravity*. What is supersymmetry? Why is this regarded by a great many physicists as a 'good thing'—to the extent that supersymmetric ideas underlie a very large number of the developments in modern fundamental theories, most importantly in string theory? Indeed, a remarkable stature has been assigned to the tenets of supersymmetry,[5] despite the fact that the predictions of this scheme of ideas seem to bear little or no relation to what has so far been observed in Nature's own scheme of things.

At this point I must again declare my bias, and provide the required statutory warning to the reader. I have found myself to be totally unconvinced of the physical relevance of the scheme of supersymmetry, at least in the form employed in particle physics and underlying theories today. As of now, observations certainly do not provide much support—and probably none at all—for the claims of supersymmetry. The attraction of the ideas comes from a much lauded mathematical elegance and from supersymmetry's undoubted value in cancelling away large batches of infinities in those QFT models that come under its umbrella. Suppose that you are a physicist interested in constructing a QFT that is to be free of uncontrollable infinities. Then your task will be made enormously easier if you take your theory to be supersymmetric!

The basic idea behind supersymmetry is that it provides a means whereby fermions and bosons can be 'paired off' according to a kind of symmetry relationship. As we have seen in §§25.5–8, the normal symmetry groups of particle physics merely 'rotate' sets of bosons among themselves and sets of fermions among themselves. They do not 'rotate' bosons into fermions or vice versa. Supersymmetry, on the other hand does just this. Recall from §26.2 that bosons satisfy commutation laws, whereas fermions satisfy anticommutation laws. An operator that sends one to the other must itself have anticommutation properties. But the operators that come from an ordinary continuous group are the group's infinitesimal generators, which form a Lie algebra; see §13.6. Ordinary Lie algebra elements satisfy commutation laws and not anticommutation laws. This means that the needed operators are not infinitesimal generators of an ordinary continuous group, but of a broader notion referred to as a *supergroup*, where the laws of a Lie algebra are extended, so that some of the generators indeed satisfy anticommutation laws as well as commutation laws.

In §§26.2,3 we already encountered such things, namely equations like

$$\boldsymbol{ab} \pm \boldsymbol{ba} = \boldsymbol{c},$$

which are satisfied by the creation, annihilation, and field operators of QFT (in equations such as the $\Psi^*\Phi \pm \Phi\Psi^* = \mathrm{i}^k\langle\psi|\phi\rangle \boldsymbol{I}$ of §26.2). In accordance with this, a *super*-Lie algebra is constructed in the same way as an ordinary Lie algebra, except that there is now the possibility of a plus (+) sign in the defining relations. In §13.6, it was noted that the defining relations of a Lie algebra have the form $[\boldsymbol{E}_\alpha, \boldsymbol{E}_\beta] = \gamma^\chi_{\alpha\beta}\boldsymbol{E}_\chi$, where $\gamma^\chi_{\alpha\beta}$ are the structure constants and $[\boldsymbol{E}_\alpha, \boldsymbol{E}_\beta] = \boldsymbol{E}_\alpha\boldsymbol{E}_\beta - \boldsymbol{E}_\beta\boldsymbol{E}_\alpha$. These relations have the form of the above displayed equation when there is the normal minus sign between $\boldsymbol{ab}$ and $\boldsymbol{ba}$. But for a super-Lie algebra we also allow the plus sign, occurring when both the $\boldsymbol{a}$ and the $\boldsymbol{b}$ are fermionic quantities (rather than both bosonic or one fermionic and one bosonic). The notation $[\boldsymbol{a}, \boldsymbol{b}]_+$ tends to be used for such anticommutators, i.e. $[\boldsymbol{a}, \boldsymbol{b}]_+ = \boldsymbol{ab} + \boldsymbol{ba}$, to supplement the usual Lie bracket notation $[\boldsymbol{a}, \boldsymbol{b}] = \boldsymbol{ab} - \boldsymbol{ba}$. This indeed requires us to go beyond the usual notion of a Lie algebra.

The generators of supergroups are normally described as being built up in a particular way. Rather than starting with ordinary real-number quantities, we take these generators to be elements of a Grassmann algebra, which, as we have seen in §11.6, involves anticommutation as well as commutation properties. We shall be catching a more detailed glimpse of how this works in §31.3.

Supergroups now form a respectable area of pure mathematics. Moreover, ideas from supersymmetry can be applied directly in mathematical arguments to obtain results that are not so easy to obtain by other means.[6] This, however, does not tell us whether supersymmetry, in the way that it has been used, has any direct relevance to physics. On the other hand, there are various instances where supersymmetry has proved useful in either motivating or establishing mathematical results that *are* of direct physical relevance.[7] But again, this does not seem to me to carry a great deal of weight for supergroups having any direct underlying relevance to particle physics or QFT.

What evidence is there that supersymmetry does have a genuine role to play in particle physics? Recall the standard model, as described in Chapter 25. Its renormalizability owes a great deal to some precise 'fine-tuning' of its parameters. These relationships can be largely understood in terms of its requirements of $\mathrm{SU}(3) \times \mathrm{SU}(2) \times \mathrm{U}(1)/\mathrm{Z}_6$ symmetry (§25.7). Yet, according to some claims,[8] the standard model requires some other very precise fine-tuning over and above those relationships. Additional symmetries could be invoked to arrange for this, and supersymmetry has been suggested as a means for achieving such fine-tuning. Accordingly, such ideas have frequently been made use of in grand unified theories (§25.8). But do we have reason to believe in such GUTs? As yet there is no observational evidence for this.

The tempting features of supersymmetry seem indeed to be that it provides a way of interrelating bosons with fermions, and that it is much easier to make supersymmetric QFTs provide finite answers than non-supersymmetric QFTs. With a supersymmetric pairing-off of bosons with fermions, the infinities of one set can be made to cancel the infinities of the other set. That makes the job of the QFT builder a good deal easier than it would be without supersymmetry. But this does not tell us that Nature herself does it this way. She may well have quite different tricks up her sleeve!

Now, the main difficulty with supersymmetry (as used today) is that it demands that every fundamental particle in Nature has what is called a 'superpartner' with a spin that differs from that of the original particle by $\frac{1}{2}\hbar$. There needs to be a 0-spin 'selectron' as partner to the electron, a 0-spin 'squark' to accompany each variety of quark, a $\frac{1}{2}$-spin 'photino' to partner the photon, a $\frac{1}{2}$-spin 'wino' and 'zino' as respective partners for the W and Z bosons, etc, etc. The trouble is that no such 'supersymmetric partner' has ever been found. The official explanation for this is that, owing to some 'supersymmetry-breaking' mechanism, the nature of which has never been adequately described, each of these putative supersymmetry partners must be enormously more massive than the particle it partners. The kind of mass that these unobserved particles are now being postulated to have is something like a thousand times that of the proton, or more. I have to say that I am far from alone in believing that this looks a little contrived.

It seems to be postulated that, of the two 'partners', the one that has the smaller spin (by $\frac{1}{2}\hbar$) is deemed to be the exceedingly more massive of the pair (except when both members of the pair are massless). Presumably, only particles considered as 'elementary' (these apparently being the photon, the graviton, the W and Z bosons, gluons, leptons and quarks) possess superpartners. Otherwise we have trouble with 0-spin particles such as pions. If there *are* 0-spin elementary particles, such as the still undiscovered Higgs boson, then they would have to count as more massive than their superpartners, on this particular reckoning (since negative spin is excluded). If this is correct, then why has the Higgs boson's superpartner not been found? Yet again, believers in both supersymmetry and inflationary cosmology must explain how the latter phenomenon's scalar φ particle (§28.4) fits into the 'superpartner' picture.

The one bit of 'positive' evidence that is now frequently cited as providing support for supersymmetry has to do with certain ideas as to how the three forces of particle physics (strong, weak, and electromagnetic) are claimed to have come together in one highly symmetrical unified scheme when the temperature of the universe had some stupendous value (about 10^{28} K), some 10^{-39} seconds (only about 10 000 Planck moments) after the Big Bang.[9] The idea is that such unification would require the interaction strengths to be all the same, at that temperature. It should be remarked that there is a factor of some 10^{13} between the strong and weak interaction strengths under ordinary conditions (although the two cannot really be directly compared). The argument is that, when renormalization effects are taken into consideration (and we recall from §26.9 how different the observed charge of a particle might be from its bare charge), then these strengths would all come together, the 'bare' values coming into their own at such enormous temperatures. (Recall the notion of 'running coupling constants', referred to at the end of §31.1.) The arguments claim that, without supersymmetry, the values do not quite come together, but just 'miss' (see Fig. 31.1); yet, when supersymmetry is brought into the picture, the curves come to one glorious coincidence, and the grand unification of particle physics can take place!

The reader may well sense my lack of conviction. (Already in §28.3 I have expressed some of my difficulties with theories for which 'symmetry restoration' occurs, when the universe's temperature is high enough.) There are enormous extrapolations involved in this particular collection of ideas claimed as observational support for supersymmetry. One of these is the presumption that nothing essentially new is to be revealed in the

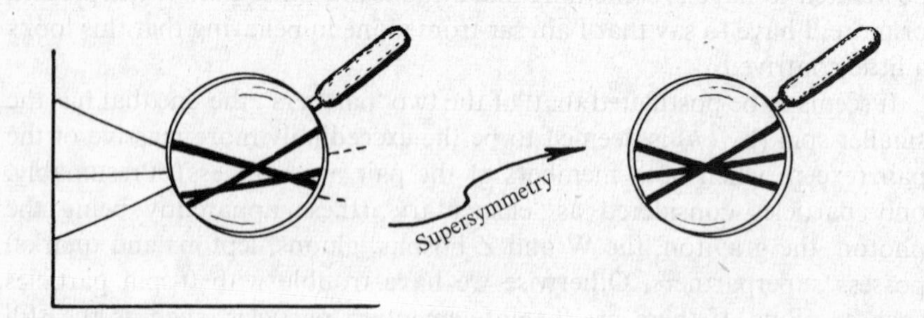

Fig. 31.1 According to a certain 'grand-unified' perspective, the coupling constants of strong, weak, and electromagnetic interactions, treated as 'running coupling constants' (see Note 26.21 and §31.1), should all attain exactly the same value at large enough temperatures, about 10^{28}K, which would have occurred at around 10 000 Planck moments after the Big Bang ($\sim 10^{-39}$s). It was found that supersymmetry is needed in order to bring all three values precisely together.

huge gap of energies (temperatures) between 10^{28} K and the roughly 10^{14} K that are accessible to present-day accelerators. This, in itself, seems an unreasonable extrapolation, and I do not see how these arguments can be regarded as providing any significant observational support for supersymmetry, except perhaps to those already committed.

31.3 The algebra and geometry of supersymmetry

Let us return to the supergravity theory with which I began this discourse. In accordance with the above, there should be a $\frac{3}{2}$-spin superpartner of the graviton, referred to as a *gravitino*. This postulated particle would be massless, like the graviton itself unless there is severe supersymmetry breaking. How is the gravitino to relate to geometry? Einstein has taught us that gravitation is described by spacetime curvature (§17.9 and §19.6). Does this imply that the gravitino ought to be playing some corresponding (super)geometrical role? In accordance with a desire for such a role, many supergravity theorists would argue that the ordinary notion of a *manifold* (as described in Chapters 10 and 12) needs to be generalized, and the concept of a *supermanifold* has accordingly been put forward. We can think of this as being defined in a very formal way, with the ordinary notion of *coordinates* being generalized to include anticommuting elements. For an ordinary manifold, the coordinates are generally real numbers (or complex numbers, if a complex manifold is being considered; see §12.9). For a supermanifold, we take them to be elements of a Grassmann algebra (§11.6).

Most supersymmetry theorists would not take such a rigorous attitude to the nature of the 'manifold' on which their supersymmetric field quantities live (even though the 'geometric' nature of standard general relativity would seem to demand this for the case of supergravity). In the following descriptions, it will not be necessary to hold rigorously to a 'supermanifold' point of view. The 'superalgebra' ideas may be considered to refer just to quantities defined on an ordinary spacetime manifold.

The simplest such algebra is obtained if we adjoin a single anticommuting element ε to the real-number system $\mathbb{R}$. The quantity ε must anticommute with itself: $\varepsilon\varepsilon = -\varepsilon\varepsilon$, whence $\varepsilon^2 = 0$. Thus, each element of the algebra has the form

$$a + \varepsilon b,$$

where a and b are real numbers, commuting with ε. Notice that the sum and product of two such numbers is given by

$$(a + \varepsilon b) + (c + \varepsilon d) = (a + c) + \varepsilon(b + d),$$
$$(a + \varepsilon b)\,(c + \varepsilon d) = ac + \varepsilon(ad + bc).$$

Notice also that, if we ignore the terms multiplying ε, then we simply get the rules of ordinary algebra back again.

This still applies if we have several different supersymmetry generators, say $\varepsilon_1, \ldots, \varepsilon_N$, which anticommute:

$$\varepsilon_i\varepsilon_j = -\varepsilon_j\varepsilon_i, \text{ whence } \varepsilon_i^2 = 0,$$

and the general element of the superalgebra has the form[31.1]

$$a + b_1\varepsilon_1 + b_2\varepsilon_2 + \cdots + b_N\varepsilon_N + c_{12}\varepsilon_1\varepsilon_2 + c_{13}\varepsilon_1\varepsilon_3 + \cdots + f_{12\ldots N}\varepsilon_1\varepsilon_2\ldots\varepsilon_N.$$

The algebra behaves in such a way that if we take the 'ordinary' part a of any element (with no εs involved) then we just get the familiar algebra of ordinary (real or complex) numbers. The 'super' part of the algebra is the remainder. It is 'nilpotent' in the sense that when any of its elements is raised to a sufficiently high power, it vanishes completely.[31.2] Sometimes the fanciful terminology 'body' and 'soul' is used for this 'ordinary' and 'super' part, respectively.

Being someone who likes to be able to have a 'picture' of what is going on, I have always found such a purely formal description of superalgebras and supermanifolds unsatisfying. It is fortunate that there is indeed a more conventional geometrical way of looking at these things. Let us, for the moment, consider the easiest case of just a single supersymmetry generator ε. Since it is to be an anticommuting entity, we might try to think of it as a 1-form $\boldsymbol{\varepsilon}$. However, it cannot be just an ordinary 1-form that refers to ordinary space—let us say the n-manifold $\mathcal{M}$—with which we are working. All the ordinary differential forms within $\mathcal{M}$ already have meanings that are taken up (see §12.4). What we must do is think of $\mathcal{M}$ as embedded as a *hypersurface* in an $(n+1)$-dimensional manifold $\mathcal{M}'$ (a 'hypersurface' being a submanifold of one dimension smaller than the ambient space—see Note 27.36), where $\boldsymbol{\varepsilon}$ is to be a 1-form that refers to the larger manifold $\mathcal{M}'$, but restricted to points of $\mathcal{M}$. We are not supposed to be interested in $\mathcal{M}'$, except just at points of $\mathcal{M}$, where $\mathcal{M}'$ supplies an additional dimension pointing away from $\mathcal{M}$. See Fig. 31.2a. (We are concerned only with what is referred to as the *first* neighbourhood of $\mathcal{M}$ in $\mathcal{M}'$. That means 'first derivatives' away from $\mathcal{M}$, so we are concerned with notions of tangent and cotangent vectors or spaces that 'point' into $\mathcal{M}'$ away from $\mathcal{M}$, but not higher-derivative notions such as *curvature* in directions away from $\mathcal{M}$.) What we are doing is still something n-dimensional, in the sense that all our quantities can be represented as functions of the n independent coordinates in the manifold

[31.1] Write down the sum and product of two such quantities when $N = 3$. What is the multiplicative inverse of such an element, where $a \neq 0$?

[31.2] Show this. What power?

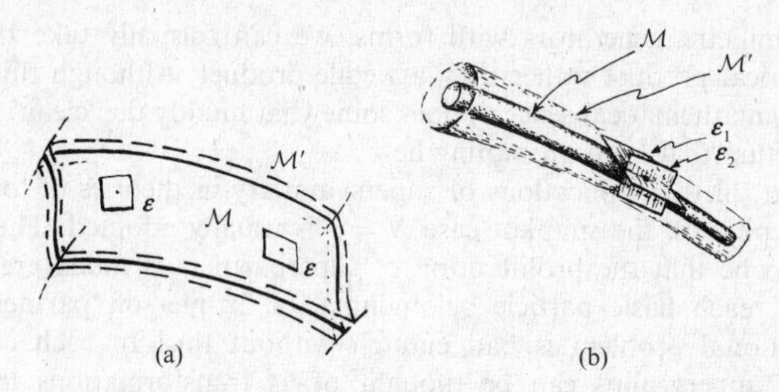

Fig. 31.2 Geometrical description of supersymmetry generators. (a) For a single generator ε, regard our n-manifold $\mathcal{M}$ as a hypersurface in an $(n+1)$-manifold $\mathcal{M}'$, where $\boldsymbol{\varepsilon}$ is a 1-form in $\mathcal{M}'$ defined at $\mathcal{M}$ (and $\boldsymbol{\varepsilon}$ defines the n-plane tangent to $\mathcal{M}$ as in Fig. 12.7, §12.3). We are interested in $\mathcal{M}'$, only to 'first order' at $\mathcal{M}$, but $\mathcal{M}'$ supplies an additional dimension pointing away from $\mathcal{M}$. (b) With N supersymmetry generators $\varepsilon_1, \ldots, \varepsilon_N$, the n-manifold $\mathcal{M}$ is now regarded as a submanifold of an $(n+N)$-manifold $\mathcal{M}'$, where again we are interested in $\mathcal{M}'$ only to first order away from $\mathcal{M}$. The N independent 1-forms $\boldsymbol{\varepsilon}_1, \ldots, \boldsymbol{\varepsilon}_N$ 'feel out' the N extra directions pointing out from $\mathcal{M}$, and into $\mathcal{M}'$.

$\mathcal{M}$. The 'soul' quantities refer to directions that point away from $\mathcal{M}$ and into $\mathcal{M}'$, whereas the 'body' quantities refer simply to directions within $\mathcal{M}$ itself.

The situation is not essentially different if we require N supersymmetry generators $\varepsilon_1, \ldots, \varepsilon_N$. We now think of our n-manifold $\mathcal{M}$ as embedded in an $(n+N)$-manifold $\mathcal{M}'$, where again we are to be interested in $\mathcal{M}'$ only in its immediate (first) neighbourhood of $\mathcal{M}$. We now require N different 1-forms $\boldsymbol{\varepsilon}_1, \ldots, \boldsymbol{\varepsilon}_N$ to probe the N extra directions[10] that point outwards away from $\mathcal{M}$ and into $\mathcal{M}'$. In my opinion, this picture (due to various people, including Abhay Ashtekar, who features strongly in Chapter 32)[11] makes the underlying ideas of supersymmetry and supermanifolds very much clearer than the formal (and rather mysterious-looking) procedures that are normally adopted. Note that the 'body' just refers to quantities that are entirely intrinsic to $\mathcal{M}$, whereas the 'soul' refers to quantities with a component 'pointing outwards' into $\mathcal{M}'$, away from $\mathcal{M}$; see Fig. 31.2b.

Even with this clear geometrical interpretation, there are oddities in the way in which 'superalgebra' is usually employed, if consistency with the geometrical picture is to be maintained. An ordinary p-form $\boldsymbol{\alpha}$ in $\mathcal{M}$, for which p is an odd number, would anticommute with a supersymmetry generator $\boldsymbol{\varepsilon}$, if we take the product of $\boldsymbol{\varepsilon}$ with $\boldsymbol{\alpha}$ to be a wedge product. However, this is not the normal convention in standard approaches to superalgebra, where $\boldsymbol{\varepsilon}$ would normally be taken to commute with $\boldsymbol{\alpha}$. This is basically a notational matter, and if we are to consider products of supersymmetry generators with forms, we can formally take this as a symmetrical product rather than a wedge product. Although this makes (formal) mathematical sense, it does somewhat muddy the 'clean' geometrical picture that I am promoting here.

In fact, in the applications of supersymmetry in theories of 'ordinary' particle physics, the simplest case $N = 1$ is usually adopted. The reason seems to be that the proliferation of superpartners is much greater for large N, each basic particle belonging to a 2^N-plet of 'partners'. The observational problem is bad enough without this! In each case, the relevant supergroups can be thought of as transformations involving 'internal symmetries' (which refer to the symmetries of the fibres of some bundle $\mathcal{B}$ over spacetime (§15.1) together with the 'rotations' that refer to the extension of $\mathcal{B}$ into the immediate neighbourhood of $\mathcal{B}$ within some space $\mathcal{B}'$ of N greater dimensions.

31.4 Higher-dimensional spacetime

Now that we have a little better idea of what supersymmetry and 'supergeometry' are really about, let us return to the matter of supergravity. The original excitement surrounding this idea, in the late 1970s, came from the hope that, unlike standard Einstein general relativity, supergravity could turn out to be renormalizable. In the Einstein vacuum theory, non-renormalizable divergences had appeared 'at the 2-loop level'—where the 'loops' refer to Feynman diagram expansions, the 'number of loops' referring to the number of cuts that would be needed in order to reduce the Feynman graph to a tree graph (see §26.8, particularly the final paragraph and Fig. 26.8, and also §§26.9,10). With matter present, however, such divergences appear already at the 1-loop level, which can be considered as a genuine disaster. In supergravity, these 1-loop divergences magically cancelled out, for the type of matter that the theory allowed, and many people had high hopes that this would continue to all loop orders. Sadly, this turned out not to be the case, and non-renormalizable divergences were found again at the 2-loop level in supergravity.[12] It was subsequently noticed that if the dimensionality of spacetime were to be increased from the standard four to eleven, then matters looked very much more promising. Despite this, a fully renormalizable version of supergravity was still not obtained and, as much more recent work has shown,[13] cannot be so obtained.

How is it that physicists could take seriously the possibility that the dimensionality of spacetime might be other than the four that we directly experience (one time and three space)? As mathematical exercises, such higher-dimensional things seem fine, but this is supposed to be a physical theory where 'spacetime' really means the combination of actual space with time. Indeed, as we shall be seeing in §31.7, string theory (as it is currently understood) requires that spacetime must indeed have more than four dimensions. In the early theory the dimension number was taken to be 26, but later innovations (which involved the ideas of supersymmetry—see §31.2) led to this spacetime dimensionality being reduced to 10.

Before we dismiss this idea as a total fantasy we must recall, from §15.1, the ingenious scheme, put forward in 1919 by the (at that time) little-known German mathematician Theodor Kaluza, and then further taken up by that same Swedish mathematical physicist Oskar Klein whom we have already encountered earlier in this chapter. Provided that the extra dimensions (in excess of 4, that is) are taken as *small* dimensions, in some appropriate sense, then we might not be directly aware of them. What does 'small' mean in this context? Recall the 'hosepipe' analogy of Fig. 15.1. When looked at from a great distance, the hosepipe appears to be 1-dimensional, but if we examine it more closely, we find a 2-dimensional surface. The idea is that some *being*, inhabiting the hosepipe universe, would not 'know' that the extra dimension wrapping around the pipe is actually 'there', provided that the physical dimensions of that being are much larger than the circumference of the hosepipe. Similar remarks would apply to a higher-dimensional 'hosepipe universe' of $4 + d$ dimensions, where d of the dimensions are 'small' and not directly perceived by a much larger being inhabiting this universe, who perceives only the 4 'large' dimensions; see Fig. 31.3.

What degree of 'smallness' is to be expected in the Kaluza–Klein model, or in modern yet-higher-dimensional versions of this idea? Klein himself came to the conclusion that the 'scale' of the tiny extra dimension ('hosepipe circumference') should be of the order of the Planck distance of 10^{-35} m. This also seems to be the most popular kind of scale (or just a little bigger than this) that is adopted in the more modern schemes, such as in higher-dimensional supergravity and string theory. It is clear that, for

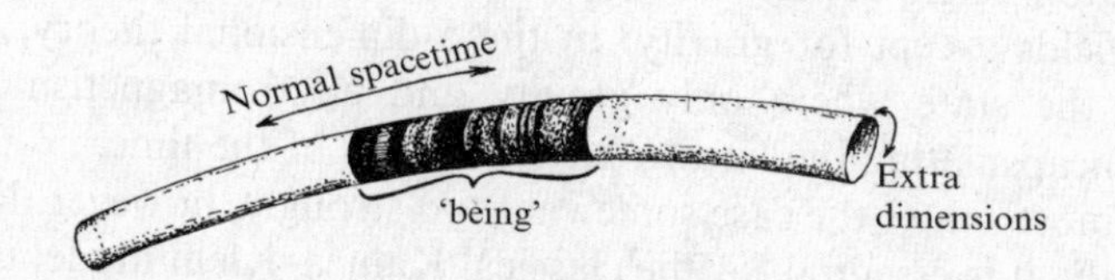

Fig. 31.3 Hosepipe model of a Kaluza–Klein-type higher-dimensional spacetime (see Fig. 15.1), where the dimension along the length of the hosepipe represents normal 4-spacetime and the dimension around the pipe represents the 'small' (perhaps Planck-scale) extra dimensions. We imagine a 'being' who inhabits this world, as straddling these 'small' extra dimensions, and so is not actually aware of them.

beings such as ourselves, this indeed counts as 'small', and it may be expected that we should have no direct experience of extra spacetime dimensions as tiny as that.

In fact, there are some recent developments (in string theory) in which the extra dimensions are *not* taken to be (that) small, but might refer to something even so 'large' as a millimetre in diameter (or perhaps not even closed up at all). One idea is that there might be observational consequences of such a scheme which show up as a modification of the inverse square law of gravitational attraction at such distances. In fact, some very delicate experiments have recently been performed in order to ascertain whether such deviations from Newton's theory might be detectable.[14] So far, no such deviation has been found, down to half a millimeter.

Whatever the status of these newer ideas, this suggestion of a higher-dimensionality for spacetime has, at this stage in our deliberations, a status no more compelling than that of a 'cute idea'—which the original Kaluza–Klein suggestion certainly was. Whatever may be the mathematical attractiveness of this idea, we have to address the question of whether there are good physical reasons for believing in such a scheme. In the case of the original Kaluza–Klein model, the reason for adopting this higher-dimensional perspective was in order to 'geometrize' electromagnetism. As we recall from §25.1, the only forces of Nature that were known (and understood) in the early part of the 20th century were the gravitational and the electromagnetic. Einstein had just shown how to incorporate the gravitational field into the curvature of 4-dimensional spacetime. It was, indeed, a very attractive and natural-looking idea to try to bring electromagnetism also into such a geometric framework. Moreover, there was something rather miraculous in the way that the very same 'vacuum Einstein equations'—namely the vanishing of the Ricci tensor ($R_{ab} = 0$; see §19.6)—apply in the Kaluza–Klein 5-dimensional theory just as in the standard 4-dimensional general relativity. In the 4-dimensional theory, this equation refers to the *vacuum* state—that is, to the absence of all physical fields except for gravity. In the 5-dimensional theory, it *almost* refers to the state where only gravity and electromagnetism operate, thereby encapsulating the known physical fields of the time.

The 'almost' states the case somewhat too strongly, however. For, most importantly, it is essential for the classical Kaluza–Klein model that there be a *symmetry* in the 'small' dimension, so that there are not infinitely too many degrees of freedom. Let us see why these extra degrees of freedom would otherwise come about. Recall the discussion of §16.7, concerning the 'size' of the infinite-dimensional space of fields on some given space. For a field that can be specified by k independent freely-chosen components on a q-dimensional initial data surface, the freedom is $\infty^{k\infty^q}$.

For standard Einstein general relativity, we have (for somewhat complicated reasons)[15] $k = 4$ and $q = 3$, so this quantity turns out to be $\infty^{4\infty^3}$ and for Maxwell theory we get exactly the same freedom. For the combined Einstein–Maxwell theory, the effective number of components per point of the initial data surface is the sum of the values for each field separately, so we have $4 + 4 = 8$ effectively independent components per point of the initial 3-surface, and the correct value for the full freedom is

$$\infty^{8\infty^3}.$$

Now, in a 5-dimensional theory subject solely to the condition of Ricci flatness (i.e. $R_{ab} = 0$; see §19.6), the initial surface is 4-dimensional (so $q = 4$), and it actually turns out that $k = 10$. This would give us an enormously greater freedom $\infty^{10\infty^4}$ for the field than the required value (as displayed above), not because 10 is greater than 8 (the k value) but because 4 is greater than 3 (the q value). There are hugely more functions of 4 variables than there are functions of 3 variables!

In the Kaluza–Klein model, we reduce the 4 back to 3 by imposing a continuous (in fact U(1), cf. §13.9) symmetry in the small dimension. There has to be a Killing vector (§14.7) expressing this symmetry and, in effect, the Kaluza–Klein 5-space is an S^1-bundle $\mathcal{B}$ over ordinary 4-dimensional spacetime $\mathcal{M}$. This does not seem so far from the conventional bundle description of electromagnetism, as described in §19.4 (and §15.8). A basic difference is that $\mathcal{B}$ itself is here assigned a Ricci-flat Lorentzian (pseudo)-metric, rather than a metric being assigned only to the spacetime $\mathcal{M}$.[16] The striking fact about the Kaluza–Klein model is that the imposition of Ricci flatness on $\mathcal{B}$ (in addition to the U(1) symmetry) is surprisingly close to providing us with the complete equations of Einstein–Maxwell theory[17] on $\mathcal{M}$. All that one needs, in addition, is that the Killing vector have a constant non-zero (in fact negative) norm. This eliminates an unwanted scalar field, and the exact 4-dimensional Einstein–Maxwell theory is thereby expressed!

Elegant as it is, the Kaluza–Klein perspective on Einstein–Maxwell theory does not provide us with a compelling picture of reality. There is certainly no strong motivation from physical directions to adopt it. Supersymmetry, for example, certainly has a stronger physical case, because of its undoubted value in reducing the problem of infinities in QFT. Why, then are higher-dimensional Kaluza–Klein-type theories so popular in modern strivings towards a deeper theory of Nature? The main reasons come from string theory which, in all comprehensive versions actively pursued today employs both supersymmetry and higher dimensions[18]—in essential ways.

31.5 The original hadronic string theory

What, then, *is* string theory? And why does it have such a powerful hold on so many of today's theorists? Again, this scheme of things gains its strongest motivation from the desire for the elimination of the infinities of QFT. In that sense, it represents a continuation of the driving ideas of Chapters 24–26. But there was also another important historical motivation, of a seminal nature, that had specifically to do with some observational curiosities in hadronic physics. Let us take a glimpse at these first.

This original issue had to do with certain relationships that were found in the particle physics of hadron scattering. In Chapter 25, it was mentioned that, among the hadrons, there are many 'particles' that are so short-lived (lasting only for about 10^{-23}s) that they just barely deserve that name, and are often referred to as *resonances*. Now, we recall that the rules of QFT (§25.2, §§26.6,8) demand that in any physical process, all the possible different activities that might take place have to be added into the total, in order to get the full quantum amplitude. All possible particles and resonances must therefore be taken into account for this. We might, for example, have a hadronic scattering process where two particles A and B come together and, after a fleeting moment, convert themselves into the pair of particles C and D. Now one way that they might do this could be for A and B to combine together to make a single particle (resonance) X which, almost at once decays into the particles C and D. There might be many such possible intermediate particles X, X′, X″, . . . , and the effect of each one would have to be added into the total. The Feynman graphs for each of these processes are indicated in Fig. 31.4a. Now, an alternative way that the transformation might happen is that a particle Y is 'exchanged' between A and B, converting A to C and B to D. Again, there might be a list of possible exchange particles Y, Y′, Y″, . . . , the Feynman graphs being shown in Fig. 31.4b. There is a third family of processes whereby the transformation might take place, which differs from this in that the outgoing particles C and D are taken the other way around, and the Feynman graphs for these are shown in Fig. 31.4c. There are other, more complicated, ways in which we might imagine that the transformation takes place, involving *closed loops* (Fig. 31.4d) but these 'higher-order' processes will be considered, for the present, as unimportant.

To obtain the total amplitude for the process whereby the pair (A, B) becomes converted to the pair (C, D), we should add up all these different contributions; but what was found, rather surprisingly, was that *each* of these three possibilities appeared to give the same answer, and that this single answer seems to be basically the correct answer. If we are to add all three answers together then we get something too big. Somehow, each of the three collections of Feynman graphs, as given in Fig. 31.4a,b,c, respectively, when summed up individually, represents physically the

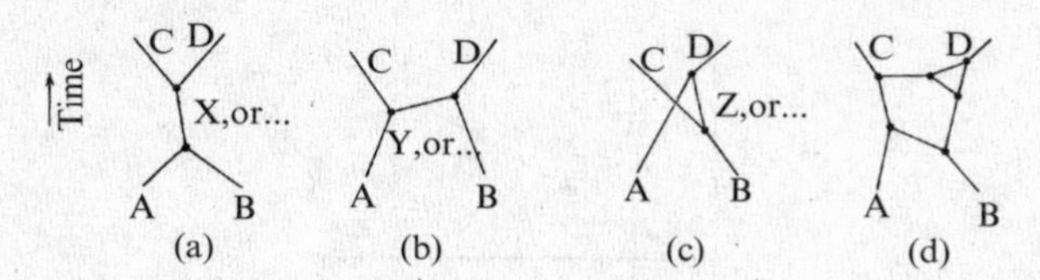

Fig. 31.4 Feynman graphs of a hadronic scattering, where two particles, A, B are converted to the pair C, D. (a) In one family of such processes, A and B combine to make a particle (resonance) which, almost at once, decays to C and D, there being many possible intermediaries X, X′, X″, ..., all contributing to the total. (b) In an alternative family of processes, a particle Y (or Y′, or Y″, or...) is 'exchanged', converting A to C and B to D, each intermediary contributing. (c) A similar 'exchange', involving Z (or Z′, or Z″, or ...), where now A is converted to D and B to C, each intermediary contributing. It turns out that to lowest order, the alternatives (a), (b), (c) are equivalent, rather than having to be added together. (d) Other ways of achieving this transformation, involve closed loops.

same thing! From the standard Feynman-graph perspective, this 'duality'[19] seems incomprehensible, but in 1970 the Japanese/American physicist Yoichiro Nambu,[20] basing his considerations on a remarkable formula[21] found by the young Italian Gabriele Veneziano in 1968, came to the conclusion that all this could be made sense of from a different perspective, whereby the individual hadronic particles were modelled by *strings* rather than point particles. A string history is a 2-dimensional surface, so the processes described in the Feynman diagrams of Fig. 31.4a,b,c,d, respectively, can now be pictured as the various alternative 'plumbings' depicted in Fig. 31.5a,b,c,d respectively. What is striking about this 'string perspective' is that the three processes (a), (b), (c), which seem so different, from the standard Feynman-diagram perspective, are now all topologically equivalent, and may be regarded as just three different ways of looking at the *same* process. Thus the 'string' picture suggests a way to make sense of a puzzling fact of hadronic physics.

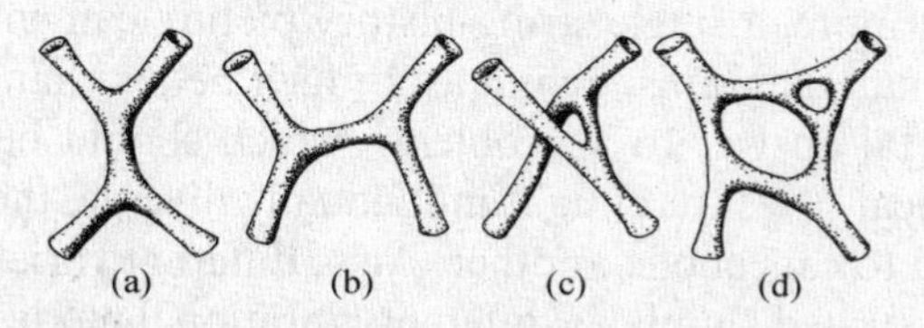

Fig. 31.5 The string-history picture of the respective processes of Fig. 31.4 provides an explanation of the equivalence between (a), (b), and (c), since they can be transformed into each other, as they are topologically the same. (d) Higher-order processes correspond to more complicated topology, where the topological genus corresponds to the number of loops (compare Fig. 8.9).

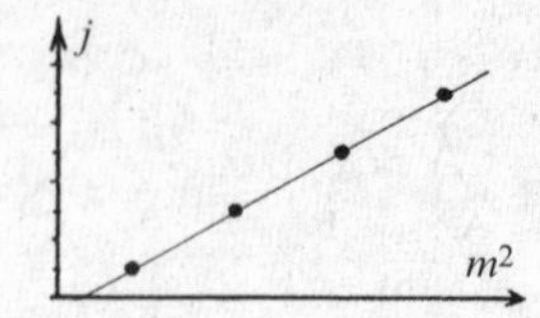

Fig. 31.6 Straight 'Regge trajectories' of particle resonances of increasing spin, plotted against mass squared. The elastic string picture provides an explanation.

This comment is just qualitative, but the string picture also presented a physical model that provided a mathematical derivation of Veneziano's formula. In addition, the string model—in which the strings behaved rather like tiny elastic bands, with a *string tension* increasing in proportion to the amount that a string is stretched—provided an explanation of another observed feature of hadronic physics, namely the straightness of *Regge trajectories*. Regge trajectories are lines that are seen when, for a particular class of hadron, we plot the value of the spin against the square of the mass. These lines turn out to be remarkably straight. An example is depicted in Fig. 31.6. As far as I am aware, there is still no complete alternative explanation of this striking observed fact concerning hadrons.[22]

Moreover, the string model gave significant and reasonable hope that a finite theory of hadronic physics might be obtainable with this picture. Roughly speaking, it served to 'smooth out' the (ultraviolet) divergences of the conventional Feynman approach (§26.8). One may think of these divergences as arising from small-distance effects when point particles get closer and closer to one another, without limit. The strings are not point particles, so this provides scope for relieving this problem. In fact, it is the issue of closed loops in the standard Feynman-graph picture that causes the divergence difficulties. In the string picture, the closed loops are simply taken over by surfaces with higher topology, as indicated in Fig. 31.5d, this being a string version of Fig. 31.4d. This ought to provide something finite rather than the divergent integrals from the Feynman graphs. Furthermore, a single string-history picture can encompass many different Feynman diagrams, giving us a much better chance to represent the physical total answer to a problem—which should be finite—rather than non-physical parts that might individually diverge, these divergences being supposed to cancel one another. Also, different varieties of particles could be incorporated simply as different vibrational modes of the strings. Finally, the 2-dimensional spacetime string histories have the remarkable additional property that they can be taken as *Riemann surfaces* which, as we recall from Chapter 8, have extraordinarily rich geometrical and analytical properties (the fact that actually underlies Veneziano's remarkable formula). Here is scope, indeed, for the complex magic that seems to be part of Nature's design at the quantum level of reality.

This is undoubtedly a most elegant mathematical picture of what could perhaps be going on at some level of physical description deeper than that of ordinary particles. When I first heard of this picture (around 1970, from Leonard Susskind, who was one of the earliest researchers in this field) I was extremely struck by the beauty and potential power of this collection of ideas. It seemed to me that here was something new which was both mathematically exciting and of apparently direct relevance to an important area of particle physics. My own primary interests at the time were in twistor theory (which we shall come to in Chapter 33), and it seemed to me that I should certainly attempt to forge some link between what I was doing and these very promising-sounding new ideas. Twistor theory makes crucial use of complex (holomorphic) structures, and with basic string theory we seem to see such structures controlling physical behaviour, via the essential use of Riemann surfaces—which are, indeed complex curves.[23]

Remarkably, some very recent work of Witten (2003) may now be realizing some of these early aspirations. I shall return to these very positive new developments, which do not employ higher dimensional spacetime, in §31.18. But as yet these do not represent a comprehensive new string theory, and my remarks in the intervening sections refer, instead, to what may be called 'main-stream' string theory.

31.6 Towards a string theory of the world

How have these remarkable initial ideas stood the test of time, after the more than 30 years that have elapsed since then? Have the developments in the subject over those years kept up or exceeded this initial promise? These are questions to which different people may give wildly differing answers. String theory has sometimes even become a highly emotionally charged subject. To its thoroughgoing supporters, string theory (with its later transmutations) *is* the physics of the 21st century, and it represents a revolution in physical thought at least comparable with, if not greater than, those of general relativity or quantum mechanics. To its most extreme detractors, it has achieved absolutely nothing, physically, so far, and has little chance to play any significant role in the physics of the future.

It would be impossible for me even to attempt to be properly dispassionate in my account of these developments, but at least I shall try to be reasonably accurate and to give reasons for the impressions that I have formed. I must, as before, provide my statutory warning to the reader that many active and exceptionally capable theoretical physicists disagree with my views. But I cannot do other than present things as I see them.

Since my viewpoint will be less than positive about a good many aspects of the current string-theory programme, I should at least give the reader scope for redressing this possible imbalance. First I present points of view of two of the most important figures in the development of the subject. In the words of Michael Green,[24] of the University of Cambridge:

> The moment you encounter string theory and realize that almost all of the major developments in physics over the last hundred years emerge—and emerge with such elegance—from such a simple starting point, you realize that this incredibly compelling theory is in a class of its own.

And Edward Witten, of the Institute for Advanced Study in Princeton, has famously remarked:[25]

> 'It is said [by Danielle Amati] that string theory is a part of twenty-first-century physics that fell by chance into the twentieth century.'

As for a popular account, which is very accessible, eloquent, and enthusiastic, and not at all critical—but which does not go deeply into the mathematical ideas—see Greene (1999).[26]

In order to present a consistent, if not necessarily fair, perspective on the subject of string theory, taken from my particular vantage point, I propose to give a roughly historical description of how the subject has itself impinged upon my own thinking. In this way, I shall try to indicate not only something of the successive developments that have taken place in the theory, but also what my own reactions to these developments have been. What makes string theory so difficult to assess dispassionately is that it gains its support and chooses its directions of development almost entirely from aesthetic judgements guided by mathematical desiderata. I believe that it is important to record each of the turnings that the theory has undergone, and to point out that almost every turn has taken us further from observationally established facts. Although string theory had its beginnings in experimentally observed features of hadronic physics, it then departed drastically from those beginnings, and subsequently has had rather little guidance from observational data concerning the physical world.

Imagine a tourist trying to locate a specific building in a vast and completely unfamiliar city. There are no street names (or at least none that make any sense to the tourist), no maps and no indication from the totally overcast sky as to which directions are north, south, or whatever. Every so often there is a fork in the road. Should the tourist turn right or left, or perhaps try that attractive little passageway hidden over to one side? The turns are frequently not right angles, and the roads are hardly ever straight. Occasionally the road is a dead-end street, so steps must be retraced and another turning made. Sometimes a route might then be spotted that had not been noticed before. There is no-one around to ask the way; in any case the local language is an unfamiliar one. At least the tourist knows that the building that is sought has a particular sublime elegance, with a supremely beautiful garden. That, after all, is one of the main reasons for looking for it. And some of the streets that the tourist chooses have a more obvious aesthetic appeal than the others, with more attractive architecture and beautiful courtyards adorned with superb shrubs and flowers—which sometimes, upon close examination may turn out to be plastic. Many choices are involved in the route to be followed and, for each choice, the tourist's only guide is the area's aesthetic appeal, together with some feeling for an overall general consistency—of style, or of some kind of imagined underlying pattern for the city.

Perhaps, for a better analogy, let us suppose that *you* are the tourist, but you are part of a group, led by a tour guide of impressive intelligence, knowledge, and sensitivity—the only trouble being that, in this case, the guide has no prior knowledge of the city and has had no prior encounter with the local language. You may well believe that the guide has better aesthetic insights than you have yourself, and certainly comes to quicker judgements about such things than you do. Occasionally, the guide's sensitivity to hidden patterns locates a building of particular sophistication in its elegance. But the criteria are not, in essence, of a kind fundamentally different from those that you might yourself choose to use. If you follow the group, then at least you will have the companionship of others, and you can talk to them about the surrounding architecture and share the excitement of the quest for your common goal. Even if you do not expect to find that goal, you enjoy the search. But perhaps, on the other hand, you prefer to go off on your own, as you become ever more suspicious that the tour guide knows no more of how to find your goal than you do yourself. Each successive choice of turn is a gamble, and on frequent occasions you may perhaps feel that a different one held more promise than the one that the guide had actually chosen...

Of course, we have witnessed several examples, in earlier chapters, where great physicists have demonstrated the power of their special insights, these insights being often of a distinctly mathematical character. One of the most impressive of these must surely be Dirac's finding of the equation for the electron, as described in §24.7. Yet the aesthetic leap was essentially just one majestic step into the unknown, from the sound body of mathematical understanding that had arisen from the experimental findings of quantum mechanics. Dirac's prediction of the electron's antiparticle involved another such leap. But it was made with great caution, and subsequently confirmed in observation. Einstein's general relativity was also partly driven by mathematical aesthetic considerations, and general relativity's strength gains, to an enormous degree, from its profoundly beautiful mathematical structure. When Einstein first formulated the theory, there was no clear demand for it on observational grounds. Yet, it can hardly be said that Einstein was simply driven by mathematical aesthetic considerations. His guidance came primarily from the physics, and lay in his conviction that the principle of equivalence (§17.4) must be central to the understanding of gravity.

In contrast, string theory has been almost entirely mathematically driven. I should first make clear that this is, in itself, not necessarily a bad thing. All successful physical theories have strong mathematical underpinnings. Mathematical consistency is indeed a crucial feature for a physical theory, if that theory is to make overall sense. And once a particular mathematical framework has been established, then rigorous mathematical developments within that framework can have powerful implications for the physical world. (The Lagrangian and Hamiltonian developments in classical physics described in Chapter 20 provide impressive examples of this.) However, difficulties arise when, in order to overcome an inconsistency, a previously believed theory must be changed, and the particular way in which some theory might be changed may depend upon the particular mathematical knowledge and aesthetic preferences of the theorist. Very often, the change will be just an idea—perhaps even a 'brilliant idea'—likely still to have failings of mathematical inconsistency, although perhaps different from those failings of the theory it replaces. Further changes might then be needed, and so on. If there are too many of these, the chance of guessing right each time may become exceedingly small.

31.7 String motivation for extra spacetime dimensions

An early inconsistency in the string-theory picture was the emergence of a serious anomaly. Recall from §30.2 that anomalies arise when the classical commutation rules, expressing a classical symmetry or invariance property, cannot be fully realized by quantum commutators, so the quantum theory loses a quality of the classical theory that might have been regarded as essential. In the case of string theory, this anomaly referred to an essential parametrization invariance in the description of the string. The presence of the anomaly led to effects that were regarded as disastrous. It was found, however[27] that increasing the number of spacetime dimensions from 4 to 26, caused the anomaly to disappear.[28] Accordingly, string theory seemed only to be quantum-mechanically consistent in a spacetime of 26 dimensions.

My own reaction to this was basically: 'there ought to be a different way around this'—although I had never looked at the problem sufficiently to appreciate the strength of the reasoning behind this '26-dimensional' conclusion. I suspect that many others reacted in a similar way, because the theory lost a good deal of its previous popularity at this point. But my own reasons for discounting a 26-dimensional universe model had an additional input, coming from twistor theory. As we shall be seeing in §§33.2,4,10, it is an essential implication of my own particular 'twistorial' perspective that spacetime indeed have the directly observed values of one time and three space dimensions (i.e. '1 + 3 dimensions').

In addition to this problem of what to do with these extra dimensions—which were presumably to be dealt with according to some Kaluza–Klein-type prescription—this relatively simple-looking string model of hadrons ran into other difficulties, such as the appearance of *tachyonic* behaviour (faster-than-light propagation). Also, the growing success of the standard model, as described in Chapter 25, led physicists to be less interested than they had been before in such 'far-out' suggestions as string models. The puzzling features of hadronic physics, referred to above, that started Veneziano, Nambu, and others on the road to strings, found a (partial) alternative QCD explanation in terms of the gluon–quark picture.

Most particularly the 'pointlike' nature of the constituents of hadrons was becoming experimentally apparent, this being consistent with the quark picture of the standard model but not with the string picture as it then was. The typical size of a loop of string would relate to the string's coupling strength, and for the original hadronic strings (with a string tension consistent with the strength of the strong-interaction coupling constant), this would give an average loop scale of some 10^{-15} m. This is hardly 'pointlike' at the scale of a proton, being comparable to the 'size' of a proton itself.

After nearly a decade during which there was little interest in string theory, a development then took place that resulted in what is sometimes referred to as 'the first superstring revolution'. In 1984, Michael Green and John Schwarz put forward a scheme (subsuming some earlier suggestions made by Schwarz and Joël Scherk) in which supersymmetry was incorporated into string theory (to provide us with 'superstrings' rather than just 'strings'), and the spacetime dimensionality[29] was thereby reduced from 26 to 10. This removed the 'tachionic problem' referred to above. Moreover, with a radical change in the scale and nature of the string tension, string theory was now to be considered as primarily a quantum-gravity theory, rather than a theory of strong interactions. It had already been recognized that there should be a massless particle/field of spin 2 arising from a vibration mode of the strings. This had been an embarrassment with the original 'hadronic' version of string theory, since there is no hadronic particle of this nature. But with the new strings, with their far greater string tension, it would indeed be appropriate to identify this massless field with gravity. Now, a typical string loop size is something of the general order of the tiny (gravitational) Planck length—about 20 orders of magnitude smaller than before, and certainly pointlike at the hadronic scale.

I should mention one further technical difference in the nature of the string tension that is introduced with the new 'gravitational-scale' strings (not normally emphasized in popular accounts). The original hadronic strings were rather like rubber bands, in that the the tension increases as the string is stretched, in proportion to the amount of the stretch.[30] However, the new gravitational-scale superstrings exert a *constant* tension $\hbar c/\alpha'$, which is thus independent of the amount of the stretch, where α' is a very small number (an area measure) referred to as the *string constant*. In this respect, the original hadronic string was much more like the type of entity that had been familiar in ordinary physics in which a classical version of it makes physical sense. (A classical version of the new superstring, with its constant tension, would almost instantly shrink away to a singularity of zero size!)

31.8 String theory as quantum gravity?

These developments completely transformed the general perception of string theory, and it rapidly gained great popularity. Frequent claims were made that string theory provided a 'complete consistent theory of quantum gravity', where the non-renormalizability of standard general relativity (see §31.1) is replaced by a completely finite string theory of quantum gravity.[31] Although, if pressed, some string theory proponents might admit that not all of the finiteness claims were completely proved, this would be regarded as a matter of little importance. As a prominent theoretical physicist and string theorist had remarked[32]

> String theory is so obviously finite that if someone were to publish a proof, I wouldn't be interested in reading it.

Moreover, the string theory of quantum gravity tended to be regarded by the string theorists as 'the only game in town', as is illustrated by the following comment on approaches to quantum gravity other than string theory, by Joseph Polchinski (1999):

> ... there are no alternatives ... all good ideas are part of string theory.

I suspect that it was the forcefulness of the early finiteness claims (although see §31.13) that supplied much of the impetus that the theory then acquired. Indeed, if this claimed discovery of the sought-for 'quantum gravity'—the missing union between the two great revolutions of 20th-century physics—were actually vindicated, then this would establish string theory as being not only one of the major intellectual achievements of that century but also a revolutionary basic framework for future progress in fundamental physics.

I believe that even many of today's *string* theorists might regard as overblown the claims that the quantum-gravity problem was completely 'solved' by string theory in the 1980s. Such people might be happier now to take this more sober position, than they would have been before, because today's string theory has moved on, and it differs appreciably from that scheme of 1984. They would, nevertheless, be likely to take the view that the 1984 string theory had at least provided the most impressive step yet towards this quantum-gravity goal.

What was my own response to these claims? Very negative, I am afraid, as was the reaction of most of my close colleagues. No doubt much of the reason for this negative reaction could be attributed to differences in cultural background between those, such as myself and my colleagues, whose outlook was rooted in a deep interest in Einstein's general relativity, and those others whose drives came more from the QFT side. The main effect of this difference in outlook was that we would have quite separate views as to the central issues to be resolved in a quantum–gravity union. Those who come from the side of QFT would tend to take renormalizability—or, more correctly, *finiteness*—as the primary aim of this union. On the other hand, we from the relativity side would take the deep conceptual conflicts between the principles of quantum mechanics and those of general relativity to be the centrally important issues that needed to be resolved, and from whose resolution we would expect to move forward to a new physics of the future. Our negative reaction to the strong claims that the string theorists were making at this time did not just arise from matters of detail or general disbelief (though these were important too), but from a frustration in the fact that the very problems that we thought were central to the whole quantum/gravity issue seemed not to be recognized by the string theorists as existing at all!

Some of these matters were touched upon in §30.11 (and another will be referred to in §33.2). It should be mentioned, however, that the issues raised in those sections barely scratch the surface of the profound conflicts that the principle of general covariance entails[33] in relation to QFT (§19.6). There is also the basic issue of what a 'quantum spacetime geometry' is to be actually like. String theory operates simply with a smooth 'classical' background spacetime, which is not even influenced directly by the presence of a string—since the basic unexcited string itself carries no energy, and so does not directly 'curve' the background spacetime. Most people in the relativity community have the expectation that the true 'quantum geometry' should take on some elements of discreteness, or should at least differ profoundly from the classical smooth-manifold picture.

These deep issues will be faced more squarely in the next two chapters, as will certain approaches to their resolution. In particular, the 'loops' that we shall encounter in the next chapter (§32.4), although very superficially resembling strings, are quite different from them in numerous respects. In particular, the spacetime geometry *is* profoundly influenced—indeed, essentially *created*—by the presence of these loops, the spatial metric being entirely concentrated along them, completely vanishing elsewhere. In string theory, on the other hand, a smooth spacetime is taken to be already present as a background for the strings, and restrictions on its metric geometry come about only from indirect influences of the strings, in a manner that we shall come to shortly (in §31.9). But for now, let us set aside this question of whether or not the really important quantum-gravity issues have been properly addressed by string theory. Instead, let us consider the string theorists' claims that they have a finite quantum theory of gravity. Do they? I shall try to address this question in the remainder of this section, and in the five that follow it.

One point of significance is perhaps contained in the wording. The claims of string theorists are that they have a 'quantum theory of gravity', not of general relativity or of Einstein's theory. What do they mean by 'gravity' if not Einstein's superbly confirmed general relativity? We recall, first, that the string theorist's spacetime is now 10-dimensional (or, as we shall be learning shortly (§31.14), *roughly* 10-dimensional—but try not to let that worry you for the moment!). What is 'gravity' in 10 dimensions? Well, the tensor calculus works just as well in 10 dimensions as it does in 4 (see §§14.4,8), so we can still construct the Ricci tensor R_{ab}, just as before. As we saw in §19.6, the condition for a *vacuum* in ordinary Einsteinian gravity is Ricci flatness, so we might guess that the string theorist's gravitational 'vacuum equation' looks the same, namely

$$R_{ab} = 0,$$

except that we are now in 10 dimensions. We might also expect, by analogy with what happens in 5-dimensional Kaluza–Klein theory where the '5-vacuum' includes both gravitation and electromagnetism, that this equation in 10 dimensions, i.e. the '10-vacuum', is also to accommodate all the non-gravitational fields as well as gravity.

Well, this *is* what the string theorists basically mean—at least roughly. Somewhat more precisely, they regard Ricci flatness as being an implication of only the first term in an infinite power series in the string constant α', the higher-order terms providing us with 'quantum corrections' to Ricci flatness. (The coefficient of $(\alpha')^r$ could involve higher derivatives of curvature tensors and polynomial expressions in such tensors.) Moreover, in addition to the metric on the 10-dimensional spacetime there are also other fields that arise in this discussion. One of these is an antisymmetric tensor field, and also there is a scalar field referred to as the *dilaton*[34] (which has to do with overall scalings) and is rather like the (unwanted) scalar field of the original Kaluza–Klein theory. (We recall that this scalar is removed by normalizing the Killing vector; see the penultimate paragraph of §31.4.) The dilaton will have some relevance to later discussions; see §31.15. Recall that the string constant is very small. It is now taken to be only very slightly more than the square of the Planck length (α' being a tiny area), with

$$\alpha' \approx 10^{-68}\mathrm{m}^2.$$

Thus, Ricci flatness, for the 10-spacetime metric, is considered to be an excellent approximation.

31.9 String dynamics

You might be wondering where these statements about spacetime curvature actually come from, since string theory is really just a theory about these little strings running about in some background spacetime (albeit of 9 spatial dimensions). In fact, I have not been specific, as yet, about the equations that control the string dynamics. Let us see to this next.

As is usual in field theory, there is a Lagrangian (§§20.5,6 and §26.6), and the string Lagrangian is defined by $1/2\alpha'$ multiplied by the surface area of the 2-surface history—the *world sheet*—that the string traces out in spacetime. The metric on the world sheet is to agree with that induced from the spacetime; and classically the dynamics would be simply that the world sheet is a kind of 'soap film', or 'minimal surface' (of appropriate metric signature) in the given spacetime background. The background is subject to no constraints, classically. The string simply flaps around according to this specified dynamics. In quantum mechanics, however, the anomaly issue looms large, and we find that even the condition that the background has 10 dimensions with supersymmetry, is now not sufficient, and the above conditions on the 10-space curvature are also needed, to provide consistency conditions on the background metric for the quantum strings.

In addition to this Einstein-equation-like consistency requirement, we recall the 'lowest mode of excitation' of a closed string, referred to in §31.7, that seemed to describe a massless particle of spin 2. Its 'spin 2' nature arises because the mode has a quadrupole (or $\ell = 2$) structure to its oscillation (see §22.11 and §32.2) and it is massless essentially because it is the lowest mode of a very 'stiff' string. Although the mode had presented a serious problem for the original hadronic strings, in its new gravitational context it was now viewed with favour because, in ordinary

(4-dimensional) physics, a *graviton* (quantum of the gravitational field) would be a massless particle of spin 2. In the conventional analysis, this comes about through an examination of perturbations in the metric field (described by a symmetric tensor 'h_{ab}', giving the infinitesimal shift in a metric from g_{ab} to $g_{ab} + \varepsilon h_{ab}$, where ε is infinitesimal see also §32.2). The viewpoint appears to be, in the new string theory—in view of the above Einstein-equation-like consistency requirement (albeit in 10 rather than 4 dimensions), and this 'graviton-like' string excitation mode—that 'string theory includes gravity'. In the words of Edward Witten (1996):

> String theory has the remarkable property of *predicting gravity*,

and Witten has further commented:[35]

> the fact that gravity is a consequence of string theory is one of the greatest theoretical insights ever.

It should be emphasized, however, that in addition to the dimensionality issue, the string theory approach is (so far, in almost all respects) restricted to being merely a perturbative theory, expressed in terms of a power series (say in the 'ε' referred to above, but most string-theory calculations refer to power series in the string constant α'). This restriction is regarded as a serious limitation by most relativity practitioners, who would not take the considerations above as sufficient to provide us with a theory having the same profound underlying principles as Einstein's general relativity.

One version of a 'string philosophy' that I have heard expressed is that we should try to think of physics as 'actually' being a *2-dimensional* QFT, with the geometrical notion of 10-dimensional spacetime being secondary to the more primitive 'reality' of the 2-dimensional string world sheet itself. Everything is to be described in terms of 'string excitations', and these are to be thought of as quantities that are merely functions of the 2 coordinates on the world-sheet. The 10 spacetime dimensions are felt out by these excitations, but everything is some kind of 'field on the 2-dimensional sheet'.

I have a great deal of difficulty with this sort of viewpoint for a theory which purports to describe gravitation, where there would be dynamical degrees of freedom in the spacetime geometry. Recall from §16.7 (and see the discussion in §31.4 above and in §§31.10–12, 15–17 below) that there are vastly more functions or fields on a larger-dimensional space than on a smaller-dimensional one, irrespective of the number of independent components that the function (field) may have at each point, provided that this number of components is finite. Moreover, for any ordinary notion of 'string excitation', this component number per point would indeed be finite (since each point of the world sheet can only be displaced in a finite number of independent directions in the ambient space). This particular 'string philosophy' would appear to be a very misleading way of looking at things. Although I doubt that it is really held to rigorously by many string theorists, the fact that some of them have been prepared to entertain such a viewpoint is perhaps indicative of the seemingly cavalier attitude of many string theorists, with regard to spacetime dimensionality. The 4-dimensionality of our *observed* spacetime, perhaps thought of as a 'low energy effect', often appears to be taken to be a matter of relatively minor importance!

In any case—even in 10 dimensions—the 'Einstein vacuum equation' of string theory is regarded as merely a consequence of the *consistency* of the string's 2-dimensional world sheet. And it is taken that Ricci flatness must continue to hold even at those spacetime locations where the string world sheet is, itself, *not* located! If the quantum theory were really describing the quantized dynamics of the coupled classical system specified by

background 9-space containing moving string,

then the consistency on the background curvature would need to hold only where the string is located. So we have to take the view that it is not this classical system that is being quantized. In fact, despite purporting to be a theory of gravity, string theory does not really properly come to terms with the problem of describing the dynamical degrees of freedom in the spacetime metric. The spacetime simply provides a fixed background, constrained in certain ways so as to allow the strings themselves to have full freedom.

31.10 Why don't we see the extra space dimensions?

If we are now going to take the full dynamics of the 10-dimensional spacetime seriously, we have to face the opposite problem of how to get the huge extra functional freedom in the 10-dimensional space down to what would be appropriate for an ordinary physical theory in four spacetime dimensions. Ricci flatness in ten dimensions allows a functional freedom $\infty^{70\infty^9}$ (see §16.7) that is enormously huger than the mere $\infty^{N\infty^3}$ that we have for an ordinary 4-space field theory, taking N independent components per point (§16.7 and §31.4). (A 10-dimensional 'Ricci-flat' theory would in fact have 70 independent functions as free data on a 9-dimensional initial surface.[36]) The extra hugeness comes from 9 being greater than 3. In comparison, the relative size of 70 and N contributes negligibly to this explosion of functional freedom.[37] An ordinary *classical* field theory in a 10-dimensional spacetime (without such a restriction as the symmetry that is decreed by the Killing vector of the original Kaluza–Klein theory—see §31.4) would certainly be in gross conflict with our observed universe, owing to this horrendous excess of functional freedom. We shall come back to this issue in §§31.12,16.

Why do string theorists seem not to be particularly troubled by this excessive functional freedom? Part of the reason would appear to be that they have significant hopes that, in an appropriately quantized string theory, there may well be additional spacetime constraints, coming from consistency requirements for the quantized strings, which effectively reduce the strings' functional freedom. We shall be catching a glimpse of these hopes in §31.16. But the main argument that is commonly voiced comes from the expectation that, if it is assumed that the six 'extra' dimensions are exceedingly 'small' (say, having a Planck scale of 10^{-35}m), then—for energies that are available in the physical world today—quantum-mechanical considerations should come to the rescue and would effectively 'kill off' the degrees of freedom that are concerned with the extra spatial dimensions.

How is this to work? As remarked above, almost all string-theoretic considerations are carried out in a perturbative framework, where merely small perturbations away from some particular basic model are examined. Here, we are to consider a basic 'spacetime' that is the *product* $\mathbb{M} \times \mathcal{Y}$ of normal Minkowski 4-space with some given compact spacelike Riemannian 6-space $\mathcal{Y}$, where the overall 'size' of this particular $\mathcal{Y}$ is very small, say of the Planck scale of 10^{-35}m. We are to look at small perturbations away from $\mathbb{M} \times \mathcal{Y}$.

First, we need to have a clearer picture of what a 'product manifold' $\mathcal{A} \times \mathcal{B}$ is, where both the m-space $\mathcal{A}$ and the n-space $\mathcal{B}$ are each taken to be (pseudo-)Riemannian manifolds. Recall from §15.2 (and Fig. 15.3a) that the points of $\mathcal{A} \times \mathcal{B}$ are described as pairs (a, b), where a belongs to $\mathcal{A}$ and b belongs to $\mathcal{B}$, so the dimension of $\mathcal{A} \times \mathcal{B}$ is $m + n$ (see Exercise [15.1]). How are we to define the (pseudo-)Riemannian metric on $\mathcal{A} \times \mathcal{B}$? This is to be the 'direct sum' of the metrics on $\mathcal{A}$ and on $\mathcal{B}$. We can use local coordinates $(x^1, \ldots, x^m, y^1, \ldots, y^n)$ for $\mathcal{A} \times \mathcal{B}$, where $(x^1, \ldots, x^m)$ and $(y^1, \ldots, y^n)$ are local coordinates for $\mathcal{A}$ and for $\mathcal{B}$, respectively. Then the metric components g_{ij} for $\mathcal{A} \times \mathcal{B}$ have 'block-diagonal form' (similar to that displayed in §13.7 for the matrices of a fully reducible representation) describing the *direct sum* of the metric components for $\mathcal{A}$ and for $\mathcal{B}$, respectively: the squared metric distance in $\mathcal{A} \times \mathcal{B}$ is the sum of those in $\mathcal{A}$ and $\mathcal{B}$ individually (Fig. 31.7).

A key fact of later relevance (see §31.14) is that if $\mathcal{A}$'s metric and $\mathcal{B}$'s metric are both Ricci-flat (vanishing Ricci tensor—see §19.6), then the direct-sum metric of $\mathcal{A} \times \mathcal{B}$ is also Ricci-flat.[31.3] The space $\mathcal{Y}$, in our product $\mathbb{M} \times \mathcal{Y}$ is taken to be indeed Ricci-flat, and $\mathbb{M}$ itself, being flat, is certainly Ricci-flat. The product $\mathbb{M} \times \mathcal{Y}$ is thus also Ricci-flat, as will be required.

[31.3] Why? *Hint:* Look at the form of the explicit expressions in Exercise [14.26] and Exercise [14.27] and in §14.7.

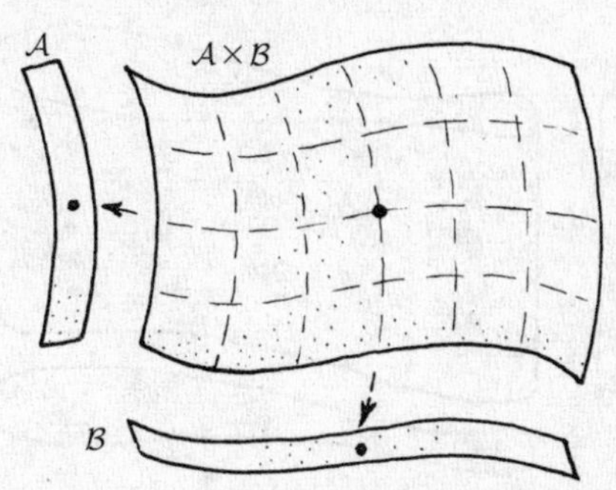

Fig. 31.7 The product manifold $\mathcal{A} \times \mathcal{B}$ (see Fig. 15.3a, §15.2) of two (pseudo-) Riemannian spaces $\mathcal{A}$ and $\mathcal{B}$ is itself (pseudo-)Riemannian. If both $\mathcal{A}$ and $\mathcal{B}$ are Ricci-flat, then so is $\mathcal{A} \times \mathcal{B}$.

We are to take the *compact* space $\mathcal{Y}$ to have an overall spatial size of the general order of the Planck scale—or maybe a little larger. (Recall the meaning of compactness, described in §12.6, Fig. 12.13.) How are perturbations away from $\mathbb{M} \times \mathcal{Y}$ to be described? These will be given by (tensor) fields on $\mathbb{M} \times \mathcal{Y}$, like the h_{ab} of §31.9, which provide us with an infinitesimal changes in the metric of $\mathbb{M} \times \mathcal{Y}$.

To study fields on $\mathbb{M} \times \mathcal{Y}$, it is useful to think in terms of an initial value problem; so we represent $\mathbb{M}$ as $\mathbb{M} = \mathbb{E}^1 \times \mathbb{E}^3$, where the Euclidean 1-space $\mathbb{E}^1$ refers to a time coordinate t, and the Euclidean 3-space $\mathbb{E}^3$ refers to space. We then analyze these fields in terms of normal modes on $\mathbb{E}^3 \times \mathcal{Y}$. See Fig. 31.8. (Recall the concept of 'normal mode', as described classically in §20.3 and in the quantum context in §§22.11,13.) What do these normal modes look like? Because of the 'product' structure of $\mathbb{E}^3 \times \mathcal{Y}$, we can represent each of these modes simply as the ordinary product of a mode on $\mathbb{E}^3$ with a mode on $\mathcal{Y}$. The modes of E^3 are just momentum states (§21.11), and they form a continuous family. As regards $\mathcal{Y}$'s normal modes, the compactness ensures that they form a *discrete* family, each characterized by some finite set of eigenvalues. (Recall the discussion at the end of §22.13.) How would one 'excite' one of these modes, so that the simple $\mathbb{E}^3 \times \mathcal{Y}$ geometry is converted to something else?

The usual string theorist's argument that we can disregard perturbations of $\mathcal{Y}$, at least at the present cosmological epoch, depends upon an expectation that the energy needed to excite any of $\mathcal{Y}$'s modes would be enormous—except for a certain particular set of modes of zero energy (which will have importance for us in §31.14) that I shall ignore for the moment. Why is this energy expected to be so large? The reasoning rests upon the very minute scale of $\mathcal{Y}$ itself. A 'standing wave' on $\mathcal{Y}$ would have a tiny wavelength, comparable with the Planck distance of $\sim 10^{-35}$m, and would therefore have something like a Planck frequency of $\sim 10^{-43}$s. The energy

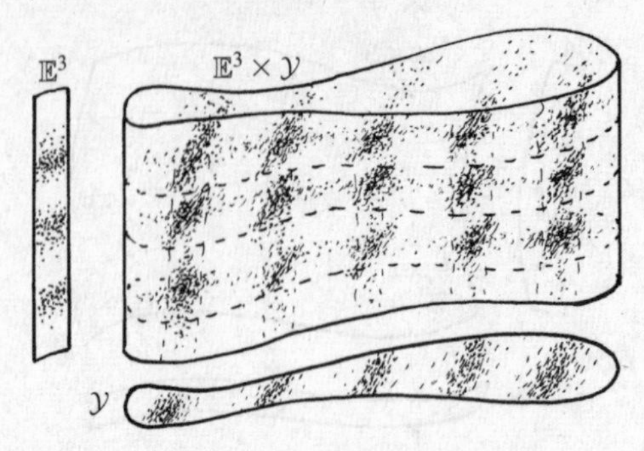

Fig. 31.8 Perturbation modes on $\mathbb{E}^3 \times \mathcal{Y}$ (for the Laplace equation) are products of modes on $\mathbb{E}^3$ with modes on $\mathcal{Y}$.

required to excite such a mode would be of the general order of a Planck energy, namely around 10^{12} joules, which is some twenty orders of magnitude larger than the largest energies involved in ordinary particle interactions! It is accordingly argued that the modes which affect $\mathcal{Y}$'s geometry will remain unexcited, in all particle-physics processes that are of relevance to the physical actions that are available today. The picture is presented that at the very early stages of the universe, six of its dimensions settled into the configuration described by a roughly Planck-scale $\mathcal{Y}$, whereas the remaining three spatial dimensions expanded outwards enormously to give the almost spatially flat picture of a 3-dimensional universe in accordance with present-day cosmology. The $\mathcal{Y}$-spaces would have remained basically undisturbed from a time not long after the first Planck moments of the universe's existence.

Let us look at this argument a little more fully. To simplify matters, consider a situation where, as with the original Kaluza–Klein theory of §31.4 and the hosepipe analogy depicted in Fig. 15.1, $\mathcal{Y}$ is just a *circle* S^1, which we take to have some very small radius ρ. We can choose a real coordinate θ for S^1 (with θ identified with $\theta + 2\pi$), where $\rho\theta$ measures actual distance round the circle. The modes of $\mathcal{Y}$ are now simply the quantities $e^{in\theta}$, where n is an integer, namely the Fourier modes that we encountered in §9.2. On $\mathbb{E}^3$ we can choose ordinary Cartesian coordinates (x, y, z). We recall from §22.11 that one way of addressing the question of finding 'modes' is to look for eigenstates of (the appropriate) Laplacian operator. In the present context, this may be regarded as an approximation (or just a 'model'). More correctly, we should be concerned with eigenstates of the Hamiltonian $\mathcal{H}$ for the evolution of the geometry. For Ricci-flat 5-spaces (our required pertubations of $\mathcal{M} \times S^1$) we should need the appropriate Hamiltonian formulation of five-dimensional general relativity, which is complicated. The leading term of this *is* essentially a Laplacian, however, and this will suffice for our present discussion.

We first encountered the 2-dimensional Laplacian $\nabla^2 = \partial^2/\partial x^2 + \partial^2/\partial y^2$ in §10.5. Here, we need the generalization to four dimensions, but the metric of our space $\mathbb{E}^3 \times \mathrm{S}^1$ is still flat, so we do not require a more elaborate expression like that of §22.11. All we need here is to increase the number of variables to four, so we can write the Laplacian as[31.4]

$$\nabla^2 = \frac{\partial^2}{\partial x^2} + \frac{\partial^2}{\partial y^2} + \frac{\partial^2}{\partial z^2} + \frac{1}{\rho^2}\frac{\partial^2}{\partial \theta^2},$$

where our fourth coordinate is the $\rho\theta$ needed for S^1. To find our 'modes', we would look for eigenstates of this ∇^2. More specifically, the procedure is merely to worry about the mode analysis for the S^1 part of $\mathbb{E}^3 \times \mathrm{S}^1$ and to leave the $\mathbb{E}^3$ part as an ordinary field. Accordingly, we split up our fields into different contributions, each having a different integer 'n', giving a dependence on θ of the specific form $\mathrm{e}^{in\theta}$, as described above. Thus, for an nth-order S^1 mode, we can write

$$\Psi = \mathrm{e}^{in\theta}\psi,$$

on our initial 4-surface $\mathbb{E}^3 \times \mathrm{S}^1$, where ψ is a function of the ordinary space coordinates x, y, z. For any such nth order mode Ψ, the term $\rho^{-2}\partial^2/\partial\theta^2$ in our above Laplacian can be replaced[31.5] simply by $-n^2/\rho^2$:

$$\frac{1}{\rho^2}\frac{\partial^2}{\partial \theta^2} \mapsto -\frac{n^2}{\rho^2}.$$

With regard to the remaining variables x, y, z, our Laplacian now reverts to the ordinary 3-space one, but we have the constant term $-n^2/\rho^2$ added to this 3-space Laplacian.

Let us recall the field equation of an ordinary (spinless) particle of mass μ, in ordinary Minkowski spacetime $\mathbb{M}$, this being the 'Klein–Gordon' wave equation (see §24.5)

$$\left(\Box + \frac{\mu^2}{\hbar^2}\right)\psi = 0,$$

where $\Box = \partial^2/\partial t^2 - \partial^2/\partial x^2 - \partial^2/\partial y^2 - \partial^2/\partial z^2$. We can think of this as a 'free incoming particle' (as would be appropriate in an S-matrix approach to QFT—see §26.8). In the 5-space $\mathbb{M} \times \mathrm{S}^1$, however, we would have an additional term $-\rho^{-2}\partial^2/\partial\theta^2$ in the wave operator $\Box$. If we take this 5-space particle to be in an n-mode eigenstate for S^1, this term gets replaced by n^2/ρ^2, as above. Accordingly, from the ordinary Minkowski 4-space point of view, our 5-space n-mode Klein–Gordon particle satisfies the 4-space equation

[31.4] Why?

[31.5] Why can we do this?

$$\left(\square + \frac{\mu^2}{\hbar^2} + \frac{n^2}{\rho^2}\right)\psi = 0.$$

This is just the Klein–Gordon equation again, but with $\mu^2/\hbar^2 + n^2/\rho^2$ in place of $\mu^2/\hbar^2$. Thus, we have the 4-space Klein–Gordon equation for a new particle, but where the mass is increased from μ to $\sqrt{(\mu^2 + \hbar^2 n^2/\rho^2)}$.

Now, any of the observed particles of Nature would have a mass μ that is enormously smaller (see §31.1) than the Planck value of roughly $\hbar/\rho$ (for our chosen value of ρ). Assuming that $n \neq 0$, this new particle would have a mass that is at least of Planck order ($\hbar n/\rho$ being much greater than μ) so it would lie far beyond the reach of presently feasible particle accelerators. It is accordingly reasoned by string theorists that no $n \neq 0$ mode can be accessed in any particle-physics process that is available at the present cosmological epoch!

Essentially the same argument would apply to the full Planck-sized compact 6-space $\mathcal{Y}$. In the relatively low-energy situation we find ourselves in today, the modes of excitation of $\mathcal{Y}$ for which $n \neq 0$ are experimentally inaccessible to us—so the string theorists maintain. It is argued, therefore, that there is no conflict between the hypothesis of extra spatial dimensions and present-day observational physics.

31.11 Should we accept the quantum-stability argument?

But is this reasoning really appropriate? I believe that there are profound reasons to question it.[38] Even if we leave aside the unanswered puzzle of why three of the spatial dimensions should behave so very differently from the remaining six, we must be very cautious about this 'particle-physics' reasoning which is relied upon for the $\mathcal{Y}$ geometry to be immunized against change during the subsequent evolution of the universe.

But before entering into the matter of Planck-energy (or higher) perturbations of $\mathcal{Y}$, I should return to the modes of $\mathcal{Y}$ of *zero energy* which I chose to ignore in §31.10. We shall see in §31.14 that these modes tend to be regarded with favour in the string-theory programme, as they offer hope for genuine contact to be made with the symmetry groups of standard particle physics (§§25.5,7). Yet, mathematically, they lead to a serious difficulty that has been referred to as the *moduli problem.* As with a Riemann surface (see §8.4) there are certain parameters referred to as *moduli* which define the specific shape of the type of $\mathcal{Y}$-space under consideration. (We shall be seeing in §31.14, that the preferred $\mathcal{Y}$s are certain complex 3-manifolds referred to as 'Calabi–Yau' spaces, whose moduli generally constitute a family of complex numbers.) The zero-energy modes refer to the varying of these moduli. We may choose to allow this variation to have a spatial $\mathbb{E}^3$-dependence, but this gives us only an acceptable $\infty^{N\infty^3}$, where N refers to the actual number of independent (real) moduli. However, it turns out that there are modes in which moduli rapidly shrink away to zero leading to a singular $\mathcal{Y}$-space. This seemingly catastrophic instability is essentially the string theorist's 'moduli problem' (see also, §31.14).[39] It appears to be unanswered; yet it is normally ignored.

Suppose that we choose to ignore it also(!). Then, are the positive-energy (Planck-scale) modes of vibration of the six extra dimensions immune from excitation? Although the Planck energy is indeed very large when compared with normal particle-physics energies, it is still not *that* big an energy, being comparable with the energy released in the explosion of about one tonne of TNT. There is, of course, enormously more energy than this available in the known universe. For example, the energy received from the Sun by the Earth in one second is some 10^8 times larger! On energy terms alone, that would be far more than sufficient to excite the $\mathcal{Y}$ space for the *entire universe*!

In the string theorist's reasoning, this energy is delivered in a local particle interaction, and we tend to imagine it as being administered in some tiny region of ordinary space. Yet the actual modes of excitation of $\mathcal{Y}$ that are supposed to be inaccessible are spread uniformly over the whole of $\mathbb{E}^3$, in our perturbations of $\mathbb{E}^3 \times \mathcal{Y}$. Recall that the modes of excitation of $\mathbb{E}^3 \times \mathcal{Y}$ are simply products of the modes on $\mathbb{E}^3$ with the modes on $\mathcal{Y}$. Those that we are considering here are just constant over $\mathbb{E}^3$. There is nothing to say that these need (or even should) be injected at a localized region in ordinary physical space.

This in itself is no argument against local particle interactions being the appropriate way to excite such modes, however. Their being spread out over the whole of $\mathbb{E}^3$ does not argue against a particle-physics perspective. We recall from the discussion of creation and annihilation operators in §26.2 and of Feynman graphs in §§26.7,8, that particles and their interactions in QFT are commonly described in terms of momentum states. Such states *are* indeed 'spread' over the whole of $\mathbb{E}^3$, as was emphasized particularly in §21.11. 'Quantum particles' need in no way be spatially localized. Perhaps a better way to think of these matters is to refer to 'quanta' rather than particles. The issue is whether or not it is reasonable to expect that a single quantum of Planck energy could be injected into a $\mathcal{Y}$ mode, by whatever means. But it does not seem to me that we need think of such 'means' as being necessarily local particle interactions, and not something else such as a non-linear disturbance of the entire spacetime geometry.

Are there reasons to believe that there ought to be some such other means? In my opinion there are indeed reasons to worry about this. Let us return to our hosepipe analogy (Fig. 31.3). Think of the hosepipe as being essentially straight in its 'large' dimension (analogous to $\mathbb{E}^3$), and with a constant S^1 cross-section (analogous to $\mathcal{Y}$), which is a circle of tiny radius ρ.

The hosepipe's modes of excitation can be composed of various waves travelling one way or the other along its length ('$\mathbb{E}^3$ modes') and of various distortions of its circular cross-sectional shape ('$\mathcal{Y}$ modes'). As we have seen, any one of these latter modes occurs simultaneously along the entire hosepipe. Quantum-mechanically, the energy in a single quantum of excitation of such a mode—an *exciton*—of vibrational frequency ν is $2\pi\hbar\nu$ (see §21.4) and independent of the hosepipe length!

For an infinite length of hosepipe, this gives a zero density of energy, for each individual exciton, so it may be less confusing if we imagine the pipe to be bent round into a very large circle, of radius R, say, where $R \gg \rho$. Now, think of a particular mode of vibration of $\mathcal{Y}$, with a particular frequency ν. The total energy $2\pi\hbar\nu$ in this exciton is indeed independent of R. This may seem puzzling, because it implies that the larger we take R to be, the smaller is the energy that exists locally in the vibration, in proportion to $1/R$. This is no inconsistency, but it tells us that the amplitude of the vibration in an exciton, for a fixed vibrational mode of $\mathcal{Y}$, is smaller, the greater the length of the pipe. If we take the limit $R \to \infty$, the energy stored locally in the mode goes to zero. We learn from this that any particular way in which the hosepipe can vibrate locally, in the limit when the hosepipe becomes infinite, must involve higher and higher numbers of quanta, the effect of each individual quantum getting less and less, so we are driven to consider that a *classical* rather than a quantum description of the behaviour of the hosepipe might become appropriate.[40]

This raises the issue of the *classical limit* of a quantum system for large quantum numbers, and the related matter of the state reduction **R** to such a classical configuration. We have seen, particularly from the discussion in Chapter 29, that the **R** issue cannot really be fully resolved within the framework of present-day quantum theory.[41] Nevertheless, a good physicist should know when a quantum description is appropriate to use, and when it makes more physical sense to use a classical one. Recall the case of ordinary angular momentum, as discussed in §22.10. A body with a very large angular momentum tends to be best treated as a classical system, so that we obtain a very well-defined rotation axis. Treating this just as a quantum system with a very large j value, we obtain a Majorana description with many spin directions, usually pointing all over the place! In practice, the classical description would be the one to use for very large angular momentum, and this provides a good picture of physical reality. More generally, classical descriptions tend to be taken as physically appropriate when quantum numbers get excessively large. In the case of angular momentum, the relevant quantum number, namely j, is measured in in terms of units of $\hbar$, so one could imagine a reasonable-looking criterion telling us when we are far away from the quantum regime: the value of j is very large in units of $\hbar$. With the hosepipe, we see that the smallness of the distance ρ is not, in itself, an appropriate measure for telling us that a 'quantum' description is more suitable than a classical one. For fixed ρ, the description of local hosepipe vibrations seems to become more and more 'classical' the larger we take R, since we need to involve larger and larger numbers of excitons and excitons involving higher and higher vibrational quantum numbers (modes of $\mathcal{Y}$).[42]

In the absence of a theory that tells us how 'large' systems become well described classically, while 'small' ones behave according to quantum rules (which in my view requires a change in the very structure of quantum mechanics along the lines of §§30.9–12), it would seem that we can come to no definitive conclusion concerning the alleged inaccessibility of excitations of $\mathcal{Y}$. (The considerations of Chapter 30 do not yet seem to provide us with any unambiguous answer, and certainly not an uncontroversial one.) Nevertheless, in view of the fact that actual perturbations of $\mathcal{Y}$ do lead us to a quantum picture of very large numbers of quanta, where each individual quantum affects the geometry of $\mathcal{Y}$ hardly at all, and to large quantum numbers, it would appear that we may well get more insights into how perturbations of an $\mathbb{M} \times \mathcal{Y}$ universe with 'small' $\mathcal{Y}$ behave if we study these *classically*, rather than quantum-mechanically. Let us consider this next.

31.12 Classical instability of extra dimensions

What can be said, if we indeed take the 10-space model as an entirely classical one? This ought, at least, to give us some guidance as to how the full quantum model will actually behave. We saw at the beginning of §31.10 that in a classical $(1+9)$-spacetime (i.e. of one time dimension and nine space dimensions), there would be an unacceptable flood of excessive degrees of freedom ($\infty^{M\infty^9} \gg \infty^{N\infty^3}$). This is serious enough, but in my opinion, things are actually much worse than this. We shall find that a classical $\mathbb{M} \times \mathcal{Y}$ universe—subject to Ricci flatness—is highly unstable against small perturbations. If $\mathcal{Y}$ is compact and of a Planck size, then spacetime singularities (§27.9) are to be expected to result within a tiny fraction of a second!

Let us first consider perturbations of $\mathbb{M} \times \mathcal{Y}$ that disturb only the $\mathcal{Y}$ geometry and which, accordingly, do not 'leak out' into the spatial $\mathbb{E}^3$. That is to say, we examine a 'generic' Ricci-flat $(1+6)$-spacetime $\mathcal{Z}$ (the perturbed evolution of $\mathcal{Y}$), the entire $(1+9)$-spacetime being $\mathcal{Z} \times \mathbb{E}^3$. We consider that $\mathcal{Z}$ is the time-evolution of some 6-space that (at some particular time) is 'close' to $\mathcal{Y}$, so $\mathcal{Z}$ starts out close to the (unchanging) 'time-evolution' $\mathbb{E}^1 \times \mathcal{Y}$ of $\mathcal{Y}$, although $\mathcal{Z}$ may deviate strongly from $\mathbb{E}^1 \times \mathcal{Y}$ at later times (Fig. 31.9). Here, I am expressing $\mathbb{M}$ as $\mathbb{M} = \mathbb{E}^1 \times \mathbb{E}^3$, as in §31.10 above (with $\mathbb{E}^1$ describing the time dimension and $\mathbb{E}^3$ the space dimensions), so we think of the spacetime $\mathbb{M} \times \mathcal{Y}$

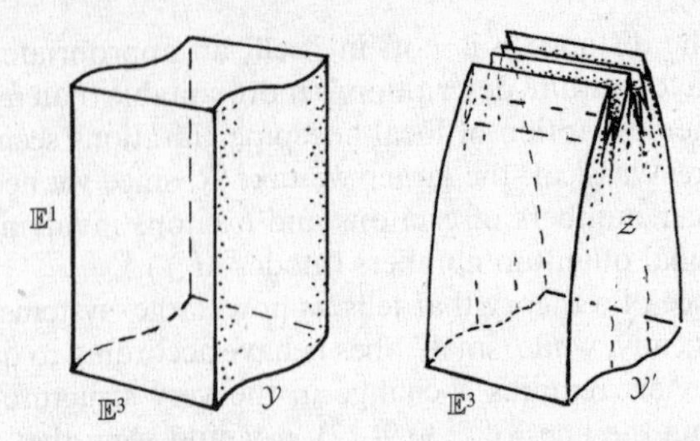

Fig. 31.9 A singularity theorem applies to perturbations of $\mathbb{M} \times \mathcal{Y}$, where $\mathcal{Y}$ is a 'small' Calabi–Yau space. (a) The non-singular canonical case $\mathbb{M} \times \mathcal{Y}$, where we express $\mathbb{M} = \mathbb{E}^1 \times \mathbb{E}^3$, with $\mathbb{E}^1$ referring to the time. (b) A general perturbation $\mathcal{Y}'$ of $\mathcal{Y}$ evolves to a space $\mathcal{Z}$ that is singular, so general perturbations of $\mathbb{E}^3 \times \mathcal{Y}$ that do not affect $\mathbb{E}^3$ evolve to spaces $\mathbb{E}^3 \times \mathcal{Z}$ that are singular.

as $(\mathbb{E}^1 \times \mathcal{Y}) \times \mathbb{E}^3$ (putting the time-evolution of $\mathcal{Y}$ first in the product); see Fig. 31.9a.

Now, in the late 1960s Stephen Hawking and I proved a singularity theorem which shows that we must expect $\mathcal{Z}$ to be singular.[43] As we explicitly stated, this theorem applies just as well to $(1 + 6)$-spacetimes (and to $(1 + 9)$-spacetimes) as it does to the conventional $(1 + 3)$-spacetimes that we originally considered. As one of this theorem's consequences, any Ricci-flat spacetime that (like $\mathbb{E}^1 \times \mathcal{Y}$ or $\mathcal{Z}$) contains a compact spacelike hypersurface, and that is 'generic' in a certain specific sense[44] (and free of closed timelike curves—see §17.9 and Fig. 17.18), must indeed be singular! The original $\mathbb{E}^1 \times \mathcal{Y}$ escapes from being singular because the generic condition fails in this case. But the generically perturbed $\mathcal{Z}$ has to be singular.

It should be noted that, in such 'singularity theorems', one does not directly establish that the curvature diverges to an infinite value, but merely that there is an obstruction of some sort to timelike or null geodesics being extendable within the spacetime to infinite length (or to infinite affine length, in the case of null geodesics—see §14.5). The normal expectation would be that this obstruction indeed arises because of the presence of diverging curvature, but the theorem does not directly show this. This theorem does, however, tell us that $\mathcal{Z}$ will become singular in this or some other way. If the perturbation away from $\mathcal{Y}$ is of the same general scale as $\mathcal{Y}$ itself (i.e. Planck scale), then we must expect the singularities in $\mathcal{Z}$ to occur in a comparable timescale ($\sim 10^{-43}$ s), but this timescale could become somewhat longer if the perturbations are of a proportionally smaller size than $\mathcal{Y}$ itself.

We conclude that if we wish to have a chance of perturbing $\mathcal{Y}$ in a generic way so that we obtain a non-singular perturbation of the full $(1 + 9)$-space $\mathbb{M} \times \mathcal{Y}$, then we must consider disturbances that significantly spill over into the $\mathbb{M}$ part of the spacetime as well. But in certain respects such disturbances are even more dangerous to our 'ordinary' picture of spacetime than those which affect $\mathcal{Y}$ alone, since the large Planck-scale curvatures[45] that are likely to be present in $\mathcal{Y}$ will spill over into ordinary space, in gross conflict with observation, and will result in spacetime singularities in very short order.[46]

Of course, unacceptable singularities in a classical theory do not necessarily tell us that such blemishes will persist in the appropriate quantum version of that theory. As we have already seen in §22.13, quantum mechanics cures the catastrophic instability of ordinary classical atoms, whereby electrons would have spiralled into the nucleus with the emission of electromagnetic radiation. However, the mere introduction of 'quantization procedures' will not necessarily ensure that classical singularities are removed. There are many examples (such as in most toy models of quantum gravity[47]) where singularities persist after quantization.

We should also take note of the fact—see §31.8—that $(1+9)$-dimensional Ricci flatness is not precisely the requirement that string theory demands. We recall that Ricci flatness is regarded merely as an excellent approximation to that requirement, coming about when terms higher than the lowest order in the string constant α' are ignored. Maybe the 'exact' requirement, involving all orders in the string constant α', could evade the above singularity theorem. However, if this requirement provides us with a condition on the Ricci tensor for which the usual local energy-positivity demands are satisfied (see especially Note 27.9, and §28.5), then the singularity theorem would still apply. On the other hand, violations of such local energy conditions can certainly occur in QFT (§24.3), so these issues are far from conclusive.

More serious, to my mind, is the fact that the full requirement, involving all orders in the string constant α', is actually an infinite system of differential equations of unbounded differential order. Accordingly, the data that would be needed on an initial 9-surface would involve derivatives of all orders in the field quantities (rather than just the first or second derivatives that are needed in ordinary field theories). The number of parameters per point needed on the 9-surface is then infinite, so we get a functional freedom *greater* than $\infty^{M\infty^9}$, for any positive integer M. This would seem to make the problem of excessive functional freedom even worse than before! I am not aware of any serious discussion of the mathematical form of this full requirement, and of what kind of initial data might be appropriate for it.

31.13 Is string QFT finite?

The kind of argument that I have been giving above illustrates why I have severe difficulty in being persuaded that string-theoretic models are likely

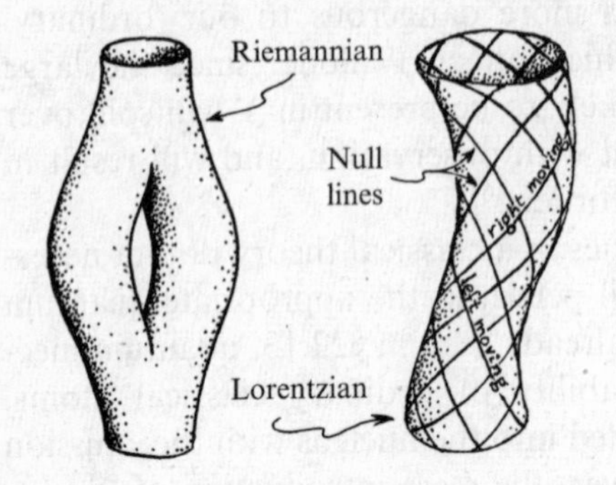

Fig. 31.10 The mathematics of string theory uses 'string histories' that are Riemann surfaces—having Riemannian (positive definite) metrics. But, physically, the string histories are Lorentzian. Passing from one to the other involves a kind of 'Wick rotation'.

to reproduce Einstein's (1 + 3)-dimensional general relativity in any kind of sensible 'classical limit'. What about the other part of string theory's claim, that it is a *consistent finite* QFT (whatever the resulting theory actually means, physically)? It seems to me that the case that a finite amplitude is obtained, for a fixed string world-sheet topology, could indeed be the strongest part of the pro-string argument, this conclusion seeming to provide a true reflection of the original virtues of the string idea. Yet, even here there are fundamental questions that must be raised.

To begin with, there is a certain worry that I have always had with regard to string theory. It is presented as a physical theory of string-like structures whose world sheets are timelike, and whose induced 'metric' is therefore a (1 + 1)-dimensional Lorentzian metric. Yet the mathematics is carried out with string world sheets that possess a (positive-definite) metric, so that the elegant ideas of Riemann surface theory can be appealed to (as in Chapter 8); see Fig. 31.10. In accordance with the former, one talks about modes of disturbance travelling along the timelike world sheet either to the left or to the right with the speed of light. (These propagate along the null curves on the Lorentzian world sheet.) In the Riemannian version of the theory, these 'left' or 'right' modes become 'holomorphic' and 'antiholomorphic' functions on the Riemann surface. The attitude seems to be that calculations are indeed done with the positive-definite Riemann-surface picture, and then a 'Wick rotation' (§28.9) is performed to get the desired Lorentzian string theory out in the end. It is certainly possible that this process is satisfactory here, but this cannot simply be assumed without specific justification. It depends critically, for example, upon approximations not being made in the computations of the amplitudes. Otherwise there could be serious question marks about the procedure, of the type that we have encountered before in relation to the Hawking approach to quantum gravity and other approaches to QFT involving analytic continuation (see §28.9). My understanding is that the Riemann-surface calculations are indeed intended to be exact, so there is some reason to be hopeful that the Wick rotation can be trusted. Nevertheless, the explicit justification for a Wick rotation depends upon the background spacetime being flat, which would certainly not be the case if we are doing serious (non-perturbative) general relativity, so it remains unclear how far this takes us in the direction of a quantum theory of actual gravity.

Even if we trust the validity of such flat-space considerations, must we go along with the forceful claims that, for each fixed Riemann-surface topology (i.e. fixed genus g, see §8.4, where g corresponds to the 'number of loops' for an ordinary Feynman graph—see §26.8 and Fig. 31.5d, §31.5), the total amplitude is indeed finite? In fact this has not been established. Despite repeated assurances, no mathematical demonstration of this claimed finiteness has yet been provided. The finiteness claims refer only to the *ultraviolet* (large momentum, small distance) divergences that quantum field theorists find themselves to be most troubled by, but even these have been established so far only at the 2-loop level. Moreover, there seems to be no argument claiming that *infrared* (small momentum, large distance) divergences (§26.9) are eliminated. Although such divergences are normally considered to be less serious than the ultraviolet ones, they certainly cannot be ignored, and need to be dealt with in some manner if the 'finite' claim is to be justified. This leaves us in some uncertainty with regard to the whole programme, as this finiteness is the linchpin of the entire string idea.[48]

Perhaps these are just irritating technicalities which will be overcome in future mathematical developments. However, even if it is accepted that we have finite amplitudes for each fixed topology, we are far from finished. The expressions have then to be *summed up*. Now there is a problem that this sum apparently actually diverges.[49] The intended *finite* theory is actually not finite after all! This particular divergence seems not to worry the string theorists, however because they take this series as an improper realization of the total amplitude. This amplitude is taken to be some analytic quantity, with the power series attempting to find an expression for it by 'expanding about the wrong point', i.e. about some point that is singular for the amplitude (a bit like trying to find a power series for $\log z$, expanded about $z = 0$, rather than expanding in terms of powers of $z - 1$; see §7.4—although in that particular case this series actually would have infinite cofficients). This could be OK, although the divergence encountered here has been shown to be of a rather uncontrollable kind ('not Borel-summable'). To make sense of the required 'Eulerian' type of reasoning (such as that which gives $1 + 2^2 + 2^4 + 2^6 + 2^8 + \cdots = -\frac{1}{3}$; see §4.3 and §26.9), some more sophisticated procedures seem to be required.[50] Moreover, if the string-theoretic (perturbational) calculations are actually expansions 'about the wrong point', then it is unclear what trust we may place in all these perturbative calculations in any case! Thus, we do not yet know whether or not string QFT is actually finite, let alone whether string theory, for all its undoubted attractions, really provides us with a quantum theory of gravity.

31.14 The magical Calabi–Yau spaces; M-theory

The particular reservations that I have had about string theory, as expressed in §§31.8–13, are not, however, the ones that seem to have worried the string theorists themselves. They have been troubled by other matters which I have not yet even referred to, namely the question of uniqueness of the theory. Originally, it had been regarded as one of the great hopes/triumphs of string theory that it might yield the one unique scheme for the universe, and much store was set on this supposed uniqueness. A clear issue had to do with the Planck-scale compact 6-manifolds $\mathcal{Y}$, into which the 10-dimensional universe is supposed to be largely curled. What are these 6-manifolds? Why does the universe tend to curl into these ones rather than some other ones? At first, it had seemed that the stringent requirements of supersymmetry, appropriate dimensionality, and Ricci flatness, together with some basic physical requirements, might lead to unique answers; but then it appeared that vast numbers of alternatives were equally possible.

Some early suggestions were that the $\mathcal{Y}$ space might be a *hypertorus* $S^1 \times S^1 \times S^1 \times S^1 \times S^1 \times S^1$, with zero curvature (see Note 31.45; recall the term 'torus' for $S^1 \times S^1$, see Figs. 8.9, 8.11, and 15.3). But it then became clear that a hypertorus-based string theory could not be made to incorporate the chiral aspects[51] of the standard model (recall §25.3), and something more sophisticated was needed. The 'stringent requirements' then led to these 6-manifolds being what are called *Calabi–Yau spaces*.[52] These are spaces of considerable pure-mathematical interest, and had been studied previously, by Eugenio Calabi and Shing-Tung Yau, for such reasons. They are examples of what are called *Kähler* manifolds, which means that they have both real Riemannian metrics and complex structures (and can therefore be interpreted as complex 3-manifolds), where these two structures are compatible (in the sense that the metric connection preserves the complex structure from which it follows that they are also *symplectic* manifolds;[31.6] see §12.9 and §§14.7,8 for the relevant concepts). Calabi–Yau spaces have additional properties, deemed essential

[31.6] Can you see why they must consequently be symplectic manifolds? *Hint*: Construct S_{ab} from the metric g_{ab} and the complex structure J^a_b, and then verify that $dS = 0$ (assume S_{ab} is non-singular).

for the string programme: they possess metrics that are Ricci-flat, and are endowed with *spinor fields* that are constant with respect to the metric connection. These constant spinor fields play necessary roles as supersymmetry generators. Without them, supersymmetry would not be possible. The various such spinor fields, for a given choice of Calabi–Yau space, can be (formally) 'rotated into each other' by a symmetry-group action. This group is then to play the kind of role that the symmetry groups of particle physics play.

It should be made clear that this symmetry does not directly apply to the Calabi–Yau spaces themselves, in the manner of a symmetry applied to the fibre $\mathcal{F}$ of a fibre bundle, as described in the discussion of Chapter 15. In fact, Calabi–Yau spaces possess no (continuous) symmetries, and we cannot regard our 10-space as a (non-trivial) fibre bundle of Calabi–Yau 6-spaces over ordinary 4-spacetime. The (internal) particle symmetries refer, instead, to the 'rotation' of the constant spinor fields among themselves. The actual Calabi–Yau spaces are not affected by the action of the symmetry.[53]

Thus, string theory leads to a very particular but unusual type of GUT theory (see §25.8 and §§28.1–3). It is intended that all of particle physics is to be found within the appropriate string-theoretic scheme. The symmetry groups that arise in this way are much larger than those of the standard model (§§25.5–7) but, as with other GUT theories (§25.8), a form of symmetry breaking is taken to be responsible for reducing the groups down to those of more direct relevance to the standard model—although this programme has not yet been successfully achieved.

What about the uniqueness issue? Unfortunately, there are tens of thousands of classes of qualitatively different possible alternatives for the Calabi–Yau spaces, so the scheme, as described, is far from unique. In fact, within a particular class of Calabi–Yau space, there are infinitely many different ones, distinguished by the values of certain parameters, called *moduli* (see §31.11), describing its shape (Fig. 31.11), just as as for a Riemann surface (§8.4, Fig. 8.11). The presence of these moduli is regarded as a good thing, because varying them provides the zero-energy modes of oscillation of the $\mathcal{Y}$ space (referred to in §31.11), which *are* taken to be physically realizable and would provide the needed route into particle

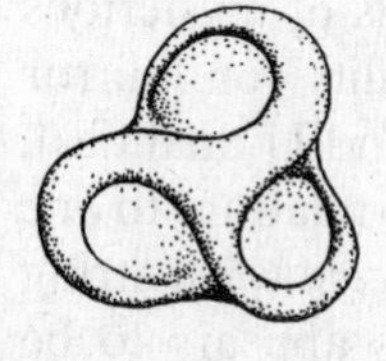
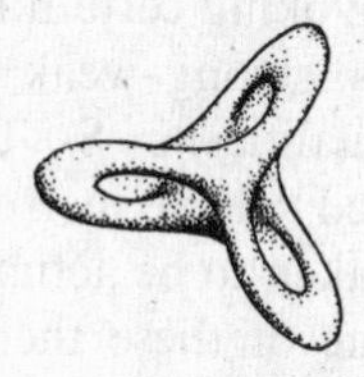

Fig. 31.11 The 'shape' of a Calabi–Yau space $\mathcal{Y}$ is described by a number of moduli (compare Figs. 8.10 and 8.11). Variations of these moduli provide zero-energy modes of oscillation of $\mathcal{Y}$.

physics and observational consequences of string theory. However, as already noted in §31.11, these can lead to instabilities.

However, there are other types of non-uniqueness that give the initial impression of being more serious even than that of the Calabi–Yau non-uniqueness. It turns out that there are five quite distinct possible overall schemes for the detailed way in which the supersymmetry interrelates the 'bosonic' and 'fermionic' modes of vibration of the string. Thus, there are five different string theories, these being referred to as Type I, Type IIA, Type IIB, Heterotic O(32), and Heterotic $E_8 \times E_8$. The groups O(32) and $E_8 \times E_8$ are those that would arise in the way outlined in the previous paragraphs of this section. (The reader may recognize the notation for these groups from §13.2, E_8 being the largest of the *exceptional* Lie groups.) The Type I theories employ open-ended strings as well as closed loops, all the others operating just with closed ones. Disturbances can travel in a right-handed or left-handed sense in all these models.[54] The Type IIA and IIB differ in how these right- and left-handed disturbances relate to each other. The Heterotic strings are particularly strange in that the left- and right-moving disturbances seem to belong to two spacetimes of different dimensionality (26 and 10, respectively). This hardly makes good geometrical sense—certainly not to me(!)—but it appears to make the appropriate formal sense. The view seems to be that the 10-dimensional picture is the geometrically appropriate one, but the left-moving disturbances behave in the same way as those of the older (non-supersymmetric) 'bosonic strings' of §§31.5,7 that were to inhabit a 26-dimensional ambient spacetime. As we have seen before, string theorists seem not to be much troubled by apparent inconsistencies in spacetime dimension, usually regarding this dimensionality as an 'energy-dependent' effect (§31.10), and therefore not of fundamental significance. We shall see more of this shortly and in §§31.15,16.

For a while, this proliferation of different string models caused many theorists to despair of being able to proceed much further. But some remarkable developments started to take place, indicating certain possible deep interrelationships between these apparently very different models. Then, in 1995, Edward Witten delivered a famous lecture,[55] initiating what has become known as 'the second superstring revolution'. In this lecture, Witten outlined a programme for the development of string theory that has completely transformed the way that the subject is to be viewed. The essential new feature is that by invoking certain kinds of mysterious 'symmetry operations' (referred to as 'strong–weak duality' or 'mirror symmetry'[56] and sometimes called S dualities, or S-, T-, and U-dualities), these different string theories are revealed to have such deep relations to one another that they may apparently be taken to be actually *equivalent* string theories. The small-scale limit of some of these theories appears to be identical (in some appropriate sense) to the large-scale limit of others, and there are other kinds of symmetry relationships of this general kind arising from a duality of Yang–Mills theory (§25.7) that is analogous to the duality between electricity and magnetism in ordinary electromagnetic theory. (Compare, also, the duality that arises in the Chan–Tsou theory, as briefly described in §25.8.) Moreover different Calabi–Yau spaces turn out also to be dual to one another in various ways. As yet, as far as my knowledge of the matter goes, not all of these relationships are proven as mathematical results.[57] But the original conjectures, coming from string theory with some impressive circumstantial evidence, have stimulated some very considerable pure-mathematical research leading to deeper understandings of Calabi–Yau manifolds and the relations between them.[58]

A particularly striking example of this 'circumstantial evidence' is worth recounting. It relates to a specific entirely mathematical problem that certain pure mathematicians (algebraic geometers) had been interested in for a number of years previously. This problem had nothing evidently to do with physics, but with counting the number of *rational* curves in certain complex 3-manifolds.[59] A rational curve is a complex curve (i.e. a Riemann surface—see Chapter 8) of *genus zero*; that is, it has the topology of a sphere S^2. These complex 3-manifolds turn out to be Calabi–Yau spaces that the demands of string theory—according to the proposals of 'the second superstring revolution'—ought, via mirror symmetry, to be related to certain other Calabi–Yau spaces. The mirror symmetry, in a certain sense, interchanges complex structure with symplectic structure; accordingly, the problem of counting rational (holomorphic) curves (which is technically a very difficult problem) is converted to a much simpler, and quite different-looking counting problem in the 'mirror' Calabi–Yau space. Two Norwegian mathematicians, Geir Ellingstrud and Stein Arilde Strømme, had developed methods for directly counting the number of rational curves (of successive orders[60] 1, 2, 3, ...) in their spaces, coming up with the successive numbers

$$2875, \quad 609\,250, \quad 2\,682\,549\,425,$$

for the first three cases. But using the assumption that the mirror-symmetry relationship holds, so that the much simpler counting procedure can be applied, Philip Candelas and his collaborators were able to come up with the numbers

$$2875, \quad 609\,250, \quad 317\,206\,375.$$

Since the mirror symmetry was only an unproved 'physicist's conjecture' at the time, it was presumed that the agreement in the first two cases was accidental and that there was no reason to accept the number 317 206 375 that Candelas and his collaborators had come up with. But then it emerged that owing to a computer-code error, the Norwegian mathematicians' number was incorrect, and the correct value was indeed exactly that obtained by the mirror-symmetry argument! Many subsequent numbers were then computed using this mirror symmetry, such as the extension of the above sequence to counting rational curves of higher orders (4, 5, 6, ..., 10):

242 467 530 000
229 305 888 887 625
248 249 742 118 022 000
295 091 050 570 845 659 250
375 632 160 937 476 603 550 000
503 840 510 416 985 243 645 106 250
704 288 164 978 454 686 113 488 249 750

This is a very remarkable example that strongly indicates that there is indeed 'something going on behind the scenes'. As things stand, this is distinctly enigmatic. There is certainly something very non-obvious to be unravelled in the mathematics, and some more recent mathematical developments appear to have gone some way towards achieving this.[61] But the more important question has to do with the physical significance of these results. Are we entitled to infer from the undoubted fact that string theory has supplied deep and previously unexpected insights into *mathematics* that it must also have a deep *physical* correctness? This answer to this enigma is far from obvious. Witten has argued that string theory, as it had been understood up until that point, was just the tip of an iceberg—or rather, it represented five tips of some mysterious yet-unknown theory that he christened 'M-theory'; see Fig. 31.12. This new theory, when it is

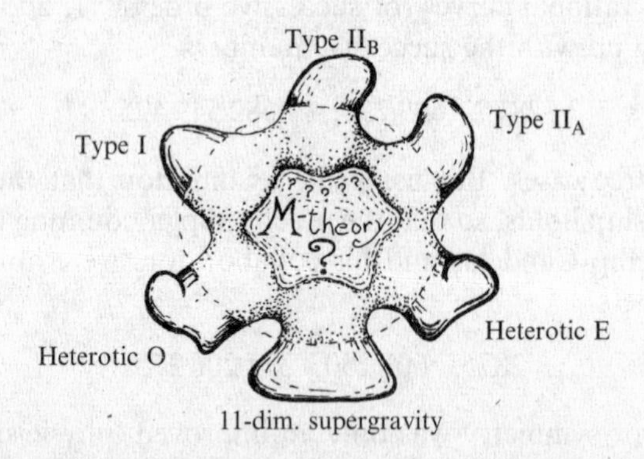

Fig. 31.12 The enigmatic 'M-theory' claim is that the five different types of string theory, all related via S-, T-, and U-dualities, and 11-dimensional supergravity, are six different aspects of one and the same yet-undiscovered structure.

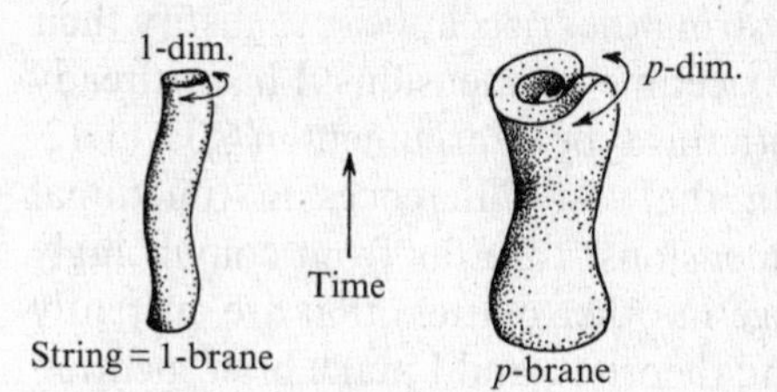

Fig. 31.13 Membranes (or p-branes, or just branes), have p spatial dimensions and 1 time dimension, the world-sheet being $(1 + p)$-dimensional. These structures are involved, together with ordinary strings (1-branes), as part of the undefined M-theory.

found, is to supplant and supersede all the different string theories that had been put forward previously.

Not only is this mysterious M-theory intended to encompass all these string theories, but it is to incorporate a number of other string-related and supersymmetry-related ideas also. Strings are now regarded as being just a special case of a more general notion which includes higher- (and even lower-) dimensional structures. These are referred to as *membranes* (or p-branes, or just branes) which have p spatial dimensions and one temporal dimension, the world sheet being $(1 + p)$-dimensional. See Fig. 31.13. Some related timelike structures, called *D-branes*, can also be involved, and I shall have something to say about these in §31.17.

In another development, M-theory is supposed also to encompass the 11-dimensional supergravity theory that we left behind earlier in this chapter (§31.4). In fact, M-theory itself seems to be thought of roughly as an 11-dimensional theory, so perhaps the dimensional mystery lies more in its relation to the various 10-dimensional string theories than to 11-dimensional supergravity. The fact that 11 dimensions seems now to be 'allowed' for a consistent string-type theory appears to be a conclusion due to Witten that, in some sense, the original argument that the 'one plus nine' dimensions needed to remove the string anomaly referred to in §31.7 are really to be regarded as an approximation (partly owing to the involvement of these higher-dimensional 'branes'), and the more correct answer is indeed 11 ($= 1 + 10$, i.e. 1 time and 10 space dimensions).[62] Yet even 11 dimensions may not satisfy the string theorists. There is some suggestion that one should move to more dimensions still, to the even more mysterious (and even more undiscovered) 'F-theory' which has $12(= 2 + 10)$ dimensions (so there are 2 time dimensions)![63]

How is it that a theory with an 11-dimensional (or perhaps 12-dimensional) 'spacetime' can be something that specializes, in certain low-energy or high-energy limits, to various theories, each (but one) of which has a 10-dimensional spacetime? Again, this discrepancy in spacetime dimensionality seems to be regarded as an 'energy effect' (§31.10), and not particularly fundamental. It can be imagined that more and more spacetime dimensions might be perceived when probed at higher and higher energies. In this kind of way, string theorists appear to justify their seemingly cavalier attitude to their spacetime's dimensions! I have already expressed my considerable unease with this type of argument in §§31.11,12. In my opinion, the difficulties with the vast differences in functional freedom in different numbers of dimensions[64] are far from convincingly addressed. This issue also looms large in other matters that are currently occupying the interests of many string theorists, and I make brief mention of these next.

31.15 Strings and black-hole entropy

Recall from §27.10 and §30.4 that the Bekenstein–Hawking formula assigns an *entropy* to a black hole, proportional to its horizon's surface area. Although several different arguments had been given in support of this conclusion, none of these unambiguously[65] equated the black-hole entropy explicitly to the logarithm of a phase-space volume, as Boltzmann's formula demands (§27.3). This would amount to a direct counting of the degrees of freedom 'lost in the hole', the counting to be carried out in accordance with the appropriate quantum gravity theory. Then in 1996, Andrew Strominger and Cumrun Vafa provided a calculation,[66] using strings and membranes, that supported an interpretation of the Bekenstein–Hawking entropy formula as 'counting degrees of freedom' in this way. This was acclaimed by the string theorists in such terms as: 'A quarter-century-old puzzle had been solved'.[67]

As appears to be usual, with such string-theoretic proclamations, this conclusion is very considerably overblown. For example, the original Strominger–Vafa calculation pertained only to black holes in a 5-dimensional spacetime. Later results do apply to ordinary 4-spacetime, but the initial excitement that led to proclamations such as the above seem to have been elicited by the original 5-dimensional calculation. Moreover, all these string-theoretic results referred only to the extremely special limiting case of an 'extremal hole' (or to perturbations away from this) for which the Hawking temperature (see §30.4) is zero—and where the hole involves additional supersymmetric Yang–Mills-type fields which have no clear justification from known physics. In addition, the actual calculations were performed in *flat space*, where there is no actual event horizon, and it is a matter of extrapolation to argue that they should also apply to a significantly curved black-hole metric.

Let me attempt some words of clarification. We have seen (§27.10) that in the ordinary general relativity theory of the vacuum, where spacetime is four-dimensional, a stationary isolated black hole is described by the Kerr metric, characterized by the values of just two (non-negative) real parameters m and a, where m is the mass and $a \times m$ the angular momentum (in natural units). For the Kerr geometry actually to describe a black hole, rather than a naked singularity (§28.8), we require $m \geq a$. (Note, for example, that this inequality is needed for the formula $A = 8\pi m[m + (m^2 - a^2)^{\frac{1}{2}}]G^2/c^4$, of §27.10, for the area A of a Kerr hole's horizon.) The *extremal* case of an ordinary black hole occurs when $m = a$, which only just qualifies it as a 'black hole'. It is really an (astrophysically unattainable) limiting case of an ordinary black hole, with zero as the value for its Hawking temperature.

This may be compared with what happens with the explicit 'black hole' considered by the string theorists, which is also 'extremal', in the sense that its Hawking temperature vanishes. But almost all of these string calculations refer to a completely different kind of animal! Instead of rotation—as is allowed for by the presence of the Kerr parameter a—additional physical fields are introduced. The string theorists' 'hole' is patterned, instead, on the Reissner–Nordstrøm solution of the Einstein equation which, unlike the Kerr solution, is spherically symmetrical. In place of Kerr's a, there is a parameter e which measures the total electric charge of the hole, the Reissner–Nordstrøm metric being a solution of the Einstein–Maxwell equations (§31.4), namely the Einstein equations with the energy–momentum tensor being that of a source-free Maxwell field.[68] The horizon's area is now given by a very similar-looking formula

$$A = 8\pi[m + (m^2 - e^2)^{\frac{1}{2}}]^2 G^2/c^4.$$

The condition for this metric to represent a black hole rather than a naked singularity is $m \geq |e|$, and the condition for the hole to be *extremal* (zero temperature) is $m = |e|$. (The modulus bars—see §6.1—simply allow for negative e.)

The type of 'black hole' that the string theorists are mainly concerned with is in essence the same as the Reissner–Nordstrøm case, but where a supersymmetric family of Yang–Mills fields (§25.7) replaces the Maxwell field. The entire solution is, in effect, a particular example of what is called a *BPS state* (where 'BPS' stands for 'Bogomoln'yi–Prasad–Sommerfeld'), in which requirements of supersymmetry, stationarity, and minimal energy determine the solution. I shall not bother with the details of what this means. Although such things have clear interest for string theorists and other people concerned with supersymmetry, there is yet no evidence of their relevance to the actual physical world (see §31.2).

What about the fact that the specific calculations of string degrees of freedom are performed in flat space, where there is no event horizon? As someone with a background in Einstein's general relativity theory, I find this to be one of the most puzzling aspects of the string theorists' claims. It is hard to see how rigorous conclusions can be reached concerning black holes, without full respect being paid to a black hole's very curved spacetime geometry, with the 'information' in its formation being captured behind an event horizon.

Let us try to get a rough idea of how the string argument proceeds.[69] To start, imagine estimating the number of physical degrees of freedom by counting the number of different possible string loops, of length l, on a Planck-scale lattice, say within a spherical volume of some fixed radius, these strings contributing a certain total mass–energy M. We are to suppose that the actual value of Newton's gravitational constant G is at our disposal. For small enough G, there will be no black hole, and in the limit as G goes to zero, the spacetime becomes actually flat. But if we invisage gradually increasing G ('cranking up Newton's constant'), we eventually come to a situation in which, according to general relativity, a black hole would form (recall Michell's expression $2MG/c^2$ for the radius of a 'Newtonian black hole'—see §27.8). In string theory, G depends on a parameter g_s, called the *string coupling constant*, and it turns out that G would get larger as g_s is made larger ($G \sim g_s^2$). In the limit of small G (small g_s), the logarithm of the count of string degrees of freedom (§27.3) gives an entropy that is the same as the Bekenstein–Hawking value for a black hole, even though there is no black hole. Scaling arguments are provided in order to show that this relation persists as G increases, so that the actual Bekenstein–Hawking formula is obtained when the black-hole stage is arrived at.

This gives only a qualitative correspondence between a count of string degrees of freedom and the Bekenstein–Hawking formula $S_{BH} = \frac{1}{4}A \times kc^3/G\hbar$, confirming a rough proportionality between string entropy and black-hole surface area A. To obtain the precise value of $\frac{1}{4}A$ in the formula, Strominger and Vafa appealed to a consideration of BPS states, where the supersymmetry requirements enable the mass to be fixed in terms of the various 'charge' values (for the supersymmetric Yang–Mills fields), and where instead of enumerating string configurations, the 'counting' now enumerates all the different BPS states[70] that contribute to the total (all BPS states with the given set of charges). This can be done explicitly enough, and the logarithm of this number gives, somewhat remarkably, the value of $\frac{1}{4}A$ precisely (in the extremal case). Not only this, but (as later work showed) perturbations away from extremality (i.e. to Hawking temperatures infinitesimally above zero) the entropy still comes out correctly, as do certain slight corrections to the purely 'black-body' nature of the Hawking radiation. Moreover, the same turns out to hold when there *is* rotation, and in 4-dimensional spacetime.

The attentive reader may be puzzled why there appears to be 'another constant' g_s in string theory when everything should have been fixed by the string constant α' (§§31.8,9), the only actual parameter that appears in the Lagrangian. The answer to this lies in the fact that things have not really been fixed, because a value needs to be settled on for the dilaton field (§31.8). The value of g_s is given by the expectation value[71] of this dilaton field, assumed to be a constant. In general, the dilaton field need not be a constant, but it is often treated as such (as in the above discussion) for convenience. Its value would depend on various things, such as the specific choice of Planck-sized $\mathcal{Y}$ space and the choice of string theory type (i.e. Type I, IIA, IIB, Heterotic O(32), or Heterotic $E_8 \times E_8$; see §31.14). In fact, this dependence underlies the strong/weak dualities of §31.14, the g_s of one theory being the reciprocal of the g_s of the 'dual' theory.

The reader will perceive that I find the above arguments to be very far from an actual string-theoretic derivation of S_{BH}, despite some remarkable agreements. I know that other general-relativists also feel considerable unease—most particularly with the fact that the essential *horizon* property of a black hole, in determining its vast entropy, does not appear to have played any real role at all. (This is in stark contrast with the loop-variable discussion of black-hole entropy that we shall come to briefly in §32.6.) Indeed, in the string-theory picture, entropy hardly increases at all at the point of formation of a black hole, giving a completely different viewpoint from the usual one (described in §27.10).

In addition, I should mention a specific technical difficulty with the string argument as it has been given.[72] This relates to the peculiar thermodynamic property of normal black holes that for small angular momentum they have a *negative specific heat* (see Note 27.2). The specific heat of a body is measured by how much its temperature rises when a small amount of heat energy is supplied to it. For an ordinary body, this is a *positive* number, our normal experience being that when heat is applied to a body, its temperature rises. But with a black hole, we tend to find that its temperature *decreases*. Heat energy supplies mass to the hole (by $E = mc^2$), so the hole gets more massive and, by Hawking's $T_{BH} = 1/8\pi m$ for a Schwarzschild hole (§30.4), its temperature decreases, so the specific heat is indeed negative. This curious negativity of a black-hole's specific heat would seem to present a difficulty for the above string-theoretic argument for black-hole entropy if it were to be applied to black holes that are not close to the extremal case, since a positive specific heat seems to be needed for the argument, and a black hole's specific heat is only positive when it is close to extremal. Indeed, we find that the specific heat is positive only when $m > a > (2\sqrt{3} - 3)^{1/2}m$ in the Kerr case, and only when $m > e > m\sqrt{3}/2$ in the Reissner–Nordstrøm case, and the same kind of thing holds for all the other Yang–Mills-charged fields.

There do appear to be some surprising relationships emerging from these string-theoretic calculations. But in my view, they fall far short of providing an independent justification of the Bekenstein–Hawking entropy formula. The loop-variable approach to this problem (briefly considered in §32.6), would seem to provide a much more impressive quantum-gravity-based attack on this problem.

31.16 The 'holographic principle'

The string arguments referred to above, like practically all string calculations, are of a perturbative nature. More recently, however, certain ideas have been proposed with the intention of providing exact results. These depend on various instances of something called the 'holographic conjecture', which has somehow got promoted to the *holographic principle*. The idea of this 'principle' seems to be that, in certain appropriate circumstances, the states of a (quantum) string field theory defined on some spacetime $\mathcal{M}$ can be put into direct 1–1 correspondence with the states of another quantum field theory, where the second QFT is defined on another spacetime $\mathcal{E}$ of lower dimension! Often, $\mathcal{E}$ is presented as though it were a (timelike) boundary of $\mathcal{M}$, or at least some conformally smooth timelike submanifold of $\mathcal{M}$, (see Fig. 31.14). However, this is not the case in the usual example, that we shall examine in a moment. The holographic principle is taken to be, in some sense, analogous to a *hologram*, where a 3-dimensional image is perceived when a (basically) 2-dimensional surface is viewed.[73]

The most familiar form of this 'holographic principle', stemmed from work of Juan Maldacena in 1998, and is sometimes referred to as the *Maldacena* conjecture, or else the ADS/CFT conjecture. Here, $\mathcal{M}$ is to be a $(1+9)$-dimensional product $\mathrm{AdS}_5 \times S^5$, where AdS_5 is the ('unwrapped') $(1+4)$-dimensional *anti-de Sitter space* (see §28.4, Figs. 28.8 and 28.9b—but here there are four space dimensions). The S^5 is a spacelike 5-sphere whose radius is of cosmological dimension, equal to $(-\Lambda')^{\frac{1}{2}}$, where Λ' is the (negative) cosmological constant of AdS_5 (§19.7). The smaller space $\mathcal{E}$ is to be the 4-dimensional 'scri' (i.e. conformal infinity—see §27.12) of AdS_5; see Fig. 31.15. We note that $\mathcal{E}$, being 4-dimensional, is certainly not the boundary of $\mathcal{M}$ in this case, since $\mathcal{M} = \mathrm{AdS}_5 \times S^5$ is 10-dimensional. Instead, the 'boundary'—i.e. the 'scri'—of $\mathcal{M}$ can be thought of (but not conformally) as $\mathcal{E} \times S^5$. The Maldacena conjecture proposes

Fig. 31.14 'Holographic principle'? A spacetime $\mathcal{E}$ is a (timelike) boundary of another spacetime $\mathcal{M}$. It is conjectured that a suitable QFT defined on $\mathcal{E}$ can be equivalent to a string QFT on $\mathcal{M}$.

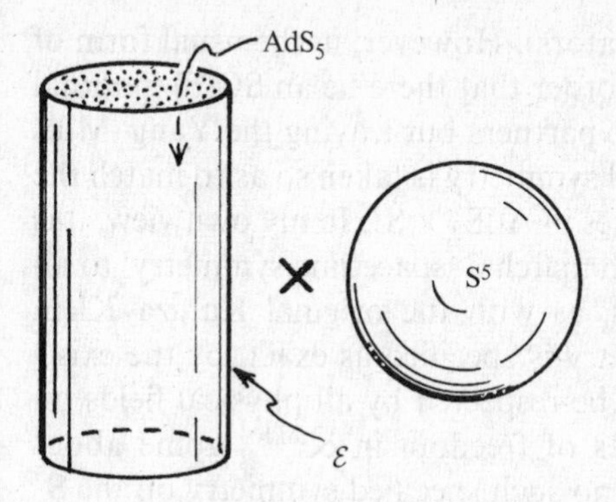

Fig. 31.15 ADS/CFT (Maldacena) conjecture. Here $\mathcal{E}$ is to be the 4-dimensional 'scri' (conformal infinity; §27.12) of anti-de Sitter 5-space AdS$_5$ (see Fig. 28.9b), rather than of the 10-dimensional $\mathcal{M} = \text{AdS}_5 \times \text{S}^5$, but the string-theory on $\mathcal{M}$ is conjectured to be equivalent to a supersymmetric Yang–Mills theory on $\mathcal{E}$.

that string-theory on $\text{AdS}_5 \times \text{S}^5$ is to be equivalent to a certain supersymmetric Yang–Mills theory on $\mathcal{E}$.

Here there is no chance of appealing to the type of 'quantum-energy' argument put forward in §31.10 for explaining away the gross discrepancy between the functional freedom of an ordinary field on $\mathcal{M}$, namely $\infty^{M\infty^9}$ and an ordinary field on $\mathcal{E}$, namely $\infty^{E\infty^3}$. Since the extra dimensions of $\mathcal{M}$ are in no way 'small'—being of cosmological scale—the flood of additional degrees of freedom, from the fields' dependence on the S^5 part of $\mathcal{M}$, would spoil any possibility of an agreement between the two field theories. The same would apply to ordinary QFTs on $\mathcal{M}$ and $\mathcal{E}$, since one-particle states are themselves described simply by 'ordinary fields' (see §26.2). The only chance of the holographic principle being actually true for these spaces is for the QFTs under consideration to be far from 'ordinary'.

In the case the string theory on $\mathcal{M}$, it is certainly conceivable that there are very strong consistency conditions which drastically reduce the $\infty^{M\infty^9}$ functional freedom. But, on the face of it, this seems very unlikely. Recall from Chapter 21, §22.8, and §16.7, that the quantum state of a single particle in $(1+n)$-dimensional spacetime has the functional freedom $\infty^{P\infty^n}$, where P is some positive integer describing the number of internal or rotational degrees of freedom (e.g. spin) of the particle. The quantum state of a single string would seem to have a much *greater* functional freedom, since a *classical* string has infinitely many degrees of freedom. If the number $\infty^{P\infty^n}$ is somehow to be reduced, then there must be huge constraints, perhaps of the type that led to the restrictions on spacetime dimension and curvature referred to in §31.7, but I am not aware of any such constraints having been suggested—which, in any case, would drastically affect the counting of string states, such as in §31.15.

The remaining possibility is to find a way of greatly increasing the functional freedom in the supersymmetric Yang–Mills fields on $\mathcal{E}$. The only way that I can see of achieving this would be have an infinite number of such fields, which could be attained by taking the limit $N \to \infty$ (N being the number of supersymmetry generators). However, in the usual form of this conjecture, one takes $N = 4$, in order that there be an SO(6) 'internal group' acting on the supersymmetric partners but leaving the Yang–Mills potentials unchanged.[74] This internal symmetry is taken so as to match the SO(6) symmetry of the S^5 that features in $AdS_5 \times S^5$. In my own view, it is fundamentally misconceived to try to match a 'spacetime symmetry' to an internal group of this kind—unless, as with the original Kaluza–Klein theory (§31.4), the spacetime symmetry is specified as exact, by the existence of Killing fields, and is also to be respected by all physical fields on the spacetime. The excessive degrees of freedom in $\infty^{M\infty^9}$ come about precisely for the reason that there is no such specified symmetry on the S^5 part of $\mathcal{M}$, which is to be respected by fields on $\mathcal{M}$.

It is my opinion that the importance this kind of discrepancy in functional freedom has been profoundly underrated. The 'sizes' of the Fock spaces (see §26.6, Note 26.12) will be completely different whenever the functional freedom in the classical fields is completely different. It should be noted that the condition of *positive frequency*, as demanded of 1-particle states in QFT, does not change the '$\infty^{M\infty^N}$' freedom for the classical fields. It simply compensates for the fact that these classical fields need to be complexified when we pass to a QFT description; see §26.3.

Why is the ADS/CFT conjecture taken so seriously? The support for it seems to come from a correspondence between BPS states on the two sides, that had been noticed by Maldacena and from a number of other correspondences. A good many of the latter can be understood purely from the correspondence between the symmetry groups (namely $SO(2, 4) \times SO(6)$) of the field theories on the two sides, but there are also some additional 'coincidences' that seem to need explaining. A reason for hoping that ADS/CFT is true appears to be that it might provide a handle on what a string theory could be like, without resorting to the usual perturbative methods, with all the severe limitations that such methods have.

Calculations on the $\mathcal{E}$ side are made easier by the fact that the space $\mathcal{E}$ is conformally flat (and sometimes referred to as 'flat', although it has no actual metric assigned to it, only a conformal metric of signature $+ - - -$). It is the universal covering space (Note 15.9) of the 'compactified Minkowski space' that we shall come to[75] in §33.3, having the topology of $S^3 \times \mathbb{E}^1$. The *metric* space $S^3 \times \mathbb{E}^1$ is sometimes called the 'Einstein cylinder' or 'Einstein universe', being the cosmological model favoured by Einstein for the period 1917–1929, when he had incorporated a cosmological constant into his field equation (§19.7)[76].

The ADS/CFT conjecture arose as another way of looking at the string 'derivation' of the Bekenstein–Hawking black-hole entropy formula (§31.15). For this, a 'black hole' is represented as a 'thermal state' on $\mathcal{E}$. This would only be of relevance to cosmological-size black holes and, at best, and provides a 'conjecture', based on some remarkable agreements between 'entropy calculations' done in different ways, rather than an actual derivation of the Bekenstein–Hawking expression.

31.17 The D-brane perspective

In various places in the above discussion, particularly in §§31.11,12,15,16, I have expressed my discomfiture with the use of higher-dimensional spacetime in string theory. One of my most fundamental difficulties with it is the enormous increase in functional freedom in higher-dimensional theories ($\infty^{P\infty^M}$, for a $(1 + M)$-dimensional spacetime), where one has to envisage some means of freezing out this extra freedom. Recall the 'hosepipe' analogy, illustrated in Fig. 15.1 and Fig. 31.3. What we perceive as a 'spacetime point' is a fully dynamical entity—here described as a circle, but generally with high dimensionality, and this is what provides us with the enormous extra freedom. We recall that Kaluza and Klein resorted to eliminating this freedom by decree, by asserting the presence of a Killing vector taking us around the 'hosepipe', in effect reducing the spacetime to a 4-dimensional one. This is, mathematically, a perfectly respectable procedure, but it appears never to have been seriously put forward in string theory. Instead, as in §31.10, string theorists seem normally to rely upon the hugeness of the energy that would be required to excite oscillations in these extra dimensions. As we have seen in §§31.11,12, there is every reason to doubt this line of argument.

Yet, the nature of the string-theory activity is such that it is hard to attack by means of specific arguments, such as those indicated in §§31.11,12,16. For the subject is always likely to metamorphose into a different form, at which point such attacks are deemed irrelevant.[77] Indeed, according to some recent ways of looking at the higher dimensions, the entire philosophy may be overturned, as far as I can see, with no public intimation that any serious change has happened at all. Although not necessarily the clear viewpoint of 'mainstream' string theorists (whatever that may be), the introduction of a certain 'D-brane philosophy' seems to be of this character.

First, what are *D-branes*? Why does string theory require them? The basic answer to the second question is that they make sense of the open strings that feature in the Type I theory, as referred to §31.14: each of the two ends of an open string must reside on some D-brane (Fig. 31.16). In response to the first question, a D-brane (or a D-q-brane) is a timelike structure of $1 + q$ spacetime dimensions (i.e. q space dimensions and 1 time dimension) that is a stable solution of 11-dimensional supergravity. (Invoking one of the M-theory dualities, we may, alternatively regard a D-brane as a solution of the equations of some other version of string/M-theory.) Basically, this is

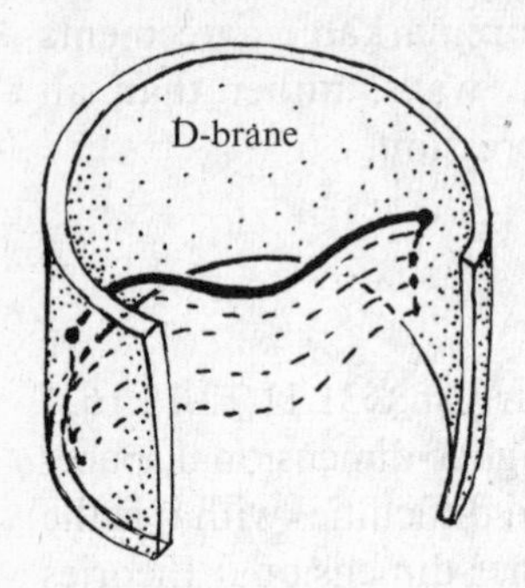

Fig. 31.16 The two ends of an open string are supposed to reside on a timelike $(q + 1)$-dimensional subspace of spacetime called a D-brane, or D-q-brane. A D-brane is an essentially classical entity (though possessing supersymmetry properties), representing a solution of 11-dimensional supergravity theory (a type of 'BPS state').

a 'brane' (as described in §31.14) of some dimension (0, 1, 2, . . . , or 9) that is a BPS state (see §§31.15,16), so it possesses some supersymmetric collection of Yang–Mills 'charges' and has minimum energy, subject to this. D-branes feature in many modern string-related discussions (for example, in black-hole entropy—see §31.15). They tend often to be treated as though they are classical objects lying within the full spacetime of $1 + 9$ (or $1 + 10$) dimensions. The D stands for 'Dirichlet', by analogy with the kind of boundary-value problem referred to as a *Dirichlet* problem, in which there is a *timelike* boundary on which data is specified (after Peter G. Lejeune Dirichlet, an eminent French mathematician who lived from 1805 to 1859—recall the 'Dirichlet series' of §7.4).

I shall make no attempt to discuss D-branes in any detail here. The only issue that I wish to raise is that, with the introduction of such 'D-branes', various theorists have provided a 'string philosophy' that seems to represent a profound shift from what had gone before. For it is not uncommonly asserted that we might 'live on' this or that D-brane, meaning that our perceived spacetime might actually lie within a D-brane. Indeed, it might even coincide with the D-brane, so that the reason that certain 'extra dimensions' are not perceived would be explained by the fact that 'our' D-brane does not itself extend into these extra dimensions.

The latter possibility would be the most economical position, of course, so 'our' D-brane (a D-3-brane) would be of $1 + 3$ dimensions. This does not *remove* the degrees of freedom in the extra dimensions, but it drastically reduces them. Why is this? Our perspective is now that we are not 'aware' of the degrees of freedom that are concerned with the deep interior of the higher-dimensional space between the D-branes, this being where the excessive functional freedom is making itself felt. We are to be aware of these extra dimensions only where they directly impinge upon the D-brane on which we 'live'. Let us return to our hosepipe analogy. Rather than the 'factor-space'[78] kind of picture that the original Kaluza–Klein analogy conjures up (Fig. 31.3), our observed spacetime now appears as a 4-dimensional *subspace* of

Fig. 31.17 An alternative viewpoint to that of Fig. 31.3, often expressed in the context of D-branes, is that a 'being' in higher dimensional space need not straddle all the extra dimensions but may be thought of as 'living' in a subspace, perhaps on a D-brane boundary

the higher-dimensional space. To visualize this, think of a strip drawn along the length of our hosepipe to represent the D-brane subspace that is now 'our observed 4-dimensional universe' (Fig. 31.17).

How much functional freedom do we now anticipate? The situation is somewhat similar to the geometrical picture that was adopted in §31.3 in order to obtain a more conventional perspective with regard to 'supergeometry'; see Fig. 31.2. Since we are now concerned with behaviour only at the D-brane (assumed to be, geometrically, an ordinary $(1+3)$-spacetime), we may imagine that our functional freedom has now become an acceptable $\infty^{M\infty^3}$, albeit for some rather large M. However, even this assumes that the restriction of the dymamics in the full 10-space (or 11-space) provides us with dynamical equations within 'our' 4-dimensional D-brane that are of the conventional type, so that initial data on some 3-space will suffice to determine behaviour throughout the 4-space. This is hardly likely, in general, so that a still excessive $\infty^{M\infty^4}$ may be expected. The problem has still not gone away!

One of the things that this attitude to D-branes has been used for is to attempt to resolve the 'hierarchy problem' referred to in §31.1. Specifically, this is the question of why gravitational interactions are so tiny in relation to the other important forces of Nature or, equivalently, the gravitationally fundamental Planck mass is so much larger than the masses of the elementary particles of Nature (by a factor of some 10^{20}). The D-brane approach to this problem seems to require the existence of more than one D-brane, one of which is 'large' and the other 'small'. There is an exponential factor involved in how the geometry stretches from one D-brane to the other, and this is regarded as helpful in addressing the 10^{40}, (or so) discrepancy between the strengths of gravitational and other forces.[79] It may be mentioned that this kind of picture of a higher-dimensional spacetime, which stretches from one D-brane boundary to another, is one of the types of geometry suggested for the 11-dimensional theories, such as M-theory, where the 11th dimension has the form of an open segment, each boundary geometry having the topological form (e.g. $\mathcal{M}\times\mathcal{Y}$) of the 10-spaces we considered earlier. In other models the 11th dimension is topologically S^1.

31.18 The physical status of string theory?

What are we to make of all this with regard to string theory's status as a physical theory of the future? The situation strikes me as having some very enigmatic and remarkable aspects to it, as well as some highly implausible-sounding ones, and it would be wrong to attempt to be completely dogmatic at this stage. Yet, many of the string theorist's claims are strongly asserted with apparent confidence. Undoubtedly, these must be watered down and taken with a sizeable heap of salt before serious consumption is contemplated. I think that it is fair to say that some of the strongest claims can be discounted altogether (such as string theory having provided a complete consistent theory of quantum gravity). But having said that, I have to admit to there being the appearance of something of genuine significance 'going on behind the scenes' in some aspects of string/M-theory. As the mathematician, Richard Thomas, of Imperial College London remarked to me, in an e-mail message

> I can't emphasise enough how deep some of these dualities are; they constantly surprise us with new predictions. They show up structure never thought possible. Mathematicians confidently predicted several times that these things weren't possible, but people like Candelas, de la Ossa, *et al.* have shown this to be wrong. Every prediction made, suitably interpreted mathematically, has turned out to be correct. And *not* for any *conceptual* maths reason so far—we have no idea why they're true, we just compute both sides independently and indeed find the same structures, symmetries and answers on both sides. To a mathematician these things cannot be coincidence, they must come from a higher reason. And that reason is the *assumption that this big mathematical theory describes nature* . . .

Yet, it still may well be that this 'something' is of purely mathematical interest, without there being any real reason to believe that it takes us any closer to Nature's secrets. I think that this is a perfectly tenable position to take, although in my actual beliefs I am prepared to accept that Nature might indeed have an interest in these matters (probably along somewhat different lines from those suggested so far). The strength of string theory's case appears to rest on a number of remarkable mathematical relationships between seemingly different 'physical situations' (where this 'physics' is usually something that would seem to be rather removed from the physics of Nature's actual world). Are these relationships 'coincidence', or is there some deeper reason behind them? It seems to me that for many of them there is indeed such a reason, in some cases as yet undiscovered, but that still does not reassure us that the string theorists are doing physics. Or, if they are, what aspect of physics are they really exploring?

I do not think that a proper assessment of these matters can be made without addressing the particular status of Edward Witten. He is generally accepted as being the figure who is most responsible for the direction of string-theory (and M-theory) research since the late 1980s. I have mentioned his role in launching the 'second superstring revolution' in 1995 (§31.14), but already by then he had established his pre-eminence in initiating several important developments in string theory, and in many other areas that have (not always obviously) some relation to string theory. String theory has had several 'tour-guides'(§31.6) throughout its over-30-year history, but Witten has been, in many respects, clearly the most impressive of these. Where Witten goes, it does not take long for the rest to follow. As an example of this, one may mention that the original Maldacena paper, which initiated much of the activity discussed in §31.16, lay essentially unnoticed on the archives by the string theory community until it was followed up by Witten in 1998. Immediately, it became the paper most cited by string theorists.[80]

It is interesting that in some very significant-looking new work,[81] Witten has reverted to considerations within a standard 4-dimensional spacetime (although supersymmetry is still involved). By combining ideas from twistor theory and string theory, Witten is able to derive some fascinating results concerning the Yang-Mills interactions of several gluons (§25.7). This work is particularly significant from my own twistor-oriented perspective (see Chapter 33), and it could well lead to some important new developments.

There is no question about the extraordinary quality of Witten's intellectual achievements. I can speak from a great deal of direct experience. There have been numerous occasions when I have attended seminars at the Mathematical Institute in Oxford (in the Geometry and Analysis series), in which a new highly original approach to some problem has been announced, and it has turned out that the seminal idea actually came wholely or at least partly from Witten. Frequently, such approaches have opened up a new field, where these unforeseen insights have shed a powerful original light on difficult mathematical problems—sometimes ones that previously seemed intractable. There is no doubt in my own mind that Witten possesses remarkable mathematical insight and understanding, of a high order. (Indeed, he won a Fields Medal in 1990, which has had the status, among mathematicians, of a Nobel Prize in the world of science. This is certainly an extraordinary achievement for a physicist.) Yet, I believe that Witten himself would deny that his abilities lie so much on the mathematical side. As I understand his own views, his successes have come from peering more deeply into the ways of Nature, gaining insights from the structure of QFT with its path integrals and infinite-dimensional function spaces, from the ideas of supersymmetry, and from the very nature of string theory and its generalizations. If he is right, then this is perhaps one of the strongest cases for accepting his contentions that supersymmetry and string theory do indeed find deep favour with Nature. On the other hand, perhaps he is a more remarkable mathematician than he admits to being!

How impressed am I that the very striking mathematical relationships that Witten and his colleagues have uncovered indicate some deep closeness to Nature? I am not totally sure how to view this issue, and I am certainly not convinced about it. Recall the remarkable achievement of the mathematician Andrew Wiles, in proving the famous 'Fermat's Last Theorem' after three and one-half centuries of failed attempts (§1.3). What Wiles actually established was that, in an important case, two very different-looking calculations actually always yield the same list of answers, the general form of this remarkable assertion being known as the Taniyama–Shimura conjecture. (In fact, Wiles's proof established only part of the full T–S conjecture—a part that was sufficient for establishing the Fermat assertion—but his methods provided an essential input for the full proof, subsequently completed by Breuil, Conrad, Diamond, and Richard Taylor.)

There is perhaps some vague similarity between this conjecture and the 'mirror-symmetry' relations of Calabi–Yau spaces referred to above (§31.14). In each case one has two infinite lists of numbers that turn out mysteriously to be the same. This kind of thing is far from unique in mathematics, and it may take a considerable number of years, in any particular case, before the underlying reasons for the equality of the lists come to light. As I understand it, many of the relations obtainable using 'mirror symmetries' have now been established by purely mathematical arguments.[82] As far as I am aware, such mysterious relationships are not normally put forward to support proposals for *scientific* (as opposed to mathematical) theories. This kind of issue will be taken up again in §34.9.

We have seen the same kind of 'coincidence' in the string-theoretic arguments for black-hole entropy, as put forward in §31.15 (and even in the much earlier 'non-string' arguments of §30.5). Are these in fact merely mathematical coincidences, or do we take these arguments as providing actual derivations? Let me end this chapter by taking another example of a striking mathematical coincidence, taken from early 20th-century physics. In 1912, Woldemar Voigt constructed a theory of spectral lines, based on an incorrect oscillator model. Fifteen years later, Heisenberg and Jordan found what we would today regard as the correct approach to this problem, and it is worth quoting from Heisenberg on his reminiscences of Voigt's work:[83]

> He was able to arrange the coupling of the oscillators with one another, and with the external field, in such a way that, in weak magnetic fields, the Paschen–Back effect was also correctly represented. For the intermediate region of moderate fields, he obtained, for the frequencies and intensities, long and complex quadratic roots; formulae, that is, which were largely

incomprehensible, but which obviously reproduced the experiments with great exactness. Fifteen years later, Jordan and I took the trouble to work out the same problem by the methods of the quantum-mechanical theory of perturbation. To our great astonishment, we came out with exactly the old Voigtian formulae, so far as both frequencies and intensities were concerned and this, too, in the complex area of the moderate fields. The reason for this we were later able to perceive; it was a purely formal and mathematical one.

I shall return, in Chapter 34, to this perplexing issue of mathematical relationships, as a driving force behind string theory and other proposals for the development of a fundamental physical theory.

Notes

Section 31.1

31.1. One should, of course, be careful not to confuse this e with the base of natural logarithms: $\mathrm{e} = 2.718281828459\ldots$; see §5.3.

31.2. This enormous discrepancy between the strengths of interactions—strong, electromagnetic, weak, and particularly gravitational, as roughly characterized by the respective coupling constants 1, $\frac{1}{137}$, $\sim 10^{-6}$, $\sim 10^{-39}$—is sometimes referred to as 'the hierarchy problem'. Georgia State University has a lovely little page explaining the finer points of comparing these couplings; see http://hyperphysics.phy-astr.gsu.edu/hbase/forces/couple.html

31.3. The relatively 'mild' renormalizaton factor for charge could come about because of the logarithmic nature of the electrodynamic divergence. The perceptive reader will, of course, notice that the puzzle of the tiny values of particle masses has not gone away, merely rephrased in terms of an absurdly tiny distance scale.

31.4. But recall 't Hooft's remark, referred to in §26.9.

Section 31.2

31.5. For a very useful collection, giving accounts of the history, personalities, and basic ideas underlying the introduction of supersymmetry, see Kane (2001) at a lay level, and Kane (1999) for something a bit more technical.

31.6. See Witten (1982); Seiberg and Witten (1994), on supersymmetric Yang-Mills theory, led to great simplifications in the Donaldson theory of 4-manifolds; see Donaldson and Kronheimer (1990). According to John Baez, the Seiberg-Witten theory shortens some proofs in Donaldson theory to 1/1000th their original length.

31.7. See Witten (1981); Deser and Teitelboim (1977) for proofs of positive energy using supersymmetry; see Gibbons (1997) for interesting black hole inequalities.

31.8. See Greene (1999), Note 5 on p. 399.

31.9. See Lawrie (1998), or, for much more detail, see Mohapatra (2002).

Section 31.3

31.10. There is a tendency for N to be a power of 2, this being the number of components of some spinor (see §11.5, §33.4). This should not be confused with the number 2^N of elements in the supersymmetry algebra. See Wess and Bagger (1992) for a discussion of supersymmetry by one of the creators!

31.11. For more information about supermanifolds, see DeWitt (1984); Rogers (1980).

Section 31.4

31.12. See the review article by Bern (2002) for an extended discussion of all this. See also Deser (1999, 2000).

31.13. See Deser (1999, 2000) on the 'last hope' for renormalizability of supergravity; also Deser and Zumino (1976).

31.14. See, for example, Hoyle *et al.* (2001), p. 1418.

31.15. For a rapid route to these counts, see Penrose and Rindler (1984), p. 389.

31.16. In a conventional bundle description, the metric on a base space $\mathcal{M}$ could be 'lifted' back into a bundle $\mathcal{B}$, over it, if desired, generally just to provide a canonical 'degenerate' metric on $\mathcal{B}$, but perhaps a non-degenerate metric if use may be made of a metric structure on the fibres. But this is not an essential aspect of the structure of a bundle.

31.17. These are the Einstein equations with Maxwell's energy–momentum tensor as source, together with the free Maxwell equations on the curved spacetime background.

31.18. There are, however, some recent applications of ideas combining string theory with twistor theory which use normal 4-dimensional spacetime. See §31.31, §33.14 and Note 31.81.

Section 31.5

31.19. Compare with Note 12.14.

31.20. See Schwarz (2001) for a general history of string theory; in particular, see Veneziano (1968); Nambu (1970); Susskind (1970); Nielsen (1970); and Goddard *et al.* (1973).

31.21. This describes things in terms of the β function, found by the great Euler in 1777. See Goddard *et al.* (1973) for the first important exposition of the duality.

31.22. Veneziano (1968) first conceived the model to explain the Regge poles. See Collins (1977) for Regge Theory in general, as well as Penrose *et al.* (1978).

31.23. Some limited success was later achieved in the bringing together of twistor ideas and those of string theory, but these were basically of a mathematical nature and did not provide a unified physical viewpoint; see Shaw and Hughston (1990) and Note 31.76.

Section 31.6

31.24. Quoted in Greene (1999), p.139, from an interview of Michael Green, by Brian Greene, 10 December, 1997.

31.25. See Witten (1996).

31.26. Authoritative works that are more detailed and technical are Green *et al.* (1987); Polchinski (1998); and Green (2000).

Section 31.7

31.27. See Green *et al.* (1987); Polchinski (1998); or Green (2000) for an argument leading to 26 dimensions.

31.28. The relevant number that anomalously appears in the quantum commutators (and must be set to zero) is $24 - \sigma$, where σ is the number of space dimensions *minus* the number of time dimensions.

31.29. With supersymmetry, the anomaly is now removed when $8 - \sigma$ is set to zero, with σ as in the previous footnote.

31.30. A hadronic string exhibits a minor difference from an ordinary rubber band in that the latter has a finite *natural* length for which the tension goes down to zero. For a hadronic string this 'natural length' would itself be zero.

Section 31.8

31.31. Many such claims are to be found in Greene (1999).

31.32. Quoted by Abhay Ashtekar in a lecture at the NSF-ITP Quantum Gravity Workshop at the University of California, Santa Barbara (1986).

31.33. Although not all of the relativity community would go the whole way with me, that the sought-for quantum/gravity union must involve a change in the rules of QFT, I continually find encouragement for this standpoint from that community. The response from the QFT community tends to be much less sympathetic!

31.34. The term 'dilaton' is not a misspelling of 'dilation', but it refers to a *quantum* version of that notion, arising from degrees of freedom available in a change of scale in the metric. Recall, from Chapter 26, that according to the rules of QFT, quantized degrees of freedom may manifest themselves as a kind of particle.

Section 31.9

31.35. Quoted in Greene (1999), p. 210 from an interview of Edward Witten by Brian Greene, 11 May 1998.

Section 31.10

31.36. The number 70 comes from the formula $n(n-3)$, for the number of independent components per point of an initial $(n-1)$-surface in a Ricci-flat n-space; See Wald (1984); Lichnerowicz (1994); Choquet-Bruhat and DeWitt-Morette (2000).

31.37. See Penrose (2003); Bryant *et al.* (1991); also Gibbons and Hartnoll (2002).

Section 31.11

31.38. See Penrose (2003).

31.39. See Dine (2000) for reflections on moduli.

31.40. One might prefer to stay within a QFT framework and use a coherent state (§26.6) instead of a classical description. This does not evade the issues being raised here, however.

31.41. Although I doubt that many string theorists would be keen on making **R** into a dynamical process, there are some notable exceptions; see Ellis *et al.* (1997a, 1997b).

31.42. The excitons behave as bosons in a quantum-field-theoretic description of the hosepipe vibrations (§22.13, §23.8, §26.2), so there can be many quanta in any one particular $\mathcal{Y}$ mode. An actual physical system for which such a quantum description can be appropriate would be long and narrow optical waveguide (e.g. an optical fibre).

Section 31.12

31.43. See Hawking and Penrose (1970).

31.44. The condition is that each timelike or null geodesic encounters 'generic' curvature, in the sense that somewhere along each one $k_{[a}R_{b]cd[e}k_{f]}k^c k^d \neq 0$, where the null vector k^c is tangent to the geodesic. A simple direct assessment of degrees of freedom shows that this condition is certainly satisfied in any 'generic' spacetime. It should be mentioned that the theorem applies in circumstances more general than Ricci flatness. We need only that the Ricci tensor satisfies an appropriate 'non-negative energy condition' (see §27.9, especially Note 27.20, and §28.5). For 'compact hypersurface', see §12.6 and Note 27.36.

31.45. There are exceptional cases of a zero-curvature $\mathcal{Y}$ with the topology of a 'hypertorus' $S^1 \times S^1 \times S^1 \times S^1 \times S^1 \times S^1$. These are not the models for $\mathcal{Y}$ favoured by today's string theorists, however (§31.14). Moreover, most perturbations of the hypertorus would not be flat.

31.46. This conclusion follows from another application of the aforementioned singularity theorem, which applies directly to the entire spacetime $\mathcal{M}$. In this application, the condition that there exist a compact spacelike hypersurface is replaced by the existence of some point p whose future light cone $\mathcal{C}$ 'curls round and meets itself' in all directions. The locus $\mathcal{C}$ is that swept out by the family of light rays ℓ (i.e. null geodesics—see §28.8), with past end point p and which extend indefinitely into the future. Technically, the required condition is satisfied if every such ℓ contains a point q for which there is a strictly timelike curve into the future from p to q. In the exact $\mathcal{M} \times \mathcal{Y}$ models just described, the condition fails (as it must, because $\mathcal{M} \times \mathcal{Y}$ can be non-singular), but it *only just* fails. Essentially what happens is that, among the 8-dimensional family of light rays ℓ, there is only a tiny 2-dimensional subfamily that fails to wander into the '$\mathcal{Y}$ part' of the spacetime and back, thereby curling into the interior of $\mathcal{C}$. It can be shown that, with a generic but small perturbation encountered by $\mathcal{C}$, this saving property will be destroyed, and the above-mentioned singularity theorem will indeed apply. Details of this argument will be presented elsewhere.

31.47. See, for example, Minassian (2002), which refers to further relevant research.

Section 31.13

31.48. See Smolin (2003) and Nicolai (2003).

31.49. See Smolin (2003); Gross and Periwal (1988); Nicolai (2003).

31.50. The series $1 + 2^2 + 2^4 + 2^6 + 2^8 + \cdots$ is not Borel-summable either, even though the 'Eulerian' value $-\frac{1}{3}$ for this sum is unambiguous, as can be seen using anaytic continuation (§7.4). I am not aware of whether such procedures have been applied to the total string amplitudes.

Section 31.14

31.51. This remark does not apply to the heterotic strings, that we shall come to shortly, for which the basic string framework is already chiral.

31.52. The most recent reference seems to be Gross *et al.* (2003). Smolin (2003) provides further references to these manifolds in string theory, and Polchinski (1998) discusses them as well.

31.53. I have a certain difficulty with this, since spinor fields actually have a geometrical interpretation. They cannot be 'rotated' (and therefore 'gauged', strictly speaking—see §§15.2,7) without this applying to the ambient space itself; see Penrose and Rindler (1984).

31.54. Applying a 'Wick rotation' to obtain a Riemann surface, the distinction is between holomorphic and antiholomorphic, as mentioned in §31.13.

31.55. Greene (1999); Smolin (2003) lists virtually all the known dualities, their status, and references.

31.56. This notion of 'mirror symmetry' is completely different from the space-reflection symmetry (parity), denoted by P, that was discussed in §25.4.

31.57. See Cox and Katz (1999), which provides an excellent coverage of such ideas.

31.58. See, for example, Kontsevich (1994); Strominger *et al.* (1996); for some of the most recent developments, see Yui and Lewis (2003).

31.59. These particular manifolds are complex 3-surfaces called 'quintics', which means that they have 'order 5'. The *order* of a complex n-surface in $\mathbb{CP}^m$ is the number of points in which it meets a general complex $(m - n)$-plane in $\mathbb{CP}^m$.

31.60. For the 'order' of a complex curve, see previous Note (31.59). Here $n = 1$.

31.61. See Cox and Katz (1999); Candelas *et al.* (1991); Kontsevich (1995).

31.62. See Smolin (2003), reference 171 particularly; Witten (1995); for a popular account, Greene (1999), p. 203.
31.63. See Vafa (1996); or Bars (2000).
31.64. See Bryant *et al.* 1991; Note 31.37.

Section 31.15

31.65. For a suggestive argument, however, see Thorne (1986).
31.66. See Strominger and Vafa (1996).
31.67. See Greene (1999), p. 340.
31.68. The reader might worry how a source-free Maxwell field might lead to a non-zero charge. There is no inconsistency here, since the black hole could have arisen from the collapse of a charged body of material, where all the charged sources have disappeared into the hole.
31.69. For a fairly readable survey of these matters, see Horowitz (1998).
31.70. The count involves structures called 'D-branes' that we shall consider in §31.17.
31.71. See Note 22.11.
31.72. Pointed out to me by Abhay Ashtekar.

Section 31.16

31.73. See Kasper and Feller (2001) for a reference on 'real' holograms.
31.74. This is a rather difficult set of ideas. For a real challenge, see Maldacena (1997) and Witten (1998).
31.75. Gary Gibbons has pointed out some intriguing geometry associated with this picture which even appears to have connections with twistor theory. Various matters of relevance to this construction are to be found in Penrose (1968a).
31.76. Einstein (1917).

Section 31.17

31.77. See Ashtekar and Das (2000) for an example of this phenomenon.
31.78. A 'factor space' is like the base space of a bundle; see §§15.1,2.
31.79. See Randall and Sundrum (1999a); see also Randall and Sundrum (1999b) for more general thoughts on these issues. Johnson (2003) is the standard reference on D-brane 'technology'. One of the more fantastical applications of this technology has been the 'ekpyrotic' model of the origin of the universe, put forward by Steinhardt and Turok (2002) in which it is proposed that the Big Bang arose from the collision of two D-branes in a previous phase of the universe. Despite invoking such exotic elements, the authors of this model make no attempt to explain the main puzzle presented by the Big Bang, namely its extraordinary specialness, as described in §27.13.

Section 31.18.

31.80. See Note 31.74.
31.81. See Nair (1988); Witten (2003); Cachazo *et al.* (2004a, 2004b, 2004c); Brandhuber *et al.* (2004).
31.82. See Notes 31.57, 31.58.
31.83. This quote is from Heisenberg's 1975 address to the German Physical Society, 'What is an elementary particle?' (I am grateful to Abhay Ashtekar for this example.) See Heisenberg (1989).

32
Einstein's narrower path; loop variables

32.1 Canonical quantum gravity

DESPITE string theory's popularity, it would be an absurdity to take the view, as some have done,[1] that it is 'the only game in town' (see §31.8). Many other interesting ideas are being pursued, these having different virtues and different difficulties. Unfortunately, it is not feasible for me to enter into a discussion of a great many of the alternative ideas for uniting quantum theory with spacetime structure here. Instead, in this chapter and the next, I shall concentrate on some of the more active areas that are closer to my own beliefs as to what lines are likely to be fruitful in the search for the true union between general relativity and quantum mechanics. As may be gathered from my comments in the previous chapter, I am of the opinion that we shall have to take a more tightly controlled position than those that allow a growth in spacetime dimensionality or ventures into supersymmetry (though I have less against the latter than the former which, as we have observed in §§31.11,12, encounters serious stability problems). Accordingly, in these two chapters, we shall be seeing some ideas that relate specifically to 4-dimensional Lorentzian spacetime, where it is intended that Einstein's actual field equation,[2] without supersymmetry, is to be addressed in some kind of genuinely quantum context. We shall see that, even here, the 'pictures of physical reality' that we encounter are still greatly removed from what is familiar, in some respects not to the extent that we saw in the previous chapter, but in other respects, more so. In this chapter, we shall witness some of the ideas behind Ashtekar variables, loop variables, and spin networks. In the following chapter, we shall make some acquaintance with twistor theory. Certain other ideas that have gained currency will also be touched upon in these two chapters, particularly discrete spacetime, q-deformed structures ('quantum groups'), and non-commutative geometry.

One of the most direct ways of approaching the quantization of Einstein's theory is to put it into a Hamiltonian form and then to try to apply the procedures of canonical quantization that were described in §§21.2,3. There are many difficulties about this, and I do not want to get mired in the details. A lot of these stem from the fact that Einstein's theory is 'generally covariant' (§19.6), so any particular coordinates that are used have no significance. Recall from the discussion of §21.2 that the standard 'quantization prescription' whereby the momentum p_a is replaced by the operator $i\hbar\partial/\partial x^a$, where x^a is the (classically) conjugate position variable, is not always correct, not even in flat spacetime if we use curvilinear coordinates. Thus, a great deal of care must be taken when carrying through this sort of quantization procedure.

Another difficulty is the complicated non-polynomial structure that is found for the standard Hamiltonian for general relativity. We should also take note of the fact that in addition to having evolution equations that take us away from an initial spacelike 3-surface $\mathcal{S}$, these being governed by the Hamiltonian, there are other equations that act *within* $\mathcal{S}$, which are called *constraints*.[3] These provide us with consistency equations for the data *on* $\mathcal{S}$, and the satisfaction of the constraint equations is necessary (and sufficient) for a satisfactory evolution of the data *away from* $\mathcal{S}$ (at least locally) and this evolution then preserves satisfaction of the constraints.

The canonical approach to quantizing general relativity has a long and distinguished history, going back to Dirac in 1932, who had to develop a whole new quantization framework, in order to handle the complicated constraints that indeed occur in Einstein's theory.[4] For many years this kind of approach was followed by a number of different researchers, with increasing sophistication,[5] but the complicated non-polynomial nature of the Hamiltonian made progress difficult. Then, in 1986, the Indian/American physicist Abhay Ashtekar made an important advance. By a subtle choice of the variables used in the theory (partly related to ideas that had been put forward earlier by Amitabha Sen),[6] whereby the constraints could be reduced to polynomial form, he was able to simplify the structure of the equations dramatically, and the awkward denominators in the Hamiltonian were eliminated, leading to a comparatively simple polynomial structure.

32.2 The chiral input to Ashtekar's variables

One of the striking features of Ashtekar's original 'new variables', as they are (still) called, is that they are asymmetrical with respect to their treatment of the right-handed and left-handed parts of the graviton (the quantum of gravitation).[7] We recall from §§22.7,9 that a (non-scalar) massless particle has two states of spin, which can be right-handed or left-handed about its direction of motion. These are referred to, respectively, as the states of positive and negative *helicity* of the particle. The graviton is to be a particle of spin 2, so its two respective helicity states should be $s=2$ and $s=-2$ (taking $\hbar = 1$), where s stands for the helicity (see also §22.12). The original Ashtekar approach treats these two states *differently*. Thus this formalism is left–right asymmetrical!

A comment is appropriate here as to why the graviton is indeed considered to be an entity of spin 2, whereas the photon has spin 1 (see §22.7, §33.7). What does this mean? The spin value of a quantum particle has to do with the symmetries (and field equations) of the field quantity describing it (and, as we shall see in §33.8, is most manifest with equations written in 2-spinor form). But it is good to have a directly *geometrical* way of seeing the difference between the spin-2 nature of gravity as opposed to the spin-1 nature of electromagnetism. Let us examine the waves appropriate to each field, namely the electromagnetic waves that constitute light on the one hand, and gravitational waves on the other.

In the case of electromagnetism, we have seen the geometric nature of the waves in Fig. 22.12, §22.9. The key point is that the electric and magnetic vectors are, indeed, *vector* quantities, so that a rotation of the wave through π (i.e. 180°) about its direction of motion sends the field quantity to its *negative*, and we need a rotation through 2π to restore it to itself. In the case of gravity, the wave would be one of spacetime distortion, as is illustrated in Figs. 17.8a, 17.9a, 28.15, and 28.16. Now if we rotate the wave through just π the distortion goes to itself, and it would be a rotation through $\frac{1}{2}\pi$ that sends it to its negative. We may note that this results from the Weyl curvature being a *quadrupole* entity, as is illustrated by the ellipse of distortion in Figs. 28.15, and we noted in §31.9 that this corresponds to a spin 2 entity. For spin value σ, a rotation through π/σ about the direction of the wave would send the field quantity to its negative, whereas a rotation through $2\pi/\sigma$ would be required to restore it to itself. (Note that this also works if σ is a half-odd integer, the field quantity being necessarily spinorial in that case; see §11.3.)

For a massless field, as is the situation here, we can go further, and think of plane waves as being composed of right-handed and left-handed circularly polarized parts, electromagnetic circular polarization being illustrated in Fig. 22.12b. For a quantized field, the relevant particles would correspondingly have positive or negative helicity (Fig. 22.13). For spin σ, these helicity values would be $\pm\sigma$ (with a corresponding description to Fig. 22.13, but with $q^{2\sigma} = z/w$ replacing $q^2 = z/w$). Thus, for gravity we indeed get the two possible helicity values $+2$ and -2.

To see how the helicity states are described, we shall need to examine the mathematics a little more explicitly. In fact, this left–right asymmetry that I have been referring to is an important feature of the twistor theory that we shall be examining in the following chapter, and the Ashtekar approach seems to have gained some of its initial inspiration from such twistorial ideas. For the moment, all that we shall require from twistor-related ideas is how this left–right asymmetry is mathematically expressed in ordinary spacetime terms. Let us recall the two tensor quantities that describe the two known massless fields of Nature, electromagnetism and gravitation. These quantities are the Maxwell field tensor $\boldsymbol{F} = F_{ab}$ (§19.2) and the Weyl conformal tensor $\boldsymbol{C} = C_{abcd}$ (§19.7). Each of these has what is called a *dual* tensor, defined in the index notation by

$$^{*}F_{ab} = \tfrac{1}{2}\varepsilon_{abpq}F^{pq} \text{ and } {}^{*}C_{abcd} = \tfrac{1}{2}\varepsilon_{abpq}C^{pq}{}_{cd},$$

(where ε_{abpq} is the antisymmetric Levi-Civita tensor, chosen here so that $\varepsilon_{0123} = 1$, in a standard right-handed orthonormal basis; see §12.7 and §19.2). We have already seen the dual Maxwell tensor $^{*}\boldsymbol{F}$ in §19.2. The dual Weyl tensor $^{*}\boldsymbol{C}$ is seen to be an analogous quantity. We might also contemplate 'dualizing' the latter pair cd of the Weyl tensor's indices instead. But this turns out to be the same as dualizing on ab.[32.1]

Recall that a 2-plane element at a point in 4-dimensional spacetime can be described by a 2-form $\boldsymbol{f}$ (or else a bivector) which is *simple* (§12.7). As with the Maxwell tensor (2-form) $\boldsymbol{F}$, we can construct its *dual* $^{*}\boldsymbol{f}$ and give meaning to the notion that, for a *complex* $\boldsymbol{f}$, it might be *self-dual*, i.e. $^{*}\boldsymbol{f} = \mathrm{i}\boldsymbol{f}$, or *anti-self-dual*, i.e. $^{*}\boldsymbol{f} = -\mathrm{i}\boldsymbol{f}$. The (complex) 2-plane element corresponding to $\boldsymbol{f}$ is called 'self-dual' or 'anti-self-dual' accordingly. This notion has importance in twistor theory (§33.6).[32.2]

Now, in the quantum theory, field quantities are allowed to take complex-number values, at least when they have the interpretation of wavefunctions. In fact there are various different, but mathematically equivalent ways of looking at these things (see Chapter 26.) It will suit our present purposes best if we indeed think of a complex Maxwell tensor, or even a complex Weyl tensor, as representing some form of wavefunction for the photon or graviton, respectively. In place of the reality condition which is characteristic of a classical field quantity, our complex wavefunctions must be of positive frequency (in accordance with the requirements specified in §24.3, §26.2). Let us not worry too much what this actually *means* in the case of the Weyl curvature. (We can, provisionally, consider that we are looking at curved spaces that are only infinitesimally different from flat, in which case $\boldsymbol{C}$ can be thought of as just a field in Minkowski space, and then the condition of positive frequency is not problematic. But it turns out that one can even do better than this, as will be indicated in §§33.10–12.[8] Now, the *right-handed* photons and

[32.1] Explain why. (You may find the diagrammatic notation and the identities of §12.8 helpful.)

[32.2] Show that, if the two indices of f_{ab}, describing a self-dual $\boldsymbol{f}$, are both contracted with any pair of indices of an anti-self-dual Weyl (or Maxwell) tensor, then the result is zero.

gravitons are described by the (positive-frequency) *self-dual* quantities ${}^{+}\boldsymbol{F}$ and ${}^{+}\boldsymbol{C}$, where ${}^{+}\boldsymbol{F}=\frac{1}{2}(\boldsymbol{F}-\mathrm{i}{}^{*}\boldsymbol{F})$ and ${}^{+}\boldsymbol{C}=\frac{1}{2}(\boldsymbol{C}-\mathrm{i}{}^{*}\boldsymbol{C})$, so that we have

$$*({}^{+}\boldsymbol{F})=\mathrm{i}\,{}^{+}\boldsymbol{F} \text{ and } *({}^{+}\boldsymbol{C})=\mathrm{i}\,{}^{+}\boldsymbol{C},$$

and the *left-handed* ones by the (positive-frequency) *anti-self-dual* quantities ${}^{-}\boldsymbol{F}=\frac{1}{2}(\boldsymbol{F}+\mathrm{i}{}^{*}\boldsymbol{F})$ and ${}^{-}\boldsymbol{C}=\frac{1}{2}(\boldsymbol{C}+\mathrm{i}{}^{*}\boldsymbol{C})$, for which

$$*({}^{-}\boldsymbol{F})=-\mathrm{i}\,{}^{-}\boldsymbol{F} \text{ and } *({}^{-}\boldsymbol{C})=-\mathrm{i}\,{}^{-}\boldsymbol{C}.$$

In the original Ashtekar framework, the self-dual and anti-self-dual parts of the Weyl curvature play different roles.

This may seem strange, from the physical point of view, because there is no evidence of any left/right asymmetry in the gravitational field, and there is certainly none in the standard Einstein theory of general relativity. As far as I can see, we can take two different attitudes to this question. On the one hand, we might regard the asymmetry as an unimporant feature of the particular mathematics that just happens to be useful in simplifying the Hamiltonian. On the other, we may take the position that there is something deeply left/right-asymmetrical about Nature, and the asymmetrical formalism is probing this in some way. In fact, we know that Nature *is* left/right asymmetrical, as is clearly manifested in weak interactions (see §25.3). In a certain sense, electromagnetism contains remnants of this asymmetry, but this is only seen indirectly, via its unification with the weak interactions in *electroweak* theory. In the absence of a (known) similar unification that relates to gravity, there is no reason to expect that gravity should itself directly or indirectly possess such asymmetrical features. Yet, as with the string theorists' perspective on their own theory, we may take the view that a quantum gravity theory is aimed at something much more than merely gravitation; it is laying the basic framework for all of physics, where the current framework of a classical spacetime is to be viewed as a convenience or approximation to something more fundamental. If that 'something' has an inbuilt chirality (as do the heterotic aspects of string theory—see §31.14—as well as the Ashtekar and twistor frameworks), then the fact that there are gross chiral asymmetries in weak interactions is far more readily comprehended.

32.3 The form of Ashtekar's variables

What then are these chiral Ashtekar variables? The chirality comes from an asymmetrical choice of one of the two kinds of 2-spinor that apply to Lorentzian 4-space. We may recall these objects from §25.2, where we saw that the electron's wavefunction ψ could be thought of as consisting of a pair of 2-complex-component entities α_A and $\beta_{A'}$, one of them having an unprimed index and the other a primed one. We noted that the unprimed index refers to the *zig* or *left-handed* (negative-helicity) part of the electron and the primed part, to the *zag* or *right-handed* (positive-helicity) part. We also saw (§25.3) that the weak interaction pays attention to the zig part α_A, but not to the zag part $\beta_{A'}$. A spacetime formalism that selects either the unprimed or the primed spinor as being 'more fundamental' than the other thus incorporates a basic chirality, and it lays down a framework that has the capability of distinguishing these two helicities at a fundamental level.

In fact, this is just what happens in the original Ashtekar (and twistor) formalisms. In the Ashtekar approach, the *canonical variables*, chosen with respect to a spacelike 3-surface $\mathcal{S}$, are basically the components of the (inverse) 3-metric $\boldsymbol{\gamma}$ *intrinsic* to $\mathcal{S}$, and the components of the (unprimed) *spin connection* $\boldsymbol{\Gamma}$ taken on $\mathcal{S}$. To be a little more precise, these are inverse metric components referred to a local spinor basis and represented as 2-forms. Moreover, the spin connection $\boldsymbol{\Gamma}$ refers to parallel transport of spinors α^A defined in the full 4-space and not just to 'spinors intrinsic to $\mathcal{S}$'.[9] Thus, $\boldsymbol{\Gamma}$ tells us how to carry an unprimed 4-space spinor α^A (a 2-spinor) parallel to itself with respect to the 4-space's metric connection (§§14.2,8) along some curve that happens to lie within the 3-space $\mathcal{S}$.[10] The field of 3-metric (density) quantities $\boldsymbol{\gamma}$ plays the role of the set of *momentum* variables and the field of connection quantities $\boldsymbol{\Gamma}$, the role of the corresponding conjugate *positions*; see Fig. 32.1. In the quantum theory, corresponding to the way that the momentum p_a is replaced by $i\hbar\partial/\partial x^a$ in the x^a-representation (§21.2), in the $\boldsymbol{\Gamma}$ representation we may take the $\boldsymbol{\gamma}$ fields to be replaced by $i\hbar\delta/\delta\boldsymbol{\Gamma}$ (where the '$\delta/\delta\boldsymbol{\Gamma}$' refers to the notion of *functional derivative* that was referred to in §20.5). Correspondingly, in the $\boldsymbol{\gamma}$ representation, $\boldsymbol{\Gamma}$ is represented by $-i\hbar\delta/\delta\boldsymbol{\gamma}$.

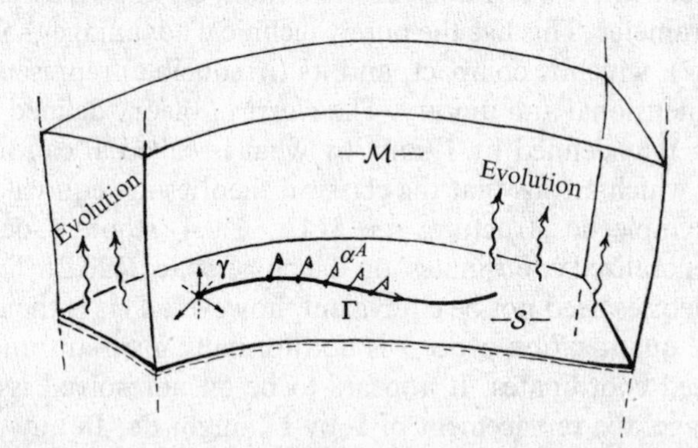

Fig. 32.1 Ashtekar's original canonical variables, defined on a spacelike 3-surface $\mathcal{S}$ in spacetime $\mathcal{M}$, takes the 'position' parameters $\boldsymbol{\Gamma}$ to be the components of the 4-space spin-connection, restricted to $\mathcal{S}$ (for spinors α^A, indicated by half-arrows). The 'momentum' parameters are basically the components of the (inverse) intrinsic metric $\boldsymbol{\gamma}$ of $\mathcal{S}$ (expressed as 2-forms and referred to an orthonormal basis at each point of $\mathcal{S}$).

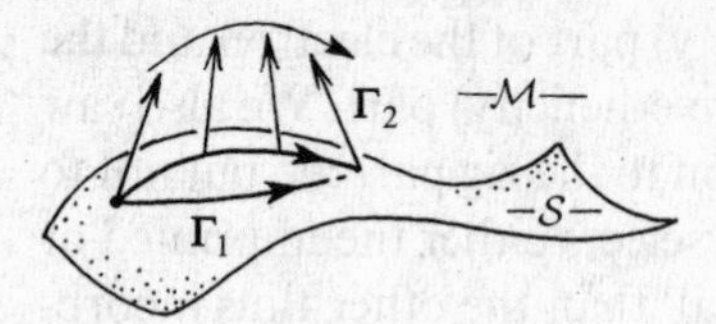

Fig. 32.2 We can express $\mathbf{\Gamma} = \mathbf{\Gamma}_1 + \mathrm{i}\mathbf{\Gamma}_2$, where $\mathbf{\Gamma}_1$ refers to the intrinsic curvature and $\mathbf{\Gamma}_2$ to the extrinsic curvature of $\mathcal{S}$ (so $\mathbf{\Gamma}_2$ measures how $\mathcal{S}$ 'bends' within $\mathcal{M}$).

The connection $\mathbf{\Gamma}$ has a part $\mathbf{\Gamma}_1$ that refers just to the *intrinsic* curvature of $\mathcal{S}$ and another part $\mathbf{\Gamma}_2$ that refers to the *extrinsic* curvature (i.e. to how $\mathcal{S}$ is 'bent' within the spacetime $\mathcal{M}$); see Fig. 32.2. The entire quantity $\mathbf{\Gamma}$ can be expressed as

$$\mathbf{\Gamma} = \mathbf{\Gamma}_1 + \mathrm{i}\mathbf{\Gamma}_2$$

(where we would have had $\mathbf{\Gamma}_1 - \mathrm{i}\mathbf{\Gamma}_2$ if we had chosen the opposite chirality for the formalism). The quantity $\mathbf{\Gamma}$ defines a bundle connection (in the sense of §15.8), where the base space is $\mathcal{S}$ and the fibre is the (unprimed) spin space $\mathbb{S}$ (a 2-dimensional complex vector space). The relevant group of the fibre is SL(2,ℂ) (see §13.3).[11]

At this point, I should mention a technical difficulty in the original Ashtekar approach. This is that the fibre group SL(2,ℂ) is non-compact and has unwanted infinite-dimensional irreducible representations, most of which are non-unitary (see §13.7). All this creates serious problems for the rigorous construction of the required quantum-gravity theory. Accordingly, in order for progress to be made, the modified connection

$$\mathbf{\Gamma}_\eta = \mathbf{\Gamma}_1 + \eta\mathbf{\Gamma}_2$$

has been used, where η is a non-zero real number, known as the Barbero-Immirzi parameter. This has the purely technical advantage that the group is now SU(2), which is compact, and its (irreducible) representations are all finite-dimensional and unitary. The *classical* theory defined by each $\mathbf{\Gamma}_\eta$ differs from that defined by $\mathbf{\Gamma}$ only by what is called a 'canonical transformation', which means that the classical theories are equivalent (having the same symplectic structure; see §14.8, §20.4), though described by different 'generalized coordinates' on the phase space (§20.2). The resulting quantum theories need not be equivalent, however. This is the issue raised in §21.2: the quantization process is not normally invariant under change of generalized coordinates. It appears to be an unresolved issue of how much 'damage' the replacement of $\mathbf{\Gamma}$ by $\mathbf{\Gamma}_\eta$ might do. In any case, much can be learnt from studying the 'easier' case of $\mathbf{\Gamma}_\eta$ first. Although one might worry that the resulting quantum gravity theory is just a 'toy model' rather than being the intended approach to quantum gravity (where values of η other than $\pm i$ or 0 seem to be without geometrical justification), the deviation from the intended quantized version of Einstein's theory is assumed not to be great.

In the case of $\boldsymbol{\Gamma}_\eta$, life is made relatively simple because the needed different irreducible representations of SU(2) are mathematically identical to the different states of *spin* (of a massive particle) in ordinary (non-relativistic) quantum mechanics. We recall from §22.8 that these different spins are labelled by the natural numbers $n = 0, 1, 2, 3, 4, 5, \ldots,$ where $\frac{1}{2}n\hbar$ is the value of the spin. (In §22.8 we also saw that the representation space, for each n, consists of symmetric spin tensors $\psi_{AB\ldots D}$, with n indices.) We shall see shortly how these different 'spin values' are to be used.

32.4 Loop variables

How do we express things in a way that makes general covariance (§19.6) more manifest, at least within the initial 3-surface $\mathcal{S}$? This is done by the ingenious device of describing our general quantum state in terms of a particularly simple family of *basic* quantum-gravity states that can be described in an essentially discrete way and for which the general covariance within $\mathcal{S}$ is taken care of very simply. (We shall come to these basic states in a moment.) The *general* state is then expressed in terms of these basic states by means of linear superposition. To understand what is needed, consider a closed loop within $\mathcal{S}$, and let us envisage the effect of using our connection $\boldsymbol{\Gamma}$ to enable us to carry an unprimed spinor α^A 'parallel to itself' all the way around such a loop. When we get back to where we started, we find that a linear transformation of the spin space $\mathbb{S}$ has been effected. This is defined, in component form, by a complex 2×2 matrix $T^A{}_B$ (see §13.3), the elements of this matrix depening upon some choice of basis in $\mathbb{S}$. However, the *trace* $T^A{}_A$ of this matrix is a basis-independent complex number,[32.3] so it is simply a property of the spin connection $\boldsymbol{\Gamma}$ in relation to the choice of loop. (This is an instance of a more general notion, referred to as a *Wilson loop*, after Kenneth Wilson, who first made use of this idea in gauge theories.[12] In 1988, Carlo Rovelli, Lee Smolin, and Ted Jacobson developed this idea in general relativity, calling these loop-dependent traces the *loop variables* for general relativity. Taking these loop variables as quantum operators, the 'basic states' referred to at the beginning of this paragraph are essentially just their *eigenstates*.

What is the geometric character of these basic quantum-gravity states? They turn out to be very peculiar from the point of view of the ordinary kind of metric geometry with which we are familiar, and very far from the 'smooth geometries' of classical general relativity. Indeed; we shall find that these basic states are very 'singular' as geometrics, analogous to the (Dirac) *delta functions* we considered §9.7 and §21.10. First, think of $\mathcal{S}$ as a featureless manifold.[13] Next, consider a family of closed loops in $\mathcal{S}$. We are

---

[32.3] Can you explain why?

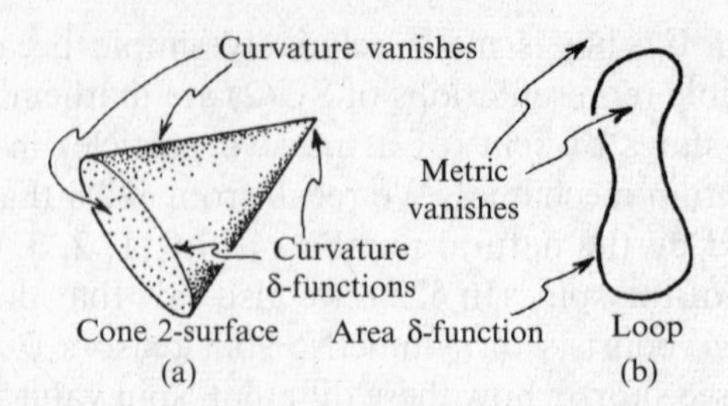

Fig. 32.3 (a) This non-smooth example of a conical 2-surface has zero curvature everywhere except at the vertex and at the rim around the base, where there are curvature δ-functions. (The Regge-calculus approach to quantum gravity operates with analogous 4-spaces, with δ-function curvature on 2-spaces; see Fig. 33.3.) (b) However, this is *not* what happens with loop quantum gravity. Here there is an *area* δ-function along the loops, the metric itself vanishing everywhere else.

to think of each loop state as having all its geometry somehow concentrated along the loop. It is not really the *curvature* that lies concentrated along the loop—which would be analogous to the geometry of the flat-based cone illustrated in Fig. 32.3, where there is a delta function (§9.7) in the curvature along the edge around the base of the cone, and also at the vertex,[32.4] this being the kind of situation arising in a different approach to quantum gravity, known as Regge calculus; see §33.1—but, instead, the entire *metric* is to be concentrated along the loop in a kind of delta function, disappearing completely outside the loop. There is a 'degree' of this concentration which is measured by a 'spin' value assigned to the loop, the different such values $j = \frac{1}{2}n$ corresponding to different irreducible representations of SU(2) (where we are using $\mathbf{\Gamma}_\eta$ for our connection, rather than the apparently more 'correct' $\mathbf{\Gamma}$).

These statements need further clarification. The notion of 'metric' that arises here is really that which assigns an *area* to any test 2-surface element that encounters the loop. In essence, there is a delta function in the area measure which is concentrated entirely along each loop in the family. What does this mean? Imagine a (not necessarily closed) 2-dimensional test surface $\mathcal{T}$ in the 3-surface $\mathcal{S}$. This may intersect various loops in a number of places. Every time $\mathcal{T}$ meets one of the loops, it clocks up a certain measure of area; there is no contribution whatever to its area *except* where it meets the loop. Thus, the 'delta-function' character of the metric is here manifested in the fact that each loop assigns a measure of area only where $\mathcal{T}$ intersects the loop. Each intersection point provides the value

$$8\pi G\eta\hbar\sqrt{j(j+1)},$$

[32.4] Can you explain? *Hint*: Use the ideas of §14.5.

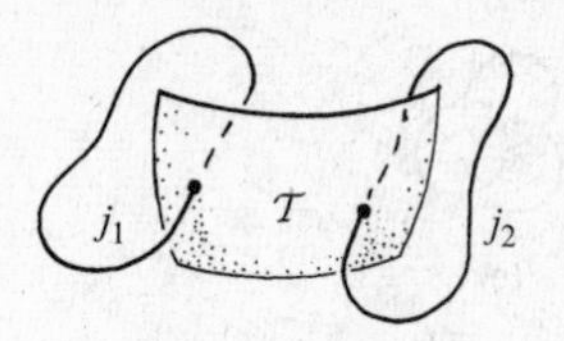

Fig. 32.4 A 2-dimensional test surface $\mathcal{T}$ in the 3-surface $\mathcal{S}$. Each intersection of $\mathcal{T}$ with a loop contributes the value $8\pi G\eta\hbar\sqrt{j(j+1)}$ of area (j being the loop's 'spin' value).

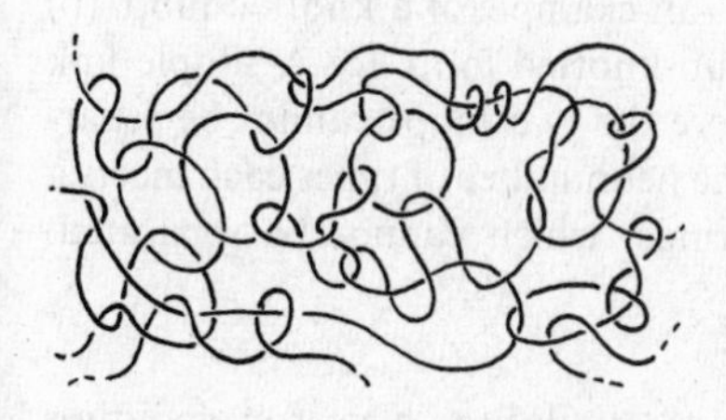

Fig. 32.5 The original loop-variable description of an approximately classical spacetime could be presented as a superposition of almost uniform spreadings of 'weaves'.

where $j = \frac{1}{2}n$ is the particular loop's 'spin' value; see Fig. 32.4. We add up these area contributions coming from all the loops of the family.

There is an interesting contrast between string theory and loop-variable theory. Whereas string theory is an almost entirely *perturbative* approach to quantum gravity, the loop-variable approach is fundamentally non-perturbative. In string theory, calculations are almost invariably performed in a background that is *flat* spacetime, i.e. a product of Minkowski space $\mathbb{M}$ with, say, some Calabi–Yau 6-space (see §31.14), and one is concerned only with weak fields in that background. The idea is that one contemplates 'perturbing away' from this weak-field limit (i.e. considering power series in some small parameter); see §26.10 and §31.9. On the other hand, in the case of the loop variables, the basic loop states (or spin-network states; see §32.6) are very far from flat (or classical), having *delta functions* in the area measure along the loops (or spin-network lines). To obtain the loop-variable description of an approximately classical spacetime, we need to consider something like an almost uniform spreading of 'weaves', as indicated in Fig. 32.5.

Notice that this is a very topological description. It makes no difference how 'close' one loop might be to another (since the notion of 'metric' makes no sense away from the loop itself). The only things that have significance are the topological 'linking' and 'knotting' (or intersecting) relations among loops, and the discrete 'spin' values that are assigned to them. Thus, general covariance (within $\mathcal{S}$) is completely taken care of, provided that we retain merely this discrete topological picture.

32.5 The mathematics of knots and links

The loop-variable picture of quantum gravity leads us into that field of mathematics which is concerned with the topology of knots and links. This is a surprisingly sophisticated subject, considering the commonplace nature of its ingredients—basically untangling bits of string! We need to

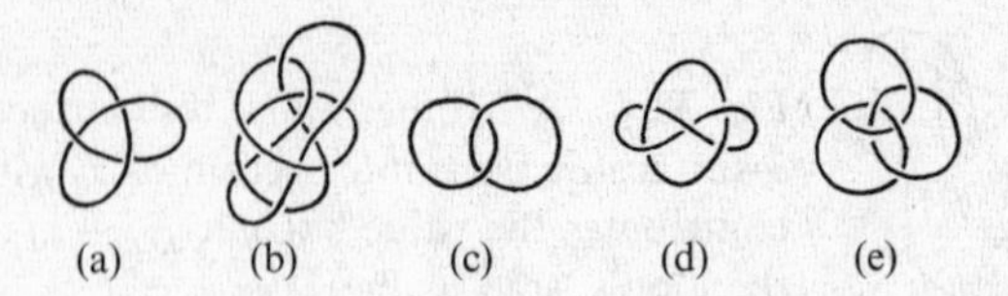

Fig. 32.6 Knots and links. (a) A trefoil knot—an example of a knotted loop. (b) An example of a (non-obviously, to the eye) un-knotted loop. (c) A simple link between two loops. (d) A Whitehead link, where the two loops cannot be separated though they have zero 'linking number' (the net number of times each meets a surface spanning the other) (e) Borromean rings, which cannot be separated, despite no pair of them being linked.

make use of mathematical criteria that are available to us for deciding whether or not a closed 'loop of string' is actually *knotted* (where 'knotted' means that it is impossible, by smooth motions within ordinary Euclidean 3-space, to deform the loop into an ordinary circle, where it is not permitted to pass stretches of the loop through each other; see Fig. 32.6). Likewise, we can ask for criteria that decide whether or not two or more distinct loops can be completely separated from one another—so they are *unlinked.* Various ingenious mathematical expressions that supply moderately complete answers to this question (such as the 'Alexander polynomial') have been known since the early 20th century, but in more recent years, a number of fascinating and more refined procedures have been found, gaining their inspiration largely from ideas coming from physics. These go under such names as 'Jones polynomial', 'HOMFLY polynomial', 'Kauffmann polynomial', etc.[14]

One way of thinking of these new mathematical structures is to consider them as arising from a kind of 'diagrammatic algebra', this being a generalization of the diagrammatic description of tensor algebra introduced in §12.8 (see Figs. 12.17 and 12.18) and used extensively in Chapter 13 (see Figs. 13.6–13.9, etc.). In this generalization, it makes a difference whether an 'index line' passes above or below another such line, whenever they cross one another in a diagram; see Fig. 32.7. There are various 'algebraic identities' that can be imposed on the algebra, such as that depicted in

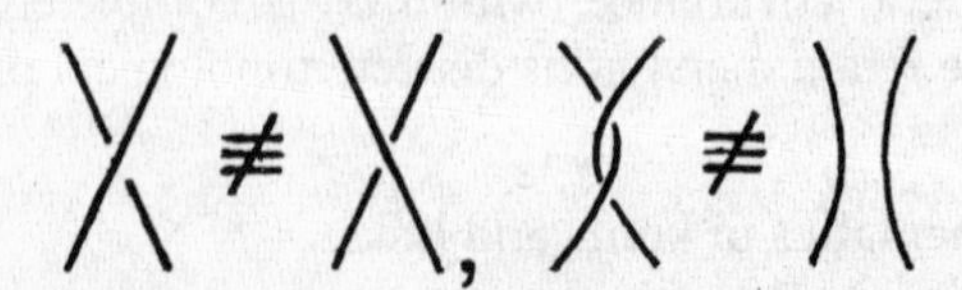

Fig. 32.7 The kind of diagrammatic tensor algebra of Figs. 12.17, 12.18 and Figs. 13.6–13.9, etc. can be generalized so as to produce an algebra for knots and links. The additional feature is that it now makes a difference whether an 'index line' passes above or below another such line when they cross.

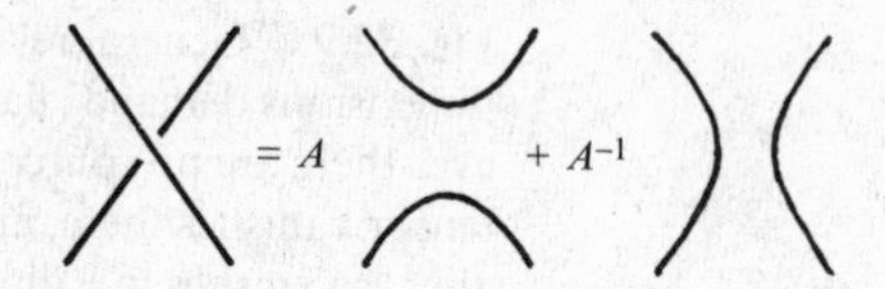

Fig. 32.8 The basic algebraic identity for the Kauffmann algebra is illustrated, where $q = A^2 = e^{i\pi/r}$, giving a q-deformed version of the 'binor' algebra underlying spin-network theory (for which $A = -1$, and the crossing issues of Fig. 32.7 do not occur).

Fig. 32.8, which serves to specify the Kauffmann algebra. These provide elegant generalizations of the combinatorial scheme that underlies the *spin-network* theory that we shall be coming to shortly.

The quantity A in Fig. 32.8 is a complex number, and sometimes A is written in terms of a quantity $q = A^2 = e^{i\pi/r}$. (The case $A = -1$ gives the 'binor calculus' that underlies spin-network theory; Penrose 1971a, 1971b.) There are analogues of the symmetrizers and antisymmetrizers of Fig. 12.17. A considerable theory of such things has been developed, sometimes referred to a *q-deformed* structures. Frequently the term 'quantum' is used, rather misleadingly, in place of 'q-deformed', such as with the notion of a 'quantum group'. There is no very clear relation between a 'quantum group' and quantum theory, however, and the existence of a significant application of quantum groups to physics at the fundamental level, though entirely possible, is largely conjectural at present.

As something of an aside, it is worth mentioning that there is another possible connection between these newly found mathematical structures (Jones polynomials, etc.) and physics, which has been developed particularly by Edward Witten.[15] This is the notion of a *topological quantum field theory*. In such a theory, the field equations disappear completely, but there is still information in global structure and in 'glitches' that can be thought of as providing 'sources' for the (locally vanishing) field. A good example would be general relativity in $1+2$ dimensions. In $1+2(=3)$ dimensions, the Weyl tensor vanishes identically, so all the curvature lies in the Ricci tensor. Thus, in 'empty space' (Ricci-flatness), the entire curvature vanishes. The gravitational field of a 'point source' is not trivial, however, because the source provides a 'glitch' which shows up in the global geometry. This is illustrated in Fig. 32.9. The geometry is very similar to the geometry of a cosmic string illustrated in Fig. 28.4, except that here the picture represents a $(1+2)$-dimensional spacetime, rather than 3-dimensional space. A segment is removed, with axis along the (timelike) world line of the source, and the two resulting planar boundaries glued together. In the classical picture, the world lines of such sources have to be straight, but a QFT based on this classical model—a *topological* QFT, since the field (here, the curvature field) vanishes—allows curved,

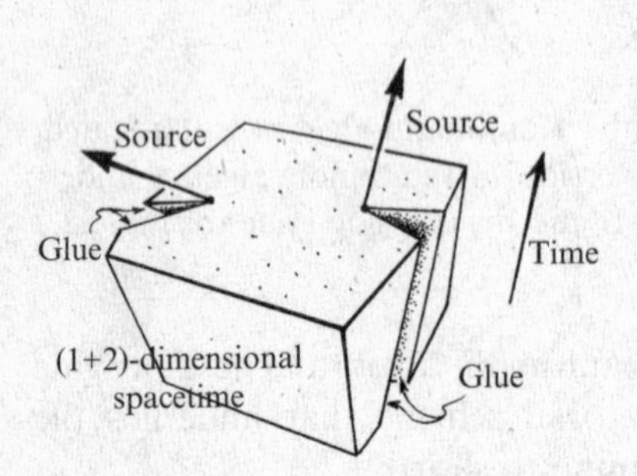

Fig. 32.9 'General relativity' in $2 + 1$ dimensions demands flat spacetime wherever there are no sources (since the Ricci tensor vanishes there, and the Weyl tensor always vanishes in 3 dimensions). But a source worldline provides a 'glitch' in the flat spacetime (a conical singularity) reminiscent of the 'cosmic string', whose spatial 3-geometry is illustrated in Fig. 28.4, §28.2. The source worldlines are always straight, classically, but this is relaxed in the quantum version of this theory—an example of a topological quantum field theory.

and indeed knotted or linked source lines. It is this that allows insights[16] to be gained concerning the mathematics of knots and links using the topological QFT idea. It will be noticed that loop variables provide a system somewhat resembling the general scheme of 'topological QFT' since the contribution to the area measure vanishes except at the 'glitches' that are the loops themselves. Nevertheless, there is a difference, because with loop variables the field equations do not disappear.

Topological QFTs are interesting as mathematical structures, but it is hard to see them playing direct roles as models of serious physical theories, owing to the complete disappearance of field equations. Most of known physics depends upon the non-triviality of such equations in order that fields propagate into the future in a controlled manner. There is, however, another distinct possibility that the ideas of topological QFT could be used in conjunction with *twistor theory*. As we shall be seeing in Chapter 33 (at the end of §33.11), in the twistor-space description, field equations do disappear locally. The application of topological QFT ideas to twistor theory has not been carried very far, as yet,[17] but it would be interesting to see what can be achieved in this area.

32.6 Spin networks

Remarkable as the loop states are, as limiting ('δ-function') configurations of 3-geometry, they still do not provide a suitable (orthonormal) *basis* for this geometry. For this, it is necessary to pass to a generalization in which loops can intersect. This leads us to consider a kind of network of 'intersecting loop lines', but we need to ask: what are we to do at these intersection points? The answer turns out to lie in certain types of structure that are formally very close to the *spin networks* that I had myself studied nearly 50 years ago, for a different but somewhat related purpose.

What are spin networks, and why had I been interested in them in the 1950s? My own particular goal had been to try to describe physics in terms of discrete combinatorial quantities, since I had, at that time, been rather strongly of the view that physics and spacetime structure should be based, at root, on *discreteness*, rather than continuity (see §3.3). A companion motivation was a form of *Mach's principle* (§28.5),[18] whereby the notion of space itself would be a *derived* one, and not initially present in the scheme. Everything was to be expressed in terms of the *relation* between objects, and not between an object and some background space.

I had come to the conclusion that the best prospect for satisfying these requirements was to consider the quantum-mechanical quantity of *total spin* of a system. 'Total spin' is defined by the scalar quantity j $(=\frac{1}{2}n)$ that measures the amount of spin as a whole, rather than a particular component of spin in some direction, measured by a quantity m. (The letters 'j' and 'm' are those commonly used in the discussion of quantum-mechanical angular momentum, taken in units of $\hbar$, where m ranges, in integer steps, between the integers or half integers $-j$ and j; see §§22.8,10,11.) The actual *magnitude* of the total spin (obtained as the square root of the sum of the squares of the m-values in three perpendicular directions) is $\hbar\sqrt{j(j+1)}$, which is the same quantity that appears in the area expression above. The allowed values of $n = 2j$ are simply natural numbers (even numbers for bosons and odd numbers for fermions; see §23.7). Moreover, though direction-independent, n is nevertheless intimately *related* to directional aspects of space. It had seemed to me that total spin, as measured by the natural number n, was an ideal quantity to fix attention upon if one were interested in building up, from scratch, some discrete combinatorial structure that leads to a notion of actual physical space. As a further ingredient, if one sets things up in the right way, one could exhibit the quantum-mechanical probabilities as being *pure probabilities*, not dependent in detail on the way in which different parts of a physical apparatus might be oriented with respect to other parts.

How does this work? Let us call a quantity of total spin $\frac{1}{2}n\hbar$ an *n-unit*. For clarity, we can think of this 'unit' as being a particle, but it need not be an elementary particle. For example, a whole hydrogen atom would do perfectly well. It simply needs to have a well-defined value for its total spin (which, in the case of a hydrogen atom would be given by $n = 0$ or 2, the case of ortho- or para-hydrogen, respectively[19]). How do we get a pure probability? We could, for example, take a couple of EPR–Bohm 1-unit pairs (A, B)

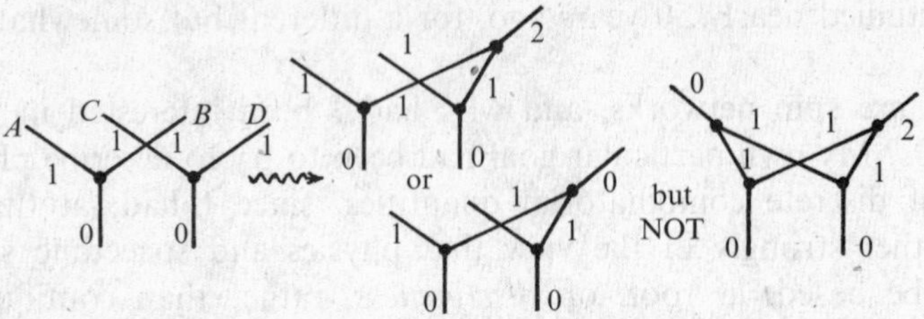

Fig. 32.10 Spin networks. Each line segment, labelled by a natural number n, represents a particle or subsystem of total spin $\frac{n}{2} \times \hbar$, called an n-unit. In this very simple example we have two EPR-Bohm 1-unit pairs, (A,B) and (C,D), each starting from 0-unit state (as in Fig. 23.2). If B and D combine to form a single unit, there are two possibilities: it can be a 0-unit or 2-unit, the respective probabilities being $\frac{1}{4}$ and $\frac{3}{4}$. The same probabilities occur if alternatively we combine A and C. But these probabilities are not independent, since we cannot have 0-unit in one case and 2-unit in other.

and (C, D), each starting from 0-unit state. (This is just a pair of arrangements, each like that of Fig. 23.3 of §23.4; see Fig. 32.10.) Now if we bring B and D together and allow them to combine into a single unit, the two possibilities are that they might result in 0-unit or a 2-unit, and the respective probabilities[32.5] are $\frac{1}{4}$ and $\frac{3}{4}$. If, alternatively, we bring A and C together, then the possible results and probabilities would be just the same. However, these two probabilities are far from independent of each other, since if we get a 0-unit in one case we cannot get a 2-unit in the other case, and vice versa.

This was the sort of idea I had for getting pure probabilities, and I had formed the opinion that any such probability had to come out as a rational number (since it would amount to Nature making a random choice of some kind between a finite number of discrete possibilities). The example just considered above is a very simple one, but it begins to illustrate the general idea. All the units in a particular spin network are imagined as being initially produced in the above way from initial 0-units (although this would not be normally expressed explicitly in the diagram), so there is no bias with regard to any particular spatial direction. Subsequently, various pairs of units may then be brought together to form single units, and the spin values for the resulting units noted. Individual units are also allowed to split into pairs of units. An example is illustrated in Fig. 32.11. We may choose to visualize all of this as happening within some spacetime. However, for the original spin-network theory, there was to be no actual background spacetime presupposed. The idea was to build up all

[32.5] Can you see why?

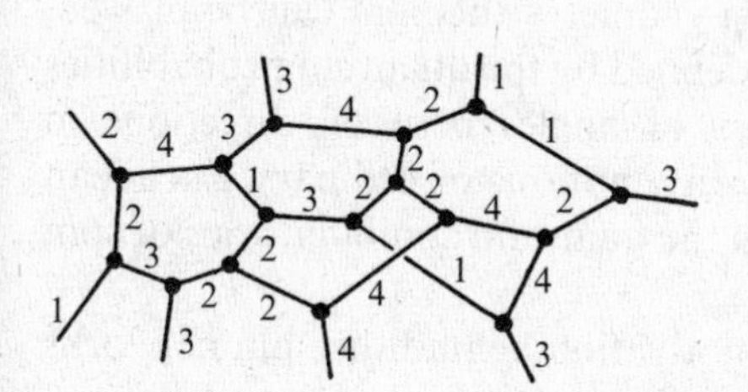

Fig. 32.11 Example of a spin-network, as originally conceived. No actual background spacetime manifold is presupposed. All spatial notions are to arise from the network of spins and from the probability values (when two units are combined to make a third). Exactly three lines meet at each vertex, uniquely specifying the connection.

the required spatial notions simply from the network of spins and from the probabilities that arise (and these can be computed using quantum-mechanical rules) when two units are brought together to make a third. A particular feature of these spin networks is that, at each such vertex, exactly three lines come together. This leads to a uniqueness in the probability calculations. The topological (graph) structure of the spin network, together with the specification of all the spin numbers on the lines, is all that is required.

I developed an entirely combinatorial ('counting') procedure for calculating the required probabilities (which, in fact, *are* all rational). The rules originally come from the standard quantum mechanics of spin, but we can then 'forget' where they come from and simply consider the spin-network system as providing a kind of 'combinatorial universe'. It is then possible to extract the notions of geometry (ordinary Euclidean 3-geometry in this case) by considering spin networks that are 'large' in an appropriate sense. The picture is that a unit of large spin might be considered to define a 'direction in space' (to be thought of as like the axis of spin of a tennis ball, for example). We can envisage measuring the 'angle between the rotation axes' of two such large units by, say, detaching a 1-unit from one and attaching it to the other. The joint probability that one spin goes up while the other goes down, in this operation, gives a measure of the *angle* between the spin axes.[32.6]

This almost works as it stands, but not quite, and a further ingredient is required. What is additionally needed is a means to distinguish the 'quantum probability'—coming from the angle of spin axes between the large units—from the 'probability through ignorance' that can come about simply because of insufficient connections between the two large units. (Recall the subtle interplay between these two notions of probability that occurs with density matrices, as discussed in §§29.3,4.) It turns out that this

[32.6] What is this angle measure, in terms of the probabilities?

'ignorance factor' can be removed by repeating the transfer of a 1-unit from the one large unit to the other, and selecting only the situations where the probabilities come out the same the second time. For families of large units for which this is the case, a geometry theorem can be proved, to the effect that the 'angle-geometry' defined by the quantum probabilities in this way is precisely the geometry of angles between directions in ordinary Euclidean 3-space.[20] In this way, notions of ordinary Euclidean geometry are seen to arise merely from the quantum combinatorics of spin networks.

It will be seen that the underlying motivation behind the spin networks that I originally had is very different from that underlying the loop-variable approach to spacetime quantization, there being no actual place for gravity in the spin networks, as originally put forward. It was therefore a considerable surprise to me to find spin networks playing such an important role in this approach to a quantum-gravity theory. Of course, there is something very much in common between the two programmes, because, in each case, one is trying to break down the notion of space into something more discrete and quantum-mechanical. There is, however, the important difference that in the loop-variable context, the quantity n is really an *area* measure, rather than the spin measure of the original spin networks. These are dimensionally different, as is reflected in the appearance of the gravitational constant G in the loop-variable expression. I shall return to this issue and its possible significance shortly.

Now, how are spin networks to feature in loop-variable quantum gravity? As I implied earlier, the spin-network nodes are, in effect, to result from the intersecting of a pair of loops. This also allows the j-value on the loop to change at such a place. Accordingly, we shall have nodes where *four* lines (or perhaps more) come together, rather than the three which occurred for my original spin networks. This results in ambiguities, since uniqueness of interpretation occurs only with the original 'trivalent' nodes. Accordingly there is an additional specification required (an 'intertwining operator') at each node. One way of expressing this specification is illustrated in Fig. 32.12, where we can represent such an 'X', where four lines

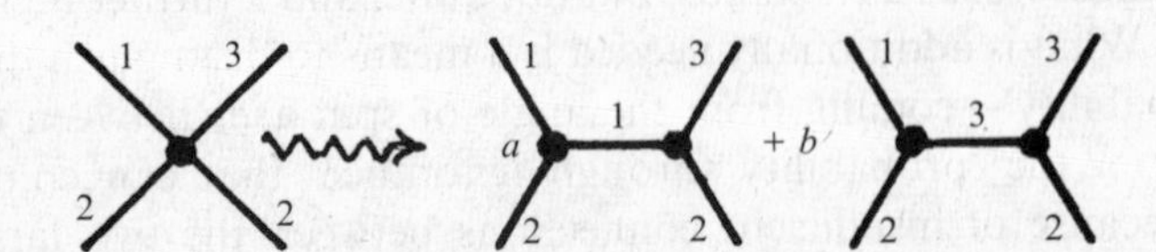

Fig. 32.12 All the nodes of spin networks in loop-variable theory are 4-valent (or more) and need extra 'intertwining' information. This can be encoded in 'X'-type vertex being expressed as a linear combination of 'H'-type 3-edge vertex pairs. The specific coefficients remove this ambiguity.

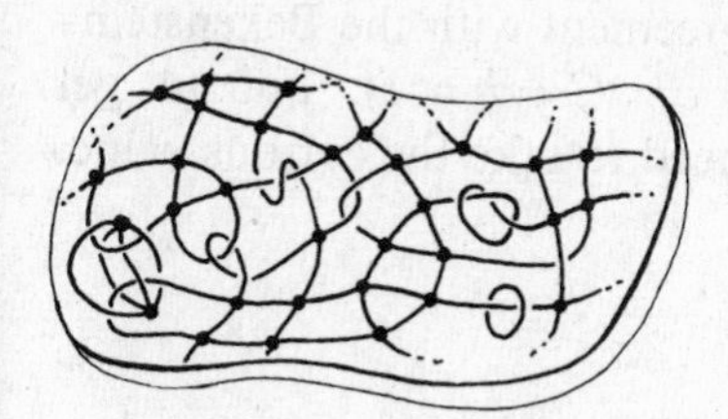

Fig. 32.13 The spin networks of the standard loop-variable approach are no longer entirely disembodied combinatorial entities, but must be embedded in a structureless (but perhaps analytic) 3-surface (like $\mathcal{S}$), their topological linking and knotting properties being of significance.

come together, as a linear combination of 'H'-type 3-edge vertex pairs. Specifying the coefficients removes the ambiguities.

There is another rather more important difference between these loop-variable spin networks and the ones that I had put forward earlier, namely that the earlier ones were entirely combinatorial structures, whereas the loop-variable networks acquire additional topological structure from their embedding in the manifold $\mathcal{S}$. The network lines could, for example, be knotted or could link one another in various ways, and this provides additional information (see Fig. 32.13). This information is still of a discrete combinatorial nature, however, being of an entirely topological character, but it is harder to express than simply specifying what goes on at the individual nodes.

So far, our loop descriptions have given us, effectively, just a *static* description, with no dynamics involved. In effect, the loops and spin networks that we have been considering have been concerned with solving the *constraint* equations of general relativity—i.e., the conditions needed to be satisfied within the surface $\mathcal{S}$—while paying full respect to Einstein's principle of general covariance. This is no mean achievement, but the formalism seems not yet to have solved the more difficult problem of the dynamical evolution away from $\mathcal{S}$ (sometimes referred to as the 'Hamiltonian constraint'), in order that the Einstein equation can be fully accommodated (see §32.1). Some significant work by Thomas Thiemann has provided one possible answer to this Hamiltonian-evolution problem, but there remains some doubt as to whether this is actually the appropriate one for Einstein's theory.[21]

Pending a fully accepted solution to these difficult dynamical issues, it has nevertheless been possible to use the loop-variable formalism to arrive at some impressive results in other directions. In particular, these spin-network ideas have proved useful in providing a much more direct and realistic-looking approach to the issue of black-hole entropy than that of string theory, as referred to in §31.15. Here the black-hole geometry is directly that of a Schwarzschild or Kerr vacuum solution of the Einstein 4-dimensional theory. The counting of gravitational quantum states can be explicitly carried out using spin networks, using appropriate approximations. When the black hole begins to get reasonably large, the answer for the entropy comes out in agreement with the Bekenstein–Hawking formula $S_{BH} = \frac{1}{4}A$ (where $k = c = G = h = 1$), but to get Hawking's precise factor $\frac{1}{4}$, one seemed to need to take the curious value

$$\eta = \frac{\log 2}{\pi\sqrt{3}}$$

for the Barbero–Immirzi parameter. Although certainly a strange value, such one choice then correctly gives the Bekenstein–Hawking entropy for all situations in which an unambiguous answer has been provided by other means, where there can be charge, rotation, and cosmological constant.[22]

In relation to this, there are various apparent numerical 'coincidences' upon which the theory seems to depend. Two separate infinite sequences, calculated in two quite different ways have to agree, term-by-term, which in fact they do. This seems to reflect a deep inner consistency of some of the ideas of quantum geometry.

Nevertheless, the black-hole results seem to have elicited something of a change in viewpoint with regard to the Barbero–Immirzi parameter. Previously, the introduction of η seems to have been just a means for making progress, where the 'geometrically correct' theory seemed to demand $\eta = \pm \mathrm{i}$. Taking a real value for η was just a mathematical convenience, in order that the compact group SU(2) would arise rather than the non-compact SL(2,$\mathbb{C}$). The impressive successes in obtaining the correct entropy values for a very wide class of horizons—this being dependent on a *single* η choice (given above as $\eta = \log 2/\pi\sqrt{3}$) for the Barbero–Immirzi parameter—has led a number proponents of the loop-variable approach to take the view that perhaps such a *real* value is actually the 'correct' one for quantum gravity, after all.

This, of course, is a possibility, though personally I find it somewhat hard to believe, as there appears to be no clear geometrical reason for such a choice. I should make the comment that with any real value for η, such as that displayed above, the *chiral* aspect of the theory that I stressed in §32.3 to introduce the subject has disappeared. The spinor transport that $\mathbf{\Gamma}_\eta$ is concerned with, with a real value of η, is a peculiar but even-handed mixture of intrinsic and extrinsic parts, whose meaning I find particularly obscure. Perhaps future work will shed light on this issue.

32.7 Status of loop quantum gravity?

I should try to give my assessment of the achievements of the Ashtekar–Rovelli–Smolin loop-variable approach to quantum gravity and its potential for future development into a full-blown theory. Again, I must alert the reader to a possible bias that may be relevant to any such assessment. In this case I must declare an interest, for not only are the people responsible for this programme all very good friends of mine, but I have also held regular visiting appointments at the two US universities (Syracuse and Penn State) where major developments in this field have taken place. To this I must add my own interest in spin-network theory; it is natural that I should find it gratifying that these old ideas should now find significant new value in this approach. Nevertheless, my own involvement in the Ashtekar-variable/loop-variable quantum-gravity programme has been somewhat tangential to the main work in this area, so I hope that I may stand back and be reasonably objective.

To begin with, I should comment that both the original Ashtekar variables and the later descriptions in terms of loop variables strike me as powerful and highly original developments in the quest for a quantum gravity theory. They are directly addressing Einstein's actual general relativity theory in the context of QFT, providing profoundly innovative ideas that are relevant to the problem at hand. In fact, I have little hesitation in saying that these developments are the most important in the canonical approach to quantum gravity since the subject itself was started roughly half a century ago, by Dirac and others. The loop states do appear to address at least some of the profound problems raised by general covariance. Moreover, these developments seem to have moved the discussion in a fascinating and perhaps not fully anticipated direction, where some gratifying elements of discreteness in spacetime structure begin to appear. Furthermore, in recent work, the original purely gravitational theory has moved in the direction of incorporating physical interactions other than just gravity, so the theory can now make claims of being an approach to fundamental physics generally.[23]

Set against this is the somewhat disturbing fact that the theory seems to have found itself necessarily deflected into adopting the $\boldsymbol{\Gamma}_\eta$ connection (with an undetermined value of η), rather than the seemingly 'geometrically correct' $\boldsymbol{\Gamma}$. In my view, a fully believable approach to quantum gravity will not come about, in this approach, until a way has been found to overcome the difficulties that seem to arise with the adoption of the original $\boldsymbol{\Gamma}$. Moreover, there is still the fundamental difficulty that the full Einstein Hamiltonian has yet to be unambiguously encompassed within the loop-variable framework, even though the constraint equations are handled by the use of spin-networks.

It strikes me as likely that these difficulties are related to another (to my mind) less-than-satisfactory feature of the Ashtekar/loop-variable theory. In common with all other conventional *canonical* approaches to quantum gravity, its formulation is directly dependent upon a 3-space description (i.e. in terms of $\mathcal{S}$), rather than being a more global spacetime one. As we have seen, the 3-space part of the 'general covariance' problem is neatly taken care of in the loop/spin-network states, but the extension of this to a full 4-space general covariance brings with is a whole 'Pandora's box' of problems. As far as I can make out, these are not very much better addressed in the loop-variable approach, as yet, than in other canonical approaches.[24]

The difficulty has to do with the issue of how time-evolution, according to the Einstein equation, is to be properly expressed in a generally covariant 4-space formalism. It is related to what is known as the 'problem of time' in quantum gravity (or, sometimes, the problem of 'frozen time'). In general relativity, one cannot distinguish time-evolution from merely a coordinate change (i.e. just replacing one time coordinate by another). A generally covariant formalism should be blind to a mere coordinate change, so the concept of time-evolution becomes profoundly problematic. My own perspective on this question, as indicated in §30.11, is that the issue is unlikely to be resolved without the problem of state-vector reduction **R** being satisfactorily addressed, and that this, in turn, will require a drastic revision of general principles.

In relation to these issues, there is another matter that I find less than satisfactory, although it is more a problem with generally covariant prescriptions *per se*, rather than with the loop-variable approach in particular. In a sense, this approach is a victim of its own success! For although the spin-network basis states individually have a pleasing coordinate-independent geometric description, it is most unclear how to interpret quantum *superpositions* of such basis states. Because of the general covariance, there is no correspondence between the 'location' of one spin network and that of another with which it is to be superposed. (This is a much more serious version of the issue addressed in §30.11, which I used to justify gravitational **OR**.) How are we expected to understand how an almost classical world is to emerge out of all this?

As the reader will have gathered by now, particularly from the discussion of Chapter 30, I regard as a necessary feature of the correct quantum–gravity union that it must depart from standard quantum mechanics in some essential way, so that **R** becomes a realistic physical process (**OR**). Is there scope for this in the loop-variable approach? Possibly so. With loop variables, the numbers $n = 2j$ on the spin-network edges refer to *area* in units of squared Planck length; but my original use of spin networks did not address such metric issues nor, in fact, any aspect of gravity at all, the spin numbers n referring to *angular momentum*. However, my original ideas demanded that each of these numbers must be, in effect, the result of an individual measurement of total spin value (action of **R** at each edge), where probabilities arise in the bringing together of two units to make a third. If **R** is an objective gravitational phenomenon, then the involvement with gravitational processes would have to enter at this stage, as is the claimed message of Chapter 30. In that case, it is not possible to separate gravity from the probability issues of spin-network theory. It may be that the full combination of loop-variable and spin-network ideas will need to incorporate state reduction into the formalism. If this proves to be the case, then it might provide a lead into an appropriate gravitational **OR** scheme, as recommended in Chapter 30. In the absence of such a formalism, however, such ideas must remain speculation.

Finally, I should comment on other work that relates to loop-variable theory. It is now not an entirely pure-gravitational theory, electromagnetism having now been addressed[25] in this formalism.[26] There are also seemingly less radical ways, than that of the previous paragraph, of making the spin networks of loop quantum gravity more '4-dimensional'. One of these involves an ingenious higher-dimensional version of spin networks referred to as *spin foams*. In these, there are 2-surfaces carrying the 'spin values' $n = 2j$, and we can picture such a spin foam as a time-evolving spin network. Such ideas, originated by Louis Crane, John Barrett, and others[27], have been further developed and modified by several others[28], but the proper connections with quantum-gravity ideas have not been fully worked out, as yet. There are also possible connections with twistor theory, and it will be interesting to see whether these can be developed more fully. In the next chapter, we shall be having a look at some of the basic notions that are indeed involved in twistor theory.

I have tried to emphasize elsewhere in this book, and in Chapter 30 most particularly, that the issue of quantum state reduction is intimately related to the structure of spacetime singularities and their asymmetry under time-reversal. It is interesting that a start has in fact been made, in examining what the loop-variable approach has to say about the effect of quantum theory on spacetime singularities.[29] I am not able to comment in any detail on this work, except to say that I see no hint of the necessary time-asymmetry arising.

Notes

Section 32.1

32.1. See Greene (1999).

32.2. There is the irony here that Einstein himself, with his venture into unified field theories in his later years, did not always stick to the narrow path that he had set out previously.

32.3. See Dirac (1964); Ashtekar (1991); Wald (1984) for the notions of constraints and Hamiltonian formulation.

32.4. See Dirac (1950, 1964); Pirani and Schild (1950); Bergmann (1956); Arnowitt *et al.* (1962), for example. See DeWitt (1967) for the Wheeler-DeWitt equation which is basically quantum gravity's Schrödinger equation on the space of compact 3-geometries.

32.5. See Isham (1975); Kuchar (1981).

32.6. See Sen (1982) and Ashtekar (1986, 1987).

Section 32.2

32.7. As we shall be seeing shortly, technical difficulties have deflected the original Ashtekar approach from this manifestly chiral description. But my understanding of the motivation behind the Ashtekar programme is that this 'deflection' may be taken as a temporary exploration of models that are close to, but not identical with, the intended quantum-gravity theory.

32.8. For the original discussion of linearized gravity, see Fierz and Pauli (1939); for futher information on linearized gravity see Sachs and Bergmann (1958). For the non-linear graviton, see Penrose (1976a and 1976b).

Section 32.3

32.9. See Ashtekar (1986 and 1987); see also Ashtekar and Lewandowski (2004) for a review, and Rovelli (2003) for a textbook!

32.10. Those wishing to understand how the 'indices' work out should note first that the inverse 3-metric is a quantity γ^{rs} with raised 3-dimensional indices. Forming the dual on *s* (§12.7), within $\mathcal{S}$, we obtain $\gamma^{r}{}_{tu}$ (a density) which is antisymmetric in *t, u*. We can read these lower indices as 2-form indices in the 4-space and take the anti-self-dual part, which provides us with a pair of lower symmetric spinor indices, giving us a quantity γ^{r}_{PQ}, or equivalently $\gamma^{r}{}_{P}{}^{Q}$ which is trace-free in *P, Q*. As for Γ, this is a quantity $\Gamma_{r}{}^{P}{}_{Q}$, as it is a 1-form matrix in spin space, which is necessarily trace-free (because it is an SL (2, ℂ) connection). Note that the index structure of Γ is opposite to that of γ, as it should be, for canonical variables. For more details, see Ashtekar (1986, 1987); Ashtekar and Lewandowski (2004); Rovelli (2003).

32.11. It was this connection that Sen introduced in 1981, reducing the constraints of general relativity to a polynomial form. It is of interest that Witten used the same connection, independently (also in 1981), in his proof procedure for positivity of total energy in general relativity; see §19.8 and Witten (1981). As far as I am aware, the first use of this connection was in the construction of the *hypersurface twistor space* for $\mathcal{S}$ (Penrose and MacCallum 1972; Penrose 1975), which is that part of twistor theory that is most closely related to Ashtekar variables.

Section 32.4

32.12. See Wilson (1976); for this technique in quantum field theory, see Zee (2003). See also Rovelli and Smolin (1990).

32.13. Even to be able to specify, its manifold structure might be considered by some to be more than we are entitled to, since this provides it with not only a topology, but also a differentiable or 'smoothness' structure—see §10.2 and §12.3. Compare also §33.1. In fact, for technical convenience, $\mathcal{S}$ is assigned an *analytic* structure, in the Ashtekar–Lewandowski approach. (Ashtekar and Lewandowski 1994), which means that it is C^{ω}, in the sense of §6.4.

Section 32.5

32.14. For a very gentle introduction to knot theory, see Adams (2000). For a more technical work, see Rolfsen (2004) and Kauffman (2001).

32.15. See Labastida and Lozano (1997) for a review article; see Witten (1988) for the original 'TQFT'.

32.16. However, topological QFTs do not directly yield rigorous mathematical theorems, owing to wild divergences.

32.17. See Penrose (1988a). It may be noted that the new twistor-string ideas put forward by Witten (2003) do provide developments of this general kind; see Note 31.81.

Section 32.6

32.18. My interest in Mach's principle was strongly stimulated by discussions with my colleague, friend, and mentor Dennis Sciama.

32.19. See Levitt (2001).

32.20. See Penrose (1971a), Moussouris (1983).

32.21. See Thiemann (1996, 1998a, 1998b, 1998c, 2001).

32.22. See Ashtekar *et al.* (1998); Ashtekar *et al.* (2000). However, it has recently been shown that this displayed value for η is in *error*, and η has now been assigned a more complicated-looking value. See Domagala and Lewandowski (2004); Dreyer, Markopoulou, and Smolin (2004). The remainder of my discussion in §§32.6,7 seems not to be affected

Section 32.7

32.23. See Thiemann (1998c).

32.24. See Hawking and Hartle (1983), and Notes 32.3–5. I can't resist referring to Smolin (2003) and to the excellent review article by Ashtekar and Lewandowski (2004), from which many of these references are taken. I owe deep thanks to Abhay Ashtekar for his help with the references in this chapter.

32.25. Varadarajan (2000).

32.26. See Varadarajan (2001).

32.27. See Barrett and Crane (1998); Baez (1998); Reisenberger (1997, 1999).

32.28. Barrett and Crane (2000); Baez (2000); Reisenberger and Rovelli (2001, 2002); Perez (2003).

32.29. Bojowald (2001); Ashtekar *et al.* (2003).

33
More radical perspectives; twistor theory

33.1 Theories where geometry has discrete elements

HAVE the theories described in the preceding chapters been sufficiently radical, in their attempts to decipher Nature's *actual* scheme whereby the quantum physics of the small is somehow united with the curved-space geometry of the large? Perhaps we should be seeking something of a character fundamentally different from the real-manifold setting of continuous spacetime which Einstein's theory and standard quantum mechanics depend upon. The question was raised in §3.3, and we must indeed ask whether the real-number spacetime continuity that is almost universally assumed in physical theories is *really* the appropriate mathematics for describing the ultimate constituents of Nature.

We have seen how the loop-variable approach to quantum gravity begins to take us away from the standard picture of a continuous and smoothly varying spacetime and towards something of a more discrete topological character. Yet, some physicists would strongly argue that a far more radical overhaul of the ideas of space and time is needed, if the appropriately deeper insights are to be gained as to the nature of a 'quantum spacetime'. The original (albeit limited) spin-network proposal of §32.6 was indeed of a completely discrete character, but the standard loop-variable picture is still dependent upon the continuous nature of the 3-surface in which the 'spin networks' are taken to be embedded. In the latter scheme, one has not really obtained the entirely discrete and manifestly 'combinatorial' framework that some would feel is necessary in order that we may come to terms with Nature's workings at the tiniest scales.

Various ideas have been proposed, quite distinct in approach from that of the initial spin-network or spin-foam schemes where the intention is to provide a completely discrete/combinatorial picture of the world. Among the more extravagant of such ideas (already referred to in §16.1) was one put forward by Ahmavaara (1965). He suggested that the real-number system, fundamental to the mathematics of conventional physics, should be replaced by some finite field $\mathbb{F}_p$, where p is some extremely large prime number. We recall (from §16.1) that $\mathbb{F}_p$ is obtained by taking the

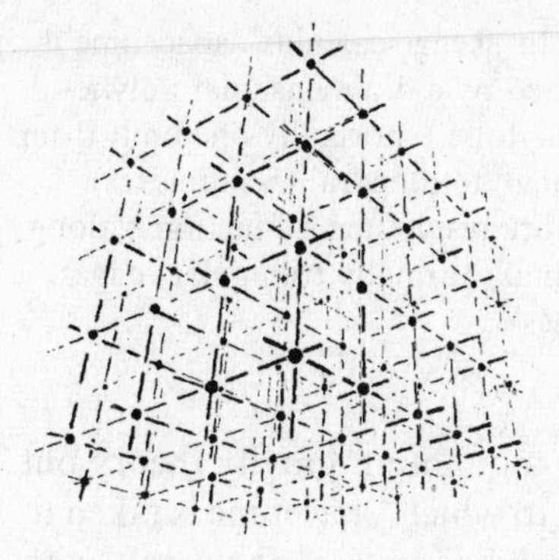

Fig. 33.1 A Snyder–Schild-type spacetime is a periodic lattice, like the vertices of a set of cubes stacked up on each other in a regular way. (There can be rather more Lorentz-invariance than one might expect!)

system of integers modulo p. Other suggestions take spacetime to have a discrete periodic *lattice* structure, like the vertices of a set of cubes stacked up on each other in a regular way[2] (Fig. 33.1). Considerably more physically plausible are schemes like Raphael Sorkin's causal-set geometry[3] (or certain closely related earlier ideas)[4] according to which spacetime is taken to consist of a discrete, possibly finite, set of points for which the notion of *causal connection* between points is taken to be the basic notion. In ordinary classical terms, this 'causal connection' refers to the possibility of sending a signal from one of the points to the other, so one of the points lies on or within the light cone of the other. See Fig. 33.2. The largely random nature of the causal connections in Sorkin's scheme enables something like the Lorentz invariance of special relativity to emerge, while there are serious difficulties for Lorentz invariance with the lattice-type structures (although there is more such symmetry than one might think at first, for the lattices—somewhat like that of Fig. 33.1). Other ideas leading to exotic spacetime structures arise from the quantum set theory or quaternionic geometry of David Finklestein,[5] the octonionic (§11.2, §16.2) physics of Corinne Manogue and Tevian Dray[6] etc.

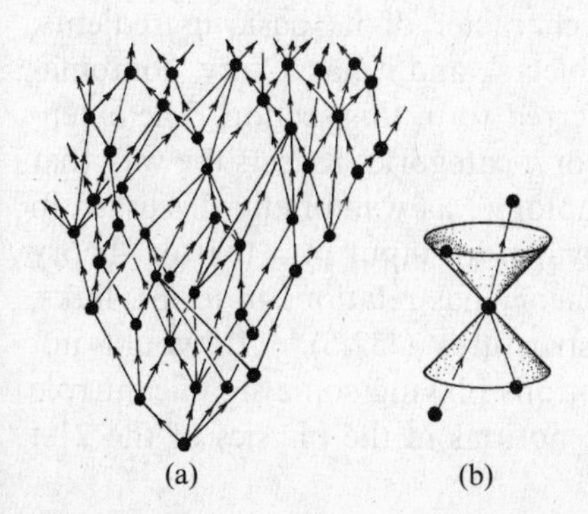

Fig. 33.2 (a) A discrete universe model, described by a causal-set geometry. (b) The relations between points are modelled on Lorentzian causality, where the arrows point either within or on the null cones.

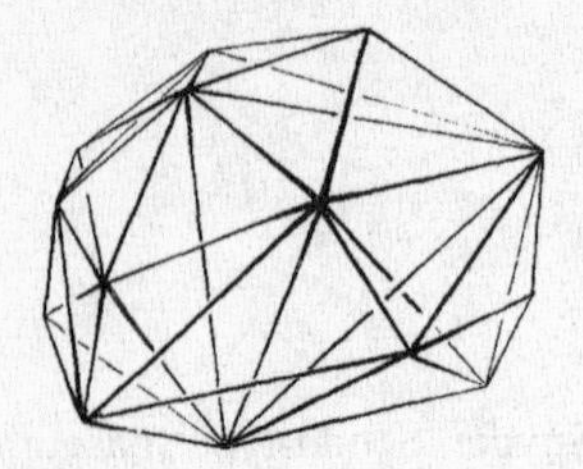

Fig. 33.3 In 'Regge calculus', spacetime is approximated by a 4-dimensional polyhedron (a 'polytope'), normally one built from 4-dimensional 'tetrahedra' (5-simplexes). The curvature resides (as δ-functions) along 2-dimensional (normally triangular) edges, called 'bones'.

There is also the interesting proposal for a quantum-gravity theory put forward by Tullio Regge in 1959, according to which, spacetime is taken to be an irregular 'tetrahedral' 4-dimensional polyhedron (or 'polytope'), with its curvature concentrated as delta functions (§9.7) along 2-dimensional 'edges'—that Regge referred to as 'bones'.[7] See Fig. 33.3 (and also Fig. 32.3a, §32.4). The quantum state is taken to be a complex-number-weighted sum of such spaces, in accordance with the 'Feynman sum over histories' described in §26.6. The description of the spaces themselves is entirely combinatorial, except for the fact that an 'angle' must be specified at each bone, to represent the strength of the curvature. In fact, the 'cosmic string' described in §28.2 (Fig. 28.4) is an example of this same type of geometry.

There have also been other intriguing radical proposals, such as those of Richard Jozsa[8] and of Christopher Isham[9] which employ *topos theory*. This is a kind of set theory[10] arising from the formalization of 'intuitionistic logic' (see Note 2.6), according to which the validity of the method of 'proof by contradiction' (§2.6, §3.1) is denied! I shall not discuss any of these schemes here, and the interested reader is referred to the literature.

Another idea that may someday find an significant role to play in physical theory is *category* theory and its generalization to n-category theory. The theory of categories, introduced in 1945 by Samuel Eilenberg and Saunders Mac Lane,[11] is an extremely general algebraic formalism (or framework) based on very primitive (but confusing) abstract notions, originally stimulated by ideas of algebraic topology. (Its procedures are often colloquially referred to as 'abstract nonsense'.) Its great power is deceptive, given the very elementary character of its basic ingredients, these being just 'arrows' connecting 'objects', and it has a very 'combinatorial' appearance, like other ideas referred to in this section. The extension of the theory of categories to that of n-categories reflects the way that 'homotopy' refines the notion of 'homology', as was briefly discussed in §7.2. Category theory has already provided an input into twistor theory (in relation to §33.9), and n-category theory has relations to loops, links, spin-foams (§32.7) and q-deformed structures (§32.5).[12] It would not altogether surprise me to find these notions playing some significant role in superseding conventional spacetime notions in the physics of the 21st century.

Rather more in line with present-day mainstream ideas is the notion of *non-commutative geometry*, developed, most particularly, by the Fields Medalist mathematician Alain Connes. What is a 'non-commutative geometry'? To appreciate the idea, think first of an ordinary smooth real manifold $\mathcal{M}$. Next, consider the family of smooth real-valued (scalar) functions on $\mathcal{M}$ (which we may take to be C^∞-smooth; see §6.3). Such functions can be added or multiplied together, and they can be multiplied by ordinary (constant) real numbers. In fact, they constitute an algebraic system $\mathcal{A}$, called a *commutative algebra over the reals* $\mathbb{R}$. (Compare §12.2 and Note 12.5.) Now, it turns out[13] that, if we know $\mathcal{A}$ only as an algebra, where no information is provided as to where this algebra came from, then we can nevertheless, reconstruct the manifold $\mathcal{M}$ simply from the algebra $\mathcal{A}$. We thus see that each of $\mathcal{M}$ and $\mathcal{A}$ can be constructed from the other, so it follows that these two mathematical structures are, in a clear sense, *equivalent* to one another.

In quantum mechanics, one frequently encounters algebras that are, on on the other hand, *non*-commutative. An example would be the algebra of x^a and p_a that satisfies the standard canonical commutation rules $p_b x^a - x^a p_b = i\hbar\delta^a_b$ of §21.2. If we try to reconstruct a 'manifold' from such an algebra, in the same kind of way that $\mathcal{M}$ would be obtained from $\mathcal{A}$ above, then we obtain what is referred to as a *non-commutative* geometry. As another example, we could take another particular case and start with the quantum-mechanical angular momentum components $\boldsymbol{L}_1$, $\boldsymbol{L}_2$, $\boldsymbol{L}_3$ of §22.8 (recall that the algebra generated by them is defined by the non-commutative laws $\boldsymbol{L}_1\boldsymbol{L}_2 - \boldsymbol{L}_2\boldsymbol{L}_1 = i\hbar\boldsymbol{L}_3$, $\boldsymbol{L}_2\boldsymbol{L}_3 - \boldsymbol{L}_3\boldsymbol{L}_2 = i\hbar\boldsymbol{L}_1$, $\boldsymbol{L}_3\boldsymbol{L}_1 - \boldsymbol{L}_1\boldsymbol{L}_3 = i\hbar\boldsymbol{L}_2$). We may think of these operators as generating the rotations of an ordinary sphere S^2. It turns out that we can get a non-commutative geometry from the algebra generated by $\boldsymbol{L}_1$, $\boldsymbol{L}_2$, $\boldsymbol{L}_3$, which we can refer to as the 'non-commutative sphere'. There are many mathematical subtleties, beautiful structures, and unexpected applications of this idea, but I cannot go into all this here. I shall return to non-commutative geometry briefly (§33.7) in relation to twistor quantization.

Connes and his colleagues have developed the idea of non-commutative geometry with a view to producing a physical theory which includes the standard model of particle physics.[14] Their model uses an algebra $\mathcal{A}$ that is a product $\mathcal{A}_1 \times \mathcal{A}_2$, where $\mathcal{A}_1$ is the (commutative) algebra of functions on spacetime (but taken to have a positive-definite metric) and where $\mathcal{A}_2$ is a non-commutative algebra arising from the internal symmetry groups of the standard model of particle physics and providing 'two copies' of spacetime. This model does not, as it stands, incorporate the Lorentzian ideas of special relativity, and certainly not general relativity. Moreover, the potential richness of the idea of non-commutative geometry does not seem to me to be at all strongly used, so far. Yet, the model makes a start, and it has intriguing features to it that entail predictions with regard to the mass of the Higgs boson.[15]

All these ideas concentrate on the construction of notions of 'spacetime' that take on aspects of discreteness or 'quantum' characteristics of some kind. In the remainder of this chapter I shall describe a quite different family of ideas, namely those of *twistor theory* (to which I have, myself, now devoted over 40 years!) in which there is no discreteness specifically imposed upon spacetime. Instead, spacetime points are deposed from their primary role in physical theory. Spacetime is taken to be a (secondary) construction from the more primitive twistor notions. Twistor theory has some relations to spin-network theory and to Ashtekar variables, and possibly to non-commutative geometry, but it does not directly lead to any notion of a 'discrete spacetime'. Instead, its departure from real-number continuity is in the opposite direction, for it calls upon the *magic of complex numbers* as a primary guiding principle for physics. According to twistor theory, there is a fundamental underlying role for complex numbers in defining spacetime structure, in addition to the well-established basic role of these same numbers in quantum mechanics. In this way, an important thread of connection is perceived, between the physics of the large and the physics of the small.

33.2 Twistors as light rays

As we have seen in Chapters 21 and 22, complex-number structure is indeed fundamental to quantum mechanics. The 'amplitudes' that appear as coefficients in the basic superposition law of quantum mechanics are complex numbers, leading to the complex Hilbert spaces of the theory. Whereas these amplitudes are commonly regarded as abstract quantities, and they play basic roles in providing probabilities when a measurement takes place, we have seen (in §22.9) that there is a strong interconnection between these complex numbers and spatial geometry. This is most manifest with the quantum mechanics of a spin $\frac{1}{2}$ particle, where the possible states of spin correspond to the different spatial directions, via the notion of a *Riemann sphere*; moreover, we saw in §22.10 that spin states for higher spin can also be described in terms of the spatial geometry of the Riemann sphere by means of the Majorana representation. Yet it is not just in quantum mechanics that we see a fundamental geometrical role for the Riemann sphere. We recall (from §18.5) that this sphere plays an important spacetime role in relativity theory, since the field of vision of an observer can also be validly regarded as a Riemann sphere. This fact has a seminal significance in twistor theory, as we shall be seeing very shortly.

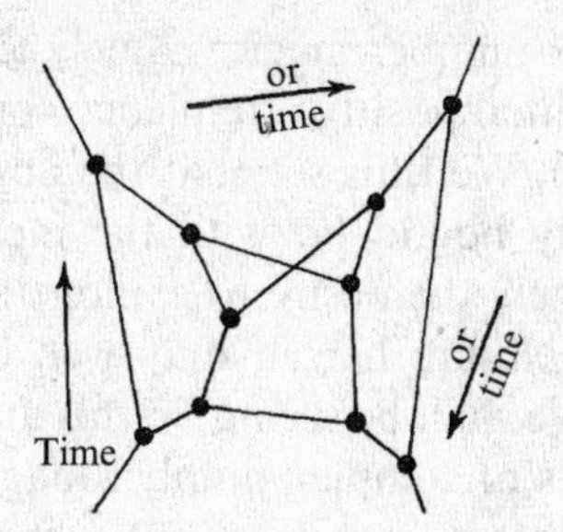

Fig. 33.4 In a spin network of the original type, the lines may be read as *quanglement* links (see §23.9 and Fig. 23.7). Any time-direction may be chosen—forwards, sideways, or backwards—and an equally valid interpretation of the spin network is obtained.

Another guiding principle behind twistor theory is quantum *non-locality*. We recall from the strange EPR effects discussed in §§23.3–6, and more specifically from the role of 'quanglement', as manifested particularly in the phenomenon of quantum teleportation, as described in §23.9, that physical behaviour cannot be fully understood in terms of entirely local influences of the normal 'causal' character. This suggests that some theory is needed in which such non-local features are incorporated at a basic level.

Some guidance for achieving this may be gained from *spin-network* theory. We recall from §32.6 that we are to regard all spin networks to be built up initially from EPR pairs. The lines of a spin network that subsequently arise can legitimately be thought of as quanglement links. The 'quantum information' that represents quanglement, can 'travel' one way or the other along a quanglement line, or spin-network line. There is no specification of a *time* in spin-network theory (and, indeed, the original spin networks could equally be read using various different time-senses—forwards, backwards, sideways, etc, see Fig. 33.4). Thus, the curious 'backwards-time' aspects of quanglement, are just reflections of this indifference to a time-flow direction that is a feature of spin networks.

It is possible to regard twistor theory as a continuation of the spin-network programme to obtain a *relativistic* scheme, in which idealized light rays (or their generalizations, with spin) appear to be, in a sense, the carriers of quanglement. Ordinary spacetime notions are not initially among the ingredients of twistor theory but are to be *constructed* from them. This has a good deal in common with the underlying philosophy behind my *original* spin networks, where spatial notions are to be constructed from the spin networks, rather than the spin networks being thought of as inhabiting a previously assigned spatial geometry.

The twistor description of spacetime indeed turns out to be a non-local one; moreover, there is a fundamentally 'holistic' character to the twistor description of physical fields that comes about via a remarkable feature of complex magic (namely holomorphic sheaf cohomology) that we have not yet properly encountered in this book—though we shall do so in §33.9 (and there was a hint of it already in the hyperfunction theory of §9.7)—

which meshes with another aspect of complex-number magic, namely the underlying holomorphic character of the essential positive-frequency condition of quantum field theory (§24.3, §33.10). We thus see that the non-local aspects of twistor theory are intimately bound up with the most important of its underlying motivations, namely the desire to exploit the magic of complex numbers in a belief that Nature herself may well be dependent upon such things at a deep level. We shall be seeing, in this and the next several sections, how all these aspects of complex-number magic begin to come together in the twistor-theoretic framework. We shall also begin to see how twistor theory finds a remarkable and unexpected deep relation to *general* relativity, and that it provides an intriguing perspective on QFT, particle physics, and the possible non-linear generalization of quantum mechanics.

How do these ideas indeed begin to come together in twistor theory? As a first step towards the understanding of twistor ideas, we may think of a twistor as representing a *light ray* in ordinary (Minkowski) spacetime $\mathbb{M}$. One can regard such a light ray as providing the primitive 'causal link' between a pair of *events* (i.e. of spacetime points). But events are themselves to be regarded as secondary constructs, these being obtained from their roles as intersections of light rays. In fact, we may characterize an event R (spacetime point R) by means of the family of light rays that pass through R; see Fig. 33.5. Thus, whereas in the normal spacetime picture a light ray Z is a locus and an event R is a point, there is a striking reversal of this in twistor space, since now the light ray is described as a *point* $\mathbf{Z}$ and an event is described as a *locus* $\mathbf{R}$.

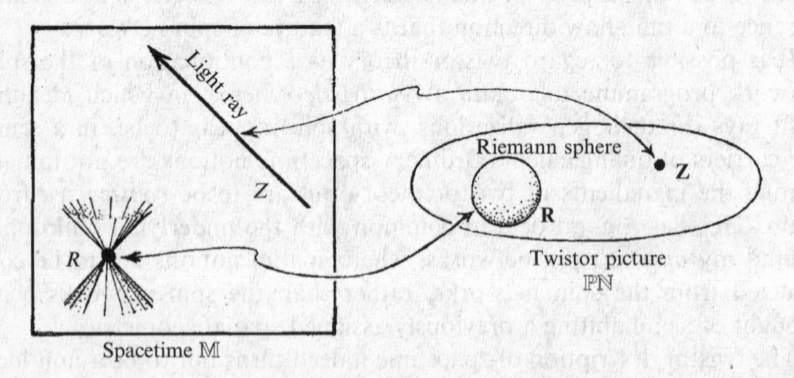

Fig. 33.5 A light ray Z in Minkowski spacetime $\mathbb{M}$ is represented as a single point $\mathbf{Z}$ in the twistor space $\mathbb{PN}$ (projective null twistor space); a single point R in $\mathbb{M}$ is represented by a Riemann sphere $\mathbf{R}$ in $\mathbb{PN}$ (this sphere representing the 'celestial sphere' of light rays at R). (For the complete correspondence this requires the compactified Minkowski space $\mathbb{M}^{\#}$ described in Fig. 33.9.)

The *twistor space* that is referred to here, whose individual points represent light rays in $\mathbb{M}$, is denoted[16] by $\mathbb{PN}$. (This notation is taken to fit in with the terminology of §33.5.) Thus, the point $\mathbf{Z}$ in $\mathbb{PN}$ corresponds to the locus Z (a light ray) inside $\mathbb{M}$ and the point R in $\mathbb{M}$ corresponds to the locus $\mathbf{R}$ (a Riemann sphere; see §18.5) inside $\mathbb{PN}$. Now, an essential part of the philosophy of twistor theory is that ordinary physical notions, which normally are described in spacetime terms, are to be translated into an equivalent (but non-locally related) description in twistor space. We see that the relationship between $\mathbb{M}$ and $\mathbb{PN}$ is indeed a non-local correspondence, rather than a point-to-point transformation. However, the space $\mathbb{PN}$ provides us merely with the beginnings of such a translation. The full richness of twistor geometry—which turns out to be quite remarkable—is revealed only gradually, as the correspondence between spacetime concepts and twistor-space geometry is developed in further detail.

This locus $\mathbf{R}$ inside $\mathbb{PN}$, describes the 'celestial sphere' (total field of vision) of an observer at R, the celestial sphere of R being regarded as the family of light rays through R. As has been noted above, this sphere is naturally a *Riemann sphere* which is a *complex 1-dimensional space* (a complex curve; see Chapter 8). Thus, we think of spacetime points as *holomorphic objects* in the twistor space $\mathbb{PN}$, in accordance with the complex-number philosophy underlying twistor theory. We shall be seeing explicitly in §§33.5,6 how this 'holomorphic philosophy' can be extended to the geometry of a more complete twistor space $\mathbb{T}$, and in §§33.8–12 how it enables us to encode, in a remarkable way, the information of linear and non-linear massless fields.

The space $\mathbb{PN}$, of light rays, does not itself immediately fit in with the 'holomorphic philosophy', however, because it is not a complex space. $\mathbb{PN}$ cannot be a complex manifold because it has *five* real dimensions[33.1] and five is an odd number, whereas any complex n-manifold must have an *even* number, $2n$, of real dimensions (see §12.9). We shall be seeing shortly (in §33.6) that if we make our 'light rays' a little more like physical massless particles, by assigning them both spin (actually helicity—see §22.7) and energy, then we get a *six*-dimensional space $\mathbb{PT}$, which actually *can* be interpreted as a complex space—of three complex dimensions. The space $\mathbb{PN}$ sits inside $\mathbb{PT}$, dividing it into two complex-manifold pieces $\mathbb{PT}^+$ and $\mathbb{PT}^-$, where $\mathbb{PT}^+$ may be thought of as representing massless particles of positive helicity and $\mathbb{PT}^-$, massless particles of negative helicity; see Fig. 33.6. However, it would not be correct to think of twistors *as* massless particles. Instead, twistors provide the variables in terms of which massless particles are to be expressed. (This is comparable with the

[33.1] Why do light rays have five degrees of freedom?

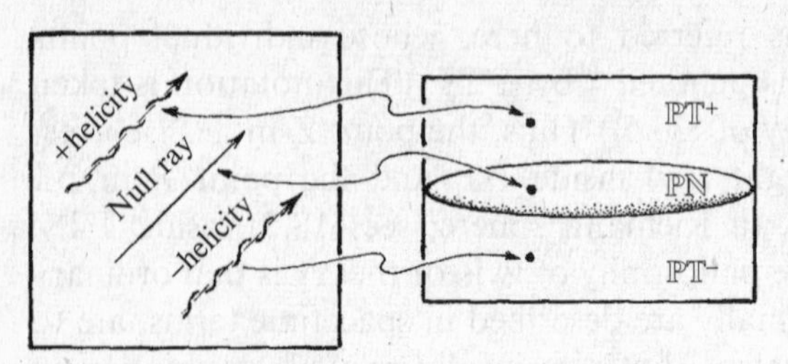

Fig. 33.6 The real 5-manifold $\mathbb{PN}$ divides projective twistor space $\mathbb{PT}$ into two complex-3-manifold pieces $\mathbb{PT}^+$ and $\mathbb{PT}^-$, these representing massless particles of positive and of negative helicity, respectively.

ordinary use of a position 3-vector **x** to label a point in space. Although a particle might occupy the point labelled by **x**, it would not be correct to identify the particle with the vector **x**.)

The twistor perspective leads us to a very different view of 'quantized spacetime' from that which is often put forward. It is quite a common 'conventional' viewpoint that the procedures of quantum (field) theory are to be applied to the metric tensor g_{ab}, this being thought of as a tensor field on the spacetime (manifold). The view is expressed that the *quantized metric* will exhibit aspects of 'fuzziness' owing to the Heisenberg uncertainty principle. One is presented with the image of some kind of four-dimensional space which possesses a 'fuzzy metric' so that, in particular, the null cones—and consequently the notion of *causality*—become subject to 'quantum uncertainties' (see Fig. 33.7a). Accordingly, there is no classically well-defined notion of whether a spacetime vector is spacelike, timelike, or null. This issue had posed foundational difficulties for any too-conventional 'quantum theory of gravity', for it is a basic feature of QFT that causality requires field operators defined at spacelike-separated events to commute. If the very notion of 'spacelike' is subject to quantum uncertainties (or has, itself, become a quantum notion), then the standard procedures of QFT—which involve the specification of commutation relations for field operators (§§26.2,3)—cannot be directly applied. Twistor theory suggests a very different picture. For now the appropriate 'quantization' procedures, whatever they may be, must be applied within twistor space rather than within the spacetime (where the latter would have been the 'conventional' viewpoint). By analogy with the way that, in the conventional approach, 'events' are left intact whereas 'null cones' become fuzzy, in a twistor-based approach it is now the 'light rays' that are left intact whereas 'events' become fuzzy (see Fig. 33.7b).

Twistor theory, as we have just seen, initially exploits a manifestation of complex number magic different from those to be found in quantum theory, namely the *classical* feature of spacetime geometry that the celestial sphere can be regarded as a Riemann sphere, which is a 1-dimensional complex manifold. The idea is that this provides us with hints as to Nature's *actual* scheme of things, which must ultimately unify spacetime

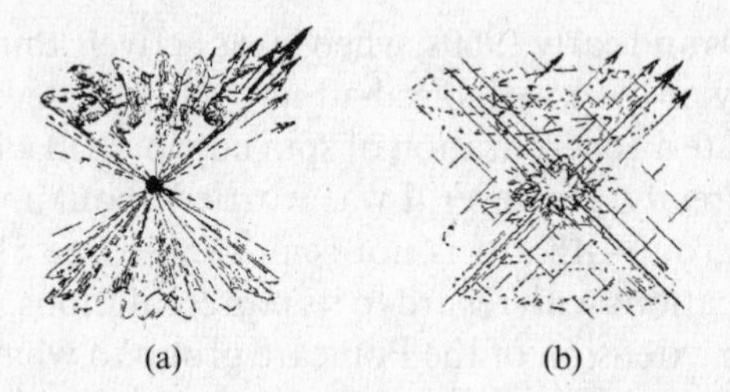

Fig. 33.7 (a) It has been a common viewpoint, with regard to the possible nature of a 'quantized spacetime', that it should be some kind of a spacetime with a 'fuzzy' metric, leading to some sort of 'fuzzy' light cone, where the notion of a direction at a point being null, timelike, or spacelike would be subject to quantum uncertainties. (b) A more 'twistorial' perspective would be to take the twistor space (in this case $\mathbb{PN}$) to retain some kind of existence (so there would still be light rays), but the condition of their intersection would become subject to quantum uncertainties. Accordingly the notion of 'spacetime point' would instead become 'fuzzy'.

structure with the procedures of quantum mechanics. It is noteworthy that this feature of spacetime geometry is specific to the particular dimension and signature actually possessed by the physical spacetime we are aware of. Indeed, the fact that the Riemann sphere plays an important role as the celestial sphere in relativity theory (§18.5) requires spacetime to be 4-dimensional and Lorentzian, in stark contrast with the underlying ideas of string theory and other Kaluza–Klein-type schemes. The full complex magic of twistor theory proper is very specific to the 4-dimensional spacetime geometry of ordinary (special) relativity theory, and does not have the same close relationship to the 'spacetime geometry' of higher dimensions (see §33.4, later).

To proceed further, let us return to the original pure spin-network picture, noting that the main thing that was missing from it was any reference to spatial displacement. In that theory, Euclidean angles arise as a kind of 'geometric limit' of pure spin-network theory; yet distances do not arise in that theory. In the loop-variable scheme, the 'distance' aspect of things is addressed by the numbers ($n = 2j$) on the lines referring to *area* rather than spin. But this is different from the interpretation in the original spin-network theory, where there is no measure of distance because spin is *angular* momentum, having to do merely with rotations and angles. We would need a corresponding role for *linear* momentum in that theory in order to be able to incorporate translational displacements and actual distances. Accordingly, it would appear that we need to move from the rotation group to the full group of Euclidean motions and, for a properly relativistic scheme, to the Poincaré group (§18.2).[17]

In the late 1950s and early 1960s, when I was actively thinking about these things, the theory of loop variables had not yet been developed, and I did indeed contemplate a generalization of spin networks in which the Poincaré group features directly. However, I was worried about an awkward aspect of the Poincaré group—that it is not semi-simple (see §13.7)—which has unpleasant implications with regard to its representations. I had the view, at that time, that an extension of the Poincaré group to what is known as the *conformal* group (which *is* semi-simple) might make the relevant analogue of the spin-network theory into a more mathematically satisfactory structure. The conformal group extends the Poincaré group by demanding merely that the light cones be preserved, rather than the Minkowski-space metric. Indeed, it turns out that the conformal group has an important place in twistor theory, as it is also the symmetry group of the space $\mathbb{PN}$ of (idealized) light rays. (The non-reflective part of the conformal group is also the symmetry group of each of the spaces $\mathbb{PT}^+$ and $\mathbb{PT}^-$, referred to above, describing massless particles with helicity and energy.) We shall see more explicitly what the role of this group is in the next two sections.

33.3 Conformal group; compactified Minkowski space

I have referred to the conformal group of spacetime above. Let us try to explore the role of this group a little more fully. It has a particular importance in physics in relation to massless fields (e.g. the Maxwell field), as it turns out that the field equations for massless fields are invariant under this larger group, not merely under the Poincaré group.[18] One may take the position that at the fundamental level, massless particles/fields are the basic ingredients, mass being something that comes in at a later stage. Indeed, this seems to be the standpoint implicit in the standard model, as described in Chapter 25, whereby mass is introduced via the Higgs boson, and is taken to come about only through a symmetry-breaking mechanism (§25.5). Be that as it may, one of the important underlying motivations behind twistor theory was indeed a belief in the basic importance of massless fields and the conformal group. We shall find (§33.8) that massless particles and fields have a remarkably concise description in twistor theory, and this fact forms one of the basic cornerstones of that theory.

What exactly is the conformal group? Strictly speaking, this group acts not on Minkowski space $\mathbb{M}$, but on a slight extension of $\mathbb{M}$ known as *compactified* Minkowski space $\mathbb{M}^\#$. The space $\mathbb{M}^\#$ is a beautifully symmetrical closed manifold, which, in many respects has a more elegant geometry than Minkowski space itself. We must not think of it as 'actual spacetime', however, but as a mathematical convenience. It is a useful intermediary to the understanding of twistor geometry and its relation to physical spacetime geometry.

A good picture to have in the back of the mind is the Riemann sphere and its relation to the complex plane. We recall, from §8.3, that the Riemann sphere is obtained from the complex plane by adjoining to it an 'infinite element', namely the point labelled ∞, and when we have done so, we obtain a geometric structure with an even greater symmetry than the plane that we started with. In a similar way, the 'compactified Minkowski space' $\mathbb{M}^{\#}$ is obtained from ordinary Minkowski space $\mathbb{M}$ by adjoining an 'infinite element' which, this time, turns out to be a complete *light cone at infinity*. The resulting space has a greater symmetry (namely the conformal group) than Minkowski space itself.

Let us see how this works. The space $\mathbb{M}^{\#}$ turns out to be a 4-dimensional real compact manifold with a Lorentzian conformal metric. Recall from §27.12 that a Lorentzian conformal metric is, in effect, just the family of null cones specified on the space. This structure is more usually phrased in terms of an equivalence class of metrics, where a metric $\boldsymbol{g}$ is considered to be equivalent to a metric $\boldsymbol{g}'$ if $\boldsymbol{g}' = \Omega^2\boldsymbol{g}$ for some smooth scalar field Ω that is everywhere positive. This rescaling indeed preserves the null cones (Fig. 33.8). Now, to pass from $\mathbb{M}$, (regarded as a conformal manifold) to the compact conformal manifold $\mathbb{M}^{\#}$, we adjoin the 3-surface $\mathscr{I}$ referred to above as the 'light cone at infinity'. Recall from §27.12 the 3-surfaces $\mathscr{I}^-$ and $\mathscr{I}^+$ (called 'scri-minus' and 'scri-plus', respectively) that represent the past and future null infinities of Minkowski space (see Fig. 27.16b). We can construct $\mathbb{M}^{\#}$ by *identifying* $\mathscr{I}^-$ with $\mathscr{I}^+$ in the way indicated in Fig. 33.9. A point of $\mathscr{I}^-$ is considered to be the same point as a corresponding point of $\mathscr{I}^+$ which is spatially antipodal to it (on the 2-sphere that most of the points on the diagram represent). The light cone of a point a^- on $\mathscr{I}^-$ comes to a focus again at the point a^+ of $\mathscr{I}^+$, and it is the points a^- and a^+ that are to be identified. In addition, all three points representing temporal and spatial infinities, i^-, i^0, i^+, are also identified as the single

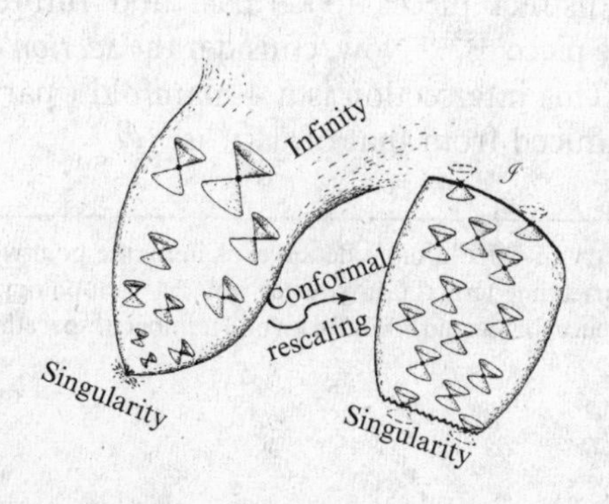

Fig. 33.8 The null-cone structure of a Lorentzian manifold $\mathcal{M}$ is equivalent to its conformal structure. A conformal rescaling of $\mathcal{M}$ affects its metric, but not its causality properties. (A metric $\boldsymbol{g}$ conformally rescales to $\boldsymbol{g}'$ if $\boldsymbol{g}' = \boldsymbol{\Omega}^2\boldsymbol{g}$, the scalar field $\boldsymbol{\Omega}$ being everywhere positive.) In favourable circumstances, such rescalings can be useful for 'bringing into view' singularities and infinite regions.

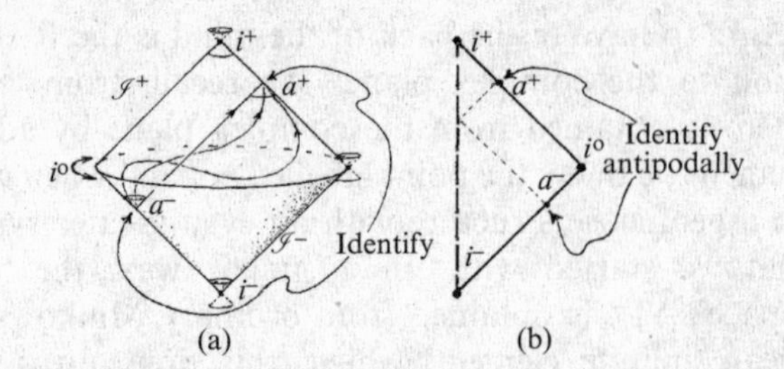

Fig. 33.9 Compactified Minkowski space $\mathbb{M}^\#$ is obtained from ordinary Minkowski space $\mathbb{M}$ by adjoining its future and past null infinities $\mathscr{I}^+$ and $\mathscr{I}^-$ and then identifying them appropriately as $\mathscr{I}$. (a) The future light cone of any point a^- on $\mathscr{I}^-$ focuses back to another vertex a^+ on $\mathscr{I}^+$ (this 'light cone' being, in ordinary terms, simply the history of a plane wavefront travelling at the speed of light), and a^- is to be identified with a^+. Spacelike infinity i^o and the past and future timelike infinities i^- and i^+ are all to be identified as a single point i. (b) This identification $\mathscr{I}$ is shown, in terms of the strict conformal diagram of Fig. 27.16b, where a^- is antipodal to a^+ on the 'S^2 of rotation' for the whole diagram.

point i.[33.2] The conformal manifold $\mathbb{M}^\#$ indeed has more symmetry than Minkowski space, having a 15-dimensional symmetry group—the *conformal group*—rather than merely the 10-dimensional Poincaré group.

There is an elegant way of describing the space $\mathbb{M}^\#$ and its group of transformations. Consider the 'light cone' $\mathcal{K}$ of the origin O in a pseudo-Euclidean 6-space $\mathbb{E}^{2,4}$, with signature $++----$. Choose standard coordinates w, t, x, y, z, v for $\mathbb{E}^{2,4}$, so that $\mathcal{K}$ is given by the equation

$$w^2 + t^2 - x^2 - y^2 - z^2 - v^2 = 0,$$

the metric $\mathrm{d}s^2$ of $\mathbb{E}^{2,4}$ being

$$\mathrm{d}s^2 = \mathrm{d}w^2 + \mathrm{d}t^2 - \mathrm{d}x^2 - \mathrm{d}y^2 - \mathrm{d}z^2 - \mathrm{d}v^2.$$

This is a 5-dimensional 'cone', with vertex O. I have done my best to depict it in Fig. 33.10, but one of the main ways that the picture is misleading is the fact that what seem to be two distinct 'pieces' to $\mathcal{K}$ ('past' and 'future') are actually connected up into 'one piece'.[33.3] Now, consider the *section* of $\mathcal{K}$ by the null 5-plane $w - v = 1$. This intersection is a 4-manifold ('paraboloid') whose intrinsic metric, induced from that of $\mathbb{E}^{2,4}$ is[33.4]

[33.2] See if you can describe the geometry of $\mathbb{M}^\#$ in more detail, explaining the pointwise identification of $\mathscr{I}^+$ with $\mathscr{I}^-$ in ordinary spacetime terms? Can you see why $\mathbb{M}^\#$'s topology is $S^1 \times S^3$? Can you think of an important difference that would occur for an odd number of spacetime dimensions?

[33.3] Can you see why?

[33.4] Why?

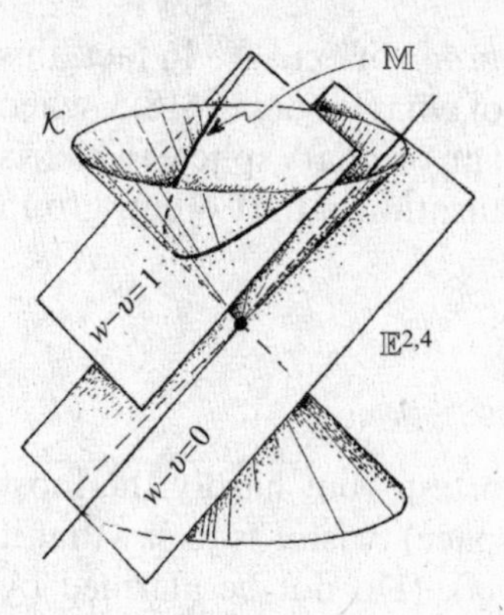

Fig. 33.10 Compactified Minkowski space $\mathbb{M}^\#$ may be identified as the space of generators of the 'light cone' $\mathcal{K}$, in pseudo-Euclidean $\mathbb{E}^{2,4}$, given by $w^2 + t^2 - x^2 - y^2 - z^2 - v^2 = 0$. The 'paraboloidal' 4-manifold section $\mathbb{M}$ of $\mathcal{K}$ by the null 5-plane $w - v = 1$, has Minkowskian intrinsic metric $ds^2 = dt^2 - dx^2 - dy^2 - dz^2$. The family of generators of $\mathcal{K}$ in $w - v = 0$ (not visible in this diagram owing to the depiction of only one 'time' dimension) are parallel to $w - v = 1$ and do not meet $\mathbb{M}$, these generators providing the points of $\mathscr{I}$.

$$ds^2 = dt^2 - dx^2 - dy^2 - dz^2.$$

We recognize this as the metric form of ordinary flat Minkowski 4-space (§18.1), so we can identify it with $\mathbb{M}$, even though it is embedded in a 'bent' way in $\mathbb{E}^{2,4}$ (with the appearance of a parabola in Fig. 33.10). How do we find $\mathbb{M}^\#$ in this picture? It is the abstract space of complete *generators* of $\mathcal{K}$ (straight lines through O that lie on $\mathcal{K}$, where the complete line through O in both directions counts as a single generator). Thus, we can think of each point of $\mathbb{M}^\#$ simply as a generator of $\mathcal{K}$ (Fig. 33.10)—so $\mathbb{M}^\#$ is the 'celestial sphere' for some 'observer' situated at the origin of $\mathbb{E}^{2,4}$!

Why does this work? Each generator that does not lie in the 5-plane $w - v = 0$ meets $\mathbb{M}$ in a unique point, so this family of generators is in a continuous 1–1 correspondence with $\mathbb{M}$. But, in addition, there are the generators that *do* lie in this 5-plane. These supply $\mathbb{M}$ with the additional points that constitute $\mathscr{I}$. The space $\mathbb{M}^\#$, defined in this way, has a conformal Lorentzian metric which is locally provided by that of any local cross-section of $\mathcal{K}$.[33.5]

The pseudo-orthogonal group O(2, 4), acting on $\mathbb{E}^{2,4}$ (see §13.8, §§18.1,2) consists of the 'rotations' that preserve the metric ds^2. This sends generators of $\mathcal{K}$ to other generators of $\mathcal{K}$, so it sends $\mathbb{M}^\#$ to itself. Moreover, it preserves $\mathbb{M}^\#$'s conformal structure.[33.6] There are exactly two elements of O(2, 4) that act as the identity on $\mathbb{M}^\#$, namely the identity element of O(2, 4) itself and the negative identity element of O(2, 4), the latter simply reversing the direction of each generator. Apart from the two-to-one nature of the correspondence arising from this reversibility of the generator directions, O(2, 4) is the conformal group. It includes a 10-dimensional subgroup preserving the

[33.5] Why is the conformal metric provided by any one local cross-section the same as that of any other? Why do the points of $\mathscr{I}$, defined in this way, agree with the definition given above? *Hint*: See §18.4 and Fig. 18.9.

[33.6] Why? You may assume the result of Exercise [33.5].

5-plane $w - v = 0$, and this gives the *Poincaré group* of $\mathbb{M}$.[33.7] In fact, this argument is just a higher-dimensional version of what we did in §18.5, when showing that the conformal transformations of an ordinary sphere (which is the compactified Euclidean plane) provide a realization of the Lorentz group O(1, 3); see E in Fig. 18.9.

33.4 Twistors as higher-dimensional spinors

How do twistors fit in with all this? The shortest—but hardly the most transparent—way to describe a (Minkowski-space) twistor is to say that it is a *reduced spinor* (or half spinor) for O(2, 4). (Do not be alarmed by the mathematical laconism of this description; I shall be giving a much more physical picture shortly!) See §11.5 for a brief mention of the notion of a reduced spinor. For a $2n$-dimensional space, on which a pseudo-orthogonal group $O(n - r, n + r)$ acts, the space of reduced spinors is 2^{n-1}-dimensional. In the present case, $n = 3$ (and $r = 1$), so we have a 4-dimensional space of reduced spinors, referred to as *twistor space*.[19]

Unfortunately, with a definition like this, we do not get a clear geometrical or physical picture of what a twistor is like. Moreover, we see that a twistor theory should exist for any even number $2(n - 1)$ of spacetime dimensions, despite what was said towards the end of §33.2. We generalize the above construction for $\mathcal{K}$ (taking it now to be of $2n - 1$ dimensions) and the compactification of the $2(n - 1)$-dimensional Minkowski spacetime works in a way analogous to that given above, where we simply introduce two new coordinates v and w as before, one with a minus sign in the metric and the other a plus sign. The 'twistor space' is now 2^{n-1}-dimensional. For an odd number $2n - 1$ of spacetime dimensions this will also work, except that we do not now have the notion of reduced spinors, and it is the entire 2^n-dimensional spin space that would have to count as our 'twistors'. An important feature of twistors is lost in the odd-dimensional case, however, namely their chiral nature (which we shall discuss more fully in §§33.7,12,14). Only when we pass to the reduced spin spaces do we achieve an essentially chiral formalism (so that left-handed and right-handed entities receive different twistorial descriptions; see §33.7), and hope may be thereby entertained that the chiral aspects of weak interactions (§25.3) may ultimately be incorporated. We shall also be seeing later why this general n-dimensional definition of a twistor misses many of the key physical (and holomorphic) properties that make twistor theory so effective.

[33.7] What is the explicit condition on a 6×6 matrix, that it represents an infinitesimal element of O(2, 4)? Which of these matrices give infinitesimal Poincaré transformations?

Since twistors refer to an active group of spacetime transformations (the conformal group) where spacetime points get sent to other spacetime points under the action of the group, twistors are seen to be entities that refer globally to the spacetime, rather than to individual points in the spacetime. Local quantities such as vectors, tensors, or ordinary spinors refer to the symmetry group that acts at a point; see §14.1—e.g. the rotation or Lorentz group (§13.8). Although this makes twistors more difficult to handle than ordinary vectors, tensors, or spinors, this globality has an advantage when we are seeking a formalism intended to replace the spacetime rather than merely to be defined in reference to a previously given spacetime manifold. As mentioned in §33.2, one of the main aims of twistor theory is indeed to obtain such a formalism. The main disadvantage of such an approach is that it is difficult to see how such a formalism can be applied to a general curved spacetime $\mathcal{M}$, when such things as the conformal group do not appear as a symmetry of $\mathcal{M}$. We shall be seeing how twistor theory comes to terms with this kind of difficulty in a remarkable way, in §§33.11,12.

The above definition of a twistor, as a reduced spinor for O(2, 4), gives us only a very limited perspective on twistor-theoretic ideas and motivation. As just stated, there is nothing specifically 4-dimensional about this approach, and it gives no clear indication of why one should be interested in twistor theory as providing us any guidance in moving forwards in our search for a deeper theory of Nature. To appreciate more fully what twistor theory is trying to do in this respect, let us recall the message of Chapters 29 and 30. Whereas it is commonly accepted that the appropriate quantum–gravity union must be a major goal in the search for a fundamentally new perspective on physics, the message of those chapters is that we should seek a development in which the very rules of quantum (field) theory are not held sacrosanct but should be bent, just as should the geometry of our conventional spacetime pictures. Nevertheless, there is clearly much truth as well as beauty in quantum-mechanical principles, and these should not simply be abandoned. In twistor theory, instead of imposing QFT rules, one looks into these rules and tries to extract features that mesh with those of Einstein's conceptions, seeking hidden harmonies between relativity and quantum mechanics. As has been stated earlier, one key element of guidance is the complex-number magic that has featured in so many places in this book. Another is an especial harmony with Einstein's theory of Lorentzian 4-space rather than with its generalizations to higher dimensions or to other signatures.

What is so special about Lorentzian 4-space, in this respect? As has been stressed in §18.5 and §33.2, the celestial sphere of an observer has a natural conformal structure and can be interpreted as a Riemann sphere. It should be borne in mind that something of this general nature actually occurs in *any* (non-zero) number of space and time dimensions, where the celestial sphere always has the structure of a conformal manifold.[33.8] What is particular about the Lorentzian 4-dimensional case, however, is that this conformal manifold can be naturally interpreted as a *complex* manifold (the Riemann sphere), a property that does not arise in any other number of space and time dimensions. What is the importance of this fact? In twistor theory, the magic of complex numbers is exploited to the full. Not only does twistor space turn out to be a complex manifold, but this complex manifold has a direct physical interpretation. In fact, general results tell us that the only cases where the 'twistor space' is a complex space of any kind[20] occur when the difference between the number of space dimensions and the number of time dimensions leaves the remainder 2 when divided by 4. It is noteworthy that this is not the case for the original Kaluza–Klein theory, nor for 10- or 11-dimensional supergravity theories, nor for the original 26-dimensional string theory, nor for 10-dimensional superstring theory, nor for 11-dimensional supergravity or M-theory, nor even for 12-dimensional F-theory (since in that case there are 2 time dimensions)!

33.5 Basic twistor geometry and coordinates

What is the physical or geometrical interpretation of a general twistor for ordinary Minkowskian 4-space? It is easiest to describe things if we use standard Minkowski coordinates t, x, y, z for a point R of $\mathbb{M}$, where we take the speed of light as unity: $c = 1$. The full twistor space $\mathbb{T}$ for $\mathbb{M}$ is a *4-dimensional complex vector space*, for which standard complex coordinates Z^0, Z^1, Z^2, Z^3 may be used. We say that a twistor $\mathbf{Z}$, with these coordinates, is *incident* with the spacetime point R—or, that R is incident with $\mathbf{Z}$—if the key matrix relation (see §13.3 for matrix notation)

$$\begin{pmatrix} Z^0 \\ Z^1 \end{pmatrix} = \frac{\mathrm{i}}{\sqrt{2}} \begin{pmatrix} t+z & x+\mathrm{i}y \\ x-\mathrm{i}y & t-z \end{pmatrix} \begin{pmatrix} Z^2 \\ Z^3 \end{pmatrix}$$

holds—from which the basics of flat-space twistor geometry all follow![33.9]

In accordance with the notation of §12.8, the (abstract-)index notation Z^α will sometimes be used to represent the twistor $\mathbf{Z}$ (where the components of $\mathbf{Z}$ in a standard frame would be Z^0, Z^1, Z^2, Z^3). Each twistor $\mathbf{Z}$, or Z^α, (an element of $\mathbb{T}$) has a complex conjugate $\bar{\mathbf{Z}}$, which is a *dual* twistor

---

[33.8] Explain why.

[33.9] Write this equation out in ordinary algebraic notation.

(element of the dual twistor space $\mathbb{T}^*$). In index form, $\bar{\mathbf{Z}}$ is written $\bar{Z}_\alpha$, with a *lower* index, and its components (in the standard frame) would be

$$(\bar{Z}_0, \bar{Z}_1, \bar{Z}_2, \bar{Z}_3) = (\overline{Z^2}, \overline{Z^3}, \overline{Z^0}, \overline{Z^1}).$$

This notation is probably a little confusing. The four quantities (complex numbers) on the left are simply the four components of the dual twistor $\bar{\mathbf{Z}}$. The four on the right are the respective complex conjugates of the complex numbers Z^2, Z^3, Z^0, Z^1. Thus, the component $\bar{Z}_0$ of $\bar{\mathbf{Z}}$ is the complex conjugate of the component Z^2 of $\mathbf{Z}$, etc. Note the interchange of the first two with the second two when forming the complex conjugation. Since $\bar{\mathbf{Z}}$ is a dual twistor, we can form its (Hermitian) scalar product (see §13.9 and §22.3) with the original twistor $\mathbf{Z}$ to obtain the (squared) *twistor norm*

$$\begin{aligned}\bar{\mathbf{Z}}\bullet\mathbf{Z} = \bar{Z}_\alpha Z^\alpha &= \bar{Z}_0 Z^0 + \bar{Z}_1 Z^1 + \bar{Z}_2 Z^2 + \bar{Z}_3 Z^3 \\ &= \overline{Z^2} Z^0 + \overline{Z^3} Z^1 + \overline{Z^0} Z^2 + \overline{Z^1} Z^3 \\ &= \tfrac{1}{2}(|Z^0 + Z^2|^2 + |Z^1 + Z^3|^2 - |Z^0 - Z^2|^2 - |Z^1 - Z^3|^2),\end{aligned}$$

where this last formula shows that the Hermitian expression $\bar{Z}_\alpha Z^\alpha$ has signature $(++--)$, in accordance with §13.9.[33.10] (The symmetry of twistor space exhibits the local equivalence, mentioned in §13.10, of the group SU(2, 2) to the O(2, 4) of §33.3.). We find, from the key incidence relation given above, that a twistor Z^α can be incident with an event in real Minkowski space $\mathbb{M}$ only if its norm *vanishes*: $\bar{Z}_\alpha Z^\alpha = 0$.[33.11] When $\bar{Z}_\alpha Z^\alpha = 0$, we say that the twistor $\mathbf{Z}$ is *null*.

To connect with the discussion of §33.2, we should first make our acquaintance with the *projective* twistor space $\mathbb{PT}$, which is the complex projective 3-space ($\mathbb{CP}^3$) constructed from the complex vector space $\mathbb{T}$. (See §15.6 for a general discusion of projective spaces.) Much of twistor geometry is most easily expressed in terms of $\mathbb{PT}$, rather than $\mathbb{T}$. The numbers, Z^0, Z^1, Z^2, Z^3 now provide *homogeneous coordinates* for $\mathbb{PT}$, so that the three independent ratios

$$Z^0 : Z^1 : Z^2 : Z^3.$$

serve to label points of $\mathbb{PT}$. The *null* projective twistors constitute the space $\mathbb{PN}$, which is the 5-real-dimensional subspace of the 6-real-dimensional space $\mathbb{PT}$ for which the twistor norm vanishes:

$$\bar{Z}_\alpha Z^\alpha = 0.$$

[33.10] Verify this final expression; explain why this tells us the signature.

[33.11] Show this; show, conversely, that such an event always exists *if* $\bar{Z}_\alpha Z^\alpha = 0$, provided that Z^2 and Z^3 do not both vanish.

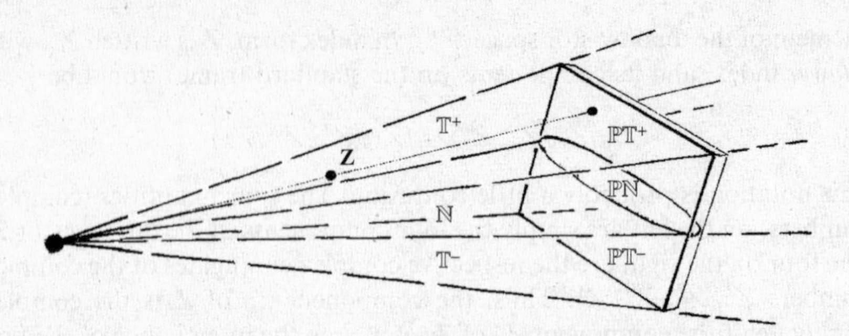

Fig. 33.11 Twistor space $\mathbb{T}$ is a complex vector space with a pseudo-Hermitian metric. The projective twistor space $\mathbb{PT}$ (a $\mathbb{CP}^3$) is the space of rays (1-dimensional subspaces) in $\mathbb{T}$. Thus, if a twistor $\mathbf{Z}$ has coordinates (Z^0, Z^1, Z^2, Z^3), the ratios $Z^0 : Z^1 : Z^2 : Z^3$ determine the corresponding point in $\mathbb{PT}$. The 7-real-dimensional subspace $\mathbb{N}$ (of null twistors: $\bar{Z}_\alpha Z^\alpha = 0$) divides twistor space $\mathbb{T}$ into the complex 4-spaces $\mathbb{T}^+$ (of positive twistors: $\bar{Z}_\alpha Z^\alpha > 0$) and $\mathbb{T}^-$ (of negative twistors: $\bar{Z}_\alpha Z^\alpha < 0$). The respective projective versions of these spaces are the 5-real-dimensional $\mathbb{PN}$ (representing light rays in $\mathbb{M}^\#$) and the two complex 3-manifolds $\mathbb{PT}^+$ (representing positive helicity massless particles) and $\mathbb{PT}^-$ (representing negative helicity massless particles).

This equation also defines the 7-real-dimensional subspace $\mathbb{N}$ of the null non-projective twistors, in the vector space $\mathbb{T}$. When $\bar{Z}_\alpha Z^\alpha > 0$, we get the space $\mathbb{T}^+$ of *positive* twistors; and when $Z_\alpha Z^\alpha < 0$, we get the space $\mathbb{T}^-$ of *negative* twistors. The projective spaces $\mathbb{PT}^+$ and $\mathbb{PT}^-$ are defined correspondingly. See Fig. 33.11 (and compare Fig. 33.6).

Let us explore the geometrical relation between $\mathbb{PN}$ and $\mathbb{M}$, depicted in Fig. 33.5, as a consequence of the key incidence relation given at the beginning of this section. It can be seen directly from this relation that two points P, R of $\mathbb{M}$ (events) that are incident with the same non-zero twistor $\mathbf{Z}$ (necessarily a null twistor) must be *null-separated* from each other (i.e. each of P and R lies on the light cone of the other). It follows that $\mathbf{Z}$ defines a *light ray*—null straight line in $\mathbb{M}$—since all the points of $\mathbb{M}$ that are incident with $\mathbf{Z}$ must be mutually null separated. See Fig. 33.12. Moreover, the twistor $\mathbf{Z}$ represents the same light ray if we replace Z^α by λZ^α, where λ is any non-zero complex number. The locus of events incident with a (non-zero) null projective twistor is indeed a light ray; but in the particular situation when $Z^2 = Z^3 = 0$, we must interpret this appropriately, for we get no actual points in $\mathbb{M}$ incident with Z^α, yet we can still regard such a null twistor as describing a *light ray at infinity* (a generator of $\mathscr{I}$, lying in $\mathbb{M}^\#$, rather than $\mathbb{M}$).[33.12]

[33.12] Demonstrate the assertions of this paragraph, explicitly.

Let us now examine the converse. Fixing the event R, with real coordinates t, x, y, z, we find that the space of twistors $\mathbf{Z}$ incident with R is defined by two linear homogeneous relations in the components Z^0, Z^1, Z^2, Z^3. Each of these linear relations defines a *plane* in $\mathbb{PT}$ and their *intersection* (the set of points in $\mathbb{PT}$ satisfying both relations) gives us a projective line $\mathbf{R}$ in $\mathbb{PT}$ (a $\mathbb{CP}^1$)—lying in $\mathbb{PN}$, in fact—which is therefore a Riemann sphere, as is required (§§15.4,6). Thus, points of $\mathbb{M}$ (events) are represented, in twistor space, by projective lines in $\mathbb{PN}$. When $Z^2 = 0 = Z^3$, we get a particular projective line in $\mathbb{PN}$ which we refer to as $\mathbf{I}$. This special line represents the point i that is the vertex of the light cone $\mathscr{I}$ at infinity. Any other point Q of $\mathscr{I}$ is represented, in $\mathbb{PN}$, by a projective line $\mathbf{Q}$ meeting $\mathbf{I}$.[33.13] The situation is illustrated in Fig. 33.12.

The way that these complex structures represent the geometry of Minkowski space (in the standard number of space and time dimensions) is very remarkable. We can re-interpret Minkowski space as the space of

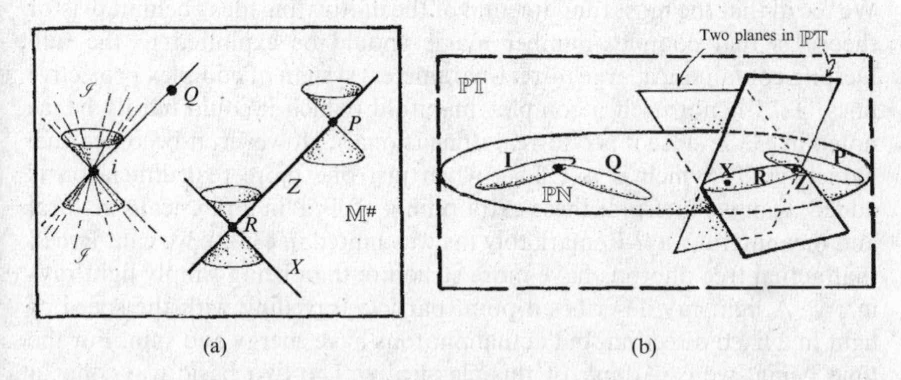

Fig. 33.12 The geometry of the basic loci in $\mathbb{M}^\#$ and $\mathbb{PN}$, given by the incidence relation of the twistor correspondence. (a) Fix a point (projective null twistor) $\mathbf{Z}$ in $\mathbb{PN}$. The points of $\mathbb{M}^\#$ (e.g. P, R) which are incident with $\mathbf{Z}$ constitute a light ray, since all such points are null separated from one another. (b) Fix a point R in $\mathbb{M}^\#$. The points of $\mathbb{PN}$ (e.g. $\mathbf{Z}, \mathbf{X}$) that are incident with R (lying on the intersection of two complex planes in $\mathbb{PT}$) constitute a complex projective line, which is a Riemann sphere. Points P and R in $\mathbb{M}^\#$ which are null separated along the light ray Z have corresponding Riemann spheres $\mathbf{P}$ and $\mathbf{R}$, which intersect in the single point $\mathbf{Z}$. (I have drawn these Riemann spheres very elongated, as a compromise with the fact that they are also projective straight lines in the projective geometry of $\mathbb{PT}$!) A particular one of these Riemann spheres is $\mathbf{I}$, which represents the point i in $\mathbb{M}^\#$. The point i specifies spacelike/timelike infinity; it is the vertex of the light cone $\mathscr{I}$ at infinity. Any other point Q of $\mathscr{I}$ is represented in $\mathbb{PN}$, by a projective line $\mathbf{Q}$ meeting $\mathbf{I}$.

[33.13] Why?

complex lines lying in $\mathbb{PN}$ (or in $\mathbb{PN}-\mathscr{I}$, if we want only the finite spacetime points), taking $\mathbb{PN}$ as the primary structure and $\mathbb{M}$ as secondary. This amounts to taking light rays as more primitive than the spacetime points themselves. *Intersection* of light rays Z and X is represented by the existence of a projective line on $\mathbb{PN}$ containing the corresponding points $\mathbf{Z}$ and $\mathbf{X}$ of $\mathbb{PN}$ and, as we have seen, the condition that two spacetime points P and R are *null-separated* is represented by the condition that the corresponding projective lines $\mathbf{P}$ and $\mathbf{R}$, in $\mathbb{PN}$, intersect (Fig. 33.12). Thus we see that twistor space provides a completely different perspective, on physical geometry, from the normal spacetime picture. Ordinary spacetime points are represented as Riemann spheres in $\mathbb{PN}$. Points of $\mathbb{PN}$ are represented as light rays in spacetime. Either way the correspondence is non-local. Yet, we can pass from one picture to the other by precise geometrical rules.

33.6 Geometry of twistors as spinning massless particles

We recall that the most fundamental of the motivating ideas behind twistor theory is that complex-number magic should be exploited to the full. Despite containing a large (4-real-parameter) system of complex projective lines, $\mathbb{PN}$ is not itself a complex manifold (which it could hardly be, as noted in §33.2, since it is odd-real-dimensional). However, it becomes one, namely $\mathbb{PT}$ (which is a $\mathbb{CP}^3$), when just one more real dimension is added. Can we interpret these extra points of $\mathbb{PT}$ in a physically natural and meaningful way? Remarkably (as was hinted in §33.2), we can. Recall that actual free photons have more structure than being simply light rays in $\mathbb{M}$. A light ray describes a point particle travelling with the speed of light in a fixed direction, but actual photons have energy and spin. For the time being, we can think of this classically. The two basic ways that a photon can spin are right-handed and left-handed about the direction of motion (positive and negative helicity, respectively, defined by right-handed and left-handed circular polarization; see §22.7). The magnitude of this helicity is just $\hbar$, in each case. It turns out that the positive-helicity classical photons can be represented as points of $\mathbb{PT}^+$, and the negative helicity ones, as points of $\mathbb{PT}^-$, where the extra dimension comes from the energy of the photon. This description holds also for any other massless particle with nonzero spin $\frac{1}{2}n\hbar$.

How does this work? This is not the place to enter into details, but the essential features can be outlined as follows. As a first step, it is helpful to realize that the first two components Z^0 and Z^1 of the twistor $\mathbf{Z}$ are really the two components of a 2-spinor $\boldsymbol{\omega}$, with index form ω^A, where $\omega^0 = Z^0$ and $\omega^1 = Z^1$ (see §22.8 and §25.2). The remaining two components Z^2 and Z^3 of $\mathbf{Z}$ are the components of a *primed* (dual) spinor $\boldsymbol{\pi}$, with index form $\pi_{A'}$, where $\pi_{0'} = Z^2$ and $\pi_{1'} = Z^3$. We sometimes write

$$\mathbf{Z} = (\boldsymbol{\omega}, \boldsymbol{\pi})$$

and refer to $\boldsymbol{\omega}$ and $\boldsymbol{\pi}$ as the *spinor parts* of the twistor $\mathbf{Z}$. The complex conjugate twistor $\bar{\mathbf{Z}}$ has its spinor parts in reverse order, i.e.

$$\bar{\mathbf{Z}} = (\bar{\boldsymbol{\pi}}, \bar{\boldsymbol{\omega}}),$$

so the twistor norm can be expressed

$$\bar{Z}_\alpha Z^\alpha = \bar{\mathbf{Z}} \bullet \mathbf{Z} = \bar{\boldsymbol{\pi}} \bullet \boldsymbol{\omega} + \bar{\boldsymbol{\omega}} \bullet \boldsymbol{\pi} = \bar{\pi}_A \omega^A + \bar{\omega}^{A'} \pi_{A'}.$$

The *incidence* relation between the twistor $\mathbf{Z}$ and the spacetime point R, with Minkowski coordinates t, x, y, z, is now written

$$\boldsymbol{\omega} = \mathrm{i}\boldsymbol{r}\boldsymbol{\pi},$$

which stands for $\omega^A = \mathrm{i}r^{AA'}\pi_{A'}$, where $\boldsymbol{r}$ (or $r^{AA'}$) has the matrix of components

$$\begin{pmatrix} r^{00'} & r^{01'} \\ r^{10'} & r^{11'} \end{pmatrix} = \frac{1}{\sqrt{2}} \begin{pmatrix} t+z & x+\mathrm{i}y \\ x-\mathrm{i}y & t-z \end{pmatrix}.$$

The spinor $\boldsymbol{\pi}$ is associated with the *momentum* of the massless particle, in the sense that the outer product $\bar{\boldsymbol{\pi}}\boldsymbol{\pi}$ (no contractions—see §14.3) describes its 4-momentum. The spinor $\boldsymbol{\omega}$ is associated with the particle's *angular momentum*, in the sense that the symmetrized product of $\boldsymbol{\omega}$ with $\bar{\boldsymbol{\pi}}$ describes the anti-self-dual part of the particle's 6-angular momentum (§18.7, §19.2, §22.12, §32.2) and the symmetrized product of $\bar{\boldsymbol{\omega}}$ with $\boldsymbol{\pi}$ describes its self-dual part.[21] Unlike the case of momentum, *angular* momentum depends upon a choice of spacetime origin O, and we sometimes refer to the angular momentum *about* O. This origin-independence/dependence is reflected in the translational behaviour of the two spinor parts $\boldsymbol{\pi}$ and $\boldsymbol{\omega}$ of a twistor $\mathbf{Z}$. Under displacement of the origin O to a new spacetime point Q, with position vector $\boldsymbol{q}$ relative to O, we find (with $\boldsymbol{q}$ in matrix form, as above) that the spinor parts undergo the transformations[33.14]

$$\boldsymbol{\pi} \mapsto \boldsymbol{\pi} \text{ and } \boldsymbol{\omega} \mapsto \boldsymbol{\omega} - \mathrm{i}\boldsymbol{q}\boldsymbol{\pi}.$$

There is also a scalar quantity that is origin-independent, which can be constructed from the momentum and angular momentum, namely the *helicity* s. It turns out that helicity is half the twistor norm:

$$s = \tfrac{1}{2}\bar{Z}_\alpha Z^\alpha = \tfrac{1}{2}\bar{\mathbf{Z}} \bullet \mathbf{Z}$$

[33.14] Show that the relation of incidence between a twistor and a spacetime point is preserved under this transformation; show that the twistor norm is preserved.

(and we note from the above that this is just the real part of $\bar{\boldsymbol{\omega}} \bullet \boldsymbol{\pi}$). In fact, twistors give a considerably more concise formalism for handling massless particles than the conventional 4-vector/tensor approach of §22.12. We now have a clear physical picture of a non-null twistor (up to the phase rescaling $\mathbf{Z} \mapsto e^{i\theta}\mathbf{Z}$, with θ real) as a classical spinning massless particle;[33.15] compare Fig. 33.6.

We still do not have a very clear *geometrical* picture of a non-null twistor. We can obtain this if we are prepared to consider *complexified* Minkowski space $\mathbb{CM}$ (or its compactification $\mathbb{CM}^{\#}$), where the spacetime coordinates $t. x, y, z$ are now taken to be complex numbers. For there is always a non-trivial 2-complex-dimensional locus of points of $\mathbb{CM}^{\#}$ incident with any (non-zero) twistor Z^{α}, called an *α-plane*, which is *self-dual* in the sense that a 2-form tangent to it is self-dual (§32.2). This α-plane represents Z^{α} up to proportionality; see Fig. 33.13. Similarly, a dual twistor W_{α} defines a *β-plane* which is an *anti*-self-dual complex 2-plane in $\mathbb{CM}^{\#}$.[33.16]

So far, this just gives us a *complex*-spacetime geometrical picture of a twistor. Can we obtain a 'real' picture that we can actually visualize? The reality structure of $\mathbb{CM}^{\#}$ is contained in its notion of *complex conjugation* (§18.1); this interchanges α-planes with β-planes, in accordance with the fact that it interchanges upper with lower twistor indices (i.e. twistors with

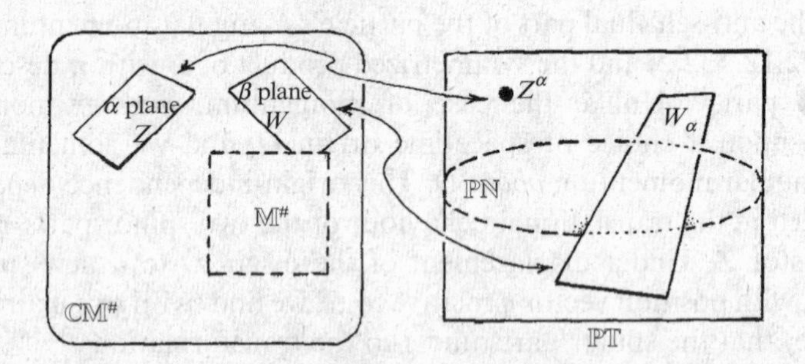

Fig. 33.13 Complex spacetime description of (generally non-null) twistors and dual twistors. For any non-zero twistor Z^{α}, there always is a 2-complex-dimensional locus in $\mathbb{CM}^{\#}$ of points incident with it, called an α plane, which is everywhere self-dual. For any non-zero dual twistor W_{α}, the points in $\mathbb{CM}^{\#}$ incident with it always constitute 2-complex-dimensional plane which is anti-self-dual, called a β plane. Only for null twistors or null dual twistors are there any real points on these loci, and then the real points constitute a light ray, in accordance with Fig. 33.12.

[33.15] Explain why there is this phase freedom and why, for a particle of given helicity $s > 0$, the particle's *energy* is encoded in the location of the point in $\mathbb{PT}^{+}$.

[33.16] Show this.

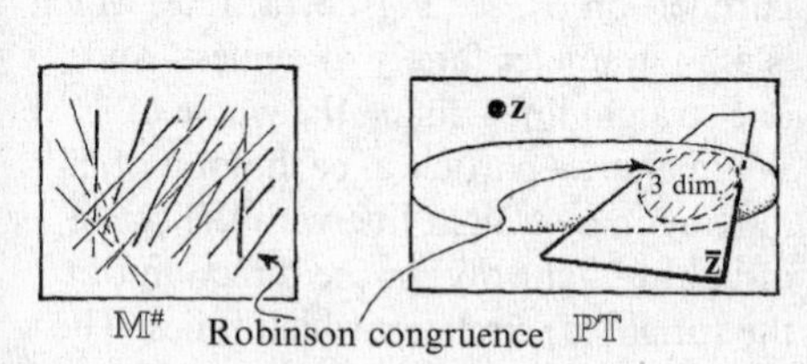

Fig. 33.14 We can obtain a 'real' picture of a non-null twistor Z^α by first passing to its complex conjugate $\bar{Z}_\alpha$, this defining a complex projective plane in $\mathbb{PT}$. This plane is fixed by its intersection with $\mathbb{PN}$, which is a 3-real-dimensional locus. This locus defines a 3-parameter family of light rays in $\mathbb{M}^\#$ called a Robinson congruence.

dual twistors) and it interchanges 'self-dual' with 'anti-self-dual'. In terms of the projective geometry of $\mathbb{PT}$, complex conjugation interchanges points with planes, since a dual twistor determines a plane in $\mathbb{PT}$.[33.17] This fact enables us to obtain a picture of a non-null projective twistor Z^α in terms of real spacetime geometry. What we do is first represent Z^α by its complex conjugate $\bar{Z}_\alpha$ which, being a dual twistor, is associated with a complex plane in $\mathbb{PT}$. This plane is fixed by its intersection with $\mathbb{PN}$, which is a real 3-dimensional locus. We can interpret this locus as providing a 3-parameter family of light rays in $\mathbb{M}$. Thus, this family of light rays geometrically represents the twistor Z^α (up to proportionality); see Fig. 33.14.

The light rays twist around in a complicated way, but it is possible to obtain a striking picture of the configuration. Consider one moment of time $\mathbb{E}^3$ (i.e. an ordinary Euclidean 3-space section—'now'—through Minkowski spacetime $\mathbb{M}$). Any light ray in $\mathbb{M}$—a point particle moving in a particular direction with the speed of light—is represented in $\mathbb{E}^3$ by a point with an 'arrow' attached to it, where the arrow determines the direction of motion. We are to picture a 3-parameter family of such light rays—called a *Robinson congruence*—to represent our single twistor **Z**. In Fig. 33.15, we see a system of oriented circles (and one straight line) filling the whole of ordinary 3-space $\mathbb{E}^3$. There will be one particle of our family at each point of $\mathbb{E}^3$, and it moves (with the speed of light) in the direction indicated by the oriented tangent to the circle that passes through that point. It is rather remarkable that, as time progresses, this entire configuration simply propagates as a whole with the speed of light in the (negative) direction of the one straight line in the picture, and this propagation represents the motion of the spinning massless particle described by the twistor. This configuration of circles is, in fact, the stereographic projection (§8.3, Fig. 8.7a), to ordinary Euclidean 3-space, of the configuration of Clifford parallels on S^3 (§15.4).

[33.17] Why?

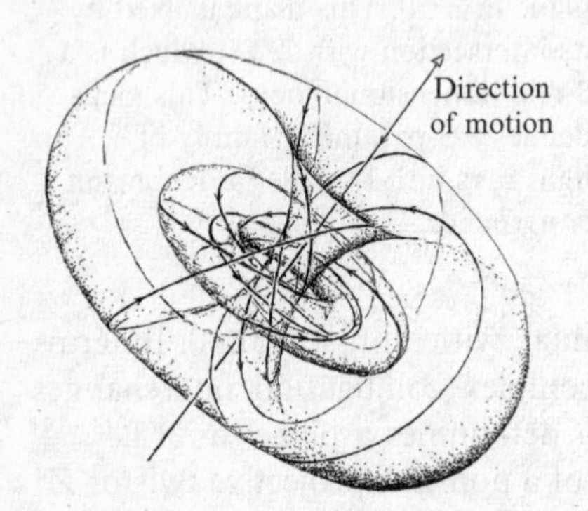

Fig. 33.15 A spatial picture giving a 'snapshot' of a Robinson congruence. (Stereographically projected Clifford parallels on S^3, see Figs. 8.7a, 15.8, which is a 3-parameter family of circles—and one straight line—filling the whole of $\mathbb{E}^3$.) We imagine a particle at each point of $\mathbb{E}^3$, moving in a straight line with the speed of light (a light ray) in the direction of the (oriented) circle on which it lies. The whole configuration propagates with the speed of light in the (negative) direction of the straight line in the figure. This represents the motion and angular momentum of the spinning massless particle described by the Z^α.

We are not to think of these 'light rays' as physical entities; they are only to provide us with a geometrical realization of a (projective) twistor. This configuration actually encodes the structure of the angular momentum of the (classical) spinning massless particle.[22] It is certainly a non-local picture. There is a smallest circle, in Fig. 33.15, whose radius is the spin divided by the energy of the particle. The centre of this circle, roughly speaking, represents the 'location' of the spinning particle (but the history of this centre cannot accurately be thought of as a light ray representing the history of the massless particle, because it does not behave appropriately under Lorentz transformations).[33.18] It was this configuration that originally provoked the name 'twistor'.[23]

33.7 Twistor quantum theory

This outlines the basic geometry of flat-space twistor theory. But some readers may be understandably impatient in wondering how such a picture, for all its geometrical prettiness, is to help us move forward in *physics*. What, indeed, does twistor theory have to say about unifying spacetime structure with quantum-mechanical principles? So far, we have merely seen some 'cute' geometrical and algebraic ways to describe massless particles, but neither quantum-mechanical nor general-relativistic ideas have yet played any roles. I had better see to this!

Let us return to the most basic idea of twistor theory. It is to regard all spacetime notions as being *subsidiary* to those of twistor space $\mathbb{T}$. Being a

[33.18] Find the centre (in the coordinates of §33.5) and show how it transforms under a general Lorentz velocity transformation.

fully complex space, $\mathbb{T}$ provides the potential to exploit complex-number magic in ways that do not readily present themselves in the standard spacetime framework. Accordingly, rather than using descriptions in terms of real spacetime coordinates, one uses the complex twistor variables Z^α instead. Now twistor variables are mixtures of position and momentum variables, and we must ask: what takes the place of the standard quantization rule (§21.2)

$$p_a \mapsto \mathrm{i}\hbar \frac{\partial}{\partial x^a}$$

(or else $x^a \mapsto -\mathrm{i}\hbar\partial/\partial p_a$)? The answer is that, in analogy with x^a and p_a being *canonical conjugate* variables, as expressed in the operator commutation law $p_b x^a - x^a p_b = \mathrm{i}\hbar\delta^a_b$ of §21.2, the twistor variables Z^α and $\bar{Z}_\alpha$ are to be taken to be canonical conjugate operators:

$$Z^\alpha \bar{Z}_\beta - \bar{Z}_\beta Z^\alpha = \hbar\delta^\alpha_\beta,$$

where, like position and momentum separately, the Z^α and $\bar{Z}_\alpha$ commute among themselves: $Z^\alpha Z^\beta - Z^\beta Z^\alpha = 0$ and $\bar{Z}_\alpha \bar{Z}_\beta - \bar{Z}_\alpha \bar{Z}_\beta = 0$.[33.19]

As an aside, it may be remarked that this *quantum* non-commutation of $\bar{Z}_\alpha$ with Z^α raises some intriguing questions with regard to the kind of 'geometry' that might arise if we take more seriously the fact that the fundamental 'coordinates' for a quantum twistor space might be such non-commuting entities. Classically, when we consider the real 8-manifold structure of twistor space $\mathbb{T}$, we can use Z^α and $\bar{Z}_\alpha$ as independent commuting variables (see §10.1). But in this quantum picture, Z^α and $\bar{Z}_\alpha$ do not commute. To attempt to use such a 'quantum' pair, Z^α and $\bar{Z}_\alpha$, as independent coordinates would lead us to the area of non-commutative geometry, which was discussed briefly above, in §33.1. It may well be interesting to pursue this line further, but I am not aware that anyone has done so.

Now, we recall that in an ordinary position-space wavefunction $\psi(\mathbf{x})$ for a particle, the momentum variables $\mathbf{p}$ do not appear, but instead momentum is represented in terms of the operator $\partial/\partial x^a$ (as above). What is the twistor analogue of this? We seem to require that our 'twistor wavefunction' $f(Z^\alpha)$ should be 'independent of $\bar{Z}_\alpha$' and that $\bar{Z}_\alpha$ should instead be represented in terms of the operator $\partial/\partial Z^\alpha$. Indeed, this is correct, but what does f being 'independent of $\bar{Z}_\alpha$' actually mean? Formally, this 'independence' would be expressed as $\partial f/\partial \bar{Z}_\alpha = 0$, which (as we recall from §10.5) are simply the Cauchy–Riemann equations asserting that $f(Z^\alpha)$ is a holomorphic function of Z^α.

[33.19] Can you see, from general considerations of Hilbert-space operators why it is appropriate that there should be no 'i' in the twistor commutator?

This is a very striking and somewhat remarkable fact. Twistor wavefunctions are indeed holomorphic entities, so they can make proper contact with the magical complex-number world. The quantum role of the complex-conjugate variables $\bar{Z}_\alpha$ is that they indeed appear as differentiation:

$$\bar{Z}_\alpha \mapsto -\hbar\frac{\partial}{\partial Z^\alpha},$$

which is a holomorphic operator, so that, at the quantum level of description, holomorphicity is preserved. It is reassuring that the interpretation of twistors in terms of momentum and angular momentum for a massless particle is consistent with the twistor commutation rules, the angular momentum and momentum commutators (§22.12) coming out correctly and being subsumed in the twistor commutators given above.[24]

A quantity of particular interest is the *helicity* s, now regarded as an operator, whose eigenvalues are the various possible half-integer values $(\ldots, -2\hbar, -\frac{3}{2}\hbar, -\hbar, -\frac{1}{2}\hbar, 0, \frac{1}{2}\hbar, \hbar, \frac{3}{2}\hbar, 2\hbar, \ldots)$ that are allowed for a massless particle. It is especially noteworthy that, with non-commutation correctly taken into account, the helicity operator becomes[25,[33.20]]

$$s = \tfrac{1}{4}(Z^\alpha\bar{Z}_\alpha + \bar{Z}_\alpha Z^\alpha) \mapsto -\tfrac{1}{2}\hbar\left(2 + Z^\alpha\frac{\partial}{\partial Z^\alpha}\right).$$

The operator

$$\boldsymbol{Y} = Z^\alpha\frac{\partial}{\partial Z^\alpha}$$

is called the *Euler homogeneity operator*. (We recall our old friend Leonhard Euler from Chapters 5, 6, 7, and 9, particularly.) As Euler showed, $\boldsymbol{Y}$ has the remarkable property that its eigenfunctions are *homogeneous*, the degree of homogeneity being the eigenvalue. That is to say, the equation

$$\boldsymbol{Y}f = uf,$$

where u is some number, is the condition for the homogeneity property

$$f(\lambda Z^\alpha) = \lambda^u f(Z^\alpha)$$

to hold.[[33.21]] It follows that a twistor wavefunction for a massless particle of a definite helicity value S (so $sf = \hbar Sf$, where s is the operator and S the eigenvalue) must be homogeneous of degree $-2S-2$ as well as being holomorphic.[[33.22]]

[33.20] Verify the equality between these two expressions for s.

[33.21] See if you can prove this.

[33.22] Why this value?

Thus, in particular, a photon's twistor wavefunction ($S = \pm 1$) would be the sum of two parts, one being homogeneous of degree 0, describing the left-handed component ($S = -1$), and one of degree -4, describing the right-handed component ($S = 1$). A neutrino, taken as a massless particle, would have a wavefunction homogeneous of degree -1 (since the helicity is $-\frac{1}{2}$), whereas a (massless) antineutrino's wavefunction would be of degree -3. A massless scalar particle's wavefunction is of homogeneity degree -2. Most important, for our deliberations here, is the case of a *graviton*, which we shall take (provisionally) to be a massless particle of spin 2 in a flat Minkowski background ($S = \pm 2$). Its left-handed part ($S = -2$) has a twistor function homogeneous of degree 2 and its right-handed part ($S = 2$), a twistor wavefunction of degree -6.

This lop-sidedness is striking, and it illustrates the essentially chiral nature of twistor theory. We shall be seeing shortly that this lop-sidedness looms particularly large when we try to bring general relativity proper under the twistor umbrella. For the moment, let us try to understand how twistor (linear) wavefunctions are to be interpreted. For these, the lop-sidedness causes no problems, and everything works very smoothly. There is, however, an important subtlety about how our wavefunction $f(Z^{\alpha})$—usually referred to as a *twistor function*—is to be interpreted. Let us come to this next.

33.8 Twistor description of massless fields

For the spacetime representation of the wavefunction of a free massless particle of general spin, the Schrödinger equation translates to a certain equation known as the *massless free-field equation*.[26] We have seen an instance of this, in the case of spin $\frac{1}{2}$, in the massless (Dirac–Weyl) neutrino equation (§25.3). It is not appropriate to go into the details here, but the equation itself is simple enough to write down once we have the 2-spinor formalism to hand, as used in §22.8 and §25.2. For negative helicity $S = -\frac{1}{2}n$, we have a quantity $\psi_{AB\ldots D}$; for positive helicity $S = \frac{1}{2}n$, there is a primed-indexed quantity $\psi_{A'B'\ldots D'}$. Each of these is completely symmetrical in all its n indices, and each of them has positive frequency, satisfying the respective equations

$$\nabla^{AA'}\psi_{AB\ldots D} = 0, \quad \nabla^{AA'}\psi_{A'B'\ldots D'} = 0,$$

where $\nabla^{AA'}$ is just the 2-spinor translation of the ordinary gradient operator ∇^{a} (written in raised-index form; see §14.3).[33.23] For spin 0, we simply

[33.23] Write these equations out explicitly, for helicity $-\frac{1}{2}n$ using the notation $\psi_r = \psi_{00\ldots011\ldots1}$, where there are $n - r$ 0s and r 1s, and translating $\nabla^{AA'}$ from ∇^{a} in just the same way that the quantity $r^{AA'}$ is translated from ordinary Minkowski coordinates t, x, y, z, as described above.

have the wave equation $\square\psi = 0$, where $\square$ is the ordinary D'Alembertian introduced in §24.5. In fact, the convenient 2-spinor notation for these equations makes light of some subtleties. When $n = 2$ (spin 1), these two equations simply become Maxwell's free-field equations in the anti-self-dual and self-dual cases, respectively.[33.24] When $n = 4$, they become the weak-field Einstein equation, split into anti-self-dual and self-dual parts, where the curvature is regarded as an infinitesimal perturbation of flat space $\mathbb{M}$.[27]

What do these equations have to do with twistor functions? It turns out, remarkably,[28] that there is an explicit contour-integral expression (§7.2) which automatically gives the general positive-frequency solution of the above massless free-field equations simply starting from the twistor function $f(Z^\alpha)$. In fact, the expression also works perfectly well without this positive-frequency requirement, though the requirement is easily ensured in the twistor formalism, as we shall be seeing in §33.10. It is inappropriate to give full details here, but the basic idea is that, in the positive-helicity case, $f(Z^\alpha)$ is first multiplied by $\boldsymbol{\pi}$ (§33.6), taken n times (this supplying n primed indices), or, in the negative-helicity case, first operated upon by $\partial/\partial\boldsymbol{\omega}$ taken n times (supplying n unprimed spinor indices); then it is multiplied by the 2-form $\boldsymbol{\tau} = \mathrm{d}\pi_{0'} \wedge \mathrm{d}\pi_{1'}$ and integrated over an appropriate 2-dimensional contour, where the incidence relation $\boldsymbol{\omega} = \mathrm{i}\boldsymbol{r}\boldsymbol{\pi}$ is first incorporated so as to eliminate $\boldsymbol{\omega}$ in favour of $\boldsymbol{\pi}$ and $\boldsymbol{r}$. This integration eliminates $\boldsymbol{\pi}$, so we end up with an indexed quantity $\psi_{\dots}$ at any chosen spacetime point R (so $\psi_{\dots}$ is a function of $\boldsymbol{r}$ alone). The contour is to lie within the locus $\boldsymbol{\omega} = \mathrm{i}\boldsymbol{r}\boldsymbol{\pi}$ (for each fixed $\boldsymbol{r}$), i.e. within (the non-projective[29] version of) the line $\mathbf{R}$, in $\mathbb{N}$, which represents the event R; see Fig. 33.16.

The positive-frequency condition is ensured by requiring that the contour integral still works when the line $\mathbf{R}$ is allowed to venture entirely into the twistor region $\mathbb{PT}^+$. Lines in $\mathbb{PT}$ correspond to 'complex spacetime points', as we have seen in §33.6, and those lying entirely in the subregion $\mathbb{PT}^+$ correspond to points of the subregion $\mathbb{M}^+$ of $\mathbb{CM}$ known as the *forward tube*.[30] We shall return to this matter in §33.10. Massless fields of mixed helicity—such as a plane-polarized photon, which is a sum of a left-handed and right-handed part—can also be described in this framework, where the twistor functions for the two different helicities are simply added.

The very existence of such an expression strikes me as being somewhat magical. The massless field equations seem to evaporate away in the twistor formalism, being converted, in effect, to 'pure holomorphicity'.

[33.24] See if you can show this, where $\psi_{00} = C_1 - \mathrm{i}C_2$, $\psi_{01} = -\mathrm{i}C_3$, and $\psi_{11} = -C_1 - \mathrm{i}C_2$, with $\mathbf{C} = 2\mathbf{E} - 2\mathrm{i}\mathbf{B}$ (see §19.2), and where corresponding expressions hold for $\psi_{A'B'}$.

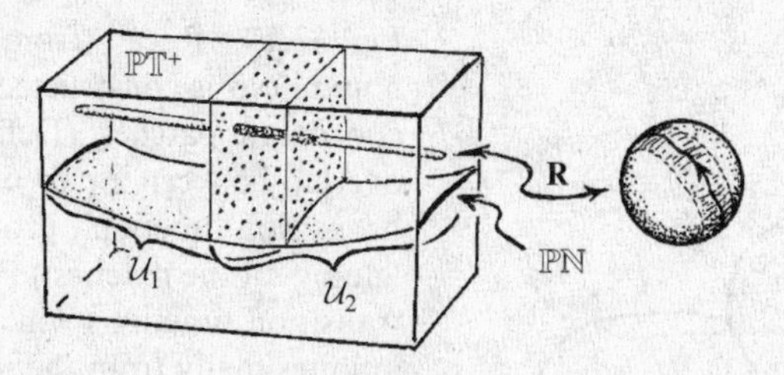

Fig. 33.16 The basic twistor contour integral. A twistor function f (for helicity $\frac{1}{2}n$), homogeneous of degree $-n-2$, is multiplied n times by $\boldsymbol{\pi}$ (n positive) or acted upon $-n$ times by $\partial/\partial\boldsymbol{\omega}$ (n negative), these supplying the indices, and then multiplied by the 2-form $\boldsymbol{\tau} = \mathrm{d}\pi_{0'} \wedge \mathrm{d}\pi_{1'}$. For any particular choice of spacetime point R, with position vector $\boldsymbol{r}$, a contour integral is then performed in the region **R** of twistor space defined by the incidence relation $\boldsymbol{\omega} = \mathrm{i}\boldsymbol{r}\boldsymbol{\pi}$. This integrates out the $\boldsymbol{\pi}$-dependence, and we are left with a solution of the massless field equations. In the case illustrated, **R** is taken in the top half of twistor space $\mathbb{PT}^+$ (or $\mathbb{T}^+$) and f is holomorphic in the intersection of $\mathcal{U}_1$ and $\mathcal{U}_2$, where the open sets $\mathcal{U}_1$ and $\mathcal{U}_2$, together cover the whole of $\mathbb{PT}^+$ (or $\mathbb{T}^+$).

When we examine this expression more carefully, we find that there is an important subtlety about how a twistor function is to be interpreted, and this relates in a striking way to the positive/negative frequency splitting of massless fields (§33.10). This subtlety is also crucial to how twistor functions manifest themselves in active ways, and provide us with curved twistor spaces. What is this subtlety? It is that twistor functions are not really to be viewed as 'functions' in the ordinary sense, but as what are called *elements of holomorphic sheaf cohomology*.[31]

33.9 Twistor sheaf cohomology

What is sheaf cohomology? The ideas are fairly sophisticated mathematically, but actually very natural. We shall be concerned here only with what is called *first* sheaf cohomology. Perhaps the easiest way to picture this notion is to think of the way in which a manifold can be constructed in terms of a number of coordinate patches, as discussed in §10.2 and §12.2 and illustrated in Fig. 12.5a. Defined on each overlap between a pair of patches is a *transition function* (providing the *gluing* of the patches). We recall from §12.2, Fig. 12.5a, that these transition functions are subject to certain consistency conditions, on triple overlaps between patches.

Now think of a manifold built up in this way but where the transition functions differ from the identity by only an infinitesimal amount. See

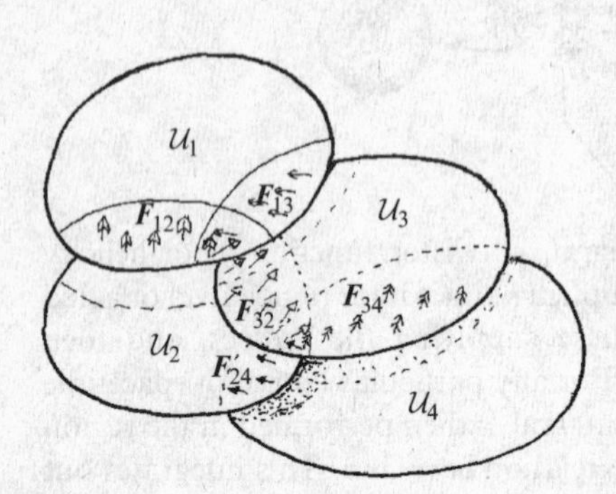

Fig. 33.17 Recall (from Fig. 12.5a) how a manifold is constructed from several coordinate patches. (Defined on each overlap between a pair of patches is a 'transition function', providing the 'gluing' of the patches.) Here, we consider transition functions which differ only infinitesimally from the identity, and therefore given by a vector field $\boldsymbol{F}_{ij}$ on each overlap of patches $\mathcal{U}_i$, $\mathcal{U}_j$, telling us how each patch is to 'shunt' relative to the others which it overlaps. (The 'patches' are open sets $\mathcal{U}_1, \mathcal{U}_2, \mathcal{U}_3, \ldots$ on the flat coordinate space.)

Fig. 33.17. This infinitesimal shift between one patch $\mathcal{U}_i$ and another patch $\mathcal{U}_j$ would be described by a vector field $\boldsymbol{F}_{ij}$ on the part of $\mathcal{U}_i$ that overlaps with $\mathcal{U}_j$, describing how the patch $\mathcal{U}_i$ is to be infinitesimally 'shunted along' with respect to $\mathcal{U}_j$. Equivalently, we can think of $\mathcal{U}_j$ to be shunted along with respect to $\mathcal{U}_i$, but in the opposite direction. This is described by the vector field $\boldsymbol{F}_{ji}$ on the part of $\mathcal{U}_j$ that overlaps with $\mathcal{U}_i$, whence on this overlap we have

$$\boldsymbol{F}_{ji} = -\boldsymbol{F}_{ij}$$

(see Fig. 33.18a). On triple overlap between patches $\mathcal{U}_i$, $\mathcal{U}_j$, and $\mathcal{U}_k$, we find (Fig. 33.18b) that the consistency relation

$$\boldsymbol{F}_{ij} + \boldsymbol{F}_{jk} = \boldsymbol{F}_{ik}$$

must hold.[33.25]

There are also 'trivial' infinitesimal deformations that arise simply from changing the coordinate system (infinitesimally) in each patch. We can think of these as being given by a vector field $\boldsymbol{H}_i$ in each particular patch $\mathcal{U}_i$, which simply 'shunts' this entire patch along over itself. This would give us a family of 'trivial' $\boldsymbol{F}_{ij}$s of the form

$$\boldsymbol{F}_{ij} = \boldsymbol{H}_i - \boldsymbol{H}_j$$

on overlaps between pairs of patches, which do not change the manifold (Fig. 33.18c).

These ideas essentially tell us the rules of first sheaf cohomology.[32] We do not need to be concerned just with vector fields, however. Ordinary

[33.25] Show that antisymmetry in F_{ij} is a consistency requirement of the triple-overlap condition.

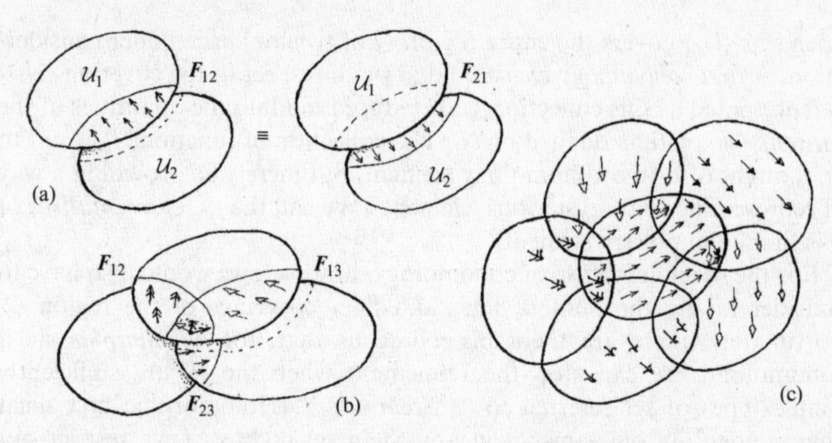

Fig. 33.18 The vector fields $\boldsymbol{F}_{ij}$ are subject to certain requirements. (a) On the intersection of $\mathcal{U}_j$ with $\mathcal{U}_i$, we have $\boldsymbol{F}_{ji} = -\boldsymbol{F}_{ij}$. This is illustrated as the movement of $\mathcal{U}_1$ over $\mathcal{U}_2$, as shown by the vector $\boldsymbol{F}_{12}$ on $\mathcal{U}_1$. But the same relative movement is achieved by the negative of this on $\mathcal{U}_2$. (b) The triple overlap condition $\boldsymbol{F}_{ij} + \boldsymbol{F}_{jk} = \boldsymbol{F}_{ik}$ is illustrated. On the triple intersection of $\mathcal{U}_1, \mathcal{U}_2$, and $\mathcal{U}_3$, the motion $\boldsymbol{F}_{12}$ of $\mathcal{U}_1$ over $\mathcal{U}_2$ is the sum of the motion $\boldsymbol{F}_{13}$ of $\mathcal{U}_1$ over $\mathcal{U}_3$ and the motion $\boldsymbol{F}_{32}$ of $\mathcal{U}_3$ over $\mathcal{U}_2$. (c) If all the patches are shunted individually in their entireties, this has no effect (except for a coordinate change in each patch). This illustrates that the overall shunt $\boldsymbol{F}_{ij} = \boldsymbol{H}_i - \boldsymbol{H}_j$ counts for nothing and must be 'factored out'.

functions f_{ij} would do as well as the vector fields $\boldsymbol{F}_{ij}$ that we have been considering. We just require that each f_{ij} be defined in the intersection of $\mathcal{U}_j$ with $\mathcal{U}_i$, that $f_{ij} = -f_{ji}$, that $f_{ij} + f_{jk} + f_{ki} = 0$ on each triple overlap, and that the whole collection $\{f_{ij}\}$ is considered to be equivalent to another such collection $\{g_{ij}\}$ if each member of the collection of corresponding *differences* $\{f_{ij} - g_{ij}\}$ has the 'trivial' form $\{h_i - h_j\}$. We say that the $\{f_{ij}\}$ are reduced *modulo* quantities of the form $\{h_i - h_j\}$, which is essentially the same sense in which the term 'modulo' was used in §16.1 (see also the notion of 'equivalence class' referred to in the Preface). In fact, the class of function (f_{ij} or h_i) that one may be concerned with in cohomology theory can be extremely general. In twistor theory, one normally deals with holomorphic functions. This gives us the notion of 'holomorphic sheaf cohomology'.

Specifically, this cohomology idea applies to twistor functions. Indeed, we are to think of a 'twistor function' as being, in general, not simply a single holomorphic function f, but as provided by a *collection* of holomorphic functions $\{f_{ij}\}$, where each individual f_{ij} is defined on the overlap between a pair of open sets $\mathcal{U}_i$ and $\mathcal{U}_j$, with $f_{ji} = -f_{ij}$, where on triple overlaps we have $f_{ij} + f_{jk} + f_{ki} = 0$, and where the whole collection of these open sets $\{\mathcal{U}_i\}$ covers the entire region $\mathcal{Q}$ of twistor space under consideration. A *first cohomology element* on $\mathcal{Q}$ (with respect to the covering $\{\mathcal{U}_i\}$) is represented as this collection $\{f_{ij}\}$, reduced modulo the quantities of the form $h_i - h_j$, with h_i defined on $\mathcal{U}_i$. The collection of functions f_{ij} is not to be thought of *as* the cohomology element, but merely as providing a way of representing that mysterious 'element'. We call the f_{ij} *representatives* of this first cohomology element.

For the strict definition of cohomology, however, we would also have to consider taking the limit of finer and finer coverings of the region $\mathcal{Q}$. Fortunately, there are theorems telling us that, for *holomorphic* sheaf cohomology, we can stop the refinement when the $\mathcal{U}_i$ are sufficiently simple types of set referred to as *Stein* sets.[33] (Holomorphic first sheaf cohomology always vanishes in any Stein set.) Thus, if we restrict our attention to coverings for which every $\mathcal{U}_i$ is a Stein set, we do not need to say 'with respect to the covering $\{\mathcal{U}_i\}$' when we refer to a cohomology element defined on $\mathcal{Q}$. The cohomology notion does not depend upon the specific choice of Stein covering. A cohomology element is a 'thing' defined on $\mathcal{Q}$, which comes out the same whichever such covering is used.[34] This remarkable fact is part of the *magic* of (holomorphic) sheaf cohomology!

How does all this apply to the twistor functions and contour integrals that we considered in §33.8? The simplest situation arises when there are just two patches $\mathcal{U}_1$ and $\mathcal{U}_2$, which together cover the region of twistor space under consideration. There is now just one function to consider, and this is the 'twistor function' of §33.8: $f(Z^\alpha) = f_{12} = -f_{21}$. According to the above rules of sheaf cohomology, we say that $f(Z^\alpha)$ is *equivalent* to $g(Z^\alpha)$ if the difference is *trivial* in the above sense, i.e. if

$$f - g = h_1 - h_2,$$

where the holomorphic function h_1 is defined globally on $\mathcal{U}_1$ and h_2 globally on $\mathcal{U}_2$. It turns out to be a simple matter to show that the appropriate contour integral applied to f is indeed the same as that applied to g whenever these functions are equivalent in this sense. Sometimes it is necessary to consider more complicated patchings, however. In essence, the above 'cohomology rules' for equivalence between twistor functions are geared to preserving the answers that the contour-integral expressions provide, but where the notion of a contour integral must now be generalized to that of a 'branched contour integral', with one branch in each overlap region. This is indicated in Fig. 33.19.[35]

An important feature of cohomology is that it is *essentially non-local*. We might have a cohomology element defined on some region $\mathcal{Q}$. It then makes sense to consider the restriction of that element down to some smaller region $\mathcal{Q}'$, contained in $\mathcal{Q}$. The non-local feature of cohomology

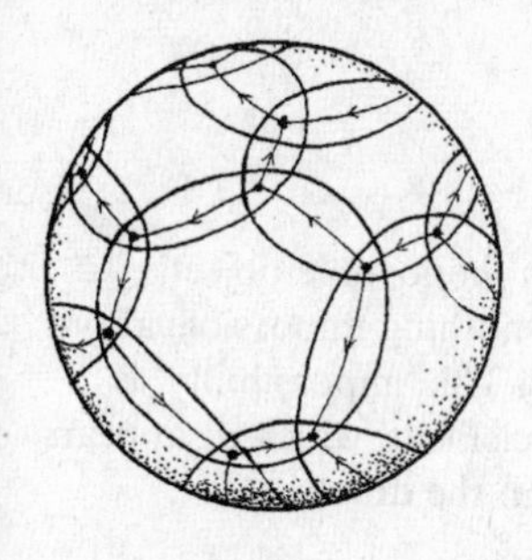

Fig. 33.19 A 'branched contour' (on the Riemann sphere), applicable to the spacetime evaluation of twistor functions for which the covering consists of more than two sets.

is manifested in the fact that, for any sufficiently small (open) subregion $\mathcal{Q}'$, in $\mathcal{Q}$, the element necessarily *vanishes* when restricted down to $\mathcal{Q}'$, in the sense that, given f_{ij} on $\mathcal{Q}'$, then hs can always be found, in $\mathcal{Q}'$, for which $f_{ij} = h_i - h_j$.

This non-locality, for twistor functions, tells us that there is no significance to be attached to the value attained by f_{ij} at some particular point. We can, indeed, restrict down to a small enough open region surrounding that point and find that the cohomology element disappears completely; see Fig. 33.20. This non-locality, as exhibited by twistor functions (regarded as first cohomology elements) is tantalizingly reminiscent of the non-local features of EPR effects and quanglement (§23.10). In my opinion, there is something important going on behind the scenes here which may someday make sense of the mysterious non-local nature of EPR phenomena, but it has yet to be fully revealed, if so.

We are to think of this 'cohomology element' as being a 'thing' defined on the space $\mathcal{Q}$, which is a bit like a function defined on $\mathcal{Q}$, but which is fundamentally non-local. One example of this kind of 'thing' is actually an entire (complex) vector bundle over $\mathcal{Q}$, as described in §§15.2,5. We recall that, in the definition of a bundle, that part of the bundle lying above a small enough region of the base space (here $\mathcal{Q}$) is 'trivial', in the sense of being just a (topological) product, cf. §15.2. (See Fig. 15.3.) This is an example of the fact that if we restrict our first cohomology element down to a small enough region, it becomes 'trivial' also; i.e. it vanishes. Thus, the

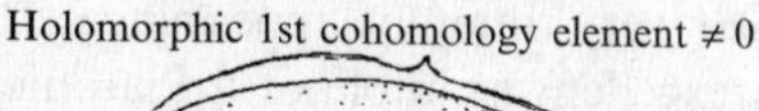

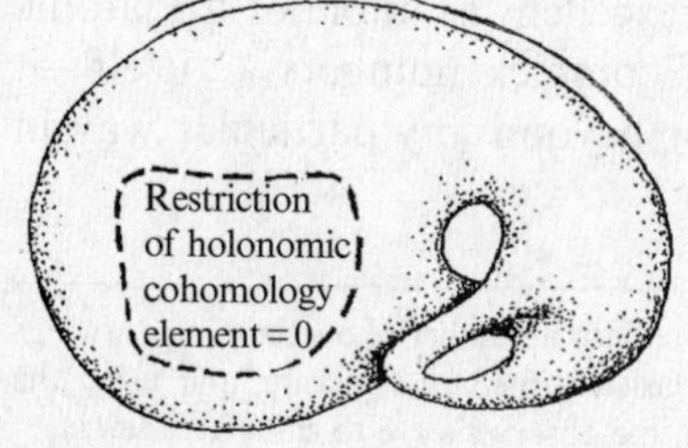

Fig. 33.20 A cohomology element can always be restricted down to a smaller region. But if this region is sufficiently small, the cohomology always disappears. This illustrates the non-locality of cohomology.

Fig. 33.21 A drawing of an 'impossible object' (a 'tribar'). Locally, there is nothing impossible about what the drawing represents. The 'impossibility' is measured by a cohomology element, which disappears in any small enough region in the drawing.

'information' expressed in a cohomology element is something of a fundamentally non-local character.

It is perhaps worthwhile to provide an elementary example that illustrates the notion of cohomology, albeit only in a simple case, in a particularly graphic way. See Fig. 33.21 for a drawing of an 'impossible object' sometimes referred to as a 'tribar'.[36] It is clear that the '3-dimensional object' which the drawing apparently depicts cannot exist in ordinary Euclidean space. Yet *locally* there is nothing impossible about the drawing. The impossibility is non-local, and disappears if one considers a small enough region in the drawing. In fact, this notion of 'impossibility' in such a drawing can be expressed as a specific cohomology element.[37,[33.26]] It is a relatively simple type of cohomology, however, where the functions $\{f_{ij}\}$ are taken to be constants.

I have only just touched upon some basic ideas of sheaf cohomology here. There are many applications of these ideas in mathematics, not all of which are concerned with holomorphicity. The 'sheaves' that one is mainly concerned with in twistor theory are those expressed in terms of holomorphic functions, and there is a special magic in cohomology theory in this particular context. (Roughly speaking, the term 'sheaf' refers to the type of function that one is concerned with, but the sheaf notion actually applies considerably more generally than just to ordinary functions.[38]) There are many other types of use of cohomology, including some that have importance in the study of the Calabi–Yau spaces that occur in string theory (§31.14), for example. Also, there are several other quite different ways of defining sheaf-cohomology elements, all of which can be shown to be mathematically equivalent, despite their very different appearances.[39] In my opinion (sheaf) cohomology is an excellent example of a Platonic notion (§1.3), where—like the system of complex numbers $\mathbb{C}$ itself—it seems to have a 'life of its own' going far beyond any particular way in which one may choose to represent it.

[33.26] See if you can do this: breaking the drawing up into a number of overlapping drawings (the $\{\mathcal{U}_i\}$), each of which individually represents a consistent 3-space structure, and using the logarithm of the distance of this 3-space structure from the observer's eye to calculate the $\{f_{ij}\}$.

33.10 Twistors and positive/negative frequency splitting

How do we incorporate the *positive-frequency* condition, so fundamental to QFT, into twistor theory? Recall, from §§9.2,3, the way in which the division of the Riemann sphere S^2 into the southern and northern hemispheres S^- and S^+ provides us with the splitting of a function, defined on the equator S^1, into its positive- and negative-frequency parts. The positive-frequency part extends into S^- and the negative-frequency part extends into S^+ (Fig. 33.22a). Projective twistor space does a corresponding thing, but in a global way that directly applies to massless fields in their entirety. It achieves this according to a direct analogy between the Riemann sphere and the projective twistor space $\mathbb{PT}$ where the analogue of a function on the Riemann sphere S^2 is a first cohomology element on $\mathbb{PT}$. The analogue of the equator S^1 is to be the space $\mathbb{PN}$, and we note that $\mathbb{PN}$ divides $\mathbb{PT}$ (which is a $\mathbb{CP}^3$) into two halves $\mathbb{PT}^+$ and $\mathbb{PT}^-$ in just the same sort of way that S^1 divides S^2 (which is a $\mathbb{CP}^1$) correspondingly[40] into two hemispheres S^- and S^+ (Fig. 33.22b).

More explicitly, the analogue of an ordinary (complex) function defined on S^1, or on S^-, or on S^+ is, respectively, a first cohomology element defined on $\mathbb{PN}$, or on $\mathbb{PT}^+$, or on $\mathbb{PT}^-$. Massless fields on $\mathbb{M}$ (strictly, on $\mathbb{M}^\#$) are represented as first cohomology elements on $\mathbb{PN}$. Each one of these can be expressed (essentially uniquely) as a sum of an element that extends into $\mathbb{PT}^+$ and an element that extends into $\mathbb{PT}^-$. The first

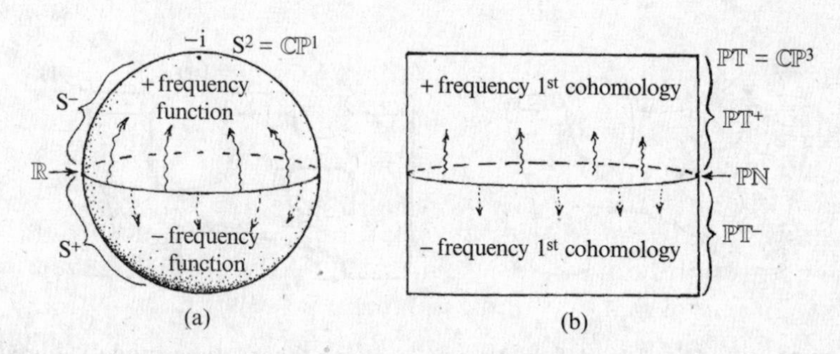

Fig. 33.22 An analogy between the Riemann sphere $S^2(=\mathbb{CP}^1)$ and projective twistor space $\mathbb{PT}(=\mathbb{CP}^3)$. (a) A complex function (i.e. a '0th cohomology element'), defined on the real axis $\mathbb{R}$ of S^2, splits into its positive frequency part, extending holomorphically into what is here depicted as the northern hemisphere S^-, and its negative frequency part, extending into the southern hemisphere S^+. (The Riemann sphere is here drawn so that $\mathbb{R}$ is its equator, but so that $-i$ is at the north pole and i at the south pole; compare Figs. 8.7 and 9.10, §9.5.) (b) A 1st cohomology element, defined on $\mathbb{PN}$ (and representing a massless field) splits into its positive frequency part, extending holomorphically into the top half $\mathbb{PT}^+$ of projective twistor space, and its negative frequency part, extending into the bottom half $\mathbb{PT}^-$.

describes a positive-frequency massless field and the second, a negative-frequency one.[41] In spacetime terms, this positive-frequency part of the field extends to be defined in the *forward tube*, which we recall from §33.8 is the region $\mathbb{M}^+$ of $\mathbb{CM}^\#$ consisting of points that are represented in twistor space by projective lines in $\mathbb{PT}^+$. In $\mathbb{CM}^+$, these are the (complex) points whose position vectors have imaginary parts that are timelike and past-pointing.[33.27]

This analogy between $\mathbb{PT}$ and the Riemann sphere leads to a possible way that twistor-theoretic ideas might find an analogy with some of those of string theory. Recall, from §§31.5,13, that Riemann surfaces are used to represent 'string histories' in that theory. The Riemann sphere ($\mathbb{CP}^1$) is the simplest such surface, but surfaces with various numbers of 'handles' (higher-genus Riemann surfaces—see §8.4) are brought in to represent more general kinds of string history. These Riemann surfaces may also have 'holes' (with S^1 boundaries), in addition to handles (see Fig. 31.5). By analogy,[42] one can consider generalizations of the space $\mathbb{PT}$, which acquire 'handles' in a corresponding way, and also 'holes' (with boundaries that are copies of $\mathbb{PN}$). These have been referred to as 'pretzel twistor spaces', and a form of QFT can be developed, based on these spaces (see Fig. 33.23). As yet, the status these ideas has not been fully ascertained.

Historically, the positive-frequency requirement—and this property that $\mathbb{PN}$ divides $\mathbb{PT}$ into two such halves—provided a key motivation in the original formulation of twistor theory, in 1963, more than 12 years before

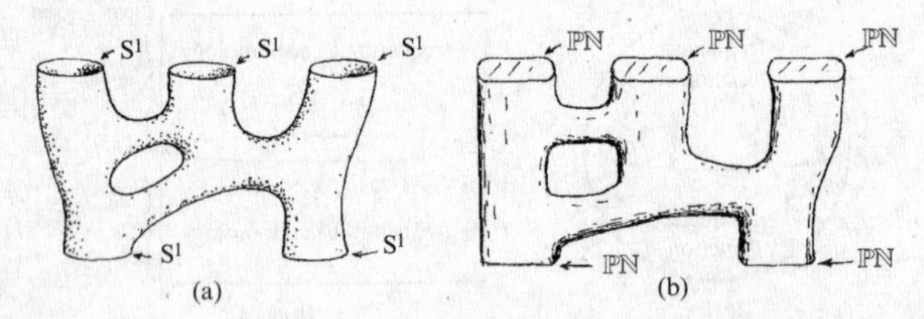

Fig. 33.23 (a) Conformal field theory (a string-theory type model), based on generalizations of the Riemann sphere to Riemann surfaces of higher genus which can have finite-sized 'holes' as well as handles (see Fig. 31.5, holes representing places where external information is fed in). (b) A twistor version which employs generalizations of $\mathbb{PT}$, acquiring 'handles' in a way corresponding to Riemann surfaces, and also 'holes' whose boundaries are copies of $\mathbb{PN}$ ('pretzel twistor spaces').

[33.27] Show this: demonstrate, from the incidence equation, that a complex position vector r^a for a point R of $\mathbb{CM}$ is represented by a projective line in $\mathbb{PT}^+$, if and only if the imaginary part of r^a is past-pointing and timelike.

the discovery that massless fields have a twistor description as holomorphic first sheaf cohomology.[43] It is striking that here again we have a property specifically arising from spacetime's four-dimensionality with Lorentzian signature. It is also specific that first cohomology elements play a role in this aspect of twistor theory, rather than ordinary functions—which are 'zeroth' cohomology elements—or second or higher cohomology elements. Higher-order notions of cohomology also exist (and have a role to play in twistor theory), but there is something unique to first cohomology, which is fundamental to twistor theory. For only these quantities find a direct role in generating *deformations* of twistor space. Let us come to this next.

33.11 The non-linear graviton

The cohomology elements (twistor functions) that we have been considering should be thought of, so far, as being entirely 'passive', in the sense that they are simply 'painted on' the (twistor) space. This corresponds to the fact that they describe spacetime fields that just reside on the spacetime, and do not influence other fields. To see how they can provide an active influence, let us think of the 'paint' on the twistor space 'drying', so that the space now becomes distorted (Fig. 33.24). To see how this can happen, we think of our previously passive twistor function f_{ij} as being associated with a vector field $\boldsymbol{F}_{ij}$ in an appropriate way. By 'sliding the patches over each other' in the direction of these vector fields by an infinitesimal amount, we begin to 'dry the paint' and construct an infinitesimally 'curved' twistor space. We can imagine that this deformation is 'exponentiated' (§14.6), until a finite deformation of twistor space is obtained (paint fully dry!).

The first situation in which this procedure was successfully applied was in the case of anti-self-dual gravity.[44] In the infinitesimal (weak-field) case, we have a massless field of helicity $S = -2$, so using the above formula

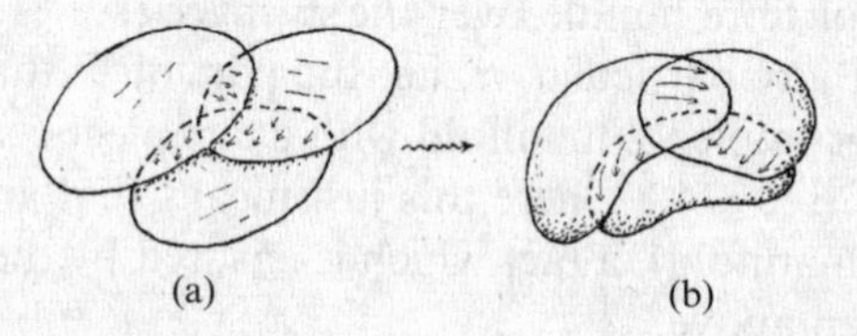

Fig. 33.24 A 1st cohomology vector field element is 'passive' (i.e., just 'painted on' the space). For it to have an active influence, think of the 'paint drying' as a result of an exponentiation of the vector field on each overlap. This results in a finite 'sliding' of one patch over another, giving a finite distortion, or 'curved space'.

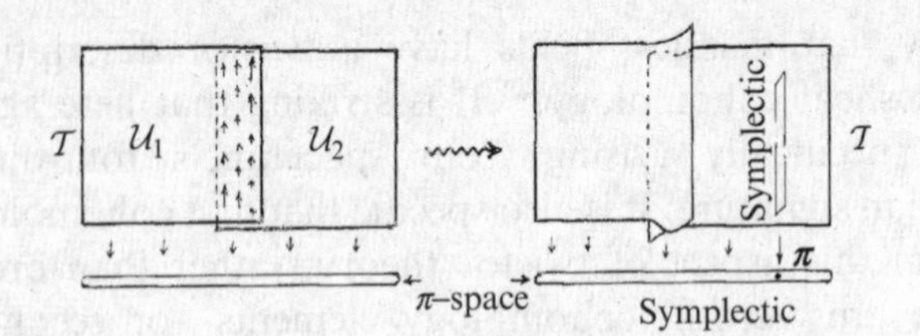

Fig. 33.25 Apply the idea of Fig. 33.24 in case of the twistor description of anti-self-dual gravity (with two patches). The vector field is $(\partial f/\partial\omega^0)\partial/\partial\omega^1 - (\partial f/\partial\omega^1)\partial/\partial\omega^0$, with f homogeneous of degree 2. We get a curved twistor space (portion) $\mathcal{T}$. There is a global projection of $\mathcal{T}$ to the $\boldsymbol{\pi}$-space. Each fibre of this projection is a complex symplectic 2-space, as is the $\boldsymbol{\pi}$-space itself.

$-2S-2$ for the homogeneity degree, we have a twistor function $f\,(=f_{ij})$ of homogeneity 2. Here we are assuming, for simplicity, that there are just two patches $\mathcal{U}_1$ and $\mathcal{U}_2$, each taken to be a portion of flat twistor space $\mathbb{T}$ with the standard coordinates of §33.5. The required vector field $\boldsymbol{F}$, constructed from f, turns out to be

$$\boldsymbol{F} = \frac{\partial f}{\partial\omega^0}\frac{\partial}{\partial\omega^1} - \frac{\partial f}{\partial\omega^1}\frac{\partial}{\partial\omega^0}.$$

Note that the homogeneity degree 2 of f exactly compensates for that of the two differential operators, to give an operator that is homogeneous of degree zero, so that is acts on the projective twistor space.[33.28]

Now, imagine exponentiating this infinitesimal shunt of one patch over the other (see Fig. 33.25). Then we obtain a *curved* twistor space (portion) $\mathcal{T}$. The absence of $\boldsymbol{\pi}$ derivatives in our infinitesimal patching relation implies that the twistor in one patch must have the same $\boldsymbol{\pi}$ part as the twistor with which it is matched in the next patch. It follows that the operation which 'projects out' the $\boldsymbol{\pi}$ spinor from the entire patched space $\mathcal{T}$ is consistent over the whole of $\mathcal{T}$. That is to say, there is a global projection of $\mathcal{T}$ down to the space of $\boldsymbol{\pi}$ spinors. Let us ignore (or preferably remove) the 'zero elements' of both $\mathcal{T}$ and the $\boldsymbol{\pi}$ space. Then we find that $\mathcal{T}$ is a kind of fibre bundle over the $\boldsymbol{\pi}$-space (see §15.2).[45] Each *fibre* (inverse image of any particular $\boldsymbol{\pi}$, i.e. the part of $\mathcal{T}$ that lies 'above' $\boldsymbol{\pi}$) turns out to be a complex 2-manifold with a symplectic structure, as does the $\boldsymbol{\pi}$ space itself (see §14.8—here this just means that an area measure is defined on the 2-manifold), a fact which is ensured by the specific form of the patching, given above.

How do we get from this curved twistor space back to some notion of 'spacetime'? The answer is that each 'spacetime point' corresponds

[33.28] Why does degree zero imply that this gives a vector field on a region in $\mathbb{PT}$? *Hint*: What's the commutator of $\boldsymbol{F}$ with $\boldsymbol{Y}$ (§33.7)?

uniquely to a holomorphic cross-section of the bundle $\mathcal{T}$. (The notion of a holomorphic cross-section was given in §15.5; here it is a map from the $\boldsymbol{\pi}$ space back into $\mathcal{T}$.) Why is this a reasonable definition? In the flat case $\mathbb{T}$, this amounts to representing the (possibly complex) spacetime point R by the map which takes $\boldsymbol{\pi}$ to $\mathbf{Z} = (\mathrm{i}\boldsymbol{r\pi}, \boldsymbol{\pi})$. In terms of the flat projective twistor space $\mathbb{PT}$, this cross-section is simply the straight line $\mathbf{R}$ (a Riemann sphere, $\mathbb{CP}^1$) in $\mathbb{PT}$ that we used in §33.5 to represent R.[33.29] It is very striking that this definition of a 'spacetime point' works just as well for the curved twistor space $\mathcal{T}$. We find[46] that there is a 4-complex-parameter family of holomorphic cross-sections, just as in the flat case. (In the projective space $\mathbb{P}\mathcal{T}$, this is a 4-complex-parameter family of lines $\mathbb{CP}^1$s.) We therefore have a 4-dimensional complex manifold $\mathcal{M}$ to represent this family. The 4-dimensionality is a remarkable fact—an instance of the complex magic of higher complex dimensions—that follows from theorems of the Japanese mathematician Kunihiko Kodaira.[47] (Experience only with *real* manifolds might have led to the expectation that there would be an *infinite*-parameter family of such things. But we have already noted in §15.5 that holomorphic cross-sections can be very restricted.)

In Fig. 33.26, this procedure is illustrated graphically (in the projective description). Start with a suitable region $\mathcal{R}$ of complex Minkowski spacetime $\mathbb{CM}$. For simplicity, let us just take $\mathcal{R}$ as some suitable (open) neighbourhood of a point R in $\mathbb{CM}$. The corresponding region $\mathcal{Q}$ of projective space $\mathbb{PT}$ is that swept out by the family of lines, each of which represents a point of $\mathcal{R}$. This will be a neighbourhood (referred to as a *tubular neighbourhood*) of the line $\mathbf{R}$, in $\mathbb{PT}$, that represents $\mathcal{R}$ (Fig. 33.26b,c). We can take it that the topology of $\mathcal{Q}$ is $S^2 \times \mathbb{R}^4$, where the S^2 comes from the topology of the line $\mathbf{R}$—or, equivalently, of the projective $\boldsymbol{\pi}$-space—and the $\mathbb{R}^4$ describes the transverse part of the immediate neighbourhood of each point of $\mathbf{R}$. We now think of S^2 (here the projective $\boldsymbol{\pi}$-space) as separated into two hemispheres, slightly extended so that there is a 'collar' of overlap, and then regard $\mathcal{Q}$ as being built up from the two overlapping (open-set) pieces $\mathcal{U}_1$ and $\mathcal{U}_2$ lying above each of these slightly extended hemispheres (Fig. 33.26d). We now 'shunt' $\mathcal{U}_2$ relative to $\mathcal{U}_1$, according to the above vector field, to get our deformed projective twistor-space region $\mathbb{P}\mathcal{T}$ (Fig. 33.26e,f).

There is still a global projection down to the $\boldsymbol{\pi}$ space (Fig. 33.26f), giving the bundle structure. However, the original 'straight lines' in $\mathcal{U}_1$ and $\mathcal{U}_2$ are now broken, so they do not give cross-sections, but Kodaira's theorem tells us that there is a *new* 4-parameter family of holomorphic curves in $\mathbb{P}\mathcal{T}$, these being the actual holomorphic cross-sections of the

[33.29] Explain the sense in which this line is a 'cross-section' of $\mathbb{PT} - \mathbf{I}$.

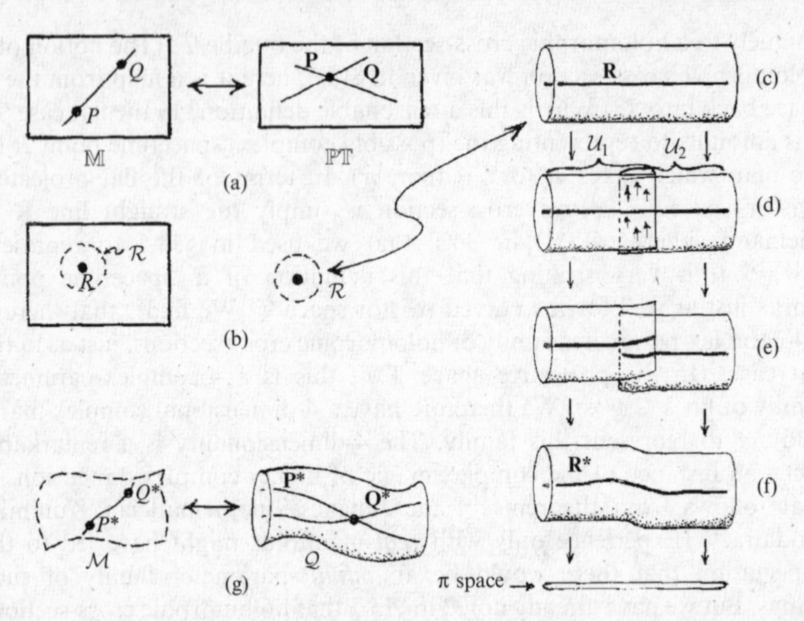

Fig. 33.26 Construction of a left-handed non-linear graviton. (a) In the standard flat-space twistor correspondence, points P and Q of $\mathbb{CM}$ are null separated whenever the corresponding lines **P** and **Q** in $\mathbb{PT}$ meet. (b) We wish to deform $\mathbb{PT}$, somehow, to a curved twistor space, but mathematical theorems tell us that this cannot be done globally. Accordingly, we take only a suitable (open) neighbourhood $\mathcal{R}$, of a point R in $\mathbb{CM}$ as our starting 'spacetime'. (c) This corresponds to a tubular neighbourhood $\mathcal{Q}$, in $\mathbb{PT}$, of the line **R**. (d) We can now apply the procedure of Fig. 33.25 to deform $\mathcal{Q}$ (considered as the union of two open sets $\mathcal{U}_1$ and $\mathcal{U}_2$. (e) However, we find that the original line **R** is now broken, and cannot be used as a sensible definition of a 'spacetime point'. (f) A theorem of Kodaira comes to our rescue, to tell us that there is a 4-parameter family of 'lines' **R*** (compact holomorphic curves, belonging to the same topological class as our original lines), which will serve this purpose. (g) The points of our sought-for 'non-linear graviton' space $\mathcal{M}$ (a complex 4-space) are given by Kodaira's curves **R***. The (complex conformal) metric of $\mathcal{M}$ is defined (as in (a)) by the condition that P^* and Q^* are null separated whenever the corresponding lines **P*** and **Q*** meet. The Weyl curvature of $\mathcal{M}$ turns out to be automatically anti-self-dual, and it is also Ricci-flat by virtue of the details of the construction.

bundle structure. The required space $\mathcal{M}$ is constructed so that each of its points corresponds to one of these cross-sections (Fig. 33.26f,g). It turns out that $\mathcal{M}$ can be assigned a metric $\boldsymbol{g}$ in a natural way, and that its Weyl curvature is anti-self-dual, and it is Ricci-flat. We can easily find the null cones of $\boldsymbol{g}$ (conformal structure) using the fact that two points P^* and Q^* of $\mathcal{M}$ are null-separated if and only if the corresponding lines **P*** and **Q*** in $\mathbb{P}\mathcal{T}$ intersect (Fig. 33.26g).

The reader may well worry what this 'spacetime' $\mathcal{M}$ actually means physically. It has turned out to be complex (and therefore 8-dimensional rather than 4-dimensional when considered as a real manifold). In the flat case, we could single out the real spacetime points (events in $\mathbb{M}$) by taking cross-sections of $\mathbb{T}$ lying in $\mathbb{N}$, and then regard our $\mathcal{M}$ as being simply the complexification $\mathbb{CM}$ of Minkowski space $\mathbb{M}$. But, in the curved case, we are not allowed such a luxury. Indeed, in this case, the 'spacetime' that we obtain by this construction is necessarily a complex manifold in its own right, and it cannot arise as a complexification of a Lorentzian real spacetime.

Why is this? A Lorentzian 4-manifold with an anti-self-dual Weyl curvature is necessarily Weyl-flat (since the complex conjugate of the zero self-dual part is the anti-self-dual part, which is therefore also zero). If it is also Ricci-flat, then it is simply flat altogether. In the complex case,[48] on the other hand, there is a vast family of non-trivial anti-self dual Ricci-flat 4-manifolds. It is a striking fact that these can *all* be obtained (at least locally) by means of the aforementioned twistor procedure!

What are we to do with this complex space $\mathcal{M}$? *Physically*, the interpretation of a complex anti-self-dual Ricci-flat complex 4-space (if it can be said to be of 'positive frequency' in some appropriate sense) is that it represents a *left-handed graviton*. In fact, it is a *non-linear* graviton, in the sense that it is a 'wavefunction' of a kind, but now it is a solution of the actual non-linear Einstein vacuum equation (Ricci-flatness), rather than of its linear approximation. The latter would have been the case if we had just taken the twistor function f as a cohomology element, rather than allowing the 'paint to dry' and thereby deform the twistor space itself. We see that twistor theory has carried us in a curious and previously unexpected direction in the unification of quantum-theoretic ideas with spacetime structure. Our twistor wavefunctions are now non-linear entities, so that deviations from standard rules of linear quantum mechanics (§§22.2–4) are beginning to appear.

There is a feature of this construction that is particularly noteworthy. If we take any point $\mathbf{Z}$ of the curved twistor space $\mathcal{T}$, we find that any sufficiently small neighbourhood of $\mathbf{Z}$ has a structure identical to that of some neighbourhood of any chosen point $\mathbf{Z}'$ of the *flat* twistor space $\mathbb{T}$ (not lying on the 'infinite' region $\mathbf{I}$—see §33.5). Accordingly, the local structure possessed by twistor space is 'floppy', in the sense in which this word was used in §14.8. Thus, all the information concerning curvature, etc., of the space $\mathcal{M}$ is stored *globally* in $\mathcal{T}$, not locally. This is a reflection of the fact, referred to above, that a cohomology element defined by a twistor function disappears completely when restricted down to a small enough region. There are no 'field equations' in twistor space. The kind of information that is normally stored in solutions of field equations in spacetime (in this case, the anti-self-dual Einstein equation) seems to be stored only non-locally in a twistor-space construction.[49]

33.12 Twistors and general relativity

This 'non-linear graviton construction' has been central to the development of twistor theory since the mid-1970s. In its initial form, it cried out for advances in two different directions. The most obvious of these was for a corresponding construction for the *right*-handed non-linear graviton, and for this to be combined with the left-handed one so that mixed polarizations states (such as plane-polarized non-linear gravitons) could be formed. This would appear to be a key part of the twistor programme. As noted above, the notion of a 'non-linear graviton' is very much in the spirit of a search for a theory, as strongly promoted in Chapter 30, in which the standard linear rules of ordinary **U**-quantum theory need to be bent in order that the correct union with Einstein's general relativity may be obtained. However, the 'graviton' that has arisen in the above construction is only 'half a graviton', in that only one of the two possible helicity states has been incorporated.

Some alert readers might venture the suggestion that if we pass to a description in terms of *dual* twistors W_α, rather than twistors Z^α, then a non-linear wavefunction for a *right*-handed graviton would be obtained, by repeating the foregoing construction in terms of dual twistors.[33.30] This way around it would be the right-handed graviton that corresponds to homogeneity of degree 2 (in W_α) and the left-handed one that corresponds to -6 homogeneity. This does not get us out of our difficulties, however, because now we lose the left-handed helicity states—and it would make no sense to use the W_α variables for the right-handed states and the Z^α variables for the left-handed ones, most immediately because we also need to describe mixed helicity states.[50]

The problem of somehow 'exponentiating' the -6 homogeneity twistor functions $f(Z^\alpha)$ to obtain a right-handed non-linear graviton has been referred to as the (gravitational) *googly* problem. (The word 'googly' is a cricketing term that describes a ball which spins in a right-handed sense about its direction of motion, though seemingly bowled with an action that would normally impart a left-handed spin.) It has taken nearly 25 years to find a plausible solution, but recent developments do appear to provide an appropriate construction for this.[51] Yet, at the time of writing, the procedures still remain conjectural in some important respects. I shall not significantly attempt to describe these developments here, and merely say that the essential new feature is that the fibres of the projection of our

---

[33.30] Why? *Hint*: Why does a spatial reflection convert twistors into dual twistors?

curved twistor space $\mathcal{T}$ down to the projective space $\mathbb{P}\mathcal{T}$ get 'twisted up' in a way that is defined by a twistor function of homogeneity degree -6. (The 'twist' is effected by exponentiating a vector field, on a pair of overlapping patches, of the deceptively simple form $Cf_{-6}Z^{\alpha}\partial/\partial Z^{\alpha}$, where C is a suitable constant, f_{-6} being a -6-homogeneity-degree twistor function.) This allows both the left-handed and right-handed parts of the graviton to be incorporated together.

At least in the case of a suitably asymptotically flat spacetime $\mathcal{M}$, there is a direct explicit construction for $\mathcal{T}$ in terms of $\mathcal{M}$. Moreover, there is a tentative proposal for obtaining $\mathcal{M}$ from a given $\mathcal{T}$, i.e. for constructing spacetime points from the purely twistorial structure of $\mathcal{T}$, which is conjectured to ensure that the required Ricci flatness (Einstein's vacuum equation) is correctly incorporated. The proposal relates, in significant ways, to a long-term research project due to Ezra T. Newman and his colleagues, for interpreting spacetime points in terms of what are called 'light-cone cuts', these being the intersections of light cones in $\mathcal{M}$ with future null infinity $\mathscr{I}^{+}$.[52] However, though apparently promising, some important aspects of this twistor construction remain unresolved at the time of writing.[53]

The other direction in which the original left-handed non-linear graviton construction (of 1975/6) cried out for advances was in generalizations from gravitational theory to other gauge fields. Very early on, in 1976–7, Richard Ward showed how the general anti-self-dual gauge fields could also be obtained using a twistor construction somewhat similar to the gravitational one. In fact, the Ward construction has led to a considerable degree of mathematical interest and development by Ward and others, particularly in the area of *integrable systems* (non-linear equations that can, in an appropriate sense, be solved in the general case). Here twistor theory has provided a powerful overview for the subject as a whole.[54] It seems likely that the above-mentioned advances towards a full solution to the gravitational mixed-helicity problem will point to a way that general (mixed-helicity) gauge fields are also to be treated within the twistor formalism.

33.13 Towards a twistor theory of particle physics

This leads us to the question of how twistor theory might develop into a full-blown physical theory—which it is not, at the moment. For this to happen, it is important that two additional areas of study in twistor theory are developed further. The first is to provide a comprehensive treatment of QFT. In fact, there has been a considerable body of activity, developed mainly by Andrew Hodges and his students in Oxford (with some initial input from me and some others in the early 1970s), which provides a perturbative approach to QFT where Feynman diagrams are replaced by constructions known as *twistor diagrams*. These involve high-dimensional contour integration, and the formalism achieves some striking success in avoiding many of the infinities that are encountered using the conventional Feynman procedures.[55] The approach is still somewhat more complicated than one would like, however, and it lacks an independent underlying guiding principle, like that of §§26.6–8, telling us exactly what contour integrals to perform, without our having to appeal to the conventional Feynman expressions as intermediaries.

The other of these areas is *twistor particle theory* which was primarily developed by Zoltan Perjés, George Sparling, Lane Hughston, Paul Tod, and Florence Tsou (Tsou Seung Tsun) from ideas that I introduced, in the mid 1970s to early 1980s, but which largely has lain dormant since that period. The basic idea here is that, whereas massless particles can be described by twistor wavefunctions of a single twistor variable, say $f(Z^\alpha)$, massive particles require more variables, e.g. $X^\alpha, \ldots, Z^\alpha$. There is an expression for the momentum and angular momentum of a massive particle that involves summing up the indvidual contributions from all these twistors, but there is now an internal symmetry *group* arising from transformations among these twistor variables and their complex conjugates that do not affect the total momentum and angular momentum. It is perhaps noteworthy that one gets groups that include, but slightly generalize, the U(2) of electroweak interactions and the SU(3) of strong interactions. A number of striking relationships with the standard classification of particles according to the standard model were noted, but the scheme stagnated for certain technical reasons. There appears to be a reasonable prospect that the recent developments on the 'googly problem'—particularly if they can be applied to gauge fields—could open up the subject again.

There is, to my mind, also a significant possibility that the Chan–Tsou proposal for a particle-physics model, briefly described in §25.8, might tie in significantly with these developments. That proposal requires there to be a dual group to each (non-Abelian) particle symmetry group, in addition to the original gauge group. Twistor theory suggests that, in accordance with Ward's construction mentioned above—together with its conjectured 'googly' version—each group should feature in anti-self-dual and self-dual versions, and this appears to require that the dual form of the gauge group should play a significant role in addition to the role of the original gauge group. Thus, through the agency of ideas from the Chan–Tsou proposal, the twistor particle programme could well play its part in a future particle physics. It is to be anticipated, moreover, that successful progress in this area should also have an important impact on the QFT programme of twistor diagram theory.

33.14 The future of twistor theory?

In my descriptions of twistor theory, I have omitted to give the reader my statutory warning that my own views on the subject do not reflect those of the community of physicists at large. In fact, since I have devoted more than half of my life span to twistor theory (on and off), it is hardly likely that my own perspective will correspond closely to the perspectives of that vast majority of physicists who have not been so devoted. I should make clear, moreover, that the community of physicists who know much about the subject is rather small, and certainly exceedingly small in relation to those who know something about string theory or supersymmetry. Twistor theory could in no way be called a 'mainstream' activity of theoretical physicists today.

Yet twistor theory, like string theory, has had a significant influence on pure mathematics, and this has been regarded as one of its greatest strengths. Twistor theory has had an important impact on the theory of integrable systems (as mentioned briefly above), on representation theory,[56] and on differential geometry. (In this last area, I should mention the work of Sergei A. Merkulov and L.J. Schwachhöfer, who were able to find a solution to what is known as the 'holonomy problem', using methods developed from those of the original non-linear graviton construction.[57] In related work, twistor theory has a significant value in the construction of what are called 'hyperkähler manifolds', 'Zoll spaces', etc.[58]) Twistor theory has been greatly guided by considerations of mathematical elegance and interest, and it gains much of its strength from its rigorous and fruitful mathematical structure.

That is all very well, the candid reader might be inclined to remark with some justification, but did I not complain, in Chapter 31, that a weakness of *string* theory was that it was itself largely mathematically driven, with too little guidance coming from the nature of the physical world? In some respects this is a valid criticism of twistor theory also. There is certainly no hard reason, coming from modern observational data, to force us into a belief that twistor theory provides the route that modern physics should follow. Also, many might well feel that the strongly chiral nature of the theory takes things too far in the direction of spatial asymmetry. There is no physical evidence, after all, that a left–right asymmetry has any role to play in gravitational physics. In Chapters 27, 28, and 30, I have been stressing the need for time-asymmetry in the appropriate quantum–gravitational union, but there is no apparent physical requirement for *space*-asymmetry (except perhaps indirectly, via the CPT theorem of QFT; see §25.4 and §30.2).

Of course, it may be the case that the space-asymmetry in the formalism will simply not translate itself into an asymmetry in physical effects. The best reason to hope that this may be true lies in the fact that the algebras generated by the pair $(Z^\alpha, -\hbar\partial/\partial Z^\alpha)$, on the one hand, and by

$(\hbar\partial/\partial\bar{Z}_\alpha, \bar{Z}_\alpha)$, on the other, are formally identical. This suggests that, whatever conclusion we might reach using a twistor description (Z^α variables) might equally be obtained using a dual-twistor description (using $W_\alpha = \bar{Z}^\alpha$ variables), and that this similarity is so complete that no left–right asymmetry in gravitation will emerge in the resulting theory. On the other hand, if the formalism is to mirror Nature, then we shall require a left–right *a*symmetry when the theory comes to describe weak interactions (§25.3). But, as twistor theory stands, in its present relatively primitive state, there is no clear reason for this difference.

The main criticism that can be levelled at twistor theory, as of now, is that it is not really a *physical* theory. It certainly makes no unambiguous physical predictions. My own (over-)optimistic perspective would be to regard twistor theory as being vaguely comparable with the Hamiltonian formalism of classical physics. Hamiltonian theory did not introduce physical changes, but it provided a different outlook on classical physics that later proved to be just what was required for the new quantum theory according to Schrödinger's prescriptions, as described in Chapters 21–23. Twistor theory, likewise, is merely a reformulation that does not necessarily introduce physical changes. The optimistic hope is that its framework might also provide a leaping-off point for some significant physical developments in the future.

There is, of course, no compulsion for the skeptic to believe that such developments will take place, and the primary case for twistor theory indeed lies, like string theory (or M-theory) in the strength of its aesthetic or mathematical appeal. The two theories are, however, mathematically incompatible as they stand, because they operate with different numbers of spacetime dimensions. One might justly (but perhaps over-harshly) say that it is a prediction of twistor theory that the aspirations of string theory are wrong—or, conversely, that it is a prediction of string theory that those of twistor theory are wrong! This incompatibility does not extend to variants or re-interpretations of string (or M-) theory in which the extra dimensions are not taken to be spacetime dimensions at all, but are regarded as 'internal' dimensions of some kind. Although such a re-interpretation appears to provide a consistent viewpoint, it is somewhat at variance with the driving force behind string theory, as normally espoused.

In this connection, I should remind the reader of certain very recent work, alluded to in §31.18, primarily by Edward Witten.[59] This points to some fascinating possibilities for a new outlook on Yang–Mills scattering amplitudes. It combines ideas of twistor theory with some of those from string theory—but now in a 4-dimensional context!

In any case, twistor theory does require some new input. Among the most important ingredients of other successful physical theories have been Lagrangians and Feynman path integrals, these providing the appropriate

QFT way of dealing with field equations (see §26.6). Twistor theory boasts the evaporation of field equations (§§33.9,11), however, so some new ideas seem to be required for the development of a full twistor QFT.[60]

Does twistor theory have any clear-cut 'predictions'? The closest to a prediction that I can think of is that the underlying motivations of the theory seem to imply that the universe ought to have negative spatial curvature, i.e. $K<0$. To see the reason for this expectation, first recall from Chapters 27 and 28 (especially §27.13) that the Big Bang seems to have had an extraordinarily uniform nature, with a very close resemblance to one of the FLRW models. These models are conformally flat (vanishing Weyl curvature) and can be described very simply in terms of a flat twistor space ($\mathbb{CP}^3$).[61] In each of the cases $K>0$, $K=0$, $K<0$, there is an exact symmetry group, but only in the case $K<0$ is this a holomorphic group. In fact, in that case, the group is precisely the one that started us off with the 'complex magic' of twistor theory, namely the Lorentz group O(1, 3), which (ignoring reflections) is the group of holomorphic transformations of the Riemann sphere. Where is this Riemann sphere? It is the 'infinity' of hyperbolic 3-space—like the bounding circle of Escher's picture, reproduced in Fig. 2.11—analogous to the celestial sphere of §18.5, as a boundary to the hyperbolic 3-space of §18.4; see Fig. 18.10.

We see that $K<0$ is not so much a prediction of twistor theory, but of the underlying holomorphic philosophy. Can we go further and say anything about the cosmological constant Λ? Presently proposed twistor constructions (see §33.12) seem to be able to accommodate the Einstein vacuum equation in a natural way only in the case $\Lambda = 0$, and it is hard to see how the present type of procedure can be modified in order to accommodate $\Lambda \neq 0$. Does this tell us that $\Lambda = 0$ is a prediction of twistor theory? It had better not (despite my own previous preference for $\Lambda = 0$)! For impressive recent observational data (see §28.10) strongly indicate $\Lambda > 0$. This simply provides twistor theory with new challenges. Clearly twistor theory will have to do a lot better than just this if it is to become respectable as a physical theory!

What about the rules of quantum theory? Does twistor theory point to any specific directions for change, in accordance with the aspirations of Chapter 30? The 'non-linear graviton' of §33.11 does begin to indicate that the twistor approach will ultimately entail a (non-linear) modification of the rules of quantum mechanics. However, there is not yet very much, within the twistor formalism, that indicates the presence of a fundamental time-asymmetry in these modifications, as would be required according to the discussions of §§30.2,3,9. However, it is a possibly suggestive feature of the particular 'googly' developments that were briefly discussed in §33.12 that they do seem to depend upon a time-asymmetric description. The strength of this possibility will have to await future developments, and the comments of the previous paragraph should be kept in mind. Accordingly, twistor theory so far says nothing useful about quantum state reduction, despite this phenomenon having provided a significant part of the initial motivating inputs behind the theory.

Finally, let us address the issue of the status of the underlying holomorphic philosophy that forms one of the main drives behind twistor theory. I think that it is fair to say that this philosophy has indeed been maintained and has provided a powerful driving force—which has in some respects exceeded expectations (such as with the twistor representations of massless fields, both linear (§§33.8–10) and non-linear (§§33.11,12). Yet, at some point, the theory will have to say something about real-number aspects of physics and non-holomorphic behaviour, such as the emergence of probability values (in accordance with the non-holomorphic squared-modulus rule $z \mapsto |z|^2$) and *real* spacetime points, where we would hope to be able to accommodate non-analytic (let alone non-holomorphic) behaviour. With regard to this last issue, some encouragement should be gained from the remarkable theory of hyperfunctions, introduced at the end of Chapter 9 (see §9.7), according to which non-analytic behaviour can be represented very elegantly within the context of holomorphic operations. The extent to which a future twistor theory will be able to address such issues is a matter for the future.

Notes

Section 33.1

33.1. See Ahmavaara (1965).

33.2. See Schild (1949); 't Hooft (1984); and Snyder (1947).

33.3. See Sorkin (1991); Rideout and Sorkin (1999); Markopoulou and Smolin (1997); one of the most important developments in this field was Markopoulou (1998).

33.4. See Kronheimer and Penrose (1967); Geroch *et al.* (1972); Hawking *et al.* (1976); Myrheim (1978); 't Hooft (1978).

33.5. See Finkelstein (1969).

33.6. See Smolin (2001); Gürsey and Tze (1996); Dixon (1994); Manogue and Schray (1993); Manogue and Dray (1999).

33.7. See Regge (1962) for the original reference. Immirzi (1997) has written an informal (and informative) review.

33.8. Jozsa developed these ideas in his D. Phil. thesis. See Jozsa (1981).

33.9. See Isham and Butterfield (2000).

33.10. See Goldblatt (1979).

33.11. See Eilenberg and Mac Lane (1945); Mac Lane (1988); Lawvere and Schanuel (1997).

33.12. See Baez and Dolan (1998); Baez (2000); Baez (2001); Chari and Pressley (1994).

33.13. See Connes and Berberian (1995).

33.14. There are also many other uses of non-commutative geometry, both in pure mathematics and as applied to physics. See Connes (1990, 1998). As an example of the latter, an elegant formalism for the comprehensive treatment of renormalization has been developed with the help of non-commutative geometry; see §26.9 and Kreimer (2000).

33.15. See Connes and Berberian (1995).

Section 33.2

33.16. For strict accuracy, we need to include a (Riemann) sphere's worth of 'light rays at infinity' to complete the definition of $\mathbb{PN}$; see §33.3.

33.17. This would need the appropriate non-directional (scalar) quantities for the Poincaré group (namely its Casimir operators; see §22.12). These are the total spin and the rest mass (squared). However, rest mass is not known to be built up as integral multiples of anything, so the combinatorial aspects of such a scheme are not so clear. This approach was nevertheless developed by John Moussouris in his Oxford D. Phil. thesis in 1983 (see Moussouris 1983). It required an additional label attached to the lines of the network as well as mass and spin.

Section 33.3

33.18. See McLennan (1965); Penrose (1963, 1964, 1965a, 1986).

Section 33.4

33.19. See Penrose and Rindler (1986), particularly the Appendix.

33.20. See Harvey (1990); Penrose and Rindler (1986); Budinich and Trautman (1988).

Section 33.6

33.21. See Penrose and Rindler (1986); Huggett and Tod (2001).

33.22. At any event $\boldsymbol{x}$ in spacetime, there are two null directions specified. There is the direction of the 'light ray' of this family through $\boldsymbol{x}$ and there is the direction of the 4-momentum of the spinning particle that the twistor represents. These two null directions are the 'principal null directions', i.e. the directions defined by the Majorana representation (see §22.10), of the (self-dual or anti-self-dual part of the) angular momentum that this particle possesses, taken about $\boldsymbol{x}$. See Wald (1984); Huggett and Tod (2001).

33.23. See Penrose (1967, 1975, 1987b); Penrose and Rindler (1986).

Section 33.7

33.24. See Penrose (1968b); Huggett and Tod (2001).

33.25. See Huggett and Tod (2001); Penrose and Rindler (1986); Hughston (1979).

Section 33.8

33.26. See Dirac (1936); Fierz (1938, 1940); Penrose (1965a).

33.27. See Fierz and Pauli (1939); Penrose and Rindler (1986); Penrose (1965a); Penrose and MacCallum (1972).

33.28. See Penrose (1968b, 1969c, 1987b); Huggett and Tod (2001); Hughston (1979); Whittaker (1903); Bateman (1904, 1944).

33.29. This is the $\mathbb{C}^2$ which represents the line $\mathcal{R}$ in $\mathbb{PN}$ (Fig. 33.11). More familiar to most twistor theorists is the entirely projective version of this contour integral, for which the 1-form $\boldsymbol{\iota} = \pi_{0'}\mathrm{d}\pi_{1'} - \pi_{1'}\mathrm{d}\pi_{0'}$ is used in place of the 2-form $\boldsymbol{\tau} = \mathrm{d}\pi_{0'} \wedge \mathrm{d}\pi_{1'}$. The contour integral is now 1-dimensional, and its relation to the 2-dimensional prescription given here is that one of these contour dimensions (given by a circle S^1) reduces the non-projective version given here to the more familiar projective version. An advantage of the present version is that it allows mixed helicity states to be described.

33.30. See Huggett and Tod (2001); Hughston (1979); Penrose and Rindler (1986).

33.31. Twistor theory owes a profound debt to Sir Michael Atiyah for an important early input to this realization. See Penrose (1979b) for the original rough arguments for twistor cohomology and Eastwood *et al.* (1981) for a thorough treatment.

Section 33.9

33.32. This is what is referred to as *Čech* cohomology. There are also many other ways of getting at the cohomology concept. See Wells (1991); Ward and Wells (1989); Griffiths and Harris (1978).

33.33. See Gunning and Rossi (1965); Ward and Wells (1989); Wells (1991); also Penrose and Rindler (1986).

33.34. See Gunning and Rossi (1965); Penrose and Rindler (1986).

33.35. Penrose and Rindler (1986).

33.36. See Penrose and Penrose (1958).

33.37. See Penrose (1991).

33.38. See Gunning and Rossi (1965); Griffiths and Harris (1978); Chern (1979); Wells (1991).

33.39. See references in Note 33.38, as well as Eastwood *et al.* (1981).

Section 33.10

33.40. There is no significance in the reversal of the + and the − here. This is just an awkward notational accident. See §9.2.

33.41. There are certain technicalities of relevance here. If the original field is not analytic (not C^ω), these fields (on $\mathbb{M}$) may turn out to be hyperfunctional in the sense of §9.7, see Bailey *et al.* (1982).

33.42. See Hodges *et al.* (1989).

33.43. See Penrose (1987b).

Section 33.11

33.44. See Penrose (1976a, 1976b); Ward (1977); Penrose and Ward (1980); Penrose and Rindler (1986).

33.45. There is a technical subtlety that it is not a *holomorphic* fibre bundle (§15.5), despite all the operations in the construction being holomorphic, since, locally in the π space, it is not strictly a holomorphic product space. $\mathcal{T}$ is referred to as a *holomorphic fibration*. See Penrose (1976b).

33.46. Under normal circumstances; see Huggett and Tod (2001), Ward and Wells (1989); Penrose and Ward (1980).

33.47. See Kodaira (1962).

33.48. Or in the real positive-definite case $(++++)$ or split signature case $(++--)$. See Penrose (1976b); Hansen *et al.* (1978); Atiyah *et al.* (1978); Dunajski (2002).

33.49. There would therefore seem to be a significant relation to the notion of a *topological QFT*, as referred to in §32.5; see Note 32.17.

Section 33.12

33.50. There is an reflection-symmetric approach to these problems using what are known as *ambitwistors*, and this has enjoyed some significant partial successes in this direction; see Penrose (1975); LeBrun (1985, 1990); Isenberg *et al.* (1978); Witten (1978). See also Penrose and Rindler (1986). A flat-space ambitwistor is basically a pair (W_α, Z^α), where $W_\alpha Z^\alpha = 0$, and it describes a complex light ray.

However, this does not fit in with the philosophy that 'twistor functions are wavefunctions', as adopted above, since an ambitwistor description is more like a classical one, in which a variable and its conjugate variable—here Z^α and W_α—both appear, rather than just one or the other as would be appropriate for a wavefunction. The ambitwistor approach also encounters some mathematical awkwardness in its description of non-linear fields.

33.51. See Penrose (2001).

33.52. See, for example, Frittelli *et al.* (1997); Bramson (1975); Penrose (1992).

33.53. See Penrose (2001).

33.54. See Mason and Woodhouse (1996).

Section 33.13

33.55. For some of the older references, see Penrose and MacCallum (1972); Penrose (1975). For more recent work, see Hodges (1982, 1985, 1990a,b, 1998).

Section 33.14

33.56. See Bailey and Baston (1990); Baston and Eastwood (1989); Mason and Woodhouse (1996) for the use of twistors in mathematics.

33.57. See Merkulov and Schwachhöfer (1998).

33.58. See Gindikin (1986, 1990); Lebrun and Mason (2002).

33.59. See Note 31.81.

33.60. Although Lagrangians have played a peripheral role in twistor theory in understanding physical interactions, they have not yet found a proper general formulation within that theory. There is perhaps some irony that the very success that twistor theory has had in presenting physical fields in a way that implicitly solves their field equations (by means of homogeneous twistor functions, in the case of a free massless fields) is what leads to a difficulty with its Lagrangian formulation. In the conventional quantum formulation, field equations come about from a 'sum over histories' (§26.6), where it must be explicitly possible to *violate* the field equations in the formulation, in order for this idea to make sense, and then the quantum corrections to the classical theory come about from the examination of this path integral in more detail. This is all lost to us if the formulation does not allow for the violation of the field equations! It seems to me that some reassessment is needed, as to what must be the true 'essence' of Lagrangians in twistor theory, and, indeed, in physical theory generally. Perhaps this relates to the worries I have expressed at the end of §26.6 and to the genuinely profound issues presented by the virtually ubiquitous divergence problem that arises with path integrals (see §26.6). But see end of §32.5 and Note 32.17.

33.61. Penrose and Rindler (1986), §9.5.

34
Where lies the road to reality?

34.1 Great theories of 20th century physics—and beyond?

Let us try to take stock of what we have learnt from our physical theories—as we begin to explore the third millennium AD—concerning the fundamental nature of this remarkable world in which we find ourselves. There is no doubt that extraordinary advances in understanding have been made, and that these have come about through careful physical observation and superb experimentation, through physical reasoning of great depth and insight, and through mathematical arguments ranging from the complicated but routine to inspirational leaps of the highest order. These have led us from the understanding of the ancient Greeks concerning the geometry of space through to Newtonian mechanics, to the magnificent structures of classical mechanics, then to Maxwell's electromagnetic theory, and to thermodynamics. More recently, the 20th century gave us special relativity, leading to Einstein's extraordinary and precisely verified general theory of relativity, and we also have the deeply mysterious yet profoundly accurate and broad-ranging quantum mechanics and its development to quantum field theory (QFT); in particular, we have the remarkably successful standard models of particle physics and cosmology.

It has been a not uncommon view among confident theoreticians that we may be 'almost there', and that a 'theory of everything' may lie not far beyond the subsequent developments of the late 20th century. Often, such comments had tended to be made with an eye on whatever had been the status of the 'string theory' that had been current at the time. It is harder to maintain such a viewpoint now that string theory has transmogrified to something (M- or F-theory) whose nature is admitted to being fundamentally unknown at present.

From my own perspective, we are much farther from a 'final theory' even than this. I have no faith at all that the developments outlined in Chapter 31 are at all close to the right lines. Various remarkable *mathematical* developments have indeed come out of string-theoretic (and related) ideas. However, I remain profoundly unconvinced that they are very much other than just striking pieces of mathematics albeit with some input from deep physical ideas. For theories whose spacetime dimensionality exceeds what we directly observe (namely $1 + 3$), I see no reason to believe (§§31.11,12) that, in themselves, they carry us much further in the direction of *physical* understanding. With regard to the other schemes that have been put forward, such as the main ones outlined in Chapters 32 and 33, with which I am much more in accord, there is no doubt in my mind that they, also, lack some important insights. It would be unwise to predict with any great confidence that even these theories are close to making the further necessary leaps that would guide us to the true road to the understanding of physical reality.

Yet, already in the 20th century, the human species had undoubtedly made some extraordinary progress towards such an understanding, and I have attempted, in this book, to convey something of what has been accomplished. Einstein's general relativity stands out, in my opinion, as that century's greatest single achievement. Quantum theory (and QFT) might well be regarded by most physicists as an even greater achievement. From my own particular perspective on the matter, I do not feel able to share that view. While it is undoubtedly the case that quantum theory has explained incomparably more than general relativity, over a vastly greater range of different phenomena, I do not regard the theory as having yet achieved the necessary coherence *as* a theory. The problem, of course, is the measurement paradox, considered at length in Chapter 29. In my opinion, quantum theory is incomplete. When it is completed—which I would anticipate happening some time in the 21st century—it will, no doubt, represent an even greater achievement than Einstein's general relativity. Indeed, as the claims of Chapter 30 would strongly suggest, such a completed quantum mechanics ought to include Einstein's theory as the limit case for large mass and distance. (And I hope that it is clear to the reader from my remarks in §31.8 that I certainly do not regard string theory as having already achieved this union, despite many claims to the contrary.)

In my view, general relativity is probably here to stay as a description of spacetime in the large-scale limit (where the presence of a cosmological constant Λ is permitted as part of Einstein's theory), although we must expect serious modifications to its descriptions at the absurdly tiny Planck distance of 10^{-35} m, or where densities may approach the Planck value of about 5×10^{93} times that of water in the vicinity of some spacetime singularities. This position on the status of general relativity must now be regarded as the conventional one. The theory's observational status, at least at the rather large end of the distance scale of orbiting neutron stars and gravitational lensing effects, and even of black holes, must be regarded as excellent. And here I mean the standard Einstein theory, without a cosmological constant.

But what about the cosmological constant? Observations over the past few years appear to favour a positive value for it. If Λ is indeed there, it is certainly very small in ordinary local terms. If we think of Λ as a curvature, then it is the reciprocal of the square of a distance, that distance being on a scale comparable with the radius of the observable universe, so Λ is certainly ignorable at all but cosmological scales. When we interpret Λ as an effective density Ω_Λ, then that density cannot be more than 2 or 3 times the tiny average matter density that our universe now has, which is about 10^{-27} kgm^{-3}—considerably less than the best artificially produced vacuum here on Earth. Again, Λ could only have relevance on cosmological scales. Yet, on the basis of the viewpoint frequently expressed by quantum field theorists, Λ is really a measure of the effective density of the vacuum, generated by 'quantum-mechanical vacuum fluctuations' (a feature of Heisenberg uncertainty in QFT, see §21.11; see also §29.6, §30.14) and, accordingly, it 'ought' to have a value (comparable to the Planck value) that is something like 10^{120} larger (or, according to some proposals, possibly only 10^{60} larger) than the upper limit of what is observed! This is regarded as a fundamental puzzle in QFT,[1] unresolved by any of the conventional approaches to quantum gravity or by string theory. My own attitude is to be less disturbed by this than many theorists appear to be. My guess (see §30.14) is that the whole issue of 'vacuum fluctuations' will need to be radically overhauled when we have a better quantum theory of gravity and, indeed, better QFT.

We must, of course, recognize the extraordinary range of phenomena that give support to existing quantum mechanics and to QFT. But I should make clear that there is no contradiction, in this, with the viewpoint that I have argued for in Chapter 30, where changes in the foundations of quantum theory are anticipated. No experiment to date seems to have yet got very close to exploring the 'quantum-gravity' level at which I expect such changes to become manifest, with state-vector reduction occurring objectively (gravitational OR). The observed quantum entanglements over distances of up to 15 kilometers[2] are completely consistent with these expected changes, since these entanglements involve only pairs of photons with energies of the order of 10^{-19}J, and spontaneous state reduction according to gravitational OR is not to be expected until the photons are actually measured (at which point OR would take place in the measuring apparatus). The present experimental situation with regard to the validity of quantum mechanics when significant mass movements are involved is best revealed in recent experiments by Anton Zeilinger and colleagues in Vienna.[3] They have performed what is basically a two-slit experiment with C_{60} (and also C_{70}) 'bucky balls'. These are *fullerenes*, where each molecule has 60 carbon atoms, in a beautifully symmetrical arrangement resembling the pattern of seams on most modern soccer balls (or else a less symmetrical arrangement involving 70 carbon atoms). These fullerene molecules are about a nanometer in diameter and they interfere with themselves after having been in a superposition of two locations separated by about 10^{-7}m, which is some 100 times the bucky ball's diameter. According to the scheme suggested in §30.11, such a superposition would last for some hundred thousand years or so before spontaneously reducing according to gravitational OR, so there is clearly no contradiction here with the Zeilinger experiment.

Of course, the situation might well be different with some future experiments. Something like the FELIX space-borne proposal of §30.13, or more likely some related experiment such as could result from Dik Bouwmeester's work in Santa Barbara, could directly test the gravitational OR scheme, and it may well be that there are other possibilities for experiments that could be performed early in the 21st century. I regard this prospect as particularly exciting, and there is the distinct possibility that such experiments could significantly change our present outlook on quantum mechanics. At the very least, they could severely limit speculations on how quantum mechanics might be modified according to some future theory.

This is in stark contrast with the present (or plausibly projected) experimental situation with regard to other attempts, such as those outlined in Chapters 31–33, at combining quantum theory with gravitation. Most considerations of experiments designed to address such quantum-gravity proposals involve particles hurled with extraordinarily high energies, absurdly far beyond the capabilities of any existing (or seriously projected) particle accelerator. (The only exceptions to this that I am aware of are experiments designed to test the possibility of the existence of 'large' extra dimensions (§31.4) which could, for example, affect the inverse square law of gravity at small distances, or some other loosely related proposals aimed at seeing whether Lorentz covariance might be violated at high energy, owing to suggested quantum–gravitational effects.[4]) There is, indeed, a profound difficulty confronting the testing of any 'conventional' quantum-gravity scheme, where the effects of this union would modify only spacetime structure (at the extraordinarily tiny Planck distances or times), leaving the standard procedures of quantum mechanics intact. We are in much better shape, experimentally, if the rules of quantum mechanics are modified by general-relativistic effects, as suggested in Chapter 30, since these proposed effects are just about within the scope of present-day technology. If such experiments are successfully performed and indicate a need for a change in the rules of quantum mechanics, then there would at last be some good physical guidance to supplement the largely mathematical desiderata driving current quantum-gravity research.

The absence of experimental data relating to the normal quantum-gravity proposals has led to a curious situation in theoretical fundamental physics research. A general consensus seems indeed to have grown up that, in order for real progress to be made in our moving beyond the standard models of particle physics (and cosmology), and thereby obtaining a deeper understanding of the basic ingredients of the universe, it will be necessary to have a quantum theory that encompasses gravity in addition to the strong, weak, and electromagnetic forces. Part of the reason for this appears to be the (no doubt physically justified) conception that a *finite* (as opposed to merely renormalizable) QFT will require divergences to be 'cut off' at the tiny Planck distance, whence gravity must necessarily be part of the picture (see §31.1). But since experiments in this area are absent, the efforts of theoreticians have been directed very much into the internal world of mathematical desiderata.

34.2 Mathematically driven fundamental physics

The interplay between mathematical ideas and physical behaviour has been a constant theme of this book. Throughout the history of physical science, progress has been made through finding the correct balance between, on the one hand, the strictures, temptations, and revelations of mathematical theory and, on the other, precise observation of the actions of the physical world, usually through carefully controlled experiment. When experimental guidance is absent, as is the case with most current fundamental research, this balance is thrown out of kilter. Mathematical coherence[5] is far from a sufficient criterion for telling us whether we are likely to be 'on the right track' (and, in many cases, even this apparently necessary requirement is thrown to the wind). We find that aesthetic mathematical values begin to take on a much larger role than they did before. Researchers often point to the successes of Dirac, of Schrödinger, of Einstein, of Feynman, and of many others, in their being guided to some considerable extent by the aesthetic attractions of the particular theoretical ideas that they put forward. It is my opinion that there is no denying the value of such aesthetic considerations, and they play a fundamentally important role in the selection of plausible proposals for new theories of fundamental physics.

Some such aesthetic judgements may sometimes merely express a clear-cut need for a mathematically coherent scheme; since mathematical beauty and coherence are indeed closely related. It seems to me that the need for such coherence, in any proposed physical model, is unarguable. Moreover, unlike many aesthetic criteria, mathematical coherence has the advantage that it is fairly clearly something objective. The difficulty with aesthetic judgements, in general, is that they tend to be very subjective.

Yet mathematical coherence need not itself be readily appreciated. Those who have worked long and hard on some collection of mathematical ideas can be in a better position to appreciate the subtle and often unexpected unity that may lie within some particular scheme. Those who come to such a scheme from the outside, on the other hand, may view it more with bewilderment, and may find it hard to appreciate why such-and-such a property should have any particular merit, or why some things in the theory should be regarded as more surprising—and, perhaps, therefore more beautiful—than others. Yet again, there may well be circumstances in which it is those from the outside who can better form objective judgements; for perhaps spending many years on a narrowly focused collection of mathematical problems arising within some particular approach results in distorted judgements!

But mathematical coherence and elegance, in the mathematics of a physical theory, despite their undoubted value, are clearly far from sufficient. Physical considerations usually have a much greater importance. But in situations where experimental guidance is lacking, mathematical qualities then assume the greater importance. I would certainly not argue that there are any simple answers to these issues. Individual researchers are right, I believe, to follow their own aesthetic drives. But they should not be surprised if they find some colleague to be completely unmoved by the alleged magnificence of the conclusions that these drives happen to lead to. I regard such aesthetic motivations to be an essential part of the development of any important new ideas in theoretical science. But without the constraints of experiment and observation, such motivations frequently carry the theory far beyond what is physically justified.

We can see many examples throughout history, where a beautiful mathematical scheme has seemed, at first, to provide a revolutionary new way to uncover Nature's secrets, yet where these initial hopes have not been realized—at least, not in the way originally anticipated. A good example must be the system of quaternions, in respect of their beautiful property of forming a division algebra. As we saw in §11.2, having been found by Hamilton in 1843, they enticed him into devoting the remaining 22 years of his life in attempts to represent Nature's laws entirely within this framework. However, this 'pure-quaternion' work (by which I mean his original actual quaternions, with their division-algebra property) had rather little direct effect on the further development of basic physical science. Hamilton's other influences on physical theory have certainly been enormous, and quite direct. For it was his own earlier researches into what we now call 'Hamiltonians', 'Hamilton's principle', and the 'Hamilton–Jacobi equation', etc.—these forming part of an exploration of the analogy between Newtonian particles and waves—that provided the jumping-off point for the 20th-century development of quantum mechanics and QFT (see §20.2 and §§21.1,2). But the influence of quaternions on physics was only distant, through generalizations in which the division-algebra property had to be thrown away.

It must have been all too easy, in the mid 19th century, to be mesmerized by the beautiful mathematical feature of quaternions that one can divide by them (§11.1). This wonderful property, enjoyed by quaternions and by comparatively few other algebras, has had a significant influence on pure mathematics, but not directly on mainstream physics. It was Clifford's generalization of quaternions to higher dimensions, together with the later ideas of Pauli and particularly of Dirac, in which the Lorentzian signature relevant to spacetime is adopted, that finally allowed enormous strides in physical theory to be made possible (§11.5, §§24.6,7). In these later developments, which were extremely important for physics, Hamilton's beautiful division property is of necessity abandoned!

I shall return to this somewhat mysterious issue of beauty in the mathematics that is successful for physics in §34.9. But these matters touch upon the important and fascinating complementary issue of mathematical 'spin-off'. Ever since ancient Greek times, theories which started close to the behaviour of the world have spawned vast areas of beautiful mathematics, initially studied for their own sake alone, but often finding applications far removed from those physical considerations from which they originated. Sometimes these applications take many centuries to be realized (such as in the case of Apollonios's study, in about 200BC, of conic sections, which played a fundamental part in the understanding of planetary motions provided by Kepler and Newton in the 16th and 17th centuries, or with Fermat's 'little theorem' of 1640, which found important applications in cryptography in the late 20th century). Mathematics—good mathematics particularly—has the habit of finding its applications in very disparate fields, which is one reason for its strength and robustness. The workings of Nature have often provided a wonderful source of such mathematical ideas. That there should be precision and reliability in such ideas stimulated by Nature is perhaps not so surprising, if we accept that Nature operates accurately in accordance with mathematical laws. More remarkable is the *subtlety* of the mathematics that seems to be involved in Nature's laws, and the habit that such mathematics seems to have in finding applications in areas far removed from its original purpose (as was the case, in particular, with the calculus of Newton and Leibniz—see Chapter 6).

But can we argue conversely that a putative physical theory which stimulates much productive research in broadly spread mathematical areas thereby gains physical believability by virtue of this? The issue has relevance particularly to the physical schemes of Chapters 31, 32, and 33. It cannot be answered in any simple way, I believe; but great caution is certainly to be recommended.

String theory, in particular, has stimulated beautiful mathematical research and gains considerable strength from this mathematical appeal.

(The same could be said of much of twistor theory and of the Ashtekar and Hawking approaches.) But it is unclear to what degree this is indicative of any underlying accord with physical reality. Yet, I have very frequently heard pure mathematicians express delight that some result that they may have found has applications in physics merely because it has a mathematical relevance to string theory! I can well understand the desire, among many pure mathematicians, that aspects of their beautiful subject should find important application to the workings of the physical world. But it should be made clear that there is (as yet) no observational reason to believe that string theory (in particular) *is* physics, although it is certainly motivated by powerful physical aspirations. String theory is also a subject that is studied by a good many physicists, but does that make it physics? This raises the issue of fashion in fundamental physical research, and I wish to address this matter next.

34.3 The role of fashion in physical theory

Let me begin by quoting a survey carried out by Carlo Rovelli, and reported in his address to the International Congress on General Relativity and Gravitation, held in Pune, India, in December 1997.[6] Rovelli is one of the originators of the loop-variable approach to quantum gravity, as described in §32.4, and he claimed no professionalism in the conducting of his survey. Yet the results he found certainly reflect what my own (unsubstantiated) expectations would have been. He made a count of articles on the subject of quantum gravity published over the previous year, as recorded in the Los Angeles Archives. The rough average of papers per month, in the various approaches to the subject, came out as follows:

String theory:	69
Loop quantum gravity:	25
QFT in curved spaces:	8
Lattice approaches:	7
Euclidean quantum gravity:	3
Non-commutative geometry:	3
Quantum cosmology:	1
Twistors:	1
Others:	6

The reader will perceive that I have only very loosely followed the demands of fashion in the space that I have devoted to these respective theories, in my accounts in this book. (I have briefly touched upon QFT in curved spaces in relation to the Hawking effect, as discussed in §30.4. Lattice approaches take discrete spacetime models in place of continuous ones—see §33.1. Euclidean quantum gravity features in the Hawking approach, as discussed in §28.9. Quantum cosmology uses simplified spacetimes where most gravitational degrees of freedom are ignored. The other approaches cited have been discussed in Chapters 31–33.) It will be noted that there were more articles in the area of string theory than in all the other areas put together. It seems to be a general view that if such a survey were repeated today, the preponderance of string theory papers would be even greater. If we were to think of scientific research as being driven by the principles of democratic government, then we would see that owing to an absolute majority being with the string theorists, all decisions as to what research should be done would be dictated by them![7]

Fortunately, the criteria of science are not those of democratic government. It is right and proper that minority activities should not suffer merely by virtue of the fact that they are in the minority. Mathematical coherence and agreement with observation are far more important. But can we ignore the whims of fashion altogether? Certainly we cannot. In addition to many less believable ideas, very fashionable in their day (such as the 11-dimensional supergravity notion of seven extra dimensions constituting a 'squashed 7-sphere'),[8] I can recall many fashions of the past that seemed—and still seem—to me to contain very significant truths (such as Regge trajectories—see §31.5—and Geoffrey Chew's analytic S matrix[9]), but which have now been out of fashion for decades. To some extent, the popularity of a theory provides a measure of its scientific plausibility—but only to *some* extent.

It is also true that, as with business concerns, it is the large ones that have a natural tendency to get larger at the expense of the smaller ones. It is not hard to see why that should be the case also with scientific fashions, particularly in the modern world of jet travel and the internet, where new scientific ideas spread rapidly across the globe, being propagated by word of mouth at scientific conferences or almost instantaneously transmitted by e-mail and on the internet in (frequently unrefereed) scientific articles. The often frantic competitiveness that this ease of communication engenders leads to 'bandwagon' effects, where researchers fear to be left behind if they do not join in. Fashion need not be so much of an issue with those theoretical ideas that continually come under experimental scrutiny. But with ideas that are as far from the possibility of experimental confirmation or refutation as are those in quantum gravity, we must be especially cautious in taking the popularity of an approach as any real indication of its validity.

Fashion also has its role to play in other areas such as with notation, or specific mathematical formalism. This is perhaps a less important issue than those discussed above, but still significant for the development of research. Let me describe one particular example, namely the highly prevalent use of Dirac's 4-component spinor formalism rather than the later 2-component one of van der Waerden (see §22.8, §24.7, §25.2). This has certain aspects of irony about it, as we shall see. In fact, in quantum electrodynamics, the 4-spinor formalism is almost universally used, whereas, as has been shown by Robert Geroch,[10] it is really a great deal simpler to use 2-spinors (briefly described in §22.8). When Dirac discovered his equation, in 1928, he used 4-spinors. Dirac's equation stimulated much interest in the importance of spinors, and a year later the distinguished Dutch mathematician Bartel L. van der Waerden formulated the powerful 2-spinor calculus.[11] However, by then, the excitement caused by Dirac's discovery of the electron's equation meant that most physicists followed Dirac's original approach, not many being even acquainted with van der Waerden's more flexible and polished formalism. Nevertheless, Dirac himself seems eventually to have appreciated the power of what van der Waerden had done. In fact, in the early 1950s I attended a lecture course of Dirac's in which he gave a beautiful introduction to the 2-spinor calculus, making the whole subject clear to me when it had been almost completely baffling in accounts that I had seen previously.

Dirac had actually used the 2-spinor approach himself, in 1936, to find generalizations of his equation for the electron to particles of higher spin.[12] But not being comfortable with the 2-spinor formalism, a number of other researchers seem to have rediscovered special cases of Dirac's higher-spin equations, which tend now to be called such things as the 'Duffin–Kemmer–Petiau' equation (1936–1939, for spins 0 and 1) or the 'Rarita–Schwinger' equation (1941, for spin $\frac{3}{2}$). It is *their* work (using tensor/4-spinor methods), in this area, rather than Dirac's earlier work, that people quote.

Dirac was no follower of fashion, and it seems that he did not even always follow the fashion that he had himself set! Nevertheless, others sometimes find themselves drawn into it even when they do not intend to be. I became acquainted with one example of this when in the mid 1970s I visited CERN to talk to Bruno Zumino, one of the originators of some of the basic ideas of supersymmetry. (His 1974 work, with Julius Wess,[13] had had a definite connection with twistor theory, and I had wanted to explore this.) He told me that he appreciated the strength of the 2-spinor formalism, and he had once written a paper in which he used 2-spinors to formulate a certain idea of his. However, a few months later, as he told me, the esteemed physicist Abdus Salam put forward the same idea, but using 4-spinors. Everyone then referred to Salam's paper and none to his own. Zumino concluded that he would not make the mistake again of using the (technically superior) 2-spinor formalism!

There is a related issue which makes it difficult for researchers, particularly young ones, to break away from the fashionable lines of research even if they wanted to. This is the sheer quantity of disparate and difficult mathematical ideas that they are confronted with in modern mathematical physics. It is hard enough to single out one small part of one particular line of work and to try to master it. To be able to make an authoritative comparative study of the overall merits of several different lines at once would certainly be beyond the capabilities of most young researchers. If they are to make a choice, they must rely on the preferences of those who are already established researchers, and this can only add to the propagation of already fashionable lines of work, at the expense of those that are less well known.

Although my remarks above have been aimed at the kind of theoretical research that is unconstrained by experimental results, the element of fashion is not unimportant in relation to experiment also, but for a somewhat different reason. This springs largely from the enormous expense that is usually involved in the setting up of experiments at the frontiers of fundamental physics. Since most experiments are indeed so expensive, they normally require government support, or the support of large commercial concerns, and there will be the need for numerous committees to decide whether to go ahead with an experiment, or whether this or that type of experiment would make a better use of limited funds. It is natural that the scientifically knowledgeable members of these committees should be those who have establised themselves for their part in developing ideas that have successfully led to the current perspectives. Thus, they would tend only to favour experiments that directly address questions that seem natural from these particular perspectives. There is therefore a significant tendency for theory to get somewhat 'locked' into particular directions. It could well be very hard to make any major change in direction for this kind of reason.

34.4 Can a wrong theory be experimentally refuted?

One might have thought that there is no real danger here, because if the direction is wrong then the experiment would disprove it, so that some new direction would be forced upon us. This is the traditional picture of how science progresses. Indeed, the well-known philosopher of science Karl Popper provided a reasonable-looking criterion[14] for the scientific admissability of a proposed theory, namely that it be *observationally refutable*. But I fear that this is too stringent a criterion, and definitely too idealistic a view of science in this modern world of 'big science'.

Let me take the example of supersymmetry in modern particle physics. It is a theoretical idea with a certain mathematical elegance and which makes the theoretician's life easier in the construction of renormalizable QFTs (§31.2). Most importantly, it is a central ingredient of string theory.

Its status among theoreticians these days is so strong that it is almost considered to be part of today's 'standard' particle-physics model. Yet, it has no (serious) experimental support (§31.2), as things stand. The theory predicts 'superpartners' for all the observed fundamental particles of Nature, but none of these has so far been observed. The reason that they have not, according to supersymmetry theorists, is that a symmetry-breaking mechanism (of unknown nature) causes the superpartners to be so massive that the energies needed to create them are still beyond the scope of present-day accelerators. With increased energy capabilities, the superpartners might be found, and a new landmark in physical theory would be thereby achieved, with important implications for the future. But suppose that still no superpartners are actually found. Would this disprove the supersymmetry idea? Not at all. It could (and probably would) be argued that there had simply been too much optimism about the smallness of the degree of the symmetry breaking, and even higher energies would be needed to find the missing superpartners.

We see that it is not so easy to dislodge a popular theoretical idea through the traditional scientific method of crucial experimentation, even if that idea happened actually to be wrong. The huge expense of high-energy experiments, also, makes it considerably harder to test a theory than it might have been otherwise. There are many other theoretical proposals, in particle physics, where predicted particles have mass–energies that are far too high for any serious possibility of refutation. Various specific versions of GUT or string theory make many such 'predictions' that are quite safe from refutation for this kind of reason.

Does the 'un-Popperian' character of such models make them unacceptable as scientific theories? I think that such a stringent Popperian judgement would be definitely too harsh. For an intriguing example, recall Dirac's argument (§28.2) that the mere existence of a single magnetic monople somewhere in the cosmos could provide an explanation for the fact that each particle in the universe has an electric charge that is an integral multiple of some fixed value (as is indeed observed). The theory which asserts that such a monopole exists *somewhere* is distinctly un-Popperian. That theory could be established by the discovery of such a particle, but it appears not to be refutable, as Popper's criterion would require; for, if the theory is wrong, no matter how long experimenters search in vain, their inability to find a monopole would not disprove the theory![15] Yet the theory is certainly a scientific one, well worthy of serious consideration.

A similar remark might be made in relation to cosmology. The region of the universe that is outside our particle horizon (§27.12) is beyond direct observation. Yet, it seems to be a reasonable scientific proposal that this region should resemble, on a broad scale, the region that is accessible to direct observation. The theory that the unobservable region does resemble the observationally accessible region—which indeed is part of the standard model of cosmology (§27.11), although not of most inflationary schemes (§28.4)—is not observationally refutable.

Moreover, even if we restrict attention to the directly observable part of the universe, we may ask whether the spatial geometry, assumed homogeneous and isotropic on a broad scale, has positive, negative, or zero curvature (the respective cases $K>0$, $K<0$, or $K=0$; see §27.11). If our theory asserts that $K=0$, then this has the character of being observationally refutable, because for any finite deviation from spatial flatness, a sufficiently precise observation could (in principle—although perhaps not in practice) discern this departure from flatness, no matter how small that spatial curvature might be. But if our theory asserts $K \neq 0$, then that theory could not be refuted if in actuality $K=0$, because there would always be some range of uncertainty in the observations that would allow for a very slight negative or positive spatial curvature. We note that the case $K>0$ could in principle be refuted if in actuality $K<0$, and $K<0$ could be refuted if in actuality $K>0$. On the other hand, $K=0$ cannot be (directly) *confirmed*,[16] whereas $K \neq 0$ could be observationally confirmed (if the universe turns out that way). Thus both assertions $K>0$ and $K<0$ are Popperian in the restricted sense that they are refutable in certain circumstances—although they are not able to be refuted if, in actuality, $K=0$—and they are also individually confirmable. We note that $K=0$ is fully Popperian, in principle, but not confirmable!

I am not sure where the Popperian perspective leaves us, in view of these various possibilities. It seems to me to be clear that each of $K>0$, $K<0$, or $K=0$ is equally 'scientific' as an assertion, despite these subtle differences with regard to Popper's criterion. And, in any case, most cosmologists would not take quite the pedantic line that I have been adopting here, namely, that '$K=0$' really means that this is to hold *precisely*. Nevertheless, a *correct* theory would be in better shape if it happens to predict either $K>0$ or <0, since then it has the chance of being observationally confirmed (and confirmation is what a scientific theory seeks, despite Popper's more negative perspective on the issue of scientific acceptability). A theory that predicts $K=0$ would have to depend upon other justifications for it to gain acceptance.

One such justification might be that $K=0$ is an implication of a particular theory that finds observational confirmation in some other way. Indeed, this is the claimed situation for the highly fashionable inflationary cosmology which was discussed in §28.4. We recall that, similarly to supersymmetry being 'almost' part of the standard model of particle physics, inflationary cosmology is frequently *almost* considered to be part of today's 'standard' model of cosmology! Let us try to examine the status of inflation with regard to Popper's refutability issue.

One might think that the situation is clear, and that inflation is indeed a Popperian theory. For over a decade, it had been consistently asserted that $K = 0$ is one of the implications of the idea of inflation,[17] and I recall having attended numerous lectures, given by supporters of inflation, in which such predictions were made.[18] Thus, if observations convincingly tell us that $K \neq 0$, then inflation is out! This seems to be as clear an adherence to Popper's principle as one could ask for. Moreover, there are some detailed predictions about the microwave background that come from inflation (together with certain other assumptions), and these seem to enjoy some considerable observational support, in a general way, particularly with regard to the continued observed *scale invariance* of the fluctuations, in agreement with most inflationary expectations. However, in the mid 1990s, evidence was beginning to mount, from various independent types of observation, that the average matter density Ω_d of the universe (baryonic plus dark) falls short of what would be required for overall spatial flatness, being at most only roughly one third of that value. (The density quantities Ω_d and Ω_Λ are taken as a fraction of the *critical* density—that being the density which would have given $K = 0$ in Einstein's theory without cosmological term; see §28.10.) Specifically, Ω_d, is about 0.3. In accordance with this trend in observations, inflation theorists began to provide inflationary models which now allowed $K \neq 0$, with $K < 0$ in fact.[19] We may note in addition that the Hawking school, in which $K > 0$ had seemed to be definite prediction (in connection with the Hartle–Hawking 'no-boundary' proposal—see §28.9), also began to perceive ways in which $K < 0$ could be accommodated within their scheme.[20]

This situation lasted until about 1998, when observations of distant supernovae (§28.10) seemed to be telling us that a positive cosmological constant needs to be incorporated into the Einstein equation, i.e. $\Lambda > 0$. This provides an effective additional density Ω_Λ which, when taken together with the matter density Ω_d, could provide the perceived critical total $\Omega_d + \Omega_\Lambda = 1$ (or, instead, the $\Omega_d + \Omega_\Lambda > 1$ that would be needed for the original Hartle–Hawking proposal). In this way, overall spatial flatness ($K = 0$) can be consistent with observation (as could overall positive spatial curvature), with $\Omega_\Lambda \approx 0.7$. In the face of this, most inflationists appear to have reverted to $K = 0$ as being a prediction of inflationary cosmology. I am not sure what Popper would have had to say about all this!

In fact, there is now an exotic inflationary proposal in which a new ingredient (a new field) is introduced, referred to as 'quintessence', which would provide an *effective* cosmological constant through a dynamical 'dark energy' of negative pressure. See Steinhardt *et al.* (1999). It has been argued that this might be signalling a *new* phase of inflation coming upon us (see §28.10)! It is to be hoped that somewhat fantastic-sounding suggestions like these will indeed find rapid ways to be convincingly settled observationally, though, in practice, matters seem rarely this clear cut.

In my opinion, we must be exceedingly cautious about claims of this kind—even if seemingly supported by high-quality experimental results. These are frequently analysed from the perspective of some fashionable theory. For example, the superb BOOMERanG[21] observations of the microwave background were originally interpreted very much from the inflationary perspective, and strong claims have been made that the observations indeed show that $K = 0$ (whence $\Lambda > 0$). Moreover, in the case of some experiments (as with BOOMERanG), with vast quantities of data and much scope for differing analyses, the raw data may not be generally released for a period of several years, in order that the people involved may (very reasonably) have 'first go' at it. For the intervening period, there is little scope for the data being analysed from a different point of view. In fact, in the case of BOOMERanG, Vahe Gurzadyan, with some members of the team, were actually able to access the data and apply his ellipticity analysis (§28.10) to it, finding the strong direct indication that $K < 0$ (later supported by his corresponding analysis of the WMAP data). Like the 'anomalous' $\ell = 2$ WMAP measurement (strangely hidden by the vertical axis in Fig. 28.19), this is not too friendly to the inflationist position. We shall have to wait until the dust thoroughly settles before coming to a clear conclusion about all this!

We see how strongly matters of scientific fashion can influence the directions of theoretical scientific research, despite the traditional protestations from scientists of the objectivity of their subject. Nevertheless, I should make it absolutely clear that the apparent lack of objectivity is not the fault of Nature herself. There is an objective physical world out there, and physicists correctly regard it as their job to find out its nature and to understand its behaviour. The apparent subjectivity that we see in the strong influences of fashion, referred to above, are simply features of our gropings for this understanding, where social pressures, funding pressures, and (understandable) human weaknesses and limitations play important parts in the somewhat chaotic and often mutually inconsistent pictures that we are presently confronted with.

34.5 Whence may we expect our next physical revolution?

I believe that, in my descriptions in this chapter, I may have presented a rather more pessimistic picture of current progress towards a fundamental understanding of physics than is often given in popular accounts. But I believe also that it is a considerably more realistic one. On the other hand,

I certainly do not wish to suggest that we have reached a stage in which fundamental progress is well-nigh impossible, as some popularizers have tried to maintain.[22] There is, indeed, an enormous quantity of observational data that still needs to be made sense of—and this would be so even if no further experiments are ever performed.

Data from modern experiments are often stored automatically, and only a small particular aspect of that stored information may be of interest to the theorists and experimenters who are directly involved. The entire mass of data would thus be likely to be analysed only in the particular way that addresses questions that they are concerned with. It is certainly possible that there are many clues to Nature's ways hidden in such data, even if we do not properly read them yet. Recall that Einstein's general relativity was crucially based on his insight (the principle of equivalence—see §17.4) which had been implicit in observational data that had been around since (and before) the time of Galileo, but not fully appreciated. There may well be other clues hidden in the immeasurably more extensive modern observations. Perhaps there are even 'obvious' ones, before our very eyes, that need to be twisted round and viewed from a different angle, so that a fundamentally new perspective may be obtained concerning the nature of physical reality.

I believe, indeed, that a new perspective is certainly needed, and that this change in viewpoint will have to address the profound issues raised by the measurement paradox of quantum mechanics and the related non-locality that is inherent in EPR effects and in the issue of 'quanglement' (Chapters 23 and 29). I have argued, in Chapter 30, that the measurement paradox must be deeply interconnected with the principles of general relativity (and, specifically, with the Galileo–Einstein principle of equivalence just referred to). Perhaps new experiments (such as that of FELIX or a more realistic ground-based alternative; §30.13) may lead the way to the needed improved understandings of quantum theory. Perhaps there will be other types of experiment shedding light on the nature of quantum gravity (such as those designed to test the possibility of higher dimensionality to space-time). Perhaps, on the other hand, it will be theoretical considerations that will take us forward.

Are the seeds of any such putative theoretical developments already to be found in some of the ideas described in the previous chapters of this book? Clearly there would be numerous differing viewpoints on this matter, and personal opinion must play a strong part in any answer. It has certainly been—and still is—my own particular hope (for over 40 years, in fact) that the framework of twistor theory might yield insights that could lead to such a change of physical viewpoint. But despite the progress that has been made (see Chapter 33), twistor theory cannot be said to have moved us significantly, as things stand, in any direction that helps us to resolve the measurement paradox.

Whatever one's stance might be, concerning the relative merits of the theories that I have described, new insights and new perspectives are definitely needed. How are these to come about? May we expect a 'new Einstein' working in a solitary way, and coming upon such revolutionary views from largely internal deliberations? Or will we find ourselves driven again by immensely puzzling experimental findings? In Albert Einstein's case, his internal insights led ultimately to general relativity, which is very largely a 'one-person theory' (despite the essential input that Einstein had from Lorentz, Poincaré, Mach, Minkowski, Grossmann, and others). Quantum theory, on the other hand, was very much a 'many-person' theory, being driven externally by the extraordinary results of a great many careful experiments. In the present climate of fundamental research, it would appear to be much harder for *individuals* to make substantial progress than it had been in Einstein's day. Teamwork, massive computer calculations, the pursuing of fashionable ideas—these are the activities that we tend to see in current research. Can we expect to see the needed fundamentally new perspectives coming out of such activities? This remains to be seen, but I am left somewhat doubtful about it. Perhaps if the new directions can be more experimentally driven, as was the case with quantum mechanics in the first third of the 20th century, then such a 'many-person' approach might work. But I perceive this happening, in the area of quantum gravity, only if there are experiments that reveal an influence of general-relativistic principles on the very structure of quantum mechanics (as I argue for in Chapter 30). Failing this, I feel that something more like the Einsteinian 'one-person' approach is likely to be needed. And for that, there is little doubt in my own mind but that mathematical aesthetics must be an important driving force in addition to physical insight.

The reason for this belief is that, the more deeply we probe the fundamentals of physical behaviour, the more we find that it is very precisely controlled by mathematics. Moreover, the mathematics that we find is not just of a direct calculational nature; it is of a profoundly sophisticated character, where there is subtlety and beauty of a kind that is not to be seen in the mathematics that is relevant to physics at a less fundamental level. In accordance with this, progress towards a deeper physical understanding, if it is not able to be guided in detail by experiment, must rely more and more heavily on an ability to appreciate the physical relevance and depth of the mathematics, and to 'sniff out' the appropriate ideas by use of a profoundly sensitive aesthetic mathematical appreciation.

It is, by the very nature of the problem, supremely difficult to lay down any sort of reliable criteria for achieving this. We have already seen, in the contrast between the approaches that have been described in the later chapters of this book, how different mathematical developments, each guided by its own set of aesthetic mathematical and physical criteria, can devlop in mutually contradictory directions. Some have argued that perhaps we should seek ways in which *all* these approaches can be brought together in some kind of synthesis, perhaps by distilling what is appropriate from the body of all of them taken together. On the other hand, it could reasonably be argued that the contradictions between the different approaches are too great, and that at most one of them can survive, all the rest having to be discarded. I suspect, myself, that the truth lies somewhere between these extremes, and that something of importance may yet be found even in many of the theories whose major tenets will eventually have to be abandoned.

Some of the theories that I have been describing, although not altogether consistent with one another, do have appreciable common ground. In particular, the loop-variable approach of Chapter 32 has significant features in common with twistor theory (Chapter 33) and I can well imagine that an appropriate combination of the ideas from each (perhaps involving spin networks, spin foams, *n*-category theory, or even noncommutative geometry) could lead to a way forward. But string theory, as it presently stands, with its dependence on extra space dimensions, strikes me as being too far from twistor or loop-variable theory for any forseeable union to emerge. Strings themselves are not a reason for incompatibility (§31.5). Even supersymmetry has been brought together with twistor ideas.[23] But string theory's insistence on higher dimensions (especially on those particular dimensions/signatures that violate twistor theory's holomorphic philosophy—see the final paragraph of §33.4) represents a fundamental conflict with both twistor and loop-variable theory. Until very recently, string theorists have shown no inclination to provide a consistent (1+3)-dimensional theory. However, as mentioned in §31.18 and §33.14, there has been a recent shift, and applications of string-theoretic ideas to ordinary (1+3)-dimensional spacetime seem now to be taken seriously.

34.6 What is reality?

As the reader will gather from all this, I do not believe that we have yet found the true 'road to reality', despite the extraordinary progress that has been made over two and one half millennia, particularly in the last few centuries. Some fundamentally new insights are certainly needed. Yet, some readers may well still take the view that the road itself may be a mirage. True—so they might argue—we have been fortunate enough to stumble upon mathematical schemes that accord with Nature in remarkable ways, but the unity of Nature as a whole with some mathematical scheme can be no more than a 'pipe dream'. Others might take the view that the very notion of a 'physical reality' with a truly objective nature, independent of how we might choose to look at it, is itself a pipe dream.

Indeed, we may well ask: what *is* physical reality? This is a question that has been posed for thousands of years, and philosophers throughout the ages have attempted various kinds of answer. Today we look back, from our vantage point of modern science, and claim to take a more sober position. Rather than attempting to answer the 'what' question, most modern scientists would try to evade it. They would try argue that the question has been wrongly posed: we should not try to ask *what* reality is; merely, *how* does it behave. 'How?' is, indeed, a fundamental question that we may consider to have been one of the main concerns of this book: how do we describe the laws that govern our universe and its contents?

Yet, many readers will no doubt feel that this is a somewhat disappointing answer—a 'cop-out', no less. To know how the contents of the universe behave does not seem to tell us very much about what it is that is doing the behaving. This 'what?' question is intimately connected with another deep and ancient question, namely 'why?'. Why do things in our universe behave in the particular ways that they do? But without knowing *what* these things are, it is hard to see *why* they should do one thing rather than another.

Modern science would be cautious in attempting answers to 'why?' questions as well as 'what?'. Yet, questions as to 'what?' and 'why?' are frequently supplied with answers. It is considered acceptable to do so provided that the questions are not asking about reality at its deepest levels. One may expect an answer to such a question as the following. 'What is a cholesterol molecule made of?'; 'why does a match burst into flame when dragged rapidly across a suitable rough surface?'; 'what is an aurora?'; 'why does the sun shine?'; 'what are the forces which hold a hydrogen atom or a hydrogen molecule together?'; and 'why is a uranium nucleus unstable?'. Yet, some other questions that one might pose could cause more embarrassment, such as 'what is an electron?' or 'why does space have just three dimensions?'. These questions can, however, find meaning within some more fundamental picture of physical reality.

It will be seen, particularly from the discussions of Chapters 31–33, that modern physicists invariably describe things in terms of mathematical models. This is irrespective of which particular family of proposals they may happen to hold to. It is as though they seek to find 'reality' within the Platonic world of mathematical ideals. Such a view would seem to be a consequence of any proposed 'theory of everything', for then physical reality would appear merely as a reflection of purely mathematical laws. As I have been arguing in this chapter, we are certainly a long way from any such theory, and it is a matter of contention whether anything resembling a 'theory of everything' will ever be found. Be that as it may, it is undoubtedly the case that the more deeply we probe Nature's secrets, the more profoundly we are driven into Plato's world of mathematical ideals as we seek our understanding. Why is this so? At present, we can only see that as a mystery. It is the first of the three deep mysteries referred to in §1.4, and illustrated in Fig. 1.3, here redrawn and embellished somewhat as Fig. 34.1.

But are mathematical notions things that really inhabit a 'world' of their own? If so, we seem to have found our ultimate reality to have its home within that entirely abstract world. Some people have difficulties with accepting Plato's mathematical world as being in any sense 'real', and would gain no comfort from a view that physical reality itself is constructed merely from abstract notions. My own position on this matter is that we should certainly take Plato's world as providing a kind of 'reality' to mathematical notions (and I tried to argue forcefully for this case in §1.3), but I might baulk at actually attempting to *identify* physical reality within the abstract reality of Plato's world. I think that Fig. 34.1 best expresses my position on this question, where each of three worlds—Platonic-mathematical, physical, and mental—has its own kind of reality, and where each is (deeply and mysteriously) founded in the one that precedes it (the worlds being taken cyclicly). I like to think that, in a sense, the Platonic world may be the most primitive of the three, since mathematics is a kind of necessity, virtually conjuring its very self into existence through logic alone. Be that as it may, there is the further mystery, or paradox, of the cyclic aspect of these worlds, where each seems to be able to encompass the succeeding one in its entirety, while itself seeming to depend only upon a small part of its predecessor.

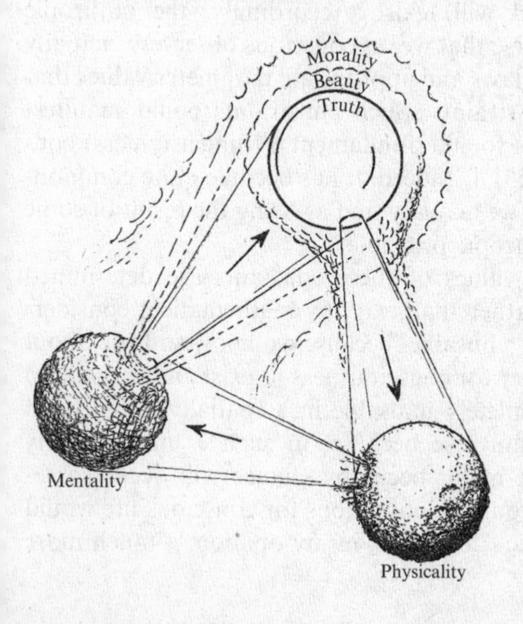

Fig. 34.1 A repeat of the diagram (Fig. 1.3) depicting 'Three Worlds and Three Mysteries', but embellished with the other 'Platonic absolutes' of Beauty and Morality, in addition to the absolute Truth that is to be found in mathematics. Beauty and Truth are intertwined, the beauty of a physical theory acting as a guide to its correctness in relation to the Physical World, whereas the whole issue of Morality is ultimately dependent upon the World of Mentality.

34.7 The roles of mentality in physical theory

We must bear in mind that each 'world' possesses its own distinctive kind of existence, different from that of the other two. Nevertheless, I do not think that, ultimately, we shall be able to consider any of these worlds properly, in isolation from the others. Since one of these is the world of mentality, this raises the issue of the role of mind in physical theory, and also of how mentality comes about in the physical structures with which it is associated (such as living, wakeful, healthy human brains, at least). I have deliberately refrained from addressing, at any great length, the question of conscious mentality in this book, despite the fact that this issue must ultimately be an important one in our quest for an understanding of physical reality. (I have discussed such matters in detail elsewhere, and I have no wish to get embroiled here in many of the contentious issues that arise.[24]) Yet, it would be inappropriate for me to try to avoid the issue of mentality altogether. Quite apart from the world of mentality having to be considered in conjunction with the other two worlds, in accordance with Fig. 34.1, there are several places in this book where the issue of consciousness has already played a significant role in physical theory, either implicitly or explicitly.

One of these is in connection with the anthropic principle, mentioned in §27.13 and discussed at some length in §§28.6,7. Any universe that can 'be observed' must, as a logical necessity, be capable of supporting conscious mentality, since consciousness is precisely what plays the ultimate role of 'observer'. This fundamental requirement could well provide constraints on the universe's physical laws, or physical parameters, in order that conscious mentality can (and will) exist. Accordingly, the anthropic principle asserts that the universe that we, as conscious observers, actually do observe, must operate with laws and appropriate parameter values that are consistent with these constraints. Such constraints could manifest themselves in particular values for the fundamental (dimensionless) constants of Nature, discussed in §31.1. Indeed, it has become quite commonplace to regard the values that we actually find as being the result of some kind of application of the anthropic principle.

Unfortunately, even if the values of these constants are determined by the anthropic principle—rather than, say, by mathematical considerations—the principle is almost unusable, because we know so little about the conditions that are necessary for consciousness to exist and actually to come about. It is almost completely unusable in a spatially infinite and essentially uniform ($K \leq 0$) universe because, in such a universe, any configuration of material that might occur by chance will occur somewhere, so that even very unfavourable conditions for conscious life would be allowed by the principle; see §28.6. It is, in my opinion, a much more optimistic possibility that these fundamental constants are actually mathematically determinable numbers. In a spatially infinite universe, this need not raise significant problems with the anthropic principle.

There is a quite separate important role played by consciousness in many interpretations of the **R** part of quantum mechanics, as discussed in Chapter 29 (particularly §§29.7,8). In fact, almost all the 'conventional' interpretations of quantum mechanics ultimately depend upon the presence of a 'perceiving being', and therefore seem to require that we know what a perceiving being actually is! We recall that the Copenhagen interpretation (viewpoint (a) in §29.1) takes the wavefunction not to be an objectively real physical entity but, in effect, to be something whose existence is 'in the observer's mind'. Moreover, at least in one of its manifestations, this interpretation requires that a measurement be an 'observation', which presumably means something ultimately observed by a conscious being—although at a more practical level of applicability, the measurement is something carried out by a 'classical' measuring apparatus. This dependence upon a classical apparatus is only a stopgap, however, since any actual piece of apparatus is still made of quantum constituents, and it would not actually behave classically—even approximately—if it adhered to the standard quantum **U** evolution. (This is simply the issue of Schrödinger's cat;—see §§29.7–9 and §§30.10–13.) The issue of environmental decoherence (viewpoint (c) in §29.1) also provides us with merely a stopgap position, since the inaccessibility of the information 'lost in the environment' does not mean that it is *actually* lost, in an objective sense. But for the loss to be subjective, we are again thrown back on the issue of 'subjectively perceived—by whom?' which returns us to the conscious-observer question.

In any case, even with environmental decoherence, if we retain rigorous adherence to **U** evolution for the 'true' quantum description of the universe, then we are driven to the many-worlds description of reality (viewpoint (b) in §29.1). The many-worlds viewpoint is manifestly dependent upon having a proper understanding of what constitutes a 'conscious observer', since each perceived 'reality' is associated with an 'observer state', so we do not know what reality states (i.e. 'worlds') are allowed until we know what observer states are allowed. Put another way, the behaviour of the seemingly objective world that is actually perceived depends upon how one's consciousness threads its way through the myriads of quantum-superposed alternatives. In the absence of an adequate theory of conscious observers, the many-worlds interpretation must necessarily remain fundamentally incomplete (see §29.8).[25]

The consistent-histories approach (viewpoint (d) in §29.2) is also explicitly dependent upon some notion of what an 'observer' might be (the notion referred to as an IGUS in the Gell-Mann–Hartle scheme[26]). The point of view suggested by Wigner (a version of viewpoint (f) in §29.2) that consciousness (or perhaps living systems generally) might *violate* **U** evolution is also one which makes explicit reference to the role of the mind (or whatever constitutes an 'observer') in the interpretation of quantum mechanics. As far as I can make out, the only interpretations that do *not* necessarily depend upon some notion of 'conscious observer' are that of de-Broglie–Bohm (viewpoint (e) in §29.2)[27] and most of those (viewpoints (f) in §29.2) that require some fundamental change in the rules of quantum mechanics, according to which **U** and **R** are both taken to be approximations to some kind of objectively real physical evolution.

As I have stated in many places in this book (particularly in Chapter 30), I am an adherent of this last view, where it is with gravitational phenomena that an objective **R** (i.e. **OR**) takes over from **U**. This gravitational **OR** would take place spontaneously, and requires no conscious observer to be part of the process. In usual circumstances, there would be frequent manifestations of **OR** occurring all the time, and these would lead to a classical world emerging on a large scale, as an excellent approximation. Accordingly, there is no need to invoke any conscious observer in order to achieve the reduction of the quantum state (**R**) when a measurement takes place.

On the other hand, I envisage that the phenomenon of consciousness—which I take to be a *real* physical process, arising 'out there' in the physical world—fundamentally *makes use* of the actual **OR** process. Thus, my own position is basically the reverse of those referred to above, in which, in one way or another, it is envisaged that consciousness is responsible for the **R** process. In my own view, it is a physically real **R** process that is (partly) responsible for consciousness!.[28]

Is this contention experimentally testable? I believe that it is. In the first place, there are specific suggestions about the structures in the brain which might be directly relevant—most particularly, A-lattice neuronal microtubules as originally suggested by Stuart Hameroff[29] (but perhaps also other structures, like synaptic clathrins,[30] whose structure closely resembles the C_{60} molecules)—and there is considerable scope for confirming/refuting these specific ideas. The scheme would require some kind of large-scale quantum coherence, acting broadly across considerable regions of the brain (having features in common with high-temperature superconductivity;[31] see §28.1), and it is envisaged that A-lattice neuronal microtubules would play an important part in this, where a conscious event would be associated with a partial state reduction (orchestrated **OR**) of this quantum system. This would normally involve many different parts of the brain together, in order to achieve sufficient coherent movement in protein molecules to effect gravitational **OR** in accordance with the proposal described in §§30.11,12.

There is an ingenious suggestion that has been put forward by Andrew Duggins, according to which such conjectured ideas—though not specific to the microtubule hypothesis—might be tested. This suggestion depends upon the fact that quite different regions in the brain are responsible for different aspects of perception (such as the visual perceptions of motion, of colour, or of shape), yet in the image that consciousness comes up with, all these different aspects come together to form a single image. This is sometimes referred to as the *binding problem*. Duggins's idea is to test to see whether there are significant violations of Bell's inequalities involved in the forming of a conscious image, indicating the presence of non-local EPR-type occurrences (§§23.3–5), which would strongly suggest that large-scale quantum effects are part of conscious perception. Preliminary results are so far inconclusive, but with some encouraging aspects.[32]

Whatever the status of these ideas, it seems to me that a 'fundamental' physical theory that lays claim to any kind of completeness at the deepest levels of physical phenomena must also have the potential to accommodate conscious mentality. Some people would try to evade (or belittle) this problem, arguing that consciousness simply 'emerges' as some sort of 'epiphenomenon'. Accordingly, so it would be claimed, there is no importance, for the emergence of consciousness, in the precise type of physics that happens to underlie the relevant (not necessarily biological) processes. A standard position is that of *computational functionalism*, according to which it is merely computational activity (of some suitable but yet unspecified nature) that gives rise to conscious mentality. I have argued strongly against this view (partly using reasoning based on Gödel's theorem and the notion of Turing computability—see §16.6), and I have indeed suggested that consciousness actually depends upon the missing (gravitational) **OR** theory.[33] My arguments demand that this missing theory must be a *non-computational* theory (i.e. its actions lie outside the scope of Turing-machine simulation, §16.6). Theoretical ideas for producing an **OR** model of this type are in a very preliminary stage, at present, but possibly there are some clues here.

34.8 Our long mathematical road to reality

I hope that it is clear, from the discussion given in the preceding sections that our road to understanding the nature of the real world is still a long way from its goal. Perhaps this goal will never be reached, or perhaps there will eventually emerge some ultimate theory, in terms of which what we call 'reality' can in principle be understood. If so, the nature of that theory must differ enormously from what we have seen in physical theories so far. The most important single insight that has emerged from our journey, of more than two and one-half millennia, is that there is a deep unity[34] between certain areas of mathematics and the workings of the physical world, this being the 'first mystery' depicted in Figs. 1.3 and 34.1. If the 'road to reality' eventually reaches its goal, then in my view there would have to be a profoundly deep underlying simplicity about that end point. I do not see this in any of the existing proposals.

This ancient Greek insight that it is *mathematics* that underscores the workings of physical reality has served us extraordinarily well, and I hope that I have made it clear that, despite our distance from our intended goal, we have come to an remarkably impressive understanding of the operations of the universe at the deepest levels that we know. Certain mathematical concepts stand out as having been particularly successful in the past. Among these are the real number system and the ideas of geometry. Initially it was the Euclidean geometry, first systematically studied by the ancient Greeks, but then the ideas developed away from the geometry of Euclid to that of Lambert, Gauss, Lobachevski, Bolyai, Riemann, Beltrami, and others. Then Minkowski told us to incorporate time with space, and Einstein presented us with his magnificent curved spacetime geometry of general relativity. The integral and differential calculus of Archimedes, Fermat, Newton, Leibniz, Euler, Cauchy, Cartan, and many others, and also the related ideas of differential equations, integral equations, and variational derivatives, have proved to be absolutely vital for the successful theories describing the workings of the world, these ideas having linked with geometry in profoundly important ways. Fundamental also, have been the statistical ideas that enable us to handle large and complicated physical systems of hugely numerous individual ingredients, as Maxwell, Boltzmann, Gibbs, Einstein, and others have taught us. Mathematics profoundly underlies quantum theory, from the matrix-theory ideas of Heisenberg to the complex Hilbert spaces, Clifford algebras, representation theory, infinite-dimensional functional analysis, etc. of Dirac, von Neumann, and many others.

I should like to single out just two particular aspects of the mathematics that underlies our understanding of the workings of the world, discussing each of these in turn, for I believe that they may hint at important but largely unaddressed questions of principle in our physical theory. The first is the role of the *complex-number* system, which we find to be so fundamental to the operations of quantum mechanics—as opposed to the real-number system, which had provided the foundation of all successful previous theories. The second is the role of *symmetry*, which has a central importance in virtually all 20th-century theories, particularly in relation to the gauge-theory formulation of physical interactions.

First, consider the complex numbers. It has been a recurring theme of this book that there is not only a special magic in the mathematics of these numbers, but that Nature herself appears to harness this magic in weaving her universe at its deepest levels. Yet, we may well question whether this is really a true feature of our world, or whether it is merely the mathematical utility of these numbers that has led to their extensive use in physical theory. Many physicists would, I believe, lean towards this second view. But, to them, there is still something of a mystery—needing some kind of explanation—as to why the role of these numbers should appear to be so universal in the framework of quantum theory, underlying, as they do, the fundamental quantum superposition principle and, in a somewhat different guise, the Schrödinger equation, the positive-frequency condition, and the infinite-dimensional 'complex structure' (§26.3) that comes about in quantum field theory. To such physicists, the real numbers seem 'natural' and the complex numbers 'mysterious'. But from a purely mathematical standpoint, there is nothing especially more 'natural' about the real numbers than the complex numbers. Indeed, in view of the somewhat magical mathematical status of the complex numbers, one might well take the opposite view and regard them as being distinctly more 'natural' or 'God-given' than the reals.

From my own peculiar standpoint, the importance of complex numbers—or, more specifically, the importance of holomorphicity (or complex analyticity)—in the basis of physics is indeed to be viewed as a 'natural' thing, and the puzzle is indeed perhaps the other way around.[35] How is it that *real* structures seem to play such an important part in physics? It should be made clear that even the standard formalism of quantum mechanics, although based on complex numbers, is not an entirely holomorphic theory. We see this in the usual requirement that quantum observables be descibed by Hermitian operators (or even normal ones, as described in §22.5) and in the unitary (rather than simply complex-linear) nature of quantum evolution—these depending upon the notion of complex conjugation ($z \mapsto \bar{z}$). Related to this, the important property of orthogonality between states is a non-holomorphic notion. The Hermitian property has to do with the usual (but not entirely necessary) demand that the results of measurements be real numbers, and the unitarity, that 'probability be conserved', i.e. that the squared-modulus rule (also to do with measurements) is maintained, whereby a complex amplitude z be converted to a probability, in accordance with the non-holomorphic operation

$$z \mapsto \bar{z}z.$$

We see that it is basically in the conversion of 'quantum information' (i.e. quanglement—see §23.10) into 'classical information' (measurement probability) where quantum holomorphicity is broken. The orthogonality of alternatives is again a crucial feature of measurement. Thus, non-holomorphicity seems to enter just at the point where measurements are introduced into quantum theory.

Of course, we see the role of real numbers also in the background spacetime within which the formalism of quantum theory is placed. If gravitational **OR** turns out to be the true basis of quantum state reduction, then we shall see the real-number (non-holomorphic) structure of actual spacetime relating to that of the operation $z \mapsto \bar{z}z$. Perhaps there is some lesson for twistor theorists here, with the theory's particular reliance on holomorphic operations? Perhaps, on the other hand, we should be seeking a role for discrete combinatorial principles somehow emerging out of complex magic, so 'spacetime' should have a discrete underlying structure rather than a real-number based one (as discussed in §§3.3,5, §32.6, and §33.1)? In any case, I believe that there are deep matters of importance, here, concerning the very mathematical basis of physical reality.

Now let us turn to the fundamental role of *symmetry* in modern physical theory. There is no doubt about the utility of this notion. Both relativity theory (in relation to the Lorentz group) and quantum theory make highly significant use of it. But are we to regard symmetry as fundamental to Nature's ways, or an incidental or approximate feature?

It seems to be a tenet of many of the modern approaches to particle physics to take symmetry as being indeed fundamental, and to regard the presently seen deviations from symmetry as a feature of symmetry breaking in the early universe. Indeed, as noted in §13.1 and §§15.2,4, exact symmetry is a necessary feature of the bundle-connection idea. Moreover, we recall from §§25.5,8 that the standard attitude to electroweak theory is to regard the U(2) symmetry as being fundamentally exact, so that it can indeed play a role as the gauge symmetry of the electroweak forces (§§15.1,8), but where it is normally envisaged that the symmetry is broken spontaneously (about 10^{-12} seconds after the Big Bang). We recall from §28.3 that there are certain difficulties with invoking the early universe to provide the needed symmetry breaking. This applies to the U(2) symmetry of electroweak theory and also to the much larger symmetries that are employed in GUT theories.

Do the large symmetry groups of GUT theories really simplify our picture of particle physics? Or would it be simpler if many of these apparent symmetries were fundamentally broken right from the start? From this second alternative perspective (which is indeed a consistent one even for electroweak theory; see §28.3), many of the symmetries that we perceive in our fundamental theories would really be only approximate at the fundamental level, and we must search more deeply for an understanding of where these apparent symmetries come from.

In ordinary quantum theory we have examples of both types of broken symmetry. There are well-understood situations in which spontaneous symmetry breaking manifestly does occur, such as with superconductivity (U(1) breaking) and other phenomena. On the other hand, there are examples where symmetry ideas can be used to provide an excellent understanding of a phenomenon, but where it is known that the symmetry is only an approximation, arising from a more exact but less symmetrical deeper underlying theory, such as in the classification of atomic spectra.[36] It remains to be seen which of these two types of situation will have greater importance in a more profound future theory of particle physics.[37]

As a related point of interest, there are circumstances where an exact symmetry group can come about even with structures where no symmetry is initially imposed. We see this with the Riemann sphere itself, which we can imagine being pieced together from patches of the complex plane in a specific way that lacks any symmetry whatsoever. But, provided that the topology of the resulting complex manifold is indeed S^2, we find that it is *equivalent* to the Riemann sphere, as a complex manifold (by a theorem of Riemann), so its symmetry group is exactly $SL(2, \mathbb{C})$, i.e. the non-reflective Lorentz group (§18.5), no matter how irregularly it is pieced together.[38]

There is a somewhat related question concerning the mysterious pure-number constants of Nature (§31.1). Are these numbers determined in the extremely early universe (such as with the Wheeler/Smolin-type proposal referred to in §28.6), in analogy with the symmetry-breaking idea? Indeed, some of these constants, like the Cabibbo and Weinberg angles (§25.7) are normally taken to arise in just this kind of way. Or might these numbers actually be mathematically determined from some deeper underlying theory? The latter would be my own personal preference, but we do not seem to be close to having a believable theory of this kind.[39]

An interesting question, in relation to this, is the *chiral* asymmetry of weak interactions (§25.3). In the normal approach to the standard model, this chiral asymmetry is built in to the framework of the theory. But neutrinos (at least most of them) are now observed to be massive particles (i.e. with non-zero rest-mass), and this fact already represents a deviation from the original standard electroweak model. One cannot simply 'blame' a left-handed neutrino for all the chiral asymmetry of weak interactions. A massive neutrino is not entirely a left-handed 'zig' particle since, with mass, it would also have a right-handed 'zag' part (see §25.2). One could imagine that, in some forms of the standard model (slightly extended so that massive neutrinos are incorporated) there was a spontaneous symmetry breaking from a previously left/right symmetric model. But in this instance it is the 'conventional' perspective that the asymmetry is there right from the start, rather than arising from a spontaneous symmetry breaking in the early universe.

We might also contemplate whether time-asymmetry (as demanded from the discussion of Chapters 27 and 28) is an issue that should be re-examined from this perspective. However, it certainly cannot arise from a conventional 'spontaneous symmetry breaking in the early universe'. The conventional picture *makes use* of the Second Law of thermodynamics; it cannot be used to derive it.

34.9 Beauty and miracles

Let us now turn to some more general and mysterious aspects of the mathematics that has been found to underlie physical theory at its deepest levels—at least at such deep a level as has been revealed to us so far. Two powerful internal driving forces have strongly influenced the direction of theoretical research, yet which usually go unmentioned in serious scientific writings—for fear, no doubt, that these influences may seem to have drifted too far from the strict rules of proper scientific procedure. The first of these is beauty, or elegance, and I have touched upon the matter in many places elsewhere in this book. The second, namely the irresistible allure of what are frequently termed 'miracles', I have only hinted at so far (in §19.8, §21.5, and §31.14); yet, as I can vouch from personal experience, these can indeed exert a powerful influence on the direction of one's research.

Before coming to the question of miracles, which are the prime concern of this section, let us first return to the issue of beauty, since the two are not unconnected. As indicated above, many of the ideas perceived to have achieved a major advance in physical theory will also be viewed as compellingly beautiful. There is the undoubted beauty of Euclidean geometry, which formed the basis of the first profoundly accurate physical theory, namely the theory of space formulated by the ancient Greeks. A millennium and a half later came the extraordinary elegance of Newtonian dynamics, with its deep and beautiful underlying symplectic geometry structure, as later revealed via the Lagrangian and Hamiltonian formalisms (§20.4). The mathematical form of Maxwell's electromagnetism, also, is indeed exquisite, and there is no doubt of the supreme mathematical beauty of Einstein's general relativity. The same can be said of the structure of quantum mechanics and many of its specific features. I would single out the extraordinary mathematical elegance of quantum-mechanical spin, of Dirac's relativistic wave equation, and of the path-integral formalism of QFT as developed by Feynman.

Yet, we may question whether the undoubted mathematical beauty in these schemes is something that would shine independently, simply as pure mathematics, had it not been for the remarkable fact that they accord so well with the workings of our universe? How would they stand comparison as just mathematical structures with some of the gems or beacons of pure mathematics? I believe that they would stand up rather well, but not overwhelmingly so. There are many bodies of pure mathematics, with no discernable relations to the physical world, whose beauty equals or even exceeds that of the physical theories that we have yet come across. (See also §16.3.)

Let us consider some deep and beautiful developments in mathematics, where the influence on physics—so far at least—has been minimal. Cantor's theory of the infinite is one noteworthy example. In my opinion, it is one of the most profoundly beautiful mathematical contributions in the whole of mathematical history. However, extraordinarily little of it seems to have relevance to the workings of the physical world as we know it (see §§16.3,4,7). The same issue arises in relation to another of the monumental achievements of mathematical understanding, a closely related descendent of Cantor's theory of the infinite, namely Gödel's famous incompleteness theorem (§16.6). Also, there are the wide-ranging and deep ideas of category theory (§33.1) that have yet seen rather little connection with physics.

In these last two areas there is some plausible indication that there might be some significant relation to a physics that could develop in the 21st century (§34.7, §33.1), but this is very speculative. It seems a good deal less plausible for some important connection with physics to transpire for the vast majority of profound and beautiful other mathematical theories that have been developed. Consider, for example, the remarkable 20th-century achievement of Andrew Wiles in establishing the truth of the assertion, of over 350 years standing, known as 'Fermat's Last Theorem'. This seems to be very remote from physical laws, as we understand them today, despite the magnificent mathematical ideas involved. Many other marvellous developments took place in the 20th century, such as the classification of simple groups, both continuous and discrete. Here there have certainly been applications to physics, but this is far from saying that the theory of simple groups provides us with a 'physical theory'. It is just that the mathematical classifications are helpful to physicists in enabling them to see what the possibilities are. Consider another example. The 19th century saw Riemann's exquisite theory of the ζ-function and its relation to the distribution of prime numbers. This seems almost equally remote from physics, despite the undoubted beauty and great mathematical importance of the still unproved Riemann hypothesis (§7.4). In fact, there *are* some intriguing connections with physics here,[40] but it would be hard to maintain that Riemann's theory provides us with anything resembling a model of the physical world.

Are we to expect that there will be a close relationship to a greater body of profound and beautiful mathematics in the physics of the future? Or are we being misled by the successes that we have so far seen in physical theory into believing that the relation between mathematics and physics is closer than it actually is? The question can be succinctly phrased in terms of Fig. 1.3. How much of the Platonic mathematical world lies at the base of the arrow that depicts the 'first mystery'?

We may also ask whether there may be any way of perceiving what kind of mathematics it is that finds a deep role in governing the behaviour of the physical world? That is an intriguing issue. Perhaps the crucial underlying factors governing the mysterious relation between mathematics and physics will be better understood at some future date.

I hope that considerations such as the above make it clear to the reader that mathematical beauty is, by itself, an ambiguous guide at best. Yet, as I have remarked at many earlier places in this book, it is hard to doubt the remarkable role that aesthetic judgements play, both mathematical and physical, in making decisions as to the most fruitful lines to follow in research into theoretical physics. Most of these are of a subtle character, and it is easy to believe that it is a highly personal matter which of several alternatives is judged to be the most attractive one to follow. Occasionally, however, something can arise, in research into mathematical theories of the physical world, which has a much more powerful impact on such choices than mere mathematical elegance, and this is what I refer to as a 'miracle'.

I can think of many examples of such things in recent history of 'quantum gravity' ideas. One of these was in supergravity theory (§31.2) where it was found that, whereas the perturbative approach to the QFT of standard Einstein general relativity theory led to non-renormalizable divergences at the second order, when supersymmetry was introduced, the divergent terms miraculously cancelled out.[41] This cancellation involved large numbers of terms, and for a time it was thought by many supergravity researchers that this apparent 'miracle' of cancellation was a signal that the theory was on the right track, so that renormalizability was therefore to be expected at all orders—and the true quantum-gravity theory would, accordingly, soon be revealed! Unfortunately for the supergravity researchers when they were able to complete the third-order calculation, non-renormalizable divergences returned. This led to higher-dimensional considerations, but things stagnated for a while. Then, in the late 1990s, supergravity was revived as part of the route leading to M-theory, as described in §§31.4,14.

I am sure that string theory and M-theory have themselves been guided by a great many such miracles. Surely one of the most important was the discovery of mirror symmetries whereby the puzzling collection of apparently quite different string theories, as described in §31.14, received strong indications that they could be united into one grand scheme referred to as 'M-theory'. These mirror symmetries acted like magic, and numbers that had previously seemed to have little to do with one another were found to be the same, such as was the case for the calculations performed by Candelas and his colleagues, as described in §31.14. This certainly qualifies as a miracle, in the sense that I am using the term here. I feel sure that when the number 317 206 375, as obtained in Candelas's calculation using mirror symmetry, was finally confirmed by the algebraic geometers, then this was hailed as a miracle, providing convincing evidence that the new string/M-theory must be on the right lines! Whether or not that turns out to be the case, this 'miracle' certainly provided excellent support for the mathematical aspects of the mirror-symmetry idea. Indeed, much pure-mathematical interest in this issue has subsequently been stimulated, and a good pure-mathematical understanding of much of what is involved has now been obtained.[42]

Are such apparent miracles really good guides to the correctness of an approach to a physical theory? This is a deep and difficult question. I can imagine that sometimes they are, but one must be exceedingly cautious about such things. It may well be that Dirac's discovery that his relativistic wave equation automatically incorporated the electron's spin seemed like such a miracle, as had Bohr's use of angular momentum quantization to obtain the correct atomic spectrum of hydrogen, and likewise Einstein's realization that his approach to gravity through the curved space of general relativity actually gave the correct answer for the perihelion motion of Mercury—which had puzzled astronomers for over 70 years previously. But these were clearly appropriate physical consequences of the theories that were being put forward, and the miracles supplied impressive confirmation of the respective theories. It is less clear what the force of the purely mathematical miracles is, such as in the case of supergravity or mirror symmetry. When finally a mathematical understanding of a miracle of this mathematical kind is obtained, there is the possibility that this may, to some degree, provide a 'debunking' of the miracle in question. Even so, this may not completely remove the psychological force of the miracle itself, which must always be viewed in its appropriate historical setting.

One thing is certain, however, and that is that such mathematical miracles cannot always be a sure guide. During the course of my own studies of twistor theory, I have come across several different pieces of encouragement that would seem to come under the heading of 'miracles' in the sense that I am using this term here. The discovery (§33.8) that homogeneous functions of single twistors generate general solutions of the massless field equations was one such, and the non-linear graviton construction of §33.11 was another. How strong an indication are these that twistor theory is 'on the right lines'? Again one must be cautious. I have no wish to make a comparison between the miracles of twistor theory with those of string theory. But they cannot both be unambiguous sign-posts, because, as pointed out in §33.14, the two theories are, as they stand, incompatible with each other!

Yet, these comments apply only to what I understand to be the state of these theories 'at the moment'. Some exciting-looking developments that have occurred only within the past several months could completely overturn the conclusion that I seem to have come to in my final remarks in the previous paragraph. These are some highly innovative applications of string-theoretic ideas in the context of twistor theory due, primarily, to Edward Witten[43] (and I have referred to these briefly in §31.18 and §33.14). In these developments, the string theory is applied to standard *four*-dimensional spacetime physics and is concerned with the kind of Yang–Mills interactions that are likely to be of direct relevance to actual particle interactions, thereby representing a substantial break with the form of string theory that I was referring to in the previous paragraph. How is this achieved? Essentially, it is done by regarding the 'target space' into which the Riemann surfaces of string theory are to be mapped as being not a Calabi–Yau complex 3-manifold (§31.14), which had been invoked to supply the 'extra spatial dimensions' of 'standard' string theory, but as the complex 3-maniford which is *projective twistor space* $\mathbb{PT}$ (a $\mathbb{CP}^3$; see §33.5). As we have seen, twistor geometry explicitly refers to ordinary 4-dimensional spacetime, and there are *no* 'extra space dimensions'! As these new ideas have been described, there is still some supersymmetry involved, and this supersymmetric version of $\mathbb{PT}$ can actually be regarded as a kind of 'Calabi–Yau space'. (This is so that a certain 'anomaly' cancels out—but it seems to me that the necessity of such anomaly cancellation may possibly be over-rated, and perhaps the supersymmetry is not really needed.) As these new ideas stand, the Riemann surfaces are taken to have genus 0 (see §8.4) i.e. they are *Riemann spheres*.[44] This enables some appreciable contact to be made with a good deal of earlier twistor theory, in which 'string' ideas had been previously involved.[45]

If string theory can become changed like this, in what appears to me to be a very substantial way, what physical relevance do the 'miracles' of that theory then have? It would be my guess that they could indeed have some significant (albeit indirect) relevance, and that some greater understanding of what is 'going on behind the scenes' (as is suggested in Richard Thomas's remarks quoted in §31.18) might possibly come more readily to light. Can one extract what is powerful in string theory and remove it from a necessary dependence on spacetime supra-dimensionality? Possibly so. What appears to be true, in essence, is that there is something deep in the idea of a quantum field theory based on the mappings of Riemann spheres into complex manifolds[46] (or perhaps also the mappings of Riemann surfaces of higher genus) and this kind of thing could still have relevance in this newer context, the complex manifold being (projective) twistor space. But exactly what is *really* going on appears still to be largely a mystery.

34.10 Deep questions answered, deeper questions posed

Issues such as those described in the preceding several sections are far from answered within present-day physical understanding, and we may hope that important light will be shed on them in a future physics in the 21st century. But if we look back to see what we had already achieved in our understanding at the end of the 20th century, the human race may justly feel some considerable pride. A great many questions that had been profoundly puzzling—and sometimes terrifying—to the ancients have found answers, and it is frequently possible to act in a positive way, in the light of these answers. Many of the terrors of disease now cause no fear, not only because of modern drugs (where scientific method has been invaluable), but where early diagnosis by use of modern technology (X rays, ultrasound, tomography, etc.) can be used, as well as sophisticated physical treatments (radiation, lasers, etc.). Often this technology depends upon deep understandings from physics that were not available to the ancients. The same type of understanding has given us many other things, such as hydroelectricity, electric lighting, modern materials that serve as protection against the elements, telecommunications such as television and mobile telephony, computer technology, the internet, modern transportation in its various forms, and numerous other aspects of our modern lives.

Many of these developments certainly depend directly upon physics in one form or another. Moreover, the basic rules of chemistry, as understood today, are also fundamentally physical ones (in principle if not in practice)—mainly coming from the rules of quantum mechanics. Biology is a good deal further from being reducible to physical laws, but we have no reason to believe (consciousness apart) that biological behaviour is not, at root, purely dependent upon physical actions that we now basically understand. Accordingly, biology seems also to be ultimately controlled by mathematics.

Consider, for example, the miraculous way in which a seed can develop into a living plant, where the superb structure of each plant is similar in great detail to each of the others that come from the same type of seed. There is deep underlying physics here, since the DNA that controls the growth of the plant is a molecule, the persistence and reliability of its structure depending crucially upon the rules of quantum mechanics (as Schrödinger famously pointed out in 1944, in his very influential little book *What is Life?*[47]). Moreover, the plant's growth is ultimately controlled by the same physical forces that govern the individual particles of which it is composed. The relevant ones are mainly electromagnetic in origin, but the strong nuclear force is vital in determining what nuclei are possible, and therefore what kinds of atoms there can be.

The weak force too, plays its role in phenomena that we see on a large scale, and it is remarkable how, despite its weakness (only about 10^{-7} of the strength of the strong force and 10^{-5} of the strength of electromagnetism), this force can result in some of the most dramatic events that have been experienced by mankind. For it is the weak force that is, through radioactive decay in the Earth's interior, largely responsible for the heating of the Earth's magma. In particular, volcanic eruptions are its legacy. There was a period of a few years in the Earth's history, starting from about 535AD when there were world-wide famines and uncharacteristically cold weather, owing to a virtually continuous cover by dust that had been thrown out in an enormous volcanic explosion. The volcano was probably the same object as Krakatoa, near Java which seems to have erupted cataclysmically in 535, and did so again (but not quite so violently) in modern times in 1883.

Possibly even more dramatic to its civilized onlookers was the volcanic explosion that destroyed the island of Thera (Santorini), which would have been easily visible from Crete, some 100 miles to the south of it, in about 1628 BC. It devastated the civilized community on Thera itself and was probably ultimately responsible for the subsequent downfall of the peaceful and cultured society of Knossos in Crete, where the famed labyrinth of Daedalus was said to be located at its Great Palace.[48] It has been persuasively argued that the destruction of Thera may well have been the source of the legend of Atlantis.[49] Perhaps we may take some comfort in the fact that some of the cataclysms of the past may have also ultimately spawned the growth of new advances that might not have taken place otherwise. (The most dramatic of these was the global annihilation of the dinosaurs, which allowed the mammalian development leading ultimately to human beings—though this seems to have been an asteroid collision rather than volcanic activity.) Did the extraordinary development of ancient Greek culture in the millennium following the destruction of Thera owe anything to that catastrophic volcanic event?

It is perhaps even more striking that the most violent explosions seen in the universe are caused by the weakest force of all—if it is fair to call it a force—namely gravitation (only about 10^{-40} of the electric force, in a hydrogen atom, and about 10^{-38} of the strength of the weak force), where black holes fuel the unbelievably powerful energy sources of quasars. But their distance from us is so great that, as seen from the Earth, the brightest quasar, 3C273 is only about 10^{-6} of the brightness of the nearby star Sirius, despite the quasar's extraordinary power. Indeed, as we examine the sky, on a clear and peaceful night, although we may feel awe at the immensity of the universe, we in fact perceive only the minutest fraction of its enormous scale. The most distant object visible to the naked eye (the Andromeda galaxy) is only a puny 10^{-3} of the distance to 3C273 and about 10^{-4} of the distance out to the edge of the observable universe!

The spacetime singularities lying at cores of black holes are among the known (or presumed) objects in the universe about which the most profound mysteries remain—and which our present-day theories are powerless to describe. As we have seen in §§34.5,7,8, particularly, there are other deeply mysterious issues about which we have very little comprehension. It is quite likely that the 21st century will reveal even more wonderful insights than those that we have been blessed with in the 20th. But for this to happen, we shall need powerful new ideas, which will take us in directions significantly different from those currently being pursued. Perhaps what we mainly need is some subtle change in perspective—something that we all have missed....

Notes

Section 34.1

34.1. See, for example, Mukohyama and Randall (2003).

34.2. See Tittel *et al.* (1998).

34.3. See Arndt *et al.* (1999).

34.4. See Amelino-Camelia *et al.* (1998); Gambini and Pullin (1999); Amelino-Camelia and Piran (2001); Sarkar (2002); or, for an alternative perspective, Magueijo and Smolin (2002).

Section 34.2

34.5. I use the word 'coherence' here to convey something that is just a little stronger than *consistency*. The word is to suggest that there is a certain economy as well as consistency in a fully *coherent* mathematical structure, where different aspects of its formalism work together in perfect accord.

Section 34.3

34.6. See Rovelli (1998).

34.7. Some of my colleagues have told me that they believe this is actually the case now!

34.8. Recall, from §15.4 that the S^7 is fibred by S^3s in analogy with Clifford parallels on S^3. In fact S^7 is what is called 'parallelizable', which means that a 7-frame of tangent vectors can be continuously assigned at all of its points. The 'squashing' of S^7 is achieved systematically along such 'parallel' directions. See Jensen (1973).

34.9. See Collins (1977) for Regge trajectories, and Chew (1962) on the S matrix.

34.10. See Geroch (University of Chicago lecture notes, unpublished).

34.11. See van der Waerden (1929); Infeld and van der Waerden (1933); Penrose and Rindler (1984, 1986); O'Donnell (2003).

34.12. See Dirac (1936); for other versions of the higher spin equations, see Corson (1953).

34.13. See Wess and Zumino (1974).

Section 34.4

34.14. See Popper (1934).

34.15. For a playful use of this kind of idea, see Aldiss and Penrose (2000).

34.16. Despite recent claims that BOOMERanG had actually provided such confirmation. See, for example, Bouchet *et al.* (2002), for a (limited) critique of these claims.

34.17. The steady-state universe, as put forward by Bondi, Gold, and Hoyle in the early 1950s (see Hoyle 1948; Bondi and Gold 1948), was emphatically Popperian, as was clearly stated by Bondi, and among its avenues of refutation would also have been the establishment of $K \neq 0$. It fell, however, because of other conflicts with observation, most notably the presence of the 2.7 K microwave background radiation, which was virtually direct evidence of the Big Bang; see §§27.7,10,11,13 and §§27.4,7,10.

34.18. See, for example, Linde (1993).

34.19. See Bucher *et al.* (1995) and Linde (1995).

34.20. See Hawking and Turok (1998).

34.21. See Lange *et al.* (2001).

Section 34.5

34.22. See, John Horgan's book *The End of Science* (1996).

34.23. See Ferber (1978); Ward and Wells (1989); Delduc *et al.* (1993); Ilyenko (1999).

Section 34.7

34.24. See Penrose (1989, 1994, 1997a, 1997b).

34.25. See Deutsch (2000); Lockwood (1989).

34.26. See Gell-Mann (1994) and Hartle (2004).

34.27. However, anyone who has ever listened to David Bohm's views on this matter would appreciate that he would not regard the issues of consciousness to be unrelated to those of quantum mechanics.

34.28. See Penrose (1989, 1994, 1997).

34.29. See Hameroff and Watt (1982); Hameroff (1987, 1998); Hameroff and Penrose (1996).

34.30. See Koruga *et al.* (1993).

34.31. See, for example, Anderson (1997).

34.32. Personal communication. As far as I am aware, Duggins's investigations are not yet complete, and no publication is available as yet.

34.33. See Penrose (1987a, 1994); Hameroff and Penrose (1996).

Section 34.8

34.34. Whether this unity is to be regarded as remarkable, and somehow mysterious, seems to be a matter of contention among experts. In a famous lecture, the highly esteemed mathematical physicist Eugene Wigner (1960) expounded upon 'The unreasonable effectiveness of mathemetics in the physical sciences'. But Andrew Gleason, a prominent mathematician, two of whose powerful theorems have featured elsewhere in this book (see Notes 13.4 and 23.4), has taken an opposing view (1990), regarding the concordance between mathematics and physics as merely a reflection of the fact that 'mathematics is the science of order'. My own personal viewpoint would be closer to that of Wigner than of Gleason. Not just the extraordinary precision, but also the subtlety and sophistication that we find in the mathematical laws operative at the foundations of physics seem to me to be much more than the mere expression of an underlying 'order' in the workings of the world.

34.35. This philosophy seems to be close to that of Geoffrey Chew, who set much store in holomorphic properties of the S matrix. See Chew (1962).
34.36. See introductory comments by Freeman Dyson, in Dyson (1966).
34.37. See Penrose (1988b).
34.38. This fact has significance in the study of the asymptotic symmetries of asymptotically flat spacetimes, in general relativity; see Sachs (1962a, 1962b); Penrose and Rindler (1986).
34.39. Compare Eddington (1946).

Section 34.9

34.40. See du Sautoy (2004) for a discussion of the connection of the Riemann hypothesis to physics; see also Berry and Keating (1999). Another relevance of the (Euler) ζ-function to physics is what is called "ζ-function regularization". A quick search of the LANL arXiv showed 142 hits at last count.
34.41. See Wess and Bagger (1992) and Note 31.13.
34.42. See Note 31.57 and Note 31.58.
34.43. See Note 31.81.
34.44. This does not in itself restrict these Riemann surfaces to be the lines of twistor space representing spacetime points, see §33.5, but they can be curves of 'higher order'. Basically the 'genus-0 condition' tells us that one is concerned with tree Yang–Mills processes, where there are no 'closed loops' involved (§26.8).
34.45. Shaw and Hughston (1990); Hodges (1985, 1990a, 1990b).
34.46. These are referred to as 'σ-models'; see Ketov (2000).

Section 34.10

34.47. Reprinted in Schrödinger (1967).
34.48. See Davies (1997), pp. 89–94 for a graphic description of this event (from which I have brrowed in the Prologue). For the legend, an old favourite is reprinted in Hamilton (1999).
34.49. See Friedrich (2000) for a recent account of this idea.

Epilogue

Antea, a postdoctoral student of physics, came from a small town in southern Italy, and she possessed remarkable artistic as well as mathematical talents. She stared at the clear night sky through a large eastward window of the Albert Einstein Institute in Golm, near Potsdam, Germany. This prestigious research institute had been set up at the end of the 20th century close to where Einstein had once owned a holiday cottage. A good part of the research was concerned with the vexed issue of 'quantum gravity' which attempts to unify the principles underlying Einstein's general relativity with those of quantum mechanics—a mystery at the very basis of the laws of the world.

This was the direction of Antea's own research, but she was a newcomer, and she had some unorthodox and not yet fully formed ideas as to how to proceed, some of which were fundamentally at variance with those of her colleagues. That night, she had continued to work well into the small hours, in the institute's upper library, at a time when all the others had long left for their beds. She had been studying some old research pertaining to gigantic energy emissions taking place at the centres of some galaxies. It is indeed fortunate, she thought to herself, that the Earth and solar system are nowhere close to any of these, else they would be, in entirety, almost instantly vaporized. The established explanation of these stupendous explosions is that each is powered by a black hole of immense proportions.

Antea knew that a black hole is a spacetime region in whose interior lies a structure known as a 'spacetime singularity'—whose scientific description was still profoundly elusive, and which depends upon the still missing theory of quantum gravity. But Antea's real interest was not so much with galactic black holes as with an even more monstrous explosion: the explosion to end all explosions—or, rather, the one that began them all—known as the 'Big Bang'. She mused that it was the origin of all things good as well as of all things bad. Yet the spacetime singularity in the big bang provided mysteries even greater than those in black holes. Antea knew that at the root of these mysteries lay the secret of how to unite Einstein's large-scale theory of space, time, and gravity with the quantum-mechanical principles of physics.

It was a peaceful night and the stars were unmistakably clear. For a while Antea stood in a pensive state, with folded arms resting on the balustrade over the staircase, staring at the patterns of the stars through the large window—she did not know for how long. She always felt awe as she contemplated, in that vast seeming hemispherical dome, the great distance of those tiny pinpricks of light, though it counted but little compared to the greater enormity of cosmological scales. Yet, she mused, if some cosmic explosion were to become visible to her *now*, no matter how far away, its little photons would have experienced no time at all in reaching her. The same would apply to the tiny gravitons produced in the explosion, some of which might be felt by the Institute's gravitational wave detector near Hannover about 250 km away. She felt moved by the thought that she would in effect be in immediate direct contact with that explosive event...

As she stood there looking to the east, she was startled by a momentary and unexpected streak of green light, just as the dawn was about to come upon her, whereupon the deep red of the Sun broke through. The phenomenon of the 'green flash' and its well-established physical explanation were known to her, but she had never actually witnessed it before and it created in her a strange emotional effect. This experience mingled with some puzzling mathematical thoughts that had been troubling her throughout the night.

Then an odd thought overtook her...

Bibliography

In addition to the great advances in physical understanding that have been achieved in the 20th century—from highly refined experiment and sophisticated mathematical theory—modern technology and innovation have vastly improved the capabilities for disseminating and retrieving information on a global scale. Specifically, there is the introduction of arXiv.org, an online repository where physicists and mathematicians, biologists and computer scientists, can publish preprints (or 'e-prints') of their work before (or even instead of!) submitting it to journals. Indeed, ArXiv.org has made it possible for scientists to communicate new ideas at an incredibly high speed, and as a consequence the pace of research activity has accelerated to an unprecedented (or, as some might consider, an alarming) degree.

I have tried to take advantage of this important new trend by supplying, wherever possible, the arXiv.org links for items in the Bibliography. Finding a paper on arXiv.org is very simple. First, use your favourite web-browser to go to www.arxiv.org. Then either search for the paper or enter 'www.arxiv.org/' followed by the identification code provided in brackets in the Bibliography. For example, to call up Lee Smolin's 2003 paper 'How far are we from the quantum theory of gravity?', one would enter as the web address:

www.arxiv.org/hep-th/0303185.

This feature will be especially helpful for those readers who have access to the World Wide Web but are far away from a university library that might stock the specialized journals where scientific papers are often published. I hope that this new feature in the Bibliography encourages readers to study the many fine papers on arXiv, whether referenced here or otherwise.

Abian, A. (1965). *The theory of sets and transfinite arithmetic*. Saunders, Philadelphia.

Abbott, B. *et al.* (2004). Detector Description and Performance for the First Coincidence Observations between LIGO and GEO. *Nucl. Instrum. Meth.* **A517**, 154–79. [gr-qc/0308043]

Adams, C. C. (2000). *The Knot Book*. Owl Books, New York.

Adams, J. F. and Atiyah, M. F. A. (1966). On K-theory and Hopf invariant. *Quarterly J. Math.* **17**, 31–8.

Adler, S. L. (1995). *Quaternionic Quantum Mechanics and Quantum Fields*. Oxford University Press, New York.

Afriat, A. (1999). The Einstein, Podolsky, and Rosen Paradox. In *Atomic, Nuclear, and Particle Physics*. Plenum Publishing Corp.

Aharonov, Y. and Albert, D. Z. (1981). Can we make sense out of the measurement process in relativistic quantum mechanics? *Phys. Rev.* **D24**, 359–70.
Aharonov, Y. and Anandan, J. (1987). Phase change during a cyclic quantum evolution. *Phys. Rev. Lett.* **58**, 1593–6.
Aharonov, Y. and Bohm, D. J. (1959). Significance of electromagnetic potentials in the quantum theory. *Phys. Rev.* **115**, 485–91.
Aharonov, Y. and Vaidman, L. (1990). Properties of a quantum system during the time interval between two measurements. *Phys. Rev.* **A41**, 11.
Aharonov, Y. and Vaidman, L. (2001). The Two-State Vector Formalism of Quantum Mechanics. In *Time in Quantum Mechanics* (ed. J. G. Muga *et al.*). Springer-Verlag.
Aharonov, Y., Bergmann, P., and Lebowitz, J. L. (1964). Time symmetry in the quantum process of measurement. In *Quantum Theory and Measurement* (ed. J. A. Wheeler and W. H. Zurek). Princeton University Press, Princeton, New Jersey, 1983; originally in *Phys. Rev.* **134B**, 1410–6.
Ahmavaara, Y. (1965). The structure of space and the formalism of relativistic quantum theory, I. *J. Math. Phys.* **6**, 87–93.
Aitchison, I. and Hey, A. (2004). *Gauge Theories in Particle Physics: A Practical Introduction*, Vols 1 and 2. Institute of Physics Publishing, Bristol.
Albert, D., Aharanov, Y., and D'Amato, S. (1985). Curious New Statistical Prediction of Quantum Mechanics. *Phys. Rev. Lett.* **54**, 5.
Aldiss, B. W. and Penrose, R. (2000). *White Mars*. St Martin's Press, London.
Alpher, Bethe, and Gamow (1948). The Origin of Chemical Elements. *Phys. Rev.* **73**, 803.
Ambjorn, J., Nielsen, J. L., Rolf, J., and Loll, R. (1999). Euclidean and Lorentzian Quantum Gravity: Lessons from Two Dimensions. *Chaos Solitons Fractals* **10** [hep-th/9805108]
Amelino-Camelia, G. and Piran, T. (2001). Planck-scale deformations of Lorentz symmetry as a solution to the UHECR and the TeV-**MATH**-gamma paradoxes. *Phys. Rev.* **D64**, 036005. [astro-ph/0008107]
Amelino-Camelia, G., *et al.* (1998). Potential Sensitivity of Gamma-Ray Burster Observations to Wave Dispersion in Vacuo. *Nature* **393**, 763–5. [astro-ph/9712103].
Anderson, P. W. (1997). *The Theory of Superconductivity in the High-T_c Cuprate Superconductors*. Princeton University Press, Princeton, New Jersey.
Anguige, K. (1999). Isotropic cosmological singularities 3: The Cauchy problem for the inhomogeneous conformal Einstein-Vlasov equations. *Annals Phys.* **282**, 395–419.
Antoci, S. (2001). The origin of the electromagnetic interaction in Einstein's unified field theory with sources. [gr-qc/018052]
Anton, H. and Busby, R. C. (2003). *Contemporary Linear Algebra*. John Wiley & Sons, Hoboken, NJ.
Apostol, T. M. (1976). *Introduction to Analytic Number Theory*. Springer-Verlag, New York.
Arfken, G. and Weber, H. (2000). *Mathematical Methods for Physicists*. Harcourt/Academic Press.
Arndt, M. *et. al.* (1999). Wave-particle duality of C_{60} molecules. *Nature* **401**, 680.
Arnol'd, V. I. (1978). *Mathematical Methods of Classical Mechanics*. Springer-Verlag, New York.
Arnowitt, R., Deser, S., and Misner, C. W. (1962). In *Gravitation: An Introduction to Current Research* (ed. L. Witten). John Wiley & Sons, Inc. New York.

Ashtekar, A. (1986). New variables for classical and quantum gravity. *Phys. Rev. Lett.* **57**, 2244–7.

Ashtekar, A. (1987). New Hamiltonian formulation of general relativity. *Phys. Rev.* **D36**, 1587–1602.

Ashtekar, A. (1991). *Lectures on Non-Perturbative Canonical Gravity* (Appendix). World Scientific, Singapore.

Ashtekar, A. and Das, S. (2000). Asymptotically anti-de Sitter space-times: Conserved quantities. *Classical and Quantum Gravity* **17**, L17–L30.

Ashtekar, A. and Lewandowski, J. (2001). Relation between polymer and Fock excitations. *Class. Quant. Grav.* **18**, L117–L127.

Ashtekar, A. and Lewandowski, J. (2004). Background Independent Quantum Gravity: A Status Report. [gr-qc/0404018].

Ashtekar, A. and Magnon, A. (1980). A geometric approach to external potential problems in quantum field theory. *General Relativity and Gravity*, vol 12, 205–223.

Ashtekar, A. and Schilling, T. A. (1998). In *On Einstein's Path* (ed. A.Harvey). Springer-Verlag, Berlin.

Ashtekar, A., and Lewandowski, J. (1994). Representation theory of analytic holonomy algebras. In *Knots and Quantum Gravity* (ed J. C. Baez Oxford University Press, Oxford.)

Ashtekar, A., Baez, J. C., and Krasnov, K. (2000). Quantum geometry of isolated horizons and black hole entropy. *Adv. Theo. Math. Phys.* **4**, 1–95.

Ashtekar, A., Baez, J. C., Corichi, A., and Krasnov, K. (1998). Quantum geometry and black hole entropy. (1998). *Phys. Rev. Lett.* **80**, 904–07.

Ashtekar, A., Bojowald, M., Lewandowski, J. (2003). Mathematical structure of loop quantum cosmology. *Adv. Theor. Math. Phys.* **7**, 233–68.

Atiyah, M. F. (1990). *The Geometry and Physics of Knots*. Cambridge University Press, Cambridge.

Atiyah, M. F. and Singer, I. M. (1963). The Index of Elliptic Operators on Compact Manifolds. *Bull. Amer. Math. Soc.* **69**, 322–433.

Atiyah, M. F., Hitchin, N. J., and Singer, I. M. (1978). Self-duality in four-dimensional Riemannian geometry. *Proc. Roy. Soc. Lond.* **A362**, 425–61.

Baez, J. C. (1998). Spin foam models. *Class. Quant. Grav.* **15**, 1827–58.

Baez, J. C. (2000). An introduction to spin foam models of quantum gravity and BF theory. *Lect. Notes Phys.* **543**, 25–94

Baez, J. C. (2001). Higher-dimensional algebra and Planck-scale physics. In *Physics Meets Philosophy at the Planck Scale* (ed. C. Callender and N. Huggett). Cambridge University Press, Cambridge. [gr-qc/9902017].

Baez, J. C. and Dolan, D. (1998). Categorification. In *Higher Category Theory* (ed. E. Getzler and M. Kapranov). Contemporary Mathematics vol. 230. AMS, Providence. RI. [Also see http://xxx.lanl.gov/abs/math.QA/9802029]

Bahcall, N., Ostriker, J. P., Perlmutter, S., and Steinhardt., P. J. (1999). The Cosmic Triangle: Revealing the State of the Universe. *Science* **284** [astro-ph/9906463].

Bailey, T. N. and Baston, R. J. (ed.) (1990). *Twistors in Mathematics and Physics*. LMS Lecture Note Series 156. Cambridge University Press, Cambridge.

Bailey, T. N., Ehrenpreis, L. and Wells, R. O., Jr. (1982). Weak solutions of the massless field equations. *Proc. Roy. Soc. Lond.* **A384**, 403–25.

Banchoff, T. (1990, 1996). *Beyond the Third Dimension*: Scientific American Library. See also http://www.faculty.fairfield.edu/jmac/cl/tb4d.htm

Bar, I. (2000). Survey of Two-Time Physics. [hep-th/0008164]

Barbour, J. B. (1989). *Absolute or Relative Motion*, Vol. I: *The Discovery of Dynamics*. Cambridge University Press, Cambridge.

Barbour, J. B. (1992). *Time and the interpretation of quantum gravity*. Syracuse University Preprint.

Barbour, J. B. (2001a). *The Discovery of Dynamics: A Study from a Machian Point of View of the Discovery and the Structure of Dynamical Theories*. Oxford University Press, Oxford.

Barbour, J. B. (2001b). *The End of Time*. Oxford University Press, Oxford.

Barbour, J. B. (2004). *Absolute or Relative Motion: The Deep Structure of General Relativity*. Oxford University Press, Oxford.

Barbour, J. B., Foster, B., and O Murchadha, N. (2002). Relativity without relativity. *Class. Quant. Grav.* **19**, 3217–48. [gr-qc/0012089]

Barrett, J. W. and Crane, L. (1998). Relativistic spin networks and quantum gravity. *J. Math. Phys.* **39**, 3296–302.

Barrett, J. W. and Crane, L. (2000). A Lorentzian signature model for quantum general relativity. *Class. Quant. Grav.* **17,** 3101–18.

Barrow, J. D. and Tipler, F. J. (1988). *The Anthropic Cosmological Principle*. Oxford University Press, Oxford.

Baston, R. J. and Eastwood, M. G. (1989). *The Penrose transform: its interaction with representation theory*. Oxford University Press, Oxford.

Bateman, H. (1904). The solution of partial differential equations by means of definite integrals. *Proc. Lond. Math. Soc.* (2) **1**, 451–8.

Bateman, H. (1944). *Partial Differential Equations of Mathematical Physics*. Dover, New York.

Becker, R. (1982). *Electromagnetic Fields and Interactions*. Dover, New York.

Begelman, M. C., Blandford, R. D., and Rees, M. J. (1984). Theory of extragalactic radio sources. *Rev. Mod. Phys.* **56**, 255.

Bekenstein, J. (1972). Black holes and the second law. *Lett. Nuovo. Cim.*, **4**, 737–40.

Belinskii, V. A., Khalatnikov, I. M., and Lifshitz, E. M. (1970). Oscilliatory approach to a singular point in the relativistic cosmology. *Usp. Fiz. Nauk* **102**, 463–500. (Engl. transl. in *Adv. in Phys.* **19**, 525–73.)

Bell, J. S. (1987). *Speakable and Unspeakable in Quantum Mechanics*. Cambridge University Press, Cambridge.

Beltrami, E. (1868). Essay on the interpretation of non-Euclidean geometry. Translated in Stillwell, J. C. (1996). Sources of Hyperbolic Geometry. *Hist. Math.,* **10**, AMS Publications.

Bennett, C. L. *et al.* (2003). First Year Wilkinson Microwave Anisotropy Probe (WMAP) Observations: Preliminary Maps and Basic Results. *Astrophys. J. Suppl.* 148, 1. [astro-ph/0302207]

Bergmann, P. G. (1956). *Helv. Phys. Acta Suppl.* **4**, 79.

Bergmann, P. G. (1957). Two-component spinors in general relativity. *Phys. Rev.* **107**, 624–9.

Bern, Z. (2002). Perturbative Quantum Gravity and its relation to Gauge Theory. *Living Rev. Relativity,* **5**. [http://relativity.livingreviews.org/Articles/lrr-2002–5/].

Berry, M. V. (1984). Quantal phase factors accompanying adiabatic changes. *Proc. Roy. Soc. Lond.* **A392**, 45–57.

Berry, M. V. (1985). Classical adiabatic angles and quantal adiabatic phase. *J. Phys. A. Math. Gen.* **18**, 15–27.

Berry, M. V. and Keating, J. P. (1999). The Riemann Zeros and Eigenvalue Asymptotics. *SIAM Review* **41**, No. 2, 236–266.

Berry, M.V. and Robbins, J.M. (1997). Indistinguishability for quantum particles: spin, statistics and the geometric phase. *Proc. R. Soc. Lond.* A **453**, 1771–1790.

Biedenharn, L. C. and Louck, J. D. (1981). *Angular Momentum in Quantum Physics.* Addison-Wesley, London.

Bilaniuk, O.-M. and Sudarshan, G. (1969). Particle beyond the light barrier. *Phys. Today* **22**, 43–51.

Birrell, N. D. and Davies, P. C. W. (1984). *Quantum Fields in Curved Space.* Cambridge University Press, Cambridge.

Bjorken, J. D. and Drell, S. D. (1965). *Relativistic Quantum Mechanics.* McGraw Hill, New York & London.

Blanchard, A., Douspis, M., Rowan-Robinson, M., and Sarkar, S. (2003). An alternative to the cosmological 'concordance model'. *Astron. Astrophys.* **412**, 35–44.

Blanford, R. D. and Znajek, R. L. (1977). Electromagnetic Extraction of Energy from Kerr Black Holes. *Monthly Notices of the Royal Astronomical Society* **179**, 433.

Bohm, D. (1951). *Quantum Theory* (Prentice–Hall, Englewood-Cliffs.) Ch. 22, sect. 15–19. *Reprinted* as: The Paradox of Einstein, Rosen and Podolsky, in *Quantum Theory and Measurement* (ed. J.A. Wheeler and W.H. Zurek) Princeton Univ. Press, Princeton, New Jersey, 1983.

Bohm, D. and Hiley, B. (1994). *The Undivided Universe.* Routledge, London.

Bojowald, M. (2001). Absence of singularity in loop quantum cosmology. *Phys. Rev. Lett.* **86**, 5227–230.

Bondi, H. (1957). Negative mass in general relativity. *Rev. Mod. Phys.* **29**, 423–8; also in *Math. Rev.* **19**, 814.

Bondi, H. (1960). Gravitational waves in general relativity. *Nature* (London) **186**, 535.

Bondi, H. (1961). *Cosmology.* Cambridge University Press, Cambridge.

Bondi, H. (1964). *Relativity and Common Sense.* Heinemann, London.

Bondi, H. (1967). *Assumption and Myth in Physical Theory.* Cambridge University Press, Cambridge.

Bondi, H. and Gold, T. (1948). The Steady-State Theory of the Expanding Universe. *Mon. Not. Roy. Astron. Soc.* **108**, 252–70.

Bondi, H., van der Burg, M. G. J., and Metzner, A. W. K. (1962). Gravitational waves in general relativity, VII. Waves from axisymmetric isolated systems. *Proc. Roy. Soc. Lond.* **A269**, 21–52.

Bonnor, W. B. and Rotenberg, M. A. (1966). Gravitational waves from isolated sources. *Proc. Roy. Soc. Lond.* **A289**, 247–74.

Börner, G. (2003). *The Early Universe.* Springer-Verlag.

Bouchet, F. R., Peter, P., Riazuelo, A. and Sakellariadou, M. (2000). Evidence against or for topological defects in the BOOMERanG data? *Phys. Rev.* **D65** (2002), 021301. [astro-ph/0005022]

Boyer, C. B. (1968). *A History of Mathematics, 2nd. ed.* John Wiley & Sons, New York.

Braginsky, V. (1977). The Detection of Gravitational Waves and Quantum Non-Distributive Measurements. In *Topics in Theoretical and Experimental Gravitation Physics* (ed. V. De Sabbata and J. Weber), pp. 105–22. Plenum Press, New York.

Bramson, B.D. (1975). The alignment of frames of reference at null infinity for asymptotically flat Einstein-Maxwell manifolds. *Proc. R. Soc. London, Ser A* **341**, 451–461.

Brandhuber, A., Spence, B., and Travaglini, G. (2004). One-Loop Gauge Theory Amplitudes in N = 4 Super Yang Mills from MHV Vertices. [hep-th/0407214].

Brauer, R. and Weyl, H. (1935). Spinors in *n* dimensions. *Am. J. Math.* **57**, 425–49.

Brekke, L. and Freund, P. G. O. (1993). *p-adic numbers in physics*. North-Holland, Amsterdam.

Bremermann, H. (1965). *Distributions, Complex Variables and Fourier Transforms*. Addison Wesley, Reading, Massachusetts.

Brody, D. C. and Hughston, L. P. (1998a). Geometric models for quantum statistical inference. In *The Geometric Universe; Science, Geometry, and the Work of Roger Penrose* (ed. S. A. Huggett, L. J. Mason, K. P. Tod, S. T. Tsou, and N. M. J. Woodhouse). Oxford University Press, Oxford.

Brody, D. C. and Hughston, L. P. (1998b). The quantum canonical ensemble. *J. Math. Phys.* **39(12)**, 6502–8.

Brody, D. C., and Hughston, L. P. (2001). Geometric Quantum Mechanics, *J. Geom. and Phys.* **38(1)**, 19–53.

Brown, J. W. and Churchill, R. V. (2004). *Complex Variables and Applications*. McGraw-Hill, New York & London.

Bryant, R. L., Chern, S. -S., Gardner, R. B., Goldschmidt, H. L., and Griffiths, P. A. (1991). *Exterior Differential Systems*. MSRI Publications, 18. Springer-Verlag, New York.

Bucher, M., Goldhaber, A., and Turok, N. (1995). An open Universe from Inflation. *Phys. Rev.* **D**52. [hep-ph/9411206]

Buckley, P. and Peat, F. D. (1996). *Glimpsing Reality*. University of Toronto Press, Toronto.

Budinich, P. and Trautman, A. (1988). *The Spinorial Chessboard*. Trieste Notes in Physics. Springer-Verlag, Berlin.

Burbidge, G. R., Burbidge, E. M., Fowler, W. A., and Hoyle, F. (1957). Synthesis of the Elements in Stars. *Revs. Mod. Phys.* **29**, 547–650.

Burkert, W. (1972). *Lore and Science in Ancient Pythagoreanism*. Harvard University Press, Harvard.

Burkill, J. C. (1962). *A First Course in Mathematical Analysis*. Cambridge University Press, Cambridge.

Byerly, W. E. (2003). *An Elementary Treatise on Fourier's Series and Spherical, Cylindrical, and Ellipsoidal Harmonics, with Applications to Problems in Mathematical Physics*. Dover, New York.

Cachazo, F., Svrcek, P., and Witten, E. (2004a). MHV Vertices and Tree Amplitudes in Gauge Theory. [hep-th/0403047]

Cachazo, F., Svrcek, P., and Witten, E. (2004b). Twistor Space Structure of One-Loop Amplitudes In Gauge Theory. [hep-th/0406177].

Cachazo, F., Svrcek, P., and Witten, E. (2004c). Gauge Theory Amplitudes In Twistor Space and Holomorphic Anomaly. [hep-th/0409245].

Candelas, P., de la Ossa, X. C., Green, P. S., and Parkes, L. (1991). A pair of Calabi-Yau manifolds as an exactly soluble superconformal theory. *Nucl. Phys.* **B359**, 21.

Cartan, É. (1923). Sur les variétés à connexion affine et la théorie de la relativité generalisée I. *Ann. École Norn. Sup.* **40**, 325–412.

Cartan, É. (1924). Sur les variétés à connexion affine et la théorie de la relativité generalisée (suite). *Ann. École Norn. Sup.* **41**, 1–45.

Cartan, É. (1925). Sur les variétés à connexion affine et la théorie de la relativité generalisée II. *Ann. École Norn. Sup.* **42**, 17–88.

Cartan, É. (1945). *Les Systèmes Différentiels Extérieurs et leurs Applications Géométriques*. Hermann, Paris.
Cartan, É. (1966). *The Theory of Spinors*. Hermann, Paris.
Carter, B. (1966a). The Complete Analytic Extension of the Reissner-Nordstrom Metric in the Special Case $e^2 = m^2$, *Phys. Letters* **21**, 423–4; Complete Analytic Extension of the Symmetry Axis of Kerr's Solution of Einstein's Equations, *Phys. Rev.* **141**, 1242–7.
Carter, B. (1966b). *Cambridge Ph.D. thesis*; the relevant pages are accessible on http://luth2.obspm.fr/~carter/Thesis/4p25A-25B.html
Carter, B. (1971). Axisymmetric Black Hole Has Only Two Degrees of Freedom. *Phys. Rev. Lett.*, **26**, 331–2.
Carter, B. (1974). Large Number Coincidences and the Anthropic Principle. In *Confrontation of Cosmological Theory with Astronomical Data* (ed. M. S. Longair), pp. 291–8. Reidel, Dordrecht. (Reprinted in Leslie 1990.)
Cercignani, C. (1999). *Ludwig Boltzmann: The Man Who Trusted Atoms*. Oxford University Press, Oxford.
Chan, H-M. and Tsou, S. T. (1993). *Some Elementary Gauge Theory Concepts*. World Scientific Lecture Notes in Physics, Vol. 47. London.
Chan, H-M. and Tsou, S. T. (2002). Fermion Generations and Mixing from Dualized Standard Model. *Acta Physica Polonica B* **12**.
Chandrasekhar, S. (1981). The maximum mass of ideal white dwarfs. *Astrophys. J.*, **74**, 81–2.
Chandrasekhar, S. (1983). *The Mathematical Theory of Black Holes*. Clarendon Press, Oxford.
Chari, V. and Pressley, A. (1994). *A Guide to Quantum Groups*. Cambridge University Press, Cambridge.
Chen, W. W. L. (2002). *Linear Functional Analysis*. [Available online: http://www.maths.mq.edu.au/~wchen/lnlfafolder/lnlfa.html]
Cheng, K. S. and Wang, J. (1999). The formation and merger of compact objects in central engine of active galactic nuclei and quasars: gamma-ray burst and gravitational radiation. *Astrophys., J.* **521**, 502.
Chern, S. S. (1979). *Complex Manifolds Without Potential Theory*. Springer-Verlag, New York.
Chernoff, P. R. and Marsden, J. E. (1974). *Properties of infinite hamiltonian systems*. Lecture Notes in Mathematics, vol. 425. Springer-Verlag, Berlin.
Chevalley, C. (1946). *Theory of Lie Groups*. Princeton University Press, Princeton.
Chevalley, C. (1954). *The Algebraic Theory of Spinors*. Columbia University Press, New York.
Chew, G. F. (1962). *S-Matrix Theory of Strong Interactions*. Pearson Benjamin Cummings.
Choquet-Bruhat, Y. and DeWitt-Morette, C. (2000). *Analysis, Manifolds, and Physics*. Parts I and II. North-Holland, Amsterdam.
Christenson, J. H., Cronin, J.W., Fitch, V.L. and Turlay, R. (1964). Evidence for the 2p decay of the K0 meson, *Phys. Rev. Lett.* **13**, 138–140.
Christian, J. (1995). Definite events in Newton–Cartan quantum gravity. Oxford preprint, submitted to *Phys. Rev. D*.
Church, A. (1936). *The calculi of lambda-conversion*. Annals of Mathematics Studies, No. 6. Princeton University Press, Princeton, NJ.
Claudel, C. M. and Newman, K. P. (1998). Isotropic Cosmological Singularities I. Polytropic Perfect Fluid Spacetimes. *Proc. R. Soc. Lond.* 454, 1073–1107.

Clauser, J. F., Horne, M. A., Shimony, A., and Holt, R. A. (1969). Proposed experiment to test local hidden-variable theories. *Phys. Rev. Lett.* **23**, 880.

Clifford, W. K. (1873). Preliminary Sketch of Biquaternions. *Proc. London Math. Soc.* **4**, 381–95.

Clifford, W. K. (1878). Applications of Grassmann's extensive algebra. *Am. J. Math.* **1**, 350–8.

Clifford, W. K. (1882). Mathematical papers by William Kingdon Clifford. (ed. R. Tucker.) London.

Cohen, P. J. (1966). *Set Theory and the Continuum Hypothesis*. W. A. Benjamin, New York.

Collins, P. D. B. (1977). *An Introduction to Regge Theory and High Energy Physics*. Cambridge University Press, Cambridge.

Colombeau, J. F. (1983). A multiplication of distributions. *J. Math. Anal. Appl.* **94**, 96–115.

Colombeau, J. F. (1985). *Elementary Introduction to New Generalized Functions*. North Holland, Amsterdam.

Connes, A. (1990). Essay on physics and non-commutative geometry. In *The Interface of Mathematics and Particle Physics* (ed. D. G. Quillen, G. B. Segal, and Tsou S. T.). Clarendon Press, Oxford.

Connes, A (1998). Noncommutative differential geometry and the structure of space-time. In *The Geometric Universe; Science, Geometry, and the Work of Roger Penrose* (ed. S. A. Huggett, L. J. Mason, K. P. Tod, S. T. Tsou, and N. M. J. Woodhouse). Oxford University Press, Oxford.

Connes, A. and Berberian, S. K. (1995). *Noncommutative Geometry*. Academic Press.

Connes, A. and Kreimer, D. (1998). Hopf Algebras, Renormalization and Noncommutative Geometry. [hep-th/9808042]

Conway, J. H. (1976). *On Numbers and Games*. Academic Press, London.

Conway, J. H. and Kochen, S. (2002). The geometry of the quantum paradoxes. In *Quantum [Un]speakables: From Bell to Quantum Information* (Eds. R. A. Bertlmann and A. Zeilinger) Springer-Verlag, Berlin.

Conway, J. H., and Norton, S. P. (1979). Monstrous Moonshine. *Bull. Lond. Math. Soc.* **11,** 308–39.

Conway, J. H. and Smith, D. A. (2003). *On Quaternions and Octonions*. A. K. Peters.

Corson, E.M. (1953). *Introduction to Tensors, Spinors, and Relativistic Wave-equations.* Blackie and Son Ltd., London.

Costa de Beauregard, O. (1995). Macroscopic retrocausation. *Found. Phy. Lett.* **8(3)**, 287–91.

Cotes, R. (1714). Logometria. *Phil. Trans. Roy. Soc. Lond.* (March).

Cox, D. A. and Katz, S. (1999). *Mirror symmetry and algebraic geometry*. Mathematical Surveys and Monographs 68. American Mathematical Society, Providence, RI.

Cramer. J. G. (1988). An overview of the transactional interpretation of quantum mechanics. *Int. J. Theor. Phys.* **27(2)**, 227–36.

Crowe, M. J. (1967). *A History of Vector Analysis: The Evolution of the Idea of a Vectorial System*. University of Notre Dame Press, Toronto. (Reprinted with additions and corrections by Dover, New York, 1985.)

Crumeyrolle, A. (1990). *Orthogonal and Symplectic Clifford Alebras Abstract: On Spinor Structures*. Kluwer [ISBN 0-7923-0541-8].

Cvitanovič, P. and Kennedy, A. D. (1982). Spinors in negative dimensions. *Phys. Scripta* **26**, 5–14.

Das, A. and Ferbel. T. (2004). *Introduction to Nuclear and Particle Physics*. World Scientific Publishing Company, Singapore.

Davenport, H. (1952). *The Higher Arithmetic: An Introduction to the Theory of Numbers*. Hutchinson's University Library, London.

Davies, M. (1997). *Europe: A History* (Oxford University Press, Oxford), pp. 89–94.

Davies, P. (2003). *How to Build a Time Machine*. Penguin, USA.

Davis, M. (1978). What is a Computation? In *Mathematics Today: Twelve Informal Essays* (ed. L. A. Steen). Springer-Verlag, New York.

Davis, M. (1988). Mathematical logic and the origin of modern computers. In *The Universal Turing Machine: A Half-Century Survey* (ed. R. Herken). Kammerer and Unverzagt, Hamburg.

Davydov, A. S. (1976). *Quantum Mechanics*. Pergamon Press, Oxford.

de Bernardis, P. *et al.* (2000). A Flat Universe from High-Resolution Maps of the Cosmic Microwave Background Radiation. *Nature* **404**, 955–9.

Delduc, F., Galperin, A., Howe, P., and Sokatchev, E. (1993). A twistor formulation of the heterotic D = 10 superstring with manifest (8,0) worldsheet supersymmetry. *Phys. Rev.* **D47,** 578–93. [hep-th/9207050]

Derbyshire, J. (2003). *Prime Obsession: Bernhard Riemann and the Greatest Unsolved Problem in Mathematics*. Joseph Henry Press, Washington, DC.

Deser, S. (1999). Nonrenormalizability of D = 11 supergravity. [hep-th/9905017]

Deser, S. (2000). Infinities in quantum gravities. *Annalen Phys.* **9**, 299–307. [gr-qc/9911073]

Deser, S. and Teitelboim, C. (1977). Supergravity Has Positive Energy. *Phys. Rev. Lett.* **39**, 248–52.

Deser, S. and Zumino, B. (1976). Consistent supergravity. *Phys. Lett.* **62B**, 335–7.

de Sitter, W. (1913). *Phys. Zeitz.*, **14**, 429. (in German).

Deutsch, D. (2000). *The Fabric of Reality*. Penguin, London.

Devlin, K. (1988). *Mathematics: The New Golden Age*. Penguin Books, London.

Devlin, K. (2002). *The Millennium Problems: The Seven Greatest Unsolved Mathematical Puzzles of Our Time*. Basic Books, London/Perseus Books, New York.

DeWitt, B. S. (1967). Quantum Theory of Gravity. I. The Canonical Theory. *Phys. Rev.* **160**, 1113.

DeWitt, B. S. (1984). *Supermanifolds*. Cambridge University Press, Cambridge.

DeWitt, B. S. and Graham, R. D. (ed). (1973). *The Many-Worlds Interpretation of Quantum Mechanics*. Princeton University Press, Princeton.

Diósi, L. (1984). Gravitation and quantum mechanical localization of macro-objects. *Phys. Lett.* **105A**, 199–202.

Diósi, L. (1989). Models for universal reduction of macroscopic quantum fluctuations. *Phys. Rev.* **A40**, 1165–74.

Dicke, R. H. (1961). Dirac's Cosmology and Mach's Principle. *Nature* **192,** 440–1.

Dicke, R. H. (1981). Interaction-free quantum measurements: A paradox? *Am. J. Phys.* **49**, 925.

Dicke, R. H., Peebles, P. J. E., Roll, P. G., and Wilkinson, D. T. (1965). Cosmic Black-Body Radiation. *Astrophys. J.* **142**, 414–19.

Dine, M. (2000). Some reflections on Moduli, their Stabilization and Cosmology. [hep-th/0001157]

Dirac, P. A. M. (1928). The quantum theory of the electron. *Proc. Roy. Soc. Lond.* **A117**, 610–24; ibid, part II, **A118**, 351–61.

Dirac, P. A. M. (1932). *Proc. Roy. Soc.* **A136**, 453.

Dirac, P. A. M. (1933). The Lagrangian in Quantum Mechanics. *Physicalische Zeitschrift der Sowjetunion*, Band 3, Heft 1.

Dirac, P.A.M. (1936). Relativistic Wave Equations. *Proc. Roy. Soc. London* **A155**, 447–59.
Dirac, P. A. M. (1937). The Cosmological Constants. *Nature* **139**, 323.
Dirac, P. A. M. (1938). A new basis for cosmology. *Proc. R. Soc. Lond.* **A165**, 199.
Dirac, P. A. M. (1950). Generalized Hamiltonian dynamics. *Can. J. Math.* **2**, 129.
Dirac, P. A. M. (1964). *Lectures on Quantum Mechanics*. Yeshiva University, New York.
Dirac, P. A. M. (1966). *Lectures in Quantum Field Theory*. Academic Press, New York.
Dirac, P. A. M. (1982a). *The Principles of Quantum Mechanics 4th edn*. Clarendon Press, Oxford.
Dirac, P. A. M. (1982b). Pretty mathematics. *Int. J. Theor. Phys.* **21**, 603–5.
Dirac, P. A. M. (1983). The Origin of Quantum Field Theory. In *The Birth of Particle Physics* (ed. Brown and Hoddeson). Cambridge University Press, New York.
Dixon, G. (1994). *Division Algebras, Quaternions, Complex Numbers and the Algebraic Design of Physics*. Kluwer Academic Publishers, Boston.
Dodelson, S. (2003). *Modern Cosmology*. Academic Press, London.
Dolan, L. (1996). Superstring twisted conformal field theory: Moonshine, the Monster, and related topics. (South Hadley, MA, 1994). *Contemp. Math.* **193**, 9–24.
Domagala, M. and Lewandowski, J. (2004). Black hole entropy from Quantum Geometry. [gr-qc/0407041].
Donaldson, S. K. and Kronheimer, P. B. (1990). *The Geometry of Four-Manifolds*. Oxford University Press, Oxford.
Douady, A. and Hubbard, J. (1985). On the dynamics of polynomial-like mappings. *Ann. Sci. Ecole Norm. Sup.* **18**, 287–343.
Dowker, F. and Kent, A. (1996). On the consistent histories approach to quantum mechanics. *J. Stat. Phys.* **82**. [gr-qc/9412067]
Drake, S. (1957). *Discoveries and Opinions of Galileo*. Doubleday, New York.
Drake, S. (trans.) (1953). *Galileo Galilei: Dialogue Concerning the Two Chief World Systems—Ptolemaic and Copernican*. University of California, Berkeley.
Dray, T. and Manogue, C. A. (1999). The Exceptional Jordan Eigenvalue Problem. *Int. J. Theor. Phys.* **38**, 2901–16, [math-ph/99110004].
Dreyer, O., Markopoulou, F., and Smolin, L. (2004). Symmetry and entropy of black hole horizons. [hep-th/0409056].
Duffin, R. J. (1938). On the characteristic matrices of covariant systems. *Phys. Rev.* **54**, 1114.
Dunajski, M. (2002). Anti-self-dual four-manifolds with a parallel real spinor. *R. Soc. Lond. Proc. Ser. A Math. Phys. Eng. Sci.* **458(2021)**, 1205–22.
du Sautoy, M. (2004). *The Music of the Primes*. Perennial, New York.
Dunham, W. (1999). *Euler: The Master of Us All*. Math. Assoc. Amer., Washington, DC.
Dyson, F. J. (1966). *Symmetry groups in nuclear and particle physics: a lecture-note and reprint volume*. W. A. Benjamin, New York.
Dyson, F. J. (1979). Time Without End: Physics and Biology in an Open Universe, *Rev. Mod. Phys*, **51**, 447–460.
Eastwood, M.G., Penrose, R., and Wells, R.O., Jr. (1981). Cohomology and massless fields. *Comm. Math. Phys.* **78**, 305–51.
Eddington, A. S. (1929a). *The Nature of the Physical World*. Cambridge University Press, Cambridge.

Eddington, A. S. (1929b). A Symmetrical Treatment of the Wave Equation. *Proc. R. Soc. Lond.* **A121**, 524–42.

Eddington, A. S. (1946). *Fundamental Theory*. Cambridge University Press, Cambridge.

Eden, R. J., Landshoff, P. V., Olive, D. I., and Polkinghorne, J. C. (2002). *The Analytic S-Matrix*. Cambridge University Press.

Edwards, C.H. and Penney, D.E. (2002). *Calculus with Analytic Geometry*. Prentice Hall; 6th edition.

Ehrenberg, W. and Siday, R. E. (1949). The refractive index in electron optics and the principles of dynamics. *Proc. Phys. Soc.* **LXIIB**, 8–21.

Ehrenfest, P. and Ehrenfest, T. (1959). *The Conceptual Foundations of the Statistical Approach in Mechanics*. Cornell University Press, Ithaca, NY.

Eilenberg, S. and Mac Lane, S. (1945). General theory of natural equivalences. *Trans. Am. Math. Soc.* **58**, 231–94.

Einstein, A. (1914), in Lorentz *et al.*, (1952).

Einstein, A. (1917). Kosmologische Betrachtungen zur allgemeinen Relativitätstheorie, *Sitzungsberichte der Preussischen Akademie der Wissenschaften*, 142–152.

Einstein, A. (1925). *S. B. Preuss. Akad. Wiss.* **22**, 414.

Einstein, A. (1945). A generalization of the relativistic theory of gravitation. *Ann. Math.* **46**, 578.

Einstein, A. (1948). A generalized theory of garitation. *Rev. Mod. Phys.* **20**, 35.

Einstein, A. (1955). Relativistic theory of the non-symmetric field. In Appendix II: *The Meaning of Relativity*, 5th ed., pp. 133–66. Princeton University Press, Princeton, NJ.

Einstein, A. and Kaufman, B. (1955). A new form of the general relativistic field equations. *Ann. Math.* **62**, 128.

Einstein, A. and Straus, E. G. (1946). A generalization of the relativistic theory of gravitation II. *Ann. Math.* **47**, 731.

Einstein, A., Podolsky, P., and Rosen, N. (1935). Can quantum-mechanical description of physical reality be considered complete? In *Quantum Theory and Measurement* (ed. J. A. Wheeler and W. H. Zurek). Princeton University Press, Princeton, New Jersey, 1983; originally in *Phys. Rev.* **47**, 777–80.

Elitzur, A. C. and Vaidman, L. (1993). Quantum mechanical interaction-free measurements. *Found. Phys.* **23**, 987–97.

Elliott, J. P., and Dawber, P. G. (1984). *Symmetry in Physics*, Vol. 1. Macmillan, London.

Ellis, J., Mavromatos, N. E., and Nanopoulos, D. V. (1997a). Vacuum fluctuations and decoherence in mesoscopic and microscopic systems. In *Symposium on Flavour-Changing Neutral Currents: Present and Future Studies*. UCLA.

Ellis, J., Mavromatos, N. E., and Nanopoulos, D. V. (1997b). Quantum decoherence in a D-foam background. *Mod. Phys. Lett.* **A12**, 2029–36.

Engelking, E. (1968). *Outline of General Topology*. North-Holland & PWN, Amsterdam.

Euler, L. (1748). *Introductio in Analysis Infinitorum.*

Everett, H. (1957). 'Relative State' formulation of quantum mechanics. In *Quantum Theory and Measurement* (ed. J. A. Wheeler and W. H. Zurek). Princeton University Press, Princeton, New Jersey, 1983; originally in Rev. Mod. Phys. **29**, 454–62.

Fauvel, J. and Gray, J. (1987). *The History of Mathematics: A Reader*. Macmillan, London.
Ferber, A. (1978). Supertwistors and conformal supersymmetry. *Nucl. Phys.* **B132** 55–64.
Fernow, R. C. (1989). *Introduction to Experimental Particle Physics* Cambridge University Press, Cambridge.
Feynman, R. P. (1948). Space-time approach to nonrelativistic quantum mechanics. *Rev. Modern Phys.* **20**, 367–87.
Feynman, R. P. (1949). The theory of positrons. *Phys. Rev.* **76**, 749.
Feynman, R. P. (1987). *Elementary Particles and the Laws of Physics: The 1986 Dirac Memorial Lectures*. Cambridge University Press, Cambridge.
Feynman, R. P. and A. Hibbs. (1965). *Quantum Mechanics and Path Integrals*. McGraw-Hill, New York.
Fierz, M. (1938). Uber die Relativitische Theorie kräftefreier Teichlen mit beliebigem Spin. *Helv. Phys. Acta* **12**, 3–37.
Fierz, M. (1940). Uber den Drehimpuls von Teichlen mit Ruhemasse null und beliebigem Spin. *Helv. Phys. Acta* **13**, 45–60.
Fierz, M. and Pauli, W. (1939). On relativistic wave equations for particles of arbitrary spin in an electromagnetic field. *Proc. Roy. Soc. Lond.* **A173**, 211–32.
Finkelstein, D. (1969). Space-time code. *Phys. Rev.* **184**, 1261–79.
Finkelstein, D. and J. Rubinstein. (1968). Connection between spin, statistics, and kinks. *J. Math. Phys.* **9**, 1972.
Flanders, H. (1963). *Differential Forms*. Academic Press. (Reissued by Dorf 1989.)
Floyd, R. M. and Penrose, R. (1971). Extraction of Rotational Energy from a Black Hole. *Nature Phys. Sci.* **229**, 177.
Fortney, L. R. (1997). *Principles of Electronics, Analog and Digital*. Harcourt Brace Jovanovich.
Frankel, T. (2001). *The Geometry of Physics*. Cambridge University Press, Cambridge.
Frenkel, A. (2000). A Tentative Expresion of the Károlyházy Uncertainty of the Space-time Structure through Vacuum Spreads in Quantum Gravity. [quant-ph/0002087]
Friedlander, F. G. (1982). *Introduction to the theory of distributions*. Cambridge University Press, Cambridge.
Friedrich, W. L. (2000). *Fire in the Sea: The Santorini Volcano: Natural History and the Legend of Atlantis*. (Trans. A.R. McBirney.) Cambridge University Press, Cambridge.
Frittelli, S., Kozameh, C. and Newman, E. T. (1997). Dynamics of light cone cuts at null infinity. *Phys. Rev.* **D56**, 8.
Fröhlich, J. and Pedrini, B. (2000). New applications of the chiral anomaly. In *Mathematical Physics 2000* (ed. A. Fokas, A. Grigoryan, T. Kibble, and B. Zegarlinski), pp. 9–47. Imperial College Press, London.
Gürsey, F. and Tze, C. -H. (1996). *On the Role of Division, Jordan, and Related Algebras in Particle Physics*. World Scientific, Singapore.
Gambini, R. and Pullin, J. (1999). Nonstandard optics from quantum spacetime. *Phys. Rev.* **D59** 124021.
Gandy, R. (1988). The confluence of ideas in 1936. In *The Universal Turing Machine: A Half-Century Survey* (ed. R. Herken). Kammerer and Unverzagt, Hamburg.
Gangui, A. (2003). Cosmology from Topological Defects. *AIP Conf. Proc.* **668**. [astro-ph/0303504]
Gardner, M. (1990). *The New Ambidextrous Universe*. W. H. Freeman, New York.
Gauss, C. F. (1900). *Werke*. Vol. VIII, pp. 357–62. Leipzig.

Gel'fand, I. and Shilov, G. (1964). *Generalized Functions*, Vol. 1. Academic Press, New York.

Gell-Mann, M. (1994). *The Quark and the Jaguar: Adventures in the Simple and the Complex*. W. H. Freeman, New York.

Gell-Mann, M. and Hartle, J. B. (1995). Strong Decoherence. In *Proceedings of the 4th Drexel Conference on Quantum Non-Integrability: The Quantum-Classical Correspondence* (ed. D.-H. Feng and B.-L. Hu). International Press of Boston, Hong Kong (1998). [gr-qc/9509054]

Gell-Mann, M. and Ne'eman, Y. (2000). *Eightfold Way*. Perseus Publishing.

Geroch, R and Hartle, J. (1986). Computability and physical theories. *Found. Phys.* **16**, 533.

Geroch, R. (1968). Spinor structure of space-times in general relativity I. *J. Math. Phys.* **9**, 1739–44.

Geroch, R. (1970). Spinor structure of space-times in general relativity II. *J. Math. Phys.* **11**, 343–8.

Geroch, R. (1984). The Everett Interpretation. *Nous*, **18**, 617–633.

Geroch, R. (unpublished). *Geometrical Quantum Mechanics*. Lecture notes given at University of Chicago.

Geroch, R., Kronheimer, E. H., and Penrose, R. (1972). Ideal points for space-times. *Proc. Roy. Soc. Lond.* **A347**, 545–67.

Ghirardi, G. C., Grassi, R., and Rimini, A. (1990). Continuous–spontaneous–reduction model involving gravity. *Phys. Rev.* **A42**, 1057–64.

Ghirardi, G. C., Rimini, A., and Weber, T. (1986). Unified dynamics for microscopic and macroscopic systems. *Phys. Rev.* **D34**, 470.

Gibbons, G. W. (1984). The isoperimetric and Bogomolny inequalities for black holes. In *Global Riemannian Geometry* (ed. T. Willmore and N. J. Hitchin). Ellis Horwood, Chichester.

Gibbons, G. W. (1997). Collapsing Shells and the Isoperimetric Inequality for Black Holes. *Class. Quant. Grav.* **14**, 2905–15. [hep-th/9701049]

Gibbons, G. W. and Hartnoll, S. A. (2002). Gravitational instability in higher dimensions. [hep-th/0206202]

Gibbons, G. W. and Hawking, S. W. (1977). *Phys. Rev.* **D15**, 2738–51.

Gibbons, G. W. and Perry, M. J. (1978). Black Holes and Thermal Green's Function. *Proc. Roy. Soc. Lond.* **A358**, 467–94.

Gibbs, J. (1960). *Elementary Principles in Statistical Mechanics*. Dover, New York.

Gindikin, S. G. (1986). On one construction of hyperkähler metrics. *Funct. Anal. Appl.* **20**, 82–3. (Russian).

Gindikin, S. G. (1990). Between integral geometry and twistors. In *Twistors in Mathematics and Physics* (ed. T. N. Bailey and R. J. Baston). LMS Lecture Note Series 156. Cambridge University Press, Cambridge.

Gisin, N. (1989). Stochastic quantum dynamics and relativity. *Helv. Phys. Acta.* **62**, 363.

Gisin, N. (1990). *Phys. Lett.* **143A**, 1.

Gisin, N., de Riedmatten, H., Scarani, V., Marcikic, I., Acin, A., Tittel, W., and Zbinden, H. (2004). Two independent photon pairs versus four-photon entangled states in parametric down conversion. J. *Mod. Opt.* **51**, 1637. [quant-ph/0310167]

Glashow, S. (1959). The renormalizability of vector meson interactions. *Nucl. Phys.* **10**, 107.

Gleason, A. M. (1957). Measures on the Closed Subspaces of a Hilbert Space. *J. Math. and Mech.* **6**, 885–893.

Gleason, A. M. (1990). In *More Mathematical People* (ed. D. J. Albers, G. L. Alexanderson, and C. Reid). Harcourt Brace Jovanovich, Boston, p. 94.

Goddard, P. *et al.* (1973). Quantum dynamics of a massless, relativistic string. *Nucl. Phys.* **B56**, 109.

Gold, T. (1962). The Arrow of Time. *Am. J. Phys.* **30**, 403.

Goldberg, J. N., Macfarlane, A. J., Newman, E. T., Rohrlich, F., and Sudarshan, E. C. G. (1967). Spin-s spherical harmonics and eth. *J. Math. Phys.* **8**, 2155–61.

Goldblatt, R. (1979). *Topoi: The Categorial Analysis of Logic*. North-Holland Publishing Company, Oxford & New York.

Goldstein, S. (1987). Stochastic mechanics and quantum theory. *J. Stat. Phys.* **47**.

Gottesman, D. and Preskill, J. (2003). Comment on 'The black hole final state'. [hep-th/0311269]

Gouvea, F. Q. (1993). *P-Adic Numbers: An Introduction.* Springer-Verlag; 2nd edition (2000), Berlin & New York.

Grassmann, H. G. (1844). *Die lineare Ausdehnungslehre*, 4th edition, Springer-Verlag.

Grassmann, H. G. (1862). *Die lineare Ausdehnungslehre Vollständig und in strenger Form bearbeitit*.

Gray, J. (1979). *Ideas of Space: Euclidean, Non-Euclidean, and Relativistic*. Oxford University Press, Oxford.

Green, M. B. (2000). Superstrings and the unification of physical forces. In *Mathematical Physics 2000* (ed. A. Fokas, T. W. B. Kibble, A. Grigouriou, and B. Zegarlinski), pp. 59–86. Imperial College Press, London.

Green, M. B., Schwarz, J. H., and Witten, E. (1978). *Superstring Theory*, Vol. I & II. Cambridge University Press, Cambridge.

Greene, B. (1999). *The Elegant Universe; Superstrings, Hidden Dimensions, and the Quest for the Ultimate Theory*. Random House, London.

Griffiths, P. and Harris, J. (1978). *Principles of Algebraic Geometry*. John Wiley & Sons, New York.

Grishchuk, L. P., *et al.* (2001). Gravitational Wave Astronomy: in Anticipation of First Sources to be Detected. Phys. Usp. **44**, 1–51. [astro-ph/0008481].

Groemer, H. (1996). *Geometric Applications of Fourier Series and Spherical Harmonics*. Cambridge University Press, Cambridge.

Gross, D. J. and Periwal, V. (1988). String Perturbation Theory Diverges. *Phys. Rev. Lett.* **60,** 2105.

Gross, M. W., Huybrechts, D., Joyce, D., and Winkler, G. D. (2003). *Calabi-Yau Manifolds and Related Geometries*. Springer-Verlag.

Grosser, M., Kunzinger, M., Oberguggenberger, M. and Steinbauer, R. (2001). *Geometric Theory of Generalized Functions with Applications to General Relativity*. Kluwer Academic Publishers, Boston and Dordrecht, The Netherlands.

Guenther, D. B., Krauss, L. M., and Demarque, P. (1998). Testing the Constancy of the Gravitational Constant Using Helioseismology. *Astrophys. J.* **498**, 871–6.

Gunning, R. C. and Rossi, H. (1965). *Analytic Functions of Several Complex Variables*. Prentice-Hall, Englewood Cliffs, New Jersey.

Gürsey, F. (1983). Quaternionic and octonionic structures in physics: episodes in the relation between physics and mathematics. *Symm. Phys. (1600–1980)*, pp. 557–92. San Feliu de Guíxols. Univ. Autònoma Barcelona, Barcelona, 1987.

Gürsey, F. and Tze, C.-H. (1996). *On the Role of Division, Jordan, and Related Algebras in Particle Physics*. World Scientific, Singapore.

Gurzadyan, V. G. *et al.* (2002). Ellipticity analysis of the BOOMERANG CMB maps. *Int. J. Mod. Phys.* **D12**, 1859–74. [astro-ph/0210021]

Gurzadyan, V. G. *et al.* (2003). Is there a common origin for the WMAP low multipole and for the ellipticity in BOOMERANG CMB maps? [astro-ph/0312305]

Gurzadyan, V. G. *et al.* (2004). WMAP confirming the ellipticity in BOOMERANG and COBE CMB maps. [astro-ph/0402399]

Gurzadyan, V. G. and Kocharyan, A. A. (1992). On the problem of isotropization of cosmic background radiation. Astron. Astrophys. **260**, 14.

Gurzadyan, V. G. and Kocharyan, A. A. (1994). *Paradigms of the Large-Scale Universe*. Gordon and Breach, Lausanne, Switzerland.

Gurzadyan, V. G and Torres, S. (1997). Testing the effect of geodesic mixing with COBE data to reveal the curvature of the universe. *Astron. and Astrophys.* **321**, 19–23. [astro-ph/9610152]

Guth, A. (1997). *The Inflationary Universe*. Jonathan Cape, London.

Haag, R. (1992). *Local Quantum Physics: Fields, Particles, Algebras*. Springer-verlag, Berlin.

Haehnelt, M. G. (2003). Joint Formation of Supermassive Black Holes and Galaxies. In *Carnegie Observatories Astrophysics Series, Vol 1: Coevolution of Black Holes and Galaxies* (ed. L. C. Ho). Cambridge University Press, Cambridge. [astro-ph/0307378]

Halverson, N. W. (2001). DASI First Results: A Measurement of the Cosmic Microwave Background Angular Power Spectrum. [astro-ph/0104489]

Halzen, F. and Martin, A. D. (1984). *Quarks and Leptons: an introductory course in modern particle physics*. John Wiley & Sons, New York.

Hannabuss, K. (1997). *An Introduction to Quantum Theory*. Oxford University Press, Oxford.

Hameroff, S. R. (1998). Funda-mental geometry: the Penrose–Hameroff 'Orch OR' model of consciousness. In *The Geometric Universe; Science, Geometry, and the Work of Roger Penrose* (ed. S. A. Huggett, L. J. Mason, K. P. Tod, S. T. Tsou, and N. M. J. Woodhouse). Oxford University Press, Oxford.

Hameroff, S. R. (1987). *Ultimate Computing. Biomolecular Consciousness and Nano-Technology*. North-Holland, Amsterdam.

Hameroff, S. R. and Penrose, R. (1996). Conscious events as orchestrated space-time selections. *J. Consc. Stud.* **3**, 36–63.

Hameroff, S. R. and Watt, R. C. (1982). Information processing in microtubules. *J. Theor. Biol.* **98**, 549–61.

Hamilton, E. (1999). *Mythology: Timeless Tales of Gods and Heroes*. Warner Books, New York.

Han, M. Y. and Nambu, Y. (1965). Three-Triplet Model with Double *SU*(3) Symmetry. *Phys. Rev.* **139**, B1006–10.

Hanany, S. *et al.* (2000). MAXIMA-1: A Measurement of the Cosmic Microwave Background Anisotropy on angular scales of 10 arcminutes to 5 degrees. *Astrophys. J.* **545**, L5.

Hanbury Brown, R. and Twiss, R. Q. (1954). A new type of interferometer for use in radio astronomy. *Phil. Mag.* **45**, 663–682.

Hanbury Brown, R. and Twiss, R. Q. (1956). Correlation between photons in 2 coherent beams of light. *Nature* **177**.

Hansen, B. M. S. and Murali, C. (1998). Gamma Ray Bursts from Stellar Collisions. [astro-ph/9806256]

Hansen, R.O., Newman, E.T., Penrose, R., and Tod, K.P. (1978). The metric and curvature properties of *H*-space. *Proc. Roy. Soc. Lond.* **A363**, 445–68.

Hardy, G. H. (1914). *A Course of Pure Mathematics,* 2nd edn. Cambridge University Press, Cambridge.

Hardy, G. H. (1940). *A Mathematician's Apology*. Cambridge University Press, Cambridge.

Hardy, G. H. (1949). *Divergent Series*. Oxford University Press, New York.

Hardy, G. H. and Wright, E. M. (1945). *An Introduction to the Theory of Numbers* (2nd edn). Clarendon Press, Oxford.

Hardy, L. (1992). Quantum mechanics, local realistic theories, and Lorentz-invariant realistic theories. Phys. Rev. Lett. **68**, 2981. [/astract/PRL/v68/i20/p2981_1]

Hardy, L. (1993). Nonlocality for two particles without inequalities for almost all entangled states. *Phys. Rev. Lett.* **71(11)**, 1665.

Hartle, J. B. (2003). *Gravity: An Introduction to Einstein's General Relativity*. Addison-Wesley, San Francisco, CA & London.

Hartle, J. B. (2004). The Physics of 'Now'. [gr-qc/0403001]

Hartle, J. B. and Hawking, S. W. (1983). The wave function of the Universe. *Phys. Rev.* **D28**, 2960.

Harvey, F. R. (1966). Hyperfunctions and linear differential equations. *Proc. Nat. Acad. Sci.* **5**, 1042–6.

Harvey, F. R. (1990). *Spinors and Calibrations*. Academic Press, San Diego, CA.

Haslehurst, L. and Penrose, R. (2001). The most general (2,2) self-dual vacuum: a googly approach. In *Further Advances in Twistor Theory, Vol.III: Curved Twistor Spaces* (ed. L. J. Mason, L. P. Hughston, P. Z. Kobak, and K. Pulvere), pp 345–9.

Hawking, S.W. (1972). Black holes in general relativity, *Commun. Math. Phys.* 25, 152–66.

Hawking, S. W. (1974). Black hole explosions. *Nature* **248**, 30.

Hawking, S. W. (1975). Particle creation by black holes. *Commun. Math. Phys.* **43**.

Hawking, S. W. (1976a). Black holes and thermodynamics. *Phys. Rev.* **D13(2)**, 191.

Hawking, S. W. (1976b). Breakdown of predictability in gravitational collapse. *Phys. Rev.* **D14**, 2460.

Hawking, S. W., King, A. R. and McCarthy, P. J. (1976). A new topology for curved space-time which incorporates the causal, differential, and conformal structures. *J. Math. Phys.* **17**, 174–81.

Hawking, S. W. and Ellis, G. F. R. (1973). *The Large-Scale Structure of Space-Time*. Cambridge University Press, Cambridge.

Hawking, S. W. and Israel, W. (ed.) (1987). *300 Years of Gravitation*. Cambridge University Press, Cambridge.

Hawking, S. W. and Penrose, R. (1970). The singularities of gravitational collapse and cosmology. *Proc. Roy. Soc. Lond.* **A314**, 529–48.

Hawking, S. W. and Penrose, R. (1996). *The Nature of Space and Time*. Princeton University Press, Princeton, New Jersey.

Hawking, S. W. and Turok, N. (1998). Open Inflation Without False Vacua. *Phys. Lett.* **B425**. [hep-th/9802030]

Hawkins, T. (1977). Weiestrass and the theory of matrices. *Arch. Hist. Exact Sci.* **17**, 119–63.

Hawkins, T. (2000). *Emergence of the theory of Lie groups*. Springer-Verlag, New York.

Heisenberg, W. (1971). *Physics and Beyond*. Addison Wesley, London.

Heisenberg, W. (1989). What is an elementary particle? In *Encounters with Einstein*. Princeton University Press, Princeton.

Helgason, S. (2001). *Differential Geometry and Symmetric Spaces*. AMS Chelsea Publishing, Providence, RI.

Hestenes, D. (1990). The *Zitterwebegung* Interpretation of Quantum Mechanics. *Found. Physics*. **20(10)**, 1213–32.

Hestenes, D. and Sobczyk, G. (1999). *Clifford Algebra to Geometric Calculus: A Unified Language for Mathematics and Physics*. Reidel, Dordrecht, Holland.

Heyting, A. (1956). *Intuitionism, Studies in Logic and the Foundations of Mathematics*. North-Holland, Amsterdam.

Heywood, P. and Redhead, M.L.G. (1983). Non-locality and the Kochen-Specker paradox, *Found. Phys.* 13 (5) 481–499.

Hicks, N. J. (1965). *Notes on Differential Geometry*. Van Nostrand, Princeton.

Hirschfeld, J. W. P. (1998). *Projective Geometries over Finite Fields* (Second Edition). Clarendon Press, Oxford.

Hodges, A. P. (1982). Twistor diagrams. *Physica*, **114A**, 157–75.

Hodges, A. P. (1985). A twistor approach to the regularization of divergences. *Proc. Roy. Soc. Lond.* **A397**, 341–74. Mass eigenstates in twistor theory, *ibid*, 375–96.

Hodges, A. P. (1990a). String Amplitudes and Twistor Diagrams: An Analogy. In *The Interface of Mathematics and Particle Physics* (ed. D. G. Quillen, G. B. Segal, and Tsou S. T.). Oxford University Press, Oxford.

Hodges, A. P. (1990b). Twistor diagrams and Feynman diagrams. In *Twistors in Mathematics and Physics*, LMS Lect. Note Ser. 156 (ed. T. N. Bailey and R. J. Baston). Cambridge University Press, Cambridge.

Hodges, A. P. (1998). The twistor diagram programme. In *The Geometric Universe; Science, Geometry, and the Work of Roger Penrose* (ed. S. A. Huggett, L. J. Mason, K. P. Tod, S. T. Tsou, and N. M. J. Woodhouse). Oxford University Press, Oxford.

Hodges, A. P., Penrose, R., and Singer, M. A. (1989). A twistor conformal field theory for four space-time dimensions. *Phys. Lett*. **B216**, 48–52.

Hollands, S. and Wald, R. M. (2001). Local Wick Polynomials and Time Ordered Products of Quantum Fields in Curved Spacetime. *Commun. Math. Phys.* 223, 289–326. [gr-qc/0103074].

Home, D. (1997). *Conceptual Foundations of Quantum Physics: An Overview from Modern Perspectives*. Plenum Press, New York & London.

Hopf, H. (1931). Über die Abbildungen der dreidimensionalen Sphäre auf die Kugelfläche. *Math. Ann.* **104**, 637.

Horgan, J. (1996). *The End of Science*. Perseus Publishing, New York.

Horowitz, G. T. (1998). Quantum states of black holes. In *Black Holes and Relativistic Stars* (ed. R. M.Wald), pp. 241–66. University of Chicago Press, Chicago.

Horowitz, G. T. and Maldacena, J. (2003). The black hole final state. [hep-th/0310281]

Horowitz, G. T. and Perry, M. J. (1982). Gravitational energy cannot become negative. *Phys. Rev. Lett*. **48**, 371–4.

Howie, J. (1989). On the SQ-universality of T(6)-groups. *Forum Math.* **1**, 251–72.

Hoyle, C. D. *et al.* (2001). Submillimeter Test of the Gravitational Inverse-Square Law: A Search for 'Large' Extra Dimensions. *Phys. Rev. Lett.* **86(8)**, 1418–21.

Hoyle, F. (1948). A New Model for the Expanding Universe. *Mon. Not. Roy. Astron. Soc.* **108**, 372.

Hoyle, F., Fowler, W. A., Burbidge, G. R., and Burbidge, E. M. (1956). Origin of the elements in stars. *Science* **124**, 611–14.

Huang, K. (1952). On the Zitterbewegung of the Dirac Electron. *Am. J. Phys.* **20(8)**, 479–484.

Huggett, S. A. and Jordon, D. (2001). *A Topological Aperitif.* Springer-Verlag, London.

Huggett, S.A. and Tod, K.P. (2001). *An Introduction to Twistor Theory.* Cambridge University Press, Cambridge.

Hughston, L. P. (1979). *Twistors and Particles.* Lecture Notes in Physics No. 97. Springer-Verlag, Berlin.

Hughston, L. P. (1995). Geometric Aspects of Quantum Mechanics. In *Twistor Theory* (ed. S. A. Huggett), pp. 59–79. Marcel Dekker, New York.

Hughston, L. P., Jozsa, R., and Wooters, W. K. (1993). A complete classification of quantum ensembles having a given density matrix. *Phys. Letts.* **A183**, 14–18.

Ilyenko, K. (1999). Twistor Description of Null Strings. Oxford D. Phil. thesis, unpublished.

Immirzi, G. (1997). Quantum Gravity and Regge Calculus. [gr-qc/9701052]

Infeld, L. and van der Waerden, B. L. (1933). Die Wellengleichung des Elektrons in der allgemeinen Relativitätstheorie. *Sitz. Ber. Preuss. Akad. Wiss. Phisik. Math.* Kl. **9**, 380–401.

Isenberg, J., Yasskin, P. B. and Green, P. S. (1978). Non-self-dual gauge fields. *Phys. Lett.* **78B**, 462–4.

Isham, C. J. (1975). *Quantum Gravity: An Oxford Symposium.* Oxford University Press, Oxford.

Isham, C. J. (1992). Canonical Quantum Gravity and the Problem of Time. [gr-qc/9210011]

Isham, C. J. and Butterfield, J. (2000). Some Possible Roles for Topos Theory in Quantum Theory and Quantum Gravity. [gr-qc/9910005]

Israel, W. (1967). Event horizons in static vacuum space-times. *Phys. Rev.* **164**, 1776–9.

Jackson, J. D. (1998). *Classical Electrodynamics.* John Wiley & Sons, New York & Chichester.

Jennewein, T., Weihs, G., Pan, J., and Zeilinger, A. (2002). Experimental Non-locality Proof of Quantum Teleportation and Entanglement Swapping, *Phys. Rev. Lett.* **88**, 017903.

Jensen, G. (1973). Einstein Metrics on Principal Fibre Bundles. *J. Diff. Geom.* **8**, 599–614.

Johnson, C. (2003). *D-Branes.* Cambridge University Press, Cambridge.

Jones, H. F. (2002). *Groups, Representations, and Physics.* Institute of Physics Publishing, Bristol.

Jozsa, R. (1981). Models in Categories and Twistor Theory. Oxford D. Phil. thesis, unpublished.

Jozsa, R. and Linden, N. (2002). On the role of entanglement in quantum computational speed-up. [quant-ph/0201143].

Jozsa, R. O. (1998). Entanglement and quantum computation. In *The Geometric Universe* (ed. S. A, Huggett, L. J. Mason, K. P. Tod, S. T. Tsou, and N. M. J. Woodhouse), pp. 369–79. Oxford University Press, Oxford.

Károlyházy, F. (1966). Gravitation and quantum mechanics of macroscopic bodies. *Nuovo Cim.* **A42**, 390.

Károlyházy, F. (1974). Gravitation and Quantum Mechanics of Macroscopic Bodies. *Magyar Fizikai Folyóirat.* **22**, 23–24. [Thesis, in Hungarian]

Károlyházy, F., Frenkel, A., and Lukács, B. (1986). On the possible role of gravity on the reduction of the wave function. In *Quantum Concepts in Space and Time* (ed. R. Penrose and C. J. Isham), pp. 109–28. Oxford University Press, Oxford.

Kahn, D. W. (1995). *Topology: An Introduction to the Point-Set and Algebraic Areas*. Dover Publications, New York.

Kaku, M. (1993). *Quantum field theory: a modern introduction*. Oxford University Press, Oxford.

Kamberov, G., *et al.* (2002). *Quaternions, Spinors, and Surfaces (Contemporary Mathematics (American Mathematical Society), v. 299.)*. American Mathematical Society.

Kane, G. (ed.) (1999). *Perspectives on Supersymmetry* (*Advanced Series on Directions in High Energy Physics*). World Scientific Pub. Co, Singapore.

Kane, G. (2001). *Supersymmetry: Unveiling the Ultimate Laws of Nature*. Perseus Publishing, New York.

Kapusta, J. I. (2001). Primordial Black Holes and Hot Matter. [astro-ph/0101515]

Kasper, J. E. and Feller, S. A. (2001). *The Complete Book of Holograms: How They Work and How to Make Them*. Dover Publications.

Kauffman, L. H. (2001). *Knots and Physics*. World Scientific Publishing, Singapore.

Kay, B. S. (1998a). Entropy defined, entropy increase and decoherence understood, and some black hole puzzles solved. [hep-th/9802172]

Kay, B. S. (1998b). Decoherence of Macroscopic Closed Systems within Newtonian Quantum Gravity. *Class. Quant. Grav.* **15**, L89–98. [hep-th/9810077]

Kay, B. S. (2000). Application of linear hyperbolic PDE to linear quantum in curved space-times: especially black holes, time machines, and a new semilocal vacuum concept. In *Journées Équations aux Dérivées Partielles, Nantes 5–9 Juin 2000*. Groupement de Recherche 1151 du CNRS. [gr-qc/0103056]

Kay, B. S. and Wald, R. M. (1991). Theorems on the uniqueness and thermal properties of stationary, nonsingular, quasifree states on space-times with a bifurcate Killing horizon. *Phys. Rept.* **207**, 49–136.

Kay, B. S., Radzikowski, M. J., and Wald, R. M. (1996). Quantum Field Theory on Spacetimes with a Compactly Generated Cauchy Horizon. Commun. Math. Phys. 183 (1997), 533–556. [gr-qc/9603012].

Kelley, J. L. (1965). *General Topology*. van Nostrand, Princeton, New Jersey.

Kemmer, N. (1938). Quantum theory of Einstein-Bose particles and nuclear interaction. *Proc. R. Soc.* **A166**, 127.

Kemmer, N. (1939). The particle aspect of meson theory *Proc. R. Soc.* **A173**, 91.

Kerr, R. P. (1963). Gravitational field of a spinning mass as an example of algebraically special metrics. *Phys. Rev. Lett.* **11**, 237–8.

Ketov, S. V. (2000). *Quantum Non-Linear Sigma-Models: From Quantum Field Theory to Supersymmetry, Conformal Field Theories, Black Holes, and Strings*. Springer-Verlag, Berlin, London.

Kibble, T. W. B. (1961). Lorentz invariance and the gravitational field. J. *Math. Phys.* **2**, 212–221.

Kibble, T. W. B. (1979). Geometrization of quantum mechanics. *Commun. Math. Phys.* **65**, 189.

Kibble, T. W. B. (1981). Is a semi-classical theory of gravity viable? In *Quantum Gravity 2: A Second Oxford Symposium* (ed. C. J. Isham, R. Penrose, and D. W. Sciama), pp. 63–80. Oxford University Press, Oxford.

Killing, W (1893). *Einfuehrung in die Grundlagen der Geometrie*. Paderborn.

Klein, F. (1898). Über den Stand der Herausgabe von Gauss' Werken. *Math. Ann.* **51**, 128–33.

Knott, C. G. (1900). Professor Klein's view of quaternions: A criticism. *Proc. Roy. Soc. Edinb.* **23**, 24–34.

Kobayashi, S. and Nomizu, K. (1963). *Foundations of Differential Geometry*. Interscience Publishers, New York & London.

Kochen, S. and Specker, E. P. (1967). The Problem of Hidden Variables in Quantum Mechanics. *Journal of Mathematics and Mechanics* **17**, 59–88.

Kodaira, K. (1962). A theorem of completeness of characteristic systems for analytic submanifolds of a complex manifold. *Ann. Math.* **75**, 146–62.

Kodaira, K. and Spencer, D. C. (1958). On deformations of complex analytic structures I, II. *Ann. Math.* **67**, 328–401, 403–66.

Kolb, E. W. and Turner, M. S. (1994). *The Early Universe*. Perseus Publishing, New York.

Komar, A. B. (1964). Undecidability of macroscopically distinguishable states in quantum field theory. *Phys. Rev.* **133B**, 542–4.

Kontsevich, M. (1994). Homological algebra of mirror symmetry. *Proceedings of the International Congress of Mathematicians, Vol. 1,2.* (Zürich, 1994). Birkhaüser, Basel.

Kontsevich, M. (1995). Enumeration of rational curves via toric actions. In *The Moduli Space of Curves* (ed. R. Dijkgraaf, C. Faber, and G. van der Geer). *Progress in Math.* **129**, 335–68 [hep-th/9405035].

Koruga, D., Hameroff, S., Withers, J., Loutfy, R., and Sundareshan, M. (1993). *Fullerene C_{60}: History, physics, nanobiology, nanotechnology*. North-Holland, Amsterdam.

Kraus, K. (1983). *States, effects and operations: fundamental notions of quantum theory*. Lecture Notes in Physics, Vol 190. Springler-Verlag, Berlin.

Krauss, L. M. (2001). *Quintessence: The Mystery of the Missing Mass*. Basic Books, New York.

Kreimer, D. (2000). *Knots and Feynman Diagrams*. Cambridge University Press, Cambridge.

Kronheimer, E.H. and Penrose, R. (1967). On the structure of causal spaces. *Proc. Camb. Phil Soc.* **63**, 481–501.

Kruskal, M. D. (1960). Maximal Extension of Schwarzschild Metric. *Phys. Rev.* **119**, 1743–45.

Kuchar, K. (1981). Canonical methods of quantization. In *Quantum Gravity 2* (ed. D. W. Sciama, R. Penrose, and C. J. Isham). Oxford University Press, Oxford.

Kuchar, K. V. (1992). Time and interpretations of quantum gravity. In *Proceedings of the 4th Canadian Conference on General Relativity and Relativistic Astrophysics* (ed. G. Kunstatter, D. Vincent and J. Williams). World Scientific, Singapore.

Labastida, J. M. F. and Lozano, C. (1998). Lectures in Topological Quantum Field Theory. [hep-th/9709192]

Landsman, N. P. (1998). *Mathematical Topics Between Classical and Quantum Mechanics*. Springer-Verlag, Berlin.

Lang, S. (1972). *Differentiable Manifolds*. Addison-Wesley, Reading, MA.

Lange, A. E. *et al.* (2001). First Estimations of Cosmological Parameters from BOOMERanG, *Phys. Rev.* **D63**, 042001. [astro-ph/0005004]

Laplace, P. S. (1799). *Allgemeine geographische Ephemeriden herausgegeben von F. von Zach*. iv Bd. 1st, 1 Abhandl., Weimar.

Laporte, O. and Uhlenbeck, G. E. (1931). Application of spinor analysis to the Maxwell and Dirac equations. *Phys. Rev.* **37**, 1380–552.

Lasenby, J., Lasenby, A. N., and Doran, C. J. L. (2000). A unified mathematical language for physics and engineering in the 21st century. *Phil. Trans. Roy. Soc. Lond.* **A358**, 21–39.

Lawrie, I. (1998). *A Unified Grand Tour of Theoretical Physics*. Institute of Physics Publishing, Bristol.

Lawson, H. B., and Michelson, M. L. (1990). *Spin Geometry*. Princeton University Press, Princeton.

Lawvere, W. and Schanuel, S. (1997). *Conceptual Mathematics: A First Introduction to Categories*. Cambridge University Press, Cambridge.

LeBrun, C. R. (1985). Ambi-twistors and Einstein's equations. *Class. and Quantum Grav.* **2**, 555–63.

LeBrun, C. R. (1990). Twistors, ambitwistors, and conformal gravity. In *Twistors in Mathematical Physics* (ed. T. N. Bailey and R. J. Baston). LMS Lecture Note Series 156. Cambridge Univ. Press, Cambridge.

Lebrun, C. and Mason, L.J. (2002). Zoll manifolds and complex surfaces. *J. Diff. Geom.* **61(3)**, 453–535.

Lefshetz, J. (1949). *Introduction to Topology*. Princeton University Press, Princeton, New Jersey.

Leggett, A. J. (2002). Testing the limits of quantum mechanics: motivation, state of play, prospects. *J. Phys.* **CM 14,** R415–451.

Lemaître, G. (1933). L'univers en expansion *Ann. Soc. Sci. Bruxelles I* **A53**, 51–85 (see p. 82).

Levitt, M. H. (2001). *Spin Dynamics: Basics of Nuclear Magnetic Resonance*. John Wiley & Sons, New York.

Lichnerowicz, A. (ed.) (1994). *Physics on Manifolds: Proceedings of the International Colloquium in Honour of Yvonne Choquet-Bruhat, Paris, June 3–5, 1992*. Kluwer Academic Publishers, Boston and Dordrecht, The Netherlands.

Liddle, A. R. (1999). *An Introduction to Modern Cosmology*. John Wiley & Sons, New York.

Liddle, A. R. and Lyth, D. H. (2000). *Cosmological Inflation and Large-Scale Structure*. Cambridge University Press, Cambridge.

Lifshitz, E. M. and Khalatnikov, I. M. (1963). Investigations in relativistic cosmology. *Adv. Phys.* **12**, 185–249.

Linde, A. (1993). Comments on Inflationary Cosmology. [astro-ph/9309043]

Linde, A. (1995). Inflation with Variable Omega. *Phys. Lett.* **B**351. [hep-th/9503097]

Littlewood, J. E. (1949). *Littlewood's miscellany*. Reprinted in 1986, Cambridge University Press, Cambridge.

Livio, M. (2000). *The Accelerating Universe*. John Wiley & Sons, New York.

Llewellyn Smith, C. H. (1973). High energy behaviour and gauge symmetry. *Phys. Lett.* **B46(2)**, 233–6. [available online]

Lockwood, M. (1989). *Mind, Brain and the Quantum; the Compound 'I'*. Basil Blackwell, Oxford.

Lorentz, H. A., Einstein, A., Minkowski, H., and Weyl, H. (1952). *The Principle of Relativity: A Collection of Original Memoirs on the Special and General Theory of Relativity*. Dover, New York.

Lounesto, P. (2001). *Clifford Algebras and Spinors*. Cambridge University Press, Cambridge.

Lüders, G. (1951). Über die Zustandsänderung durch den Messprozess. *Ann. Physik* **8**, 322–8.

Ludvigsen, M. (1999). *General Relativity: A Geometric Approach*. Cambridge University Press, Cambridge.

Ludvigsen, M. and Vickers, J. A. G. (1982). A simple proof of the positivity of the Bondi mass. *J. Phys.* **A15**, L67–70.

Luminet, J.-P. *et al.* (2003). Dodecahedral space topology as an explanation for weak wide-angle temperature correlations in the cosmic microwave background. *Nature* **425**, 593–95.

Lyttleton, R. A. and Bondi, H. (1959). *Proc. Roy. Soc.* (London) **A252**, 313.

MacDuffee, C. C. (1933). *The theory of matrices*. Springer-Verlag, Berlin. (Reprinted by Chelsea).

MacLane, S. (1988). *Categories for the Working Mathematician*. Springer-Verlag, Berlin.

McLennan, J. A., Jr. (1956). Conformal invariance and conservation laws for relativistic wave equations for zero rest mass. *Nuovo. Cim.*, 3, 1360–79.

MACRO Collaboration (2002). Search for massive rare particles with MACRO. *Nucl. Phys. Proc. Suppl.* **110**, 186–8. [hep-ex/0009002]

Magueijo, J. (2003). *Faster Than the Speed of Light: The Story of a Scientific Speculation*. Perseus Publishing, New York.

Magueijo, J. and Smolin, L. (2002). Lorentz invariance with an invariant energy scale. [gr-qc/0112090]

Mahler (1981). *P-Adic Numbers and Functions*. Cambridge University Press, Cambridge.

Majorana, E. (1932). Teoria relativistica di particelle con momento intrinsico arbitrario. *Nuovo Cimento,* **9**, 335–44.

Majorana, E. (1937). Teoria asimmetrica dell' elettrone e del positrone. Nuovo cimento **14**, 171–84.

Maldacena, J. (1997). The Large N Limit of Superconformal Field Theories and Supergravity. [hep-th/9711200]

Manogue, C. A. and Dray, T. (1999). Dimensional Reduction. *Mod. Phys. Lett.* **A14**, 93–7. [hep-th/9807044]

Manogue, C. A. and Schray, J. (1993). Finite Lorentz transformations, automorphisms, and division algebras. *J. Math.Phys.* **34**, 3746–67.

Markopoulou, F. (1997). Dual formulation of spin network evolution. [gr-qc/970401]

Markopoulou, F. (1998). The internal description of a causal set: What the universe looks like from the inside. *Commun. Math. Phys.* **211**, 559–83. [gr-qc/9811053]

Markopoulou, F. and Smolin, L. (1997). Causal evolution of spin networks. *Nucl. Phys.* **B508**, 409–30. [gr-qc/9702025]

Marsden, J. E. and Tromba, A. J. (1996). *Vector Calculus*. W. H. Freeman & Co., New York. [new edn 2004]

Marshall, W., Simon, C., Penrose, R., and Bouwmeester, D. (2003). Towards Quantum Superpositions of a Mirror. *Phys. Rev. Lett.* **91**, 13.

Mason, L. J. and Woodhouse, N. M. J. (1996). *Integrability, Self-Duality, and Twistor Theory*. Oxford University Press, Oxford.

Mattuck, R. D. (1976). *A Guide to Feynman Diagrams in the Many-Body Problem*. Dover, New York.
Merkulov, S. A. and Schwachhöfer, L. J. (1998). Twistor solution of the holonomy problem. In *The Geometric Universe: Science, Geometry, and the Work of Roger Penrose* (ed. S. A. Huggett, L. J. Mason, K. P. Tod, S. T. Tsou, and N. M. J. Woodhouse). Oxford University Press, Oxford.
Michell, J. (1784). On the means of discovering the distance, magnitude, etc., of the fixed stars, in consequence of the diminution of their light, in case such a diminution should be found to take place in any of them, and such other data should be procured from observations, as would be further necessary for that purpose. *Phil. Trans. Roy. Soc. Lond.* **74**, 35–57.
Mielnik, B. (1974). Generalized Quantum Mechanics. *Commun. Math. Phys.* **37**, 221.
Milgrom, M. (1994). Dynamics with a non-standard inertia-acceleration relation: an alternative to dark matter. *Annals Phys.* **229**. [astro-ph/9303012]
Miller, A. (2003). Erotica, Aesthetics, and Schröedinger's Wave Equation. In *It Must Be Beautiful* (ed. G. Farmelo). Granta, London.
Minassian, E. (2002). Spacetime singularities in (2+1)-dimensional quantum gravity. *Class. Quant. Grav.* **19,** 5877–900.
Minkowski, H. (1952), in Lorentz *et al.*, (1952).
Misner, C. W. (1969). Mixmaster Universe. *Phys. Rev. Lett.* **22**, 1071–4.
Misner, C. W., Thorne, K. S., and Wheeler, J. A. (1973). *Gravitation*. Freeman, San Francisco.
Mohapatra, R. N. (2002). *Unification and Supersymmetry*. Springer-Verlag, Berlin & London.
Montgomery, D. and Zippin, L. (1955). *Topological Transformation Groups*. Interscience, New York & London.
Moore, A. W. (1990). *The Infinite*. Routledge, London & New York.
Moroz, I. M., Penrose, R., and Tod, K. P. (1998). Spherically-symmetric solutions of the Schrödinger–Newton equations. *Class. Quant. Grav.* **15**, 2733–42.
Mott, N. F. (1929). The wave mechanics of α-ray tracks. *Proc. Roy. Soc. Lond.* **A126**, 79–84. *Reprinted* in *Quantum Theory and Measurement* (ed. J. A. Wheeler and W. H. Zurek). Princeton Univ. Press, Princeton, New Jersey, 1983.
Moussouris, J. P. (1983). Quantum models of space-time based on recoupling theory. Oxford D. Phil. thesis, unpublished.
Mukohyama, S. and Randall, L. (2003). A Dynamical Approach to the Cosmological Constant. *Phys. Rev. Lett.* **92** (2004) 211302. [hep-th/0306108]
Munkres, J. R. (1954). *Elementary Differential Topology*. Annals of Mathematics Studies, 54. Princeton University Press, Princeton, New Jersey.
Myrheim, J. (1978). Statistical geometry. CERN preprint, TH-2538, unpublished.
Nahin, P. J. (1998). *An Imaginary Tale: The Story of* $\sqrt{-1}$. Princeton Univ. Press, Princeton.
Nair, V. (1988). A Current Algebra For Some Gauge Theory Amplitudes. *Phys. Lett.* **B214**, 215.
Nambu, Y. (1970). *Proceedings of the International Conference on Symmetries and Quark Models*. Wayne State Uniersity, p. 269. Gordon and Breach Publishers.
Narayan, R. *et al.* (2003). Evidence for the Black Hole Event Horizon. *Astronomy & Geophysics,* **44(6)**, 6.22–6.26.
Nayfeh, Ali H. (1993). *Methods of Normal Forms*, John Wiley & Sons, New York.

Needham, T. (1997). *Visual Complex Analysis*. Clarendon Press, Oxford University Press, Oxford.

Negrepontis, S. (2000). The Anthyphairetic Nature of Plato's Dialectics. In *Interdisciplinary Approach to Mathematics and their Teaching, Volume 5*, pp. 15–77. University-Gutenberg, Athens. (In Greek).

Nester, J. M. (1981). A new gravitational energy expression, with a simple positivity proof. *Phys. Lett.* **83A**, 241–2.

Netterfield, C. B. *et al.* (2002). A measurement by BOOMERANG of multiple peaks in the angular power spectrum of the cosmic microwave background *Astrophys. J.* **571**, 604–614 [astro-ph/0104460].

Newlander, A., and Nirenberg, L. (1957). Complex Analytic Coordinates in Almost Complex Manifolds. *Ann. of Math.* **65**, 391–404.

Newman, E. T. (2002). On a Classical, Geometric Origin of Magnetic Moments, Spin-Angular Momentum and the Dirac Gyromagnetic Ratio. *Phys. Rev.* **D65** 104005. [gr-qc/0201055].

Newman, E. T. and Penrose, R. (1962). An approach to gravitational radiation by a method of spin coefficients. *J. Math. Phys.* **3**, 896–902; errata (1963), **4**, 998.

Newman, E. T. and Penrose, R. (1966). Note on the Bondi–Metzner–Sachs group. *J. Math. Phys.* **7**, 863–70.

Newman, E. T. and Unti, T. W. J. (1962). Behavior of asymptotically flat empty space. *J. Math. Phys.* **3**, 891–901.

Newman, E. T., Couch, E., Chinnapared, K., Exton, A., Prakash, A., and Torrence, R. (1965). Metric of a rotating charged mass. *J. Math. Phys.* **6**, 918–9.

Newman, M. H. A. (1942). On a string problem of Dirac, *J. Lond. Math. Soc.* **17**, 173–7.

Newman, R. P. A. C. (1993). On the Structure of Conformal Singularities in Classical General Relativity. *Proc. R. Soc. Lond.* **A443**, 473.

Newton, I. (1687). *The Principia: Mathematical Principles of Natural Philosophy*. Reprinted by University of California Press, 1999.

Newton, I. (1730). *Opticks*. Dover, 1952.

Ng, Y. J. (2004). Quantum Foam. [gr-qc/0401015]

Nicolai, H. (2003). Remarks at AEI Symposium "Strings meet Loops", 29–31 October 2003. http://alphaserv3.aei.mpg.de/events/stringloop.html

Nielsen, H. B. (1970). Submitted to *Proc. of the XV Int. Conf. on High Energy Physics, Kiev* (unpublished).

Nielsen, M. A. and Chuang, I. L. (2000). *Quantum Computation and Quantum Information*. Cambridge University Press, Cambridge.

Nomizu, K. (1956). *Lie Groups and Differential Geometry*. The Mathematical Society of Japan, Tokyo.

Novikov, I. D. (2001). *The River of Time*. Cambridge University Press, Cambridge.

O'Donnell, P. (2003). *Introduction to 2-Spinors in General Relativity*. World Scientific, Singapore.

O'Neill, B. (1983). *Semi-Riemannian Geometry: With Applications to Relativity*. Academic Press, New York.

Oppenheimer, J. R. (1930). On the theory of electrons and protons. *Phys. Rev.* **35**, 562–3.

Ozsvath, I. and Schucking, E. (1962). *Nature* **193**, 1168.

Ozsvath, I. and Schucking, E. (1969). *Ann. Phys.* **55**.

Page, D. N. (1976). Dirac equation around a charged, rotating black hole. *Phys. Rev. D.* **14**, 1509–10.

Page, D. N. (1987). Geometrical description of Berry's phase. *Phys. Rev.* **A36**, 3479–81.

Page, D. N. (1995). Sensible Quantum Mechanics: Are Only Perceptions Probabilistic? [quant-ph/9506010]

Pais, A. (1982). *'Subtle is the Lord'... 'The Science and the Life of Albert Einstein'.* Clarendon Press, Oxford.

Pais, A. (1986). *Inward Bound: Of Matter and Forces in the Physical World.* Clarendon Press, Oxford.

Parker, T. and Taubes, C. H. (1982). On Witten's proof of the positive energy theorem. *Comm. Math. Phys.* **84**, 223–38.

Pars, L. A. (1968). *A Treatise on Analytical Dynamics.* Reprinted in 1981, Ox Bow Press.

Pearle, P. (1985). Models for reduction. In *Quantum Concepts in Space and Time* (ed. C. J. Isham and R. Penrose), pp. 84–108. Oxford University Press, Oxford.

Pearle, P. and Squires, E. J. (1995). Gravity, energy conservation and parameter values in collapse models. *Durham University preprint*. DTP/95/13.

Peitgen, H.-O. and Richter, P. H. (1986). *The Beauty of Fractals: Images of Complex Dynamical Systems.* Springer-Verlag, Berlin & Heidelberg.

Peitgen, H.-O. and Saupe, D. (1988). *The Science of Fractal Images.* Springer-Verlag, Berlin.

Penrose, L.S. and Penrose, R. (1958). Impossible Objects: A Special Type of Visual Illusion *Brit. J. Psych.* **49**, 31–3.

Penrose, R. (1959). The apparent shape of relativistically moving sphere. *Proc. Camb. Phil. Soc.* **55**, 137–9.

Penrose, R. (1960). A spinor approach to general relativity, *Ann. Phys.* (New York) **10**, 171–201.

Penrose, R. (1962). The Light Cone at Infinity. In *Proceedings of the 1962 Conference on Relativistic Theories of Gravitation Warsaw.* Polish Academy of Sciences, Warsaw. (Published 1965.)

Penrose, R. (1963). Asymptotic properties of fields and space-times. *Phys. Rev. Lett.* **10**, 66–8.

Penrose, R. (1964). Conformal approach to infinity. In *Relativity, Groups and Topology: The 1963 Les Houches Lectures* (ed. B. S. DeWitt and C. M. DeWitt). Gordon and Breach, New York.

Penrose, R. (1965a). Zero rest-mass fields including gravitation: asymptotic behaviour. *Proc. R. Soc. Lond.* **A284**, 159–203.

Penrose, R. (1965b). Gravitational collapse and space-time singularities. *Phys. Rev. Lett.* **14**, 57–59.

Penrose, R. (1966). *An analysis of the structure of space-time.* Adams Prize Essay, Cambridge University, Cambridge (unpublished; but much of it is in Penrose 1968a).

Penrose, R. (1967). Twistor algebra. J. *Math. Phys.* **8**, 345–66.

Penrose, R. (1968a). Structure of space-time. In *Battelle Rencontres, 1967* (ed. C. M. DeWitt and J. A. Wheeler). Lectures in Mathematics and Physics. Benjamin, New York.

Penrose, R. (1968b). Twistor quantization and curved space-time. *Int. J. Theor. Phys.* **1**, 61–99.

Penrose, R. (1969a). Gravitational collapse: the role of general relativity. *Rivista del Nuovo Cimento*; Serie I, Vol. 1; *Numero speciale*, 252–76.

Penrose, R. (1969b). Solutions of the zero rest-mass equations, *J. Math. Phys.* **10**, 38–9.
Penrose, R. (1971a). Angular momentum: an approach to combinatorial space-time. In *Quantum Theory and Beyond* (ed. T. Bastin). Cambridge University Press, Cambridge.
Penrose, R. (1971b). Applications of negative dimensional tensors. In *Combinatorial Mathematics and its Applications* (ed. D. J. A. Welsh). Academic Press, London.
Penrose, R. (1975). Twistor theory: its aims and achievements. In *Quantum Gravity, an Oxford Symposium* (ed. C. J. Isham, R. Penrose, and D. W. Sciama). Oxford University Press, Oxford.
Penrose, R. (1976a). The non-linear graviton. *Gen. Rel. Grav.* **7**, 171–6.
Penrose, R. (1976b). Non-linear gravitons and curved twistor theory. *Gen. Rel. Grav.* **7**, 31–52.
Penrose, R. (1978). Gravitational collapse: a Review. In *Physics and Astrophysics of Neutron Stars and Black Holes, LXV Corso.* Soc. Italiana di Fisica, Bologna, Italy, pp. 566–82.
Penrose, R. (1979a). Singularities and time-asymmetry. In *General Relativiy: An Einstein Centenary* (ed. S. W. Hawking and W. Israel). Cambridge University Press, Cambridge.
Penrose, R. (1979b). On the twistor description of massless fields. In *Complex Manifold Techniques in Theoretical Physics* (eds. D. E. Lerner and P. D. Sommers). Pitman, San Francisco. See also various articles in L. P. Hughston and R. S. Ward (Editors) (1979) *Advances in Twistor Theory*. Pitman Advanced Publishing Program, San Francisco.
Penrose, R. (1980). On Schwarzschild causality—a problem for 'Lorentz-covariant' general relativity. In *Essays in General Relativity* (A. Taub Festschrift) (ed. F. J. Tipler), pp. 1–12. Academic Press, New York.
Penrose, R. (1982). Quasi-local mass and angular momentum in general relativity. *Proc. Roy. Soc. Lond.* **A381**, 53–63.
Penrose, R. (1986). Gravity and state-vector reduction. In *Quantum Concepts in Space and time* (ed. R. Penrose and C. J. Isham), pp. 129–46. Oxford University Press, Oxford.
Penrose, R. (1987a). Quantum physics and conscious thought. In *Quantum Implications: Essays in Honour of David Bohm* (ed. B. J. Hiley and F. D. Peat). Routledge and Kegan Paul, London & New York.
Penrose, R. (1987b). On the origins of twistor theory. In *Gravitation and Geometry: a volume in honour of I. Robinson.* (ed. W. Rindler and A. Trautman). Bibliopolis, Naples.
Penrose, R. (1987c). Newton, quantum theory, and reality. In *300 years of Gravity* (ed. S. W. Hawking and W. Israel), pp. 17–49. Cambridge University Press, Cambridge.
Penrose, R. (1988a). Topological QFT and Twistors: Holomorphic Linking; Holomorphic Linking: Postscript. *Twistor Newsletter* **27**, 1–4.
Penrose, R. (1988b). Fundamental asymmetry in physical laws. *Proceedings of Symposia in Pure Mathematics* **48**. American Mathematical Society, pp. 317–328.
Penrose, R. (1989). *The Emperor's New Mind: Concerning Computers, Minds, and the Laws of Physics.* Oxford University Press, Oxford.
Penrose, R. (1991). On the cohomology of impossible figures [La cohomologie des figures impossibles] *Structural Topology [Topologie structurale]* **17**, 11–16.

Penrose, R. (1992). $\mathcal{H}$-space and Twistors. In *Recent Advances in General Relativity*. Einstein Studies, Vol. 4 (ed. A. I. Janis and J. R. Porter), pp. 6–25. Birkhäuser, Boston.

Penrose, R. (1994). *Shadows of the Mind: An Approach to the Missing Science of Consciousness*. Oxford University Press, Oxford.

Penrose, R. (1996). On gravity's role in quantum state reduction. *Gen. Rel. Grav.* **28**, 581–600.

Penrose, R. (1997a). *The Large, the Small and the Human Mind*. Cambridge University Press, Cambridge. Canto edition (2000).

Penrose, R. (1997b). On understanding understanding. *Internat. Stud. Philos. Sci.* **11**, 7–20.

Penrose, R. (1998a). Quantum computation, entanglement and state-reduction. *Phil. Trans. Roy. Soc. Lond.* **A356**, 1927–39.

Penrose, R. (1998b). The question of cosmic censorship. In *Black Holes and Relativistic Stars* (ed. R. M. Wald). University of Chicago Press, Chicago, Illinois. Reprinted in *J. Astrophys. Astr.* **20**, 233–48 (1999).

Penrose, R. (2000a). Wavefunction collapse as a real gravitational effect. In *Mathematical Physics 2000* (ed. A. Fokas, T. W. B. Kibble, A. Grigouriou, and B. Zegarlinski), pp. 266–82. Imperial College Press, London.

Penrose, R. (2000b). On Bell non-locality without probabilities: some curious geometry. In *Quantum Reflections*. (Eds. J. Ellis and D. Amati). Cambridge Univ. Press, Cambridge, 1–27.

Penrose, R. (2001). Towards a twistor description of general space-times; introductory comments. In *Further Advances in Twistor Theory, Vol.III: Curved Twistor Spaces* (ed. L.J. Mason, L.P. Hughston, P.Z. Kobak, and K. Pulverer). Chapman & Hall/CRC Research Notes in Mathematics 424, London. 239–55.

Penrose, R. (2002). John Bell, State Reduction, and Quanglement. In *Quantum [Un]speakables: From Bell to Quantum Information* (ed. R. A. Bertlmann and A. Zeilinger). Springer-Verlag, Berlin.

Penrose, R. (2003). On the instability of extra space dimensions. *The Future of Theoretical Physics and Cosmology, Celebrating Stephen Hawking's 60th Birthday* (ed. G. W. Gibbons, E. P. S. Shellard, S. J. Rankin), Cambridge University Press, Cambridge.

Penrose, R. and MacCallum, M.A.H. (1972). Twistor theory: an approach to the quantization of fields and space-time. *Phys. Repts.* **6C**, 241–315.

Penrose, R. and Rindler, W. (1984). *Spinors and Space-Time*, Vol. I: *Two-Spinor Calculus and Relativistic Fields*. Cambridge University Press, Cambridge.

Penrose, R. and Rindler, W. (1986). *Spinors and Space-Time*, Vol. II: *Spinor and Twistor Methods in Space-Time Geometry*. Cambridge University Press, Cambridge.

Penrose, R., Robinson, I., and Tafel, J. (1997). Andrzej Mariusz Trautman. *Class. Quan. Grav.* **14**, A1–A8.

Penrose, R., Sparling, G. A. J., and Tsou, S. T. (1978). Extended Regge Trajectories. *J. Phys. A. Math. Gen.* **11**, L231–L235.

Penzias, A. A. and Wilson, R. W. (1965). A Measurement of Excess Antenna Temperature at 4080 Mc/s. *Astrophys. J.* **142**, 419.

Percival, I. C. (1994). Primary state diffusion. *Proc. R. Soc. Lond.* A**447**, 189–209.

Percival, I. C. (1995). Quantum space-time fluctuations and primary state diffusion. [quant-ph/9508021]

Peres, A. (1991). Two Simple Proofs of the Kochen-Specker Theorem. *Journal of Physics A: Mathematical and General* **24**, L175–L178.
Peres, A. (1995). Generalized Kochen-Specker Theorem. [quant-ph/9510018].
Peres, A. (2000). Delayed choice for entanglement swapping, *J.Mod.Opt.* **47**, 531. [quant-ph/9904042]
Perez, A. (2001). Finiteness of a spinfoam model for Euclidean quantum general relativity. *Nucl. Phys.* **B599**, 427–34.
Perez, A. (2003). Spin foam models for quantum gravity. *Class. Quant. Grav.* **20**, R43–R104.
Perlmutter, S. *et. al.* (1998). Cosmology from Type Ia Supernovae. *Bull. Am. Astron. Soc.* **29**. [astro-ph/9812473]
Peskin, M. E. and Schröder, P. V. (1995). *Introduction to Quantum Field Theory.* Westview Press, Reading, MA & Wokingham.
Petiau, G. (1936). Contribution à la théorie des équations d'onde corpusculaires. *Adad. Roy. Belgique.* (Cl. Sci. Mem. Collect. **16** No 2)
Pirani, F. A. E. and Schild, A. (1950). On the Quantization of Einstein's Gravitational Field Equations. *Phys. Rev.* **79**, 986–91.
Pitkaenen, M. (1994). p-Adic description of Higgs mechanism I: p-Adic square root and p-adic light cone. [hep-th/9410058]
Polchinski, J. (1998). *String Theory.* Cambridge University Press, Cambridge.
Polkinghorne, J. (2002). *Quantum Theory, A Very Short Introduction.* Oxford University Press, Oxford.
Popper, K. (1934). *The Logic of Scientific Discovery.* Routledge; New Ed edition (March 2002).
Pound, R.V. and Rebka, G. A. (1960). *Phys. Rev. Lett.* **4**, 337.
Preskill, J. (1992). Do black holes destroy information? [hep-th/9209058]
Priestley, H. A. (2003). *Introduction to Complex Analysis.* Oxford University Press, Oxford.
Rae, A. I. M. (1994). *Quantum Mechanics.* Institute of Physics Publishing, 4th edn. 2002.
Raine, D. J. (1975). Mach's principle in General Relativity. *Monthly Notices RAS* **171**, 507–528.
Randall, L. and Sundrum, R. (1999a). A Large Mass Hierarchy from a Small Extra Dimension. *Phys. Rev. Lett.* **83**, 3370–3. [hep-ph/9905221]
Randall, L. and Sundrum, R. (1999b). An Alternative to Compactification. *Phys. Rev. Lett.* **83**, 4690–3. [hep-th/9906064]
Rarita, W. and Schwinger, J. (1941). On the theory of particles with half-integer spin. *Phys. Rev.* **60**, 61.
Redhead, M. L. G. (1987). *Incompleteness, Nonlocality, and Realism.* Clarendon Press, Oxford.
Reed, M. and Simon, B. (1972). *Methods of Mathematical Physics Vol. 1: Functional Analysis.* Academic Press, New York & London.
Reeves, J. N. *et al.* (2002). The signature of supernova ejecta in the X-ray afterglow of the gamma-ray burst 011211. *Nature* **416,** 512–15.
Regge, T. (1961). General Relativity without Coordinates. *Nuovo Cimento A* **19**, 558–571.
Reisenberger, M. P. (1997). A lattice worldsheet sum for 4-d Euclidean general relativity. [gr-qc/9711052]
Reisenberger, M. P. (1999). On relativistic spin network vertices. *J. Math. Phys.*, **40**, 2046–054.

Reisenberger, M. P. and Rovelli, C. (2001). Spacetime as a Feynman diagram: the connection formulation. *Class. Quant. Grav.* **18**, 121–40.
Reisenberger, M. P. and Rovelli, C. (2002). Spacetime states and covariant quantum theory. *Phys. Rev.* **D65**, 125016.
Reula, O. and Tod, K. P. (1984). Positivity of the Bondi energy. *J. Math. Phys.* **25**, 1004–8.
Riemann, G. B. F. (1854). Über die Hypothesen, welche der Geometrie zu Grunde liegen (Habilitationsschrift, Göttingen); see *Collected Works of Bernhardt Riemann*, Ed. Heinrich Weber, 2nd edn. (Dover, New York, 1953), pp. 272–287.
Rindler, W. (1977). *Essential Relativity*. Springer-Verlag, New York.
Rindler, W. (1982). *Introduction to Special Relativity*. Clarendon Press, Oxford.
Rindler, W. (2001). *Relativity: Special, General, and Cosmological*. Oxford University Press, Oxford.
Rizzi, A. (1998). Angular momentum in general relativity: A new definition. *Phys. Rev. Lett.* **81(6)**, 1150.
Robinson, D. C. (1975). Uniqueness of the Kerr Black Hole. *Phys. Rev. Lett.*, **34**, 905–6.
Rogers, A. (1980). A global theory of supermanifolds. *J. Math. Phys.* **21**, 1352–65.
Rolfsen, D. (2004). *Knots and Links*. American Mathematical Society, Providence, RI.
Roseveare, N. T. (1982). *Mercury's Perihelion from Le Verrier to Einstein*. Clarendon Press, Oxford.
Rovelli, C. (1991). Quantum mechanics without time: A model. *Phys. Rev.* **D42**, 2638.
Rovelli, C. (1998). Strings, loops and others: a critical survey of the present approaches to quantum gravity. In *Gravity and Relativity: At the turn of the Millennium* (15th International Conference on General Relativity and Gravitation, eds. N. Dadhich and J. Narlikar, Inter-University Centre for Astronomy and Astrophysics, Pune, India), 281–331.
Rovelli, C. (2003). *Quantum Gravity*. http://www.cpt.univ-mrs.fr/~rovelli/book.pdf
Rovelli, C. and Smolin, L. (1990). Loop representation for quantum general relativity. *Nucl. Phys.* **B331**, 80–152.
Runde, V. (2002). The Banach-Tarski paradox–or–What mathematics and religion have in common. *Pi in the Sky* **2** (2000), 13–15. [math.GM/0202309]
Russell, B. (1903). *Principles of Mathematics*. Most recent republication by W. W. Norton & Company, 1996.
Russell, B. (1927). *The Analysis of Matter*. Allen and Unwin; reprinted 1954, Dover, New York.
Ryder, L. H. (1996). *Quantum Field Theory*. Cambridge University Press, Cambridge.
Sabbagh, K. (2003). The Riemann Hypothesis: The Greatest Unsolved Problem in Mathematics. Farrar, Straus and Giroux.
Saccheri, G. (1733). *Euclides ab Omni Naevo Vindicatus*. Translation in Halsted, G. B. (1920). *Euclid Freed from Every Flaw*. Open Court, La Salle, Illinois.
Sachs, R. and Bergmann, P. G. (1958). Structure of Particles in Linearized Gravitational Theory. *Phys. Rev.* **112**, 674–680.
Sachs, R. K. (1961). Gravitational waves in general relativity, VI: the outgoing radiation condition. *Proc. Roy. Soc. Lond.* **A264**, 309–38.
Sachs, R. K. (1962a). Gravitational waves in general relativity, VIII: waves in asymptoticaly flat space-time. *Proc. Roy. Soc. Lond.* **A270**, 103–26.

Sachs, R. K. (1962b). Asymptotic symmetries in gravitational theory. *Phys. Rev.* **128**, 2851–64.
Sakellariadou, M. (2002). The role of topological defects in cosmology. Invited lectures in NATO ASI / COSLAB (ESF) School 'Patterns of Symmetry Breaking', September 2002 (Cracow). [hep-ph/0212365]
Salam, A. (1980). Gauge Unification of Fundamental Forces. *Rev. Mod. Phys.* **52(3)**, 515–23.
Salam, A. and Ward, J. C. (1959). Weak and electromagnetic interaction. *Nuovo Cimento* **11**, 568.
Sarkar, S. (2002). Possible astrophysical probes of quantum gravity. *Mod. Phys. Lett.* **A17**, 1025–1036. [gr-qc/0204092]
Sato, M. (1958). On the generalization of the concept of a function. *Proc. Japan Acad.* **34**, 126–30.
Sato, M. (1959). Theory of hyperfunctions I. *J. Fac. Sci. Univ. Tokyo*, Sect. I, **8**, 139–93.
Sato, M. (1960). Theory of hyperfunctions II. *J. Fac. Sci. Univ. Tokyo*, Sect. I, **8**, 387–437.
Schild, A. (1949). Discrete space-time and integral Lorentz transformations. *Can. J. Math.* **1**, 29–47.
Schoen, R. and Yau, S. –T. (1979). On the proof of the positive mass conjecture in the general relativity. *Comm. Math. Phys.* **65**, 45–76.
Schoen, R. and Yau, S. –T. (1982). Proof that Bondi mass is positive. *Phys. Rev. Lett.* **48**, 369–71.
Schouten, J. A. (1954). *Ricci-Calculus*. Springer, Berlin.
Schrödinger, E. (1930). *Sitzungber. Preuss. Akad. Wiss. Phys.-Math. Kl.* **24**, 418.
Schrödinger, E. (1935). Probability relations between separated systems. *Proc. Camb. Phil. Soc.* **31**, 555–63.
Schrödinger, E. (1950). *Space-Time Structure*. Cambridge University Press, Cambridge.
Schrödinger, E. (1952). *Science and Humanism: Physics in Our Time*. (Cambridge Univ. Press, Cambridge).
Schrödinger, E. (1956). *Expanding Universes*. Cambridge University Press, Cambridge.
Schrödinger, E. (1967). *'What is Life?' and 'Mind and Matter'*. Cambridge Univ. Press, Cambridge.
Schutz, B. (2003) *Gravity from the ground up: an introductory guide to gravity and general relativity*. Cambridge University Press, Cambridge.
Schutz, J. W. (1997). *Independent Axioms for Minkowski Space-Time*. Addison Wesley Longman Ltd., Harlow, Essex.
Schwartz, L. (1966). *Thèorie des distributions*. Hermann, Paris.
Schwarz, J. H. (2001). String Theory. *Curr. Sci.* **81(12)**, 1547–53.
Schwarzschild, K. (1916). Über das Gravitationsfeld eines Massenpunktes nach der Einsteinschen Theorie. *Sitzber. Deut. Akad. Wiss. Berlin Math.-Phys. Tech.* Kl. 189–96.
Schwinger, J. (1951). *Proc. Nat. Acad. Sci.* **37**, 452.
Schwinger, J. (ed.) (1958). *Quantum Electrodynamics*. Dover.
Sciama, D. W. (1959). *The Unity of the Universe*. Doubleday & Company, Inc., New York.
Sciama, D. W. (1962). On the analogy between charge and spin in general relativity, in *Recent Developments in General Relativity*. Pergamon & PWN, Oxford.

Sciama, D. W. (1972). *The Physical Foundations of General Relativity*. Heinemann, London.

Sciama, D. W. (1998). Decaying neutrinos and the geometry of the universe. In *The Geometric Universe: Science, Geometry, and the Work of Roger Penrose* (ed. S. A. Huggett, L. J. Mason, K. P. Tod, S. T. Tsou, and N. M. J. Woodhouse). Oxford University Press, Oxford.

Seiberg, N. and Witten, E. (1994). Electric-magnetic duality, monopole condensation, and confinement in $N = 2$ supersymmetric Yang-Mills theory. *Nucl. Phys.* **B426**. [hep-th/9407087]

Sen, A. (1982). Gravity as a spin system. *Phys. Lett.* **B119**, 89–91.

Shankar, R. (1994). *Principles of Quantum Mechanics*, 2nd edn. Plenum Press, New York & London.

Shapiro, I. I. *et al.* (1971). *Phys. Rev. Lett.* **13**, 789.

Shaw, W. T. and Hughston, L. P. (1990). Twistors and strings. In *Twistors in Mathematics and Physics* (ed. T. N. Bailey and R. J. Baston). *London Mathematical Society Lecture Notes Series*, 156. Cambridge University Press, Cambridge.

Shawhan, P. (2001). The Search for Gravitational Waves with LIGO: Status and Plans. *Intl. J. Mod. Phys. A* **16**, supp. 01C, 1028–30.

Shih, Y. H. *et al.* (1995). Optical Imaging by Means of Two-Photon Entanglement. *Phys. Rev. A, Rapid Comm.* **52**, R3429.

Shimony, A. (1998). Implications of transience for spacetime structures. In *The Geometric Universe: Science, Geometry, and the Work of Roger Penrose* (ed. S. A. Huggett, L. J. Mason, K. P. Tod, S. T. Tsou, and N. M. J. Woodhouse). Oxford University Press, Oxford.

Shrock, R. (2003). *Neutrinos and Implications for Physics Beyond the Standard Model.* World Scientific Pub. Co., Singapore.

Silk, J. and Rees, M. (1998). Quasars and galaxy formation. *Astronomy and Astrophysics*, v. 331, p. L1–L4.

Simmonds, J. G. and Mann, J. E. (1998). *A First Look at Perturbation Theory*, Dover Publications, New York.

Simon, B. (1983). Holonomy, the quantum adiabatic theorem, and Berry's phase. *Phys. Rev. Lett.* **51**, 2160–70.

Singh, S. (1997). *Fermat's Last Theorem.* Fourth Estate, London.

Slipher, V. A. (1917). Nebulae. *Proc. Am. Phil. Soc.* **56**, 403.

Smolin, L. (1991). Space and time in the quantum universe. In *Conceptual Problems in Quantum Gravity* (ed. A. Ashtekar and J. Stachel). Birkhauser, Boston.

Smolin, L. (1997). *The Life of the Cosmos.* Oxford University Press, Oxford.

Smolin, L. (1998). The physics of spin networks. In *The Geometric Universe: Science, Geometry, and the Work of Roger Penrose* (ed. S. A. Huggett, L. J. Mason, K. P. Tod, S. T. Tsou, and N. M. J. Woodhouse). Oxford University Press, Oxford.

Smolin, L. (2001). The exceptional Jordan algebra and the matrix string. (hep-th/0104050)

Smolin, L. (2002). *Three Roads To Quantum Gravity.* Basic Books, New York.

Smolin, L. (2003). How far are we from the quantum theory of gravity? [hep-th/0303185]

Smoot, G. F. *et al.* (1991). Preliminary results from the COBE differential microwave radiometers: large-angular-scale isotropy of the Cosmic Microwave Background. *Astrophys. J.* **371**, L1.

Snyder, H. S. (1947). Quantized space-time. *Physical Review,* **71**, 38–41.

Sorabji, R. J. (1984). *Time, Creation, and the Continuum*. Cornell University Press.
Sorabji, R. J. (1988). *Matter Space and Motion*. Duckworth Publishing.
Sorkin, R. D. (1991). Spacetime and Causal Sets. In *Relativity and Gravitation: Classical and Quantum* (ed. J. C. D'Olivo *et al*). World Scientific, Singapore.
Sorkin, R. D. (1994). Quantum Measure Theory and its Interpretation. In *Proceedings of 4th Drexel Symposium on Quantum Nonintegrability*, 8–11 Sep., Philadelphia, PA. [gr-qc/9507057]
Spergel, D. N. (2003). First Year Wilkinson Microwave Anisotropy Probe Observations: Determination of Cosmological Parameters. *Astrophys. J. Suppl.* **148**, 175.
Stachel, J. (1995). History of relativity. In *History of 20th Century Physics* (ed. L. Brown, A. Pais, and B. Pippard), Chapter 4. American Institute of Physics (AIP) and British Institute of Physics (BIP).
Stairs, A. (1983) Quantum logic, realism and value-definiteness. Phil. Sci. 50(4), 578–602.
Stapp, H. P. (1971). S-matrix Interpretation of Quantum Mechanics. *Phys. Rev.* **D3**, 1303–20
Stapp, H. P. (1979). Whiteheadian Approach to Quantum Theory and the Generalized Bell's Theorem. *Found. Phys.* **9**, 1–25.
Steenrod, N. E. (1951). *The Topology of Fibre Bundles*. Princeton University Press, Princeton.
Steinhardt, P. J. and Turok, N. (2002). A Cyclic Model of the Universe. *Science* **296(5572)** 1436–39. [hep-th/0111030]
Stoney, G. J. (1881). On the Physical Units of Nature. *Philosophical Magazine*, vol. 11, 381.
Strauss, W. (1992). *Partial Differential Equations: An Introduction*. John Wiley and Sons.
Strominger, A. and Vafa, C. (1996). Microscopic Origin of the Bekenstein-Hawking Entropy. *Phys. Lett.* **B379**, 99–104.
Strominger, A., Yau, S-T., and Zaslow, E. (1996). Mirror symmetry is T-duality. *Nucl. Phys.* **B479**, 1–2, 243–59.
Struik, D.J. (1954). *A Concise History of Mathematics*. Dover, New York.
Sudarshan, G. and Dhar, J. (1968). Quantum Field Theory of Interacting Tachyons. *Phys. Rev.* **174**, 1808.
Sudbery, A. (1987). Division algebras, (pseudo) orthogonal groups and spinors. *J. Phys.* **A17**, 939–55.
Susskind, L. (1970). Structure of Hadrons Implied by Duality. *Nuovo Cimento* **A69**, 457.
Susskind, L. (2003). 'Twenty Years of debate with Stephen.' In *The Future of Theoretical Physics and Cosmology* (ed. G. W. Gibbons, P. Shellard, and S. Rankin). Cambridge University Press, Cambridge.
Susskind, L., Thorlacius, L., and Uglum, J. (1993). The stretched horizon and black hole complementarity. *Phys. Rev.* **48**, 3743. [hep-th/9306069]
Sutherland, W. A. (1975). *Introduction to Topology*. Oxford University Press, Oxford.
Swain, J. (2004). The Majorana representations of spins and the relation between $SU(\infty)$ *and S Diff* (S^2). hep-th/0405004.
Synge, J. L. (1950). The gravitational field of a particle. *Proc. Irish Acad.* **A53**, 83–114.
Synge, J. L. (1956). *Relativity: The Special Theory*. North-Holland, Amsterdam.

Synge, J. L. (1960). *Relativity: The General Theory*, North-Holland Publ. Co., Amsterdam.

Szekeres, G. (1960). On the Singularities of a Riemannian Manifold. *Publ. Mat. Debrecen,* **7**, 285–301.

't Hooft, G. (1978a). On the phase transition towards permanent quark confinement. *Nucl. Phys.* **B138**, 1.

't Hooft, G. (1978b). Quantum gravity: a fundamental problem and some radical ideas. In *Recent Developments in Gravitation* (ed. M. Levy and S. Deser). Plenum, New York.

Tait, P. G. (1900). On the claim recently made for Gauss to the invention (not the discovery) of quaternions. *Proc. Roy. Soc. Edinb.* **23**, 17–23.

Taylor, E. F. and Wheeler, J. A. (1963). *Spacetime Physics.* W.H. Freeman, San Francisco.

Terrell, J. (1959). Invisibility of the Lorentz contraction. *Phys. Rev.* **116**, 1041–5.

Thiele, R. (1982). *Leonhard Euler*. Leipzig (in German).

Thiemann, T. (1996). Anomaly-free formulation of non-perturbative, four-dimensional Lorentzian quantum gravity. *Phys. Lett.* **B380**, 257–64.

Thiemann, T. (1998a). Quantum spin dynamics (QSD). *Class. Quant. Grav.* **15**, 839–73.

Thiemann, T. (1998b). QSD III: Quantum constraint algebra and physical scalar product in quantum general relativity. *Class. Quant. Grav.* **15**, 1207–47.

Thiemann, T. (1998c). QSD V: Quantum gravity as the natural regulator of matter quantum field theories. *Class Quant. Grav.* **15**, 1281–314.

Thiemann, T. (2001). QSD VII: Symplectic Structures and Continuum Lattice Formulations of Gauge Field Theories. *Class. Quant. Grav.* **18**, 3293–338.

Thirring, W. E. (1983). *A Course in Mathematical Physics: Quantum Mechanics of Large Systems*. Springer-Verlag, Berlin & London.

Thomas, I. (1939). *Selections Illustrating the History of Greek Mathematics*, Vol. I: *From Thales to Euclid.* The Loeb Classical Library, Heinemann, London.

Thorne, K. (1986). *Black Holes: The Membrane Paradigm.* Yale University Press, New Haven.

Thorne, K. (1995a). *Black Holes and Time Warps.* W. W. Norton & Company.

Thorne, K. (1995b). Gravitational Waves. [gr-qc/9506086].

Tipler, F. J. (1997). *The Physics of Immortality*. Anchor.

Tipler, F. J., Clarke, C. J. S., and Ellis, G. F. R. (1980). Singularities and horizons—a review article. In *General Relativity and Gravitation*, Vol. II (ed. A. Held), pp. 97–206. Plenum Press, New York.

Tittel, W., Brendel, J., Zbinden, H., and Gisin, N. (1998). Violation of Bell Inequalities by Photons More Than 10 km Apart. *Phys. Rev. Lett.* **81**, 3563.

Tod, K. P. and Anguige, K. (1999a). Isotropic cosmological singularities 1: Polytropic perfect fluid spacetimes. *Annals Phys.* **276**, 257–93. [gr-qc/9903008]

Tod, K. P. and Anguige, K. (1999b). Isotropic cosmological singularities 2: The Einstein-Vlasov system. *Annals Phys.* **276**, 294–320. [gr-qc/9903009]

Tolman, R. C. (1934). *Relativity, Thermodynamics, and Cosmoogy.* Clarendon Press, Oxford.

Tonomura, A., Matsuda, T., Suzuki, R., Fukuhara, A., Osakabe, N., Umezaki, H., Endo, J., Shinagawa, K., Sugita, Y., and Fujiwara, F. (1982). Observation of Aharonov–Bohm effect with magnetic field completely shielded from the electronic wave. *Phys. Rev. Lett.* **48**, 1443.

Tonomura, A., Osakabe, N., Matsuda, T., Kawasaki, T., Endo, J., Yano, S., and Yamada (1986). Evidence for Aharonov–Bohm effect with magnetic field completely shielded from electron wave. *Phys. Rev. Lett*. **56**, 792–5.

Trautman, A. (1958). Radiation and boundary conditions in the theory of gravitation. *Bull. Acad. Polon. Sci. Sér. Sci. -Math., Astr. Phys*. **6**, 407–12.

Trautman, A. (1962) Conservation laws in general relativity, in *Gravitation: An Introduction to Current Research* (ed. L. Witten) Wiley, New York.

Trautman, A. (1965) in Trautman, A., Pirani, F. A. E., and Bondi, H. (1965) Lectures on General Relativity. *Brandeis 1964 Summer Institute on Theoretical Physics*, vol. I. Prentice-Hall, Englewood Cliffs, N.J.

Trautman, A. (1970). Fibre bundles associated with space-time. *Rep. Math. Phys. (Torún)* **1**, 29–34.

Trautman, A. (1972, 1973). On the Einstein-Cartan equations I-IV. *Bull. Acad. Pol. Sci.*, Ser. Sci. Math. Astron. Phys. **20**, 185–90; 503–6; 895–6; **21**, 345–6.

Trautman, A. (1997). Clifford and the 'Square Root' Ideas. In *Contemporary Mathematics* **203**.

Trèves F. (1967). *Topological Vector Spaces, Distributions and Kernels*, (Academic Press, New York).

Turing, A. M. (1937). On computable numbers, with an application to the Entscheidungsproblem. *Proc. Lond. Math. Soc*. **42(2)**, 230–65; a correction (1937), **43**, 544–6.

Unruh, W. G. (1976). Notes on black hole evaporation. *Phys. Rev*. **D14**, 870.

Vafa, C. (1996). Evidence for F-theory. *Nucl. Phys*. **B469**, 403.

Valentini, A. (2002). Signal-Locality and Subquantum Information in Deterministic Hidden-Variables Theories. In *Non-Locality and Modality* (ed. T. Placek and J. Butterfield). Kluwer. [quant-ph/0112151]

van der Waerden, B. L. (1929). Spinoranalyse. *Nachr. Akad. Wiss. Götting. Math.-Physik,* **Kl.** 100–9.

van der Waerden, B. L. (1985). *A History of Algebra: From al-Khwrizmi to Emmy Noether*, pp. 166–74. Springer-Verlag, Berlin.

van Heijenoort, J. (ed.) (1967). *From Frege to Gödel: A Source Book in Mathematical Logic, 1879–1931*. Harvard University Press, Cambridge, MA.

van Kerkwijk, M. H. (2000). Neutron Star Mass Determinations. [astro-ph/0001077]

Varadarajan, M. (2000). Fock representations from U(1) holonomy algebras. *Phys. Rev*. **D61**, 104001.

Varadarajan, M. (2001). M. Photons from quantized electric flux representations. *Phys. Rev*. **D64**, 104003.

Veneziano, G. (1968). *Nuovo Cimento*. **57A**, 190.

Vilenkin, A. (2000). *Cosmic Strings and Other Topological Defects*. Cambridge University Press, Cambridge.

Vladimirov, V. S. and Volovich, I.V. (1994). *P-Adic Analysis and Mathematical Physics*. World Scientific Publishing Company, Inc.

von Neumann, J. (1955). *Mathematical Foundations of Quantum Mechanics*. Princeton University Press, Princeton, New Jersey.

Wald, R. M. (1984). *General Relativity*. University of Chicago Press, Chicago.

Wald, R. M. (1994). *Quantum Field Theory in Curved Spacetime and Black Hole Thermodynamics*. University of Chicago Press, Chicago.

Ward, R. S. (1977). On self-dual gauge fields, *Phys. Lett*. **61A**, 81-2.

Ward, R.S. and Wells, R.O., Jr. (1989). *Twistor Geometry and Field Theory*. Cambridge University Press, Cambridge.

Weinberg, S. (1967). A model of leptons. *Phys. Rev. Lett.* **19**, 1264–66.
Weinberg, S. (1972). Gravitation and Cosmology: Principles and Applications of the General Theory of Relativity (Wiley, New York).
Weinberg, S. (1989). Precision Tests of Quantum Mechanics. *Phys. Rev. Lett.* **62**, 485–8
Weinberg, S. (1992). *Dreams of a Final Theory: The Scientists Search for the Ultimate Laws of Nature*. Pantheon Books, New York.
Wells, R. O. (1991). *Differential analysis on complex manifolds*. Prentice Hall, Englewood Cliffs.
Werbos, P. (1989). Bell's theorem: the forgotten loophole and how to exploit. In *Bell's Theorem, Quantum Theory, and Conceptions of the Universe* (ed. M. Kafatos). Kluwer, Dordrecht, The Netherlands.
Werbos, P. J. and Dolmatova, L. (2000). The Backwards-Time Interpretation of Quantum Mechanics: Revisited With Experiment. [http://arxiv.org/ftp/quant-ph/papers/0008/0008036.pdf]
Wess, J. and Bagger, J. (1992). *Supersymmetry and Supergravity*. Princeton University Press, Princeton.
Wess, J. and Zumino, B. (1974). Supergauge transformations in four dimansions, *Nucl. Phys.* **70**, 39–50.
Weyl, H. (1918), in Lorentz *et al.*, (1952).
Weyl, H. (1928). *Gruppentheorie und Quantenmechanik*. Hirzel, Leipzig; English translation of 2nd edn, *The Theory of Groups and Quantum Mechanics*. Dover, New York.
Weyl, H. (1929). Elektron und Gravitation I. *Z. Phys.* **56**, 330–52.
Wheeler, J. A. (1957). Assessment of Everett's 'Relative State' Formulation of Quantum Theory. *Rev. Mod. Phys.* **29**, 463–65.
Wheeler, J. A. (1960). *Neutrinos, Gravitation and Geometry: contribution to Rendiconti della Scuola Internazionale di Fisica' Enrico Fermi-XI, Corso, July 1959*. Zanichelli, Bologna. (Reprinted in 1982.)
Wheeler, J. A. (1965). Geometrodynamics and the issue of the final state. In *Relativity, Groups and Topology* (ed. B. S. and C. M. DeWitt). Gordon and Breach, New York.
Wheeler, J. A. (1973). From Relativity to Mutability. In *The Physicist's Conception of Nature* (ed. J. Mehra). D. Reidel, Boston, pp. 202–247.
Wheeler, J.A. (1983). Law without law, in *Quantum Theory and Measurement* (eds. J. A. Wheeler and W. H. Zurek). Princeton Univ. Press, Princeton, 182–213.
Whittaker, E. T. (1903). On the partial differential equations of mathematical physics. *Math. Ann.* **57**, 333–55.
Wick, G. C. (1956). Spectrum of the Bethe-Salpeter equation. *Phys Rev.* **101**, 1830.
Wigner, E. P. (1960). The Unreasonable Effectiveness of Mathematics in the Physical Sciences. *Commun. Pure Appl. Math.* **13**, 1–14.
Wilder, R. L. (1965). *Introduction to the foundations of mathematics*. John Wiley & Sons, New York.
Wiles, A. (1995). Modular elliptic curves and Fermat's Last Theorem. *Ann. Maths* **142**, 443–551.
Williams, R.K. (1995). Extracting X-rays, γ-rays, and Relativistic e^-e^+ Pairs from Supermassive Kerr Black Holes Using the Penrose Mechanism. *Phys. Rev.* **D51**, 5387.
Williams, R.K. (2002). Production of the High Energy–Momentum Spectra of Quasars 3C279 and 3C273 Using the Penrose Mechanism. [astro-ph/0306135]. Accepted for publication in Astrophysical Journal, 2004.

Williams, R.K. (2004). Collimated Escaping Vortical Polar e^-e^+ Jets Intrinsically Produced by Rotating Black Holes and Penrose Processes. [astro-ph/0404135].

Willmore, T. J. (1959). *An Introduction to Differential Geometry*. Clarendon Press, Oxford.

Wilson, K. (1975). *Phys. Reps.* **23**, 331.

Wilson, K. (1976). Quarks on a lattice, or the colored string, model. *Phys. Rep.* **23(3)**, 331–47.

Winicour, J. (1980). Angular momentum in general relativity. In *General Relativity and Gravitation* Vol. 2 (ed. A. Held), pp. 71–96. Plenum Press, New York.

Witten, E. (1978). An interpretation of classical Yang–Mills theory. *Phys. Lett.* **77B**, 394–8.

Witten, E. (1981). A new proof of the positive energy theorem. *Comm. Math. Phys.* **80**, 381–402.

Witten, E. (1982). Supersymmetry and Morse theory. *J. Diff. Geom.* **17**, 661–92.

Witten, E. (1988). Topological quantum field theory. *Commun. Math. Phys.* **118**, 411.

Witten, E. (1995). String theory in various dimensions. *Nucl. Phys.* **B443**, 85.

Witten, E. (1996). Reflections on the Fate of Spacetime. *Phys. Today*, April 1996.

Witten, E. (1998). Anti de Sitter Space and Holography. [hep-th/9802150]

Witten, E. (2003). Perturbative Gauge Theory as a String Theory in Twistor Space. [hep-th/0312171]

Witten, L. (1959). Invariants of general relativity and the classification of spaces. *Phys. Rev.* **113**, 357–62.

Wolf, J. (1974). *Spaces of Constant Curvature*. Publish or Perish Press, Boston, MA.

Woodhouse, N. M. J. (1991). *Geometric Quantization* , 2nd edn. Clarendon Press, Oxford.

Woodin, W. H. (2001). The Continuum Hypothesis Part I *Notices Amer. Math. Soc.* **48**, 567–76, Part II, *ibid.* 681–90; also (Part I) online at http://www.ams.org/notices/200106/fea-woodin.pdf and (Part II) ditto, but with /200107/.

Wooters, W. K. and Zurek, W. H. (1982). A single quantum cannot be cloned. *Nature* **299**, 802–3.

Wykes, A. (1969). *Doctor Cardano: Physician Extraordinary*. Frederick Muller, London.

Yang, C. N. and Mills, R. L. (1954). Conservation of Isotopic Spin and Isotopic Gauge Invariance. *Phys. Rev.* **96**, 191–5.

Yui, N. and Lewis, J. D. (2003). *Calabi-Yau Varieties and Mirror Symmetry*. Fields Institute Communications, V. 38. American Mathematical Society, Providence, RI.

Zee, A. (2003). *Quantum Field Theory in a Nutshell*. Princeton University Press, Princeton.

Zeilinger, A., Gaehler, R., Shull, C. G., and Mampe, W. (1988). Single and double slit diffraction of neutrons. *Rev. Mod. Phys.* **60**, 1067.

Zel'dovich Ya, B. (1966). Number of quanta as an invariant of the classical electromagnetic field. *Soviet Phys.-Doklady* **10**, 771–2.

Zimba, J. and Penrose, R. (1993). On Bell non-locality without probabilities: more curious geometry. *Stud. Hist. Phil. Sci.* **24**, 697–720.

Zinn-Justin, J. (1996). *Quantum Field Theory and Critical Phenomena*. Oxford University Press, Oxford.

Index

Page numbers in italic, e.g. *921*, refer to figures.